D1172178

*The microbial world*

**Roger Y. Stanier**

*Professor of Bacteriology, University of California, Berkeley*

**Michael Doudoroff**

*Professor of Bacteriology, University of California, Berkeley*

**Edward A. Adelberg**

*Professor of Microbiology, Yale University*

*Englewood Cliffs, New Jersey*

# *The microbial world*

*third edition*

**Prentice-Hall, Inc.**

*The microbial world*

*third edition*

**Roger Y. Stanier**
**Michael Doudoroff**
**Edward A. Adelberg**



C-13-581017-5

Library of Congress
Catalog Card Number 70-110090

Current Printing (last digit):
16 15 14 13 12 11
10 9 8 7 6 5 4 3 2

**Printed in the United States of America**

*Prentice-Hall International, Inc., London*
*Prentice-Hall of Australia, Pty., Ltd., Sydney*
*Prentice-Hall of Canada, Ltd., Toronto*
*Prentice-Hall of India Private Ltd., New Delhi*
*Prentice-Hall of Japan, Inc., Tokyo*

# *Preface to the third edition*

*About 100 years ago,* Pasteur revealed both the role of microorganisms in nature and their importance to human welfare. His successors, more concerned with the second point than with the first, turned microbiology into a flourishing field of applied science which evolved in almost complete isolation from the rest of biology. When the first edition of *The Microbial World* appeared, this long isolation was coming to an end. Our book had a frankly propagandist purpose—that of accelerating this change by presenting microbiology in the framework of the facts and concepts of general biology. Since then biology has undergone a revolution, only the dim future outlines of which could be discerned when the first edition of *The Microbial World* was written. Bacteria and viruses were then the material objects through which the molecular basis of cell function was discovered. Today, *Escherichia coli* and its phages serve as primary reference points for charting biological phenomena on the cellular and molecular levels.

The impacts of this revolution are already evident in university instruction. College freshmen (and even high-school students) now learn the central dogmas of molecular biology; they may not understand very clearly the general anatomy of a vertebrate, but the genetic code is firmly implanted in their minds. These changes in the intellectual climate of biology have forced us to reappraise both the *scope* and the *level* of the material to be included in the third edition of *The Microbial World.* In consequence, the book has been almost entirely rewritten and substantially reorganized.

With respect to the level of presentation, we have operated on the premise that students who today begin the study of microbiology have a considerably more solid scientific background than their predecessors of twelve years ago. We have therefore assumed that they possess some knowledge of the rudiments of organic chemistry, general biochemistry, genetics, and cell physiology. Where necessary, we have provided general references to the literature in these fields, but we have not started from first principles when introducing material dependent on these subjects.

The scope of the book has been changed in a number of respects. Material dealing with microbial diseases has been incorporated into a much broader description of symbiotic relationships that involve microorganisms. This is an important aspect of microbial ecology that to our knowledge has never been adequately presented in a textbook for student use. The treatment of the principal constituent groups of bacteria has been greatly enlarged, since this is also an area in which information is not easy to obtain and users of previous editions have suggested that such expansion would be desirable. Finally, it is impossible today to avoid discussion of the central concepts of molecular biology in a book that deals largely with bacteria. We have tried to do this without losing sight of our primary goal—to describe microorganisms and their behavior, not the properties and behavior of molecules.

We have also given considerable thought to the problem of reference material. For many readers the book would be more useful if facts were fully documented by citations of the original literature. However, this would have greatly increased its length. Instead, we have included at the end of each chapter an annotated bibliography of books and reviews, through which the interested reader can find references to substantiate most of the factual material presented in the text. This solution is not an entirely satisfactory one, since a good deal of searching may be required to pin down the origin of any given statement, but it is the best that we have been able to devise.

As a result of all these changes, *The Microbial World* has really become a new book rather than a revision of the last edition. The authors hope that it will continue to meet the needs of both teachers and students in its present reincarnation. They will welcome comments and criticisms, as well as corrections of any factual errors.

We should like to take this opportunity to express our thanks to the many individuals who helped us in the preparation of this edition. Once again we are grateful to the many colleagues, too numerous to list, who responded generously to requests for new illustrative material. We are indebted to Dr. Norberto Palleroni for skillfully undertaking the onerous task of preparing the index. Dr. Barbara Bachmann read the entire manuscript and made many constructive criticisms with respect to both style and content. Drs. Frank Black, Lawrence Freedman, S. Mark Henry, John Ingraham, P. T. Magee, Raphael Martinez, N. J. Palleroni, and H. P. Treffers read and criticized specific portions of the text; their suggestions were of great value. Finally, we should like to acknowledge the help of our secretaries, Mrs. Lis Nissen and Mrs. Beatrice L. Jacobson.

*Roger Y. Stanier*
*Michael Doudoroff*
*Edward A. Adelberg*

# Contents

## Chapter one The beginnings of microbiology 1



## Chapter two The nature of the microbial world 24

# *The microbial world*

# *Chapter one*
# *The beginnings of microbiology*

*Microbiology is the study of organisms* that are too small to be clearly perceived by the unaided human eye, called *microorganisms.* If an object has a diameter of less than 0.1 mm, the eye cannot perceive it at all, and very little detail can be perceived in an object with a diameter of 1 mm. Roughly speaking, therefore, organisms with a diameter of 1 mm or less are microorganisms, and fall into the broad domain of microbiology. Microorganisms have a wide taxonomic distribution; they include some metazoan animals, protozoa, many algae and fungi, bacteria, and viruses. The very existence of this microbial world was unknown to mankind until the invention of microscopes, optical instruments that serve to magnify objects so small that they cannot be clearly seen by the unaided human eye. Microscopes, invented at the beginning of the seventeenth century, opened the biological realm of the very small to systematic scientific exploration.

Early microscopes were of two kinds. The first were *simple microscopes* with a single lens of very short focal length, consequently capable of a high magnification; such instruments do not differ in optical principle from ordinary magnifying glasses able to increase an image severalfold, which had been known since antiquity. The second were *compound microscopes* with a double lens system consisting of an ocular and objective. The compound microscope has the greater intrinsic power of magnification and eventually displaced completely the simple instrument; all our contemporary microscopes are of the compound type. However, nearly all the great original microscopic discoveries were made with simple microscopes.

## The discovery of the microbial world

*Figure 1.1 Antony van Leeuwenhoek (1632–1723). In this portrait, he is holding one of his microscopes. Courtesy of the Rijksmuseum, Amsterdam.*

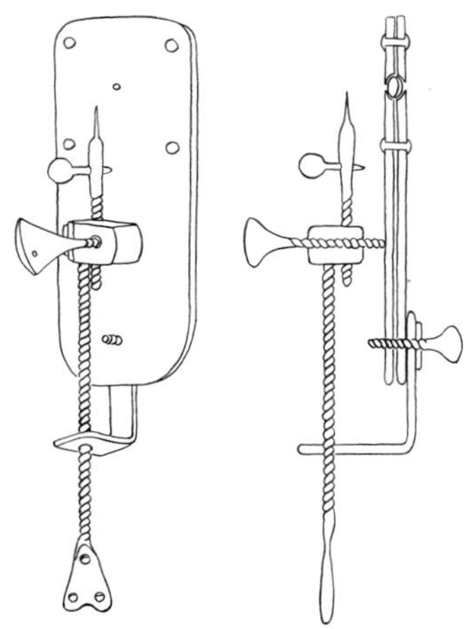

*Figure 1.2 A drawing to show the construction of one of Leeuwenhoek's microscopes. After C. E. Dobell, Antony van Leeuwenhoek and His Little Animals. New York: Russell and Russell, Inc., 1932.*

The discoverer of the microbial world was a Dutchman, Antony van Leeuwenhoek (1632-1723). Leeuwenhoek was a merchant in the small town of Delft, and his scientific activities were fitted into a life well filled with business affairs and civic duties. In this, he was no exception for his time; professional scientists did not exist in the seventeenth and eighteenth centuries. The great discoveries of this period in all fields of science were made by amateurs, who earned their living in other ways, or who were freed from the necessity of earning a living because of their personal wealth. However, Leeuwenhoek differed from his scientific contemporaries in one respect: he had little formal education and never attended a university. This was probably no disadvantage scientifically, since the scientific training then available would have provided little basis for his life's work; more serious handicaps, insofar as the communication of his discoveries went, were his lack of connections in the learned world and his ignorance of any language except Dutch. Nevertheless, through a fortunate chance, his work became widely known in his own lifetime, and its importance was immediately recognized. About the time that Leeuwenhoek began his observations, the Royal Society had been established in England for the communication and publication of scientific work. The Society invited Leeuwenhoek to communicate his observations to its members and a few years later (1680) elected him as a Fellow. For almost 50 years, until his death in 1723, Leeuwenhoek transmitted his discoveries to the Royal Society in the form of a long series of letters written in Dutch. These letters were largely translated and published in English in the *Proceedings of the Royal Society*, so becoming quickly and widely disseminated (Figure 1.1).

Leeuwenhoek's microscopes (Figure 1.2) bore little resemblance to the instruments with which we are familiar. The single, minute, almost spherical lens was mounted between two small metal plates. The specimen was placed on the point of a blunt pin attached to the back plate and was brought into focus by manipulating two screws which varied the position of the pin relative to the lens. During this operation, the observer held the instrument with its other face very close to his eye and squinted through the lens. No change of magnification was possible, the magnifying power of each microscope being an intrinsic property of its simple lens. Despite the simplicity of their construction, Leeuwenhoek's microscopes were able to give clear images at magnifications which ranged, depending on the focal length of the lens, from about 50 to nearly 300 diameters. The highest magnification that he could obtain was consequently somewhat less than one third of the highest magnification that is obtainable with a modern compound light microscope. Leeuwenhoek constructed hundreds of such instruments, a few of which survive today.

Leeuwenhoek's place in scientific history depends not so much on his skill as a microscope maker, essential though this was, as on the extraordinary range and skill of his microscopic observations. He was endowed with an unusual degree of curiosity and studied almost every conceivable object that could be looked at through a microscope. He made magnificent observations on the microscopic structure of the seeds and embryos of plants and on small invertebrate animals. He discovered the existence of spermatozoa and of red blood cells and was thus the founder of animal histology. By discovering and describing capillary circulation he completed the work on the circulation of blood begun by Harvey half a century before. Indeed, it would be easy to fill a page with a mere list of his major discoveries about the structure of higher plants and animals. His greatest claim to fame rests, however, on his discovery of the microbial world: the world of "animalcules," or little animals, as he and his contemporaries called them. A new dimension was thus added to biology. All the main kinds of unicellular microorganisms that we know today—protozoa, algae, yeasts, and bacteria—were first described by Leeuwenhoek, often with such accuracy that it is possible to identify individual species from his accounts of them. In addition to the diversity of this microbial world, Leeuwenhoek emphasized its incredible abundance. For example, in one letter describing for the first time the characteristic bacteria of the human mouth, he wrote:

*I have had several gentlewomen in my house, who were keen on seeing the little eels in vinegar: but some of them were so disgusted at the spectacle, that they vowed they'd never use vinegar again. But what if one should tell such people in future that there are more animals living in the scum on the teeth in a man's mouth, than there are men in a whole kingdom?*

Although Leeuwenhoek's contemporaries marveled at his scientific discoveries, the microscopic exploration of the microbial world which he had so brilliantly begun was not appreciably extended for over a century after his death. The principal reasons for this long delay seem to have been technical ones. Simple microscopes of high magnification are both difficult and tiring to use, and the manufacture of the very small lenses is an operation that requires great skill. Consequently, most of Leeuwenhoek's contemporaries and immediate successors used compound microscopes. Despite the intrinsic superiority of compound microscopes, the ones available in the seventeenth and eighteenth centuries suffered from serious optical defects, which made them less effective working instruments than Leeuwenhoek's simple microscopes. This is clearly shown by the fact that Leeuwenhoek's English contemporary, Robert Hooke, who was a very capable and careful observer, could not repeat with his own compound microscope many of the finer observations reported by Leeuwenhoek.

The major optical improvements which were eventually to lead to compound microscopes of the quality that we use today began about 1820 and extended through the succeeding half century. These improvements were closely followed by resumed exploration of the microbial world and resulted, by the end of the nineteenth century, in a detailed knowledge of its constituent groups. In the meantime, however, the science of microbiology had been developing in other ways, which led to the discovery of the roles the microorganisms play in the transformations of matter and in the causation of disease.

## *The controversy over spontaneous generation*

After Leeuwenhoek had revealed the vast numbers of microscopic creatures present in nature, scientists began to wonder about the origin of these forms. From the beginning, there were two schools of thought. Some believed that the animalcules were formed spontaneously from nonliving materials, whereas others (Leeuwenhoek included) believed that they were formed from the "seeds" or "germs" of these animalcules, which were always present in the air. The belief in the spontaneous formation of living beings from nonliving matter is known as the doctrine of *spontaneous generation*, or *abiogenesis*, and has had a long existence. In ancient times, it was considered self-evident that many plants and animals can be generated spontaneously under special conditions. The doctrine of spontaneous generation was accepted without question until the Renaissance.

As knowledge of living organisms accumulated, it gradually became evident that the spontaneous generation of plants and animals simply does not occur. A decisive step in the abandonment of the doctrine as applied to animals took place as the result of experiments performed about 1665 by an Italian physician, Francesco Redi. He showed that the maggots that develop in putrefying meat are the larval stages of flies and will never appear if the meat is protected by placing it in a vessel closed with fine gauze so that flies are unable to deposit their eggs on it. By such experiments, Redi destroyed the myth that maggots develop spontaneously from meat. Consequently, the doctrine of spontaneous generation was already being greatly weakened by exact studies on the development and life cycles of plants and animals at the time when Leeuwenhoek discovered the microbial world. For technical reasons, it is far more difficult to show that microorganisms are not generated spontaneously, and so as time went on the proponents of the doctrine came to center their claims more and more on the mysterious appearance of these simplest forms of life in organic infusions. Those who did not believe in the spontaneous generation of microorganisms were in the position, always difficult, of having to prove a negative point; in fact, it was not until the middle of the nineteenth century that the cumulative negative evidence became sufficiently abundant to lead to the general abandonment of this ancient but erroneous doctrine.

One of the first to provide strong evidence that microorganisms do not arise spontaneously in organic infusions was the Italian naturalist Lazzaro Spallanzani, who conducted a long series of experiments on this problem in the middle of the eighteenth century. He could show repeatedly that heating can prevent the appearance of animalcules in infusions, although the duration of the heating necessary is variable. From various observations, Spallanzani concluded that animalcules can be carried into infusions by air and that this is the explanation for their supposed spontaneous generation in well-heated infusions. Earlier workers had closed their flasks with corks, but Spallanzani was not satisfied that any mechanical plug could completely exclude air, and he resorted to hermetic sealing. He observed that after sealed infusions had remained barren for a long time a tiny crack in the glass would be followed by the development of animalcules. His final conclusion was that, to render an infusion *permanently* barren, it must be sealed hermetically and boiled. Animalcules could never appear unless new air somehow entered the flask and came in contact with the infusion.

One might have thought that Spallanzani's beautiful experiments, which showed so clearly all the difficulties of work of this kind, would settle the issue once and for all. Had later investigators fully adopted the precautions outlined by Spallanzani, this might well have been the case. However, faulty experiments continued to be performed, and the results continued to be brought forward as evidence for the occurrence of spontaneous generation. In the meantime, however, an interesting *practical* application of Spallanzani's discoveries had been made. His experiments had shown that even very perishable plant or animal infusions do not undergo putrefaction or fermentation when they have been rendered free of animalcules, from which it seemed probable that these chemical changes were in some way connected with the development of microbes. In the beginning of the nineteenth century, François

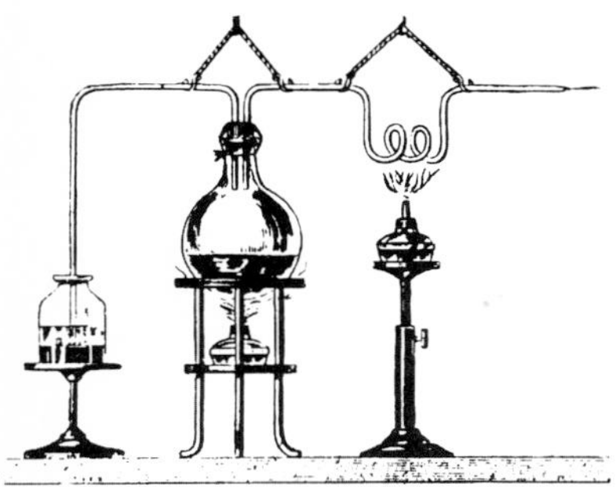

*Figure 1.3 The apparatus used by Schwann in his experiments on spontaneous generation. The central flask, which is being rendered sterile by heating, contains the organic infusion. Air was passed into it through the heated coiled tube to the right. The vessel at the left is a mercury trap, through which the air escaped after passing through the central flask.*

Appert found that one can preserve foods by enclosing them in airtight containers and heating the containers. He was able in this way to preserve highly perishable foodstuffs indefinitely, and "appertization," as this original canning process was called, came into extensive use for the preservation of foods long before the scientific issue had been finally settled.

In the late eighteenth century, the work of Joseph Priestley, Henry Cavendish, and Antoine Lavoisier laid the foundations of the chemistry of gases. One of the gases first discovered was oxygen, which soon was recognized to be essential for the life of animals. In the light of this knowledge, it seemed possible that the hermetic sealing recommended by Spallanzani and practiced by Appert was effective in preventing the appearance of microbes and the decomposition of organic matter, not because it excluded air carrying germs but because it excluded oxygen, required both for microbial growth and for the initiation of fermentation or putrefaction. Consequently, the influence of oxygen on these processes was a matter of much discussion in the early nineteenth century. It was finally shown that neither growth nor decomposition will occur in an infusion that has been properly heated, even when it is exposed to air, provided that the air entering the infusion has been previously treated so as to remove any germs that it contains. The first successful demonstration of this fact was made in 1837 by Theodor Schwann, who passed air into a boiled infusion through a coiled tube heated almost to the melting point and found that the infusion remained sterile. Schwann's apparatus is illustrated in Figure 1.3. In 1854 similar experiments were performed by H. Schroeder and T. von Dusch, who substituted a long tube filled with cotton wool for the heated coil.

### The experiments of Pasteur

By 1860 some scientists had begun to realize that there is a *causal relationship* between the development of microorganisms in organic infusions and the chemical changes that take place in these infusions; *microorganisms are the agents that bring about the chemical changes.* The great pioneer in these studies was Louis Pasteur (Figure 1.4). However, the acceptance of this important concept was conditional on the demonstration that spontaneous generation does not occur. Stung by the continued claims of adherents to the doctrine of spontaneous generation, Pasteur finally turned his attention to this problem. His work on the subject was published in 1861 as a *Memoir on the Organized Bodies Which Exist in the Atmosphere.*

Pasteur first demonstrated that air really does contain microscopically observable "organized bodies." He aspirated large quantities of air through a tube that contained a plug of guncotton to serve as a filter. The guncotton was then removed and dissolved in a mixture of alcohol and ether, and the sediment was examined microscopically. In addition to inorganic matter, it contained considerable numbers of small round or oval bodies, indistinguishable from microorganisms. Pasteur next proceeded to repeat the experiments of Schwann and confirmed the fact that heated air can be supplied to a boiled infusion without giving rise to microbial development. Having established this point, he went on to show that in a closed system the addition of a piece of germ-laden guncotton to a sterile infusion invariably provoked microbial growth. These experiments showed Pasteur how germs can enter infusions and led him to what was perhaps his most elegant experiment on the subject. This was the demonstration that infusions will remain sterile indefinitely in open flasks, provided that the neck of the flask is drawn out and bent down in such a way that the germs from the air cannot ascend it. Pasteur's bent-necked flasks are illustrated in Figure 1.5. If the neck of such a flask was broken off, the infusion rapidly became populated with microbes. The same thing happened if the sterile liquid in the flask was poured into the exposed portion of the bent neck and then poured back. This type of experiment finally disposed of all criticisms based on possible effects of air itself as an activating agent for the development of life in organic infusions.

*Figure 1.4 Louis Pasteur (1822–1895). Courtesy of the Institut Pasteur, Paris.*

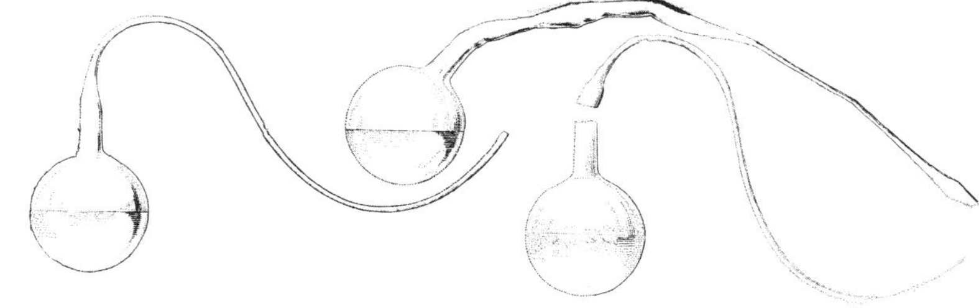

*Figure 1.5 The famous swan-neck flask used by Pasteur during his studies on spontaneous generation. The construction of the neck permitted free access of air to the flask contents but prevented entry of microorganisms present in the air.*

Pasteur rounded out his study by determining in semiquantitative fashion the distribution of microorganisms in the air and by showing that these living organisms are by no means evenly distributed through the atmosphere.

**The experiments of Tyndall**

With these experiments, Pasteur won the battle over spontaneous generation as far as his French contemporaries were concerned, and attempts to refute his work were no longer taken seriously in France. In England, however, there appeared a new advocate of the doctrine of

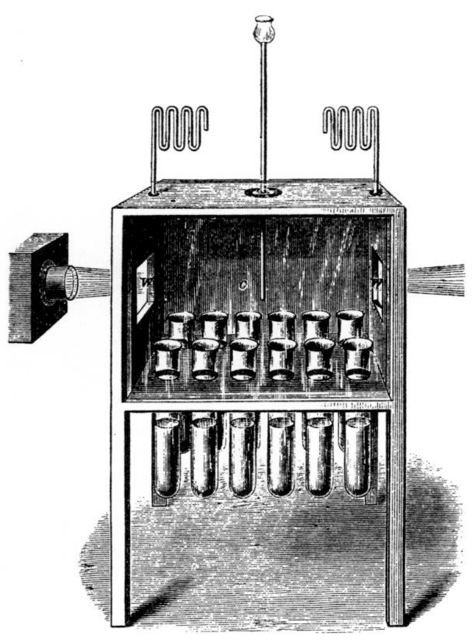

*Figure 1.6 The apparatus used by Tyndall for his experiments on spontaneous generation.*

spontaneous generation, Bastian, who in 1872 published a large book on the subject. The challenge was taken up by John Tyndall, an English physicist who was an ardent supporter of Pasteur and his work. Tyndall had spent many years studying the relation of radiant heat to gases, and in these experiments he had been impressed with the difficulty of removing particles that float in the atmosphere. He found that such particles, even though invisible to the naked eye, could be made apparent when a powerful condensed beam of light was passed through air in a darkened room. The path of the beam became visible when observed at right angles because part of the light was scattered by the floating matter. Air that had been made dust-free no longer scattered light in this fashion, and the path of the beam was no longer visible. In Tyndall's terminology, such air was "optically empty." One day he examined a flask that had been standing closed and untouched for a long time and found that the air in it had become "optically empty" as a result of the deposition of the dust on the bottom and sides of the flask. This led him to construct a special chamber in which experiments could be carried out (Figure 1.6). The front consisted of glass and the remaining walls of wood. Two panes of glass were let in on the sides to permit passage of a beam of light, and at the back was a door to allow the introduction of removal of material. The bottom was pierced with holes to accommodate test tubes that contained the various solutions under study, and through the top there passed a movable pipette to be used for adding solutions to the test tubes. Also inserted in the top were two narrow tubes, bent up and down several times on the principle of Pasteur's bent-necked flasks, to permit a dustproof connection between the air inside the chamber and the atmosphere.

When prepared for an experiment on spontaneous generation, the chamber was closed and left untouched until a beam of light passed through the windows showed that the air had become "optically empty." The test tubes were then filled with infusions through the pipette in the top and lowered into a bath of boiling brine for 5 minutes to sterilize them. Tyndall found that boiled infusions in this chamber would remain germ-free for months. However, in experimenting with infusions of dried hay, he observed that these were much more difficult to sterilize than the infusions that he had previously used; 5 minutes of boiling would no longer do the job. For these experiments, Tyndall had brought a bale of hay into his laboratory. When he turned again to the other kinds of infusions, with which he had previously had such striking successes, he found that now they too could no longer be successfully sterilized by boiling, even for an hour or more. After many experiments, Tyndall finally realized what had happened. The dried hay contained spores of bacteria that were many times more resistant to heat than any microbes with which he had previously been dealing, and as a result of the presence of the hay in his laboratory, the laboratory air had become thoroughly infected with these spores. Once he had grasped this point, he proceeded to test the actual limits of heat resistance of the spores of hay bacteria and found that boiling infusions for even as long as 5½ hours would not render them sterile with certainty. From these results, he concluded that bacteria have phases, one relatively thermolabile (destroyed by boiling for 5 minutes) and one thermoresistant to an almost incredible extent. These conclusions were almost immediately confirmed by a German botanist, Ferdinand Cohn, who demonstrated that the hay bacteria can produce microscopically distinguishable resting bodies (*endospores*), which are highly resistant to heat.

Tyndall then proceeded to develop a method of sterilization by *discontinuous heating*, later called *tyndallization*, which could be used to kill *all* bacteria in infusions. Since growing bacteria are easily killed by brief boiling, all that is necessary is to allow the infusion to stand for a certain period before applying heat to permit germination of the spores with a consequent loss of their heat resistance. A very brief period of boiling can then be used, and repeated, if need be, several times at intervals to catch any spores late in germination. Tyndall found that discontinuous boiling for 1 minute on five successive occasions would make an infusion sterile, whereas a single continuous boiling for 1 hour would not.

With the publication of Tyndall's findings, the scientific world acknowledged the demise of the doctrine of spontaneous generation. Thus the final overthrow of this incorrect biological theory was the joint achievement of a chemist (Pasteur) and a physicist (Tyndall). Of the two, Tyndall perhaps deserves the greater credit, since the recognition of the tremendous heat resistance of bacterial spores was essential to the development of adequate procedures for sterilization.

It is often stated that the work of Pasteur and Tyndall "disproved" the possibility of spontaneous generation, and their experimental findings have been used to support the contention that spontaneous generation has never occurred in the past. This is an unjustifiable extension of their actual findings. The conclusion that we may safely draw is a much more limited one: that at the present time microorganisms do not arise spontaneously in properly sterilized organic infusions. It is probable that the primary origin of life on earth did involve a kind of spontaneous generation, although a far more gradual and subtle one than that envisaged by the proponents of the doctrine during the eighteenth and nineteenth centuries. Like so many biological statements, the statement "spontaneous generation does not occur" is strictly valid only within temporal limits and should really be accompanied by the phrase "at the present stage in the history of the earth."

## *The discovery of the role of microorganisms in transformations of organic matter*

During the long controversy over spontaneous generation, a correlation between the growth of microorganisms in organic infusions and the onset of chemical changes in the infusion itself was frequently observed. These chemical changes were designated as fermentation and putrefaction. Putrefaction, a process of decomposition that results in the formation of ill-smelling products, occurs characteristically in meat and is a consequence of the breakdown of proteins, the principal organic constituents in such natural materials. Fermentation, a process that results in the formation of alcohols or organic acids, occurs characteristically in plant materials as a consequence of the breakdown of carbohydrates, the predominant organic compounds in plant tissues.

**Fermentation as a biological process**

In 1837, three men, C. Cagniard-Latour, Th. Schwann, and F. Kützing, independently proposed that the yeast which appears during alcoholic fermentation is a microscopic plant and that the conversion of sugars to ethyl alcohol and carbon dioxide characteristic of the alcoholic fermentation is a physiological function of the yeast cell. This theory was bitterly attacked by such leading chemists of the time as J. J. Berzelius, J. Liebig, and F. Wohler, who held the view that fermentation and putrefaction are purely chemical processes. To understand the attitude of the chemists, one must realize that the science of chemistry had made great advances during the first decades of the nineteenth century and that in 1828 the whole field of synthetic organic chemistry had been opened up by the first synthesis of an organic compound, urea, from inorganic materials. With the demonstration that organic compounds, until that time known exclusively as products of living activity, could be made in the laboratory, the chemists rightly felt that a large body of natural phenomena had now become amenable to analysis in physicochemical terms. The conversion of sugars to alcohol and carbon dioxide appeared to be a relatively simple chemical process. Accordingly, the chemists did not look with favor on the attempt to interpret this process as the result of the action of a living organism.

Ironically enough it was Pasteur, himself a chemist by training, who eventually convinced the scientific world that *all fermentative proc-*

*esses are the results of microbial activity.* Pasteur's work on fermentation extended with minor interruptions from 1857 to 1876. This work had a practical origin. The distillers of Lille, where the manufacture of alcohol from beet sugar was an important local industry, had encountered difficulties and called on Pasteur for assistance. Pasteur found that their troubles were caused by the fact that the alcoholic fermentation had been in part replaced by another kind of fermentative process, which resulted in the conversion of the sugar to lactic acid. When he examined microscopically the contents of fermentation vats in which lactic acid was being formed, he found that the cells of yeast characteristic of the alcoholic fermentation had been replaced by much smaller rods and spheres. If a trace of this material was placed in a sugar solution containing some chalk, a vigorous lactic fermentation ensued, and eventually a grayish deposit was formed, which again proved on microscopic examination to consist of the small spherical and rod-shaped organisms. Successive transfers of minute amounts of material to fresh flasks of the same medium always resulted in the production of a lactic fermentation and an increase in the amount of the formed bodies. Pasteur argued that the active agent, or new "yeast," was a microorganism that specifically converted sugar to lactic acid during its growth.*

Using similar methods, Pasteur studied a considerable number of fermentative processes during the following 20 years. He was able to show that fermentation is invariably accompanied by the development of microorganisms. Furthermore, he showed that each particular chemical type of fermentation, as defined by its principal organic end products (for example, the lactic, alcoholic, and butyric fermentations), is accompanied by the development of a *specific type of microorganism.* Many of these specific microbial types could be recognized and differentiated microscopically by their characteristic size and shape. In addition they could be distinguished by the specific environmental conditions which favored their development. To cite one example of such physiological specificity, Pasteur observed very early that whereas the agent of alcoholic fermentation can flourish in an acid medium, the agents of the lactic fermentation grow best in a neutral medium. It was for this reason that he added chalk (calcium carbonate) to his medium for the cultivation of the lactic organisms; this substance serves as a neutralizing agent and prevents too strong an acidification of the medium that would otherwise occur as a result of the formation of lactic acid.

*The agents of the lactic acid fermentation which Pasteur studied were in fact bacteria, but at that time knowledge of microorganisms was still sketchy, and so Pasteur called them a new type of yeast.

**The discovery of anaerobic life**

During his studies on the butyric fermentation, Pasteur discovered another fundamental biological phenomenon: the *existence of forms of life which can live only in the absence of free oxygen.* Unlike the agents of the alcoholic and lactic fermentations, the bacteria that cause the butyric fermentation are motile organisms. While examining microscopically fluids that were undergoing a butyric fermentation, Pasteur observed that the bacteria at the margin of a flattened drop, in close contact with the air, became immotile, whereas those in the center of the drop remained motile. This observation suggested that air had an inhibitory effect on the microorganisms in question, an inference which Pasteur quickly confirmed by showing that passage of a current of air through the fermenting fluid could retard, and sometimes completely arrest, the butyric fermentation. He thus concluded that some microorganisms can live only in the absence of oxygen, a gas previously considered essential for the maintenance of all life. He introduced the terms *aerobic* and *anaerobic* to designate, respectively, life in the presence and in the absence of oxygen.

**The physiological significance of fermentation**

The discovery of the anaerobic nature of the butyric fermentation provided Pasteur with an important clue for the understanding of the role that fermentations play in the life of the microorganisms that bring them about. Free oxygen is essential for most organisms as an agent for the oxidation of organic compounds to carbon dioxide. Such oxygen-linked biological oxidations, known collectively as *aerobic respirations,* provide the energy that is required for maintenance and growth.

Pasteur was the first to realize that the breakdown of organic compounds in the absence of oxygen can also be used by some organisms as a means of obtaining energy; as he put it, *fermentation is life without air.* Some strictly anaerobic microorganisms, such as the butyric acid bacteria, are dependent on fermentative mechanisms to obtain energy. Many other microorganisms, including certain yeasts, are *facultative anaerobes,* which have two alternative energy-yielding mechanisms at their disposal. In the presence of oxygen, they employ aerobic respiration, but they can employ fermentation if no free oxygen is present in their environment. This was beautifully demonstrated by Pasteur, who showed that sugar is converted to alcohol and carbon dioxide by yeast in the absence of air but that in the presence of air, little or no alcohol is formed; carbon dioxide is the principal end-product of this aerobic reaction.

The amount of growth which can occur at the expense of an organic compound is determined primarily by the amount of energy that can be obtained by the breakdown of that compound. Fermentation is a less efficient energy-yielding process than aerobic respiration, because part of the energy present in the substance decomposed is still present in the organic end-products (for example, alcohol or lactic acid) characteristically formed by fermentative processes. As Pasteur was the first to show, the breakdown of a given weight of sugar results in substantially less growth of yeast under anaerobic conditions than under aerobic ones, thus establishing the relative inefficiency of fermentation as an energy source.

Pasteur's work showed beyond doubt that fermentations are "vital processes," which play a role of basic physiological importance in the life of many cells. The further development of knowledge about the nature of fermentation resulted from an accidental observation made in 1897 by H. Buchner. In attempting to preserve an extract of yeast, prepared by grinding yeast cells with sand, Buchner added a large quantity of sugar to it and was surprised to observe an evolution of carbon dioxide accompanied by the formation of alcohol. A soluble enzymatic preparation, able to carry out alcoholic fermentation, was thus discovered. Buchner's discovery inaugurated the development of modern biochemistry; the detailed analysis of the mechanism of cell-free alcoholic fermentation was eventually to show that this complex metabolic process can be interpreted as resulting from a succession of chemically intelligible reactions, each catalyzed by a specific enzyme. Today, the belief that even the most complex physiological process can be similarly understood in physicochemical terms is accepted as a matter of course by all biologists. In this sense, the intuition of the nineteenth-century chemists who battled against the biological theory of fermentation has proved to be a correct one.

## *The discovery of the role of microorganisms in the causation of disease*

During his studies on fermentation, Pasteur, who was ever conscious of the practical applications of his scientific work, devoted considerable attention to the spoilage of beer and wine, which he showed to be caused by the growth of undesirable microorganisms. Pasteur used a peculiar and significant term to describe these microbially induced spoilage processes; he called them "diseases" of beer and wine. In fact, he was already considering the possibility that microorganisms may act as agents of infectious disease in higher organisms. A certain amount of evidence in support of this hypothesis already existed. It had been shown in 1813 that specific fungi can cause diseases of wheat and rye, and in 1845 M. J. Berkeley had proved that the great Potato Blight of Ireland, a natural disaster which deeply influenced Irish history, was caused by a fungus. The first recognition that fungi may be specifically associated with a disease of animals came in 1836 through the work of A. Bassi in Italy on a fungal disease of silkworms. A few years later, J. L. Schönlein showed that certain skin diseases of man are caused by fungal infections. Despite these indications, very few medical scientists were willing to entertain the notion that the major infectious diseases of man could be caused by microorganisms, and fewer still believed that organisms as small and apparently simple as the bacteria could act as agents of disease.

**Surgical antisepsis**

The introduction of anesthesia about 1840 had made possible a very rapid development of surgical methods. Speed was no longer a primary consideration, and the surgeon was able to undertake operations of a length and complexity that would have been unthinkable previously. However, with the elaboration of surgical technique, a problem that had always existed became more and more serious: *surgical sepsis*, or the infections that followed surgical intervention and often resulted in the death of the patient. Pasteur's studies on the problem of spontaneous generation had shown the presence of microorganisms in the air and at the same time indicated various ways in which their access to and development in organic infusions could be prevented. A young British surgeon, Joseph Lister, who was deeply impressed by Pasteur's work, reasoned that surgical sepsis might well result from microbial infection of the human tissues exposed during operation. He decided to develop methods for preventing the access of microorganisms to surgical wounds. By the scrupulous sterilization of surgical instruments, by the use of disinfectant dressings, and by the conduct of surgery under a spray of disinfectant to prevent airborne infection, he succeeded in greatly reducing the incidence of surgical sepsis. Lister's procedures of antiseptic surgery, developed about 1864, were initially greeted with considerable scepticism but, as their striking success in the prevention of surgical sepsis was recognized, gradually became common practice. This work provided powerful *indirect* evidence for the germ theory of disease, even though it did not cast any light on the possible microbial causation of specific human diseases. Just as in the case of Appert's development of canning as a means of food preservation half a century before, so with Lister's introduction of surgical antisepsis, practice had run ahead of theory.

The discovery that bacteria can act as specific agents of infectious disease in higher animals was made through the study of anthrax, a serious infection of domestic animals that is transmissible to man. In the terminal stages of a generalized anthrax infection, the rod-shaped bacteria responsible for the disease occur in enormous numbers in the bloodstream. These objects were first observed as early as 1850, and their presence in the blood of infected animals was reported by a series of investigators during the following 15 years. Particularly careful and detailed studies were carried out between 1863 and 1868 by C. J. Davaine, who showed that the rods are invariably present in diseased animals but are undetectable in healthy ones and that the disease can be transmitted to healthy animals by inoculation with blood containing these rod-shaped elements.

**The bacterial etiology of anthrax**

The conclusive demonstration of the bacterial causation, or *etiology*, of anthrax was provided in 1876 by Robert Koch (Figure 1.7), then a young German country doctor. He had no laboratory, and his experiments were conducted in his home, using very primitive improvised equipment and small experimental animals. He showed that mice could be infected with material from a diseased domestic animal. He transmitted the disease through a series of 20 mice by successive inoculation; at each transfer, the characteristic symptoms were observed. He then proceeded to cultivate the causative bacterium by introducing minute, heavily infected particles of spleen from a diseased animal into drops of sterile serum. Observing hour after hour the growth of the organisms in this culture medium, he saw the rods change into long filaments within which ovoid, refractile bodies eventually appeared. He showed that these bodies were spores, which had not been seen by previous workers. When spore-containing material was transferred to a fresh drop of sterile serum, the spores germinated and gave rise once more to typical rods. In this fashion, he transferred cultures of the bacterium eight successive times. The final culture of the series, injected into a healthy animal, again produced the characteristic disease, and from this animal, the organisms could again be isolated in culture (Figure 1.8).

*Figure 1.7 Robert Koch (1843–1910). Courtesy of VEB George Thieme, Leipzig.*

This series of experiments fulfilled the criteria which had been laid down 36 years before by J. Henle as logically necessary to establish the causal relationship between a specific microorganism and a specific disease. In generalized form, these criteria are: (1) the microorganism must be present in every case of the disease; (2) the microorganism must be isolated from the diseased host and grown in pure culture; (3) the specific disease must be reproduced when a pure culture of the micro-

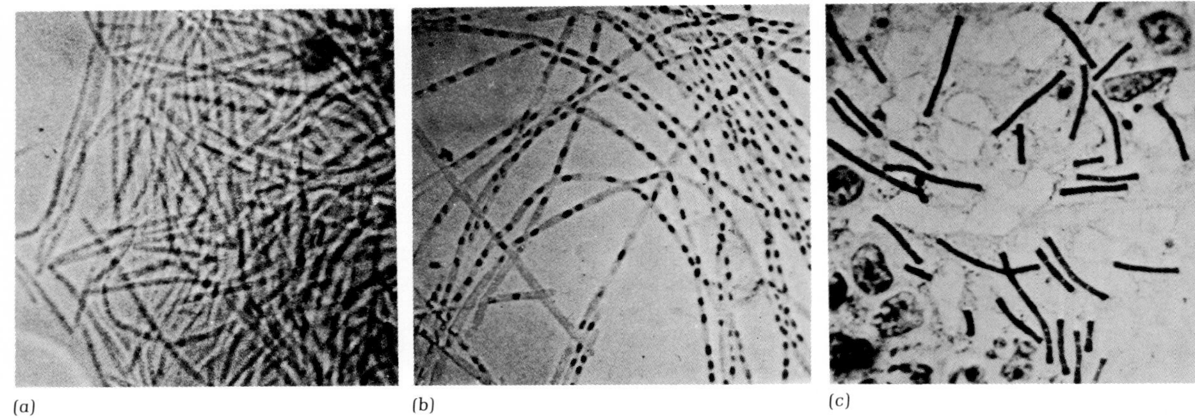


*Figure 1.8 The first photomicrographs of bacteria, taken by Robert Koch in 1877. (a) Unstained chains of vegetative cells of Bacillus anthracis. (b) Unstained chains of B. anthracis; the cells contain refractile spores. (c) A stained smear of B. anthracis from the spleen of an infected animal. Note the rod-shaped bacilli and the larger tissue cells.*

organism is inoculated into a healthy susceptible host; and (4) the microorganism must be recoverable once again from the experimentally infected host. Since Koch was the first to apply these criteria experimentally, they are now generally known as *Koch's postulates.*

Koch carried out another series of experiments which was of cardinal importance in demonstrating the *biological specificity* of the disease agent. He showed that another spore-forming bacterium, the hay bacillus, does not cause anthrax upon injection, and he also differentiated bacteria that cause other infections from the anthrax organism. From these studies he concluded that "only one kind of bacillus is able to cause this specific disease process, while other bacteria either do not produce disease following inoculation, or give rise to other kinds of disease."

In the meantime, Pasteur had found a collaborator, J. Joubert, with a knowledge of medical problems. Unaware of Koch's work, Pasteur and Joubert undertook the study of anthrax. They did not add anything new to the conclusions reached by Koch, but they confirmed his work and provided additional demonstrations that the bacillus, and not some other agent, was the specific cause of the disease.

## The rise of medical bacteriology

This work on anthrax abruptly ushered in the golden age of medical bacteriology, during which newly established institutes, created in Paris and in Berlin for Pasteur and Koch, respectively, became the world centers of bacteriological science. The German school, led by Koch, concentrated its efforts primarily on the isolation, cultivation, and characterization of the causative agents for the major infectious diseases of man. The French school, under the leadership of Pasteur, turned almost immediately to a more subtle and complex problem:

the experimental analysis of how infectious disease takes place in the animal body and how recovery and immunity are brought about. Within 25 years, most of the major bacterial agents of human disease had been discovered and described, and methods for the prevention of many of these diseases, either by artificial immunization or by the application of hygienic measures, had been developed. It was by far the greatest medical revolution in all human history.

**The discovery of filterable viruses**

One of the early technical contributions from Pasteur's new institute was the development of filters able to retain bacterial cells and thus to yield bacteria-free filtrates. Infectious fluids were often tested for the presence of disease-producing bacteria by passing them through such filters; if the filtrate was no longer able to produce infection, the presence of a bacterial agent in the original fluid was indicated. In 1892 a Russian scientist, D. Iwanowsky, applied this test using an infectious extract from tobacco plants infected with mosaic disease. He found to his surprise that the filtrate was fully infectious when applied to healthy plants. His specific discovery was soon confirmed, and within a few years other workers found that many major plant and animal diseases are caused by similar, filter-passing, submicroscopic agents. A whole class of infectious entities, much smaller than any previously known organisms, was thus discovered. The true nature of these *viruses*, as they came to be known, remained obscure for many decades, but eventually it was established that they are a distinctive group of biological objects entirely different in structure and mode of development from all cellular organisms (see Chapter 11).

## *The development of pure culture methods*

Pasteur possessed an intuitive skill in the handling of microorganisms and was able to reach correct conclusions about the specificity of fermentative processes, even though he undoubtedly always worked with cultures that contained a considerable mixture of microbial forms. The classical studies of Koch and Pasteur on anthrax, which firmly established the germ theory of animal disease, were conducted under experimental conditions which did not really permit certainty that rigorously pure cultures of the causative organism had been obtained. There are pitfalls for the unwary in working with mixed microbial populations, and not all the scientists who began to study microorganisms in the middle of the nineteenth century were as wary as Pasteur and Koch. It was frequently claimed that microorganisms had a very large capacity for variation with respect both to their *morphological form* and to their *physiological function*. This belief became known as the doctrine of *pleomorphism*, while the opposing belief, that microorganisms show a constancy and specificity of form and function, became known as the doctrine of *monomorphism*.

**The origin of the belief in pleomorphism**

Let us consider what happens when a nutrient solution is inoculated with a mixed microbial population. The principle of natural selection at once begins to operate, and the microbe that can grow most rapidly under the conditions provided soon predominates. As a result of its growth and chemical activities, the composition of the medium changes, and after some time, conditions no longer permit growth of the originally predominant form. The environment may now be favorable for the growth of a second kind of microorganism, also originally introduced into the medium but hitherto unable to develop. As a consequence, the second kind of organism gradually replaces the first as the predominant form in the culture. In this fashion one may obtain the *successive development of many different microbial types* in a single culture flask seeded with a mixed population. It is often possible to maintain the predominance of the form that first develops only by repeated transfer of the mixed population at short intervals into a fresh medium of the same composition; this was essentially the trick used by Pasteur in his studies on fermentation.

If one does not recognize the possibility of such microbial successions, it is easy to conclude that the chemical and morphological changes observable over the course of time in a single culture inoculated with a mixed population reflect *transformations undergone by a single kind of microorganism*. Many scientists fell into this intellectual trap, and between 1865 and 1885, claims for the extreme variability of microorganisms, based on such observations, were frequently made.

The term *pleomorphism* (derived from the Greek, meaning "doctrine of many shapes") implies that its proponents were concerned primarily with the possibilities of morphological variation. In fact, this was often not the case. Many pleomorphists insisted equally on the variability of function. For them, there was no such thing as a specific microbial agent for alcoholic fermentation or for a given disease; they considered that it is essentially the nature of the environment that determines both form and function. The widespread persistence of such beliefs represented a threat to the whole development of microbiology as a science, and they were opposed by such leaders of the new discipline as Pasteur, Koch, and Cohn; they upheld the doctrine of monomorphism, insisting on the constancy of microbial form (and function).

Around 1870 it began to be realized that a sound understanding of the form and function of microorganisms could be obtained only if the complications inherent in the study of mixed microbial populations were avoided by the use of pure cultures. *A pure culture is one that contains only a single kind of microorganism.* The leading advocates of the use of pure cultures were two great mycologists (students of fungi), A. de Bary and O. Brefeld.

**The first pure cultures**

Much of the pioneering work on pure culture techniques was done by Brefeld, working with fungi. He introduced the practice of isolating single cells, as well as the cultivation of fungi on solid media, for which purpose he added gelatin to his culture liquids. His methods of obtaining pure cultures worked admirably for the fungi but were found to be unsuitable when applied to the smaller bacteria. Other methods had, therefore, to be devised for bacteria. One of the first to be proposed was the *dilution method.* A fluid containing a mixture of bacteria was diluted with sterile medium in the hope that ultimately a growth could be obtained that took its origin from a single cell. In practice, the method is tedious, difficult, and uncertain; it also has the obvious disadvantage that at best one can only isolate in pure form the particular microorganism that predominates in the original mixture. The first man to use it for the isolation of bacteria was Lister, who in 1878 isolated Pasteur's "lactic ferment" by this means and thereby obtained for the first time a pure bacterial culture.

Koch realized very early that the development of simple methods for obtaining pure cultures of bacteria was a vital requirement for the growth of the new science. The dilution method, used successfully by Lister, was obviously too tedious and uncertain for routine use. A more promising approach had already been suggested by the earlier observations of J. Schroeter, who had noted that on such solid substrates as potato, starch paste, bread, and egg albumen, isolated bacterial growths, or *colonies,* arise. The colonies differed from one another, but within each colony the bacteria were of one type. At first Koch experimented with the use of sterile, cut surfaces of potatoes, which he placed in sterile, covered glass vessels and then inoculated with bacteria. However, potatoes have obvious disadvantages: the cut surface is moist, which allows motile bacteria to spread freely over it; the substrate is opaque, and hence it is often difficult to see the colonies; and most important of all, the potato is not a good nutrient medium for many bacteria. Koch perceived that it would be far better if one could solidify a well-tried liquid medium with some clear substance. In this fashion, a translucent gel could be prepared on which developing bacterial colonies would be clearly visible. At the same time, the varying nutritional requirements of different bacteria could be met by modifying the composition of the liquid base. With this in mind, he added gelatin as a hardening agent. Once set, the gelatin surface was seeded by picking up a minute quantity of bacterial cells (the *inoculum*) on a platinum needle, previously sterilized by passage through a flame, and drawing it several times rapidly and lightly across the surface of the jelly. Different bacterial colonies soon appeared, each of which could be purified by a repetition of the streaking process. This became known as the *streak method* for isolating bacteria. The pure cultures were transferred to tubes containing sterile nutrient gelatin that had been plugged with cotton wool and set in slanted position. Such cultures became known as *slant cultures*. Shortly thereafter, Koch discovered that instead of streaking the bacteria over the surface of the already solidified gelatin, he could mix them with the melted gelatin. When the gelatin set, the bacteria were immobilized in the jelly and there developed into isolated colonies. This became known as the *pour plate method* for isolating bacteria.

Gelatin, the first solidifying agent used by Koch, has several disadvantages. It is a protein highly susceptible to microbial digestion and liquefaction. Furthermore, it changes from a gel to a liquid at temperatures above 28°C. A new solidifying agent, *agar*, was soon introduced. Agar is a complex polysaccharide, extracted from red algae. A temperature of 100°C is required to melt an agar gel, so it remains solid throughout the entire temperature range over which bacteria are cultivated. However, once melted, it remains a liquid until the temperature falls to about 44°C, a fact that makes possible its use for the preparation of cultures by the pour plate method. It produces a stiff and transparent gel. Finally, it is a complex carbohydrate that is attacked by relatively few bacteria, so the problem of its liquefaction rarely arises. For these reasons, agar rapidly replaced gelatin as the hardening agent of choice for bacteriological work. All modern attempts to find an equally satisfactory synthetic substitute for agar have failed.

### The development of culture media by Koch and his school

Pasteur had developed simple, transparent liquid media of known chemical composition for the selective cultivation of fermentative microorganisms. For the isolation of the microbial agents of disease, different types of culture media were required, and this was the second major technical problem to which Koch and his collaborators devoted their attention. Disease-producing bacteria develop normally within the tissues of an infected host, so it seemed logical that their cultivation outside the animal body would succeed best if the medium resembled as much as possible the environment of the host tissues. This line of reasoning led Koch to adopt *meat infusions* and *meat extracts* as the basic ingredients in his culture media. *Nutrient broth* and its solid counterpart, *nutrient agar,* which are still the most widely used media in general bacteriological work, were the outcome of Koch's experiments along these lines. Nutrient broth contains 0.5 percent peptone, an enzymatic digest of meat; 0.3 percent meat extract, a concentrate of the water-soluble components of meat; and 0.8 percent NaCl, to provide roughly the same total salt concentration as that found in tissues. For the cultivation of more fastidious disease-producing organisms, this basal medium can be supplemented in various ways (e.g., with sugar, blood, or serum). Considering the specific purposes for which these media were designed, the choice of ingredients may be considered logical, although there is no evidence that the traditional inclusion of NaCl has any real value, for most bacteria are insensitive to changes in the salt concentration of their environment over a very wide range. As time went on, however, many bacteriologists came to consider that these media were universal ones, suitable for the cultivation of nearly all bacteria. This is untrue; as we shall see later, bacteria vary greatly in their nutritional requirements, and no single medium is capable of supporting the growth of more than a very small fraction of the bacteria that exist in nature.

*Figure 1.9 Sergius Winogradsky (1856–1953). Courtesy of Masson et Cie., Paris. Reprinted with the permission of the Annales de l'Institut Pasteur.*

## *Microorganisms as geochemical agents*

The study of the role played by microorganisms as agents of infectious disease became the central field of microbiological interest in the last decades of the nineteenth century. Nevertheless, some scientists carried forward the work which had been initiated by Pasteur through his early investigations on the role of microorganisms in the production of fermentations. This work had clearly shown that microorganisms can serve as specific agents for large-scale chemical transformations and indicated that the microbial world as a whole might well be responsible for a wide variety of other geochemical changes.

The establishment of the cardinal roles that microorganisms play in the biologically important cycles of matter on earth—the cycles of carbon, nitrogen, and sulfur—was largely the work of two men, S. Winogradsky and M. W. Beijerinck. In contrast to plants and animals, microorganisms show an extraordinarily wide range of physiological diversity. Many groups are specialized for carrying out chemical transformations that cannot be performed at all by plants and animals, and thus play vital parts in the turnover of matter on earth.

*Figure 1.10 Martinus Willem Beijerinck (1851–1931). Courtesy of Martinus Nijhoff, The Hague. Reprinted with permission.*

One of the most striking examples of microbial physiological specialization is provided by the autotrophic bacteria, discovered by Winogradsky. These bacteria can grow in completely inorganic environments, obtaining the energy necessary for their growth by the oxidation of reduced inorganic compounds, and using carbon dioxide as the source of their cellular carbon. Winogradsky found that there are several physiologically distinct groups among the autotrophic bacteria, each characterized by the ability to use a particular inorganic energy source; for example, the sulfur bacteria oxidize inorganic sulfur compounds, the nitrifying bacteria, inorganic nitrogen compounds.

Another discovery, to which both Winogradsky and Beijerinck contributed, was the role that microorganisms play in the fixation of atmospheric nitrogen, which cannot be used as a nitrogen source by most living organisms. They showed that certain bacteria and blue-green algae, some symbiotic in higher plants, others free-living, can use gaseous nitrogen for the synthesis of their cell constituents. These microorganisms accordingly help to maintain the supply of combined nitrogen, upon which all other forms of life are dependent.

## Enrichment culture methods

For the isolation and study of the various physiological types of microorganisms that exist in nature, Winogradsky and Beijerinck developed a new and profoundly important technique: the technique of the *enrichment culture.* It is essentially an application on a microscale of the principle of natural selection. The investigator devises a culture medium of a particular defined chemical composition, inoculates it with a mixed microbial population, such as can be found in a small amount of soil or mud, and then ascertains by examination what kinds of microorganisms come to predominate. Their predominance is caused by their ability to grow more rapidly than any of the other organisms present in the inoculum, hence the term *enrichment medium.* To take a specific example, if we wish to discover microorganisms that can use atmospheric nitrogen, $N_2$, as the only source of the element nitrogen, we prepare a medium that is *free of combined nitrogen* but which contains all the other nutrients—an energy source, a carbon source, minerals—necessary for growth. This is then inoculated with soil, placed in contact with $N_2$, and incubated under any desired set of physical conditions. Since nitrogen is an essential constituent of every living cell, the only organisms of all those present in the original inoculum that will be able to multiply in such a medium are those that can fix atmospheric nitrogen. Provided that such types are present in the soil sample, they will grow. Such experiments can be varied in innumerable ways: by modifying such factors as the carbon source, the energy supply, the temperature, and the hydrogen ion concentration. For each particular set of conditions, a particular kind of microorganism will come to predominance, provided that there are any organisms existing in the inoculum that can grow under such conditions. The enrichment culture method is thus one of the most powerful experimental tools available to the microbiologist; by its use he can isolate microorganisms with any desired set of nutrient requirements, provided that such organisms exist in nature.

## Further reading


**Books**

Bulloch, W., *The History of Bacteriology.* New York: Oxford University Press, 1960.

Brock, T. D. (editor and translator), *Milestones in Microbiology.* Englewood Cliffs, N.J.: Prentice-Hall, 1961.

Conant, J. B., *Harvard Case Histories in Experimental Science,* Vol. 2. New York: Dover, 1960.

Dobell, C., *Antony van Leeuwenhoek and His "Little Animals."* London: Staples Press, 1932.

Dubos, R., *Louis Pasteur, Free Lance of Science.* Boston: Little, Brown, 1950.

Large, E. C., *Advance of the Fungi.* London: Jonathan Cape, 1940.

**Reviews**

van Niel, C. B., "Natural Selection in the Microbial World." *J. Gen. Microbiol.* **13,** 201 (1955).

# *Chapter two*
# *The nature of the microbial world*

*The term "microorganism"* is applicable to any organism of microscopic dimensions. It does not have the precise taxonomic significance of such terms as "vertebrate" or "flowering plant," each of which defines a restricted biological group with common structural and functional properties. Microorganisms occur in many taxonomic groups, some of which (for example, the algae) also contain members too large to fall into the microbial category. In this chapter, we shall characterize the major taxonomic groups that consist, in whole or in part, of microorganisms. First, however, it may be useful to review briefly some general biological principles.

## *The common properties of biological systems*

A mammal, a flowering plant, and a bacterium are all organisms, but superficial examination does not suggest that they have much in common. It is only when we start to examine the living world in terms of its *chemical* and *microscopic organization* that the initial impression of diversity gives way to a realization of its fundamental unity. The superficial diversity becomes understandable as the consequence of the operation of the evolutionary process on a complex structure, the *cell,* which probably had a single origin and which still preserves many common features.

What are the common features of biological systems? First, all organisms share a *common chemical composition,* as shown by the invariable presence of three types of complex organic macromolecules: *protein, deoxyribonucleic acid,* and *ribonucleic acid.* Deoxyribonucleic acid (DNA) is the cellular substance that carries in coded form all the information required to determine the specific properties of the organism. The information embodied in this *genetic message* is translated, through the intermediacy of ribonucleic acid (RNA), into specific patterns of protein synthesis. The proteins serve as the catalysts or *enzymes* responsible for all the varied operations of the cell, and it is the specific pattern of protein synthesis characteristic of an organism that ultimately determines its distinctive gross properties.

In addition, organisms all perform certain *common chemical activities,* known collectively as *metabolism* or *metabolic activities.* There are, naturally, many differences in metabolic detail among the different kinds of organisms. However, all organisms are obliged to synthesize the universal constituents of living matter from external chemical building blocks and to generate the energy necessary for such synthetic activities. With respect to these fundamental features of metabolism, the resemblances between organisms are much more marked than are the differences.

Finally, organisms share a *common physical structure,* being organized into microscopic subunits known as *cells.* As a consequence of their cellular organization, growth results in *cell division,* with an increase in the total number of cells.

These generalizations apply to all biological groups except viruses, a class of very small entities in which the cell is not the basic unit of structure. The special properties of viruses will be taken up at the end of this chapter.

## *The cell*

**The common properties of cells**

The region of a cell in which the genetic information is principally localized is called the *nucleus.* The surrounding region is called *cytoplasm.* The nucleus, which is surrounded by the cytoplasm, contains all or most of the DNA of the cell. It accordingly serves as the carrier of the genetic plan and as the information center for direction of cellular synthesis. The cytoplasm contains most of the RNA and protein of the cell. It is separated from the external environment by a delicate membrane, known as the *cytoplasmic,* or *cell, membrane.* This membrane, composed of a mosaic of proteins and lipids, is the barrier through which all materials from the external environment must pass in order to enter the cell. It is endowed with the property of *semipermeability;* that is, it allows the passage of some substances into the cell and excludes others. At the same time, it allows the exit of the waste products of cellular metabolism, while keeping within the cell the various kinds of molecules, large and small, that are needed for the maintenance of cellular function.

In some cells, the membrane is the only bounding structure; this is true of animals and of many microorganisms. In plants and in certain groups of microorganisms, the cytoplasmic membrane is surrounded by a much thicker and sturdier structure with a different chemical composition, known as the *cell wall.* Cell walls are not endowed with semipermeability and thus take no part in regulating the selective transport of materials into or out of the cell. They also appear to be devoid of enzymatic function. Their primary role appears to be a protective one; they provide protection from mechanical injury, and in particular they prevent osmotic rupture of the membrane which is likely to occur when the cell is placed in an environment with a high water content, owing to the free movement of water through the semipermeable membrane. In the case of cells that are surrounded by walls, the enclosed functional unit is often referred to as a *protoplast.*

**The origin of cells**

A cell arises only by the *divison of a preexisting cell* or, in the special case of *sexual reproduction,* by the *fusion of two preexisting cells.*

**Types of cellular organization**

The simplest organisms are those in which the whole organism consists of a single cell. Since cells are of microscopic dimensions, such *unicellular organisms* are necessarily small and fall in the general category of *microorganisms.* Unicellular organization is common, though not universal, in the groups known as *bacteria, protozoa,* and *algae;* it also occurs, though more rarely, among the *fungi.* There are very considerable differences between the different groups of unicellular micro-

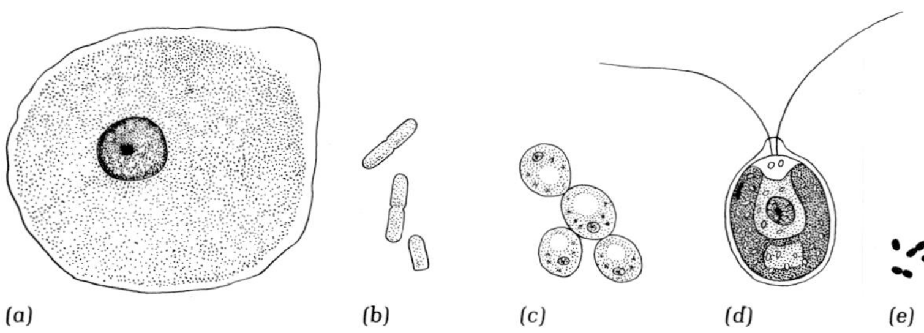


*Figure 2.1 Drawings of several unicellular microorganisms on the same relative scale. (a) An ameba; (b) a large bacterium; (c) a yeast; (d) a flagellate alga; (e) a small bacterium (×1,000).*

organisms; such differences are based entirely on variations in the *size*, *form*, and *internal structure* of the individual cell which here serves as the complete organism. Some typical unicellular organisms, drawn to the same scale, are shown in Figure 2.1.

A more complex type of biological organization is *multicellularity*. Although multicellular organisms usually arise from a single cell, they consist in the mature state of many cells, permanently attached to one another in a characteristic way, which confers on the organism its external form. When a multicellular organism contains a relatively small total number of cells, it may still be of microscopic dimensions; many examples exist among the bacteria and the algae. These simple multicellular organisms are commonly composed of similar cells, arranged in the form of a unidimensional filament, and show little increase in structural complexity over unicellular forms.

However, when the number of cells in a multicellular organism becomes sufficiently great, the organism acquires a certain degree of structural complexity, derived simply from the manner in which the similar structural units are linked together. The best examples of this type of multicellular organization can be found among the more highly developed algae such as the seaweeds, which have a characteristic plantlike macroscopic form (Figure 2.2). However, microscopic examination shows that in many seaweeds with a quite complex superficial form

*Figure 2.2 A simple multicellular organism, the brown alga Dictyosiphon. Despite the superficially plantlike structure, this organism has essentially no internal cellular differentiation, as can be seen in the cross section at lower left. After G. M. Smith, Cryptogamic Botany, Vol. 1. New York: McGraw-Hill, 1938.*

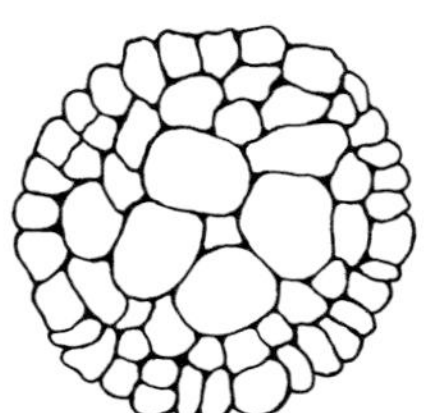

there is little or no specialization of the component cells; the form is conferred by a specific pattern of arrangement of like structural units.

A much greater intrinsic structural complexity within the framework of multicellularity is found in higher plants and animals, owing to the development of distinct *tissue regions* that differ from one another with respect to the *kinds of cells* of which they are composed. A further level of internal complexity may be achieved by the combination of different tissues into a specialized local structure known as an *organ*. The liver is an animal organ, the leaf a plant organ; both consist of a number of different tissues, and both are differentiated by their whole form and cellular structure from other local regions of the organism in which they occur (Figure 2.3). The structural complexity of a higher

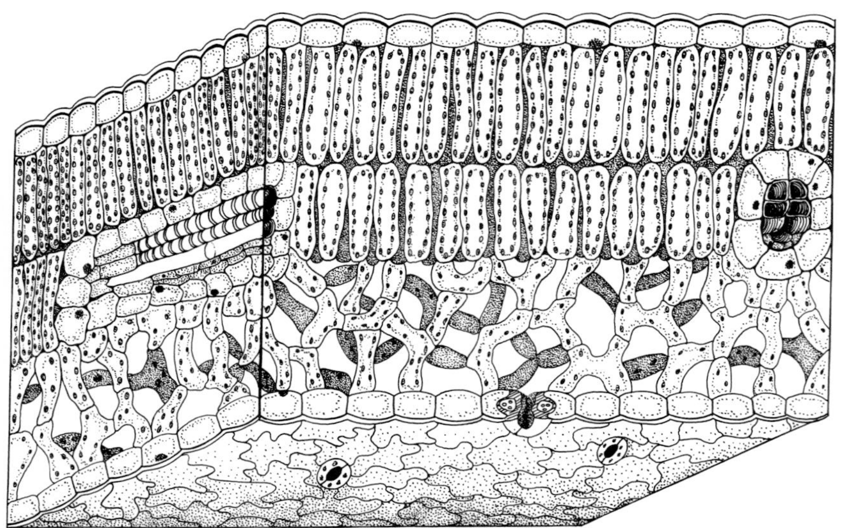

*Figure 2.3 A magnified cross section of a plant organ, the leaf, illustrating multicellular construction based on the internal differentiation of a variety of cell types, which is characteristic of higher plants and animals. After C. L. Wilson and W. E. Loomis, General Botany. New York: Dryden Press, 1957.*

plant or animal thus proves on microscopic analysis to be vastly greater than that of such a large multicellular organism as a seaweed, which is not highly differentiated. Higher organisms develop from a single cell (the fertilized ovum); the ultimate complexity of the adult form arises from the *differentiation of numerous cell types* during growth and development.

In a few groups of organisms, there is a third type of organization, known as *coenocytic structure*, which at first sight seems to contradict the axiom that organisms are composed of cells. In the customary description of a cell, there is generally the statement that it contains a

Figure 2.4 *A coenocytic organism, the green alga Bryopsis. After G. M. Smith, Cryptogamic Botany, Vol. 1. New York: McGraw-Hill, 1938. Despite its plant-like form (a) this organism is not constructed from separate cells. As is shown by the magnified cross section of the apical portion (b), the cytoplasm is continuous.*

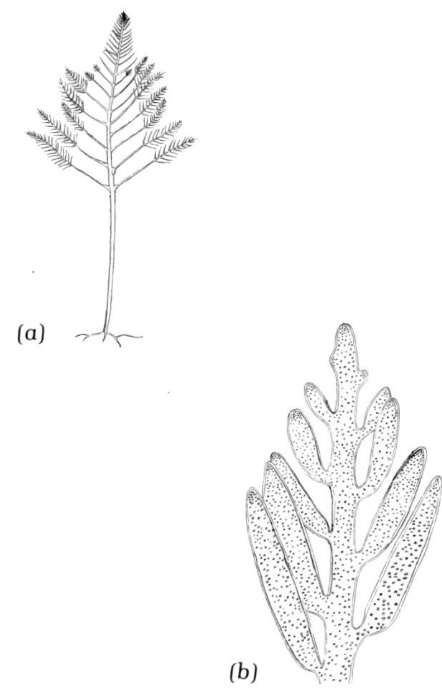


(single) nucleus. Coenocytic organisms are *multinucleate*. When examined microscopically, they are found to consist of a large mass of cytoplasm containing many nuclei and are not separated into regular uninucleate cellular units by means of a system of bounding membranes as is a multicellular organism (Figure 2.4). However, if we follow the *development* of a coenocytic organism, we invariably find that at some stage of its life history it presents a truly cellular structure, the coenocytic organization arising secondarily. One road to coenocytic organization, known as *plasmodial development*, is most clearly exemplified by a group known as the slime molds. A slime mold starts out as a single cell, but instead of undergoing successive cellular divisions, the cell simply increases in bulk, and repeated nuclear divisions take place. A similar mode of development is found in the true fungi and in some seaweeds.

**Primary nutritional categories**

Organisms can be subdivided into two major groups in terms of the energy source that they use to support biosynthesis and growth. Ultimately, of course, the maintenance of all life on earth is dependent on the steady flux of energy that our planet receives from the sun. However, only *photosynthetic organisms* can directly use this solar radiant energy. The utilization of radiant energy, the process of *photosynthesis*, involves two sequential events. The first is the trapping or absorption of certain wavelengths of light, which requires an appropriate array of cellular pigments, known as the *photosynthetic pigment system;* at least one pigment of the chemical class known as *chlorophylls* is an invariable component of this system. The second is the conversion of the absorbed radiant energy into a form of chemical bond energy that can be used to drive the synthetic activities of the cell. Most photosynthetic organisms have simple nutrient requirements, which can be met by the absorption of all required elements in inorganic form. The requirement for carbon (the major element of living matter) is met by carbon dioxide, the most oxidized form of carbon; such photosynthetic organisms are termed *photoautotrophs*. Some of the photosynthetic bacteria use organic compounds, rather than $CO_2$, as the principal carbon source; these are termed *photoheterotrophs*.

All other organisms are dependent on the availability of a suitable *chemical* energy source to meet their energetic needs. A few bacteria, the *chemoautotrophs,* can obtain energy by the oxidation of reduced *inorganic* compounds; like photoautotrophs, they have simple nutrient requirements, purely inorganic. With this minor exception, however, nonphotosynthetic organisms are *chemoheterotrophs,* which require *organic compounds as energy sources.* Since photosynthetic organisms (and chemoautotrophs) are the only net producers of organic matter on earth, it is they who ultimately provide, either directly or indirectly, the organic forms of energy required by all other organisms.

Nonphotosynthetic organisms that require organic nutrients can be further subdivided on the basis of *the way in which these nutrients are obtained from the environment.* Fungi and bacteria absorb all their nutrients in dissolved form: this is termed *osmotrophic nutrition.* Most protozoa ingest their food in the form of solid particles; the ingested material is subsequently made soluble by internal digestion. This is termed *phagotrophic nutrition.*

## *The major divisions of the living world*

It is a judgement of common sense as old as mankind that the earth is inhabited by two different kinds of organisms, plants and animals. Early in the history of biology, this intuitive judgement was scientifically formalized by the division of the living world into two *kingdoms,* plants and animals. The distinction between the two kingdoms can be based on many characters, both anatomical and functional (Table 2.1). It is satisfactory as long as one deals with the more highly differentiated, multicellular groups. Even such oddities as nonphotosynthetic flowering plants are not difficult to place, since their many other "plantlike" characters leave no doubt as to their correct assignment.

*Table 2.1 Principal differences between higher animals and higher plants*

| | | **Plants** | **Animals** |
|---|---|---|---|
| *Physiological differences* | *Energy source* | *Photosynthesis* | *Organic materials* |
| | *Chlorophyll* | *Present* | *Absent* |
| | *Principal reserve food* | *Starch* | *Glycogen, fat* |
| | *Active movement* | *Absent* | *Present* |
| *Structural differences* | *Cell walls* | *Present* | *Absent* |
| | *Mode of growth* | *Open*[a] | *Closed*[a] |

[a]In animals the individual achieves a more or less fixed size and form as an adult. In most plants growth continues throughout the life of the individual, and the final size and form are much less rigidly fixed.

## The place of microorganisms

When exploration of the microbial world got under way in the eighteenth and nineteenth centuries, there seemed no reason to doubt that these simple organisms could be distributed between the plant and animal kingdoms. In practice, the assignment was usually made on the basis of the most easily determinable differences between plants and animals: the power of active movement and the ability to photosynthesize. Multicellular algae, which are immotile, photosynthetic, and in some cases plantlike in form, found a natural place in the plant kingdom. Although they are all nonphotosynthetic, the coenocytic fungi were also placed in the plant kingdom on the basis of their general immotility. Microscopic motile forms were lumped together as one group of animals, the Infusoria (Table 2.2).

*Table 2.2 Early attempts (about 1800) to allocate microorganisms to the plant and animal kingdoms*

| Plants | Animals |
|---|---|
| *Algae (immotile, photosynthetic)* | *Infusoria (motile)* |
| *Fungi (immotile, nonphotosynthetic)* | |

Following the enunciation and acceptance of the cell theory, in about 1840, biologists perceived that the Infusoria were, in fact, a very heterogeneous group in terms of their cellular organization. Some of these microscopic forms (e.g., the rotifers) are invertebrate animals, with a body plan based on differentiation during multicellular development. Furthermore, the strictly unicellular representatives can be subdivided into two groups: *protozoa*, with relatively large and complex cells, and *bacteria*, with much smaller and simpler cells. The old Infusoria was thus split three ways. Some of its component groups were classified as metazoan (multicellular) invertebrate animals. Others, the protozoa, were kept in the animal kingdom, but differentiated from all other animals on the basis of their unicellular structure. Finally, the bacteria were transferred to the plant kingdom, despite their generally nonphotosynthetic nature, as a result of the discovery that one algal group, the blue-green algae, was characterized by cells with a comparably simple structure.

However, subsequent experience showed that the treatment of the protozoa (a large and complex microbial group) as unicellular animals led to considerable difficulties. Such protozoa as the ciliates and amebae, phagotrophic organisms devoid of cell walls, could be fitted quite

*Table 2.3 Final effort (about 1860) to allocate microorganisms to the plant and animal kingdoms*

| **Plants** | **Contested groups** | **Animals** |
|---|---|---|
| | | *Small metazoans* |
| | | *Rotifers* |
| | | *Nematodes (some)* |
| | | *Arthropods (some)* |
| *Algae (photosynthetic)* | | *Protozoa* |
| | | *Ciliates* |
| *Immotile forms* ← | *Photosynthetic flagellates* | → *Nonphotosynthetic flagellates* |
| *Fungi (nonphotosynthetic)* | | |
| *True fungi* ← | *Slime molds* | → *Ameboid protozoa* |
| *Bacteria* | | |

satisfactorily into the confines of the animal kingdom, but other protozoa could not. On closer study, the flagellate protozoa proved to be a very odd assortment of creatures, in some of which motility by means of flagella was the only "animallike" character. Some possessed cell walls, others did not. Some were phagotrophic, some osmotrophic, some photosynthetic. Accordingly, this one microbial group shows all possible combinations of plantlike and animallike characters, often manifested by a single species. The problem of the placement of the flagellates became even more acute when it was recognized that in terms of cellular properties many of the photosynthetic flagellates resembled very closely certain of the multicellular, immotile algae. Another protozoan group, the slime molds, also presented difficulties. In the vegetative state, these organisms are phagotrophic and ameboid, but they can also form complex fruiting structures, similar in size and form to those characteristic of the true fungi. Should the slime molds be classified with the fungi, as plants, or with the protozoa, as animals?

Consequently, as knowledge of the properties of the various microbial groups deepened, it became apparent that at this biological level a division of the living world into two kingdoms cannot really be maintained on a logical and consistent basis. Some groups (notably the photosynthetic flagellates and the slime molds) were claimed both by botanists as plants and by zoologists as animals (Table 2.3). The problem is easy enough to understand in evolutionary terms. The major microbial groups can be regarded as the descendants of very ancient evolutionary lines, which antedated the emergence of the two great lines that eventually led to the development of the more complex plants and animals. Hence most microbial groups cannot be pigeonholed in terms of the properties that define these two more advanced evolutionary groups.

**The kingdom of protists**

Dissatisfaction with existing classification, coupled with a clear understanding of the root of the trouble, led one of Darwin's disciples, E. H. Haeckel, to propose the obvious way out. In 1866 he suggested that logical difficulties could be avoided by the recognition of a *third* kingdom, the *protists,* to include protozoa, algae, fungi, and bacteria. The protists accordingly include both photosynthetic and nonphotosynthetic organisms, some plantlike, some animallike, some sharing properties specific to both the traditional kingdoms. What distinguishes all protists from plants and animals is their *relatively simple biological organization.* Many protists are unicellular or coenocytic; and even the multicellular protists (e.g., the larger algae) lack the internal differentiation into separate cell types and tissue regions characteristic of plants and animals. A primary division in the biological world can accordingly be best made in terms of *level of biological organization;* this can then be followed, for the more highly organized forms, by a secondary division on the basis of the properties long used to separate plants from animals (Table 2.4).

*Table 2.4 Component groups of the three kingdoms of organisms*

| | **Plants** | **Animals** |
|---|---|---|
| *Multicellular, showing extensive differentiation* | *Seed plants*<br>*Ferns*<br>*Mosses and liverworts* | *Vertebrates*<br>*Invertebrates* |
| | **Protists** | |
| *Unicellular, coenocytic, or multicellular; without differentiation* | *Algae*<br>*Protozoa*<br>*Fungi*<br>*Bacteria* | |

**The internal division of the protists: eucaryotic and procaryotic groups**

Within the past 20 years, the perfection of the electron microscope as an instrument for studying fine details of cellular structure has revolutionized our knowledge about the way cells are organized. This has led to the recognition that there are two different kinds of cells among organisms. The more highly differentiated kind, the *eucaryotic cell,* is the unit of structure in plants, animals, protozoa, fungi, and most groups of algae. The less differentiated *procaryotic cell* is the unit of structure in bacteria and blue-green algae. The protists can accordingly be subdivided into *eucaryotic protists*, with a cellular ground plan similar to that of plants and animals, and *procaryotic protists*, with a different cellular ground plan. In the following paragraphs, we shall summarize the principal differences between the eucaryotic and procaryotic cells.

## *Ground plans of eucaryotic and procaryotic cells*

Figures 2.5 and 2.6 show electron micrographs of thin sections of a typical eucaryotic cell and a typical procaryotic cell. The relatively high degree of internal complexity of the eucaryotic cell is immediately apparent.

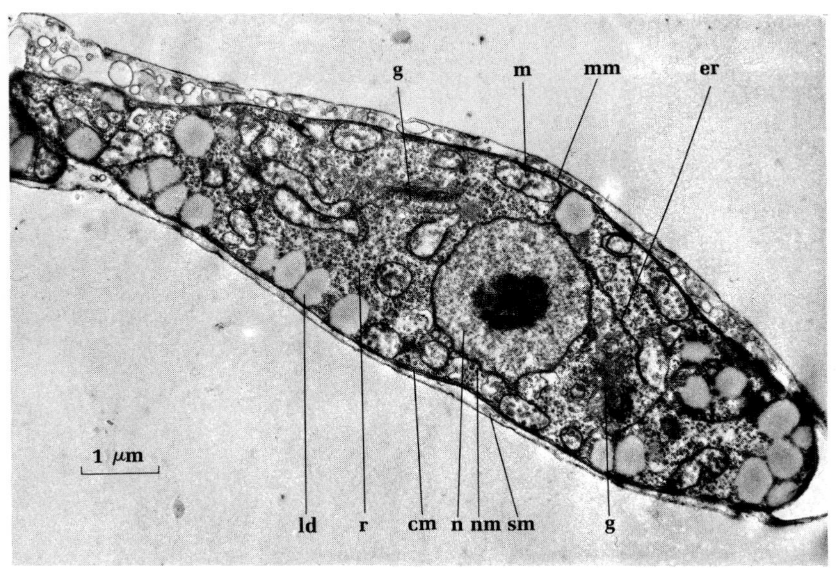


*Figure 2.5 Electron micrograph of a thin section of a nonphotosynthetic eucaryotic protist, Labyrinthula, to show the typical ground plan of a relatively undifferentiated eucaryotic cell (×10,660). The cell is devoid of a wall, but it is surrounded by a loose extracellular slim matrix (sm): er, endoplasmic reticulum; g, Golgi bodies; m, mitochondrion; mm, outer mitochondrial membranes; n, resting nucleus; nm, nuclear membrane; r, ribosomes; cm, cell membrane; ld, large lipid droplets. Courtesy of John Waterbury and Stanley Watson.*

Both kinds of cells are bounded by cytoplasmic membranes with similar profiles, resolvable at high magnification as triple-layered structures about 80 A* thick. Membranes with this fine structure and thickness, which are universal components of cells, are known by the general name *unit membranes*. In eucaryotic cells, certain internal regions are also bounded by unit membranes, separated from the outer cyto-

*The angstrom unit (A) is the unit of length most commonly used in description of biological fine structure. It is equal to $10^{-4}$ micrometer (μm), or $10^{-8}$ cm.

plasmic membrane. The regions include the *nucleus*, in its resting state; the *mitochondria*, organelles that are the sites of cellular respiration; the *chloroplasts*, sites of photosynthesis in eucaryotic photosynthetic organisms; *lysosomes*, which contain various hydrolytic or digestive enzymes; and *vacuoles*, liquid-filled bodies which have various functions, to be described later. In procaryotic cells, there are no such unit-membrane-bounded internal compartments; *the cytoplasmic membrane is the only bounding membrane of the cell.* In the procaryotic cell there are, accordingly, only two major internal regions, the *nucleus* and the *cytoplasm*, each relatively uniform in fine structure and not separated from one another by a membrane.

The principal structural elements observable with the electron microscope in the cytoplasmic region of a procaryotic cell are ribonucleoprotein particles, about 100 A in diameter, known as *ribosomes*. They are the sites of cellular protein synthesis, and their structure and role in this process will be described later. The ribosomes of the procaryotic cell have a particle weight of approximately 2.7 million and are termed *70-S ribosomes*, a name derived from their characteristic sedimentation rate in the ultracentrifuge.

Leaving aside for the moment its membrane-bounded inclusions, the cytoplasmic region of a eucaryotic cell shows a much more varied fine structure when examined in thin section with the electron microscope. It is traversed by a membrane system, known as the *endoplasmic reticulum*, which is continuous with the cell membrane. Many ribosomes, which are typically less abundant than in a procaryotic cell, tend to be aligned on the surfaces of the endoplasmic reticulum. Although eucaryotic cytoplasmic ribosomes resemble procaryotic ribosomes in composition and structure, they are about 50 percent larger, and hence tend

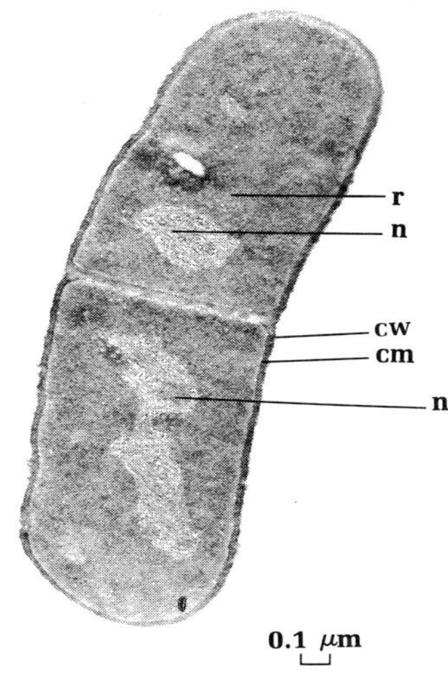


*Figure 2.6 Electron micrograph of a thin section of a nonphotosynthetic procaryotic protist, Bacillus subtilis (×22,200). The dividing cell is surrounded by a relatively dense wall (cw) overlying the cell membrane (cm). Within the cell, portions of the nonmembrane-bound nucleus (n) are distinguishable by their fibrillar structure. The cytoplasmic region is densely filled with ribosomes (r). Courtesy of C. F. Robinow.*

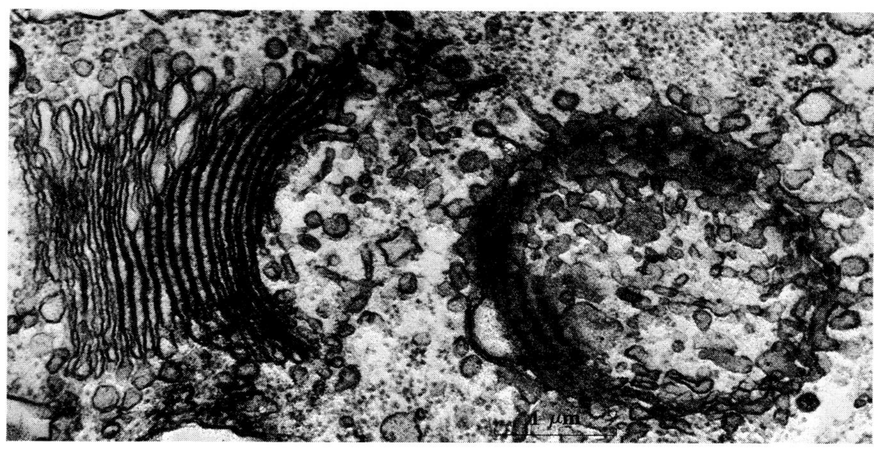

*Figure 2.7 The Golgi apparatus as seen in an electron micrograph of a thin section of Euglena gracilis (×28,800). Two adjacent Golgi bodies have been sectioned in different planes. At left, vertical section through the membrane stack; at right, section parallel to the stack. Courtesy of Gordon F. Leedale.*

to sediment more rapidly in the ultracentrifuge. For this reason, they are termed *80-S ribosomes.*

Another distinctive structure in the cytoplasm of eucaryotic cells, which never occurs in procaryotic cells, is a complex membranous body known as the *Golgi apparatus*, or *dictyosome* (Figure 2.7). It is a disc-shaped body, the numerous membranes of which lie closer to one another than those of the endoplasmic reticulum; and, at one surface, vacuoles are produced. The functions of the Golgi apparatus were unknown until recently, but it now appears to serve as a packaging structure, within the vacuoles of which various kinds of cellular products are enclosed and transported. Lysosomes appear to arise from the Golgi apparatus. In a number of specialized eucaryotic cells, other kinds of materials are formed and transported to the cell surface in Golgi vacuoles. These include the nematocysts (or stinging organelles) of coelenterates, the surface scales which cover some algae, and various secreted cellular products.

## The structural basis of respiration and photosynthesis

Chloroplasts and mitochondria, the respective sites of photosynthetic and respiratory function in eucaryotic cells, have relatively elaborate fine structures (Figures 2.8 and 2.9). Each kind of organelle contains a characteristically arranged system of *internal membranes*, dispersed

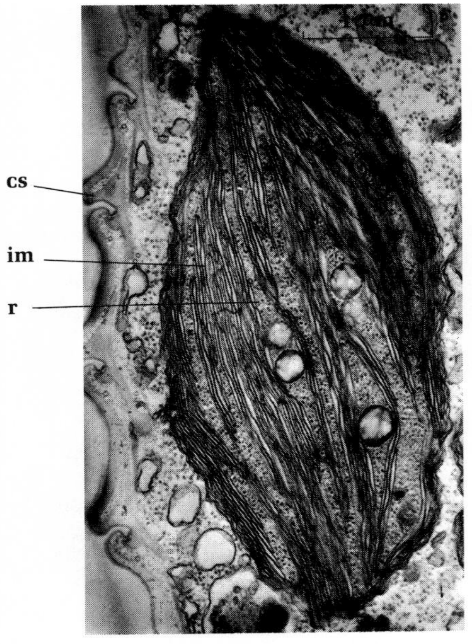


(a)

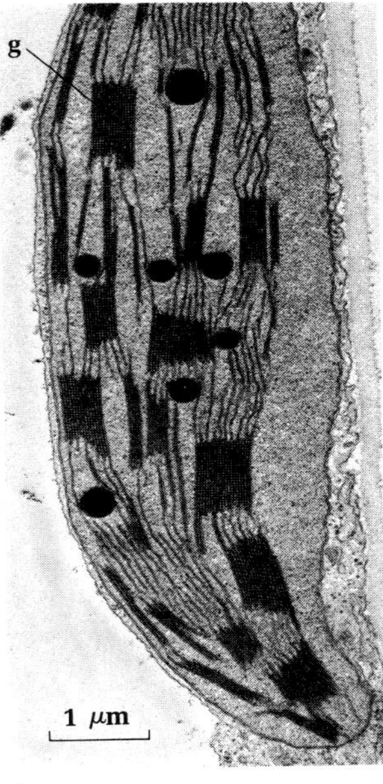


(b)

*Figure 2.8 The structure of chloroplasts as revealed in electron micrographs of thin sections of eucaryotic cells. (a) Chloroplast of the unicellular alga Euglena (×21,200). The internal membranes (im) are arranged in irregular parallel groups and run in the long axis of the chloroplast. Ribosomes (r) are scattered between the lamellae. The chloroplast lies just below the sculptured cell surface (cs). From G. F. Leedale, B. J. D. Meeuse, and E. G. Pringsheim, "Structure and physiology of Euglena spirogyra," Arch. Mikrobiol. 50, 68 (1965). (b) Chloroplast of a sugar beet leaf (×14,840). The internal membranes tend to be arranged in dense, regular stacks, termed grana (g), in the chloroplasts of plants. Courtesy of W. M. Laetsch.*

through a less-structured internal region. Both photosynthesis and respiration are very complex metabolic processes and involve the operation of many different enzymes. Biochemical analyses made on isolated chloroplasts and mitochondria have shown that their internal

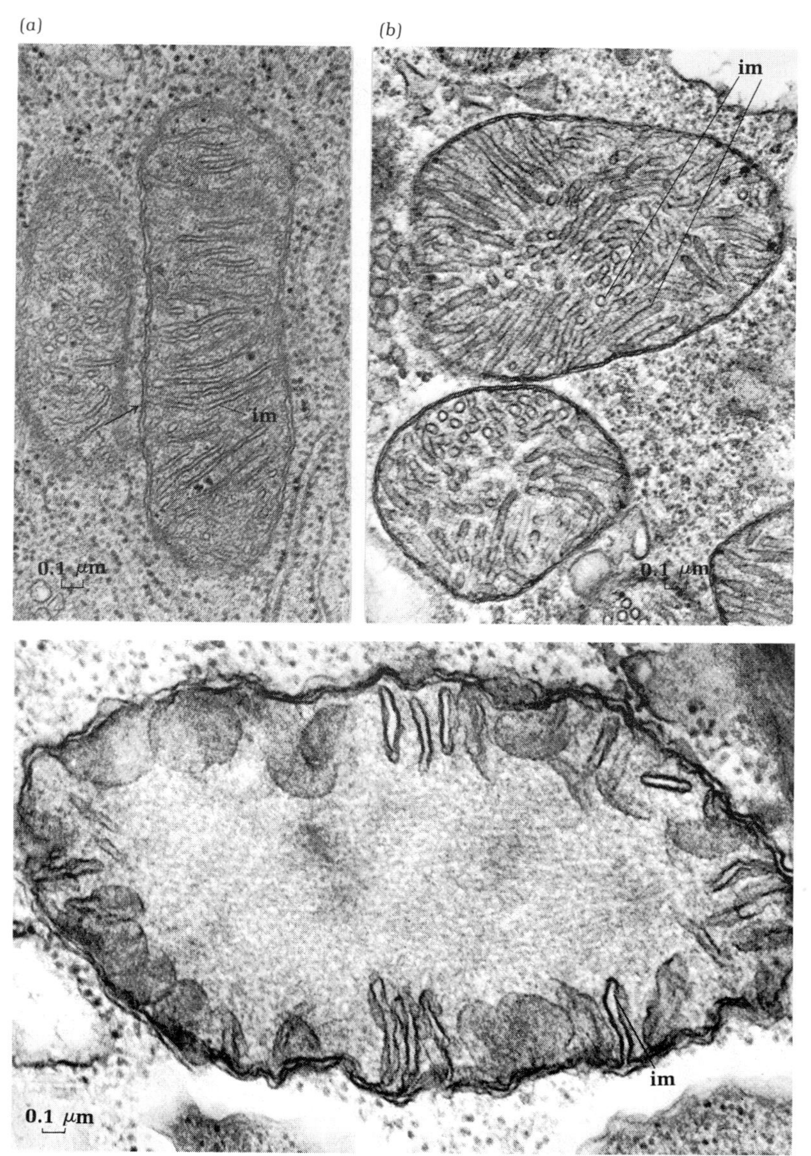


*Figure 2.9 The structure of mitochondria as seen in electron micrographs of thin sections of eucaryotic cells. (a) Mitochondria in a mammary gland cell of the mouse (×51,000). Numerous flattened internal membranes (im) arise by invagination from the inner enclosing membrane of the organelle (arrow). Courtesy of Dorothy Pitelka. (b) Mitochondria of a ciliate, Condylostoma (×51,000). The internal membranes (im) are tubular in cross section, and are very abundant. Courtesy of Dorothy Pitelka. (c) Mitochondrion of a photosynthetic flagellate, Euglena (×59,500). The flattened internal membranes (im) are less numerous and less extensively intruded than in (a) and (b). Courtesy of Gordon F. Leedale.*

structural differentiation reflects an internal differentiation of function, certain components of the overall metabolic machinery being specifically located on the internal membranes of the organelle. In mitochondria, the membranes carry the enzymes responsible for the transport of electrons from the organic substrates to oxygen and for the generation of energy; in chloroplasts, they bear the photosynthetic pigments, together with the enzymes responsible for energy conversion. Enzymes that mediate other parts of the metabolic processes (conversion of $CO_2$ to sugars in photosynthesis, conversion of organic substrates to $CO_2$ in respiration) are not physically linked to the membrane systems and are presumably localized in the nonmembranous internal portions of the organelles.

Chloroplasts and mitochondria are not simply the cellular sites of metabolic functions. Each carries in addition a small amount of genetic material (DNA), which specifies some of the properties of the organelle, together with a protein-synthesizing system for the translation of this genetic message. There are accordingly ribosomes within chloroplasts and mitochondria as well as in the surrounding cytoplasm. The ribosomes of chloroplasts and mitochondria are of the 70-S, or procaryotic, type, thus differing physically from the 80-S cytoplasmic ribosomes. The existence of special organelles for the performance of respiration and photosynthesis in eucaryotic cells thus also implies a dispersion of the genetic system and of the machinery for protein synthesis.

*Figure 2.10 The mitotic apparatus in a living cell. First mitotic division in a fertilized starfish egg, enclosed within the fertilization membrane. Photomicrograph taken with polarized light, which reveals the spindle (bright area), as a result of the birefringence of its constituent microtubules. Courtesy of S. Inoue.*

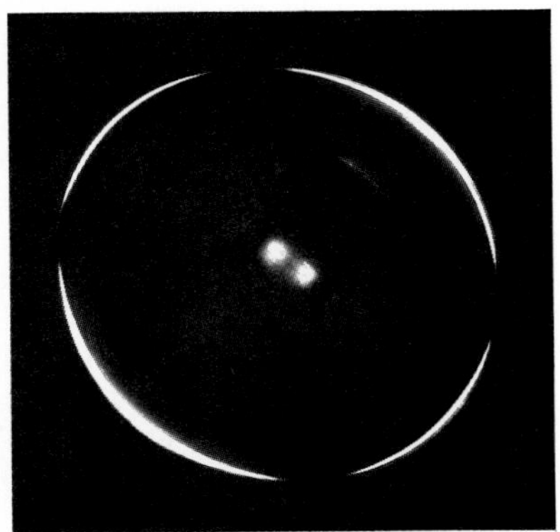

Chloroplasts and mitochondria are never assembled from their constituent parts within the eucaryotic cell: they always arise by the division of a preexisting organelle of the same type. Their maintenance in a cell line is therefore dependent on the transmission of the intact organelle to the daughter cells at the time of cell division. Accordingly, chloroplasts and mitochondria are endowed with a certain degree of genetic autonomy and continuity. The loss of the chloroplast in an originally photosynthetic cell line is completely irreversible and results in a permanent loss of photosynthetic ability. Such a loss is, however, not necessarily lethal, provided that the organism in which it has taken place can use organic compounds as chemical energy sources; nonphotosynthetic flowering plants illustrate the survival of biological lines that have undergone this loss. The same phenomenon has occurred in many groups of photosynthetic flagellates, giving rise to nonphotosynthetic counterparts. This is a point to which we shall return in Chapter 4. The loss of mitochondria seems to have been a much rarer evolutionary event; these organelles are almost universally present in eucaryotic cells, both photosynthetic and nonphotosynthetic, since respiration is in general an important cell function. The only eucaryotic organisms in which mitochondria appear to be absent are certain anaerobic protozoa and fungi which obtain energy exclusively by the fermentation of organic compounds.

The structural basis of respiratory and photosynthetic function in procaryotic cells is entirely different. As a general rule, the membrane-associated machinery of both metabolic processes is housed in the cytoplasmic membrane or in infoldings derived from it. The enzymes that mediate other parts of each metabolic process are located in the cytoplasm. Hence, the procaryotic cytoplasmic membrane is functionally much less specialized than the eucaryotic cytoplasmic membrane. It controls the passage of materials into and out of the cell, but, in addition, it performs certain metabolic functions which, in eucaryotic cells, are internalized within chloroplasts and mitochondria.

**The genetic organization of the eucaryotic cell**

The resting, or *interphase*, nucleus of a eucaryotic cell is enclosed by a *nuclear envelope*, consisting of a unit membrane in which there are numerous pores about 400 to 500 A in diameter. The genetic material is carried within this envelope on a number of different structural subunits, the *chromosomes*. Each chromosome in the interphase nucleus is a very long thread, 200 to 300 A wide, and consists of DNA linked to a special kind of basic protein, known as a histone. The orderly transmission of the genetic message at the time of cell division requires that each chromosome be replicated, so that a complete chromosome set can pass to each daughter cell. This is achieved by a complex process of nuclear division known as *mitosis*. The events of mitosis show many special variations in different eucaryotic groups, which cannot be described in the very brief general account given below.

The replication of the individual chromosomes occurs in the resting nucleus, before the onset of division. As mitosis starts, the nuclear membrane usually dissolves, and a spindle-shaped system consisting of numerous microtubules, the *mitotic apparatus* (Figure 2.10), is rapidly synthesized in the nuclear region. In many eucaryotic cells, the mitotic spindle is organized between two cylindrical bodies about 5,000 A long and 1,500 A wide, known as *centrioles* (Figure 2.11). Each microtubule is a hollow structure about 200 A wide. At the same time, the chromosomes shorten and thicken as a result of extensive coiling, so that they become individually visible with the light microscope (Figure 2.12), and become aligned in the equatorial plane of the mitotic spindle. Thereafter, one set of daughter chromosomes is withdrawn to each pole of the spindle, and in the polar regions, two new interphase nuclei are reorganized, through dissolution of the spindle microtubules, uncoiling of the chromosomes, and the elaboration of a new nuclear envelope.

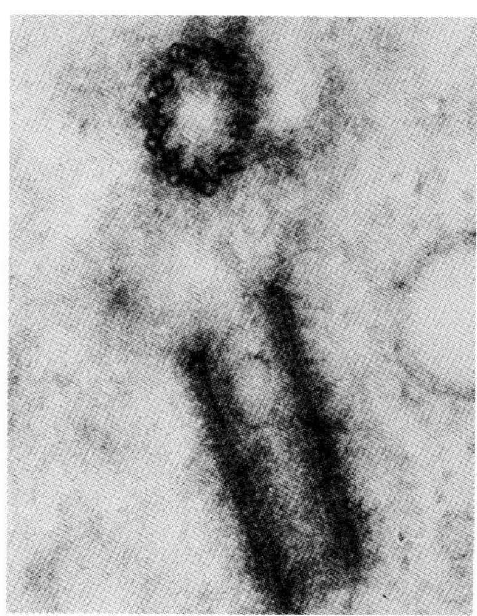

*Figure 2.11 Electron micrograph of a pair of centrioles in a dividing human lymphosarcoma cell (×88,350). One is sectioned transversely, the other longitudinally, revealing the typical hollow cylindrical structure of this organelle. Courtesy of G. Bernhard, Institut de Recherches sur le Cancer, Villejuif, France.*

Figure 2.12 Photomicrographs illustrating three successive phases of a mitotic nuclear division. (a) Early organization of the spindle; the nuclear membrane has disappeared, and the chromosomes are already visible in the region of the organizing spindle. (b) The spindle is now fully developed, and the chromosomes are regularly aligned in its equatorial plane; this stage is often referred to as the metaphase of mitosis. (c) Separation of the two daughter sets of chromosomes has occurred, and each set is being withdrawn toward one pole of the spindle.

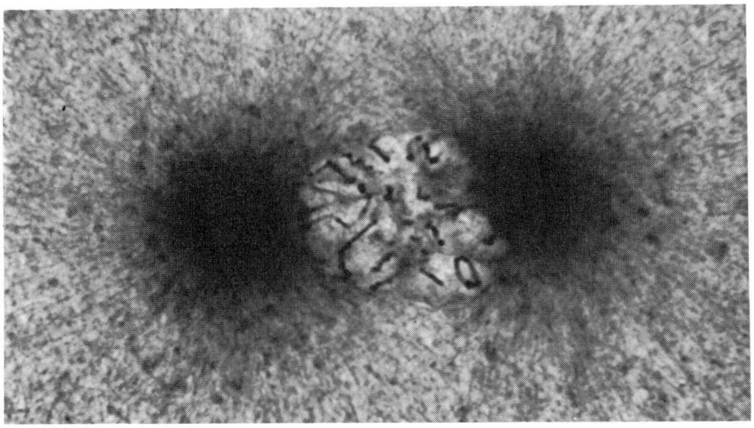

(a)

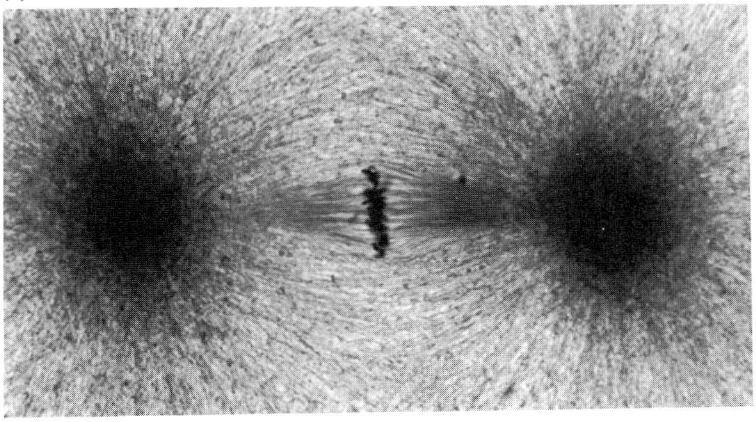

(b)

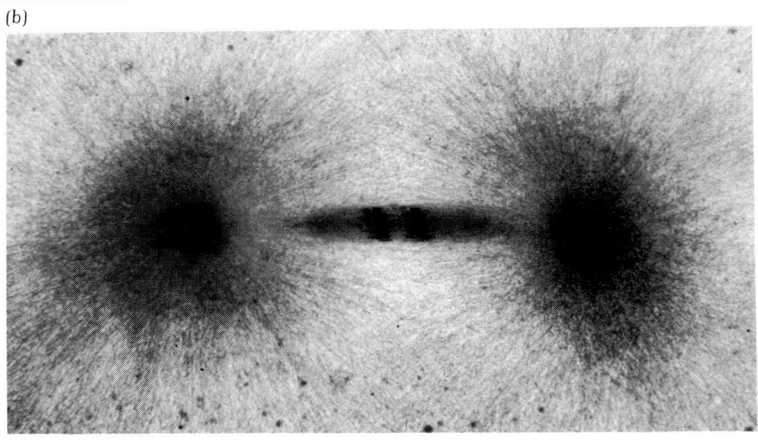

(c)

Except in coenocytic eucaryotic organisms, nuclear division is very closely linked with the ensuing process of cell division; in fact, the position and plane of cell division are determined by the mitotic apparatus, in the sense that the equatorial plane of the spindle is the site at which the subsequent cell division takes place. As a result of this close linkage between nuclear and cell division, each cell of a eucaryotic organism characteristically has a single nucleus.

**Eucaryotic sexual processes**

In the definition of the cell it was stated that cells can arise not only by division of a preexisting cell but by the fusion of two preexisting cells. Such cellular fusion is the first step in the process of *sexual reproduction*. The two cells that participate are known as *gametes* and the resulting fusion cell as a *zygote*. In all eucaryotic organisms, gametic fusion is followed by nuclear fusion, with the result that the zygote nucleus contains *two complete sets of genetic determinants*, one derived from each gametic nucleus.

Sexual reproduction is common in the life cycle of plants and animals. In vertebrates and many invertebrates, it is the *only* method for the production of a new individual. Plants can also be propagated asexually (e.g., by cuttings), and asexual modes of reproduction exist in many groups of invertebrates. Among the eucaryotic protists, sexual reproduction is rarely an obligatory event in the life cycle. Many of these organisms completely lack a sexual stage in their life cycles, and even in species where sexuality does exist, sexual reproduction may occur infrequently, the formation of new individuals taking place principally by asexual means (for example, by binary fission or the formation of spores).

Sexual fusion results in a *doubling of the number of chromosomes*, since the nuclei of the gametes, each containing $N$ chromosomes, fuse to form the nucleus of the zygote, which consequently contains $2N$ chromosomes. Hence in passing from one sexual generation to the next, there must at some stage be a *halving of the number of chromosomes*, if the chromosome content of the nucleus is not to increase indefinitely. In fact, the halving of the chromosome number is a universal accompaniment of sexuality. It is brought about by a special process of nuclear division termed *meiosis* (Figure 2.13). In animals, *meiosis takes place immediately prior to the formation of gametes*. In other words, each individual of the species has $2N$ chromosomes in its cells through most of the life cycle. Such an organism is termed *diploid*. This state of affairs is, however, by no means universal among sexually reproducing eucaryotic organisms. In many eucaryotic protists, *meiosis takes place immediately after zygote formation*, with the consequence that the or-

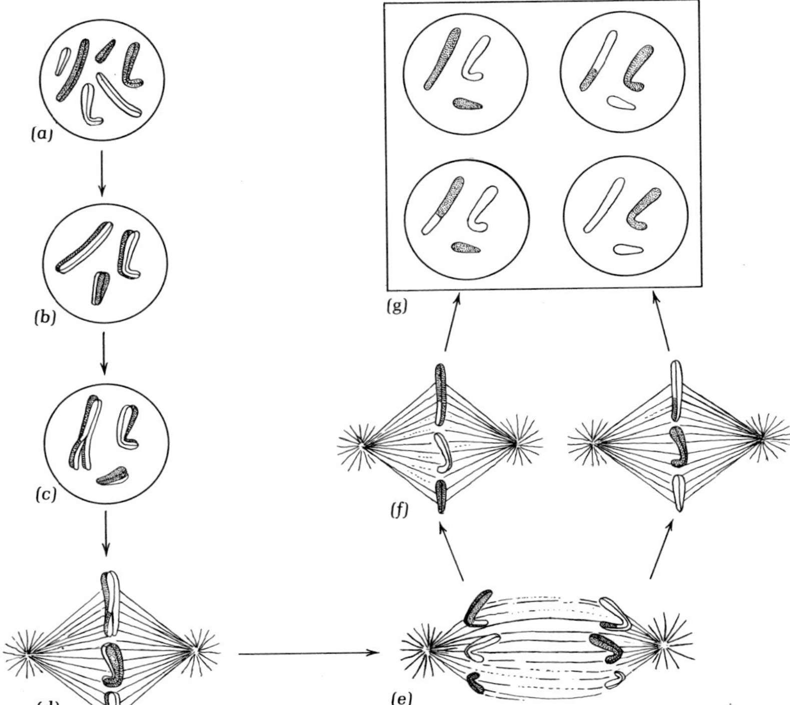


*Figure 2.13 Meiosis in a diploid plant cell (a) with three pairs of chromosomes. (b) Homologous chromosomes pair. (c) An exchange of segments (crossing over) takes place (shown for one chromosome only). (d) The chromosomes are shown at "first metaphase." (e) The chromosomes of each pair separate. (f) Two nuclei have formed, and within each a metaphase spindle forms. This time, however, the sister chromatids that make up each single chromosome separate. This phase, called the "second metaphase," is thus analogous to a mitotic division. (g) The four haploid nuclei that result from meiosis are shown.*

ganisms have N chromosomes through most of the life cycle. Such organisms are termed *haploid*. In many algae and plants, as well as in some fungi and protozoa, there is a well-marked *alternation of haploid and diploid generations*. In this type of life cycle, the diploid zygote gives rise to a diploid individual, which forms, by meiosis, haploid *asexual* reproductive cells. Each such haploid cell gives rise to a haploid individual, which eventually forms haploid gametes; gametic fusion, with the formation of a diploid zygote once again, completes the cycle.

**The genetic organization of the procaryotic cell**

In the nucleus of a procaryotic cell, a very different kind of organization of the genetic material has been revealed, although detailed studies have so far been made with relatively few organisms. Most of our present information has been obtained with the bacterium *Escherichia coli*. The nuclear region of *E. coli* contains a single molecule of DNA, unlinked to protein. This molecule is only 25 A thick but has an extraordinary total length, slightly more than 1 mm. It suffices to specify all the essential properties of the bacterium. A genetic system of this extreme simplicity is possible because much less information is required to specify the properties of a procaryotic cell than is required to specify the properties of a eucaryotic cell, as illustrated by some typical data on cellular DNA contents assembled in Table 2.5.

Nuclear division in *E. coli* thus requires the replication of a single, very long DNA molecule, followed by separation of the daughter molecules. The DNA remains permanently in an extended molecular state, never undergoing the shortening and coiling characteristic of the eucaryotic chromosomes during the mitotic divisional process. Nor is there formation of a mitotic apparatus, as a framework for the separation of the daughter molecules. The mechanism of this separation is not yet fully elucidated; current interpretations of it will be described in Chapter 10.

Although *E. coli* is the procaryotic organism in which nuclear organization has been most intensively investigated, the less complete information that we possess about the nuclei of other procaryotic organisms is in no way incompatible with the description given above for *E. coli*. It is, accordingly, probable that *all* procaryotic organisms share this distinctive nuclear organization.

In procaryotic cells, nuclear division and cell division are not as closely geared to one another as in eucaryotic cells. During the growth of unicellular procaryotic organisms, nuclear division typically runs ahead of cell division: each daughter cell, immediately after a cell division has occurred, may contain two or even more already separate nuclei. As a general rule, procaryotic cells only become uninucleate after growth has ceased and the organism has entered the resting state.

*Table 2.5 Relative amounts of genetic material in bacteria and in some eucaryotic organisms*

| **Bacteria** | **Absolute amount, grams** | **Relative amount** |
|---|---|---|
| *Escherichia coli*, DNA/vegetative nucleus | $5 \times 10^{-15}$ | 1 |
| *Bacillus subtilis*, DNA/spore | $5 \times 10^{-15}$ | 1 |
| Recorded ranges for some eucaryotic organisms, DNA/sperm cell | | |
| Mollusks | $0.9–2.5 \times 10^{-12}$ | 180–500 |
| Echinoderms | $0.6–1.5 \times 10^{-12}$ | 120–300 |
| Fishes | $0.5–1.8 \times 10^{-12}$ | 100–360 |
| Mammals | $\sim 3 \times 10^{-12}$ | 600 |

**Genetic transfer in procaryotic organisms**

Genetic studies indicate that most, if not all, procaryotic organisms are normally haploid; the prevalence of the diploid state during the life cycle, with all its profound genetic and evolutionary consequences, appears to be confined to certain eucaryotic groups.

Although transfer of genetic material between individuals has now been demonstrated in many procaryotic groups, such events are probably relatively rare in nature; for the most part, procaryotic organisms maintain themselves by asexual reproduction. Genetic transfer can be effected by three mechanisms: direct cell contact (conjugation), passage through a particle of a bacterial virus or bacteriophage (transduction), and passage through the medium as free DNA (transformation). In conjugation, transformation, and transduction, DNA transfer is always unidirectional, from a *donor cell* to a *recipient cell*, and with very rare exceptions in the case of conjugation, *only part of the donor genome* (frequently a very small part) *is transferred*. Hence, the recipient cell does not become completely diploid, like a eucaryotic zygote is; it is a *partial diploid* after completion of DNA transfer from the donor. As a result, the ensuing genetic recombination involves exchanges between the *complete set of haploid genes of the recipient and the partial set received from the donor*. The haploid state is usually rapidly restored, by elimination of that portion of the DNA which is not incorporated into the recombinant genome; under special circumstances, partial diploidy can persist for some time in the progeny of the recipient cell, but such partial diploids tend to be extremely unstable. The return to the haploid state does not, accordingly, involve a regular reduction division equivalent to meiosis in a eucaryotic organism.

## *The physical state of the cytoplasm in eucaryotic and procaryotic cells*

There are important differences in the physical state of the cytoplasm between eucaryotic and procaryotic cells, differences that are quickly evident when one watches living cells with the light microscope. In the cells of eucaryotic organisms, the cytoplasm is frequently in a *state of motion*, as revealed by the continuous displacement of particulate inclusions such as nuclei, mitochondria and chloroplasts; this phenomenon is known as *cytoplasmic streaming*. Furthermore, the cytoplasm of eucaryotic cells almost always contains *vacuoles*, sacs of liquid enclosed within unit membranes. These vacuoles vary greatly in size, form, and function; they are typically evanescent structures, which can appear and disappear or undergo marked changes of shape or position in the cell through the course of time.

The abilities to undergo cytoplasmic streaming and to form vacuoles are associated in eucaryotic cells with a series of functional attributes wholly lacking in procaryotic cells. These attributes receive their fullest expression in *eucaryotic cells devoid of walls*, characteristic of many protozoa and of animals. They include the ability to ingest solid food materials, a phenomenon known as *phagocytosis;* the ability to take droplets of liquid into the cell, known as *pinocytosis;* the ability to use directed cytoplasmic streaming as a means of cellular locomotion, known as *ameboid movement;* and the ability to control the water relations of the cell by a device known as a *contractile vacuole*.

**Phagocytosis and pinocytosis**

Among protists, phagocytosis is characteristic of the ameboid and ciliate protozoa, the slime molds, and some flagellates. It underlies the phagotrophic mode of nutrition. This ability has been conserved in animals.

Phagocytosis involves an active extension of the surface of the cell around a solid particle; the particle finally becomes completely enclosed within the cell membrane and enters the cell in a *food vacuole*. In many phagotrophic protists, as well as in phagocytic cells of higher animals, the intake of particles can occur at any point on the cell surface. In certain protozoa, there is a special, funnel-shaped region at the surface of the cell, known as a gullet, or cytostome, into which food particles are swept. The formation of food vacuoles occurs at the base of the gullet.

Once a food vacuole has been formed, its contents are subjected to *intracellular digestion*, by means of the hydrolytic (digestive) enzymes contained in *lysosomes*. The digestive enzymes of the lysosomes, being enclosed by the bounding unit membrane of the organelle, cannot act upon the components of the surrounding cytoplasm, but when a lysosome comes into contact with a food vacuole, its membrane dissolves at the point of contact, and its enzymes are discharged into the vacuole, where they proceed to digest the enclosed food particle. The products

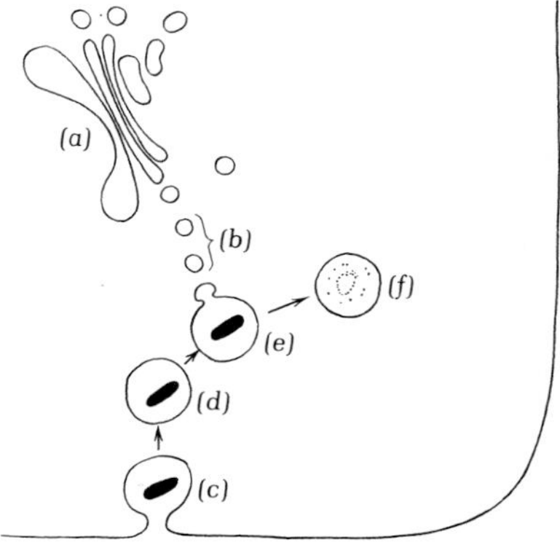


*Figure 2.14 A diagrammatic representation of the events of intracellular digestion: (a) Golgi apparatus; (b) lysosomes produced from the Golgi apparatus; (c) phagocytic capture of a food particle (a bacterium) at the surface of the cell, during which the particle is almost completely surrounded by the cell membrane; (d) newly formed food vacuole; (e) coalescence of the food vacuole with a lysosome; (f) digestion of the vacuolar contents by hydrolytic enzymes released from the lysosome. Modified from N. Novikoff, E. Essner and N. Quintana, Federation Proc.* **23,** *1011 (1964).*

of digestion then diffuse through the membrane of the food vacuole into the surrounding cytoplasm. This method of digestion is not confined to phagotrophic protists; it also occurs in invertebrate animals that do not have a well-defined gastrointestinal tract. A schematic diagram of the events of intracellular digestion, including the generation of lysosomes from the Golgi apparatus, is shown in Figure 2.14.

In the more highly developed animal phyla, the digestion of solid food takes place through the action of secreted enzymes in a specialized gastrointestinal tract, and the soluble products of digestion are absorbed through the wall of the gut. Digestion is accordingly extracellular. In such animal groups, the capacity for phagocytosis has nevertheless been retained by some specialized cells of the body. However, the function of phagocytosis in animals that perform extracellular digestion has changed. Its role is no longer nutritional but *protective:* it provides a means for the engulfment and destruction of invading microorganisms, as discussed in Chapter 26.

Pinocytosis is a process of cellular drinking, mechanistically analogous to the formation of a food vacuole, by means of which the cell can ingest a considerable volume of liquid from its surroundings. From a raised area formed at the cell surface, a long, narrow channel is formed, through which liquid is drawn into a vacuole deep within the cell.

**Ameboid movement**

Eucaryotic cells not completely enclosed by walls may be able, through directed cytoplasmic streaming, to move over a solid surface; since this

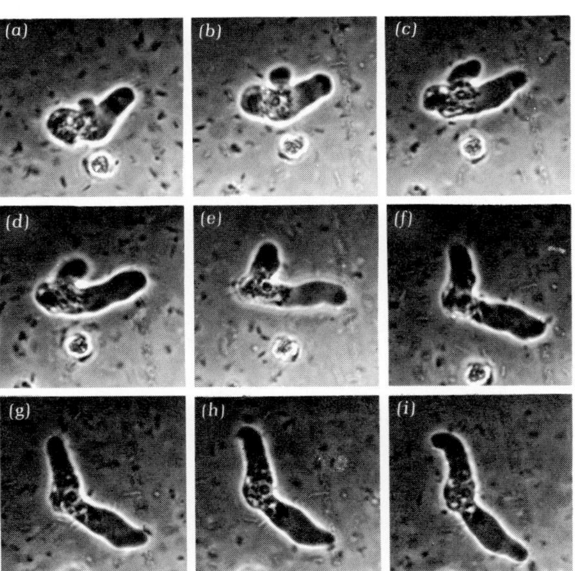


*Figure 2.15 Ameboid movement. A series of successive photomicrographs by phase contrast, taken at intervals of 15 seconds, of a small ameba, Tetramitus (×9,380). Courtesy of Jeanne Stove Poindexter.*

is the characteristic method of movement in ameboid protozoa, it has received the name *ameboid movement* (Figure 2.15). It is also manifested by slime molds and by many of the phagocytic cells in the bodies of animals. Cells that are completely enclosed by walls are incapable of ameboid movement, since the necessary contact between the mobile cell surface and a solid substrate cannot be achieved. Thus, the higher fungi, in which the coenocyte is completely enclosed within a branching wall, are immotile, even though vigorous cytoplasmic streaming can often be observed in these organisms.

**Osmoregulation**

Most free-living microorganisms inhabit *hypotonic* environments, in which the concentration of water is higher than it is within the cell. Since the cell membrane is freely permeable to water but not to many solutes, there is a tendency for water to enter the cell, to equalize the internal and external water concentrations. Unless this tendency is counterbalanced in some manner, the cell will swell and eventually undergo *osmotic lysis*. In many eucaryotic and nearly all procaryotic microbial groups, the danger of osmotic lysis is mechanically prevented by enclosure of the cell within a rigid wall of sufficient mechanical strength to counterbalance water pressure and thus prevent swelling of the enclosed protoplast. In protozoa that do not possess such rigid walls, a special type of vacuole, the *contractile vacuole*, functions as a cellular pump to eliminate water and thus to maintain osmotic balance. The contractile vacuole is, accordingly an *active* mechanism of osmoregulation, in contrast to the *passive* mechanism represented by the cell wall. Water is collected from the surrounding cytoplasm by the contractile vacuole, which thus progressively increases in volume; periodically, its contents are discharged through the cell membrane to the exterior.

## *The surface layers of the cell in eucaryotic protists*

Among eucaryotic protists there are considerable differences with respect to the nature of the surface layer of the cell, and these differences have important functional consequences with respect to movement and food intake.

In fungi and in several of the major algal groups, the protoplast is completely enclosed by a rigid wall consisting in large part of polysaccharide. Such an outer structure confers a fixed form on the cell and protects it from mechanical and osmotic damage. It precludes the possibility of ameboid movement; the cells of organisms with complete walls, if motile, move by other mechanisms. Phagocytosis is also precluded, and these organisms absorb their nutrients in dissolved form. Digestion of insoluble food materials, if it occurs, is extracellular and is mediated by the secretion of hydrolytic enzymes.

A few groups of eucaryotic protists characterized by the presence of a cell wall have managed to avoid these functional restrictions by the device of synthesizing a wall that does not completely enclose the protoplast. Motile diatoms, which synthesize a rigid wall impregnated with silica, have a wall with an open slot, through which contact between the protoplast and the substrate can be effected, permitting a modified form of ameboid movement. In some phagotrophic dinoflagellates, the wall consists of a series of plates between which pseudopodia can be extended.

In euglenid algae and many nonphotosynthetic flagellate and ciliate protozoa, the protoplast is covered by a thin and flexible outer layer, probably composed of protein, which is termed a *pellicle*. The cell has a relatively fixed form, conferred by this semirigid pellicle, but its shape can be distorted under mechanical stress (or, in euglenoid algae, by internal movements of the protoplast). Organisms with pellicles cannot perform ameboid movement and move by other mechanisms. Furthermore, phagocytosis cannot take place through the pellicular layer. The many phagotrophic protists with this kind of surface structure ingest solid food through a gullet, and phagocytic engulfment takes place at a small specialized area in the base of the gullet.

## *The surface layers of the cell in procaryotic protists*

Virtually all procaryotic protists have cells completely enclosed by rigid walls of distinctive chemical composition, which will be described in Chapter 5. The quasi-universality of walls in procaryotic protists is no doubt a reflection of the fact that these organisms do not possess an active mechanism of osmoregulation and can exist in hypotonic environments only when protected passively by enclosure in a wall. This interpretation is confirmed by the few exceptions to the rule. Rigid walls are absent from extreme halophiles, which grow in saturated brines, and from the *Mycoplasma* group, which, as parasites of animals, live in an isotonic environment. The extreme halophiles undergo immediate osmotic lysis in solutions that contain less than 12 percent salt, and the mycoplasmas can be cultivated only in media sufficiently rich in solutes to preserve isosmotic conditions. The digestion of insoluble nutrients by procaryotic organisms is invariably extracellular, mediated by the secretion of hydrolytic enzymes.

## *The mechanisms of eucaryotic and procaryotic cellular movement*

It has already been pointed out that the phenomenon of cytoplasmic streaming, one of the distinctive properties of eucaryotic cells, can serve as the basis for active movement on solid surfaces (ameboid movement). However, an indispensable precondition is that the cell not be completely enclosed within a wall. Some eucaryotic and procaryotic microorganisms can move actively by different mechanisms, which do not involve cytoplasmic streaming and which can operate even when the cell is completely enclosed by a wall. The most widespread of these mechanisms is *flagellar movement*, which permits movement in a liquid medium and is mediated by special threadlike organelles extending from the cell surface, termed flagella or cilia.*

Many protozoa, unicellular algae, and bacteria move by means of flagella. Aquatic fungi, multicellular algae, and lower plants, although completely immotile in the mature state, may produce motile flagellated reproductive cells (asexual spores, gametes, or zygotes). In the animal kingdom, specialized cells are motile by means of flagella; this is almost universally true of the male gametes or sperm cells. Flagella are also born by some *fixed* cells of the mature animal, their function being the creation of currents in the overlying liquid that bathes them. It should be noted that among protists ameboid and flagellar movement are not necessarily mutually exclusive; both types of locomotion may be possible for cells without walls (e.g., certain protozoa, spores of aquatic fungi).

Although there are minor variations in the structure of eucaryotic flagella or cilia, these organelles show a remarkable basic constancy of form throughout the enormous range of organisms that produce them.

A eucaryotic flagellum always arises from a cylindrical anchoring structure within the cytoplasm, known as a *basal body*. Basal bodies are homologous in structure with the centrioles, which act as the organizing centers for the mitotic apparatus. The external part of the

*The distinction between these two kinds of organelle is based on their relative length: a cilium is much shorter than a flagellum. Their fine structures as revealed by electron microscopy are, however, completely homologous. It would therefore be useful to have a single collective name for both flagella and cilia, but none has yet come into general use. For this reason, we shall use both names with their traditional meanings; the reader should keep in mind, however, that there are no fundamental differences, and that a cilium is merely a short flagellum.

*Figure 2.16 The fine structure of eucaryotic flagella and cilia, as revealed by electron micrographs of thin sections. (a) Longitudinal section through the cell of Bodo, a nonphotosynthetic flagellate (×40,050): bb, cylindrical basal body; outer microtubules (om); inner microtubules (im). Underlying the basal body is a specialized mitochondrion (m). At left (arrow), transverse section of a flagellum external to the cell. Note enclosure by an extension of the cell membrane (cm). (b) Section through the body surface of a ciliate, Didinium (×53,400). Within the cell (lower left), basal bodies (bb) have been sectioned transversely; their walls are composed of nine triple rows of microtubules. Just above the cell surface, several cilia (c) have been sectioned transversely; note the nine outer pairs of microtubules and the absence of the inner pair of microtubules. (c) Insert at upper right: section through two cilia at a point some distance from the cell surface. Note the inner pair of microtubules, the nine outer pairs, and the enclosing membrane. Courtesy of Dorothy Pitelka.*

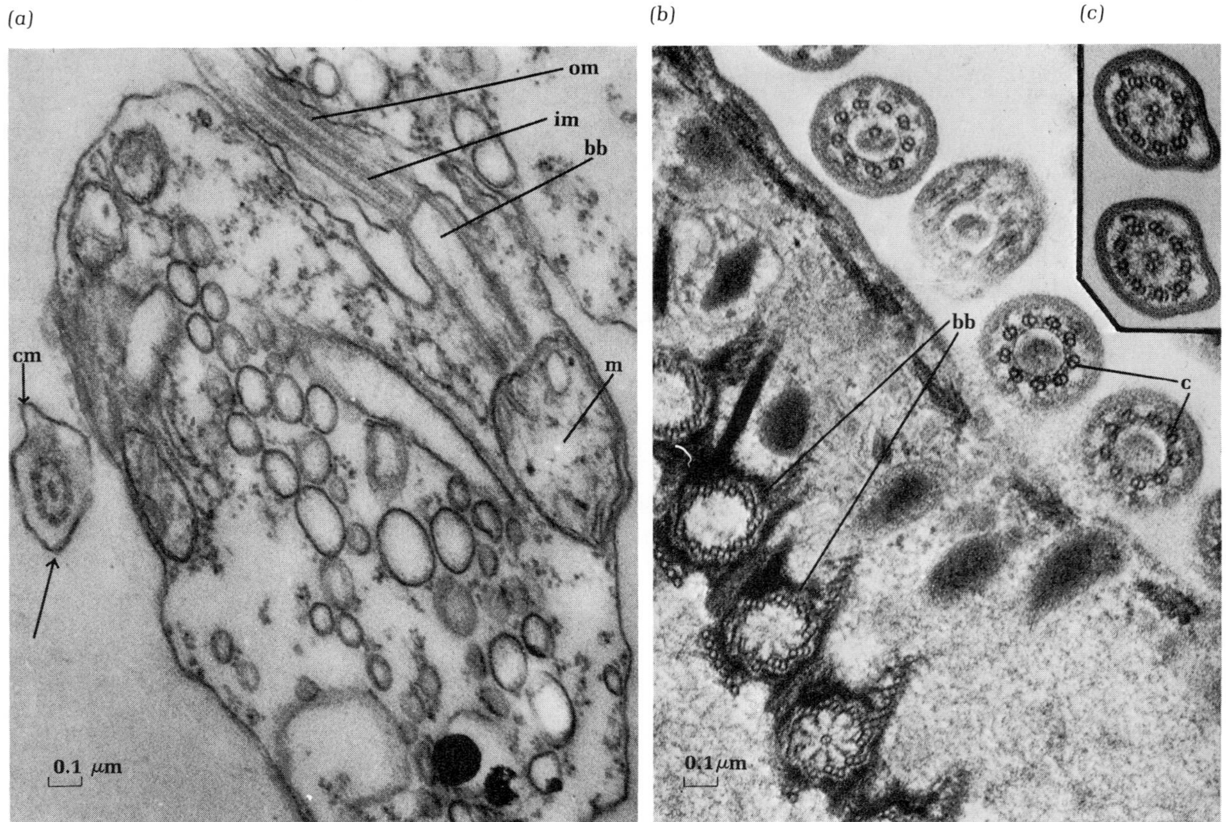

flagellum is surrounded by an extension of the cell membrane, which encloses a system of longitudinal microtubules that has a very regular arrangement. There are nine outer pairs of microtubules, arranged in a circle around a central inner pair. The two central microtubules arise from a plate in the flagellum near the surface of the cell, whereas the outer nine pairs are derived from the microtubular structure present in the basal body. Electron micrographs that illustrate various features of eucaryotic flagellar structure are shown in Figure 2.16.

Although it bears the same name, a *bacterial flagellum* shows a completely different fine structure and organization. It consists of a *single helical fibril*, far thinner than a eucaryotic flagellum, having approximately the dimensions of one of the individual microtubules within a eucaryotic flagellum. With rare exceptions, it is not enclosed by an extension of the cytoplasmic membrane, but protrudes through it. It is anchored in a basal organelle that is much smaller than a basal body and has a different structure. A slightly modified version of the bacterial flagellum is responsible for the movement of the bacteria known as *spirochetes*. These organisms possess an *axial filament*, consisting of two sets of flagellalike fibrils anchored at the two poles of the cell and enclosed within the outermost layer of the wall. The structure of a bacterial flagellum is shown in Figure 2.17.

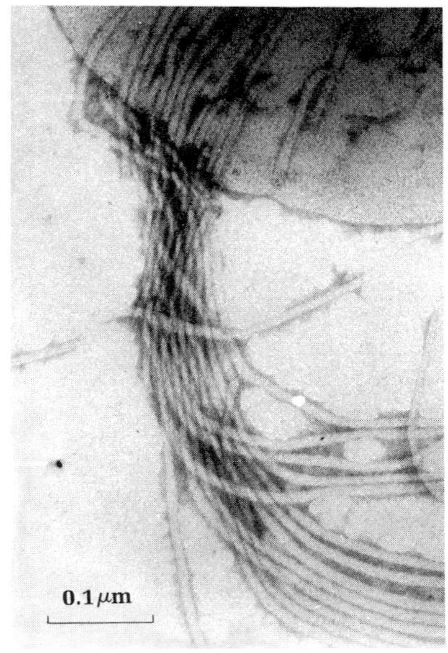


*Figure 2.17 The fine structure of bacterial flagella. Part of the pole of a cell of Spirillum serpens, showing numerous bacterial flagella emerging at various points from the cell surface. Electron micrograph of phosphotungstate-stained material (×85,500). Courtesy of R. G. E. Murray.*

Some microorganisms that are enclosed within walls can move slowly over solid surfaces, even though no locomotor organelles can be detected. This type of movement, known as *gliding movement*, is the sole method of locomotion in blue-green algae and in myxobacteria. Gliding movement also occurs in a few eucaryotic protists. In all these cases, the mechanism of movement is still unknown.

## *The biological significance of the two levels of cellular structure*

The numerous and fundamental differences in organization and function between the eucaryotic and procaryotic cell which have been outlined above have become fully appreciated only in recent years. It is now evident that the biological gap which separates bacteria and blue-green algae from all other cellular organisms represents one of the largest evolutionary discontinuities in the present-day living world. In these two microbial groups, the universal functions of cells—biosynthesis, growth, respiration, photosynthesis, movement—have a unique mechanistic basis, being conducted within a material framework far simpler than that of other cellular organisms. This fact poses some major evolutionary questions. Does the procaryotic cell, still perpetu-

*Table 2.6 Major differences between eucaryotic and procaryotic cells*

| | **Procaryotic** | **Eucaryotic** |
|---|---|---|
| *Genetic structure* | | |
| *Chromosome number* | 1 | >1 |
| *Mitotic nuclear division* | − | + |
| *Nuclear membrane* | − | + |
| *Nuclear DNA bound to histones* | − | + |
| *DNA in organelles* | − | + |
| *Cytoplasmic structure* | | |
| *Nature of cytoplasmic ribosomes* | *70 S* | *80 S* |
| *Nature of organellar ribosomes* | *Absent* | *70 S* |
| *Mitochondria* | − | + |
| *Chloroplasts* | − | + or − |
| *Golgi apparatus* | − | + |
| *Pinocytosis* | − | + or − |
| *Phagocytosis* | − | + or − |
| *Ameboid movement* | − | + or − |
| *Cytoplasmic streaming* | − | + or − |

ated in bacteria and blue-green algae, represent a stage in the evolution of the more complex eucaryotic cell, or did these two kinds of cells have completely different evolutionary origins? The discovery of DNA and 70 S ribosomes in mitochondria and chloroplasts has suggested another fascinating evolutionary possibility—that these organelles might have had an origin different from the other components of the eucaryotic cell, being derived from originally free-living procaryotic organisms, which entered an endosymbiotic relationship with a more advanced cell type and gradually became so closely integrated with the function of the host organism that they lost the ability to lead an independent existence.

The major differences between eucaryotic and procaryotic cells are summarized in Table 2.6.

## The general properties of viruses

The viruses are the smallest of all biological entities. They differ radically from all cellular organisms with respect to structure, chemical composition, and mode of development.

Viruses can develop only within the cells of host organisms. They are transmitted from cell to cell in the form of infectious particles known as *virions*. These particles, which range in diameter from 200 to 3,000 A, have a relatively simple chemical composition: they consist typically of a protein coat, or capsid, enclosing a *single kind of nucleic acid,* either RNA or DNA, depending on the kind of virus in question. The viral nucleic acid contains the genetic information requisite for intracellular development and formation of virions.

The protein components of some virions may contain an enzyme or enzymes specifically involved in the penetration of the host cell, but with this exception, *they are devoid of enzymatic activity.* Hence, intracellular development of a virus is dependent on enzymes formed within the host cell. In part, viral growth utilizes the preexisting enzymatic machinery of the host cell; in part, it depends on newly formed enzymes, the synthesis of which by the cell is directed by the viral nucleic acid itself. Following viral infection, accordingly, the synthetic and metabolic activities of the host cell are redirected into manufacturing and assembling the component parts of virions rather than into making normal cell constituents. This process usually culminates in the death of the infected cell. The newly synthesized virions, after their liberation from the host cell, can once again initiate the infective cycle when they encounter a suitable host. The major differences between viruses and cellular organisms have been summarized in Table 2.7.

*Table 2.7 Major differences between viruses and cellular organisms*

| | | **Viruses** | **Cellular organisms** |
|---|---|---|---|
| *Unit of structure* | | *Virion* | *Cell* |
| *Composition* | *Nucleic acids* | *One only (either DNA or RNA)* | *Always both DNA and RNA* |
| | *Proteins* | + | + |
| | *Lipids and polysaccharides* | –[a] | + |
| | *Enzymes* | *Absent, or restricted to a few* | *Many hundreds* |
| | *Ability to generate energy* | – | + |
| *Nature of growth* | *Reproduced exclusively from genetic material* | + | – |
| | *Independent synthesis of parts, subsequent assembly* | + | – |
| | *Direct formation from similar, preexisting structural elements* | – | + |

[a]Except in a few animal viruses, which apparently incorporate these compounds from the cells of the host.

## *Further reading*


**Books**

DuPraw, E. J., *Cell and Molecular Biology*. New York: Academic Press, 1968.

Gunsalus, I. C., and R. Y. Stanier (editors), *The Bacteria*, Vol. I. New York: Academic Press, 1960.

Loewy, A. G., and P. Siekevitz, *Cell Structure and Function*. New York: Holt, Rinehart and Winston, 1963.

Knight, B. C. J. G., and H. P. Charles (editors), *Organization and Control in Prokaryotic and Eukaryotic Cells*. New York: Cambridge University Press, 1970.

Pollock, M. R., and M. H. Richmond (editors), *Function and Structure in Microorganisms*. New York: Cambridge University Press, 1965.

Stern, H., and D. L. Nanney, *The Biology of Cells*. New York: Wiley, 1965.

**Reviews**

Lwoff, A., "The Concept of Virus." *J. Gen. Microbiol.* **17,** 239 (1957).

Stanier, R. Y., and C. B. van Niel, "The Concept of a Bacterium." *Arch. Mikrobiol.* **42,** 17 (1962).

# Chapter three
# The methods of microbiology

*As a result* of the small size of microorganisms, the amount of information that can be obtained about their properties from the examination of *individuals* is limited; for the most part, the microbiologist studies *populations*, containing thousands or millions of individuals. Such populations are obtained by growing microorganisms under more or less well-defined conditions, as *cultures*. A culture that contains only one kind of microorganism is known as a *pure* or *axenic culture*. A culture that contains more than one kind of microorganism is known as a *mixed culture;* in the special case when it contains only two kinds of microorganisms, deliberately maintained in association with one another, it is known as a *two-membered culture*.

At the very heart of microbiology, there accordingly lie two kinds of operations: *isolation*, the separation of a particular microorganism from the mixed populations that exist in nature; and *cultivation*, the growth of microbial populations in artificial environments (culture media) under laboratory conditions. These two operations come into play irrespective of the kind of microorganism with which the microbiologist deals; they are basic alike to the study of viruses, bacteria, fungi, algae, protozoa, and even small invertebrate animals. Furthermore, they have been extended in recent years to the study in isolation of cell or tissue lines derived from higher plants and animals *(tissue culture)*. The unity of microbiology as a science, despite the biological diversity of the organisms with which it deals, is derived from this common operational base.

## *Pure culture technique*

Microorganisms are ubiquitous, so the preparation of a pure culture involves not only the isolation of a given microorganism from a mixed natural microbial population, but also the maintenance of the isolated individual and its progeny in an artificial environment to which the access of other microorganisms is prevented. Microorganisms do not require much space for development; hence an artifical environment can be created within the confines of a test tube, a flask, or a petri dish, the three kinds of containers most commonly used to cultivate microorganisms. The culture vessel must be rendered initially *sterile* (free of any living microorganism) and, after the introduction of the desired type of microorganism, it must be protected from subsequent external contamination. The primary source of external contamination is the atmosphere, which always contains floating microorganisms. The form of a petri dish, with its overlapping lid, is specifically designed to prevent atmospheric contamination. Contamination of tubes and flasks is prevented by closure of their orifices with an appropriate stopper. This is usually a plug of cotton wool, although metal caps or plastic screw caps are now often employed, particularly for test tubes.

The external surface of a culture vessel is, of course, subject to contamination, and the interior of a flask or tube can become contaminated when it is opened to introduce or withdraw material. This danger is minimized by passing the orifice through the flame of a burner, immediately after the stopper has been removed and again just before it is replaced.

The *inoculum* (i.e., the microbial material used to seed or inoculate a culture vessel) is commonly introduced on a metal wire or loop, which is rapidly sterilized just before its use by heating in the flame of a burner. Transfers of liquid cultures can also be made by pipette. For this purpose, the mouth end of the pipette is plugged with cotton wool, and the pipette is sterilized in a paper wrapping or in a glass or metal container, which keeps both inner and outer surfaces free of contamination until the time of use.

The risks of accidental contamination may be further reduced by transferring cultures in a small closed room, the atmosphere of which is rendered more or less free of microorganisms by exposure to ultraviolet light or other appropriate treatment before use.

**The isolation of pure cultures by plating methods**

If one is dealing with microorganisms that form discrete colonies on solid media (e.g., most bacteria, yeasts, many fungi and unicellular algae), pure cultures may be most simply obtained by one of the modifications of the plating method. This method involves the separation and immobilization of individual organisms on or in a nutrient medium solidified with agar or some other appropriate jelling agent. Each viable organism gives rise, through growth, to a colony from which transfers can be readily made.

The *streaked plate* is in general the most useful plating method. A sterilized bent wire is dipped into a suitable diluted suspension of organisms and is then used to make a series of parallel, nonoverlapping streaks on the surface of an already solidified agar plate. The inoculum is progressively diluted with each successive streak, so that even if the initial streaks yield confluent growth, well-isolated colonies develop along the lines of later streaks (Figure 3.1). Alternatively, isolations can be made with *poured plates:* successive dilutions of the inoculum are placed in sterile petri dishes and mixed with the molten but cooled agar medium, which is then allowed to solidify. Colonies subsequently develop embedded in the agar.

The isolation of strictly anaerobic bacteria by plating methods poses special problems. Provided that the organisms in question are not rapidly killed by exposure to oxygen, plates may be prepared in the usual manner and then incubated in closed containers, from which the oxygen is removed either by chemical absorption or combustion. For the more oxygen-sensitive anaerobes, a modification of the pour plate method, known as the *dilution shake culture*, is to be preferred. A tube of melted and cooled agar is inoculated and mixed, and approximately one tenth of its contents is transferred to a second tube, which is then mixed and used to inoculate a third tube in a similar fashion. After 6 to 10 successive dilutions have been prepared, the tubes are rapidly cooled and sealed, by pouring a layer of sterile petroleum jelly and paraffin on the surface, thus preventing access of air to the agar column. In shake cultures, the colonies develop deep in the agar column (Figure 3.2), and are thus not easily accessible for transfer. To make a transfer, the petroleum jelly–paraffin seal is removed with a sterile needle, and the agar column is gently blown out of the tube into a sterile petri dish by passing oxygen-free gas through a capillary pipette inserted between the tube wall and the agar. The column is sectioned into discs with a sterile knife to permit examination and transfer of colonies.

*Figure 3.1 Isolation of a pure culture by the streak method. A Petri dish containing nutrient agar was streaked with a suspension of bacterial cells. As a result of subsequent growth, each cell has given rise to a macroscopically visible colony.*

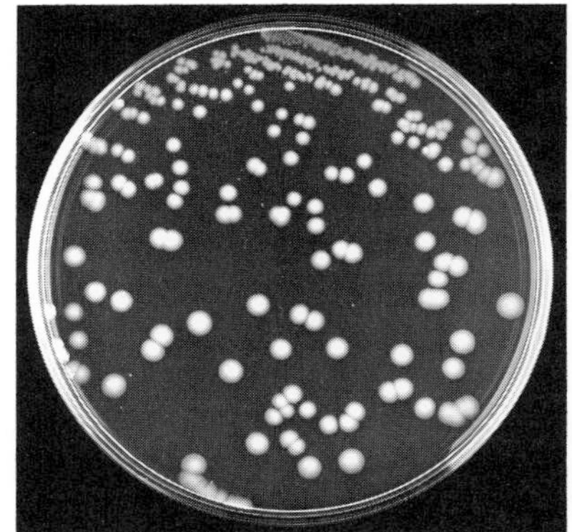

*Figure 3.2 Isolation of a pure culture of anaerobic bacteria by the dilution shake method. A complete series of dilution shakes is shown. Note the confluent growth in the more densely seeded tubes (at right), and the well-isolated colonies in the two final tubes of the series (at left). After the agar had solidified, each tube was sealed with a mixture of sterile vaseline and paraffin to prevent the access of atmospheric oxygen, which inhibits the growth of anaerobic bacteria.*

In isolating from a mixed natural population it is often possible, provided one's technique is good, to prepare a first plate or dilution shake series in which many of the colonies that develop are well separated from one another. Can one then pick material from such a colony, transfer it to an appropriate medium, and call it a pure culture? Although this is often done, a culture so isolated may be far from pure. Microorganisms vary greatly in their nutritional requirements, and consequently no single medium and set of growth conditions will permit the growth of all the microorganisms present in a natural population. Indeed, it is probable that only a very small fraction of the microorganisms initially present will be able to form colonies on any given medium. Hence, for every visible colony on a first plate, there may be a thousand cells of other kinds of microorganisms which were also deposited on the agar surface but have failed to give macroscopically visible growth, although they may still be viable. The probability is high that some of these organisms will be picked up and carried over when a transfer is made. *One should never pick from a first plate for the preparation of a pure culture.* Instead, a second plate should be streaked from a cell suspension prepared from a well-isolated colony. If all the colonies on this second plate appear identical, a well-isolated colony can be used to establish a pure culture.

Not all microorganisms able to grow on solid media necessarily give rise to well-isolated colonies. Certain motile flagellated bacteria (*Pro-*

*teus, Pseudomonas)* can rapidly spread over the slightly moist surface of a freshly poured plate. This can be prevented by the use of plates with well-dried surfaces, on which the cells are immobilized. Spirochetes and organisms that show gliding movement (myxobacteria, blue-green algae) can move over or through an agar gel, even when its surface is well dried. In such cases, the movement of the organisms in question is an aid to their purification, since they can move away from other kinds of microorganisms immobilized on the agar. Thus, purification can often be achieved by allowing migration to occur and transferring repeatedly to fresh plates from the advancing edge of the migrating population.

The incorporation into the medium of selectively inhibitory substances is also sometimes helpful in making isolations from nature. Because of their biological specificity, certain antibiotics are particularly useful in this respect. Bacteria vary greatly in their sensitivity to the antibiotic penicillin, which can consequently be used at low concentrations to prevent the development of sensitive bacteria in the initial population. At higher concentrations, penicillin is generally toxic for procaryotic organisms but not for eucaryotic ones. It is thus a very useful agent for the purification of protozoa, fungi, and eucaryotic algae that are contaminated by bacteria. Conversely, procaryotic organisms are insensitive to polyene antibiotics such as nystatin, which are generally toxic for eucaryotic organisms. The incorporation of this kind of antibiotic into the isolation medium can sometimes be used to advantage in the purification of bacteria heavily contaminated by fungi or amebae. Many other variations on the theme of selective toxicity can be used to facilitate isolations by the plating method.

## The isolation of pure cultures in liquid media

Plating methods are in general satisfactory for the isolation of bacteria, blue-green algae, and fungi, because the great majority of the representatives of these groups can grow well on solid media. However, some of the larger-celled bacteria have not yet been successfully cultivated on solid media, and many protozoa and algae are also cultivable only in a liquid medium. Although plating methods for the isolation of viruses have been greatly extended in recent years, many of these organisms are most easily isolated by the use of liquid media. In the case of viruses, of course, a pure culture is never obtainable, since these organisms are obligate intracellular parasites; a two-membered culture, consisting of a specific virus and its biological host, represents the goal of purification for this microbial group.

The simplest procedure of isolation in liquid media is the *dilution method.* The inoculum is subjected to serial dilution in a sterile medium, and a large number of tubes of medium are inoculated with aliquots of each successive dilution. The goal of this operation is to inoculate a series of tubes with a microbial suspension so dilute that the probability of introducing even one individual into a given tube is very small: a probability of the order of 0.05. When a large number of tubes is seeded with an inoculum of this size, it can be calculated from probability theory that the fraction of tubes which receives no organisms is 0.95; the fraction receiving one organism is 0.048; the fraction receiving two organisms is 0.0012; the fraction receiving three organisms is 0.00002. As a result, if a tube shows *any* subsequent growth, there is a very high probability that this growth has resulted from the introduction of a *single* organism. The probability is

$$\frac{0.048}{0.048 + 0.0012 + 0.00002} = 0.975$$

The probability that growth has originated from a single organism declines very rapidly as the mean number of organisms in the inoculum increases. It is, therefore, essential to isolate from a series of tubes the great majority of which show *no* growth.

The dilution method was the technique used to isolate the first bacterial pure culture: it was applied by Lister to obtain in pure form the bacterium responsible for the souring of milk. It has, however, one major disadvantage—that it can be used only to isolate the *numerically predominant* member of a mixed microbial population. In the isolation of viruses, this is usually no problem, because the size of the population of virions in infected material is characteristically extremely large. It can almost never be effectively used for the isolation of larger microorganisms that are incapable of developing on solid media (e.g., protozoa, algae), because in nature these microorganisms are as a rule greatly outnumbered by bacteria. Hence, the usefulness of the dilution method is limited.

When neither plating nor dilution methods can be applied, the only alternative is to resort to the *microscopically controlled isolation of a single cell or organism from the mixed population,* a technique known by the name of *single-cell isolation.* The technical difficulty of single-cell isolation is inversely related to the *size* of the organism which one wishes to isolate: it is relatively easy to use with large-celled microorganisms, such as algae and protozoa, but becomes much harder with bacteria.

In the case of large microorganisms, purification involves the capture of a single individual in a fine capillary pipette and the subsequent transfer of this individual through several washings in relatively large volumes of sterile medium to eliminate microbial contaminants of smaller size. The successive operations can be performed manually, with control by direct microscopic observation at a relatively low magnification, such as that provided by a dissecting microscope. Provided that the operator possesses the necessary dexterity, this method is an exceedingly effective one for the isolation of protozoa and unicellular algae.

The technique of the capillary pipette can no longer be applied if the organism which one wishes to isolate is so small that it cannot be readily observed at a magnification of 100 times or less, because one cannot achieve the necessary fineness of control to manipulate a capillary pipette directly at higher magnifications. In this event, a mechanical device known as a *micromanipulator* must be used, in conjunction with specially prepared, very fine glass operating instruments. The essential purpose of a micromanipulator is to *gear down* manual control, so that very slight and precisely controlled movements of the operating instruments can be effected in a small operating area (a microdrop) under continuous microscopic observation at high magnifications (500 to 1,000×).

**Two-membered cultures**

The goal of isolation is normally to obtain a pure culture. However, there are certain situations where this cannot be achieved or where achievement is so difficult as to be impractical. Under such circumstances, the alternative is to obtain the next best degree of purification, in the shape of a *two-membered culture,* which contains only two kinds of microorganisms. As already mentioned, a two-membered culture is in principle the only possible way to maintain viruses, since these organisms are all obligate intracellular parasites of cellular organisms. Obligate intracellular parasitism is also characteristic of several groups of cellular microorganisms: the bacteria of the rickettsia and psittacosis groups, agents of animal disease; certain bacteria of the *Bdellovibrio* group, which parasitize other bacteria; some protozoa that cause animal diseases (e.g., the agent of malaria); and a number of parasitic fungi. In all these instances, a two-membered culture represents the nearest approach to cultivation under controlled laboratory conditions that can be achieved.

Many of thé phagotrophic protozoa, which feed in nature on smaller microorganisms, are also most easily maintained in the laboratory as two-membered cultures in association with their smaller microbial prey. This is true, for example, of ciliates, amebae, and slime molds. In such instances, the association is probably never an *obligate* one, since careful nutritional studies on a few representatives of these groups have shown that they can be grown in pure culture; however, the nutritional requirements of phagotrophic protozoa are typically extremely complex, so that the preparation of media for the maintenance of pure cultures is both difficult and laborious. For purposes of routine maintenance, and also for many experimental purposes, two-membered cultures are satisfactory.

The establishment of a two-membered culture is an operation that is conducted in two phases. First, it is necessary to establish a pure culture of the food organism (the host in the cases of viruses and obligate intracellular parasites, the prey in the case of phagotrophic protozoa). Once this has been achieved, the parasite or predator can be isolated by any one of a variety of methods (plating on solid media in the presence of the food organism, dilution in a liquid medium, single-cell isolation) and introduced into the pure culture of the food organism.

The successful maintenance of two-membered cultures requires considerable art, as a reasonably stable biological balance between the two components is essential. The medium must be one that permits sufficient growth of the food organism to meet the needs of the parasite or predator but should not be so rich that the food organism can outgrow its associate or produce metabolic products that are deleterious to it.

*Figure 3.3 Exponential (logarithmic) order of death of bacteria. The same data are plotted semilogarithmically in (a) and arithmetically in (b); N is the number of surviving bacteria.*

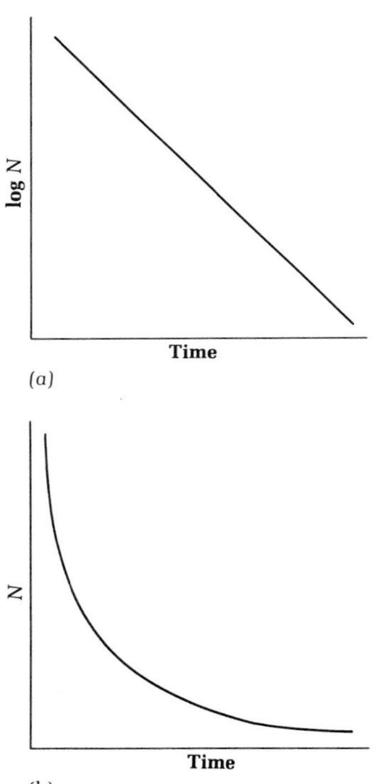


## *The theory and practice of sterilization*

Sterilization is a treatment that *frees the treated object of all living organisms.* It can be achieved by exposure to lethal physical or chemical agents or, in the special case of solutions, by filtration.

To understand the basis of sterilization by lethal agents, it is necessary to describe briefly the kinetics of death in a microbial population. The only valid criterion of death in the case of a microorganism is *irreversible loss of the ability to reproduce;* this is usually determined by quantitative plating methods, survivors being detected by colony formation. When a pure microbial population is exposed to a lethal agent, the kinetics of death are nearly always *exponential:* the number of survivors decreases geometrically with time. This reflects the fact that all the members of the population are of similar sensitivity; probability alone determines the actual time of death of any given individual. If the logarithm of the number of survivors is plotted as a function of the time of exposure, a straight line is obtained (Figure 3.3); its negative slope defines the *death rate.*

The death rate tells one only what *fraction* of the initial population survives a given period of treatment. To determine the *actual number* of survivors, one must also know the *initial population size,* as illustrated graphically in Figure 3.4. Accordingly, for the establishment of procedures of sterilization, two factors have to be taken into account: the death rate and the initial population size.

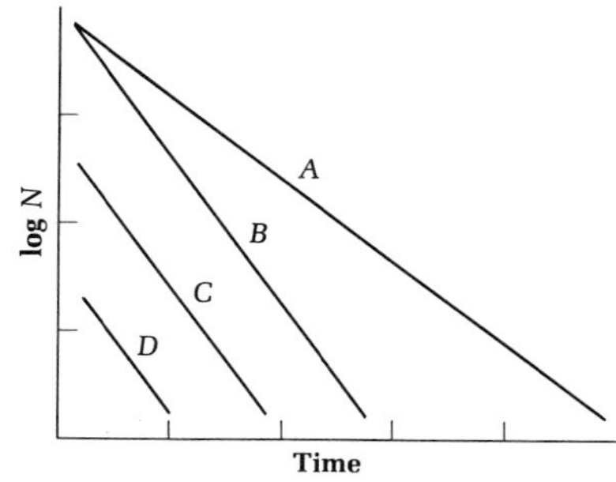


*Figure 3.4 Relationship of death rate and population size to the time required for the destruction of bacterial cultures; N is the number of surviving bacteria. Cultures B, C, and D have identical death rates. Culture A has a lower death rate.*

In the practice of sterilization, the microbial population to be destroyed is almost always a *mixed* one. Since microorganisms differ widely in their resistance to lethal agents, the significant factors become the initial population size and the death rate of the *most resistant* members of the mixed population. These are almost always the highly resistant endospores of certain bacteria. Consequently, spore suspensions of known resistance are the objects commonly used to assess the reliability of sterilization methods.

Taking into account the kinetics of microbial death, we can formulate the practical goal of sterilization by a lethal agent in a slightly more refined way: *the probability that the object treated contains even one survivor should be infinitesimally small.* For example, if we wish to sterilize a liter of a culture medium, this goal will be achieved for all practical purposes if the treatment is one which will leave no more than one survivor in $10^6$ liters; under such circumstances, the probability of failure is very small indeed. Procedures of routine sterilization are always designed to provide a very wide margin of safety.

**Sterilization by heat**

Heat is the most widely used lethal agent for purposes of sterilization. Objects may be sterilized by dry heat, applied in an oven in an atmosphere of air, or by moist heat, provided by wet steam. Of the two methods, sterilization by dry heat requires a much greater duration and intensity. Heat conduction is less rapid in air than in steam. In addition, bacteria can survive in a completely desiccated state and, in this state, the intrinsic heat resistance of vegetative bacterial cells is greatly increased, almost to the level characteristic of spores. Consequently, the death rate is much lower for dry cells than for moist ones.

Dry heat is used principally to sterilize glassware or other heat-stable solid materials. The objects are wrapped in paper or otherwise protected from subsequent contamination and exposed to a temperature of 170°C for 90 minutes in an oven.

Steam must be used for the heat sterilization of aqueous solutions. Treatment is usually carried out in a metal vessel known as an *autoclave,* which can be filled with steam at a pressure greater than atmospheric. Sterilization can thus be achieved at temperatures considerably above the boiling point of water; laboratory autoclaves are commonly operated at a steam pressure of 15 lb/in.$^2$ above atmospheric pressure, which corresponds to a temperature of 120°C. Even bacterial spores that survive several hours of boiling are rapidly killed at 120°C. Small volumes of liquid (up to about 3 liters) can be sterilized by exposure for 20 minutes; if larger volumes are to be sterilized, the time of treatment must be extended.

A temperature of 120°C within the autoclave will be attained under a pressure of 15 lb/in.$^2$ *only if the atmosphere consists entirely of steam.* At the start of the operation, accordingly, all the air originally in the chamber must be expelled and replaced by steam; this is achieved by the use of a steam trap, which remains open as long as air is being passed through it but closes when the atmosphere consists of steam. If some air remains in the sterilization chamber, the partial pressure of steam will be lower than that indicated on the pressure gauge, and the temperature will be correspondingly lower. For this reason an autoclave should always be equipped with both a temperature and a pressure gauge. The temperature within the sterilization chamber can be monitored by including, among the objects to be sterilized, special indicator papers, which change color if the heat treatment has been adequate.

Sterilization by autoclaving cannot be applied to solutions containing substances that are easily destroyed by heat. If, as is the case for many sugars, these substances are not seriously affected by heating at 100°C, fractional sterilization *(tyndallization)* can be used. It involves exposure to flowing steam at atmospheric pressure for 30 minutes on 3 successive days. This is essentially the procedure of sterilization originally devised by Tyndall in his studies on the problem of spontaneous generation, described in Chapter 1. The success of tyndallization is dependent on the germination of spores that survive the initial heating. Since the spores of some bacteria require nutrients in order to germinate, tyndallization can be safely used only for the sterilization of culture media, not for that of nonnutrient solutions (pure sugar solutions, salt solutions).

**Pasteurization**

Although not, strictly speaking, a method of sterilization, the process of heat treatment termed *pasteurization* deserves brief mention. Originally devised by Pasteur as a means of destroying microorganisms in wine and beer that would otherwise cause spoilage of these beverages, it is now used primarily to make beverages and foods *safe* for consumption, by killing any disease-producing microorganisms that they might contain. The most widespread application is the pasteurization of milk, commonly performed by heating to 62°C for 30 minutes. This treatment is sufficient to kill disease-producing bacteria of the kinds commonly transmitted through milk. It leaves many harmless bacteria alive and does not affect the taste of the product.

**Sterilization by chemical treatment**

Many of the substances used in preparing culture media are too heat labile to be sterilized by autoclaving or tyndallization. For such substances, a reliable method of chemical sterilization would be extremely useful. The essential requirement for a chemical sterilizing agent is that it should be *volatile* as well as *toxic*, so that it can be readily eliminated from the object sterilized after treatment. The best available candidate is *ethylene oxide*, a liquid that boils at 10.7°C. It can be added to solutions in liquid form (final concentration of approximately 0.5 to 1.0 percent) at a temperature of 0 to 4°C, or used as a sterilizing gas at temperatures above the boiling point. It is chemically unstable, decomposing in aqueous solution to ethylene glycol, which is nonvolatile and may have undesirable effects. Furthermore, ethylene oxide is both explosive and toxic for humans, so special precautions must be taken in its handling. For these reasons, ethylene oxide sterilization has not become a routine laboratory procedure. It is, however, used industrially, for the sterilization of plastic petri dishes and other plastic objects which melt at temperatures greater than 100°C.

**Chemical disinfection**

Although they cannot be used for preparative sterilization, a variety of nonvolatile toxic chemicals are used in the laboratory to destroy microorganisms on discarded objects used in microbiological work (e.g., pipettes, centrifuge tubes, cultures that are no longer required). Since the primary goal of such treatment is usually to kill potentially dangerous infectious microorganisms that have been used in experiments, these chemical agents are commonly termed *disinfectants*. The use of disinfectants is sometimes advisable even when dealing with harmless microorganisms, to reduce contamination of the laboratory atmosphere (e.g., from the spores in discarded cultures of fungi). Disinfectant solutions may consist of phenolic chemicals such as Lysol, hypochlorite ($OCl^-$), mercuric chloride ($HgCl_2$), or organic mercurials.

**Sterilization by filtration**

The principal laboratory method used to sterilize solutions of heat-labile materials is *filtration* through filters capable of retaining microorganisms. The action of such filters is almost always complex. Mi-

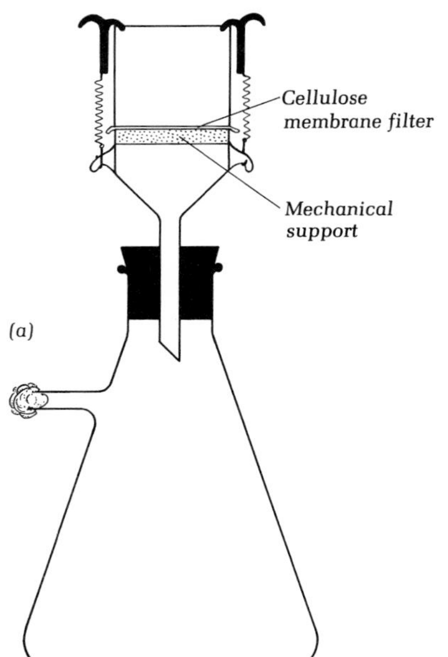


croorganisms are retained in part by the small size of the filter pores and in part by adsorption on the pore walls during their passage through the filter. The importance of adsorption is indicated by the fact that a filter may effectively retain microorganisms even when the average diameter of its pores is somewhat greater than the mean size of the cells that are retained. Sterilization by filtration is subject to one major theoretical limitation. Since the viruses range down in size to the dimensions of large protein molecules, they are not necessarily retained by filters that can hold back even the smallest of cellular microorganisms. Consequently, it is never possible to be certain that filtration procedures which render a solution bacteria-free will also free it of viruses. In practice, this limitation of the method seldom leads to difficulties.

The most commonly used sterilizing filters (Figure 3.5) are the following:

(1) *membrane filters,* consisting of porous discs of cellulose esters (made in a wide range of diameters and pore sizes);

(2) *Seitz filters,* consisting of discs of an asbestos–cellulose mixture;

(3) *sintered glass filters,* prepared by fusing together fine glass fragments;

(4) *candle filters,* made of unglazed ceramic.

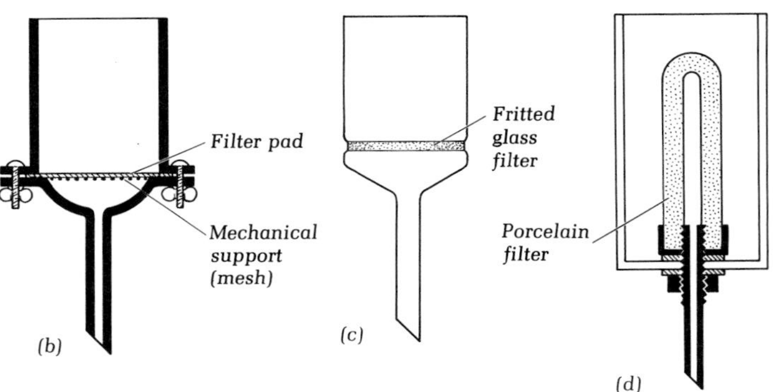


*Figure 3.5 Diagrammatic drawings of several kinds of filters that are used to remove bacteria from liquids: (a) cellulose membrane filter shown mounted and ready for use on a sterile suction flask; (b) Seitz filter, unmounted; (c) fritted glass filter, unmounted; (d) porcelain candle filter, unmounted.*

The filter is appropriately mounted, and the mounting is inserted into a receiving flask. The assembly is sterilized by heat. After the solution has been poured onto the filter, its passage is accelerated either by suction on the receiving flask or by pressure on the unfiltered liquid. Pressure filtration is more convenient, as it does not lead to foaming of solutions that contain proteins and allows a change of the receiving flask without disturbing the filter.

## *The principles of microbial nutrition*

To grow, organisms must draw from the environment all the substances which they require for the synthesis of their cell materials and for the generation of energy. These substances are termed *nutrients*. A culture medium must therefore contain, in quantities appropriate to the specific requirements of the microorganism for which it is designed, all necessary nutrients. However, microorganisms are extraordinarily diverse in their specific physiological properties, and correspondingly in their specific nutrient requirements. Literally thousands of different media have been proposed for their cultivation, and in the descriptions of these media, the reasons for the presence of the various components are often not clearly stated. Nevertheless, the design of a culture medium can and should be based on scientific principles, the *principles of nutrition*, which we shall outline as a preliminary to the description of culture media.

The chemical composition of cells, broadly constant throughout the living world, indicates the major material requirements for growth. *Water* accounts for some 80 to 90 percent of the total weight of cells and is always therefore the major essential nutrient, in quantitative terms. The solid matter of cells (Table 3.1) contains, in addition to

*Table 3.1 Approximate elementary composition of the microbial cell*[a]

| Element | Percentage of dry weight |
|---|---|
| *Carbon* | 50 |
| *Oxygen* | 20 |
| *Nitrogen* | 14 |
| *Hydrogen* | 8 |
| *Phosphorus* | 3 |
| *Sulfur* | 1 |
| *Potassium* | 1 |
| *Sodium* | 1 |
| *Calcium* | 0.5 |
| *Magnesium* | 0.5 |
| *Chlorine* | 0.5 |
| *Iron* | 0.2 |
| *All others* | ~0.3 |

[a]Data for a bacterium, *Escherichia coli*, assembled by S. E. Luria, in *The Bacteria* (I. C. Gunsalus and R. Y. Stanier, editors), Vol. I, Chap. 1. New York: Academic Press, 1960.

*Table 3.2 General physiological functions of the principal elements*

| **Element** | **Physiological functions** |
|---|---|
| Hydrogen | Constituent of cellular water, organic cell materials |
| Oxygen | Constituent of cellular water, organic cell materials; as $O_2$, electron acceptor in respiration of aerobes |
| Carbon | Constituent of organic cell materials |
| Nitrogen | Constituent of proteins, nucleic acids, coenzymes |
| Sulfur | Constituent of proteins (as amino acids cysteine and methionine); of some coenzymes (e.g., CoA, cocarboxylase) |
| Phosphorus | Constituent of nucleic acids, phospholipids, coenzymes |
| Potassium | One of the principal inorganic cations in cells, cofactor for some enzymes |
| Magnesium | Important cellular cation; inorganic cofactor for very many enzymatic reactions, including those involving ATP; functions in binding enzymes to substrates; constituent of chlorophylls |
| Manganese | Inorganic cofactor for some enzymes, sometimes replacing Mg |
| Calcium | Important cellular cation; cofactor for some enzymes (for example, proteinases) |
| Iron | Constituent of cytochromes and other heme or nonheme proteins; cofactor for a number of enzymes |
| Cobalt | Constituent of vitamin $B_{12}$ and its coenzyme derivatives |
| Copper, zinc, molybdenum | Inorganic constituents of special enzymes |

*hydrogen* and *oxygen* (derivable metabolically from water), *carbon, nitrogen, phosphorus,* and *sulfur,* in order of decreasing abundance. These six elements account for about 95 percent of the cellular dry weight. Many other elements are included in the residual fraction. Nutritional studies show that *potassium, magnesium, calcium, iron, manganese, cobalt, copper, molybdenum,* and *zinc* are required by nearly all organisms. The known functions in the cell of these 15 elements are summarized in Table 3.2.

All the required metallic elements can be supplied as nutrients in the form of the *cations of inorganic salts.* Potassium, magnesium, calcium, and iron are required in relatively large amounts and should always be included as salts in culture media. The quantitative requirements for manganese, cobalt, copper, molybdenum, and zinc are very small—so small, in fact, that it is often technically difficult to demonstrate their essentiality, since they are present in adequate amounts as contaminants of the major inorganic constituents of media. They are often referred to as *trace elements* or *micronutrients.* One nonmetallic element, phosphorus, can also be used as a nutrient when provided in inorganic form, as phosphate salts.

It should be noted that some biological groups have additional, specific mineral requirements; for example, diatoms and certain other algae synthesize cell walls that are heavily impregnated with silica and consequently have a specific *silicon* requirement, supplied as silicate. Although a requirement for *sodium* cannot be demonstrated for most microorganisms, it is required at relatively high concentrations by certain marine bacteria, by blue-green algae, and by photosynthetic bacteria. In these groups, it cannot be replaced by other monovalent cations.

The needs for *carbon, nitrogen, sulfur,* and *oxygen* cannot be so simply described, because organisms differ with respect to the *specific chemical form* under which these elements must be provided as nutrients.

### The requirements for carbon

Organisms that perform photosynthesis, and bacteria that obtain energy from the oxidation of inorganic compounds, typically use the most oxidized form of carbon, $CO_2$, as the sole or principal source of cellular carbon. The conversion of $CO_2$ to organic cell constituents is a reductive process, which requires a net input of energy. In these physiological groups, accordingly, a considerable part of the energy derived from light or from the oxidation of reduced inorganic compounds must be expended for the reduction of $CO_2$ to the level of organic matter.

All other organisms obtain carbon principally from organic nutrients. Since organic substrates are at the same general oxidation level as organic cell constituents, they do not have to undergo a primary reduction to serve as sources of cell carbon. In addition to meeting the biosynthetic needs of the cell for carbon, organic substrates must supply the energetic requirements of the cell. Consequently, much of the carbon present in the organic substrate enters the pathways of energy-yielding metabolism and is eventually excreted again from the cell, as $CO_2$ (the major product of energy-yielding respiratory metabolism) or as a mixture of $CO_2$ and organic compounds (the typical end-products of fermentative metabolism). Organic substrates thus usually have a *dual nutritional role:* they serve at the same time as a source of carbon and as a source of energy. Many microorganisms can use *a single organic compound* to supply completely both these nutritional needs. Others,

*Table 3.3 Organic compounds capable of serving as the sole source of carbon and energy for the bacterium Pseudomonas multivorans*

1. Carbohydrates and carbohydrate derivatives (sugar acids and polyalcohols)
   - Ribose
   - Xylose
   - Arabinose
   - Fucose
   - Rhamnose
   - Glucose
   - Mannose
   - Galactose
   - Fructose
   - Sucrose
   - Trehalose
   - Cellobiose
   - Salicin
   - Gluconate
   - 2-Ketogluconate
   - Saccharate
   - Mucate
   - Mannitol
   - Sorbitol
   - Inositol
   - Adonitol
   - Glycerol
   - Butylene glycol
2. Fatty acids
   - Acetate
   - Propionate
   - Butyrate
   - Isobutyrate
   - Valerate
   - Isovalerate
   - Caproate
   - Heptanoate
   - Caprylate
   - Pelargonate
   - Caprate
3. Dicarboxylic acids
   - Malonate
   - Succinate
   - Fumarate
   - Glutarate
   - Adipate
   - Pimelate
   - Suberate
   - Azelate
   - Sebacate
4. Other organic acids
   - Citrate
   - α-Ketoglutarate
   - Pyruvate
   - Aconitate
   - Citraconate
   - Levulinate
   - Glycolate
   - Malate
   - Tartrate
   - Hydroxybutyrate
   - Lactate
   - Glycerate
   - Hydroxymethylglutarate
5. Primary alcohols
   - Ethanol
   - Propanol
   - Butanol
6. Amino acids
   - Alanine
   - Serine
   - Threonine
   - Aspartate
   - Glutamate
   - Lysine
   - Arginine
   - Histidine
   - Proline
   - Tyrosine
   - Phenylalanine
   - Tryptophan
   - Kynurenine
   - Kynurenate
7. Other nitrogenous compounds
   - Anthranilate
   - Benzylamine
   - Putrescine
   - Spermine
   - Tryptamine
   - Butylamine
   - Amylamine
   - Betaine
   - Sarcosine
   - Hippurate
   - Acetamide
   - Nicotinate
   - Trigonelline
8. Nitrogen-free ring compounds
   - Benzoylformate
   - Benzoate
   - o-Hydroxybenzoate
   - m-Hydroxybenzoate
   - p-Hydroxybenzoate
   - Phenylacetate
   - Phenol
   - Quinate
   - Testosterone

however, cannot grow when provided with only one organic compound, and need a variable number of additional organic compounds as nutrients. These additional organic nutrients have a purely biosynthetic function, being required as precursors of certain organic cell constituents that the organism is unable to synthesize. They are termed *organic growth factors*, and their roles are described in greater detail below.

Microorganisms are extraordinarily diverse with respect to both the *kind* and the *number* of organic compounds that they can use as a principal source of carbon and energy. This diversity is shown by the fact that *there is no naturally occurring organic compound that cannot be used as a source of carbon and energy by some microorganism.* Hence, it is impossible to describe concisely the chemical nature of organic carbon sources for microorganisms. This extraordinary variation with respect to carbon requirements is one of the most fascinating physiological aspects of microbiology.

When the organic carbon requirements of *individual* microorganisms are examined, some show a high degree of versatility, whereas others are extremely specialized. Certain bacteria of the *Pseudomonas* group, for example, can use any one of over 90 different organic compounds as sole carbon and energy source (Table 3.3). At the other end of the spectrum are methane-oxidizing bacteria, which can use only two organic substrates, methane and methanol, and certain cellulose-decomposing bacteria, which can use only cellulose.

Most (and probably all) organisms that depend on organic carbon sources also require $CO_2$ as a nutrient in very small amounts, because this compound is utilized in a few biosynthetic reactions. However, as $CO_2$ is normally produced in large quantities by organisms that use organic compounds, the biosynthetic requirement can be met through the metabolism of the organic carbon and energy source. Nevertheless, the complete removal of $CO_2$ often either delays or prevents the growth of microorganisms in organic media; and a few bacteria and fungi require a relatively high concentration of $CO_2$ in the atmosphere (5 to 10 percent) for satisfactory growth in organic media.

### The requirements for nitrogen and sulfur

Nitrogen and sulfur occur in the organic compounds of the cell principally in reduced form, as amino and sulfhydryl groups, respectively. Most photosynthetic organisms assimilate these two elements in the oxidized inorganic state, as nitrates and sulfates; their biosynthetic utilization thus involves a preliminary reduction. Many nonphotosynthetic bacteria and fungi can also meet the needs for nitrogen and sulfur from nitrates and sulfates. Some microorganisms are unable to bring about a reduction of one or both of these anions and must be supplied with the elements in a *reduced form.* The requirement for a reduced nitrogen source is relatively common and can be met by the provision of nitrogen as ammonium salts. A requirement for reduced sulfur is rarer; it can be met by the provision of sulfide or of an organic compound that contains a sulfhydryl group (e.g., cysteine).

The nitrogen and sulfur requirements can often also be met by organic nutrients that contain these two elements in reduced organic combination (amino acids or more complex protein degradation products, such as peptones). Such compounds may also, of course, provide organic carbon and energy sources, meeting simultaneously the cellular requirements for carbon, nitrogen, sulfur, and energy.

Several procaryotic groups can also utilize the most abundant natural nitrogen source, $N_2$, which is unavailable to eucaryotic organisms. This process of nitrogen assimilation is termed *nitrogen fixation* and involves a preliminary reduction of $N_2$ to ammonia.

**Organic growth factors**

Any organic compound that an organism requires as a precursor or constituent of its organic cell material, but which it cannot synthesize from simpler carbon sources, must be provided as a nutrient. Organic nutrients of this type are known collectively as *growth factors.* They fall into three classes, in terms of chemical structure and metabolic function:

*Table 3.4 Relation of some water-soluble vitamins to coenzymes*

| Vitamin | Coenzyme | Enzymatic reactions involving the coenzyme form |
|---|---|---|
| *Nicotinic acid (niacin)* | *Pyridine nucleotide coenzymes (NAD and NADP)* | *Dehydrogenations* |
| *Riboflavin (vitamin $B_2$)* | *Flavin nucleotides (FAD and FMN)* | *Some dehydrogenations, electron transport* |
| *Thiamin (vitamin $B_1$)* | *Thiamin pyrophosphate (cocarboxylase)* | *Decarboxylations and some group-transfer reactions* |
| *Pyridoxine (vitamin $B_6$)* | *Pyridoxal phosphate* | *Amino acid metabolism<br>Transamination<br>Deamination<br>Decarboxylation* |
| *Pantothenic acid* | *Coenzyme A* | *Keto-acid oxidation, fatty acid metabolism* |
| *Folic acid* | *Tetrahydrofolic acid* | *Transfer of one-carbon units* |
| *Biotin* | *Prosthetic group of biotin enzymes* | *$CO_2$ fixation, carboxyl transfer* |
| *Cobamide (vitamin $B_{12}$)* | *Cobamide coenzymes* | *Molecular rearrangement reactions* |

*Table 3.5 Concentrations of several water-soluble vitamins[a] in the cells of bacteria (in parts per million of dry weight)*

| **Vitamin** | *Aerobacter aerogenes* | *Pseudomonas fluorescens* | *Clostridium butyricum* |
|---|---|---|---|
| Nicotinic acid | 240 | 210 | 250 |
| Riboflavin | 44 | 67 | 55 |
| Thiamin | 11 | 26 | 9 |
| Pyridoxine | 7 | 6 | 6 |
| Pantothenic acid | 140 | 91 | 93 |
| Folic acid | 14 | 9 | 3 |
| Biotin | 4 | 7 | Required for growth |

[a]In the cell these substances are present in the coenzyme form, but since their quantitative measurement after extraction is dependent on conversion to the corresponding vitamins, the data are presented in this form. Taken from R. C. Thompson, *Texas Univ. Publ.* **4237**, 87 (1942).

(1) *amino acids,* required as constituents of proteins;

(2) *purines* and *pyrimidines,* required as constituents of nucleic acids;

(3) *vitamins,* a diverse collection of organic compounds that form parts of the prosthetic groups or active centers of certain enzymes (Table 3.4).

Because growth factors fulfill specific needs in biosynthesis, they are required in only small amounts, relative to the principal cellular carbon source, which must serve as a general precursor of cell carbon. Some 20 different amino acids enter into the composition of proteins, so the need for any specific amino acid that the cell is unable to synthesize is obviously not large. The same argument applies to specific need for a purine or a pyrimidine: five different compounds of these classes enter into the structure of nucleic acids. The quantitative requirements for vitamins are even smaller, since the various coenzymes of which they are precursors have catalytic roles and consequently are present at levels of a few parts per million in the cell, as shown in Table 3.5.

The biosynthesis of amino acids, purines, pyrimidines, and coenzymes typically involves complex series of individual step reactions, which will be discussed in Chapter 7. The inability to perform *any one* of these step reactions makes an organism dependent on the provision of the end-product as a growth factor. However, the growth factor itself may not be absolutely essential; if the blocked reaction occurs at an early stage in its biosynthesis, organic precursors that follow the blocked step may be able to satisfy the needs of the cell as specific nutrients. A close analysis of a particular growth-factor requirement shown by a number of different microorganisms usually reveals that they differ in the particular chemical form or forms of the growth factor which they require. This can be illustrated by considering the rather common requirement for vitamin $B_1$ (thiamin), which has the following structure:

```
           NH2                CH3
           |                  |
           C                  C==C—CH2—CH2OH
         //  \               /   |
        N     C—CH2—N+       |
        |     ||      \\     |
  CH3—C       CH        C—S
         \\  /          |
           N            H
```

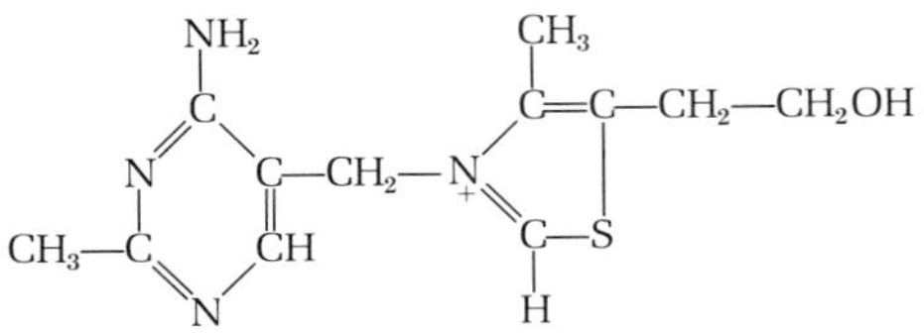

*pyrimidine*            *thiazole*

Some microorganisms require the entire molecule as a growth factor. There are, however, some microorganisms that, given the two halves of the molecule as nutrients, can put them together. Others require only the pyrimidine portion because they can synthesize thiazole. Still others need only the thiazole portion, because they can make and add the pyrimidine portion. For each type of organism described above, the *minimal* growth-factor requirement is different. Yet, in every case, what the organism must eventually have is the entire thiamin molecule, and if this compound is provided as a nutrient, it can be used as a growth factor by all the types described. Even the entire thiamin molecule, however, is not the compound that organisms must eventually make as an essential component of their cells. The functional compound is the coenzyme cocarboxylase, which acts as a prosthetic group in several enzymatic reactions. This coenzyme is thiamin pyrophosphate and has the following structure:

```
           NH2                CH3                  O       O
           |                  |                    ||      ||
           C                  C==C—CH2—CH2O—P—O—P—OH
         //  \               /   |                 |       |
        N     C—CH2—N+       |                 OH      OH
        |     ||      \\     |
  CH3—C       CH        C—S
         \\  /          |
           N            H
```

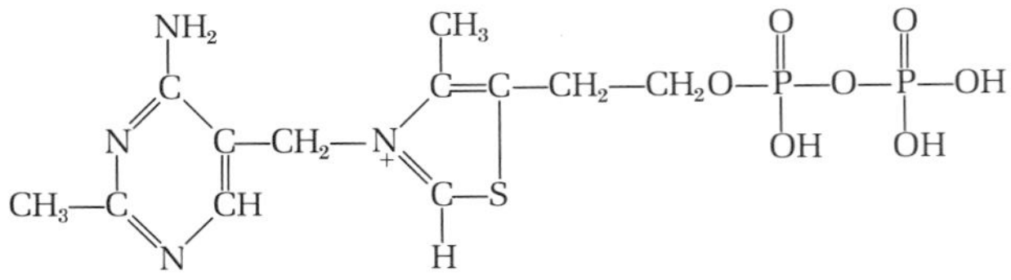

### The roles of oxygen in nutrition

As an elemental constituent of water and of organic compounds, oxygen is a universal component of cells and is always provided in large amounts in the major nutrient, water. However, many organisms also require *molecular oxygen* ($O_2$). These are organisms that are dependent on aerobic respiration for the fulfillment of their energetic needs and for which molecular oxygen functions as a terminal oxidizing agent. Such organisms are termed *obligately aerobic.*

At the other physiological extreme are those microorganisms which obtain energy by means of reactions that do not involve the utilization of molecular oxygen and for which this chemical form of the element is not a nutrient. Indeed, for many of these physiological groups, molecular oxygen is a toxic substance, which either kills them or inhibits their growth. Such organisms are *obligately anaerobic.* The means by which oxygen inhibits obligate anaerobes are probably complex and are not fully understood. In some cases it has been shown that obligate anaerobes contain enzymes that must remain in the reduced state to function, and the inhibition by $O_2$ probably reflects the inactivation of these enzymes. Most obligate anaerobes lack the enzyme catalase, which splits hydrogen peroxide to water and oxygen, but do possess enzymes capable of generating hydrogen peroxide from molecular oxygen. It has been suggested that death of such organisms following exposure to $O_2$ is caused by the intracellular formation and accumulation of hydrogen peroxide, which is a highly toxic compound.

Some microorganisms are *facultative anaerobes,* which can grow either in the presence or in the absence of molecular oxygen. In metabolic terms, facultative anaerobes fall into two subgroups. Some, like the lactic acid bacteria, have an exclusively fermentative energy-yielding metabolism but are not sensitive to the presence of oxygen. Others (e.g., many yeasts, coliform bacteria) can shift from a respiratory to a fermentative mode of metabolism. Such facultative anaerobes use $O_2$ as a terminal oxidizing agent when it is available but can also obtain energy in its absence by fermentative reactions.

Among microorganisms that are obligate aerobes, some grow best at partial pressures of oxygen considerably below that (0.2 atm) present in air. They are termed *microaerophilic.* This probably reflects the possession of enzymes that are inactivated under strongly oxidizing conditions and can thus be maintained in a functional state only at low partial pressures of $O_2$. Many bacteria that obtain energy by the oxidation of molecular hydrogen show this behavior, and it is known that hydrogenase, the enzyme involved in hydrogen utilization, is readily inactivated by oxygen.

## Nutritional categories among microorganisms

Originally, biologists recognized two principal nutritional classes among organisms: the *autotrophs,* exemplified by plants, which can use completely inorganic nutrients; and the *heterotrophs,* exemplified by animals, which require organic nutrients. Today, these two simple categories are insufficient to encompass the variety of nutritional patterns known to exist in the living world, and various attempts to construct more elaborate systems of nutritional classification have been made. The principal disadvantage of the simple dichotomy between autotrophs and heterotrophs is that it does not take into account the subtleties of carbon requirements. There is a very important nutritional difference between a bacterium which uses glucose as its principal carbon source, and a green alga which can grow in an otherwise completely inorganic medium, provided that it is supplied with vitamin $B_{12}$ at a concentration of 1 $\mu$g/liter. Yet both these organisms "require organic nutrients" and are thus, technically speaking, heterotrophs.

The most useful relatively simple primary classification of nutritional categories is one that takes into account two parameters: the *nature of the energy source* and the *nature of the principal carbon source,* disregarding requirements for specific growth factors. Organisms that use light as an energy source are termed *phototrophic* (photosynthetic); organisms that use chemical energy sources are termed *chemotrophic.* Organisms that use $CO_2$ as the principal carbon source are defined as *autotrophic;* organisms that use organic compounds as the principal carbon source are defined as *heterotrophic.* A combination of these two criteria leads, accordingly, to the establishment of four principal categories:

(1) *Photoautotrophic* organisms, dependent on light as an energy source and employing $CO_2$ as the principal carbon source. This category includes higher plants, eucaryotic algae, blue-green algae, and certain photosynthetic bacteria (the purple and green sulfur bacteria).

(2) *Photoheterotrophic* organisms, dependent on light as an energy source and employing organic compounds as the principal carbon source. The principal representatives of this category are a group of photosynthetic bacteria known as the purple nonsulfur bacteria; a few eucaryotic algae also belong to it.

(3) *Chemoautotrophic* organisms, dependent on chemical energy sources and employing $CO_2$ as the principal carbon source. The use of $CO_2$ as a principal carbon source by chemotrophs is always associated with the ability to use *reduced inorganic compounds* as energy sources. This ability is confined to the bacteria and occurs in a number of specialized groups that can use reduced nitrogen compounds ($NH_3$, $NO_2^-$), ferrous iron, reduced sulfur compounds ($H_2S$, S, $S_2O_3^{2-}$), or $H_2$ as oxidizable energy sources.

(4) *Chemoheterotrophic* organisms, dependent on chemical energy sources and employing organic compounds as the principal carbon source. It is characteristic of this category that both energy and carbon requirements are supplied at the expense of an organic compound. Its members are numerous and diverse, including higher animals, protozoa, fungi, and the great majority of the bacteria.

It must be emphasized that this scheme of nutritional categories is based on certain arbitrarily selected metabolic properties and does not provide a *complete* description of the nutritional needs of an organism. Specifically, it leaves out of consideration organic-growth-factor requirements, which may occur in any one of the four principal categories. It also ignores the important fact that microorganisms are nutritionally versatile, so that a given species may be formally assignable to more than one category. Many algae, for example, although capable of photoautotrophic development in the light, can *also* grow in the dark as chemoheterotrophs. Chemoheterotrophy also represents a nutritional alternative for certain of the photoheterotrophic purple bacteria and for the chemoautotrophic, hydrogen-oxidizing bacteria.

The chemoheterotrophs represent a very complex category, which may be subdivided with respect to certain secondary nutritional characters. One possible means of subdivision has already been discussed in Chapter 2—subdivision with respect to the *physical state in which organic nutrients enter the cell.* Organisms that take them up in dissolved form are *osmotrophic,* whereas organisms that engulf them in particulate form are *phagotrophic.* An alternative subdivision may be made on the basis of the *chemical manner in which the energy-yielding organic substrate is degraded.* This viewpoint allows a primary differentiation between *respiratory organisms,* which couple the oxidation of organic substrates with the reduction of an inorganic oxidizing agent (electron acceptor), usually $O_2$, and *fermentative organisms,* in which the energy-yielding metabolism of organic substrates is not so coupled.

## Nutritional interactions among organisms: syntrophy

When two or more different microorganisms are placed together in a medium, their combined metabolic activities may differ either quantitatively or qualitatively from the sum of the activities of the individual members, growing in isolation in the same medium. Such phenomena result from nutritional or metabolic interactions and are collectively termed *syntrophic,* or *synergistic, effects.* For example, the obligately anaerobic, methane-producing bacteria cannot use glucose as a substrate, but they can grow readily at the expense of the end-products formed from glucose by anaerobic bacteria able to ferment this compound.

Particularly interesting examples of syntrophy are encountered in mixed cultures one or more of whose members require growth factors for their development. Microorganisms that can synthesize their own cell materials from simple organic constituents of the medium often excrete small amounts of vitamins or amino acids that are essential for

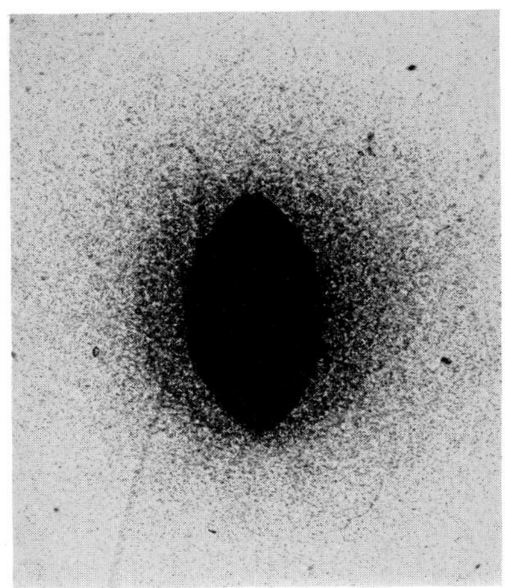

*Figure 3.6 A halo of satellite colonies surrounding a single colony on which they depend for growth factors. A magnified photograph of a portion of agar pour plate in which glucose was provided as the sole carbon nutrient. Before incubation, the agar was heavily seeded with a strain of* Escherichia coli *that required the amino acids isoleucine and valine as growth factors. It was simultaneously inoculated with a few cells of another strain of* E. coli *that requires no growth factors and excretes isoleucine and valine into the medium. Microcolonies of the first strain have developed only in the immediate vicinity of the growing colony of the excreter.*

the growth of others. This permits the development of those types that have more exigent nutritional requirements and are said to be "fed" by the excreters. The phenomenon of *satellite colony formation* (Figure 3.6) is often observed on solid media, where colonies of one type of organism may develop only in the vicinity of those of another type which provide them with essential growth factors.

*Cross-feeding* is an even more complex nutritional relationship between organisms in which each type depends on the other for some essential nutrient in an environment that is nutritionally deficient for either one alone. It will be recalled that some microorganisms cannot synthesize the pyrimidine portion of thiamin, while others cannot synthesize the thiazole portion. In certain instances such organisms can be grown together in a medium containing neither growth factor, since they can supplement each other's dietary needs. Cross-feeding is a simple form of the symbiotic relationship known as *mutualism*. The nutritional interdependence of different organisms has led, in the course of evolution, to a variety of symbiotic associations, some of which will be discussed in Chapter 23.

## The construction of culture media

In constructing a culture medium for any microorganism, the primary goal is to provide a balanced mixture of the required nutrients, at concentrations that will permit good growth. It might seem at first sight reasonable to make the medium as rich as possible, by providing all nutrients in great excess. However, this approach is not a wise one. In the first place, many nutrients become growth inhibitory or toxic as the concentration is raised. This is true of organic substrates, such as salts of fatty acids (e.g., acetate) and even of sugars, if the concentration is high enough. Some inorganic constituents may also become inhibitory if provided in excess; many algae are very sensitive to the concentration of inorganic phosphate. Second, even if growth can occur in a concentrated medium, the metabolic activities of the growing microbial population will eventually change the nature of the environment to the point where it becomes highly unfavorable and the population becomes physiologically abnormal or dies. This may be brought about by a drastic change in the hydrogen ion concentration (pH), by the accumulation of toxic organic metabolites, or, in the case of strict aerobes, by the depletion of oxygen. Since the usual goal of the microbiologist is to study the properties and behavior of *healthy* microorganisms, it is wise to limit the total growth of cultures by providing a limiting quantity of one nutrient; in the case of chemoheterotrophs, the principal carbon source is usually selected for this purpose. Examples of the appropriate concentrations of nutrients will be provided in the various media described below.

The rational point of departure for the preparation of media is to compound a *mineral base*, which provides all those nutrients which can be supplied to any organism in inorganic form. This base can then be supplemented, as required, with a carbon source, an energy source, a nitrogen source, and any required growth factors; these supplements will, of course, vary with the nutritional properties of the particular organism that one wishes to grow. A medium composed entirely of chemically defined nutrients is termed a *synthetic medium*. One that contains ingredients of unknown chemical composition is termed a *complex medium*.

We may illustrate these principles by considering the composition of four media of increasing chemical complexity, each of which is suitable for the cultivation of certain kinds of chemotrophic bacteria (Table 3.6). All four media share a common mineral base. Medium 1 is supplemented with $NH_4Cl$ at a concentration of 1 g/liter but has no added source of carbon. However, if it is incubated aerobically, the $CO_2$ of the atmosphere will be available as a carbon source. In the dark, the only organisms that can grow in this medium are chemoautotrophic nitrifying bacteria, such as *Nitrosomonas*, which obtain carbon from $CO_2$ and energy from the aerobic oxidation of ammonia; the ammonia also provides them with a nitrogen source.

Medium 2 is additionally supplemented with glucose at a concentration of 5 g/liter. Under aerobic conditions, it will support the growth of many bacteria and fungi, since glucose can commonly be used as a

*Table 3.6 Four media of increasing complexity*

| Common ingredients | Additional ingredients | | | |
|---|---|---|---|---|
| | **Medium 1** | **Medium 2** | **Medium 3** | **Medium 4** |
| *Water, 1 liter* | $NH_4Cl$, *1 g* | *Glucose,*[a] *5 g* | *Glucose, 5 g* | *Glucose, 5 g* |
| $K_2HPO_4$, *1 g* | | $NH_4Cl$, *1 g* | $NH_4Cl$, *1 g* | *Yeast extract, 5 g* |
| $MgSO_4 \cdot 7\,H_2O$, *200 mg* | | | *Nicotinic acid, 0.1 mg* | |
| $FeSO_4 \cdot 7\,H_2O$, *10 mg* | | | | |
| $CaCl_2$, *10 mg* | | | | |
| *Trace elements* (Mn, Mo, Cu, Co, Zn) *as inorganic salts, 0.02–0.5 mg of each* | | | | |

[a]If the media are sterilized by autoclaving, the glucose should be sterilized separately and added aseptically. When sugars are heated in the presence of other ingredients, especially phosphates, they are partially decomposed to substances that are very toxic to some microorganisms.

carbon and energy source for aerobic growth. If incubated in the absence of oxygen, it can also support the development of many facultatively or strictly anaerobic bacteria, able to derive carbon and energy from the fermentation of glucose. Note, however, that this medium is not a suitable one for any microorganism that requires growth factors; it contains only a single carbon compound.

Medium 3 is additionally supplemented with one vitamin, nicotinic acid. It can therefore support the growth of all those organisms able to develop in medium 2, together with others, such as the bacterium *Proteus vulgaris*, which require nicotinic acid as a growth factor.

For the three media so far described, the chemical nature of every ingredient is known; thus, they are good examples of synthetic media. Medium 4 is a complex medium, in which the $NH_4Cl$ and nicotinic acid of medium 3 have been replaced by a nutrient of unknown composition, yeast extract, at a concentration of 5 g/liter. It can support the growth of a great many chemoheterotrophic microorganisms, both aerobic and anaerobic, having no growth-factor requirements, relatively simple ones, or highly complex ones. The yeast extract provides a variety of organic nitrogenous constituents (partial breakdown products of proteins) which can fulfill the general nitrogen requirements, and it also contains most of the organic growth factors likely to be required by microorganisms.

Complex media are, accordingly, useful for the cultivation of a wide range of microorganisms, including ones whose precise growth-factor requirements are not known. Even when the growth-factor requirements of a microorganism have been precisely determined, it is often more convenient to grow the organism in a complex medium, particularly if the growth factor requirements are numerous. This point is illustrated in Table 3.7, which describes the composition of a synthetic medium that will support growth of the lactic acid bacterium *Leuconostoc mesenteroides*. This bacterium can also be cultivated satisfactorily in the complex medium 4. In this particular instance, accordingly, the yeast extract of medium 4 must furnish the following requirements: the organic acid, acetate; 19 amino acids; 4 purines and pyrimidines; and 10 vitamins.

The media described in Table 3.6 can support the development of microorganisms only if certain other requirements for growth are also met. These include a suitable temperature of incubation, favorable osmotic conditions, and a hydrogen ion concentration within the range tolerated by the organism in question. Suitable chemical adjustments may be required to accommodate the osmotic conditions and hydrogen ion concentration to the needs of some microorganisms for which these media are satisfactory with respect to their content of nutrients.

*Table 3.7 Medium for Leuconostoc mesenteroides*[a]

| | | | |
|---|---|---|---|
| *Water* | *1 liter* | | |
| *Energy source:* | | | |
| *Glucose* | *25 g* | | |
| *Nitrogen source:* | | | |
| $NH_4Cl$ | *3 g* | | |
| *Minerals:* | | | |
| $KH_2PO_4$ | *600 mg* | $FeSO_4 \cdot 7H_2O$ | *10 mg* |
| $K_2HPO_4$ | *600 mg* | $MnSO_4 \cdot 4H_2O$ | *20 mg* |
| $MgSO_4 \cdot 7H_2O$ | *200 mg* | NaCl | *10 mg* |
| *Organic acid:* | | | |
| *Sodium acetate* | *20 g* | | |
| *Amino acids:* | | | |
| DL-*α-Alanine* | *200 mg* | L-*Lysine* · HCl | *250 mg* |
| L-*Arginine* | *242 mg* | DL-*Methionine* | *100 mg* |
| L-*Asparagine* | *400 mg* | DL-*Phenylalanine* | *100 mg* |
| L-*Aspartic acid* | *100 mg* | L-*Proline* | *100 mg* |
| L-*Cysteine* | *50 mg* | DL-*Serine* | *50 mg* |
| L-*Glutamic acid* | *300 mg* | DL-*Threonine* | *200 mg* |
| *Glycine* | *100 mg* | DL-*Tryptophan* | *40 mg* |
| L-*Histidine* · HCl | *62 mg* | L-*Tyrosine* | *100 mg* |
| DL-*Isoleucine* | *250 mg* | DL-*Valine* | *250 mg* |
| DL-*Leucine* | *250 mg* | | |
| *Purines and pyrimidines:* | | | |
| *Adenine sulfate* · $H_2O$ | *10 mg* | *Uracil* | *10 mg* |
| *Guanine* · HCl · $2H_2O$ | *10 mg* | *Xanthine* · HCl | *10 mg* |
| *Vitamins:* | | | |
| *Thiamine* · HCl | *0.5 mg* | *Riboflavin* | *0.5 mg* |
| *Pyridoxine* · HCl | *1.0 mg* | *Nicotinic acid* | *1.0 mg* |
| *Pyridoxamine* · HCl | *0.3 mg* | *p-Aminobenzoic acid* | *0.1 mg* |
| *Pyridoxal* · HCl | *0.3 mg* | *Biotin* | *0.001 mg* |
| *Calcium pantothenate* | *0.5 mg* | *Folic acid* | *0.01 mg* |

[a]From H. E. Sauberlich and C. A. Baumann, "A Factor Required for the Growth of *Leuconostoc citrovorum*," *J. Biol. Chem.*, **176**, 166 (1948).

**The control of pH***

Although a given medium may be suitable for the *initiation* of growth, the subsequent development of a bacterial population may be severely limited by chemical changes that are brought about by the growth and metabolism of the organisms themselves. For example, in glucose-containing media, organic acids that may be produced as a result of fermentation may become inhibitory to growth.

*The pH scale (Figure 3.7) provides a convenient method of expressing hydrogen ion concentrations in aqueous solutions and is almost invariably used by biologists and biochemists. The pH value of a given solution is the *logarithm of the reciprocal of the hydrogen ion concentration* (expressed in moles per liter), or

$$pH = \log \frac{1}{[H^+]}$$

For example, a solution of acid, which is 0.1 *N* with respect to hydrogen ions, has a pH value of 1.0. This follows from the fact that log (1/0.1) equals the logarithm of 10, which is 1.0.

Figure 3.7 *The pH scale. Acidity and alkalinity are expressed in normality of completely dissociated acid and base, respectively.*

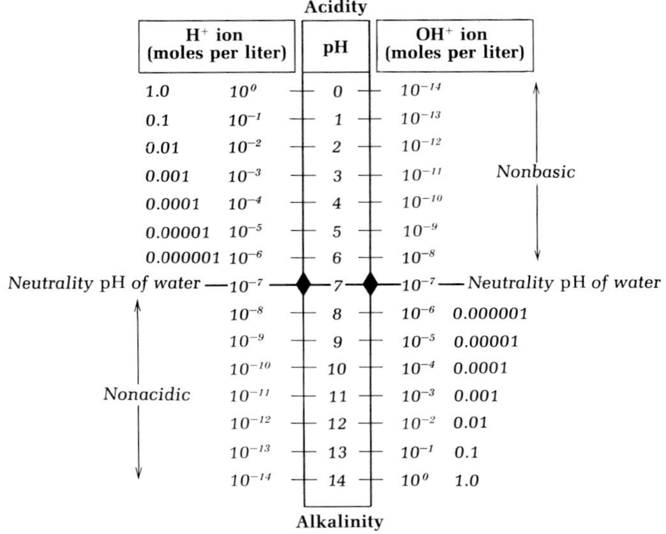


In contrast, the microbial decomposition or utilization of anionic components of a medium tends to make the medium more alkaline. For example, the oxidation of a molecule of sodium succinate liberates two sodium ions in the form of the very alkaline salt, sodium carbonate. The decomposition of proteins and amino acids may also make a medium alkaline as a result of ammonia production.

To prevent excessive changes in hydrogen ion concentration either *buffers* or *insoluble carbonates* are often added to the medium.

The phosphate buffers, which consist of mixtures of monohydrogen and dihydrogen phosphates (e.g., $K_2HPO_4$ and $KH_2PO_4$), are the most useful ones. $KH_2PO_4$ is a weakly acidic salt, whereas $K_2HPO_4$ is slightly basic, so that an equimolar solution of the two is very nearly neutral, having a pH of 6.8. If a limited amount of strong acid is added to such a solution, part of the basic salt is converted to the weakly acidic one:

$$K_2HPO_4 + HCl \rightarrow KH_2PO_4 + KCl$$

If, however, a strong base is added, the opposite conversion occurs:

$$KH_2PO_4 + KOH \rightarrow K_2HPO_4 + H_2O$$

Thus, the solution acts as a buffer in that it resists radical changes in the hydrogen ion concentration when acid or alkali is produced in the medium. By using different ratios of acidic and basic phosphates, different pH values may be established, ranging from approximately 6.0 to 7.6. Good buffering action, however, is obtained only in the narrower range of pH 6.4 to 7.2 because the capacity of a buffer solution is limited by the amounts of its basic and acidic ingredients. Hence, the more acidic the initial buffer, the less is its ability to prevent increases in hydrogen ion concentration (decreases in pH) and the greater its capacity for reacting with alkali. Conversely, the more alkaline the initial buffer, the less is its ability to prevent increases in pH and the greater its ability to prevent acidification.

The phosphates are used widely in the preparation of media because they are the only inorganic agents that buffer in the physiologically important range around neutrality and that are relatively nontoxic to microorganisms. In addition, they provide a source of phosphorus, which is an essential element for growth. In high concentrations, phosphate becomes inhibitory, so the amount of phosphate buffer that can be used in a medium is limited by the tolerance of the particular organism being cultivated. Generally, about 5 g of potassium phosphates per liter of medium can be tolerated by bacteria and fungi.

When a great deal of acid is produced by a culture, the limited amounts of phosphate buffer that may be used become insufficient for the maintenance of a suitable pH. In such cases, carbonates may be added to media as "reserve alkali" to neutralize the acids as they are formed. In the presence of hydrogen ions, carbonate is transformed to bicarbonate, and bicarbonate is converted further to carbonic acid, which decomposes spontaneously to $CO_2$ and water. This sequence of reactions, all of which are freely reversible, can be summarized:

$$CO_3^{2-} \underset{-H^+}{\overset{+H^+}{\rightleftharpoons}} HCO_3^- \underset{-H^+}{\overset{+H^+}{\rightleftharpoons}} H_2CO_3 \rightleftharpoons CO_2 + H_2O$$

Because $H_2CO_3$ is an extremely weak acid and because it decomposes with the loss of $CO_2$ to the atmosphere, the addition of carbonates prevents the accumulation of hydrogen ions and hence of free acids in a medium. The soluble carbonates, such as $Na_2CO_3$, are strongly alkaline and are therefore not suitable for use in culture media. In contrast, *insoluble carbonates* are very useful ingredients for many culture media. Of these insoluble carbonates, finely powdered chalk ($CaCO_3$) is the most generally employed. Because of its insolubility, calcium carbonate does not create strongly alkaline conditions in the medium, especially if it is used in conjunction with other buffers. When, however, the pH of the liquid drops below approximately 7.0, the carbonate is decomposed with the evolution of $CO_2$. It thus acts as a neutralizing agent for any acids that may appear in a culture by converting them to their calcium salts.

The addition of $CaCO_3$ to agar media used for the isolation and cultivation of acid-forming bacteria helps to preserve neutral conditions.

Furthermore, since the acid-forming colonies dissolve the precipitated chalk and become surrounded by clear zones, they can be easily recognized against the opaque background of the medium.

In some instances, neither buffers nor insoluble carbonates can be used to maintain a relatively constant pH in a culture medium. Special problems arise, for example, when very large amounts of acid are formed in a medium in which the presence of calcium carbonate is not desired. Even more serious difficulties are encountered in controlling the pH of slightly alkaline media in which basic substances are produced as a result of bacterial growth. This is due to the fact that phosphate buffers are not effective in the pH range 7.2 to 8.5, and no other suitable buffers in this range are available. In certain cases, therefore, it is necessary to adjust the pH of the culture, either periodically or continuously, by the aseptic addition of strong acids or bases. In some laboratories and in industrial plants, elaborate mechanical devices are used for this purpose. With their aid, a continuous titration of the medium is feasible and the pH is kept nearly constant.

The media described in Table 3.6 are all slightly alkaline at the beginning because they contain the alkaline salt $K_2HPO_4$. Many organisms prefer neutral or slightly acidic conditions, which can be achieved by the use of appropriate buffers.

**The avoidance of mineral precipitates: chelating agents**

A troublesome problem often encountered in the preparation of synthetic media is the formation of a precipitate upon sterilization, particularly if the medium has a relatively high phosphate concentration. This results from the formation of insoluble complexes between phosphates and certain cations, particularly calcium and iron. Although it usually does not affect the nutrient value of the medium, it may make the observation or quantitation of microbial growth difficult. The problem can be avoided by sterilizing separately the calcium and iron salts in concentrated solution and adding them to the sterilized and cooled medium. Alternatively, one can incorporate in the medium a small amount of a chelating agent, which will form a soluble complex with these metals and thus prevent them from forming an insoluble complex with phosphates. The chelating agent most commonly used for this purpose is ethylenediaminetetraacetic acid (EDTA), used at a concentration of approximately 0.01 percent.

**The control of oxygen concentration**

Oxygen is an essential nutrient for the obligately aerobic bacteria. Aerobic microorganisms can be grown easily on the surface of agar plates and in shallow layers of liquid medium. In unshaken liquid cultures, growth usually occurs at the surface. Below the surface, however, conditions become anaerobic, and growth is impossible. To obtain large populations in liquid cultures, *it is therefore necessary to aerate the medium.* Various types of shaking machines that constantly agitate, and thus aerate, the medium are available for laboratory use. Another method of aeration is the continuous passage of a stream of air through a culture. To ensure a large surface of contact between gas and liquid, the air may be introduced through a porous "sparger," which delivers it in the form of very fine bubbles.

For successful cultivation and handling of obligate anaerobes, the organisms must be exposed to oxygen as little as possible. Ground-glass-stoppered bottles completely filled with medium are suitable containers for anaerobic cultures, provided that the medium is first deoxygenated by boiling. The cultivation of anaerobic bacteria on solid media can be achieved either on the surface of ordinary agar plates incubated in an oxygen-free atmosphere or within a solidified deoxygenated agar medium. Jars of various types are used for the anaerobic incubation of plate cultures. Vacuum desiccators, which can be evacuated and refilled with pure nitrogen, hydrogen, or a mixture of either gas with $CO_2$, are commonly employed. Sometimes, air is replaced with hydrogen, and the last traces of oxygen are removed by combustion, which is effected with the aid of a built-in, electrically heated catalyst. Oxygen can also be removed chemically from the air within a container by using a mixture of concentrated solutions of pyrogallic acid and sodium carbonate, which are kept separate until after the container is closed.*

Cultivation of anaerobic bacteria has been greatly facilitated by the discovery that they can be protected from the toxic effects of oxygen by the addition of strong reducing agents to the medium. The inclusion of ascorbate, sodium thioglycolate, cysteine, or traces of sodium sulfide in culture media makes it possible to handle anaerobes even under conditions where oxygen is not rigorously excluded. In a medium containing sodium thioglycolate (1 g/liter), strongly reducing conditions can be established by boiling or autoclaving. The medium can then be protected from contact with air by covering the surface with a layer of mineral oil. A small amount of agar (0.5 to 2.0 g/liter) is commonly incorporated in such media to increase their viscosity. This prevents the free circulation of the medium by mechanical agitation and convection currents and thus helps to preserve anaerobic conditions. Sodium sulfide, which becomes converted to $H_2S$ in neutral solutions,

*Sodium carbonate is preferable to the strongly alkaline hydroxides despite the fact that the latter are just as effective in reacting with pyrogallic acid to remove oxygen. Many bacteria require traces of $CO_2$ for growth. Hydroxides remove $CO_2$ as well as oxygen from the atmosphere. Sodium carbonate, in contrast, actually enriches the atmosphere with $CO_2$ as a result of its reaction with pyrogallic acid.

establishes strongly reducing conditions very rapidly upon its addition to media. Since $H_2S$, even in relatively low concentrations, is toxic to bacteria, the amount of sodium sulfide that can safely be added is 0.1 g/liter or less.

**The provision of carbon dioxide**

A problem frequently encountered in the cultivation of photoautotrophs and chemoautotrophs is the provision of carbon dioxide in sufficient amounts. Although the diffusion of carbon dioxide from the atmosphere into the culture medium will permit growth to occur, the carbon dioxide concentration in the atmosphere is very low (0.03 percent in the open atmosphere, somewhat higher inside a building), and the growth rates of autotrophs are often limited by the availability of carbon dioxide under these conditions. The solution is to gas the cultures with air that has been artificially enriched with carbon dioxide and contains from 1 to 5 percent of this gas. The control of pH becomes a problem, for reasons already discussed (p. 81), and if this solution is adopted, care must be taken to modify the buffer composition of the medium. In the case of autotrophs that can be grown under anaerobic conditions in stoppered bottles (e.g., the purple and green sulfur bacteria) the requirement for $CO_2$ can be met by the incorporation of $NaHCO_3$ in the medium. Soluble bicarbonates cannot be used in media exposed to air because the rapid loss of $CO_2$ to the atmosphere causes the medium to become extremely alkaline.

**The provision of light**

For the cultivation of phototrophic microorganisms (eucaryotic algae, blue-green algae, photosynthetic bacteria), *light* is an essential requirement. The provision of adequate illumination combined with control of temperature is not a simple matter. In the cultivation of nonphotosynthetic organisms, temperature control is provided by the use of incubators, maintained by a thermostatic device at the desired value; however, most models commercially available are not designed with a system of internal illumination and cannot be used for the cultivation of phototrophic organisms.

A relatively uncontrolled and discontinuous illumination may be obtained by the exposure of cultures to daylight, in a north window. Direct exposure to sunlight should be avoided, because the intensity may be too high, and the temperature may rise to a point where growth is prevented. Most phototrophic microorganisms can tolerate continuous illumination, and their growth is much more rapid under these conditions, so artificial light sources are advantageous. The *emission spectrum* of the lamp employed is important. Fluorescent light sources have the practical advantage of producing relatively little heat, so maintenance of a suitable temperature is not difficult. However, their emission spectra are deficient, compared to sunlight, in the longer wavelengths of the visible spectrum and the near infrared region. They are satisfactory for the cultivation of the eucaryotic and blue-green algae, which perform photosynthesis with light of wavelengths shorter than 700 nm but provide little or no photosynthetically effective light for purple and green bacteria, which use wavelengths in the range 750 to 1,000 nm. The only suitable artificial light sources for photosynthetic bacteria are incandescent lamps, and if high intensities are used, the dissipation of heat becomes a problem. The easiest solution is to immerse culture vessels in a glass or plastic water bath which can be subjected to lateral illumination and maintained at the desired temperature by the circulation of cold water. The other solution is to construct a light cabinet with internal incandescent illumination, in which the temperature can be controlled by ventilation or refrigeration.

**The routine maintenance of laboratory cultures**

The maintenance of a culture collection is a tedious task, for the cultures must be transferred at regular intervals to prevent their dying. The media and the conditions of cultivation for this purpose are designed to ensure the maximal survival of the organisms. Although good growth must be obtained in each culture, the massive accumulation of metabolic products in the medium must be avoided because these often accelerate the death of the organisms.

To prevent a rapid senescence of "stock cultures," they are maintained, whenever possible, at relatively low temperatures. Incubation at elevated temperatures is carried out only for a minimal period of time. The cultures are then kept at room temperature or in a refrigerator.

Cultures of most bacteria and fungi can also be preserved, often for many years, in a dry state in a vacuum. The dehydration, known as *lyophilization*, is achieved by subjecting frozen suspensions of the organisms to a high vacuum.

## *Selective media*

It is clear that no single medium or set of conditions will support the growth of all the different types of organisms that occur in nature. Conversely, any medium that is suitable for the growth of a specific organism is, to some extent, *selective* for it. In a medium inoculated with a variety of organisms, only those that can grow in it will reproduce, and all others will be suppressed. Further, if the nutritional requirements of an organism are known, experimental conditions can be set up that will favor the development of one particular organism over that of others. Thus, if specially designed selective media are used, desired types of microorganisms can be isolated from natural environments such as soil, in which many different kinds of organisms are present. Selective media can be used to obtain microorganisms from nature either by *direct isolation* or by *enrichment*.

**Direct isolation**

Solid media are employed for direct isolation. When a mixed inoculum containing a variety of different organisms is spread directly on the surface of a *selective solid medium*, all the bacteria that can grow on it will produce colonies. Dispersal of the organisms over the medium eliminates, to a large extent, a competition for nutrients; hence, slow-growing bacteria are able to produce colonies in the same environment with those that grow rapidly. If the inoculum is not diluted too much before or during the spreading, not only the predominant organisms but also those that are present in relatively small numbers will have a chance to develop discrete colonies.

Direct isolation is the most effective technique to use when the goal is to isolate the *maximum possible number of types* of microorganisms that are able to grow in a particular environment. For example, to obtain a wide variety of aerobic bacteria that can utilize starch as their only organic nutrient, an agar medium containing starch and all essential mineral salts is used. The medium is inoculated by spreading evenly over its surface a bit of soil or leaf mold suspended in a drop of water. After incubation under aerobic conditions, many different types of colonies develop, and from these, different strains can be isolated on fresh plates of the same medium.

**Enrichment**

As opposed to the direct isolation of organisms on solid selective media, in which competition between different microbial types is avoided, the enrichment culture technique is based on free competition among different organisms in *liquid media*. If a liquid selective medium is inoculated with a variety of organisms and incubated under any chosen set of conditions, those organisms for which the environment is most suitable will outgrow all others and eventually become predominant. A gram of garden soil generally serves as an excellent source of a variety of microorganisms for the inoculation of enrichment media. In certain cases, other sources are used either because they are known to contain large quantities of a desired organism or because the purpose of enrichment is the detection of an organism in a given environment.

Two important factors are involved in the preparation of an enrichment culture for any given organism: (1) the conditions of cultivation must be made suitable for the organism to be enriched; (2) the environment must be made as unsuitable as possible for the growth of other organisms. In a few cases, a medium can be prepared that will support the development of only one type of organism. In such a medium the desired organisms become enriched simply because they have no competitors. For example, if a medium containing no organic compounds and no oxidizable substrates other than ammonium salts (medium 1 in Table 3.6) is incubated under aerobic conditions in the dark, only ammonia-oxidizing chemoautotrophs (e.g., *Nitrosomonas*) can develop. In practice, however, most enrichment media cannot be designed to be so highly selective; they can support the growth of more than one type of organism. Their effectiveness depends, therefore, on rather subtle advantages that they may offer to one or another physiological group of microorganisms.

In a mixed culture, the ultimate predominance of one type of organism depends on its ability to grow more rapidly or more extensively than its competitors. The enrichment of various organisms can be achieved by the manipulation of such environmental factors as the nutrient composition of the medium, aeration, pH, temperature, osmotic pressure, and surface tension. For example, the closely related genera *Escherichia* and *Aerobacter* can both be grown and enriched in similar media (medium 2 of Table 3.6). However, while the former is a normal inhabitant of the intestines of warm-blooded animals, the latter is a common soil organism. Hence, *Escherichia* is naturally adapted to grow at somewhat higher temperatures than is *Aerobacter*. It is also relatively tolerant to bile salts that occur in the digestive tract and which are toxic to *Aerobacter*. The selective enrichment for *Escherichia* rather than for *Aerobacter* can therefore be obtained either by incubating the medium at 45°C or by adding bile salts to it.

Another situation in which temperature selection can be used with great effectiveness is in the isolation of blue-green algae. These organisms share with many eucaryotic algae the same general nutrient requirements: they grow well in a simple mineral medium when provided with light. However, competition from eucaryotic algae can be almost entirely eliminated by incubating such enrichment cultures at a temperature of 35°C, which is nearly optimal for the growth of most blue-green algae but too high to permit the development of eucaryotic algae, for most of which 30°C is the temperature maximum.

To isolate endospore-forming bacteria, the inoculum is pasteurized by brief exposure to high temperatures (e.g., 2 to 5 minutes at 80°C). Thus, competition from nonsporulating bacteria is eliminated. If the pasteurized cultures are incubated in aerobic conditions, members of the genus *Bacillus* are obtained. In the absence of oxygen, various species of anaerobes of the genus *Clostridium* develop.

Environmental changes resulting from the growth of the organisms themselves may create strongly selective conditions that favor some organisms and suppress others. In special cases, it is possible to take advantage of this phenomenon for purposes of enrichment.

**Enrichment methods for some specialized physiological groups**

The tool of the enrichment culture is one of the most powerful techniques available to the microbiologist. An almost infinite number of permutations and combinations of the different environment variables, nutritional and physical, can be developed for the specific isolation of microorganisms from nature. Enrichment techniques provide a means for isolating known microbial types at will from nature, by taking advantage of their specific requirements, and can also be indefinitely elaborated, as a means of obtaining for study hitherto undescribed organisms capable of growing in the environments devised by the scientist. Here we shall attempt to summarize a few of the enrichment procedures that can be used to isolate major physiological groups of microorganisms, principally bacteria, from nature.

*Table 3.8 Primary environmental factors that determine the outcome of enrichment procedures for chemoheterotrophic bacteria with the use of synthetic media*

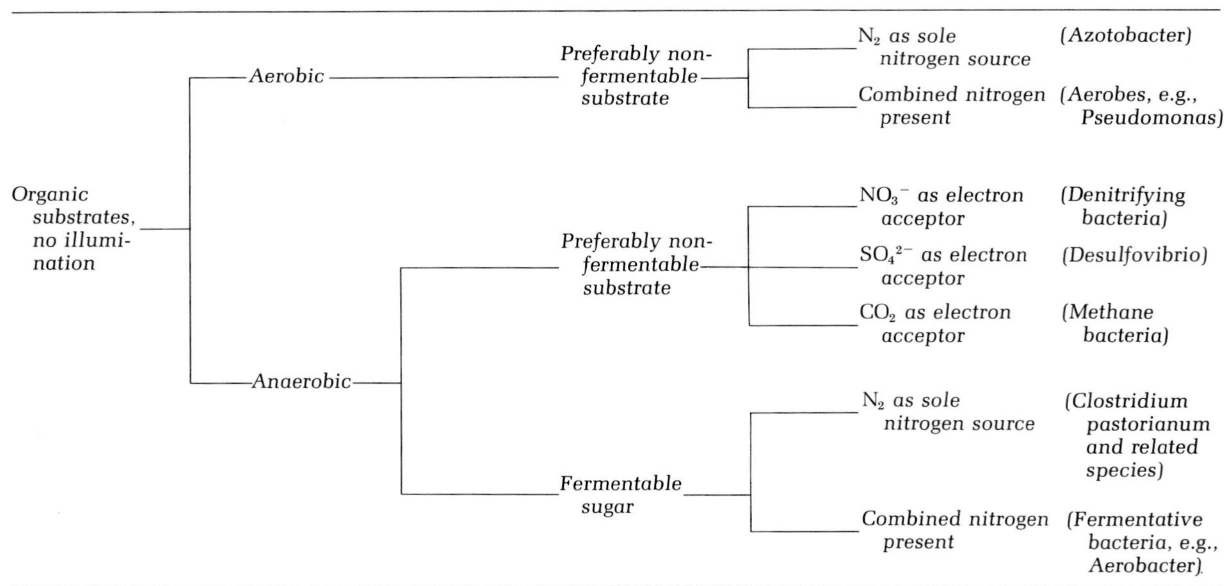

| | | | | |
|---|---|---|---|---|
| *Organic substrates, no illumination* | *Aerobic* | *Preferably nonfermentable substrate* | $N_2$ *as sole nitrogen source* | *(Azotobacter)* |
| | | | *Combined nitrogen present* | *(Aerobes, e.g., Pseudomonas)* |
| | *Anaerobic* | *Preferably nonfermentable substrate* | $NO_3^-$ *as electron acceptor* | *(Denitrifying bacteria)* |
| | | | $SO_4^{2-}$ *as electron acceptor* | *(Desulfovibrio)* |
| | | | $CO_2$ *as electron acceptor* | *(Methane bacteria)* |
| | | *Fermentable sugar* | $N_2$ *as sole nitrogen source* | *(Clostridium pastorianum and related species)* |
| | | | *Combined nitrogen present* | *(Fermentative bacteria, e.g., Aerobacter)* |

*Table 3.9 Enrichment conditions for chemoheterotrophic bacteria*[a,b]

| Additions to basal medium | | Special environmental conditions | | |
|---|---|---|---|---|
| **Organic** | **Inorganic** | **Atmosphere** | **pH** | **Organisms enriched** |
| *Ethyl alcohol, 4.0* | *None* | *Air* | *7.0* | *Azotobacter* |
| *Ethyl alcohol, 4.0* | $NH_4Cl$, *1.0* | *Air* | *7.0* | *Aerobes (e.g., Pseudomonas)* |
| *Ethyl alcohol, 4.0* | $NaNO_3$, *3.0* | *None (stoppered bottle)* | *7.0* | *Denitrifying bacteria (e.g., some species of Pseudomonas)* |
| *Ethyl alcohol, 4.0* | $NH_4Cl$, *1.0*<br>$Na_2SO_4$, *5.0* | *None (stoppered bottle)* | *7.2* | *Desulfovibrio* |
| *Ethyl alcohol, 4.0* | $NH_4Cl$, *1.0*<br>$NaHCO_3$, *1.0*<br>$CaCO_3$, *5.0* | *None (stoppered bottle)* | *7.4* | *Methane-producing bacteria* |
| *Glucose, 10.0* | $CaCO_3$, *5.0* | Pure $N_2$ | *7.0* | *Clostridium pastorianum and related species* |
| *Glucose, 10.0* | $NH_4Cl$, *1.0* | *None (stoppered bottle)* | *7.0* | *Fermentative bacteria (e.g., Aerobacter)* |

[a]The components of the medium are given in grams per liter.
[b]Incubated at 25–30°C without illumination. *Basal medium:* $MgSO_4 \cdot 7H_2O$, *0.2;* $K_2HPO_4$, *1.0;* $FeSO_4 \cdot 7H_2O$, *0.05;* $CaCl_2$, *0.02;* $MnCl_2 \cdot 4H_2O$, *0.002;* $NaMoO_4 \cdot 2H_2O$, *0.001.*

**Synthetic enrichment media for chemoheterotrophs**

The nutritional and environmental conditions necessary for the enrichment of various groups of chemoheterotrophs in synthetic media are outlined in Table 3.8, and the detailed composition of each enrichment medium is described in Table 3.9.

For the enrichment of fermentative organisms, the chemical nature of the organic substrate is important; to be attacked by fermentation it must be neither too oxidized nor too reduced. Sugars are excellent fermentative substrates, but many other classes of organic compounds on the same approximate oxidation level can also be fermented. The enrichment cultures must be incubated anaerobically, not only because some fermentative organisms are obligate anaerobes, but to prevent competition from aerobic forms. Nitrate should not be used as a nitrogen source, since it will also allow growth of denitrifying bacteria (see below). Calcium carbonate may be added if the organisms being enriched produce acid and are themselves sensitive to it.

Three special physiological groups of chemoheterotrophic bacteria, which can use inorganic compounds other than molecular oxygen as terminal oxidants for respiratory metabolism, may also be enriched in synthetic media under anaerobic conditions. In these cases, readily fermentable organic substrates such as sugars should be avoided. For the denitrifying bacteria, nitrate is included as a terminal oxidant, and acetate, butyrate, or ethyl alcohol as a carbon and energy source. For the sulfate-reducing bacteria, a relatively large amount of sulfate is included, since it is the specific terminal oxidant; lactate or malate, but not acetate, provides the best source of carbon and energy. For the carbonate-reducing (methane-producing) bacteria, $CO_2$ must be present as an oxidant and such compounds as formate as the source of carbon and energy. It should also be noted here that in enrichments for sulfate- and carbonate-reducing bacteria, the use of ammonia as a nitrogen source is desirable; addition of nitrate will favor development of the nitrate reducers. Similarly, in enrichments for carbonate and nitrate reducers, the concentration of sulfate should be kept to a minimum to prevent overgrowth by sulfate reducers.

Obviously, aerobic conditions are essential for the enrichment of microorganisms that can obtain energy from respiration. Either fermentable or nonfermentable substrates are satisfactory organic nutrients. If a fermentable substrate is used, however, the culture must be aerated thoroughly and continuously, because the depletion of oxygen by respiration favors the enrichment of anaerobes. Readily fermentable compounds generally stimulate the growth of facultative anaerobes. Thus, when glucose or other sugars are included in enrichment media, *Aerobacter* appears as one of the principal organisms under aerobic as well as under anaerobic conditions.

The use of nonfermentable substrates usually results in the enrichment of organisms that are obligate aerobes, most commonly members of the genus *Pseudomonas*. For instance, benzoate-oxidizing strains of *P. putida* can be obtained by using sodium benzoate (1 g/liter) as the sole organic nutrient. If asparagine (2 g/liter) is used as the sole source of both carbon and nitrogen in the medium, asparagine-oxidizing strains of this or other related species become predominant.

Ammonium salts are generally used as a nitrogen source in synthetic enrichment media for aerobes. If no combined nitrogenous compounds are provided, cultures of the aerobic nitrogen-fixing bacteria of the genus *Azotobacter* can be obtained. Species of *Azotobacter* can use a great variety of organic substrates, including alcohol, butyrate, benzoate, and glucose, as their only organic nutrients.

### The enrichment of chemoautotrophic and photosynthetic organisms

For the enrichment of chemoautotrophic and photoautotrophic organisms, organic compounds must be omitted from the medium, and $CO_2$ or bicarbonate must be used as the only source of carbon (Tables 3.10 to 3.13). Photosynthetic forms require light, whereas the chemolithotrophs should be cultivated in the dark to prevent the development of photosynthetic types. During incubation of cultures, either aerobic or

*Table 3.10 Primary environmental factors that determine the outcome of enrichment procedures for some chemoautotrophic bacteria*

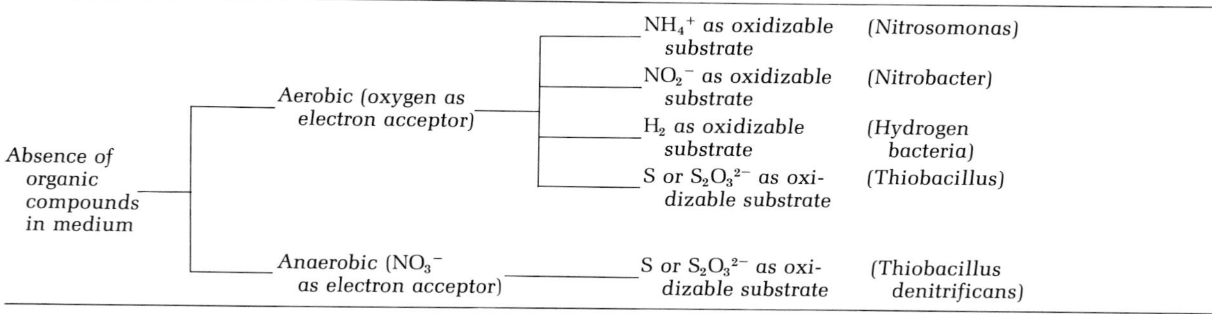


anaerobic conditions must be maintained, depending on whether or not the organisms need oxygen. An exception to this rule is found in the selective cultivation of algae. Since these organisms *produce* oxygen in their metabolism, it makes virtually no difference whether the enrichment cultures are incubated under aerobic or anaerobic conditions.

Media devised for the enrichment and propagation of photosynthetic organism should contain sodium because this element is known to be required by blue-green algae and photosynthetic bacteria.

For the enrichment of purple nonsulfur bacteria, the medium should

*Table 3.11 Enrichment conditions for some chemoautotrophic bacteria*[a]

| | Special environmental features | | |
|---|---|---|---|
| **Additions to medium** | **Atmosphere** | **pH** | **Organism enriched** |
| $NH_4Cl$, 1.5; $CaCO_3$, 5.0 | Air | 8.5 | *Nitrosomonas* |
| $NaNO_2$, 3.0 | Air | 8.5 | *Nitrobacter* |
| $NH_4Cl$, 1.0 | 85% $H_2$, 10% $O_2$, 5% $CO_2$ | 7.0 | Hydrogen bacteria |
| $NH_4Cl$, 1.0; $Na_2S_2O_3 \cdot 7\,H_2O$, 7.0 | Air | 7.0 | *Thiobacillus* |
| $NH_4NO_3$, 3.0; $Na_2S_2O_3 \cdot 7\,H_2O$, 7.0; $NaHCO_3$, 5.0 | None (stoppered bottle) | 7.0 | *Thiobacillus denitrificans* |

[a]The components of the medium are given in grams per liter. *Basal medium:* $MgSO_4 \cdot 7\,H_2O$, 0.2; $K_2HPO_4$, 1.0; $FeSO_4 \cdot 7\,H_2O$, 0.05; $CaCl_2$, 0.02; $MnCl_2 \cdot 4\,H_2O$, 0.002; $NaMoO_4 \cdot 2\,H_2O$, 0.001. *Environment:* in the dark; temperature, 25°–30°C.

*Table 3.12 Primary environmental factors that determine the outcome of enrichment procedures for photosynthetic microorganisms*

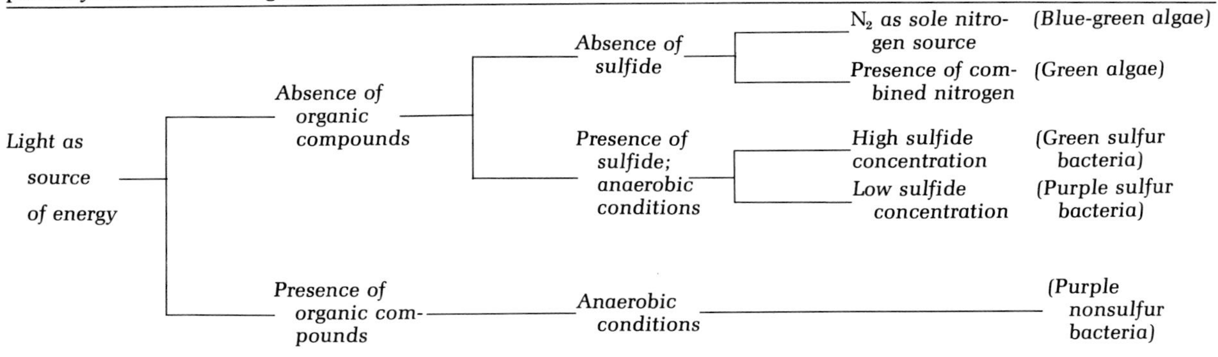


include a suitable organic substrate and, in some instances, bicarbonate. The substrate should not be a readily fermentable one; acetate, butyrate, or malate is customarily used. Bicarbonate must be added if the substrate (e.g., butyrate) is more reduced than cell material because photosynthesis is accompanied by a net consumption of $CO_2$. With substrates such as malate, which are metabolized with a net production of $CO_2$, the addition of bicarbonate is unnecessary. Because the photoheterotrophic bacteria require various growth factors, a small amount of yeast extract is generally added to the enrichment medium.

*Table 3.13 Enrichment conditions for photosynthetic microorganisms*[a]

| **Additions to medium** | | **Special environmental features** | | |
|---|---|---|---|---|
| **Organic** | **Inorganic** | **Atmosphere** | **pH** | **Organisms enriched** |
| None | None | Air, or air + 5% $CO_2$ | 6.0–8.0 | Blue-green algae |
| None | $NaNO_3$ or $NH_4Cl$, 1.0 | Air, or air + 5% $CO_2$ | 6.0–8.0 | Green algae |
| None | $NH_4Cl$, 1.0; $Na_2S \cdot 9\,H_2O$, 2.0; $NaHCO_3$, 5.0 | None (stoppered bottle) | 7.5 | Green sulfur bacteria |
| None | $NH_4Cl$, 1.0; $Na_2S \cdot 9\,H_2O$, 1.0; $NaHCO_3$, 5.0 | None (stoppered bottle) | 8.0–8.5 | Purple sulfur bacteria |
| Sodium malate, 5.0; yeast extract, 0.5 | $NH_4Cl$, 1.0 | None (stoppered bottle) | 7.0–7.5 | Purple nonsulfur bacteria |

[a]The components of the medium are given in grams per liter. *Basal medium:* $MgSO_4 \cdot 7\,H_2O$, 0.2; $K_2HPO_4$, 1.0; $FeSO_4 \cdot 7\,H_2O$, 0.01; $CaCl_2$, 0.02; $MnCl_2 \cdot 4\,H_2O$, 0.002; $NaMoO_4 \cdot 2\,H_2O$, 0.001; NaCl, 0.5. *Environment:* constant illumination, temperature, 25–30°C.

**The use of complex media for enrichment**

Some bacteria cannot be enriched in defined media because they have extremely complex nutritional requirements. Nevertheless, such organisms may, in some cases, be obtained from nature by the use of specially designed complex media. The lactic acid bacteria illustrate this point. These organisms are characterized by their remarkable resistance to lactic acid, which they themselves produce in the fermentation of sugar. To enrich for lactic acid bacteria, a poorly buffered medium containing glucose and a rich source of growth factors is used (e.g., 20 g of glucose and 10 g of yeast extract per liter). After inoculation, preferably with natural materials that are rich in lactic acid bacteria (e.g., vegetable matter, raw milk, sewage), the medium is incubated under anaerobic conditions. The first organisms to develop are usually bacteria such as *Aerobacter* and *Escherichia*. However, as lactic acid gradually accumulates, conditions become less and less favorable for these bacteria, whereas the lactic acid bacteria continue to grow. Eventually, the acidity of the medium becomes so high that the lactic acid bacteria predominate and most other organisms are destroyed.

Another good example of a complex medium that is quite selective for a specific group of organisms is one designed for the enrichment of the propionic acid bacteria. These organisms produce propionic acid, acetic acid, and $CO_2$ in fermentation. Although they can ferment glucose readily, they cannot compete either with *Aerobacter* or with the lactic acid bacteria in glucose media because they grow relatively slowly and do not tolerate acidic conditions. However, the propionic acid bacteria can also ferment lactic acid, which is not a suitable substrate for most other fermentative organisms. This capacity is the key to their enrichment. If a neutral medium containing 20 g of sodium lactate and 10 g of yeast extract per liter is inoculated with natural materials containing propionic acid bacteria and incubated at 30°C under anaerobic conditions, an enrichment of these organisms is obtained. Swiss cheese is the best inoculum for the cultures because propionic acid bacteria are the principal agents in its ripening.

Complex media can be used successfully for the selective cultivation of the acetic acid bacteria. These bacteria are especially adapted to environments that contain high concentrations of alcohol. They are also far less susceptible than other bacteria to inhibition by acetic acid, which they produce from alcohol in respiration. To enrich for them, a complex medium containing alcohol is inoculated with materials containing these bacteria and then incubated under aerobic conditions. Fruits, flowers, and unpasteurized (draught) beer are good sources of inoculum. A medium containing 40 ml of alcohol and 10 g of yeast

*Table 3.14 Complex media for enrichment of chemoheterotrophs*[a]

| Additions | Special environmental conditions | Preferred choice of inoculum | Organisms enriched |
|---|---|---|---|
| *None* | pH *7.0; aerobic* | *Soil* | *Aerobic amino acid oxidizers* |
| *None* | pH *7.0; aerobic* | *Pasteurized soil* | *Bacillus spp.* |
| *None* | pH *7.0; anaerobic* | *Pasteurized soil* | *Amino acid–fermenting clostridia* |
| *Urea, 50.0* | pH *8.5; aerobic* | *Pasteurized soil* | *Alkali-tolerant urea-decomposing bacilli (Bacillus pasteurii)* |
| *Glucose, 20.0* | pH *2.0–3.0; anaerobic* | *Soil* | *Anaerobic Sarcina spp.* |
| *Glucose, 20.0* | pH *6.5; anaerobic* | *Plant materials, milk* | *Lactic acid bacteria* |
| *Glucose, 20.0;* $CaCO_3$, *20.0* | pH *7.0; aerobic or anaerobic* | *Soil or sewage* | *Coliform bacteria* |
| *Glucose, 20.0;* $CaCO_3$, *20.0* | pH *7.0; anaerobic* | *Pasteurized soil* | *Sugar-fermenting clostridia* |
| *Sodium lactate, 20.0* | pH *7.0; anaerobic* | *Swiss cheese* | *Propionic acid bacteria* |
| *Ethanol, 40.0* | pH *6.0; aerobic* | *Fruits, unpasteurized beer* | *Acetic acid bacteria* |

[a]Common components are yeast extract, 10.0; $KH_2PO_4$ or $K_2HPO_4$, 1.0; $MgSO_4$, 0.2. (All components are given in grams per liter.) Incubation is generally at 30°C.

extract per liter, adjusted to pH 6.0, can be used for enrichment. Beer and hard cider are also excellent enrichment media, for these fermented beverages closely resemble the natural environment in which acetic acid bacteria predominate. Although a large surface of contact with air must be ensured, the inoculated medium is usually not aerated vigorously because many acetic acid bacteria grow best in a pellicle that they form at the surface.

Examples of complex media and environmental conditions used for the enrichment of selected organotrophs are given in Table 3.14.

## *Further reading*


**Books**

Meynell, G. G., and E. Meynell, *Theory and Practice in Experimental Bacteriology*. New York: Cambridge University Press, 1965.

Schlegel, H. G. (editor), *Anreicherungskultur und Mutantenauslese* (Enrichment Culture and Mutant Selection). Stuttgart: Fischer, 1965. The only systematic description of enrichment methods; many of the articles are in English.

**Review**

Luria, S. E., "The Bacterial Protoplasm: Composition and Organization," in *The Bacteria* (I. C. Gunsalus and R. Y. Stanier, editors), Vol. I. p. 1. New York: Academic Press, 1960.

# Chapter four
# The eucaryotic protists

*Among protists* that possess a eucaryotic cell structure, three major groups can be recognized: *algae, protozoa,* and *fungi.* Each of these groups is very large and internally diverse. The more highly specialized representatives—for example, a seaweed, a ciliate, and a mushroom—can be readily assigned to the algae, the protozoa, and the fungi, respectively. However, there are many eucaryotic protists for which the assignment is arbitrary: numerous transitions exist between algae and protozoa and between protozoa and fungi. For this reason, the three major groups of eucaryotic protists cannot be sharply distinguished in terms of simple sets of clear-cut differences. Broadly speaking, the eucaryotic algae may be defined as organisms that perform oxygen-evolving photosynthesis and possess chloroplasts. Some of them are unicellular microorganisms; some are filamentous, colonial, or coenocytic; and some have a plantlike structure that is formed through extensive multicellular development, with little or no differentiation of cells and tissues. In organismal terms, accordingly, the algae are highly diverse, and by no means all fall into the category of microorganisms. The brown algae known as kelps may attain a total length of as much as 50 m. The protozoa and fungi are nonphotosynthetic organisms, and the difference between them is essentially one of organismal structure; protozoa are predominantly unicellular, whereas fungi are predominantly coenocytic and grow in the form of a filamentous, branched structure known as a *mycelium.*

For historical reasons that have been discussed in Chapter 2, the

*Table 4.1 Major groups of eucaryotic algae*

| **Group name** | **Pigment system** | | **Composition of cell wall** | **Nature of reserve materials** | **Number and type of flagella** | **Range of structure** |
|---|---|---|---|---|---|---|
| | **Chlorophylls** | **Other special pigments** | | | | |
| *Green algae:* division Chlorophyta | a + b | — | *Cellulose* | *Starch* | *Generally two identical flagella per cell* | *Unicellular, coenocytic, filamentous; plant-like multicellular forms* |
| *Euglenids:* division Euglenophyta | a + b | — | *No wall* | *Paramylum and fats* | *One, two, or three flagella per cell* | *All unicellular* |
| *Dinoflagellates and related forms:* division Pyrrophyta | a + c | *Special carotenoids* | *Cellulose* | *Starch and oils* | *Two flagella, dissimilar in form and position on cell* | *Mostly unicellular, a few filamentous forms* |
| *Chrysophytes and diatoms:* division Chrysophyta | a ± c | *Special carotenoids* | *Wall composed of two overlapping halves, often containing silica (some have no walls)* | *Leucosin and oils* | *Two flagella, arrangement variable* | *Unicellular, coenocytic, filamentous* |
| *Brown algae:* division Phaeophyta | a + c | *Special carotenoids* | *Cellulose and algin* | *Laminarin and fats* | *Two flagella, of unequal length* | *Plantlike multicellular forms* |
| *Red algae:* division Rhodophyta | a | *Phycobilins* | *Cellulose* | *Starch* | *No flagella* | *Unicellular; plantlike multicellular forms* |

algae and fungi were traditionally regarded as "plants" and have been largely studied by botanists, while the protozoa were traditionally regarded as "animals" and have been largely studied by zoologists. As a result of this specialization, the many interconnections between the three groups have tended to be overlooked. We shall attempt in this chapter to provide a unified account of the properties of eucaryotic protists that emphasizes possible evolutionary interrelationships.

## *The eucaryotic algae*

The primary classification of algae is based on cellular, not organismal, properties: the chemical nature of the wall, if present; the organic reserve materials produced by the cell; the nature of the photosynthetic pigments; and the nature and arrangement of the flagella borne by motile cells. In terms of these characters, the algae are arranged in a series of divisions, summarized in Table 4.1.

The divisions are not equivalent to one another in terms of the range of organismal structure of their members. For example, the Euglenophyta (euglenid algae) consist entirely of unicellular or simple colonial organisms, while the Phaeophyta (brown algae) consist only of plantlike, multicellular organisms. The largest and most varied group, the Chlorophyta (green algae), from which the higher plants probably originated, spans the full range of organismal diversity, from unicellular organisms to multicellular representatives with a plantlike structure.

The common cellular properties of each algal division suggest that its members, however varied their organismal structure may be, are representatives of a single major evolutionary line. Evolution among the algae thus in general appears to have involved *a progressive increase in organismal complexity in the framework of a particular variety of eucaryotic cellular organization.* Although it is possible to perceive these evolutionary progressions *within* each algal division, the relationships *between* divisions are completely obscure. The primary origin of the eucaryotic algae as a whole is accordingly an unsolved (and no doubt insoluble) problem.

### The photosynthetic flagellates

In many algal divisions, the simplest representatives are motile, unicellular organisms, known collectively as *flagellates.* The cell of a typical flagellate, illustrated by *Euglena* (Figure 4.1), has a very marked polarity: it is elongated and leaf-shaped, the flagella usually being inserted at the anterior end. In the Euglenophyta, to which *Euglena* belongs, there are two flagella of unequal length, which originate from a

Figure 4.1 *Euglena gracilis. (a) Photomicrograph of fixed cell (×1,000). Courtesy of Gordon F. Leedale. (b) Schematic drawing of the same cell, to show principal structural features: n, nucleus; c, chloroplast; m, mitochondrion; e, eyespot; $f_1$, $f_2$, the two flagella of unequal length, originating within a small cavity of the anterior end of the cell.*

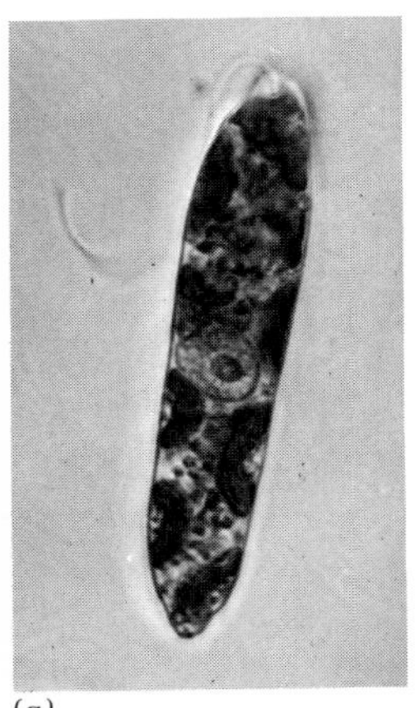

(a)

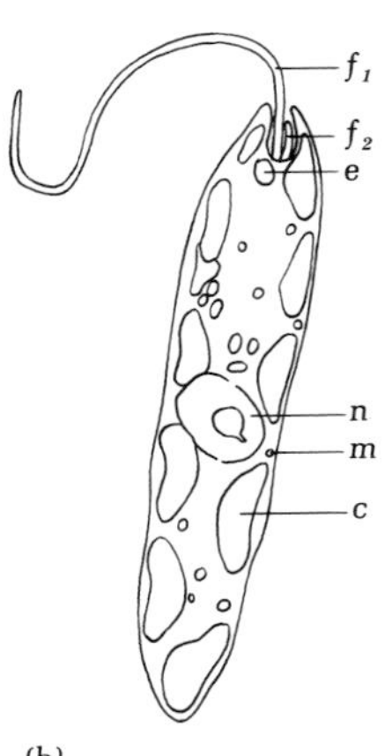


(b)

small cavity at the anterior end of the cell. Many chloroplasts and mitochondria are dispersed throughout the cytoplasm. Near the base of the flagellar apparatus is a specialized organelle, the *eyespot*, which is red, owing to its content of special carotenoid pigments; the eyespot serves as a photoreceptor to govern the active movement of the cell in response to the direction and intensity of illumination. The cell of *Euglena*, unlike that of many other flagellates, is not enclosed within a rigid wall; its outer layer is an elastic *pellicle*, which permits considerable changes of shape. Cell division occurs by *longitudinal fission* [Figure 4.2(*a*)]. About the time of the onset of mitosis, there is a duplication of the anterior organelles of the cell, including the flagella and their basal apparatus; cleavage subsequently occurs through the long axis, so that the duplicated organelles are equally partitioned between the two daughter cells. This mode of cell division is characteristic of all flagellates except those belonging to the Chlorophyta, such as *Chlamydomonas*, where each cell undergoes *two or more multiple fissions* to produce four smaller daughter cells, liberated by rupture of the parental cell wall [Figure 4.2(*b*)]. Even in such cases, however, the internal divisions take place in the longitudinal plane. As we shall see in a subsequent section, longitudinal division also occurs in the nonphotosynthetic flagellate protozoa and is one of the primary characters that distinguish these organisms from the other major group of protozoa that possess flagellalike locomotor organelles, the ciliates.

In terms of their cellular properties, the photosynthetic flagellates are a highly diverse collection of organisms and these cellular differences are of great importance in classification of these algae.

Most multicellular algae are immotile in the mature state. However, their reproduction frequently involves the formation and liberation of

Figure 4.2 *Longitudinal and multiple fission in flagellate algae. (a, b) Two cells of Euglena gracilis in the course of longitudinal fission (phase contrast, ×1,740). [Reproduced from G. F. Leedale, in The Biology of Euglena (D. E. Buetow, ed. New York: Academic Press, 1968.] In (a), division of the nucleus and of the locomotor apparatus at the anterior end of the cell is complete. In (b), cell cleavage has begun. (c, d, e, f) Four steps in the cellular life cycle of Chlorogonium tetragamum, a green alga that reproduces by multiple fission (phase contrast, ×2,000). (c) newly liberated daughter cell; (d) two-celled stage; (e) four-celled stage; (f) four daughter cells just after liberation from the mother cell. Original photomicrographs of material provided by Paul Kugrens, Department of Botany, University of California, Berkeley.*

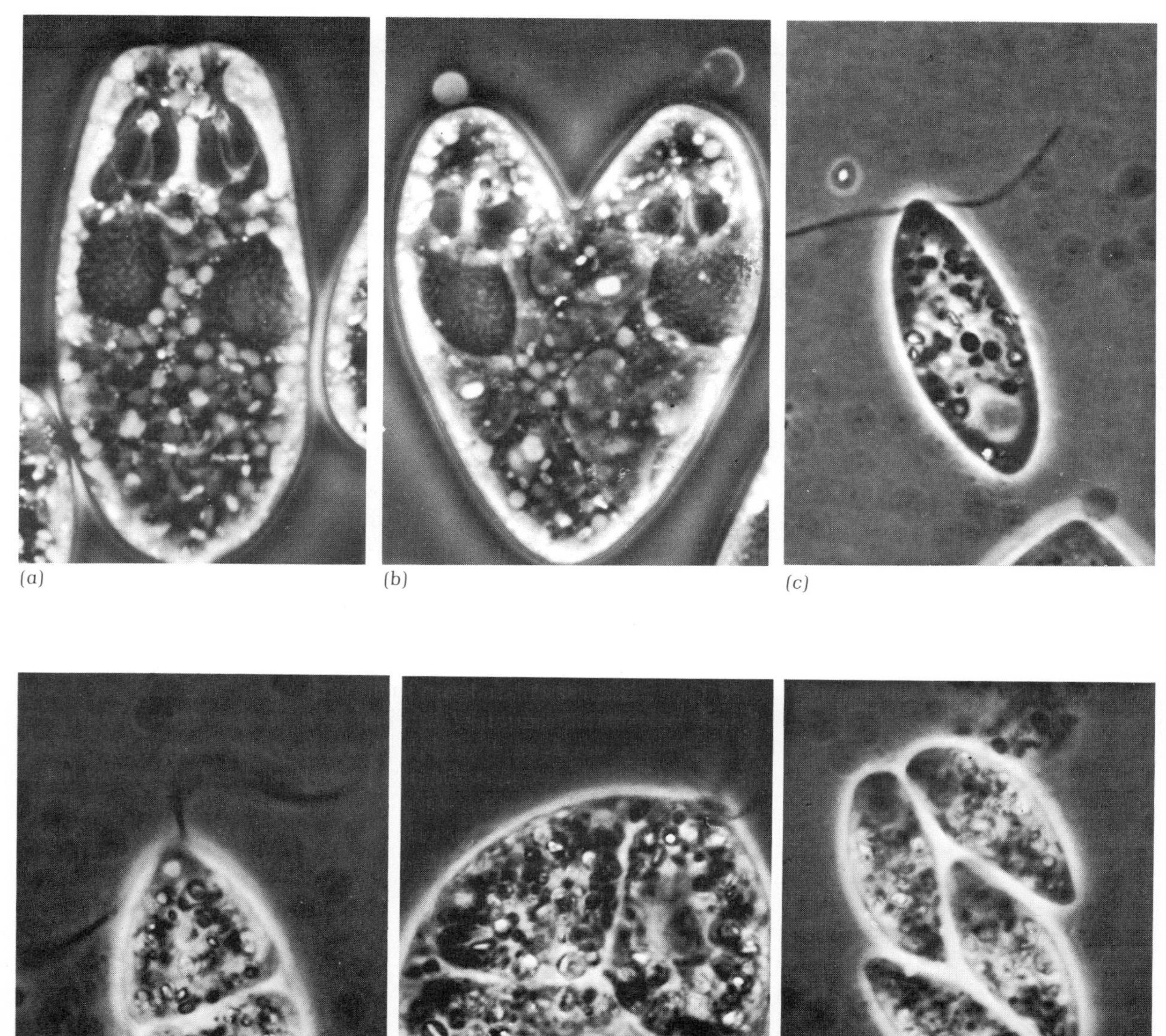
(a)
(b)
(c)
(d)
(e)
(f)

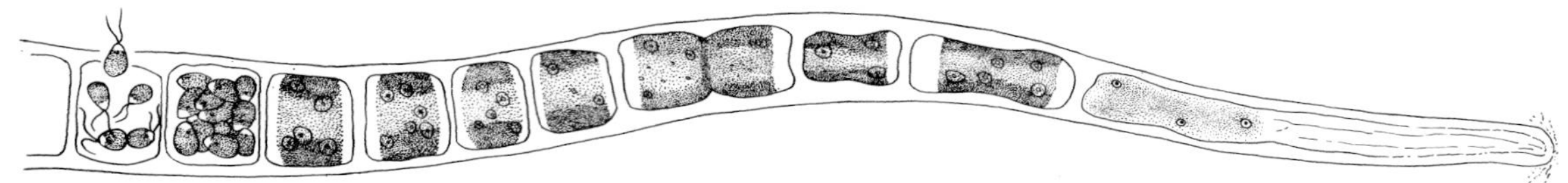

Figure 4.3 *The filamentous green alga, Ulothrix (×1,250). At left, the formation and liberation of biflagellate zoospores.*

motile cells, either asexual reproductive cells *(zoospores)* or gametes. Figure 4.3 shows the liberation of zoospores from a cell of a filamentous member of the Chlorophyta, *Ulothrix;* it can be seen that these zoospores have a structure very similar to that of the *Chlamydomonas* cell, illustrated in Figure 4.2*(b)*. The structure of the motile reproductive cells of multicellular algae thus often reveals their relatedness to a particular group of unicellular flagellates.

### The nonflagellate unicellular algae

By no means all unicellular algae are flagellates; several algal divisions also contain unicellular members which are either immotile or possess other means of movement. Many of these unicellular nonflagellate algae possess strikingly specialized and elaborate cells, which may be illustrated by considering two groups, the *desmids* and the *diatoms.*

The desmids, members of the Chlorophyta, have flattened, relatively large cells, with a characteristic bilateral symmetry (Figure 4.4). Asexual reproduction involves the synthesis of two new half cells in the equatorial plane, followed by cleavage between the new half cells to produce two bilaterally symmetrical daughters, each of which has a cell consisting of an "old" and a "new" half.

Figure 4.4 *Phase contrast photomicrographs of living desmid cells: (a) Micrasterias (×390); (b) Cosmarium (×1,704). Material from the collection of algae of the Department of Botany, University of California, Berkeley, provided by R. Berman.*

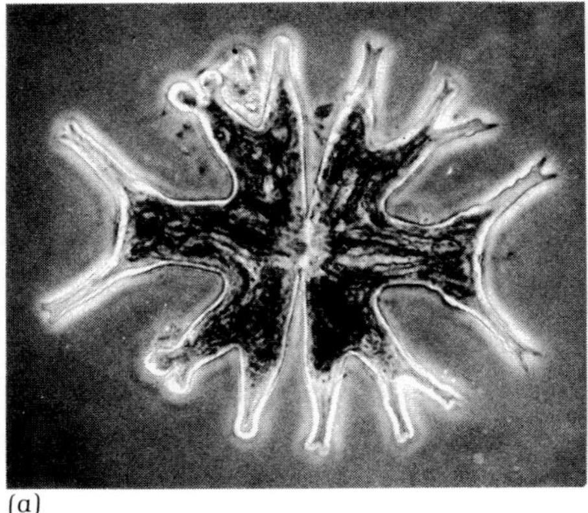

(a)

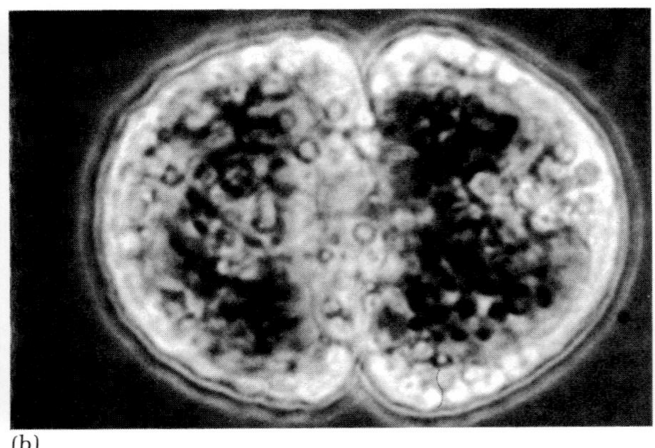

(b)

The diatoms (Figure 4.5), members of the Chrysophyta, have organic walls impregnated with silica. The architecture of the diatom wall is exceedingly complex; it always consists of two overlapping halves, like the halves of a petri dish. Division is longitudinal, each daughter cell retaining half of the old wall and synthesizing a new half.

Although devoid of flagella, some desmids and diatoms can move slowly over solid substrates. The mechanism of desmid locomotion is not known. The locomotion of diatoms is accomplished by a special modification of ameboid movement. In motile diatoms, there is a narrow longitudinal slot in the wall, known as a *raphe*, through which

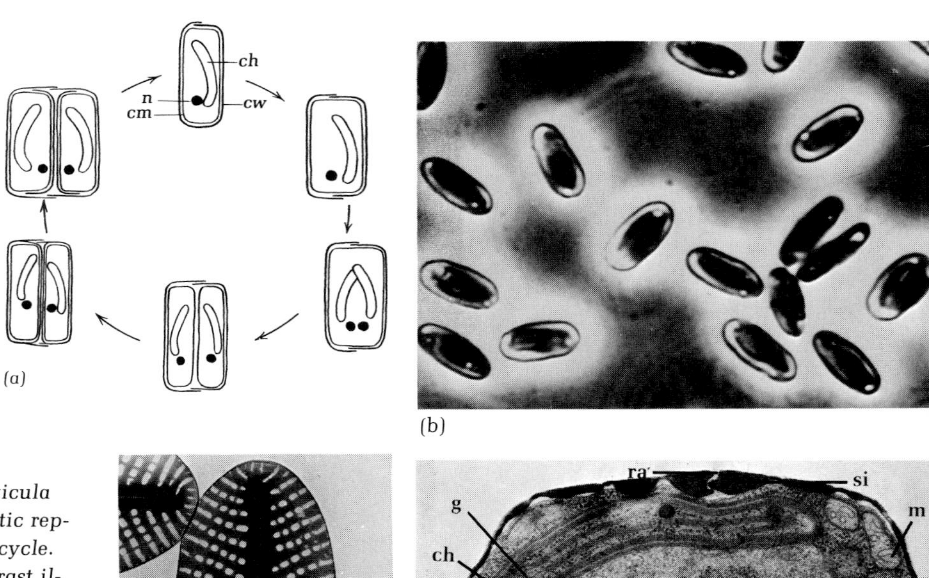


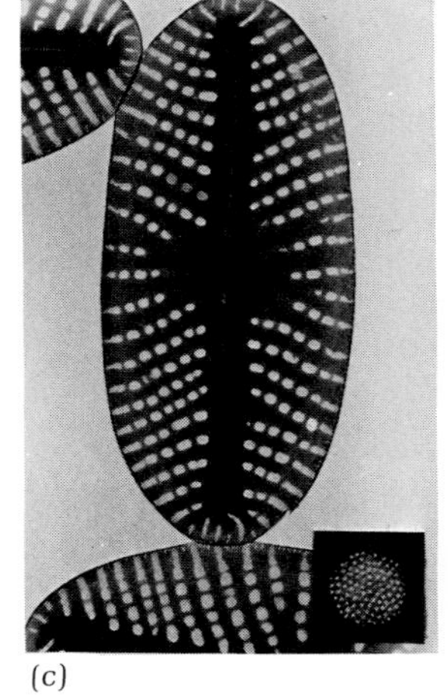


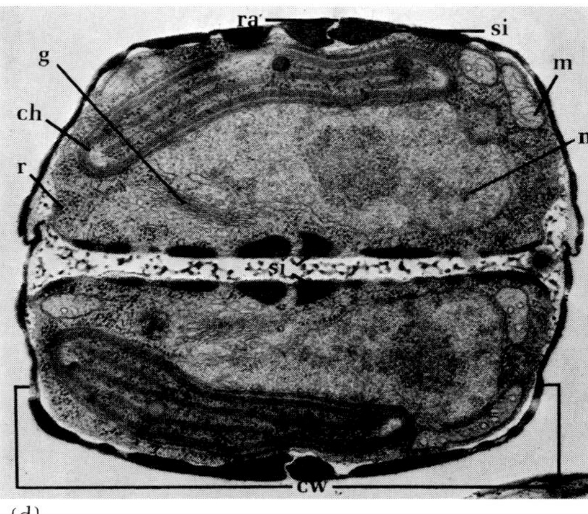


*Figure 4.5 The diatom Navicula pelliculosa. (a) Diagrammatic representation of the division cycle. (b) Living cells, phase contrast illumination (×1,323). (c) Electron micrograph of the wall (×9,800). Insert depicts fine structure of one of the wall pores (×56,000). (d) Transverse section of a dividing cell (×23,850): ch, chloroplast; g, Golgi apparatus; n, nucleus; m, mitochondrion; r, ribosomes; ra, raphe; si, silica in wall; cw, cell wall; cm, cell membrane. Courtesy of M. L. Chiappino and B. E. Volcani, University of California, San Diego.*

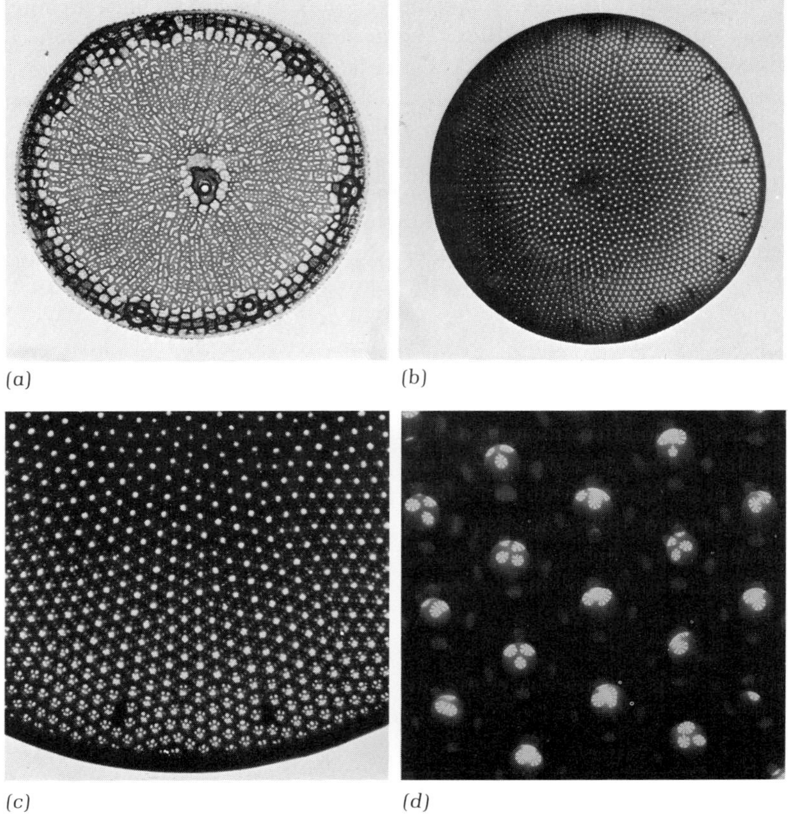

Figure 4.6 *Structural complexity of the wall in diatoms, illustrated by electron micrographs of isolated walls. (a) Wall of Cyclotella nana (×10,000). (b) Wall of Coscinodiscus granii (×1,133). (c) Part of (b) at a higher magnification (×3,000). (d) Part of B at still higher magnification, showing detailed fine structure of the pores (×18,667). Courtesy of M. L. Chiappino and B. E. Volcani, University of California, San Diego.*

the protoplast can make direct contact with the substrate. Movement is brought about by directed cytoplasmic streaming in the canal of the raphe, which pushes the cell over the substrate.

Many fossil diatoms are known, because the siliceous skeleton of the wall (Figure 4.6) is practically indestructible, and as diatoms are one of the major groups of algae in the oceans, large fossil deposits of diatom walls have accumulated in many areas. These deposits, known as *diatomaceous earth*, have industrial uses as abrasives and filtering agents.

## The natural distribution of algae

Most algae are aquatic organisms that inhabit either fresh water or the oceans. These aquatic forms are principally free-living, but certain unicellular marine algae have established durable symbiotic relationships with specific marine invertebrate animals (e.g., sponges, corals, various groups of marine worms) and grow within the cells of the host animal. Some terrestrial algae grow in soil or on the bark of trees. Others have established symbiotic relationships with fungi, to produce the curious, two-membered natural associations termed *lichens,* which form slowly growing colonies in many arid and inhospitable environments, notably on the surface of rocks. Many of the symbiotic relationships into which algae have entered will be described in Chapters 24 and 25.

The marine algae play a very important role in the cycles of matter on earth, since their total mass (and consequently their gross photosynthetic activity) is at least equal to that of all land plants combined and is probably much greater. This role is by no means evident, because the most conspicuous of marine algae, the seaweeds, occupy a very limited area of the oceans, being attached to rocks in the intertidal zone and the shallow coastal waters of the continental shelves. The great bulk of marine algae are unicellular floating *(planktonic)* organisms, predominantly diatoms and dinoflagellates, distributed through the surface waters of the oceans. Although they sometimes become abundant enough to impart a definite brown or red color to local areas of the sea, their density is usually so low that there is no gross sign of their presence. It is the enormous total volume of the earth's oceans which they occupy that makes them the most abundant of all photosynthetic organisms.

**Nutritional versatility of algae**

The ability to perform photosynthesis confers on many algae very simple nutrient requirements; in the light, they can grow in a completely inorganic medium. This is not always true, however, because many algae have *specific vitamin requirements,* a requirement for vitamin $B_{12}$ being particularly common. In nature the source of these vitamins is probably bacteria that inhabit the same environment. The ability to perform photosynthesis does not necessarily preclude the utilization of organic compounds as the principal source of carbon and energy, and many algae have *a mixed type of metabolism.*

Even when growing in the light, certain algae (e.g., the green alga *Chlamydobotrys*) cannot use $CO_2$ as their principal carbon source and are therefore dependent on the presence of acetate or some other suitable organic compound to fulfill their carbon requirements. This is caused by a defective photosynthetic machinery: although these algae can obtain energy from their photosynthetic activity, they cannot obtain the reducing power to convert $CO_2$ to organic cell materials.

Many algae that perform normal photosynthesis in the light, using $CO_2$ as the carbon source, can grow well in the dark at the expense of a variety of organic compounds; such forms can thus shift from photosynthetic to respiratory metabolism, the shift being determined primarily by the presence or absence of light. Algae completely enclosed by cell walls are osmotrophic and dependent on dissolved organic substrates as energy sources for dark growth. However, a considerable number of unicellular algae which lack a cell wall, or are not completely enclosed by it, can phagocytize bacteria or other smaller microorganisms and thus employ a phagotrophic mode of nutrition as well. It is not correct, accordingly, to regard the algae as an *exclusively* photosynthetic group; on the contrary, many of their unicellular members possess and can use the nutritional capacities characteristic of the two major subgroups of nonphotosynthetic eucaryotic protists, the protozoa and fungi.

**The leucophytic algae**

As mentioned in Chapter 2, loss of the chloroplast from a eucaryotic cell is an *irreversible event,* which results in a *permanent loss of photosynthetic ability.* Such a change appears to have taken place many times among unicellular algal groups with a mixed mode of nutrition, to yield nonpigmented counterparts, which can be clearly recognized on the basis of other cellular characters as *nonphotosynthetic derivatives of algae.* Such organisms, known collectively as *leucophytes,* exist in many flagellate groups and also in diatoms and in nonmotile groups among the green algae. The recognition of leucophytes is often easy, since they may have preserved a virtually complete structural identity with a particular photosynthetic counterpart. In some cases, this structural near-identity may include the preservation of vestigial, nonpigmented chloroplasts, as well as a pigmented eyespot. There can be little doubt accordingly that these nonphotosynthetic organisms are close relatives of their structural counterparts among the algae and have arisen from them by a loss of photosynthetic ability in the recent evolutionary past. Indeed, the transition can be demonstrated experimentally in certain strains of *Euglena,* which yield stable, colorless races when treated with the antibiotic streptomycin or when exposed to small doses of ultraviolet irradiation or to high temperatures (Figure 4.7). These colorless races cannot be distinguished from the naturally oc-

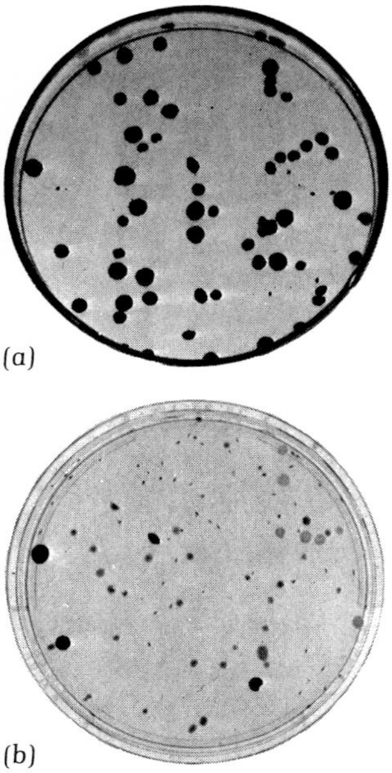

(a)

(b)

*Figure 4.7 The loss of chloroplasts in Euglena gracilis as a result of ultraviolet irradiation. (a) A light-grown plate culture of E. gracilis. (b) A light-grown culture of the same organism, after exposure to brief ultraviolet irradiation. Most of the cells have given rise to clones devoid of chloroplasts (pale colonies). Courtesy of Jerome A. Schiff.*

curring nonphotosynthetic euglenid flagellates of the genus *Astasia*.

The classification of the leucophytes raises a difficult problem. In terms of cell structure, they can be easily assigned to a particular division of algae, as nonphotosynthetic representatives, and this classification is no doubt the most satisfactory one. However, since they are nonphotosynthetic unicellular eucaryotic protists, they can alternatively be regarded as protozoa, and they are, in fact, included among the protozoa by zoologists. The leucophytes accordingly provide the first and by far the most striking case of a group, or rather a whole series of groups, which are clearly transitional between two major assemblages among the eucaryotic protists.

## The origins of the protozoa

The protozoa are a highly diverse group of unicellular, nonphotosynthetic eucaryotic protists, most of which show no obvious resemblances to the various divisions of algae. Nevertheless, the various kinds of leucophytes, which are recognizably of algal origin, provide a plausible clue concerning the evolutionary origin of many groups among the protozoa. The loss of photosynthetic function abruptly reduces the nutritional potentialities of an organism; leucophytes are therefore immediately confined to a more restricted range of environments than their photosynthetic ancestors. Specific features of cellular construction which possessed adaptive value in the context of photosynthetic metabolism become superfluous; the eyespot is the most obvious example. Hence one could expect that loss of photosynthetic ability would be followed by a series of evolutionary changes in the structure of the cell which better fit the organism for an osmotrophic or phagotrophic mode of life. Beyond a certain point, these changes would make the algal origin of the organism unrecognizable, and it would then be classified without question as a protozoon.

One group of eucaryotic protists, the dinoflagellates, has several features of cell structure that permit the biologist to recognize a dinoflagellate origin even in organisms which have evolved very far from the typical unicellular photosynthetic flagellate members of this algal group (Figure 4.8). The motile cell of a dinoflagellate has two flagella, which differ in structure and arrangement. One lies in a groove or girdle around the equator of the cell; the other extends away from the cell in a posterior direction. The dinoflagellate nucleus is also unusual; its

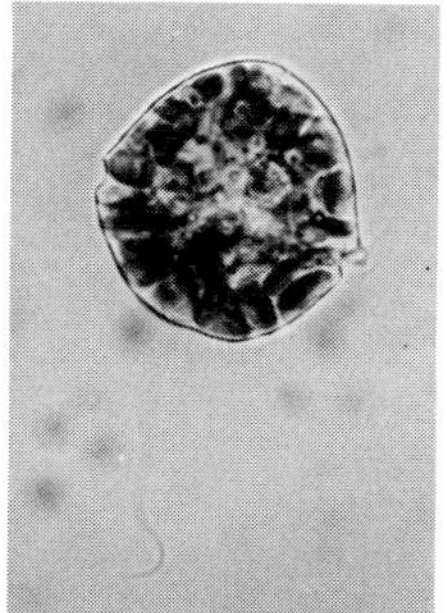

(a)

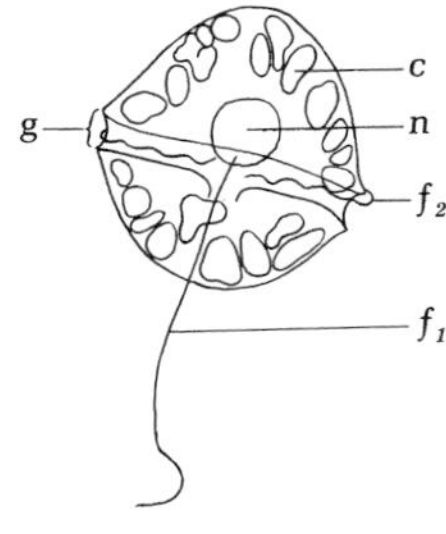


(b)

*Figure 4.8 A photosynthetic dinoflagellate, Glenodinium foliaceum. (a) A living cell (×1,000). (b) A diagrammatic drawing of the cell: c, chloroplast; n, nucleus; g, girdle; $f_1$ and $f_2$, flagella.*

division is highly specialized, and the chromosomes often remain visible in interphase.

Most photosynthetic dinoflagellates are unicellular planktonic organisms, widely distributed in the oceans, and characteristically brown or yellow in color as a result of the possession of a distinctive set of photosynthetic pigments. Many (the so-called "armored" dinoflagellates) possess very elaborate cell walls, composed of a series of plates, which do not completely enclose the protoplast. There is a very pronounced tendency to phagotrophic nutrition among these photosynthetic members of the group, because the wall structure permits pseudopodial extension and the engulfment of small prey. A few filamentous algae, completely enclosed by walls, can be recognized as of dinoflagellate origin, since they form zoospores with the characteristic flagellar arrangement.

A much more extensive series of specialized forms can be traced among the nonphotosynthetic members of this flagellate group. Many of the free-living unicellular dinoflagellates are nonphotosynthetic phagotrophic organisms. Some preserve close structural similarities to photosynthetic members of the group; others, such as the large marine organism, *Noctiluca*, have a highly specialized cellular organization

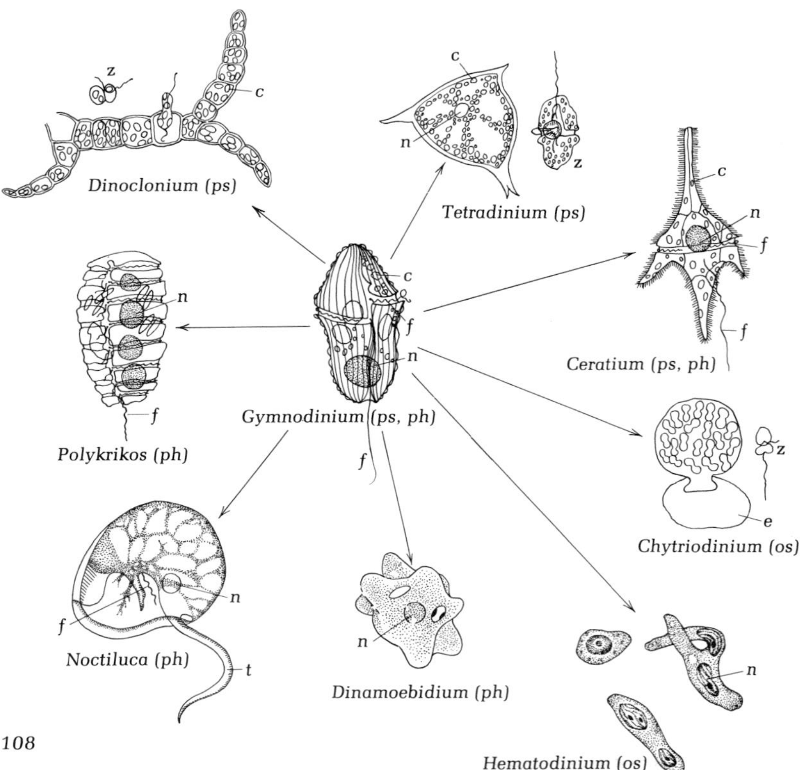


*Figure 4.9 The different evolutionary trends that are represented among dinoflagellates. Gymnodinium is a relatively unspecialized photosynthetic dinoflagellate, which is both photosynthetic (ps) and phagotrophic (ph). Ceratium is a more specialized photosynthetic dinoflagellate, characterized by a very complex wall with spiny extensions, comprised of many plates. Tetradinium and Dinoclonium are nonmotile, strictly photosynthetic organisms, which reproduce by multiple cleavage to form typical dinoflagellate zoospores. Polykrikos, Noctiluca, and Dinamoebidium are three free-living phagotrophic dinoflagellates. Polykrikos is a coenocytic, multinucleate organism, the cell of which bears a series of pairs of flagella. Noctiluca has one small flagellum, and bears a large and conspicuous tentacle. Dinamoebidium is an ameboid organism. Chytriodinium and Hematodinium are parasitic dinoflagellates whose nutrition is osmotrophic (os). Chytriodinium parasitizes invertebrate eggs and reproduces by cleavage of a large saclike structure into dinoflagellate zoospores. Hematodinium is a blood parasite in crabs: n, nucleus; f, flagellum; c, chloroplast; z, zoospore; e, parasitized invertebrate egg; t, tentacle.*

not found in any photosynthetic member of the group. However, the most far-reaching modifications of cell structure within the dinoflagellates are to be found among its parasitic members, most of which occur in marine invertebrates. *Hematodinium*, which occurs in the blood of certain crabs, is completely devoid of flagella. *Chytriodinium*, which parasitizes the eggs of copepods, develops as a large, saclike structure within the egg, subsequently giving rise by multiple internal cleavage to numerous motile spores with a typical dinoflagellate structure. Were it not for the retention of the distinctive nuclear organization (and, in the case of *Chytriodinium*, the flagellar structure of the spores), neither of these parasitic protists could be recognized as belonging to the same group as the photosynthetic dinoflagellates. *Hematodinium* could be classified with the sporozoan protozoa and *Chytriodinium* with the primitive group of fungi known as chytrids.

Accordingly, within this one small flagellate group, it is possible to reconstruct some major patterns of evolutionary radiation that were probably characteristic of eucaryotic protists as a whole (Figure 4.9).

## The protozoa

In the light of the preceding discussion, the protozoa can best be regarded as comprising a number of groups of nonphotosynthetic, typically motile, eucaryotic unicellular protists, which have probably derived at various times in the evolutionary past from one or another group among the unicellular algae (see Table 4.2).

*Table 4.2 Primary subdivisions of the protozoa*

I. *Class Mastigophora. The flagellate protozoa. Motile by means of one or more flagella. Cell division always longitudinal.*
*Included in this class are the "phytoflagellates" (i.e., unicellular motile representatives of the various algal divisions) as well as the "zooflagellates," nonphotosynthetic organisms not recognizable as leucophytes. These forms are in the main osmophilic.*

II. *Class Rhizopoda. The ameboid protozoa. Motile by means of pseudopodia. It should be noted that the distinction from class I on the basis of locomotion is not absolute, since many of the Rhizopoda can also form flagella. Reproduction by binary fission. Phagotrophic.*

III. *Class Sporozoa. A very diverse group of parasitic protozoa. Immotile or showing gliding movement. Reproduction by multiple fission. Osmophilic.*

IV. *Class Ciliata. The ciliates. Motile by means of numerous cilia, organized into a coordinated locomotor system. The cell has two nuclei, differing in structure and function. Division always transverse. Phagotrophic.*

**The flagellate protozoa: the mastigophora**

The Mastigophora are protozoa that always bear flagella as the locomotor organelles. In contrast to the ciliates, in which cell division is transverse, flagellate protozoa undergo longitudinal division, preceded by duplication of the flagellar apparatus at the anterior end of the cell. This mode of division has already been described for a photosynthetic flagellate, *Euglena*. In addition to leucophytes, this protozoan group includes many representatives that show no resemblance to photosynthetic flagellates and are for the most part parasites of animals.

The trypanosomes are frequently parasitic in vertebrates, where they develop in the bloodstream, being transmitted from host to host by the bite of insects. They include important agents of disease, such as the agent of African sleeping sickness, transmitted by the tsetse fly. The cell is slender and leaf-shaped, its single flagellum being directed posteriorly and attached through part of its length to the body of the cell, to form an undulating membrane [Figure 4.10(*a*)]. The trypanosomes are osmotrophic protozoa, which absorb their nutrients from the blood of the host.

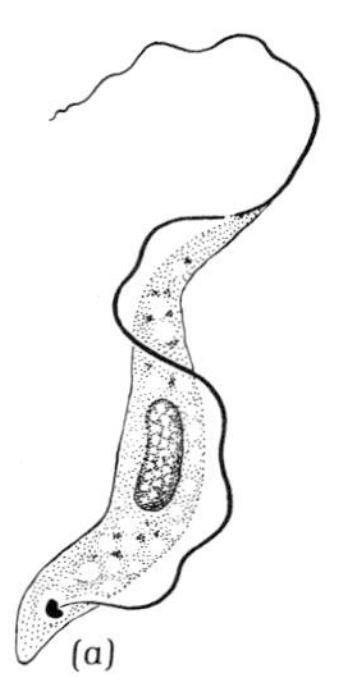


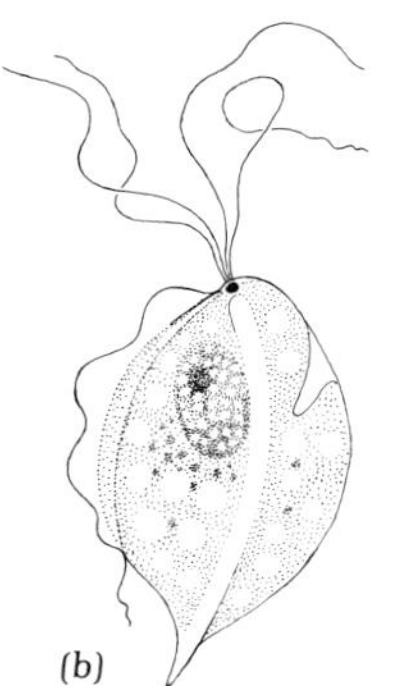


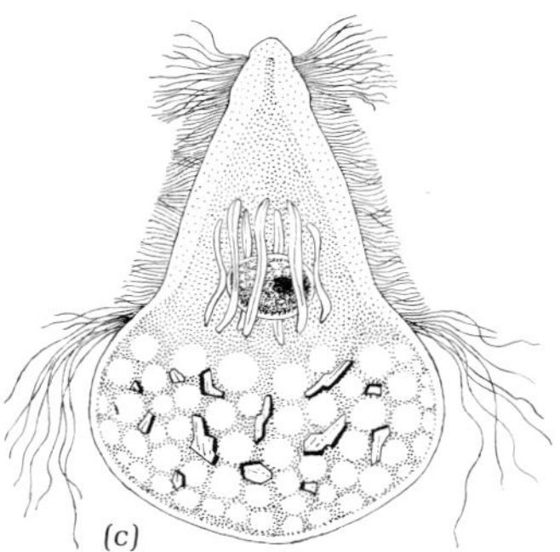


*Figure 4.10 Some nonphotosynthetic flagellate protozoa (Mastigophora). (a) A trypanosome. The leaf-shaped cell and the long undulating membrane, to which the flagellum is attached, are characteristic of this organism. (b) Trichomonas. (c) Trichonympha.*

Other parasitic flagellates inhabit the gut of vertebrates or invertebrates. The trichomonads, which have 4 to 6 flagella [Figure 4.10(*b*)] are harmless inhabitants of the gut of vertebrates. Several very highly specialized groups of flagellate protozoa inhabit the gut of termites; one of the most striking of these organisms, *Trichonympha*, is illustrated in Figure 4.10(*c*). The relationship between termites and their intestinal protozoa is a symbiotic one, described in Chapter 25. Among these termite protozoa, the flagellar apparatus is extremely elaborate, and the cell may bear several hundred individual flagella.

**The ameboid protozoa: the rhizopoda**

The Rhizopoda are protozoa in which ameboid locomotion is the predominant mode of cell movement, although some of them are able to produce flagella as well. The simplest members of this group are amebas, which have characteristically amorphous cells as a result of the continuous changes of shape brought about by the extension of pseudopodia. Most amebas are free-living soil or water organisms which phagocytize smaller prey. A few inhabit the animal gut, including forms that cause disease (amebic dysentery). Other members of the Rhizopoda have a well-defined cell form, as the result of the formation of an exoskeleton or shell (typical of the foraminifera) or an endoskeleton (typical of the heliozoa and radiolaria). Several members of the Rhizopoda are illustrated in Figure 4.11.

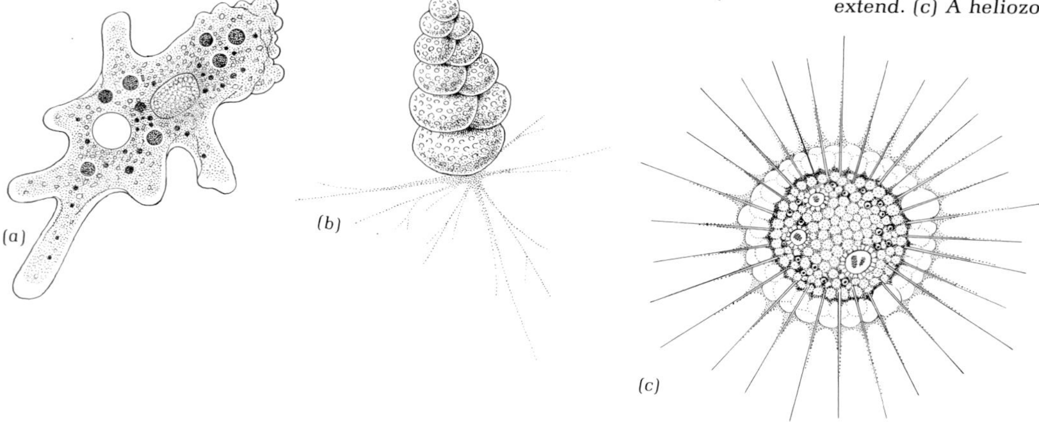


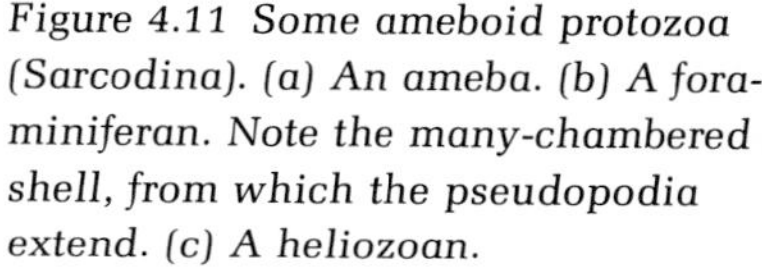

*Figure 4.11 Some ameboid protozoa (Sarcodina). (a) An ameba. (b) A foraminiferan. Note the many-chambered shell, from which the pseudopodia extend. (c) A heliozoan.*

**The ciliate protozoa: the ciliophora**

The ciliate protozoa are a very large and varied group of aquatic, holozoic organisms that are particularly widely distributed in fresh water. The ciliates share a number of fundamental cellular characters which distinguish them sharply from all other eucaryotic protists. This suggests that despite the very great internal diversity of this group, it is the one class of protozoa that may have had a single common evolutionary origin.

The common characters of ciliates can be summarized as follows:

(1) At some time in the life history, the cell is motile by means of numerous short, hairlike projections, structurally homologous with flagella, which are termed *cilia*.

(2) Each cilium arises from a basal structure, the kinetosome, which

is homologous with the kinetosome of a flagellum; however, in ciliates the kinetosomes are interconnected by rows of fibrils called kinetodesmata to form very elaborate compound locomotor structures termed *kineties*. This internal system persists, even when the cell is devoid of cilia.

(3) Cell division is transverse, not longitudinal, as in flagellates. Ciliates show a marked polarity, with posterior and anterior differentiation of the cell, so the transverse mode of cell division necessarily entails an elaborate process of morphogenesis each time division occurs, during which the anterior daughter cell resynthesizes posterior structures, while the posterior daughter cell resynthesizes anterior structures. The morphogenetic transformations are generally almost complete when the two daughter cells separate.

(4) Each individual contains two dissimilar nuclei, a large *macronucleus* and a much smaller *micronucleus,* which differ in function as well as in structure.

We may illustrate the distinctive character of the ciliates by considering the properties of a simple member of the group, *Tetrahymena pyriformis* (Figure 4.12). It has a pear-shaped body about 50 $\mu$m long,

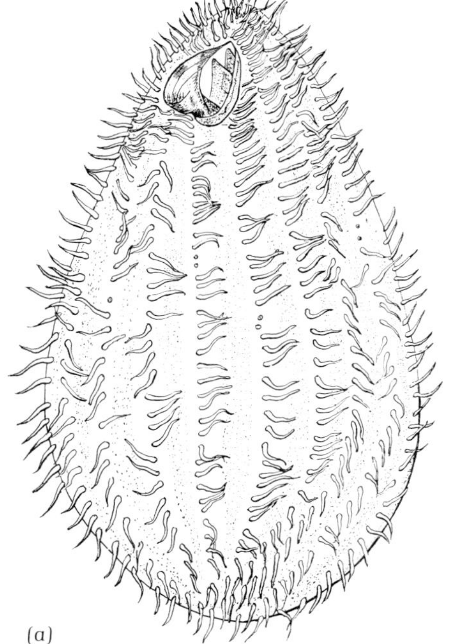


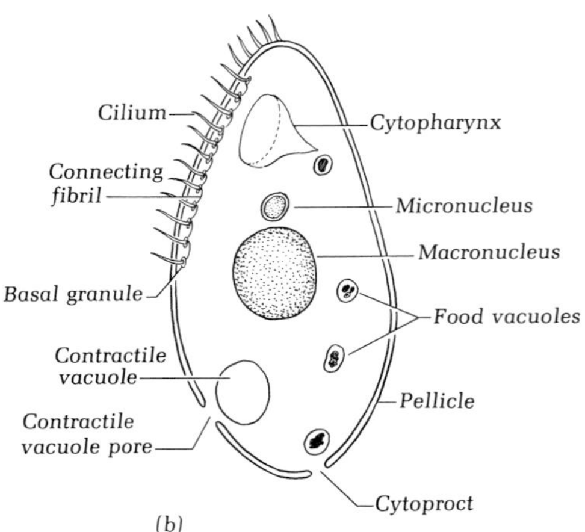


*Figure 4.12 The ciliate protozoon, Tetrahymena. (a) General view, showing external appearance. (b) Diagrammatic cross section, showing main structural features of the cell.*

enclosed by a semirigid pellicle. The surface is covered with hundreds of cilia, arranged in longitudinal rows of kineties. The beating of the cilia, which propels the organism, is rhythmic and coordinated.

Near the narrow anterior end of the cell is the mouth or *cytostome*. It consists of an oral aperture, a mouth cavity that extends some distance into the cell, an undulating membrane, and three membranelles. The undulating membrane and membranelles are composed of specialized, adherent cilia, the movements of which sweep food particles into the mouth cavity. Captured food enters the cytoplasm by being enclosed in food vacuoles which are formed in succession at the base of the mouth cavity. These food vacuoles then circulate within the cell as a result of cytoplasmic streaming until the food material has been digested and the soluble products absorbed; undigested material is ejected from the cell by a posteriorly located pore known as the *cytoproct*. In nature, *Tetrahymena* is normally a predator and feeds on smaller microorganisms. However, in the laboratory it can be grown in pure culture on a medium that contains only soluble nutrients. Under such conditions, the liquid nutrients must still be taken in through the mouth, in the form of vacuoles.

Although its natural environment is a dilute one, with an osmotic pressure far below that of the contents of the cell, *Tetrahymena* is able to maintain water balance by the operation of a *contractile vacuole*. This structure, located near the posterior end of the cell, is formed by the coalescence of smaller vacuoles in the cytoplasm; when it reaches a certain critical size, it discharges its liquid contents into the environment through a pore in the pellicle and then starts to grow in volume again. As mentioned above, typical ciliates have two dissimilar nuclei in the cell. The larger *macronucleus*, which is polyploid, is necessary for normal cell division and growth and is therefore sometimes referred to as the "vegetative nucleus." Some strains of *Tetrahymena* have only this kind of nucleus; they can reproduce indefinitely by binary fission but cannot undergo sexual reproduction. Other strains possess also a small, diploid *micronucleus*, which plays an essential role in sexual reproduction. As will be described presently, the macronucleus can be derived after conjugation from a micronucleus; hence strains of *Tetrahymena*, having only a macronucleus can be regarded as deficient cell lines which have probably lost their micronucleus by an accident of vegetative growth. In *Tetrahymena*, the first step in cell division is an elongation of the macronucleus parallel to the long axis of the cell. At the same time, a structural reorganization of the cytoplasm begins. Its principal feature is the formation of *a second cytostome* just posterior to the future plane of cell division. A furrow then develops across the center of the cell, which becomes dumbbell-shaped. If a micronucleus is present, it divides mitotically, and the two daughter nuclei migrate respectively to the anterior and posterior portions of the cell. Finally, the elongated macronucleus divides, and the two daughter cells separate.

Sexual reproduction in *Tetrahymena* (Figure 4.13) takes place only between cells of two different strains of compatible mating type. The mating cells fuse in the region of the cytostome, and their micronuclei undergo meiotic division, each giving rise to four haploid nuclei. Three of the nuclei in each cell disintegrate, while the fourth undergoes another, mitotic division. A nuclear exchange between the mating part-

*Figure 4.13 Stages in the conjugation of Tetrahymena.*

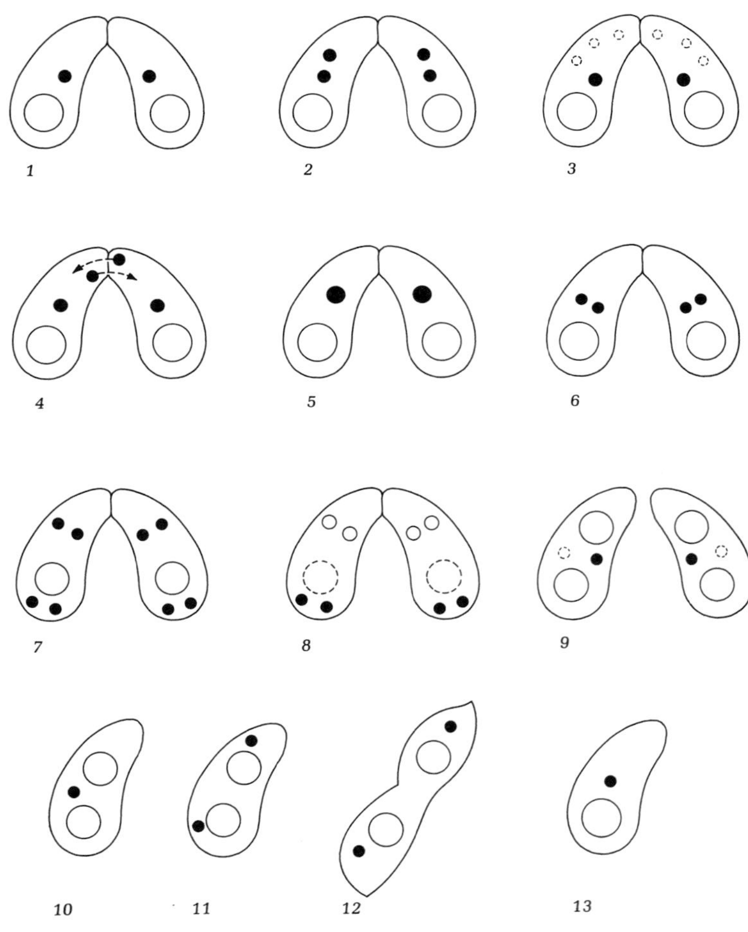

ners then takes place, each partner receiving one of the two haploid nuclei derived from the micronucleus of the other. This nuclear exchange is followed, in each cell of the pair, by a fusion between the haploid nucleus that was not transferred and the haploid nucleus received from the partner. The resulting diploid zygote nuclei undergo two successive mitotic divisions, producing four diploid nuclei in each cell. Two of these then proceed to develop into macronuclei, a process that involves polyploidization. The two others remain as micronuclei. At the same time, the old macronucleus in each cell disintegrates. The two partners then separate from one another, each now containing two new macronuclei and two new micronuclei, all derived from the zygote nucleus. One of the micronuclei in each cell disintegrates, and the remaining one divides mitotically, after which the cell undergoes binary fission to yield two daughter cells, each containing a single macronucleus and a single micronucleus. The organisms continue to reproduce by binary fission until the next cycle of conjugation.

A curious feature of sexuality in ciliates is that the disintegration of three of the four haploid daughter nuclei in each cell which follows meiotic division of the micronucleus *results in the complete elimination from the cell of half its previous diploid genome.* When the remaining haploid nucleus divides, each member of the conjugating pair thus contains two genetically identical haploid nuclei. The following reciprocal nuclear exchange and zygotic nuclear fusion therefore makes the two mating cells into *identical diploid twins;* after their separation from one another, they transfer identical genotypes to all their respective progeny during subsequent vegetative growth.

*Tetrahymena* is among the simplest of ciliates. The foregoing account suffices to show what an extraordinarily elaborate and complex biological organization has been evolved in this protozoan group within the framework of unicellularity. The ciliates represent the apex of biological differentiation on the unicellular level, but they appear to be a terminal evolutionary group. The development of more complex biological systems took place through the establishment of multicellularity and involved the differentiation of specialized cell types during the growth of the individual organism, characteristic of all plants and animals.

## *The fungi*

Like the protozoa, the fungi are nonphotosynthetic. Although some of the more primitive aquatic fungi show resemblances to flagellate protozoa, the fungi as a whole have developed a highly distinctive biological organization which can be regarded as an adaptation to life in their most common habitat, the soil. We shall start out by considering the main features of this type of biological organization.

Most fungi are coenocytic organisms and have a vegetative structure known as a *mycelium* (Figure 4.14). The mycelium consists of a multi-

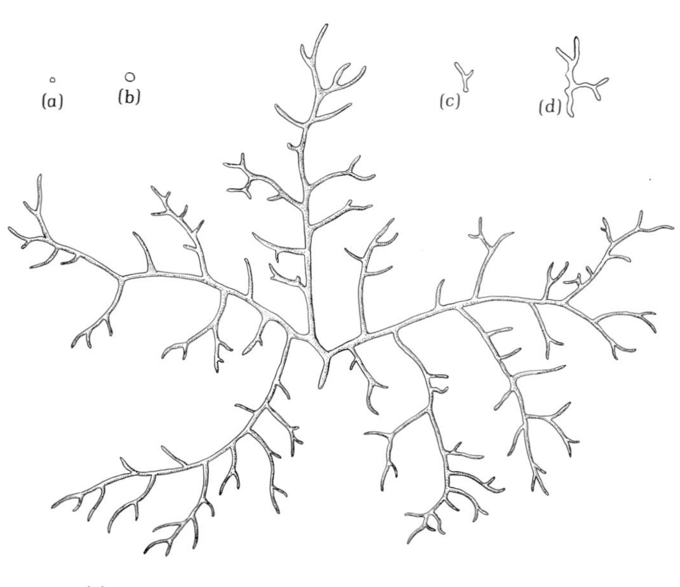


*Figure 4.14 Successive stages in the development of a fungal mycelium from a reproductive cell or conidium (×85). After C. T. Ingold, The Biology of Fungi. London: Hutchinson, 1961.*

nucleate mass of cytoplasm enclosed within a rigid, much-branched system of tubes, which are fairly uniform in diameter. The enclosing tubes represent a protective structure that is homologous with the cell wall of a unicellular organism. A mycelium normally arises by the germination and outgrowth of a single reproductive cell, or spore. Upon germination, the fungal spore puts out a long thread, or *hypha*, which branches repeatedly as it elongates to form a ramifying system of hyphae which constitutes the mycelium. Fungal growth is characteristically confined to the tips of the hyphae; as the mycelium extends, the cytoplasmic contents may disappear from the older, central regions. The size of a single mycelium is not fixed; as long as nutrients are available, outward growth by hyphal extension can continue, and in some of the basidiomycetes, a single mycelium may be as much as 50 ft in diameter. Usually, asexual reproduction occurs by the formation of uninucleate or multinucleate spores which are pinched off at the tips of the hyphae. Neither the spores nor the mycelium of higher fungi is capable of movement. However, the internal contents of a mycelium show streaming movements, which cannot be translated into progression over the substrate because the cytoplasm is completely enclosed within its wall. In fact, the simplest brief definition of the structure of a higher fungus is: *a multinucleate mass of cytoplasm, mobile within a much-branched enclosing system of tubes.*

Since a mycelium is capable of almost indefinite growth, it frequently attains macroscopic dimensions. In nature, however, the vegetative mycelium of fungi is rarely seen, because it is normally embedded in soil or other opaque substrates. Many fungi (the mushrooms) form specialized, spore-bearing fruiting structures, however, which project above soil level and are readily visible as macroscopic objects. Such structures were known long before the beginning of scientific biology, although their nature and mode of formation were not clearly understood until the nineteenth century. The superficial resemblance of these fruiting structures to plants was undoubtedly a very important factor in the decision of the early biologists to assign fungi to the plant kingdom, despite their nonphotosynthetic nature.

Since fungi are always enclosed by a rigid wall, they are unable to engulf smaller microorganisms. Most fungi are free-living in soil or water and obtain their energy by the respiration or fermentation of soluble organic materials present in these environments. Some are parasitic on plants or animals. A number of soil forms are predators and have developed ingenious traps and snares, composed of specialized hyphae, which permit them to capture and kill protozoa and small invertebrate animals such as the soil-inhabiting nematode worms. After the death of their prey, such fungi invade the body of the animal by hyphal growth and absorb the nutrients contained in it.

**The primitive fungi: aquatic phycomycetes**

Although soil is by far the most common habitat of the fungi as a whole, many of the primitive fungal groups are aquatic. These fungi are known collectively as *water molds* or *aquatic phycomycetes*. They occur on the surface of decaying plant or animal materials in ponds and streams; some are parasitic and attack algae or protozoa. It is these fungi which show the closest resemblances to protozoa; they produce motile spores or gametes, furnished with flagella, and in the simpler forms the vegetative structure is not mycelial. This description applies, for example, to many of the fungi known as *chytrids*.

The developmental cycle of a typical simple chytrid, which occurs in ponds on decaying leaves, is shown in Figure 4.15. The mature vegetative structure consists of a sac about 100 $\mu$m in diameter which is anchored to the solid substrate by a number of fine, branched threads known as *rhizoids*. The sac is a *sporangium*, within which reproductive

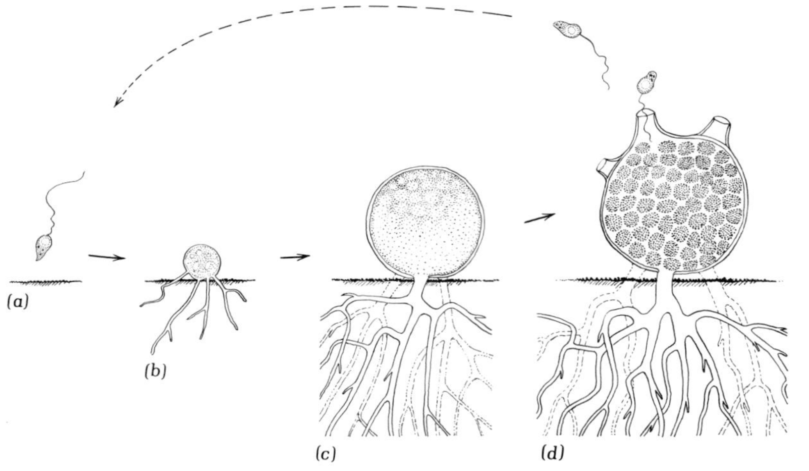


*Figure 4.15 The life cycle of a primitive fungus, a chytrid. The flagellated zoospore (a) settles down on a solid surface. As development begins (b), a branching system of rhizoids is formed, anchoring the fungus to the surface. Growth results in the formation of a spherical zoosporangium, which cleaves internally to produce many zoospores (c). The zoosporangium ruptures to liberate a fresh crop of zoospores (d).*

cells, or spores, are produced. The enclosed cytoplasm contains many nuclei, formed by repeated nuclear division during vegetative growth. Each nucleus eventually becomes surrounded with a distinct volume of cytoplasm, bounded by a membrane. The sporangium then ruptures to release uninucleate flagellated zoospores, each of which can settle down and grow into a new organism. The rhizoids are not reproductive structures; they serve purely to anchor the developing sporangium to the substrate and to absorb the nutrients required for its growth.

The aquatic phycomycetes are a very varied group with respect to their mechanisms of reproduction and life cycles. The range of this variation can be well illustrated by comparing a chytrid with another aquatic phycomycete, *Allomyces*. *Allomyces* shows a well-marked alternation of haploid and diploid generations. We will describe first the diploid *sporophyte*. When mature it looks like a microscopic tree, with a basal system of anchoring rhizoids from which springs a much-branched mycelium bearing two different kinds of sporangia. The mitosporangia have thin, smooth, colorless walls, whereas the meiosporangia have thick, dark-pitted walls. Upon maturation, both kinds of sporangia liberate flagellated spores, but the subsequent development of these spores is very different. The mitospores derived from mitosporangia are diploid and germinate into sporophytic individuals. The meiospores derived from meiosporangia are haploid, because meiosis takes place during the maturation of the meiosporangium; they give rise to haploid or *gametophytic* individuals.

The gametophyte is grossly similar in structure to the sporophyte, but instead of bearing meio- and mitosporangia, it produces male and female gametangia, which are generally borne in pairs. The female gametangium looks very much like a mitosporangium, whereas the male gametangium is distinguished by its brilliant orange color. The gametangia rupture to liberate male and female gametes, a considerable number arising from each gametangium. Both male and female gametes are motile, moving by means of flagella, but they can be readily distinguished from one another by size and color. The female gamete is larger than the male and colorless, and the male has an orange oil droplet at the anterior end. The gametes fuse in pairs to form biflagellate zygotes which eventually settle down and develop once more into sporophytes. The whole life cycle is illustrated in Figure 4.16.

*Figure 4.16 The life cycle of Allomyces, an aquatic phycomycete with a well-marked alternation of haploid and diploid generations. From a drawing made by Raphael Rodriguez and reprinted by permission of Arthur T. Brice.*

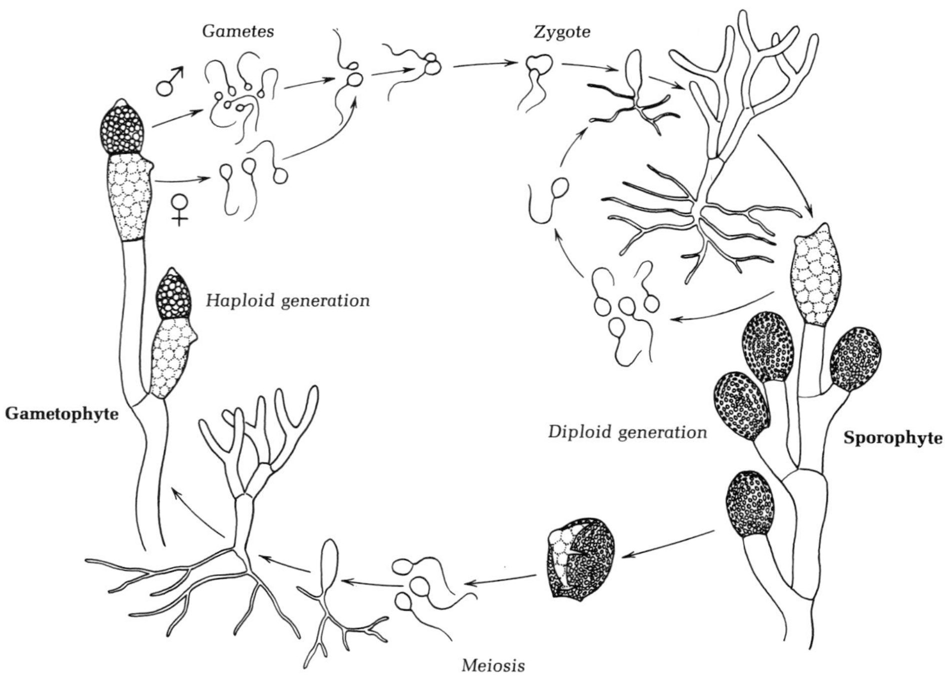


**The terrestrial phycomycetes**

The phycomycetes also include a group known as the *terrestrial phycomycetes*, which are inhabitants of soil. These organisms differ from all aquatic phycomycetes in not possessing motile flagellated reproduc-

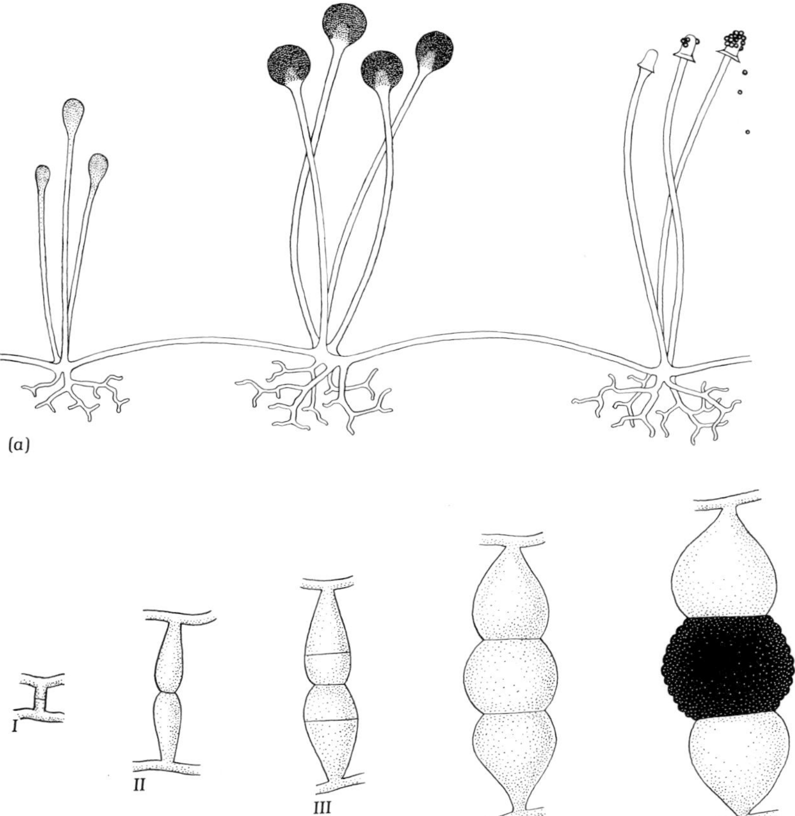


*Figure 4.17 (a) The vegetative stage of Rhizopus, a terrestrial phycomycete. (b) Sexuality in Rhizopus. Successive stages of sexual fusion and the formation of a zygospore.*

tive cells. *They are thus permanently immotile.* This is a property they share with all the higher groups of fungi. The absence of motility characteristic of the higher fungi is understandable in terms of their ecology: motile reproductive cells are of value only when dispersion occurs through water. The reproductive cells of soil-inhabiting fungi are dispersed in the main through the air.

As a typical example of a terrestrial phycomycete, we may take *Rhizopus*. The mycelium is differentiated into branched rhizoids that penetrate the substrate, horizontal hyphae known as stolons that spread over the surface of the substrate, bending down at intervals to form tufts of rhizoids, and erect sporangiophores that emerge from the stolons in tufts [Figure 4.17(*a*)]. The unbranched sporangiophore enlarges at the tip to form a rounded sporangium which becomes separated from the rest of the sporangiophore by a cross wall. Within this sporangium, large numbers of spherical spores are formed. These asexual *sporangiospores* are eventually released by rupture of the surrounding wall and are dispersed by air currents. They give rise on germination to a new vegetative mycelium.

*Rhizopus* also reproduces sexually, but sexual reproduction can occur only when two mycelia of opposite sex come into contact with one another. Fungi that show this phenomenon are known as *heterothallic fungi* in contrast to *homothallic fungi* (such as *Allomyces*) that can produce both kinds of sex cells on a single mycelium. In *Rhizopus*, the two kinds of mycelia between which sexual reproduction can take place are known as + and − strains because there are no morphological indications of maleness and femaleness. As the hyphae from a + and a − mycelium meet, each produces a short side branch at the point of contact. This side branch then divides to form a special cell, the *gametangium*. The two gametangia, which are in direct contact with one another, fuse to form a large zygospore, surrounded by a thick, dark wall. The whole sequence of events is shown in Figure 4.17(*b*). It can be seen that the behavior of both partners in the sexual act is identical; hence, there is no basis for a designation as "male" or "female." Upon germination of the zygospore, meiosis occurs, and a hypha emerges and produces a sporangium. The haploid spores from this sporangium in turn develop into the typical vegetative mycelium.

**Distinctions between phycomycetes and higher fungi**

Despite the considerable differences among them, all phycomycetes share two properties that readily distinguish them from the remaining classes of fungi (Ascomycetes, Basidiomycetes, and Fungi imperfecti). First, *their asexual spores are always endogenous*, formed inside a saclike structure, the zoosporangium of the aquatic types or the sporangium containing immotile sporangiospores of the terrestrial types. In the other groups of fungi, the asexual spores are always exogenous, being formed free at the tips of hyphae (Figure 4.18). Second,

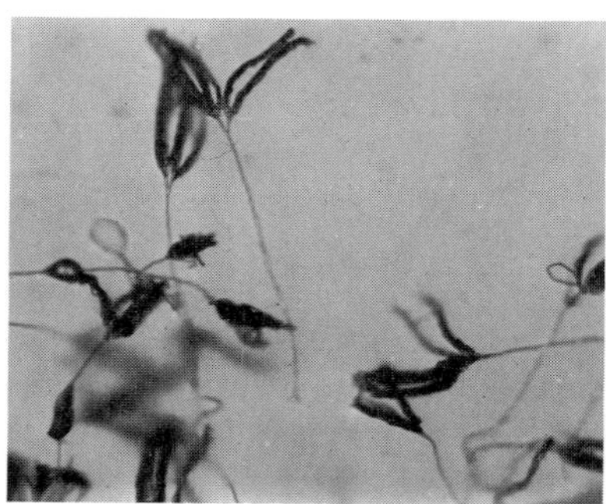

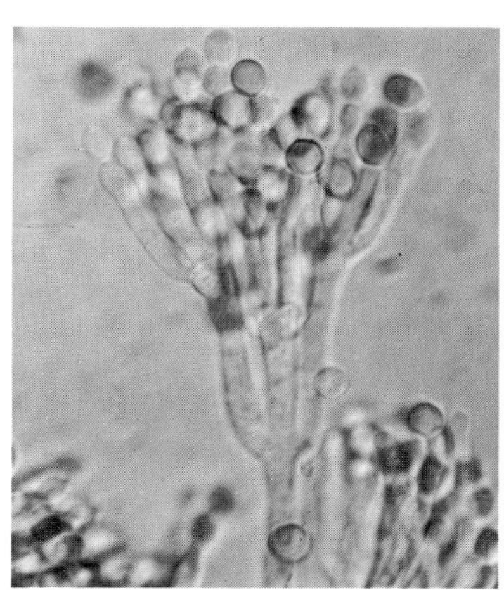

*Figure 4.18 Penicillium. Left, edge of a colony at relatively low magnification, showing spore heads. Right, conidiophore at high magnification, showing branched structure and terminal chains of spherical conidia. Courtesy of D. K. B. Raper.*

*the mycelium in phycomycetes shows no cross walls* except in regions where a specialized cell, such as a sporangium or gametangium, is formed from a hyphal tip. Such a mycelium is known as a *nonseptate mycelium*. In the remaining groups of fungi, distinct cross walls occur at regular intervals along the hyphae. Thus, on the basis of these two simple criteria, one can readily distinguish a phycomycete from any other type of fungus.

Since the mycelium of phycomycetes is nonseptate, it is clear that these organisms are coenocytic. The regular occurrence of cross walls in the mycelium of other groups of fungi suggests, in contrast, that they are cellular organisms. This is not true, however. The cross walls do not divide the cytoplasm into a number of separate cells: each cross wall has a central pore, through which both cytoplasm and nuclei can move freely. There is thus just as much cytoplasmic continuity in the septate fungi as in the phycomycetes, and both groups are, in fact, coenocytic.

## The ascomycetes and basidiomycetes

The higher fungi, with septate mycelia and exogenous asexual spores, are broadly classified into two groups, ascomycetes and basidiomycetes, on the basis of their sexual development. Following zygote formation in these fungi, there is an immediate reduction division followed by the formation of four or eight haploid sexual spores, which are borne in or on structures known as *asci* and *basidia*. The formation of an ascus is characteristic of fungi of the class Ascomycetes, and the formation of a basidium is characteristic of fungi of the class Basidiomycetes. In ascomycetes, the zygote develops into a saclike structure, the ascus, while the nucleus undergoes two meiotic divisions, often followed by one or more mitotic divisions. A wall is formed around each daughter nucleus and the neighboring cytoplasm to produce four, eight, or more ascospores within the ascus (Figure 4.19). Eventually the ascus ruptures, and the enclosed spores are liberated.

In basidiomycetes, the zygote enlarges to form a club-shaped cell, the basidium; at the same time, the diploid nucleus undergoes meiosis. The subsequent course of events is strikingly different from that which occurs in an ascus. No spores are formed within the basidium; instead, a slender projection known as a sterigma develops at its upper end, and a nucleus migrates into this sterigma as the latter enlarges. Eventually,

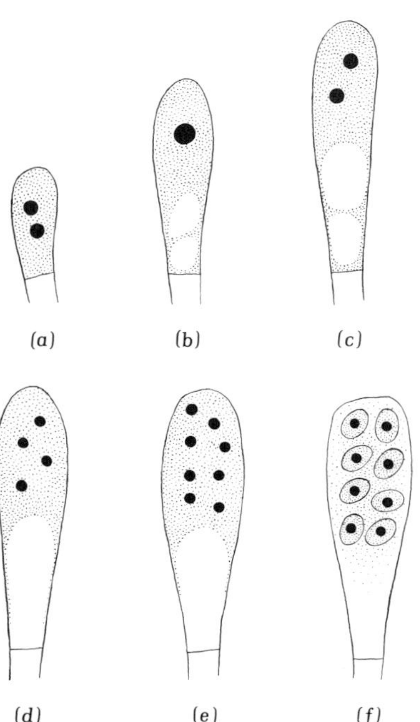


*Figure 4.19 Successive stages in the formation of an ascus. (a) Binucleate fusion cell; (b) nuclear fusion; (c), (d), (e), nuclear divisions; (f) ascospore formation.*

a cross wall is formed near the base of the sterigma, the cell thus cut off being a basidiospore. The same process is repeated for the remaining three nuclei in the basidium, so that a mature basidium bears on its surface four basidiospores (Figure 4.20). Basidiospore discharge occurs

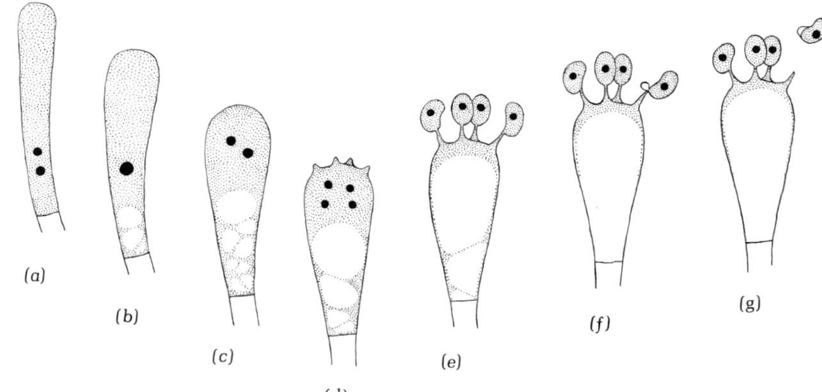


*Figure 4.20 Successive stages in basidium formation and basidiospore discharge: (a) binucleate cell; (b) nuclear fusion; (c), (d) nuclear division; (e) formation of basidiospores; (f), (g) basidiospore discharge.*

by a remarkable mechanism. After the basidiospore has matured, a minute droplet of liquid appears at the point of its attachment to the basidium. This droplet grows rapidly until it is about one fifth the size of the spore, and then, quite suddenly, both spore and droplet are shot away from the basidium.

**The Fungi Imperfecti**

The classification of the septate fungi into ascomycetes and basidiomycetes has one practical disadvantage. Obviously, the assignment of a fungus to its correct class is possible only if one has observed the sexual stage of its life cycle. If one happens to deal with a fungus that is incapable of sexual reproduction, or in which the sexual stage is unknown, it cannot be assigned either to the ascomycetes or to the basidiomycetes. Since heterothallism is very common in the higher fungi, it often happens that a single isolate of an ascomycete or basidiomycete will never undergo sexual reproduction, which requires the presence of another strain of opposite mating type. Accordingly, it has been necessary to create a third class, the Fungi Imperfecti, for those kinds in which a sexual stage has not so far been observed. It should be realized that the fungi imperfecti is essentially a provisional taxonomic group; from time to time the sexual stage is discovered in a fungus originally assigned to this group, and the organism in question is then transferred to either the ascomycetes or the basidiomycetes.

**The development of an ascomycete**

As a typical ascomycete, we may consider a mold of the genus *Neurospora*. The vegetative stage of *Neurospora* consists of a mycelium, on the surface of which there develop special hyphae, the *conidiophores*, carrying chains of exogenous asexual spores, the *conidia*. The conidia are pigmented and are responsible for the characteristic pink to orange color of a *Neurospora* colony. When mature, conidia are easily dislodged and float through the air. When they come into contact with a substrate favorable for development, they germinate and give rise once more to the development of a vegetative mycelium.

*Neurospora* is a heterothallic ascomycete, which also has a sexual reproductive cycle. The haploid mycelia form immature fruiting bodies, termed *protoperithecia*, each consisting of a coiled hypha lying within a hollow sphere formed from a compact mass of ordinary hyphae. When a hypha from a mycelium of opposite mating type comes into contact with a protoperithecium, it fuses with the coiled hypha within the protoperithecium, and the nuclei of the two haploid strains mingle in a common cytoplasm. Each type of nucleus divides repeatedly, giving rise to many haploid nuclei of opposite mating types; these eventually fuse in pairs, to form many diploid nuclei, which then undergo immediate meiosis. An ascus develops at the site of each meiosis, around the four haploid nuclei. Meanwhile, the wall of the protoperithecium thickens and becomes pigmented, to form a mature *perithecium* that contains several dozen asci. The maturation of the asci is completed by the formation of ascospores, each delimited within the

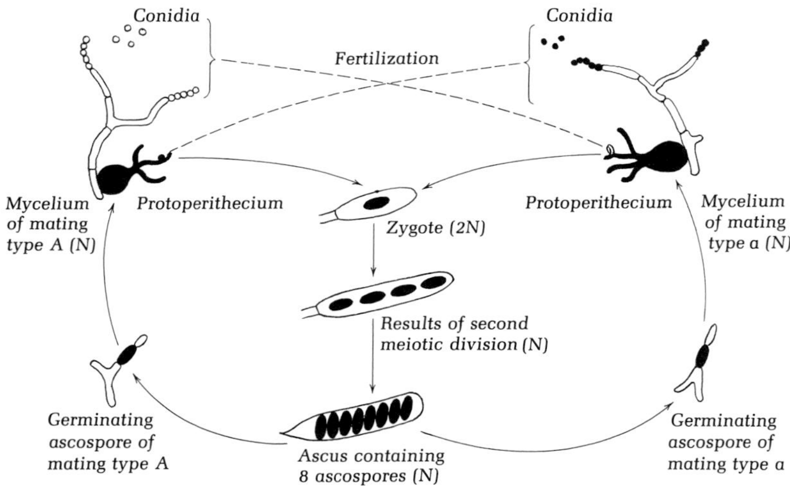


*Figure 4.21 The life cycle of Neurospora, a heterothallic ascomycete. Asexual reproduction occurs by the formation of conidia from a haploid mycelium of each mating type, A and a. These mycelia also bear protoperithecia which, when fertilized by conidia or hyphae of the opposite mating type, develop into perithecia, within which numerous zygotes are formed. Each zygote undergoes two meiotic and one mitotic division to form an ascus containing 8 ascospores, four of mating type A and four of mating type a. Germination of the ascospores gives rise once more to haploid mycelia.*

ascus by a resistant spore wall that surrounds one of the haploid nuclei and the adjacent cytoplasm. A mature ascus may contain four or eight ascospores, the number depending on whether or not meiosis is followed by a further mitotic division of the four haploid nuclei.

The mature perithecium is roughly spherical in shape, with a short protruding neck. At maturity, a pore forms at the tip of the neck, through which the ascospores are violently discharged. Upon germination, ascospores (like conidia) produce a haploid mycelium. This life cycle is shown in Figure 4.21.

**The development of a basidiomycete**

The most conspicuous members of the basidiomycetes are the mushrooms. The portion of a mushroom that is seen by the casual observer is a small part of the whole organism, being the specialized fruiting structure that bears the basidia. The vegetative portion of the organism is entirely concealed from view and consists of a loose mycelium spreading, often for many yards, under the soil (Figure 4.22). The vegetative mycelium grows more or less continuously. When conditions are favorable (generally following a spell of wet, warm weather), fruiting bodies are formed at various points on its surface and push up through the soil to become the parts of the fungus visible to the observer.

In the common field mushroom (Figure 4.22) the fruiting body consists of a stalk surmounted by a cap, both composed of closely packed hyphae. The underside of the cap consists of rows of radiating gills, each gill being lined with thousands of basidia. The basidia project horizontally from the vertical walls of the gill, and consequently when the basidiospores are ejected, they pass into the air space between adjacent gills, and from there fall to the ground below. When a mushroom is mature, basidiospore discharge is a massive phenomenon. If one places a ripe cap on a piece of paper for a few hours, a "negative" of the gill structure will be formed on the paper by the deposition of millions of basidiospores.

*Figure 4.22 Cross section of a mushroom, showing the subterranean vegetative mycelium and the fruiting structure. In the mature fruiting body at left, large numbers of basidiospores are being discharged from the gills on the underside of the cap, and are being dispersed by the wind. A second, immature fruiting body is shown at right, just emerging from the soil. From A. H. R. Buller, Researches on Fungi, Vol. 1, p. 219. New York: Longmans Green, 1909.*

Heterothallism is widespread among basidiomycetes. Consequently, isolated haploid basidiospores give rise on germination to haploid mycelia that are incapable of fructifying. However, if two such haploid mycelia of compatible mating types come into contact, hyphal fusion followed by nuclear exchange takes place, to produce a still haploid *dicaryon;* the two kinds of nuclei become associated in a very regular fashion, one pair occurring in each compartment of the septate mycelium. During growth of the dicaryon, the two kinds of nuclei divide synchronously.

A dicaryotic mycelium may continue to grow vegetatively for a long time, and during such vegetative growth, fusion between the paired nuclei never occurs. Fusion takes place only at the time of fructification, when the basidia are produced and is followed in each basidium by immediate meiosis to produce the four haploid nuclei destined to enter the basidiospores.

### Heterocaryosis and parasexuality in mycelial fungi

Among mycelial fungi, hyphal fusions are not restricted to heterothallic individuals of opposite mating types; they can occur between any two strains belonging to the same species. If the two strains belong to the same mating type, so that eventual completion of the sexual cycle cannot follow hyphal fusion, the result is the *formation of a single organism containing nuclei derived from both parental strains; such an organism is a heterocaryon.* The nuclei derived from each parental mycelium migrate into the cytoplasm of the other; because the mycelia are coenocytic, migration, coupled with nuclear division, eventually leads to a complete mingling of the two types of nuclei.

In most cases, the two types of nuclei do not fuse but multiply by mitosis independently of one another. The phenotypic properties of a heterocaryon are determined by both types of nuclei, so that heterocaryons have the properties of hybrids between the two parental strains, even though all their nuclei may be haploid. The two types of nuclei may become dissociated if uninucleate conidia are produced. However, since many fungi produce multinucleate conidia, heterocaryosis may be perpetuated through the asexual life cycle.

On rare occasions other than when fruiting bodies are being formed, nuclear fusion may occur within a heterocaryotic mycelium, to form a true diploid nucleus derived from the two types of haploid nuclei. This was first detected in *Aspergillus* through the use of mutant strains which produce conidia with altered pigmentation. Mutants forming yellow and white conidia, respectively, instead of the normal green conidia of the wild type, were used to prepare a heterocaryon. As the conidia of *Aspergillus* are uninucleate, each spore head of this heterocaryon normally produces chains of both yellow and white spores. The presence of diploid nuclei is revealed by the occasional appearance of a chain of diploid spores, easily detected as a result of their green color, characteristic of the wild type. Such spores can be isolated and used to propagate diploid mycelia. Haploidy is never reestablished by meiosis but can be attained by the successive losses of one member of each chromosome pair. The resulting haploids are recombinants between the two parental forms, since the individual chromosome may be derived from either parent and crossovers between chromosome pairs occur during mitosis in the diploid state. This phenomenon is termed *parasexuality:* it has the genetic consequences of a sexual process but does not involve the usual sequence of gametic fusion and meiosis. It provides a mechanism for genetic recombination in imperfect fungi but is probably of little importance in fungi with a sexual stage.

**The yeasts**

Among the ascomycetes, basidiomycetes, and fungi imperfecti, the characteristic vegetative structure is the coenocytic mycelium. Nonetheless, there are a few groups in these classes that have largely lost the mycelial habit of growth and become unicellular. Such organisms are known collectively as *yeasts*. A typical yeast consists of small, oval cells that multiply by forming buds. The buds enlarge until they are almost equal in size to the mother cell, nuclear division occurs, and then a cross wall is formed between the two cells (Figure 4.23). Although the yeasts constitute a minor branch of the higher fungi in terms

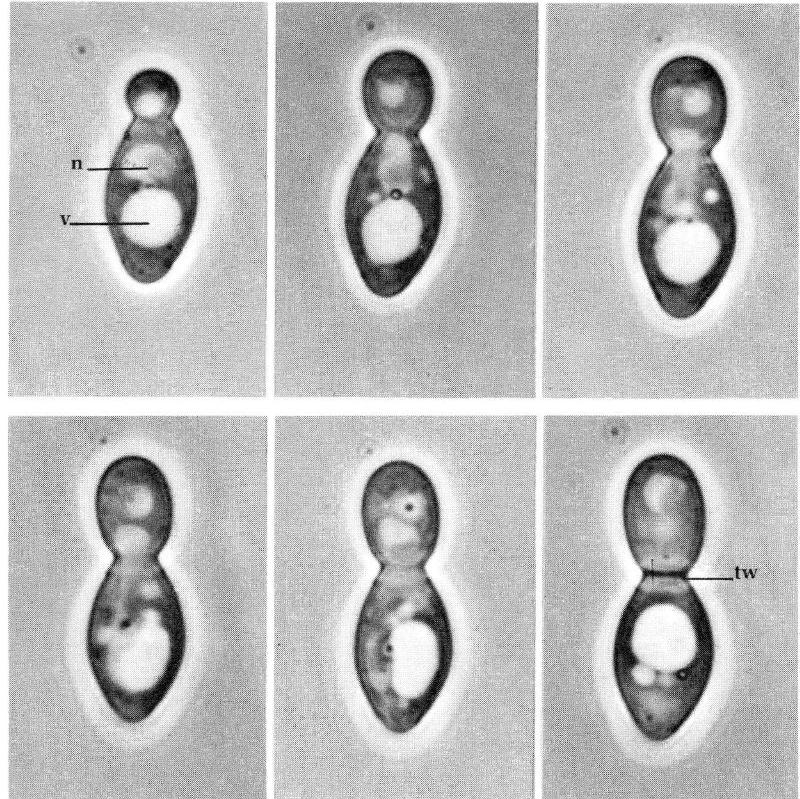


*Figure 4.23 A sequence of photomicrographs of a budding cell of the ascomycetous yeast, Wickerhamia, showing nuclear division and transverse wall formation (phase contrast, ×2,295): nucleus, n; vacuole, v; transverse wall, tw. From P. Matile, H. Moor, and C. F. Robinow, The Yeasts, Vol. 1. New York: Academic Press, 1969.*

of number of species, they are very important microbiologically. Most yeasts do not live in soil but have instead become adapted to environments with a high sugar content, such as the nectar of flowers and the surface of fruits. Many yeasts (the fermentative yeasts) perform an alcoholic fermentation of sugars and have been long exploited by man (see Chapter 29).

Yeasts are classified in all three classes of higher fungi: ascomycetes, basidiomycetes, and fungi imperfecti. The principal agent of alcoholic fermentation, *Saccharomyces cerevisiae,* is an ascomycetous yeast. Budding ceases at a certain stage of its growth, and the vegetative cells become transformed into asci, each containing four ascospores. For a long time it was believed that ascospore formation in *S. cerevisiae* was not preceded by zygote formation because pairing of vegetative cells prior to the formation of ascospores could never be observed. Eventually, however, it was discovered that zygote formation takes place at an unexpected stage of life cycle—immediately after the germination of the haploid ascospores. Pairs of germinating ascospores, or the first vegetative cells produced from them, fuse to form diploid vegetative cells. Diploidy is then maintained throughout the entire subsequent period of vegetative development, and meiosis occurs immediately prior to the formation of ascospores. Thus, *S. cerevisiae* exists predominantly in the diplophase. Other ascomycetous yeasts do not share this pattern of behavior but form zygotes by fusion between vegetative cells immediately before ascospore formation. The germinating ascospores then give rise to haploid vegetative progeny.

Although budding is the predominant mode of multiplication in yeasts, there are a few that multiply by binary fission, much like bacteria; these are placed in a special genus, *Schizosaccharomyces.*

In ascomycetous yeasts, the vegetative cell or zygote becomes entirely transformed into an ascus at the time of ascospore formation. Yeasts of the genus *Sporobolomyces* form basidiospores, and in this case the entire vegetative cell becomes transformed into a basidium. Just as in the mushrooms, basidiospore discharge in *Sporobolomyces* is a violent process, and the colonies of this yeast are readily detectable

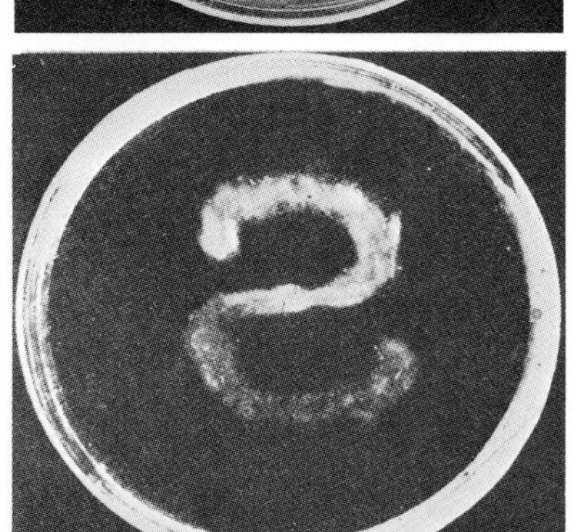

*Figure 4.24 The formation of a mirror image of a colony of Sporobolomyces by basidiospore discharge in a Petri dish incubated in the inverted position. Top, the colony on the agar surface, streaked in the form of an S. Bottom, the deposit of basidiospores formed on the lid of the Petri dish as a result of spore discharge from the colony. From A. H. R. Buller, Researches on Fungi, Vol. 5, p. 175. New York: Longmans Green, 1933.*

on plates that have been incubated in an inverted position because the portion of the glass cover underlying a *Sporobolomyces* colony becomes covered with a deposit of discharged spores that form a mirror image of the colony above (Figure 4.24).

## The slime molds

We shall conclude this survey of the higher protists by discussing the *slime molds*, which are not classified as true fungi, although they possess certain characteristics that resemble those of the fungi. The best-known representatives of the slime molds are the myxomycetes, organisms that are found most commonly growing on decaying logs and stumps in damp woods. The vegetative structure, known as a *plasmodium*, is a multinucleate mass of cytoplasm unbounded by rigid walls, which flows in ameboid fashion over the surface of the substrate, ingesting smaller microorganisms and fragments of decaying plant material. An actively moving plasmodium is characteristically fan-shaped, with thickened ridges of cytoplasm running back from the edge of the fan; it resembles a spreading layer of thin, colored slime (Figure 4.25). As long as conditions are favorable for vegetative development, the plasmodium continues to increase in bulk with accompanying repeated nuclear divisions. Eventually, the organism may become a mass of cytoplasm containing thousands of nuclei and weighing several hundred grams. Fruiting occurs when a plasmodium migrates to a relatively dry region of the substrate. Out of the undifferentiated plasmodium there is then produced a fruiting structure that is often of remarkable complexity and beauty [as illustrated by the case of *Ceratiomyxa* (Figure 4.26)]. As this fruiting body develops, small, uninucleate sections of the plasmodium become surrounded by walls to form large numbers of uninucleate spores, borne on the fruiting structure. After liberation, the spores germinate to produce uniflagellate ameboid gametes which fuse in pairs to form biflagellate zygotes. After some time, a zygote loses its flagella and develops into a new plasmodium. The vegetative nuclei in a growing plasmodium are diploid, meiosis taking place just prior to the formation of spores in the fruiting body.

*Figure 4.25 The plasmodium of a myxomycete, Didymium, growing at the expense of bacteria on the surface of an agar plate. Courtesy of Dr. K. B. Raper.*

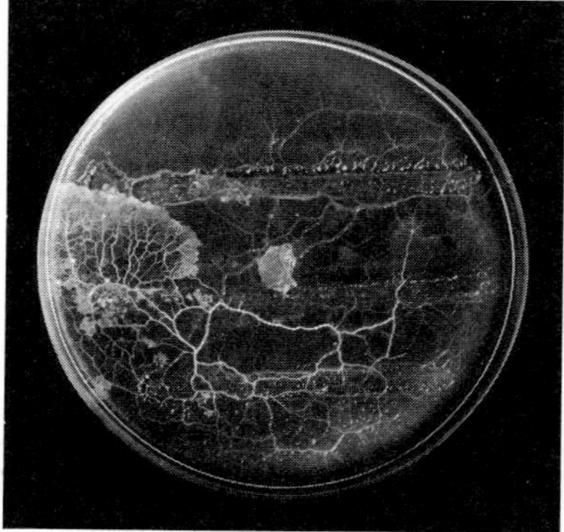

*Figure 4.26 Fruiting bodies of a myxomycete, Ceratiomyxa, on a piece of wood. From C. M. Wilson and I. K. Ross, "Meiosis in the Myxomycetes," Am. J. Botan. 42, 743 (1955).*

It is, of course, the fruiting stage of a myxomycete that at once reminds one of a true fungus; at first sight, the amorphous, plasmodial vegetative stage appears to resemble little, if at all, the branched, mycelial vegetative stage of the fungi but suggests, rather, a relationship to the ameboid protozoa. In fact, the plasmodium and the mycelium are basically similar structures. Both are coenocytic, and in both the cytoplasm can flow, although in the mycelium cytoplasmic streaming is confined within the walls of branched tubes. The superficial difference between a plasmodium and a mycelium is essentially caused by the fact that in a plasmodium the cytoplasm is not bounded by rigid walls and is thus free to flow in any direction.

The slime molds also include a small group, the Acrasieae (Figure 4.27), which show far greater resemblances to the unicellular ameboid protozoa than do the true myxomycetes. The vegetative stage of an acrasian consists of small, uninucleate amebas, which multiply by binary fission and can in no way be distinguished, at this stage of their life history, from other small ameboid protozoa. Nevertheless, when conditions are favorable, thousands of these isolated amebas are capable of aggregating and cooperating, without ever losing their cellular distinctness, in the construction of an elaborate fruiting body. The first sign of approaching fructification is the aggregation of the vegetative cells to form a macroscopically visible heap. This heap of cells gradually differentiates into a tall stalked structure that bears a rounded head of asexual spores. At all stages in the formation of this fruiting body, the cells remain separate; some individuals form the stalk, which is surrounded and given rigidity by a cellulose sheath, while others are carried up the outside of the rising stalk to form the spore head. As this

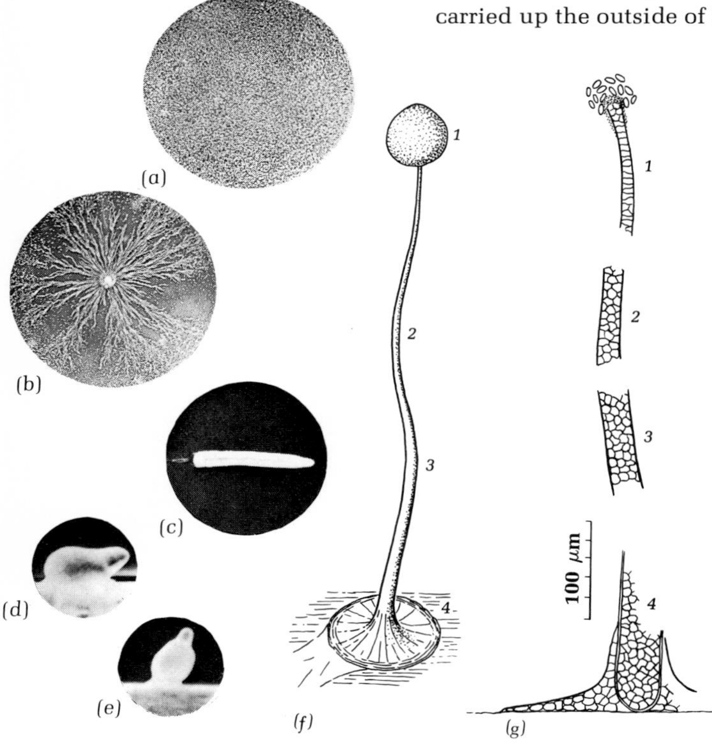


*Figure 4.27 The life cycle of Dictyostelium, a representative of the Acrasieae. (a) A uniform mass of vegetative amoebae. (b) Aggregation of the amoebae to a fruiting center. (c) Motile mass of aggregated cells. (d), (e) Early stages in formation of the fruiting body. (f) A mature fruiting body. (g) Magnified sections through various regions of the fruiting body. Courtesy of K. B. Raper and Quart. Rev. Biol. Reprinted with permission.*

matures, each ameba in it rounds up and becomes surrounded by a wall. These spores, following their release, germinate and give rise to individual ameboid vegetative cells once more. This remarkable kind of life cycle, where a communal process of fructification is imposed on a unicellular phase of vegetative development, occurs in one procaryotic group, the myxobacteria (described in Chapter 5).

## *The eucaryotic protists: summing up*

It is not possible in one chapter to do justice to the extraordinary profusion and biological variety of the eucaryotic protists; only a few representatives of each major subgroup have been somewhat summarily described. For more detailed information about these organisms, the reader should consult specialized books dealing with the algae, the protozoa, or the fungi (see the bibliography at the end of the chapter). There is, unfortunately, no single book that provides a more extended survey of the entire biological group. Comprehension of the comparative biology of eucaryotic protists is further impeded by major terminological difficulties, because botanists and zoologists have applied entirely different names to structures common to all three subgroups. Moreover, the taxonomic treatments adopted for flagellate algae and leucophytes by zoologists and botanists are widely different. Here we have tried to bridge these differences and to provide a broader account than is customary of the eucaryotic protists, in terms of their possible evolutionary interrelationships. These suggested interrelationships are summarized in Figure 4.28.

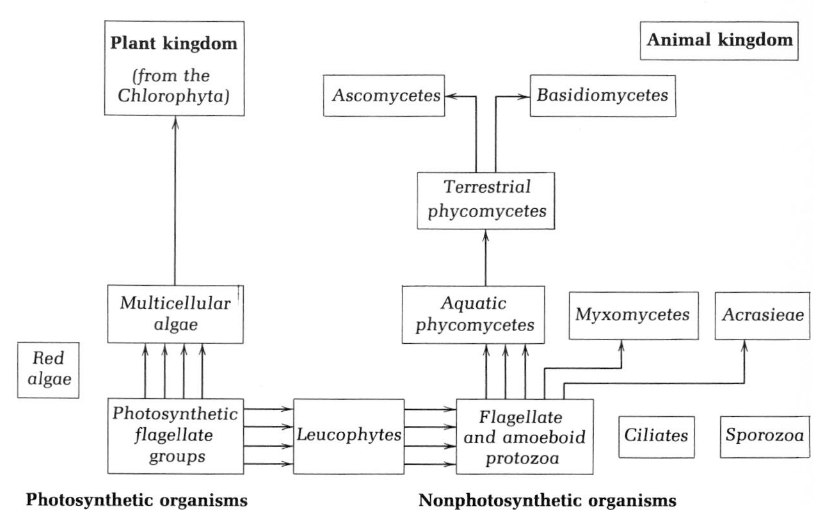


*Figure 4.28 Possible evolutionary relationships between eucaryotic biological groups. The arrows connecting boxes indicate directions of evolution; where several arrows connect two boxes, this indicates probable multiple evolutionary lines connecting the two groups in question. Isolated boxes contain groups of which the evolutionary relationships are uncertain.*

## *Further reading*

**Books**

*Eucaryotic algae:*

Fritsch, F. E., *Structure and Reproduction of the Algae.* New York: Cambridge University Press, 1935, Vol I; 1945, Vol II.

Lewin, R. (editor), *Physiology and Biochemistry of Algae.* New York: Academic Press, 1962.

Smith, G. M., *Cryptogamic Botany,* 2nd ed., Vol. I. New York: McGraw-Hill, 1955.

*Protozoa:*

Hall, R. P., *Protozoology.* Englewood Cliffs, N.J.: Prentice-Hall, 1953.

Mackinnon, D. L., and R. S. J. Hawes, *An Introduction to the Study of Protozoa.* New York: Oxford University Press, 1961.

*Fungi:*

Alexopoulos, C. J., *Introductory Mycology,* 2nd ed. New York: Wiley, 1962.

Burnett, J. H., *Fundamentals of Mycology.* New York: St. Martin's Press, 1968.

# Chapter five
# The procaryotic protists

*The procaryotic protists* (bacteria and blue-green algae) are defined primarily by the structure of their cells, as discussed in Chapter 2. These organisms show a considerably narrower range of structural variation than the eucaryotic protists; there are, for example, no representatives that have achieved the size and organismal complexity characteristic of seaweeds or mushrooms, or the extraordinarily specialized unicellular condition characteristic of ciliates. Evolutionary specialization among procaryotic protists is expressed in metabolic rather than in structural terms; no other major biological group is as varied in this respect. Representatives of all four primary nutritional categories—photoautotrophs, photoheterotrophs, chemoautotrophs, and chemoheterotrophs—occur among the procaryotic protists. Indeed, almost every type of metabolism found in eucaryotic organisms can also be found in procaryotic ones, and many specialized types of metabolism exist *uniquely* in procaryotic organisms. The blue-green algae are physiologically uniform, consisting overwhelmingly of photoautotrophs; the extraordinary spectrum of physiological diversity is principally expressed among the bacteria. The many important roles that bacteria play in nature as agents in the cycles of matter (see Chapter 22) are a reflection of their physiological diversity and explain both the numerical abundance and the ubiquity of these organisms throughout the biosphere.

The bacteria and blue-green algae possess several distinctive general properties, in addition to the procaryotic organization of their cells, which will be discussed as a preliminary to the description of their constituent groups.

## *The range of cell size*

Among both eucaryotic and procaryotic protists, the dimensions of the cell show a very substantial spread if we take all representatives of each group into consideration. The largest unicellular bacteria and blue-green algae are somewhat larger than the smallest unicellular eucaryotic protists, so that size alone cannot be used as an absolute differential character to distinguish the two great microbial assemblages from one another. Nevertheless, it is certainly true to say that the *average cell size* of the procaryotic protists is considerably less than that of the eucaryotic protists. Furthermore, the smallest unicellular bacteria are far below the lower limit of size found among eucaryotic groups and are, in fact, only barely resolvable with a light microscope. These bacteria overlap in size with the largest viruses.

Because cells can differ so greatly in shape, the only satisfactory basis on which their size can be defined is *cellular volume*. In Table 5.1, data on the volume of the cell for a number of protists, as well as for the structural unit (virion) of some representative viruses, have been assembled. From the individual data, a *size range* for each of the three groups can be established; these values are also given in Table 5.1.

*Table 5.1 Size of the unit of structure of unicellular protists and of viruses*

| Nature of structural unit | Biological group | Volume of structural unit, $\mu m^3$ | | |
|---|---|---|---|---|
| | | **Normal range** | **Extreme limits** | **Limits for major group** |
| *Eucaryotic cell* | *Unicellular algae* | *5,000–15,000* | *5–100,000* | *5–150,000,000* |
| | *Protozoa* | *10,000–50,000* | *20–150,000,000* | |
| | *Yeasts* | *20–50* | *20–50* | |
| *Procaryotic cell* | *Eubacteria* | *1–5* | *0.10–5,000* | *0.01–5,000* |
| | *Spirochetes* | *0.1–2* | *0.05–1,000* | |
| | *Myxobacteria* | *1–5* | *0.50–20* | |
| | *Blue-green algae* | *5–50* | *0.10–5,000* | |
| | *Rickettsias* | *0.01–0.03* | *0.01–0.03* | |
| *Virion* | *Pox group* | *0.01* | *Fixed for each group* | *0.00001–0.01* |
| | *Rabies virus* | *0.0015* | | |
| | *Flu virus* | *0.0005* | | |
| | *Polio virus* | *0.00001* | | |

The smallest known eucaryotic protist is a flagellate alga, *Micromonas*. This interesting organism has a cell that contains a single, small chloroplast and one mitochondrion. Together with the nucleus, they occupy a large part of the total volume of the cell. Therefore, any further reduction of size in such an organism could be achieved only *by elimination of a cellular organelle that is vital to metabolic function.* The lower size limit possible for this kind of cell is accordingly set by *structural limitations.*

In principle, the lower size for a procaryotic cell is set by *molecular limitations.* To reproduce itself, any cell requires a large number of *different enzymes* and hence of *different kinds of proteins.* We do not know the precise minimum, but the number is probably of the order of several hundred. Furthermore, *many molecules* of most enzymes are needed. To these must be added the cellular nucleic acids and the various other organic constituents of the cell such as lipids and carbohydrates. All in all, it seems likely that the smallest bacteria, which can just be seen with a light microscope, have a size very close to this *molecular limit* for the maintenance of cellular function.

The minimal size of the unit of structure of a virus, the virion, is also determined by molecular considerations. Virions are composed of only one kind of nucleic acid and of one (or a few) kinds of protein; their minimal size is, therefore, defined by the volume occupied by a small number of molecules of protein and nucleic acid.

The factors that operate to determine minimal size adequately explain why the cells of some bacteria can be two orders of magnitude smaller than the smallest eucaryotic cell. It is not immediately obvious, however, why the cells of bacteria and blue-green algae cannot attain the largest cell sizes found among protozoa and eucaryotic algae, although the data assembled in Table 5.1 clearly show that they do not. Several factors probably operate to keep procaryotic cells small. In the first place, it is a general biological rule, valid at all levels of biological complexity, that *the rate of metabolism is inversely related to the size of the organism.* Since the growth rate is primarily determined by the overall rate of metabolism, small size is an essential condition for rapid growth. Many bacteria are characterized by extraordinarily rapid growth, with doubling times of less than an hour under optimal conditions. Such growth rates are far higher than those in most eucaryotic groups. This property undoubtedly confers a decisive biological advantage in many situations where bacteria compete for nutrients with other kinds of microorganisms and is probably the most important single factor in the survival of many bacteria in nature.

A second factor can be surmised from the properties of very large bacteria, such as the chemoheterotroph *Spirillum volutans* and the photoautotrophs *Chromatium okenii* and *Thiospirillum jenense*. In contrast to the smaller-celled members of the same physiological groups, these giant bacteria are extremely difficult to cultivate, primarily because the physicochemical conditions necessary for their development are very narrowly defined. In other words, they lack adaptive flexibility and cannot readily accommodate to minor changes in the environment. This suggests that the power to *regulate metabolic function*, essential to achieve adaptive flexibility, is very limited in such organisms. The upper size limit for the procaryotic cell may, accordingly, be determined by the difficulty of maintaining a satisfactory regulation and coordination of metabolic activity in a large cell with the procaryotic type of organization.

## The chemical nature of the cell wall

With a few exceptions to be discussed later, the cells of procaryotic organisms are enclosed within walls. In itself, this is not a distinctive group character, as cell walls also exist in many eucaryotic protists and in plants. However, analytical studies have shown that the *procaryotic cell wall has a unique chemical composition*.

In all biological groups, the cell wall is chemically complex, being typically composed of a number of different kinds of macromolecules, associated either in a molecular mosaic or in adjacent layers. However, the tensile strength which enables the wall to withstand the osmotic pressure of the enclosed protoplast and hence prevents osmotic lysis in a hypotonic environment is largely contributed by one particular component of the wall, which forms a molecular mesh, or *sacculus*, enclosing and constraining the protoplast and conferring on it the characteristic external form of the cell. The nature of this component can often be inferred by the effect of exposing cells to hydrolytic enzymes, known to destroy one specific macromolecular constituent of the wall. If such treatment results in swelling and osmotic lysis of the cell, it is clear that the constituent in question is essential for the structural integrity of the wall. Thus, the cells of higher plants undergo osmotic lysis upon treatment with the enzyme *cellulase*, pinpointing *cellulose* as the constituent primarily responsible for the structural integrity of the wall. Similarly, the polysaccharide *chitin* can be identified as the main strengthening element of the wall in many fungi.

In all procaryotic groups, the sacculus, which confers tensile strength on the wall and defines the shape of the cell, is composed of a unique kind of organic polymer, known as a *murein*. Mureins are *heteropolymers*, composed of several different kinds of subunits, some of which are carbohydrates and some amino acids. The two carbohydrates in mureins are the amino sugars N-acetylglucosamine (AG) and N-acetylmuramic acid (AMA) (Figure 5.1). Although AG occurs as a constituent

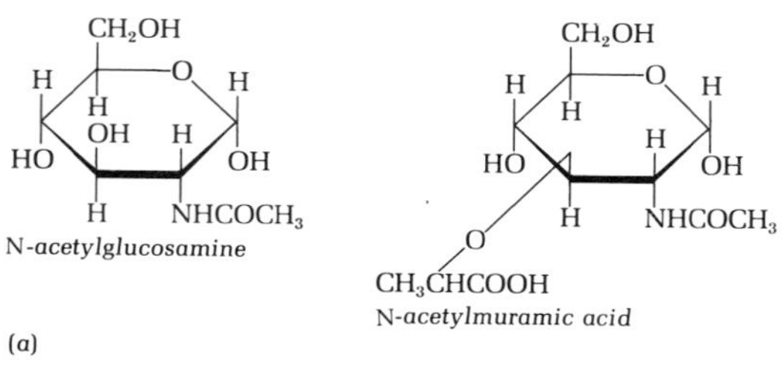


L-alanine

D-alanine

D-glutamic acid

diaminopimelic acid

or

L-lysine

(b)

Figure 5.1 *The chemical constituents of the mureins: (a) amino sugars; (b) amino acids. N-Acetylmuramic acid and diaminopimelic acid are particularly distinctive constituents of this type of cell-wall polymer and are not known to occur in any other types of biological polymers.*

of many other biological polymers (for example, chitin), AMA is found only in mureins. The amino acid constituents (Figure 5.1) are somewhat more variable but always include as a minimum three different amino acids: *alanine*, *glutamic acid*, and either *diaminopimelic acid* or a structurally related basic amino acid, *lysine*. Alanine, glutamic acid, and lysine are also universal constituents of proteins; however, half the alanine residues and all the glutamic acid residues in mureins are in the form of the "unnatural," or D, isomers, whereas in proteins all the amino acid residues have the L configuration. Diaminopimelic acid never occurs in proteins.

Bacterial cell walls are often very complex in structure, containing other polymeric constituents in addition to murein (See Chapter 10 for a more detailed discussion). In fact, the murein component may be a minor constituent in quantitative terms. The importance of mureins as a taxonomic marker of the procaryotic cell rests on their almost universal presence in the walls of such cells, coupled with their complete absence as structural constituents of eucaryotic cells.

## The absence of sterols

A second chemical peculiarity of nearly all procaryotic cells is a negative one: the *absence of sterols*. The sterols are a special class of lipids, all of which share a distinctive condensed ring structure derived biosynthetically from an open-chain unsaturated hydrocarbon with 30 carbon atoms, squalene (Figure 5.2). One or more members of this class

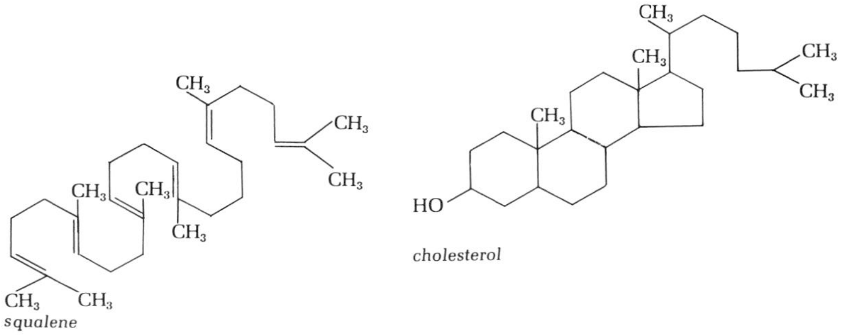

*Figure 5.2 The structures of the $C_{30}$ hydrocarbon, squalene, which is the biosynthetic precursor of sterols, and of a typical sterol, cholesterol.*

of lipids are universally present in eucaryotic cells, where they appear to be built into the structure of the cell membrane and possibly of the endoplasmic reticulum. Small amounts of sterols have recently been detected in blue-green algae. Sterols do not occur in bacteria, however, except in members of the mycoplasma group cultivated on media containing a large amount of blood serum; in this special case, the sterols are derived from the medium. It seems clear, therefore, that an *inability to synthesize sterols* is a general characteristic of bacteria, even though they can synthesize many other lipid substances (for example, carotenoid pigments) which have the same primary biosynthetic origin as sterols. Among the bacteria, only certain mycoplasmas require sterols as growth factors.

A number of eucaryotic organisms also lack the ability to synthesize sterols, but in all such cases, *sterols are essential growth factors*, which must be supplied in the medium if the organisms in question are to grow. These facts indicate that the sterols have universal and very important functions in the eucaryotic cell, probably concerned, at least in part, with the maintenance of the integrity of the cell membrane. The absence of sterols in most procaryotic organisms probably points to a basically different molecular organization of the procaryotic cell membrane, even though its fine structure as seen in the electron microscope cannot be distinguished from that of a eucaryotic cell membrane (see Chapter 2).

## *The structural bases for the selective toxicity of antibiotics*

Antibiotics are organic compounds of low molecular weight and of varied and often unusual chemical structure. They are biological products, which are formed and excreted by certain microorganisms, and are either growth inhibitory or toxic for other microorganisms. A characteristic feature of the antibiotics, which underlies their therapeutic value in the treatment of infectious diseases, is their *selective toxicity*. Some of these compounds are selectively toxic for procaryotic organisms, others for eucaryotic ones. In a number of cases, the analysis of the mode of action of such antibiotics has shown that the target is a structural component peculiar either to the eucaryotic or to the procaryotic cell (Table 5.2).

The penicillins and several other antibiotics are selectively toxic for procaryotic organisms, because they inhibit specific steps involved in the synthesis of the murein component of the cell wall. The polyene antibiotics, such as nystatin and filipinin, are selectively toxic for eucaryotic organisms, in which they cause disruption of the cell membrane. Since their effect can be antagonized by the presence of sterols

*Table 5.2 Interpretation in structural terms of the selective toxicity of certain antibiotics for procaryotic and eucaryotic organisms*

| | Active against: | | | |
|---|---|---|---|---|
| **Antibiotics** | **Procaryotes** | **Eucaryotes** | **Structural target in cell** | **Mode of action** |
| *Penicillins, oxamycin, vancomycin* | + | − | *Murein component of procaryotic cell wall* | *Inhibit steps in murein synthesis* |
| *Polyene antibiotics: filipinin, nystatin* | − | + | *Eucaryotic cell membrane* | *Interfere with sterol incorporation or function* |
| *Streptomycin, gentamycin, kanamycin, neomycin, paromycin* | + | −(+)[a] | *70-S ribosome* | *Cause misreading of genetic message during protein synthesis* |
| *Chloramphenicol* | + | −(+)[a] | *70-S ribosome* | *Interferes with protein synthesis by preventing peptide bond formation* |

[a]See the discussion in the text.

in the medium, it is probable that they interfere in some way with the sterol component of the eucaryotic membrane.

Finally, there are numerous antibiotics that specifically interfere with protein synthesis in procaryotic organisms and for which the structural target is the 70-S ribosome. Streptomycin and a number of related antibiotics combine with 70-S ribosomes in such a way as to interfere with normal protein synthesis; chloramphenicol prevents protein assembly on 70-S ribosomes by blocking peptide bond formation. Although relatively nontoxic for eucaryotic organisms, very high concentrations of all these antibiotics can have deleterious effects on eucaryotic cells, affecting the synthesis either of chloroplasts or of mitochondria. Thus, streptomycin treatment destroys the chloroplasts of plants and of the alga *Euglena*, while chloramphenicol treatment of yeast causes the formation of structurally abnormal, and completely nonfunctional, mitochondria. It will be recalled (Chapter 2) that protein synthesis within these organelles is mediated by 70-S ribosomes, whereas cytoplasmic protein synthesis in eucaryotic cells is mediated by 80-S ribosomes. Thus, *organellar* protein synthesis in eucaryotes, like the total machinery of protein synthesis in procaryotes, is susceptible to the action of these antibiotics; this is symbolized in Table 5.2 by indicating their toxicity for eucaryotes as "—(+)."

## *The major subdivisions of the procaryotic protists*

The procaryotic protists are customarily divided into two groups: bacteria and blue-green algae. This division is a convenient one, because the blue-green algae are readily distinguishable on physiological grounds. They perform photosynthesis accompanied by oxygen evolution, and they have a photosynthetic pigment system that includes chlorophyll *a* and $\beta$-carotene, which are invariably associated with oxygen-evolving photosynthesis. Not all blue-green algae are motile, but motility, when it exists, is of the gliding type.

The bacteria are a large and very varied group of procaryotic organisms. They do not really share any common properties that justify treating them as a single group distinct from, and equivalent to, the blue-green algae. They cannot be characterized as nonphotosynthetic, because photosynthetic ability, although not widespread, is possessed by the purple and green bacteria. The pigment systems of these bacteria are different from that of blue-green algae, and their photosynthesis never results in oxygen formation. The type of motility cannot serve to distinguish bacteria from blue-green algae, since many nonphotosynthetic bacteria are capable of gliding movement resembling that of blue-green algae. Some of these organisms are probably leucophytes, in the sense that they are nonphotosynthetic structural counterparts of particular blue-green algae; others show no close resemblances to known blue-green algae. Instead of attempting to define the bacteria as a whole, it is more convenient for purposes of description to recognize three principal subgroups among them, which can be distinguished from one another and from blue-green algae by a combination of structural and physiological characters. These are *eubacteria, myxobacteria,* and *spirochetes,* into which can be fitted the great majority of organisms customarily included among the bacteria. The principal characters of the four subgroups among the procaryotic protists are summarized in Table 5.3.

The eubacteria, the myxobacteria, and the spirochetes are distinguished primarily by differences with respect to the *mode of movement* and the *structure of the cell wall.* The myxobacteria comprise all those unicellular bacteria which show gliding movement. Their cell wall is thin and delicate; in the few cases where it has been analyzed chemically, it appears to consist largely of murein. As a result of their wall structure, the rod-shaped cells of myxobacteria are readily deformable,

*Table 5.3 Four major subgroups of procaryotic organisms*

| | **Blue-green algae** | **Myxobacteria** | **Spirochetes** | **Eubacteria** |
|---|---|---|---|---|
| *Mechanism of cellular movement* | *Gliding or immotile* | *Gliding* | *Axial filament* | *Flagella or immotile* |
| *Cell wall* | *Thick, rigid* | *Thin, flexible* | *Thin, flexible* | *Thick, rigid* |
| *Range of organismal structure* | | | | |
| *Unicellular* | + | + | + | + |
| *Filamentous* | + | − | − | + |
| *Coenocytic (mycelial)* | − | − | − | + |
| *Shapes of cells (unicellular forms)* | | | | |
| *Rods* | + | + | − | + |
| *Spheres* | + | − | − | + |
| *Spirals* | − | − | + | + |
| *Resting cells (some members only of each group)* | *Akinetes* | *Microcysts* | *None* | *Endospores, cysts* |
| *Nutritional categories* | | | | |
| *Photoautotrophic* | + | − | − | + |
| *Photoheterotrophic* | − | − | − | + |
| *Chemoautotrophic* | +[a] | − | − | + |
| *Chemoheterotrophic* | +[a] | + | + | + |

[a]Filamentous gliding organisms (e.g., *Beggiatoa*) which can be regarded for taxonomic purposes either as nonphotosynthetic representatives of the blue-green algae or as a special group of filamentous, gliding bacteria.

able to bend into arcs or circles during the course of gliding movement. They are also characteristically of low refractility, less readily visible in the living state than the thicker-walled cells of unicellular eubacteria or unicellular blue-green algae. The spirochetes constitute a small group of unicellular organisms, motile by means of an *axial filament*, which is composed of two sets of fibrils, anchored at the ends of the helical cell and wound around it, underneath the outer layer of the cell wall. The wall is delicate and thin, and the cells of spirochetes, like those of myxobacteria, are readily deformable. By far the largest and most varied group of bacteria are the eubacteria. When motile, these organisms swim by means of bacterial flagella, fine fibrils that extend through the cell wall into the surrounding medium. However, permanent immotility coupled with an absence of flagella is characteristic of many types of eubacteria. In such cases, differentiation from myxobacteria and spirochetes can still be made on the basis of the wall structure. Eubacterial cell walls are relatively thick and rigid, conferring on the organism a fixed form and a considerable refractility. The eubacteria include unicellular, filamentous, and coenocytic organisms, and the cell may have a variety of different shapes. All photosynthetic bacteria are unicellular members of the eubacteria; as already mentioned, they can be distinguished from blue-green algae of similar form by their pigment systems and their mechanism of photosynthesis.

## The blue-green algae

The blue-green algae are photoautotrophs, and in most cases obligate ones, completely unable to grow in the dark. The mechanism of photosynthesis is identical with that in eucaryotic algae. The pigment system is, however, distinctive. In addition to chlorophyll a and $\beta$-carotene, common to all oxygen-evolving phototrophs, it contains group-specific carotenoids (echinone, myxoxanthophyll) which occur

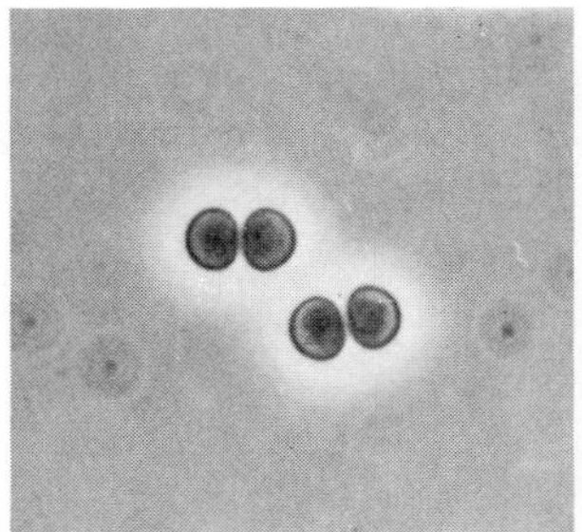

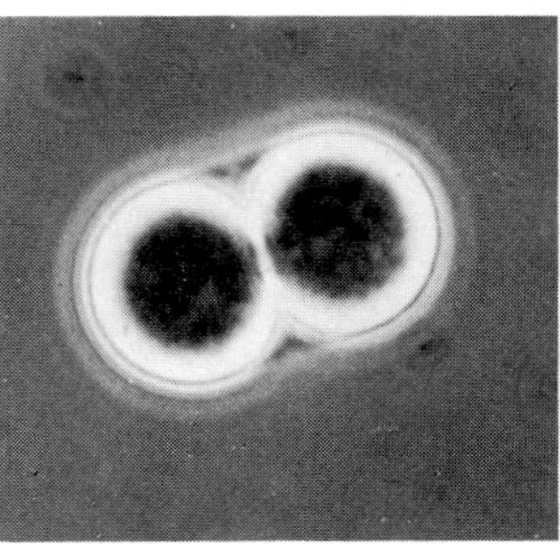

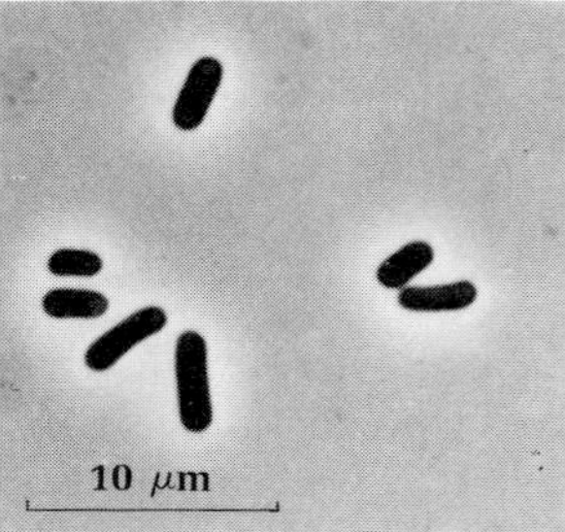



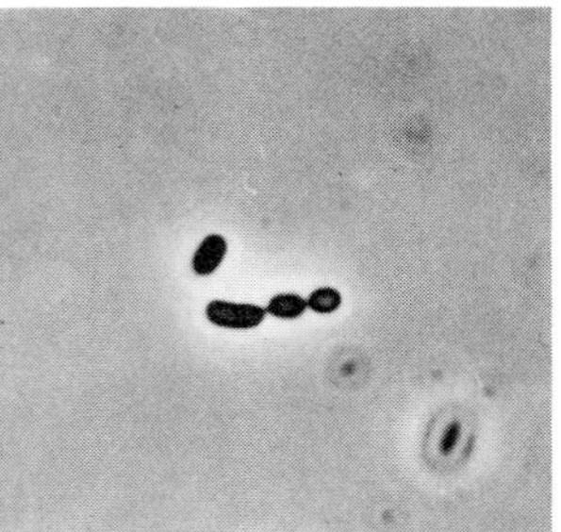

*Figure 5.3 Some unicellular blue-green algae. Courtesy of Rosmarie Rippka and Riyo Kunisawa.*

rarely, if ever, in other groups of oxygen-evolving phototrophs, as well as colored proteins conjugated with linear tetrapyrroles, the *phycobiliproteins.* Phycobiliproteins may be either blue (phycocyanins) or red (phycoerythrins), and depending on the particular combination of pigments belonging to this class that are present in the cell, the color of blue-green algae can vary widely. Species that contain only phycocyanin have the characteristic blue-green color; when phycoerythrin is present in addition to (or in place of) phycocyanin, the organisms may appear red, brown, brownish-purple, or almost black. Phycobiliprotein pigments are not confined to blue-green algae; they are always formed by one group of eucaryotic algae, the Rhodophyta (red algae), and occur sporadically in a few unicellular eucaryotic algae belonging to other groups.

Many blue-green algae are unicellular organisms, either rod-shaped or spherical, which multiply by binary fission. These representatives of the group resemble closely in form certain of the eubacteria; and they are usually immotile. Recognition of blue-green algae rests, accordingly, on their photosynthetic properties and pigment system. The color of these forms frequently cannot be detected when the small individual cells are examined in the microscope; it is evident only when a mass of cells is observed. Many of the spherical forms produce colonies of characteristic shape, in which the cells are held together by a common sheath or slime layer external to the wall. Some typical representatives are shown in Figures 5.3 and 5.4.

The blue-green algae also include several groups of multicellular organisms, consisting of filaments of cells, held together by a common outer wall. The filament increases in length by repeated divisions of its component cells in one plane, and reproduction occurs by the liberation of short elements, containing a few cells, which are termed *hormogonia.* Filamentous blue-green algae of the *Oscillatoria* type are straight filaments of even width, sometimes tapering near the tip; a similar form is *Spirulina,* in which the filament is helicoid. In the *Nostoc* group, the cells in the filament are more rounded and barrel-shaped, so that the filament has a beaded appearance. In organisms of the *Nostoc* type, the filament often bears at intervals paler cells, known as heterocysts; breakage of the filament with formation of hormogonia tends to occur at these points. Members of this filamentous group also

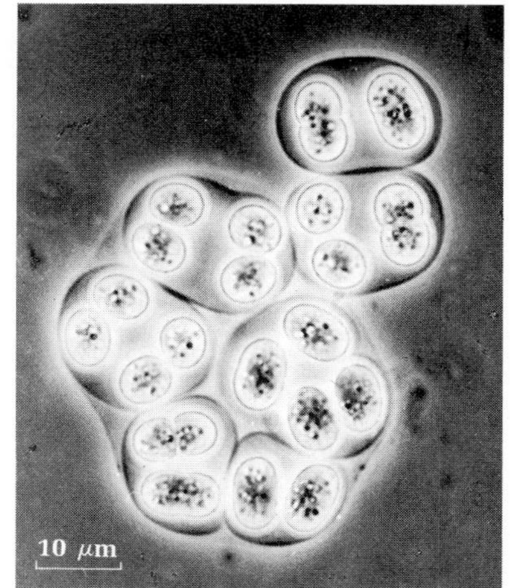


*Figure 5.4 A large unicellular blue-green alga, the cells of which are held together in groups by a multilayered sheath (phase contrast ×800). Courtesy of Rosmarie Rippka and Riyo Kunisawa.*

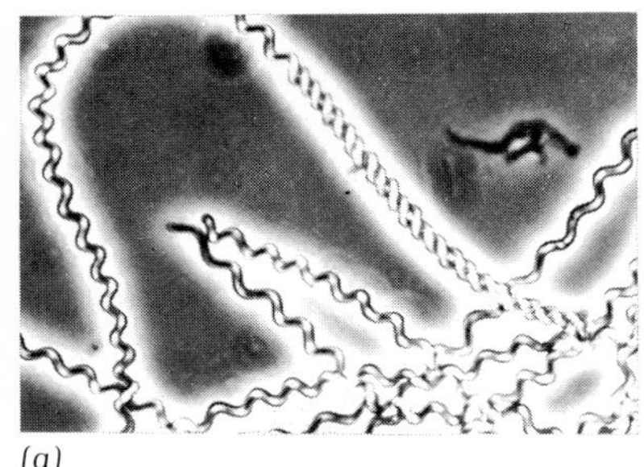

(a)

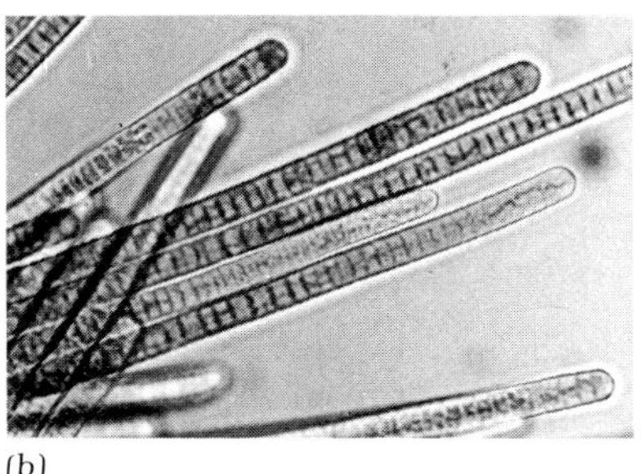

(b)

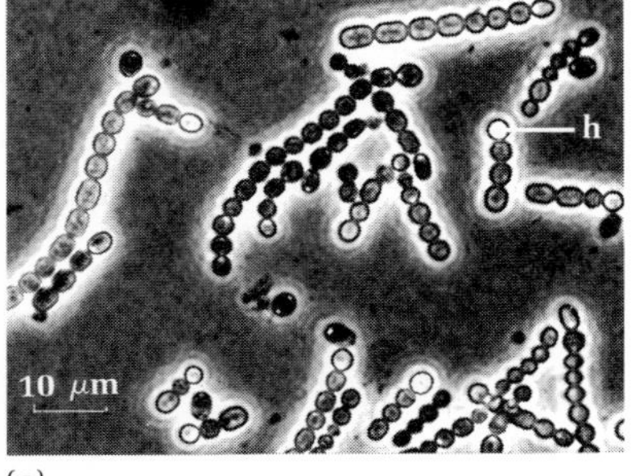


(c)

*Figure 5.5 Some filamentous blue-green algae. (a) Spirulina, in which the filament of cells is helical (phase contrast ×584). (b) Oscillatoria, in which the filament of cells is straight (ordinary illumination, ×584). (c) Anabaena, in which there is a marked constriction between adjacent cells on the filament. Specialized cells known as heterocysts (h) occur in the filament (phase contrast, ×584). Courtesy of Rosmarie Rippka.*

form, adjacent to heterocysts, larger cells known as *akinetes,* which serve as resting cells. All these filamentous blue-green algae are motile, the entire filament being capable of gliding movement. There is frequently formation of a copious slime layer. Some typical representatives are illustrated in Figure 5.5. In filamentous blue-green algae of the type of *Rivularia,* the filaments taper from the base to the tip and are often anchored to solid substrates through a basal heterocyst. In such cases, the filament itself is, of course, immotile; however, the hormogonia produced by detachment at the tip are capable of gliding movement.

The blue-green algae have an exceptionally wide natural distribution, occurring in soil, fresh water, and the oceans in all parts of the world. Some have the ability to grow at very high temperatures (above 70°C); these so-called *thermophilic* forms are characteristic inhabitants of hot springs. Certain blue-green algae, alone of all organisms that perform plant photosynthesis, are able to fulfill their nitrogen requirements by fixing atmospheric nitrogen, a biochemical peculiarity also found in a wide variety of bacteria. Coupled with the ability to perform plant photosynthesis, the capacity for nitrogen fixation confers on these blue-green algae *the simplest nutritional requirements of any known organisms.* They can flourish in an environment that provides, in addition to $N_2$ and $CO_2$, which are always present in the atmosphere, only water, light, and minerals. It is this fact which explains the occurrence of blue-green algae in very bleak natural environments which cannot support the development of other biological groups.

It has long been known that among the filamentous blue-green algae, the capacity for nitrogen fixation is confined to types that form heterocysts, such as *Nostoc* and *Rivularia.* Recent work has shown that the heterocyst is the actual site of nitrogen fixation. Furthermore, the heterocyst is unable to assimilate $CO_2$, as a result of a modification of its photosynthetic machinery. In such blue-green algae, accordingly, the two kinds of cells in the filament – heterocysts and vegetative cells – differ basically in metabolic function: the vegetative cells are responsible for $CO_2$ assimilation; the heterocysts, for nitrogen fixation.

**Nonphotosynthetic organisms related to filamentous blue-green algae**

The classical representative of this group is *Beggiatoa*, a form that can be found in nature whenever hydrogen sulfide is present in contact with the atmosphere. It is very common near sulfur springs and on the surface of black mud. *Beggiatoa* is a chemoautotroph that oxidizes hydrogen sulfide. Elemental sulfur is stored within the cells as highly refractile droplets; these inclusions confer on *Beggiatoa* filaments a characteristic and unmistakable appearance (Figure 5.6) The filaments reproduce by the liberation of hormogonia.

*Figure 5.6 Beggiatoa. Filaments containing many sulfur droplets (×850). Courtesy of E. J. Ordal.*

The taxonomic placement of *Beggiatoa* and related filamentous forms is essentially arbitrary. Structurally, these organisms are counterparts of filamentous blue-green algae and on structural grounds could be assigned a position close to their specific photosynthetic counterparts. If, however, one wishes to define the blue-green algae in functional terms as an exclusively photosynthetic group, these nonphotosynthetic, filamentous gliding organisms become "bacteria" by definition. In fact, they have traditionally been regarded as a special group of bacteria. Their properties will be discussed at greater length in Chapter 16.

## The myxobacteria

The myxobacteria are unicellular organisms with small, rod-shaped cells that reproduce by binary transverse fission. They are endowed with the capacity for gliding movement and consequently form flat, spreading colonies on solid media. Movement is particularly active at the edge of a colony, which has a lacy, irregular structure, as a result

Figure 5.7 *Expanding edge of myxobacterial colony on agar, photographed at intervals as follows: (a) initial shape; (b) after 7 minutes; (c) after 15 minutes; (d) after 25 minutes. From R. Y. Stanier, "Studies of Nonfruiting Myxobacteria, 1. Cytophaga johnsonae, n. sp., a Chitin-decomposing Myxobactium." J. Bacteriol.* **53,** *310 (1947).*

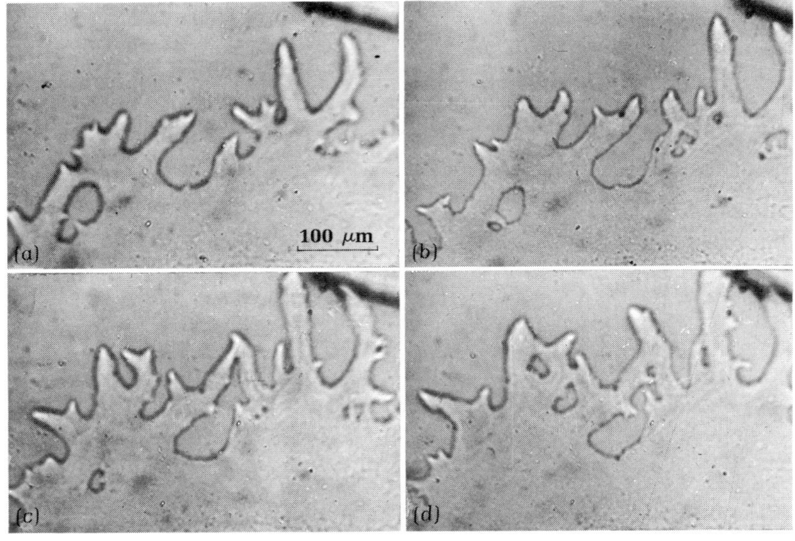

of the advance of flame-shaped groups of cells (Figure 5.7). The individual cells are weakly refractile and flexible, as a result of their thin and delicate cell wall, which has a fine structure similar to that of the cytoplasmic membrane. The myxobacteria are all classified as chemoheterotrophs.

Two principal subgroups may be recognized; the fruiting myxobacteria and the cytophagas. The fruiting myxobacteria have a remarkable cycle of development. During vegetative growth the individual cells tend to be held in loose association by the formation of a thin, underlying slime layer produced as the colony expands. Within certain regions of the colony, large numbers of cells proceed to *aggregate*, and the initial aggregates then differentiate into many-celled *fruiting bodies*, which are often of considerable size and complexity. The simplest types of fruiting bodies, characteristic of the genus *Myxococcus*, are glistening, brightly colored droplets, sometimes as much as 2 mm in diameter, which project above the surface of the substrate (Figure 5.8). They consist of masses of small spherical, thick-walled cells known as

Figure 5.8 *A fruiting myxobacterium, Myxococcus. Mature fruiting bodies on a dung particle (×25). From A. T. Henrici, and E. J. Ordal, The Biology of Bacteria, 3rd ed. Boston: Heath, 1948. Courtesy of Mrs. N. A . Woods, E. J. Ordal, and the publisher.*

Figure 5.9 *A fruiting myxobacterium, Myxococcus. Various stages in the conversion of the long, thin, rod-shaped vegetative cells to the spherical microcysts of which the fruiting body is composed (×1,150). From A. T. Henrici, and E. J. Ordal, The Biology of Bacteria, 3rd ed. Boston: Heath, 1948. Courtesy of Mrs. N. A. Woods, E. J. Ordal, and the publisher.*

microcysts, each formed by the shortening and rounding up of an individual vegetative cell (Figure 5.9).

The life cycle of the fruiting myxobacteria is analogous to that of the cellular slime molds (Acrasieae), described in Chapter 4. In the cellular slime molds, however, the vegetative cell is a small, eucaryotic ameba. The differences between the two groups in cellular organization indicate that there is no evolutionary relationship between them, so the striking parallelism with respect to their developmental cycles is best regarded as an example of *evolutionary convergence:* acquisition of superficially similar biological properties through the action of natural selection on two initially entirely distinct biological groups.

The cytophagas possess vegetative cells similar in structure to those of fruiting myxobacteria but do not undergo aggregation and fruiting body formation.

## *The eubacteria*

The eubacteria constitute by far the largest and most diversified group among the procaryotic protists and cannot be adequately characterized in a few pages. Many of their constituent groups will be discussed in detail in later chapters (Chapters 17 to 21). Here we shall discuss the range of structure and development that occurs in these organisms.

### The structural properties of unicellular eubacteria

Most eubacteria are unicellular organisms, which multiply by binary transverse fission. In terms of cell shape, three principal types of unicellular eubacteria can be recognized: *cocci* (singular *coccus*), with spherical or ovoid cells; *rods,* with cylindrical cells; and *spirilla* (singular *spirillum*), with helical cells (Figure 5.10). In many unicellular eubacteria, separation of the daughter cells takes place immediately after division has been completed. In some, however, the daughter cells remain more or less loosely attached to one another after division, so that many-celled aggregates occur. Since cell division always takes place at right angles to the long axis of the cell, the only possible type

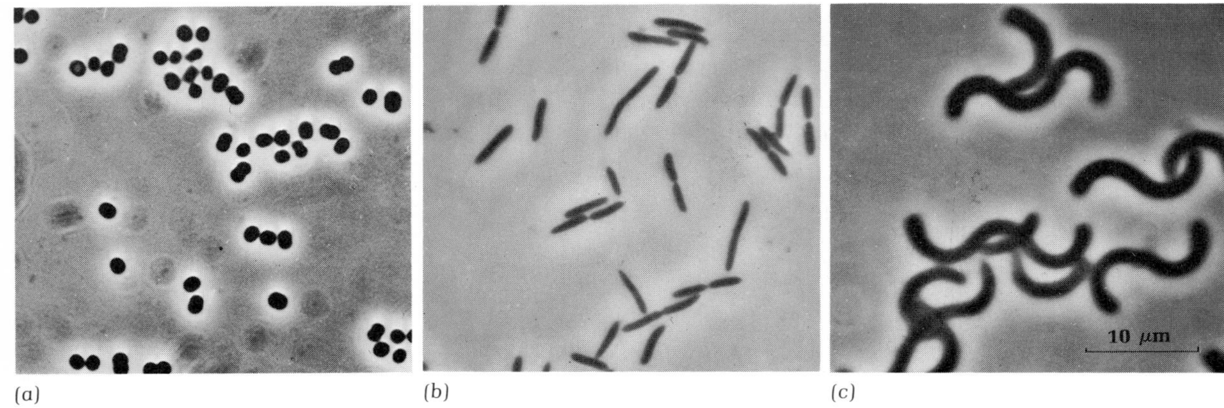


Figure 5.10 *The three cell shapes that occur among unicellular true bacteria. (a) coccus; (b) rod; (c) spiral (phase contrast, ×1,500).*

of aggregate in rod-shaped and helical eubacteria is a *chain of cells*. The formation of such chains is common in certain groups of rod-shaped bacteria, notably members of the genera *Bacillus* (Figure 5.11) and *Lactobacillus*. It should be specifically noted that these chains of cells are *not* structurally equivalent to the multicellular filaments of such organisms as *Oscillatoria* or *Beggiatoa*, in which the individual cells are permanently associated through a continuous, common outer wall. In the chain-forming bacilli or lactobacilli, each individual cell is a separate organism which can be mechanically detached from its neighbors (e.g., by mechanical agitation) and remain viable.

The possibilities for aggregate formation by adherence of daughter cells after division are more varied among the spherical eubacteria: several geometrical arrangements can exist, each of which is determined by the *planes of successive divisions*. In some cocci, successive divisions always take place in the same plane, giving rise to a chain of cells. This is the cell arrangement characteristic of the spherical bac-

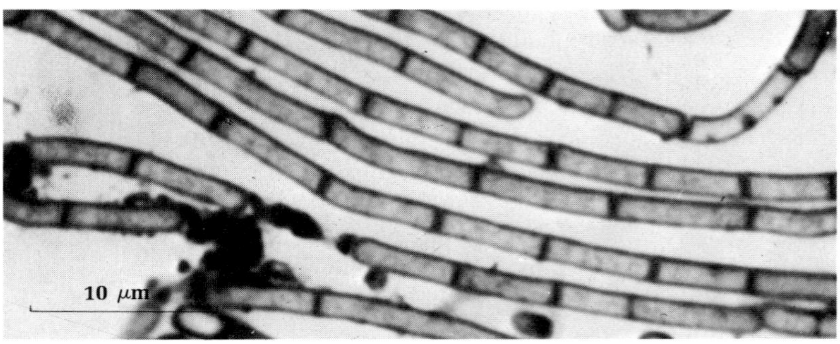


Figure 5.11 *A chain-forming Bacillus, stained with tannic acid and crystal violet to reveal cross walls (×2,350). Courtesy of C. F. Robinow.*

Figure 5.12 *Different cell arrangements of true bacteria with spherical cells. (a) A chain-forming coccus, Streptococcus (×2,200). Courtesy of Daisy Kuhn and Patricia Edlemann (b) A coccus that forms flat, rectangular plates of cells, Lampropedia (×1,500). Courtesy of Daisy Kuhn. (c) A coccus that forms cubical packets of cells, Sarcina (×540). From E. Canale-Perola and R. S. Wolfe, "Studies on Sarcina ventriculi." J. Bacteriol.* **79**, *887 (1960).*

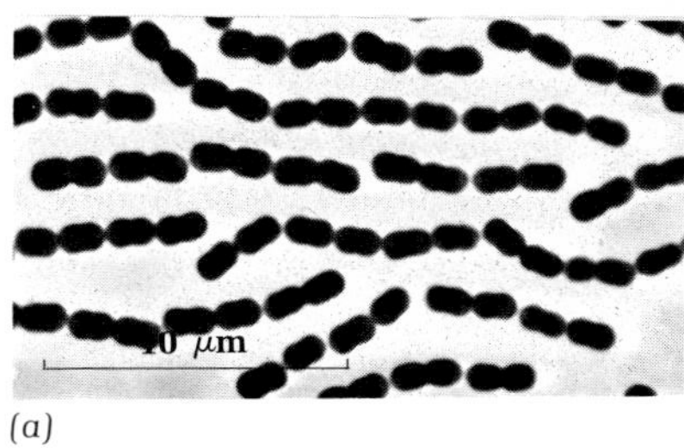

(a)

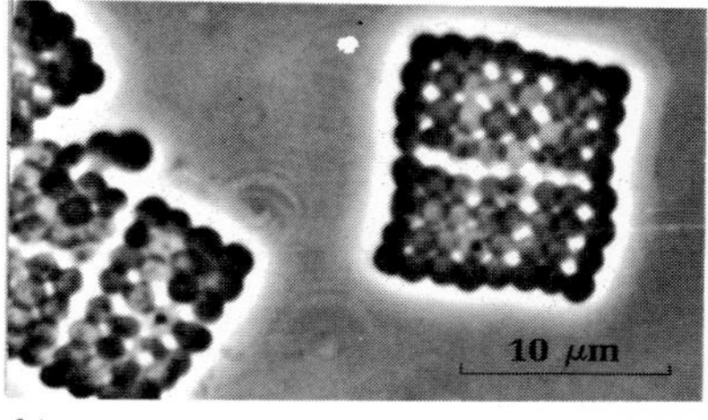

(b)

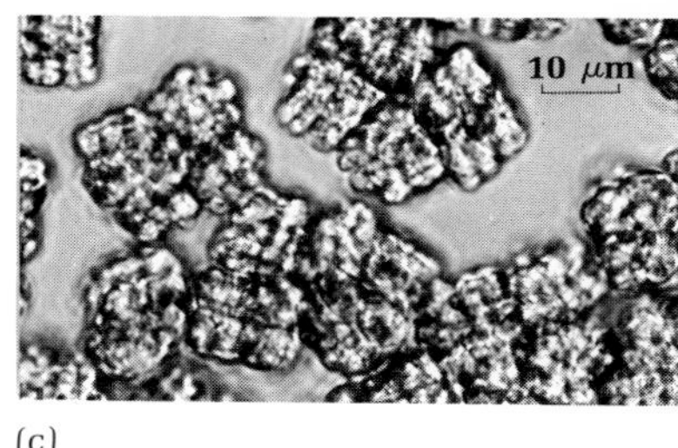

(c)

teria known as streptococci (Figure 5.12). In other cocci, successive divisions take place regularly in two planes at right angles to one another, to produce four-celled tetrads. The most striking example of this arrangement occurs in the large coccus *Lampropedia*, where continued adherence of the cells after division gives rise to flat tablets, containing as many as 64 regularly arranged individuals (Figure 5.12). Finally, certain cocci divide regularly in three successive planes at right angles to one another, to form regular cubical packets consisting of eight cells; this is characteristic of the spherical true bacteria known as sarcinae illustrated in Figure 5.12.

The simple geometry of the eubacterial cell is modified in certain groups by the presence of cellular extensions, which are enclosed by the wall; they are known by the general name *prosthecae*. The presence of a kind of prostheca commonly termed a "stalk" is characteristic of one group of rod-shaped bacteria, the caulobacters (Figure 5.13). The caulobacter stalk is a short, filiform appendage extending from a polar site on the cell and usually terminated by an extracellular adhesive disc, or *holdfast*, which anchors the cell to solid substrates. Through

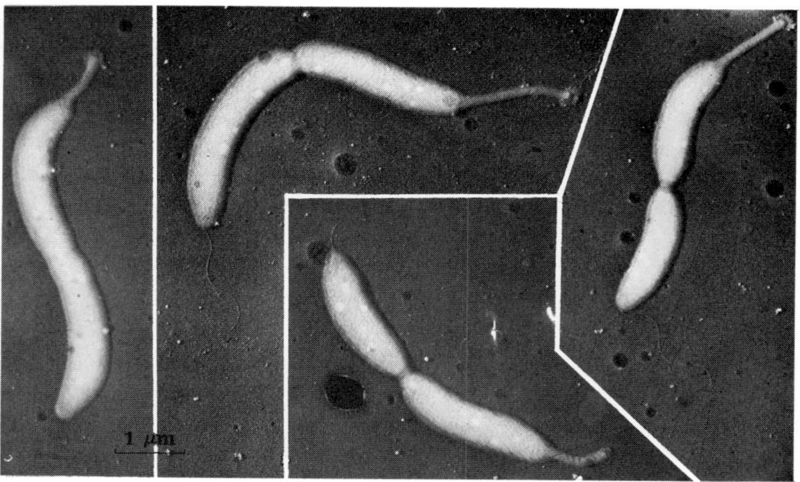


Figure 5.13 *Metal-shadowed electron micrograph of stalked cells of Caulobacter, several of which are dividing (×8,850). Note the flagellum on the distal daughter cell in the divisional stages and the continuity between the body of the cell and the stalk. From A. E. Houwink, "Caulobacter, Its Morphogenesis, Taxonomy, and Parasitism." Antonie van Leeuwenhoek* **21**, *1, 54 (1955).*

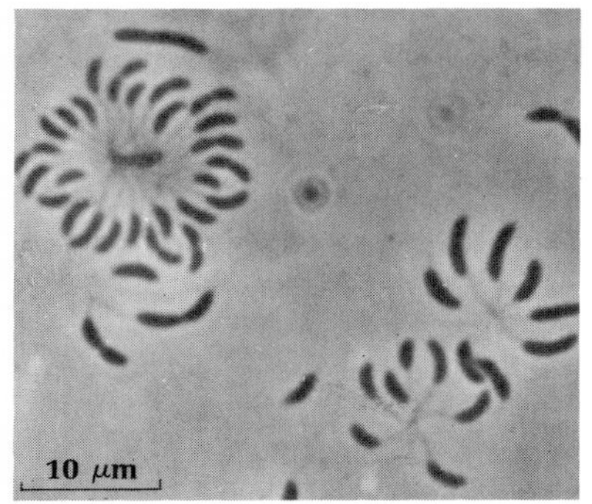


*Figure 5.14 Caulobacter rosettes formed by the adherence of cells to one another at the base of the stalks (phase contrast, ×1,308). Courtesy of Jeanne Stove Poindexter.*

adhesion of their holdfasts, characteristic rosette-shaped groups of cells can be formed (Figure 5.14). The geometry of the caulobacter cell results in the formation of *structurally dissimilar cells* as a result of binary transverse fission: the basal daughter cell receives the stalk, while the apical cell bears a single flagellum. Following its liberation, the apical cell is briefly motile, but it rapidly loses its flagellum and develops a stalk at the former site of flagellar insertion.

### The flagella of unicellular eubacteria

Many groups of unicellular eubacteria are permanently immotile: this is true of nearly all cocci and of a large number of rod-shaped forms. When motility does occur in eubacteria, it is always mediated by *bacterial flagella,* very fine threads of protein with a characteristic helical structure which extend out from the cytoplasm through the cell wall. The diameter of an individual flagellum is approximately 100 A, although these organelles may be as long as 10 μm. Hence, single bacterial flagella cannot be perceived with the light microscope and cannot be seen on living cells unless the cells bear a large number of flagella, which become intertwined during movement to produce a larger bundle (Figure 5.15).

The classical method for demonstrating flagella involves special staining procedures, which make them individually visible with the

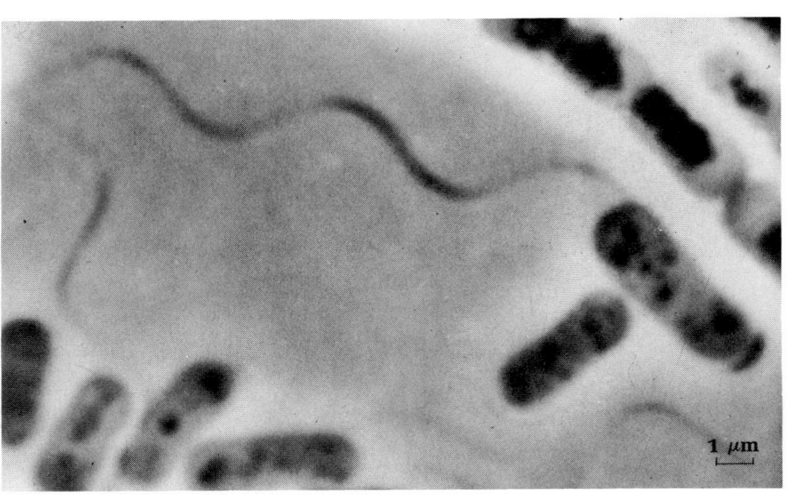


*Figure 5.15 Polar tufts of flagella on living cells of Sphaerotilus natans, demonstrated by phase contrast illumination (×5,250). From J. L. Stokes, "Studies on the Filamentous Sheathed Iron Bacterium Sphaerotilus natans." J. Bacteriol.* **67**, *285 (1954).*

Figure 5.16 *Stained Flagella of a Pseudomonas that has a single polar flagellum (×2,520). Courtesy of George Hageage and C. F. Robinow.*

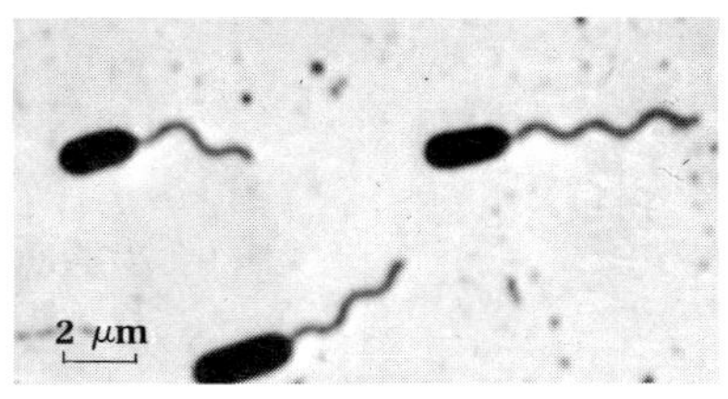


light microscope. Although these procedures vary greatly in detail, they all involve the same principle. The fixed cells are first treated with a *mordant:* an unstable colloidal solution that precipitates a thick layer of stainable material over the surfaces of the cells and their flagella. By a simultaneous or subsequent application of a suitable dye, this precipitated material can be visualized with the light microscope as a stained thread, which encloses and reproduces the form of the flagellum itself (Figure 5.16). In the electron microscope, the flagella of bacteria can be visualized directly, particularly in metal-shadowed preparations (Figure 5.17), and this is the best method of determining their number and arrangement on the cell.

Figure 5.17 *Electron micrograph of a metal-shadowed Pseudomonas cell bearing two polar flagella (×10,944). Courtesy of A. E. Houwink and Technical Physics Department, Technical Highschool, Delft, Netherlands.*

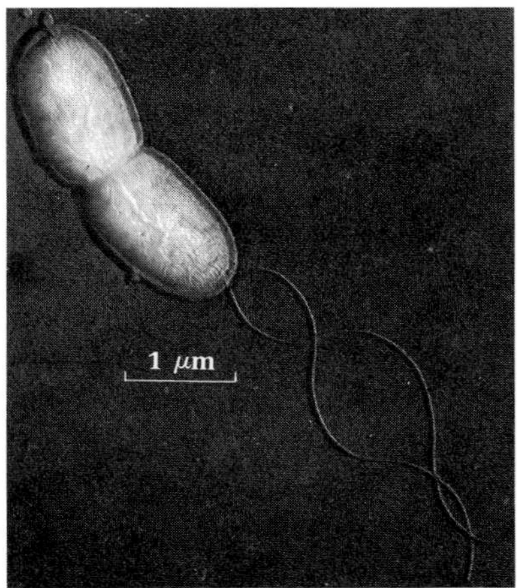


The manner of insertion of flagella on the cell is a constant group character among the motile unicellular eubacteria and has an important place in the classification of these organisms. There are two principal modes of insertion. In some groups the flagella are inserted only at the pole or poles of the cell, an arrangement termed *polar flagellation.* In certain cases, notably the spirilla, a large number of individual flagella arise from one or both poles; this arrangement is termed polar *multitrichous* flagellation. Other polarly flagellated bacteria, notably the rod-shaped pseudomonads and the caulobacters, have only a single polar flagellum; this arrangement is termed polar *monotrichous flagellation.* In other groups, for example the rod-shaped bacilli and the coliform bacteria, the flagella are inserted at many points along the sides of the cell, an arrangement termed *peritrichous flagellation.* Finally, it should be noted that in certain groups the mode of flagellar insertion is difficult to determine and describe with certainty. These are sparsely flagellated organisms, which bear from one to four flagella per cell, generally inserted a slight distance away from the pole. This arrangement, which occurs in bacteria of the genera *Agrobacterium* and *Rhizobium,* is sometimes termed *degenerately peritrichous flagellation.*

**The resting cells of unicellular eubacteria**

Certain unicellular eubacteria can form a special kind of resting cell, known as an *endospore* (Figure 5.18); this property is found in a number of rod-shaped bacteria and in one coccus. As its name implies, the endospore is formed within a vegetative cell, by a special process of division that will be described in detail in Chapter 20. As a rule, only one spore is formed in each cell and is liberated by the lysis of the sur-

Figure 5.18 *Bacterial endospores. Stained wet mount of an anaerobic spore-former (Clostridium) in the course of sporulation. The rod-shaped vegetative cells are swollen at one end by the presence of oval, highly refractile spores. Courtesy of C. F. Robinow.*

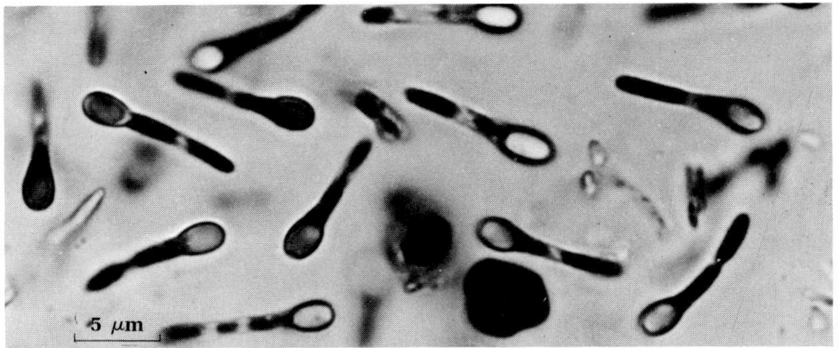


rounding mother cell. Endospores are extremely refractile because of their very low water content and can be stained only with difficulty. They are extremely resistant to heat and to other treatments lethal for vegetative cells and can remain dormant for long periods of time (years, in some species). Each spore gives rise on germination to a vegetative cell.

A few unicellular eubacteria, notably the members of the genus *Azotobacter*, can form a different kind of resting cell, known as a *cyst*, which is similar in structure and mode of formation to the microcysts characteristic of certain myxobacteria. Each cyst is formed by the shortening of a rod-shaped vegetative cell, accompanied by the secretion of an unusually thick enclosing wall (Figure 5.19). Cysts do not, however, possess the striking physiological properties of endospores; they are, for example, not significantly more heat resistant than the vegetative cells from which they are derived.

Only a minority of unicellular eubacteria can form either endospores or cysts. In the great majority of these organisms, no structurally differentiated resting cells are known, and survival is essentially dependent on the persistence, in a nonproliferating state, of unmodified vegetative cells.

Figure 5.19 *Stained preparation of vegetative cells and cysts of Azotobacter (×1,535). The vegetative cells are oval, deeply stained rods; the cysts are spherical, surrounded by thick, lightly stained walls. From S. Winogradsky, Microbiologie du sol, p. 780. Paris: Masson, 1949. Reprinted with permission of M. Manigault and the publishers.*

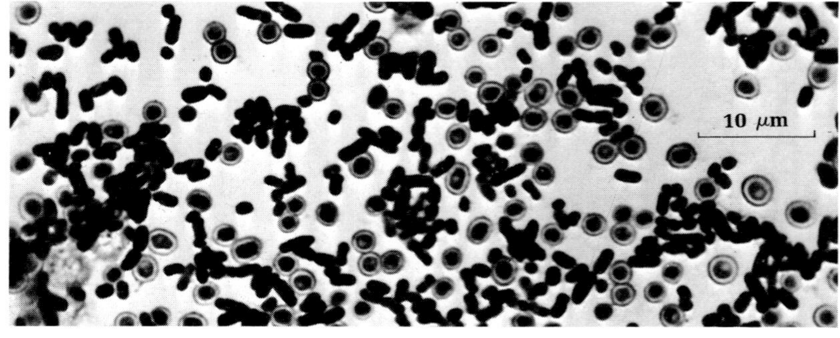

**The filamentous eubacteria**

A truly filamentous, multicellular organization, so common among the blue-green algae, occurs very rarely among eubacteria. The only well-studied example is the large organism *Caryophanon* (Figure 5.20). The plump filament, which contains from 10 to 30 cells, and can attain dimensions of 4 by 40 $\mu$m, is actively motile by means of numerous lateral flagella. Reproduction takes place by *binary fission of the entire filament*, and the daughter filaments subsequently elongate by internal cell divisions, which take place at right angles to the long axis, much as in filamentous blue-green algae.

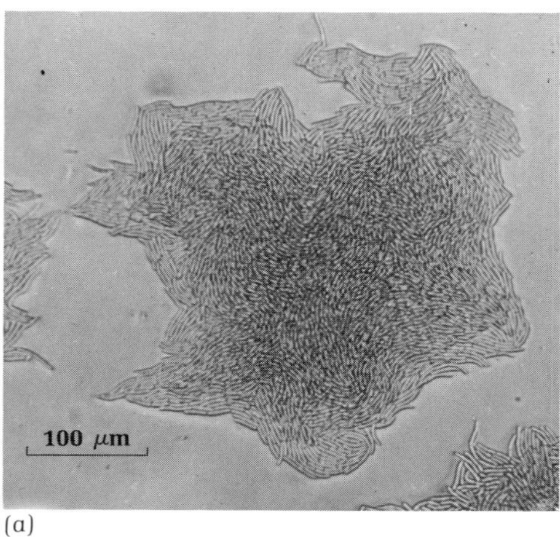

(a)

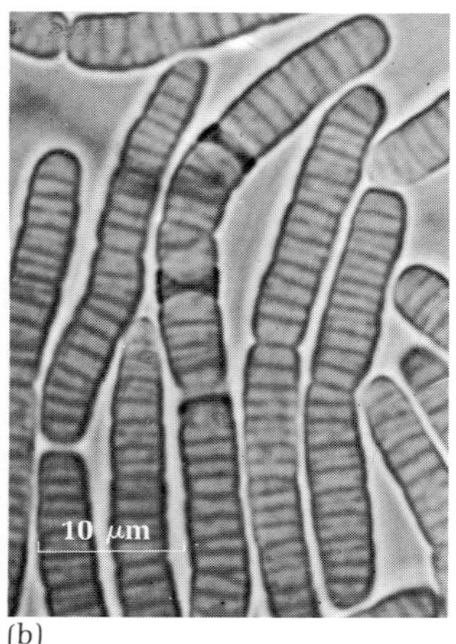

(b)

*Figure 5.20 Caryophanon. (a) A small colony on agar (×138). (b) Living unstained filaments (×1,380). From E. G. Pringsheim and C. F. Robinow, "Observations on Two Very Large Bacteria, Caryophanon latum Peshkoff and Lineola longa (nomen provisorium." J. Gen. Microbiol.* **1***, 278 (1947).*

**The mycelial eubacteria: the actinomycetes**

One large group of eubacteria, the actinomycetes, is characterized by a mycelial vegetative structure, analogous to that which occurs in the fungi. The simplest representatives of this group are the mycobacteria and proactinomycetes, in which there is formation of a transient mycelium (Figure 5.21), which breaks up into short, branched, rod-shaped or coccoid elements, often indistinguishable from the cells of unicellular eubacteria. These elements can grow out to form a new mycelium.

Other actinomycetes have a *permanently mycelial vegetative structure* and reproduce by the formation of special cells known as conidiospores or sporangiospores. This mode of development may be illus-

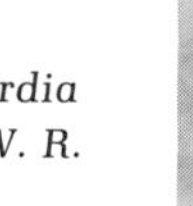

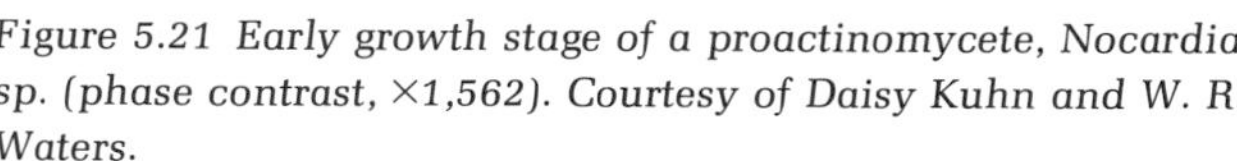

*Figure 5.21 Early growth stage of a proactinomycete, Nocardia sp. (phase contrast, ×1,562). Courtesy of Daisy Kuhn and W. R. Waters.*

trated by considering the properties of a typical member of the genus *Streptomyces.* Outgrowth of a conidiospore leads to the formation of a dense, much-branched *substrate mycelium;* an early stage of growth is shown in Figure 5.22. The structure is largely coenocytic, although

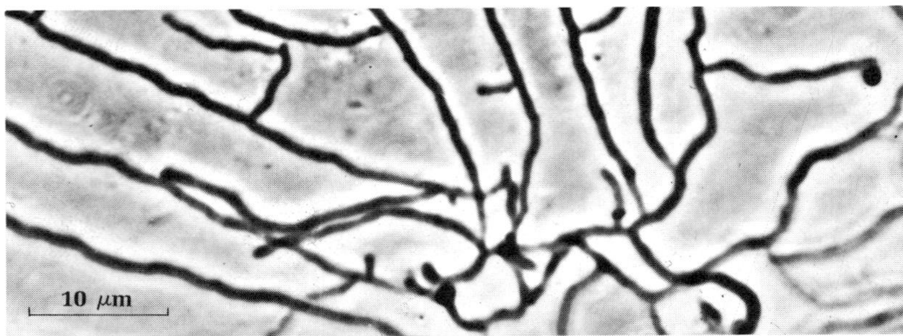


*Figure 5.22 The mycelium of an actinomycete, Streptomyces (phase contrast, ×1,500). From D. A. Hopwood, "Phase contrast observations on Streptomyces coelicolor." J. Gen. Microbiol.* **22,** *295 (1960).*

cross walls occur rarely. After formation of the substrate mycelium, a somewhat looser *aerial mycelium,* composed of thicker and more refractile hyphae, develops upon its surface. Chains of conidiospores are then produced by multiple divisions in the tips of the aerial hyphae

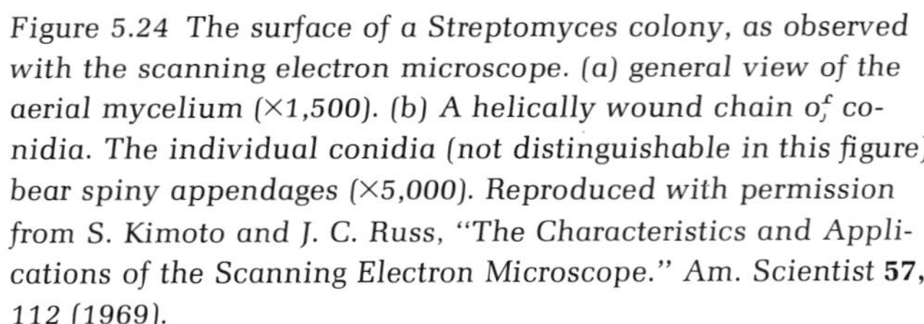

*Figure 5.24 The surface of a Streptomyces colony, as observed with the scanning electron microscope. (a) general view of the aerial mycelium (×1,500). (b) A helically wound chain of conidia. The individual conidia (not distinguishable in this figure) bear spiny appendages (×5,000). Reproduced with permission from S. Kimoto and J. C. Russ, "The Characteristics and Applications of the Scanning Electron Microscope." Am. Scientist* **57,** *112 (1969).*

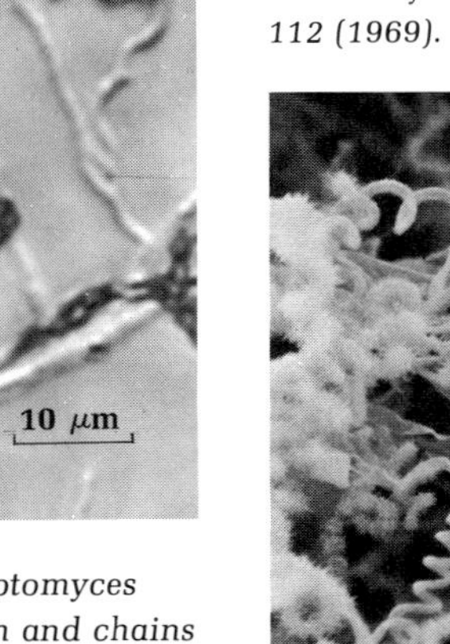


*Figure 5.23 Part of a Streptomyces colony, showing mycelium and chains of spherical conidia (living material, ×1,024).*

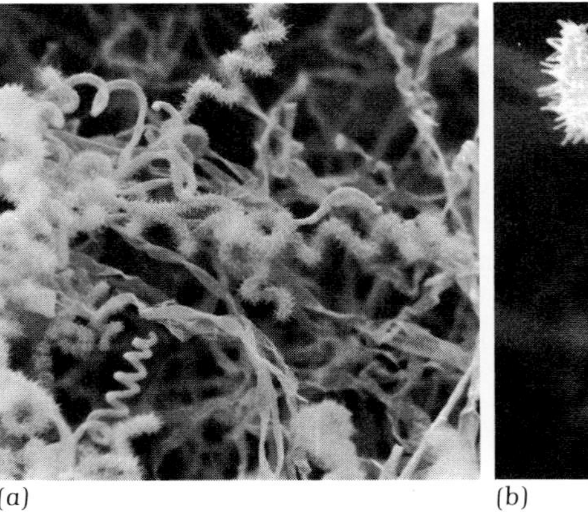

(a)

(b)

(Figure 5.23). The mature *Streptomyces* colony (Figure 5.24) develops a powdery surface, much like that of a mold colony, as a result of conidiospore formation. When mature, the individual conidiospores readily detach from the mycelium; each individual can give rise on germination to a new mycelial colony.

Most actinomycetes are permanently immotile; however, the sporangiospores in some genera have flagella (see Chapter 21).

**The Gram stain: its cytochemical and taxonomic significance**

A differential staining procedure of great value in the identification of eubacteria is the *Gram stain,* first developed in 1884 by a Danish physician, Christian Gram. Many modifications exist, but they all embody the same essential steps. A heat-fixed smear of bacteria is stained successively with a solution of crystal violet (or a related basic dye) and with a dilute solution of iodine. The preparation is then treated with an organic solvent, such as alcohol or acetone. The cells of some bacteria, known as *Gram-negative bacteria,* are rapidly and completely decolorized by the organic solvent; the cells of others, known as *Gram-positive bacteria,* resist decolorization. The Gram reaction can be performed without the use of a microscope, in which case one simply examines the slide after treatment with the organic solvent to determine whether the smear of cells is still colored or not. The customary procedure is, however, to subject the preparation to microscopic examination. In this event, treatment by the organic solvent is followed by *counterstaining* with a red dye such as safranin. Gram-positive cells retain the deep purple color conferred on them by the initial staining with crystal violet and iodine, whereas Gram-negative cells, which have been decolorized, exhibit the red color of the counterstain. As a result of this difference in color, Gram-positive and Gram-negative cells can be readily distinguished from one another under the microscope.

In determining the Gram reaction of a bacterium, it is essential to perform the stain on cells derived from a young culture, since certain bacteria are Gram positive only during the course of active growth and lose the capacity to retain the crystal violet–iodine complex after active growth ceases. This is particularly true of many sporeformers, which are strongly Gram positive when examined in young cultures but which later become Gram variable or Gram negative. The attribute of Gram positiveness is thus to some extent conditioned by the physiological state of the cells.

The Gram stain was originally developed as an empirical procedure for demonstrating the presence of Gram-positive bacteria in animal tissues, which are Gram negative. Its present importance does not, however, rest on its use as a simple aid in the observation of bacteria.

In the course of time, it became apparent that the outcome of the Gram reaction is *correlated with other structural properties of the eubacterial cell.* Consequently, the division of the eubacteria into two groups, based on the outcome of this staining procedure, constitutes a division of profound taxonomic importance.*

The obvious value of the Gram reaction as an index of internal relationships among eubacteria suggested that it must be indicative of some *basic chemical difference* between the cells of Gram-positive and Gram-negative forms. This conclusion was further strengthened by the discovery that certain *physiological* properties of eubacteria are related to their Gram reaction. Thus, most Gram-positive bacteria are much more sensitive than Gram-negative ones to the toxic effects of basic aniline dyes (e.g., crystal violet) and of certain antibiotics (e.g., penicillin). However, for 75 years after its discovery, the mechanism of the Gram reaction defied explanation in chemical terms.

The first real clue to the meaning of the Gram reaction emerged about 15 years ago, when comparative chemical analyses of the composition of eubacterial cell walls began to accumulate. These show that there are *consistent and major compositional differences between the walls of Gram-positive and Gram-negative eubacteria.* Furthermore, *these chemical differences are systematically correlated with a difference between the two groups in the fine structure of the wall,* as revealed by the examination of its profile in the electron microscope. In the last analysis, accordingly, the Gram reaction has proved to be a rough-and-ready method, not invariably reliable, for distinguishing between two major subgroups of eubacteria in which the walls have a different structure and chemical composition.

The discussion of wall structure has taken us far from the simple laboratory procedure of the Gram stain, to which we must now return. The role played by the cell wall in the Gram reaction has been partially elucidated. The walls of some Gram-positive bacteria can be completely dissolved by the enzyme lysozyme. If such bacteria are Gram stained and then treated with lysozyme under appropriate conditions, the naked cells or *protoplasts* are liberated; these protoplasts still retain the stain, showing that the cell itself, and not the wall, is the site of the Gram reaction. However, the protoplasts can be *immediately decolorized with alcohol* and have thus lost their Gram positiveness together with their walls. It is accordingly evident that the wall of a Gram-positive bacterium interposes a barrier which prevents the leaching out of the dye complex.

*It should be specifically noted that the Gram reaction possesses this taxonomic value *only for eubacteria;* it is not a useful taxonomic character for other microbial groups.

Finally, the broad correlations between eubacterial wall structure, as revealed by the Gram stain, and other structural characters in the group must be described. Eubacterial groups with polar flagellation are almost invariably Gram negative. This generalization holds for the following polarly flagellated groups: the rod-shaped pseudomonads; the vibrios, with curved, rod-shaped cells; the spirilla; the caulobacters; and the entire group of photosynthetic bacteria. Peritrichously flagellated eubacteria, on the other hand, include both Gram-positive and Gram-negative groups. The major groups of Gram-positive eubacteria include the actinomycetes in their entirety; several large groups of immotile, rod-shaped, nonsporeforming bacteria (e.g., the lactobacilli and the propionic acid bacteria); all rod-shaped bacteria that form endospores; and the great majority of cocci.

**The Mycoplasma group and bacterial L forms**

A bacterial group whose nature was for a long time obscure consists of the members of the genus *Mycoplasma*, sometimes also referred to as the "pleuropneumonia-like organisms," or PPLO group. This common name derives from the fact that the first member to be described was the agent of a disease of cattle known as bovine pleuropneumonia. Similar organisms were subsequently shown to cause a number of other diseases of animals and humans; they also exist (apparently as harmless parasites) on mucous membranes of man and a number of other animals.

The cells of these bacteria are nonmotile and very delicate and plastic, because *they lack a cell wall.* It is difficult to describe their form, which varies with the method of cultivation and can also be affected by the manner of handling during microscopic examination. On a solid medium, the undisturbed cells are flattened. In liquid cultures, irregular forms with branches and threadlike extensions often occur (Figure 5.25). The small colonies on solid media have a characteristic

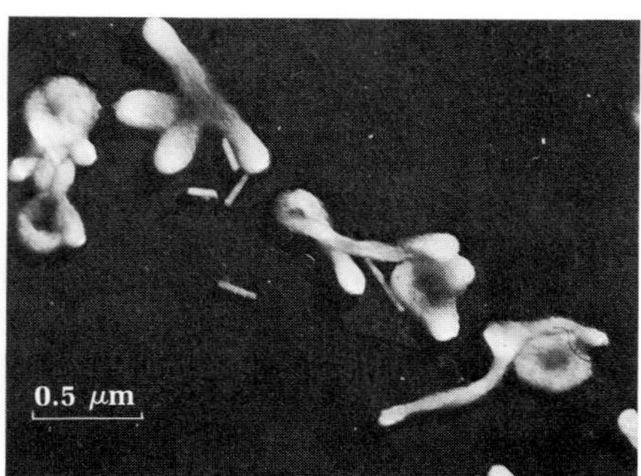


*Figure 5.25 Electron micrograph of cells of a member of the Mycoplasma group, the agent of bronchopneumonia in the rat (×1,960). From E. Klieneberger-Nobel and F. W. Cuckow, "A study of organisms of the pleuropneumonia group by electron microscopy." J. Gen. Microbiol.* **12**, *99 (1955).*

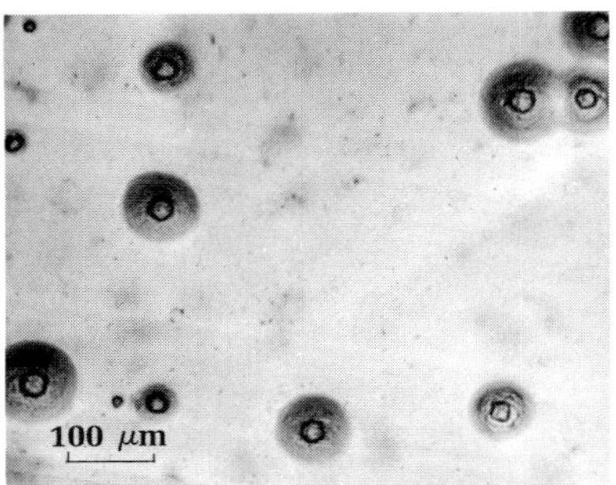


*Figure 5.26 Characteristic colony structure of organisms of the pleuropneumonia group (×68). Courtesy of M. Shifrine.*

"fried-egg" appearance (Figure 5.26). The isolation and cultivation of these bacteria at first proved difficult, and many of them could be grown only on very rich complex media, supplemented with a high concentration of blood serum. Since the recognition that they lack a cell wall, and are hence susceptible to osmotic lysis, the cultivation of these bacteria has become easier: the complexity of the medium can be considerably reduced, provided that it contains sufficient amounts of solutes to be isotonic with the cells.

A remarkable property of the mycoplasma group, recognized early in their study, is their ability to pass through filters fine enough to retain ordinary bacterial cells. Since the volume of the cells of the mycoplasma group is not significantly less than the volume of ordinary unicellular eubacteria, this filter-passing ability was at first interpreted as evidence for the occurrence of very small, filterable reproductive elements. Subsequent work has shown this to be incorrect; despite their size, the cells of these organisms are so plastic that they can be sucked mechanically through the pores of a filter that retains rigid bacterial cells of the same volume.

About 1935 it was discovered that certain unicellular eubacteria can give rise spontaneously to so-called "L forms," which have a structure and developmental character very similar to that of the organisms of the mycoplasma group. At first, some workers believed that L forms were independent biological entities, parasitic on the species of eubacteria from which they could be isolated. However, this interpretation had to be abandoned when it was discovered that *cultures of L forms can sometimes revert to the eubacteria from which they have been derived.* Consequently, they must represent a growth form of eubacteria, despite the very different structural properties of these two groups.

The existence of bacterial L forms has an important bearing on the problem of the biological nature of the mycoplasma group. These bacteria could well be *naturally occurring L forms* which have arisen by some genetic or physiological accident from eubacteria in an environment that allows them to persist and grow despite their failure to synthesize a cell wall. The crucial test of this hypothesis would be the demonstration that a member of the mycoplasma group can revert to eubacterial form, but this has not yet been shown.

If this interpretation of the nature of the mycoplasma group is a correct one, the organisms have no doubt arisen, at different places and times, from a variety of different eubacteria. The only common property which marks all members of this group is their inability to synthesize a cell wall, and their peculiar structure would seem to be a consequence of this defect.

## The spirochetes

The spirochetes, a small group of chemoheterotrophic bacteria, are unicellular organisms with a distinctive form. The cell, always very long relative to its width, is helicoidal in shape, surrounded by a delicate, flexible wall. External to the protoplast, but also enclosed within the wall, is a longitudinal fibrillar structure, the *axial filament*, spirally wound about the protoplast. Spirochetes can swim actively in a liquid medium; during this movement, the cells often bend into broad loops or coils, while retaining at all times their finely wound helicoid shape. They multiply by binary fission, and no resting stages are known. The characteristic form of some of the large spirochetes is shown in Figure 5.27. Such organisms can be easily seen by ordinary light microscopy. However, the smaller members of the group have cells so narrow that they are barely visible with the light microscope and can be best observed in the living state by dark-field or phase-contrast microscopy.

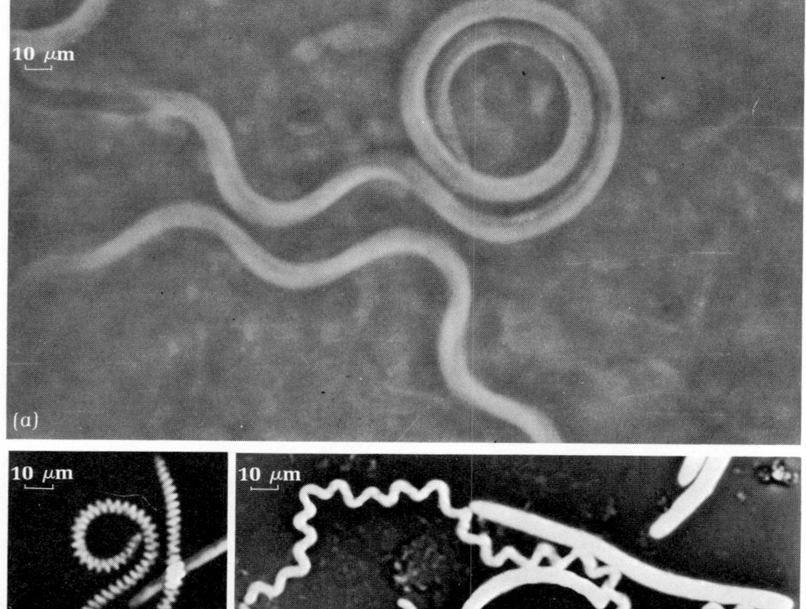


*Figure 5.27 Some of the larger spirochetes. (a) Phase contrast of living* Cristispira, *a form found in the crystalline style of clams (×352). Courtesy of S. W. Watson. (b) Nigrosin mount of an unidentified spirochete from water (×316). (c) Nigrosin mount of* Spirochaeta plicatilis, *a large spirochete common in water (×316). The preparation also contains cells of rod-shaped bacteria. (b) and (c) courtesy of C. F. Robinow.*

The organization of the spirochetal cell cannot be determined by light microscopy, and the topological relationships between the wall and the protoplast and axial filament have been elucidated only by use of the electron microscope (Figure 5.28). Studies have shown that the wall always completely encloses both protoplast and axial filament; however, since the wall is easily ruptured, the fibrils of the axial filament frequently fray out from the surface of the protoplast in poorly fixed or dead cells. The axial filament is not a continuous structure; it is composed of two overlapping sets of fibrils. The fibrils are anchored through hooklike bases in the two poles of the cell and extend toward the equator of the cell between the wall and the membrane. The overlap in the equatorial region creates the illusion of a continuous filament. These fibrils closely resemble eubacterial flagella in their dimensions, their helicoid form, and their structure. Indeed, the cell of a spirochete is analogous to that of a spirillum, with its two polarly inserted flagellar tufts; the basic difference is that in a spirochete the polarly inserted fibrils lie *within* the wall, whereas in a spirillum they extend *through* it. The movement of the spirochetal cell is presumably mediated by the fibrils of the axial filament, but the mechanism is not yet understood.

The number of fibrils that compose the axial filament varies widely in the different species of the group, in more or less direct relationship

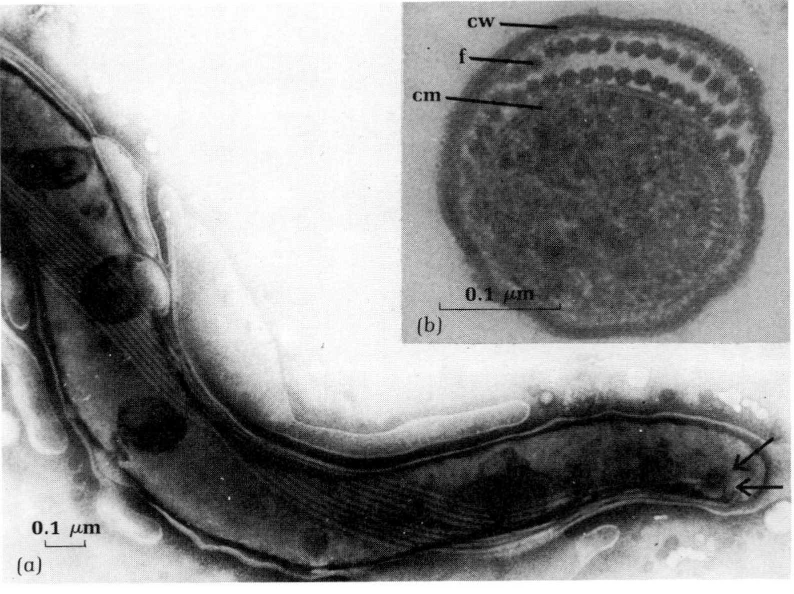


*Figure 5.28 The structure of the spirochetal cell as shown by electron micrographs of spirochetes from the mouth. (a) End of a cell, negatively stained with phosphotungstic acid, showing the relationship of the multifibrillar axial filament to the protoplast (×43,800). The insertion points of two fibrils from the axial filament are just visible at the pole of the cell (arrows). (b) Cross section of a large spirochete, showing the location of the fibrils (f) of the axial filament between the cell membrane (cm) and the cell wall (cw) (×157,680). From M. A. Listgarten and S. S. Socransky, "Electron microscopy of axial fibrils, outer envelope and cell division of certain oral spirochetes." J. Bacteriol.* **88,** *1087 (1964).*

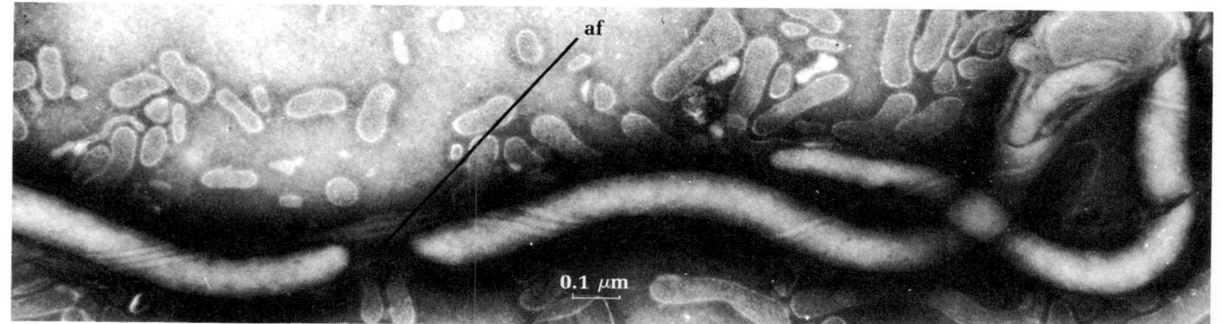


*Figure 5.29 Electron micrograph of negatively stained dividing cell of the spirochete Treponema microdentium (×59,940). The daughter cells are beginning to separate, but are still connected by fibrils of the axial filament (af). From M. A. Listgarten and S. S. Socransky, "Electron microscopy of axial fibrils, outer envelope and cell division of certain oral spirochetes." J. Bacteriol.* **88,** *1087 (1964).*

with the size of the cell. In the smallest spirochetes there is only a single fibril inserted at each pole. In the very large forms, such as *Cristispira*, there are several hundred. In fact, this organism owes its name to the fact that the thick bundle of fibrils composing its axial filament forms an elevated longitudinal ridge, or *crista*, which spirals around the cell and is visible with the light microscope (see Figure 5.27). Cell division in spirochetes (Figure 5.29) involves a primary division of the protoplast within the enclosing wall. This is followed by division of the wall and separation of the two daughter individuals, each individual bearing one of the two sets of fibrils that compose the axial filament. A new initiation point and a new set of fibrils are synthesized at the newly formed pole of each daughter cell.

Free-living spirochetes are inhabitants of mud and water; most of these organisms appear to be anaerobes, found characteristically in the oxygen-free regions of these environments. The members of the genus *Cristispira* are parasites of clams and other mollusks. They occur, frequently in enormous numbers, in a special region of the host, a gelatinous rod known as the crystalline style, which lies in a sac connected to the digestive tract. Many of the smaller spirochetes are parasites of man or of other vertebrates and include agents responsible for several human diseases: syphilis and yaws *(Treponema)*, relapsing fever *(Borrelia)*, and one type of infectious jaundice *(Leptospira)*. Few spirochetes, free-living or pathogenic, have been grown in pure culture, apart from the members of the *Leptospira* group. These organisms, in contrast to other spirochetes, are strict aerobes, although they grow best at oxygen tensions less than atmospheric, thus belonging to the category of microaerophils. Some free-living members of the genus *Treponema* have also been successfully cultivated; they have proved to be strict anaerobes.

## *Obligate intracellular parasites of uncertain taxonomic position*

Two microbial groups that are clearly bacterial in nature, the rickettsias and the bedsonias, cannot be assigned with certainty to any of the major bacterial assemblages described so far. Both groups have extremely small cells and appear to be obligate intracellular parasites of animals. Thus far, all attempts to cultivate them, or even to maintain them for some time in a viable state outside the host cell, have failed. Hence, it has been difficult to obtain information about their properties; the bacterial nature of the bedsonias has been established only recently. In the past it was suggested frequently that these organisms might be either large viruses or forms in some sense transitional between bacteria and viruses. This view of their relationships rested primarily on their smallness and obligate intracellular parasitism; it had to be abandoned when, in 1957, A. Lwoff first clearly recognized and defined the distinctive noncellular biological properties of the viruses. Our present knowledge identifies both the rickettsias and the bedsonias as cellular, procaryotic organisms, albeit ones of an unusual character.

### The rickettsias

The rickettsias are obligate intracellular parasites in certain groups of arthropods (notably fleas, lice, and ticks). They do not appear to produce symptoms of disease in their arthropod hosts, but if they are transmitted by bite to a vertebrate host, a severe and often fatal infection may result. The major rickettsial disease of man is epidemic typhus, the biography of which has been recounted in popular form in one of the few literary classics written by a microbiologist, Hans Zinsser's *Rats, Lice and History*. Other human rickettsial diseases of minor importance are Rocky Mountain spotted fever, transmitted by ticks, and scrub typhus, normally transmitted by mites to field mice, but also transmissible to man. The laboratory study of the rickettsias has been greatly facilitated by the discovery that they can be grown in the yolk sacs of chick embryos, a much more convenient laboratory material than the animal hosts.

The rickettsias are short rods, about 0.3 by 1.0 $\mu$m (Figure 5.30), which multiply by binary transverse fission. Their cellular nature is clearly established by the fact that they contain both DNA and RNA, are enclosed by a cell wall, and can perform certain metabolic activities. The structure of the cell, as revealed by the electron microscopy of thin sections, is clearly procaryotic, and this has been confirmed chemi-

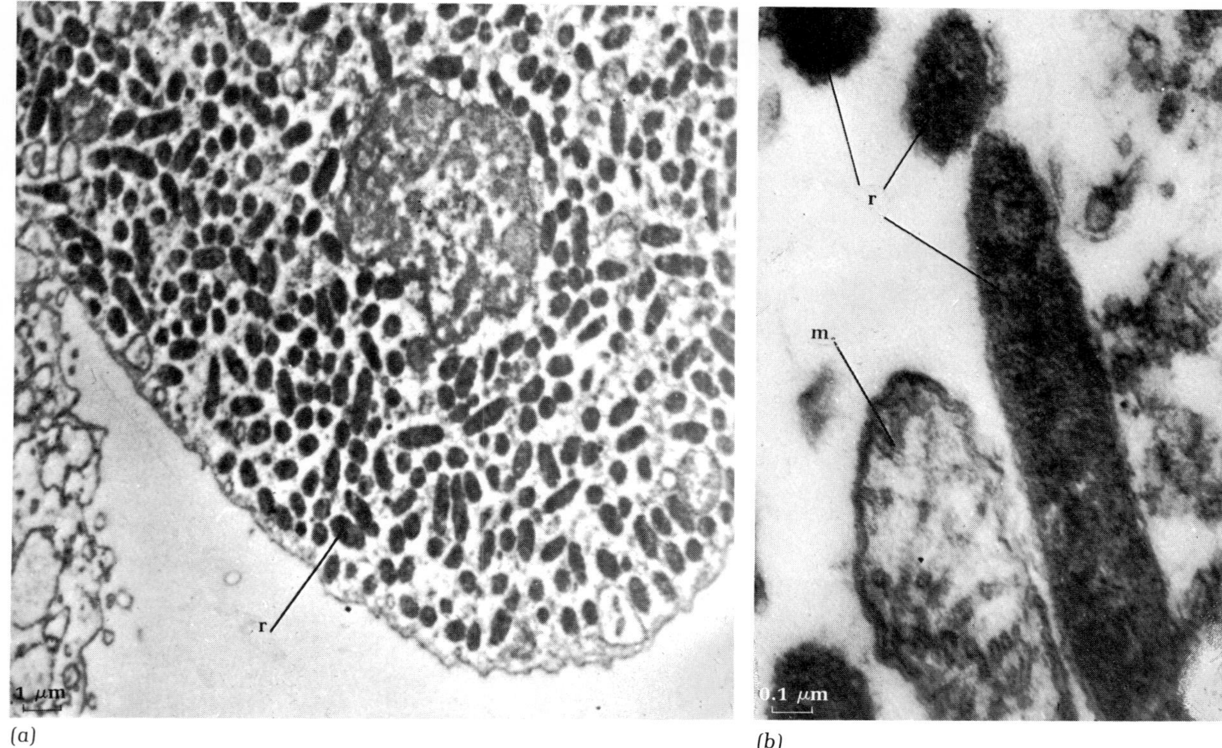


*Figure 5.30 Electron micrographs of thin sections through the tissues of chicken embryos infected with Rickettsia. (a) A single embryonic cell in the later stage of rickettsial infection (×10,026). The whole cytoplasmic region is filled with the rod-shaped rickettsial cells (r); the nucleus (n), although abnormal in appearance, is not infected. (b) Portion of an infected embryonic cell at a high magnification (×57,031). Several rickettsial cells (r) are to be seen in longitudinal or transverse section; also a mitochondrion (m), which is roughly of the same dimensions as the rickettsial cells. From S. L. Wissig, L. G. Caro, E. B. Jackson, and J. E. Smadel, "Electron Microscopic Observations on Intracellular Rickettsiae." Am. J. Pathol.* **32,** *1117 (1956).*

cally, by the detection of components of murein as cell constituents.

The explanation of their obligate intracellular parasitism is still far from clear. Isolated cell suspensions can oxidize the amino acid glutamate through the reactions of the tricarboxylic acid cycle, which suggests that rickettsias have an autonomous system of energy-yielding metabolism. However, isolated cells rapidly become inviable. The loss of viability can be retarded by adding certain coenzymes to the suspending medium. Since these are substances to which intact cells are normally completely impermeable, it has been suggested that the adaptation of the rickettsias to an intracellular mode of life has caused far-reaching changes in the properties of the cell membrane, which enable them to absorb and use coenzymes or other chemically complex metabolites produced by the host. Such changes in membrane properties would, however, make the rickettsias extremely vulnerable in an extracellular environment.

**The bedsonias**

The bedsonias, often referred to by the clumsy name "psittacosis–lymphogranuloma venereum–trachoma group," are the agents of a number of diseases of birds and of mammals, including man. In contrast to rickettsias, they are not transmitted through invertebrate hosts, but pass directly between their vertebrate hosts.

The cells of bedsonias (Figure 5.31) are more or less spherical and slightly smaller than those of rickettsias, being 0.2 to 0.7 $\mu$m in diameter.

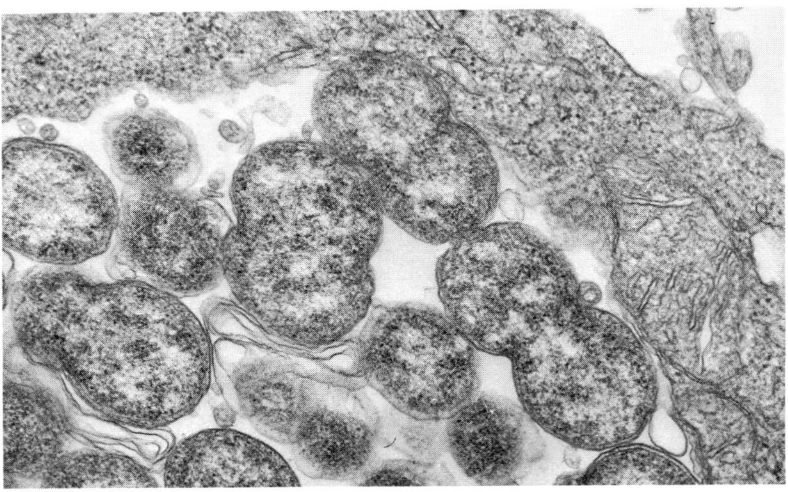

*Figure 5.31 Electron micrograph of a thin section of part of a mammalian cell growing in tissue culture, and infected by many Bedsonia cells (×28,350). Each of the small round-to-ovoid Bedsonia cells is surrounded by a unit membrane; many are in the course of binary fission. Courtesy of R. R. Friis, Department of Microbiology, University of Chicago.*

As a result of their very small size, it is difficult to establish with certainty the mode of growth within the host cell; and there is some dispute about whether multiplication occurs by binary fission or by budding. Chemical analyses have established the presence in the cells of both DNA and RNA, as well as components of murein. Although biochemical studies show that the bedsonias possess extensive enzymatic capacities, it has not yet been demonstrated that they are capable of performing energy-yielding metabolism. This has led to the interesting hypothesis that they may be "energy parasites," the growth of which is strictly dependent on the provision of energy-rich materials derived from the energy-yielding metabolism of the host cell.

## *Some physiological and biochemical attributes peculiar to procaryotes*

This survey of procaryotic organisms has dealt primarily with their structural attributes. In conclusion, we shall describe briefly certain physiological and biochemical properties that occur exclusively among procaryotic organisms. Unlike the presence of murein in the cell wall, or the inability to synthesize sterols, these properties are sporadically distributed among procaryotic groups and cannot therefore serve as taxonomic markers of the procaryotes as a whole. However, when one of these properties is displayed by an organism, it provides an almost infallible indication of bacterial or blue-green algal affinities.

**Nitrogen fixation**

The ability to use $N_2$ as a nitrogen source for the synthesis of cell materials is a procaryotic character. This statement may appear paradoxical, in view of the fact that several groups of plants, such as legumes, are able to grow in the absence of combined nitrogen, but analysis shows that in all such cases the ability is conferred by a symbiotic relationship with bacteria or blue-green algae. Many blue-green algae, as well as bacteria belonging to a number of different groups, can use $N_2$ as a nitrogen source to support free-living growth.

**The ability to grow at very high temperatures**

Extreme thermophily, which we may define as the ability to grow at temperatures in excess of 55°C, coupled with the failure to grow at temperatures below 40°C, is a physiological character entirely confined to procaryotic organisms; it exists among the blue-green algae, the endospore-forming unicellular bacteria, and the actinomycetes.

**Chemoautotrophy**

The ability to use reduced inorganic compounds as sole energy sources for growth is confined to certain unicellular eubacteria and to colorless, filamentous organisms that are structurally related to the blue-green algae.

**Poly-β-hydroxybutyrate as a cellular reserve material**

The most common cellular reserve materials of procaryotic organisms are glycogenlike polysaccharides (polyglucoses) and poly-β-hydroxybutyrate, an insoluble polyester of β-hydroxybutyric acid. Polyglucoses also occur widely as reserve materials in eucaryotic organisms, but poly-β-hydroxybutyrate does not; it seems to be a uniquely bacterial reserve material. It is the principal or sole organic reserve material in many groups of eubacteria.

**Obligate anaerobiosis**

Obligate anaerobiosis is not *strictly* confined to procaryotic organisms. A small number of obligate anaerobes (all symbionts in the digestive tracts of animals) are known among the protozoa, and one well- authenticated case of an obligately anaerobic fungus has been described. However, this property is widespread in bacteria, being characteristic of many groups among the eubacteria and spirochetes. Furthermore, the modes of energy-yielding metabolism displayed by the obligately anaerobic bacteria are exceptionally varied with respect both to the substrates that can be used and to the end-products formed. The obligately anaerobic eucaryotes are all dependent on the fermentation of sugars.

Our knowledge of the geochemical history of the earth shows that the formation of free oxygen as a major atmospheric constituent was a relatively late event in the earth's history and was probably of biological origin, a consequence of the evolution of photosynthesis of the plant type. Hence, much of the early evolution of life on earth must have occurred under strictly anaerobic conditions. It is conceivable that the numerous and varied groups of contemporary anaerobic bacteria represent surviving evolutionary lines whose physiological properties had already been fixed during the early, anaerobic period of biological evolution. The strictly anaerobic eucaryotic microorganisms, on the other hand, are organisms in which this physiological character has been secondarily acquired, as a result of evolutionary adaptation to specialized oxygen-free habitats.

## *Further reading*

**Books**

Fritsch, F. E., *The Structure and Reproduction of the Algae*, Vol. 2. New York: Cambridge University Press, 1945. Contains the best general account of blue-green algae.

Gunsalus, I. C., and R. Y. Stanier (editors), *The Bacteria*, Vol. I. New York: Academic Press, 1960.

Moulder, J. W., *The Biochemistry of Intracellular Parasitism*. Chicago: University of Chicago Press, 1962.

Moulder, J. W., *The Psittacosis Group as Bacteria*. New York: Wiley, 1964.

Salton, M. R. J., *The Bacterial Cell Wall*. Amsterdam: Elsevier, 1964.

**Reviews**

Lwoff, A., "The Concept of Virus." *J. Gen. Microbiol.* **17,** 239 (1957).

Murray, R. G. E., "Fine Structure and Taxonomy of Bacteria," in *Microbial Classification*, p. 119. New York: Cambridge University Press, 1962.

Stanier, R. Y., "Toward a Definition of the Bacteria," in *The Bacteria*, Vol. V, p. 445. New York: Academic Press, 1964.

# Chapter six
# Microbial metabolism: generation and transfer of energy

*The metabolism* of any organism can be subdivided into two major categories: *energy-generating or degradative pathways*, sometimes collectively referred to as *catabolism*, and *energy-consuming* or *biosynthetic pathways*, sometimes collectively referred to as *anabolism*. Growth is the result of an intimate linkage between the processes of catabolism and anabolism, part of the free energy derived from degradations being used to drive the processes of biosynthesis.

## Thermodynamic properties of living systems

The second law of thermodynamics states that in any energy-yielding reaction, the total energy liberated, $\Delta H$, is never fully available for the performance of work, since part of it is lost as an increase of entropy. The free energy potentially available for the performance of work, $\triangle F$, is given by the relation

$$\Delta H = \Delta F + T\Delta S$$

where $T\Delta S$ expresses the entropy increase. Living systems are subject to this physical law; consequently, it is the free energy, $\Delta F$, liberated by energy-yielding metabolic processes that is potentially available to drive the energy-consuming reactions of biosynthesis in the cell. Purely thermodynamic considerations accordingly determine the *maximal* amount of energy available to the cell. However, the *actual* amount of the energy derived from catabolism that can be applied to energy-consuming metabolic processes is always considerably less. In the alcoholic fermentation of sugars, for example, the free energy released by the conversion of 1 mole of carbohydrate to 2 moles of ethanol and 2 moles of $CO_2$, according to the equation

$$C_6H_{12}O_6 \rightarrow 2CO_2 + 2C_2H_5OH$$

is 56 kcal. However, the amount that is in fact available to drive energy-consuming reactions is only about 16 kcal. The real thermodynamic efficiency of alcoholic fermentation is accordingly about 29 percent, and the rest of the free energy produced is liberated as heat. This is a feature of all energy-yielding metabolic processes, and the magnitude of heat production, which can be measured in a calorimeter, *provides an indication of the thermodynamic inefficiency of the coupling between catabolism and anabolism.*

## *The chemical links between energy-yielding metabolism and biosynthesis*

To understand why only part of the free energy made available through energy-yielding metabolic processes can be applied to the biosynthetic, energy-consuming processes of the cell, it is necessary to consider the *nature of the chemical coupling* between the two components of metabolism. The essential chemical function of all energy-yielding metabolic processes is to produce certain organic compounds that contain a high level of potential energy in the form of *energy-rich bonds*. These compounds serve as *energy carriers*, in the sense that their potential-energy content can be used, in coupled reactions, to drive the various energy-requiring step reactions involved in the biosynthesis of cell material. Since each of these molecular energy carriers has a fixed content of potential energy, the usable energy derived from catabolic processes is made available in discrete packets, each having a certain potential-energy content. The compounds that serve as energy carriers, linking catabolic and biosynthetic phases of metabolism, are *adenosine triphosphate* and *reduced pyridine nucleotides*.

**Energetic coupling through adenosine triphosphate**

The chemical structure of adenosine triphosphate (ATP) is shown in Figure 6.1. It is a derivative of the nucleotide adenosine monophosphate (AMP) or adenylic acid, to which two additional phosphate groups are linked through pyrophosphate bonds. These two bonds (indicated by the symbol ~) are energy rich, in the sense that their removal by hydrolysis to yield successively adenosine diphosphate (ADP) and adenosine monophosphate (AMP) liberates a substantial

Figure 6.1 *The structure of ATP (adenosine triphosphate), showing the various components of the molecule that can be obtained by hydrolysis.*

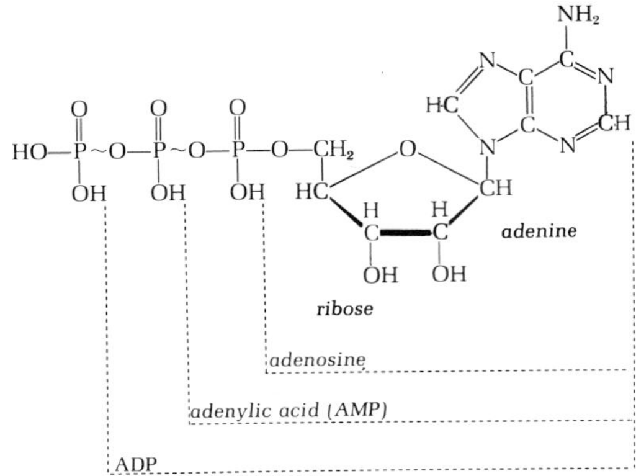


amount of energy as heat (over 8 kcal/mole for each bond broken). In a biological system, however, the potential energy contained in the two terminal phosphate bonds of ATP is not normally liberated as heat by simple hydrolysis. Instead, *energetic coupling is achieved by the transfer of one or both of the terminal phosphate groups of adenosine monophosphate to an acceptor molecule, part of the bond energy being preserved in the newly formed molecule.* Roughly speaking, such transfer reactions can preserve up to 8 kcal of phosphate bond energy, so that the two terminal phosphate bonds of ATP represent together a potential-energy supply of about 16 kcal.

In such transfer reactions, the terminal phosphate bond is often transferred, with the formation of ADP. An example is the activation of glucose:

glucose + ATP → glucose-6-phosphate + ADP

Sometimes the transfer involves addition of the adenosine monophosphate residue to the acceptor molecule, with liberation of the two terminal phosphate groups as inorganic pyrophosphate (P—P). Such a reaction occurs in the activation step in the conversion of amino acids to proteins, which precedes the actual polymerization:

amino acid + ATP → AMP–amino acid + P—P

When AMP is produced through energy-transfer reactions, it can be converted to ADP by the following enzymatic reaction:

AMP + ATP → 2ADP

Grossly speaking, therefore, the utilization of the phosphate bond energy of ATP in coupled reactions can be regarded as representing a conversion of ATP to ADP. One of the essential functions of energy-yielding metabolism is *to regenerate the cellular pool of ATP, by resynthesis of this compound from ADP*. Several different biochemical mechanisms associated with energy-yielding metabolism can achieve this. As an illustration, we may take one of the two ATP-generating reactions that occur during the course of the alcoholic fermentation and many other fermentative processes, the conversion of 2-phosphoglyceric acid to pyruvic acid:

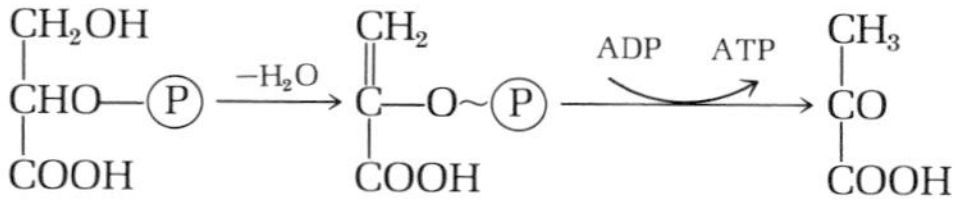

$$\begin{array}{c} CH_2OH \\ | \\ CHO\text{—}(P) \\ | \\ COOH \end{array} \xrightarrow{-H_2O} \begin{array}{c} CH_2 \\ \| \\ C\text{—}O\sim(P) \\ | \\ COOH \end{array} \xrightarrow{ADP \curvearrowright ATP} \begin{array}{c} CH_3 \\ | \\ CO \\ | \\ COOH \end{array}$$

*2-phosphoglyceric acid* — *phosphoenol-pyruvic acid* — *pyruvic acid*

In this reaction sequence, the phosphate ester bond of phosphoglyceric acid, which has a low energy content, is converted by a molecular rearrangement to a phosphate bond of high energy content (indicated by a wavy bond) in the intermediate phosphoenolpyruvic acid; the phosphate group can then be donated to ADP, to form ATP. In this and in subsequent reactions, we shall symbolize esterified phosphate groups in organic structural formulae as (P) and use the symbols $P_i$ and P—P to indicate inorganic phosphate and pyrophosphate, respectively.

Another type of ATP-generating reaction, common to many fermentative and respiratory energy-yielding processes, occurs in association with the oxidation of $\alpha$-keto acids, such as pyruvic and $\alpha$-ketoglutaric acids, and involves the intermediacy of a coenzyme known as *coenzyme A*. This compound is a derivative of adenosine diphosphate. There is attached to ADP, through the terminal phosphate bond, a molecule of pantothenic acid, which carries a terminal thiol (—SH) group:

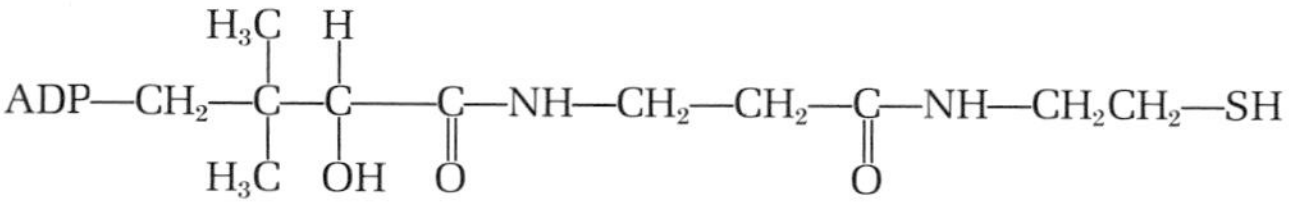

$$ADP\text{—}CH_2\text{—}\underset{H_3C}{\overset{H_3C}{C}}\text{—}\underset{OH}{\overset{H}{C}}\text{—}\underset{O}{\overset{}{\underset{\|}{C}}}\text{—}NH\text{—}CH_2\text{—}CH_2\text{—}\underset{O}{\underset{\|}{C}}\text{—}NH\text{—}CH_2CH_2\text{—}SH$$

The molecule can be symbolized as CoA—SH.

In the oxidation of $\alpha$-keto acids, coenzyme A becomes attached, through formation of a thioester linkage, to the carboxyl group of the oxidized product. The oxidation of pyruvic acid can be represented as shown here:

$$CH_3COCOOH + CoA\text{-}SH \xrightarrow{-2H} CO_2 + CH_3\text{—}CO\text{—}S\text{—}CoA$$

*pyruvic acid* — *acetyl coenzyme A*

The mechanism of this oxidation, which is complex, results in a conservation of part of the energy of oxidation in the thioester bond between coenzyme A and the oxidized product. This bond energy can be used subsequently to synthesize ATP from ADP and inorganic phosphate

$$\underset{\text{acetyl coenzyme A}}{CH_3\text{—}CO\text{—}S\text{—}CoA} + ADP + P_i \rightarrow \underset{\text{acetic acid}}{CH_3COOH} + CoA\text{-}SH + ATP$$

The general role of ATP as a coupling agent between catabolism and biosynthesis is schematized in Figure 6.2. As this figure shows, the

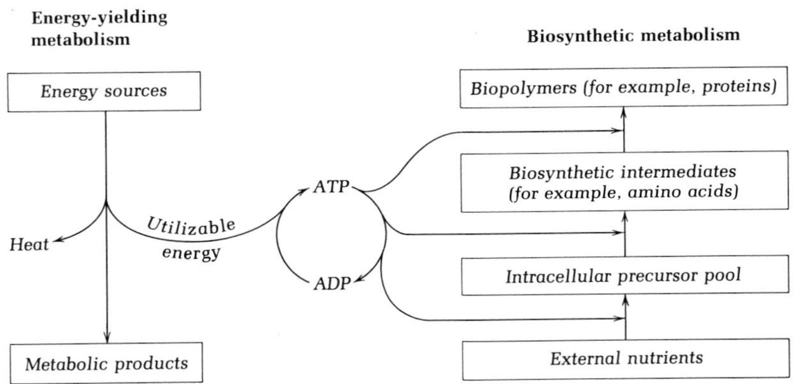


Figure 6.2 A schematic diagram showing the role of ATP in the coupling of energy-yielding metabolism and biosynthesis.

phosphate bond energy of ATP intervenes at a number of different stages of the overall biosynthetic process. First, it is required to drive the entry of nutrients into the cell, almost always an active process that requires a net input of energy. Second, it is used to convert these nutrients, once inside the cell, to intermediary metabolites of low molecular weight: amino acids, sugar phosphates, nucleotides, fatty acids, and others. Finally, it is used to achieve the polymerization of these intermediates into the various classes of biopolymers that are the major constituents of cell material: proteins, polysaccharides, nucleic acids, and lipids.

## Oxidations and reductions of organic compounds

Like all oxidations, the oxidation of an organic compound represents a removal of electrons. In most cases each step in the oxidation of an organic compound involves the removal of the two electrons and the simultaneous loss of two protons; this is equivalent to the removal of two hydrogen atoms and is termed a *dehydrogenation.* Conversely, the reduction of an organic compound usually involves an addition of two electrons and two protons and can therefore be considered a *hydrogenation.* In describing the oxidations and reductions of organic compounds, it is convenient to represent the reactions as additions or removals of *pairs of hydrogen atoms,* symbolized as 2H. For example, the oxidation of lactic acid to pyruvic acid and the reduction of pyruvic acid to lactic acid, can be expressed

$$\begin{matrix} CH_3 & & CH_3 \\ | & & | \\ CHOH & \underset{+2H}{\overset{-2H}{\rightleftharpoons}} & C{=}O \\ | & & | \\ COOH & & COOH \end{matrix}$$

*lactic acid* *pyruvic acid*

**Energetic coupling through pyridine nucleotides**

Biosynthetic processes invariably require a major net input of energy derived from the energy-rich bonds of ATP. In addition, they often require a net input of *reducing power,* which is also derived from the catabolic reaction sequences. The *net* requirement for reducing power is highly variable; *it reflects the difference in oxidation state between the nutrients that are used for biosynthesis and the final products of biosynthesis.* The variability of this requirement is caused by the differences between organisms with respect to the oxidation state of their major nutrients, specifically, the nutrients that serve as sources of *carbon, nitrogen,* and *sulfur.* Carbon contributes much more on a weight basis to cell material than nitrogen, and nitrogen more than sulfur, so the net requirements for reducing power are most substantial for the assimilation of oxidized forms of carbon and proportionately less for the assimilation of oxidized forms of nitrogen and sulfur.

With respect to carbon, the overall state of oxidation of the organic components of the cell is approximately that of carbohydrate; it can be symbolized as $(CH_2O)$. Hence, chemoheterotrophic organisms that use glucose or other carbon sources on this oxidation level for the synthesis of cell material require little or no net input of reducing power for the assimilation of carbon. On the other hand, an organism that uses $CO_2$ as its sole source of carbon (e.g., a plant or a chemoautotrophic bacterium) must have available a large amount of reducing power for the purposes of carbon assimilation. Roughly speaking, a net input of four hydrogen atoms is required for the assimilation of each molecule of $CO_2$, to reduce it to the oxidation level of cell material:

$$CO_2 + 4H \rightarrow (CH_2O) + H_2O$$

Most of the nitrogen of the cell material is present in the form of amino groups ($R—NH_2$), which are on the same oxidation level as ammonia. Hence, no net input of reducing power is required for nitrogen assimilation when nitrogen is furnished to the cell either as ammonia or as complex organic nitrogen compounds (amino acids, peptones). However, if nitrogen is furnished as nitrate, considerable reducing power is needed to reduce it to the level of ammonia, as a preliminary to the synthesis of nitrogen-containing organic cell materials:

$$NO_3^- + 8H \rightarrow NH_3 + 2H_2O + OH^-$$

The sulfur of cell material is principally present as sulfhydryl groups (R—SH), on the same oxidation level as sulfide. If sulfate is the sulfur source, it must accordingly be reduced to sulfide prior to incorporation into sulfur-containing materials of the cell:

$$SO_4^{2-} + 8H \rightarrow S^{2-} + 4H_2O$$

The transfer of reducing power from energy-yielding metabolism to reductive biosynthetic step reactions is mediated by two carrier molecules known as *pyridine nucleotides*. Nicotinamide adenine dinucleotide (NAD)* has the structure shown in Figure 6.3. It is composed of

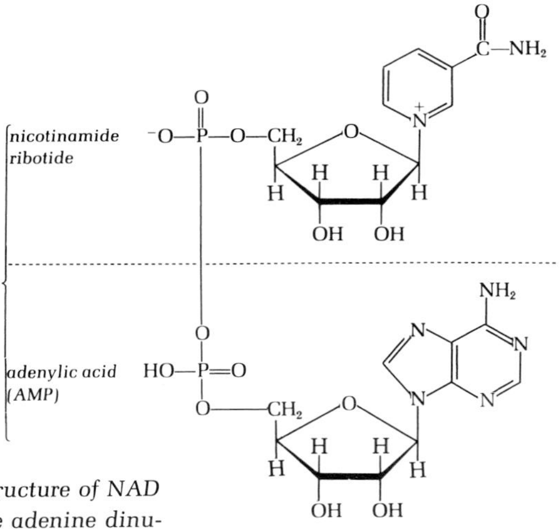

*nicotinamide adenine dinucleotide (NAD) or diphosphopyridine nucleotide (DPN)*

*nicotinamide ribotide*

*adenylic acid (AMP)*

*Figure 6.3 Structure of NAD (nicotinamide adenine dinucleotide).*

two nucleotides, adenylic acid (AMP) and nicotinamide ribotide, linked through their phosphate groups. The structure can thus be expressed as nicotinamide–ribose–Ⓟ—Ⓟ–ribose–adenine. Nicotin-

*NAD was formerly referred to as DPN (diphosphopyridine nucleotide) and NADP as TPN (triphosphopyridine nucleotide).

amide adenine dinucleotide phosphate (NADP) has an identical structure except that it contains an additional phospahte group, linked to one of the ribose residues.

Both pyridine nucleotides can readily undergo reversible oxidation and reduction. The site of these oxidation–reductions is the nicotinamide group, and the nature of the reversible reaction is shown in Figure 6.4. It can be seen that the oxidized form of a pyridine nucleotide car-

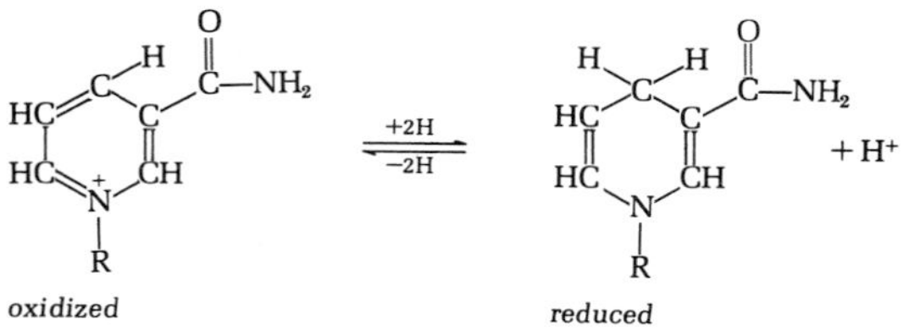

oxidized $\underset{-2H}{\overset{+2H}{\rightleftharpoons}}$ reduced $+ H^+$

*Figure 6.4 Oxidized and reduced forms of the nicotinamide moiety of pyridine nucleotides.*

ries one hydrogen atom less than the reduced form and in addition has a positive charge on the nitrogen atom, which enables it to accept a second electron upon reduction. The reversible oxidation–reduction of NAD can thus be symbolized

$$NAD^+ + 2H \rightleftharpoons NADH + H^+$$

For convenience, we will hereafter use the symbols NAD and $NADH_2$ to designate the oxidized and reduced forms, respectively, of NAD. The oxidized and reduced forms of NADP will be written NADP and $NADPH_2$.

*The reductive reactions of biosynthesis are mediated by a transfer of electrons from the reduced forms of the pyridine nucleotides.* An example is the reductive amination of the keto acid $\alpha$-ketoglutaric acid to form the amino acid, glutamic acid:

$$HOOCCH_2CH_2COCOOH + NH_3 + NADH_2 \rightarrow HOOCCH_2CH_2CHNH_2COOH + NAD + H_2O$$

Many enzymatic reactions that involve an oxidation or a reduction of pyridine nucleotides function specifically with either NAD or NADP. However, in terms of the overall metabolism of the cell, these two electron carriers are often equivalent, since a special enzyme, transhydrogenase, present in many organisms, can mediate a reversible electron transfer between them:

$$NADPH_2 + NAD \rightleftharpoons NADP + NADH_2$$

The generation of reducing power during energy-yielding metabolism is achieved by electron transfers from oxidizable compounds to one of the pyridine nucleotides. A typical example is the oxidation of ethanol to acetaldehyde, mediated by the enzyme alcohol dehydrogenase:

$$CH_3CH_2OH + NAD \rightarrow CH_3CHO + NADH_2$$

Not infrequently, a particular catabolic reaction sequence may give rise to the formation both of ATP and of reduced pyridine nucleotide. This can occur, for example, during the oxidation of pyruvic acid to acetic acid, a reaction central to the oxidative breakdown of many organic substrates. Without entering into the biochemical details of this oxidation, we may express its net consequences, insofar as the provision of materials for biosynthetic reactions is concerned, as

$$CH_3COCOOH + NAD + H_2O + ADP + P_i \rightarrow CH_3COOH + CO_2 + NADH_2 + ATP$$

In respiratory metabolism, much of the reduced pyridine nucleotide formed by the oxidation of organic substrates is used to generate ATP, as a result of the reoxidation of $NADH_2$ or $NADPH_2$ through an electron transport chain linked to oxygen. The reoxidation of a molecule of pyridine nucleotide yields sufficient energy to permit the formation, under optimal conditions, of three molecules of ATP from ADP. Roughly speaking, therefore, the energy of a molecule of reduced pyridine nucleotide is biologically equivalent to that contained in three energy-rich phosphate bonds. Insofar as reduced pyridine nucleotides are reoxidized in reductive biosynthetic reactions (such as the conversion of $\alpha$-ketoglutarate to glutamate), their bond energy is conserved in the reduced product of the reaction and cannot at the same time be directly used to generate energy-rich phosphate bonds. It follows, accordingly, that the reducing power produced during energy-yielding metabolism as $NADH_2$ or $NADPH_2$ can be used in two possible ways: to mediate reductive biosynthetic reactions or, in the case of respiratory metabolism, to produce additional ATP through reactions associated with electron transport to oxygen.

## *Modes of energy-yielding metabolism*

**The definition and nature of fermentation**

Of the three principal modes of energy-yielding metabolism, the simplest in terms of mechanism is *fermentation*. It can be defined as *an energy-yielding metabolic process in which organic compounds serve both as electron donors and as electron acceptors.* The compounds that perform these two functions are usually two different metabolites derived from a fermentable substrate (such as from a sugar). In fermentations, the substrate gives rise to a mixture of end-products, some of which are more oxidized than the substrate and others more reduced.

There are, accordingly, limits to the kinds of organic compounds that can be degraded by fermentation: they must be neither highly oxidized nor highly reduced. *Carbohydrates* are the principal substrates of fermentation. Among the bacteria, some compounds belonging to other chemical classes can also be fermented: organic acids, amino acids, purines, and pyrimidines.

Since fermentable substrates are on roughly the same oxidation level as cell material and also usually serve as the principal source of carbon for biosynthesis, the net biosynthetic requirement for reducing power during fermentative growth is negligible. *The major or sole energetic contribution of a fermentation is accordingly the provision of ATP.* ATP is formed by the transfer to ADP of high-energy phosphate groups from phosphorylated intermediates that arise during substrate breakdown. This mode of ATP synthesis, the only one possible as a result of fermentation, is known as *substrate-level phosphorylation.*

Pasteur, who first recognized the physiological function of fermentation, called it "the consequence of life without air." This statement is still correct, because all fermentations can proceed under strictly anaerobic conditions. Many of the organisms that obtain energy by fermentation are strict anaerobes. Others are facultative anaerobes, however, able to grow either in the presence or in the absence of air. As a rule, facultatively anaerobic fermentative organisms *change their mode of energy-yielding metabolism upon exposure to air:* the presence of molecular oxygen induces a metabolic shift from fermentation to respiration. However, one facultatively anaerobic bacterial group, the lactic acid bacteria, provides a notable exception to this rule; oxygen does not modify their mode of energy-yielding metabolism and fermentation continues even when they are grown in air.

Fermentative processes always maintain a strict oxidation–reduction balance, in the sense that the *average oxidation level of the end-products is identical with that of the substrate fermented.* In other words, the number of molar equivalents of carbon, hydrogen, and oxygen in the end-products is the same as in the substrate. This can be readily seen in the case of alcoholic fermentation:

$$\underset{\textit{substrate}}{C_6H_{12}O_6} = \underset{\textit{end-products}}{2CO_2 + 2C_2H_5OH}$$

**The definition and nature of respiration**

*Respiration* can be defined as *an energy-yielding metabolic process in which either organic compounds or reduced inorganic compounds serve as electron donors and molecular oxygen as the ultimate electron acceptor.* This definition covers all modes of respiratory metabolism, with the exception of a special class of respiratory processes characteristic of a few bacteria, the *anaerobic respirations. In anaerobic respiration, an oxidized inorganic compound other than oxygen serves as ultimate electron acceptor.* The compounds that can so act are sulfates, nitrates, and carbonates. To distinguish oxygen-linked respiratory processes from these anaerobic respirations, it is useful to qualify the former as *aerobic respiration.*

A distinctive feature of most respiratory processes is the presence in the cell of a special set of carrier enzymes that comprise the *respiratory electron transport chain.* Electrons removed from the oxidizable substrate enter the chain and pass through a series of intermediate carriers to the ultimate inorganic electron acceptor.

The coupling of substrate oxidation with an external electron acceptor characteristic of respiratory metabolism usually permits a complete oxidation of organic substrates to $CO_2$. The free-energy change for the complete oxidation of an organic compound is very much greater than that for its fermentation. For example, the complete oxidation of 1 mole of glucose liberates 688 kcal, whereas most fermentations of this sugar liberate about one tenth as much energy. The substrate-level phosphorylations characteristic of the fermentation of organic compounds also operate in the respiration of organic compounds, but a considerably greater quantity of ATP is generated as a result of the passage of electrons through the respiratory transport chain, a process known as *oxidative phosphorylation.* The gross yield of ATP per mole of substrate respired is accordingly much greater than that obtainable by a fermentative metabolism of the same compound.

The fraction of the energy of respiration that must be supplied as reducing power is extremely variable, since the oxidation state of the nutrients assimilated (used for biosynthesis) during respiratory metabolism varies widely. If the respiratory substrate is an organic compound either more reduced than or at approximately the same oxidation level as cell material, little or no reducing power will be required, since the substrate also serves as the cellular carbon source. However, if the organic substrate is highly oxidized, a considerable net input of reducing power, generated during respiration through the reduction of the pyridine nucleotides, is required for its assimilation. In the extreme case of the chemoautotrophic bacteria, where the respiratory substrate is a reduced inorganic compound, the sole source of cellular carbon is $CO_2$. The generation of reduced pyridine nucleotides for the reductive assimilation of $CO_2$ becomes a very important function of respiratory metabolism in these organisms.

Many microorganisms that perform aerobic respiratory metabolism are strict aerobes. Some, however, may also be able to grow under anaerobic conditions, using as a source of energy either fermentation or anaerobic respiration with nitrate as terminal acceptor. These forms are, accordingly, *facultative* anaerobes. The bacteria that perform anaerobic respiration with sulfate or carbonate as the terminal acceptor are *strict* anaerobes; they cannot use aerobic respiration as an alternative means of energy-yielding metabolism.

**The definition and nature of photosynthesis**

The third (and mechanistically most complex) mode of energy-yielding metabolism is *photosynthesis:* the use of light as a source of energy. Historically, the term "photosynthesis" was used to describe the overall metabolism of plants, eucaryotic algae, and blue-green algae, represented by the following reaction:

$$CO_2 + H_2O \xrightarrow{\text{light}} (CH_2O) + O_2$$

It is obvious that this equation does not describe an *energy-yielding* process but rather its *major biosynthetic consequence,* the light-mediated conversion of $CO_2$ to organic cell materials. Indeed, the energy-yielding reactions of photosynthesis have been discovered only recently. They are initiated by the absorption of light by the photosynthetic pigment system, followed by a process of *energy conversion.* The net result of energy conversion is the *transformation of part of the absorbed light energy into chemical bond energy which can be directly applied to biosynthesis—that of ATP and reduced pyridine nucleotides.* The light-dependent generation of ATP, which is the common denominator of all photosynthetic processes, is known as *photophosphorylation.* It is mechanistically analogous to oxidative phosphorylation: a special electron transport chain is intimately associated with the photopigment system, and ATP is formed during the passage through this chain of electrons generated during the primary photochemical reaction. Some electrons generated through the photochemical reaction may be used to reduce pyridine nucleotides. Substrate-level phosphorylations, which provide all the ATP in fermentations and some ATP in respirations, do not occur in photosynthetic energy-yielding metabolism.

The biosynthetic requirement for reducing power by photosynthetic organisms varies widely. In organisms that perform photosynthesis of the plant type, it is always great, because $CO_2$ is the principal or sole source of carbon. Many of the photosynthetic bacteria use organic compounds as the principal source of carbon; in such cases, the primary and sometimes the sole function of photosynthesis is the provision of ATP.

Molecular oxygen does not play a role in the energy-generating reactions of photosynthesis; consequently, all photosyntheses can in principle occur under strictly anaerobic conditions. However, in plant photosynthesis, the *oxidation of water* is a special feature of the photochemical reactions, and molecular oxygen is thus a *product* of photosynthesis. All organisms that use this photosynthetic mechanism are thus *oxygen producers*, and hence aerobes, in the sense that they must be able to tolerate the presence of oxygen. The fundamentally anaerobic nature of photosynthetic energy-yielding reactions is much more clearly evident in photosynthetic bacteria, where the oxidation of water (and hence the production of oxygen) does not occur. Most of these organisms are strict anaerobes; and in the minority that are facultative aerobes, photosynthetic metabolism is suppressed by the presence of oxygen, being replaced as a source of energy by respiration.

## *The biochemistry of fermentations*

### The fermentations of carbohydrates

The pathways employed for the fermentative breakdown of carbohydrates vary widely in different microbial groups, as do the end-products formed. Practically every carbohydrate or carbohydrate derivative can be used as a fermentable energy source by some microorganism. The list includes polysaccharides such as starch, cellulose, and chitin; disaccharides such as lactose, maltose, and sucrose; hexoses such as glucose, fructose, and galactose; pentoses such as arabinose and xylose; sugar acids such as gluconic and glucuronic acids; and polyalcohols such as mannitol and glycerol. In some groups, the particular range of carbohydrates that can be fermented is an important taxonomic character, used for the characterization of species or genera. We shall not develop this aspect of the problem here; it will be taken up in later chapters, devoted to the characteristics of special bacterial groups. For the purpose of general discussion, we shall instead consider the possible modes of fermentation of *glucose*, a sugar that is fermented by practically all organisms capable of fermenting any carbohydrate.

The fermentation of glucose is always initiated by a phosphorylation at the expense of ATP, to yield glucose-6-phosphate. This is converted, through one of several pathways, to the keto acid *pyruvic acid*, which is a central metabolic intermediate in the fermentative metabolism of all carbohydrates. The reactions that intervene between the initial phosphorylation of glucose and the formation of pyruvic acid are of such a nature that they permit the generation of ATP, in molar excess over the initial input required to phosphorylate the primary substrate; *part, and sometimes all, of the net energy yield of fermentations is the result of reactions that precede the formation of pyruvic acid*. Pyruvic acid is more oxidized than glucose, and its formation from glucose is accompanied by a reduction of pyridine nucleotides. *For the fermenta-*

*tion to achieve a final oxidation–reduction balance, the reduced pyridine nucleotide must be reoxidized. This reoxidation characteristically occurs in the terminal step reactions, which result in the conversion of pyruvic acid to the metabolic end-products.* Sometimes pyruvic acid is itself reduced to lactic acid. In other cases, it is further metabolized, and the reoxidation of pyridine nucleotide is accompanied by a reduction of a product derived from pyruvic acid.

**The alcoholic and homolactic fermentations**

These general principles can be illustrated by the most familiar and widely distributed fermentations of glucose, the alcoholic and homolactic fermentations. They are closely similar in biochemical mechanism, differing only with respect to the terminal fate of pyruvic acid. The homolactic fermentation occurs in animal tissues, in a number of

(*a*) Preliminary phosphorylations conversion of glucose to fructose-1,6-diphosphate

*glucose (pyranose form)* —ATP → ADP→ *glucose-6-phosphate (pyranose form)* → *fructose-6-phosphate (furanose form)* —ATP → ADP→ *fructose-1,6-diphosphate (furanose form)*

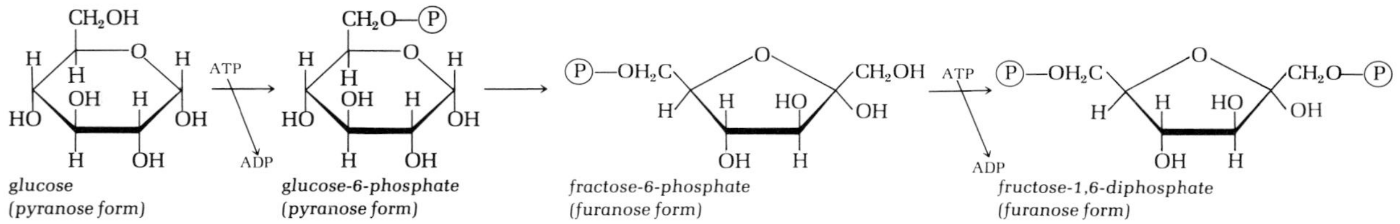

(*b*) Cleavage of fructose-1,6-diphosphate and formation of 1,3-diphosphoglyceric acid: esterification of inorganic phosphate and reduction of NAD

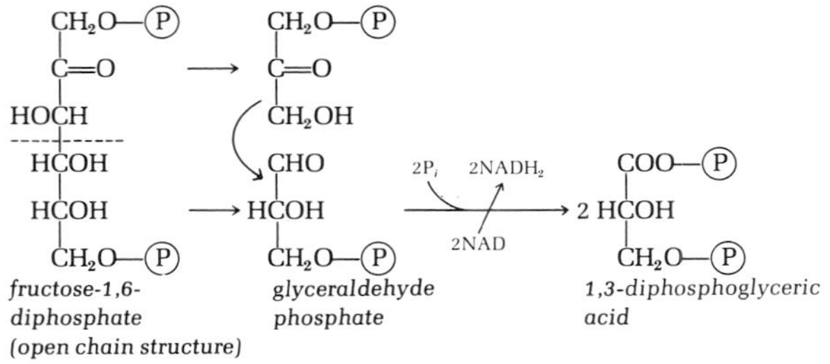

*fructose-1,6-diphosphate (open chain structure)* → *glyceraldehyde phosphate* —2P$_i$, 2NAD → 2NADH$_2$→ 2 *1,3-diphosphoglyceric acid*

(*c*) Conversion of the two molecules of 1,3-diphosphaglyceric acid formed in (*b*) to pyruvic acid: regeneration of ATP

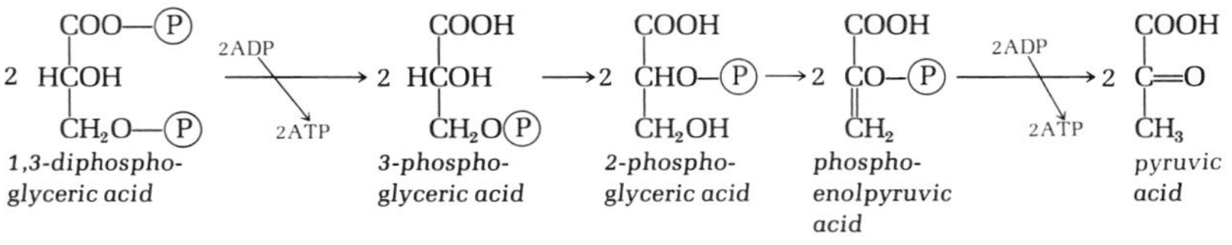

2 *1,3-diphospho-glyceric acid* —2ADP → 2ATP→ 2 *3-phospho-glyceric acid* → 2 *2-phospho-glyceric acid* → 2 *phospho-enolpyruvic acid* —2ADP → 2ATP→ 2 *pyruvic acid*

Overall reaction: glucose + 2ADP + 2P$_i$ + 2NAD → 2 pyruvic acid + 2ATP + 2NADH$_2$

*Figure 6.5 The Embden-Meyerhof pathway: conversion of glucose to pyruvic acid.*

eucaryotic organisms belonging to the fungi and the protozoa, and in several bacterial groups, notably the lactic acid bacteria. The alcoholic fermentation is found in plant tissues and some fungi, notably yeasts.

The conversion of glucose to pyruvic acid in the alcoholic and homolactic fermentations involves a metabolic sequence known as the Embden-Meyerhof pathway, shown in Figure 6.5. Two molecules of ATP are expended in the initial reactions, to produce fructose-1,6-diphosphate. This is cleaved to yield triose phosphates, glyceraldehyde phosphate and dihydroxyacetone phosphate, which are freely interconvertible. The oxidation of triose phosphate, coupled with a reduction of NAD, is accompanied by an esterification of inorganic phosphate, to yield, from each $C_3$ moiety, a molecule of 1,3-diphosphoglyceric acid. The subsequent steps in the conversion of this compound to pyruvic acid permit a transfer of both phosphate groups to ADP, so that a total of 4 moles of ATP are formed per mole of glucose used. Since 2 moles of ATP are expended in the initial activation steps, the net ATP yield is 2 moles per mole of glucose fermented. A metabolic divergence between the two fermentations occurs at the level of

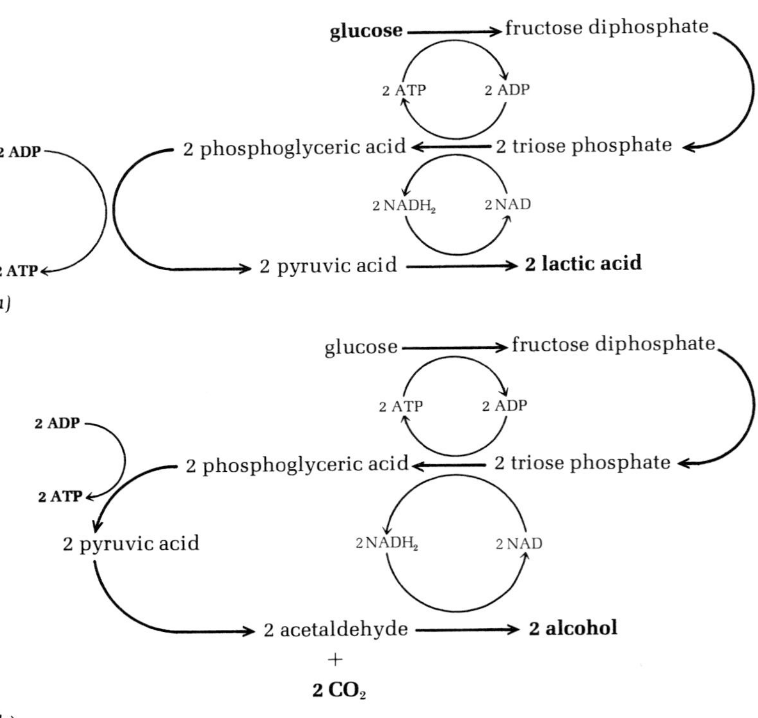


*Figure 6.6 A comparison between (a) lactic acid and (b) alcoholic fermentations.*

pyruvic acid. In the homolactic fermentation, $NADH_2$ is reoxidized by a direct reduction of pyruvic acid to lactic acid; in the alcoholic fermentation, pyruvic acid is first decarboxylated to $CO_2$ and acetaldehyde, and $NADH_2$ is reoxidized by a reduction of acetaldehyde to ethanol. Flow patterns of both fermentations are compared in Figure 6.6.

**Other carbohydrate fermentations using the Embden–Meyerhof pathway**

The Embden–Meyerhof pathway is the most widespread mechanism for the fermentative conversion of glucose to pyruvic acid, and it is employed by many groups of bacteria that produce fermentative end-products different from those characteristic of the alcoholic and homo-

*Table 6.1 Bacterial sugar fermentations which proceed through the Embden–Meyerhof pathway*

| Class of fermentations | Principal products from pyruvic acid | Bacterial group(s) where found |
|---|---|---|
| *(1) Homolactic* | $CH_3CHOHCOOH$<br>*(lactic acid)* | *Lactic acid bacteria of genera Streptococcus, Pediococcus, Lactobacillus (some species)* |
| *(2) Mixed acid* | $CH_3CHOHCOOH$<br>*(lactic acid)*<br>$CH_3COOH$ *(acetic acid)*<br>$HOOCOH_2CH_2COOH$<br>*(succinic acid)*<br>HCOOH *(formic acid)*<br>or $CO_2$ *and* $H_2$<br>$CH_3CH_2OH$ *(ethanol)* | *Many enteric bacteria (e.g., Escherichia, Salmonella, Shigella, Proteus, Yersinia)* |
| *(2a) Butanediol* | *As in 2, but also*<br>$CH_3(CHOH)_2CH_3$<br>*(2,3-butanediol)* | *Aerobacter, Serratia, Aeromonas, Bacillus polymyxa* |
| *(3) Butyric acid* | $CH_3CH_2CH_2COOH$<br>*(butyric acid)*<br>$CH_3COOH$ *(acetic acid)*<br>$CO_2$ *and* $H_2$ | *Many anaerobic sporeformers (Clostridium); some non-spore-forming anaerobes (Butyribacterium)* |
| *(3a) Butanol–acetone* | *As in (3), but also*<br>$CH_3CH_2CH_2CH_2OH$<br>*(butanol)*<br>$CH_3CH_2OH$ *(ethanol)*<br>$CH_3COCH_3$ *(acetone)*<br>$CH_3CHOHCH_3$<br>*(isopropanol)* | *Certain anaerobic sporeformers (Clostridium spp.)* |
| *(4) Propionic acid* | $CH_3CH_2COOH$<br>*(propionic acid)*<br>$CH_3COOH$ *(acetic acid)*<br>$HOOCCH_2CH_2COOH$<br>*(succinic acid)*<br>$CO_2$ | *Propionibacterium, Veillonella* |

Figure 6.7 *Derivations of some major end-products of the bacterial fermentations of sugars from pyruvic acid. The end-products are shown in boldface.*

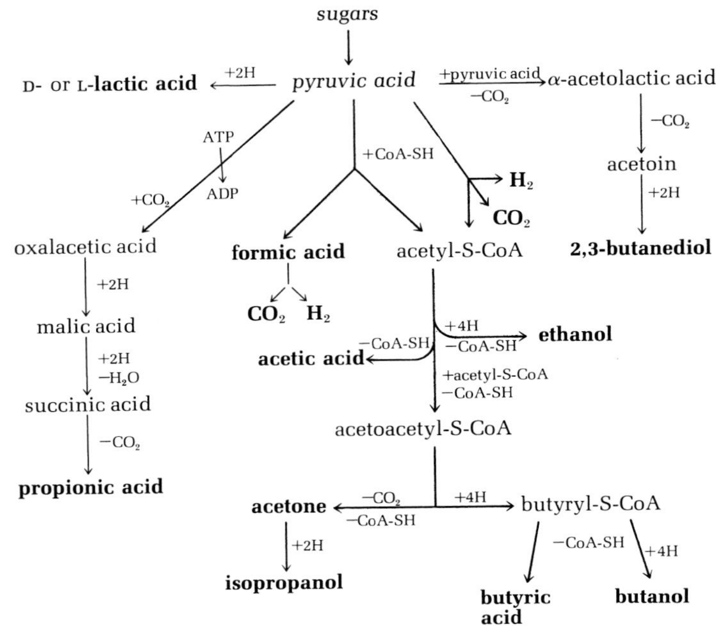


lactic fermentations (Table 6.1). *These differences reflect exclusively differences with respect to pyruvic acid metabolism* (Figure 6.7).

Many facultatively anaerobic bacteria, including *Escherichia, Salmonella, Shigella, Proteus, Yersinia,* and *Vibrio,* carry out a *mixed-acid fermentation.* In the case of *Escherichia coli,* the overall reaction can be approximately formulated as

$$2\ \text{glucose} + H_2O \rightarrow 2\ \text{lactate} + \text{acetate} + \text{ethanol} + 2CO_2 + 2H_2$$

In certain cases (e.g., *Shigella*), there is no formation of $CO_2$ and $H_2$, these end-products being replaced by an equivalent amount of an acidic end-product, formic acid (HCOOH). A characteristic feature of this fermentation is a *cleavage of pyruvic acid* involving CoA to yield acetyl CoA and formate. The acetyl CoA is converted to acetyl phosphate, half of which is used for the generation of additional ATP through the following reaction:

$$\text{acetyl-S-CoA} + P_i \rightarrow \text{CoA-SH} + \text{acetyl-}\text{Ⓟ} \xrightarrow{\text{ADP} \rightarrow \text{ATP}} \text{acetate}$$

The other half of the acetyl phosphate is reduced to ethanol:

$$\text{acetyl-}\text{Ⓟ} \xrightarrow{\text{NADH}_2 \rightarrow \text{NAD}} \text{acetaldehyde} \xrightarrow{\text{NADH}_2 \rightarrow \text{NAD}} \text{ethanol}$$

In organisms such as *E. coli* the formic acid produced by pyruvate cleavage is converted to $CO_2$ and $H_2$; in bacteria that do not possess formic hydrogenlyase, the enzyme responsible for this cleavage, formic acid is an end-product.

The conversion of glucose to acetic acid, ethanol, and formate (or $CO_2$ and $H_2$) is a balanced reaction, so these products could, in theory, be the sole end-products of this type of fermentation. However, all organisms that perform such a mixed-acid fermentation are also able to reduce pyruvic acid to lactic acid, and a substantial part of the pyruvic acid is always disposed of in this manner. The molar ratio of lactic acid to the other end-products can vary considerably, depending on the conditions of cultivation. A small amount of succinic acid is also often produced. Consequently, the quantitative relationships between the end-products of the mixed-acid fermentation are not fixed as they are in the alcoholic and homolactic fermentations.

Closely related in mechanism to the mixed-acid fermentation is the *butanediol fermentation*, carried out by such facultatively anaerobic bacteria as *Aerobacter*, *Serratia*, *Erwinia*, and certain species of *Bacillus* and *Aeromonas*. Besides the typical products of the mixed-acid fermentation, an additional end-product, 2,3-butanediol, is formed from pyruvate by a special series of reactions (Figure 6.8). The diversion

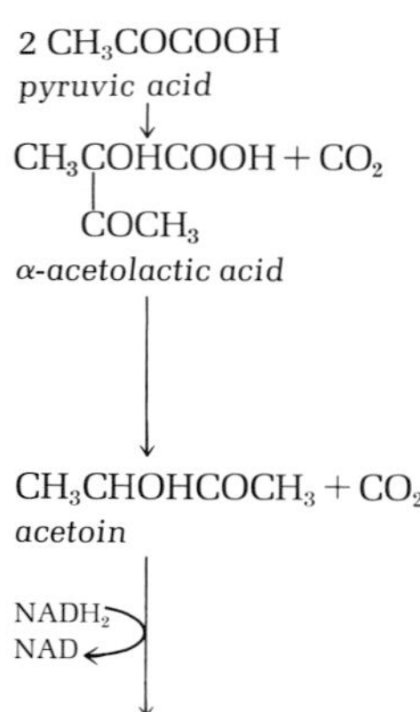


*Figure 6.8 Formation of 2,3-butanediol from pyruvic acid in bacterial fermentations of carbohydrates.*

of part of the pyruvate to 2,3-butanediol greatly reduces the total amount of acid produced relative to that produced in a mixed-acid fermentation. This reflects in part the fact that 2,3-butanediol is a neutral end-product, and in part the fact that its formation from 2 moles of pyruvic acid is accompanied by the reoxidation of only 1 mole of

$NADH_2$, so that the maintenance of the oxidation–reduction balance requires a greater reduction of acetyl phosphate to ethanol, with a consequent diminution in the amount of acetic acid formed.

Many of the anaerobic spore-forming bacteria of the genus *Clostridium* carry out a fermentation in which acetic and butyric acids, $CO_2$, and $H_2$ are the major end-products, the *butyric acid fermentation.* This fermentation results from a cleavage of pyruvate to yield directly acetyl-S-CoA, $H_2$, and $CO_2$:

$$\text{pyruvic acid} + \text{CoA-SH} \rightarrow \text{acetyl-S-CoA} + CO_2 + H_2$$

Part of the acetyl-S-CoA is converted to acetic acid with accompanying formation of ATP, while part is condensed to acetoacetyl-S-CoA, which is subsequently reduced to butyryl-S-CoA. The conversion of butyryl-S-CoA to butyric acid also permits ATP formation. In a butyric acid fermentation, accordingly, a considerable net generation of ATP, over and above that obtainable through the initial reactions of the Embden–Meyerhof pathway, is possible.

The clostridia and other strictly anaerobic bacteria contain a special electron carrier, *ferredoxin,* an iron-containing protein that has a very low redox potential and can thus mediate the utilization of molecular hydrogen. The presence of this carrier, which does not occur in facultative anaerobes, permits the clostridia to reutilize part of the $H_2$ formed

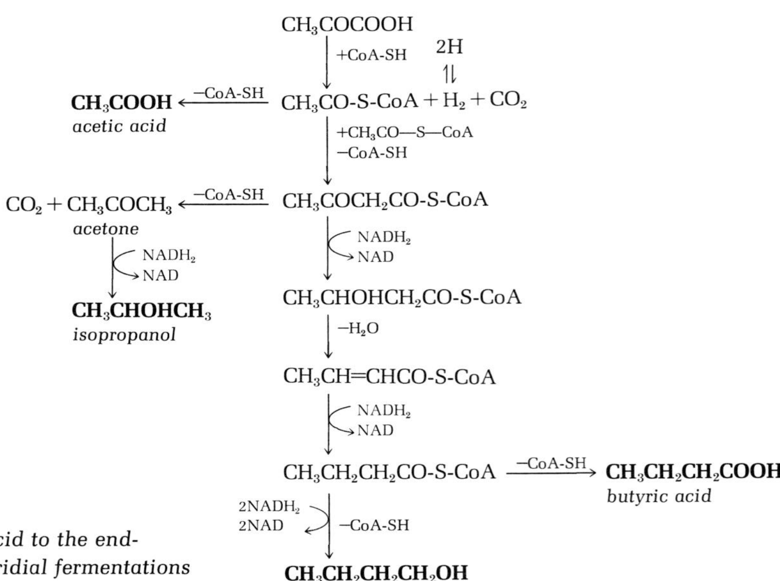


*Figure 6.9 Conversion of pyruvic acid to the end-products (boldface) formed in clostridial fermentations of the butyric acid and acetone–butanol types.*

in the initial cleavage of pyruvic acid. This results in the reduction of the primary acidic products of the butyric acid fermentation to neutral end-products: butanol, acetone, isopropanol, and ethanol (Figure 6.9). Oxidation–reduction balance is thus maintained by the use of molecular hydrogen as a reducing agent. As a result of the alternative biochemical possibilities, the end-products of these clostridial fermentations can be very numerous, and the ratios between them can vary widely. A clostridial fermentation in which neutral end-products predominate is sometimes referred to as an *acetone–butanol fermentation;* however, it is mechanistically only a minor variation on the butyric acid fermentation.

The *propionic acid fermentation* of sugars is carried out principally by a group of non-spore-forming bacteria, the members of the genus *Propionibacterium. Most members of this group can also ferment lactic acid, which is an end-product of other bacterial fermentations.* The principal products, either from sugars or from lactic acid, are propionic acid, acetic acid, and $CO_2$. The ability of these bacteria to grow fermentatively at the expense of lactate is a significant property, both physiologically and ecologically. It shows that the propionic acid fermentation permits a net generation of ATP through the breakdown of lactic acid, over and above that obtainable by the fermentation of glucose through the reactions of the Embden–Meyerhof pathway.

In the fermentation of lactic acid, the substrate is initially oxidized to pyruvic acid. Part of the pyruvic acid is further oxidized to $CO_2$ and acetyl-S-CoA. The energy-rich bond of acetyl-S-CoA can be used for the synthesis of ATP, by the mechanism already described in the context of the mixed-acid fermentation. The reduced pyridine nucleotide generated by the oxidation of 1 mole of lactic acid to acetic acid and $CO_2$ is reoxidized through the reduction of 2 moles of lactic acid to propionic acid, as shown in Figure 6.10.

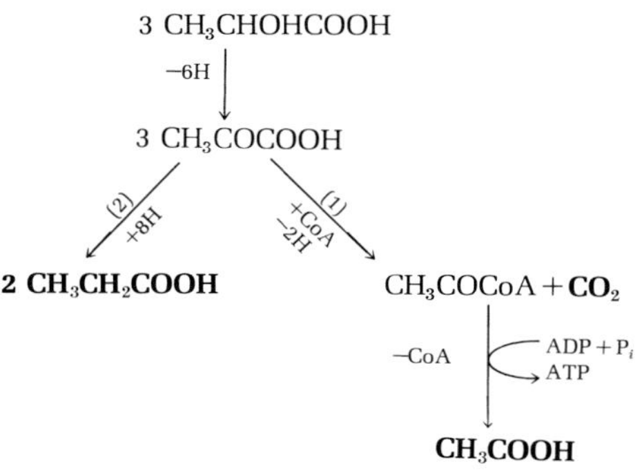


*Figure 6.10 A diagrammatic representation of the propionic acid fermentation of lactic acid. The detailed nature of the reactions involved in the conversion of pyruvic acid to propionic acid is shown in Figure 6.11. (End-products shown in boldface.)*

The formation of propionic acid from lactic acid is an indirect and complex process, involving six reactions, in two of which a biotin-containing enzyme capable of a reversible binding of $CO_2$ participates:

lactic acid → pyruvic acid + 2H
pyruvic acid + enzyme-biotin-$CO_2$ → oxalacetic acid + enzyme-biotin
oxalacetic acid + 4H → succinic acid + $H_2O$
succinic acid + propionyl-S-CoA → succinyl-S-CoA + propionic acid
succinyl-S-CoA → methylmalonyl-S-CoA
methylmalonyl-S-CoA + enzyme-biotin →
propionyl-S-CoA + enzyme-biotin-$CO_2$

The sum of these reactions gives the deceptively simple equation

lactic acid + 2H → propionic acid + $H_2O$

The reactions that lead to propionic acid formation are cyclic, as shown in Figure 6.11.

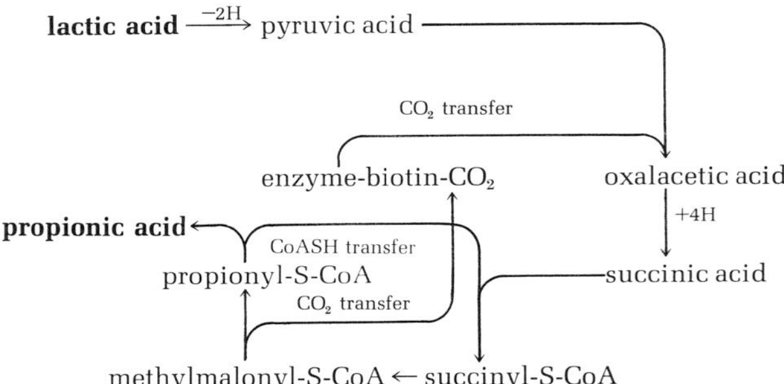


*Figure 6.11 The reactions responsible for the formation of propionic acid from lactic acid, showing the cyclic transfers of* CoA-SH *and* $CO_2$.

With glucose as a substrate, the products of the propionic acid fermentation are the same, since glucose and lactic acid are on the same oxidation level. The bacteria obtain the meager yield of 1 mole of ATP for every 3 moles of lactic acid fermented; the ATP yield resulting from glucose fermentation is, of course, considerably greater, as a result of ATP formation through the reactions of the Embden–Meyerhof pathway.

**Alternative pathways: the hexose monophosphate shunt**

In certain bacterial fermentations, the initial phosphorylation of glucose is followed immediately by an *oxidation of glucose-6-phosphate* coupled with a reduction of NAD, to yield 6-phosphogluconic acid:

glucose-6-phosphate + NAD → 6-phosphogluconate + $NADH_2$

Although pyruvate is ultimately formed in these fermentations, the mechanism of the reactions that lead to the production of $C_3$ intermediates is different from that in the Embden–Meyerhof pathway. This mechanistic difference has an important effect on energetic yield. *When glucose is fermented through the hexose monophosphate pathway, the net yield of ATP is half that characteristic of the Embden–Meyerhof pathway: 1 mole per mole of glucose converted to pyruvate.*

The most common bacterial fermentation of this type is the *heterolactic fermentation*, characteristic of many lactic acid bacteria. The products formed from glucose are lactic acid, ethanol, and $CO_2$, in a strict molar ratio of 1 : 1 : 1.

glucose → lactic acid + ethanol + $CO_2$

As shown in Figure 6.12, the molar equivalence of the three end-products is a consequence of the mechanism of 6-phosphogluconate

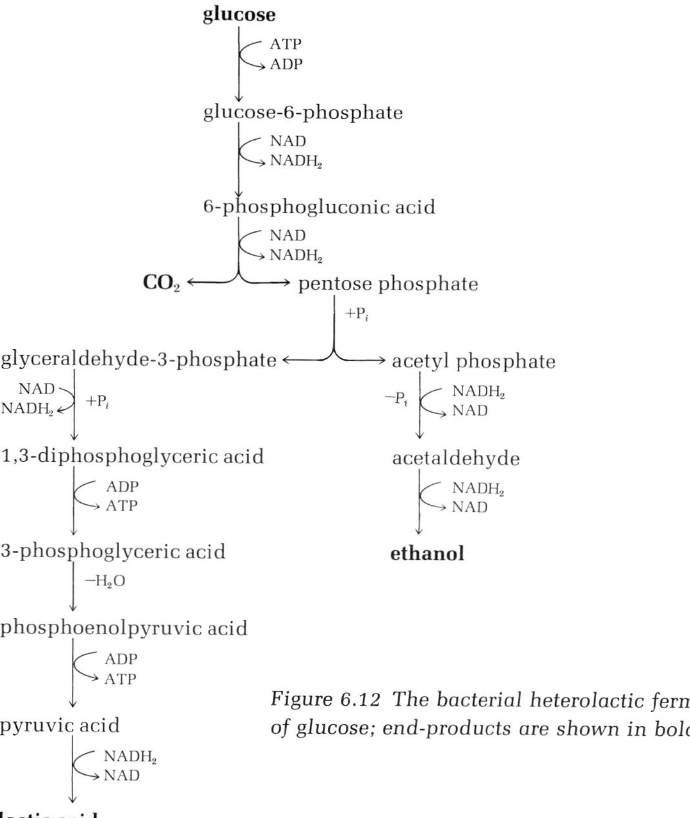


*Figure 6.12 The bacterial heterolactic fermentation of glucose; end-products are shown in boldface.*

breakdown. The $CO_2$ is derived from an oxidative decarboxylation of this intermediate, which yields pentose phosphate. The pentose phosphate is cleaved, with an uptake of inorganic phosphate, to acetyl phosphate and glyceraldehyde-3-phosphate. Ethanol arises from acetyl phosphate by two successive reductions, which balance the two oxidations involved in the conversion of glucose-6-phosphate to pentose phosphate and $CO_2$. The glyceraldehyde-3-phosphate is converted through later reactions of the Embden-Meyerhof pathway to pyruvic acid and thence to lactic acid. Although two energy-rich phosphate bonds are generated in this sequence, the net yield of ATP is only 1 mole per mole of glucose, because the energy-rich bond of acetyl phosphate is lost during the reduction of this intermediate to ethanol.

A different variation of the hexose monophosphate pathway, known as the Entner–Doudoroff pathway, occurs in the alcoholic fermentation of glucose performed by the bacterium *Zymomonas lindneri* (Figure 6.13). In this case, 6-phosphogluconate is dehydrated and then cleaved

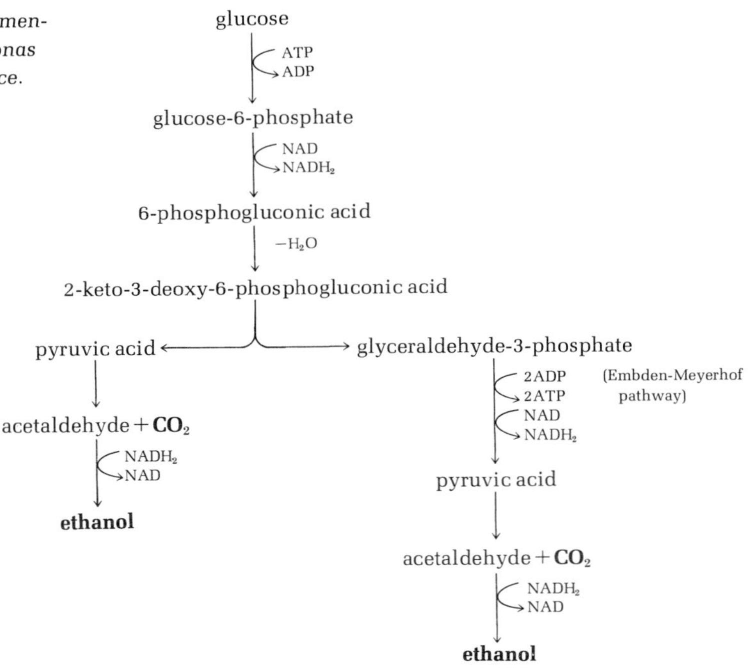


*Figure 6.13 Mechanism of the alcoholic fermentation of glucose by the bacterium Zymomonas lindneri; end-products are shown in boldface.*

to yield one molecule of glyceraldehyde-3-phosphate and one molecule of pyruvate, from which ethanol and $CO_2$ are formed through the same series of reactions as in alcoholic fermentation by yeast.

**Alternative pathways: the $C_2$–$C_4$ split**

The bacteria of the genus *Bifidobacterium* ferment glucose with the formation of a mixture of lactic and acetic acids. The mechanism of this fermentation has only recently been clarified (Figure 6.14). Following

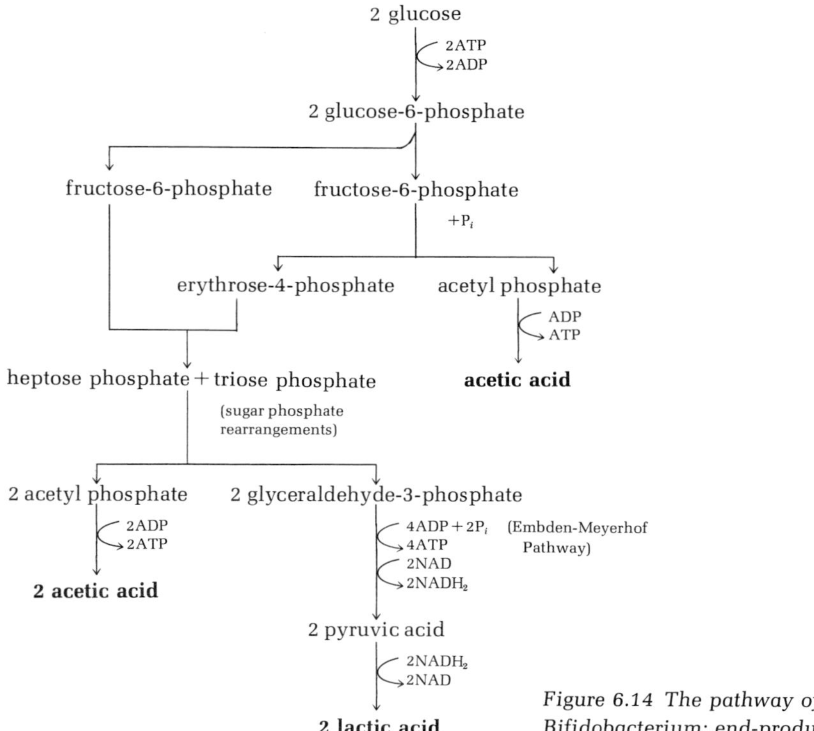


*Figure 6.14 The pathway of glucose fermentation by Bifidobacterium; end-products are shown in boldface.*

phosphorylation, glucose is converted to fructose-6-phosphate, as in the Embden–Meyerhof pathway; this intermediate is then immediately cleaved, with an uptake of inorganic phosphate, to yield acetyl phosphate and a 4-carbon sugar, erythrose-4-phosphate. Through a complex series of rearrangements initiated by the reaction of erythrose-4-phosphate with another molecule of fructose-6-phosphate, glyceraldehyde-3-phosphate and more acetyl phosphate are formed. The glyceraldehyde-3-phosphate is converted to lactic acid and the acetyl phosphate to acetic acid. The overall equation for this fermentation is

2 glucose → 3 acetic acid + 2 lactic acid

Energetically, it is slightly more favorable than the homolactic fermentation, since a total of 5 moles of ATP can be formed for every 2 moles of glucose fermented.

### The fermentations of nitrogenous organic compounds

The anaerobic spore-forming bacteria of the genus *Clostridium* display a large number of variations on the theme of fermentative metabolism. In addition to the butyric acid fermentation of sugars, already discussed, many clostridia can ferment amino acids, these fermentations being used either as the exclusive means of obtaining energy or in some species as an alternative to sugar fermentation.

The most widespread type of amino acid fermentation among clostridia is the *Stickland reaction*, which can be defined as a *coupled oxidation–reduction involving a pair of amino acids, one of which serves as electron donor and the other as electron acceptor*. A typical example is the fermentation of alanine and glycine to yield ammonia, acetic acid, and $CO_2$:

$$\text{alanine} + 2\ \text{glycine} + 2H_2O \rightarrow 3\ \text{acetic acid} + 3NH_3 + CO_2$$

In this particular case, alanine serves as the electron donor, being oxidized via pyruvic acid to $CO_2$ and acetyl CoA, from which ATP can be formed by the mechanism already described. This reaction is coupled with a transfer of electrons to NAD. The $NADH_2$ is reoxidized by reduction of two molecules of glycine to acetic acid and ammonia (Figure 6.15).

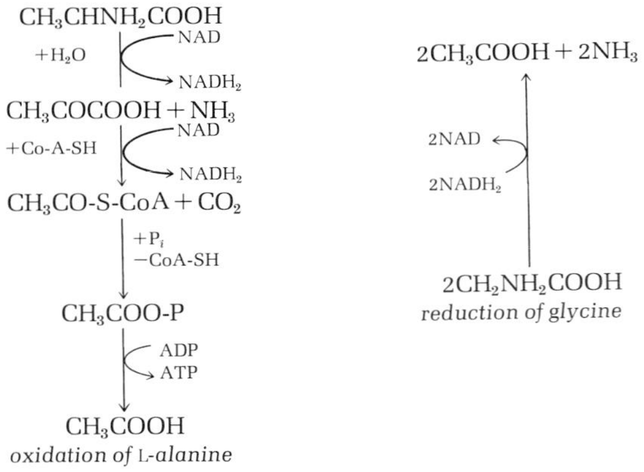

$CH_3CHNH_2COOH$
$+H_2O$ ↓ NAD → $NADH_2$
$CH_3COCOOH + NH_3$
$+$Co-A-SH ↓ NAD → $NADH_2$
$CH_3CO\text{-}S\text{-}CoA + CO_2$
↓ $+P_i$, $-$CoA-SH
$CH_3COO\text{-}P$
↓ ADP → ATP
$CH_3COOH$
*oxidation of* L-*alanine*

$2CH_2NH_2COOH$
↑ $2NADH_2$ → 2NAD
$2CH_3COOH + 2NH_3$
*reduction of glycine*

*Figure 6.15 Mechanism of the Stickland reaction as a mechanism of fermentative energy-yielding metabolism in clostridia. In this illustration,* L*-alanine serves as the electron donor and glycine as the electron acceptor; both are converted to acetic acid.*

Clostridia can also ferment single amino acids. Such reactions are exemplified by the fermentation of threonine, to yield propionic acid:

$$CH_3CHOHCHNH_2COOH + H_2O \rightarrow CH_3CH_2COOH + H_2 + CO_2 + NH_3$$

and by the fermentation of glutamate, to yield butyric and acetic acids:

$$2HOOCCH_2CH_2CHNH_2COOH + 2H_2O \rightarrow CH_3CH_2CH_2COOH + 2CH_3COOH + 2CO_2 + 2NH_3$$

Certain clostridia and other anaerobic bacteria can obtain energy by the fermentation of ring compounds, such as uric acid, uracil, allantoin, and nicotinic acid.

**The ethanol-acetic acid fermentation**

The most remarkable clostridial fermentation is that performed by *Clostridium kluyveri,* which obtains the energy for growth by a synthesis of fatty acids at the expense of ethanol and acetic acid. When acetic acid is in excess, butyric acid is the main product:

$$CH_3CH_2OH + CH_3COOH \rightarrow CH_3CH_2CH_2COOH + H_2O$$

When ethanol is in excess, the main end-product is caproic acid:

$$2CH_3CH_2OH + CH_3COOH \rightarrow CH_3CH_2CH_2CH_2CH_2COOH + 2H_2O$$

The biochemical mechanism of butyric acid formation is illustrated in Figure 6.16.

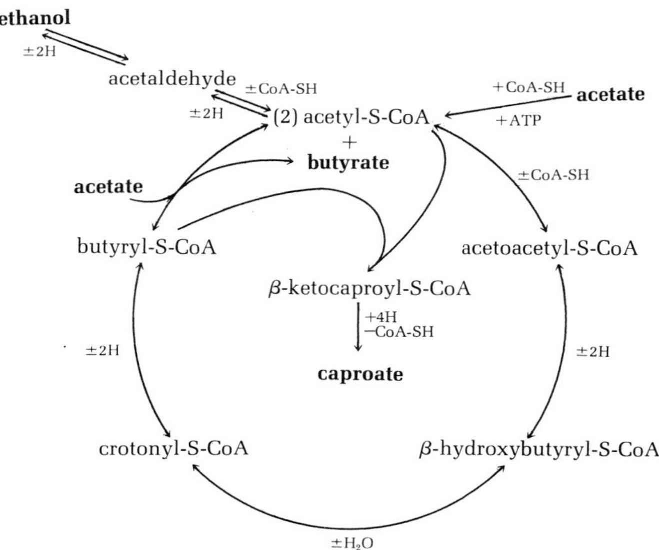


*Figure 6.16 Fatty acid synthesis in Clostridium kluyveri.*

## *The biochemistry of aerobic respiration*

Although the oxidation of the substrate is coupled with a reduction of molecular oxygen in aerobic respiration, such coupling is very rarely a *direct* one. Instead, electrons removed from the substrate are passed through a series of intermediate carriers, capable of undergoing freely reversible oxidation and reduction, until the last carrier of the series reacts in its reduced state with oxygen, a reaction mediated by a terminal oxidase. As a result of this terminal oxidation, which is irreversible, the whole chain of carriers is reoxidized, and oxygen is reduced to water. The sequence of carriers that mediates between the oxidation

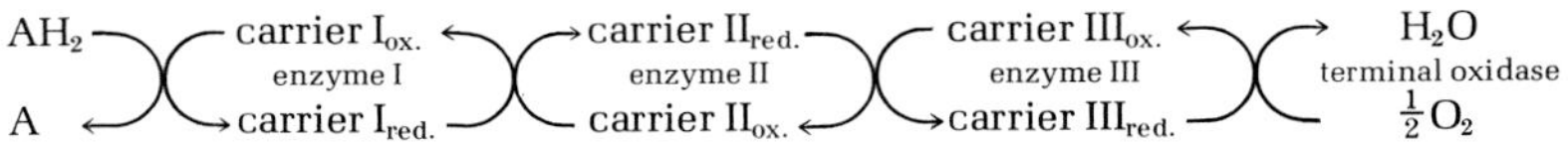


*Figure 6.17 A schematic representation of a respiratory electron-transport system, linking the dehydrogenation of an oxidizable substrate, $AH_2$, with the reduction of molecular oxygen to water.*

of the substrate and the reduction of water is termed the *respiratory electron transport chain.* The operation of such a transport chain is shown in Figure 6.17.

**Intracellular location of the respiratory electron transport chain**

Respiratory electron transport chains are very complex. They comprise a large number of distinct enzymes and electron carriers, physically associated in a rigid matrix having a high lipid content. This multicomponent system is contained in unit membranes. In eucaryotic cells, these unit membranes are located within mitochondria; in procaryotic cells, they are located in the cell membrane or its invaginations (see Chapter 10, p. 351). When the cells of an aerobic bacterium are disrupted, the fragments of the originally continuous membrane system can then be isolated as a particulate fraction by centrifugation of the extract.

The biochemical analysis of respiratory electron transport chains is difficult because of their physical structure. Since most of their components are insoluble in water, they cannot be easily separated from one another in a functional state. Most of our present knowledge on this subject has been derived from work with the mitochondrial electron transport chain, which is remarkably uniform in all eucaryotic groups. The nature of the electron transport chains found in different bacteria is much more varied, and the components of these chains often differ markedly from those characteristic of mitochondrial systems, but none of the bacterial systems has as yet been really well characterized.

### The physiological function of electron transport chains

When the removal of a pair of electrons or hydrogen atoms from an organic substrate is coupled with the reduction of oxygen to water, there is a large free-energy change, often approximately equivalent to that resulting from the combustion of a molecule of hydrogen gas. *The passage of these electrons through the transport chain permits a stepwise release of this energy, some of which can be converted into the bond energy of ATP.* To amplify this statement it is necessary to introduce the concept of *oxidation–reduction potentials.*

The tendency of a substance to donate or accept electrons (i.e., to undergo oxidation or reduction) can be described quantitatively in terms of its *electrode potential* or *oxidation–reduction potential.* This is measured relative to a *standard potential,* that of the hydrogen electrode,

$$\tfrac{1}{2}H_2 \rightleftharpoons H^+ + e^-$$

assigned an arbitrary value of zero volts at a pH of zero. At pH 7, near which most biological oxidation–reductions take place, the standard potential of the hydrogen electrode is −0.42 V. The symbol $E'_0$ designates an electrode potential in volts, measured at pH 7.

When the $E'_0$ values of two half reactions are known, the free-energy change resulting from the coupled reaction can be calculated from the following relationship:

$$\text{free-energy change} = -nF\Delta E'_0$$

where $n$ is the number of electrons transferred; $F$ the faraday, a physical constant equal to 23,000 cal/V; and $\Delta E'_0$ the algebraic difference between the potentials of the two half reactions. Thus the combustion of hydrogen involves the two half reactions.

$$H_2 \rightleftharpoons 2H^+ + 2e^- \qquad (E'_0 = -0.42)$$

and

$$\tfrac{1}{2}O_2 \rightleftharpoons O^{2-} - 2e^- \qquad (E'_0 = +0.82)$$

It can be readily calculated from the above equation that the free-energy change involved is −57,000 cal.

For a typical biological oxidation, the oxygen-linked oxidation of malate to pyruvate and $CO_2$ ($E'_0$ = −0.33), the analogous calculation shows a free-energy change of −53,000 cal, not significantly different from that for the oxidation of hydrogen.

*The carriers in a respiratory electron transport chain participate in a series of reactions of gradually increasing $E'_0$ values, interposed be-*

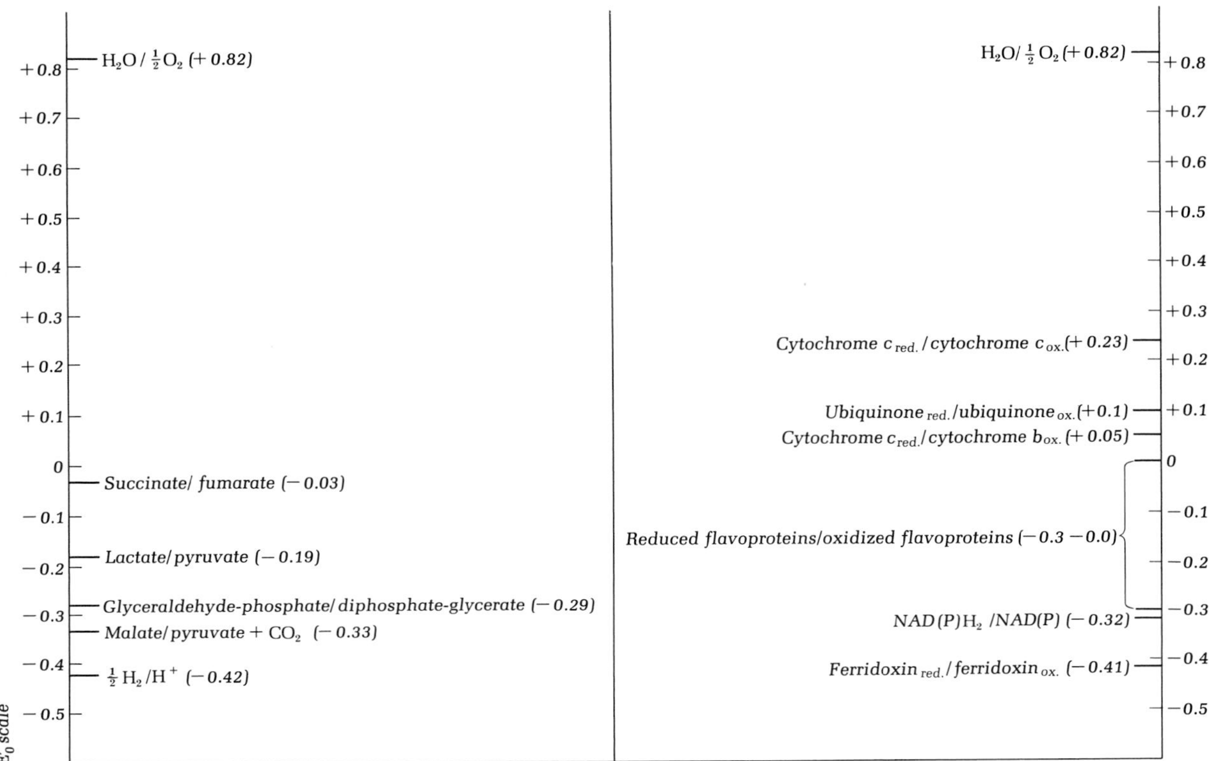


*Figure 6.18 The $E_0'$ values for various half-reactions involved in substrate dehydrogenations (left) and the electron transport (right).*

*tween the initial substrate dehydrogenation, which has as a rule a negative $E'_0$ value, and the terminal reduction of oxygen, which has a highly positive $E'_0$ value.* The position on the $E'_0$ scale of several typical respiratory carriers, relative to substrate dehydrogenations and oxygen reduction, is shown in Figure 6.18.

**The major components of the transport chain**

Every respiratory transport chain so far examined contains three different classes of molecules that participate in the transport process. Two classes consist of enzymes with firmly bound prosthetic groups capable of undergoing oxidation and reduction: the *flavoproteins* and the *cytochromes*. The third class consists of nonprotein carriers of relatively low molecular weight, the *quinones*.

Flavoproteins are enzymes that contain a yellow prosthetic group, derived biosynthetically from the vitamin riboflavin [Figure 6.19(*a*)]. The prosthetic group may be either flavin mononucleotide (FMN) or flavin adenine dinucleotide (FAD); both possess the same active site,

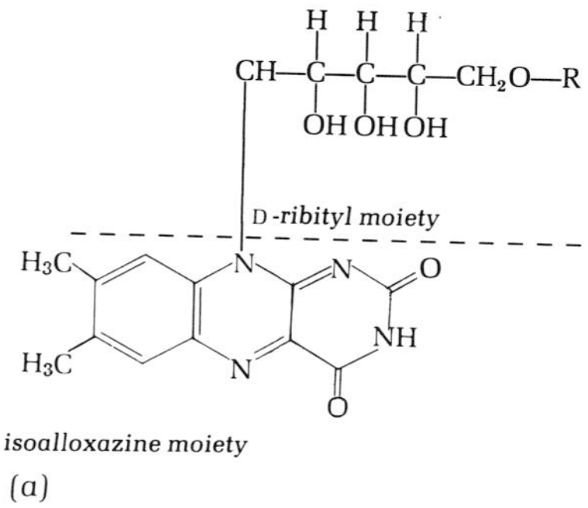

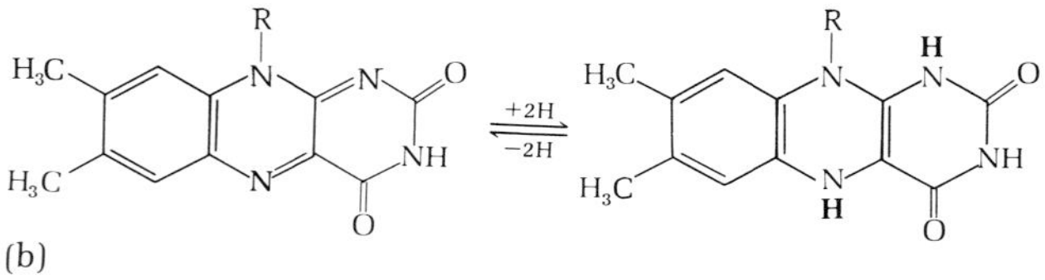

Figure 6.19 *(a) Structures of the vitamin riboflavin (R is H) and of its two coenzyme derivatives, FMN (R is $PO_3H_2$) and FAD (R is ADP). (b) The reversible oxidation and reduction of the ring structure of FMN and FAD.*

capable of undergoing reversible oxidation and reduction [Figure 6.19(*b*)]. The flavoproteins are a large class of enzymes, which differ widely with respect to their $E'_0$ values, and mediate many diverse oxidation–reductions. Some are active in the primary dehydrogenation of organic substrates; an example is succinic dehydrogenase, which mediates the oxidation of succinic acid to fumaric acid. Others act by accepting electrons from reduced pyridine nucleotides formed in a primary substrate dehydrogenation. The acceptors for the reoxidation of these enzymes also vary. Some reduced flavoproteins are autoxidizable by molecular oxygen, a reaction accompanied by the formation of hydrogen peroxide:

$$\text{reduced flavoprotein} + O_2 \rightarrow \text{oxidized flavoprotein} + H_2O_2$$

Hydrogen peroxide is a highly toxic substance, and most aerobic organisms contain an enzyme, *catalase*, which decomposes the peroxide formed through such reactions to water and oxygen:

$$H_2O_2 \rightarrow H_2O + \tfrac{1}{2}O_2$$

The flavoproteins associated with the respiratory electron transport chain do not, however, transfer electrons directly to oxygen; their reoxidation is mediated by subsequent carriers of the chain, either quinones or cytochromes.

In the mitochondrial electron transport system, the major quinone is ubiquinone (Figure 6.20); a diversity of other quinones occur in bac-

Figure 6.20 *(a) Structure of ubiquinone. The number of isoprenoid units*

$$—(CH_2—CH{=}C(CH_3)—CH_2)—$$

*in the side chain varies in different organisms from 6 to 10. (b) Reversible oxidation and reduction of a quinone.*

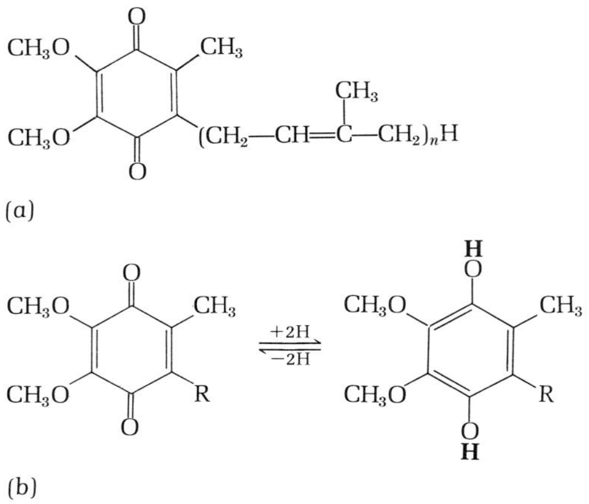

terial electron transport systems. The roles of the quinones as carriers in the chain are in most cases not precisely determined, but in some systems they appear to act as carriers intermediate between the flavoproteins and the cytochromes.

The cytochromes are enzymes that belong to the class of heme proteins, which also includes hemoglobin and catalase. All heme proteins have a prosthetic group derived from *heme,* a cyclic tetrapyrrole with an atom of iron chelated within the ring system (Figure 6.21). Electron

Figure 6.21 *Structure of heme.*

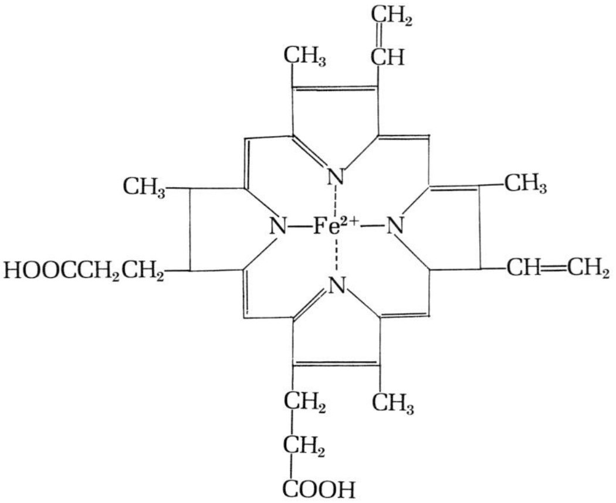

transfer by the cytochromes involves a reversible oxidation and reduction of this iron atom:

$$Fe^{2+} \rightleftharpoons Fe^{3+} + e^-$$

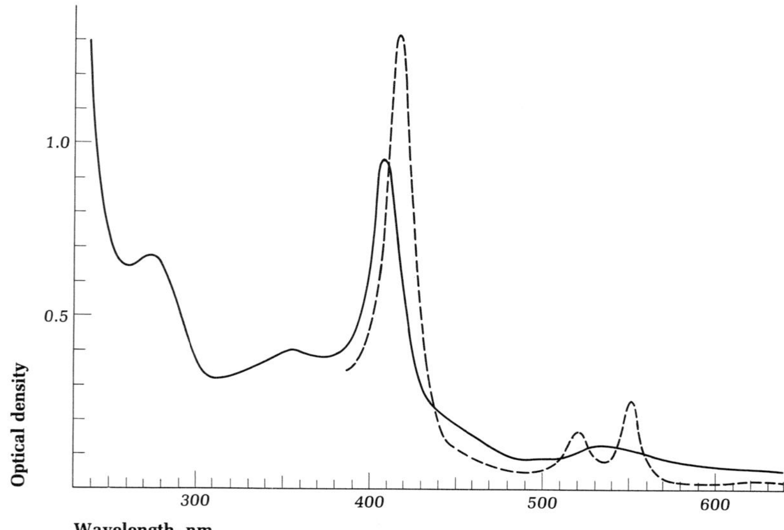


*Figure 6.22 The absorption spectrum of cytochrome c. Solid line: oxidized cytochrome c; dotted line; reduced cytochrome c.*

The cytochromes have characteristic absorption bands in the reduced state (Figure 6.22), and their spectral and functional properties permit the recognition of several different members of the class in electron transport chains, distinguished by a terminal letter (e.g., cytochrome c). Typically, the transport sequence is

$$\text{cytochrome b} \rightarrow \text{cytochrome c} \rightarrow \text{cytochrome a} \xrightarrow[\text{oxidase}]{\text{cytochrome}} O_2$$

The carrier chain of a respiratory electron transport system can accordingly be schematized as shown in Figure 6.23. It must be emphasized, however, that many variations on this scheme exist among the bacteria.

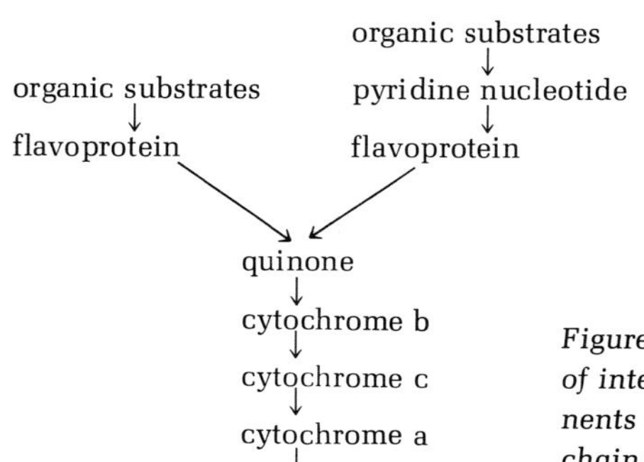


*Figure 6.23 A diagrammatic representation of interrelationships among the components of a respiratory electron transport chain. The arrows denote the flow of electrons.*

**Oxidative phosphorylation**

The generation of ATP associated with the transport of electrons through the carrier chain to oxygen is termed *oxidative phosphorylation*, to distinguish it from the *substrate-level phosphorylations*, which are the only mechanisms of ATP synthesis in fermentative metabolic processes, and from the *photophosphorylations* characteristic of photosynthetic energy generation. The precise mechanism and location of oxidative phosphorylations still remain unknown. However, the evidence suggests that there are three sites in the respiratory transport chain at which ATP synthesis can occur: in the reoxidation of pyridine nucleotides by flavoproteins, in electron transfer from flavoproteins to cytochrome c, and in electron transfer from cytochrome c to oxygen. The transfer of a pair of electrons from an organic substrate to oxygen can, accordingly, give rise to two molecules of ATP if the initial dehydrogenation is effected by a flavoprotein, and to three molecules of ATP if the dehydrogenation is mediated by a pyridine nucleotide-linked dehydrogenase.

**The uncoupling of respiration and oxidative phosphorylation**

Certain synthetic dyes can serve as *artificial electron acceptors*. When such compounds are added to a suspension of respiring microorganisms, electrons derived from the oxidizable substrate are shunted from the normal transport chain to the artificial acceptor, and oxidative phosphorylation is therefore reduced. An example of an artificial electron acceptor is methylene blue, which is rapidly reduced in the absence of air by suspensions of cells to its colorless, hydrogenated form:

$$MB + 2H \rightarrow MBH_2$$

Reduced methylene blue is autoxidizable by molecular oxygen, as shown here:

$$MBH_2 + O_2 \rightarrow MB + H_2O_2$$

With this dye, the shunt occurs at the level of the flavoproteins. The addition of methylene blue to a suspension of respiring cells thus creates an *artificial bypass* of the cytochrome system (Figure 6.24).

Another group of compounds, represented by *dinitrophenol*, can interfere with oxidative phosphorylation in a different manner. In the

*Figure 6.24 The effect of methylene blue in diverting the flow of electrons through the respiratory electron transport system from cytochromes. Arrows denote directions of electron flow.*

cytochromes $\rightarrow O_2 \rightarrow H_2O$
↗
substrate $\rightarrow$ DPN $\rightarrow$ flavoproteins
↘
methylene blue $\rightarrow O_2 \rightarrow H_2O_2$

presence of one of these compounds, electron transport still proceeds through the normal carrier chain, but the specific reactions responsible for ATP synthesis are inhibited.

When organic substrates are oxidized by intact microbial cells, the ATP generated during electron transport is immediately used for biosynthesis; consequently, a substantial fraction of the organic substrate carbon is assimilated and converted to cell material. The fraction of the substrate so assimilated during microbial respiration ranges between 20 and 50 percent, varying both with the organism and the substrate examined. As a result, only part of the substrate carbon is liberated in the form of carbon dioxide or other oxidized products. However, when oxidative phosphorylation is prevented by one of the artificial agents described above, the ATP required for assimilation is no longer produced, and the organic substrate may be almost completely oxidized to $CO_2$. Treatment of cells with these agents therefore greatly increases the total oxygen consumption at the expense of the organic substrate provided, as shown for dinitrophenol in Figure 6.25.

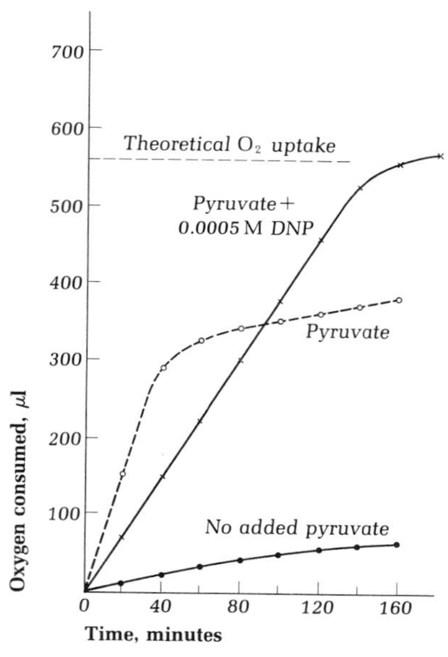


*Figure 6.25 The effect of 0.0005 M dinitrophenol on the extent of oxygen uptake by resting cells of Escherichia coli respiring 10 micromoles of pyruvate. In the absence of dinitrophenol, the total oxygen uptake is only about 60 percent of the theoretical amount (dashed line) required for the complete oxidation of this quantity of pyruvate, since much of the pyruvate is assimilated by the cells. The addition of 0.0005 M dinitrophenol increases total oxygen uptake to the theoretical level, since assimilation is prevented as a result of the inhibition of oxidative phosphorylation. Redrawn from C. E. Clifton and W. A. Logan, J. Bacteriol.* **37**, *523 (1939).*

**Bacterial cytochrome systems**

The remarkable diversity of bacterial respiratory electron transport systems, in contrast to the essential uniformity of mitochondrial ones, is revealed by the comparison of the cytochrome absorption spectra of

*Table 6.2 Wavelengths (nm) of the main cytochrome absorption bands in mitochondrial and bacterial electron transport systems*

| | α bands | | | | | Principal | Principal |
|---|---|---|---|---|---|---|---|
| | a[a] | | | b[a] | c[a] | β bands | Soret bands |
| *Mitochondrial systems* | | | | | | | |
| *Yeast* | 605 | — | — | 563 | 552 | 525 | 426 |
| *Mammalian heart muscle* | 605 | — | — | 563 | 552 | 525 | 426 |
| *Bacterial systems* | | | | | | | |
| *Bacillus subtilis* | 604 | — | — | 564 | 552 | 523 | 422 |
| *Micrococcus luteus* | 605 | — | — | 562 | 552 | 523 | 430 |
| *Micrococcus lysodeikticus* | — | 600 | — | — | 552 | 520 | 432 |
| *Pseudomonas acidovorans* | — | 600 | — | — | 553 | 530 | 425 |
| *Pseudomonas aeruginosa* | — | — | — | 560 | 552 | 530 | 428 |
| *Pseudomonas putida* | — | — | — | 560 | 552 | 530 | 427 |
| *Gluconobacter suboxydans* | — | — | — | — | 554 | 523 | 428 |
| *Acetobacter pasteurianum* | — | — | 588 | — | 554 | 523 | 428 |
| *Aerobacter aerogenes* | 628 | — | 592 | 560 | — | 530 | 430 |
| *Escherichia coli* | 630 | — | 593 | 560 | — | 533 | 433 |
| *Proteus vulgaris* | 630 | — | 595 | 560 | — | 533 | 430 |
| *Pseudomonas maltophilia* | 628 | — | 597 | 558 | — | 530 | 430 |

[a]Designates type of cytochrome in mitochondrial system to which each α band corresponds.

intact bacterial cells and of mitochondria (Table 6.2). By measuring the absorption spectrum of the reduced system (anaerobic suspension) relative to that of the oxidized system (aerobic suspension), it is possible to identify the characteristic absorption bands of the various cytochrome components in the reduced states. The so-called α bands, at longest wavelengths, are particularly characteristic; the shorter β bands and Soret bands are generally composite ones, to which several pigments contribute.

As Table 6.2 shows, the mitochondria of mammalian heart muscle and of yeast (at opposite ends of the scale of eucaryotic biological complexity) have identical cytochrome spectra. The cytochrome spectra of a few bacteria (e.g., *Bacillus subtilis* and *Micrococcus luteus*) are similar to mitochondrial ones. Even in these cases, however, there are functional differences, since the cell-free bacterial systems are unable to oxidize effectively mammalian cytochrome c. In most aerobic bacteria, few if any of the main cytochrome peaks coincide with those of the mitochondrial system.

It will be noted that a number of the bacterial species for which data are given in Table 6.2 show no peak at 552 to 554 nm, characteristic of cytochromes of the c type. The lack of a c-type cytochrome in such organisms is correlated with the outcome of an empirical procedure, the *oxidase test*, which has considerable diagnostic importance in the identification of aerobic bacteria. The test is performed by putting a patch of bacteria on a piece of filter paper soaked in a solution of dichlorophenol indophenol or N,N-dimethyl-*para*-phenylenediamine, which are rapidly converted to oxidized colored products by "oxidase-positive" species (containing a cytochrome c) but not by "oxidase-negative" species (lacking a cytochrome c). Certain groups of aerobic bacteria, notably the enteric group (*Escherichia coli* and related forms), are characteristically oxidase negative. In other groups (e.g., the *Pseudomonas* and *Moraxella* groups) some species are oxidase positive and others are oxidase negative.

## Bioluminescence

*Bioluminescence*, the ability of an organism to emit light, is a property of wide but sporadic occurrence in the living world. It is possessed by certain bacteria, fungi, and dinoflagellates among protists; by many different kinds of invertebrates; and by some fishes. Light emission by animals may be an intrinsic property of the species or may be caused by symbiotic luminous bacteria, harbored in special organs (see Chapter 24). Since the light produced is confined to a fairly narrow band of wavelengths in the visible spectrum, it is always colored, the color being characteristic for the emitting organism: thus, fungi produce a yellow-green luminescence (maximum near 530 nm), whereas bacterial luminescence is blue-green (maximum near 490 nm). Bacterial and fungal light emission is always continuous, whereas in dinoflagellates and in some animals it occurs in flashes.

Some simple organic molecules can be treated in a manner that causes them to emit light, a phenomenon termed *chemiluminescence*. Study of these model systems shows that the light-emitting molecule, A, must be brought into an electronically excited state, with a high potential energy, $A^*$; part of the excitation energy is then liberated as a photon ($h\nu$), accompanied by a return of the molecule to its ground state:

*Figure 6.26 Luminous bacteria photographed by their own light: left, a streaked plate of Photobacterium phosphoreum; right, two flasks containing a suspension of the same organism in a sugar medium. A stream of air was passed continuously through the flask on the right during the photographic exposure. The bacteria in the unaerated flask on the left had exhausted the dissolved oxygen and had ceased to luminesce except at the surface, where organisms were exposed to the air.*

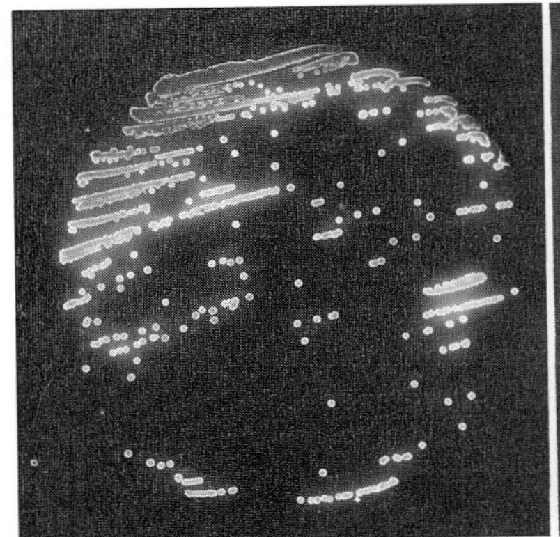

$$A \xrightarrow{\text{excitation}} A^*$$
$$A^* \longrightarrow A + h\nu$$

The same principle underlies bioluminescence. The biochemical study of cell-free systems from many different organisms has shown, however, that the mechanisms of light emission are extremely varied; it is probable that this striking functional property has been evolved independently in the many different biological groups where it exists. In bacteria and fungi, it is always associated with a consumption of molecular oxygen; and light emission accordingly occurs only under aerobic conditions (Figure 6.26). In fact, light emission by a suspension of luminous bacteria provides one of the most sensitive known methods for detection of traces of dissolved oxygen. We shall describe here the mechanism of bacterial luminescence (Figure 6.27).

$$\begin{array}{l} FMNH_2 + \text{enzyme} \\ \downarrow \\ FMNH_2 \cdot \text{enzyme} \\ \quad \downarrow + \tfrac{1}{2}O_2 \\ \text{enzyme}^* + FMN \\ \quad \downarrow \text{long-chain aldehydes} \\ \text{enzyme} + \text{photon} \end{array}$$

*Figure 6.27 A schematic diagram of the series of reactions involved in bacterial luminescence.*

The specific sequence of reactions is initiated by a reduction of free (not enzyme-bound) flavin mononucleotide by $NADH_2$. Reduced FMN then reacts with an enzyme, *luciferase*. The resulting complex is oxidized by molecular oxygen with formation of oxidized FMN and excited luciferase: in other words, some of the energy of the reduced FMN is not liberated as heat but is conserved in the excited luciferase molecule. The return of luciferase to its ground state is accompanied by emission of a photon. A peculiar feature of this return reaction, as studied in bacterial cell-free extracts, is the dependence of light emission on the presence of a long-chain aldehyde, such as palmityl aldehyde. If this component is absent, the return of excited luciferase to the ground state gives a light yield some thousand times less, most of the energy of excitation being liberated as heat. It is not known whether long-chain aldehydes play such a role in the intact bacterial cell, or whether some other chemical substance is the natural promoter of bacterial light emission. The aldehydes do not appear to be substrates for luciferase; their function has been interpreted as the maintenance of the excited enzyme in a conformation that is highly favorable for light emission.

Bacterial luminescence can be regarded in physiological terms as a *minor natural bypass of the normal bacterial respiratory electron transport chain*, through which a small fraction of the electrons derived from the substrate are passed to oxygen. Using arrows to denote the paths of electron flow, this may be schematized as follows:

$$\begin{array}{l} \text{substrate} \rightarrow \text{NAD} \rightarrow \text{flavoproteins} \rightarrow \text{cytochrome system} \rightarrow O_2 \\ \qquad\qquad\quad \llcorner\!\!\rightarrow \text{FMN} \xrightarrow{\text{luciferase}} O_2 \end{array}$$

The physiological link between the two chains is shown by the effect of cyanide. When added to a suspension of luminous bacteria, this inhibitor of normal electron transport via the cytochromes causes an abrupt enhancement of luminescence. No ATP is formed when electrons reach oxygen through the bypass, since most of the energy is emitted as light.

### The conversion of organic substrates to carbon dioxide

The availability of oxygen as a terminal electron acceptor in respiratory metabolic processes makes possible, in principle, a complete oxidation of organic compounds to $CO_2$. As already mentioned, a certain fraction of the substrate carbon is assimilated and used for the synthesis of cell material, but in aerobic respiration, the rest of the substrate carbon is usually entirely converted to $CO_2$.

Some of the biochemical pathways that serve for the conversion of carbohydrates to pyruvate, and thence to acetyl-S-CoA, have already been described in the context of fermentation. These pathways also operate in the respiratory metabolism of carbohydrates; the only difference is that the reduced pyridine nucleotides formed are reoxidized via the electron transport chain, instead of (as in fermentations) by the reduction of pyruvic acid or of products derived from it. *Metabolic reactions peculiar to respiratory metabolism, which play no part in fermentation, first intervene at the level of acetyl-S-CoA.*

In respiratory metabolism, acetyl-S-CoA is oxidized to $CO_2$ through a special cyclic series of reactions known as the tricarboxylic acid (TCA) cycle, or Krebs cycle (Figure 6.28). The cycle is initiated by the

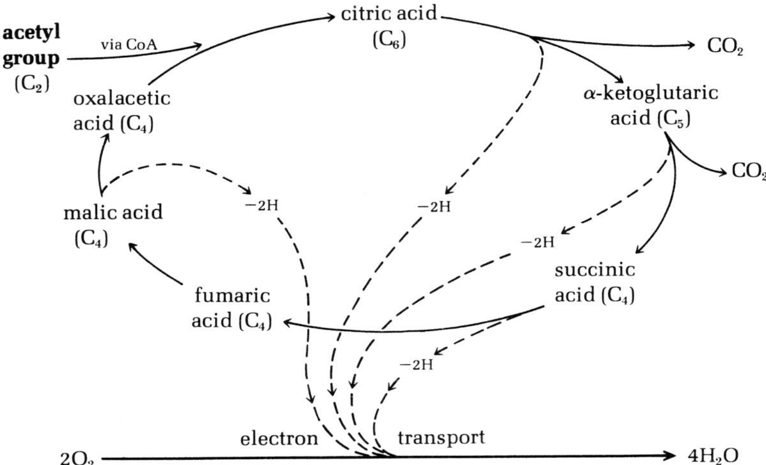


*Figure 6.28 The citric acid cycle.*

condensation of acetyl-S-CoA with oxalacetate to form citric acid. The ensuing reactions result in the conversion of citric acid to oxalacetate, with a concomitant liberation of two carbon atoms as $CO_2$ and the production of eight electrons which enter the electron transport chain. One complete turn of the cycle therefore achieves a complete oxidation of the equivalent of one molecule of acetyl-S-CoA, with the regeneration of one molecule of oxalacetate.

If the TCA cycle operated exclusively for the terminal oxidation of acetyl residues derived from primary substrates, it could be maintained without a net input of oxalacetate, the role of this compound being purely catalytic. However, the TCA cycle also has a very important biosynthetic role, since several of its intermediates, notably oxalacetic, α-ketoglutaric and succinic acids, are precursors for the synthesis of amino acids and other cell constituents.* Hence, in a grow-

*Metabolic pathways which, like the TCA cycle, lie at the juncture between catabolic and anabolic reaction sequences and have a dual role in energy-yielding metabolism and in biosynthesis are termed *amphibolic pathways*. Their significance is discussed at greater length in Chapter 7.

ing organism, the cycle is never in fact closed, and its maintenance requires a considerable net synthesis of oxalacetic acid. This is formed by the carboxylation of either pyruvate or phosphoenolpyruvic acid:

$$\begin{array}{l} CH_3 \\ | \\ C{=}O + CO_2 + ATP \rightarrow \\ | \\ COOH \end{array} \begin{array}{l} COOH \\ | \\ CH_2 + ADP + P_i \\ | \\ C{=}O \\ | \\ COOH \end{array}$$

$$\begin{array}{l} CH_2 \\ \| \\ C{-}O \sim \text{(P)} + CO_2 \longrightarrow \\ | \\ COOH \end{array} \begin{array}{l} COOH \\ | \\ CH_2 + P_i \\ | \\ C{=}O \\ | \\ COOH \end{array}$$

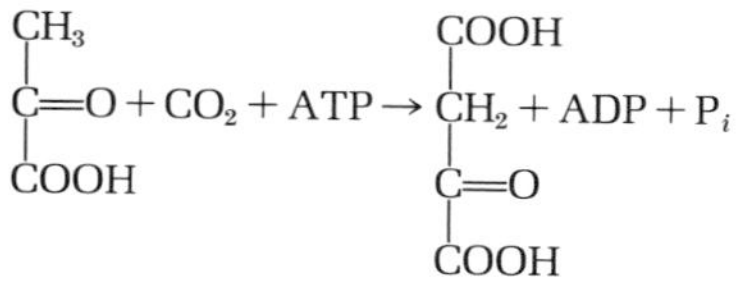

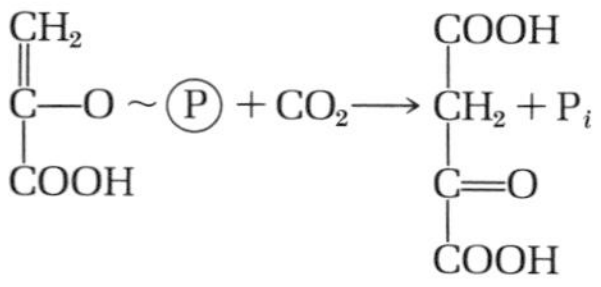

As a result, carbon from pyruvic acid enters the cycle by two routes: via oxalacetic acid and via acetyl-S-CoA.

**The role of the glyoxylate cycle in acetic acid oxidation**

A special modification of the TCA cycle, known as the glyoxylate cycle, comes into play during oxidation of acetic acid or of primary substrates (such as higher fatty acids), which are converted to acetyl-S-CoA without the intermediate formation of pyruvic acid. Under these circumstances, oxalacetic acid cannot be generated by the carboxylation of pyruvic or phosphoenolpyruvic acid, since in aerobic microorganisms there is no mechanism for synthesizing pyruvic acid directly from acetic acid: the oxidation of pyruvic acid to acetyl-S-CoA and $CO_2$ is a completely irreversible reaction.

The oxalacetic acid required for acetic acid oxidation is replenished by the oxidation of succinic and malic acids, which are produced through a sequence of two reactions. In the first reaction, isocitric acid, which is a normal intermediate of the TCA cycle, is cleaved to yield succinic and glyoxylic acids:

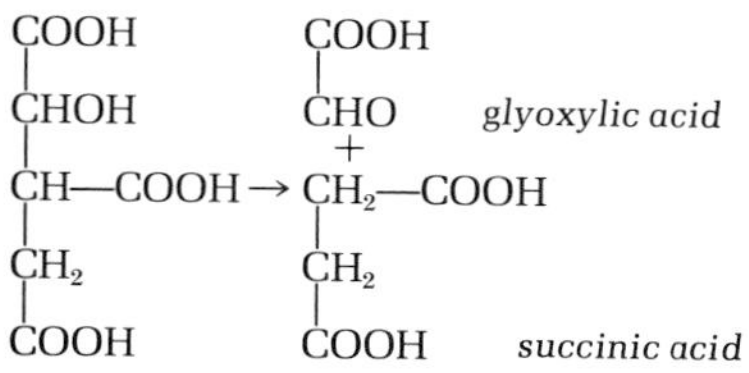

$$\begin{array}{l} COOH \\ | \\ CHOH \\ | \\ CH{-}COOH \rightarrow \\ | \\ CH_2 \\ | \\ COOH \end{array} \begin{array}{ll} COOH & \\ | & \\ CHO & \textit{glyoxylic acid} \\ \;+ & \\ CH_2{-}COOH & \\ | & \\ CH_2 & \\ | & \\ COOH & \textit{succinic acid} \end{array}$$

In the second reaction, acetyl-S-CoA is condensed with glyoxylic acid to yield malic acid:

$$COOH{-}CHO + CH_3{-}CO{-}S{-}CoA \rightarrow COOH{-}CHOH{-}CH_2{-}COOH + CoASH$$

In combination, these two reactions constitute a bypass of some of the reactions of the TCA cycle and result in the net conversion of two acetyl residues to oxalacetic acid, permitting the normal TCA cycle to operate and, in addition, providing a source of pyruvic acid from the oxidation of the $C_4$ acids. The cyclic process, which does not result in acetate conversion to $CO_2$, is known as the "glyoxylate cycle" (Figure 6.29). It does not operate during the attack on primary substrates that

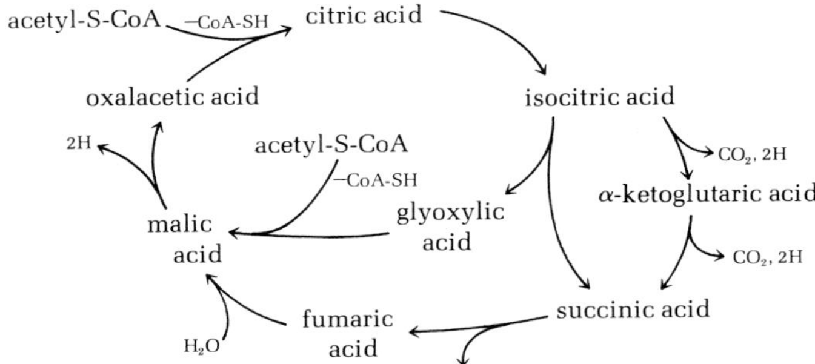


*Figure 6.29 The glyoxylate bypass and its relation to the reactions of the citric acid cycle.*

are decomposed through pyruvic acid, when oxalacetic acid can be formed by carboxylation. The two enzymes of the bypass are synthesized during growth on acetate or its direct metabolic precursors.

**Special pathways for the primary attack on organic compounds by microorganisms**

There is probably no naturally occurring organic compound which cannot be used as a substrate for respiratory metabolism by some microorganism. However complex the structure of the primary substrate may be, its utilization as a source of energy always involves the same basic principle: *a stepwise degradation to yield eventually one or more small aliphatic fragments capable of entering the terminal reactions of the TCA cycle*. The preliminary reaction sequences in the utilization of such substrates, frequently long and complex, therefore represent physiological adjuncts to the TCA cycle. As a specific illustration of such specialized microbial metabolic pathways, we shall describe pathways involved in the bacterial utilization of aromatic compounds.

**Oxidation of aromatic compounds through the β-ketoadipate pathway**

Most aerobic bacteria that use aromatic compounds as respiratory substrates attack them through one or other of the two convergent branches of the β-ketoadipate pathway (Figure 6.30). Through these reactions, the six carbon atoms of the aromatic nucleus in the primary substrate are converted to the six carbon atoms of an aliphatic acid, β-ketoadipic

p-hydroxy-benzoate → protocatechuate $\xrightarrow{+O_2}$ β-carboxv-cis,cis-muconate → γ-carboxv-muconolactone $\xrightarrow{-CO_2}$ β-ketoadipate enol lactone → β-ketoadipate

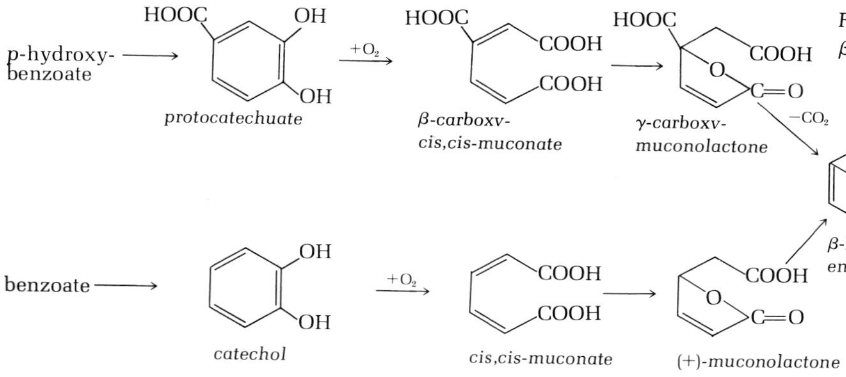

benzoate → catechol $\xrightarrow{+O_2}$ cis,cis-muconate → (+)-muconolactone → β-ketoadipate enol lactone

β-ketoadipate → β-ketoadipyl-CoA → acetyl-CoA + succinate

*Figure 6.30 The chemistry of the β-ketoadipate pathway.*

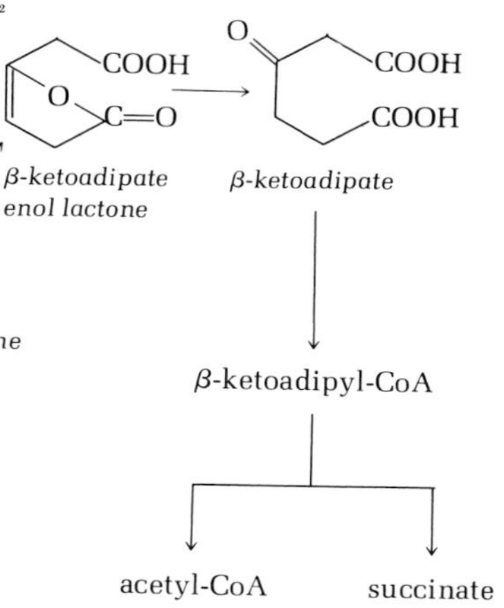

acid. This is in turn cleaved to acetyl-S-CoA and succinic acid, both of which can immediately enter the TCA cycle.

An interesting feature of the attack on aromatic substrates is the important role of *oxygenases,* enzymes that mediate a direct addition of molecular oxygen to the substrate. Oxygenases function both in the addition of hydroxyl groups to aromatic substrates and in the ring cleavage of diphenolic intermediates; two specific examples of such reactions are

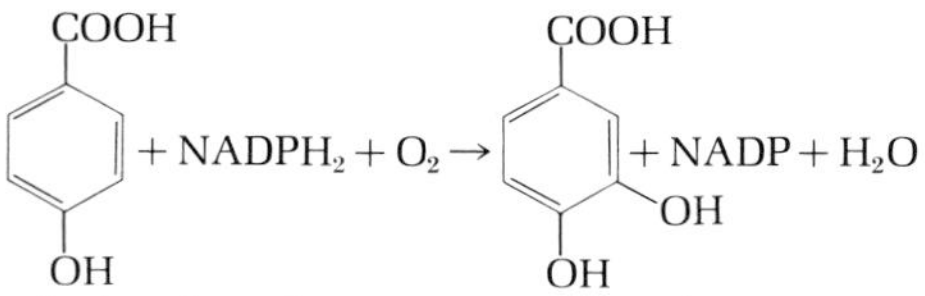

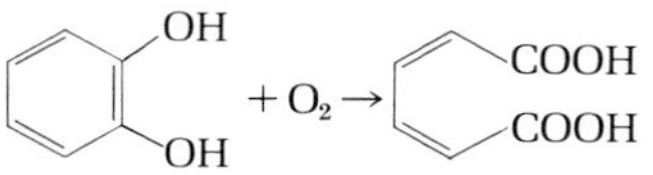

*p-hydroxybenzoic acid* + $NADPH_2 + O_2 \rightarrow$ *protocatechuic acid* + $NADP + H_2O$

*catechol* + $O_2 \rightarrow$ *cis,cis-muconic acid*

When the oxidation of a substrate is brought about by oxygenation, rather than by dehydrogenation, no transfer of electrons to the respiratory transport chain is possible; in fact, the hydroxylation of aromatic compounds actually dissipates a considerable amount of potential energy, since it is accompanied by an oxidation of reduced pyridine nucleotide. For these reasons, neither reducing power nor ATP can be derived from the oxidative step reactions involved in the initial attack on aromatic substrates. The energetic gain to the cell resulting from the respiratory metabolism of such substrates is made not in the initial reaction sequences but in the subsequent terminal oxidation of the aliphatic products through the TCA cycle, during which electrons are transferred to the respiratory transport chain.

**Incomplete oxidations of organic compounds**

The respiration of organic substrates by microorganisms normally results in a total conversion of substrate carbon to assimilatory products and $CO_2$. In certain cases, however, oxidation of the organic substrate is incomplete, and partially oxidized organic end-products are formed. This often reflects a derangement of the terminal reactions of substrate oxidation, produced by abnormal physiological conditions: high substrate concentrations, oxygen limitation, or an unfavorable pH. Some fungi excrete large quantities of intermediates of the tricarboxylic acid cycle (citric acid, $\alpha$-ketoglutaric acid, fumaric acid) when grown with very high sugar concentrations, and aerobic pseudomonads may produce $\alpha$-ketoglutaric acid under similar conditions. Growth of aerobic bacteria with high concentrations of glucose may also lead to the accumulation and excretion of oxidized derivatives of glucose, such as gluconic acid.

Among aerobic bacteria, the tendency to oxidize organic substrates incompletely is particularly marked among the acetic acid bacteria (genera *Acetobacter* and *Gluconobacter*). These organisms owe their common name to their characteristic ability to oxidize ethanol with the accumulation of acetic acid, exploited industrially for vinegar manufacture (see Chapter 29). The *overoxidizing acetic acid bacteria* of the genus *Acetobacter* carry out an oxidation of ethanol in two stages; initially, the substrate is almost stoichiometrically converted to acetic acid, but after the alcohol has been exhausted, the organisms oxidize the acetic acid that has accumulated to $CO_2$. Such organisms accordingly possess the potential capacity to carry out complete oxidations through the reactions of the tricarboxylic acid cycle together with the glyoxylate bypass.

The *underoxidizing acetic acid bacteria* of the genus *Gluconobacter* are completely unable to oxidize acetic acid and therefore form it in stoichiometric yields from ethanol. Experiments with one species, *Gluconobacter suboxydans*, have shown that it lacks the enzymatic machinery of the TCA cycle, being wholly unable to oxidize $\alpha$-ketoglutarate, succinate, fumarate, and malate; oxalacetic acid is converted via pyruvic acid stoichiometrically to acetic acid. In this particular case, the performance of incomplete oxidations is genetically, not environmentally, determined.

The acetic acid bacteria oxidize incompletely many other substrates (primary and secondary alcohols, carbohydrates) in addition to ethanol; each substrate yields a characteristic oxidation product, with the same carbon skeleton, as shown by the typical reactions listed in Table 6.3.

The incomplete oxidations performed by acetic acid bacteria can provide the cell with ATP, formed through oxidative phosphorylation; but the underoxidizers cannot derive assimilable carbon from most of the substrates that they attack and hence must grow at the expense of other organic nutrients furnished in the medium.

*Table 6.3 Some incomplete oxidations of organic compounds carried out by acetic acid bacteria*

$$\underset{\textit{ethanol}}{CH_3CH_2OH} + O_2 \rightarrow \underset{\textit{acetic acid}}{CH_3COOH} + H_2O$$

$$\underset{\textit{propanol}}{CH_3CH_2CH_2OH} + O_2 \rightarrow \underset{\textit{propionic acid}}{CH_3CH_2COOH} + H_2O$$

$$\underset{\textit{isopropanol}}{\begin{matrix} H_3C \\ H_3C \end{matrix}\!\!>\!CHOH} + \tfrac{1}{2}O_2 \rightarrow \underset{\textit{acetone}}{\begin{matrix} H_3C \\ H_3C \end{matrix}\!\!>\!C{=}O} + H_2O$$

$$\underset{\textit{glycerol}}{\begin{array}{l} CH_2OH \\ | \\ CHOH \\ | \\ CH_2OH \end{array}} + \tfrac{1}{2}O_2 \rightarrow \underset{\textit{dihydroxyacetone}}{\begin{array}{l} CH_2OH \\ | \\ C{=}O \\ | \\ CH_2OH \end{array}} + H_2O$$

$$\underset{\textit{2,3-butanediol}}{\begin{array}{l} CH_3 \\ | \\ (CHOH)_2 \\ | \\ CH_3 \end{array}} + \tfrac{1}{2}O_2 \rightarrow \underset{\textit{acetoin}}{\begin{array}{l} CH_3 \\ | \\ CHOH \\ | \\ C{=}O \\ | \\ CH_3 \end{array}} + H_2O$$

$$\underset{\textit{mannitol}}{HOCH_2{-}\begin{array}{c} H \\ | \\ C \\ | \\ OH \end{array}{-}\begin{array}{c} H \\ | \\ C \\ | \\ OH \end{array}{-}\begin{array}{c} OH \\ | \\ C \\ | \\ H \end{array}{-}\begin{array}{c} OH \\ | \\ C \\ | \\ H \end{array}{-}CH_2OH} + \tfrac{1}{2}O_2 \rightarrow \underset{\textit{fructose}}{HOCH_2{-}\begin{array}{c} \\ \\ C \\ \| \\ O \end{array}{-}\begin{array}{c} H \\ | \\ C \\ | \\ OH \end{array}{-}\begin{array}{c} OH \\ | \\ C \\ | \\ H \end{array}{-}\begin{array}{c} OH \\ | \\ C \\ | \\ H \end{array}{-}CH_2OH} + H_2O$$

$$\underset{\textit{glucose}}{HOCH_2{-}\begin{array}{c} H \\ | \\ C \\ | \\ OH \end{array}{-}\begin{array}{c} H \\ | \\ C \\ | \\ OH \end{array}{-}\begin{array}{c} OH \\ | \\ C \\ | \\ H \end{array}{-}\begin{array}{c} H \\ | \\ C \\ | \\ OH \end{array}{-}CHO} + \tfrac{1}{2}O_2 \rightarrow \underset{\textit{gluconic acid}}{HOCH_2{-}\begin{array}{c} H \\ | \\ C \\ | \\ OH \end{array}{-}\begin{array}{c} H \\ | \\ C \\ | \\ OH \end{array}{-}\begin{array}{c} OH \\ | \\ C \\ | \\ H \end{array}{-}\begin{array}{c} H \\ | \\ C \\ | \\ OH \end{array}{-}COOH} + H_2O$$

$$\underset{\textit{gluconic acid}}{HOCH_2{-}\begin{array}{c} H \\ | \\ C \\ | \\ OH \end{array}{-}\begin{array}{c} H \\ | \\ C \\ | \\ OH \end{array}{-}\begin{array}{c} OH \\ | \\ C \\ | \\ H \end{array}{-}\begin{array}{c} H \\ | \\ C \\ | \\ OH \end{array}{-}COOH} + \tfrac{1}{2}O_2 \rightarrow \underset{\textit{5-ketogluconic acid}}{HOCH_2{-}\begin{array}{c} \\ \\ C \\ \| \\ O \end{array}{-}\begin{array}{c} H \\ | \\ C \\ | \\ OH \end{array}{-}\begin{array}{c} OH \\ | \\ C \\ | \\ H \end{array}{-}\begin{array}{c} H \\ | \\ C \\ | \\ OH \end{array}{-}COOH} + H_2O$$

## The oxidation of inorganic compounds

The use of reduced inorganic compounds as substrates for respiratory metabolism is confined to bacteria, being characteristic of a number of special physiological groups, known collectively as the *chemoautotrophs*. The substances that can serve as energy sources are $H_2$, CO, $NH_3$, $NO_2^-$, $Fe^{2+}$, and reduced sulfur compounds ($H_2S$, S, $S_2O_3^{2-}$). In this mode of respiratory metabolism, the sole function of substrate oxidation is to provide reducing power and ATP, the cellular carbon being derived from $CO_2$, assimilated through the reactions of the Calvin cycle (see Chapter 7).

The use of $H_2$ as an energy source occurs in a number of different groups of aerobic bacteria; these organisms are without exception facultative chemoautotrophs, also able to respire and grow at the expense of a wide variety of organic substrates. With respect to the respiratory electron transport system, the use of hydrogen as an energy source is not markedly different from the use of an organic substrate. Provided that a bacterium contains the enzyme hydrogenase, which activates molecular hydrogen:

$$H_2 \rightleftharpoons 2H$$

the oxidation can be coupled with a reduction of pyridine nucleotide, and thence with the normal respiratory electron transport system (Figure 6.31).

$$H_2 \rightarrow \text{pyridine nucleotide} \rightarrow \text{flavoprotein} \rightarrow \text{cytochrome } c \rightarrow \text{cytochrome } a \rightarrow O_2$$

$$\left.\begin{matrix} NH_3 \\ NO_2^- \\ H_2S \end{matrix}\right\} \rightarrow \text{cytochrome } c \rightarrow \text{cytochrome } a \rightarrow O_2$$

*Figure 6.31 The paths of electron flow from $H_2$ and other reduced inorganic energy sources to oxygen.*

The situation is not the same for the two physiological groups of nitrifying bacteria, which obtain energy by one of the following oxidations:

$$2NH_3 + 3O_2 \rightarrow 2NO_2^- + 2H^+ + 2H_2O$$

and

$$2NO_2^- + O_2 \rightarrow 2NO_3^-$$

It can be calculated that the $E'_0$ values for these two oxidations are too positive to permit a coupling with the reduction of pyridine nucleotides. This thermodynamic problem exists also for the sulfur-oxidizing chemoautotrophs, which obtain energy by the following reactions:

$$2H_2S + O_2 \rightarrow 2S + 2H_2O$$
$$2S + 2H_2O + 3O_2 \rightarrow 2SO_4^{2-} + 4H^+$$
$$S_2O_3^{2-} + H_2O + 2O_2 \rightarrow 2SO_4^{2-} + 2H^+$$

For some of these organisms, there is direct evidence that the electrons derived from the oxidation of the inorganic substrate enter the transport chain at the level of cytochrome c (Figure 6.31). As already discussed, a site for oxidative phosphorylation exists between cytochrome c and oxygen; hence these oxidations can yield ATP directly. However, the primary reactions of substrate oxidation provide no mechanism for the generation of reduced pyridine nucleotide, required in large amounts by chemoautotrophs for the assimilation of $CO_2$. Biochemical evidence suggests that part of the ATP derived from oxidative phosphorylation is expended to mediate pyridine nucleotide reduction, by "driving" electrons from a higher to a lower redox potential (Figure 6.32). These particular groups of chemoautotrophs are physiologically

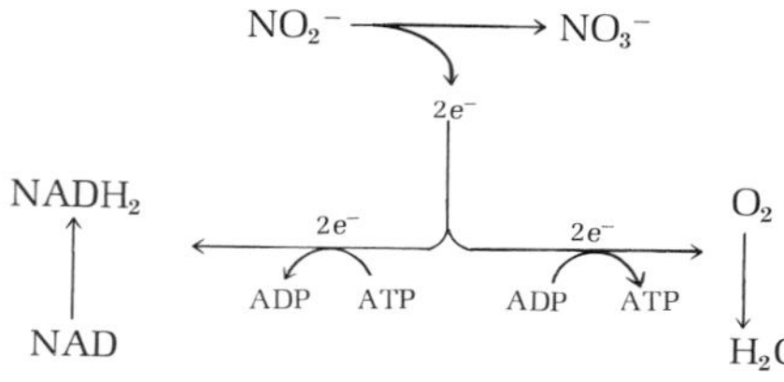


*Figure 6.32 A schematic diagram illustrating a possible mechanism for the reduction of pyridine nucleotides in Nitrobacter, by expenditure of ATP derived from oxidative phosphorylation.*

highly specialized organisms; in contrast to the hydrogen-oxidizing bacteria, most of them are incapable of using organic substrates as energy sources.

## *Anaerobic respirations*

Some bacteria are capable of performing respiratory metabolism under completely anaerobic conditions, by using nitrate, sulfate, or carbonate as a terminal inorganic electron acceptor. This is a variation on the theme of respiratory metabolism that does not occur in other biological groups.

**The use of nitrate as an electron acceptor**

Many aerobic bacteria contain nitrate reductases, enzymes that mediate the reaction

$$NO_3^- + 2e^- + 2H^+ \rightarrow NO_2^- + H_2O$$

The ability to reduce nitrate to nitrite does not, however, permit normal growth under anaerobic conditions, since a large amount of nitrate must be reduced to oxidize a small amount of substrate, and the reduction product, nitrite, is highly toxic. A few normally aerobic bacteria (principally *Pseudomonas* and *Bacillus* species) can use nitrate as a physiologically useful terminal electron acceptor by reducing it beyond the level of nitrite, to molecular nitrogen. The ability to form $N_2$ from nitrate is termed *denitrification*. For each molecule of nitrate reduced, five electrons can be accepted, as shown by the equation

$$2NO_3^- + 10e^- + 12H^+ \rightarrow N_2 + 6H_2O$$

Under these conditions, nitrate serves as an extremely effective electron acceptor; furthermore, the reduction product is an inert gas, which is nontoxic.

Denitrification involves the operation of special enzymes, linked to the normal aerobic respiratory electron transport chain, which enable electrons to be transferred to nitrate and its partial reduction products. It is, therefore, always an *alternative* mode of respiratory energy-yielding metabolism, used by denitrifying bacteria to support growth in the absence of $O_2$. In the presence of air, even when nitrate is present, respiration proceeds entirely through the aerobic electron transport chain, since the enzymes responsible for denitrification are not synthesized under these conditions.

The oxidation of organic substrates under denitrifying conditions is complete, $CO_2$ being the only oxidized end-product. Denitrifying bacteria can oxidize the same range of organic substrates through aerobic respiration and through denitrification, with the exception of compounds (notably aromatic substances), the metabolism of which involves the participation of oxygenases. Such compounds can be attacked only if free oxygen is present; they cannot be oxidized anaerobically by denitrifying bacteria.

Denitrification can be used for the respiration of inorganic compounds: some bacteria that oxidize $H_2$ couple its oxidation with the reduction of nitrate to $N_2$, and certain chemoautotrophs of the genus *Thiobacillus* can similarly oxidize sulfur anaerobically. The anaerobic oxidation of elemental sulfur can be represented as

$$5S + 6NO_3^- + 2H_2O \rightarrow 5SO_4^{2-} + 3N_2 + 4H^+$$

The detailed chemistry of the steps in denitrification between nitrite and $N_2$ is still unclear.

**The use of sulfate as an electron acceptor**

The ability to use sulfate as an electron acceptor is a rare property among bacteria, being confined to a small group of vibrioid organisms (the genus *Desulfovibrio*) and to a few anaerobic sporeforming rods (members of the genus *Clostridium*). *All these bacteria are obligate anaerobes; sulfate reduction is not, therefore, an alternative to normal respiratory metabolism, as is denitrification, but a completely independent respiratory process.*

In sulfate reduction, the electron acceptor is reduced to sulfide; each molecule of sulfate can therefore accept eight electrons:

$$SO_4^{2-} + 8e^- + 8H^+ \rightarrow S^{2-} + 4H_2O$$

Sulfite ($SO_3^{2-}$) is the only intermediate in the reduction whose role is clearly established; the biochemical mechanism for the conversion of sulfite to sulfide remains unclear.

Sulfate-reducing bacteria are able to attack a relatively limited range of organic compounds, the principal substrates being lactic acid and $C_4$ dicarboxylic acids. These substrates are oxidized only to the level of acetate, since the sulfate reducers do not possess the enzymes of the tricarboxylic acid cycle (Figure 6.33). Some sulfate reducers can also oxidize molecular hydrogen.

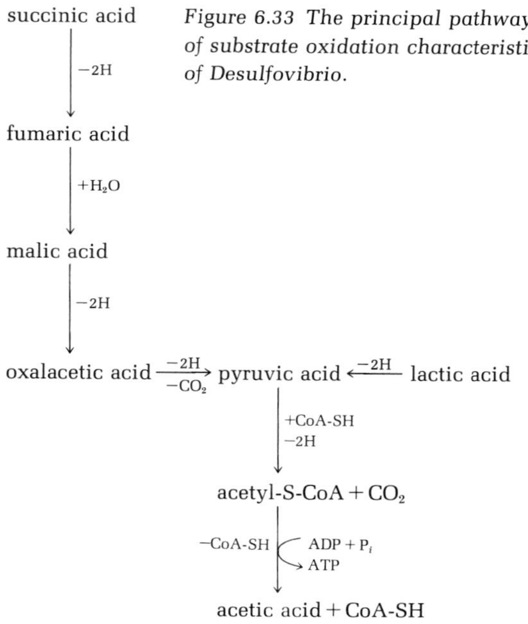


*Figure 6.33 The principal pathways of substrate oxidation characteristic of Desulfovibrio.*

The nature of the anaerobic electron transport system that couples these substrate oxidations with the reduction of sulfate is not yet established; however, the sulfate reducers are unique among anaerobic nonphotosynthetic bacteria in containing large amounts of special cytochromes of the c type. They also contain ferredoxin and other nonheme iron proteins. Both these components are presumed to play a role in the anaerobic transport system.

**The use of carbonate as an electron acceptor**

A small group of strictly anaerobic bacteria, the methane-producing bacteria, use carbonate as an electron acceptor: its reduction results in formation of the gas, *methane* ($CH_4$). The simplest and most clear-cut example of this mode of anaerobic respiration is the oxidation of molecular hydrogen, an energy-yielding reaction performed by all these bacteria:

$$4H_2 + CO_2 \rightarrow CH_4 + 2H_2O$$

In this particular case, methane can arise only by the reduction of $CO_2$. However, these organisms can also produce methane from organic substrates, such as methanol and acetic acid.

In such cases, methane does not necessarily arise through carbonate reduction; in fact, labeling experiments with acetate marked either in the first or the second carbon atom with $^{14}C$ have shown that methane arises *exclusively* from the methyl carbon of acetic acid:

$$\begin{array}{l} CH_3 \rightarrow CH_4 \\ | \\ COOH \rightarrow CO_2 \end{array}$$

Accordingly, the methane-producing bacteria, although capable of forming this gas by the reduction of $CO_2$, can also form it from partially reduced forms of carbon, contained in organic substrates. Such reactions can be formally regarded as fermentations, whereas methane production from hydrogen and $CO_2$ is clearly classifiable as anaerobic respiration.

The biochemistry of methane formation is complex and not yet clearly elucidated. Methane-producing bacteria do not contain cytochromes, but have a very high content of vitamin $B_{12}$ and of folic acid. These two vitamins, in coenzyme form, play a role in the reduction process.

## *Photosynthetic energy conversion*

Photosynthetic energy conversion is a process of considerable complexity and occurs in a membrane system containing pigments, electron carriers, lipids, and proteins. This is termed the *photosynthetic apparatus.* In eucaryotic organisms, it is contained within the chloroplast. In procaryotic organisms, it is usually localized in extensions of the cell membrane (see Chapter 10, p. 351).

Radiant energy is always transferred in discrete packets known as *quanta;* the energy content of a quantum is inversely related to its wavelength (Figure 6.34). When radiant energy is absorbed by matter,

**Light absorption**

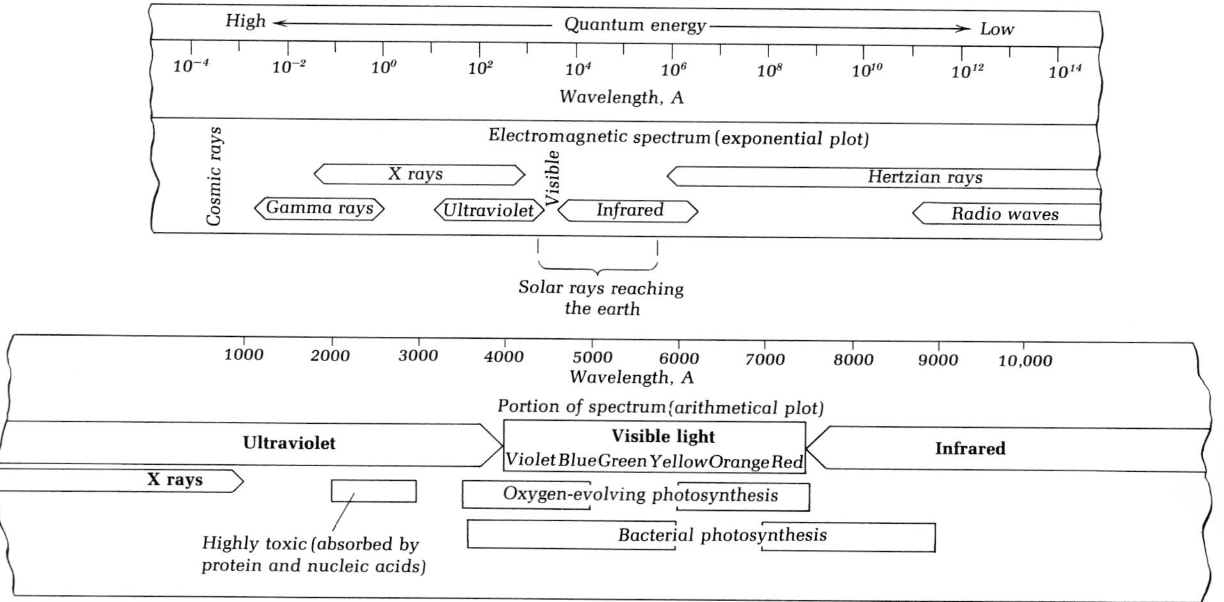


Figure 6.34 The electromagnetic spectrum: above, the entire spectrum is plotted on an exponential scale; below, the ultraviolet, visible, and near-infrared regions are greatly expanded and plotted on an arithmetic scale.

its possible effects are a function of the energy content of the quantum and hence of the wavelength of the radiation. Infrared light of wavelength longer than 1200 nm has an energy content so small that the absorbed energy in this spectral range is immediately converted to heat; it cannot mediate chemical change. So-called "ionizing" radiations of very short wavelength (X rays, $\alpha$ particles, cosmic rays) have such a high energy content that molecules in their path are immediately ionized. Between these two extremes, radiations ranging in wavelength from 200 to 1,200 nm (ultraviolet, visible, and near infrared light) have an energy content such that their absorption is capable of producing a chemical change in the absorbing molecule; it is this portion of the electromagnetic spectrum which can serve for the performance of photosynthesis.

## The spectral quality of sunlight

Solar radiation is profoundly modified by its passage through the atmosphere, which effectively filters out much of the shorter, high-energy radiation. At sea level, about 75 percent of the total energy of sunlight is contained in light of wavelengths between 400 and 1,100 nm (the visible and near infrared portions of the spectrum), and it is within these limits that the pigments responsible for light capture in photosynthesis have their effective absorption bands.

## Photosynthetic pigment systems

In all photosynthetic organisms, the pigment system responsible for light capture is a complex one, containing pigments belonging to at least two different chemical classes. Pigments of the classes of *chlorophylls* and *carotenoids* are universal components of photosynthetic pigment systems; in some algae, these are supplemented by pigments of a third class, the *phycobiliproteins*.

The chlorophylls, of which at least seven kinds occur in the various groups of photosynthetic organisms, have a common molecular ground plan, shown in Figure 6.35. These compounds are related structurally

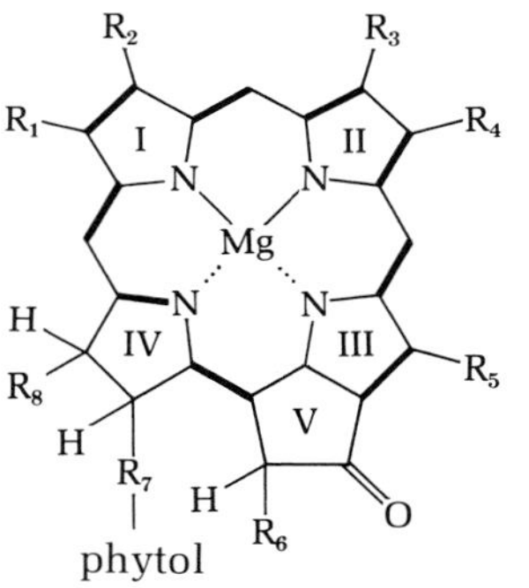

*Figure 6.35 The molecular ground plan of the chlorophylls. Double bonds between carbon atoms are indicated by heavy lines. The tetrapyrrolic nucleus (rings I, II, III, and IV) has the same derivation as that of the hemes, but it is chelated with magnesium. In chlorophylls, one or more of the pyrrole rings are reduced; in this diagram, ring IV is shown reduced, as is characteristic of chlorophyll a; $R_1$, $R_2$, and so forth, designate aliphatic side chains attached to the tetrapyrrolic nucleus. The presence of ring V, the pentanone ring, and the substitution of $R_7$ by a long-chain alcohol (phytol or farnesol) are characteristic features of the chlorophylls that do not occur in hemes.*

and biosynthetically to the hemes, which serve as the prosthetic groups in cytochromes and many other respiratory enzymes; both hemes and chlorophylls have a central tetrapyrrolic nucleus, within which a metal is chelated. In hemes this metal is iron; in chlorophylls, magnesium. Two additional features distinguish chlorophylls chemically from

*Figure 6.36 The absorption spectrum of pure chlorophyll a dissolved in ether. The presence of two strong absorption bands, one in the blue-violet region of the spectrum and one in the red region of the spectrum, is characteristic of all chlorophylls.*

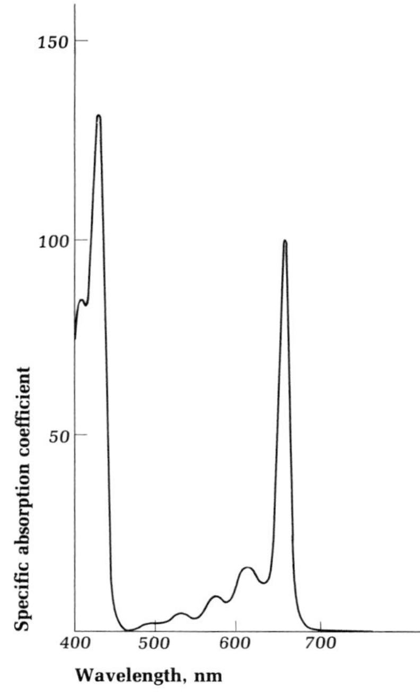


hemes: the existence of a fifth ring, the pentanone ring, and the esterification of one side chain on the tetrapyrrolic nucleus with an alcohol, phytol, or farnesol. As shown for chlorophyll a in Figure 6.36, chlorophylls absorb light intensely in two spectral regions: the violet, around 400 nm, and the red or near infrared, around 600 to 800 nm. The position of this peak at longer wavelength varies considerably with the chlorophyll species. In the photosynthetic apparatus, chlorophyll is intimately associated with the other cell constituents, which may considerably modify its spectral properties.

The carotenoids, of which a large number of different kinds occur in photosynthetic organisms, have the basic structure of a long, unsaturated hydrocarbon with projecting methyl groups, containing a total of 40 carbon atoms (Figure 6.37). In particular members of the class, this basic structure can be modified in several ways: by terminal ring closure, to form six-membered alicyclic or aromatic rings, and by the addition of oxygenated substituents, notably hydroxyl, methoxyl, or keto groups. Carotenoids have a single broad region of light absorption, between about 450 and 550 nm.

The phycobiliproteins are conjugated proteins, containing linear tetrapyrroles (bile pigments). Red phycobiliproteins (phycoerythrins) have a major absorption band around 550 nm; blue phycobiliproteins (phycocyanins), around 625 nm.

The particular combination of pigments present in the photosynthetic apparatus determines the gross absorption spectrum of any given organism and the spectral quality of the light that it can use for the per-

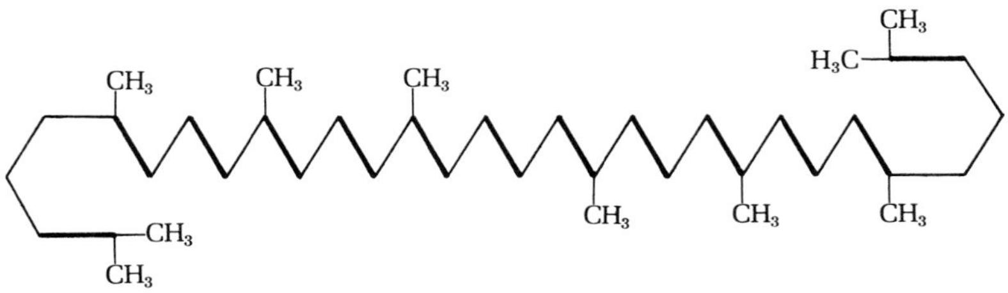

*Figure 6.37 The molecular ground plan of the carotenoids, illustrated by an open-chain carotenoid which does not contain oxygen. Double bonds between carbon atoms are indicated by heavy lines. This basic structure may be modified, in the different kinds of carotenoids, by terminal ring closure at one or both ends of the molecule and by the introduction of hydroxyl (—OH), methoxyl (—$OCH_3$), or ketone (=O) groups.*

formance of photosynthesis. Figure 6.38 shows the absorption spectra of several different algae and photosynthetic bacteria and indicates for each organism roughly the contributions to the total absorption of the three major classes of photosynthetic pigments. No single type of photosynthetic organism possesses a pigment system able to absorb with equal effectiveness throughout the available spectral range of sunlight, but in aggregate, the different kinds of photosynthetic organisms make effective use of practically all wavelengths between 400 and 1,100 nm.

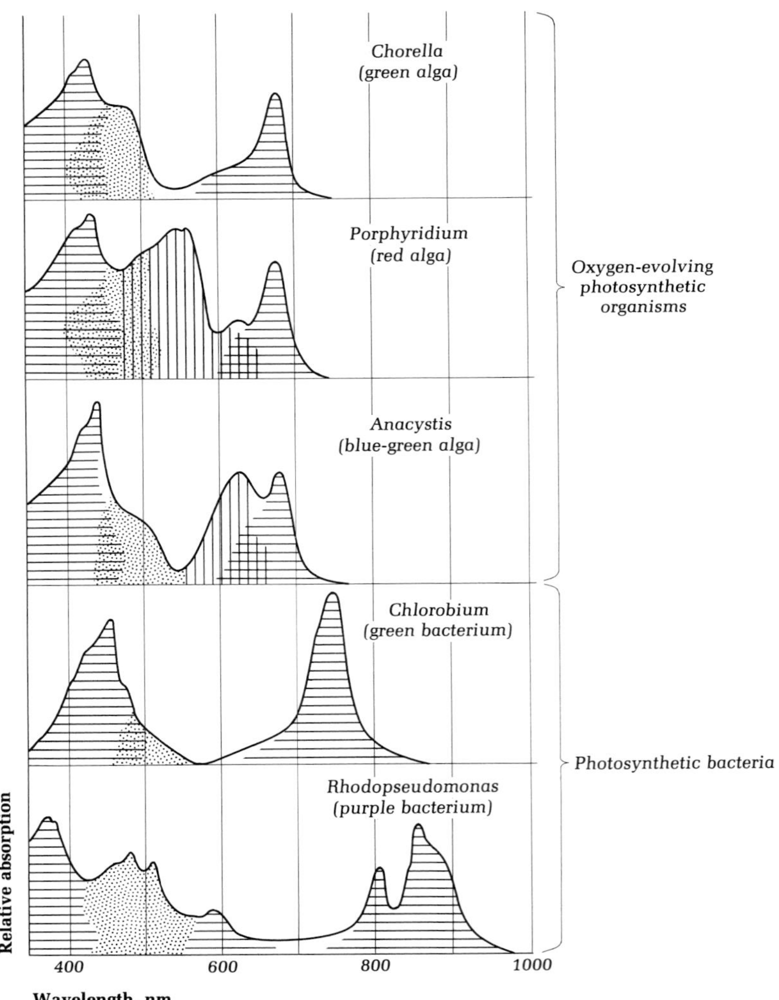


*Figure 6.38 The absorption spectra of some photosynthetic organisms. The contributions by the various classes of photosynthetic pigments are approximately indicated as follows: chlorophylls, horizontal hatching; carotenoids, stippling; phycobilins, vertical hatching.*

**Photochemical reaction centers**

The energy contained in the light quanta absorbed by the photosynthetic pigment system is transferred to special *reaction centers* within the photosynthetic apparatus. The reaction centers contain chlorophyll molecules in a special state, closely associated with the components of a photosynthetic electron transport chain. The first identified electron acceptor in this chain is the nonheme iron protein, ferredoxin, which has a very low redox potential and which also plays roles in the metabolism of strictly anaerobic nonphotosynthetic bacteria. Also present in the photosynthetic electron transport chains are components of the classes found in respiratory electron transport chains: flavoproteins, quinones, and cytochromes.

The process of photosynthetic energy conversion is initiated when transferred light energy is absorbed by a molecule of reaction center chlorophyll. This chlorophyll molecule is oxidized by the ejection of an electron, which is accepted by ferredoxin:

$$\text{chlorophyll} + h\nu \rightarrow \text{chlorophyll}^{+} + e^{-}$$

$$\text{ferredoxin} + e^{-} \rightarrow \text{reduced ferredoxin}$$

Since the $E'_0$ of the oxidized ferredoxin–reduced ferredoxin half reaction is about −0.4 V, the reoxidation of reduced ferredoxin makes available approximately as much energy as the oxidation of molecular hydrogen. This energy can then be harnessed through the intermediacy of other electron carriers of the photosynthetic electron transport system in order to generate biologically useful forms of chemical bond energy.

*Chlorophyll thus plays a dual role in photosynthetic energy conversion, as a light-gathering pigment and as the site of the initial photochemical event.* Carotenoids and phycobiliproteins, on the other hand, function uniquely as light-gathering pigments, passing the light energy which they absorb to the reaction center, that is to say, to the chlorophyll.

In all photosynthetic organisms, the energy of the electrons ejected from chlorophyll can be used to generate ATP by the process of *cyclic photophosphorylation,* the mechanism of which is shown schematically in Figure 6.39. The electrons are passed through the photosynthetic electron transport chain, the terminal component of which reduces oxidized chlorophyll; the electrons flow through a closed circuit, the flow being triggered by the absorption of light energy, part of which is captured in the transport chain by the synthesis of ATP.

The electrons ejected by light from reaction center chlorophyll can

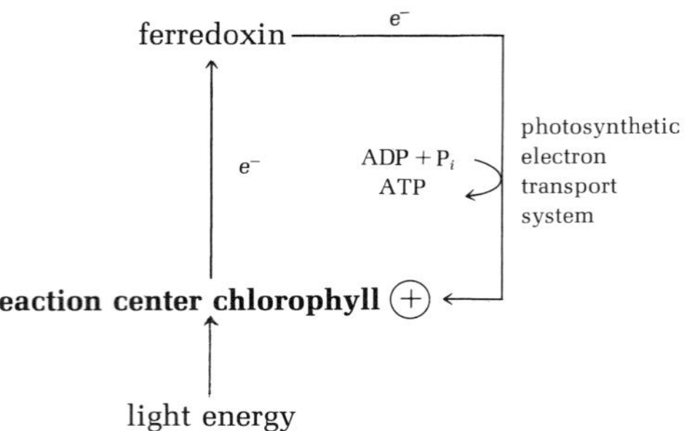


*Figure 6.39 A schematic diagram to show the mechanism of noncyclic photophosphorylation.*

alternatively be used for the reduction of pyridine nucleotide (Figure 6.40). In this event, the flow of electrons becomes open, or *noncyclic*,

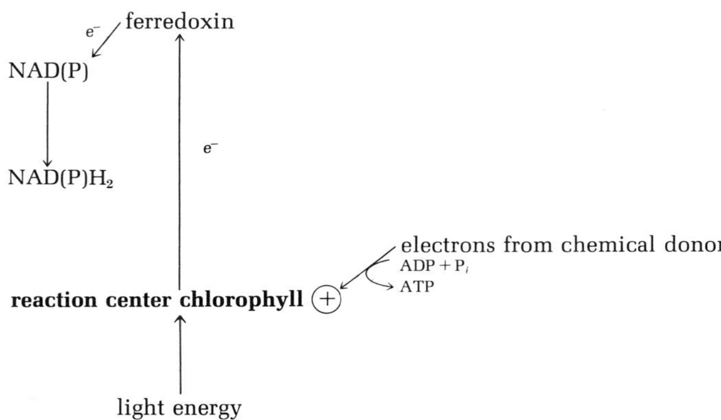


*Figure 6.40 The photosynthetic reduction of pyridine nucleotides through the reactions of noncyclic photophosphorylation.*

because chlorophyll$^+$ must be reduced by electrons derived from an appropriate chemical donor, via the transport system of the photosynthetic apparatus. Since this transfer of electrons can also give rise to ATP formation, the process is termed *noncyclic photophosphorylation.*

**The difference between plant photosynthesis and bacterial photosynthesis**

The most obvious gross difference between photosynthesis of the plant and of the bacterial types is the absence of oxygen evolution in bacterial photosynthesis. In plants, the formation of oxygen results from the oxidation of water, which is coupled, through the reactions of noncyclic photophosphorylation, with the reduction of NADP:

$$2NADP + 2H_2O + 2ADP + 2P_i \rightarrow 2NADPH_2 + O_2 + 2ATP$$

The oxidation of water to molecular oxygen is not an energy-yielding reaction: on the contrary, it requires a considerable net input of energy. How, then, are plants and algae able to use this energetically unfavorable electron donor? A partial answer to the question can now be given, as a result of the discovery in recent years that the photosynthetic apparatus of plants and algae contains *two distinct kinds of photochemical reaction centers,* which can be distinguished by their responses to certain wavelengths of light and to inhibitors of photosynthesis. Reaction centers of type I mediate both cyclic photophosphorylation and the photochemical reduction of NADP. Light absorption by these reaction centers alone is, however, insufficient to couple the reduction of NADP with the oxidation of water. *Noncyclic photophosphorylation in plant photosynthesis requires simultaneous light absorption by reaction centers of type II, at which the photochemically driven oxidation of water takes place.* The electrons derived from this oxidation pass through the photosynthetic electron transport system and reduce the oxidized chlorophyll formed through passage of electrons to NADP at reaction centers of type I. This coupling of the two light reactions characteristic of plant photosynthesis is shown in Figure 6.41.

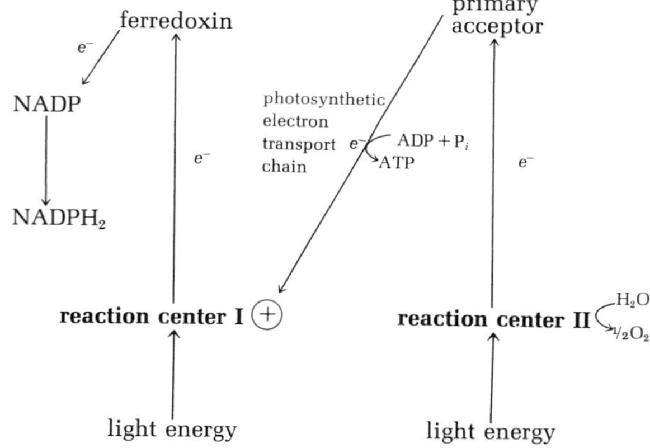


*Figure 6.41 The coupling of the two light reactions characteristic of plant photosynthesis, accompanied by the oxidation of water to molecular oxygen.*

**The coupling of photosynthetic energy generation and biosynthesis**

All present evidence suggests that the inability of bacteria to bring about a photosynthetic production of oxygen is a consequence of the *absence of type II reaction centers from the bacterial photosynthetic apparatus.* The bacterial photochemical reaction centers, exclusively of type I, can mediate cyclic photophosphorylation, and they can also mediate the reduction of pyridine nucleotide, provided that a suitable electron donor (e.g., $H_2S$, $H_2$, an organic compound) is available for the reduction of oxidized chlorophyll molecules in the type I reaction centers. Hence, noncyclic electron flow in bacterial photosynthesis involves only *one* type of photochemical reaction, mediated by type I reaction centers. In terms of its photochemistry, bacterial photosynthesis is a simpler process than plant photosynthesis.

We have described how light absorbed by the pigments of the photosynthetic apparatus is used to mediate the synthesis of ATP and reduced pyridine nucleotides, providing the forms of energy essential for biosynthesis. It is these reactions of *energy conversion* (i.e., the conversion of light energy into chemical bond energy) that are unique to photosynthesis. The particular manner in which the chemical bond energy so produced is used for the performance of biosynthetic reactions is neither unique to photosynthesis nor, strictly speaking, a part of the photosynthetic process. The ensuing biosynthetic processes involve so-called "dark reactions," catalyzed by enzymes in the absence of light. It may nevertheless be useful to terminate this section by a brief discussion of the coupling between photosynthetic energy generation and the assimilation of $CO_2$, the primary carbon source for most photosynthetic organisms.

In all organisms, photosynthetic and nonphotosynthetic, which use $CO_2$ as principal carbon source, $CO_2$ assimilation occurs through a special cyclic series of reactions, discussed in Chapter 7, which result in conversion of $CO_2$ to carbohydrate ($(CH_2O)$). Grossly, this conversion involves an expenditure of two molecules of reduced pyridine nucleotide and three molecules of ATP:

$$CO_2 \xrightarrow[2PNH_2 \;\to\; 2PN]{3ATP \;\to\; 3ADP + 3P_i} (CH_2O)$$

The reactions of noncyclic photophosphorylation in plant photosynthesis yield equimolar quantities of reduced pyridine nucleotide and of ATP. Consequently, noncyclic photophosphorylation cannot supply all the ATP needed for $CO_2$ assimilation and must be supplemented by an additional synthesis of ATP, derived from the reactions of cyclic photophosphorylation. Both cyclic and noncyclic photochemical processes must therefore operate in plants to permit $CO_2$ assimilation.

The situation in bacterial photosynthesis is less clear, since at least some of the inorganic electron donors used by these organisms (e.g., $H_2$) can serve directly for the generation of reduced pyridine nucleotide by "dark" enzymatic reactions, thus making unnecessary the intervention of a photochemical reaction for the production of reducing power. It seems likely, therefore, that noncyclic photophosphorylation plays a much less important role in bacterial photosynthesis than in plant photosynthesis. The most important function of the photochemical events in bacterial photosynthesis is probably the provision of ATP through the reactions of cyclic photophosphorylation.

## Further reading

**Books**


Barker, H. A., *Bacterial Fermentations.* New York: Wiley, 1957.

Doelle, H. W., *Bacterial Metabolism.* New York: Academic Press, 1969.

Gunsalus, I. C., and R. Y. Stanier (editors), *The Bacteria,* Vol. II. New York: Academic Press, 1961.

Lehninger, A. L., *Bioenergetics.* New York: Benjamin, 1965.

Sokatch, J. R., *Bacterial Physiology and Metabolism.* New York: Academic Press, 1969.

# Chapter seven
# Microbial metabolism: biosynthesis

*The many different ways* in which microorganisms can obtain the energy required for biosynthesis have been outlined in Chapter 6. Despite their mechanistic diversity, all modes of energy-yielding metabolism have the same ultimate function: the provision of ATP (and of reduced pyridine nucleotides, when there is a net requirement for reducing power) to drive the reactions of biosynthesis. In this sense, there is a fundamental unity underlying the superficial diversity of energy-yielding metabolism. This *unity of biochemistry*, a concept first emphasized by the microbiologist A. J. Kluyver in 1926, becomes even more evident when we start to analyze the ways in which the ATP derived from energy-yielding metabolism is employed for the performance of biosynthesis. In every cell, the major end-products of biosynthesis are proteins and nucleic acids, and the biochemical reactions that underlie their formation show little variation from group to group. This is also true of the synthesis of the diverse types of organic molecules—pyridine nucleotides, porphyrin derivatives, flavins, and so on—which serve as coenzymes of prosthetic groups of enzymes. There is, accordingly, a *central core of biosynthetic reactions that are performed in similar fashion by all organisms.* A greater degree of diversity occurs in the synthesis of other classes of cell constituents, in particular polysaccharides and lipids, since the chemical composition of these substances is often group specific.

In this chapter we shall focus attention on the reactions of biosynthesis that are common to most or all organisms. Some more specialized biosynthetic processes, distinctive of procaryotic organisms, will be discussed in Chapter 10.

## Methods of studying biosynthesis

Biochemistry was initially concerned with elucidation of the pathways of energy-yielding metabolism, many of which (e.g., the fermentations) are chemically fairly simple and are mediated by enzyme systems present at high levels in the cell. The elucidation of biosynthetic mechanisms is more recent. Work on this problem could not even be initiated until the role of ATP as an energetic coupling agent between energy-yielding metabolism and biosynthesis had been recognized. Furthermore, the unraveling of biosynthetic pathways required the development of new techniques which, although helpful, are rarely essential for the analysis of energy-yielding metabolism. The most important ones are the *use of isotopic labeling* and the *use of mutants.*

When a biosynthetic intermediate (for example, an amino acid) is furnished in the medium to a growing population of cells, it will often prevent its own endogenous synthesis (the mechanisms by which this control is effected are discussed in Chapter 8). The exogenously furnished compound is therefore preferentially incorporated by the cell into biosynthetic end-products. If the exogenously furnished compound is labeled with a radioisotope, chemical fractionation of the labeled cells can then reveal the ultimate location of radioactivity in various cell constituents. Such experiments show, for example, that $^{14}C$-labeled glutamic acid is incorporated into protein not only as glutamic acid residues but also as residues of two other amino acids, arginine and proline. This result demonstrates that the carbon skeleton of glutamic acid serves as a biosynthetic precursor of the carbon skeletons of arginine and proline.

Another valuable technique using radioisotopes is *pulse labeling.* A growing culture is briefly exposed to a highly radioactive biosynthetic precursor. During this exposure, a small quantity of the precursor enters the cell and starts to be distributed, through the various biosynthetic pathways, into biosynthetic end-products. If aliquots of the cell material are subjected to chemical fractionation at various times after pulse labeling, the appearance of radioactivity in biosynthetic intermediates can be followed. Such an experiment reveals the sequence of the intermediates leading to end-products. The pathway for the conversion of $CO_2$ to organic compounds by photosynthetic organisms and chemoautotrophs was largely established by experiments of this kind.

Radioisotopic methods are also valuable for the detection of biosynthetic reactions in cell-free systems, where the reaction rates are often too low to permit detection of the end-products by ordinary chemical analysis. The use of radioactive precursors permits radiochemical detection and identification of the end-products. These methods were indispensable for the study of the synthesis of proteins and nucleic acids in cell-free systems.

Biochemical mutants became an important tool for the study of biosynthesis after the demonstration in 1940 by G. Beadle and E. Tatum that it is possible to isolate so-called *auxotrophic* mutants. Such mutants require as growth factors biosynthetic intermediates which the parental strain can synthesize de novo. This requirement is caused by the inactivation of one enzyme, mediating a specific step in the specific biosynthetic pathway. If a series of mutants affected at different points in a particular biosynthetic pathway are isolated, the general nature of the pathway can often be deduced from the specific chemical nature of the substances which each of the mutants requires for growth. Furthermore, mutants of this type frequently synthesize and excrete the substrate for the specifically blocked reaction, which can be isolated from the medium and identified chemically.

## *The assimilation of inorganic carbon, nitrogen, and sulfur*

Many microorganisms assimilate carbon, nitrogen, and sulfur in inorganic form. The use of $CO_2$ as the sole carbon source is characteristic of algae, many photosynthetic bacteria, and chemoautotrophic bacteria. Ammonia can be used as a sole nitrogen source by many organisms, belonging to all nutritional categories. Some of these organisms (but not all of them) can likewise use nitrate as a nitrogen source. The ability to use $N_2$ as a nitrogen source (nitrogen fixation) is much rarer, being confined to certain groups of procaryotic protists. Most microorganisms can use sulfate as a source of sulfur.

The ability to use inorganic sources of C, N, and S as precursors of the organic components of the cell involves special series of biosynthetic reactions, required to incorporate these elements into compounds which serve as points of departure for general biosynthesis.

### The assimilation of carbon dioxide

The mechanisms by which chemoheterotrophs can fix $CO_2$ in the carboxyl group of oxalacetate, a reaction essential for the maintenance of the citric acid cycle, have been described in Chapter 6. In such organisms, $CO_2$ fixation is a minor reaction, most cell carbon being derived from organic sources. Photoautotrophs and chemoautotrophs, for which $CO_2$ serves as the sole or principal source of cellular carbon, fix $CO_2$ by a different mechanism, cyclic in nature. This mechanism was first elucidated in a green alga, *Chlorella*, by M. Calvin, A. Benson, and J. A. Bassham, and is sometimes called the *Calvin cycle*. Subsequent work has shown that it is universal in organisms, both eucaryotic and procaryotic, which depend on $CO_2$ as the principal carbon source.

The main features of this cycle are shown, in highly diagrammatic form, in Figure 7.1. The net result is the conversion of $CO_2$ to hexose

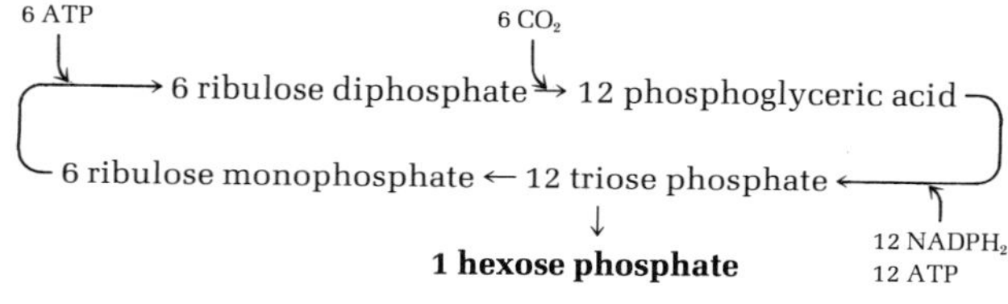


*Figure 7.1 The cyclic mechanism for the synthesis of hexose phosphate from $CO_2$ by photoautotrophs and chemoautotrophs.*

monophosphate and requires an expenditure of 18 moles of ATP (converted to ADP) and 12 moles of $NADPH_2$ (converted to NADP) for every mole of hexose phosphate formed. The biochemical events of the cycle can be divided into three categories:

*(a) The carboxylation step.* This involves the synthesis from a common intermediary metabolite, ribulose-5-phosphate, of a special $CO_2$ acceptor, ribulose-1,5-diphosphate. This pentose derivative then reacts with $CO_2$ to yield two molecules of a glycolytic intermediate, 3-phosphoglyceric acid:

$$\text{ribulose-5-phosphate} + \text{ATP} \rightarrow \text{ribulose-1,5-diphosphate} + \text{ADP}$$
$$\text{ribulose-1,5-diphosphate} + CO_2 \rightarrow 2\ \text{3-phosphoglyceric acid}$$

*(b) Conversion of 3-phosphoglyceric acid to fructose-6-phosphate.* This occurs through a reversal of the reactions of the Embden–Meyerhof pathway, the mechanism of which was discussed in Chapter 6 (see Figure 6.5). The conversion of hexose phosphate to 3-phosphoglyceric acid is an energy-yielding oxidation, so its reversal requires an expenditure of ATP and of reduced pyridine nucleotide, as shown in the following:

$$2\ \text{3-phosphoglyceric acid} + 2\text{ATP} + 2\text{NADPH}_2 \rightarrow \text{fructose-6-phosphate} + 2\text{ADP} + 2\text{P}_i + 2\text{NADP}$$

*(c) The regeneration of ribulose-5-phosphate.* This is brought about by a complex series of rearrangements of phosphorylated sugars formed through the reactions described in category (*b*). The overall reaction is

the conversion of 1 mole of fructose-6-phosphate and 3 moles of triose phosphate to 3 moles of pentose phosphate:

fructose-6-phosphate + 2 glyceraldehyde-3-phosphate +
dihydroxyacetone phosphate → 3 ribulose-5-phosphate

Schematically,

$C_6 + 3C_3 \rightarrow 3C_5$

A total of six separate reactions is involved:

fructose-6-phosphate + glyceraldehyde-3-phosphate ⇌
erythrose-4-phosphate + xylulose-5-phosphate
erythrose-4-phosphate + dihydroxyacetone phosphate ⇌
sedoheptulose-1,7-diphosphate
sedoheptulose-1,7-diphosphate + $H_2O$ → sedoheptulose-7-phosphate + $P_i$
sedoheptulose-7-phosphate + glyceraldehyde-3-phosphate ⇌
ribose-5-phosphate + xylulose-5-phosphate
xylulose-5-phosphate ⇌ ribulose-5-phosphate
ribose-5-phosphate ⇌ ribulose-5-phosphate

The way in which these six reactions are integrated to effect the overall conversion is shown in Figure 7.2, and the integration of the three

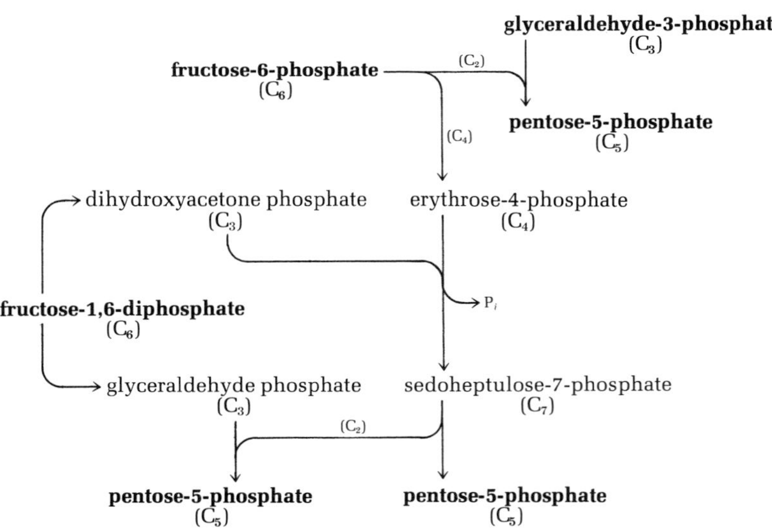


*Figure 7.2 The interconversion of sugar phosphates, which results in the formation of 3 moles of pentose phosphate from 1 mole of glyceraldehyde phosphate, 1 mole of fructose monophosphate, and 1 mole of fructose diphosphate.*

component parts of the cycle (*a*, *b*, and *c*) is schematized in Figure 7.3.

It is important to note that of the 13 individual steps that compose this cycle, a total of 11 (those listed under *b* and *c*) occur in most or all organisms: the reactions of the Embden–Meyerhof pathway (*b*) are of

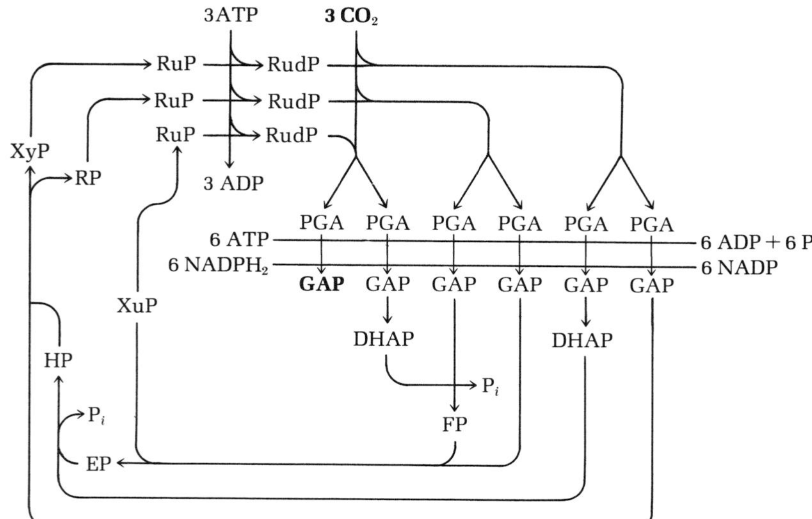


*Figure 7.3 A schematic representation of the reactions involved in the conversion of 3 moles of $CO_2$ to triose phosphate: RuP, ribulose-5-phosphate; RudP, ribulose-1,5-diphosphate; PGA, phosphoglyceric acid; GAP, glyceraldehyde phosphate; DHAP, dihydroxyacetone phosphate; FP, fructose-6-phosphate; EP, erythrose phosphate; HP, heptose phosphate; XyP, xylose phosphate; XuP, xylulose phosphate; RP, ribose phosphate. After H. R. Mahler and E. H. Cordes, Biological Chemistry. New York: Harper & Row, 1966.*

wide biological distribution, and the sugar phosphate rearrangements (c) always play an important role in the central core of biosynthetic reactions, since they are involved in the formation of pentose phosphate, required for nucleic acid synthesis. *Only two reactions, those listed under a, are specific to organisms that use $CO_2$ as sole carbon source. It is, consequently, the possession of the two enzymes mediating these reactions that distinguishes the carbon metabolism of photoautotrophs and chemoautotrophs from that of heterotrophic organisms.*

**The assimilation of ammonia**

The nitrogen atom of ammonia (valence of −3) is at the same oxidation level as the nitrogen atoms in the organic constituents of the cell (R—$NH_2$, R═NH, R—N═R). The assimilation of ammonia does not, therefore, necessitate a change of valence. Ammonia is directly converted to organic form by the reductive amination of an intermediate of the TCA cycle, $\alpha$-ketoglutaric acid:

$$HOOC\text{—}(CH_2)_2\text{—}CO\text{—}COOH + NH_3 + NADH_2 \rightarrow HOOC\text{—}(CH_2)_2CHNH_2COOH + NAD + H_2O$$

The product, L-glutamic acid, can itself act as the acceptor of additional ammonia in an ATP-dependent reaction, which results in the formation of an amide, glutamine:

$$HOOC\text{—}(CH_2)_2CHNH_2COOH + ATP + NH_3 \rightarrow HOOC\ (CH_2)_2CHNH_2CONH_2 + ADP + P_i + H_2O$$

Glutamic acid and glutamine are, accordingly, the primary organic nitrogenous compounds formed through the assimilation of ammonia. Other nitrogenous compounds are formed from them by transfer of an amino or amide group. For example, other amino acids can be formed by *transamination* between glutamic acid and nonnitrogenous intermediary metabolites:

$$\text{L-glutamate} + \text{pyruvate} \rightleftharpoons \alpha\text{-ketoglutarate} + \text{L-alanine}$$
$$\text{L-glutamate} + \text{oxalacetate} \rightleftharpoons \alpha\text{-ketoglutarate} + \text{L-aspartate}$$
$$\text{L-glutamate} + \text{glyoxylate} \rightleftharpoons \alpha\text{-ketoglutarate} + \text{glycine}$$

**The assimilation of nitrate and molecular nitrogen**

The two other inorganic forms of nitrogen utilizable for biosynthesis by many microorganisms are $NO_3^-$ and $N_2$, in which the valence of the nitrogen atom is, respectively, +5 and 0. The assimilation of these nitrogen sources involves a *reduction of nitrogen to the valence of −3, by preliminary conversion to ammonia.*

Many organisms that can use nitrate as a nitrogen source cannot use it as a terminal electron acceptor for anaerobic respiration (see Chapter 6, p. 213). The reduction of nitrate as a source of cell nitrogen is therefore termed *assimilatory nitrate reduction,* to distinguish it from *dissimilatory nitrate reduction* (or denitrification) which was described in the context of energy-yielding metabolism in Chapter 6. Although the first step is chemically identical in both cases (reduction of nitrate to nitrite), it is mediated by different enzymes.

The first step in assimilatory nitrate reduction is mediated by nitrate reductase, a flavoprotein containing molybdenum. This enzyme is reduced by $NADPH_2$, and the electrons are transferred via FAD to molybdenum and thence to nitrate. The path of electron transfer can be schematized as

$$\begin{matrix} NADPH_2 & & FAD & & 2Mo^{5+} & & NO_3^- \\ & \times & & \times & & \times & \\ NADP & & FADH_2 & & 2Mo^{6+} & & NO_2^- \end{matrix}$$

*nitrate reductase*

The mechanism for the further reduction of nitrite to ammonia is not clear. It is presumed to involve three successive two-electron transfers:

$$\begin{matrix} & NO_2^- & \rightarrow & (x) & \rightarrow & NH_2OH & \rightarrow & NH_3 \\ \text{Valence of N:} & +3 & & +1 & & -1 & & -3 \end{matrix}$$

There is good evidence for the occurrence of hydroxylamine ($NH_2OH$) as an intermediate, but the nature of the intermediate of valence +1 is obscure. It could be one of the highly unstable compounds nitroxyl (HNO), nitramide ($NO_2NH_2$), or hyponitrous acid ($H_2N_2O_2$).

The reduction of nitrate to ammonia is thus a relatively complex process, involving several enzymes. This explains why many microorganisms that grow readily with ammonia cannot use nitrate as an alternative inorganic nitrogen source. It should be noted that molybdenum plays an essential role in assimilatory nitrate reduction. This fact had been suspected long before the biochemical mechanism of nitrate reduction to nitrite was known, as a result of the nutritional finding that molybdenum is essential for microbial growth with nitrate but not with ammonia. This element also plays an important role in the fixation of $N_2$ and is therefore an essential nutrient for all organisms growing at the expense of an oxidized form of inorganic N.

The mechanism of biological nitrogen fixation long resisted attempts at elucidation, primarily because of the difficulty of obtaining active cell-free extracts from nitrogen-fixing microorganisms. Even the product of the fixation process was unclear until recently; hydroxylamine and ammonia both appeared to be possible products. It is now established that $N_2$ is always reduced to ammonia, prior to incorporation into organic form. The experimental analysis of this problem became possible when methods were devised for obtaining nitrogen fixation in cell-free extracts; this has now been achieved both with blue-green algae and with a variety of $N_2$-fixing bacteria. Most of the experimental work has been done with two bacteria, the anaerobe *Clostridium pasteurianum* and the aerobe *Azotobacter vinelandii*. The essential requirements for activity of the nitrogen-fixing enzyme system, or *nitrogenase*, are ATP and a source of reducing power, which can be supplied in various ways. In clostridia, the natural reductant appears to be reduced ferredoxin; but this nonheme iron compound does not occur in aerobic bacteria and therefore cannot function in *Azotobacter*. In *Azotobacter* extracts, the necessary reducing power can be supplied by an artificial donor, sodium dithionite ($Na_2S_2O_4$). No inorganic intermediates between $N_2$ and $NH_3$ have so far been detected, and it is presumed that reduction takes place directly on the surface of the enzyme, as shown schematically in Figure 7.4.

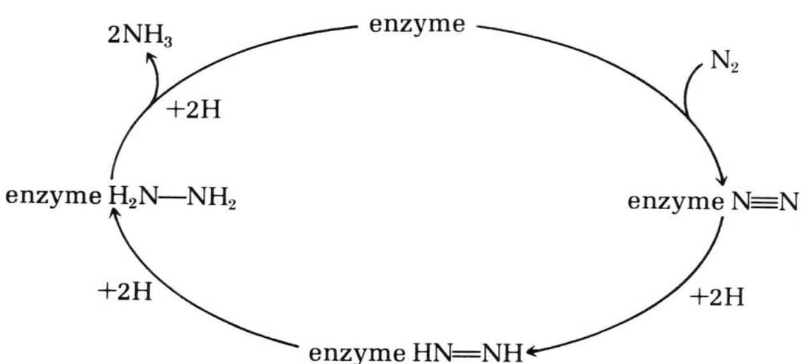


*Figure 7.4 A schematic diagram of the mechanism of biological nitrogen fixation.*

Nitrogenase appears to comprise at least two components: a protein that contains both iron and molybdenum and one that contains iron alone. An interesting feature of the system is that it can reduce other substrates as well as $N_2$. They include $N_2O$, cyanide, azide, and acetylene. These reductions will occur both in extracts and in the cell; the recently discovered reduction of acetylene, which yields ethylene,

$$CH{\equiv}CH \xrightarrow{+2H} CH_2{=}CH_2$$

provides the simplest method for detecting and measuring nitrogen-fixing capacity, both in vitro and in vivo. If no suitable electron acceptor is provided, nitrogenase uses the electrons derived from the donors to form $H_2$.

**The assimilation of sulfate**

The great majority of microorganisms can fulfill their sulfur requirements from sulfate. Sulfur has a valence of −2 in the sulfur-containing organic compounds of the cell, whereas it has a valence of +6 in sulfate. Accordingly, sulfate assimilation involves *a reduction of sulfate to sulfide*, prior to its incorporation into organic compounds. Chemically, this is equivalent to the reduction of sulfate by the sulfate-reducing bacteria, which use this anion as an electron acceptor in anaerobic respiration, as discussed in Chapter 6. The enzymatic mechanisms are different, however; and the reduction of sulfate for use as a sulfur source is termed *assimilatory sulfate reduction* (by analogy with assimilatory nitrate reduction) to distinguish it from the mechanistically and physiologically different process of *dissimilatory sulfate reduction:* the use of sulfate as a terminal electron acceptor.

Sulfide is incorporated into organic form as the sulfur atom of the amino acid cysteine. Serine (or a derivative of it) serves as the sulfide acceptor. In plants and eucaryotic microorganisms, the sulfide acceptor is serine itself:

$$CH_2OH{-}CHNH_2{-}COOH + H_2S \rightarrow CH_2SH{-}CHNH_2{-}COOH + H_2O$$

In bacteria that have been examined, the acceptor is O-acetylserine, formed by an acetylation of serine with acetyl CoA. The reaction then proceeds:

$$\text{O-acetylserine} + H_2S \rightarrow \text{cysteine} + \text{acetic acid} + H_2O$$

## *The strategy of biosynthesis*

On a weight basis, most of the organic matter of the cell consists of relatively large molecules which belong to four classes: nucleic acids, proteins, polysaccharides, and complex lipids. These substances are

*Table 7.1 Classes of large organic molecules of the cell and their building blocks*

| Class | Chemical nature of subunits | Number of different kinds used |
|---|---|---|
| *Nucleic acids* | *Nucleotides* | 8 |
| *Deoxyribonucleic acids* | *Deoxyribonucleotides* | 4 |
| *Ribonucleic acids* | *Ribonucleotides* | 4 |
| *Proteins* | *Amino acids* | 20 |
| *Polysaccharides* | *Monosaccharides* | ~15[a] |
| *Complex lipids* | *Variable* | ~20[a] |

[a]The number of building blocks used for the synthesis of any given member of these classes is usually much smaller.

assembled stepwise from organic precursors of lower molecular weight. Each class is defined by the specific nature of its precursors: nucleotides in the case of nucleic acids, amino acids in the case of proteins, and simple sugars in the case of polysaccharides. Complex lipids are more variable and heterogeneous in composition; their precursors may include fatty acids, polyalcohols, simple sugars, amines, and amino acids. As shown in Table 7.1, approximately 70 different precursors are required for the synthesis of all four classes. In addition, the cell contains a number of organic constituents which do not serve as precursors of larger molecules. These include about 20 coenzymes and carriers, which play a variety of catalytic roles in association with specific enzymes. In all, the minimal number of organic compounds of relatively low molecular weight that must be synthesized by any cell as precursors of larger molecules or as cell constituents in their own right is probably between 100 and 200. All these compounds are synthesized from a considerably smaller number of organic substances, which comprise a dynamic pool: the *central intermediary metabolites* of the cell. The most important of these central intermediary metabolites are sugar phosphates, pyruvate, acetate, oxalacetate, succinate, and α-ketoglutarate. They contribute the greater part of the carbon skeletons of the many different constituents of the cell.

The number of primary biosynthetic pathways that lead from this pool of central intermediary metabolites is considerably smaller than the number of products formed, as a result of the fact that most biosynthetic pathways are *branched* ones, several different biosynthetic products being formed by terminal divergence from a single primary biosynthetic reaction sequence.

In the following pages we shall trace the biosynthetic origins from the central intermediary metabolic pool of a number of organic substances which either serve as precursors for the large organic molecules of the cell or play specific roles themselves in cellular function.

The concluding section of the chapter will discuss the nature of the stepwise assembly processes which lead to the formation of the large organic molecules of the cell.

## *The synthesis of nucleotides*

The monomeric precursors for the synthesis of nucleic acids are purine and pyrimidine nucleotides. One example has already been given in Chapter 6: adenylic acid(also called adenosine monophosphate, AMP). All nucleotides have the same general structure: the purine or pyrimidine base is attached through one of its nitrogen atoms by a glycosidic bond to pentose phosphate. The chemical parent compound, consisting of a purine or pyrimidine base attached to a nonphosphorylated pentose molecule, is termed a *nucleoside.* Free nucleosides do not play a role in biosynthesis; they are usually phosphorylated, and may bear one, two, or three phosphate groups. A nucleotide with one phosphate residue is termed a nucleoside monophosphate and may be abbreviated NMP, where N stands for the specific purine or pyrimidine base. Similarly, the nucleoside diphosphates and triphosphates may be designated NDP and NTP, respectively. The nucleotides derived from the nucleoside adenosine, namely, adenosine monophosphate, diphosphate and triphosphate, are thus abbreviated AMP, ADP, and ATP, respectively.

In *all* nucleotides, the bond between pentose and the first phosphate group is not energy rich: NMPs are therefore never energy-rich compounds. The pyrophosphate bonds between the first and succeeding phosphates are both energy rich, so that NDPs and NTPs contain one and two energy-rich phosphate bonds, respectively. The general structure of nucleotides is shown in Figure 7.5.

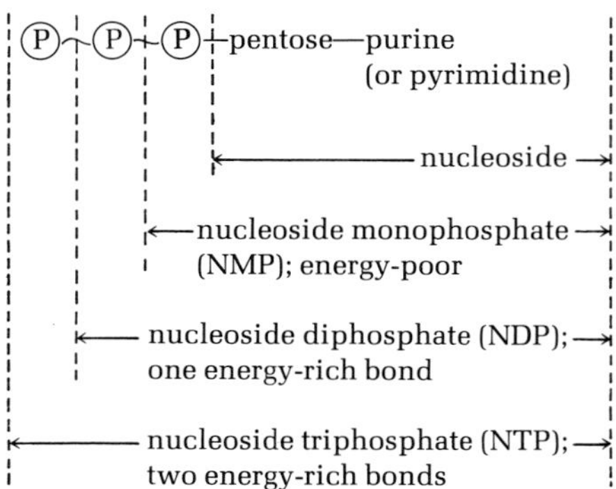


*Figure 7.5 The structure of purine and pyrimidine nucleotides.*

The nucleotides that serve as precursors of RNA contain ribose as the constituent sugar; those that serve as precursors of DNA contain a reduced pentose derived from ribose, deoxyribose, as the constituent sugar. Biosynthetically, the deoxyribonucleotides are all derived from ribonucleotides, which therefore serve as *indirect* precursors of DNA, in addition to serving as *direct* precursors of RNA. It should also be noted that the free ribonucleotides themselves play important direct roles in metabolism, as coenzymes, as carriers of energy-rich phosphate bonds, and as agents of *group transfer* in biosynthesis (e.g., in transferring sugar or amino acid residues into polymers). Free deoxyribonucleotides do not have these metabolic functions; they serve uniquely as precursors for DNA synthesis. The general patterns of nucleotide biosynthesis are shown schematically in Figure 7.6.

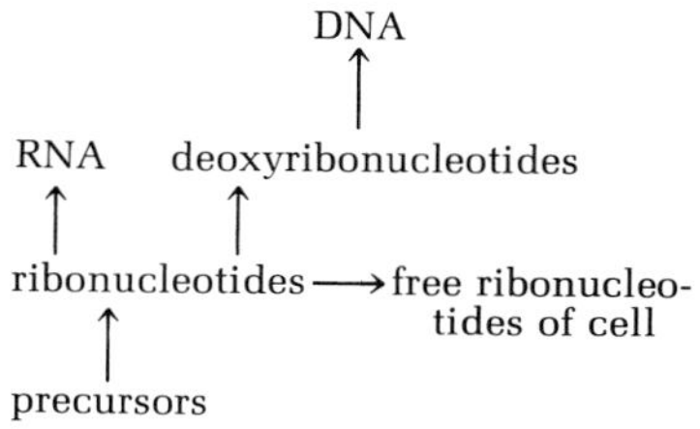

DNA
↑
RNA deoxyribonucleotides
↑ ↑
ribonucleotides ⟶ free ribonucleotides of cell
↑
precursors

*Figure 7.6 General pathway of nucleotide and nucleic acid synthesis.*

**The synthesis of ribonucleotides**

The synthesis of any nucleotide involves two distinct biosynthetic sequences: synthesis of the sugar moiety and synthesis of the attached purine or pyrimidine base. The sugar phosphate residue for all ribonucleotides is derived from the same precursor, ribose-5-phosphate, which is in all cases initially subjected to an additional phosphorylation, through transfer of two phosphate groups in pyrophosphate linkage to the 1 position on the sugar phosphate molecule. The donor is ATP:

ribose-5-phosphate + ATP → 5-phosphoribosyl-1-pyrophosphate + AMP

The product, 5-phosphoribosyl-1-pyrophosphate (PRPP), has the structure shown in Figure 7.7.

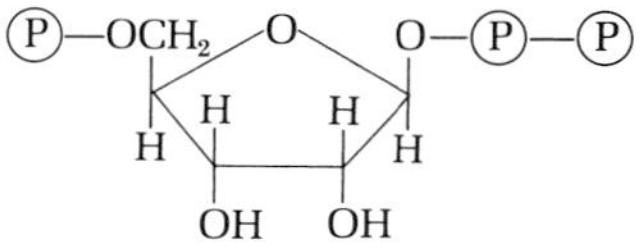

Ⓟ—$OCH_2$ O O—Ⓟ—Ⓟ
H H H H
OH OH

*Figure 7.7 Structure of 5-phosphoribosyl-1-pyrophosphate:* Ⓟ *denotes a phosphate group.*

This ribose derivative represents the starting point for the assembly of both purine and pyrimidine ribonucleotides. The two classes of ribonucleotides are formed through two distinct biosynthetic pathways, involving completely different chemical mechanisms. The purine nucleotides are synthesized through a stepwise assembly of the purine ring, in attachment to pentose phosphate; the pyrimidine nucleotides are synthesized by the separate assembly of the pyrimidine nucleus, which is attached to PRPP after its assembly.

In the purine pathway (Figure 7.8) PRPP is aminated by transfer of an amino group from glutamine, which replaces the pyrophosphate group at position 1, thus establishing immediately the bond between the sugar moiety and the nitrogen atom of the eventual purine ring. The product then serves as an acceptor for subsequent additions, yielding eventually inosinic acid. This intermediate lies at a metabolic branch point: divergent reactions lead to the formation of the two major purine mononucleotides, AMP and GMP (guanosine monophosphate,

Figure 7.8 *Biosynthesis of purine nucleotides.*

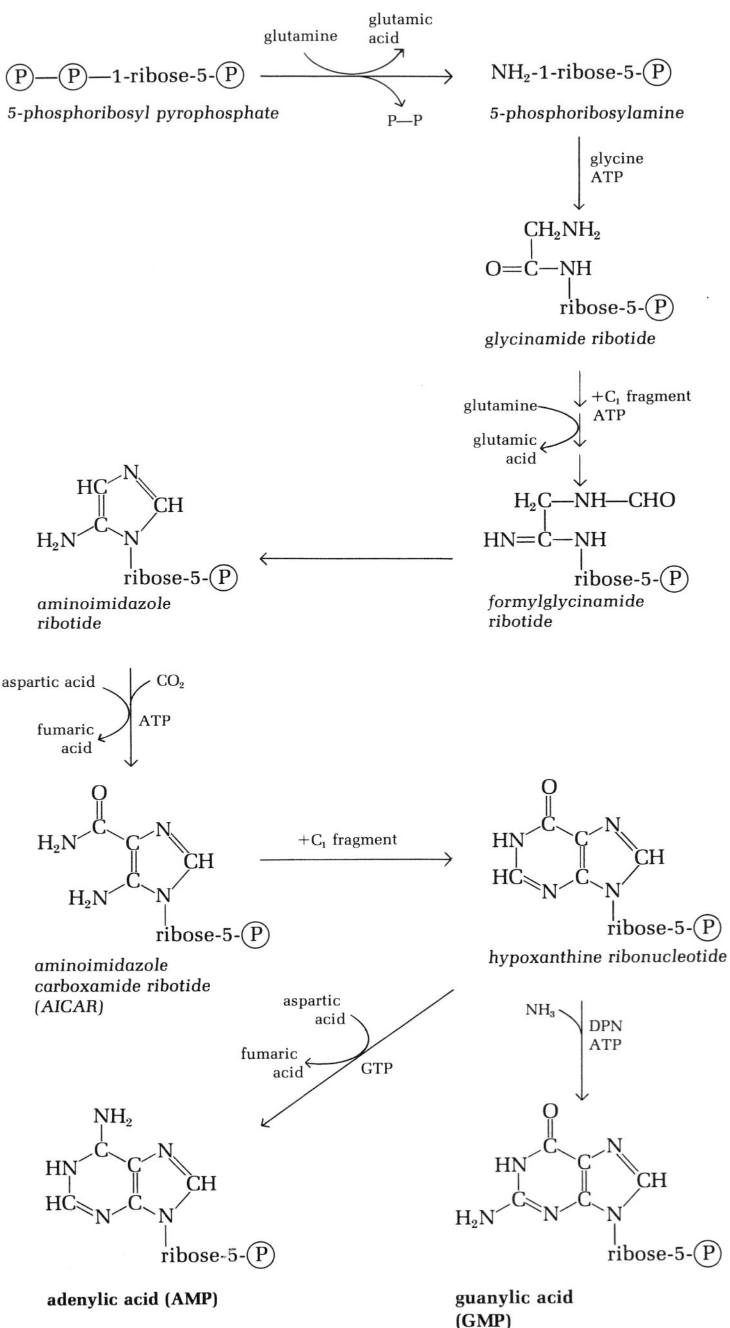

or guanylic acid). By additional phosphorylations these mononucleotides are converted to ATP and GTP, respectively, the immediate precursors of RNA.

In the pyrimidine pathway (Figure 7.9) the formation of the pyrimidine ring is initiated by a condensation between an amino acid, aspartate, and carbamyl phosphate; ring closure occurs by dehydration to

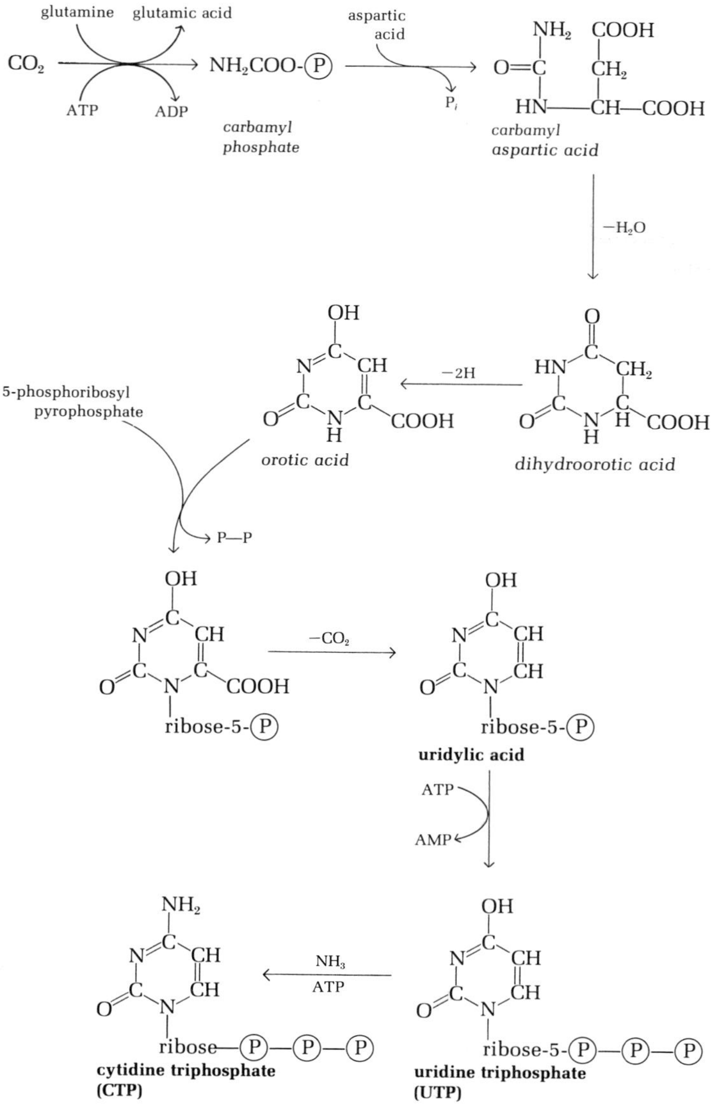

glutamine glutamic acid

$CO_2$ → $NH_2COO$-(P)

ATP ADP

*carbamyl phosphate*

aspartic acid → $P_i$

$NH_2$ COOH
O=C $CH_2$
HN—CH—COOH

*carbamyl aspartic acid*

$-H_2O$

*dihydroorotic acid*

−2H

*orotic acid*

5-phosphoribosyl pyrophosphate

P—P

ribose-5-(P)

$-CO_2$

ribose-5-(P)

**uridylic acid**

ATP

AMP

ribose-5-(P)—(P)—(P)

**uridine triphosphate (UTP)**

$NH_3$

ATP

ribose—(P)—(P)—(P)

**cytidine triphosphate (CTP)**

*Figure 7.9 Biosynthesis of pyrimidine nucleotides.*

yield dihydroorotate, which is reduced to a pyrimidine, orotate. Orotate is then coupled to PRPP, with elimination of pyrophosphate, to form a precursor from which there arises the pyrimidine mononucleotide, uridylic acid (also called uridine monophosphate, UMP). Successive phosphorylations lead to the formation of uridine triphosphate (UTP); this is aminated to form cytidine triphosphate (CTP): UTP and CTP are the immediate precursors of RNA. As already mentioned, all four ribonucleoside triphosphates, ATP, GTP, UTP, and CTP, also perform important metabolic functions in the free state.

### The synthesis of deoxyribonucleotides

The metabolic precursors of DNA are four deoxyribonucleotides. If the symbol dNTP is used to denote a deoxyribonucleoside triphosphate, they can be symbolized as dATP, dGTP, dCTP, and dTTP and contain the nucleosides adenosine, guanosine, cytidine, and thymidine, respectively. The bases in dATP, dGTP, and dCTP are, accordingly, identical with the bases in the corresponding ribonucleotides, ATP, GTP, and CTP. These three deoxyribonucleotides are synthesized by direct reduction of the pentose moiety in the corresponding ribonucleotides; the reductions occur at the level of the diphosphates [e.g., ADP → dADP; Figure 7.10(*b*)]. The ultimate reductant is $NADPH_2$, from which electrons are transferred via a protein carrier.

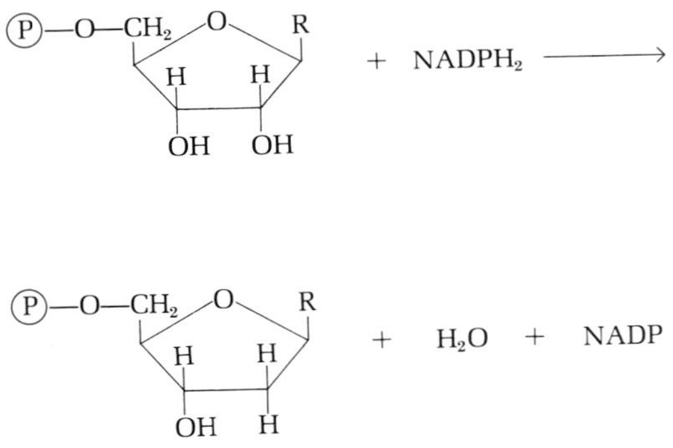

Ⓟ—O—$CH_2$ (ribose ring: O, R; H, H; OH, OH) + $NADPH_2$ ⟶

Ⓟ—O—$CH_2$ (deoxyribose ring: O, R; H, H; OH, H) + $H_2O$ + NADP

*Figure 7.10 Formation of deoxyribonucleotides from the corresponding nucleotides:* R *denotes the purine or pyrimidine residue.*

The formation of thymidine deoxyribonucleotide involves an additional sequence of reactions, in which the cytidine residue of dCDP is converted to the thymidine residue of dTDP [Figure 7.10(*b*)]. The deoxynucleoside diphosphates are then phosphorylated to yield the corresponding triphosphates.

## The synthesis of amino acids

Eighteen amino acids and two amides are required for the synthesis of proteins. Only one amino acid, histidine, has a completely isolated biosynthetic origin. The other 19 compounds are derived, through branched biosynthetic pathways, from a relatively small number of metabolic precursors; they can be grouped, in terms of biosynthetic origin, into a total of five "families," as shown in Table 7.2. We shall sketch in a summary manner the pathways involved.

We have already discussed the origin of two members of the glutamate family, glutamate itself and the corresponding amide, glutamine. Both are synthesized from $\alpha$-ketoglutarate, by two successive additions of ammonia. The other two members of the glutamate family, L-proline and L-arginine, are derived from glutamate through divergent pathways (Figure 7.11).

The parent amino acid of the aspartate family, L-aspartate, arises by transamination of oxalacetate and can be further aminated to yield the amide asparagine, in a reaction analogous to the formation of glutamine from glutamate. The other amino acids belonging to this family are

*Table 7.2 Biosynthetic derivations of the amino acids*

| Precursors | Products | |
|---|---|---|
| *α-Ketoglutarate → glutamate* → | *glutamine, arginine, proline* | *"Glutamate family"* |
| *Oxalacetate → aspartate* → | *asparagine, methionine, threonine (→ isoleucine), isoleucine (in part), lysine (in part)*[a] | *"Aspartate family"* |
| *Phosphoenolpyruvate + erythrose-4-phosphate* → | *phenylalanine (in part), tyrosine (in part), tryptophan (in part)* | *"Aromatic family"* |
| *3-Phosphoglycerate → serine* → | *glycine, cysteine* (serine also → *tryptophan*) | *"Serine family"* |
| *Pyruvate* → | *alanine, valine, leucine (in part)* | *"Pyruvate family"* |
| *Phosphoribosyl pyrophosphate + ATP* → | *histidine (in part)* | |

[a] In all procaryotic and some eucaryotic protists, lysine has an alternative derivation (see the text).

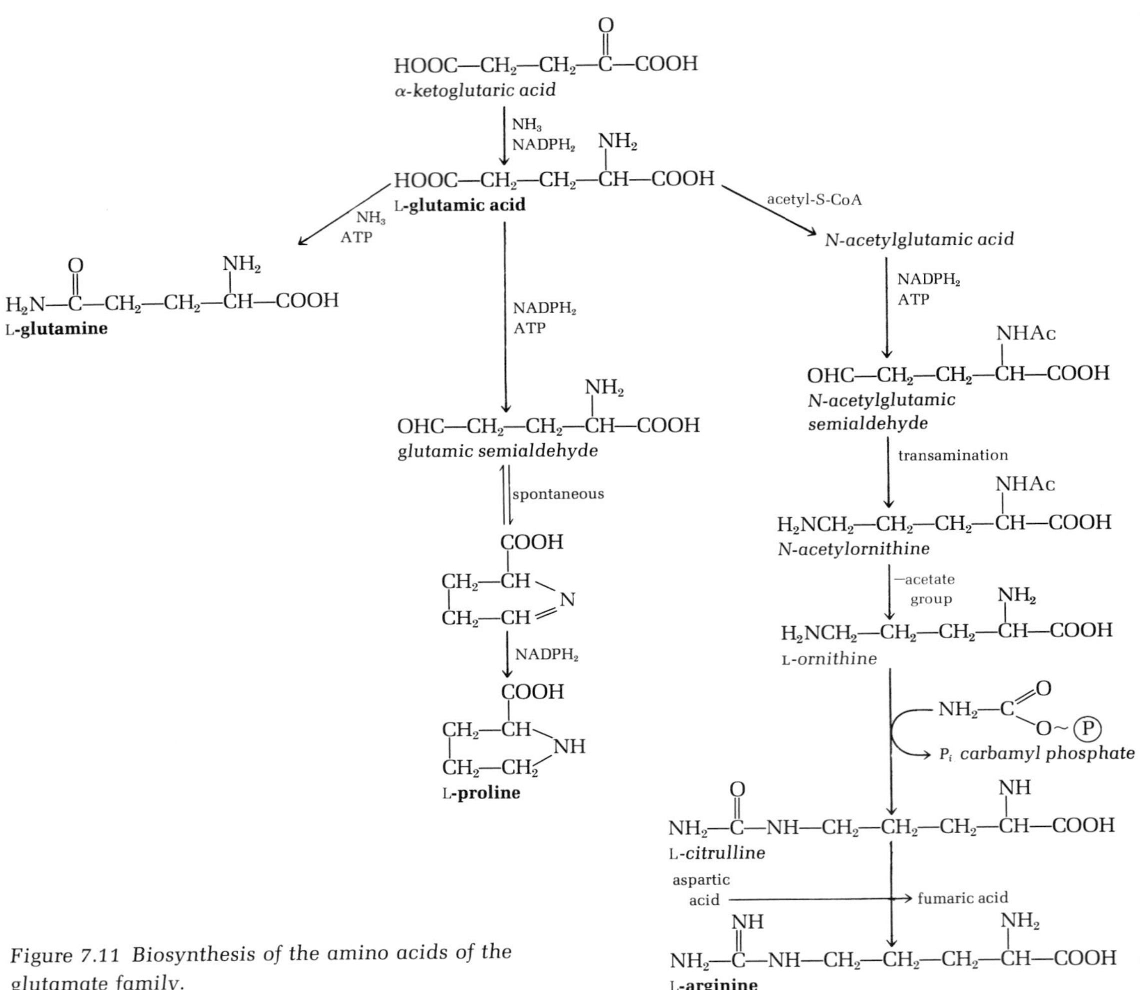


*Figure 7.11 Biosynthesis of the amino acids of the glutamate family.*

formed through a branched pathway (Figure 7.12). The initial common steps are the phosphorylation of aspartate and the reduction of the product, aspartyl phosphate, to aspartic semialdehyde. In many organisms, the first divergence occurs at this point: one branch leads to L-lysine, the other to L-methionine and L-threonine; L-threonine serves also (in part) for the synthesis of L-isoleucine.

The sequence of reactions (Figure 7.13) that leads from aspartic semialdehyde to L-lysine is characteristic of all procaryotic protists, certain aquatic phycomycetes, green algae, and plants. However, some eucaryotic protists (most fungi and the euglenid algae) synthesize L-

Figure 7.12 *Biosynthesis of the amino acids of the aspartate family.*

$HOOC-CH_2-C(=O)-COOH$
*oxalacetic acid*

↓ transamination

$HOOC-CH_2-CH(NH_2)-COOH$
L-**aspartic acid** $\xrightarrow{+NH_3,\ ATP}$ $H_2N-C(=O)-CH_2-CH(NH_2)-COOH$
L-**asparagine**

↓ ATP

↓ $NADPH_2$

$OCH-CH_2-CH(NH_2)-COOH$ → → → → L-**lysine**
*aspartic semialdehyde*

↓ $NADPH_2$

$HOCH_2-CH_2-CH(NH_2)-COOH$
*homoserine* → → $CH_3-S-CH_2-CH_2-CH(NH_2)-COOH$
L-**methionine**

↓ ATP

Ⓟ$-OCH_2-CH_2-CH(NH_2)-COOH$
*phosphohomoserine*

↓ $-P_i$

$CH_3-CHOH-CH(NH_2)-COOH$
L-**threonine**

↓ $-NH_3$

$CH_3-CH_2-C(=O)-COOH$
*α-ketobutyric acid*

↓

$CH_3CH_2-CH(CH_3)-CH(NH_2)-COOH$
L-**isoleucine**

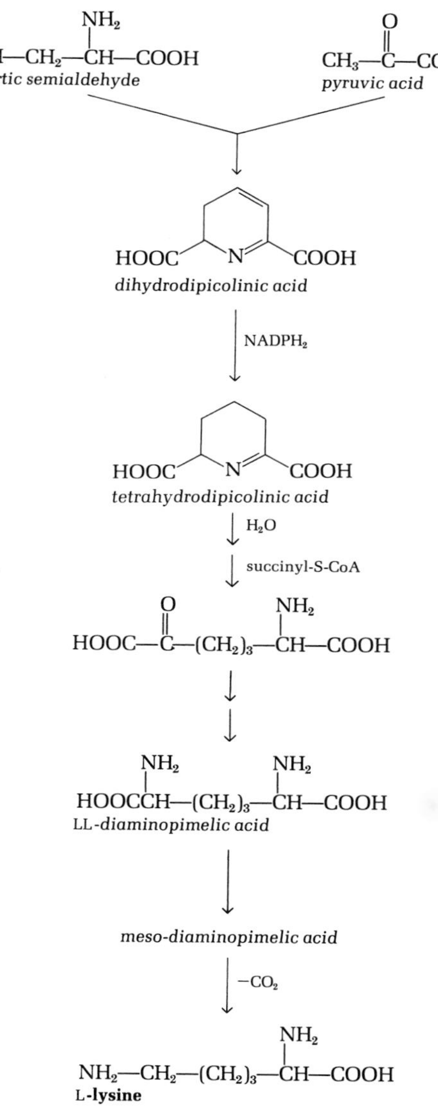


7.13 *The lysine branch of the aspartate pathway.*

lysine by a completely independent pathway (Figure 7.14), so that in these organisms lysine is not a member of the aspartate family. This is the most significant exception to the essential unity of biosynthetic core reactions that has so far been discovered.

In the biosynthesis of lysine from aspartic semialdehyde, the LL and meso isomers of diaminopimelic acid serve as penultimate intermediates. Although these two amino acids are never incorporated into proteins, they serve, in most procaryotic organisms, as *precursors of the murein component of the cell wall.* In green algae and plants, they serve

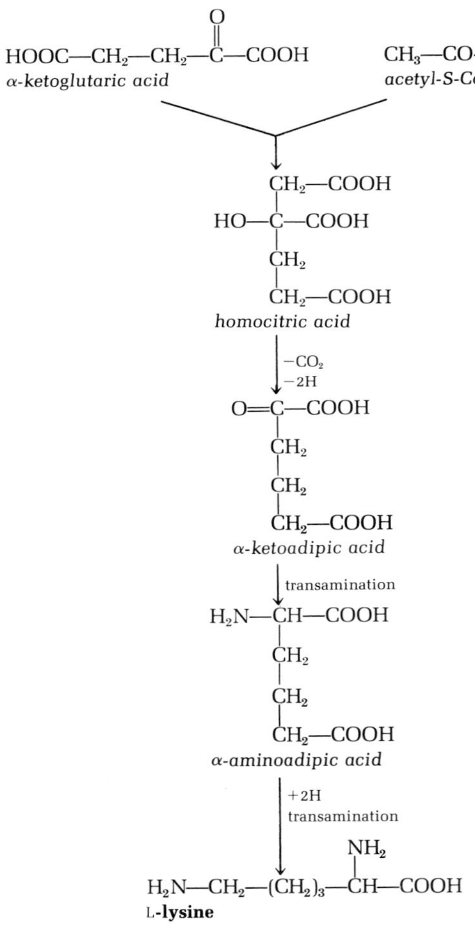

$$HOOC{-}CH_2{-}CH_2{-}\overset{\displaystyle O}{\overset{\|}{C}}{-}COOH \qquad CH_3{-}CO\text{-}S\text{-}CoA$$

*α-ketoglutaric acid* *acetyl-S-CoA*

↓

$$\begin{array}{l} \quad\;\; CH_2{-}COOH \\ \quad\;\;\; | \\ HO{-}C{-}COOH \\ \quad\;\;\; | \\ \quad\;\; CH_2 \\ \quad\;\;\; | \\ \quad\;\; CH_2{-}COOH \end{array}$$

*homocitric acid*

↓ $-CO_2$, $-2H$

$$\begin{array}{l} O{=}C{-}COOH \\ \quad\;\; | \\ \quad\; CH_2 \\ \quad\;\; | \\ \quad\; CH_2 \\ \quad\;\; | \\ \quad\; CH_2{-}COOH \end{array}$$

*α-ketoadipic acid*

↓ transamination

$$\begin{array}{l} H_2N{-}CH{-}COOH \\ \qquad\;\; | \\ \qquad CH_2 \\ \qquad\;\; | \\ \qquad CH_2 \\ \qquad\;\; | \\ \qquad CH_2{-}COOH \end{array}$$

*α-aminoadipic acid*

↓ $+2H$ transamination

$$H_2N{-}CH_2{-}(CH_2)_3{-}\overset{\displaystyle NH_2}{\overset{|}{C}H}{-}COOH$$

L-**lysine**

*Figure 7.14 The alternate pathway for the synthesis of lysine.*

uniquely as biosynthetic intermediates. One other substance, found as a cell constituent only in certain procaryotic organisms, is derived from this pathway: dipicolinic acid, which is a major constituent of the endospore in spore-forming eubacteria (see Chapter 20). Among procaryotic organisms, accordingly, the lysine pathway has biosynthetic roles in addition to its role in lysine synthesis (Figure 7.15).

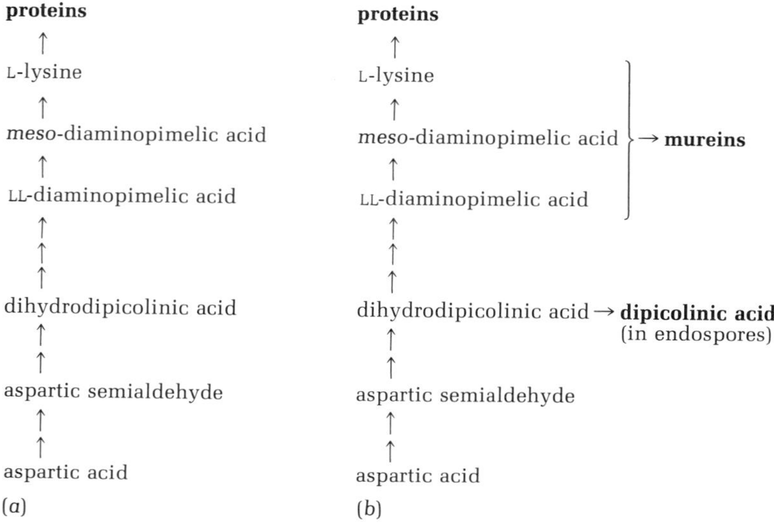


*Figure 7.15 Differences in the biosynthetic role of the lysine branch of the aspartate pathway in (a) eucaryotic groups (green algae and plants) and (b) procaryotic groups (blue-green algae and bacteria).*

The third major family, the aromatic family, includes as products the three aromatic amino acids L-tyrosine, L-phenylalanine, and L-tryptophan (Figure 7.16). Their synthesis is initiated by a condensation between phosphoenolpyruvate and erythrose-4-phosphate to yield a 7-carbon phosphorylated intermediate, which undergoes cyclization to form 5-dehydroquinic acid; the ring carbons of this intermediate eventually became the aromatic ring carbons of the end-products. Addition of another molecule of phosphoenolpyruvate to a later hydroaromatic intermediate, 5-phosphoshikimic acid, provides the carbon atoms for the side chains of tyrosine and phenylalanine. The product of this second condensation, chorismic acid, is the site of the first metabolic divergence: it can be converted either to prephenic acid, the precursor of tyrosine and phenylalanine, or (by loss of the side chain) to

Figure 7.16 Biosynthesis of the amino acids of the aromatic family.

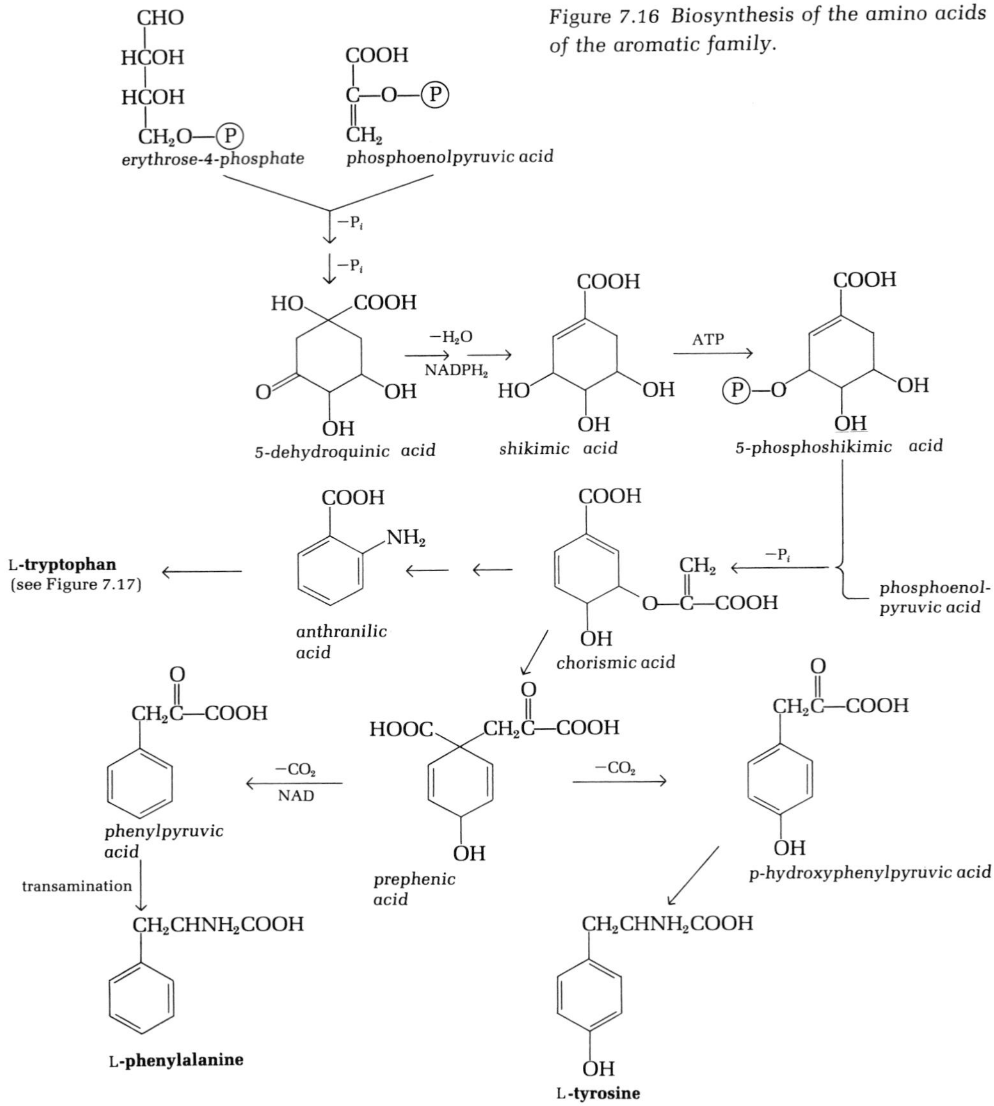

erythrose-4-phosphate

phosphoenolpyruvic acid

$-P_i$

$-P_i$

5-dehydroquinic acid

$-H_2O$

$NADPH_2$

shikimic acid

ATP

5-phosphoshikimic acid

phosphoenol-pyruvic acid

$-P_i$

chorismic acid

anthranilic acid

L-**tryptophan** (see Figure 7.17)

prephenic acid

$-CO_2$

NAD

phenylpyruvic acid

transamination

L-**phenylalanine**

p-hydroxyphenylpyruvic acid

L-**tyrosine**

anthranilic acid, the precursor of tryptophan. A further series of additions and rearrangements (Figure 7.17) results in the conversion of anthranilic acid to tryptophan. The aromatic pathway also furnishes, via chorismate, certain aromatic products that do not enter protein synthesis: *p*-aminobenzoic acid, a precursor of the coenzyme folic acid; and *p*-hydroxybenzoic acid, a precursor of quinones, such as ubiquinone.

Figure 7.17 *Terminal steps in the conversion of anthranilic acid to L-tryptophan.*

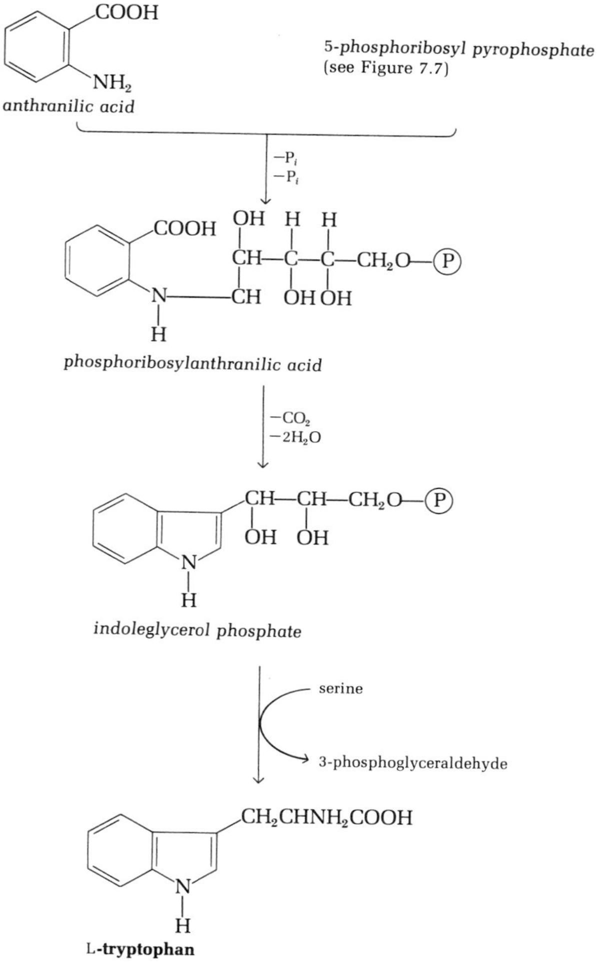

anthranilic acid + 5-phosphoribosyl pyrophosphate (see Figure 7.7)

$-P_i$, $-P_i$

phosphoribosylanthranilic acid

$-CO_2$, $-2H_2O$

indoleglycerol phosphate

serine → 3-phosphoglyceraldehyde

$CH_2CHNH_2COOH$

**L-tryptophan**

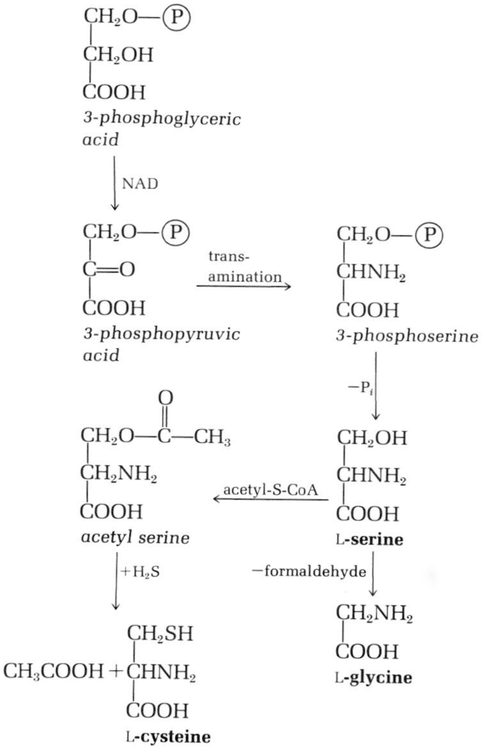

3-phosphoglyceric acid ($CH_2O$—Ⓟ, $CH_2OH$, COOH)

NAD

3-phosphopyruvic acid ($CH_2O$—Ⓟ, C=O, COOH) —trans-amination→ 3-phosphoserine ($CH_2O$—Ⓟ, $CHNH_2$, COOH)

$-P_i$

**L-serine** ($CH_2OH$, $CHNH_2$, COOH) —acetyl-S-CoA→ acetyl serine ($CH_2O$—C(=O)—$CH_3$, $CH_2NH_2$, COOH)

$+H_2S$

$CH_3COOH$ + **L-cysteine** ($CH_2SH$, $CHNH_2$, COOH)

L-serine —formaldehyde→ **L-glycine** ($CH_2NH_2$, COOH)

Figure 7.18 *Biosynthesis of the amino acids of the serine family.*

The pathways for formation of the amino acids of the serine family (serine, glycine, and cysteine) and of the pyruvate family (alanine, valine, and leucine) are shown in Figures 7.18 and 7.19, respectively.

Finally, there is the isolated pathway of L-histidine biosynthesis (Figure 7.20). The chain of five carbon atoms in the skeleton of this amino acid is derived from PRPP, already encountered as a precursor of nucleotides and of tryptophan. Two of these atoms contribute to the

*Figure 7.19 Biosynthesis of the amino acids of the pyruvate family.*

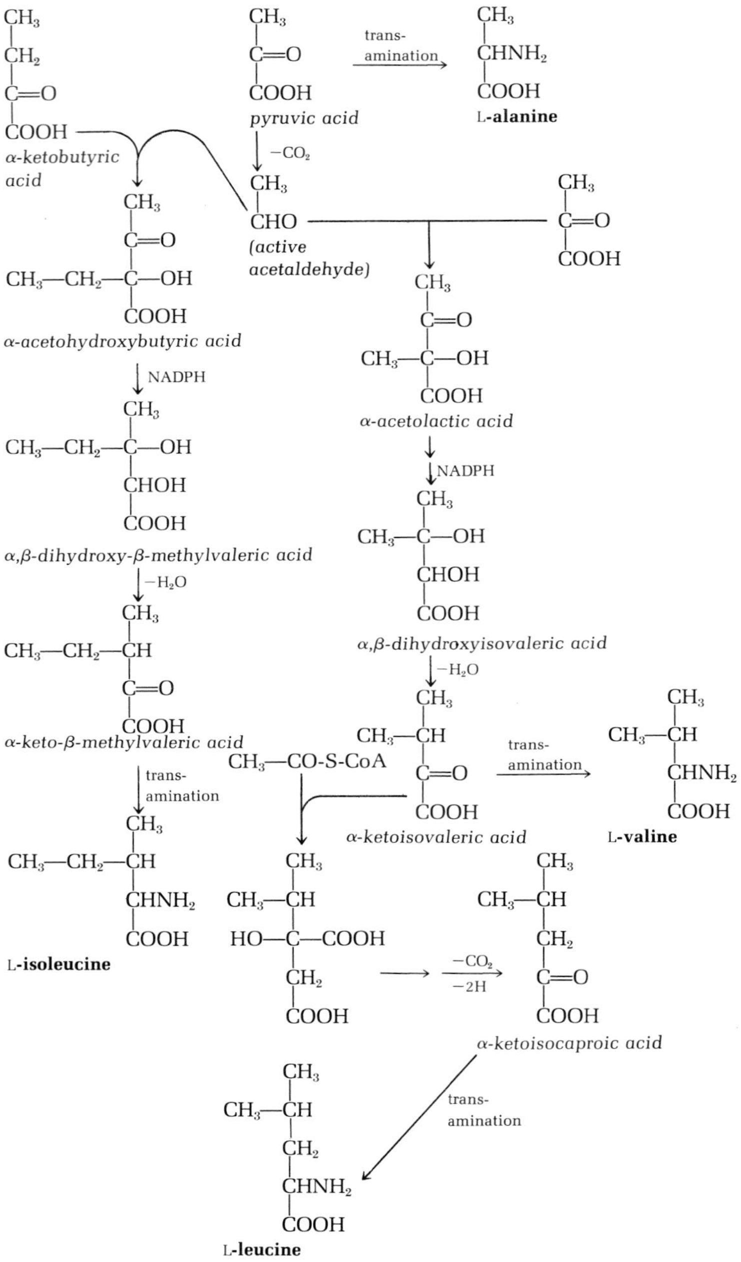

$CH_3$—$CH_2$—C(=O)—COOH
*α-ketobutyric acid*

$CH_3$—C(=O)—COOH
*pyruvic acid*

trans-amination →

$CH_3$—$CHNH_2$—COOH
**L-alanine**

↓ $-CO_2$

$CH_3$—CHO
*(active acetaldehyde)*

$CH_3$—C(=O)—COOH

$CH_3$—$CH_2$—C(OH)(COOH)—C(=O)—$CH_3$
*α-acetohydroxybutyric acid*

↓ NADPH

$CH_3$—$CH_2$—C($CH_3$)(OH)—CHOH—COOH
*α,β-dihydroxy-β-methylvaleric acid*

↓ $-H_2O$

$CH_3$—$CH_2$—CH($CH_3$)—C(=O)—COOH
*α-keto-β-methylvaleric acid*

↓ trans-amination

$CH_3$—$CH_2$—CH($CH_3$)—$CHNH_2$—COOH
**L-isoleucine**

$CH_3$—C(OH)(COOH)—C(=O)—$CH_3$
*α-acetolactic acid*

↓ NADPH

$CH_3$—C($CH_3$)(OH)—CHOH—COOH
*α,β-dihydroxyisovaleric acid*

↓ $-H_2O$

$CH_3$—CH($CH_3$)—C(=O)—COOH
*α-ketoisovaleric acid*

trans-amination →

$CH_3$—CH($CH_3$)—$CHNH_2$—COOH
**L-valine**

$CH_3$—CO-S-CoA

$CH_3$—CH($CH_3$)—C(OH)(COOH)—$CH_2$—COOH

→ $\xrightarrow[-2H]{-CO_2}$

$CH_3$—CH($CH_3$)—$CH_2$—C(=O)—COOH
*α-ketoisocaproic acid*

trans-amination

$CH_3$—CH($CH_3$)—$CH_2$—$CHNH_2$—COOH
**L-leucine**

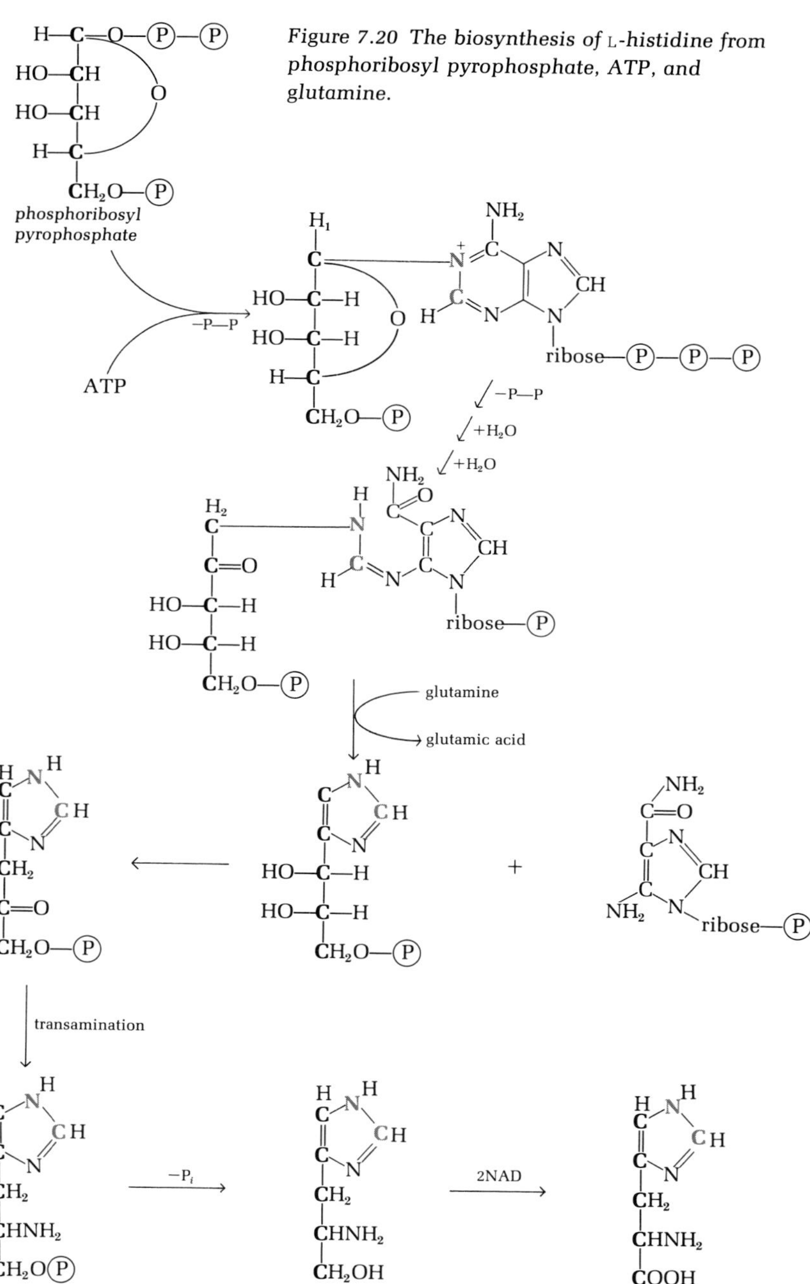


*Figure 7.20 The biosynthesis of* L*-histidine from phosphoribosyl pyrophosphate, ATP, and glutamine.*

five-membered imidazole ring. The remaining three atoms of the imidazole ring have a curious derivation: a C—N— fragment is contributed from the purine nucleus of ATP and an additional nitrogen atom from glutamine. This metabolic utilization of ATP, as a donor of two atoms from the purine nucleus, is unique. At first sight, it

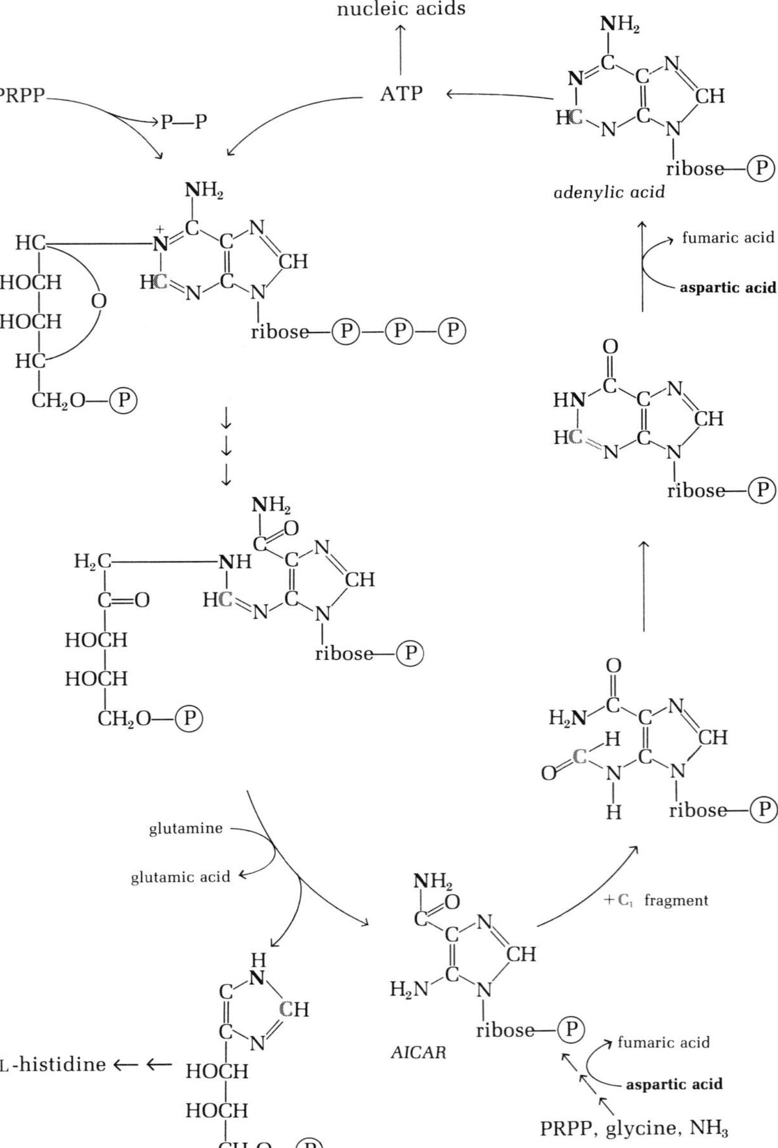


Figure 7.21 The intermeshing of histidine and purine biosynthesis.

seems an extremely uneconomical way of performing the required operation. The physiological rationale becomes apparent when one realizes that the product formed from ATP in this process, aminoimidazole carboxamide ribotide, is *itself an intermediate in purine biosynthesis.* There is, accordingly, an intimate connection between the biosynthesis of histidine and of the purine nucleotides: ATP is destroyed in the course of histidine synthesis, but the product of its breakdown can reenter the path of purine synthesis, at a relatively late stage (Figure 7.21).

## *The synthesis of porphyrins*

Each of the many different organic molecules that serve as coenzymes or as prosthetic groups of enzymes is synthesized through a special pathway. Space does not permit individual accounts of all these pathways. As one illustration, we shall describe the synthesis of *porphyrins.* The biosynthetic end-products fall into two major groups: the iron-containing *hemes,* which serve as the prosthetic groups of cytochromes and many other enzymes, known collectively as *hemoproteins;* and the magnesium-containing *chlorophylls.*

The synthesis of porphyrins is initiated by a condensation of the amino acid glycine with succinyl CoA; this gives rise in three steps to porphobilinogen, the precursor of the entire porphyrin molecule (Figure 7.22). The condensation of four molecules of this intermediate

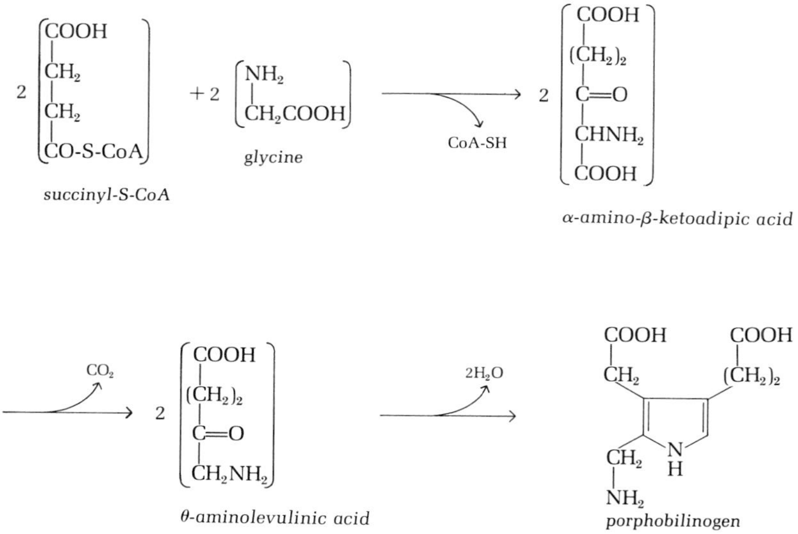

2 [COOH–$CH_2$–$CH_2$–CO-S-CoA] (*succinyl-S-CoA*) + 2 [$NH_2$–$CH_2COOH$] (*glycine*) → (−CoA-SH) 2 [COOH–$(CH_2)_2$–C=O–$CHNH_2$–COOH] (*α-amino-β-ketoadipic acid*)

→ (−$CO_2$) 2 [COOH–$(CH_2)_2$–C=O–$CH_2NH_2$] (*θ-aminolevulinic acid*) → (−$2H_2O$) porphobilinogen (pyrrole ring N–H bearing COOH–$CH_2$–, COOH–$(CH_2)_2$– and $NH_2$–$CH_2$– substituents)

Figure 7.22 *The synthesis of porphobilinogen, the monopyrrolic precursor of hemes and chlorophylls, from succinyl-S-CoA and glycine.*

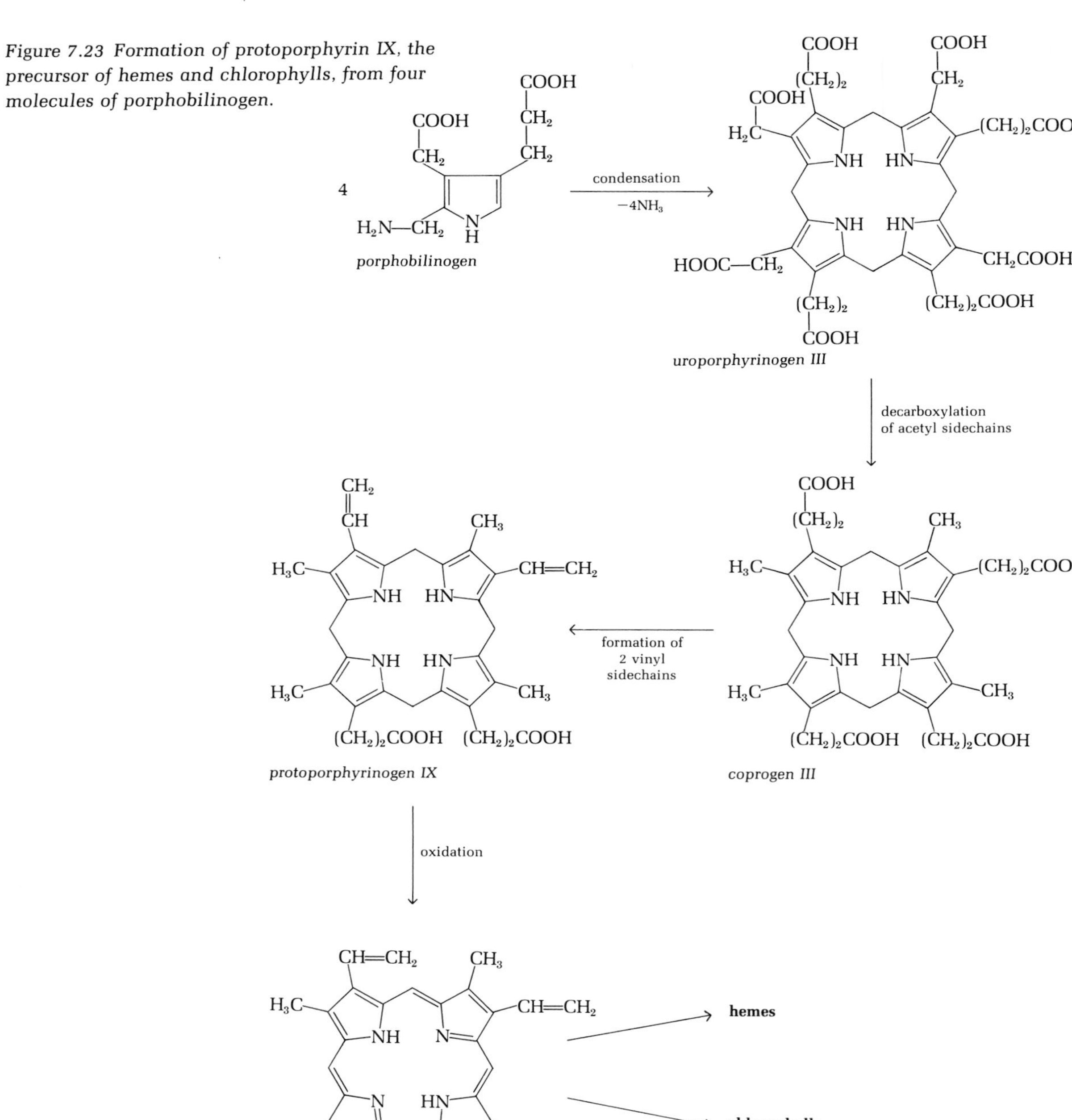

Figure 7.23 Formation of protoporphyrin IX, the precursor of hemes and chlorophylls, from four molecules of porphobilinogen.

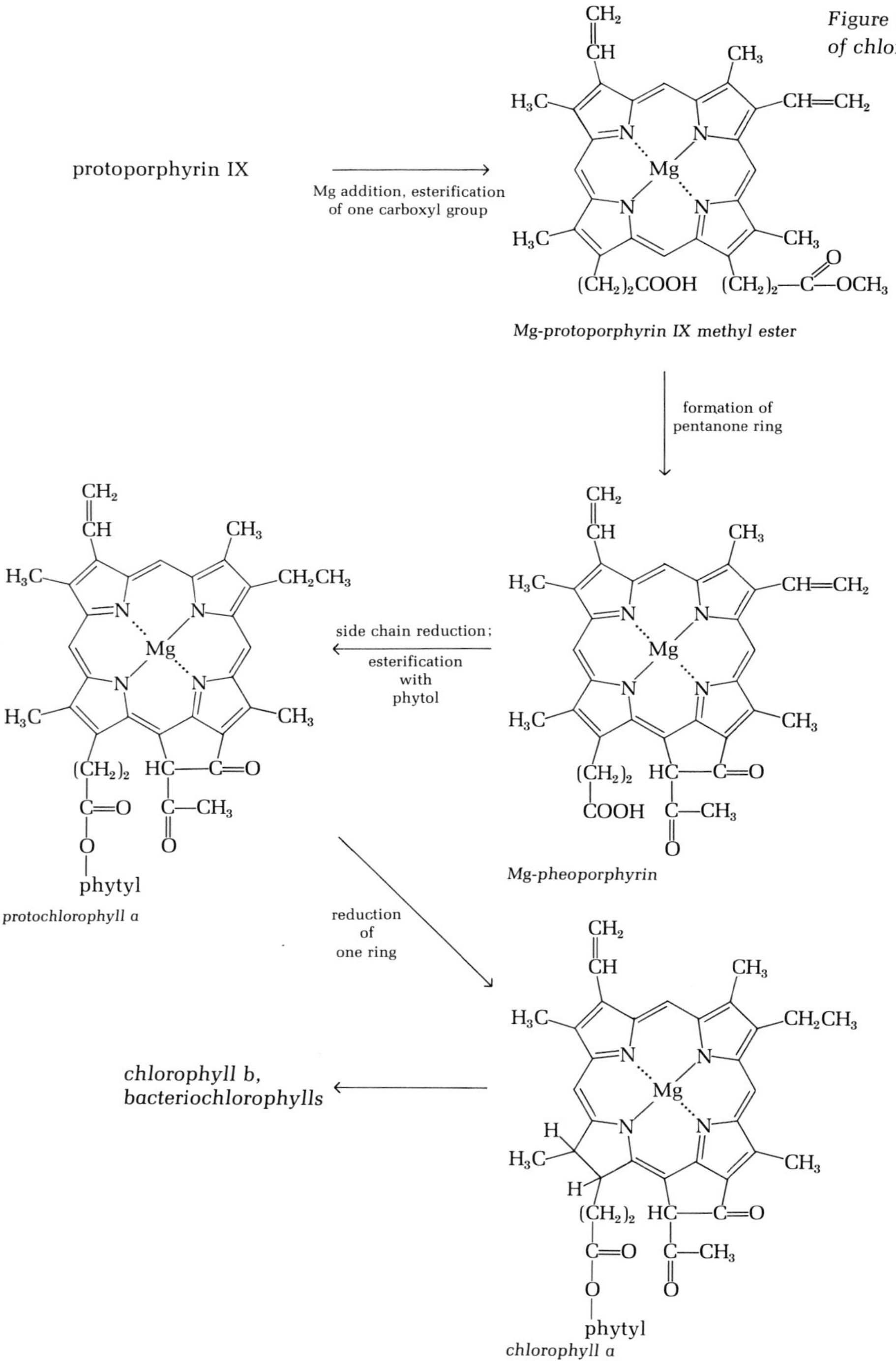

*Figure 7.24 Steps in the biosynthesis of chlorophylls.*

protoporphyrin IX

Mg addition, esterification of one carboxyl group

*Mg-protoporphyrin IX methyl ester*

formation of pentanone ring

*Mg-pheoporphyrin*

side chain reduction; esterification with phytol

*protochlorophyll a*

reduction of one ring

*chlorophyll a*

*chlorophyll b, bacteriochlorophylls*

forms the tetrapyrrolic nucleus of uroporphyrinogen III; subsequent modifications of the side chains and eventual oxidation yield protoporphyrin IX (Figure 7.23). The insertion of iron as a chelate in this molecule leads directly to the formation of a heme. If, alternatively, the metal magnesium is chelated with the ring, the product is Mg protoprophyrin, which is the precursor of chlorophylls. A long series of subsequent steps leads to the formation of the chlorophylls characteristic of the various groups of photosynthetic organisms. Most of these steps in chlorophyll synthesis are common ones: the divergences that give rise to the various specific plant chlorophylls and the bacteriochlorophylls occur near the end of the biosynthetic sequence (Figure 7.24).

### *The synthesis of lipid constituents from acetate*

Among the classes of large molecules in the cell, the lipids occupy an anomalous position. In the first place, even the most complex lipids are compounds of relatively low molecular weight, compared to nucleic acid, proteins, and polysaccharides. The molecular weights of lipids rarely exceed 1,000, whereas the molecular weights of the members of the other three classes are one or more orders of magnitude greater (in the range $10^4$ to $10^8$). Second, lipids are not made up of simpler building blocks that all belong to one chemical class. This point is illustrated in Figure 7.25 by the structure of several complex lipids. The *phospholipids*, which are universal membrane constituents, are derivatives of glycerol phosphate. Various classes of compounds,

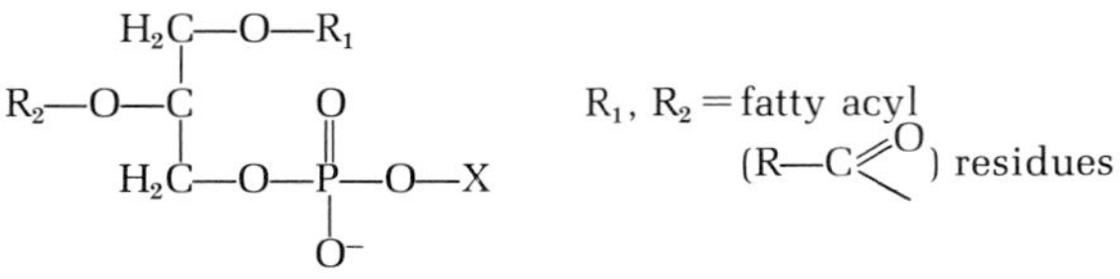

$$\begin{array}{l} \quad\;\; H_2C{-}O{-}R_1 \\ \quad\;\;\;\;\;| \\ R_2{-}O{-}C \qquad\quad O \\ \quad\;\;\;\;\;| \qquad\quad\;\; \| \\ \quad\;\; H_2C{-}O{-}P{-}O{-}X \\ \qquad\qquad\quad\;\;\; | \\ \qquad\qquad\quad\;\; O^- \end{array}$$

$R_1$, $R_2$ = fatty acyl ($R{-}C{\lessgtr}^{O}$) residues

*Basic structure of a phospholipid:*

| | |
|---|---|
| $X = {-}CH_2{-}CH_2{-}N^+{-}(CH_3)_3$ | *lecithin* |
| $X = {-}CH_2{-}CH_2{-}NH_2$ | *phosphatidyl ethanolamine* |
| $X = {-}CH_2{-}CHNH_2{-}COOH$ | *phosphatidyl serine* |
| $X = {-}CH_2{-}CHOH{-}CH_2OH$ | *phosphatidyl glycerol* |

*Basic structure of a glycolipid:*

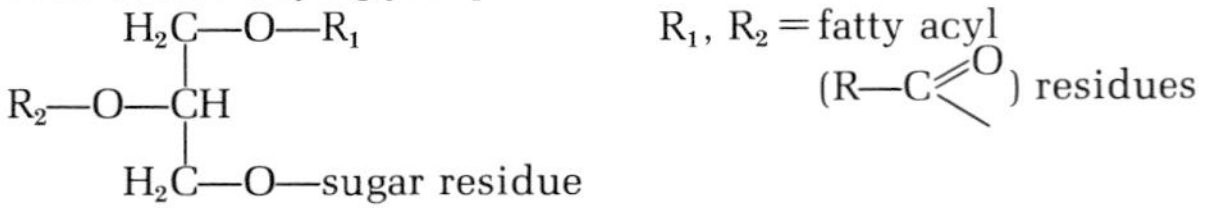

$$\begin{array}{l} \quad\;\; H_2C{-}O{-}R_1 \\ \quad\;\;\;\;\;| \\ R_2{-}O{-}CH \\ \quad\;\;\;\;\;| \\ \quad\;\; H_2C{-}O{-}\text{sugar residue} \end{array}$$

$R_1$, $R_2$ = fatty acyl ($R{-}C{\lessgtr}^{O}$) residues

Figure 7.25 General structure of phospholipids and glycolipids.

notably amines and amino acids, may be attached to the phosphate group, while the free hydroxyl groups on the other two carbon atoms of the glycerol residue are linked to fatty acids, aldehydes, or hydrocarbons. In the *glycolipids* a sugar residue replaces the phosphate group and its substituents.

It is not possible here to give a detailed account of how all the possible building blocks of lipids are synthesized; this section will be confined to a discussion of the way in which acetate serves as a precursor for the synthesis of fatty acids and other lipid materials.

## The synthesis of saturated fatty acids

The fatty acids of the cell are predominantly of considerable chain length ($C_{12}$ to $C_{18}$), unbranched, and contain an even number of carbon atoms. They may be completely saturated, or partially desaturated, containing one or more double bonds. These compounds are synthesized from acetyl units, but at no stage of their biosynthesis does acetate or any other free fatty acid participate in the process. All the biosynthetic intermediates are acyl derivatives

$$R—\overset{\diagup}{C}{=}O$$

which are attached by a —C—S— bond to a coenzyme or carrier protein.

In *Escherichia coli*, a special protein known as *acyl carrier protein* (ACP) plays a crucial part in the successive transformations (Figure 7.26). The formation of long-chain fatty acids starts with the transfer of an acetyl group from acetyl CoA to ACP (reaction 1). This complex serves as an acceptor, to which successive $C_2$ units are transferred, each transfer being followed by a reduction. The vehicle for each successive transfer is malonyl ACP, formed from acetyl CoA by reactions 2 and 3 (Figure 7.26). Each successive addition of an acetyl unit is accompanied by the liberation of the free carboxyl group of malonyl ACP as $CO_2$, which drives the reaction in the direction of synthesis (reaction 4; Figure 7.26). The product of this reaction is a $\beta$-ketoacyl ACP, which is reduced to the corresponding acyl ACP by a sequence of three reactions (reactions 5, 6, and 7; Figure 7.26). Repetitions of reactions 4, 5, 6, and 7 result in the successive formation of progressively longer acyl ACPs. The overall reaction can be schematized as

$$CH_3—CO—ACP + n\overset{\text{COOH}}{\overset{|}{C}}H_2—CO—ACP + 2nNADPH_2 + 2nH^+ \rightarrow$$
$$CH_3(CH_2—CH_2)_n—CO—ACP + nCO_2 + nACP + 2nNADP + nH_2O$$

In this bacterial system, the enzymes that catalyze reactions 1 to 7 (Figure 7.26) and ACP, the carrier, can be readily separated from one

Figure 7.26 *Mechanism for the synthesis of saturated fatty acids from acetyl-S-CoA in E. coli. The reactions leading to the synthesis of hexanoyl-($C_6$)-S-ACP are shown. By further transfers of acetyl units from malonyl-S-ACP and subsequent reductions (repetitions of reaction steps 4 to 7) unbranched fatty acids of progressively greater chain length containing even numbers of carbon atoms are formed.*

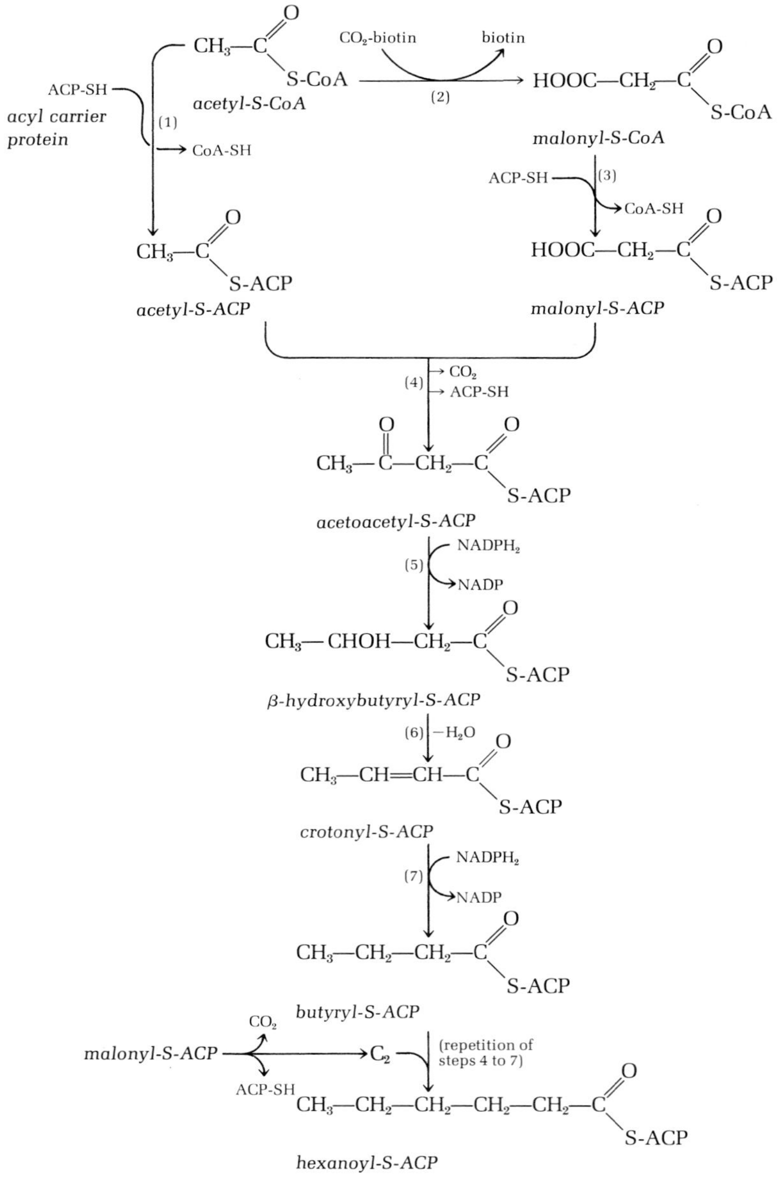

$$\text{(1)}\quad CH_3\text{—}CO\text{—S-CoA (acetyl-S-CoA)} + \text{ACP-SH (acyl carrier protein)} \rightarrow CH_3\text{—}CO\text{—S-ACP (acetyl-S-ACP)} + \text{CoA-SH}$$

$$\text{(2)}\quad CH_3\text{—}CO\text{—S-CoA} + CO_2\text{-biotin} \rightarrow HOOC\text{—}CH_2\text{—}CO\text{—S-CoA (malonyl-S-CoA)} + \text{biotin}$$

$$\text{(3)}\quad HOOC\text{—}CH_2\text{—}CO\text{—S-CoA} + \text{ACP-SH} \rightarrow HOOC\text{—}CH_2\text{—}CO\text{—S-ACP (malonyl-S-ACP)} + \text{CoA-SH}$$

$$\text{(4)}\quad \text{acetyl-S-ACP} + \text{malonyl-S-ACP} \rightarrow CH_3\text{—}CO\text{—}CH_2\text{—}CO\text{—S-ACP (acetoacetyl-S-ACP)} + CO_2 + \text{ACP-SH}$$

$$\text{(5)}\quad \text{acetoacetyl-S-ACP} + NADPH_2 \rightarrow CH_3\text{—}CHOH\text{—}CH_2\text{—}CO\text{—S-ACP } (\beta\text{-hydroxybutyryl-S-ACP}) + NADP$$

$$\text{(6)}\quad \beta\text{-hydroxybutyryl-S-ACP} - H_2O \rightarrow CH_3\text{—}CH{=}CH\text{—}CO\text{—S-ACP (crotonyl-S-ACP)}$$

$$\text{(7)}\quad \text{crotonyl-S-ACP} + NADPH_2 \rightarrow CH_3\text{—}CH_2\text{—}CH_2\text{—}CO\text{—S-ACP (butyryl-S-ACP)} + NADP$$

$$\text{malonyl-S-ACP} \rightarrow C_2 + CO_2 + \text{ACP-SH}; \quad \text{butyryl-S-ACP} + C_2 \xrightarrow{\text{(repetition of steps 4 to 7)}} CH_3\text{—}CH_2\text{—}CH_2\text{—}CH_2\text{—}CH_2\text{—}CO\text{—S-ACP (hexanoyl-S-ACP)}$$

another. In eucaryotic microorganisms (yeast) and in animal tissues, on the other hand, the entire sequence of reactions is catalyzed by a single multienzyme complex ("fatty acid synthetase") which exists as a particle of very high molecular weight (approximately 2.3 million) and cannot be dissociated into active subunits.

**The synthesis of unsaturated fatty acids**

Monounsaturated fatty acids, containing one double bond in the carbon chain, are universal cellular constituents. Components of this class include the derivative of palmitic acid ($C_{16}$), palmitoleic acid; and the derivatives of stearic acid ($C_{18}$), oleic and *cis*-vaccenic acid. The most common biochemical mechanism for the formation of monounsaturated fatty acids involves an oxygenative desaturation of the corresponding fatty acid; the double bond is always introduced between the ninth and tenth carbon atoms from the carboxyl end, so that palmitic acid yields palmitoleic acid and stearic acid yields oleic acid (Figure 7.27). Because of the obligatory involvement of molecular oxygen in this mechanism of desaturation, it is known as the *aerobic pathway*. Many bacteria (including aerobes as well as anaerobes) form monounsaturated fatty acids by a completely different mechanism, which does not involve the participation of molecular oxygen and is hence known as the *anaerobic pathway*. Its mechanism is outlined in Figure 7.28 and involves the introduction of a special branch from the normal chain of reactions of fatty acids biosynthesis at the $C_{10}$ level. The $C_{10}$ hydroxyacyl intermediate, $\beta$-OH decanoyl ACP, can undergo either normal $\alpha,\beta$ desaturation, leading to the formation of longer chain saturated fatty acids; or a $\beta,\gamma$ dehydration, leading to the homologous monounsaturated fatty acids. Note that in the anaerobic pathway, the position of the double bond in the carbon chain of the eventual end-products is determined by the point in biosynthesis at which it is introduced. Subsequent chain elongation leads to its location between carbon atoms 9 and 10 in the $C_{16}$ product (palmitoleic acid). In the $C_{18}$ product, however, the double bond becomes located between carbon atoms 11 and 12. Hence bacteria that employ the anaerobic pathway contain *cis*-vaccenic acid as their monounsaturated $C_{18}$ fatty acid, not oleic acid, the product of direct desaturation of stearic acid by the aerobic pathway. The biological distribution of the two pathways for the formation of monounsaturated fatty acids is shown in Table 7.3.

$$\underset{\textit{palmitic acid}}{CH_3(CH_2)_{14}COOH} + \tfrac{1}{2}O_2 \rightarrow \underset{\textit{palmitoleic acid}}{CH_3(CH_2)_5CH{=}CH(CH_2)_7COOH} + H_2O$$

$$\underset{\textit{stearic acid}}{CH_3(CH_2)_{16}COOH} + \tfrac{1}{2}O_2 \rightarrow \underset{\textit{oleic acid}}{CH_3(CH_2)_7CH{=}CH(CH_2)_7COOH} + H_2O$$

*Figure 7.27 The formation of monounsaturated fatty acids from the corresponding saturated fatty acids.*

Figure 7.28 *The anaerobic pathway to monounsaturated fatty acids, characteristic of many bacteria, showing its relationship to the pathway for saturated fatty acid synthesis.*

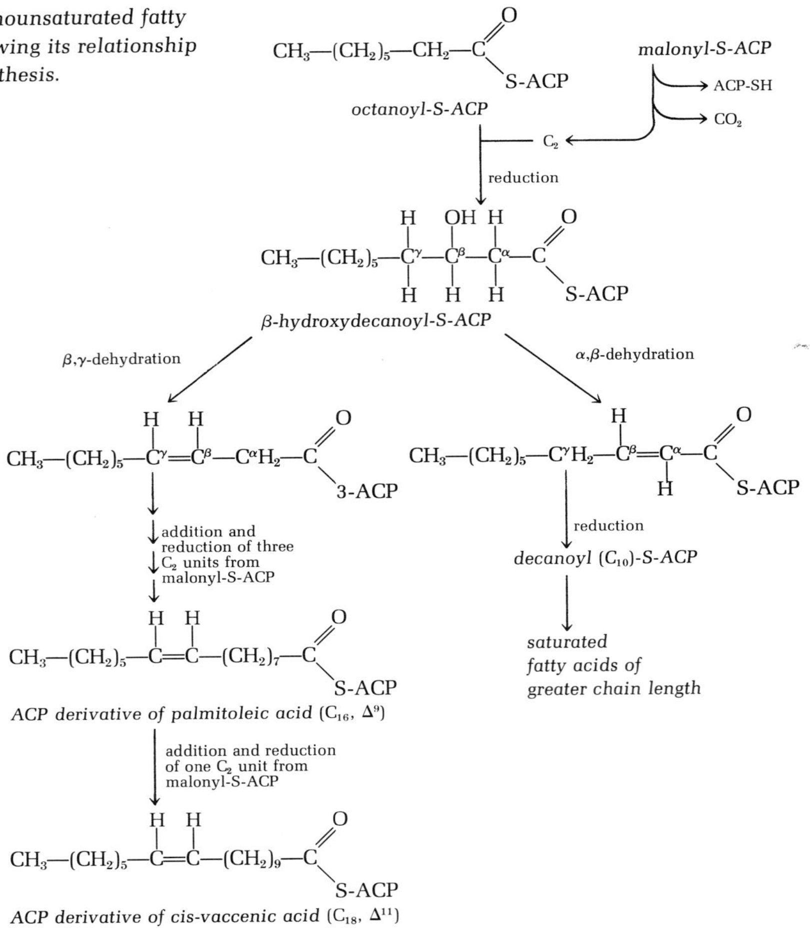

$CH_3$—$(CH_2)_5$—$CH_2$—C(=O)—S-ACP
*octanoyl-S-ACP*

*malonyl-S-ACP* → ACP-SH, $CO_2$, $C_2$

reduction

$CH_3$—$(CH_2)_5$—$C^{\gamma}H_2$—$C^{\beta}H(OH)$—$C^{\alpha}H_2$—C(=O)—S-ACP
*β-hydroxydecanoyl-S-ACP*

β,γ-dehydration

α,β-dehydration

$CH_3$—$(CH_2)_5$—$C^{\gamma}H$=$C^{\beta}H$—$C^{\alpha}H_2$—C(=O)—S-ACP

$CH_3$—$(CH_2)_5$—$C^{\gamma}H_2$—$C^{\beta}H$=$C^{\alpha}H$—C(=O)—S-ACP

addition and reduction of three $C_2$ units from malonyl-S-ACP

reduction

*decanoyl* ($C_{10}$)*-S-ACP*

*saturated fatty acids of greater chain length*

$CH_3$—$(CH_2)_5$—CH=CH—$(CH_2)_7$—C(=O)—S-ACP
*ACP derivative of palmitoleic acid* ($C_{16}$, $\Delta^9$)

addition and reduction of one $C_2$ unit from malonyl-S-ACP

$CH_3$—$(CH_2)_5$—CH=CH—$(CH_2)_9$—C(=O)—S-ACP
*ACP derivative of cis-vaccenic acid* ($C_{18}$, $\Delta^{11}$)

Table 7.3 *Biological distribution of mechanisms for the synthesis of monounsaturated fatty acids*

| **Anaerobic pathway** | **Aerobic pathway** |
|---|---|
| *Clostridium* spp. | *Mycobacterium* spp. |
| *Lactobacillus* spp. | *Corynebacterium* spp. |
| *Escherichia coli* | *Micrococcus lysodeikticus* |
| *Pseudomonas* spp. | *Bacillus megaterium* |
| Photosynthetic bacteria | Fungi |
| Blue-green algae | Protozoa |
| | Animals |

### The synthesis of polyisoprenoid compounds

A large number of different cell constituents have carbon skeletons which consist (in a formal chemical sense) of repeating $C_5$ units with the structure of isoprene:

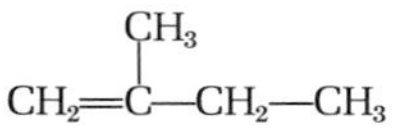

$$CH_2{=}C(CH_3){-}CH_2{-}CH_3$$

Like the fatty acids, these *polyisoprenoid compounds* are synthesized exclusively from acetyl units; however, the mechanism of chain elongation differs markedly from that characteristic of fatty acid synthesis, diverging at the $C_4$ level (Figure 7.29). Acetoacetyl CoA undergoes a "head-to-head" condensation with acetyl CoA, to yield, after rearrangement, *mevalonic acid*, a branched $C_6$ acid. This is in turn con-

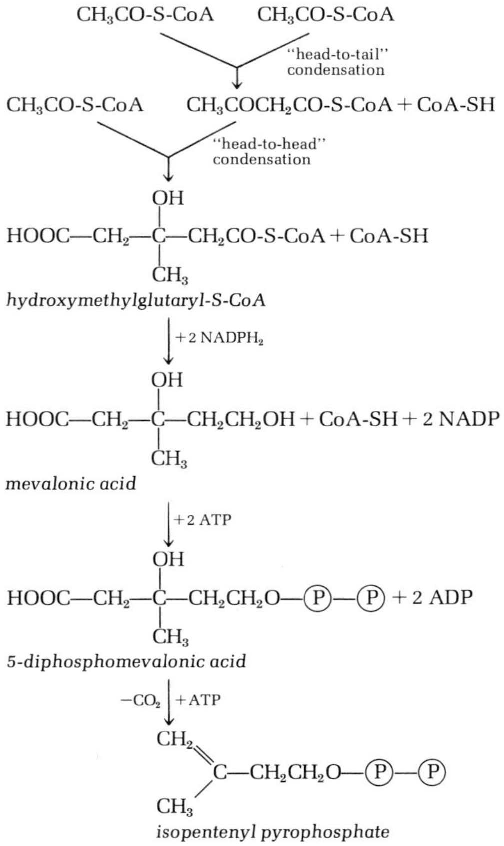

$CH_3CO\text{-}S\text{-}CoA$ + $CH_3CO\text{-}S\text{-}CoA$

↓ "head-to-tail" condensation

$CH_3CO\text{-}S\text{-}CoA$ + $CH_3COCH_2CO\text{-}S\text{-}CoA + CoA\text{-}SH$

↓ "head-to-head" condensation

$HOOC{-}CH_2{-}C(OH)(CH_3){-}CH_2CO\text{-}S\text{-}CoA + CoA\text{-}SH$

*hydroxymethylglutaryl-S-CoA*

↓ $+2\ NADPH_2$

$HOOC{-}CH_2{-}C(OH)(CH_3){-}CH_2CH_2OH + CoA\text{-}SH + 2\ NADP$

*mevalonic acid*

↓ $+2\ ATP$

$HOOC{-}CH_2{-}C(OH)(CH_3){-}CH_2CH_2O{-}Ⓟ{-}Ⓟ + 2\ ADP$

*5-diphosphomevalonic acid*

↓ $-CO_2$, $+ATP$

$CH_2{=}C(CH_3){-}CH_2CH_2O{-}Ⓟ{-}Ⓟ$

*isopentenyl pyrophosphate*

*Figure 7.29 Synthesis of isopentenylpyrophosphate, the precursor of all polyisoprenoid compounds, from acetyl-S-CoA.*

verted, by two successive phosphorylations and decarboxylation, to *isopentenyl pyrophosphate*, the activated $C_5$ compound from which polyisoprenoid compounds are synthesized. The successive steps by which $C_{15}$ and $C_{20}$ derivatives are synthesized from this intermediate are shown in Figure 7.30. The tail-to-tail condensation of two mole-

*Figure 7.30 Chain elongation in polyisoprenoid biosynthesis.*

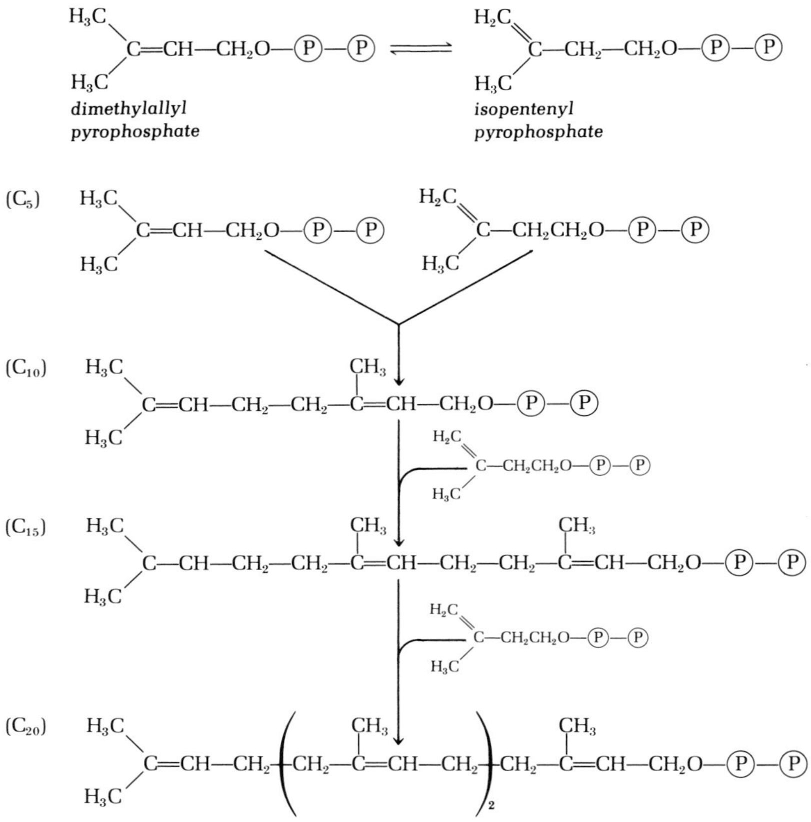

$(H_3C)_2C{=}CH{-}CH_2O{-}Ⓟ{-}Ⓟ \rightleftharpoons H_2C{=}C(H_3C){-}CH_2{-}CH_2O{-}Ⓟ{-}Ⓟ$

*dimethylallyl pyrophosphate* — *isopentenyl pyrophosphate*

($C_5$) $(H_3C)_2C{=}CH{-}CH_2O{-}Ⓟ{-}Ⓟ$ + $H_2C{=}C(H_3C){-}CH_2CH_2O{-}Ⓟ{-}Ⓟ$

($C_{10}$) $(H_3C)_2C{=}CH{-}CH_2{-}CH_2{-}C(CH_3){=}CH{-}CH_2O{-}Ⓟ{-}Ⓟ$

+ $H_2C{=}C(H_3C){-}CH_2CH_2O{-}Ⓟ{-}Ⓟ$

($C_{15}$) $(H_3C)_2C{-}CH{-}CH_2{-}CH_2{-}C(CH_3){=}CH{-}CH_2{-}CH_2{-}C(CH_3){=}CH{-}CH_2O{-}Ⓟ{-}Ⓟ$

+ $H_2C{=}C(H_3C){-}CH_2CH_2O{-}Ⓟ{-}Ⓟ$

($C_{20}$) $(H_3C)_2C{=}CH{-}CH_2{-}(CH_2{-}C(CH_3){=}CH{-}CH_2)_2{-}CH_2{-}C(CH_3){=}CH{-}CH_2O{-}Ⓟ{-}Ⓟ$

cules of the $C_{15}$ derivative, farnesyl pyrophosphate, yields squalene, a precursor of sterols. The analogous condensation of two molecules of the $C_{20}$ derivative yields phytoene, the precursor of carotenoids. The $C_{15}$ and $C_{20}$ polyisoprenoid alcohols, farnesol and phytol, are components of the chlorophylls. Further chain elongation by head-to-tail condensation yields polyisoprenoid compounds containing from 50 to 60 carbon atoms. These occur in the structure of quinones, such as ubiquinone.

### Amphibolic pathways

We have now surveyed the special pathways involved in the synthesis of many biosynthetic intermediates of the cell. Before continuing, it may be useful to summarize and draw together these metabolic details, in terms of a metabolic map (Figure 7.31), which shows the interrelationships of some of the pathways that have been discussed. This makes more evident a point already previously alluded to—that the special

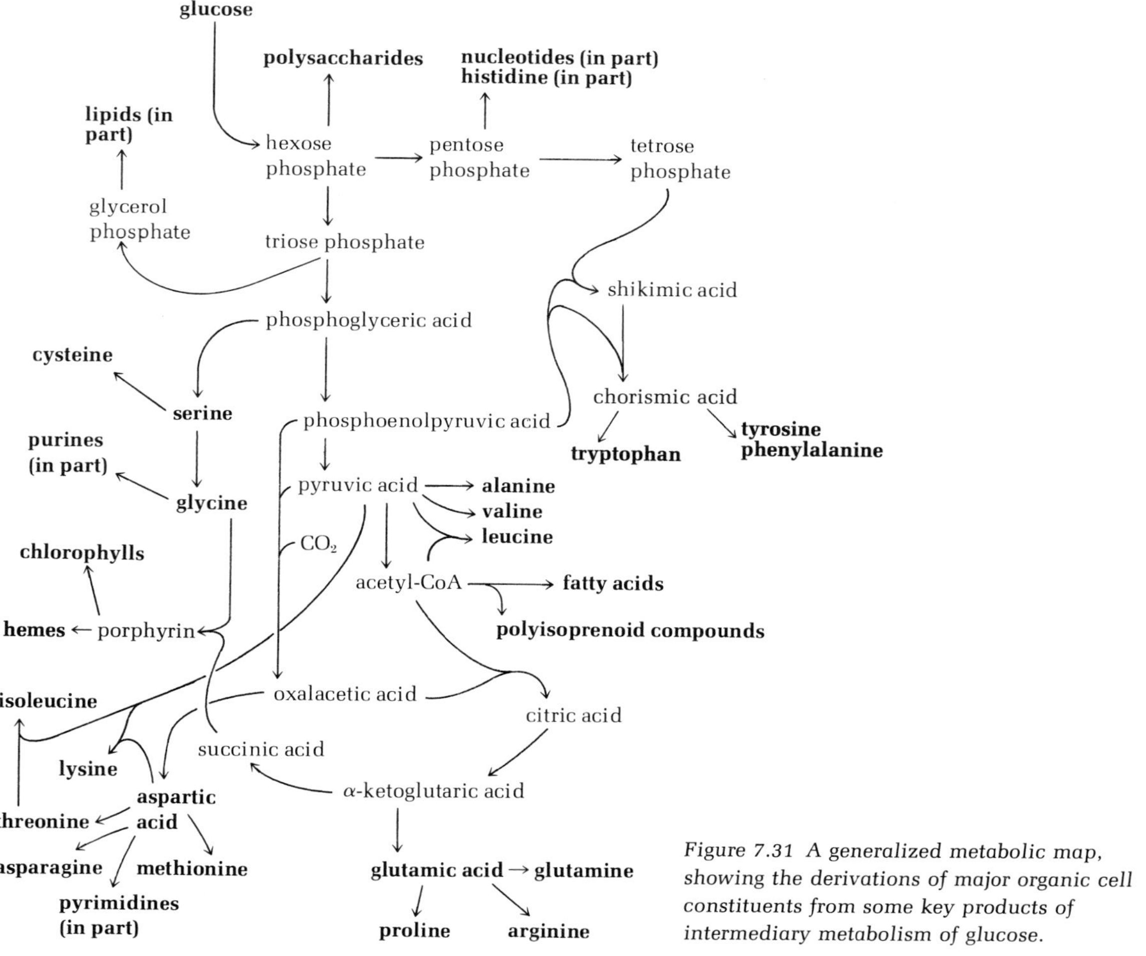


*Figure 7.31* *A generalized metabolic map, showing the derivations of major organic cell constituents from some key products of intermediary metabolism of glucose.*

pathways of biosynthesis are interconnected through reaction sequences (for example, the reactions of the Embden–Meyerhof pathway and of the tricarboxylic acid cycle) which also play important roles in the energy-yielding metabolism of organic substrates.

In heterotrophic organisms there is, accordingly, a *dual flow of carbon from the exogenous organic substrate;* part of this carbon emerges in the catabolic end-products, and part of it is drawn off from intermediates in the catabolic pathways for the purposes of biosynthesis. In autotrophic organisms, on the other hand, the carbon flow is unidirectional: it proceeds from $CO_2$, via sugar phosphates, to the various building blocks.

Pathways that perform both catabolic and anabolic functions, such as the TCA cycle, are known as amphibolic (Greek *amphi-*, both) pathways. The functions of amphibolic pathways in the formation of biosynthetic precursors mean that they must operate, in whole or in part, *even in organisms that do not utilize them for the purpose of energy generation.* An interesting illustration of this principle is seen in the use made of certain reactions of the tricarboxylic acid cycle by obligately chemoautotrophic bacteria and by obligate photoautotrophs, such as the blue-green algae. These organisms cannot oxidize exogenous organic substrates, and consequently the tricarboxylic acid cycle never serves a *catabolic* function for them. Nevertheless, *all the enzymes of the cycle with the exception of α-ketoglutarate dehydrogenase are present in their cells.* The absence of this enzyme destroys the *cyclic* nature of the system, but it does not affect the universal *biosynthetic* functions of the reactions of the TCA cycle, which are to produce α-ketoglutarate and $C_4$ dicarboxylic acids. Figure 7.32 shows how

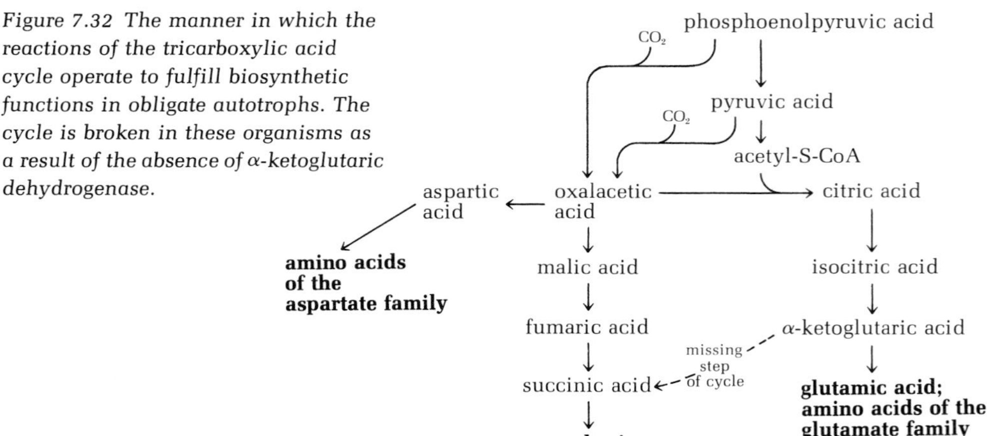


*Figure 7.32 The manner in which the reactions of the tricarboxylic acid cycle operate to fulfill biosynthetic functions in obligate autotrophs. The cycle is broken in these organisms as a result of the absence of α-ketoglutaric dehydrogenase.*

$\alpha$-ketoglutarate and $C_4$ dicarboxylic acids are formed by the autotrophs. The "top" reactions of the cycle, operating in the usual cyclic sense, furnish $\alpha$-ketoglutarate; the "bottom" reactions, operating in the reverse of the usual cyclic sense, furnish $C_4$ dicarboxylic acids.

In facultative anaerobes, such as *Escherichia coli*, it can be shown that during fermentative growth at the expense of sugars, the reactions of the cycle operate in the same manner: the cycle is broken at the $\alpha$-ketoglutarate-succinate step, the top half being used to generate $\alpha$-ketoglutarate, and the bottom half, in the reverse of the cyclic direction, to generate $C_4$ dicarboxylic acids. *Escherichia coli* is, however, capable of oxidizing organic compounds aerobically. Unlike the autotrophs, it can synthesize $\alpha$-ketoglutarate dehydrogenase, essential for the operation of the cycle as an energy-generating system. Hence, when this bacterium grows aerobically, the full cycle operates, serving at one time a respiratory and a biosynthetic function.

Withdrawal of carbon from the cycle for biosynthetic purposes makes it essential for the cell to synthesize oxalacetate in substantial quantities, to maintain the cycle as a respiratory device. In Chapter 6 we discussed two ways in which organisms can achieve this: by the carboxylation of pyruvate or phosphoenolpyruvate, when a pyruvate precursor is the oxidizable substrate; and by a synthesis of succinate from two molecules of acetyl CoA, when acetate or a direct acetate precursor is the oxidizable substrate, via the glyoxylate cycle.

## *The assembly of biopolymers: general principles*

Polysaccharides, proteins, and nucleic acids are polymers composed of subunits (monomers), linked to one another in linear sequence by a particular type of bond. The sequence of monomer units in nucleic acids and proteins is always strictly linear; a few polysaccharides (e.g., glycogen) are branched molecules, but such branching is a secondary biosynthetic event, superimposed on a primary linear arrangement of the subunits. Some polysaccharides are *homopolymers*, consisting of a single, chemically identical repeating subunit. Many polysaccharides, all proteins, and all nucleic acids are *heteropolymers*, consisting of chemically similar but nonidentical subunits. The polysaccharides that contain more than one kind of subunit show a regular arrangement of the subunits, whereas in proteins and nucleic acids, the sequences of subunits are irregular. These variations on the theme of polymer structure are illustrated in Figure 7.33.

The subunits of which a biopolymer is composed can be liberated

*Figure 7.33 The modes of arrangement of subunits in biopolymers.*

| | | |
|---|---|---|
| polysaccharides | homopolymers | —A—A—A—A—A—A— |
| | regular heteropolymers | —A—B—A—B—A—B— |
| nucleic acids | irregular heteropolymers | —A—A—B—C—D—D—B—C—A— |
| proteins | irregular heteropolymers | —A—B—C—C—D—E—B—F—G— |

in free form by subjecting it to *hydrolysis*. This fact shows that the subunits are linked together through *anhydride bonds;* in other words, the assembly of the polymer chain involves coupling of the subunits through reactions that are, in a formal chemical sense, equivalent to *dehydrations*. The nature of the bonds that connect subunits in each major class of biopolymers is shown in Figure 7.34.

The readiness with which the biopolymers can be hydrolyzed to their free subunits, by either chemical or enzymatic means, shows that in an aqueous solution their formation cannot occur simply through anhydride bond formation between the subunits, with the elimination

| **Type of polymer** | **Designation of bond** | **Structure of bond** |
|---|---|---|
| *protein* | *peptide* | 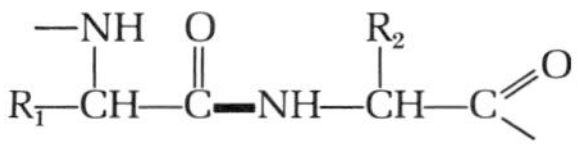 —NH, O, $R_2$; $R_1$—CH—C—NH—CH—C=O |
| *polysaccharide* | *glycoside* | 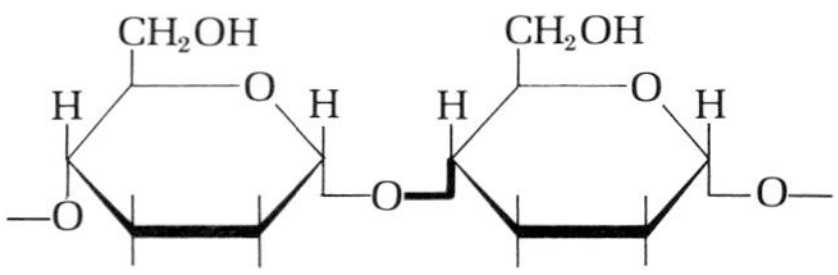 $CH_2OH$, O, H, H; —O ... O— ... O— |
| *nucleic acid* | *phosphodiester* | 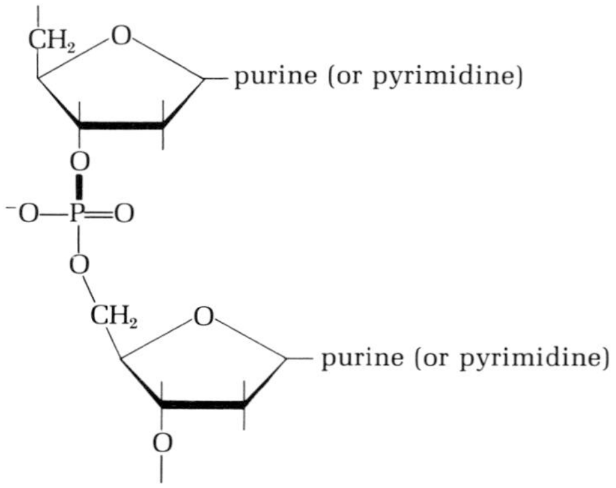 $CH_2$, O — purine (or pyrimidine); O; $^-O$—P=O; O; $CH_2$, O — purine (or pyrimidine); O |

*Figure 7.34 Nature of the bonds that link together the subunits in the major classes of biological polymers.*

*Table 7.4 Biopolymers and their monomeric constituents, showing activating agents and activated forms of the monomers*

| Polymer class | Constituent monomers (product of hydrolysis) | Activating agents | Activated forms of monomers |
|---|---|---|---|
| *Polysaccharides* | *Sugars* | *Nucleoside triphosphates (NTP)* | *Sugar-NDP* |
| *Proteins* | *Amino acids* | *ATP, tRNA* | *Aminoacyl AMP→ aminoacyl tRNA* |
| *Nucleic acids* | *Nucleoside monophosphates* | *ATP* | *Nucleoside triphosphates* |

of water; such a process is thermodynamically most unfavorable. In fact, *polymerization is always preceded by an activation of the monomers.* Activation results in the attachment of the monomers to carrier molecules and requires the expenditure of at least one molecule of ATP (or its equivalent) for the activation of each monomer. *Such activation effectively converts the monomers to their anhydrides, attached to the carrier molecules by labile (energy-rich) bonds.* As a consequence, each monomeric unit can be readily incorporated into a growing polymer chain, its transfer being thermodynamically favored. The monomers of each major class of biopolymer are activated in a characteristic chemical fashion (Table 7.4).

Monosaccharides occur in the cell principally as phosphate esters. The sugar phosphates are activated by reaction with one of the ribonucleotide triphosphates, ATP, UTP, GTP, or CTP. The nature of such activations can be illustrated by the activation of glucose-6-phosphate, which precedes the incorporation of glucose residues into the polysaccharide glycogen in bacteria:

$$\text{glucose-6-phosphate} + \text{ATP} \rightarrow \text{ADP-glucose} + \text{P—P}$$

These reactions always lead to the formation of inorganic pyrophosphate.

Amino acids are activated in two steps, both catalyzed by specific amino acid–activating enzymes. The first step involves a reaction with ATP:

$$\text{amino acid} + \text{ATP} \rightarrow \text{aminoacyl AMP} + \text{P—P}$$

Although water is not a product of the reaction, the amino acid is in effect dehydrated by its transfer to AMP:

$$\text{R—CHNH}_2\text{COOH} + \text{adenosine—}\underset{\text{OH}}{\overset{\text{O}}{\overset{\|}{\underset{|}{\text{P}}}}}\text{—}\underset{\text{OH}}{\overset{\text{O}}{\overset{\|}{\underset{|}{\text{P}}}}}\text{—}\underset{\text{OH}}{\overset{\text{O}}{\overset{\|}{\underset{|}{\text{P}}}}}\text{—OH} \rightarrow \text{R—CHNH}_2\text{—}\overset{\text{O}}{\overset{\|}{\text{C}}}\text{—O—}\underset{\text{OH}}{\overset{\text{O}}{\overset{\|}{\underset{|}{\text{P}}}}}\text{—adenosine} + \text{HO—}\underset{\text{OH}}{\overset{\text{O}}{\overset{\|}{\underset{|}{\text{P}}}}}\text{—}\underset{\text{OH}}{\overset{\text{O}}{\overset{\|}{\underset{|}{\text{P}}}}}\text{—OH}$$

*Table 7.5 Polysaccharide-synthesizing systems*

| Polysaccharide | Repeating unit | Precursor |
|---|---|---|
| *Glycogen* | *α-D-Glucose (1→4)* | *UDP-glucose (animals), ADP-glucose (bacteria)* |
| *Cellulose* | *β-D-Glucose (1→4)* | *GDP-glucose* |
| *Xylan* | *β-D-Xylose (1→4)* | *UDP-xylose* |
| *Pneumococcus type III capsular polysaccharide* | *β-D-Glucuronic acid (1→4), β-D-glucose (1→3)* | *UDP-glucuronic acid, UDP-glucose* |

The aminoacyl residues are then transferred to specific transfer RNA (tRNA) molecules to form aminoacyl RNAs:

aminoacyl AMP + tRNA → aminoacyl tRNA + AMP

The nucleoside monophosphates, the monomers of nucleic acids, are all activated prior to the polymerization by conversion to the corresponding triphosphates.

The assembly of a biopolymer always occurs by the sequential addition of new subunits to one end of the growing polymer chain. Polymer growth is, therefore, unidirectional.

## *The synthesis of polysaccharides*

The properties of several systems for the synthesis of polysaccharides are described in Table 7.5. A characteristic feature of polysaccharide synthesis is the requirement for a *primer*—a short segment of the poly-

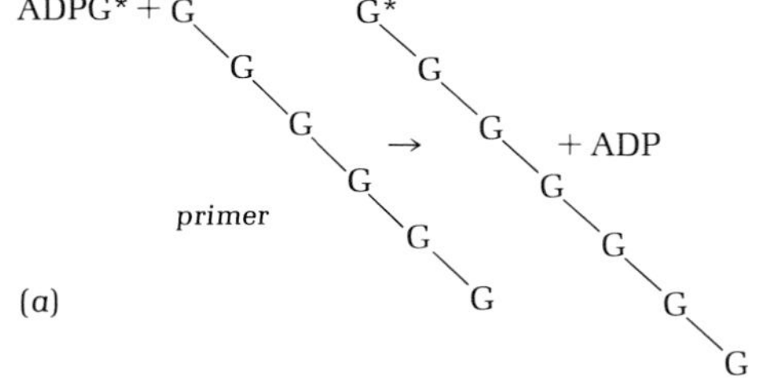


*Figure 7.35 Chain elongation and branching in the enzymatic synthesis of the polysaccharide glycogen: G denotes glucosyl units. (a) Transfer of a glycosyl unit from UDPG to a primer molecule in glycogen synthesis. (b) Reaction catalyzed by the branching enzyme in glycogen synthesis.*

saccharide in question—to act as an acceptor of the monomer units. In the synthesis of glycogen, where the function of the primer has been studied in detail, it has been found that the primer must contain more than four sugar units to function effectively (Figure 7.35). The molecular branching characteristic of glycogen is produced by a special *branching enzyme* that cleaves off small fragments from the end of the 1,4-linked linear polysaccharide chain and reinserts them in 1,6 linkage at another point (Figure 7.35).

## *The synthesis of DNA*

The DNA molecule is a double helix, consisting of a complementary pair of polynucleotide chains held together by hydrogen bonds between pairs of purines and pyrimidines. When these bases are present in their energetically most probable forms (the keto, rather than the enol, form of the oxygenated bases, and the amino, rather than the imino, form of the aminated bases), the only possible base pairs permitted by hydrogen-bonding distances are adenine with thymine (AT) and guanine with cytosine (GC). The entire molecule can thus be described as a linear sequence of nucleotide pairs (Figure 7.36); the exact order of these pairs constitutes a *genetic message* which contains all the information necessary to determine the specific structures and functions of the cell.

The details of the mechanism by which the genetic message is dupli-

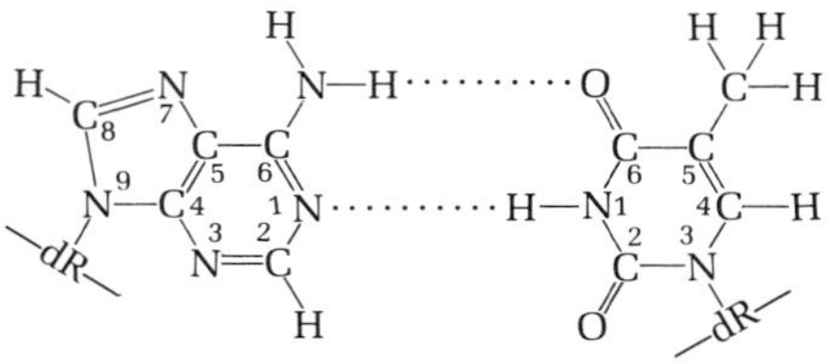

*adenine* *thymine*

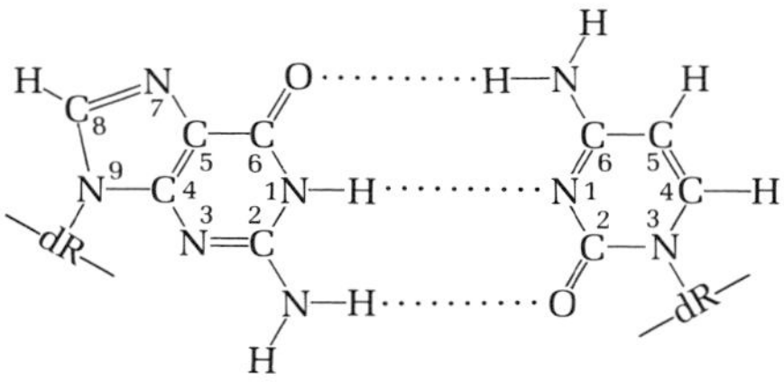

*guanine* *cytosine*

*Figure 7.36 The pairing of adenine with thymine and guanine with cytosine by hydrogen-bonding. The symbol —dR— represents the deoxyribose moieties of the sugar-phosphate backbones of the double helix. Hydrogen bonds are shown as dotted lines.*

cated and transmitted to every cell of the organism are only gradually being elucidated. It is known that the complementary strands separate, each strand acting as a *template* on which is assembled a new complementary strand by the enzymatic polymerization of deoxyribonucleoside triphosphates. The sequence of bases in the new strand is rigidly dictated by the hydrogen-bonding constraints described above (i.e., wherever the template carries adenine the new strand will acquire a thymine, and so on). Replication thus leads to the formation of two new double helices, each identical with the original double helix.

**The antiparallel structure of the DNA double helix**

Each strand of the double helix is a polarized structure; its polarity results from the sequential linkage of polarized subunits, the deoxyribonucleotides. As shown in Figure 7.37, each nucleotide has a 5′-

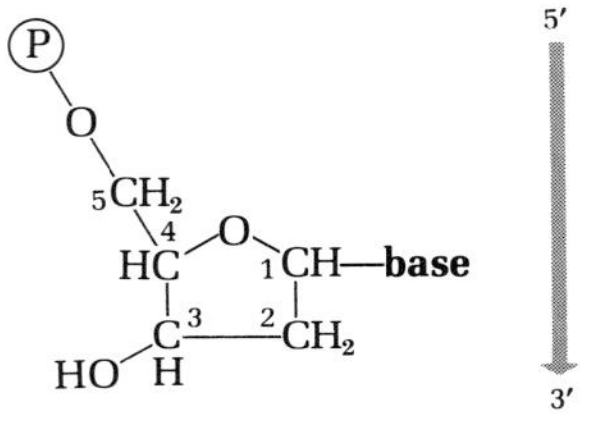

*Figure 7.37 A deoxyribonucleotide. The molecule is polarized, having a 3′-hydroxyl group at one end and a 5′-phosphate group at the other.*

phosphate end and a 3′-hydroxyl end; when a series of nucleotides are connected by phosphodiester linkages, a polarized chain is formed which also has a 5′-phosphate end and a 3′-hydroxyl end.

The two strands of the double helix are *antiparallel* (i.e., they have *opposite polarity*). This is shown in Figure 7.38; if we scan the diagram from top to bottom we see that the left-hand strand has 5′→3′ polarity, whereas the right-hand strand has 3′→5′ polarity. The significance of the antiparallel structure of DNA will become apparent when we consider the process of DNA replication.

**The polymerization of the deoxyribonucleoside triphosphates**

In 1958, A. Kornberg and his coworkers succeeded in isolating from extracts of *Escherichia coli* cells an enzyme that catalyzes the polymerization of deoxyribonucleoside triphosphates to form a polynucleotide chain. The enzyme, called *DNA polymerase*, requires a mixture of the four triphosphates (containing adenine, guanine, cytosine, and thymine), as well as some *primer DNA*.

The mechanism of action of DNA polymerase is best illustrated by its activity when presented with a primer consisting of DNA containing single-stranded regions. With such a primer, *DNA polymerase catalyzes the sequential addition of nucleotides to the free, 3′-OH ends,*

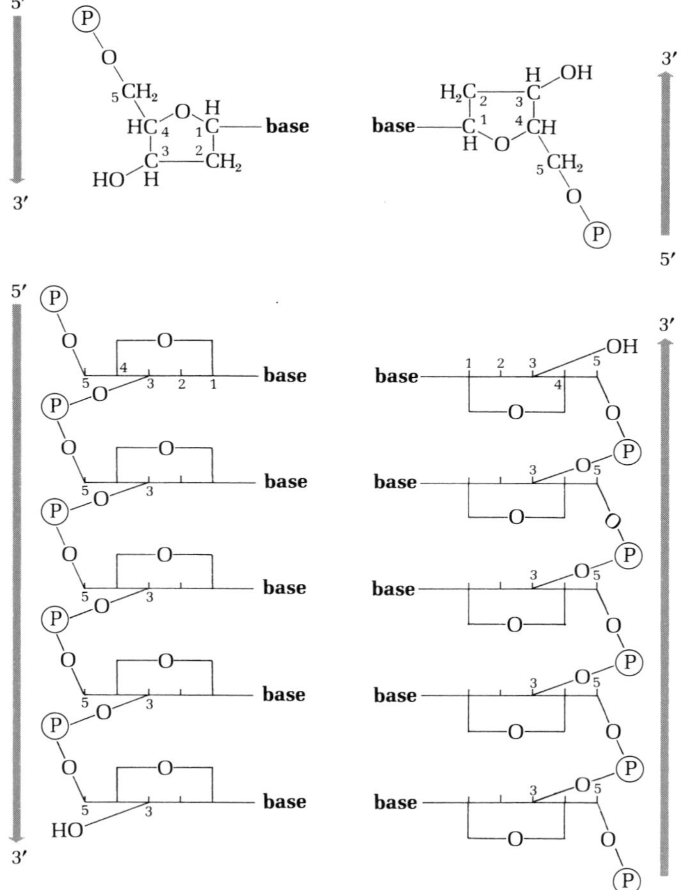


*Figure 7.38 The antiparallel nature of the double helix. Above, the complete structural formula of one base pair is shown. Below, a segment of duplex is shown diagrammatically. Note that the left-hand strand runs from 5′ to 3′, reading from top to bottom, while the right-hand strand runs from 3′ to 5′.*

*using the opposite, intact strand as template* (Figure 7.39). The reaction is driven by the energy-rich bonds of the triphosphates; inorganic pyrophosphate (P—P) is split off in the process.

As we will discuss in Chapter 10, the replication of DNA may be pictured as occurring at a *fork* in the molecule, formed by the separation of the two complementary strands (Figure 7.40). For the purposes of this discussion, the fork may be considered to move along the DNA molecule, each branch of the fork constituting a growing double helix. Since the double helix is an antiparallel structure, replication in one branch of the fork must take place by the sequential addition of triphosphates to the 3′-OH end of a growing chain, while in the other branch of the fork replication must take place by the sequential addition of triphosphates to the 5′-phosphate end of a growing chain.

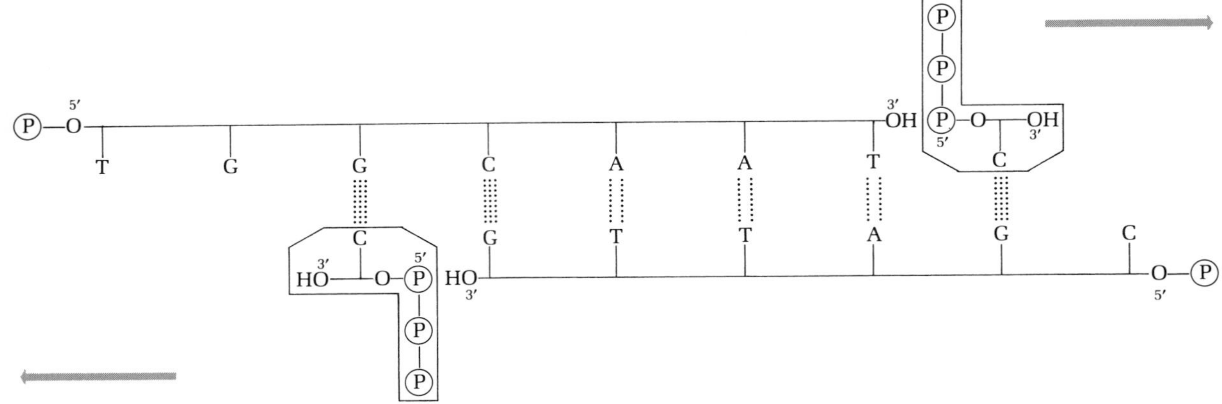


Figure 7.39 *Diagram showing a short fragment of double-stranded DNA, with single-stranded regions at each end. The polynucleotides are in the process of being lengthened by the action of DNA polymerase, which catalyzes the addition of deoxyribonucleotide triphosphates to the 3′-hydroxyl ends of the chains. The triphosphate that is added is governed by the base in the opposite chain, which thus serves as template. The arrows show the direction of sequential addition of deoxyribonucleotides; inorganic pyrophosphate* (P—P) *is split off in the process.*

This requirement presents a problem that has not yet been completely solved. As we have shown, the DNA polymerase isolated by Kornberg is capable of only one of the two chain-lengthening activities required for DNA replication; it can only add triphosphates to the 3′-hydroxyl

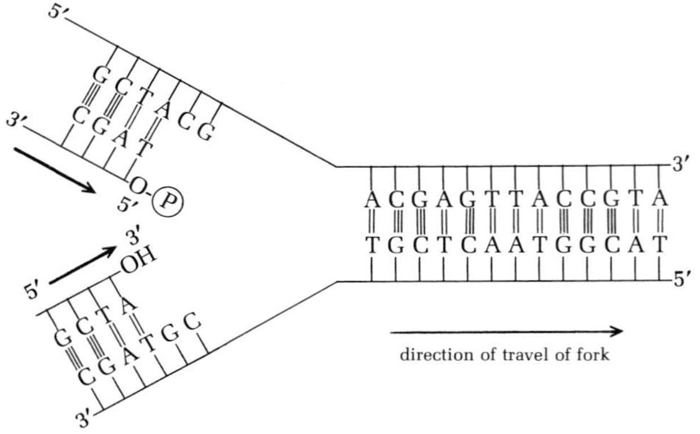


Figure 7.40 *A replicating fork in a DNA molecule. The double helix must separate progressively from left to right, the separated strands serving as templates for the sequential growth of complementary strands. The direction of growth of the new strands is indicated by the arrows.*

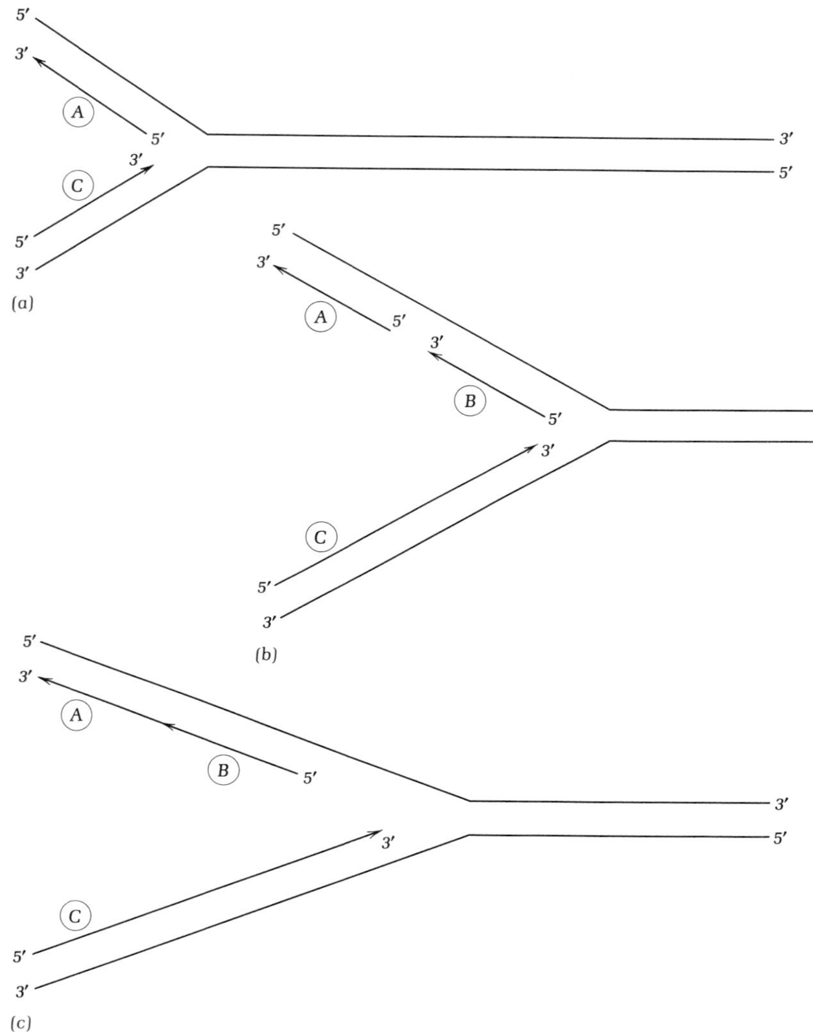


*Figure 7.41* Replication by the combined action of DNA polymerase and polynucleotide ligase. (a) A fork has been formed by separation of the two strands. Two complementary strands, labeled A and C, have been synthesized by DNA polymerase. Chain lengthening takes place at the 3′-hydroxyl ends in both cases. (b) The fork has progressed farther. A second oligonucleotide, labeled B, has been synthesized along the upper branch of the fork. New strand C has been lengthened. (c) Oligonucleotides A and B have been joined by polynucleotide ligase. Strand C continues to lengthen at the 3′ end. The arrowheads indicate the direction of chain lengthening by DNA polymerase.

end of a growing chain. The mechanism by which the chain terminating in a 5′-phosphate is elongated is still not clear. One proposal, shown in Figure 7.41, has recently received strong support from the discovery of an enzyme, *polynucleotide ligase*, which is capable of catalyzing the joining of 3′-hydroxyl and 5′-phosphate termini. The ligase also plays an essential role in the repair of damaged DNA, as described in chapter 12.

The replication of DNA will be discussed further in Chapter 10, in the context of the replication of the bacterial chromosome.

## *The synthesis of RNA*

RNA is synthesized in the cell by a DNA-dependent *RNA polymerase.* This enzyme brings about the polymerization of ribonucleoside triphosphates, to form a polynucleotide strand that is *complementary in base sequence to one of the two strands of DNA.* This process is called *transcription.* The ribonucleoside triphosphates are linked together by $5' \rightarrow 3'$ phosphodiester linkages analogous to those described above for DNA.

In vitro, the transcribing enzyme requires the presence of the ribonucleoside triphosphates of adenine, guanine, cytosine, and uracil, as well as a DNA template. When it is presented with fragments of double-stranded DNA as template, it synthesizes RNA molecules which are complementary to both strands of the DNA. When, however, the enzyme is presented with circular double-stranded DNA, such as the replicative form of the single-stranded DNA virus, $\phi$X174, it copies only one of the two strands. Thus, intact DNA appears to possess specific sites for the initiation of RNA synthesis; these sites orient the polymerase so that it produces copies of one specific strand. The fragmentation of DNA appears to create artificial initiation sites, with the result that both strands are copied.

Three classes of RNA are formed in the cell by the process of transcription: ribosomal RNA, transfer RNA (tRNA), and messenger RNA (mRNA). The sequence of bases in each mRNA molecule is *translated* into a sequence of amino acids in a specific polypeptide, through the process of protein synthesis. The process requires the participation of ribosomal RNA (as a component of ribosomes) and of tRNA.

## *The synthesis of proteins*

**Polypeptide synthesis**

Protein synthesis is carried out on the surfaces of the *ribosomes.* Ribosomes are complex particles composed of two subunits; each subunit consists of a molecule of ribosomal RNA combined with a number of different proteins. The two subunits are of different sizes: in procaryotic cells, the larger, 50-S subunit, contains an RNA molecule with a molecular weight of about 1 million; the smaller, 30-S subunit, contains an RNA molecule half that size. Figure 7.42 shows bacterial ribosomes as they appear in the electron microscope.

The synthesis of a specific protein begins when a 30-S and a 50-S ribosomal subunit come together at one end of an mRNA molecule,

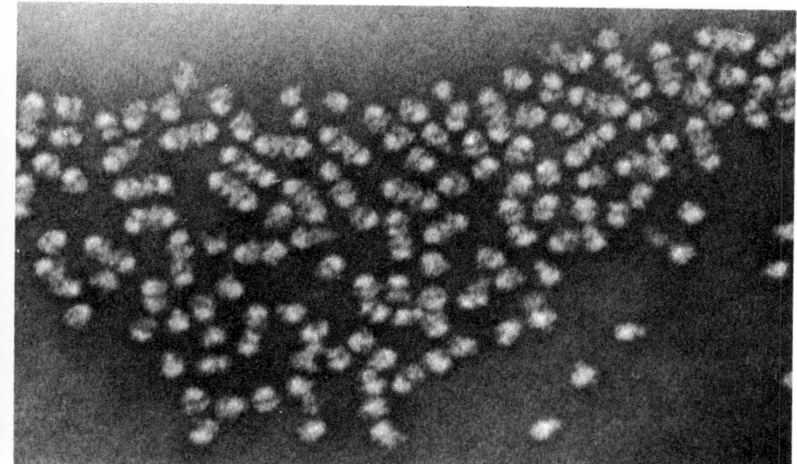

*Figure 7.42* *Preparation of 70-S particles, negatively stained with phosphotungstic acid. Each particle is composed of two unequal subunits, which can be separated to yield 50-S and 30-S particles. Some of the 70-S particles in this picture are present as 100-S dimers. From H. E. Huxley, and A. Zubay, "Electron microscope observations on the structure of microsomal particles from E. coli." J. Mol. Biol.* **2,** *10(1960).*

forming a 70-S ribosome–mRNA complex. This complex has the unique ability to bind tRNA molecules; the tRNA molecule which is first bound is that which carries in its structure a triplet of bases, or *anticodon*, complementary to the first triplet of bases, or *codon*, on the mRNA. It is bound to one of two binding sites on the 50-S subunit, which we will call "site A" (Figure 7.43, top left).

Before binding to the ribosome–mRNA complex, tRNA molecules are enzymatically "charged" with their respective amino acids. A tRNA molecule having a particular anticodon is always charged with a particular amino acid; for example, the tRNA whose anticodon is complementary to the triplet UUU (uracil-uracil-uracil) is always coupled to a phenylalanine molecule. For each of the 20 amino acids of proteins the cell contains one or more tRNA molecules, each with a specific anticodon; the amino acids are coupled to the tRNA molecule through their carboxyl groups.

Having bound the first charged tRNA molecule, the 70-S ribosome binds a second tRNA, charged with its specific amino acid, at "site B" (Figure 7.43, top right). A series of enzymatic reactions then takes place, the end result of which is the *transfer* of the amino acid on the first tRNA (let us call it $AA_1$) to the amino acid on the second tRNA ($AA_2$). The two amino acids, now coupled by a *peptide bond* between the carboxyl group of $AA_1$ and the amino group of $AA_2$, form the chain $AA_1$-$AA_2$ attached to the tRNA of $AA_2$ (Figure 7.43, bottom right).

The ribosome now moves a distance of one codon along the mRNA molecule; the first tRNA—discharged of its amino acid—is displaced, and the second tRNA, with its attached polypeptide chain, moves from

*Figure 7.43* *Four stages in the lengthening of a polypeptide chain on the surface of a 70 S ribosome. (a) A tRNA molecule, bearing the anticodon complementary to codon 1 at one end and $AA_1$ at the other, binds to site A: $AA_1$ is attached to the tRNA through its carboxyl group; its amino nitrogen bears a formyl group (symbolized Ⓕ). (b) A tRNA molecule, bearing $AA_2$, binds to site B; its anticodon is complementary to codon 2. (c) An enzyme complex catalyses the transfer of $AA_1$ to the amino group of $AA_2$, forming a peptide bond. (Note that transfer in the opposite direction is blocked by the prior formylation of the amino group of $AA_1$). (d) The ribosome moves to the right, so that sites A and B are now opposite codons 2 and 3; in the process, $tRNA_1$ is displaced and $tRNA_2$ moves to site A. Site B is again vacant, and is ready to accept $tRNA_3$, bearing $AA_3$. (When the polypeptide is completed and released, the formyl group is enzymatically removed.)*

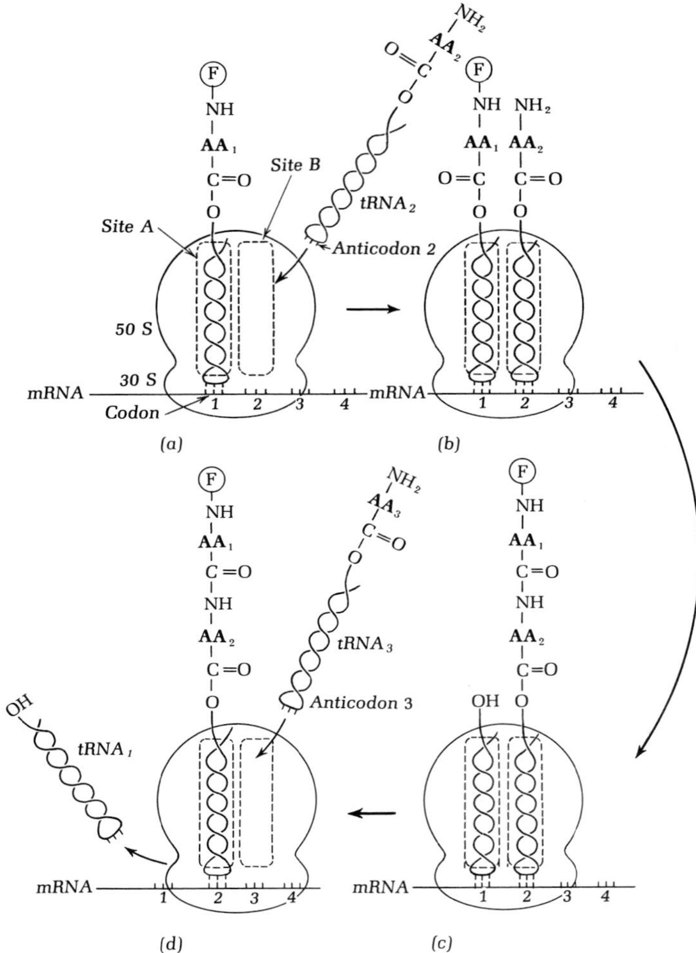


site B to site A (Figure 7.43, bottom left). A third charged tRNA is now bound at site B, and the transpeptidation series of reactions again takes place: $AA_1$-$AA_2$ is transferred to the amino group of $AA_3$, forming the chain $AA_1$-$AA_2$-$AA_3$. This cycle is repeated as the ribosome passes each codon of the mRNA, the growing polypeptide chain always terminating with the amino group of $AA_1$ and remaining attached to ribosome-bound tRNA through the carboxyl group of the most recently added amino acid.

The process is completed when the ribosome reaches one of the *chain-terminating codons* such as UAA (uracil-adenine-adenine), at which point the polypeptide chain is released from the tRNA-ribosome-mRNA complex. When the ribosome arrives at the end of the mRNA molecule it detaches, and the free 70-S ribosome eventually dissociates into 30-S and 50-S subunits which can reinitiate the translation process.

The correct reading of the triplet code is ensured by a mechanism which prevents the ribosomal subunits from attaching to mRNA anywhere except at the beginning of the molecule. Apparently there is a specific site at that position which is necessary for proper binding. Once the ribosome is bound to the initiation site, the mechanism by which the tRNA molecules are aligned determines that the message will be "read" as a sequence of correct triplets. The consequences of shifting the "reading frame" will be discussed in Chapter 12, in the section on mutations that result from the insertion or deletion of base pairs in DNA.

The genetic code can now be written by listing the 64 possible triplets, along with the amino acid which each triplet specifies through the process of translation. The code, which is summarized in Table 7.6, appears to be universal: the translation components from any organism

*Table 7.6 The genetic code*

| | **Second letter** | | | | | | | |
|---|---|---|---|---|---|---|---|---|
| **First letter** | U | | C | | A | | G | |
| U | UUU | phe[a] | UCU | ser | UAU | tyr | UGU | cys |
| | UUC | phe | UCC | ser | UAC | tyr | UGC | cys |
| | UUA | leu | UCA | ser | UAA | (none)[b] | UGA | (none)[b] |
| | UUG | leu | UCG | ser | UAG | (none)[b] | UGG | try |
| C | CUU | leu | CCU | pro | CAU | his | CGU | arg |
| | CUC | leu | CCC | pro | CAC | his | CGC | arg |
| | CUA | leu | CCA | pro | CAA | glu-N | CGA | arg |
| | CUG | leu | CCG | pro | CAG | glu-N | CGG | arg |
| A | AUU | ileu | ACU | thr | AAU | asp-N | AGU | ser |
| | AUC | ileu | ACC | thr | AAC | asp-N | AGC | ser |
| | AUA | ileu | ACA | thr | AAA | lys | AGA | arg |
| | AUG | met | ACG | thr | AAG | lys | AGG | arg |
| G | GUU | val | GCU | ala | GAU | asp | GGU | gly |
| | GUC | val | GCC | ala | GAC | asp | GGC | gly |
| | GUA | val | GCA | ala | GAA | glu | GGA | gly |
| | GUG | val | GCG | ala | GAG | glu | GGG | gly |

[a]Amino acids are abbreviated as the first three letters in each case, except for glutamine (glu-N), asparagine (asp-N), and isoleucine (ileu).
[b]The codons UAA, UAG, and UGA are nonsense codons (see Chapter 12); UAA and UAG are called the ochre codon and the amber codon, respectively.

can be used with a given mRNA without altering the sequence of amino acids that is specified.

The code presented in Table 7.6 was determined by the use of two kinds of in vitro systems. In one, synthetic polymers, each containing a mixture of two different nucleotides, were used as messengers for the synthesis of polypeptides in vitro; by analyzing the ability of different polymers to stimulate the incorporation of different amino acids, many codons could be deciphered. In the other system, synthetic nucleotide triplets were used to stimulate the binding of specific tRNA molecules to ribosomes in vitro. The "breaking of the genetic code," representing the work of several groups over the remarkably short time of 5 years, ranks as a major achievement of molecular biology and indeed of science in general.

The genetic code is *degenerate:* more than one codon can specify the same amino acid. Leucine, for example, is specified by six different codons. This is made possible by the existence of several different tRNA species, all of which are charged with leucine but each of which has a different anticodon. In some cases, moreover, a given anticodon can match either of two codons which differ only in the third nucleotide (e.g., UUU and UUC).

**The secondary and tertiary structure of proteins**

Even before the nascent polypeptide chain detaches from the ribosome, it begins to fold into a compact, three-dimensional mass. First, parts of the polypeptide become coiled into a regular, helical structure called an *α-helix:* this is designated as the *secondary structure* of the protein. Next, the entire molecule, including those regions which have the α-helical configuration, folds on itself to assume a specific three-dimensional shape, called the *tertiary structure* of the protein.

*Both the secondary and tertiary configurations of a protein are determined solely by its primary structure;* a polypeptide with a given amino acid sequence will ultimately assume one particular form which represents its most stable state. This can be demonstrated experimentally. If a protein, such as an enzyme, is *denatured* (i.e., caused to unfold) under special conditions, removal of the denaturing agent permits the protein to refold into its native state, regaining full enzymatic activity.

The fraction of the molecule which exists in the α-helical configuration varies from zero to nearly 100 percent in different proteins. The α-helix is maintained by hydrogen bonds between the carboxyl oxygen of one peptide bond and the amide nitrogen of another peptide bond three residues farther along the chain. The resulting configuration is a helix in which a complete 360° turn is made once every 3.6 residues.

*Figure 7.44 Photograph of a three-dimensional model of the protein ribonuclease S. The polypeptide backbone of the molecule is represented by the continuous black tube, the amino acyl side chains are represented by thinner stick models. Sulfur atoms are represented by large balls; four disulfide bridges are visible. Courtesy of F. Richards.*

The tertiary structures of several different proteins have been completely elucidated by X-ray diffraction analysis. The first to be determined were those of myoglobin, the oxygen-carrying heme protein of muscle, and hemoglobin, the oxygen-carrying heme protein of blood. The complete solutions of these structures required many years of work. The complete three-dimensional structures of two enzymes, lysozyme and ribonuclease, have since been determined. A model of ribonuclease is shown in Figure 7.44.

Figure 7.45 The primary structure of hen egg-white lysozyme, showing the four disulfide bridges between cysteine residues. The amino acids are abbreviated as the first three letters in all cases except isoleucine (ILEU), asparagine (ASN), and glutamine (GLN). From R. E. Canfield and A. L. Liu, "The disulfide bonds of egg white lysozyme (muramidase)." J. Biol. Chem. **240,** 1997 (1965).

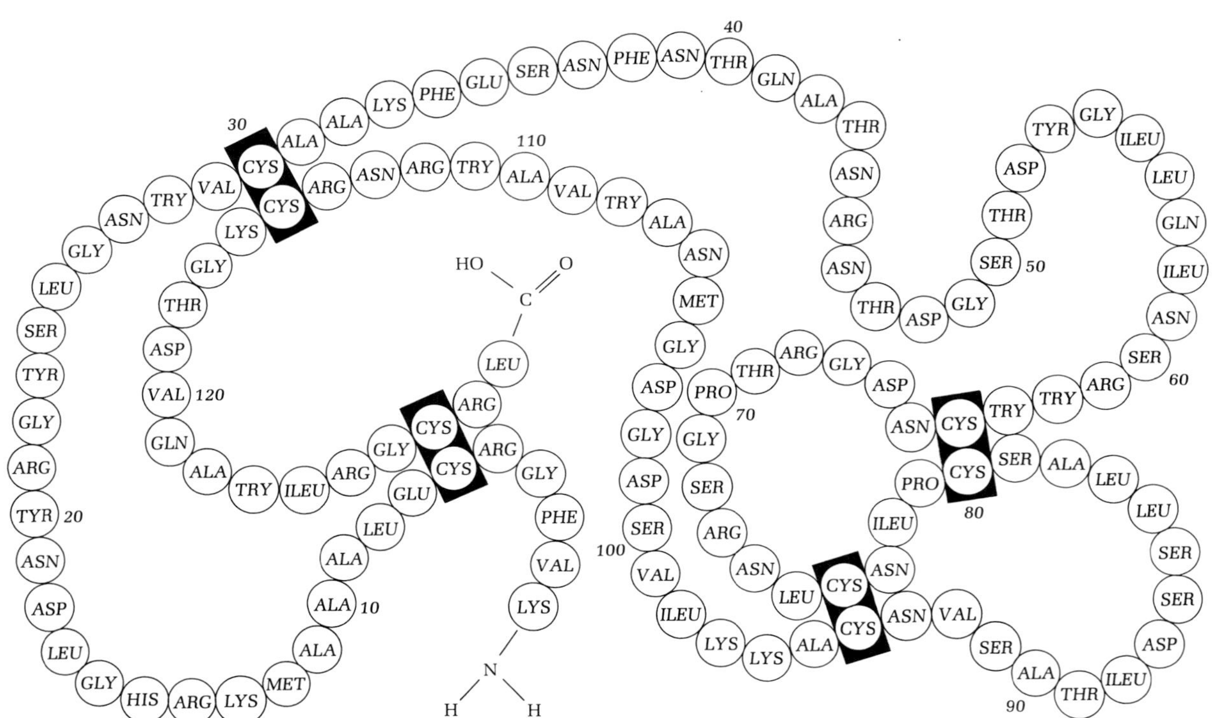


The tertiary structure of most proteins is maintained by several types of bonds, of which the most important are *disulfide bridges* and *hydrophobic bonds.* Lysozyme, for example, contains four disulfide bridges; these are shown more clearly in Figure 7.45. Disulfide bridges are also of major importance in maintaining the structure of antibody molecules, as shown in Figure 28.1. Hydrophobic bonds result in the folding of the molecule such that the hydrophobic (nonpolar) amino acid side chains are closely packed together on the inside of the three-dimensional mass, most of the polarized groups projecting toward the outside. The configuration of the molecule is also maintained by interpeptide hydrogen bonds, similar to those of the $\alpha$-helix; by side-chain hydrogen bonds, such as a hydrogen bond between a tyrosine hydroxyl group and a free carboxyl group; and ionic bonds, between free carboxyl and amino groups of acidic and basic amino acids, respectively.

**The quaternary structure of proteins**

The quaternary structure of a protein is formed by the noncovalent association of several polypeptides, each of which has its own primary, secondary, and tertiary structure. Glutamic dehydrogenase (GDH), for example, is composed of between 24 and 30 identical subunits, each with a molecular weight of about 40,000. These subunits show two levels of aggregation: first, they associate to form monomers with a molecular weight of 250,000 to 350,000; second, four of these monomers associate to form an enzyme "molecule" (oligomer). The oligomer and monomer, but not the subunit, are catalytically active.

An extremely important property of many oligomeric enzymes is their susceptibility to regulation by *effectors;* small molecules that are unrelated to their substrates. Such enzymes are *allosteric proteins:* in addition to its catalytic sites, an allosteric protein possesses one or more binding sites for specific effectors. *The binding of an effector by an allosteric enzyme changes its affinity for its substrate,* so that its catalytic action is either stimulated or inhibited.

## Further reading

**Books**


Bernhard, S., *The Structure and Function of Enzymes.* New York: Benjamin, 1968.

Cohen, G. N., *Biosynthesis of Small Molecules.* New York: Harper & Row, 1967.

Hartman, P., and S. Suskind, *Gene Action,* 2nd ed. Englewood Cliffs, N.J.: Prentice-Hall, 1969.

Haynes, R., and P. Hanawalt (editors), *The Molecular Basic of Life—An Introduction to Molecular Biology. Readings from Scientific American.* San Francisco: Freeman, 1968.

Ingram, V. M., *The Biosynthesis of Macromolecules.* New York: Benjamin, 1965.

Kornberg, A., *Enzymatic Synthesis of DNA.* New York: Wiley, 1962.

Mandelstam, J., and K. McQuillen (editors), *The Biochemistry of Bacterial Growth.* New York: Wiley, 1968.

"Replication of DNA in Microorganisms." *Cold Spring Harbor Symp. Quant. Biol.* **33** (1968).

Sokatch, J. R., *Bacterial Physiology and Metabolism.* New York: Academic Press, 1969.

"Synthesis and Structure of Macromolecules." *Cold Spring Harbor Symp. Quant. Biol.* **28** (1963).

Watson, J. D., *Molecular Biology of the Gene.* New York: Benjamin, 1965.

**Reviews**

Kornberg, A., "Active Center of DNA Polymerase." *Science* **163,** 1410 (1969).

Lipmann, F., "Polypeptide Chain Elongation in Protein Biosynthesis." *Science* **164,** 1024 (1969).

# Chapter eight
# Microbial metabolism: regulation

*The overall metabolic activity* of a growing microbial cell reflects the simultaneous operation of a large number of interconnected pathways, both energy yielding and biosynthetic. Each specific pathway comprises a number of individual reactions, catalyzed by specific enzymes. Unless the cell has been deranged by mutation or by exposure to adverse environmental conditions, a close balance is maintained between the component parts of this extremely complex network of reactions; there is an orderly increase in the quantities of all cell constituents. The cell must therefore possess appropriate means of balancing the rates of the constituent reactions in each metabolic pathway, as well as the overall rates of flow through different pathways. The *regulation* of metabolic activity is indispensable for the maintenance of cellular function. Two basically different kinds of regulatory mechanism operate in the cell: the specific regulation of *enzyme synthesis* and the specific regulation of *enzyme activity*. Both are mediated by compounds of low molecular weight, which are either formed in the cell as intermediary metabolites or enter it from the environment.

## The regulation of enzyme synthesis

**The induction of enzyme synthesis**

Microorganisms may possess the potential ability to perform many metabolic activities which are not obligatory for the maintenance of cellular function and which come into play only under certain special environmental conditions. Such activities are typically concerned with energy-yielding metabolism. For example, many bacteria can use a wide range of different organic compounds as carbon and energy sources, but in any given environment, only one of these compounds may be available to the organism. The enzymatic machinery for the performance of these facultative metabolic activities is usually synthesized by cells only in response to a specific chemical signal from the environment. The necessary genetic information is always present, but its phenotypic expression is environmentally determined.

As a specific illustration of this phenomenon, we shall analyze the utilization of a disaccharide, lactose, as a carbon and energy source by *Escherichia coli*. Although *E. coli* contains the enzymes necessary for the metabolism of glucose under all growth conditions, cells that have been grown in the absence of lactose are unable to metabolize this sugar immediately when it is presented to them; an appreciable lag occurs before it can be attacked at a high rate. Lactose-grown cells, on the other hand, can metabolize this sugar at almost the same rate as they metabolize glucose.

A comparison of the enzymatic constitution of glucose- and lactose-grown cells shows that a number of enzymes, barely detectable in glucose-grown cells, are present at high levels in lactose-grown ones. They include *galactoside permease*, an enzyme that mediates entry of lactose into the cells; *β-galactosidase*, which hydrolyzes lactose to its constituent monosaccharides, glucose and galactose; and a sequence of enzymes that convert galactose to a glucose derivative able to enter the reactions of the Embden–Meyerhof pathway. Hence, exposure to one compound, lactose, accordingly elicits a substantial change in the enzymatic constitution of the cells. Lactose itself induces only two of these enzymes, galactoside permease and β-galactosidase. As a result of their action on lactose, galactose is formed within the cell, and this in turn induces the sequence of enzymes responsible for galactose metabolism. Such a complex inductive event is known as a *sequential induction*, since it involves the formation, within the cell, through the metabolism of the primary substrate inducer, of one or more metabolites with inductive properties different from those of the compound supplied in the environment (Figure 8.1).

Since the study of β-galactosidase formation by *E. coli* has provided much of our basic information concerning induced enzyme synthesis, we shall discuss the physiological aspects of induction in the context of this enzyme; the genetic aspects will be treated in Chapter 13.

Since lactose is a substrate for the enzymes that it induces, it seemed possible that its combination with preexisting enzyme molecules in the cell might play a role in the inductive process. An examination of the effects of chemical analogs of lactose (Table 8.1) soon showed that

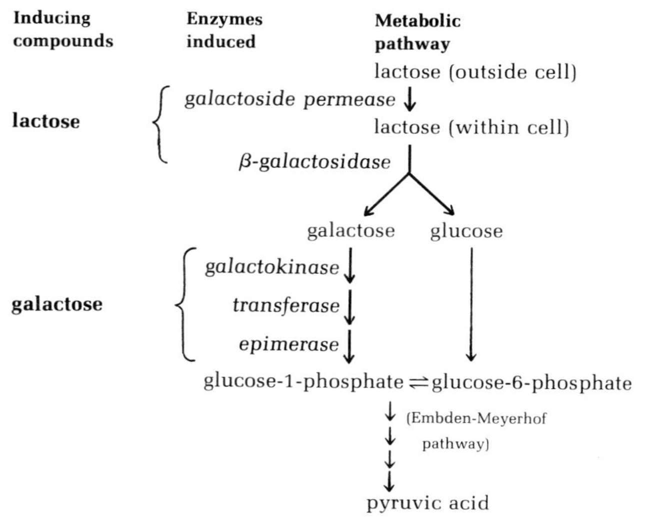


*Figure 8.1 The changes in the enzymatic composition of Escherichia coli that result from exposure to the substrate-inducer, lactose. Boldface arrows indicate reactions mediated by inducible enzymes, formed either directly or indirectly in response to lactose; lightface arrows indicate reactions mediated by constitutive enzymes. Two inducible enzymes are synthesized coordinately (braces) in direct response to lactose; three inducible enzymes are synthesized coordinately in response to galactose, a metabolic intermediate formed within the cells through the hydrolysis of lactose.*

this was not true: some analogs (e.g., isopropylthiogalactoside) that are not substrates for β-galactosidase are excellent inducers; others (e.g., phenylgalactoside), which are split by the enzyme, possess no inductive capacity. The process of induction is therefore mechanistically entirely distinct from enzyme function. Although many naturally occurring inducers are either substrates or products of enzymes they induce, the roles of inducer and substrate are not obligately linked.

If an inducer of β-galactosidase is added to a culture of *E. coli* growing at the expense of another carbon source (e.g., glycerol), the kinetics of induced enzyme synthesis follows a characteristic course (Figure

*Table 8.1. Specificity of lactose analogs as inducers and substrates for the β-galactosidase of Escherichia coli*

| Class and compound | Inducer | Substrate |
|---|---|---|
| *β-Galactosides* | | |
| *Glucose- (lactose)* | + | + |
| *Methyl-* | + | + |
| *Phenyl-* | − | + |
| *β-Thiogalactosides* | | |
| *Isopropyl-* | + | − |
| *Phenyl-* | − | − |
| *Methyl-* | + | − |
| *α-Galactosides* | | |
| *Glucose- (melibiose)* | + | − |

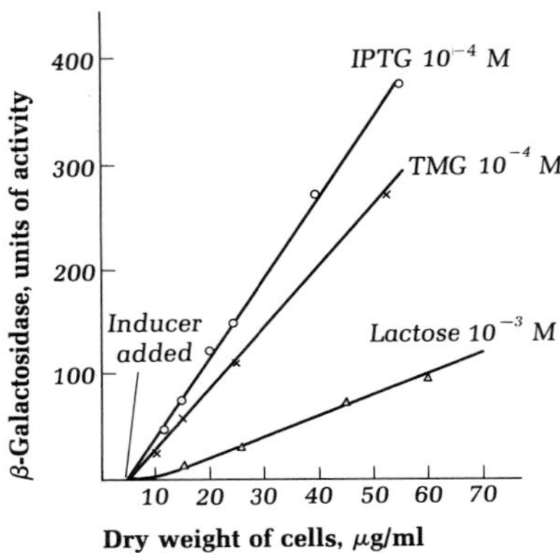


*Figure 8.2 The kinetics of induction of β-galactosidase in growing cultures of E. coli exposed to three different inducers: the natural substrate, lactose, and two lactose analogs, isopropylthiogalactoside (IPTG) and thiomethylgalactoside (TMG). Note that IPTG and TMG, neither of which can be hydrolyzed by β-galactosidase, are both considerably more effective as inducers than is lactose.*

8.2). When the increase in the amount of enzyme is plotted as a function of the increase in cell mass or of total cellular protein, there is a very brief lag, after which the relationship becomes linear. This implies that the primary events leading to induction are rapid ones, and that after they have taken place, the cells immediately acquire the ability to synthesize β-galactosidase at a fixed, maximal rate. The slope of the curve in this plot expresses the differential rate of induced enzyme synthesis, relative to the rate of total protein synthesis by the cell. For any given enzyme, each inducer elicits a characteristic differential rate of synthesis, and the effectiveness of different inducers can thus be compared in quantitative terms (Figure 8.2).

Any compound capable of inducing β-galactosidase also induces the synthesis of galactoside permease, and the relative rate of synthesis of these two enzymes is always constant. When the synthesis of two or more enzymes shows such a tight physiological linkage, it is said to be *coordinate*. Coordinacy is usually, but not always, the consequence of a close association on the chromosome of the structural genes governing the enzymes in question, the phenotypic expression of these genes being subject to a common regulatory control (see Chapter 13).

It is often convenient for practical reasons to differentiate between those enzymes of a given organism that are inducible and those that are synthesized irrespective of the cellular environment, the so-called *constitutive enzymes*. Inducibility and constitutivity are not, however, intrinsic properties of an enzyme but rather of the enzyme-forming system. By selection, one can obtain mutants of *E. coli* that synthesize β-galactosidase constitutively (in the absence of an inducer) at rates as high as the rate of induced synthesis in the parental strain.

### Catabolite repression

Many of the enzymes of catabolic pathways are subject to a type of regulation called *catabolite repression*. If the cell is provided with a rapidly metabolizable energy source, the resulting increase in the intracellular concentration of ATP leads to the repression of enzymes that degrade less rapidly metabolized energy sources. The synthesis of β-galactosidase, for example, is repressed in cells of *E. coli* that are actively metabolizing glucose. This is the basis of *diauxic growth*, a peculiar growth response to a mixture of substrates (cf. Chapter 9).

The manner in which an increase in ATP concentration leads to the repression of specific enzymes is not yet fully understood, but it has been found that such repression may be overcome by the addition to the medium of 3′,5′-cyclic AMP. Cyclic AMP is normally present in bacterial cells and may serve as an *activator* of the synthesis of many catabolic enzymes. It has been suggested that the accumulation of ATP leads to a decrease in the intracellular concentration of cyclic AMP; if so, this might account for the failure of many catabolic enzymes to be synthesized when the ATP concentration is high.

It should be noted that the effectiveness with which an energy source can act as a catabolite repressor depends entirely on the rate at which it can be metabolized by the cell, not on its specific chemical structure. Although glucose is a highly effective catabolite repressor in *E. coli*, it is not in all other bacteria. For example, *Pseudomonas putida* grows more slowly with glucose than with succinate, and in this species succinate is accordingly a much more effective repressor than glucose.

**End-product repression**

A second and much more specific mode of repression can affect the synthesis of enzymes, particularly those operative in biosynthetic pathways. In many of these pathways, the enzymes that catalyze the reaction sequence are subject to repression by the end-product of the pathway. For example, in *E. coli* and related bacteria the coordinate set of enzymes that mediates the synthesis of the amino acid L-histidine is subject to repression by histidine. This phenomenon results in the regulation of the rate of a specific biosynthetic sequence by the size of the pool of its end-product within the cell. If the end-product is not immediately used in further biosynthetic reactions (protein synthesis, in the case of an amino acid), its pool size increases, and the rate of its synthesis is thereby diminished; when the pool size declines, the rate of synthesis is accelerated. End-product repression also provides an effective mechanism for cutting off further endogenous synthesis of biosynthetic intermediates when these are available from an external source, thus making possible a substantial economy in the utilization of the principal carbon source for biosynthesis. Like inducibility, repressibility by an end-product is not an inherent property of an enzyme but a reflection of the way in which its synthesis is controlled, and it can be changed by mutation. So-called "derepressed" mutants are ones in which the ability of a specific metabolite to influence the rate of synthesis of the enzymes responsible for its formation has been abolished. One might expect that such mutants would overproduce and excrete the metabolite in question. However, they rarely do so, because a different mechanism for regulating the rate of synthesis of metabolites by the cell, end-product inhibition, operates effectively. This control mechanism will be described below.

## The regulation of enzymatic activity

Mechanisms of regulation that act at the level of enzyme synthesis all operate with a measurable time lag. Even though the initiation of induced enzyme synthesis following exposure of a population of cells to the inducer is typically very rapid, several generations of growth elapse before the newly synthesized enzyme attains its maximal level in the cells. Similarly, when repression begins to operate, several generations of growth are required for the repressed enzyme to fall, through dilution, to a minimal level in the cell. The generation times of bacteria are rarely less than 20 minutes, and frequently much longer, so the establishment of a steady level for an enzyme after the onset of induction or repression is of the order of hours.

The other principal kind of regulatory mechanism, the regulation of enzyme *activity*, operates without a significant time lag and therefore makes possible a very rapid change of metabolic rates. By no means all the enzymes of the cell are subject to this mode of control; the susceptible enzymes are typically those that occupy strategic branch points in the complex network of intermediary metabolic sequences, so that a reduction or acceleration of their activity immediately affects the flow rate through an entire pathway, the first step of which is catalyzed by the affected enzyme. Although these enzymes catalyze a wide variety of different chemical reactions, they all share certain common properties, structural and kinetic, which distinguish them collectively from the other enzymes of the cell. They are termed *allosteric enzymes*.

## The properties of allosteric enzymes

Many enzymes can be inhibited by compounds that are not substrates but which have a sufficiently close steric similarity to the substrate to permit their attachment to the catalytic site on the enzyme surface. Such attachment impedes the combination of the enzyme with its substrate and thus reduces the rate of the reaction. A classical example is the inhibition of the enzyme succinic dehydrogenase by malonic acid, a chemical analog of the substrate containing one less carbon atom in the chain (Figure 8.3). Inhibitors of this type, which may affect any enzyme, are termed *isosteric* ("same shape"), since they owe their inhibitory action to their steric resemblance to the substrate.

*Figure 8.3 The structures of malonic and succinic acids. Malonic acid is a competitive inhibitor of succinic dehydrogenase since it resembles succinic acid sufficiently in structure to combine with the catalytic site of this enzyme.*

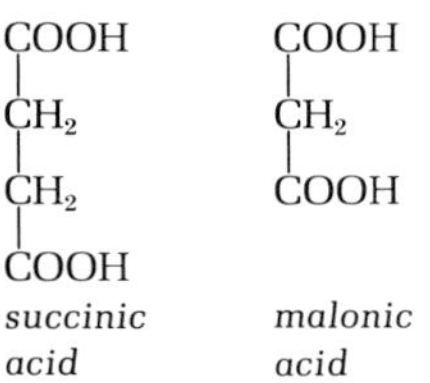


The distinctive property of allosteric enzymes is their ability to be either inhibited or activated by compounds of low molecular weight that bear no close steric relationship to the enzyme substrate, the *allosteric* ("other shape") effectors. An allosteric enzyme always has at least two *different kinds* of combining sites for the attachment of small molecules: the catalytic sites, occupied by the substrate, and the allosteric sites, occupied by its allosteric effectors. The extreme structural differences that often exist between the substrates and effectors of an allosteric enzyme can be illustrated by the example of the aspartate transcarbamylase of *E. coli*, the first enzyme operative in the pathway of pyrimidine biosynthesis. The substrates for this enzyme are carbamyl phosphate + aspartic acid —(aspartic transcarbamylase)→ carbamyl aspartic acid + $P_i$

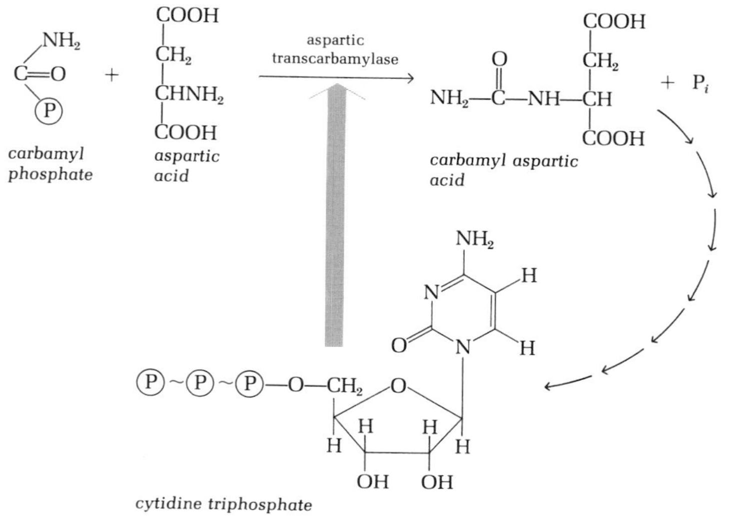

cytidine triphosphate

Figure 8.4 *The allosteric control of the first step in pyrimidine biosynthesis (condensation of carbamyl phosphate and aspartic acid to form carbamyl aspartic acid). The enzyme responsible, aspartic transcarbamylase, is allosterically inhibited (bold arrow) by cytidine triphosphate, the eventual product of the biosynthetic sequence.*

carbamyl phosphate and aspartate. It is allosterically *inhibited* by cytidine triphosphate, the end-product of the specific biosynthetic sequence, and allosterically *activated* by the end-product of a parallel and distinct biosynthetic pathway, adenosine triphosphate (ATP). As shown in Figure 8.4, the allosteric inhibitor may have absolutely no steric resemblance to either of the substrate molecules, and this is also true of the allosteric activator.

When the rate of reaction of an allosteric enzyme as a function of substrate concentration is determined, a sigmoid curve is often obtained, instead of the hyperbolic curve characteristic of nonallosteric enzymes (Figure 8.5). The effect of the allosteric inhibitor, also shown in Figure 8.5, is to reduce the rate of reaction at low substrate concentrations; this inhibition is largely reversed by increasing the concentration of the substrate. The sigmoid nature of the curve relating activity to substrate concentration shows that the enzyme contains more than one catalytic site and that the complexing of one site with the substrate increases the ability of the enzyme to complex with the substrate at additional catalytic sites; there is a cooperative interaction of the substrate molecules with the enzyme. Precisely the same relationship obtains for effector molecules, which can also interact cooperatively with the enzyme.

Many allosteric enzymes can be readily desensitized to their allosteric effectors by relatively mild treatments (for example, exposure to a low temperature, treatment with organic mercurials) which do not

Figure 8.5 *The rate of reaction of aspartic transcarbamylase as a function of the concentration of one of its substrates, aspartic acid. Note sigmoid nature of the curve. The effect of the allosteric inhibitor, CTP, on aspartic transcarbamylase activity is also shown. Redrawn from J. C. Gerhart and A. B. Pardee, "The enzymology of control by feedback inhibition." J. Biol. Chem.* **237,** *891 (1962).*

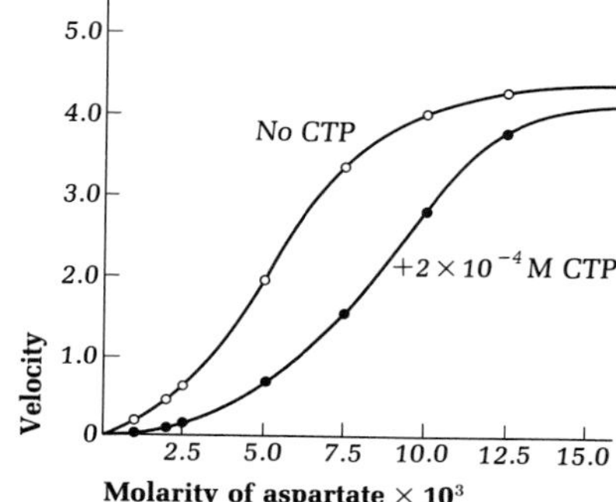

abolish their catalytic function. After such treatment, however, the interaction with the substrate is modified, and the curve relating activity to substrate concentration changes from a sigmoid to a hyperbolic one. This change implies that the desensitized enzyme no longer reacts cooperatively with substrate molecules: the catalytic sites have become independent, so that attachment of substrate to one site does not influence the ability of other sites to attach to substrate molecules.

Finally, it should be noted that allosteric enzymes are always proteins of relatively high molecular weight, composed of multiple subunits. In some cases, these subunits are *identical*, and each therefore carries both catalytic and allosteric sites; in others (notably aspartic transcarbamylase) the enzyme molecule consists of two kinds of subunits, one bearing the catalytic site and the other the regulatory one.

The many peculiar properties of allosteric enzymes suggest a general model for their structure. The molecule is flexible, and its shape can be modified by combination either with the substrate or with the allosteric effector, such modifications of shape in turn influencing the structure of the catalytic site and hence the catalytic activity of the enzyme. Complexing of one catalytic site with substrate promotes an "active" configuration at additional sites and thus increases the activity of the enzyme. Complexing of an allosteric site with an activator has a similar effect, whereas complexing of an allosteric site with an inhibitor promotes a configuration of the enzyme with reduced activity.

## The role of allosteric enzymes in end-product inhibition

Allosteric enzymes, as already mentioned, tend to be located at strategic branch points in metabolic pathways, where their activity can control the rate of flow of material through an entire branch. In biosynthetic pathways, the enzymes that mediate the first step of a specific biosynthetic sequence are characteristically allosteric. Their activity is subject to *end-product inhibition* (sometimes called *feedback inhibition*) by the ultimate product of that branch. A typical example has already been presented in the specific case of aspartate transcarbamylase. A more complex illustration of the same principle, which involves a series of different end-product inhibitions at a number of specific sites in a branched biosynthetic pathway, is portrayed in Figure 8.6. This shows the way in which the activity of enzymes responsible for the synthesis of the amino acids of the aspartate family is regulated in *E. coli*. Each enzyme subject to regulation is so located in the pathway that its inhibition by the end-product of one branch does not interfere with the flow of carbon into other branches.

The regulation of the biosynthesis of the amino acids of the aspartate family in *E. coli* is made even more fine by the existence of three

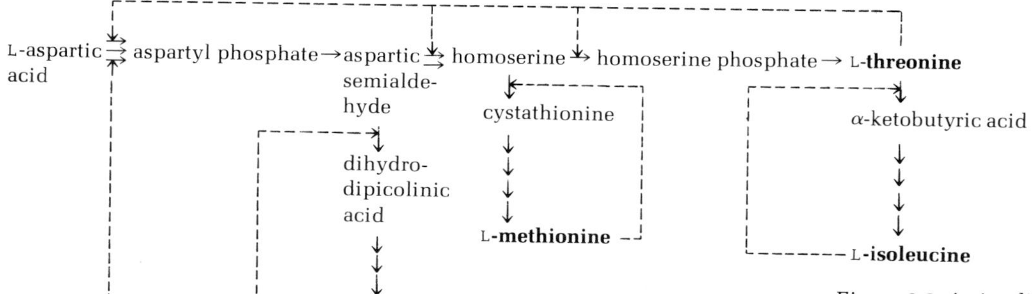


*Figure 8.6 A simplified diagram of the aspartate pathway in E. coli. Each solid arrow designates a reaction catalyzed by one enzyme. The biosynthetic products of the pathway (in boldface) are all allosteric inhibitors of one or more reactions. The dashed arrows leading from each boldface compound indicate the specific reactions subject to allosteric inhibition. Careful study of this diagram reveals that the targets of each allosteric inhibitor are enzymes not essential to the biosynthesis of the other end-products of the pathway. The regulatory significance of the mediation of certain reactions by more than one enzyme (aspartic acid →aspartylphosphate; aspartic semialdehyde→homoserine) is also evident.*

different enzymes with aspartokinase function, mediating the initial step, which governs the entire flow into the pathway (Table 8.2). One of these enzymes is allosterically inhibited by lysine and another by threonine. Consequently, if a pool of either lysine or threonine accumulates within the cell, only a fraction of the total aspartokinase activity is shut off, since the activity of the other aspartokinases, not subject to inhibition by the amino acid that has accumulated, remains unaffected. If *both* these amino acids accumulate, however, most of the aspartokinase activity is cut off.

It should be noted that the control of a biosynthetic pathway by end-product inhibition does not necessarily preclude a parallel control through end-product repression. The lysine-sensitive aspartokinase of *E. coli* is also *repressed* by lysine, and the threonine-sensitive aspartokinase by threonine. These two enzymes are thus subject to dual regulation, both modes of regulation being effected by the same small molecule (Table 8.2).

End-product inhibition can be abolished by mutation. Such a mutation does not affect the synthesis of the enzyme (as do mutations resulting in derepression) but rather the *structure of the enzyme molecule itself.* The enzyme is modified in such a manner that it no longer

*Table 8.2 Control of the first step of the aspartate pathway, mediated by three different aspartokinases, in the bacterium Escherichia coli*

| **Enzyme** | **Repressed by** | **Inhibited allosterically by** |
|---|---|---|
| *Aspartokinase I* | *Threonine and isoleucine* | *Threonine* |
| *Aspartokinase II* | *Methionine* | *No allosteric control* |
| *Aspartokinase III* | *Lysine* | *Lysine* |

exhibits allosteric properties. Mutants in which end-product inhibition has been abolished typically overproduce and excrete the biosynthetic end-product.

**The allosteric regulation of catabolic reaction sequences**

Although allosteric regulation always plays an important role in regulating biosynthetic pathways, it is by no means confined to this portion of the cellular metabolic machinery. This kind of control operates in catabolic sequences, particularly ones that are not subject to regulation by induction. The sites and nature of allosteric activations and inhibitions known to affect the metabolism of glucose and glycogen in *E. coli* are shown in Figure 8.7.

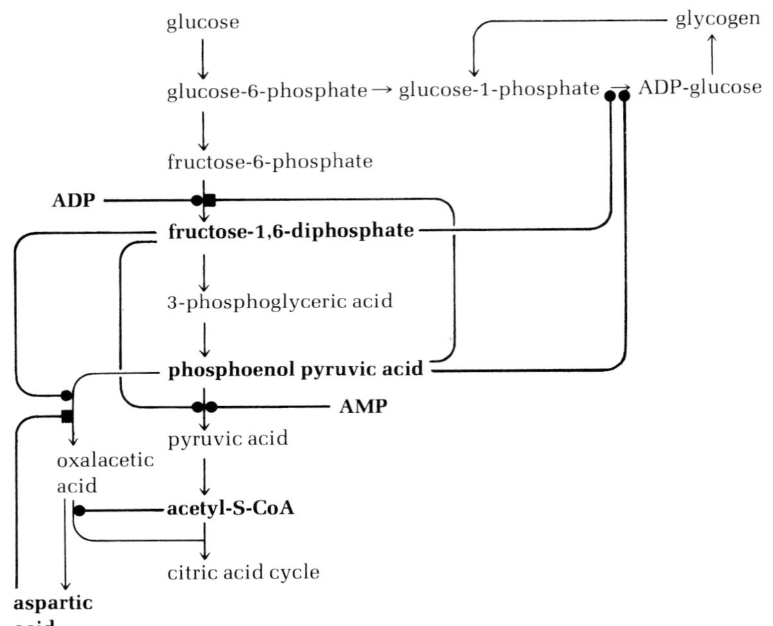


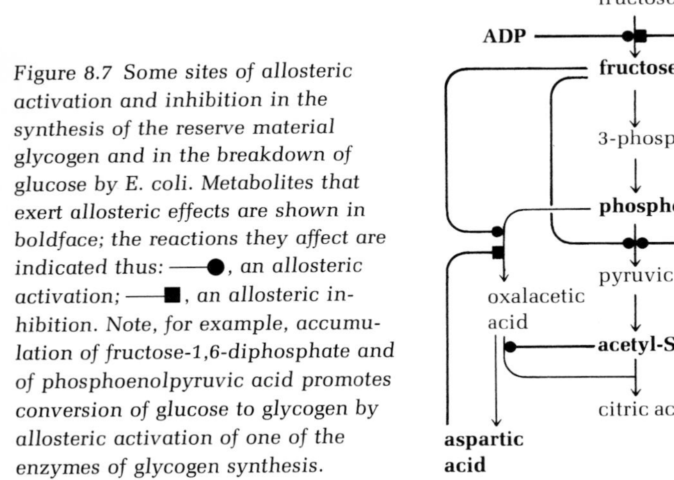

*Figure 8.7 Some sites of allosteric activation and inhibition in the synthesis of the reserve material glycogen and in the breakdown of glucose by E. coli. Metabolites that exert allosteric effects are shown in boldface; the reactions they affect are indicated thus: ——●, an allosteric activation; ——■, an allosteric inhibition. Note, for example, accumulation of fructose-1,6-diphosphate and of phosphoenolpyruvic acid promotes conversion of glucose to glycogen by allosteric activation of one of the enzymes of glycogen synthesis.*

## *The diversity of bacterial regulatory mechanisms*

Many metabolic pathways are of widespread occurrence in different groups of bacteria. A given pathway may be *metabolically* identical in two bacterial groups but subject to markedly different modes of regulation, each group specific.

In all bacteria that belong to the enteric group, the primary flow of carbon into the amino acids of the aspartate family is regulated by the device of multiple aspartokinases subject to independent end-product control, as discussed above in the specific case of *E. coli*. Among bacteria as a whole, however, this mode of control is rare. Most bacteria belonging to other groups that have been examined synthesize only one aspartokinase, subject to a more complex kind of end-product inhibition. The activity of this enzyme is little affected by either lysine or threonine individually, but in the presence of *both* amino acids, a severe inhibition of activity occurs. Evidently the two effectors can cause a conformational change of the enzyme protein that neither can produce alone. This phenomenon is known as *concerted end-product inhibition* (Figure 8.8). At the physiological level, concerted end-product inhibition and independent end-product inhibitions affecting isofunctional enzymes have much the same effect: both mechanisms prevent the end-product of one branch of the pathway from shutting off the flow of carbon into other branches.

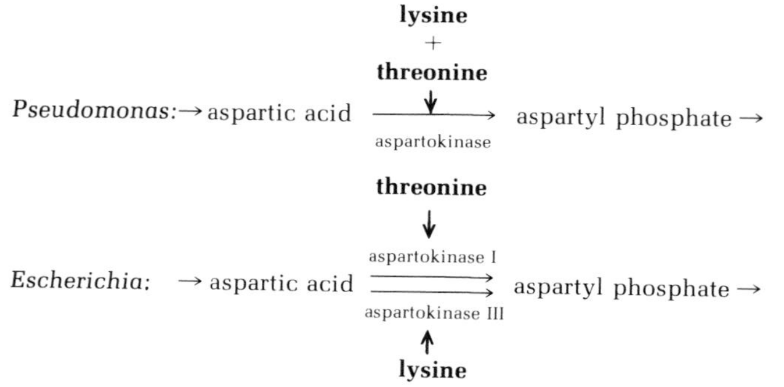


*Figure 8.8 Two different modes of allosteric regulation of the initial step in the biosynthesis of the amino acids of the aspartate family. In Pseudomonas, this reaction is mediated by one type of aspartokinase, subject to concerted feedback inhibition by lysine and threonine. In Escherichia, two of its three aspartokinases are subject to separate feedback inhibition, by lysine and threonine, respectively.*

The $\beta$-ketoadipate pathway for the oxidation of aromatic substrates was discussed in Chapter 6. It occurs in many groups of aerobic bacteria and is always subject to regulation by induction. Although the pathway is metabolically identical in all bacteria, the way in which it is regulated differs from group to group, being constant within a given group. The specific patterns of regulation characteristic of the two bacterial genera *Pseudomonas* and *Acinetobacter* are shown in Figures 8.9 and 8.10. A comparison of these two figures shows major differences in the chemical nature of the inducer substrates, in the degree of coordinate regulation, and in the use made of the device of isofunctional enzymes subject to independent regulatory control.

*Figure 8.9 Regulation of the β-ketoadipate pathway in Pseudomonas. Enzymes in braces are coordinately controlled.*

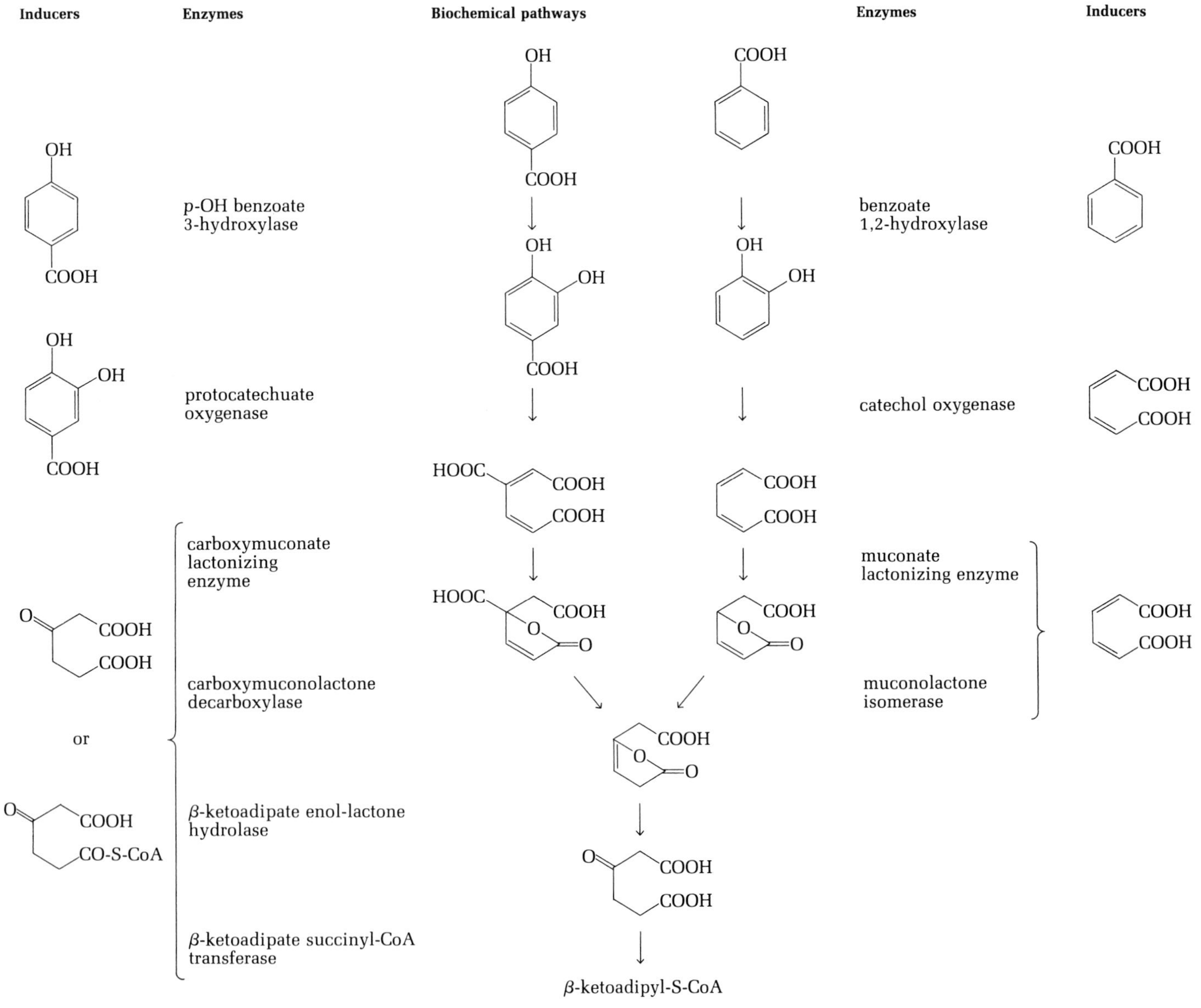

**Inducers** | **Enzymes** | **Biochemical pathways** | **Enzymes** | **Inducers**

OH
COOH
p-OH benzoate
3-hydroxylase
OH
COOH
OH
OH
COOH
COOH
OH
OH
benzoate
1,2-hydroxylase
COOH

OH
OH
COOH
protocatechuate
oxygenase
catechol oxygenase
COOH
COOH

HOOC
COOH
COOH
COOH
COOH

carboxymuconate
lactonizing
enzyme
HOOC
COOH
O
O
COOH
O
O
muconate
lactonizing enzyme
COOH
COOH

O
COOH
COOH
carboxymuconolactone
decarboxylase
muconolactone
isomerase

or

COOH
O
O

O
COOH
CO-S-CoA
β-ketoadipate enol-lactone
hydrolase
O
COOH
COOH

β-ketoadipate succinyl-CoA
transferase

β-ketoadipyl-S-CoA

*Figure 8.10 Regulation of the β-ketoadipate pathway in Acinetobacter. Enzymes in braces are coordinately controlled.*

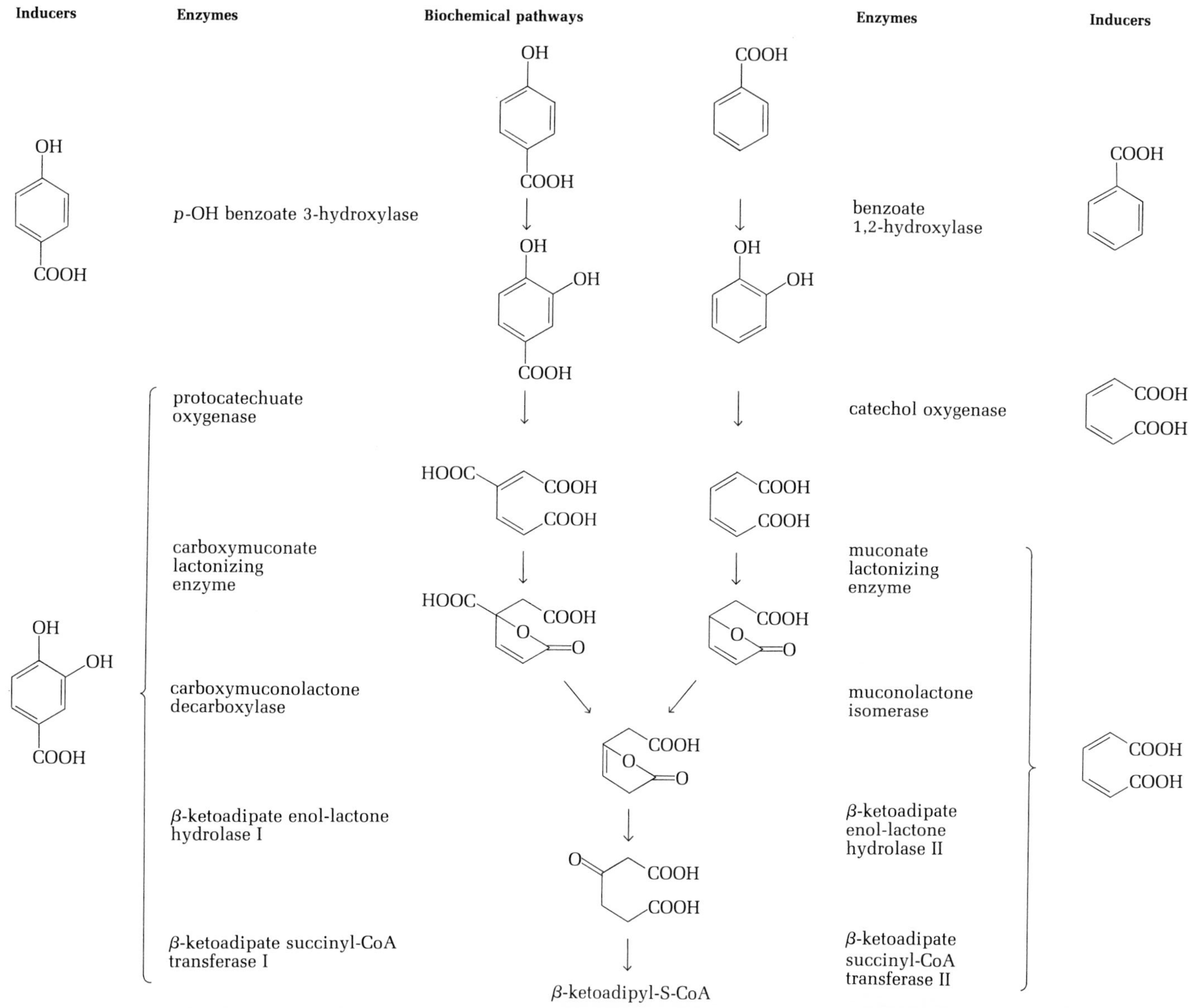

| Inducers | Enzymes | Biochemical pathways | | Enzymes | Inducers |
|---|---|---|---|---|---|
| OH, COOH | | OH, COOH | COOH | | |
| | *p*-OH benzoate 3-hydroxylase | ↓ | ↓ | benzoate 1,2-hydroxylase | COOH |
| | | OH, OH, COOH | OH, OH | | |
| | protocatechuate oxygenase | ↓ | ↓ | catechol oxygenase | COOH, COOH |
| | | HOOC, COOH, COOH | COOH, COOH | | |
| | carboxymuconate lactonizing enzyme | ↓ | ↓ | muconate lactonizing enzyme | |
| OH, OH, COOH | | HOOC, COOH, O, =O | COOH, O, =O | | |
| | carboxymuconolactone decarboxylase | ↘ | ↙ | muconolactone isomerase | COOH, COOH |
| | | COOH, O, =O | | | |
| | β-ketoadipate enol-lactone hydrolase I | ↓ | | β-ketoadipate enol-lactone hydrolase II | |
| | | O, COOH, COOH | | | |
| | β-ketoadipate succinyl-CoA transferase I | ↓ | | β-ketoadipate succinyl-CoA transferase II | |
| | | β-ketoadipyl-S-CoA | | | |

## *The regulation of the synthesis of nucleic acids*

**The regulation of DNA synthesis**

In Chapter 7 the synthesis of DNA was described as a process in which the two strands of the double helix separate, each strand serving as a template for the polymerization of deoxyribonucleoside triphosphates. It will be recalled that the replication process creates a *fork* in the DNA molecule; as the DNA unwinds the fork progresses along its length, and this is accompanied by replication of the two branches (Figure 7.10).

The rate of progression of the fork, and hence of the polymerization reaction, is constant under a wide range of conditions. In *E. coli*, for example, the time required for a complete doubling of the DNA (i.e., the time required for a single fork to travel the entire length of the molecule) is approximately 40 minutes at 37°C. If the cycle of growth and cell division takes longer than 40 minutes, as a result of a limited rate of protein synthesis or RNA synthesis, the cycle of DNA synthesis still occupies only 40 minutes; during the remainder of the division cycle, the DNA is "resting." In a nonsynchronized culture, this alternation between the replicating and resting states of the DNA of individual cells produces a continuous increase in DNA in the total population; the rate of this increase is directly proportional to the fraction of time that the division cycle is occupied by the replication process.

In a rich medium, permitting a division cycle shorter than 40 minutes, the rate of DNA synthesis must increase. This increase is not achieved by a faster rate of polymerization at the replication fork; rather, it is achieved *by the initiation of new cycles of replication prior to the completion of the previous cycle* (Figure 8.11). The maximal growth rate of *E. coli* is attained when every chromosome contains three replication forks; at this point, four copies of the chromosome are produced every 40 minutes, instead of two, and a division cycle as short as 20 minutes is maintained.

DNA synthesis is thus regulated by a mechanism that controls the *initiation* of the replication cycle. It has been shown that the initiation of replication, but not its continuation, requires the synthesis of a specific protein, or *initiator:* if protein synthesis is blocked, DNA synthesis continues until all replication cycles already in progress are completed. New cycles are not initiated until the block is removed.

The nature of the initiator is not known. In Chapter 10 we shall discuss the relationship between DNA replication and cell division; the nature of this relationship suggests that the initiation of DNA synthesis requires a signal produced at the end of the division process. This sig-

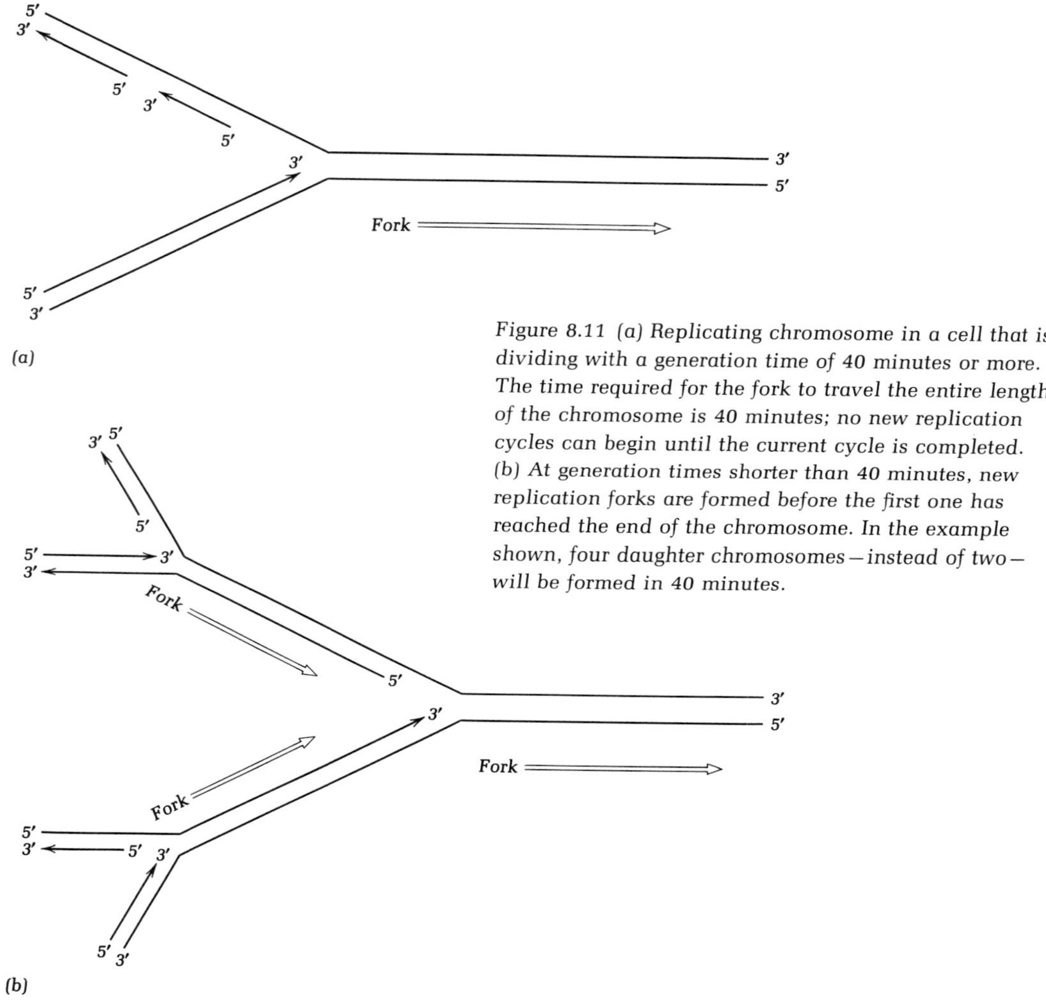


Figure 8.11 *(a) Replicating chromosome in a cell that is dividing with a generation time of 40 minutes or more. The time required for the fork to travel the entire length of the chromosome is 40 minutes; no new replication cycles can begin until the current cycle is completed. (b) At generation times shorter than 40 minutes, new replication forks are formed before the first one has reached the end of the chromosome. In the example shown, four daughter chromosomes—instead of two—will be formed in 40 minutes.*

nal would appear to be the synthesis of the initiator protein.

Jacob and Brenner have proposed a general model for the control of DNA replication; according to their model, each unit of replication, or *replicon*, possesses a genetic locus for production of the initiator and a locus at which replication begins. The latter locus has been designated the replicator. Replication begins when a molecule of the initiator binds to the replicator and continues until the cycle of replication is completed. Production of the initiator by the initiator locus is governed by some step in the cell-division process, as yet unknown.

**The regulation of RNA synthesis**

When cells that are in a steady state of growth are transferred to a medium that will support a higher rate of division ("shift-up"), the rate of RNA synthesis increases abruptly. In contrast, the rates of protein and DNA synthesis increase very gradually, and only after a considerable lag period. The sudden rise in the rate of RNA synthesis reflects the synthesis of new ribosomal RNA (rRNA). The rate of protein synthesis in the cell appears to be limited by the number of ribosomes; hence, protein synthesis cannot increase until more ribosomes are made.

Conversely, when cells are transferred to a medium that can support only a lower rate of cell division ("shift-down"), protein and DNA synthesis slow to the new steady-state rate, but the synthesis of ribosomal RNA *stops completely*. The synthesis of ribosomes is not resumed until the number of ribosomes per cell has been diluted by growth to the level appropriate to the new growth rate.

What is the nature of the chemical signal to which the synthesis of ribosomal RNA responds? In searching for an answer to this question, investigators have concentrated their attention on the products that accumulate when protein synthesis slows down and tend to disappear when the rate of protein synthesis is maximal. Such a product, by acting as a specific inhibitor of rRNA synthesis, could account for the sudden decrease in rRNA synthesis after shift-down and for the sudden increase in rRNA synthesis after shift-up.

Two products that might act in this manner are (1) uncharged tRNA and (2) free ribosomes not attached to mRNA. Both of these products tend to accumulate in the cell when protein synthesis slows down, and to decrease in concentration when protein synthesis is maximal. If either of these products mediates the regulation of rRNA synthesis, it should be possible to correlate its concentration in the cell with the rate of synthesis of rRNA. Although attempts to find such a correlation for uncharged tRNA have been unsuccessful, the concentration of free ribosomes does show the predicted correlation. It thus appears possible that free ribosomes regulate rRNA synthesis by acting as repressors of the genes that code for rRNA.

## *Further reading*

**Books**

"Cellular Regulatory Mechanisms." *Cold Spring Harbor Symp. Quant. Biol.* **26** (1961).

Cohen, G. N., *The Regulation of Cell Metabolism.* New York: Holt, Rinehart, and Winston, 1968.

Maaløe, O., and N. Kjeldgaard, *Control of Macromolecular Synthesis*. New York: Benjamin, 1965.

**Reviews**

Ames, B. N., and R. G. Martin, "Biochemical Aspects of Genetics: The Operon." *Ann. Rev. Biochem.* **33,** 235 (1964).

Cohen, G. N., "Regulation of Enzyme Activity in Microorganisms." *Ann. Rev. Microbiol.* **19,** 105 (1965).

Monod, J., J. Changeux, and F. Jacob, "Allosteric Proteins and Cellular Control Systems." *J. Mol. Biol.* **6,** 306 (1963).

# *Chapter nine*
# *Microbial growth*

*This chapter* will describe the methods used for the measurement of microbial growth, together with the various modes and phases of growth and their mathematical expression and interpretation. The discussion will be centered on the growth of unicellular bacterial populations, which are ideal objects for the study of the growth process and have largely served for the elucidation of its nature.

## *The definition of growth*

In any biological system, growth can be defined as an *orderly increase of all the chemical constituents of an organism.* An increase in total mass is not necessarily a reflection of growth; it may result from the synthesis and accumulation of a cellular reserve material (for example, glycogen or poly-$\beta$-hydroxybutyrate), unaccompanied by synthesis of the major biopolymers (proteins and nucleic acids). Growth normally results in cellular multiplication, except in the special case of coenocytic organisms. In a multicellular organism, cellular multiplication leads to an increase in the size of the individual; in a unicellular organism, it leads to an increase in the number of individuals.

## *The measurement of growth*

To follow the course of growth, it is necessary to make quantitative measurements. For a unicellular population, growth can be measured in terms of two different parameters: *cell mass* and *cell number*. Both types of measurement are expressed relative to a fixed volume of medium (most commonly, a milliliter). It should be emphasized that cell mass and cell number are not necessarily equivalent. In the first place, the mass of the individual cell may vary. In the second place, cell mass increases continuously with time, whereas the increase in cell number is discontinuous, occurring as a result of successive cell divisions at fixed time intervals. Their difference is important in the case of synchronous growth, when all cells of the population divide in phase (p. 322). Usually, however, multiplication in a large microbial population is nonsynchronous; under these circumstances, the increases in the cell mass and in the number of cells are equivalent for all practical purposes.

**The measurement of cell mass**

The only direct way to measure cell mass is to determine the dry weight of cell material in a fixed volume of culture. Provided that the chemical composition of the cell material remains constant during growth, cell mass may also be determined somewhat more indirectly, by measuring some chemical component of the cell material: its carbon or nitrogen content or its protein content. These are practically the only methods that can be used to follow the growth of mycelial or filamentous organisms. They are very rarely used for unicellular bacteria, however, because of the relative insensitivity of gravimetric or chemical analysis. It is difficult to weigh with accuracy less than 1 mg; yet this represents the dry weight of from 1 to 5 billion bacteria of ordinary size. Somewhat more sensitive indirect methods for estimating cell mass are the determinations of the amount of a particular enzyme, or the overall rate of a metabolic process, such as respiration or fermentation. The rate of acid production resulting from the fermentation of sugars is sometimes used to follow the growth of the lactic acid bacteria.

Cell mass can also be indirectly measured with greater sensitivity by determining the amount of radioactive precursor incorporated into cell material (e.g., the incorporation of an exogenously furnished amino acid into protein).

The method of choice for measuring the cell mass of unicellular microorganisms is an optical one: the *determination of the amount of light scattered by a suspension of cells*. This technique is based on the fact that when small particulate objects are evenly suspended in a liquid, their power to scatter light is proportional, within certain limits, to their concentration. When a beam of light is passed through such a suspension, the reduction in the amount of light transmitted affords measure of the cell density.

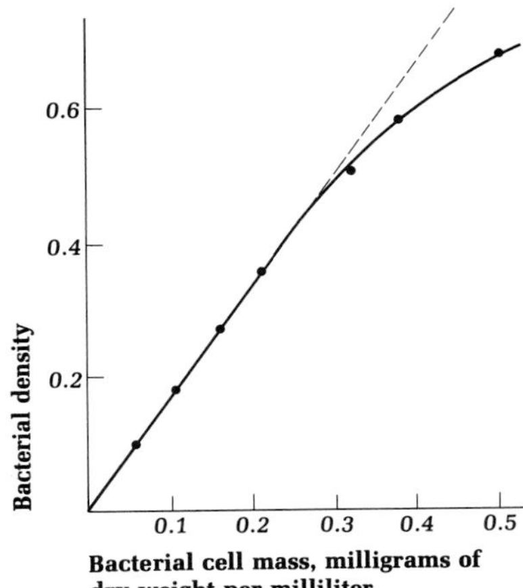


*Figure 9.1 The relationship between the optical density of a suspension of bacteria and bacterial cell mass. Note that proportionality is strict at low optical-density values but deviates from strict proportionality (dashed line) at high optical-density values.*

The relationship between the mass of a cell suspension and its optical density must be determined empirically for any given organism and measuring instrument, as well as for cells of that organism in different phases of growth (since cell size varies with the growth phase). This is done by direct measurement of the dry weight of cells in a sample suspension of a given optical density. It is also necessary to determine the region over which linearity between optical density and mass is maintained (Figure 9.1), and in making subsequent measurements, to dilute dense suspensions to an optical density that falls on the linear portion of the curve. When these precautions are observed, measurements of optical density with a colorimeter or spectrophotometer provide the most accurate and easy method for determining the cell mass of unicellular microorganisms. The lower limit of sensitivity of the method is reached with bacterial suspensions that contain about 10 million cells per milliliter.

**The measurement of cell number**

The number of unicellular organisms in a suspension can be determined microscopically, by counting the individual cells in an accurately determined, very small volume. Such counting is usually done with the aid of special slides known as *counting chambers*. These are ruled with squares of known area and are so constructed that a film of liquid of known depth can be introduced between the slide and the cover slip. Consequently, the volume of liquid overlying each square is accurately known. Such a direct count is known as a *total cell count*. It includes both viable and nonviable cells, since, at least in the case of bacteria, these cannot be distinguished from one another by microscopic examination.

The principal limitation of the direct microscopic enumeration of bacterial populations is that relatively high concentrations of cells must be present in the suspension. The high magnification required for seeing the bacteria limits the volume of liquid that can be examined carefully with the microscope, yet a sufficient number of cells must be found in a known volume to make the count statistically significant. As a consequence, only suspensions that contain 10 million or more cells per milliliter can be counted with any degree of accuracy in a counting chamber designed for bacteria.

An electronic instrument, known as the *Coulter counter,* can be used for the direct enumeration of cells in a suspension. A portion of the suspension is passed through a very fine orifice; during its passage, the cells move through an electric current path in the suspending fluid; detection is based upon differences in electrical conductivity between the cell and the suspending fluid. The number of cells per unit volume is thus registered and recorded. The instrument can also be used to register and record the size distribution of the cells in the population examined. While the Coulter counter is satisfactory for the enumeration of large cells (e.g., the cells of protozoa and algae), the enumeration of very small cells (e.g., bacteria) is more difficult. For bacterial measurements, a very small orifice must be used, and the suspending medium must be completely free of small inanimate particles (e.g., dust particles), which cannot be distinguished by the registering system from bacterial cells.

The enumeration of unicellular organisms can also be made by *plate count,* because single viable cells, separated from one another in space by dispersion on or in an agar medium, give rise through growth to separate, macroscopically visible colonies. Hence, by preparing appropriate dilutions of a bacterial population and using them to seed a suitable medium, one can ascertain the number of viable cells in the initial population by counting the number of colonies that develop after incubation of the plates and multiplying this figure by the dilution factor. This method of enumeration is often termed a *viable count:* in contrast to direct microscopic enumeration, it measures *only those cells that are capable of growth on the plating medium employed.* The viable count is by far the most sensitive method for estimating bacterial number, since even one viable cell in a suspension can be detected. Its accuracy is dependent on the observation of certain precautions. Two or three replicate plates should be prepared from each dilution, to minimize the sampling error. Since the sampling error decreases with increasing sample size, the highest accuracy is obtainable when relatively large numbers of colonies are present on each plate. The practical limit is reached with colony numbers of 300 to 400 per plate. When larger numbers of cells are plated, not all the viable cells may form colonies.

Both direct microscopic enumeration and plating are necessary if one wishes to determine the fraction of a population that is viable: the former method reveals the total number of cells in the population, and the latter reveals the number of viable cells. It is also possible to measure the viable fraction directly, by a technique that combines microscopic examination and growth. A sample of the population, appropriately diluted, is spread over a thin layer of agar medium

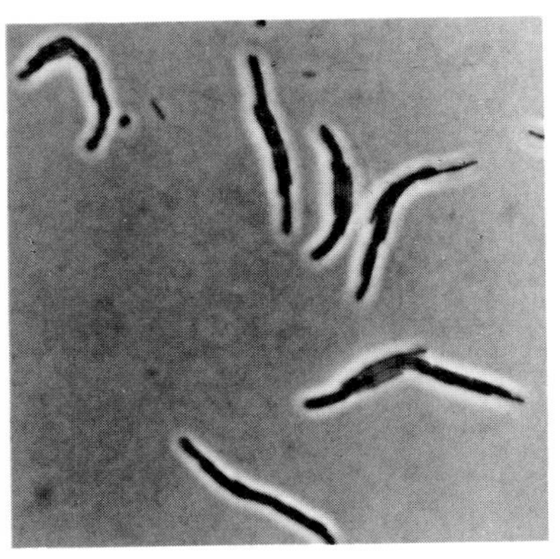

Figure 9.2 *Determination of viability by a combination of growth and microscopic examination. A sample of a culture has been spread on a thin layer of agar medium, and was photographed after incubation for 3.5 hours. Viable cells have formed microcolonies. From J. R. Postgate, J. E. Crumpton, and J. R. Hunter, "The measurement of bacterial viabilities by slide culture." J. Gen. Microbiol.* **24,** *15 (1961).*

poured onto the surface of a sterile glass slide. The inoculated slide is protected by placing a cover slip on its surface, incubated for a period sufficient to permit the occurrence of several cell divisions, and examined microscopically with phase-contrast illumination. Under these conditions, the viable cells can be easily identified as a result of the fact that they have developed into microcolonies, and their number relative to nonviable cells, which remain single, can be precisely determined (Figure 9.2).

## *The mathematical nature and expression of growth*

Under favorable conditions, a growing unicellular microbial population doubles at regular intervals, because each of the two daughter cells produced by a division has the same potential for growth as the parent cell. If the population is started from a single cell, the first few divisions are reasonably synchronous: every cell divides more or less simultaneously, so that the number of cells in the population increases in a stepwise manner. However, as a result of minor differences in growth rate between the individual members of the population, the times of division soon become random, and the number of cells in the population then increases in a continuous fashion. This situation is always encountered in large unicellular populations, unless they have been specially treated to make division synchronous. When the time of division is random, both cell mass and cell number increase continuously with time, doubling at fixed intervals. The time required for a doubling of mass or number is known as the *mean doubling time.*

For any initial population size, $N_0$, we can determine the population size in successive generations as follows:

after 1 generation: $N_1 = 2N_0$
after 2 generations: $N_2 = 2 \times 2N_0 = 2^2 N_0$
after 3 generations: $N_3 = 2 \times 2^2 N_0 = 2^3 N_0$
after *n* generations: $N_n = 2^n N_0$

The exponent of 2 corresponding to any given number is the logarithm of that number to the base 2 ($\log_2$). It is accordingly evident that *the logarithm of the number of cells or of cell mass increases in direct proportion to time.* Such growth is termed *exponential* or *logarithmic growth.* The exponential growth of a population, starting from one cell and proceeding through 10 generations, is shown in Table 9.1. In the last columns of the table, the logarithm of the number of cells has been calculated to the bases of 2 and 10. It can be seen that, *irrespective of*

*Table 9.1 Exponential growth of a unicellular organism with a generation time of 20 minutes*

| Time, minutes | Number of divisions | Number of organisms, expressed as: Arithmetic number | $\log_2$ | $\log_{10}$ |
|---|---|---|---|---|
| 0 | 0 | **1** | 0 | 0.000 |
| 20 | 1 | **2** | 1 | 0.301 |
| 40 | 2 | **4** | 2 | 0.602 |
| 60 | 3 | **8** | 3 | 0.903 |
| 80 | 4 | **16** | 4 | 1.204 |
| 100 | 5 | **32** | 5 | 1.505 |
| 120 | 6 | **64** | 6 | 1.806 |
| 140 | 7 | **128** | 7 | 2.107 |
| 160 | 8 | **256** | 8 | 2.408 |
| 180 | 9 | **512** | 9 | 2.709 |
| 200 | 10 | **1024** | 10 | 3.010 |

*the logarithmic base chosen*, the logarithm of the number of cells increases in direct proportion to time.

**The graphical expression of growth**

*When the logarithm of the number of cells to any base is plotted as a function of time, a straight line will result if the population is in the course of exponential growth.* This is shown in Figure 9.3(*a*), which presents graphically the data from Table 9.1. Only the scale of the ordinate is changed by a change of the logarithmic base, not the shape of the curve. This kind of graphical representation is termed a *semilogarithmic plot*, since the ordinate (number of cells or cell mass) is

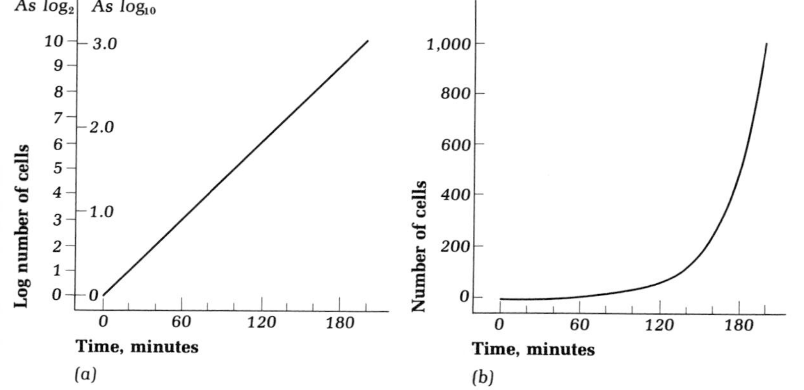


Figure 9.3 Exponential growth of unicellular organisms. (*a*) Logarithms of the numbers of cells plotted against time. (*b*) Numbers of cells plotted against time.

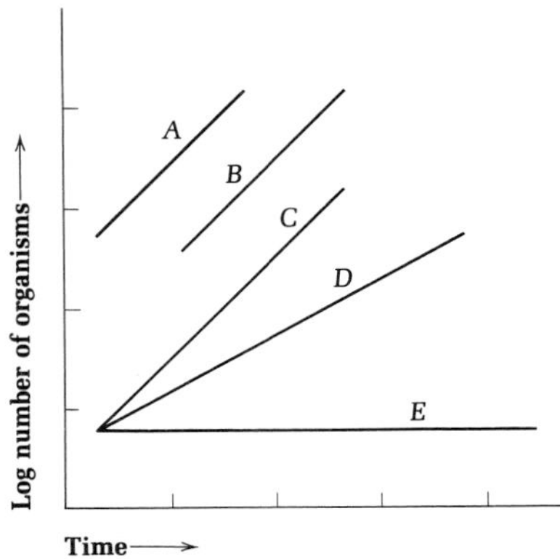


*Figure 9.4 Relation between growth rates and the slopes of the semilogarithmic plots of the number of cells against time. Cultures A, B, and C have identical growth rates; culture D has a lower growth rate; while in culture E the rate is zero.*

expressed in logarithmic numbers and the abscissa (time) in arithmetic numbers. Such plots have several advantages over a purely arithmetic plot for the representation of growth. In the first place, with an arithmetic ordinate it is not possible to illustrate the reproductive behavior of both a large and a small population on the same graph. This is illustrated in Figure 9.3(*b*) by the arithmetic plot of the data from Table 9.1; the early growth behavior of the population is not discernible, and only the last two or three generations of growth are apparent. Second, the slope of the straight line in a semilogarithmic plot corresponds to the growth rate of the population: the steeper the slope, the more rapid the rate of growth. The slope is independent of population size. These points are illustrated in Figure 9.4. Finally, a semilogarithmic plot of growth data immediately shows whether the population is growing at a fixed or at a changing rate. This is shown by the three semilogarithmic plots of bacterial growth in Figure 9.5. Only culture B in this figure is growing at a constant rate. In culture A the growth rate is increasing, and in culture C it is declining. Such differences in reproductive behavior are not always apparent when an arithmetic plot of growth data is made.

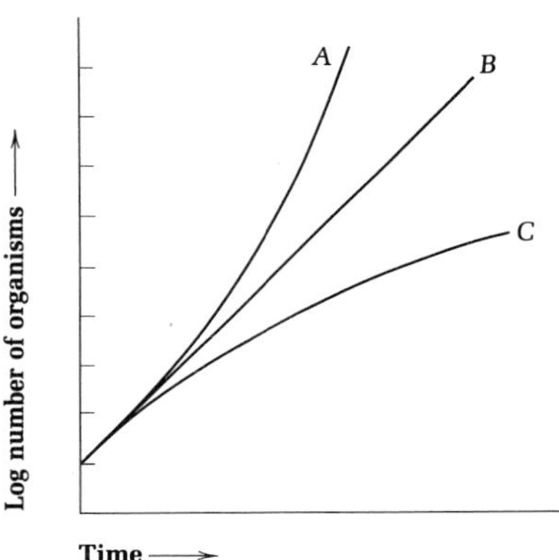


*Figure 9.5 Semilogarithmic growth plots to illustrate the reproductive behavior of three bacterial cultures. In culture A, the growth rate is increasing; culture B is growing at a constant rate; culture C has a declining growth rate.*

In making semilogarithmic plots of growth data, it is convenient to express cell number or mass in terms of $\log_2$ rather than the more familiar $\log_{10}$ values; as shown in Figure 9.3(*a*), each unit on the ordinate then corresponds to a twofold increase in number or mass, so that the doubling time of the population can be read directly from the graph. $\text{Log}_2$ values can be derived from $\log_{10}$ values as shown in the following equation:

$$\log_2 n = \frac{\log_{10} n}{\log_{10} 2} = \frac{\log_{10} n}{0.301}$$

**The calculation of growth-rate constants**

If $N_0$ is the population size at a certain time and $N_t$ its size at a subsequent time, *t*, the number of generations that has occurred can be calculated from the following relationship:

$$N_t = 2^{kt} N_0$$

where *k* is the *exponential growth-rate constant,* defined as the number of doublings per unit time and usually expressed as the *number of doublings per hour*. The logarithmic form of this equation is

$$\log_2 \frac{N_t}{N_0} = kt$$

from which the exponential growth rate, $k$, can be derived as follows:

$$k = \frac{\log_2 N_t - \log_2 N_0}{t}$$

The number of doublings per hour of a growing population can thus be calculated by subtracting the $\log_2$ of the initial population from the $\log_2$ of the final population and dividing this figure by the elapsed time in hours. If tables of logarithms to the base 2 are not available, the calculation can be made by taking the logarithms of the two numbers to base 10, and dividing the expression by $\log_{10}2$, which is approximately 0.301:

$$k = \frac{\log_{10} N_t - \log_{10} N_0}{0.301t}$$

The use of this equation may be illustrated by a specific example. A bacterial population increases from 100 ($10^2$) cells to a billion ($10^9$) cells in 10 hours. What is the exponential growth rate?

$$k = \frac{\log_{10} 10^9 - \log_{10} 10^2}{0.301 \times 10} = \frac{9-2}{3.01} = 2.33 \text{ generations per hour}$$

It is often convenient to express the growth rate as the time required for the population to double, or *mean doubling time*. This is the reciprocal of the exponential growth rate: $1/k$. In the specific example considered above, the mean doubling time is, accordingly, 1/2.33 hours, or approximately 26 minutes.

For certain purposes, particularly calculations of growth kinetics in continuous cultures (p. 000), it is necessary to use a different growth-rate constant, $\mu$, known as the *instantaneous growth-rate constant*. This constant is derived as follows: $dN$, the change in $N$ over the small increment of time, $dt$, is proportional to the number of cells, $N$, as well as to $\mu$. Thus,

$$\frac{dN}{dt} = \mu N$$

Integration of this expression between the limits $N_t$ and $N_0$ yields the growth equation,

$$N_t = e^{\mu t} N_0$$

which becomes, in logarithmic form,

$$\ln \frac{N_t}{N_0} = \mu t$$

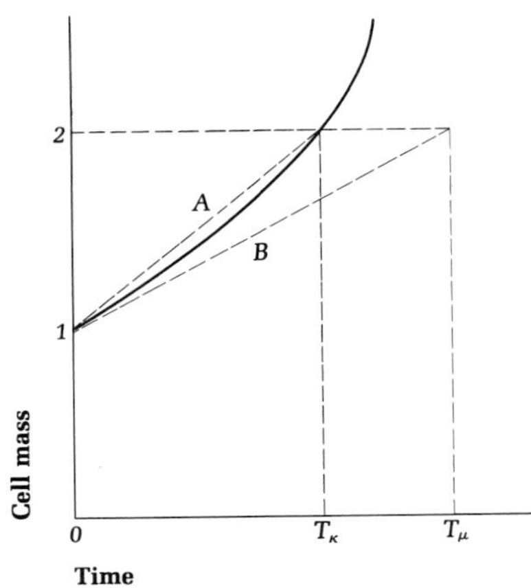


Figure 9.6 An arithmetic plot of exponential growth for one generation (solid line), showing in graphical terms the relationships between the growth rate constants $k$ and $\mu$. The slope of line A expresses $k$, and the slope of line B, $\mu$. $T_\mu$, the instantaneous generation time, equals 1.45 $T_k$, the mean doubling time.

The instantaneous growth-rate constant expresses the relative population increase per unit time; its reciprocal, $1/\mu$, the instantaneous generation time, represents the time that would be required for the population to double if the rate of increase obtaining at zero time remained unchanged. The relationships between $k$ and $\mu$ are best shown graphically, as slopes derived from an arithmetic plot of exponential growth through one generation (Figure 9.6). The numerical relationship between the two growth-rate constants can be calculated from the two equations for growth just presented:

$$N_t = 2^{kt} N_0 \quad \text{and} \quad N_t = e^{\mu t} N_0$$

By combining these two equations and solving for $\mu$ it is found that

$$\mu = k(\ln 2) = 0.69k$$

Hence, the instantaneous generation time, $1/\mu$, is $1/0.69k$, or 1.45 times the mean generation time.

## The growth curve

Microbial populations seldom maintain exponential growth at high rates for long. The reason is obvious if one considers the consequences of exponential growth. A single bacterium with a generation time of 20 minutes would produce, after 48 hours of exponential growth at this rate, a progeny of $2^{144}$ cells, or approximately $2.2 \times 10^{43}$ cells. Although a bacterial cell of average size weighs only $10^{-12}$ g, the total weight of such a population would be $2.2 \times 10^{31}$ g, or roughly 4,000 times the weight of the earth.

The growth of microbial populations is normally limited either by the exhaustion of available nutrients or by the accumulation of toxic metabolic products. Since these changes in the environment are produced by the microorganisms themselves, the development of microbial populations is typically *self-limited*, and after growth has ceased, the population starts to decline as a result of the death of its individual members. These are the factors that determine the typical shape of the microbial growth curve in a culture where the nutrients are not renewed, shown in Figure 9.7. Four principal phases in the history of such a culture, each characterized by a certain rate of population change, can be recognized. These are the *lag phase*, in which the growth rate is zero; the *phase of exponential growth (logarithmic phase)*, in which the growth rate reaches a constant maximal value; the *stationary phase*, in which the growth rate again falls to zero; and the *death phase*, in which the growth rate has a negative value. These four phases are,

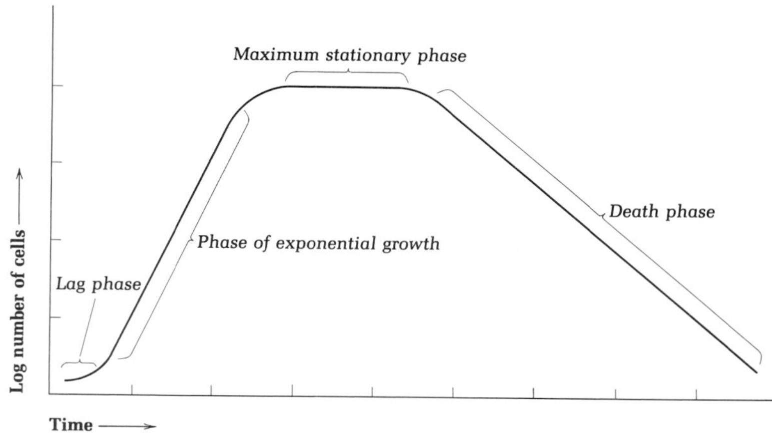


Figure 9.7 Generalized growth curve of a bacterial culture.

of course, interconnected by transitional periods during which the growth rate changes continuously. It will be noted that in Figure 9.7, the growth curve is plotted in terms of the logarithm of the *number of viable cells* in the population. Had cell mass been plotted, the same curve would have been obtained during the exponential phase, but it would have deviated somewhat during the other three phases of growth; we shall discuss in due course the reasons for these differences between cell number and cell mass at certain stages of growth.

**The lag phase**

A lag phase does not always occur in the growth of a bacterial population, and when it does occur, its duration can vary considerably. Lags are in general caused either by the use of an inoculum taken from an old culture, in which the population is in the stationary or death phase, or by a transfer from a chemically different medium. If, on the other hand, a population in the course of exponential growth is transferred into a fresh medium of the same chemical composition, exponential growth will be maintained at the same rate.

The lag that results from the use of an inoculum in the stationary phase is a reflection of the fact that, after the cessation of growth, cells may undergo considerable changes in chemical composition; notably, the cellular population of ribosomes declines and must be restored to a high level before protein synthesis can resume at a high rate. Furthermore, cells in the stationary phase are smaller than those in the exponential phase, and the size of the individual cells increases to that characteristic of the exponential phase before cell division begins. For this reason, the lag as measured by viable count is typically somewhat longer than that measured by cell mass. If an inoculum is taken from a culture in the death phase, many of the cells in the inoculum will be nonviable but still capable of scattering light; under these circumstances, growth as measured by optical density may not become noticeable for some time after the small fraction of viable cells has started growing exponentially. Here, accordingly, the lag measured in terms of viable count will be somewhat shorter than that measured in terms of cell mass.

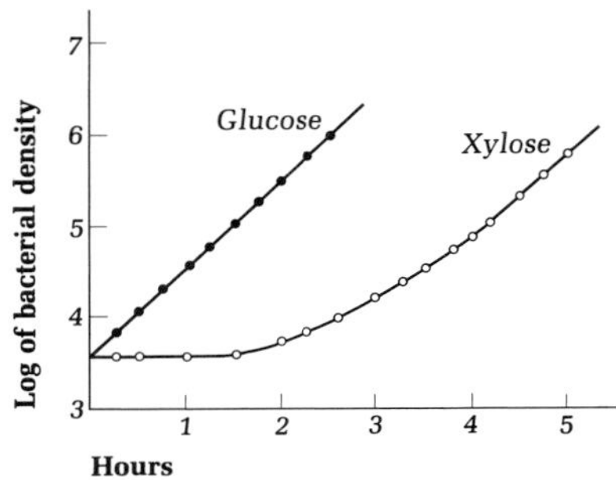


Figure 9.8 The growth of E. coli in synthetic media containing glucose and xylose, respectively, as sole carbon sources. The inoculum was taken from a culture in the course of exponential growth with arabinose.

The kinds of lag described in the preceding paragraph are eliminated if the inoculum is taken from a culture in the course of exponential growth. Nevertheless, even with an exponentially growing inoculum, a lag is frequently observed if the transfer is made into a medium with a different carbon or nitrogen source. Such lags are caused by the fact that the cells do not initially have high levels of the enzymes required for the utilization of the new nutrient, and a certain time elapses before these inducible enzymes reach levels in the cell sufficient to support growth at a maximal rate. A typical example of this phenomenon is shown in Figure 9.8. *Escherichia coli* growing exponentially with the sugar arabinose continues exponential growth upon transfer to a medium containing glucose, for the metabolism of which the cells already possess the necessary enzymes, but the same inoculum shows a lag of about 2.5 hours at 37°C upon transfer to a medium containing xylose, for the metabolism of which the necessary enzymes must be newly synthesized (see Chapter 8, p. 281).

*Diauxic growth (diauxy)* is sometimes observed when a bacterial culture is grown in the presence of two different carbon sources. Diauxy is manifested by two phases of exponential growth, separated by a transient lag, and is caused by catabolite repression of induced enzyme synthesis (see Chapter 8, p. 284). Thus, when *E. coli* is grown with a mixture of glucose and of another sugar that is attacked by inducible enzymes, glucose will first be utilized. Only after the disappearance of glucose from the medium (which results in a transient cessation of growth) is catabolite repression alleviated. This permits the synthesis of the enzymes necessary for growth at the expense of the second sugar (Figure 9.9).

The lag time can be determined graphically on a semilogarithmic

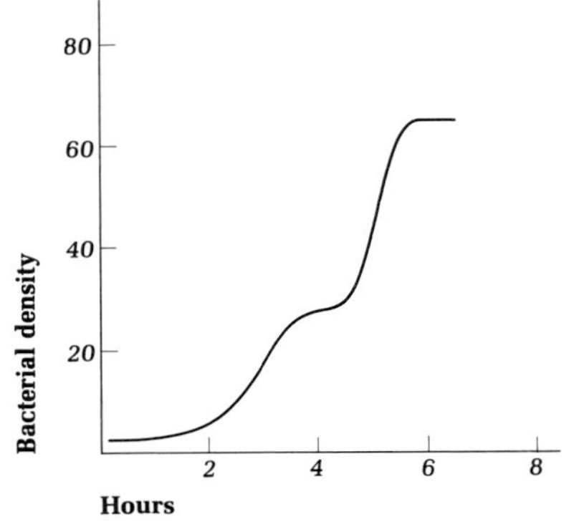


Figure 9.9 Diauxic growth (solid line) of E. coli on a medium supplying equal quantities of glucose and xylose. The transient cessation of growth after about 4 hours reflects the complete utilization of the amount of glucose furnished. After J. Monod, La croissance des cultures bacteriennes. Paris: Hermann, 1942.

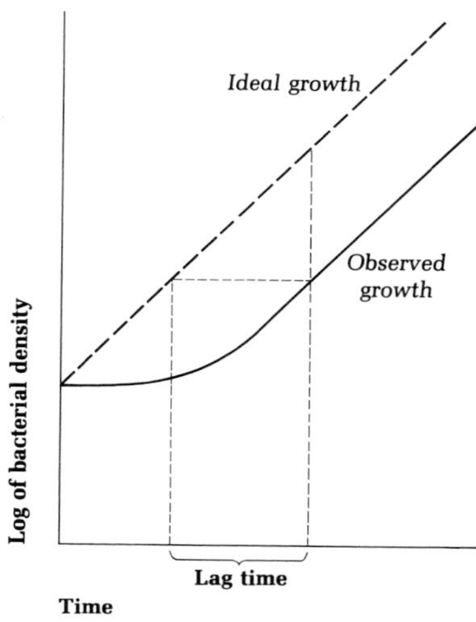


*Figure 9.10 Graphical determination of the lag time. Solid line: observed growth of a culture. Dashed line: ideal growth of the same culture, had growth proceeded without a lag. The horizontal distance between these two lines represents the lag time.*

plot by drawing a line parallel to the eventual exponential portion of the growth curve and passing through the point of origin of the growth curve on the ordinate; the distance between these two parallel lines, measured on the abscissa, is the lag time (Figure 9.10).

Occasionally, a phenomenon distinct from enzyme induction may underlie delayed development of a culture that has been transferred to a chemically different medium. This is *the selection of a mutant population.* It can happen that the bulk of the population is unable to utilize the available form of carbon or nitrogen in the new medium but contains a very small fraction of mutant cells which do possess this capacity. In such a case, the inoculum is a mixture of two genetically different kinds of cells, of which only the minority population can grow under the conditions imposed. A very long lag, sometimes accompanied by death of part of the nonmutant population, will result, and the cell mass will begin to increase only when the rare mutant cells have multiplied sufficiently to become a significant fraction of the population. The lag in this case is apparent, not real, since that small fraction of cells able to grow under the circumstances provided begin to multiply exponentially, long before their presence is manifested by a turbidity increase. The possibility that a lag reflects the growth of a mutant population should always be suspected when the duration of the lag phase corresponds to the time required for many generations of exponential growth.

## The exponential phase

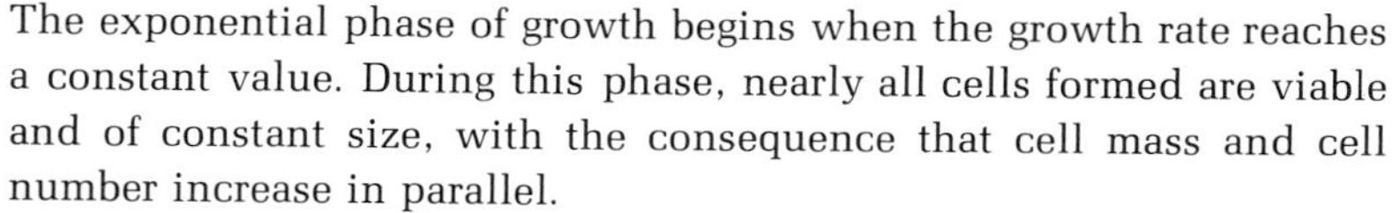

The exponential phase of growth begins when the growth rate reaches a constant value. During this phase, nearly all cells formed are viable and of constant size, with the consequence that cell mass and cell number increase in parallel.

The numerical value of the growth-rate constant is influenced by both genetic and environmental factors. The maximal reproductive potential of any given microorganism in an optimal environment is an inherent property of the species and differs widely from species to species (Table 9.2). Subject to this genetically determined upper limit, the actual growth rate of any given microorganism is determined by the

*Table 9.2 Maximal recorded growth rates for certain bacteria, measured at or near the temperature optimum, in complex media unless otherwise noted*

| Organism | Temperature, °C | Generation time, hr |
|---|---|---|
| *Pseudomonas natriegens* | 37 | 0.16 |
| *Bacillus stearothermophilus* | 60 | 0.18 |
| *Escherichia coli* | 40 | 0.35 |
| *Bacillus subtilis* | 40 | 0.43 |
| *Pseudomonas putida* | 30 | 0.75[a] |
| *Vibrio marinus* | 15 | 1.35 |
| *Rhodopseudomonas spheroides* | 30 | 2.2 |
| *Mycobacterium tuberculosis* | 37 | ~6 |
| *Nitrobacter agilis* | 27 | ~20[a] |

[a]Grown in a synthetic medium.

environment. The factors that affect growth rate include the nature and concentrations of nutrients, the temperature, the pH, and the ionic strength of the medium. Some of them will be discussed later in this chapter (pp. 314–318).

**The maximum stationary phase**

In a culture where nutrients are not renewed, exponential growth continues for only a few generations, and the growth rate then starts to decline as a consequence of the accumulation of toxic metabolic products or the approaching exhaustion of a nutrient. Eventually, the culture enters the *maximum stationary phase*, characterized by a growth rate of zero. The character and duration of this phase are both influenced by the factor that has limited growth. If the population enters the stationary phase as a result of the exhaustion of an essential nutrient (e.g., the carbon and energy source), the viable count, the total count, and the cell mass all become stationary at approximately the same moment, and the population may remain unchanged in size for many hours. The transition to the stationary phase is much less sharp when growth is limited by the accumulation of toxic materials, as commonly happens when a culture is grown in a rich, complex medium. As a result of slight differences in the susceptibility of the cells, the population may have reached the stationary phase, as judged by viable count, but still show a slow increase of both cell mass and total count. Under these circumstances, the stationary phase is a statistical phenomenon, marginal growth by some members of the population being counterbalanced by the death of others, as shown in Figure 9.11.

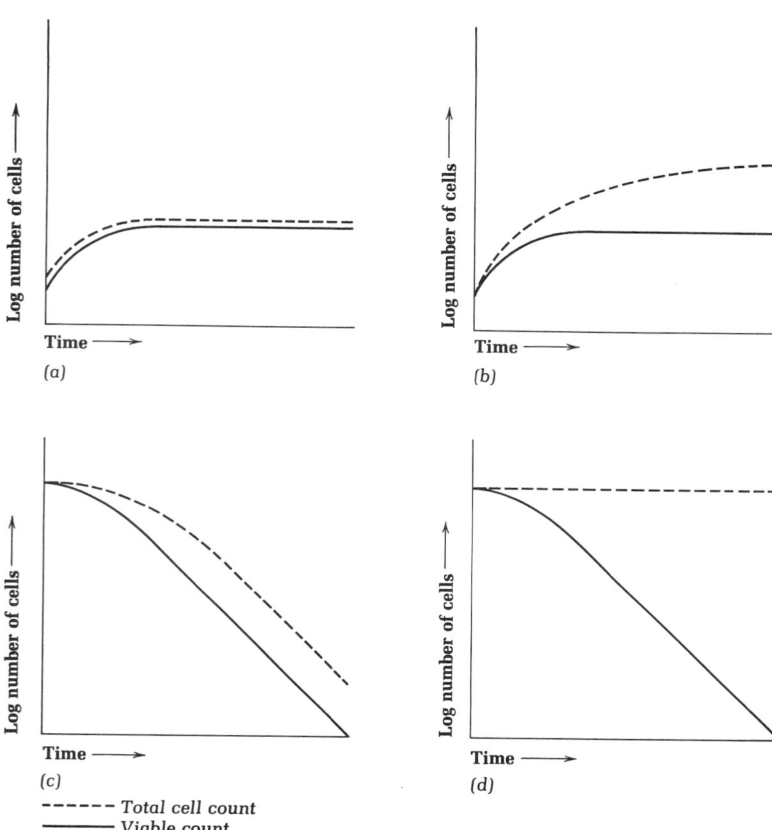


*Figure 9.11 Relationship between viable and total cell counts of cultures in the maximum stationary phase and death phase: (a) growth of a culture limited by the complete exhaustion of a source of energy; (b) growth limited by the accumulation of toxic products or by the partial exhaustion of nutrients; (c) death of culture accompanied by rapid autolysis; (d) death without autolysis of cells.*

**The death phase**

The maximum stationary phase is eventually followed by the death phase, in which the viable population declines. The kinetics of death of bacterial populations is exponential, as discussed in Chapter 3. In bacterial populations, death of an individual cell is often unaccompanied by lysis, so that the cell mass may remain constant or show only a slight decline during this phase, even though a large fraction of the population has become nonviable. Death is sometimes closely followed by lysis; in this event, cell mass declines together with viable count (Figure 9.11).

**Arithmetic (linear) growth**

Under certain special conditions, the kinetics of bacterial growth becomes arithmetic: population density increases as a direct function of time. This is a reflection of physiological imbalance and implies that the cells are unable to synthesize one or more essential catalysts. Two examples will be described.

If a bacterium that requires nicotinic acid as a growth factor is deprived of this nutrient, it is unable to synthesize pyridine nucleotides, of which nicotinic acid is a specific biosynthetic precursor. As a result of the functions of pyridine nucleotides in electron transport, their levels in the cell determine the overall rate of metabolism and hence of growth. When no further synthesis of these compounds can occur, the growth rate of the population becomes directly proportional to the supply of pyridine nucleotides in the cells: the total catalytic power of the population can no longer increase, and at each successive doubling the catalytic power of the individual cells (and hence their growth potential) is approximately halved. Growth therefore becomes directly proportional to time, i.e., arithmetic.

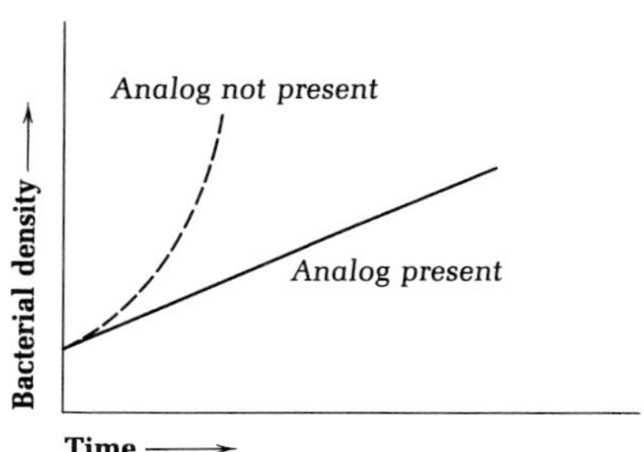


*Figure 9.12 Linear growth of E. coli in a growth medium containing an amino acid analog, p-fluorophenylalanine (solid line). The dashed line shows the growth of a parallel culture not containing p-fluorophenylalanine.*

When microorganisms are grown in the presence of certain chemical analogs of the natural amino acids, such as *p*-fluorophenylalanine, the analog may be taken up by the cells, activated, and incorporated into newly synthesized protein in place of the corresponding natural amino acid. The "false" proteins so produced are often devoid of catalytic power; as a result, total catalytic power of the population is fixed at the level that it had attained when the analog was added to the culture, and growth becomes arithmetic (Figure 9.12).

## *The efficiency of growth: growth yields*

The total growth of a culture represents the difference between the initial cell mass (or number of cells) and the final cell mass (or number of cells) formed when the population enters the stationary phase. When growth is limited by the concentration of a particular nutrient, there is a fixed relationship between total growth and the initial concentration of that nutrient in the medium, as shown in Figure 9.13. The mass of cell material produced per unit mass of nutrient furnished is accordingly a constant, known as the *growth yield*.

In the case of chemoheterotrophic microorganisms, the growth yield measured in terms of the organic substrate used provides an index of the efficiency with which the substrate supports biosynthesis. The data in Figure 9.13 were obtained with an aerobic, chemoheterotrophic pseudomonad, growing in a synthetic medium that contained fructose as the sole source of carbon and energy. Inspection of the graph shows that approximately 0.4 g (dry weight) of cell material is produced per gram of fructose used. The carbon content of fructose and of cell material is, respectively, 40 and 50 percent. When this difference is taken into ac-

*Figure 9.13 The relationship between the growth yield of an aerobic bacterium (Pseudomonas sp.) and the initial concentration of fructose, the growth-limiting nutrient. The experiment was performed in a synthetic medium containing fructose as the sole source of carbon and energy.*

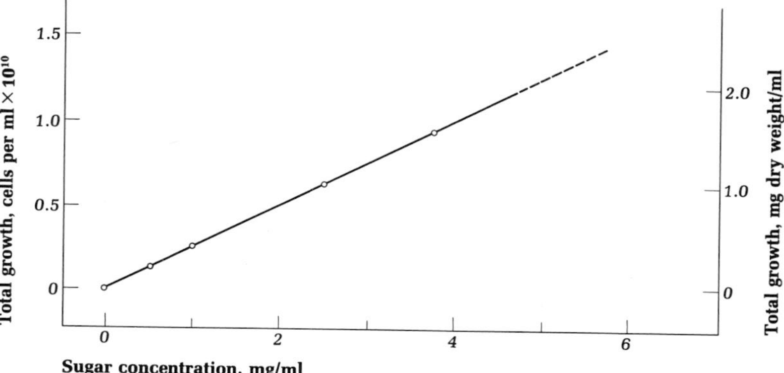


count, the growth yield permits a calculation of the fraction of fructose carbon converted to cell carbon: it is 0.5. This microorganism accordingly uses half the fructose carbon to make cell material and oxidizes the other half to $CO_2$. Analogous experiments with other aerobic chemoheterotrophic bacteria growing with sugars as sole carbon sources show that the fraction of the substrate converted to cell material may range, depending on the organism, from 20 to 50 percent. The differences between aerobes with respect to the growth yield at the expense of a given carbon source probably reflect differences in the efficiency with which the substrate can be used to generate ATP; evidence in support of this inference is described below.

When fermentative organisms are grown anaerobically in a complex medium, it can be shown by radioisotopic measurements that the fermentable substrate is converted almost entirely to end-products, the synthesis of cell material taking place at the expense of other organic nutrients in the medium. Under these conditions, the fermentable substrate serves exclusively to provide ATP, and since the amount of ATP produced through many fermentations is known, the growth yield measured as a function of the concentration of the fermentable substrate affords a means of determining directly the efficiency with which ATP is used for the synthesis of cell material. Such measurements have been made with yeast and with a number of fermentative bacteria (Table 9.3). The growth yields as a function of ATP formed are remarkably constant; they approximate 10 g of cell material per gram-mole of ATP. It should be noted that growth yields calculated per gram-mole of sugar fermented vary, as a result of the differing yields of ATP characteristic of different fermentative pathways (see Chapter 6). The alcoholic fermentation of glucose via the Entner–Doudoroff pathway characteristic of *Zymomonas mobilis* yields only 1 mole of ATP per mole of substrate.

*Table 9.3 Growth yields of fermentative microorganisms, measured in terms of glucose fermented and of ATP produced*

| Organism | Fermentation and pathway | ATP formed per mole of glucose fermented | Molar growth yields expressed as grams of cell material per mole of: | |
|---|---|---|---|---|
| | | | Glucose fermented | ATP produced |
| *Yeast (Saccharomyces cerevisiae)* | *Alcoholic, Embden–Meyerhof* | 2 | 21 | 10.5 |
| *Streptococcus fecalis* | *Homolactic, Embden–Meyerhof* | 2 | 22 | 11 |
| *Lactobacillus delbrueckii* | *Homolactic, Embden–Meyerhof* | 2 | 21 | 10.5 |
| *Zymomonas mobilis* | *Alcoholic, Entner–Doudoroff* | 1 | 8.6 | 8.6 |

On the other hand, the alcoholic fermentation of glucose via the Embden–Meyerhof pathway performed by yeast, and the mechanistically similar homofermentative lactic acid fermentation performed by *Streptococcus fecalis* and *Lactobacillus delbrueckii*, yield 2 moles of ATP per mole of substrate. As shown in Table 9.3, the growth yield of *Z. mobilis* is approximately half that of yeast or of homofermentative lactic acid bacteria when calculated on the basis of the amount of sugar fermented, but is similar when calculated on the basis of the ATP made available.

## *Some effects of environment on growth rate and growth yield*

**The effect of nutrient concentration on growth rate**

As a general rule, the growth rate of microorganisms remains unaffected by the concentrations of its nutrients until these have fallen to very low values. At these low concentrations, growth rate is closely proportional to concentration, but as the concentration increases, the growth rate rises rapidly to a maximal value, which is maintained until the nutrient concentration reaches an inhibitory level, at which point the growth rate begins to fall again. As shown in Figure 9.14, the same type of hyperbolic curve relating these two parameters is obtained with either a carbon source or an essential growth factor as the rate-limiting nutrient.

Figure 9.14 *The effect of nutrient concentration on the growth rate of E. coli. (a) Effect of glucose concentration; (b) effect of tryptophan concentration (for a tryptophan-requiring mutant). In each case, the vertical dashed line indicates the concentration of the nutrient that supports growth at a half-maximal rate. This value is 0.22 × $10^{-4}$ M for glucose, and 1.1 ng/ml for tryptophan.*

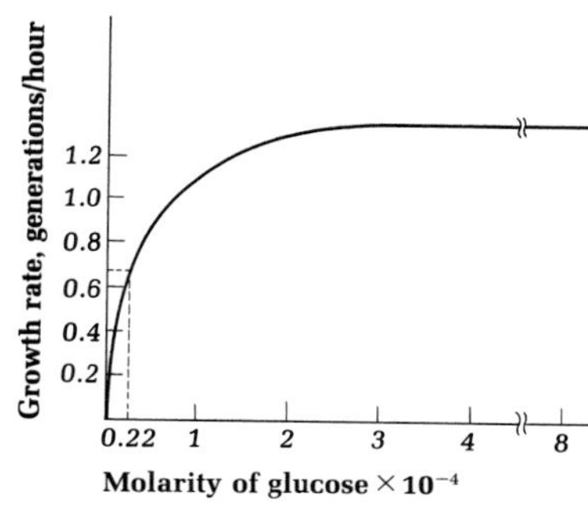


(a)

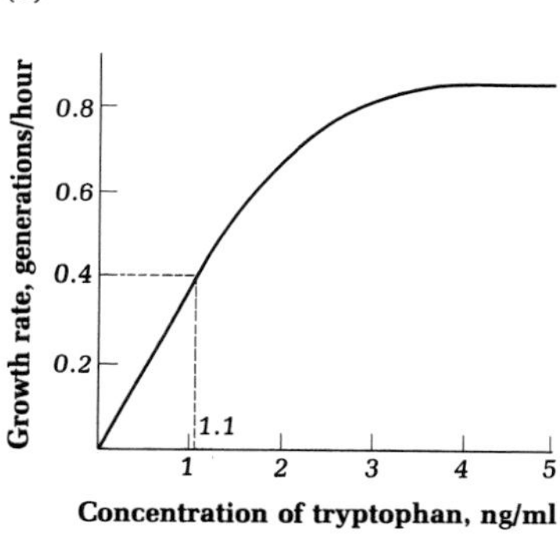


(b)

The effects of different nutrients on growth rate can best be compared in terms of the concentrations that support a half-maximal rate of growth. For carbon and energy sources, this concentration is usually of the order of $10^{-5}$ to $10^{-6}$ *M*, which corresponds for glucose to a concentration of between 20 and 200 mg/liter. For other nutrients (e.g., minerals, growth factors) the half-saturating concentration is usually even lower. The microbial cell accordingly has a very high affinity for its nutrients, of the same order of magnitude as the affinities of enzymes for their substrates. As will be discussed in Chapter 10, the entry of many nutrients into the cell is mediated by specific permeases, and the observed dependence of growth rate on nutrient concentration is probably determined in most cases by the affinity of the specific permease for the nutrient in question. The profound importance of permeases in facilitating bacterial growth is clearly evident from the growth behavior of permeaseless mutants. When a bacterium no longer makes a specific permease as a result of mutation, and is hence dependent on entry of the corresponding nutrient into the cell by diffusion, an external concentration of this nutrient 1,000 times as great as that which supports a maximal growth rate of the parental strain may still not allow the permeaseless mutant to grow at a comparable rate.

## The effect of temperature on growth rate

The total temperature span within which organisms can grow is a narrow one, extending from about −5 to +80°C. The upper temperature limit of biological growth is determined by the thermolability of the cellular proteins. The lower temperature limit is set by the freezing point of water, several degrees below zero when it contains appreciable quantities of solutes. It should be noted that biological *survival* is possible at temperatures far below the minimal temperature that permits growth. Microorganisms can retain viability for long periods when kept at low temperatures in a frozen state, under conditions where all metabolic activity is arrested; in fact, the maintenance of bacterial strains at the temperature of liquid nitrogen (−196°C) is an excellent means of preserving them.

For any specific microorganism, the temperature that permits growth seldom exceeds 35°C. When growth rate is plotted as a function of tem-

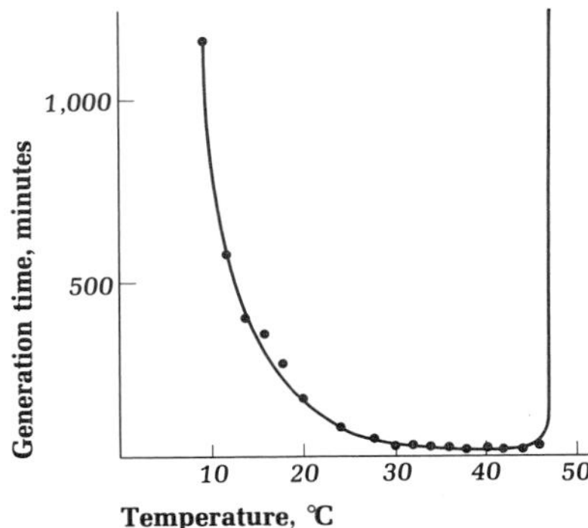


*Figure 9.15 The effect of temperature on the growth rate of E. coli, a typical mesophile. After J. L. Ingraham, "Growth of psychrophilic bacteria." J. Bacteriol.* **76,** *75 (1958).*

perature, a curve of characteristic shape is obtained, illustrated in Figure 9.15 for the bacterium *E. coli*.

Among procaryotic microorganisms, differences with respect to the temperature range that supports growth are particularly marked. In terms of this property, it is possible to distinguish three major physiological groups, *thermophiles, mesophiles,* and *psychrophiles* (Table 9.4); *E. coli* is a typical member of the mesophilic group. The temperature optimum of mesophiles lies between 30 and 45°C and the minimum between 10 and 15°C. In thermophiles, the whole curve relating growth rate to temperature is shifted to a higher range: the optimum lies between 55 and 75°C; the minimum is never lower than 35 to 40°C and may be somewhat higher. Among the psychrophiles, two subgroups can be distinguished. *Facultative psychrophiles* possess temperature optima and maxima in the customary range of mesophiles, but unlike mesophiles, they can grow, albeit slowly, at temperatures near or below 0°C. *Obligate psychrophiles,* whose existence has been demonstrated only recently, are rapidly killed at a temperature of 20°C or above; their temperature minimum is similar to that of the facultative group, but the temperature optimum lies at 15 to 18°C.

In interpreting the influence of temperature on microbial growth rate, it is necessary to consider the effect of temperature on the rate of chemical reactions. The rate of most chemical reactions approximately doubles for a temperature increment of 10°C. The relation of reaction rate to temperature is described mathematically by the law of Arrhenius:

$$\log_{10} v = \frac{-\Delta H}{2.303RT} + C$$

where $v$ represents the reaction rate, $-\Delta H$ the activation energy of the reaction, $R$ the gas constant, and $T$ the absolute temperature in degrees

*Table 9.4 Principal physiological categories of bacteria in terms of the relationship between growth rate and temperature*

| | Temperature, °C | | |
|---|---|---|---|
| **Group** | **Minimum** | **Optimum** | **Maximum** |
| *Thermophiles* | 40–45 | 55–75 | 60–80 |
| *Mesophiles* | 10–15 | 30–45 | 35–47 |
| *Psychrophiles* | | | |
| *Obligate* | (−5)–(+5) | 15–18 | 19–22 |
| *Facultative* | (−5)–(+5) | 25–30 | 30–35 |

Kelvin. Since $-\Delta H$, $R$, and $C$ are constants, it follows that there is a linear relationship between the logarithm of the reaction rate and the inverse of the absolute temperature. When microbial growth rates as a function of temperature are plotted in terms of the Arrhenius equation (Figure 9.16), one finds a linear relationship between rate and the inverse of the absolute temperature over a certain part of the temperature range. However, there is a substantial deviation from linearity at temperatures that approach both the maximum and the minimum. How can these deviations at the two extremes of the temperature range be explained?

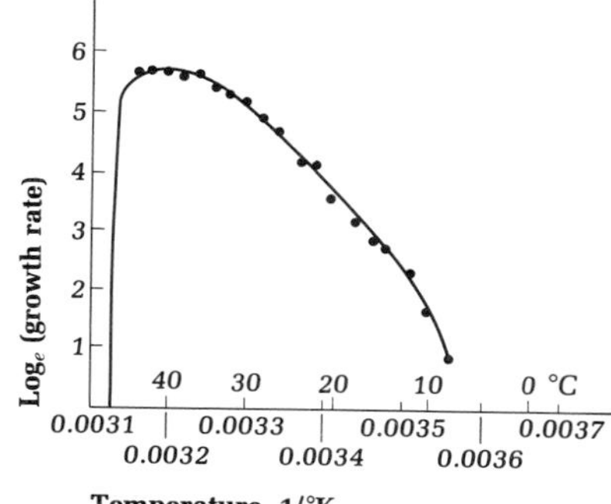


*Figure 9.16 An Arrhenius plot of the relationship between growth rate and temperature for E. coli, calculated from the data of Figure 9.15. Growth rate is in terms of generations per hour. After J. L. Ingraham, "Growth of psychrophilic bacteria." J. Bacteriol.* **76,** *75 (1958).*

The deviation near the temperature maximum is easily interpreted. As temperature increases, microbial growth is eventually arrested by the thermal denaturation of the cell proteins; above the temperature optimum, this becomes a factor of rapidly increasing importance, causing an abrupt decline of the growth rate. It should be noted that growth will cease when the *most thermolabile essential protein* of the cell is destroyed. This is clearly shown by the properties of certain mutants, known as *conditional lethal mutants* (see Chapter 13). In such mutants, an enzyme which plays an essential cellular role has undergone a change that does not impair its catalytic function but renders it much more susceptible to heat denaturation than the enzyme of the parental type. Such mutants can grow normally at low temperatures, where their behavior is indistinguishable from that of the parent, but at higher temperatures, which still permit normal growth of the wild type, their growth ceases as a result of the thermal inactivation of the mutated enzyme. It is thus possible to change the temperature maximum of an organism to a very considerable extent by mutation of a single gene.

In the light of this analysis, it follows that thermophilic microorganisms must possess a *complete complement of cellular proteins with an unusually high degree of thermal stability*. This inference has been confirmed experimentally, by comparative studies on certain enzymes isolated from thermophiles and mesophiles; in every case examined so far, an enzyme with a given catalytic function isolated from a thermophile is considerably more resistant to heat denaturation than the corresponding enzyme that has been isolated from mesophilic bacteria.

The deviation from linearity of the Arrhenius plot at low temperatures is less easily interpreted. In terms of the properties of a chemical reaction, one would have predicted that all bacteria would continue to grow (although at progressively lower rates) as the temperature was reduced, until the whole system froze solid. Until recently, there has

*Table 9.5 Influence of temperature on the growth rate and growth yield of Aerobacter aerogenes, growing aerobically at the expense of glucose in a synthetic medium*[a]

| Temperature, °C | Growth rate, $\mu$ | Growth yield, grams (dry weight) of cells per gram of glucose |
|---|---|---|
| 23 | 0.552 | 0.354 |
| 27 | 0.751 | 0.356 |
| 32 | 0.967 | 0.336 |
| 37 | 1.305 | 0.324 |
| 38.8 | 0.833 | 0.217 |
| 39.7 | 0.600 | 0.172 |
| 40.8 | 0.400 | 0.095 |
| 42 | 0 | 0 |

[a]From J. Senez, *Bacteriol. Rev.* **26**, 104 (1962).

been no satisfactory explanation of the fact that most mesophiles stop growing at a temperature of 10 to 15°C and most thermophiles at a temperature as high as 40 to 45°C. It is now known, however, that *the temperature minimum is determined by the regulatory machinery of the cell.* The sensitivity to low temperatures of the feedback control systems is considerably greater than that of the metabolic sequences which they control. Consequently, as the temperature is lowered, end-product inhibition may become total at a temperature that would still permit the activity of an unregulated enzyme, and growth ceases. In functional terms, psychrophiles can be interpreted as organisms whose control systems are far less susceptible to derangement at low temperatures than those of mesophiles.

As shown in Table 9.5, the growth yield of a microorganism remains practically constant with increasing temperature until the temperature optimum has been reached but declines very rapidly between the optimum and the maximum. This reflects the fact that at near-maximal temperatures the coupling between energy-yielding metabolism and biosynthesis becomes impaired.

## *The continuous cultivation of microorganisms*

As we have already discussed, microbial growth in a culture where the nutrients are not renewed remains exponential for only a few generations. The population can, of course, be maintained in the exponential phase by repeated transfers into fresh medium at short time intervals. However, the maintenance of a microbial population in a state of expo-

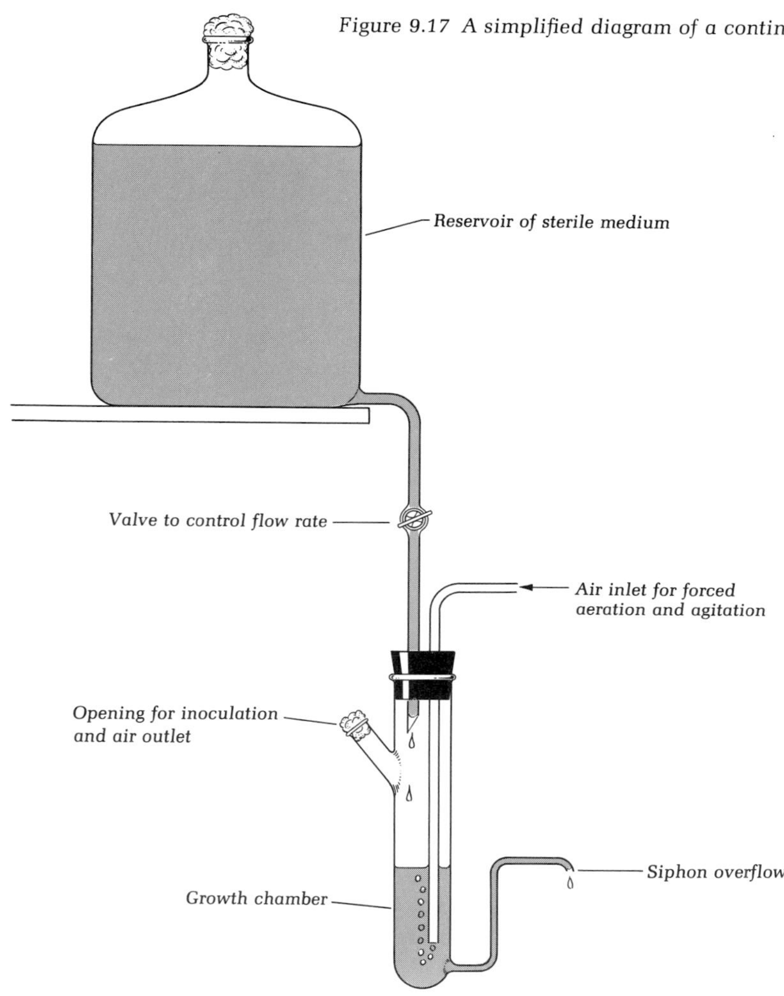


*Figure 9.17 A simplified diagram of a continuous culture system.*

nential growth over a long period can be achieved more easily and reliably by using a system of *continuous culture* (Figure 9.17). The growth chamber is connected to a reservoir of sterile medium. Once growth has been initiated, fresh nutrients are continuously supplied from the reservoir. The volume of liquid in the growth chamber is maintained constant, the excess population being removed continuously through a siphon overflow. In such an open system, exponential growth can be maintained for days, or even weeks.

Continuous culture has many advantages, both practical and theoretical. In practical terms, it can provide a continuous supply of cells that are both uniform and in an optimal physiological state for experimental uses. Many of the parameters that affect growth can be studied with greater ease and precision in continuous cultures than in cultures where the medium is not renewed.

**The mathematical theory of continuous growth**

In a continuous culture system, the siphon overflow maintains the culture in the growth chamber at a fixed volume, $V$ ml. Fresh medium enters at a flow rate of $w$ ml/hr, homogeneous conditions being maintained by rapid mixing of the inflowing medium with the culture. The hourly rate of dilution of the culture, equal to $w/V$, will be termed $D$.

In the growth chamber, two opposing factors influence the size of the population. The population increases continuously as a result of growth; at the same time, it is continuously diminished by washing out through the siphon overflow. The rate of these two opposing processes can be described mathematically.

By derivation from the growth equation

$$N_t = e^{\mu t} N$$

the instantaneous growth rate of the population is

$$\frac{dN}{dt} = \mu N$$

The rate of loss of cells by washing out can be expressed as

$$\frac{dN}{dt} = -DN$$

where $N$ is the initial population size and $D$ the dilution rate.

The net rate of change of population size in the growth chamber is accordingly determined by the algebraic sum of these two partial derivatives:

$$\frac{dN}{dt} = \mu N - DN = (\mu - D)N$$

We can now ask how $N$, the population size, will vary with time. For any given microorganism growing under a fixed set of environmental conditions, the growth-rate constant, $\mu$, cannot exceed a certain maximal value, which we shall call $\mu_{max}$. If the dilution rate, $D$, is *greater* than $\mu_{max}$, the expression $dN/dt$ has a negative value. Under these circumstances, the population in the growth chamber will diminish until it has been completely washed out.

If the dilution rate, $D$, is *less than* the maximal growth rate, $\mu_{max}$, the value of $dN/dt$ is positive, and the size of the population will increase.

*This increase will continue until the rate of growth begins to be limited by the concentration of the limiting nutrient in the inflowing medium.* At this point, the growth rate, $\mu$, falls to a lower value. When $\mu$ becomes equal to $D$, the size of the population in the growth chamber becomes constant. Under these circumstances, accordingly, *growth is self-regulating.* This is the principle of the continuous culture device known as the *chemostat.*

**The chemostat**

The constancy of population size in a chemostat is achieved by internal feedback control. The growth rate is never maximal, being maintained at a submaximal value that is determined by the steady-state concentration of the limiting nutrient in the growth chamber. This concentration is, in turn, determined by $D$, the rate of inflow of fresh medium. If the flow rate is increased, the growth rate increases; if the flow rate decreases, the growth rate decreases. In this manner, the growth rate of the population in a chemostat can be varied at will over a very wide range (from a value approaching zero to about $0.5\mu_{max}$), simply by changing the flow rate.

Under steady-state conditions in a chemostat, the instantaneous growth rate, $\mu$, is equal to the flow rate, $w/V$, which is easily measurable. The mean generation time, $G$, can be derived from the flow rate by means of the following relation:

$$G=\frac{1}{k}=\frac{1}{0.69\mu}=\frac{V}{0.69w}$$

As this relation shows, the population in the growth chamber doubles every time a volume equal to $0.69V$ flows through it.

The steady-state size of the population in the growth chamber is directly proportional to the concentration of the growth-limiting nutrient in the inflowing medium.

## *The influence of growth rate on size and composition of cells*

By use of a chemostat, the microbial growth rate can be varied widely by changes of flow rate, while all other environmental conditions remain constant. One can therefore ascertain the specific effect of growth rate on the size and chemical composition of the cell. Contrary, perhaps, to intuitive expectation, the growth rate has significant effects, as shown in Table 9.6. The size of the cells is directly related to their growth rate (column 1). The cellular contents of DNA and of protein

*Table 9.6 Influence of growth rate on the size and macromolecular composition of the cells of Salmonella typhimurium*[a]

| Growth rate, $\mu$ | Number of cells ($\times 10^{-12}$) per gram dry weight | Composition of cells, mg/g dry weight | | |
|---|---|---|---|---|
| | | DNA | Protein | RNA |
| 2.4 | 1.3 | 30 | 670 | 310 |
| 1.2 | 3.1 | 35 | 740 | 220 |
| 0.6 | 4.8 | 37 | 780 | 180 |
| 0.2 | 6.3 | 40 | 830 | 120 |

[a]From O. Maaløe and N. O. Kjeldgaard, *Control of Macromolecular Synthesis*. New York: Benjamin, 1966.

are inversely related to growth rate, being highest in slowly growing cells (columns 2 and 3). The most significant compositional changes concern the RNA content (column 4), which is directly related to the growth rate and increases between two- and threefold with a tenfold change in growth rate. This change in RNA content reflects almost entirely a change in *ribosomal* RNA: rapidly growing cells have a far higher ribosomal content than slowly growing ones. The mechanism whereby these changes of ribosomal content are mediated when growth conditions change was discussed in Chapter 8.

## Synchronous growth

In any large unicellular microbial population, the times of division of the individual members of the population are random. By certain methods, however, it is possible to impose synchronous division, lasting two or three generations, on a relatively large population of initially nonsynchronous cells. The principal means of accomplishing this are: subjection of the population to cyclic changes of temperature; separation by physical means of cells in the population that are all at the same stage of the cellular life cycle; and attachment of growing cells to a membrane, from which one product of cell division is shed.

When a culture of *Salmonella typhimurium* is subjected at fixed intervals to alternate cycles of growth at 25 and 37°C, the entire population divides synchronously (Figure 9.18). As shown by viable counts, division occurs during each short exposure to 37°C, the population level remaining fixed during the intermediate periods of cultivation at 25°C. Cell mass increases at both temperatures, although its rate of increase is less at 25°C than at 37°C because the latter temperature is closer to the optimal one. When the culture is placed at its optimum temperature, after such cyclic treatment, synchrony persists for several

Figure 9.18 *The synchronization of a broth culture of S. typhimurium by cyclic changes of temperature. After O. Maaløe and K. G. Lark, in Recent Developments in Cell Physiology (J. A. Kitching, editor), p. 159. London: Butterworth, 1954.*

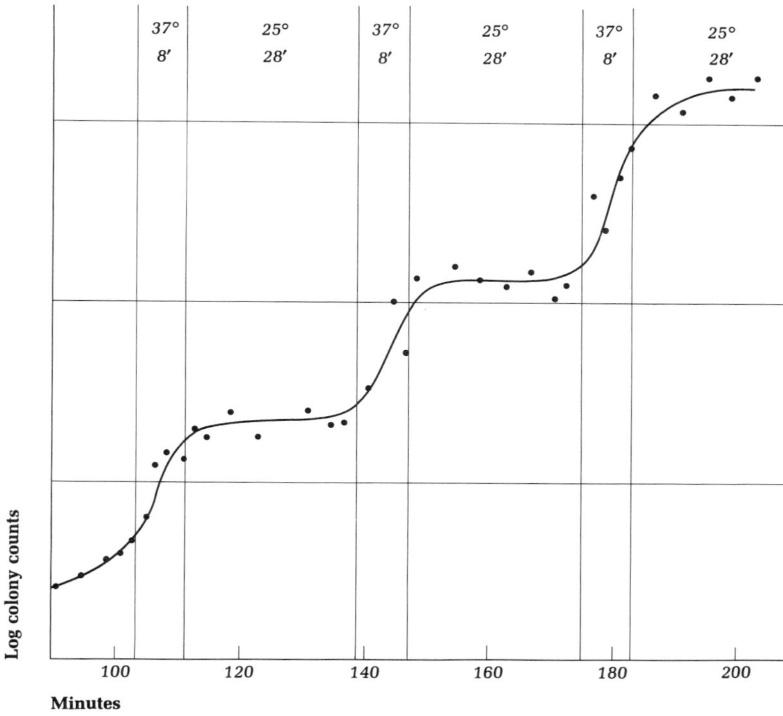


generations. The temperature cycle necessary to provoke synchronous growth must be determined empirically for each organism studied. The mechanism of synchronization can be explained as follows. One of the steps in cell division is particularly temperature dependent. Consequently, at the lower temperature, the growth of cells tends to be arrested at this point in the division cycle, which gives a chance for the more tardy members of the population to catch up with others that have already reached this point. When the temperature is suddenly raised, the barrier to completion of division is lifted, and virtually the entire population proceeds to divide. The necessity of several repetitions of the cycle results from the fact that arrest of the temperature-sensitive stage of division is not perfectly efficient.

Another method for achieving synchrony is to filter a growing population through a filter paper pad, which permits selective passage of the smallest cells. Since these members of the population represent the products of a recent cell division, they are all at more or less the same stage of the cellular life cycle and will subsequently divide in synchrony. This method has the advantage of yielding a synchronous population that has not been subjected to physiological manipulation.

## *Further reading*


**Books**

Maaløe, O., and N. O. Kjeldgaard, *Control of Macromolecular Synthesis.* New York: Benjamin, 1966.

Mandelstam, J., and K. McQuillen, *The Biochemistry of Bacterial Growth.* New York: Wiley, 1968.

**Reviews**

Farrell, J. and A. Rose, "Temperature Effects on Microorganisms." *Ann. Rev. Microbiol.* **21,** 101 (1967).

Monod, J., "The Growth of Bacterial Cultures." *Ann. Rev. Microbiol.* **3,** 371 (1949).

Novick, A., "Growth of Bacteria." *Ann. Rev. Microbiol.* **9,** 97 (1955).

Scherbaum, O. H., "Synchronous Division of Microorganisms." *Ann. Rev. Microbiol.* **14,** 283 (1960).

Senez, J. C., "Some Considerations on the Energetics of Bacterial Growth." *Bacteriol. Rev.* **26,** 95 (1962).

# Chapter ten
# Relations between structure and function in procaryotic cells

*The structural features* that distinguish procaryotic cells from eucaryotic cells have been outlined in Chapter 2, and further information about the properties of procaryotic cells was presented in Chapter 5. As a result of their relatively simple organization, procaryotic cells have provided the most favorable model systems for the study of the interrelations among the biosynthesis, the structure, and the function of the component parts of cells, the topic to be discussed in this chapter.

## *The surface layers of procaryotic cells*

The procaryotic cell may be surrounded by as many as three surface layers, which differ in fine structure, chemical composition, and

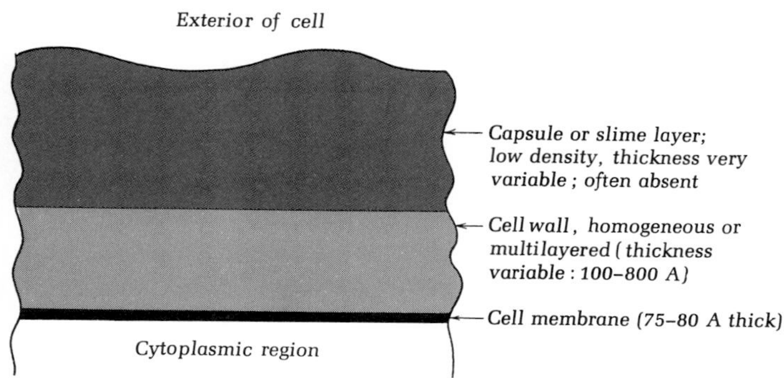


*Figure 10.1 Schematic diagram of the surface layers of the procaryotic cell.*

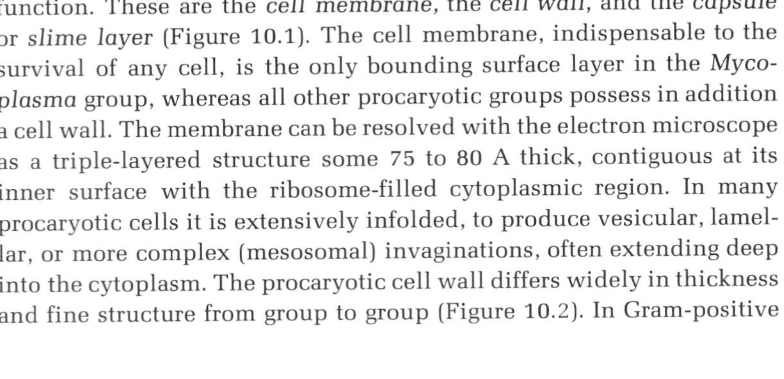

function. These are the *cell membrane*, the *cell wall*, and the *capsule* or *slime layer* (Figure 10.1). The cell membrane, indispensable to the survival of any cell, is the only bounding surface layer in the *Mycoplasma* group, whereas all other procaryotic groups possess in addition a cell wall. The membrane can be resolved with the electron microscope as a triple-layered structure some 75 to 80 A thick, contiguous at its inner surface with the ribosome-filled cytoplasmic region. In many procaryotic cells it is extensively infolded, to produce vesicular, lamellar, or more complex (mesosomal) invaginations, often extending deep into the cytoplasm. The procaryotic cell wall differs widely in thickness and fine structure from group to group (Figure 10.2). In Gram-positive

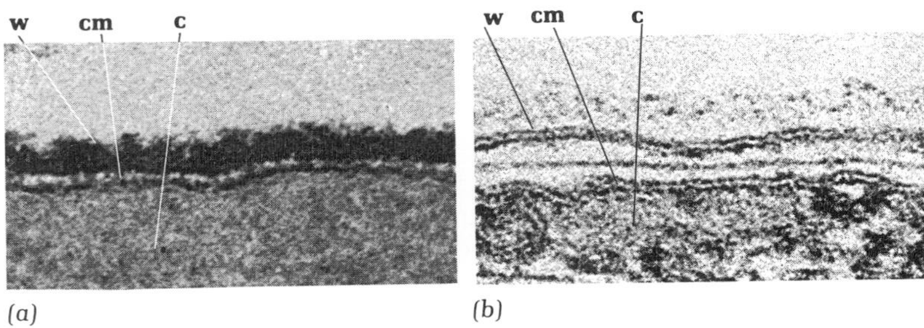


*Figure 10.2 Electron micrographs of sections of the surface layers of (a) a Gram-positive bacterium, and (b) a Gram-negative bacterium, illustrating the differences in cell-wall profiles: c, cytoplasm; cm, cell membrane; w, wall. In the Gram-positive bacterium the wall consists of a single, thick, continuous layer. In the Gram-negative bacterium it is multilayered; in this particular organism there is a thin inner (murein) layer; an intermediate layer, similar in profile to the cell membrane; and a loose outer layer.*

bacteria, it is usually a dense, uniform layer, from 200 to 800 A in thickness. In other procaryotic groups (Gram-negative true bacteria, blue-green algae) it is clearly multilayered, and consists of a thin, electron-dense inner layer (as little as 20 to 30 A thick in most Gram-negative bacteria), covered by one or more less dense outer layers, some 100 to 300 A thick. Capsules or slime layers, although widespread among procaryotic organisms, are by no means of universal occurrence. This layer is typically thicker and less dense than the cell wall, and is as a rule poorly visible with the electron microscope. It can be visualized most clearly by examining a suspension of cells in India ink with the light microscope. The carbon particles of the ink cannot penetrate the capsular layer; they thus reveal more or less sharply its contour (Figure 10.3).

Two different kinds of filiform appendages may extend out through the surface layers of the cell. *Flagella*, the locomotor organelles of true bacteria, are helicoid threads with a diameter of 120 to 185 A.

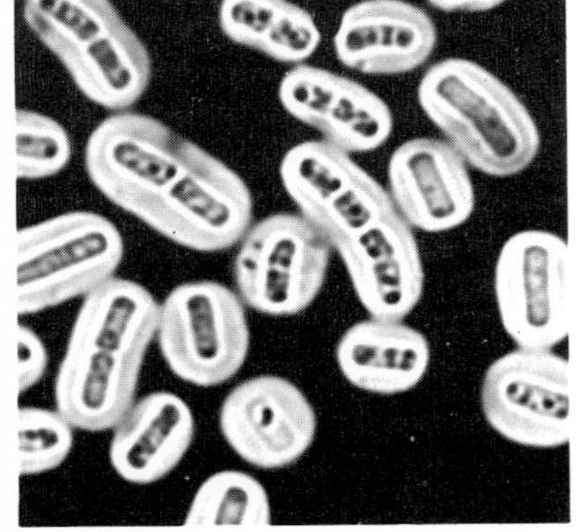

*Figure 10.3 Bacterial capsules, demonstrated by dispersing the cells in India ink. The organism shown is Bacillus megaterium (× 2,700). Courtesy of C. F. Robinow.*

*Pili* or *fimbriae* are stiff, straight rods, which vary in diameter from 40 to as much as 350 A.

**The chemical composition of the surface layers**

In the past 20 years, techniques of separation and purification have been developed which permit the isolation and chemical analysis of the various surface structures just described. The cell membrane, like the membranes of eucaryotic cells, consists largely of protein (50 to 75% of the dry weight) and lipid (20 to 35% of the dry weight). The lipid is predominantly phospholipid (e.g., phosphatidylethanolamine), and sterols are absent. The membrane proteins are tightly bound to the phospholipids but can be dissociated by treatment of the membrane fraction with detergents. Electrophoresis of detergent-treated membranes shows that a considerable number of different proteins, separable by their charges, are present. The cell membranes of aerobic bacteria contain in addition all the components of the respiratory electron transport system, together with certain substrate dehydrogenases (succinic and malic dehydrogenases). In purple bacteria, the pigment system and the associated machinery of photosynthetic electron transport are incorporated into the membrane or its extensions.

As mentioned in Chapter 5, procaryotic cell walls are notable for their chemical complexity and diversity. They always contain mureins, a unique class of heteropolymers that provide the principal strengthening elements of the wall, and these may be accompanied by polymers of other classes. Polysaccharides, teichoic acids, and (more rarely) proteins occur in the walls of Gram-positive bacteria; lipoproteins, lipopolysaccharides, and proteins occur in the walls of other groups.

Capsules or slime layers are as a rule of comparatively simple composition: in most cases, they consist of a single polysaccharide or a polypeptide containing a single amino acid. Occasionally, "mixed" capsules are formed, containing two different polymeric constituents. The flagella and the pili are both composed of protein.

**The roles of surface structures as antigens and as phage receptors**

It has long been known that bacterial cells possess a large number of different *antigenic determinants* and hence display a very high degree of *antigenic specificity* (see Chapter 28). This specificity is so great that detailed antigenic analyses, performed on a large number of strains belonging to a single species, permit their subdivision into many immunologically specific groups, or *serotypes*. In the course of time, it has become evident that the determinants which confer such antigenic specificity reside in the various surface components of the cell:

capsule, pili, flagella, and cell wall. Since the capsule, the pili, and the flagella are each of relatively uniform composition, they generally each carry only one or a few determinants; but the cell wall is often the site of a large number of different antigenic determinants (*somatic, or O, antigens*), and hence makes a major contribution to the antigenic complexity and specificity of the bacterial cell. Of the wall components, proteins, teichoic acids, and lipopolysaccharides have all been implicated as determinants of bacterial antigenic specificity. This question will be discussed at greater length in connection with the lipopolysaccharides of Gram-negative bacteria (p. 342).

For a bacterium to undergo infection by bacterial viruses (phages), the virus particles must first adsorb to the bacterial surface (see Chapter 11). Adsorption is a highly specific phenomenon, which is dependent on the existence of an appropriate *receptor site* on the surface of the bacterial cell. As a general rule, such receptor sites are contained in the cell wall. Detailed studies on the receptor sites of *Escherichia coli* for phages of the T series have shown that the specific receptors for certain of these phages lie in the lipopolysaccharide layer of the wall and, for others, in the lipoprotein layer. For certain phages, the specific receptor sites lie in the filiform appendages of the cell (pili or flagella). Thus, the male-specific phages of *E. coli* adsorb specifically to a certain type of pilus, the F pilus, carried only by $F^+$ or Hfr cells (see Chapter 14). Other phages attach specifically to the flagellum; the most remarkable of these is a phage specific for *Bacillus subtilis*, which attaches to flagella by means of a pair of helical tail fibers (Figure 10.4).

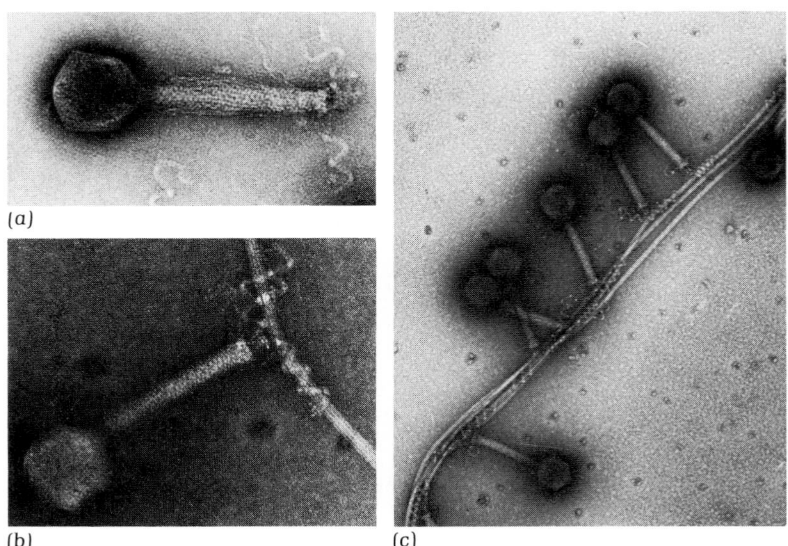


*Figure 10.4 Electron micrographs of a bacteriophage of Bacillus subtilis which adsorbs on the bacterial flagellum. (a) A free phage particle with helical tail fibers (× 120,000). (b) A phage particle that has adsorbed to a bacterial flagellum, around which the tail fibers are wrapped (× 120,000). (c) A group of phage particles attached to several flagella (× 61,000). From L. M. Raimondo, N. P. Lundh, and R. J. Martinez, "Primary adsorption site of phage PBSI: the flagellum of Bacillus subtilis." J. Virol.* **2**, *256 (1968).*

Attachment to these filiform appendages is presumably then followed by a transfer of the phage particle to the cell surface at the base of the appendage, to permit introduction of the phage nucleic acid into the cytoplasm; however, the details of this transfer are not yet clearly understood.

**The functions of wall and membrane**

The respective functions of the wall and the membrane were first clearly demonstrated by C. Weibull in *Bacillus megaterium*. This Gram-positive bacterium has a cell wall which can be completely destroyed by treatment with lysozyme. When cells are exposed to lysozyme in a hypotonic solution, they undergo rapid osmotic lysis. However, if the lysozyme treatment is conducted in a medium isotonic with the cell contents (e.g., one containing a high concentration of sucrose), destruction of the wall results in conversion of the rod-shaped cells to spherical *protoplasts*, which remain stable for some hours in isotonic media (Figure 10.5). The protoplasts possess an osmotic barrier; they retain the full respiratory activity of the cells from which they are derived; they are capable of synthesizing proteins and nucleic acids and can even support phage growth, if infection took place prior to removal of the specific phage receptors of the wall by lysozyme treatment. However, they are unable to synthesize a new cell wall after the removal of lysozyme: reversion to the rod form has never been observed. Furthermore, although they can increase in dry weight, and may possibly divide, they cannot multiply indefinitely. These observations define the major functions of the cell wall. It has a vital *mechanical* role in protecting the enclosed cell from osmotic lysis in a hypotonic environment, it confers specific shape on the cell, and it seems to play an important part in normal cell division. In addition, it cannot be re-formed once it has been completely removed from

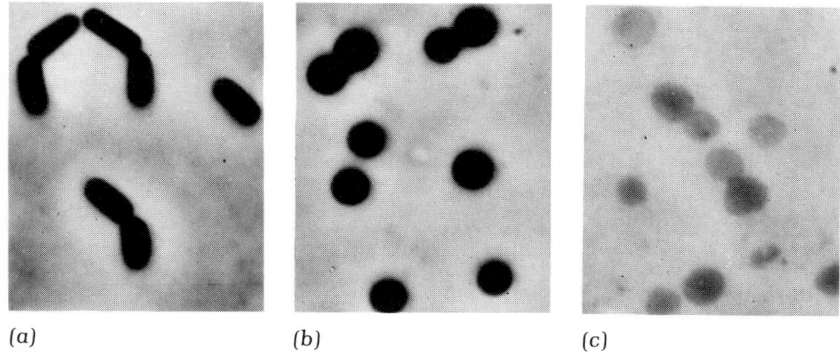


*Figure 10.5 Bacillus megaterium (phase contrast, × 2,600): (a) the intact cells; (b) the spherical protoplasts, formed by enzymatic dissolution of the cell wall with lysosyme in an isotonic medium; (c) the ghosts (empty cytoplasmic membranes), formed by osmotic rupture in a hypotonic medium. Courtesy of C. Weibull.*

the cell. On the other hand, the wall does not itself contribute to the osmotic barrier of the cell, a function exclusively fulfilled by the membrane; and it probably does not contain proteins endowed with enzymatic activity, with the possible exception of enzymes involved in synthesis or breakdown of the wall polymers.

The membrane, on the other hand, is clearly the site of a number of different and highly important enzymatic activities. These include multienzyme systems involved in respiratory and photosynthetic function, which have already been mentioned. In addition, it has been shown to contain ATPase, enzymes and carriers operative in the late stages of the synthesis of wall polymers, and enzyme systems that operate to facilitate the passage of nutrients into the cell (permease systems). Indeed, considering the multiplicity of catalytic functions now associated with membranes, it seems probable that all the membrane proteins possess specific catalytic activities.

### The possible functions of capsules and slime layers

The capsule or slime layer is clearly a dispensable structure for the procaryotic cell. Furthermore, it is variable (even in one species) in chemical composition (Table 10.1). Many bacteria do not possess such a surface layer, and those which do can lose it without any apparent effect on their viability or rate of growth. In certain cases, the formation

*Table 10.1 Chemical composition of the capsule in certain bacteria*

| Organism | Nature of capsule | Chemical subunits |
|---|---|---|
| *Bacillus anthracis* | *Polypeptide* | D-*Glutamic acid* |
| *Leuconostoc mesenteroides* | *Dextran* | *Glucose* |
| *Streptococcus pneumoniae (pneumococcus)* | *Complex polysaccharides (many types), e.g.,* | |
| | *Type II* | *Rhamnose, glucose, glucuronic acid* |
| | *Type III* | *Glucose, glucuronic acid* |
| | *Type VI* | *Galactose, glucose, rhamnose* |
| | *Type XIV* | *Galactose, glucose,* N-*acetylglucosamine* |
| | *Type XVIII* | *Rhamnose, glucose* |
| *Streptococcus spp.* | *Hyaluronic acid* | N-*Acetylglucosamine, glucuronic acid* |
| *Streptococcus salivarius* | *Levan* | *Fructose* |
| *Acetobacter xylinum* | *Cellulose* | *Glucose* |
| *Aerobacter aerogenes* | *Complex polysaccharide* | *Glucose, fucose, glucuronic acid* |

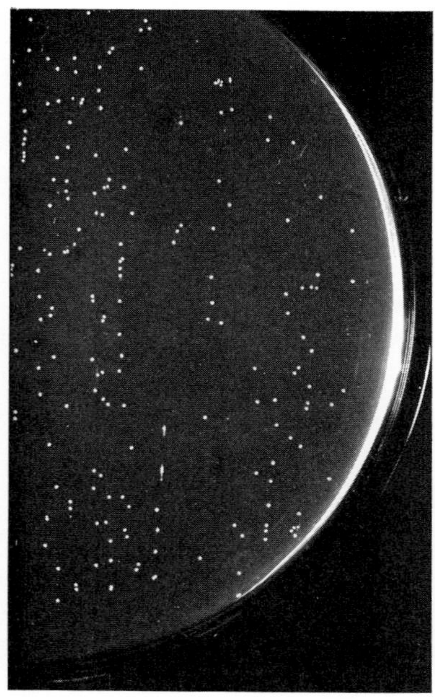

(a)

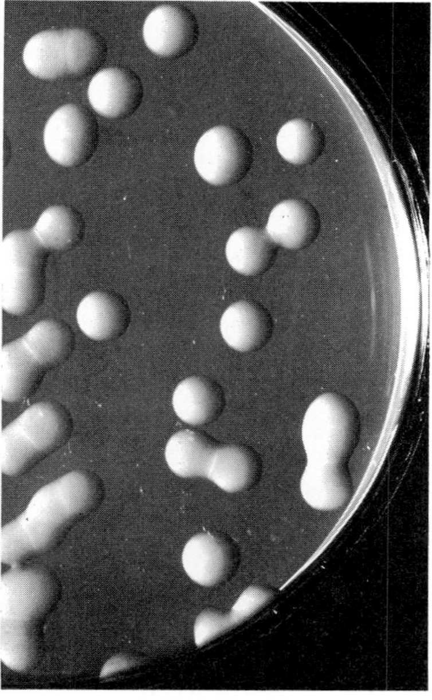

(b)

*Figure 10.6 The formation of extracellular polysaccharides by bacteria. Two plates of Leuconostoc mesenteroides, streaked on glucose medium (a) and sucrose medium (b). The large size and mucoid appearance of the colonies on sucrose are caused by the massive synthesis and deposition around the cells of dextran.*

of a slime layer is environmentally determined. This is true of the bacteria that form extracellular dextrans (polyglucoses) or levans (polyfructoses). These particular polysaccharides can be synthesized only from the disaccharide sucrose, not from other sugars. Hence the cells are surrounded by a copious slime layer when grown at the expense of sucrose but not when grown at the expense of other sugars (Figure 10.6). In bacteria that normally form a capsule under all conditions of growth, the property can be lost by spontaneous mutation. This is often termed an *S→ R mutation*, because it changes the surface of the colonies from a smooth, glistening texture (S) to a relatively rough one (R). When such a mutation occurs in certain normally capsulated pathogenic bacteria, such as *Streptococcus pneumoniae*, it results in a *loss of virulence for the normal susceptible animal host.* It has been shown that capsule-free mutants are much more readily engulfed and destroyed by phagocytes than are encapsulated cells of the wild type, and it is evidently this increased susceptibility to phagocytic destruction in the animal body that accounts for their failure to establish a successful infection. By extension from this finding, it may be supposed that a capsule is also advantageous in natural environments to nonpathogenic bacteria, since it could well confer resistance to phagocytic engulfment by phagotrophic protozoa. Just as in the case of pathogenic bacteria, the survival value of the capsule would not be expressed under conditions of pure culture in the laboratory.

## *The structure and synthesis of the cell wall*

**The murein layer**

The chemical nature of bacterial cell walls remained unknown until 1952, when Salton first developed methods for their isolation and purification (Figure 10.7). Despite the complexity and diversity of wall

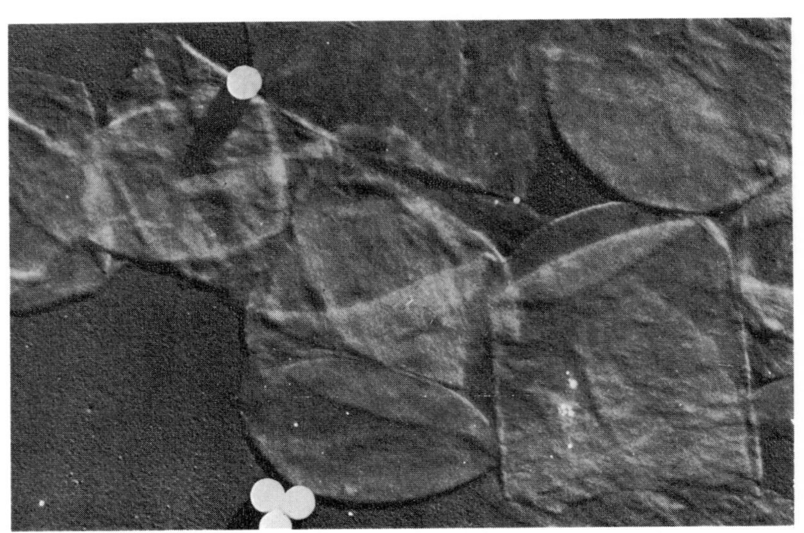

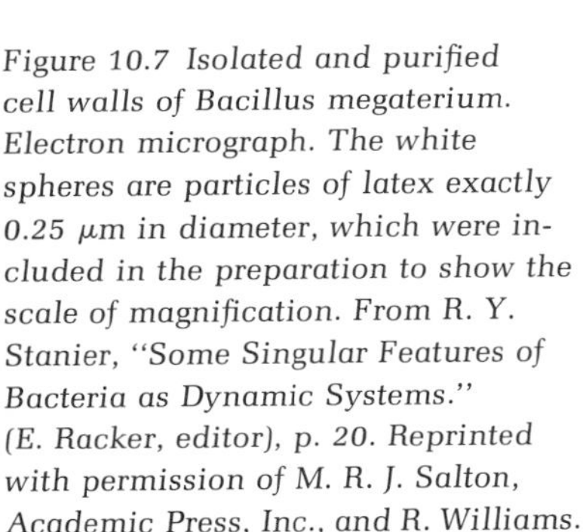

*Figure 10.7 Isolated and purified cell walls of Bacillus megaterium. Electron micrograph. The white spheres are particles of latex exactly 0.25 μm in diameter, which were included in the preparation to show the scale of magnification. From R. Y. Stanier, "Some Singular Features of Bacteria as Dynamic Systems." (E. Racker, editor), p. 20. Reprinted with permission of M. R. J. Salton, Academic Press, Inc., and R. Williams.*

composition revealed by the analyses of purified preparations, it has become evident that all procaryotic walls have one common chemical denominator: presence of *mureins* (also referred to as a *peptidoglycans* or *mucopeptides*) as the principal strengthening and shape-determining constituents of the wall. The mureins constitute a special class of biological heteropolymers. The nature of the typical subunits of a murein and the characteristic molecular ratios in which they occur are shown in Table 10.2. In Gram-positive unicellular bacteria and actinomycetes, murein accounts for at least 50 percent (and often more) of the total dry weight of the wall, but in other procaryotic groups, such as Gram-negative bacteria, it may constitute less than 10 percent of the dry weight of the wall. These quantitative differences are illustrated by some data on the relative amino sugar contents of walls from

*Table 10.2 Molecular composition of mureins: constant features*

| Mureins | Molecular ratio |
|---|---|
| *Amino sugars* | |
| N-*Acetylglucosamine* | 1 |
| N-*Acetylmuramic acid* | 1 |
| *Amino acids* | |
| L-*Alanine* | 1 |
| D-*Glutamic acid* | 1 |
| *meso-Diaminopimelic acid or* LL-*Diaminopimelic acid or* L-*Lysine* | 1 |
| D-*Alanine* | 1 |

Gram-positive and Gram-negative bacteria (Table 10.3). The general importance of mureins as the strengthening element of the procaryotic cell wall has been demonstrated primarily through studies on the mode of action of two antibacterial agents: the enzyme lysozyme and the antibiotic penicillin. Both agents destroy the integrity of the murein layer, but do so in entirely different ways: lysozyme hydrolyzes glycosidic bonds in mureins, whereas penicillin inhibits a terminal step in the biosynthesis of mureins. As a consequence, lysozyme lyses resting cells, whereas penicillin lyses only actively growing ones. The gross effects of both agents are, however, similar; as the murein layer is weakened the cell first loses its specific shape, eventually becoming spherical, and soon after undergoes osmotic lysis. Most procaryotic organisms (except the *Mycoplasma* group) are susceptible to penicillin lysis, differing only in their relative sensitivity to this antibiotic. Many are resistant to lysozyme lysis, because the murein layer, overlain by other wall layers, is not directly accessible to the

*Table 10.3 Amino sugar content (glucosamine plus muramic acid) of the purified walls of some eubacteria, expressed as a percentage of the dry weight of the wall*

| Gram-positive organisms | Range | Gram-negative organisms | Range |
|---|---|---|---|
| *Bacillus (8 spp.)* | 7–32 | *Escherichia (2 spp.)* | 3–5 |
| *Lactobacillus (1 sp.)* | 12 | *Proteus (2 spp.)* | 4–5 |
| *Streptococcus (1 sp.)* | 22 | *Aerobacter (1 sp.)* | 2 |
| *Micrococcus (7 spp.)* | 10–22 | *Salmonella (1 sp.)* | 3.5 |
| *Staphylococcus (4 spp.)* | 10–17 | *Serratia (1 sp.)* | 2 |
| *Corynebacterium (1 sp.)* | 14 | *Vibrio (1 sp.)* | 2 |
| | | *Spirillum (1 sp.)* | 7 |

enzyme; in such cases, however, the simultaneous addition of a chelating agent, ethylenediaminetetraacetic acid (EDTA), will often provoke lysis.

In Gram-positive unicellular bacteria and actinomycetes in which murein is a major wall component, it occurs throughout the wall, forming a complex matrix with other wall polymers (polysaccharides and teichoic acids). In Gram-negative bacteria and blue-green algae, the murein is localized in the innermost layer of the wall, where it occurs in practically pure state. By appropriate treatments, it is possible to strip off successively the outer layers of the isolated cell walls of Gram-negative bacteria and to obtain finally the pure murein sac, which, although extremely thin, still retains the original shape of the cell (Figure 10.8). Such preparations provided an essential clue to the

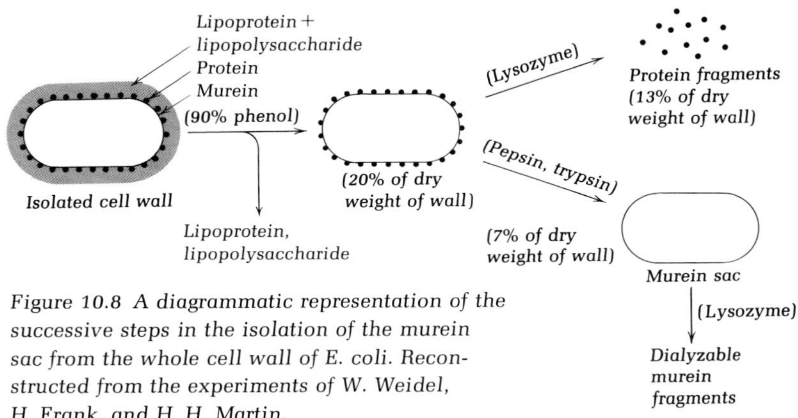


*Figure 10.8 A diagrammatic representation of the successive steps in the isolation of the murein sac from the whole cell wall of E. coli. Reconstructed from the experiments of W. Weidel, H. Frank, and H. H. Martin.*

manner in which the murein molecule is organized, since they can be completely hydrolyzed to yield principally the two products shown in Figure 10.9. Since lysozyme is known to attack specifically $\beta$-1,4-glycosidic bonds between muramic acid and glucosaminyl residues, the structure of the intact molecule could be readily deduced (Figure 10.10). The murein sac is essentially a single gigantic and rigid molecule, in which the polysaccharide chains, which consist of alternating N-acetylmuramic acid and N-acetylglucosamine residues, are periodically *cross-linked* by peptide bonding between amino acids in some of the short peptide chains attached to the muramic acid residues. Cross linking takes place between the carboxyl group of the terminal D-alanine of one peptide chain and the free $\epsilon$-$NH_2$ group of the diaminopimelic acid of a second chain. The extreme thinness of the murein layer in the wall of Gram-negative bacteria suggests that the murein sac

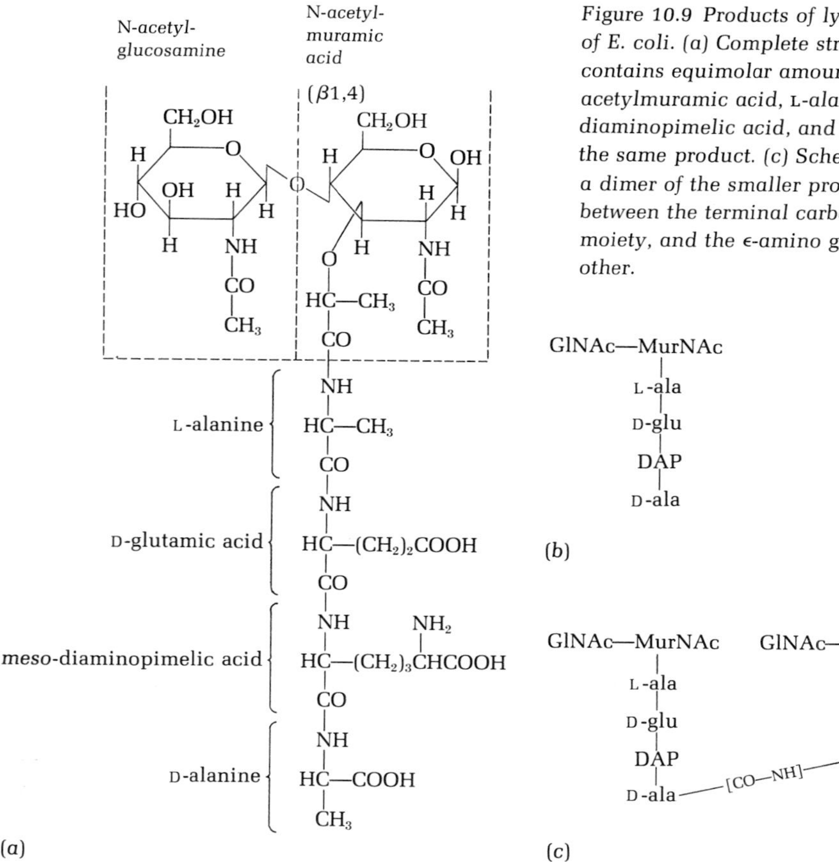


Figure 10.9 Products of lysozyme hydrolysis of the murein sac of E. coli. (a) Complete structure of the smaller product, which contains equimolar amounts of N-acetylglucosamine, N-acetylmuramic acid, L-alanine, D-glutamic acid, meso-diaminopimelic acid, and D-alanine. (b) Schematic structure of the same product. (c) Schematic structure of the larger product, a dimer of the smaller product, held together by a peptide bond between the terminal carboxyl group of D-alanine on one moiety, and the ϵ-amino group of diaminopimelic acid on the other.

GlNAc—MurNAc
L-ala
D-glu
DAP
D-ala

(b)

GlNAc—MurNAc GlNAc—MurNAc
L-ala L-ala
D-glu D-glu
DAP —[CO—NH]— DAP
D-ala D-ala

(c)

characteristic of these groups may be a *two-dimensional mesh*, or molecular monolayer.

The much thicker murein layers in the walls of Gram-positive bacteria, although they may be susceptible to disintegration by lysozyme, are not hydrolyzed completely to monomers and dimers of the basic

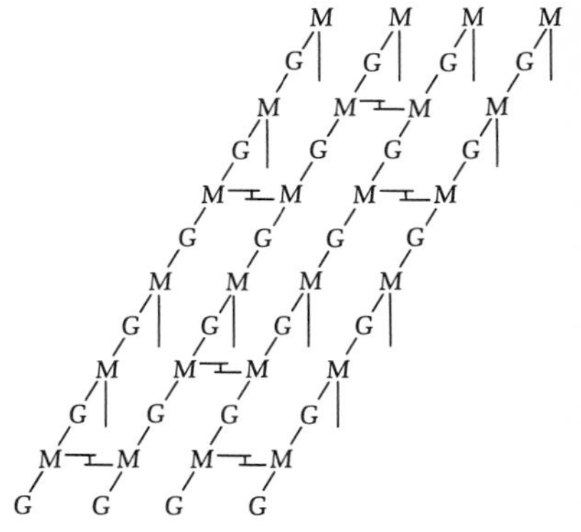

Figure 10.10 A schematic representation of the organization of the intact murein sac of E. coli: G and M designate residues of N-acetylglucosamine and N-acetylmuramic acid, respectively, joined (diagonal lines) by β-1,4-glycosidic bonds. The vertical lines represent free tetrapeptide side chains, attached to muramic acid residues. The symbol —I— represents cross-linked tetrapeptide sidechains. From J. M. Ghuysen, "Bacteriolytic enzymes in the determination of wall structure." Bacteriol. Rev. **32,** 425 (1968).

*Table 10.4 Some variations in the structure of bacterial mureins*

| Bacterial group | Diamino acid in tetrapeptide chain | Amino acids in peptide bridge between tetrapeptide chains |
|---|---|---|
| *Gram-negative bacteria* | *meso-DAP* | *None* |
| *Gram-positive bacteria* | | |
| *Staphylococcus aureus* | L-*lys* | *-(gly-gly-gly-gly-gly)-* |
| *Micrococcus lysodeikticus, other Micrococcus spp.* | L-*lys* | *Repeats of tetrapeptide* |
| *Streptococcus spp.* | L-*lys* | -(L-*ala*-L-*ala*)- *or* -(L-*ala*-L-*ala*-L-*ala*)- |
| *Streptomyces spp.* | LL-*DAP* | *Not established (gly?)* |

murein repeating unit. This reflects the fact that cross linking by peptide bonds, which are not susceptible to lysozyme, is far more extensive. If such cross linking takes place in two planes, rather than in one, a continuous, three-dimensional murein layer of considerable thickness can be formed; and this appears to be the molecular structure of the murein polymer in the walls of many Gram-positive organisms.

**Variations in murein structure**

In all mureins the composition of the poly(amino sugar) moiety is the same: it consists of alternating N-acetylmuramyl and N-acetylglycosaminyl residues, attached by $\beta$-1,4 linkages. There is, however, considerable variation in the composition of the peptide chains attached to the muramyl residues. The most common pattern, characteristic of Gram-negative bacteria, myxobacteria, blue-green algae, and certain Gram-positive bacteria, is that exemplified by the murein of *E. coli* (Figure 10.9). Other patterns occur, as far as is known, only among Gram-positive unicellular bacteria and actinomycetes. The differences concern principally the nature of the diamino acid at position 3 in the peptide chain and the presence and nature of supplementary amino acids, which serve as a *peptide bridge* between cross-linked peptide side chains. Murein structure is a character of considerable taxonomic significance among Gram-positive unicellular eubacteria and actinomycetes, since these structural variations are often either group or species specific (Table 10.4).

**The biosynthesis of mureins**

The biosynthesis of mureins has been elucidated by J. T. Park, J. L. Strominger, and their collaborators, working largely with a Gram-positive coccus, *Staphylococcus aureus*. The murein of this species has peptide chains of the composition: L-ala-D-glu-L-lys-D-ala, cross-linked through pentapeptide bridges consisting of glycine residues

Figure 10.11 *A schematic representation of the murein of Staphylococcus aureus. In this species, the tetrapeptide chains (-*L*-ala-*D*-glu-*L*-lys-*D*-ala), symbolized by vertical rows of four dots attached to muramic acid residues (M), are almost completely cross-linked to one another by pentaglycine bridges, symbolized by horizontal rows of five dots. From J. M. Ghuysen, "Bacteriolytic enzymes in the determination of wall structure."* Bacteriol. Rev. **32,** *425 (1968).*

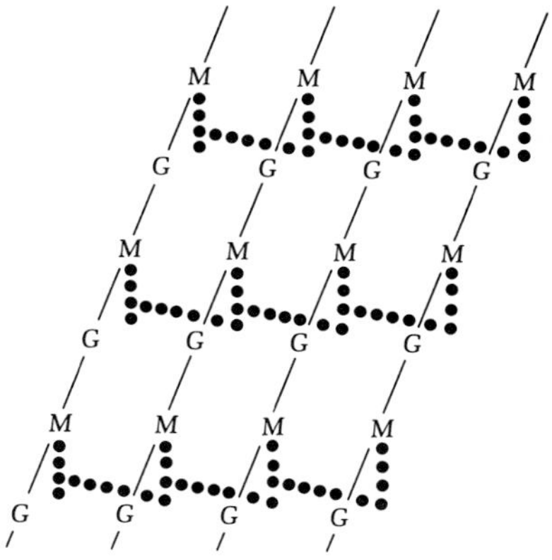

(Figure 10.11). A clue to the way in which biosynthesis proceeds was obtained through studies on the effect of penicillin. When growing in the presence of a sublethal penicillin concentration, *S. aureus* excretes unusual nucleotides, of which the most complex is known as the *Park nucleotide* (Figure 10.12). It contains, attached to UDP, a residue of muramic acid with a chain of five amino acids:

L-ala-D-glu-L-lys-D-ala-D-ala

As subsequent work showed, the Park nucleotide is a biosynthetic intermediate in murein synthesis.

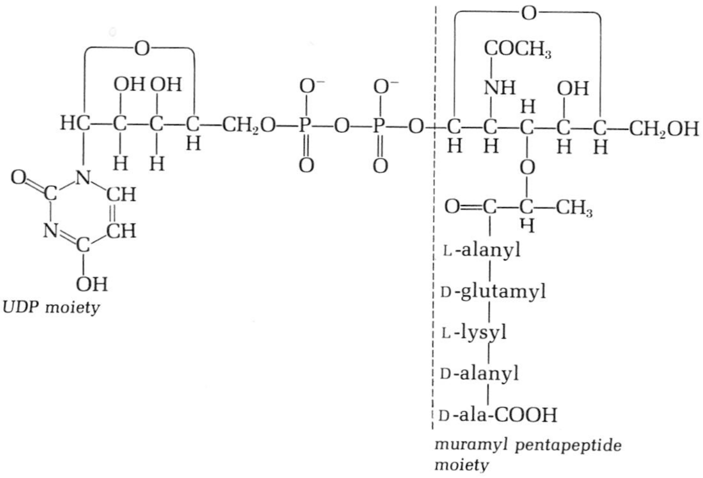

Figure 10.12 *Structure of the Park nucleotide, an intermediate in murein synthesis.*

The process of murein synthesis may be divided into three stages:

(1) The sequential assembly, in attachment to UDP, of the Park nucleotide. The step reactions of stage 1 are mediated by soluble enzymes in the cytoplasm.

(2) The transfer of the incomplete murein subunit from UDP to a lipid constituent of the cell membrane, where the terminal steps in subunit assembly take place.

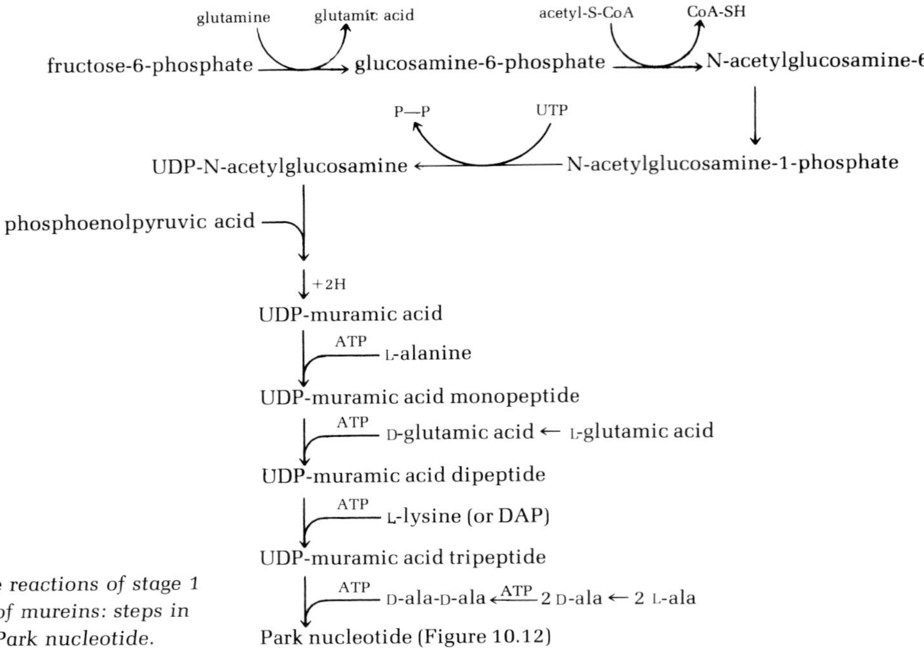


*Figure 10.13 The reactions of stage 1 in the synthesis of mureins: steps in assembly of the Park nucleotide.*

(3) Transfer of the completed murein subunit from the membrane to a growing point of the murein layer in the cell wall, where the final step in biosynthesis (cross linking of the peptide chains) takes place.

The reactions of stage 1 are schematized in Figure 10.13. UDP-N-acetylglucosamine is initially converted to UDP-N-acetylmuramic acid. The amino acids of the side chain are then added sequentially to this unit: the reactions do not involve the usual machinery of protein synthesis (see Chapter 7) and can be represented as

$$\text{acceptor} + \text{AA} + \text{ATP} \rightarrow \text{acceptor} - \text{AA} + \text{ADP} + \text{P}_i$$

The two terminal D-alanine residues are added simultaneously, in the form of a dipeptide, D-alanyl-D-alanine. This is synthesized from L-alanine by the following sequence of reactions:

$$\text{L-alanine} \rightleftharpoons \text{D-alanine}$$
$$2\text{D-alanine} + \text{ATP} \rightarrow \text{D-alanyl-D-alanine} + \text{ADP} + \text{P}_i$$

The nature of the diamino acid (DAP or lysine) in the peptide chain is determined by the specificity of the enzyme that adds this component to the growing subunit; thus, bacteria which contain a lysine residue have a specific lysine-transferring enzyme that is inactive with DAP.

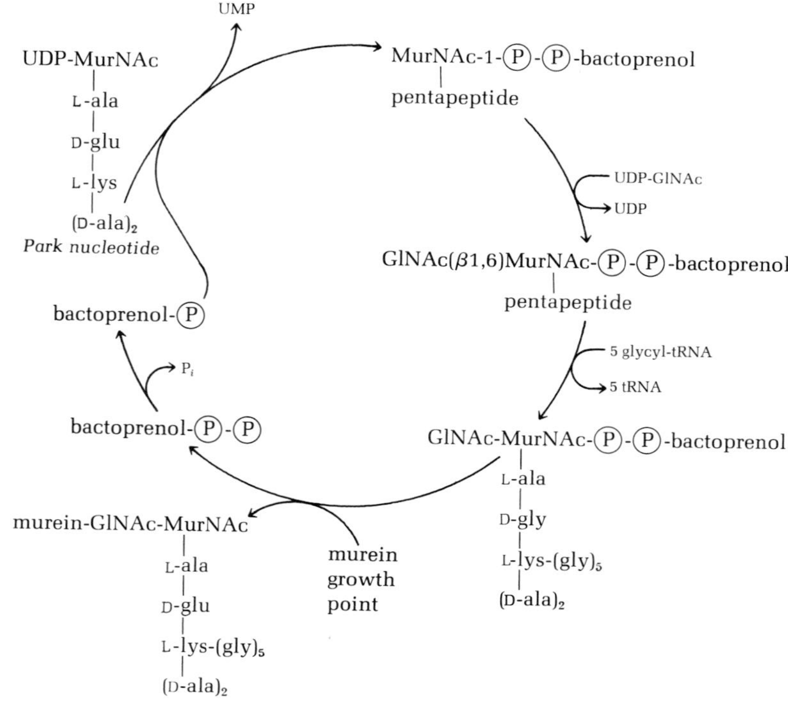


*Figure 10.14 The reactions of stage 2 in the synthesis of the murein of S. aureus: MurNAc, N-acetylmuramic acid; GlNAc, N-acetylglucosamine. These reactions, which take place in the cell membrane, involve the conversion of the Park nucleotide to a complete murein subunit and its transfer to a growing point in the murein layer of the cell wall.*

The reactions of stage 2 are summarized in Figure 10.14. They are cyclic in nature and are initiated by the transfer of the murein precursor, in conjunction with one phosphate residue, from the nucleotide to a phosphorylated lipid carrier in the membrane. This substance is a polyisoprenoid compound known as a bactoprenol (Figure 10.15).

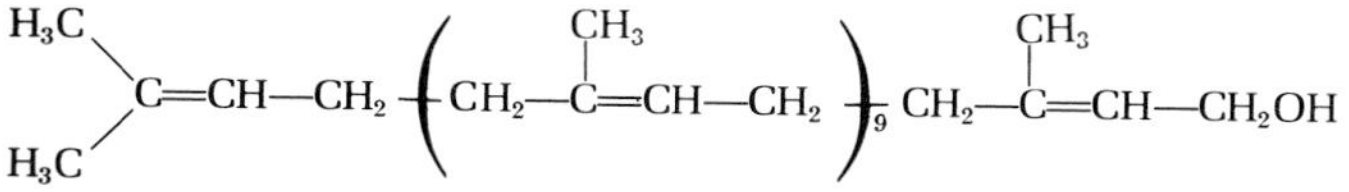

$$(H_3C)_2C{=}CH{-}CH_2{-}\left(CH_2{-}\overset{CH_3}{\overset{|}{C}}{=}CH{-}CH_2\right)_9{-}CH_2{-}\overset{CH_3}{\overset{|}{C}}{=}CH{-}CH_2OH$$

*Figure 10.15 The structure of bactoprenol.*

The next step is the addition of N-acetylglucosamine to the muramic acid moiety, by transfer from UDP-N-acetylglucosamine. Finally, the pentaglycine bridge is attached to the ϵ-amino group of lysine. The mechanism is different from that of the primary assembly of the peptide chain, since the glycine subunits are transferred from glycyl tRNA; the reaction does not, however, involve participation of ribosomes. Finally, the complete subunit is added to a poly(amino sugar) chain in the cell wall, with the liberation of bactoprenol, bearing a pyrophosphate residue. After removal of the terminal phosphate residue by hydrolysis, the bactoprenol monophosphate can repeat its role in the cycle.

The new murein subunit is now incorporated into the wall fabric but is not yet cross linked. It still bears an extra terminal D-alanine, which does not exist in the finished polymer, and cross linking of the peptide chains has not yet occurred. Cross linkage involves the formation of a covalent peptide bond, a reaction that normally requires an input of bond energy from ATP. However, this energy source is not available external to the membrane of the cell. Instead, the *peptide bond of the terminal D-ala-D-ala sequence is used to achieve cross linking by transpeptidation,* as shown in Figure 10.16. This reaction is medi-

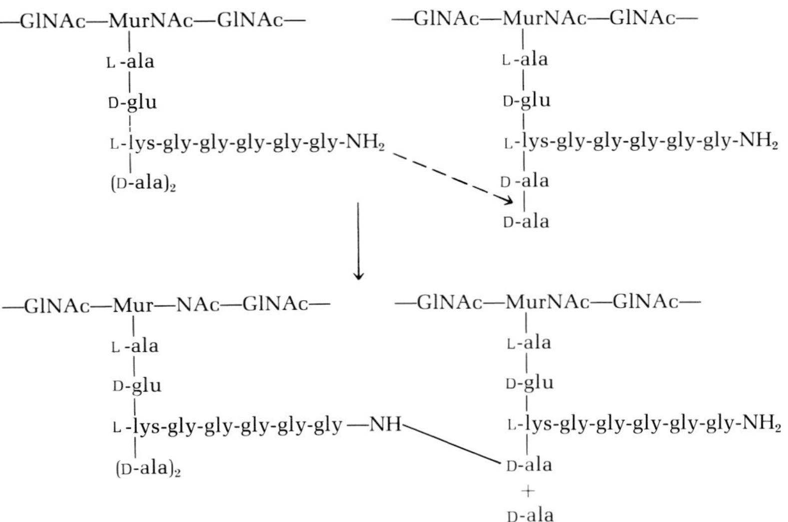


*Figure 10.16 The cross-linking of the peptide chains in the murein layer of the S. aureus wall, by transpeptidation of the amino group of glycine on one chain to the peptide bond between the two terminal D-alanine residues on a second chain, with the elimination of D-alanine.*

ated by a specific transpeptidase, with liberation of the terminal D-alanine residue. A second enzyme, D-alanine carboxypeptidase, cleaves off the terminal D-alanine residues from any peptide chains that do not engage in cross linking. *It is these two terminal reactions which are specifically inhibited by penicillin.* As a consequence, growth in the presence of penicillin leads to the formation of new murein chains which are not cross-linked and lack tensile strength. Weak points therefore develop in the growing wall, resulting in osmotic lysis. Certain other antibiotics owe their antibacterial activity to interference at other specific points with the process of murein synthesis (Table 10.5). The most characteristic wall polymers of Gram-positive bacteria,

*Table 10.5 Sites of action of antibacterial antibiotics that affect the synthesis of mureins*

| Antibiotic | Reaction(s) inhibited |
|---|---|
| D-*Cycloserine* | L-*ala* ⇄ D-*ala*; 2D-*ala* → D-*ala*-D-*ala* |
| *Bacitracin* | *Bactoprenol*-Ⓟ—Ⓟ → *bactoprenol*-Ⓟ + $P_i$ |
| *Vancomycin* } *Ristocetin* | *Undetermined stage 2 reactions* |
| *Penicillins* } *Cephalosporins* | *Terminal transpeptidation in growing wall* |

**Other wall polymers of Gram-positive bacteria**

apart from mureins, are the *teichoic acids*. They are present in many (and possibly all) Gram-positive bacteria and may comprise as much as 50 percent of the dry weight of the cell wall. There are two classes of teichoic acids; glycerol teichoic acids and ribitol teichoic acids (Figure 10.17). The former are partially associated with the cell membrane,

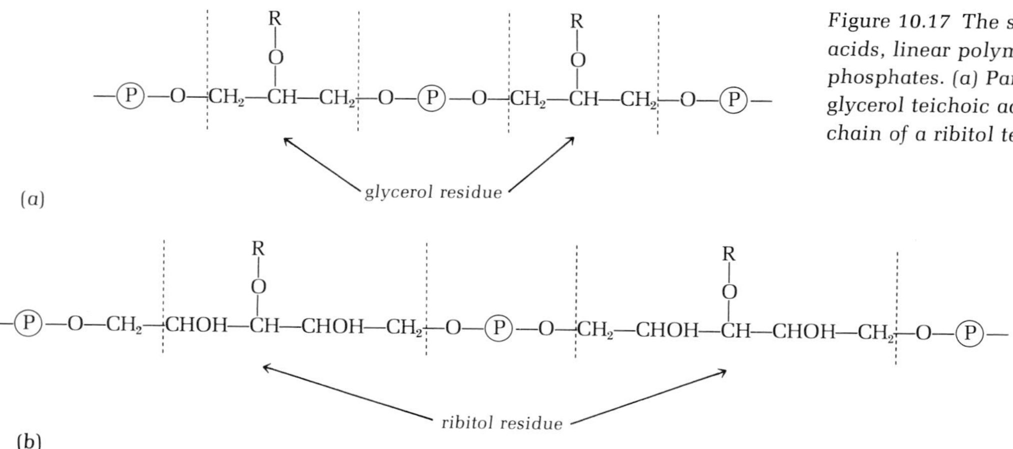


*Figure 10.17 The structure of teichoic acids, linear polymers of polyalcohol phosphates. (a) Part of the chain of a glycerol teichoic acid. (b) Part of the chain of a ribitol teichoic acid.*

whereas the latter seem to occur exclusively as constituents of cell walls.

The teichoic acids are linear polymers composed of alternating polyalcohol and phosphate groups, the two ends of the polyalcohol being linked through ester bonds to phosphate groups. This basic structure may be modified by substitutions of the hydroxyl groups on the glycerol or ribitol residues in the polymer chain. These substituents (designated —R in Figure 10.17) may consist of residues of either D-alanine or sugars, such as glucose and N-acetylglucosamine.

Teichoic acids can be extracted from the wall by relatively mild

*Table 10.6 Monosaccharide constituents of wall polysaccharides in Gram-positive bacteria*

| Genera | Monosaccharides |
|---|---|
| *Propionibacterium, Streptococcus, Lactobacillus, Clostridium* | **Rhamnose,** *glucose, galactose, mannose* |
| *Corynebacterium, Mycobacterium, Nocardia* | **Arabinose,** *glucose, galactose, mannose* |
| *Arthrobacter, Bacillus, Micrococcus, Staphylococcus* | *Glucose, galactose, mannose* |

chemical treatment, without affecting its fabric. Recent work suggests that they are covalently linked to the murein, through phosphodiester bonding to the $—CH_2OH$ group of the muramyl residues.

Their biosynthesis has not been studied in detail, but it is known that glycerol teichoic acids are assembled by transfer of glycerophosphate residues from CDP-glycerol to the growing chain, as shown in the following equations:

$$\text{glycerophosphate} + \text{CTP} \rightarrow \text{CDP-glycerol} + \text{P—P}$$
$$\text{CDP-glycerol} + (\text{glycerophosphate})_n \rightarrow \text{CMP} + (\text{glycerophosphate})_{n+1}$$

A wide variety of polysaccharides may also occur in the fabric of Gram-positive cell walls. In many cases these polysaccharides are group specific, as indicated by the presence of particular constellations of sugars as components of the isolated walls. The sugars found in the walls of different groups of Gram-positive bacteria are listed in Table 10.6; here, the characteristic group-specific sugars are shown in boldface type.

**The lipopolysaccharides in the walls of Gram-negative bacteria**

The outer layers of the wall of Gram-negative true bacteria, which may account for as much as 80 percent of the dry weight of the wall, consist principally of lipoprotein and lipopolysaccharide. The lipopolysaccharide component has been much studied for two reasons. In the first place, this wall component is primarily responsible for the somatic (O) antigenic specificity of Gram-negative bacteria. Among bacteria of the enteric group, this specificity is exceedingly high, so it is often possible to distinguish, among otherwise not readily differentiable strains of a single bacterial species, a large number of *serotypes*, which differ with respect to the structure and antigenic specificity of their lipopolysaccharides. In the genus *Salmonella*, for example, over 1,000 specific serotypes are known, although only about half a dozen species can be differentiated on other phenotypic grounds. This genus has been studied in particular depth because its members are important agents of food-borne infection, and the built-in label represented by serotypic specificity often permits the epidemiologist to trace the chain of transmission of an outbreak of disease.

The lipopolysaccharide fraction of the cell wall also has *endotoxic* activity; it is responsible for the symptoms that characteristically occur when the cells of Gram-negative bacteria are introduced into the bloodstream of an animal; these include fever and, in severe cases, shock and internal hemorrhage.

The lipopolysaccharide of the wall can be extracted by a variety of methods and fractionated into a lipid moiety ("lipid A") and a polysaccharide moiety. Neither moiety alone carries endotoxic activity. The polysaccharide moiety bears all the antigenic determinants. The polysaccharides characteristic of the wall of enteric bacteria may contain as many as 12 different constituent sugars. Comparative analyses of many bacterial strains belonging to a given species of enteric bacterium show that five are almost always present; the remainder are variable. The strains of a particular species of enteric bacterium can be grouped, in terms of these characters, into a number of different wall chemotypes (Table 10.7); any given chemotype may in turn contain several different wall serotypes.

These facts suggest that the polysaccharides consist of a constant and common *core region*, containing glucose, galactose, N-acetylglu-

*Table 10.7 Some chemotypes of the species Escherichia coli, defined by the nature of the sugars present in the wall lipopolysaccharide*

| | **Wall sugars** | | | | | | | | | | |
|---|---|---|---|---|---|---|---|---|---|---|---|
| **Chemotype** | **Glucose** | **Galactose** | **Heptose** | **Glucosamine** | **2-Keto-3-deoxyoctonate** | **Mannose** | **Fucose** | **Rhamnose** | **6-Deoxytalose** | **Colitose** | **Galactosamine** |
| I | + | + | + | + | + | − | − | − | − | − | − |
| II | + | + | + | + | + | − | − | − | − | − | + |
| III | + | + | + | + | + | + | − | − | − | − | − |
| IV | + | + | + | + | + | + | − | − | − | − | + |
| V | + | + | + | + | + | − | + | − | − | − | − |
| VI | + | + | + | + | + | − | + | − | − | − | + |
| VII | + | + | + | + | + | − | − | + | − | − | − |
| VIII | + | + | + | + | + | − | − | + | − | − | + |

Figure 10.18 Structures of the $C_7$ and $C_8$ constituents of the lipopolysaccharide core region.

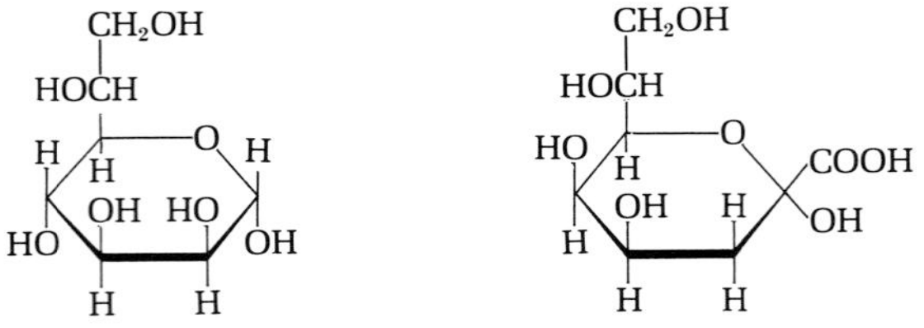

heptose

2-keto-3-deoxyoctonic acid

cosamine, 2-keto-3-deoxyoctonate, and heptose (Figure 10.18) together with a more variable *side-chain region*, containing the sugars associated with serological specificity. This interpretation is supported by the changes in wall polysaccharide structure associated with an S→R mutation in the enteric groups. The R mutants, recognizable by the relatively rough surfaces of their colonies, have invariably lost the antigenic specificity of the serotype within which they arose, and their wall polysaccharides retain only the five sugars of the core region. The gross structure of the wall lipopolysaccharides can, accordingly, be represented as shown in Figure 10.19.

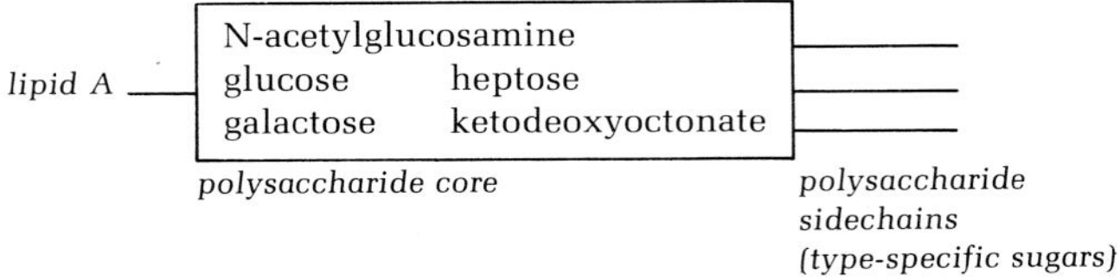


Figure 10.19 Schematic diagram of the structure of the wall lipopolysaccharide in Gram-negative bacteria.

The analysis of the biosynthesis of wall polysaccharides has been greatly facilitated by the study of mutants, both spontaneous mutants of the R type and mutants that have undergone specific derangements of carbohydrate metabolism. The use of mutants is possible precisely because the wall polysaccharides are not, like mureins, indispensable structural constituents; their nature can be substantially modified by mutation without affecting viability. We shall illustrate such work by considering the specific case of wall polysaccharide synthesis in *Salmonella typhimurium*. In addition to the five "core" compounds, the wall polysaccharide of this species normally contains D-mannose; the monodeoxyhexose L-rhamnose; and a dideoxyhexose, abequose.

Figure 10.20 shows the metabolic interconversions of hexoses in *S. typhimurium;* the immediate precursors for wall polysaccharide synthesis (all sugar nucleotides) are underlined, and the sites of mutational lesions in carbohydrate metabolism that have proved useful in the analysis of wall synthesis are numbered. Table 10.8 shows the

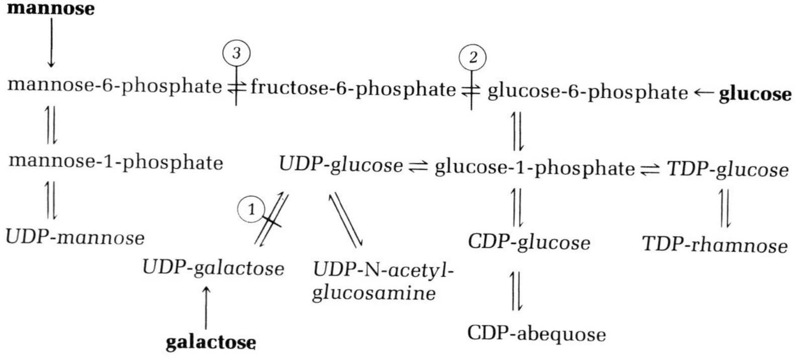


*Figure 10.20 Interconversions of sugar phosphates and nucleotides in Salmonella typhimurium. Italicized compounds are the substrates for wall polysaccharide synthesis. In the wild type, all these sugar derivatives can be synthesized from the intermediary metabolic pool of the cell via fructose-6-phosphate. They can also be formed from glucose, mannose, or galactose (boldface), if one of these sugars is furnished as a nutrient. The sites of three known mutational lesions are also indicated: (1) UDP-galactose epimerase; (2) phosphoglucose isomerase; (3) phosphomannose isomerase.*

composition of the wall polysaccharide in the wild type and mutants grown in nutrient broth not containing any sugar as a nutrient: in some cases, the effect of adding a specific sugar to the medium is also shown.

Considering first the properties of the mutant that lacks UDP-gal epimerase, it can be seen from Figure 10.20 that the loss of this enzyme specifically prevents the cell from synthesizing UDP-gal from the intermediary metabolic pool. Such a mutant cannot, therefore, incorporate galactose residues into its lipopolysaccharide when grown in the

*Table 10.8 Composition of the wall polysaccharide in the wild type of Salmonella typhimurium and in certain mutants derived from it, which lack enzymes involved in the synthesis of wall polysaccharides*

| Strain | Carbohydrate additions to growth medium | Constituent sugars in wall polysaccharide[a] KDO | hep | glu | gal | GNAC | man | rha | abe |
|---|---|---|---|---|---|---|---|---|---|
| *Wild type* | *None* | + | + | + | + | + | + | + | + |
| *Mutants* | | | | | | | | | |
| *UDP-galactose epimerase negative (Enzyme 1)* | *None* | + | + | + | 0 | 0 | 0 | 0 | 0 |
| | *+ Galactose* | + | + | + | + | + | + | + | + |
| *Phosphoglucose isomerase negative (Enzyme 2)* | *None* | + | + | 0 | 0 | 0 | 0 | 0 | 0 |
| | *+ Glucose* | + | + | + | + | + | + | + | + |
| *Phosphomannose isomerase negative (Enzyme 3)* | *None* | + | + | + | + | + | 0 | 0 | 0 |
| | *+ Mannose* | + | + | + | + | + | + | + | + |

[a]KDO, 2-keto-3-deoxyoctulosonic acid; hep, heptose; glu, glucose; gal, galactose; GNAC, N-acetylglucosamine; man, mannose; rha, rhamnose; abe, abequose.

absence of galactose. *At the same time, the incorporation of N-acetylglucosamine, mannose, rhamnose, and abequose is prevented.* If this strain is furnished with galactose (which can be directly converted to UDP-gal, thus circumventing the mutational block, as shown in Figure 10.20) a normal polysaccharide is formed. Similarly, a mutant that lacks phosphomannose isomerase cannot synthesize UDP-man from the intermediary pool; it fails to incorporate mannose, rhamnose, and abequose into the polysaccharide. Again, the wild-type phenotype is restored by provision of an external supply of mannose. Finally, the phosphoglucose isomeraseless mutant makes a wall polysaccharide in which KDO and heptose are the only sugars; however, a normal polysaccharide is formed when it is supplied with glucose. The properties of these mutants show immediately that the wall polysaccharide is synthesized by a sequential addition of specific sugars to a core that contains KDO and heptose; if the addition of any given sugar is prevented, the addition of other sugars is likewise prevented. One can, therefore, reconstruct the biosynthetic sequence as

$$\underbrace{[\text{KDO-heptose}] \leftarrow \text{glucose} \leftarrow \text{galactose} \leftarrow \text{GNAC}}_{\text{“core” region}} \leftarrow \underbrace{(\text{mannose, rhamnose, abequose}}_{\text{specific side chain}}$$

Biochemical experiments, using the highly incomplete lipopolysaccharide made by the phosphoglucose isomeraseless mutant as a substrate, reveal the terminal steps in the assembly of the core region (Figure 10.21).

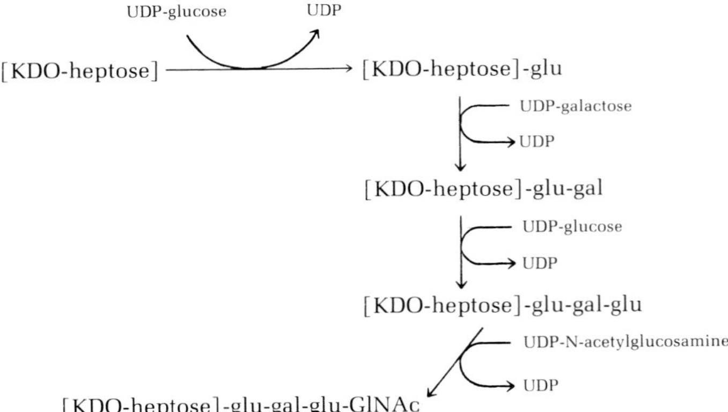


*Figure 10.21 Terminal steps in the assembly of the core region of Salmonella lipopolysaccharide.*

Structural analyses have shown that the serologically specific side chains are made up of repeating units containing the constituent monomers; in the case of *S. typhimurium*, the repeating unit is

```
(-rhamnose-mannose-galactose-)n
              |
           abequose
```

These side chains are first polymerized and then attached to the completed core portion of the polysaccharide. The assembly of the side chain has been studied in particular detail in *Salmonella newington*, where the repeating unit is

(-mannose-rhamnose-galactose-)$_n$

As shown in Figure 10.22, the assembly of this molecule, and its

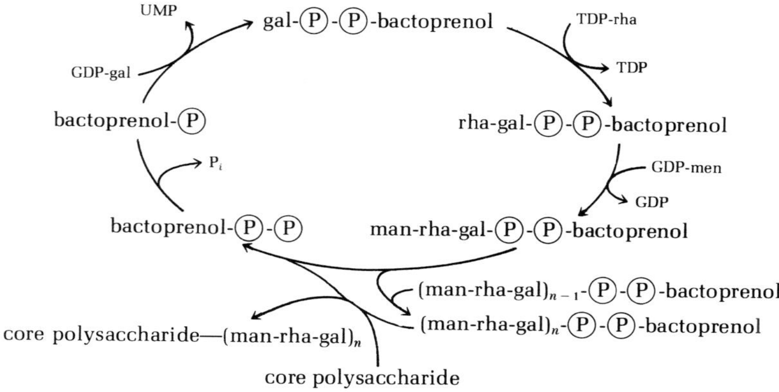


*Figure 10.22 Mechanism for the synthesis of the repeating unit of the polysaccharide side chain of Salmonella newington, and its transfer to the lipopolysaccharide core of the cell wall. (Compare with Figure 10.14.)*

subsequent attachment to the core portion of the lipopolysaccharide, involves a cyclic series of reactions, strikingly analogous to those which operate in the assembly of mureins. There is a sequential construction of the repeating subunits, initiated by attachment to a $C_{55}$ polyisoprenoid carrier (bactoprenol) in the membrane; after completion of the side-chain polysaccharide, it is transferred to the core, with a liberation of bactoprenol pyrophosphate. It thus appears probable that assembly of subunits on a lipid carrier in the membrane, with subsequent transfer to "open" ends of the growing polymer fabric of the wall, is a general feature of cell wall biosynthesis in procaryotic organisms.

**The surface topology of the wall**

As might be expected from their molecular complexity, bacterial cell walls do not often show a regular surface topology. The walls of Gram-negative bacteria as a rule have a highly convoluted surface structure (Figure 10.23). In a few bacteria, both Gram positive and Gram negative, electron microscopic examination of isolated walls does, however,

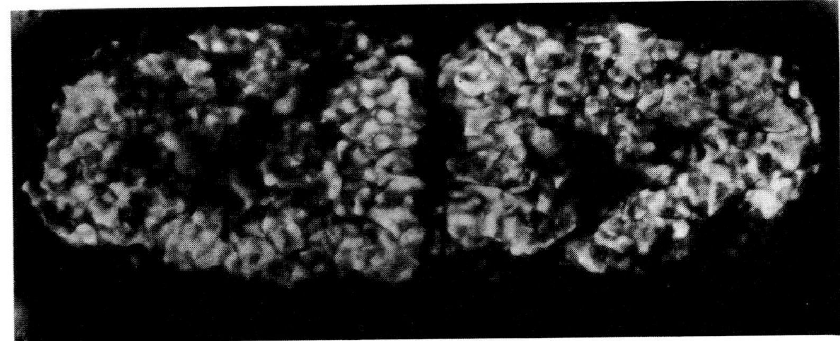

Figure 10.23 *Electron micrograph of a cell of E. coli, negatively stained with silicotungstate to reveal the highly regular surface structure of the outer layer of the cell wall (× 62,000). From M. E. Bayer and T. F. Anderson, "The surface structure of Escherichia coli." Proc. Natl. Acad. Sci. U.S.* **54,** *1592 (1965).*

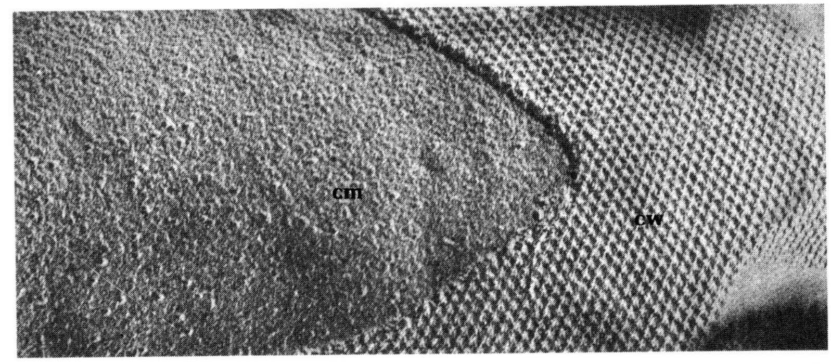


Figure 10.24 *Electron micrograph of a freeze-etched preparation of the cell surface of a marine Nitrosomonas, cut tangentially so as to reveal the regular surface structure of the cell wall (cw) and the surface structure of the underlying cell membrane (cm) (× 130,000). Courtesy of S. W. Watson.*

reveal a very regular arrangement of molecular subunits (Figure 10.24). These arrangements cannot yet be clearly interpreted in chemical terms.

**The manner of wall growth**

The manner of wall growth can be analyzed experimentally by the use of specific antibodies, directed against wall constituents and labeled by conjugation with a fluorescent dye. When living cells are coated with such an antibody, the cell walls become intensely fluorescent, and because antibody molecules adhere to the specific sites of their attachment, the areas of new wall synthesis can be determined by direct observation of the growing cells. In the Gram-positive bacterium *Streptococcus pyogenes*, such experiments, conducted with two different antibody preparations, each directed against a specific wall polymer, have shown that new wall material is laid down at a highly localized growing point, in a narrow equatorial band around the spherical cell. As shown in Figure 10.25, the cell caps retain their fluorescent intensity through several generations, while new "dark" areas are progressively inserted between the preformed wall material at each succes-

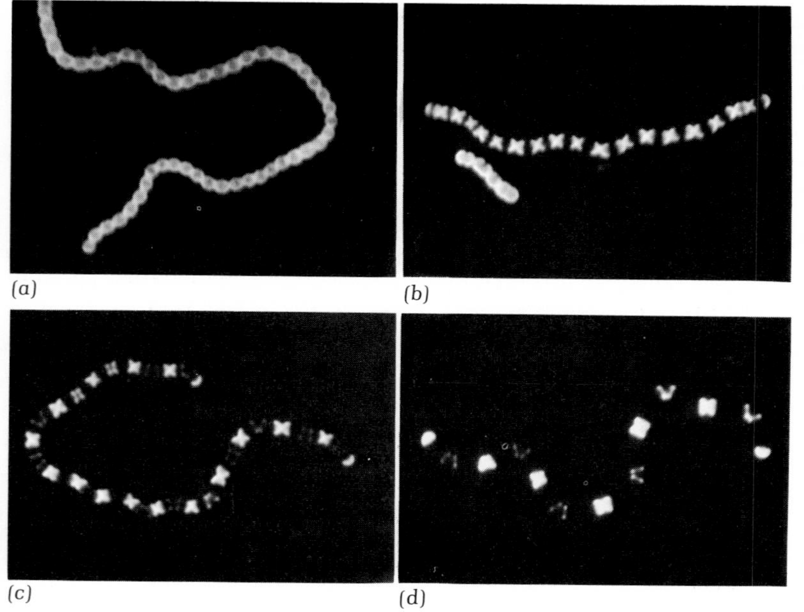


*Figure 10.25 Growth of the wall of Streptococcus pyogenes, followed by ultraviolet photomicrography of growing chains of cells, in which the wall had been initially coated with a fluorescent antibody. (a) Immediately after antibody treatment; the cells are evenly fluorescent. (b) After 15 minutes of growth. New (nonfluorescent) wall material has been formed around the equator of each cell; the polar caps of the cells, previously labeled with fluorescent antibody, remain fluorescent. (c), (d) The appearance of cell chains after 30 and 60 minutes of growth, respectively. From R. M. Cole and J. J. Hahn, "Cell wall replication in Streptococcus pyogenes." Science* **135,** *722 (1962).*

sive generation. Analogous experiments with Gram-negative bacteria, on the other hand, show a gradual overall weakening of fluorescent intensity as growth of the wall proceeds, which suggests that new material is inserted at a large number of growing points over the surface of the wall. Bacteria may, accordingly, differ with respect to the sites of wall growth.

**The origin and nature of bacterial L forms**

In 1935 it was observed that the bacterium *Streptobacillus moniliformis* can give rise on rich media to atypical colonies, in which the normal rod-shaped cells are replaced by irregularly shaped, often globular growth forms. It was found that these so-called *L forms* could be propagated indefinitely on serum-enriched complex media. At first, it was believed that the L forms, so different in cell structure from *S. moniliformis,* were symbionts or parasites of the bacterium, but this interpretation had to be abandoned when it was found that they could, on rare occasions, revert to rods. Subsequent observations on L forms of other bacteria showed that the phenomenon was by no means confined to *S. moniliformis.*

After the discovery of penicillin, it was found that L forms are *peni-*

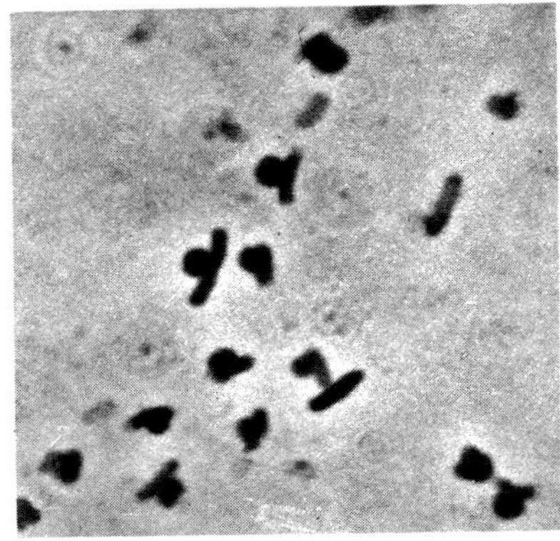

*Figure 10.26 The genesis of L forms in Proteus vulgaris: conversion of the rods to spherical bodies during growth in the presence of penicillin (phase contrast, ×2,450). Courtesy of E. Kellenberger and K. Leibermeister.*

*cillin resistant* and can in fact be selected by cultivation of bacteria on an osmotically buffered penicillin-containing medium (Figure 10.26). As a rule, removal of penicillin after short exposure leads to immediate reversion to the normal cell form. After longer periods of growth with penicillin, some bacteria may continue to grow in the L form when penicillin is removed. Both the duration of the penicillin exposure necessary to maintain L forms in the absence of penicillin, and the fraction of the population so converted, are variable. Furthermore, the stability of the resulting L forms, even those derived from a single bacterial species, is variable: some ("stable L forms") have never been observed to revert; others ("unstable L forms") do occasionally revert.

From the known mode of action of penicillin as an inhibitor of the terminal, cross-linking step in murein synthesis, L forms can be interpreted as bacteria in which the synthesis of the murein layer has been severely deranged but which can continue to grow, even though with aberrant cell form, in media sufficiently concentrated to prevent their osmotic lysis.

In certain stable L forms, the components of the murein layer are completely absent from the cells. All unstable L forms, and certain stable ones, still contain muramic acid, but the concentration is relatively low (about 10 to 15 percent of its concentration in normal cells). Furthermore, the muramic acid is in an unusual chemical state, being readily extractable with dilute acid, whereas the muramic acid in a normal cell wall is not. From these facts, the following general interpretation of the nature of L forms can be derived. *They are bacteria in which the murein primer for cell wall synthesis has been either eliminated or modified by penicillin treatment.* In the first case, no new murein material can be deposited in the wall; in the second, new subunits are incorporated in an irregular and uncoordinated manner, so that the formation of the normal continuous murein layer is prevented. On rare occasions, L forms of the latter type may succeed in resynthesizing a continuous murein layer and will then revert to the normal bacterial form.

L forms can be obtained from both Gram-positive and Gram-negative bacteria. In the latter, the lipopolysaccharide and lipoprotein layers of the cell wall continue to be synthesized in an apparently normal fashion; this is indicated both by electron microscopic examination of the wall layer, and by the fact that such L forms retain O antigens and are still susceptible to infection by phages for which the receptors are contained in the outer wall layers.

## *The bacterial cell membrane*

We have already alluded to the functional versatility of the bacterial cell membrane: It is not simply the surface layer that regulates the entry of nutrients to the cell, but it plays other important roles as well, in energy-yielding metabolism and in biosynthesis.

**The topology of the membrane**

The cell membrane is too thin to be resolved by light microscopy, and the elucidation of its structure and arrangement in bacteria have become possible only since the introduction of electron microscopy.

In some bacteria, the membrane appears to have a simple shape, and its contour follows that of the inner layer of the wall over the whole surface of the protoplast. Often, however, its shape is more complex, as the result of the occurrence of intrusions into the cytoplasmic region. Such intrusions may be few or numerous, and relatively superficial or penetrating deeply into the central regions of the cell.

A type of intrusion that occurs in many different bacterial groups, particularly near the sites of transverse wall formation, is known as a *mesosome* (Figure 10.27). Mesosomes are extremely complex infoldings, which in thin sections often appear to be separated from the cell

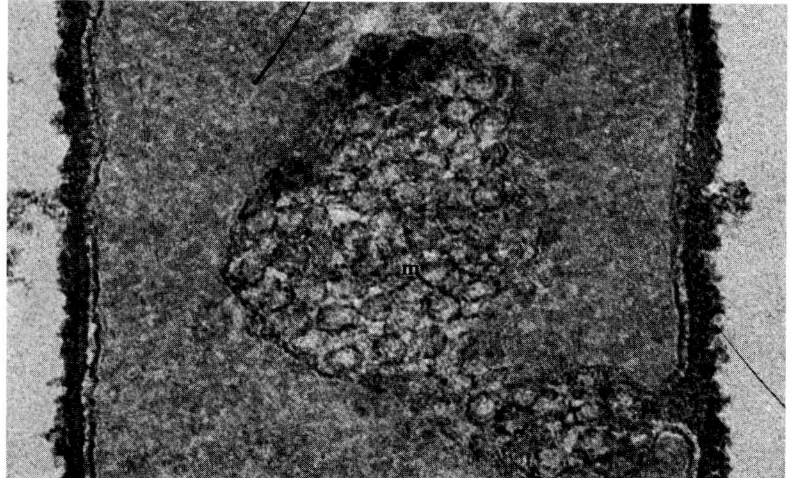


*Figure 10.27 Electron micrograph of a thin section of part of a cell of Bacillus subtilis, showing a mesosome (m) and its connection to the cell membrane. Courtesy of W. van Iterson.*

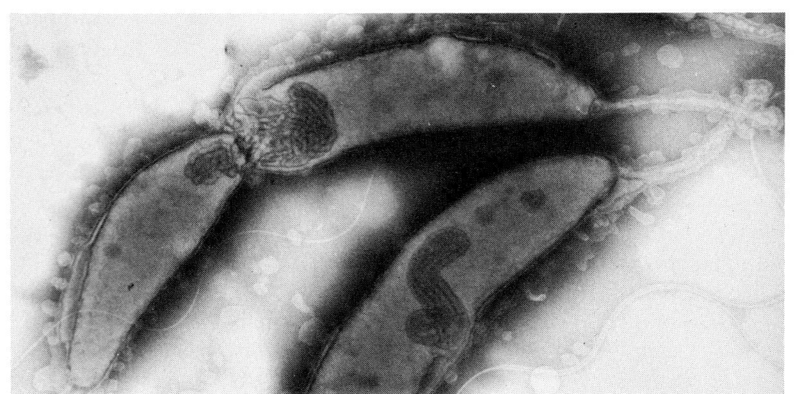

Figure 10.28 Mesosomes of *Caulobacter crescentus*: electron micrograph of whole cells negatively stained with phosphotungstate, which has penetrated the mesosomal involutions of the cell membrane, clearly outlining their positions (×28,160). Courtesy of Germaine Cohen-Bazire.

membrane itself. However, their topological relationship to the cell membrane can often be visualized in electron micrographs of whole cells, negatively stained with a solution of an electron-dense inorganic compound, such as phosphotungstic acid (Figure 10.28). Phosphotungstic acid cannot penetrate into the cytoplasm but frequently passes through the cell wall, filling the space between the wall and the surface of the membrane. In such preparations, the negative stain penetrates deeply into the mesosome, showing that this structure is, in

(a)

(b)

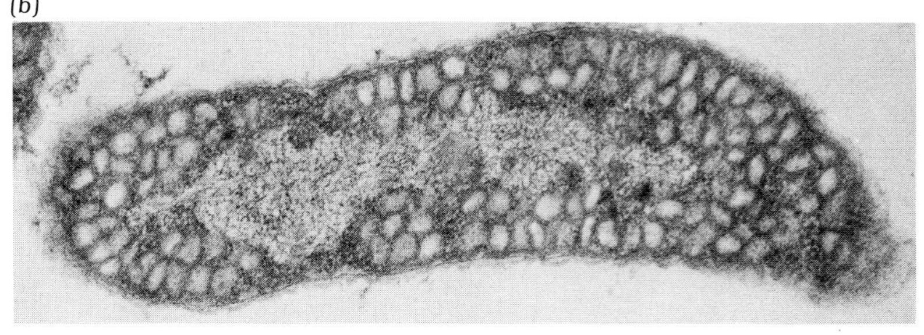

Figure 10.29 Electron micrographs of longitudinal sections of the photosynthetic bacterium, *Rhodospirillum rubrum* (×46,200), showing the effect of the environment on the extent of the intrusion of the cytoplasmic membrane. (a) Cell from a culture grown in bright light (1,000 foot-candles), and having a relatively low chlorophyll content. (b) Cell from a culture grown in dim light (50 foot-candles) and having a high chlorophyll content. Courtesy of Germaine Cohen-Bazire.

fact, a much-folded portion of the cell membrane. Different functions have been proposed for the mesosome: a localized center of respiratory activity, an organelle operative in transverse wall formation, and a device for separating nuclei after their replication. In support of the latter hypothesis, it has been shown for a few bacteria that the bacterial chromosome is attached to the mesosome. However, the role or roles of mesosomes in cell function are still not clearly established, and the existence of many bacterial species that seem to lack mesosomes suggests that this organelle is not a universal structural element of the bacterial cell.

In purple bacteria, where the photosynthetic pigment system is located in the membrane, the whole cytoplasmic region is characteristically filled with vesicular or lamellar membranous intrusions, the particular form of which is characteristic for each species. Among purple nonsulfur bacteria, the photosynthetic pigment content of the cells can vary over a wide range in response to environmental factors, and such variations are more or less directly connected with the extent of intrusion of the membrane (Figure 10.29). In this particular group, accordingly, an increase in the photosynthetic activity of the cell involves an increase in the total area and also in the degree of infolding of the membrane, upon which are borne the centers of photochemical function.

The extensive membrane intrusions—either vesicular or lamellar—characteristic of certain aerobic bacteria that have a high content of respiratory pigments probably have an analogous function. For example, in *Azotobacter*, which has the highest known respiratory rate of any bacterium, vesicular membrane intrusions, structurally similar to those of many purple bacteria, are very abundant (Figure 10.30). The nitrifying bacteria, which also have a high content of respiratory pigments, typically contain lamellar membranous intrusions, as shown for *N. europea* in Figure 10.31.

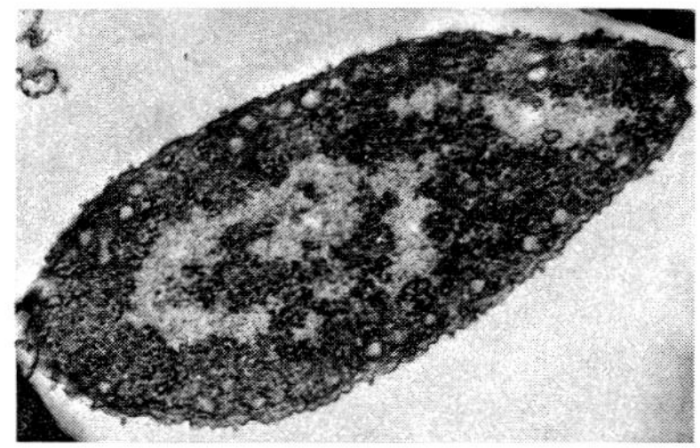

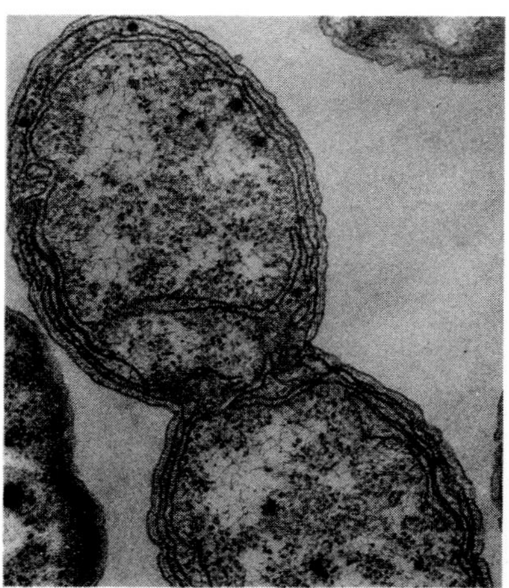

*Figure 10.30 (Above) Electron micrograph of thin section of the nitrogen-fixing bacterium, Azotobacter vinelandii, showing vesicular intrusions of the membrane similar in structure to those of certain photosynthetic bacteria (see Figure 10.29). Courtesy of J. Pangborn and A. G. Marr.*

*Figure 10.31 Electron micrograph of a thin section of Nitrosomonas europaea, an obligate chemoautotroph, showing membrane intrusions (×39,100). Courtesy of S. W. Watson.*

**The role of the membrane in transport of nutrients**

The existence of specific permeases, enzymatic systems that facilitate the entry of nutrients into the bacterial cell, was demonstrated in 1957 by G. N. Cohen and J. Monod. Recent work on sugar transport in *Escherichia coli* has shown that sugar permeases are complex enzymatic systems, which include both specific and nonspecific components. These are probably bound to the membrane, although it has been possible to obtain some components in soluble form in cell-free extracts.

The transport process requires an energy input, which is provided specifically by the energy-rich phosphate bond of phosphoenolpyruvate (PEP), not by ATP. The nonspecific component of the transport system involves an enzymatically catalyzed transfer of phosphate to a carrier protein:

PEP + protein → pyruvate + protein phosphate

The nonspecific role of this part of the system is shown by the existence of mutants that have lost the ability to utilize many sugars, as a result of derangement either of the transferring enzyme or of the carrier protein.

The activated carrier protein serves to phosphorylate a sugar, in a second enzymatically catalyzed reaction, mediated by enzymes that are specific for the sugar in question:

protein phosphate + sugar → sugar-6-phosphate + protein

It is this reaction which results in the transfer of the sugar across the membrane, as shown in Figure 10.32.

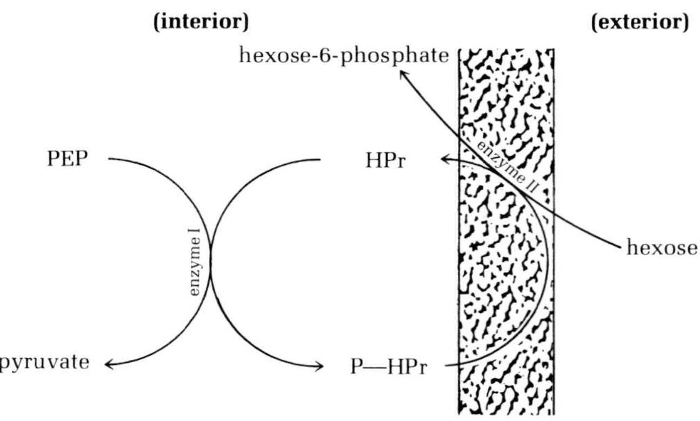


*Figure 10.32 A diagrammatic representation of the biochemical events associated with hexose transport into the bacterial cell. The activation of a carrier protein, HPr, is mediated by phosphate transfer from PEP to form P—HPr. The activated carrier then reacts with free hexose at the level of the cell membrane, under the influence of a membrane-bound enzyme (enzyme II), bringing hexose into the cell as hexose-6-phosphate.*

## The pili

The existence of pili, which are filiform appendages on many bacteria, was not suspected until they were revealed by electron microscopic examination of bacteria (Figure 10.33). Piliation is very common among enteric bacteria, where these structures have been studied in detail. It has also been reported in Gram-negative bacteria belonging to other groups but is not known to occur among Gram-positive forms.

Among enteric bacteria, a number of different types of pili can be distinguished, primarily by their width. They range in thickness from 30 A (type I pili) to 250 A (type V pili); in other groups even thicker pili (300 A) have been reported. Most pili are, however, considerably thinner than flagella, which probably explains why they were not detected prior to the introduction of the electron microscope. The number of pili of a particular type on each cell can range from one to several hundred.

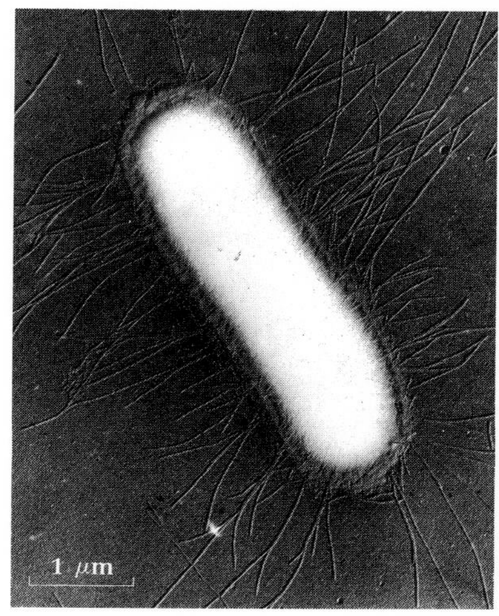


*Figure 10.33 Electron micrograph of a metal-shadowed preparation of a cell of E. coli surrounded by pili (×13,530). Courtesy of C. Brinton and E. Kellenberger.*

Any given bacterial strain can exist in either a piliated or an unpiliated state and, if piliated, may bear simultaneously more than one type of pilus. The character of piliation is genetically determined. The mutation rates of the genes in question are often unusually high. The only gross property conferred on a strain by pili is stickiness: the cells tend to adhere to one another and typically form coherent surface pellicles when grown in liquid media not subjected to agitation.

Pili can be mechanically removed from cells by agitation in a blender. Chemical analysis of purified preparations of type I pili reveal that they consist of protein. Isolated type I pili can be disaggregated by heat or treatment with acid to yield protein subunits ("pilin") with a molecular weight of about 17,000; at neutrality in the cold, the subunits will reaggregate to form structures indistinguishable from the original pili. The native pilus is a structure of some complexity, consisting of a helically wound fiber, with a central "hole" about 20 A wide (Figure 10.34).

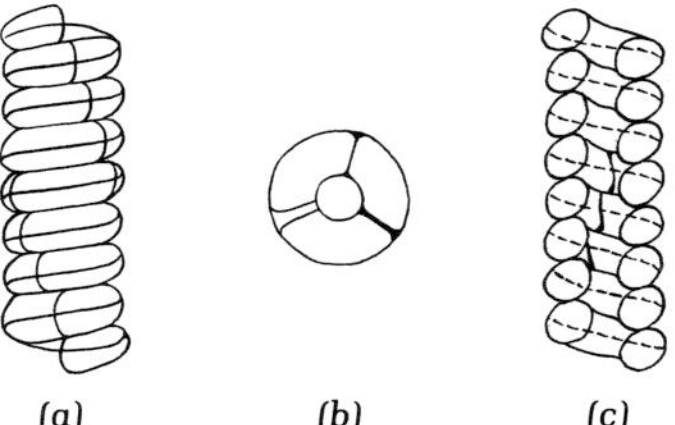


*Figure 10.34 A diagrammatic interpretation of the construction of the type I pilus of E. coli illustrated in Figure 10.33: (a) side view; (b) end view; (c) longitudinal section. This filiform structure is composed of identical protein subunits with a molecular weight of 17,000, which are aggregated to form a shallow helix. There are $3\frac{1}{8}$ subunits in each turn of the helix, as is shown in (b). The outer diameter of the helix is 70 A, and it has a hollow core with a diameter of 20 to 25 A. After C. C. Brinton, Jr.*

Among enteric bacteria, particular interest has centered on the *F pili*, a special class of pili whose formation is determined by certain transmissible, extrachromosomal factors (F and R factors), which also confer on cells that contain them the ability to act as genetic donors in bacterial conjugation. Generally, cells that contain such factors make only one or a few F pili; however, this organelle can be readily distinguished, even if the cell also carries pili of other kinds, as a result of its ability to serve as a specific receptor for certain phages.

Although experiments show that the presence of an F pilus is essential in order for a cell to act as a genetic donor, the role played by F pili in bacterial conjugation is still controversial (see Chapter 14).

## *The bacterial flagella and flagellar movement*

Although other mechanisms of movement occur among procaryotic organisms (see Chapter 5), flagellar movement, characteristic of true bacteria, has been studied in the greatest detail and is the only aspect that will be discussed in the present chapter.

### The anatomy of bacterial flagella

The external part of a bacterial flagellum is of variable length, ranging up to 10 times the diameter of the cell to which it is attached. It has an intrinsically helical form, and both the pitch and the wavelength of this helix are constant for a given bacterial strain. Flagellar width ranges in most bacteria between 120 and 185 A. In a few groups (*Vibrio, Campylobacter*) the flagellum is notably thicker (approximately 350 A), being surrounded by a distinct sheath, some 75 A thick, which appears to originate either from the wall or from the membrane.

The precise mode of attachment of flagella to the cell is difficult to study and is still not entirely clear. At or near its base, the other-

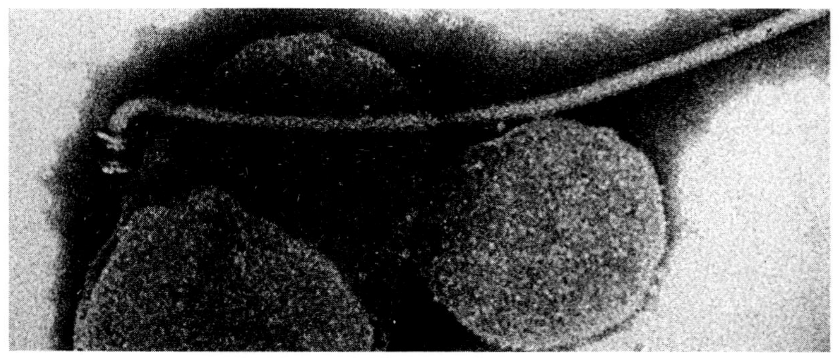

*Figure 10.35 Electron micrograph of a negatively stained lysate of the purple bacterium, Rhodospirillum molischianum, showing the basal structure of an isolated flagellum (×151,200). Note the basal hook and the attached paired discs. The other objects in the field are fragments of the photosynthetic membrane system. From G. Cohen-Bazire and J. London, "Basal organelles of bacterial flagella." J. Bacteriol.* **94**, *458 (1967).*

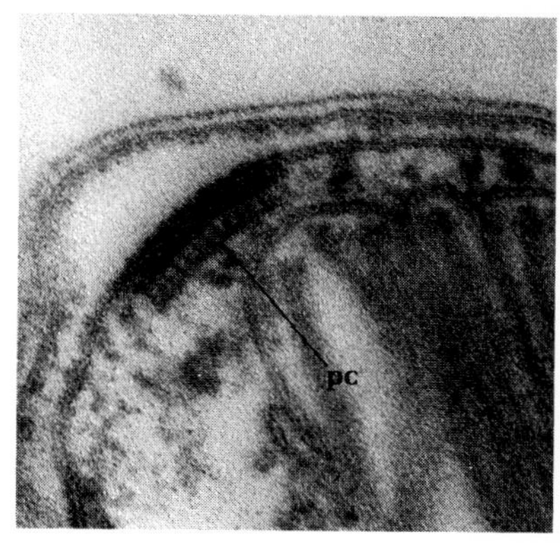


*Figure 10.36 The polar cap (pc) of a purple sulfur bacterium, Ectothiorhodospira mobilis (×143,500). This structure occurs near the flagellated cell pole in bacteria that bear a polar tuft of flagella. Courtesy of S. W. Watson. From C. C. Remsen, S. W. Watson, J. B. Waterbury and H. G. Trüper, "Fine structure of Ectothiorhodospira mobilis," J. Bacteriol.* **95,** *2374 (1968).*

wise uniform structure of the flagellum changes into a slightly thickened basal hook; and in favorable preparations from lysed cells, a double spindle-shaped structure, attached to the end of the hook by a short collar, has been detected (Figure 10.35). In spirilla and a few other bacteria that bear a large tuft of flagella inserted in a small area at one pole of the cell, a specialized polar membrane system, immediately underlying the cell membrane, has been observed in thin sections (Figure 10.36). A possible reconstruction of the basal region of flagella, relative to other surface regions of the cell, is shown in Figure 10.37.

**The chemistry of flagella**

Preparations of bacterial flagella can be obtained by mechanical shearing from the cells, followed by differential centrifugation; such methods of preparation yield, of course, only the external portions of the flagella, not the structurally more complex anchoring elements within the cell. Sheared flagella, which consist exclusively (or almost exclusively) of protein, can be purified by the customary procedures of protein fractionation. The purified flagella retain their characteristic form. If a suspension of purified flagella is brought to a pH of 3 to 4, or treated with detergents, the threads disintegrate, to yield a soluble protein ("flagellin") with a molecular weight of 30,000 to 40,000. The flagella of a given bacterial strain appear thus to be constructed from a single kind of molecular subunit. Since the serological specificity of bacteria is in part determined by their flagella, it is evident that strains or species differ with respect to the chemical nature of their flagella (and flagellins); this is confirmed by amino acid analyses, which show systematic differences between the flagella from different species. When a solution of flagellin, prepared by acid treatment, is readjusted to pH 5.3 to 6.0, the subunits reaggregate spontaneously, to form helical fibrils grossly indistinguishable from the original flagella. The rate of reaggregation is strongly dependent on flagellin concentration but is almost instantaneous in concentrated solutions.

The external part of a bacterial flagellum is, accordingly, built up from a single type of protein subunit, of relatively low molecular weight, capable of spontaneous polymerization to form a helical thread of fixed diameter and characteristic wavelength. The way in which the

*Figure 10.37 A diagrammatic reconstruction of the basal part of the flagellum of a Spirillum cell, showing its possible relationship to the cell wall and cell membrane: lp, lipoprotein layer of wall; m, murein layer of wall; cm, cell membrane. Courtesy of R. G. E. Murray.*

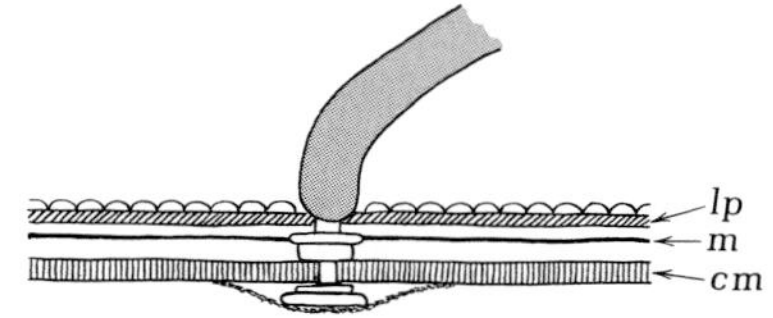

spherical flagellin subunits, each about 45 A in diameter, are arranged in a flagellum is not precisely known.

Bacterial flagella illustrate a principle of widespread importance in the creation of biological structure: that of *spontaneous self-assembly from identical subunits.* In the case of flagellins, the capacity to undergo such assembly is clearly determined by the primary structure of their polypeptide chains. Two observations demonstrate this important fact. If a *Salmonella* strain is fed *p*-fluorophenylalanine, an analog of tyrosine, this compound is incorporated in place of tyrosine into the flagellar protein. The flagella become nonfunctional, and their wavelength is shortened to half its normal value. In another bacterium, a mutational change that results in the formation of straight, nonfunctional flagella has been traced to a single amino acid substitution in the constituent flagellin.

**The synthesis of flagella**

Cells can be mechanically deflagellated, without loss of viability, by a variety of mechanical or chemical treatments. Regrowth of the flagella is relatively rapid, the normal number and length being restored in about one generation. Such experiments provide definitive evidence of the functional role of flagella; the deflagellated cells are rendered completely immotile and gradually regain the power of movement during flagellar regrowth. At first, such movement is rotatory, not transitional, which suggests that flagella must attain a critical length before they can propel the cell through a liquid medium.

**The nature of flagellar movement**

Owing to the extreme thinness of individual bacterial flagella, few bacteria provide favorable objects for the observation of flagellar movement. Observations have been made largely with spirilla, the polar flagellar tufts of which are sufficiently thick to be visualized by phase-contrast microscopy. On a moving *Spirillum* cell with two flagellar tufts, the tufts form cones of revolution with the same orientations; reversal of the direction of movement can occur by a flipping over of both cones (Figure 10.38). When, as sometimes happens, the cones of revolution of the two tufts are oppositely oriented, translational movement is replaced by a slow rotation of the cell. The functional equivalence of the flagella at each pole with respect to translational movement is shown by the behavior of cells that bear only one flagellar tuft (Figure 10.38). During active movement, the flagellar tuft rotates at approximately 40 rpm, while the body of the cell rotates in the opposite direction, at approximately 12 rpm. The speed of movement can attain as much as 50 $\mu$m/sec.

*Figure 10.38 A diagrammatic drawing showing the cones of rotation of the flagellar tufts on a Spirillum cell relative to the direction of movement.*

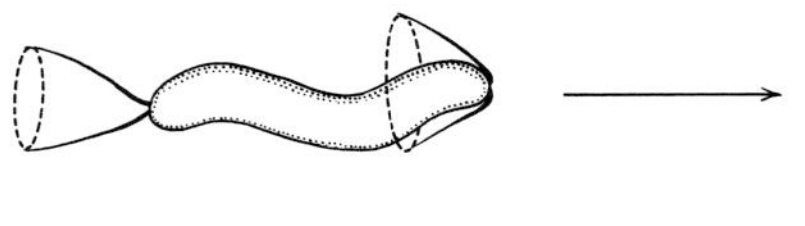

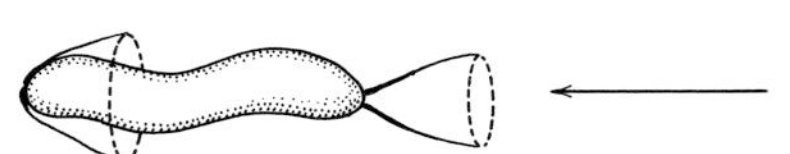

Observations on peritrichously flagellated bacteria show that during movement the separately inserted flagella coalesce, to form a rotating "tail" behind the moving cell; they do not have independent, oarlike motion.

It has been suggested that flagella are contractile proteins, capable, like myosin (the protein of muscle cells), of undergoing chemically mediated modifications of shape. However, there is no direct evidence for this view; attempts to demonstrate ATPase activity in bacterial flagella, comparable to the ATPase activity of myosin, have consistently failed. Hence, the mechanochemical basis of bacterial flagellar movement remains unknown. The one point that may be considered established is that flagellar movement is mediated, probably indirectly, by ATP. The clearest evidence for this has been obtained by experiments with fluorescent pseudomonads. Like other strictly aerobic bacteria, these organisms normally become immotile under anaerobic conditions. However, motility can be maintained in anaerobic preparations if they are provided with a particular amino acid, L-arginine. The duration of movement is directly related to arginine concentration, for a given density of bacterial suspension. The arginine-induced anaerobic movement of fluorescent pseudomonads is made possible by a special mode of metabolism of arginine:

$$\text{arginine} + 2H_2O \rightarrow \text{ornithine} + NH_3 + CO_2$$

The significance of the reaction becomes apparent when its mechanism is examined (Figure 10.39); it is a nonoxidative process, which permits the formation of ATP.

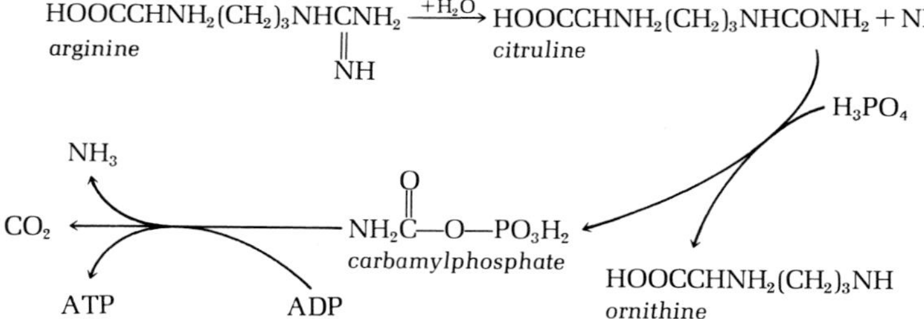


*Figure 10.39 The mechanism of the so-called "arginine dihydrolase" reaction, which permits the generation of ATP by substrate-level phosphorylation.*

**The tactic responses of flagellated bacteria**

In a suspension of flagellated bacteria, the cells are normally in a state of continuous, but random, active movement. However, if such a suspension is exposed to certain kinds of physical or chemical gradients, the cells will accumulate in that portion of the gradient that provides optimal conditions for their energy-yielding metabolism. This is known as a *tactic response*. If the stimulus is provided by a chemical substance (e.g., of an energy source or, in aerobes, of oxygen) the response is known as *chemotaxis*. The chemotactic response to oxygen is often referred to as *aerotaxis*. Motile photosynthetic bacteria will also respond to a gradient of light intensity, a phenomenon termed *phototaxis*.

To understand the mechanism of such tactic responses, it is necessary to consider the behavior of the individual cells in an evenly dispersed suspension, following the imposition of an appropriate gradient of light intensity or nutrient concentration. In the absence of the gradient, there is an equal probability that a cell will move in any direction. Once a gradient has been established, however, cells that move away from the physiologically optimal region of the gradient show a special type of behavior, known as a *shock movement*. Such cells momentarily cease to swim and then move off again in a direction different from that which they had previously followed. In the case of spirilla, the direction of movement is precisely reversed (i.e., it changes by 180 degrees); in peritrichously flagellated bacteria, the change of direction is less regular. For a shock movement to occur at all, a cell must experience a relatively sharp decline in the concentration of the triggering agent; in other words, there is a definite *threshold* required for the response.

Essentially, therefore, bacterial chemotactic and phototactic responses are negative ones: a cell is not directed in its movement *toward* the most physiologically favorable region of the gradient; instead, it is prevented by repeated shock movements from moving into an unfavorable region and thus arrives by trial and error in the favorable region. The mechanism of phototactic response in motile eucaryotic algae is much more complex and mechanistically different; these organisms have a special, light-perceiving organelle, the eyespot, which enables them to perceive the direction of a light source, and move directly toward it, a phenomenon termed *topotaxis*. In procaryotic organisms, the response is never topotactic. This is particularly clearly shown by the accumulation of motile photosynthetic bacteria in a *light trap* (Figure 10.40), produced by projecting a narrow spot of light on an otherwise dimly illuminated suspension of cells. The moving cells enter the lighted spot by random movement; once within it, however, they are prevented from leaving again by the shock movements that occur every time they penetrate the sharp gradient of light intensity which separates the light spot from the surrounding, dimly illuminated area. As a result, the bulk of the cells in the suspension are eventually captured and held within the light trap. This occurs quite rapidly; within 10 to 15 minutes a large fraction of the population of an actively motile suspension of photosynthetic spirilla may be captured in the illuminated spot.

*Figure 10.40 A diagrammatic illustration of the operation of a "light trap." The dashed line indicates the trajectory of a purple bacterium in a darkened field that contains a single illuminated area (white circle).*

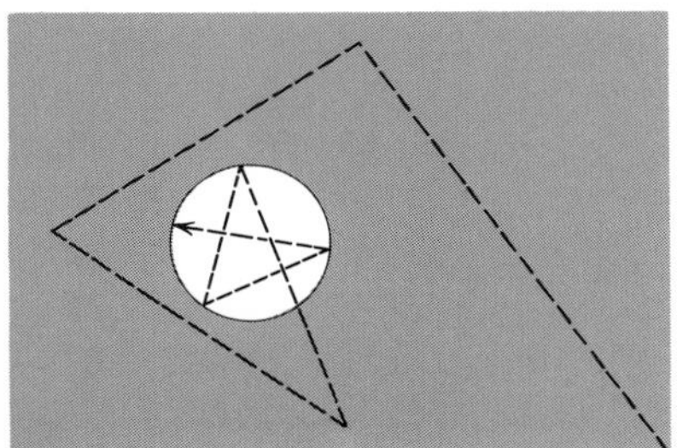

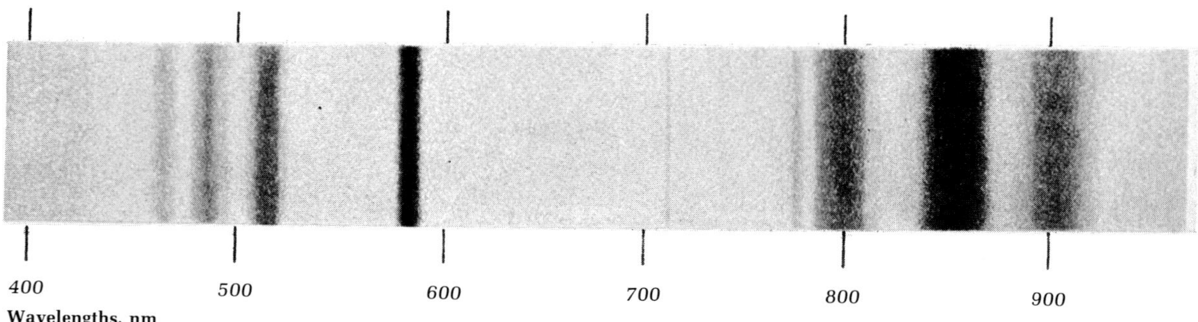


*Figure 10.41 The pattern of phototactic accumulation of motile purple bacteria in a wet mount, which has been exposed to illumination in a spectrum. The cells accumulate massively at wavelengths corresponding to the absorption bands of their chlorophyll and carotenoids, which are therefore photosynthetically effective. The relatively weak accumulation around 500 nm corresponds to the positions of carotenoid absorption bands; the accumulations at 590, 800, 850 and 900 nm correspond to the positions of chlorophyll absorption bands. After J. Buder, "Zur Biologie des Bakteriopurpurins und der Purpurbakterien." Jahrb. wiss. Botan.* **58,** *525 (1919).*

If a wet mount of motile purple bacteria is illuminated not with white light but with a spectrum produced by focusing light that has passed through a prism on the preparation, the bacteria rapidly accumulate in a series of bands corresponding to the principal absorption bands of their photosynthetic pigment system (Figure 10.41). A careful quantitative study of the relative effectiveness of different wavelengths in mediating phototaxis has shown that the action spectrum for phototaxis by purple bacteria corresponds exactly to the action spectrum for the performance of photosynthesis.

The responses of motile nonphotosynthetic bacteria to an oxygen gradient similarly serve to bring the cells into the physiologically most favorable region. As first shown by Beijerinck, such bacteria can be classified into three groups in terms of their patterns of accumulation in a wet mount, where oxygen penetrates only by diffusion from the edge of the cover slip (Figure 10.42). Strict aerobes accumulate in a band at the very edge of the cover slip, while strict anaerobes accumulate in the center of the preparation, thus minimizing contact with oxygen. Certain spirilla, which grow best at partial pressures of oxygen less than atmospheric, accumulate in a band some distance in from the edge of the cover slip.

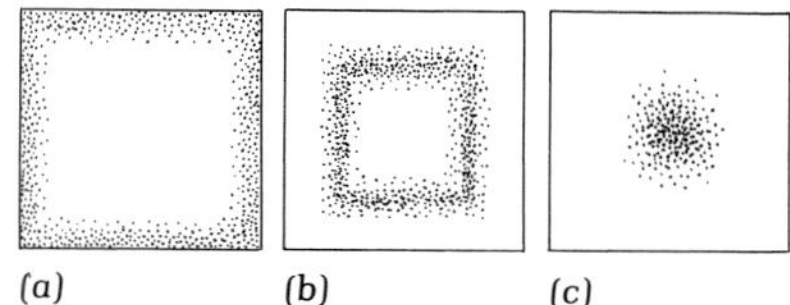


*Figure 10.42 Aerotactic responses of motile bacteria (after Beijerinck). Suspensions of various bacteria were placed on slides under coverslips. (a) Aerobic bacteria accumulate near the edges of the coverslip, where oxygen concentration is greatest. (b) Microaerophilic bacteria accumulate at some distance from the edge. (c) Obligate anaerobes accumulate in the centralmost, anaerobic region.*

The bacterium *E. coli* responds chemotactically to gradients not only of oxygen but also of many different organic compounds, each of which serves as a potent attractant. Two hypotheses could account for this versatility: either the shock movement is triggered by a common product arising from the metabolism of the different attractants (such as a catabolic intermediate, or ATP) or it is triggered by the binding of each attractant to its specific receptor.

The latter hypothesis has been shown to be correct. Many compounds that are good energy sources for *E. coli* are not attractants. Pyruvate, for example, is not an attractant although it is the first catabolic product of L-serine, which itself is a strong attractant. Secondly, many compounds that are active as attractants are not metabolizable. Fucose, for example, which is a nonmetabolizable analog of galactose, is as strong an attractant as galactose itself. In fact, bacteria will migrate away from a good energy source, such as pyruvate, in response to a nonmetabolizable attractant such as fucose.

The specificities of the chemotactic receptors have been identified by competition experiments as well as by the isolation of mutants, each of which has lost a separate receptor. These studies have shown that *E. coli* possesses a large number of receptors with narrow specificities: there is one receptor for glutamate and aspartate, for example, there is one receptor for serine, and there is one for each of many sugars. The receptor for galactose also binds fucose, its nonmetabolizable analog.

The mechanism by which the shock movement is triggered by the binding of any one of a series of different compounds to their specific receptors is not known. The shock movement in *E. coli* consists of a tumbling action, lasting for a fraction of a second, after which the cell resumes swimming in a randomly chosen direction. It appears as though a sudden decrease in the binding of an attractant causes an effect that propagates through the cell membrane, temporarily disorienting the flagellar movement.

### Motile aerobic bacteria as indicators of photosynthetic oxygen production

W. Engelmann, who made the pioneering observations on bacterial phototaxis and chemotaxis in the late nineteenth century, also used the aerotactic response of aerobic bacteria as a tool to investigate plant photosynthesis. One of his classical experiments, which provided the first direct evidence that the chloroplast is the site of oxygen production in eucaryotic photosynthetic organisms, is illustrated and explained in Figure 10.43.

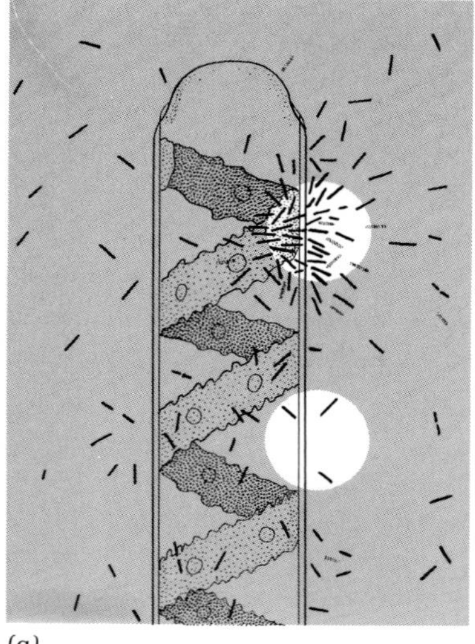

(a)

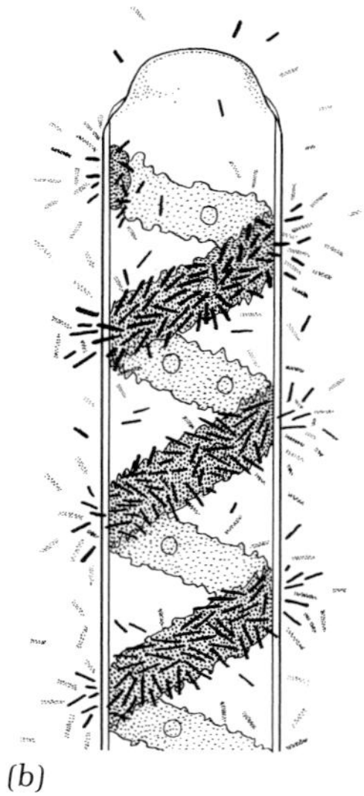

(b)

*Figure 10.43 Microscopic detection of oxygen production by the use of the aerotactic response of motile aerobic bacteria (after Engelmann). A suspension of bacteria was placed under a coverslip together with cells of the filamentous green alga, Spirogyra, which possesses long spiral chloroplasts. Oxygen that is produced when the chloroplasts are illuminated attracts the bacteria. (a) Small circular areas on the slide were illuminated. Note that the bacteria accumulate around the algal cell only where the chloroplast is in the light beam. (b) The entire field is illuminated, and the bacteria can be seen to cluster around those portions of the cell in which the chloroplast is located.*

## *Procaryotic organelles bounded by nonunit membranes*

In most procaryotic cells, the cytoplasmic region is more or less uniformly filled with ribosomes, except where local accumulations of cellular reserve materials (to be discussed in the next section) occur. However, two types of specialized cytoplasmic organelles, separated from the ribosomal region by nonunit membranes, are found in certain groups. The *chlorobium vesicle* is uniquely characteristic of green bacteria. The *gas vacuole* is of wider, but sporadic, occurrence in a number of procaryotic groups.

### The chlorobium vesicles

In all purple bacteria, the photosynthetic apparatus is located in the topologically complex cytoplasmic membrane, which is extensively intruded into the cytoplasmic region, in the form of vesicles or lamellae. It is probable that the lamellar membrane system that carries the photosynthetic apparatus of blue-green algae also originates from the cell membrane, although the evidence is less clear than in the case of purple bacteria. In green bacteria, however, the photosynthetic apparatus has a different intracellular location: it is contained in a series of cigar-shaped vesicles, which immeditely underlie, and are distinct from, the cell membrane (Figure 10.44).

The vesicles, detectable only by electron microscopy, are each about

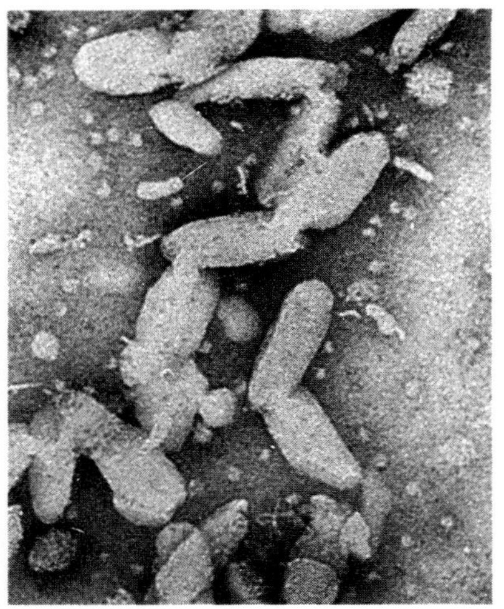

*Figure 10.44 Electron micrograph of a partly purified preparation of chlorobium vesicles, negatively stained with phosphotungstate (×100,800). The small particles are contaminating ribosomes. From G. Cohen-Bazire, N. Pfennig, and R. Kunisawa, "The fine structure of green bacteria." J. Cell Biol.* **22,** *207 (1964).*

1,000 to 1,500 A long and 500 A wide; each is enclosed by a single-layered membrane some 50 A thick. The entire photosynthetic pigment system of the cell is localized in these structures. The contents of each vesicle appear homogeneous; there are no clear-cut indications of internal structure. Both in dimensions and in structure, chlorobium vesicles differ markedly from the chloroplasts of eucaryotic photosynthetic organisms. Nevertheless, they represent the unique example of a localization of photosynthetic function within a specialized organelle in the procaryotic cell.

## The gas vacuoles

As a rule, cells are slightly more dense than water and will therefore sink, at a rate dependent on their size, in an aqueous medium. Certain procaryotic organisms have cells that are buoyant and will rise in an aqueous medium. This property is correlated with the possession of a special gas-filled organelle, known as a *gas vacuole*. Gas vacuoles are sufficiently large to be detectable with the light microscope; they appear as densely refractile bodies of irregular contour (Figure 10.45). Cells that contain gas vacuoles also have striking gross optical properties; they appear dark by transmitted light and pale by reflected light, as a result of the scattering of light by the gas-filled intracellular structures.

Gas vacuoles occur sporadically in members of all three major photosynthetic procaryotic groups and have also been detected in a few

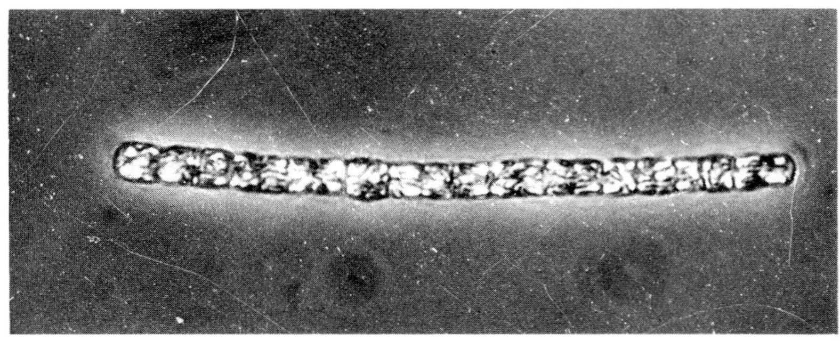

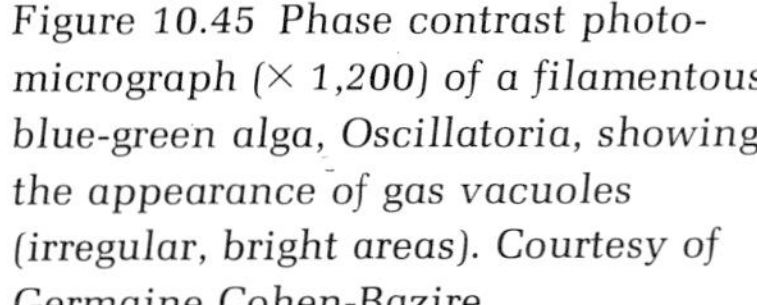

*Figure 10.45 Phase contrast photomicrograph (× 1,200) of a filamentous blue-green alga, Oscillatoria, showing the appearance of gas vacuoles (irregular, bright areas). Courtesy of Germaine Cohen-Bazire.*

Table 10.9 Procaryotic groups in which gas vacuoles occur

| Group | Occurrence |
|---|---|
| *Blue-green algae* | *Certain species belonging to many genera, both unicellular and filamentous (e.g., Oscillatoria, Spirulina, Aphanizomenon)* |
| *Purple bacteria* | *Lamprocystis, Amoebobacter, Thiodictyon* |
| *Green bacteria* | *Pelodictyon* |
| *Nonphotosynthetic eubacteria* | *Halobacterium (some strains), Ankalomicrobium* |

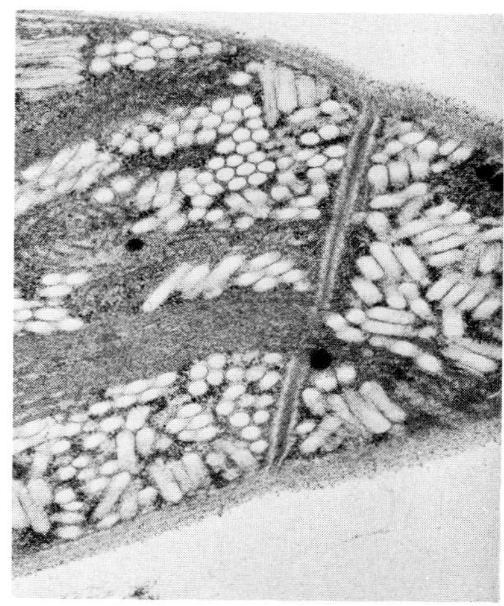

*Figure 10.46 Electron micrograph of a thin section of the Oscillatoria shown in Figure 10.45, showing the intracellular arrangement of the cylindrical gas vesicles which compose gas vacuoles (×25,800). Courtesy of Germaine Cohen-Bazire.*

groups of nonphotosynthetic bacteria (Table 10.9). The organisms that contain them are aquatic in habitat and (with a few exceptions) nonmotile. Comparative electron microscopic studies have shown that the gas vacuoles of all procaryotic organisms have a very similar fine structure. They are composite organelles, made up of a number of cylindrical *gas vesicles* with conical tips; particularly in blue-green algae, these vesicles are arranged in an extremely regular manner (Figure 10.46). The wall of each vesicle consists of a single-layered membrane about 50 A thick, with a very regular fine structure, particularly evident in isolated vesicles (Figure 10.47).

When a suspension of cells containing gas vacuoles is subjected to a sudden increase of pressure, the vesicles collapse, and the cells lose their distinctive optical properties, together with their buoyancy.

The gas vacuole can be interpreted as an organelle of flotation, which enables cells to maintain a more or less fixed vertical position in an aqueous medium. Aerobic gas-vacuole-containing organisms (e.g., blue-green algae, *Halobacterium*) are carried to the surface by the buoyancy of their cells. This is not true, however, of the anaerobic,

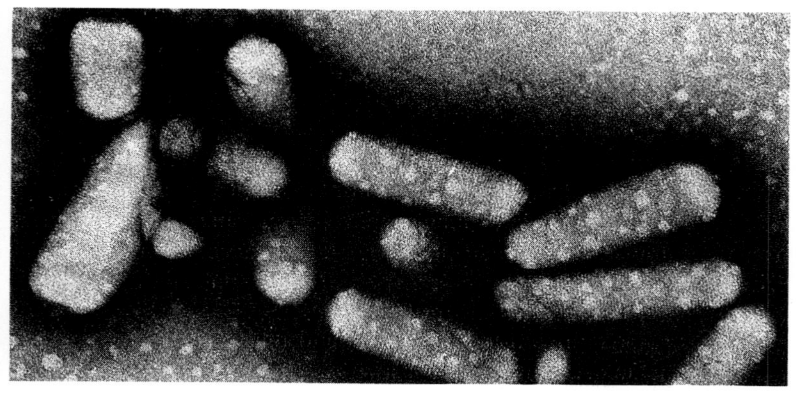

*Figure 10.47 Electron micrograph of purified gas vesicles from Oscillatoria, negatively stained with uranyl acetate (×102,600). The vesicles are still inflated. Note the regular, banded fine structure of the vesicle wall. Courtesy of Germaine Cohen-Bazire.*

gas-vacuole-containing purple or green sulfur bacteria; in nature, these organisms characteristically accumulate in a relatively narrow, horizontal plane in the deeper, anaerobic layers of lakes (see Chapter 17).

The most dramatic natural manifestation of flotation caused by gas vacuoles is the phenomenon known as a *water bloom*. Water blooms, which can occur both on lakes and in oceans, consist of massive surface accumulations of gas-vacuole-containing blue-green algae. The biological precondition for a water bloom is an abundant local development of the species in question. However, the bloom itself is precipitated by meteorological factors: a period of absolute calm which permits the cells, previously dispersed through the upper layer of the water by wave action, to float to the surface.

## The procaryotic cellular reserve materials

A variety of cellular reserve materials may occur in procaryotic organisms; they are frequently detectable as granular cytoplasmic inclusions.

**Organic reserve materials**

Two chemically different kinds of organic reserve materials, each of which can provide an intracellular store of carbon or energy, are widespread among procaryotic organisms. They are glucose-containing polysaccharides such as starch and glycogen, and a polyester of $\beta$-hydroxybutyric acid, poly-$\beta$-hydroxybutyric acid. The former class of substances also occur as reserve materials in many eucaryotic organisms. Poly-$\beta$-hydroxybutyric acid is, however, uniquely found in procaryotic groups. Procaryotic organisms do not store neutral fats, which commonly occur as reserve materials in eucaryotic organisms: poly-$\beta$-hydroxybutyric acid could, accordingly, be considered the procaryotic equivalent of this type of storage material.

As a general rule, only one kind of reserve material is formed by a given species or group. Thus, the bacteria of the enteric group and the anaerobic sporeformers *(Clostridium)* synthesize only glycogen or starch as a reserve material, whereas many *Pseudomonas, Azotobacter, Spirillum,* and *Bacillus* species synthesize only poly-$\beta$-hydroxybutyrate. Certain bacteria can, however, synthesize both types of reserve material; this is characteristic of the purple bacteria. Finally, it should be noted that a few bacteria (e.g., the fluorescent species of the genus *Pseudomonas*) do not synthesize any specific organic reserve material.

Bacterial polysaccharide reserves are usually deposited more or less evenly throughout the cytoplasm, in areas detectable with the electron

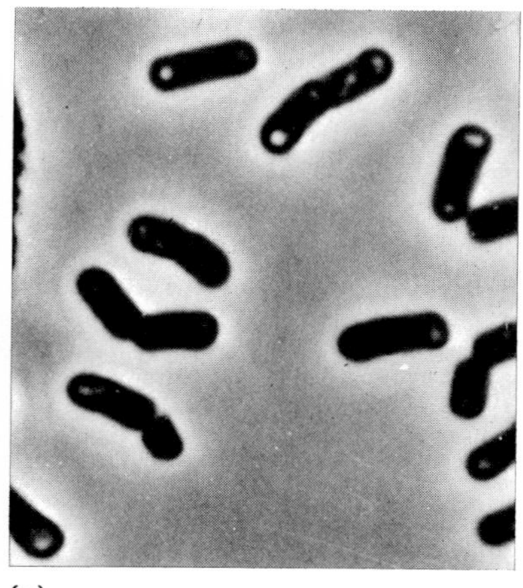

(a)

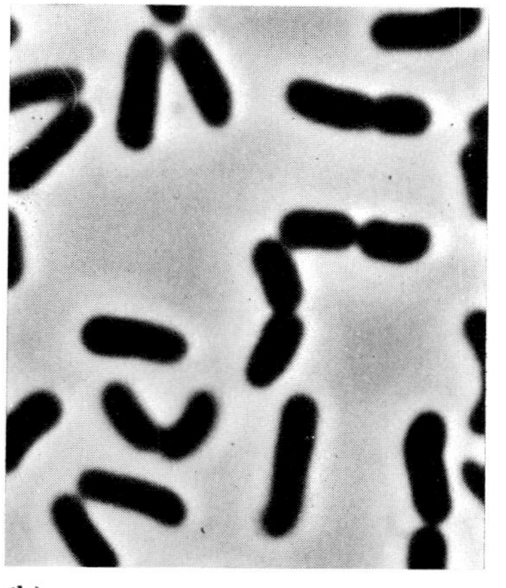

(b)

Figure 10.48 The formation and utilization of poly-β-hydroxybutyric acid in *Bacillus megaterium* (phase contrast, × 2,700). (a) Cells grown with a high concentration of glucose and acetate. All cells contain one or more granules of poly-β-hydroxybutyric acid (light areas). (b) Cells from the same culture after incubation for 24 hours with a nitrogen source, in the absence of an external carbon source. Almost all the polymer granules have disappeared. Courtesy of J. F. Wilkinson.

microscope but not visible with the light microscope. The presence of large amounts of such reserves in cells can often be detected by treatment with a solution of iodine in potassium iodide, which stains unbranched polyglucoses such as starch dark blue and branched polyglucoses such as glycogen reddish brown. Deposits of poly-β-hydroxybutyrate, on the other hand, are readily visible with the light microscope, occurring as refractile granules of variable size scattered through the cell. They are specifically stainable with Sudan black, a property also shown, in other groups, by neutral fat deposits. For this reason, bacterial poly-β-hydroxybutyrate granules have sometimes been incorrectly identified as lipid reserves.

As a general rule, the cellular content of these reserve materials is relatively low in actively growing cells: however, they accumulate massively when cells are limited in nitrogen but still have a carbon energy source available. Under such circumstances, where nucleic acid and protein synthesis are impeded, much of the assimilated carbon is stored by the cells, and reserve materials may accumulate until they represent as much as 50 percent of the cellular dry weight. If such cells are then deprived of an external carbon source and furnished with an appropriate nitrogen source (e.g., $NH_4Cl$), the reserve materials can be used for the synthesis of cellular nucleic acid and protein (Figure 10.48). Essentially, the synthesis of polyglucoses of poly-β-hydroxybutyrate represents a device for accumulating a carbon store in a form that is *osmotically inert*. In the case of poly-β-hydroxybutyrate, polymer synthesis also represents a method of *neutralizing an acidic metabolite,* since the free carboxyl group of β-hydroxybutyric acid is eliminated through the formation of the ester bonds between the subunits of the polymer (Figure 10.49). The cell can thus accommodate a very large store of such materials, whereas an equivalent intracellular accumulation of free glucose or β-hydroxybutyric acid could have catastrophic physiological consequences.

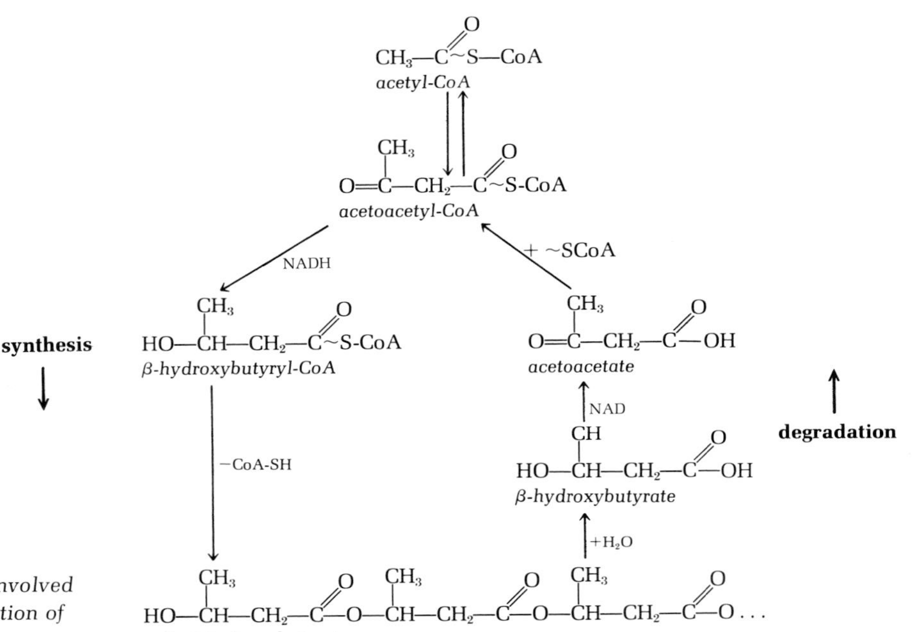


*Figure 10.49 The reactions involved in the synthesis and degradation of poly-β-hydroxybutyrate.*

The accumulation and subsequent reutilization of carbon reserves is mediated by special enzymatic machinery, under close regulatory control. Some idea of the nature of these systems has been obtained through studies on the bacterial metabolism of poly-β-hydroxybutyrate. This substance is formed through a side path on the metabolic route of fatty acid synthesis (Figure 10.50). The native polymer granules into which it is incorporated have associated with them a complex system for degradation of the polymer. The polymer in native granules cannot be attacked until the granules have been "activated" by an enzyme that requires $Ca^{2+}$ ions. This enzyme may be proteolytic, since its effect can be mimicked by trypsin. The activated granules become subject to the action of a depolymerase, which hydrolyzes the polymer to a dimeric ester, which can in turn be converted to free β-hydroxybutyric acid by a specific dimerase. A remarkable feature of this system is that *the depolymerase cannot hydrolyze the chemically purified*

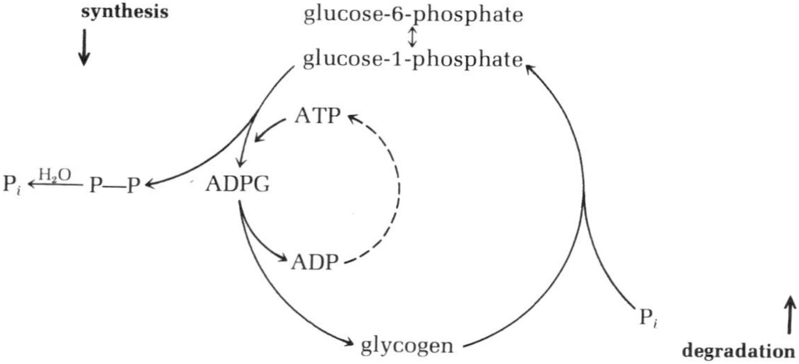


*Figure 10.50 The reactions involved in the bacterial synthesis and degradation of glycogen.*

*polymer;* the only substrate that it can attack is the activated polymer granule. Even relatively mild treatments of the native polymer granules (e.g., freezing and thawing) may render them unutilizable by the intracellular polymer-degrading system.

The bacterial synthesis of glycogen occurs at the expense of glucose-1-phosphate and ATP, which are converted to ADP-glucose and pyrophosphate; ADP-glucose is then converted to glycogen, with the liberation of ADP. The degradation of glycogen occurs through the action of a phosphorylase to yield glucose-1-phosphate.

**Volutin granules**

Many microorganisms, both procaryotic and eucaryotic, may accumulate *volutin granules,* which are stainable with basic dyes such as methylene blue (Figure 10.51). These bodies are also sometimes termed *metachromatic granules,* as a result of the fact that they exhibit a *metachromatic effect,* appearing red when stained with a blue dye. In electron micrographs of bacteria, they appear as extremely electron-dense bodies. About 20 years ago it was demonstrated that volutin granules owe their metachromatic behavior to the presence of large amounts of inorganic polyphosphate. The polyphosphates are linear polymers of orthophosphate, of varying chain lengths.

The conditions for volutin accumulation in bacteria have been studied in some detail. In general, starvation of the cells for almost any nutrient leads to volutin formation. Sulfate starvation is particularly effective and leads to a rapid and massive accumulation of polyphosphate. When cells that have built up a polyphosphate store are again

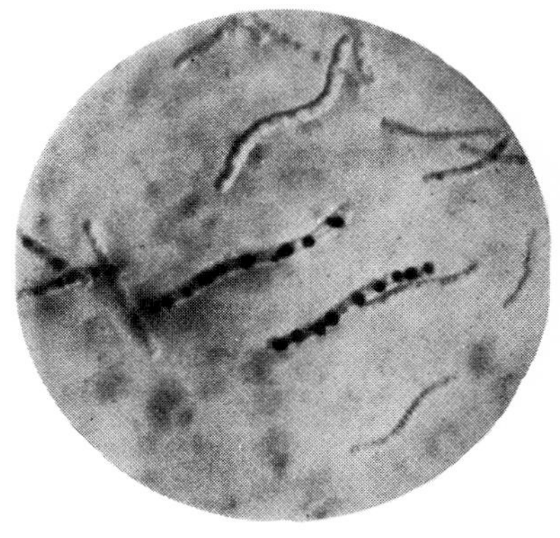

*Figure 10.51 Volutin ("metachromatic granules") in the cells of a Spirillum, demonstrated by staining with methylene blue (×1,200). From George Giesberger, "Beitrage zur Kenntnis der Gattung Spirillum Ehbg." (1936), p. 46.*

furnished with sulfate, the polyphosphate rapidly disappears, and tracer experiments with $^{32}P$ show that the phosphate is incorporated into nucleic acids. The volutin granules therefore appear to function primarily as an intracellular phosphate reserve, formed under a variety of conditions when nucleic acid synthesis is impeded. The formation of polyphosphate occurs by the sequential addition of phosphate residues to pyrophosphate, ATP serving as the donor:

$$P{-}P + ATP \rightarrow P{-}P{-}P + ADP$$
$$({-}P{-})_n + ATP \rightarrow ({-}P{-})_{n+1} + ADP$$

It is conceivable that the degradation of polyphosphate, if it occurs by the reversal of this reaction, can also provide a source of ATP for the cell.

**Sulfur inclusions**

Inclusions of inorganic sulfur may occur in two physiological groups: the purple sulfur bacteria, which use $H_2S$ as a photosynthetic electron donor, and the filamentous, nonphotosynthetic organisms, such as *Beggiatoa* and *Thiothrix*, which use $H_2S$ as an oxidizable energy source. In both these groups, the accumulation of sulfur is transitory and takes place when the medium contains sulfide; after the sulfide in the medium has been utilized completely, the stored sulfur is further oxidized to sulfate.

## *The structure and replication of the nucleus*

**The recognition and demonstration of bacterial nuclei**

When bacteria are treated with a chemical fixative and stained with a basic dye such as is used to reveal the chromosomes of a eucaryotic cell, the dye is taken up evenly throughout the cell. This behavior is due to the abundant presence of ribosomes, which are many times more numerous in bacterial cells than in eucaryotic ones and which are strongly basophilic as a result of their high content of RNA. Before it was known that the cytoplasm of bacterial cells is unusually rich in RNA, the uniform staining of the cell by basic dyes was interpreted by many bacteriologists to mean that bacteria have dispersed genetic material rather than discrete nuclei.

If, however, fixed bacterial cells are first treated with an agent that destroys RNA, such as HCl or ribonuclease, staining with a basic dye reveals discrete basophilic bodies, or *nuclei* (Figure 10.52). The presence in these bodies of genetic material is confirmed by their ability to stain with the Feulgen reagent, which is specific for DNA.

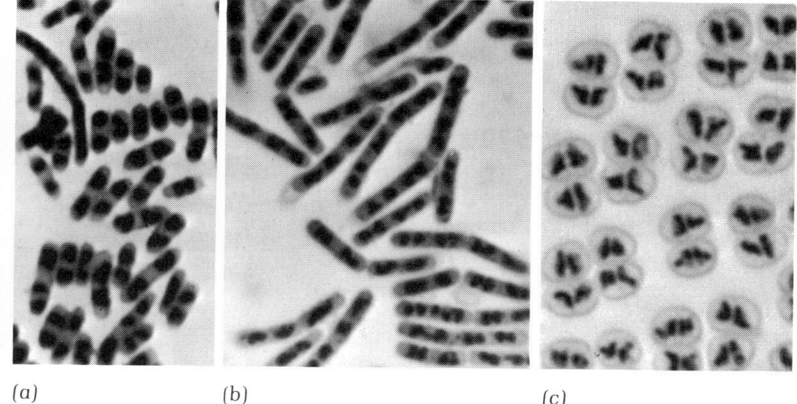


*Figure 10.52 Stained preparations photographed through the light microscope of the nuclear structures of bacteria: (a) Proteus vulgaris (×620); (b) Bacillus mycoides (×540); (c) Unidentified coccus (×730). Courtesy of C. F. Robinow.*

The bacterial nuclei can also be observed in *living* cells by phase-contrast microscopy, if the cells are suspended in a medium of appropriate contrast. In such preparations, the division of the nuclei can be followed. Figure 10.53 presents a series of phase contrast photomicrographs of *E. coli*, showing two successive nuclear divisions.

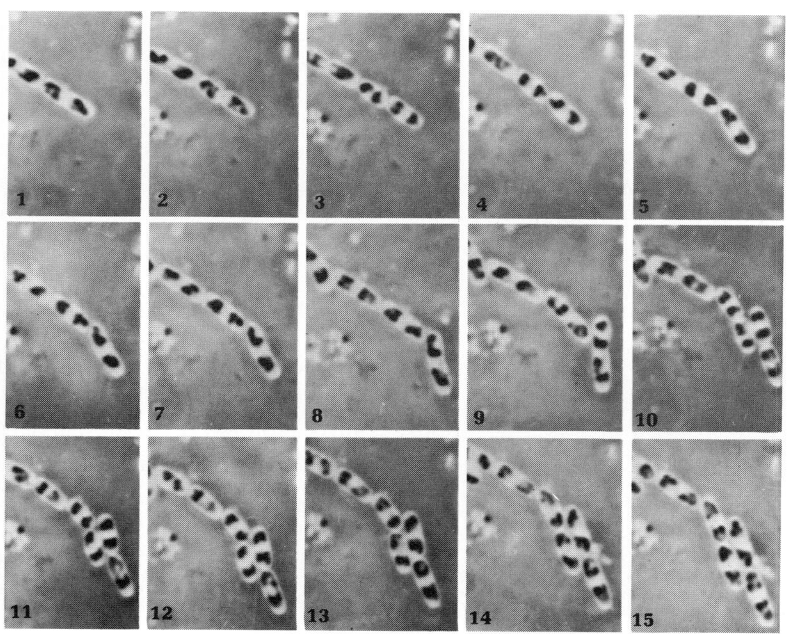


*Figure 10.53 Successive photomicrographs of growth and nuclear division in a single group of E. coli cells suspended in a concentrated protein solution to enhance the contrast between nuclear and cytoplasmic regions (phase contrast, ×825). The sequence was taken over a total period of 78 minutes, equivalent to 2.5 bacterial divisions. Courtesy of D. J. Mason and D. Powelson.*

## The structure and chemical composition of the bacterial nucleus

With the rapid development of electron microscopic techniques in the 1950s, it became possible to resolve structures far smaller than the bacterial nucleus, the size of which is just at the limit of resolving power of the light microscope. Good electron micrographs were not obtained, however, until methods had been developed that would permit the fixing of the material without coagulating the DNA. When thin sections of bacteria are properly fixed, the nucleus is revealed as a region of the cell that is packed with fibrillar DNA (Figure 10.54). There is no bounding membrane, and no substructures within the nucleus are visible.

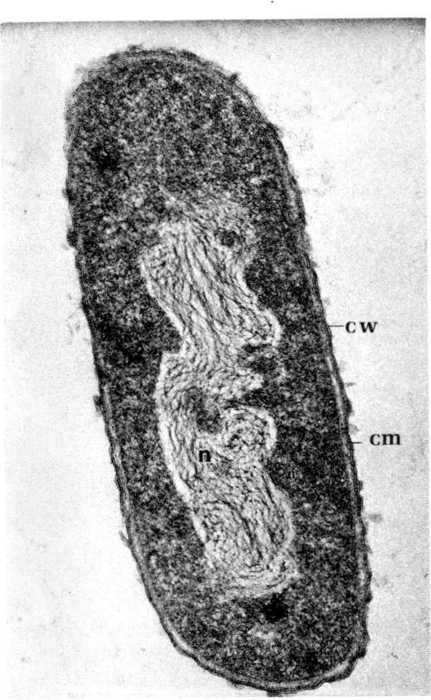


*Figure 10.54 Thin section of a dividing cell of a unicellular procaryotic organism, Bacillus subtilis (×28,200): n, nucleus; cm, cytoplasmic membrane; cw, cell wall. Courtesy of C. F. Robinow.*

By the time this cytological picture of the bacterial nucleus had emerged, genetic analysis had revealed that *E. coli* and related enteric bacteria contain only one genetic linkage group. Hence, each nucleus should contain only one element of structure, or *chromosome*. Furthermore, the genetic data yielded a closed ("circular") genetic map (see Chapter 14). Although a circular genetic map does not necessarily mean that the physical chromosome is a closed, circular structure, it seemed possible that in bacteria this might be the case. If so, then the fibrillar DNA seen in thin sections of the bacterial nucleus might represent one continuous, circular structure, folded and packed into a compact mass.

In 1963 J. Cairns succeeded in extracting bacterial DNA under conditions that minimize its shearing or degradation. Cells were grown with $^3$H-labeled thymine, so that only their DNA would be radioactive, and were placed in a chamber sealed at one end with glass and at the other end with a membrane filter. The cells were then gently lysed by dialysis against a solution of a detergent such as sodium lauryl sulfate or against a solution of lysozyme and ethylenediaminetetraacetic acid (EDTA). Further treatments were then carried out to dissociate any proteins from the DNA; all the treatments were done by allowing the reagents to *diffuse* into and out of the chamber through the membrane filter, so that the material was never subjected to mechanical agitation. Finally, the membrane was punctured, the chamber was drained, and the membrane was mounted on a slide for radioautography.

During the final draining of the chamber, the DNA of individual bacterial cells was spread out on the surface of the membrane filter. Examination of the developed radioautographs showed that the DNA was present as extremely long threads, the longest of which were slightly more than 1 mm in length. Furthermore, a few of the threads were *circular* (Figure 10.55). These threads are contained within cells which have an average length of approximately 2 $\mu$m.

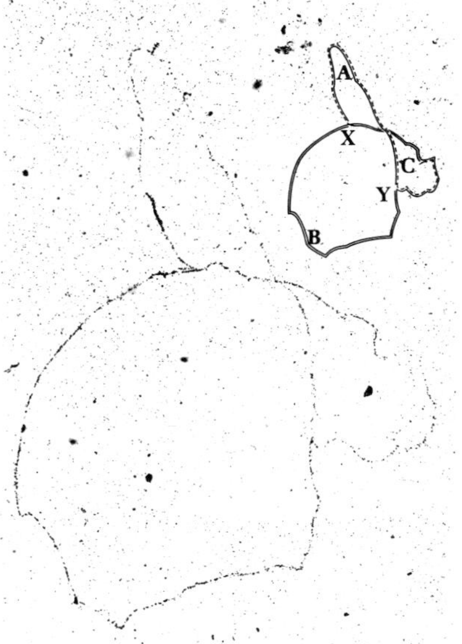


Figure 10.55 *Autoradiograph of the chromosome of E. coli strain K12, labeled with tritiated thymidine for two generations and extracted as described in the text. The scale at the bottom represents 100 μm. Inset: A diagram of the same structure, showing regions (A, B, C) in which both strands contain tritium (double solid lines) and in which only one strand is labeled (one solid and one dashed line); X indicates the starting point for replication (the replicator); Y indicates the replicating fork. From J. Cairns, "The chromosome of Escherichia coli." Cold Spring Harbor Symp. Quant. Biol.* **28,** *43 (1964).*

The length of 1 mm for the DNA thread agrees well with the amount of DNA per bacterial nucleus as determined chemically, assuming that the radioactive structure in Cairn's pictures is an extended double helix. This amount of DNA represents approximately $5 \times 10^6$ base pairs, with a molecular weight of about $3 \times 10^9$. The fact that an intact structure is obtained after treatments that degrade lipids, proteins, and RNA suggests that *the bacterial chromosome consists of a single circular molecule of DNA, about $5 \times 10^6$ base pairs in length.*

The bacterial chromosome is a highly charged molecule, since for each base pair there is present a phosphate with one ionized hydroxyl group; the negative charges must be balanced in the cell by an equal number of cationic groups. In eucaryotes, this is achieved by a coupling of DNA to basic proteins, or histones; histones have not been detected in bacteria, however, which contain instead a high concentration of small polyamines such as spermine and spermidine (Figure 10.56). It is possible that all the DNA of the bacterial cell is neutralized by small polyamines such as these.

$$\underset{\displaystyle NH_2}{CH_2}—(CH_2)_3—NH—(CH_2)_2—\underset{\displaystyle NH_2}{CH_2}$$

*spermidine*

$$\underset{\displaystyle NH_2}{CH_2}—(CH_2)_3—NH—(CH_2)_4—NH—(CH_2)_2—\underset{\displaystyle NH_2}{CH_2}$$

*spermine*

Figure 10.56 *The structures of the polyamines spermidine and spermine.*

The bacterial nucleus may be defined as the region of the cell in which is packed the single molecule of DNA representing the bacterial chromosome. Since there is no nuclear membrane and no visible internal structure other than fibrillar DNA, the nucleus and chromosome can be considered to be equivalent.

## The replication of a circular double helix

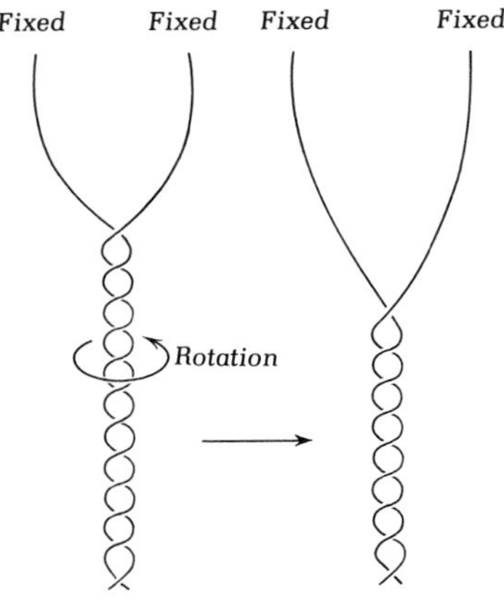


*Figure 10.57 The unwinding of a double helix by rotation.*

In Chapters 7 and 8 DNA replication was described as occurring at the *replication fork* in the molecule. The replication fork represents a "growing point"; it moves progressively along the length of the molecule, the two branches of the fork acting as templates for the replication enzyme system (Figure 7.40). For this to occur the double helix must unwind; if the two branches of the fork are thought of as remaining stationary, unwinding requires that the original duplex structure rotate along its axis (Figure 10.57).

The replication of a *circular* double helix may be considered to begin by the separation of the two strands at a fixed point on the circle; this point, the *replicator*, has been defined in Chapter 8 as the site of initiation of DNA synthesis. The replicating enzyme system attaches to the DNA at this site and begins to move along the circle as the double helix rotates and unwinds. Such unwinding would be impossible, however, if both strands of the duplex molecule formed intact, covalent circles. It must therefore be presumed that *one strand of the double helix is broken at the replicator site, the unbroken phosphodiester bond of the other strand serving as a swivel to permit rotation and unwinding.* Replication of the circular structure can then proceed as shown in Figure 10.58.

## The attachment of the chromosome to the cell membrane: chromosomal separation

Phase-contrast photomicrographs of dividing nuclei (Figure 10.53) show that the newly replicated sister chromosomes gradually separate from each other, so that the nuclei become more or less equally spaced apart in the cell. In eucaryotic cells, such movement of the chromosomes is brought about by the formation and action of a mitotic spindle; in bacteria, however, no mitotic apparatus exists. What, then, is the mechanism of chromosomal separation in bacteria?

To answer this question, F. Jacob and S. Brenner proposed a model* for chromosomal replication. Their model, in somewhat modified form, can be summarized by presenting a sequence of hypothetical events:

*The Jacob–Brenner model also accounts for the *transfer* of one replica of the chromosome to a recipient cell in bacterial conjugation (see Chapter 14).

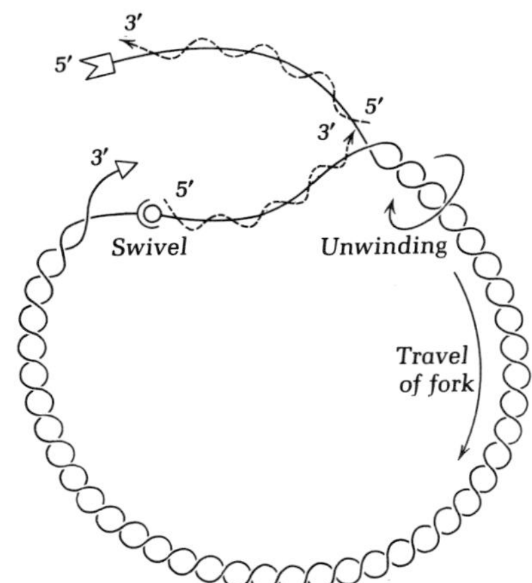


*Figure 10.58 The replication of a circular double helix. One strand of the duplex is broken; the other then provides a swivel, permitting unwinding at the replication fork. Newly synthesized complementary strands are shown in dashed lines.*

(1) At the start of the replication cycle, *the chromosomal replicator is attached to the cell membrane. The replicating enzyme system is localized in the membrane, at the site of chromosomal attachment.* One strand of the double helix, which has served as template during the preceding cycle of replication, is a closed circle; the other, newly synthesized strand, is a broken circle (Figure 10.59).

(2) Immediately following completion of the previous replication cycle, a new attachment site had been formed in the cell membrane, adjacent to the old attachment site. One free end of the recently synthesized strand of DNA attaches to this new membrane site.

(3) Replication of the chromosome begins. As it progresses, *the replicating enzyme system remains fixed in the membrane, and the chromosome moves past it.* The replicating fork thus remains at the membrane attachment site at all times; its movement along the chromosome results from the movement of the chromosome past the attachment site.

(4) During replication, the synthesis of new cell membrane takes place in an annular region between the two membrane attachment sites. *The two daughter chromosomes are thus separated from each other,*

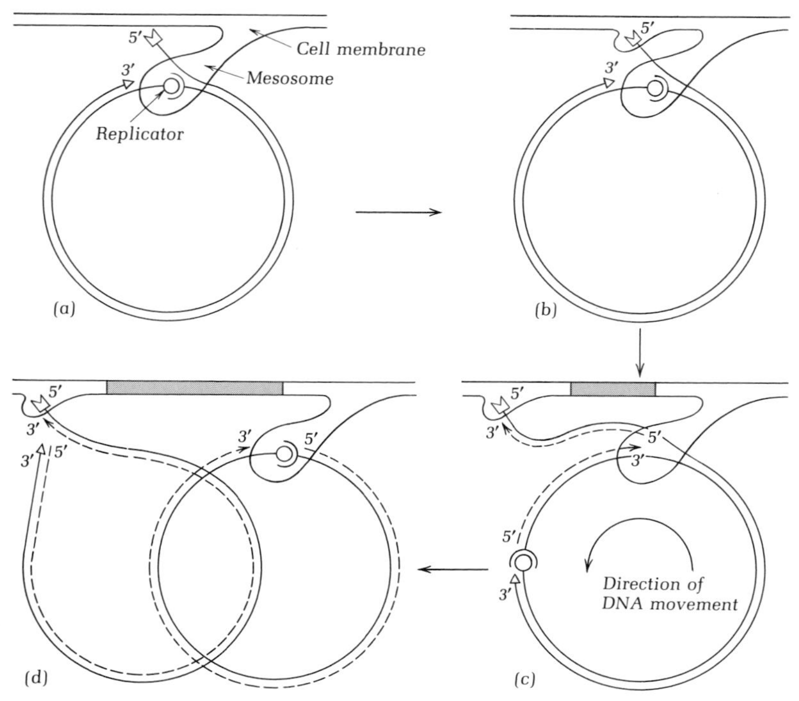


*Figure 10.59 Replication of the bacterial chromosome, according to the model of Jacob and Brenner. (a) The chromosome is attached to a mesosome at the replicator site, which serves as a swivel. One of the strands is broken. (b) The 5′ end of the broken strand attaches to a new site in the membrane. (c) The chromosome rotates counterclockwise past the mesosomal attachment site, at which is fixed the replicating enzyme system. Newly synthesized strands are shown as dashed lines. The attachment sites are separated by localized membrane synthesis (shown by shaded area). (d) The cycle of replication has been completed. The final step will be the joining of the free ends of one strand of the new chromosome (solid line).*

Figure 10.60 *Electron micrographs of thin sections showing attachment of bacterial DNA to the cell membrane. (a) A B. subtilis cell, showing the DNA in contact with a mesosome. From A. Ryter and F. Jacob, "Membrane et segregation nucléaire chez les bactéries." Protides of the Biological Fluids* **15,** *267 (1967). Courtesy of Elsevier, Amsterdam. (b) A protoplast of B. subtilis prepared by lysozyme treatment. A mesosome has been extruded, and the DNA of the cell has been pulled up against the cell membrane at that site. From A. Ryter, "Association of the nucleus and the membrane of bacteria: a morphological study." Bacteriol. Rev.* **32,** *39 (1968).*

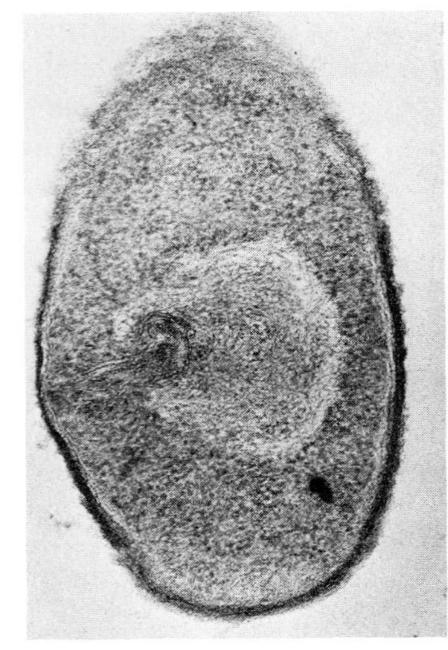
(a)

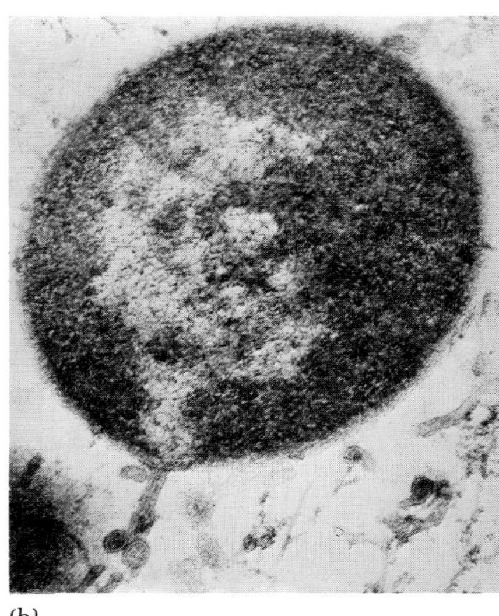
(b)

*as the attachment sites are spread farther and farther apart.* The cell wall also elongates during this process; the synthesis of new cell wall material may or may not take place adjacent to the region of membrane growth.

(5) When replication is complete, one daughter chromosome again contains one circular and one broken strand. The other daughter chromosome, which is completely broken, must undergo ring closure of one strand to complete the cycle. The strand that becomes closed is the one that acted as template during the replication process.

The details of the above model are mainly conjectural. The attachment of the bacterial chromosome to the cell membrane, however, has been confirmed by direct electron microscopic examination [Figure 10.60(*a*)]. Thin sections of bacteria show that a *mesosome* is usually present at the attachment site. If the cells are placed in a hypertonic medium, so that the mesosome is expelled into the space between the membrane and wall, the chromosome is pulled up against the flattened cell membrane [Figure 10.60(*b*)].

## Chromosomal replication and cell division

Following the separation of the two daughter chromosomes, the cell initiates synthesis of a transverse septum midway between them. This process begins with the inward growth of cell wall; the growing points are often found to be associated with mesosomes (Figure 10.61). New cell wall material is deposited between the invaginating membranes of

Figure 10.61 *Electron micrograph of a thin section of a dividing bacterium, showing mesosomes associated with the inwardly growing septum. Courtesy of A. Ryter.*

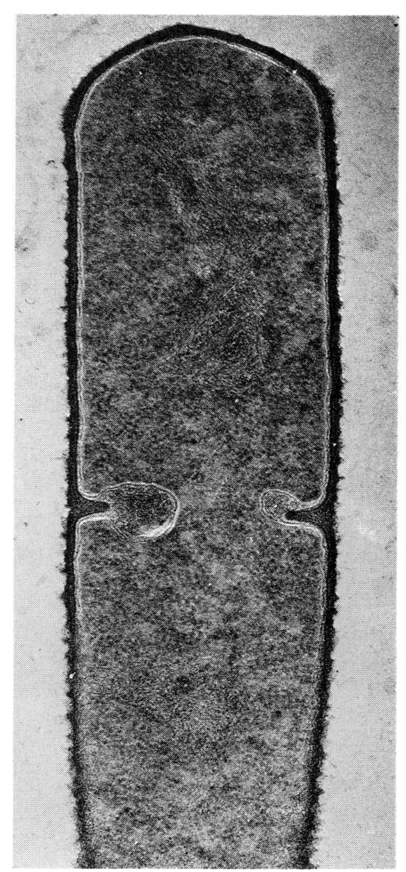

the septum and ultimately cleaves to form two daughter cells.

In normal cells, the time relationship between replication of the chromosome and synthesis of the transverse septum is closely regulated. The completion of replication appears to be a signal for the initiation of septum formation; the completion of septum formation appears to be a signal for another cycle of chromosomal replication. The nature of these signals is still unknown; it can be shown, however, by the isolation of mutants in which *replication and cell division are uncoupled,* that specific gene products are involved.

Mutants have been obtained in which the regulation of cell division is disrupted in a variety of ways. These mutants are of the type called "conditional lethal mutants" and will be discussed in Chapter 13. In this case, the mutants grow normally at 30°C but are unable to form colonies at 42°C. When the different types of cell-division mutants are incubated at 42°C and examined with the light microscope, several types of aberrant growth are observed. In one type, nuclear division

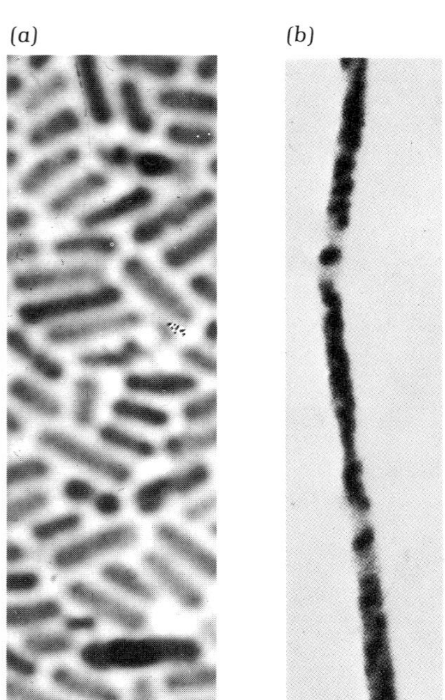


Figure 10.62 *Temperature-sensitive cell-division mutants. (a) DNA-free cells of E. coli, formed by a mutant that continues to elongate and divide at 40°C without replicating its DNA. The DNA-free cells have been separated from the normal cells by passage through a millipore filter of appropriate pore size. The photograph shows a suspension that has been stained for DNA: only two cells show nuclei. From Y. Hirota and F. Jacob, "Production de bactéries sans DNA," Compt. Rend. Acad. Sci. (Paris)* **263,** *1619 (1966). (b) Multinucleate filament of E. coli, formed by a mutant in which DNA replication and segregation continue at 40°C, without septum formation. From M. Kohiyama, A. Cousin, A. Ryter, and F. Jacob, "Mutants thermo-sensibles d'Escherichia coli K12—Isolement et caractérisation rapide," Ann. Inst. Pasteur* **110,** *465 (1966).*

ceases but cell elongation and septum formation continue for a limited period of time. The result is the formation of anucleated (DNA-free) cells, which are incapable of further growth [Figure 10.62(*a*)]. In another type, nuclear division continues but septum formation does not occur; the result is the formation of extremely long, multinucleated filaments [Figure 10.62(*b*)].

## Further reading

**Books**

Gunsalus, I. C., and R. Y. Stanier (editors), *The Bacteria*, Vol. I. New York: Academic Press, 1960.

Pollock, M. R., and M. H. Richmond (editors), *Function and Structure in Micro-organisms*. New York: Cambridge University Press, 1965.

Salton, M. R. J., *The Bacterial Cell Wall*. Amsterdam: Elsevier, 1964.

**Reviews**

Adler, J., "Chemotaxis in Bacteria." *Science* **153,** 708 (1966).

Cairns, J., "The Chromosome of *Escherichia coli*." *Cold Spring Harbor Symp. Quant. Biol.* **28,** 43 (1963).

Cohen, G. N., and J. Monod, "Bacterial Permeases." *Bacteriol. Rev.* **21,** 169 (1957).

Cohen-Bazire, G., and W. R. Sistrom, "The Procaryotic Photosynthetic Apparatus." In *The Chlorophylls* (L. P. Vernon and G. R. Seely, editors). New York: Academic Press, 1966.

Doetsch, R. N., and G. J. Hageage, "Motility in Procaryotic Organisms." *Biol. Rev.* **43,** 317 (1968).

Horecker, B. L., "Biosynthesis of Bacterial Polysaccharides." *Ann. Rev. Microbiol.* **20,** 253 (1966).

Ryter, A., "Association of the Nucleus and the Membrane of Bacteria: a Morphological Study." *Bacteriol. Rev.* **32,** 39 (1968).

Salton, M. R. J., "Structure and Function of Bacterial Cell Membranes." *Ann. Rev. Microbiol.* **21,** 417 (1967).

"Symposium on the Fine Structure and Replication of Bacteria and Their Parts." *Bacteriol. Rev.* **29,** 277 (1965).

# Chapter eleven
# The viruses

*Long before the discovery* of the microbial world, the term *virus* was used to denote any agent capable of producing disease. The word is a Latin one and originally meant "venom" or "poisonous fluid." Ideas concerning the causation of infectious disease were necessarily vague and abstract until the nineteenth century, when specific microbial agents of disease were first recognized. In the early days of microbiology, these microbial agents, whether bacteria, fungi, or protozoa, were often indiscriminately referred to as "viruses." The term is no longer used in this general sense.

## *The discovery of filterable viruses*

In 1892 D. J. Ivanowsky found that an infectious extract from tobacco plants with mosaic disease retained its infectivity after passage through a filter able to prevent the passage of bacteria. He assumed that the infectious agent was a small microorganism. During the following two or three decades, it was shown that many major diseases of both plants and animals are caused by similar infectious agents, so small that they cannot be seen with the light microscope. Since the basic criterion that served to differentiate these forms from the more familiar microbial agents of disease was their ability to pass through filters with pores sufficiently fine to retain even very small bacteria, they came to be known collectively as "filterable viruses." With the passage of time, the adjective "filterable" was gradually dropped, and the word *virus* became a *specific* group designation for these ultramicroscopic, filter-passing infectious agents.

Most of the scientists who studied viruses during the first decades of the twentieth century assumed that they were simply another class of microorganisms which differed in size but not in any really fundamental biological respect from the better-known kinds of microorganisms. Studies of their behavior in the laboratory led to the conclusion that they were *obligate intracellular parasites,* able to multiply only within the host cells. However, since obligate intracellular parasitism is also a character of some protists, this could not be considered a distinctive property of viruses.

The first indications that the viruses might be different *in nature* from cellular organisms had, however, been obtained not long after Ivanowski's discovery of the filterability of the tobacco mosaic virus. In the course of confirming Ivanowski's observations, M. W. Beijerinck discovered that the virus of tobacco mosaic disease could be precipitated from a suspension by alcohol without losing its infectious power and was capable of diffusing through an agar gel. These are properties never shown by a living organism, and Beijerinck accordingly concluded that the virus was not a living organism but rather a "fluid infectious principle." Nearly 40 years passed, however, before this brilliant intuition was confirmed. In 1935 W. M. Stanley showed that the infectious principle of the same virus could be crystallized and that the crystals consisted largely of protein. At first, this was taken to mean that the virus was a protein molecule. The first assumption proved oversimplified; a few years later, purified tobacco mosaic virus was found to contain, in addition to protein, a much smaller but constant amount of ribonucleic acid. The infectious principle is therefore not a protein molecule but a molecular complex, built up from two different kinds of macromolecule: protein and nucleic acid. The nucleic acid is specific to the virus and may be either DNA or RNA.

The first viruses to be described were agents of disease for higher plants and animals. About 1915 F. W. Twort and F. d'Herelle independently discovered that bacteria are susceptible to infection by ultramicroscopic, filter-passing agents, designated as *bacteriophages* (i.e., eaters of bacteria). This designation is often shortened to *phages.* Although d'Herelle very early stressed fundamental similarities between bacteriophages and plant and animal viruses, it was some time before the bacteriophages were universally recognized to be a subgroup of the group of viruses.

## The general properties of viruses

A virus alternates in its life cycle between two phases, one extracellular and the other intracellular. In its *extracellular phase*, it exists as an inert, infectious particle, or *virion*. The virion consists of one or more molecules of nucleic acid, either DNA or RNA, contained within a protein coat, or *capsid;* this "nucleocapsid" may, as in the case of certain animal viruses, be enclosed within a membranous *envelope*.

In its *intracellular phase*, a virus exists in the form of replicating nucleic acid, either DNA or RNA. During the intracellular phase, the genetic material of the virus is not only replicated by the host cell but also serves as a genetic message for the synthesis by the cell of specific *viral proteins*. These proteins include the subunits, or *capsomers*, from which the capsid of the mature virus particle is assembled.

### The nucleic acid of the mature virion

The mature virion contains a *single kind of nucleic acid*. As will be shown in Table 11.1 (page 389), the type of nucleic acid present in the virion varies from one virus to another. Most viruses contain either double-stranded DNA or single-stranded RNA, but single-stranded DNA is found in some viruses (e.g., coliphages fd and $\phi$X174, and minute virus of mice), and double-stranded RNA is found in the virions of reoviruses and of certain plant viruses.

When nucleic acid is extracted from virions under conditions that minimize shearing, it is generally found—with the exception of the large RNA viruses—that each virion contains a *single molecule of nucleic acid*. The length of the molecule is constant for a given virus but varies from a few thousand nucleotides (or nucleotide pairs) to as many as 250,000 nucleotides (or nucleotide pairs) in different viruses. If we take 1,000 nucleotides to be the size of an average gene (see Chapter 12), the smallest viruses contain fewer than 10 genes while the largest viruses contain several hundred.

The DNA of several bacterial and animal viruses is *cyclic*, or "circular" (Figure 11.1). In other DNA viruses, as in all RNA viruses, the nucleic acid of the virion is noncyclic, or "linear." In one case, that of bacteriophage $\lambda$, the DNA of the virion is linear but becomes circular immediately after it penetrates the host cell (see Figure 11.16). Circularity is also characteristic of bacterial chromosomes (e.g., *Escherichia coli*) and of certain viruslike genetic elements found in bacteria (e.g., the sex factor of *E. coli;* see Chapter 14).

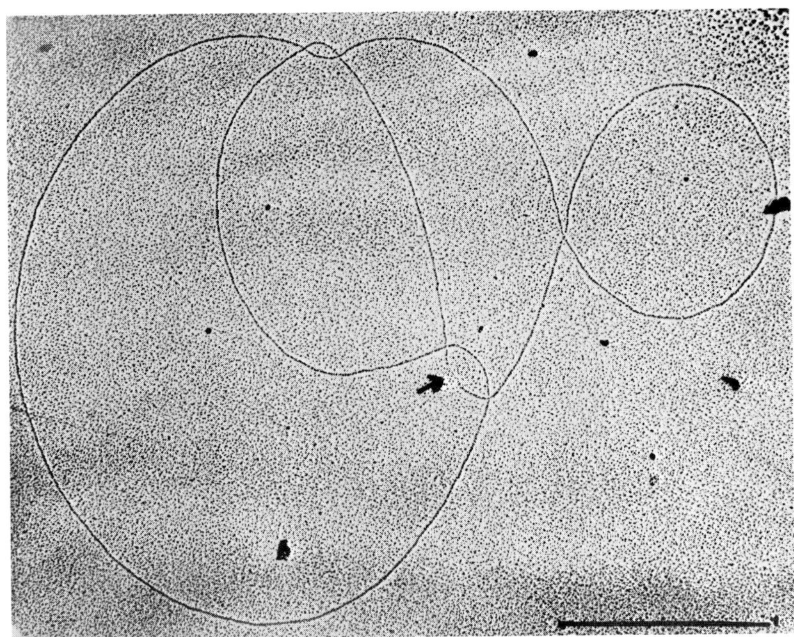

*Figure 11.1 Lambda phage DNA. The scale marker represents 1 μm; the length of the DNA molecule is 16.3 μm. The arrow points to a region of discontinuity, believed to contain the "sticky ends" described in the text. From H. Ris and B. L. Chandler, "The ultrastructure of genetic systems in prokaryotes and eukaryotes." Cold Spring Harbor Symp. Quant. Biol.* **28,** *1 (1963).*

The replication of the various types of viral nucleic acid will be discussed in a later section.

### The architecture of the nucleocapsid

Most virions fall into two classes with respect to the architecture of their nucleocapsids: they are either *helical* or *polyhedral.* These basic structures may be further modified: in many bacteriophages a polyhedral head is joined to a helically constructed tail, while in many animal viruses the nucleocapsid is enclosed in a membranous envelope.

The most thoroughly studied helical virion is that of the tobacco mosaic virus (Figure 11.2). The single-stranded RNA molecule lies in a groove formed in the helically arranged capsid; the complete virion forms a rod-shaped structure containing about 2,000 identical cap-

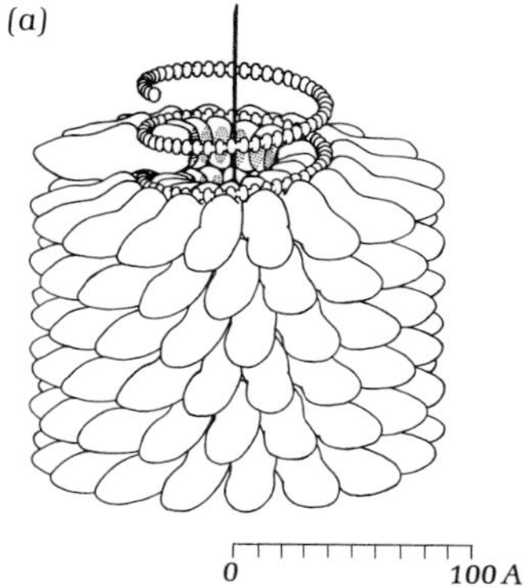


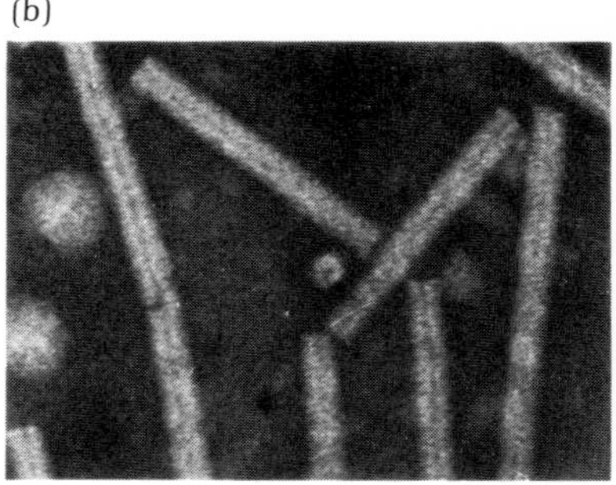


Figure 11.2 (a) A drawing of the structure of tobacco mosaic virus. For clarity, part of the ribonucleic acid chain is shown without its supporting framework of protein. From A. Klug, and D. C. D. Caspar, "The structure of small viruses," Adv. Virus Res. **1**, 225 (1960). (b) Electron micrograph of tobacco mosaic virus particles in phosphotungstic acid. From S. Brenner and R. W. Horne, "A negative staining method for high resolution electron microscopy of viruses." Biochim. Biophys. Acta **34**, 103 (1959). Courtesy of R. W. Horne.

somers. The virion of the tobacco mosaic virus is typical of many plant viruses; some of the animal viruses also form helical nucleocapsids, but these are always irregularly coiled within an envelope. Some examples are shown in Figure 11.3.

In the polyhedral virions, the nucleic acid is packed in an unknown manner within a hollow, polyhedral head. In many of the polyhedral plant and animal virions, the capsid is an *icosahedron*—a regular polyhedron with 12 corners, 20 triangular faces, and 30 edges.

The capsid of such a virion is composed of two types of capsomers: *pentamers*, which are situated at the corners, and *hexamers*, which fill

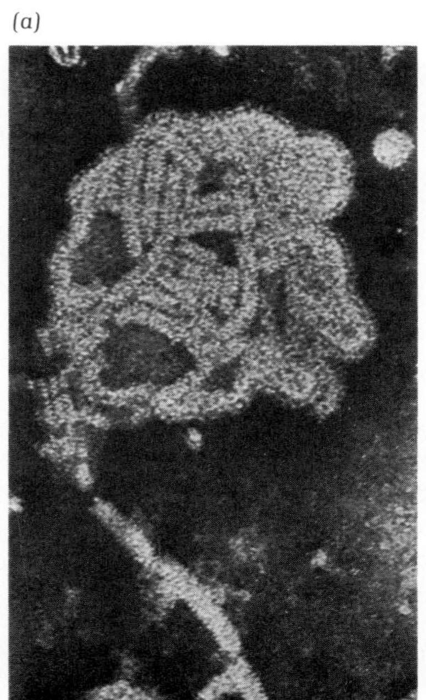


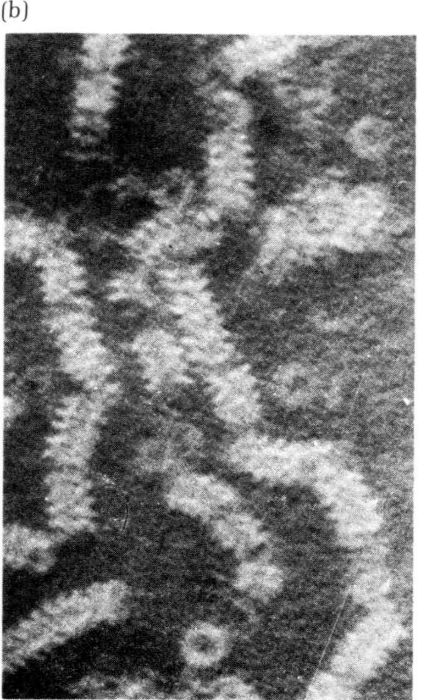


Figure 11.3 The nucleocapsid of enveloped viruses. (a) A partially disrupted particle of Newcastle disease virus, releasing the internal component (nucleocapsid). (b) High-magnification electron micrograph of the nucleocapsid released from particles of mumps virus. From R. W. Horne et al., "The structure and composition of myxoviruses. I. Electron microscope studies of the structure of myxovirus particles by negative staining techniques." Virology **11**, 79 (1960).

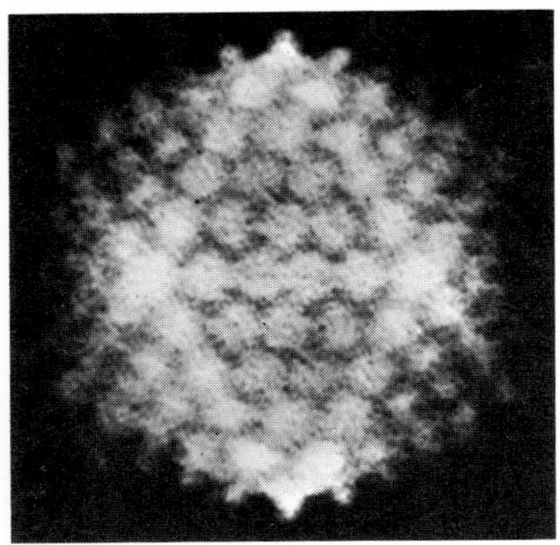

*Figure 11.4 Adenovirus virion, showing the icosahedral array of capsomers. The capsomers at the vertices are surrounded by five nearest neighbors, all the others by six. From R. C. Valentine and H. G. Pereira, "Antigens and structure of the adenovirus." J. Mol. Biol.* **13,** *13 (1965).*

the triangular faces (Figure 11.4). Each capsomer is a multimeric protein, the pentamer containing five identical subunits and the hexamer six. A subunit may consist of a single polypeptide chain; in one case, however, the subunit has been found to contain two different polypeptide chains.

The size of an icosahedral virus is determined by the number of capsomers it contains. This number follows the laws of crystallography: only certain numbers are possible in an icosahedral structure. For example, the smallest possible icosahedral capsid would have 12 pentamers and no hexamers, the next smallest would have 12 pentamers and 20 hexamers, and so on. The largest known icosahedral virion, that of an insect virus, contains 812 capsomers.

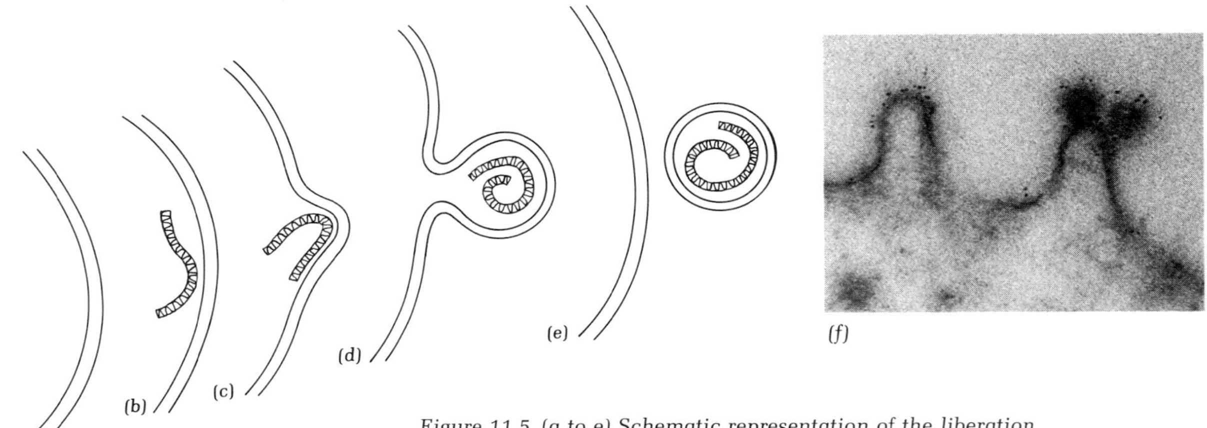


*Figure 11.5 (a to e) Schematic representation of the liberation of an enveloped virus. Modified from B. Davis et al., Microbiology. New York: Harper and Row, 1967. (f) Electron micrograph of a thin section through an animal cell that is liberating influenza virus particles. The material was treated with antibody against influenza virions; the antibody molecules were coupled to ferritin, which is extremely electron-dense due to bound iron atoms. Micrograph from C. Morgan et al., "The application of ferritin-conjugated antibody to electron microscopic studies of influenza virus in infected cells." J. Exptl. Med.* **114,** *825 (1961).*

**The envelope of certain animal viruses**

As shown in Figure 11.5, some animal viruses are liberated from the host cell by an extrusion process which coats the nucleocapsid with a layer of cell membrane. Viruses that form nucleocapsids in the nucleus of the host cell may be similarly coated as they pass through the nuclear membrane into the cytoplasm.

Not all the materials of the envelope are derived from normal cell components, however. In many cases, the envelope contains proteins that are determined by viral genes. The finished envelope is often a complex, highly organized structure containing several layers (Figure 11.6).

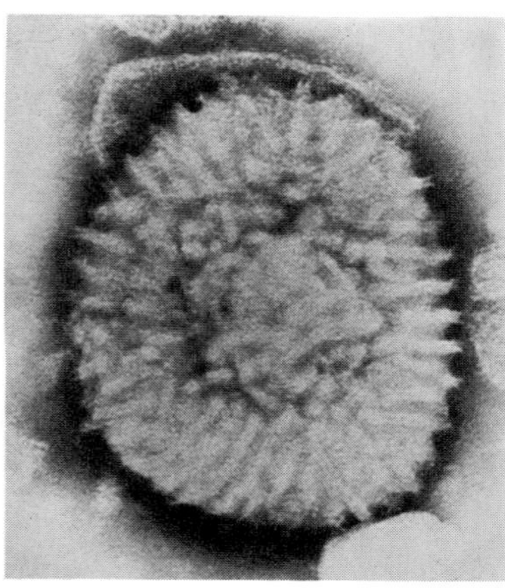

(a)

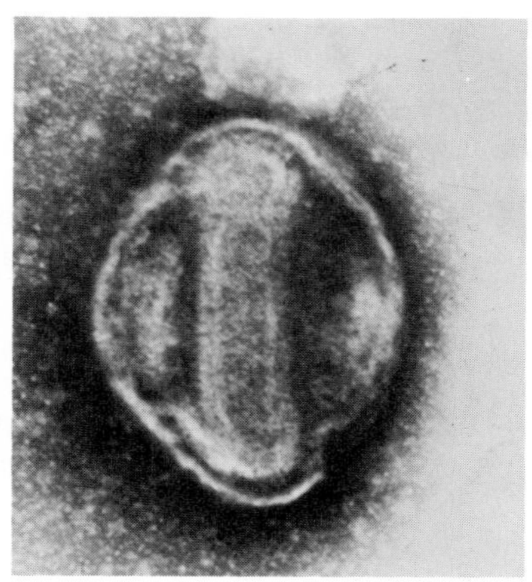

(b)

*Figure 11.6 (a) A whole vaccinia particle, negatively stained with phosphotungstic acid. The ridges on the surface may be long rodlets or tubules. (b) A negatively stained vaccinia particle that has been centrifuged in a sucrose gradient. The particle has been partially disrupted, and has lost its outer membrane. The remaining structure includes a biconcave inner core, containing the nucleic acid, two elliptical bodies, and a surrounding membrane. From S. Dales, "The uptake and development of vaccinia virus in strain L cells followed with labeled viral deoxyribonucleic acid." J. Cell Biol.* **18,** *51 (1963).*

**The reproduction of viruses**

The process of viral reproduction can be considered to take place in five stages: penetration of the host cell, synthesis of enzymes needed for viral nucleic acid replication, synthesis of viral constituents, assembly of the constituents to form mature virions, and release of the mature virions from the host cell.

The penetration process differs in bacterial, plant, and animal viruses. Bacterial and plant viruses must penetrate the cell wall of the host, whereas animal viruses may adsorb directly to the host cell membrane. The penetration of the cell wall by some, but not all, bacterial viruses is accomplished through injection: the protein coat of the virus remains adsorbed to the outside of the cell, and the viral nucleic acid is injected through the wall. Plant viruses, on the other hand, do not appear to be equipped with a specific device for penetration of the host cell wall: experimental infection requires mechanical injury of the plant cells. In nature, plant viruses are usually transmitted from one host plant to another by vectors such as insects, which inject the virus particles through the cell walls of the plant.

Animal viruses adsorb to the host cell membrane and may then be taken into the cell by phagocytosis. As a result of phagocytosis, the ingested virion is trapped within a membrane-bounded food vacuole (see Figure 26.5); it still must penetrate the cell membrane to reach the cytoplasm or nucleus, where it will be reproduced. In the case of viruses that possess an envelope partially derived from the cell membrane of its previous host, penetration is probably facilitated by fusion of the viral envelope with the membrane of the new host cell. In viruses lacking envelopes, the manner in which the nucleocapsid penetrates the cell membrane is not known.

Only the viral nucleic acid of most bacterial viruses reaches the cytoplasm of the host cell. In plant and animal viruses, on the other hand, the entire nucleocapsid may reach the cytoplasm; consequently, the final step in the penetration process is the removal of the protein capsid, presumably by the action of proteolytic enzymes. Thus, penetration culminates with the appearance in the interior of the cell of free viral nucleic acid.

The liberation of viral nucleic acid within the host cell initiates two distinct processes: the synthesis of virus-specified proteins and the replication of the viral nucleic acid itself. If the viral nucleic acid is RNA, it serves directly as messenger RNA; if the viral nucleic acid is DNA, it is first transcribed by a DNA-dependent RNA polymerase to form viral messenger RNA. In both cases, the viral messenger is translated by the host cell ribosomes to form enzymes necessary for viral replication as well as subunits of the viral capsid.

The viral nucleic acid serves as template for its own replication, complementary strands being synthesized by specific polymerases. The details of the mechanism differ, however, with the nature of the nucleic acid: DNA or RNA, single- or double-stranded, cyclic or linear. Some different modes of replication will be discussed in later sections of this chapter.

The processes of protein synthesis and replication lead to the accumulation in the host cell of numerous molecules of viral nucleic acid together with numerous subunits of the viral capsid. These combine with each other spontaneously to form complete nucleocapsids. The final stage in the reproductive process is the liberation of mature virions from the host cell. In animal cells this is often accomplished by extrusion through the cell membrane, whereas in bacteria it is usually accomplished by lysis of the host cell through the action of a viral enzyme.

**Infectious nucleic acid**

We have seen that the host cell can synthesize complete virions from free viral nucleic acid liberated within the cytoplasm. Indeed, it is possible to infect host cells with free nucleic acid extracted from virions; the *efficiency* of this infection, however, is usually one-thousandth to one-millionth that of the intact virion. The capsid thus facilitates the penetration of the host cell by the virus by providing a mechanism for *adsorption* of the virion to the host cell surface and—in the case of some bacteriophages—by providing a mechanism for the injection of viral nucleic into the cell. The capsid also serves to protect the viral nucleic acid from enzymatic degradation or chemical denaturation.

The host-range specificity of a virus is determined largely by the specificity of its binding to receptor sites on the host cell. This binding involves specific adsorption sites on the surface of the viral capsid; consequently, the host range of free viral nucleic acid is much wider than that of the intact virion. In nature, however, the low efficiency of infection by free nucleic acid prevents the wider host range from being expressed.

**The differences between viruses and cellular organisms**

From the foregoing account, it is clear that viruses represent a unique class of biological entities, different from cellular organisms. The differences between viruses and cells may be summarized as follows.

(1) In its extracellular phase, the virion, a virus consists of a single kind of nucleic acid—DNA or RNA—combined with a protein capsid. The virion may include one or a few enzymes which play a role in the initiation of an infection, but the enzymatic complement is far from sufficient to reproduce another virion. In contrast, a cell always contains both DNA and RNA, along with an elaborate complement of enzymes capable of catalyzing all the reactions that are necessary for the reproduction of the cell.

(2) A virus is reproduced by the assembly of nucleic acid replicas and capsid subunits, independently synthesized by the host cell. Growth of a cell, on the other hand, consists in the orderly increase in the amount of each of its constituent parts, during which the integrity of the whole is continuously maintained. Thus, the virion never arises directly from a preexisting virion, whereas the cell always arises directly from a preexisting cell. Cellular growth culminates in an increase in cell number by the process of cell division.

The nature of viral reproduction and the nature of cellular growth and division are thus completely different. At one time it was believed that certain infectious agents which were intermediate in *size* between typical cells and typical viruses also represented intermediate forms of life, but the criteria listed above now permit the unequivocal identification of these agents as cellular organisms in certain cases and as true viruses in others. Thus, the agents of psittacosis, lymphogranuloma venereum, and trachoma are clearly cellular, whereas the agents of vaccinia and of smallpox—which are equally large—are clearly viral.

**The classification of viruses**

The most widely used system for the classification of viruses is shown in Table 11.1. According to this system, introduced by A. Lwoff and his colleagues in 1962, viruses are grouped according to the properties of their virions: type of nucleic acid, capsid architecture, presence or absence of an envelope, and capsid size. Further subdivisions are based on other features of the virion, such as the number of nucleic acid strands (one or two); on features of viral development, such as the site of viral synthesis in the cell; and on host–virus interactions, as exemplified by host range.

This system is not intended to represent a natural, or phylogenetic, classification; that is, it does not attempt to express the evolutionary relationships between viruses, relationships that are completely obscure. Instead, it groups viruses according to common sets of chemical and structural features, which are constant properties that can be determined with precision.

It will be noted that host range plays only a minor role in this classification. The different methodologies required for the study of viruses in widely different hosts, however, have led to the general practice of grouping them according to whether their normal hosts are bacteria, animals, or plants.*

## *The bacterial viruses*

Nearly every species of bacterium investigated so far has been found to serve as host to one or more viruses or bacteriophages. Most of the research on bacteriophages, however, has been done with the phages that attack *Escherichia coli;* our discussion will therefore be mainly confined to results obtained with this group.

*For a long time, the only protists known to serve as hosts to viruses were the bacteria. Recently, however, viruses have been found in some members of the blue-green algae, and particles resembling virions have been reported in one species of fungus.

Table 11.1 *System for the classification of viruses*

| Nucleic acid | Capsid symmetry | Naked or enveloped | Size of capsid, A[a] | Number of capsomers | Special features | Examples: Bacterial | Examples: Animal | Examples: Plant |
|---|---|---|---|---|---|---|---|---|
| RNA | *Helical* | *Naked* | *175 × 3,000* | | | | | *Tobacco mosaic virus* |
| | | *Enveloped* | *90* | | | | *Myxoviruses* | |
| | | | *180* | | | | *Paramyxoviruses* | |
| | *Polyhedral* | *Naked* | *200–250* | | | *Coliphage f2* | | |
| | | | *280* | *32* | | | *Picornaviruses* | *Bushy stunt virus* |
| | | | *700* | *92* | *Double-stranded RNA* | | *Reoviruses* | |
| DNA | *Helical* | *Naked* | *50 × 8,000* | | *Single-stranded DNA* | *Coliphage fd* | | |
| | | *Enveloped* | *90–100* | | | | *Poxviruses* | |
| | *Polyhedral* | *Naked* | *220* | *12* | *Single-stranded DNA* | *Coliphage $\phi$X174* | | |
| | | | *450–550* | *72* | | | *Polyoma, papilloma* | |
| | | | *600–900* | *252* | | | *Adenoviruses* | |
| | | | *1,400* | *812* | | | *Tipula, insect viruses* | |
| | | *Enveloped* | | *162* | | | *Herpesviruses* | |
| | *Binal (polyhedral "heads," helical "tails")* | *Naked* | *Head: 950 × 650*<br>*Tail: 170 × 1,150* | | | *Coliphages T2, T4, T6* | | |

[a]Diameter, in case of polyhedral virion.

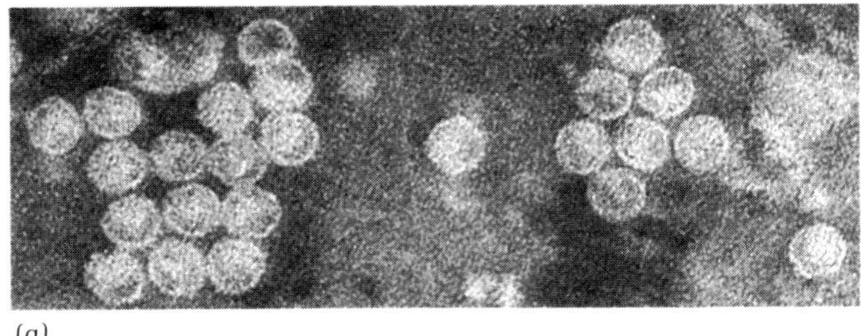
(a)

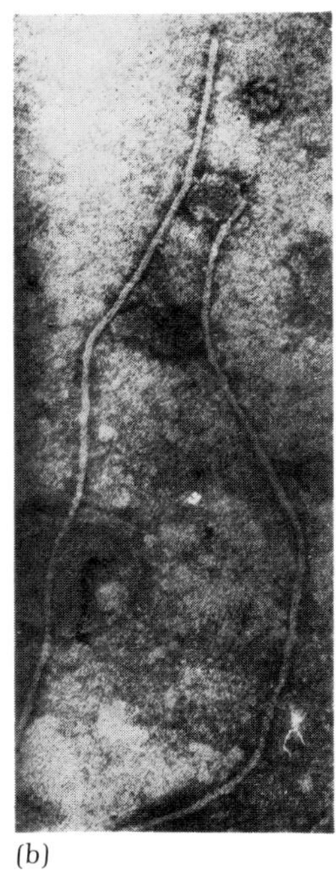
(b)

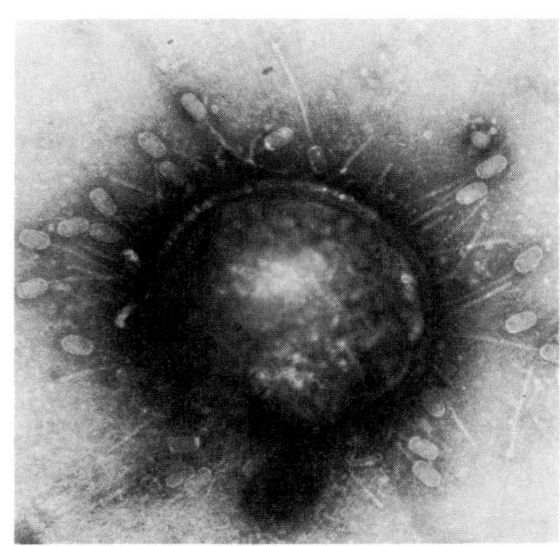
(c)

Figure 11.7 *Bacteriophage virions: (a) icosahedral; (b) filamentous (helical symmetry); (c) complex (polyhedral head, helically symmetrical tail). Micrographs (a) and (b) from D. E. Bradley, "The structure of some bacteriophages associated with male strains of Escherichia coli." J. Gen. Microbiol.* **35,** *471 (1964); (c) from D. E. Bradley, "The morphology and physiology of bacteriophages as revealed by the electron microscope." J. Roy. Microscop. Soc.* **84,** *257 (1965); courtesy of Cambridge University Press, New York.*

Both DNA and RNA phages are known. In most of the DNA phages the nucleic acid of the virion is double-stranded, although in a few cases it is single-stranded. The RNA phages discovered so far have single-stranded RNA in their virions. In all but one group of bacteriophages, the nucleic acid of the virion is contained within a polyhedral capsid; in many cases this is joined to a helical protein structure, or "tail," which serves as an adsorption organ (Figure 11.7). The exception to the polyhedral structure are the filamentous phages, of which the phage called fd is typical; fd is a rod-shaped structure containing single-stranded DNA (Figure 11.7).

## The detection and enumeration of bacteriophage particles

When a small number of virulent bacteriophage particles is added to a growing liquid culture of susceptible bacteria, some of the bacterial cells become infected. The infected cells show no obvious changes for a certain period, which commonly extends from 15 minutes to 1 hour or more, depending on the nature of the bacterium and phage in question and the conditions of cultivation. Then, quite suddenly, the infected cells undergo *lysis*. The lysis of the infected cell liberates a large number of new phage particles. These particles can in turn infect other cells in the population, with a repetition of the same cycle. Consequently, even if the number of infectious phage particles originally introduced is small relative to the number of bacterial cells, practically the entire bacterial population may be destroyed in a few hours.

*Figure 11.8 Plaques formed by bacteriophage T2. Courtesy of G. S. Stent.*

It is very easy to determine the number of phage particles or infected bacterial cells in a suspension by spreading an appropriate dilution of this suspension over the surface of an agar plate that is evenly inoculated with a thin suspension of susceptible bacteria. After appropriate incubation, the surface of the plate shows a confluent layer of bacterial growth, except at those points where a phage particle or an infected cell has been deposited. Around such sites of infection, clear zones of lysis, or *plaques*, are formed as a result of the localized destruction of the film of bacteria by successive cycles of phage growth (Figure 11.8). By appropriate methods (e.g., differential centrifugation) one can separate infected cells from free phage particles in a mixture that contains both, and enumerate each separately by this *plaque-counting* method. If only the phage particles in a suspension are to be counted, the bacterial cells that are present can be killed by shaking an aliquot of the suspension with chloroform.

## The isolation and purification of bacteriophages

The plaque method may be used to isolate from nature phages that are capable of attacking a particular species or strain of bacterium. A sample of material from the natural habitat of the bacterium is shaken with water and filtered; the filtrate is then sterilized with chloroform and aliquots are mixed with suspensions of the bacterium and plated on agar. Any plaques that appear represent phage particles that were present in the natural material. The phage from a single plaque can be isolated by stabbing the plaque with a sterile inoculating needle and suspending the adhering material in a small volume of sterile diluent; such a suspension usually contains between $10^4$ and $10^6$ phage particles. Purification of the phage is then achieved by repeated serial isolations from single plaques.

To prepare large batches of phage for chemical or physical analysis, a culture of bacteria growing exponentially in liquid medium is inoculated with phage particles. The series of cycles of phage growth which ensues results in the lysis of most or all the cells in the culture; the culture is then freed of remaining cells and cellular debris by low-speed centrifugation and sterilized either by filtration or by treatment with chloroform. The resulting *sterile lysate* usually contains between $10^9$ and $10^{12}$ bacteriophage particles per milliliter, together with soluble and particulate material liberated from the lysed cells.

The final purification of the bacteriophage is usually accomplished by ultracentrifugation: even the smallest phage particles will sediment at centrifugal forces exceeding 100,000 × g (i.e., 100,000 times the force of gravity). As a preliminary step, it is often possible to precipitate the viral particles by treatment with ammonium sulfate or other agents.

## *The DNA bacteriophages: the lytic cycle of infection*

The essential features of the lytic cycle of infection are shown by a fundamental quantitative experiment known as the *one-step growth experiment*. A culture of sensitive bacteria is mixed with a suspension of phage particles and incubated for a short time to permit the phage particles to become adsorbed on the bacterial cells. If the bacteria are in considerable excess, almost all the phage particles are adsorbed. The culture is then greatly diluted and allowed to grow. The evolution of the number of phage particles and infected bacteria is determined by periodic plaque counts. Figure 11.9 shows the typical result. After infection, the plaque counts remain constant for some time, since each infected bacterium serves as a center for the formation of a single plaque, not matter how many particles it may contain at the time that it is plated. This period is known as the *latent period*. It ends abruptly as the infected bacteria begin to lyse and liberate new phage particles. At this time, the plaque count rises very rapidly until all infected cells have lysed; this is known as the *burst period*. After lysis is complete, the plaque count remains more or less constant, even if there are uninfected cells left in the population, because the initial dilution of the culture largely prevents adsorption of the newly liberated phage particles on the remaining uninfected bacteria. The average number of new phage particles liberated by each infected cell is known as the *burst size;* in the actual experiment portrayed, it was about 150, but the value can vary greatly depending on the host–phage system studied and the conditions of the experiment.

*Figure 11.9 A typical one-step growth curve for a bacteriophage. The sudden increase in plaque count indicates that the host cells are lysing and liberating free phage.*

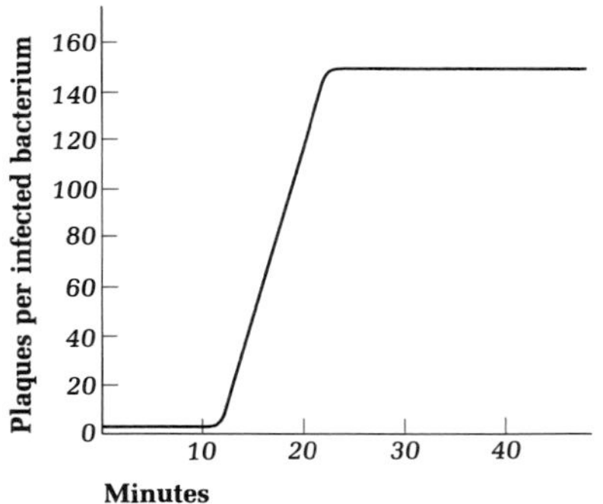

Let us now consider in more detail the events that take place during the lytic cycle of infection. Most of the experiments on which this account is based have been done with a small group of phages that attack *Escherichia coli*. Certain phages, numbered T1 through T7 (T standing for "Type"), were arbitrarily chosen by the group of investigators that began the experimental attack on phage reproduction in 1939. Three of the phages, T2, T4, and T6, proved to be closely related and to have a number of features that make them especially suitable for experimental purposes. For example, the DNA of the "T-even" phages contains a unique base, 5-hydroxymethylcytosine, in place of cytosine; using a specific chemical assay for hydroxymethylcytosine it is thus possible to follow the synthesis of viral DNA in infected cells in the presence of an excess of bacterial DNA.

The T-even phages were for many years assumed to be typical of phages in general. Electron microscopy, however, has revealed that the virions of the T-even phages are much more complex than those of other phages, and that their mechanisms of adsorption to and penetration of the host cell are quite specialized. The events which follow penetration, however, appear to be essentially the same in the T-even phages as in other DNA phages.

**Adsorption and penetration**

The lytic cycle of infection begins when a bacteriophage particle undergoes a chance collision with a host cell. If the virion possesses an adsorption site that is chemically complementary to a specific *receptor site* on the bacterial cell wall, irreversible *adsorption* occurs. For some phages that attack *E. coli* the receptor sites are present on the lipoprotein layer of the wall, whereas for others they are on the lipopolysaccharide layer. The adsorption sites of the virions also differ from one phage to another. The T-even phages are unique in possessing specialized "tail fibers" which behave as adsorption organs (Figure 11.10); in general, for those phages that possess "tails," the tails serve as adsorption organs.

Following adsorption, the bacteriophage particle *injects* its DNA into the bacterial cell. This process has been investigated in the T-even phages, which have been shown to possess contractile tail sheaths. After the tail fibers become adsorbed, the sheath contracts; the tail *core* is thus driven through the cell wall (Figure 11.11). When the tip of the core reaches the cell membrane the DNA contents of the phage "head" are injected into the cell. The syringelike mechanism of the T-even phages is unique; it is not known how the other phages penetrate the cell walls of their hosts, nor how the DNA of any of the phages penetrates the cell membrane.

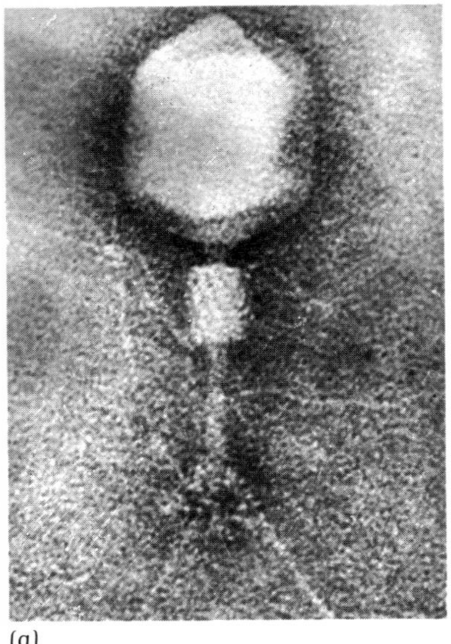

(a)

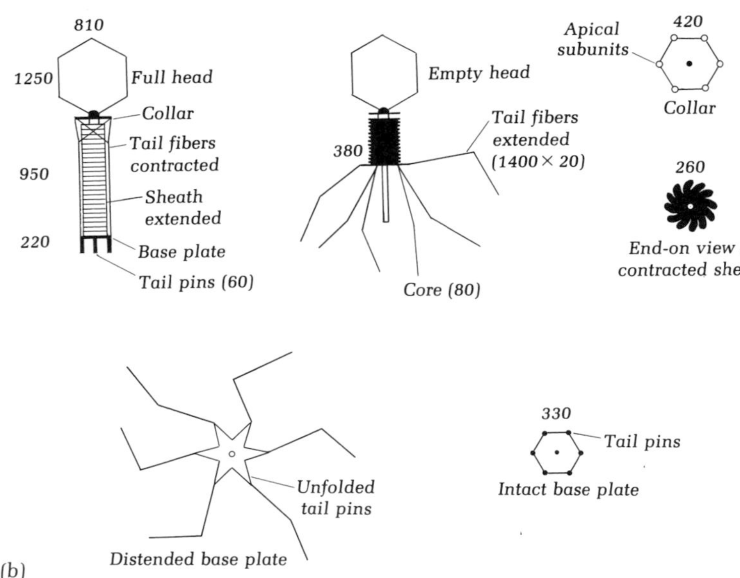


(b)

Figure 11.10 (a) An isolated particle of one of the T-even bacteriophages embedded in phosphotungstic acid. Note the filled head, contracted sheath, core, and tail fibers. From S. Brenner et al., "Structural components of bacteriophage." J. Mol. Biol. **1,** 281 (1959). (b) T-even phage components, with dimensions indicated in angstrom units. From D. E. Bradley, "Ultrastructure of bacteriophages and bacteriocins." Bacteriol. Revs. **31,** 230 (1967).

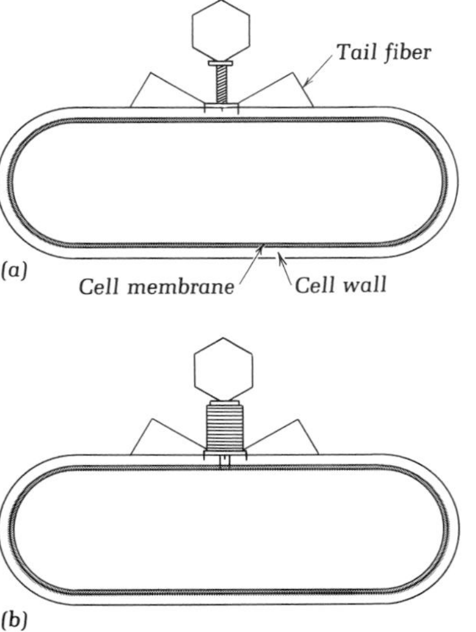


The penetration process discovered in the T-even phages, in which the protein capsid is left outside the cell, has been assumed to be characteristic of bacteriophages in general. For some phages, however, such as those which contain RNA, the penetration mechanism may be quite different. In one case—that of the single-stranded DNA phage fd—the injection method is not used. In this phage, the protein of the helical capsid enters the cell along with the DNA.

Figure 11.11 The syringe-like action of the T-even phages. (a) Phage adsorbed to bacterial cell wall; the sheath is extended. (b) The sheath has contracted, driving the tail core through the cell wall.

**The formation of "early" proteins**

Once the phage DNA has reached the host cell cytoplasm, part of it is immediately transcribed by host cell RNA polymerase to form "early" viral messenger RNA. The preexisting ribosomes of the host then translate this viral mRNA to form a complement of new enzymes, including all those necessary for the *replication* of the phage DNA. In the case of the T-even phages, as many as 11 new enzymes are formed by the translation of viral messenger; these include enzymes that are necessary for the formation of 5-hydroxymethylcytosine and its triphosphate, as well as a new DNA polymerase and a deoxyribonuclease that destroys the host DNA.

It seems probable that all viruses carry the genetic information for at least some of the enzymes necessary for the replication of viral nucleic acid. Many viruses, however, unlike the T-even phages, rely on the host DNA to provide some of the essential enzymes; in these viruses the host DNA is not broken down but rather continues to function in protein synthesis.

**The replication of bacteriophage DNA**

The replication of bacteriophage DNA proceeds according to the general mechanism described in Chapter 7; the details of the mechanism differ according to whether the phage DNA replicates as a cyclic or linear molecule. In those phages in which the virion contains a molecule of single-stranded DNA, the injected DNA is rapidly converted to the double-stranded form by bacterial DNA polymerase, following which replication proceeds in the usual semiconservative manner. Normal replication soon ceases, however, and is replaced by a new enzymatic process in which only one type of complementary strand is made from the double-stranded template. These new single strands are finally incorporated into progeny phage particles.

**Maturation**

Soon after the replication of phage DNA begins, that part not already transcribed is used as template for the synthesis of "late" viral messenger RNA. The translation of the late messenger results in the formation of a second set of viral-specified proteins, among them the subunits of the viral capsid. Simultaneously with the accumulation of capsid subunits, the viral DNA molecules undergo *condensation:* each molecule assumes a tightly packed, polyhedral shape as a result of its combination with a specific viral protein or "condensing principle" (Figure 11.12). The capsid subunits then crystallize on the surface of the condensates to form mature phage heads; in the T-even phages maturation

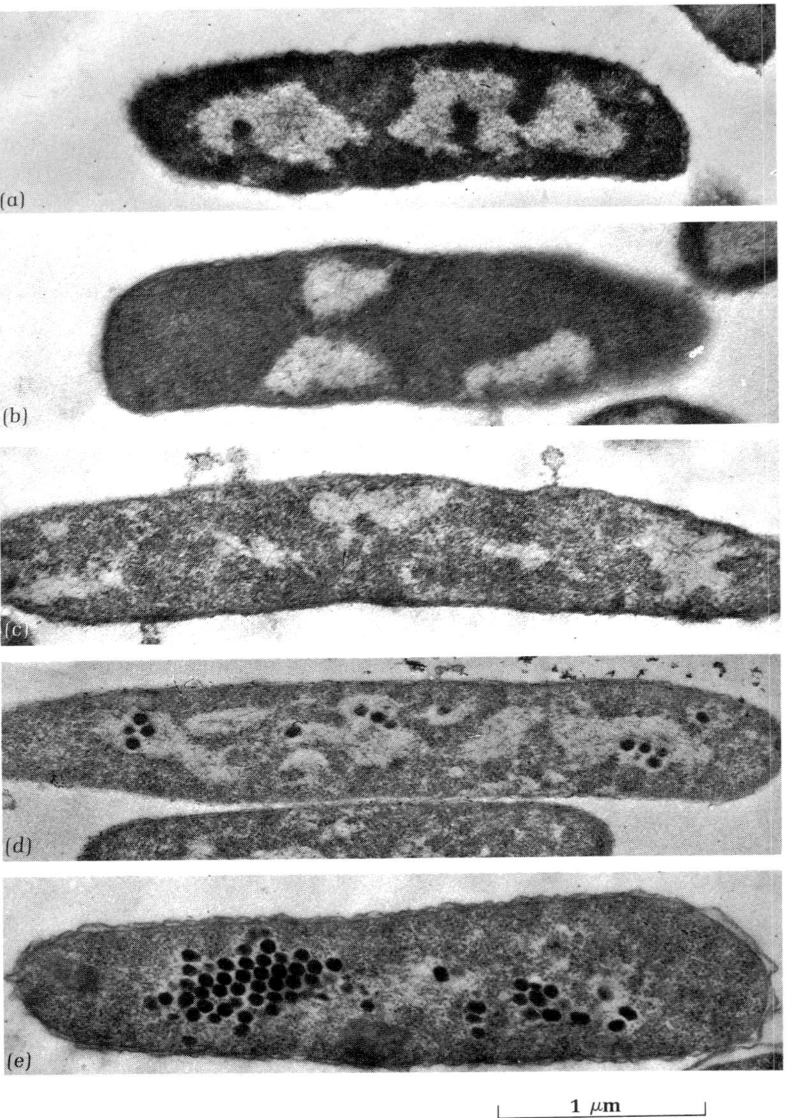


Figure 11.12 The course of phage infection in the bacterium *E. coli*, illustrated by electron micrographs of successive thin sections of the cells (×33,200). (a) Uninfected cell. (b) Four minutes after infection. Note change in structure of the DNA-containing region (light areas of cell). (c) Ten minutes after infection. (d) Twelve minutes after infection. The first new phage bodies, or DNA condensates (dark spots), are developing within the DNA-containing regions of the cell. (e) Thirty minutes after infection (shortly before lysis). Many new phage bodies are evident in the infected cell. Courtesy of E. Kellenberger, E. Boy de la Tour, J. Sechaud, and A. Ryter.

is completed by a process in which other subunits polymerize to form the tail, followed by assembly of tails and heads into complete virions.

The process of *self-assembly* is guided in an unknown manner by the products of certain viral genes. These "morphopoietic" gene products govern, for example, the positioning of special capsomers to form the corners of polyhedral phage heads. In phages containing

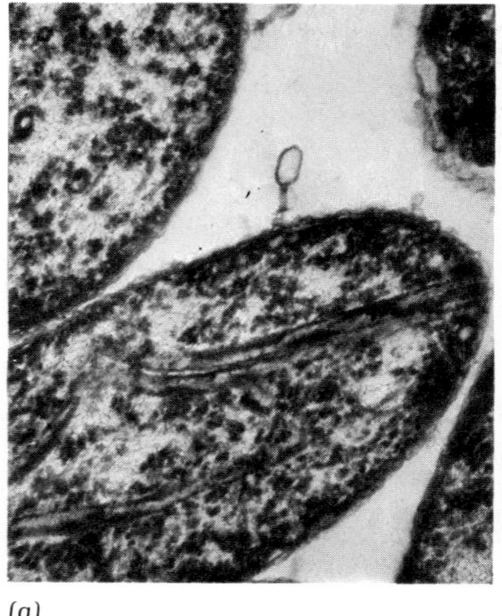

(a)

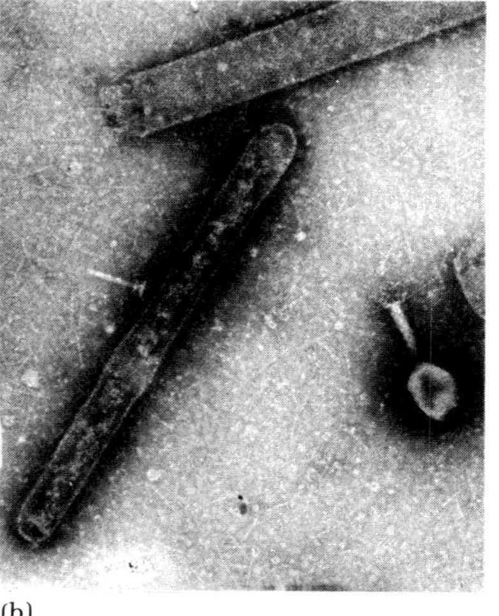

(b)

*Figure 11.13 Polyheads of phage T4. (a) Intracellular polyheads in a partially lyzed cell. Note the adsorbed phage particle. (b) Formaldehyde-fixed polyheads prepared in phosphotunstate. The preparation includes an intact phage particle for comparison. From R. Favre, E. Boy de la Tour, N. Segré, and E. Kellenberger, "Studies on the morphopoiesis of the head of phage T-even. I. Morphological, immunological, and genetic characterization of polyheads." J. Ultrastructure Res.* **13**, *318 (1965).*

mutant morphopoietic genes,* the capsomers polymerize incorrectly to form bizarre structures such as "polyheads" or "polysheaths" (Figure 11.13).

**The liberation of mature virions**

At the end of the latent period of lytic infection, another viral-specified "late" protein appears in the cell: phage lysozyme. This enzyme attacks the murein layer of bacterial cell walls, hydrolyzing the linkages between the sugar residues of the backbone chains. The wall is thus progressively weakened until it is ruptured by the internal osmotic pressure of the cell, and the progeny phage are liberated into the environment along with the other contents of the cell (see Figure 11.7c).

An exception to the general rule that phages are liberated by lysis of the host cell is provided by the helical ("filamentous") phages, such as fd. As this phage matures, the filamentous particles are extruded through the cell wall of the host, which remains viable throughout the liberation process (Figure 11.14).

*Such mutations are lethal and can thus be studied only in "conditional lethal mutants" in which the mutation is expressed in one host but not in another (see Chapter 13).

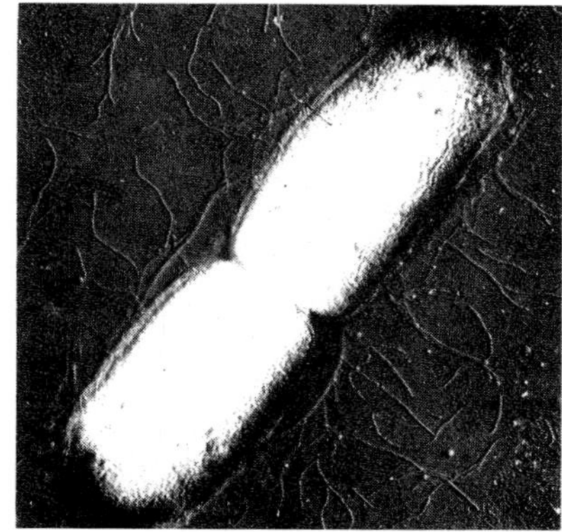

*Figure 11.14 The liberation of filamentous phage by extrusion from living bacterial cells. This micrograph was made 30 minutes after the cells had been infected and washed free of unadsorbed phage particles. From P. H. Hofschneider and A. Preuss, "M13 bacteriophage liberation from intact bacteria as revealed by electron microscopy." J. Mol. Biol.* **7**, *450 (1963).*

## *Lysogeny*

In the lytic cycle of infection, described above for bacteriophages, every infected cell eventually lyses and liberates a crop of progeny virus particles. Many bacteriophages, however, are capable of an alternative interaction with their host: following penetration, the viral genome may reproduce in synchrony with that of the host, which survives and undergoes normal cell division to produce a clone of infected cells. In most of the progeny cells no viral structural proteins are formed, most viral genes having become *repressed.* In an occasional cell, however, derepression occurs spontaneously and the viral genome initiates a lytic cycle of development; the cell in which this occurs lyses and liberates mature virions.

This virus–host relationship is called *lysogeny;* infected cells that possess the latent capacity to produce mature phage particles are said to be *lysogenic.* Bacteriophages capable of entering this relationship are called *temperate,* and the viral genome present in the cells of a lysogenic culture is called *prophage.*

Thus, when a temperate phage particle infects a susceptible host, the entering phage genome has two alternative fates: it may commence rapid vegetative multiplication, culminating in the formation of mature virions and lysis of the host cell, or it may enter the prophage state, giving rise to a clone of lysogenic cells. When a suspension of phage particles is added to a culture of a susceptible host strain, so that every bacterium becomes infected, some of the cells undergo "productive" infection and lyse, while the remainder of the cells are lysogenized. The fraction that encounters each fate is determined by the genetic constitution of the host, the genetic constitution of the virus, and also by the environment; by experimenting with each of these variables, it has been possible to investigate the factors that decide the fate of the individual cell.

From such experiments, the following picture has emerged. In every newly infected cell, three processes take place more or less simultaneously: (1) the viral genome is transcribed and translated to form the "early proteins" of infection, including one or more types of *repressor* molecules which inhibit both viral replication and viral gene expression; (2) the viral genome commences replication; and (3) one or more of the replicated genomes enters the prophage state. The fate of the cell in which these three processes have begun then depends on the outcome of a "race" between repressor production and viral maturation: if mature virions and lysozyme are produced before repressor action can interfere, the cell will be lysed; if repressor molecules accumulate in time to shut off viral replication and gene expression before mature virions and phage lysozyme have been produced, the cell will be lysogenized. In the lysogenized cell, phage repressor molecules continue to be produced by the prophage and prevent both vegetative replication and maturation of the phage. Repressor molecules do not interfere, however, with the replication of prophage in synchrony with the cell. *Prophage replication* and *vegetative replication* are thus different processes; they will be discussed in the following sections.

## Lysogeny of the λ type

Most of the research on the nature of prophage and the lysogenic state has been done with the DNA bacteriophage called *lambda* (λ), which was first discovered as a prophage present in strain K12 of *E. coli*.* When a culture of strain K12 is grown in the laboratory, about one cell in $10^4$ lyses and liberates mature λ particles. As an even more rare event, an occasional cell is formed that fails to receive a prophage; such a cell is said to have been "cured." When a cured (nonlysogenic) cell is isolated and cultivated, the λ-free cells can be reinfected and again lysogenized; if a genetic mutant of λ is used for this purpose, every lysogenic cell inherits the capacity to produce λ particles of the mutant type.

Such observations tell us that λ prophage consists of one or more copies of the λ genome. Where is the λ genome located in the lysogenic cell, and what mechanism ensures its synchronous replication and the near-perfect segregation of its replicas to daughter cells?

The answers to these questions were provided by experiments in which lysogenic bacteria were allowed to *recombine* genetically with nonlysogenic (cured) bacteria or with lysogenic cells carrying prophages derived from different mutants of λ. As we shall discuss in Chapter 14, *E. coli* strain K12 is capable of recombination by either of two processes: conjugation and transduction. In both processes, segments of the bacterial chromosome are transferred from one cell (the genetic donor) to another (the genetic recipient). The recipient thus becomes a partial diploid, or *zygote;* recombination events (or "crossovers") in the zygote produce recombinant chromosomes which are finally segregated to produce recombinant daughter cells.

As we will see in Chapter 14, recombination experiments have permitted the *mapping* of genes on the bacterial chromosome. If the donor and recipient differ in several genetic traits, the recombinants can be

*Strain K12 is one of a collection of *E. coli* strains that had been isolated from clinical specimens at the Stanford University Hospital. It was arbitrarily chosen for use in a pioneering project on bacterial genetics and has since become a "standard" strain for genetical research.

Figure 11.15 *The formation of λ prophage. (a) Adsorption of the virus. (b) Injection of viral DNA. (c) Circularization of the viral genome. (d) Pairing of homologous regions on the viral and bacterial genomes. (e) A crossover event occurs within the region of pairing. (f) The two genomes have been integrated, forming a single circle. Note that the attachment site for λ is at a specific location, between the loci* gal *and* bio *(see Figure 14.12). The specific attachment site on λ is indicated by four short vertical lines on the DNA strand.*

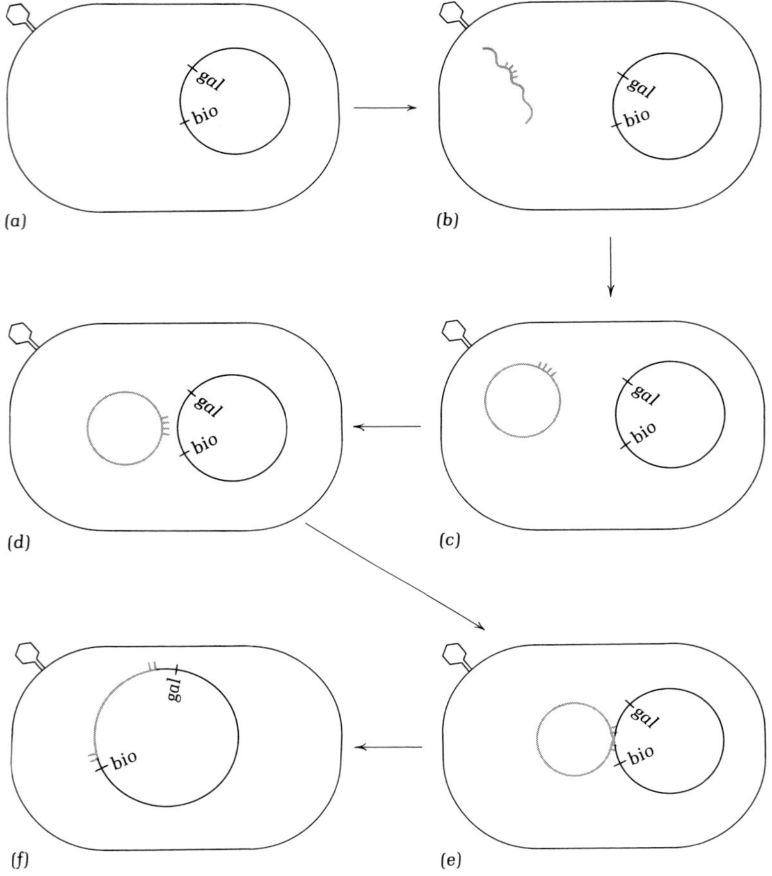


analyzed to determine which genes they have inherited from the donor parent and which from the recipient. The closer together that two genes lie on the donor chromosome, the higher the frequency with which they will be inherited together by the same recombinant cell; such genes are said to be closely *linked.* When nonlysogenic donor cells were recombined with recipient cells harboring λ prophage, *the prophage was found to be inherited as though it occupied a discrete site on the bacterial chromosome,* closely linked to the *gal* genes, which govern the enzymes of galactose utilization. Thus, the λ prophage is attached to the bacterial chromosome and replicates in synchrony with it. At cell division, every daughter cell receives a bacterial chromosome and with it the attached λ prophage.

The nature of the attachment between prophage and chromosome remained a mystery for almost 10 years after its existence was discovered. In 1962, A. Campbell suggested that attachment occurs by the process shown in Figure 11.15. According to the "Campbell model," the λ genome circularizes immediately after its penetration of the host cell. The λ genome possesses a region which specifically pairs with a homologous region on the bacterial chromosome, adjacent to the *gal* loci; *a single crossover event within the region of pairing inserts the λ genome into the continuity of the bacterial chromosome.* The inserted genome constitutes the λ prophage; it is henceforth replicated as a part of the bacterial chromosome, even after phage repressor has completely inhibited the replication of free λ DNA.

The Campbell model has been confirmed by a variety of experiments. The λ genome which is present in the mature virion has been shown to possess *cohesive ends,* which consist of single-stranded regions with complementary base sequences (Figure 11.16). Hydrogen bonding in these regions leads to circularization of the λ DNA; the circles are then *covalently* joined by the action of polynucleotide ligase.

The crossover event that integrates λ with the bacterial chromosome is brought about by a set of recombination enzymes which cut the DNA molecules and rejoin the broken ends in new combinations (see Chapter 14). The same enzymes, operating on the integrated structure, can bring about the *detachment* of the prophage, and such detachment has been shown to be the first step in the process that leads to the production of mature virus by an occasional lysogenic cell.

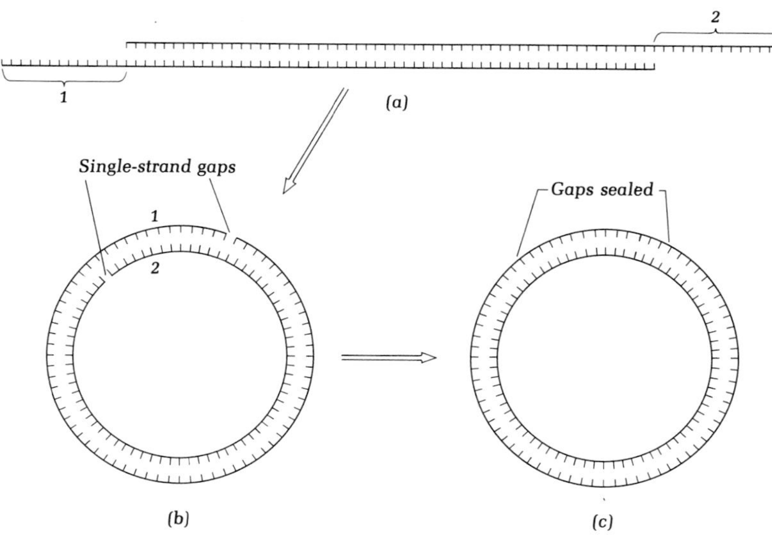


*Figure 11.16 The conversion of the λ genome from the linear form to the circular form. (a) The linear genome has single-stranded regions at each end that are complementary to each other. (b) The single-stranded regions are joined by hydrogen bonds between complementary bases. (c) The sugar-phosphate backbones are joined by polynucleotide ligase, forming a fully covalent circle.*

This finding has posed an apparent paradox: since *E. coli* strain K12 contains a full set of recombination enzymes, what prevents the detachment of λ in the great majority of the cells of a lysogenic culture? The paradox was solved by the discovery that the integration of λ with the bacterial chromosome requires the action of *specific* recombination enzymes, governed by λ genes: the integration of λ occurs before λ repressor has had time to appear; once it has appeared, however, it represses all λ genes, including those for the specific recombination enzymes. The detachment of λ is thus prevented, except in the rare cell in which the repressor is inactivated. The inactivation of repressor, leading to the detachment of λ and its entry into the lytic cycle, will be discussed in a later section.

The events described above have been found to occur in the formation of prophage by a number of other phages closely related to λ [e.g., phage φ80, which undergoes prophage insertion in the chromosome next to the *trp* loci (the genes governing the enzymes of tryptophan biosynthesis)]. The attachment sites of a number of prophages are shown on the genetic map of *E. coli* strain K12, which is reproduced in Figure 14.12. Until recently it was believed that all temperate phages produced lysogenic states of the λ type, involving attachment of the phage genome to the host chromosome. Now, however, it is clear that a completely different lysogenic state occurs with phages of the P1 type.

## Lysogeny of the P1 type

A number of bacteriophages, of which phage P1 is the best studied, differ from the λ group of phages in two important respects. First, recombination experiments between lysogenic and nonlysogenic cells fail to reveal a discrete chromosomal location for P1 or the other phages of this type. The transfer of P1 prophage is occasionally observed, but no linkage to known bacterial genes is evident. Second, P1 and λ differ in the manner in which they form *transducing particles*, virions that contain fragments of host DNA. In Chapter 14 it will be seen that transducing particles of the λ type contain a segment of λ genome covalently linked to a segment of the host chromosome derived from the region adjacent to the prophage attachment site. As shown in Figure 14.20, a small error in the normal process of prophage detachment creates a transducing particle. Transducing particles of the P1 type, on the other hand, contain only host (bacterial) DNA: if any phage DNA is present, the amount is too small to be detected. Furthermore, phages of the P1 type are capable of transducing *any* segment of the bacterial chromosome, in contrast with λ, which can transduce only those genes which are immediately adjacent to its attachment site.

Thus, both genetic mapping experiments and the analysis of transducing particles have failed to reveal a chromosomal location for prophages of the P1 type. Recently, P1 DNA has been extracted from lysogenic cells and shown to consist of circular DNA structures of the same size as the DNA from P1 virions; it shows no association with bacterial DNA.

As an alternative to chromosomal attachment, it is very possible that prophages of the P1 type occupy attachment sites in the *cell membrane*, since (as will also be discussed in Chapter 14) such attachment sites account for the orderly replication and segregation of the extrachromosomal genetic elements known as *plasmids* and *episomes*. For example, the cell membrane of *E. coli* strain K12 possesses attachment sites not only for the bacterial chromosome but also for F, an episome that behaves as a "sex factor" in *E. coli*. As discussed in Chapter 14, both chromosome and F are replicated and segregated by a mechanism that involves their attachment to the cell membrane.

In summary, a phage of the P1 type can replicate in harmony with the cell as a prophage, presumably attached to the cell membrane, or can undergo uncontrolled vegetative replication as part of a lytic cycle. The latter process is prevented in the majority of lysogenic cells by the action of phage repressor, just as in the λ type of lysogeny.

## Repression and induction

The mechanism of phage repression, which blocks the lytic cycle and thus permits the perpetuation of the lysogenic state, has been most extensively studied in the *E. coli* strain K12 (λ) system. The λ genome has been shown to possess a gene (the $C_I$ gene) which produces the repressor protein; the λ genome also possesses a receptor site, to which the repressor can specifically bind. When the receptor site is unoccupied, λ DNA is replicated and its genes are expressed through the formation of viral proteins. When a genome binds a repressor molecule, both replication and gene expression are inhibited. An exception is the $C_I$ gene itself, which continues to produce repressor protein after all other λ functions have been shut off.

*Virulent mutants* of temperate phage often occur. Such mutants are unable to lysogenize their hosts, although they retain the ability to produce the lytic cycle of infection. They are easily recognized by the fact that they produce *clear plaques* on a lawn of sensitive bacteria; in contrast, the parental temperate phage particles produce *cloudy plaques* as a result of the fact that many cells in the region of the plaque become lysogenized and produce minute colonies (Figure 11.17). When the virulent mutants are isolated, they prove to be of two different types: those in which the $C_I$ gene has mutated so that repressor

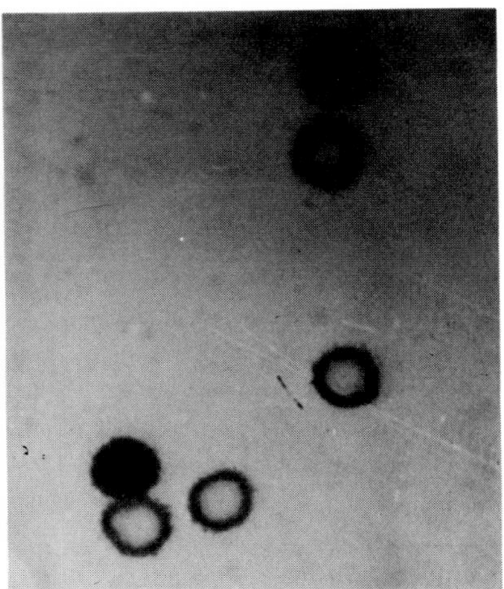

*Figure 11.17 Plaques formed by the wild type and by a virulent mutant of the bacteriophage, lambda. The wild-type particles form cloudy plaques as a result of the growth of lysogenized cells; the mutant particles, which are unable to lysogenize the bacterial host, form clear plaques with sharp edges. Courtesy of C. Radding.*

is no longer produced, and those in which the receptor is altered such that the repressor can no longer act.

Certain phages, including λ, are *inducible:* their prophages are caused to enter the lytic cycle of development by treatment of the host cell with any of a number of agents that affect DNA replication—ultraviolet irradiation, temporary thymine starvation, treatment with mitomycin C, or treatment with alkylating agents. Such treatments lead to the *inactivation of phage repressor;* the lytic cycle of development is then initiated, and every cell in the treated culture undergoes lysis. The inducing agents seem to act indirectly, by blocking the synthesis of host DNA. Cells that are so blocked appear to accumulate a metabolite that serves as an endogenous inducer, perhaps combining with repressor and inactivating it. The rare cells in a lysogenic culture which spontaneously lyse and liberate progeny phage are apparently those cells in which the repressor mechanism fails; this may reflect the production of endogenous inducer or the failure of repressor to be produced.

The massive induction of a λ-lysogenic culture can be brought about in two other ways. One way is to permit the cells to transfer their chromosomes (including the inserted λ prophage) to nonlysogenic recipients, by the process of conjugation (see Chapter 14). Since the cytoplasm of the recipient is free of repressor, the transferred prophages are immediately induced. This process is called *zygotic induction.* The other mode of induction occurs with a strain of λ in which the $C_I$ gene has mutated so that the repressor is thermolabile. The mutant repressor functions normally at 37°C but is denatured at 43°C, unlike the wild-type repressor, which is stable at both temperatures. Thus, a lysogenic culture containing the mutant λ can be grown at 37°C and then massively induced by shifting the culture to 43°C.

## Immunity

If the cells of a lysogenic culture are allowed to adsorb particles of a different phage, they are said to be *superinfected.* The DNA of the second phage will penetrate the lysogenic cells, but its ability to reproduce will depend upon its sensitivity to the repressor produced by the resident prophage. If the superinfecting phage is resistant to the repressor, it will multiply and lyse the cells. If it is sensitive, the superinfection will be *abortive;* the DNA of the superinfecting phage will be neither replicated nor expressed and will be diluted out during subsequent growth of the cells. In that case the lysogenic cells are said to be *immune* to the second phage.

Two phages that share sensitivity to the same repressor are termed *coimmune.* For example, if a cell harbors one strain of λ as prophage, it will be immune to superinfection with a second strain of λ. The exception, of course, will be the virulent mutant that owes its virulence to its resistance to its own repressor; such a virulent mutant will successfully superinfect any λ-lysogenic cell.

It is easy to demonstrate that immunity to a second phage is due to the same repressor that is responsible for the maintenance of lysogeny. For example, irradiating a culture that is lysogenic for an inducible phage will simultaneously induce the prophage and break the immunity barrier. If such a culture is superinfected immediately after irradiation, every cell will lyse and liberate a mixture of phage progeny arising from both the prophage and the superinfecting phage.

Immunity must be distinguished from the genetically controlled property of *phage resistance.* Phage resistance results from a mutation in the bacterial chromosome, resulting in the loss of a phage receptor on the cell surface. In contrast to an immune cell, a genetically resistant cell does not adsorb the phage.

**Defective prophages**

From time to time, prophages undergo gene mutations or deletions (the loss of a segment of DNA spanning several genes) which render them incapable of normal development. For example, a mutation may cause the loss of ability to replicate in the vegetative state or to synthesize an essential viral protein. Such prophages are termed *defective;* their presence in the cell is revealed by the fact that the *cell retains its immunity to superinfection* by coimmune phages. Furthermore, if the defective prophage arose from an inducible phage, treatment of the cells with inducing agents may still derepress the functional genes and bring about lysis of the cell by phage lysozyme.

The existence of defective prophages in bacteria isolated from nature may thus be tested for by treatment of the cells with inducing agents. Lysis of the cells suggests the presence of an inducible, defective prophage; confirmation may be obtained by electron microscopic examination of the lysate, in which incomplete capsids may be observed. In some cases it has been possible to demonstrate the presence of a defective prophage by superinfection of the cells with a related phage and recovery in the resulting lysate of *recombinant* phages that have inherited some genes from the defective prophage and others from the superinfecting phage. Alternatively, the lysate may contain a *mixture* of normal and defective phages, the latter having been enabled to reproduce by the superinfecting phage that provided the missing function.

## The RNA bacteriophages

In a number of bacteriophages, the genetic material in the virion consists of a molecule of single-stranded RNA.

When a molecule of viral RNA reaches the cytoplasm of the host cell, it is immediately recognized as messenger RNA by the ribosomes, which bind to it and initiate its translation into viral proteins. One such viral protein is a complex enzyme, *RNA synthetase*. This enzyme brings about the replication of the viral RNA; it polymerizes the ribonucleoside triphosphates of adenine, guanine, cytosine, and uracil, using viral RNA as a template.

The first step in the process of RNA replication is the formation of a *double-stranded intermediate*, in which the entering viral RNA strand (called the "plus" strand) is used by the synthetase as a template to form the complementary, or "minus," strand (Figure 11.18). The synthetase now uses the double-stranded molecule as a template for the repeated synthesis of new plus strands, each new plus strand displacing the previous one from the replication intermediate (Figure 11.18). The original plus strand, however, is not released; it appears to be permanently bound to the minus strand through their attachment to a cell structure. This structure may be the host cell membrane, since labeling studies suggest that the viral RNA is attached to the membrane during replication.

As the newly synthesized plus strands are released from the replica-

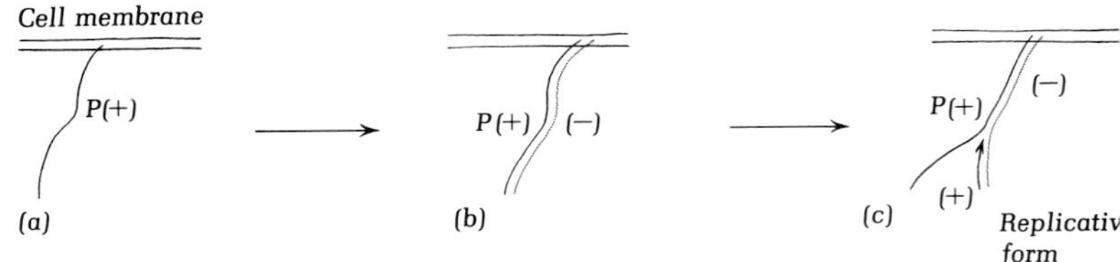


*Figure 11.18 The replication of an RNA virus. (a) The infecting parental strand is labeled P(+). (b) An RNA synthetase uses the P(+) strand as a template to form a complementary (−) strand. (c to f) A different RNA synthetase then uses the (−) strand as a template, forming new (+) progeny strands repeatedly. Modified from B. Davis et al., Microbiology. New York: Harper and Row, 1967.*

tive intermediate, they are either used by the synthetase to form a new double-stranded intermediate or are assembled into mature virions by the attachment of capsomers.

## *The animal viruses*

**Some important groups of animal viruses**

As the study of animal viruses has progressed, a number of different categories, each of which groups together viruses possessing many characteristics in common, have been recognized. In some cases, the category is defined strictly on the basis of virion structure: the picornaviruses, for example, include all small RNA viruses ("picorna" is a composite of pico-, "small," and RNA). In other cases, the category is defined ecologically: the arboviruses, for example, include all arthropod-borne viruses. Some important groups of animal viruses are described in Table 11.2.

**The detection and enumeration of animal virus particles**

The virions of many animal viruses, including both enveloped and naked types, bind strongly to specific receptor sites on the surface of red blood cells. Since a virion has several adsorption sites, it can bind to two red blood cells simultaneously and form bridges between them. Thus, if a sufficient number of virions are mixed with a suspension of red blood cells, the latter are agglutinated.

*Hemagglutination,* as this phenomenon is called, provides a rapid assay for those viruses which adsorb to red blood cells. In practice, serial dilutions of the virus preparation are tested for hemagglutinating activity, and the titer of the preparation is defined as the reciprocal of the highest dilution that has detectable activity. For example, if a preparation of virus shows hemagglutination at a dilution of 1 : 320 but not at 1 : 640, its titer is 320.

The hemmagglutination assay described above is an example of the general method called *end-point titration,* in which one measures the highest dilution of a preparation to exhibit a given viral activity. Other activities that can be used for the end-point tritration of viruses include killing of host animals and pathological effects on tissue cultures.

Several methods for the enumeration of animal viruses are completely analogous to the plaque method described earlier for the enumeration of bacteriophages. In each of these methods, virus particles are placed on the surface of a layer of host cells; each particle initiates the infection of a single cell, which becomes a focus of infection for neighboring cells.

*Table 11.2 Principal groups of animal viruses capable of causing infections in mammals*

| Group | Physicochemical properties | Other characteristics | Examples |
|---|---|---|---|
| *Adenoviruses* | *Naked, icosahedral capsid containing double-stranded DNA* | *Ubiquitous agents of latent infections of lymphoid tissue; many produce tumors in experimental animals* | *Members of this group are designated as numbered "types"* |
| *Other DNA-containing tumor viruses* | *Naked, icosahedral capsid containing double-stranded DNA* | *Production of tumors in animals* | *Polyoma, papilloma, SV-40* |
| *Herpesviruses* | *Enveloped, icosahedral capsid containing double-stranded DNA* | *Tendency to cause latent infections* | *Herpes simplex, varicella (herpes zoster), pseudorabies* |
| *Poxviruses* | *Enveloped, helical (?) capsid containing double-stranded DNA* | *Very large; brick-shaped to ovoid; predilection for epidermal cells; two members (myxoma and fibroma viruses) cause tumors* | *Smallpox, vaccinia, cowpox* |
| *Picornaviruses* | *Naked, icosahedral capsid containing single-stranded RNA* | *Small size; cause enteric and/or respiratory infections* | *Poliovirus, Coxsackievirus, echovirus, rhinovirus* |

In the *pock method,* the virus particles are placed on the surface of the chorioallantoic membrane of a chick embryo; the suspension of particles is introduced through a small hole drilled in the egg shell, which is then carefully sealed. The eggs are incubated until each virus has produced a visible lesion, or pock, on the membrane. The shell is then cut away and the pocks are counted.

In the *plaque method,* the virus particles are placed on the surface of a sheet of cells growing on the inner surface of a flat glass bottle. After allowing the virus particles to adsorb to the cells, the fluid medium is removed and the cells are covered with an overlay of soft agar. Virus particles released from the initially infected cells can spread to neighboring cells but are prevented from wider diffusion by the agar. After a period of incubation, each primary infected cell gives rise to a *plaque,* or zone of infected cells. Methods for visualizing the plaques vary for different viruses. Some viruses produce a *cytopathic effect:*

Table 11.2 (Continued)

| Group | Physicochemical properties | Other characteristics | Examples |
|---|---|---|---|
| *Reoviruses* | *Naked, icosahedral capsid containing double-stranded RNA* | *Large size; ubiquitous infectious agents but overt disease not produced* | *Members of this group are designated as numbered "types"* |
| *Arboviruses* | *Enveloped, probably icosahedral capsid containing single-stranded RNA* | *Arthropod-borne (mosquitos, ticks)* | *Yellow fever, equine encephalitis, dengue, Colorado tick fever, sindbis* |
| *Myxoviruses* | *Enveloped, helical capsid containing single-stranded RNA* | *Multiply in the nucleus; filamentous virions commonly formed* | *Influenza* |
| *Paramyxoviruses* | *Enveloped, helical capsid containing single-stranded RNA* | *Multiply in the cytoplasm; possess a hemolysin* | *Parainfluenza, mumps, measles, Newcastle disease, canine distemper* |
| *RNA-containing tumor viruses* | *Enveloped capsid of uncertain architecture, containing single-stranded RNA* | *Production of tumors in animals* | *Murine leukemia viruses, mouse mammary tumor virus, avian leucosis viruses, Rous sarcoma virus* |

they kill their host cells. For these viruses, the plaques are revealed by flooding the cell layer with dyes that differentially stain live and dead cells (Figure 11.19). Other viruses do not kill their host cells, and their plaques are detected by flooding the cell layer with reagents that are specific for viral substances. For example, cells infected with some hemagglutinating viruses will adsorb red blood cells, which will thus adhere specifically to plaques. Fluorescent antibodies (see Chapter 28) can also be used to reveal the cells that contain viral antigens.

Any of the above methods can be used to assay the fractions produced at each step during the purification of a virus. When the preparation is sufficiently pure, the number of viral particles it contains can be directly counted in the electron microscope. When this is done, it is usually found that the number of *infectious units* is considerably lower than the number of morphologically recognizable particles. The ratio

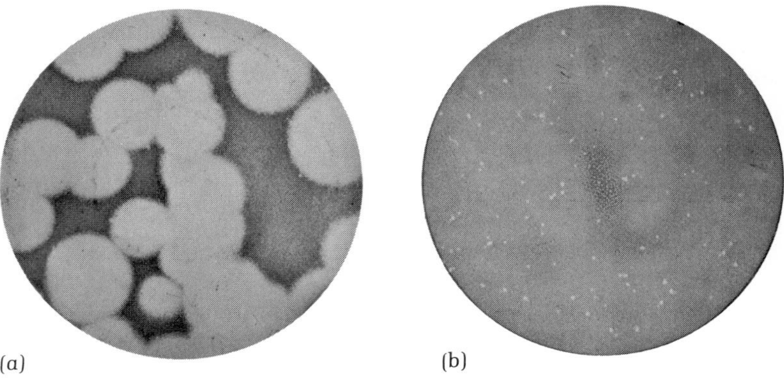


*Figure 11.19 Plaques formed by encephalomyocarditis virus on a layer of animal cells. Two genetically distinct plaque types are shown; the large plaques (a) are 10 to 12 mm diameter, and the small plaques (b) are 0.5 mm diameter. Courtesy of H. Liebhaber.*

of infectious units to visible particles varies from $10^{-1}$ for some viruses to as low as $10^{-4}$ or $10^{-5}$ for others.

## *The reproduction of animal viruses*

**Adsorption and penetration**

The adsorption of an animal virus to a host cell begins with the formation of noncovalent bonds between sites on the virion surface and *receptor sites* on the cell surface.

The adsorption sites on the virion vary in nature from one virus to another. In adenoviruses, for example, each pentamer at the corners of the icosahedral virion bears a small fiber that acts as an adsorption organ. In many of the enveloped viruses, the adsorption organs are the numerous spikes which stud the surface of the envelope (Figure 11.20).

The receptor sites on the host cells are also varied. For the myxoviruses, for example, they consist of mucoproteins, while for poliovirus they consist of lipoproteins. Little is known, however, about the molecular reactions involved in the adsorption process.

Following adsorption, a series of events occurs which includes the *penetration* of the cell membrane and the *uncoating* of the virion; the result is the liberation within the host cytoplasm of free viral nucleic acid. The exact sequence of events is still obscure, and it seems to vary from one virus to another. In one case (poliovirus), uncoating appears to begin while the virion is still attached to the membrane; in other cases, such as the poxviruses, uncoating is partially accomplished by the action of lysosomal enzymes on virions contained within phagocytic vacuoles; in still other cases, uncoating may take place entirely within the cytoplasm.

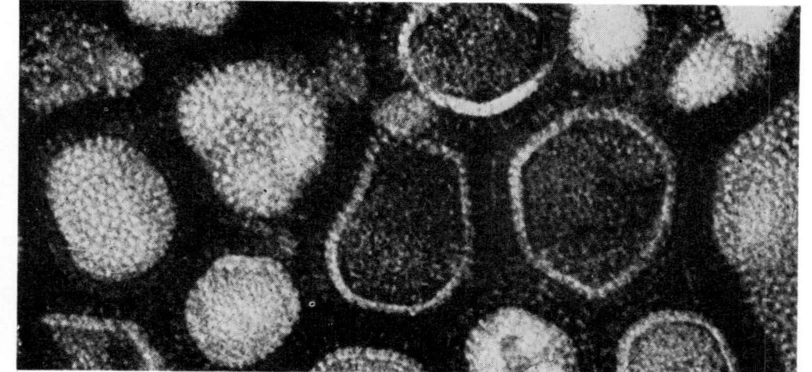

*Figure 11.20 Particles of influenza virus, showing surface projections or "spikes." From R. W. Horne et al., "The structure and composition of the myxoviruses. I. Electron microscope studies of the structure of myxovirus particles by negative staining techniques." Virology* **11,** *79 (1960).*

**Replication and maturation**

The synthesis of certain viral proteins must precede the replication of viral nucleic acid. In general, the enzymes and ribosomes of the host are utilized for the transcription of viral DNA into viral RNA as well as for the synthesis of proteins from viral RNA. In a number of the poxviruses, however, the mature virion actually contains a small amount of DNA-dependent RNA polymerase; this enzyme guarantees that the viral DNA will be transcribed, regardless of the availability of host polymerase.

The mechanisms of viral DNA and RNA replication are those already described for the bacteriophages. DNA replication is semiconservative; RNA replication proceeds through the repeated synthesis of single strands of the type found in mature virions, using a double-stranded replicative intermediate. With the exception of the poxviruses, DNA viruses replicate in the host nucleus; those RNA viruses which have been thoroughly investigated replicate in the host cytoplasm. Maturation occurs, as in bacteriophage, by the self-assembly of capsid proteins and nucleic acid molecules to form nucleocapsids.

A number of animal viruses have been shown to undergo *genetic recombination:* mixed infection with two genetically marked viruses can yield progeny that inherit markers from both parents. Mutant characters employed as markers in these studies include altered plaque morphologies, resistance to inhibitory agents, and altered host range. Recombination has been shown to occur in several groups of DNA viruses; what is more remarkable is that recombination has also been reported in two RNA viruses, influenza (a myxovirus) and polio (a picornavirus). In the case of influenza virus, recombination presumably reflects the random assortment of parental RNA molecules at the time that they are assembled into mature virions. In the case of poliovirus, a mechanism involving the pairing of RNA molecules, followed by breakage and reunion, has been postulated by analogy with the mechanism of DNA recombination, but direct evidence is still lacking.

**The release of virions from the cell**

There are two general modes of virion release among the animal viruses. In some, the nucleocapsid is extruded from the cell in such a manner that the virus particle acquires an outer envelope derived from the cell membrane (Figure 11.5). Prior to the extrusion, a number of viral proteins are incorporated into the cell membrane, so that the envelope contains both viral and host material. All the lipids of the envelope, however, are derived from the host membrane.

Other viruses are released through ruptures in the membranes; the free virions are naked nucleocapsids. This type of release usually follows death of the host cell and is characterized by the sudden appearance of a large number of free virions. In contrast, the release of virions by "budding off" of the membrane does not immediately kill the host cell, and the release process continues for many hours. In some cases, the host cells continue to multiply during the process and survive indefinitely. This situation leads to a *steady-state infection:* a clone of cells arises, all of which are continually liberating virus particles. Reinfection of the cells by extracellular virus is not necessary to maintain the steady state, since each daughter cell acquires intracytoplasmic virus particles at cell division.

## The tumor viruses

**The discovery of viruses as etiological agents of tumors**

The characteristic tissues of animals are formed by the *regulated, limited growth* of their component cells. As a rare event, a cell may escape the normal regulatory processes and divide without restraint, forming an abnormal mass of tissue. Such masses are called *tumors* or *neoplasms;* the development of a tumor is called *neoplasia.*

Some tumors, such as most papillomas (warts) are *benign:* they remain localized and the animal is unharmed. Other tumors are *malignant:* their growth is invasive, so that the organ in which it occurs is damaged and the animal dies. Often, unregulated cells are released from the tumor and establish new neoplastic foci in other parts of the body. The development of malignant tumors is called *cancer.*

Tumors are usually named by appending the suffix *-oma* to the name of the tissue in which the tumor has arisen. Thus, a lymphoma is a cancer of the lymphoid tissue; a sarcoma is a cancer of fleshy, nonepithelial tissue; an adenocarcinoma is a cancer of glandular tissue;

and so on. An important exception to this system of nomenclature is *leukemia*, or cancer of the white blood cells.

The first evidence of a relationship between viruses and cancer was obtained as early as 1908, when V. Ellerman and O. Bang demonstrated that certain chicken leukemias could be transmitted to healthy birds by cell-free filtrates of diseased blood. Some years later, P. Rous was able similarly to transmit a chicken sarcoma. These discoveries received little attention at the time. Only in 1932, when R. E. Shope demonstrated a viral origin of rabbit papillomas, was importance attached to viruses as tumor agents.

An important step was taken when a *natural route of transmission* was found for a virus-induced cancer. In 1936, J. J. Bittner showed that mammary tumors of mice are caused by a virus that is transmitted from mother to offspring through the milk. Bittner's work led to an understanding of several important aspects of virus-induced cancer. First, an animal that is infected with a tumor virus at the time of birth may not develop a tumor until it has become an adult. Second, the ability of a virus to induce a tumor depends on certain environmental factors, such as the physiology of the host; mammary tumors of mice, for example, occur at high frequency only in animals undergoing hormonal stimulation characteristic of pregnancy. Even male mice will develop mammary tumors if injected with the hormone estradiol over a long period of time, provided that they are infected with the virus.

**Types of tumor viruses**

Both RNA and DNA viruses are found among those capable of inducing tumors in their hosts. The RNA viruses include the agents of mouse leukemias, the Rous sarcoma virus, and the mouse mammary tumor viruses. The DNA viruses include the Shope papilloma virus and the polyoma virus, which causes a variety of tumor types in mice.

The most significant discovery with respect to a possible role of viruses in human cancer has been that certain viruses which are widespread in monkeys and in man can induce tumors when injected into hamsters, mice, or rats. One of these viruses is called SV-40 (simian virus 40); it was discovered as a common contaminant of rhesus monkey kidney-cell cultures, which were used in the large-scale commercial production of polio vaccine. SV-40 was detected by its ability to form cytopathic plaques in tissue cultures derived from a different monkey species.

Among the human viruses that have been found to induce neoplasias in experimental animals are some of the *adenoviruses*. These viruses were discovered accidentally as agents of cytopathic changes in tissue cultures derived from human adenoids; at the same time, they were recognized as the cause of massive respiratory disease epidemics in military populations. Related viruses have since been isolated from a variety of animals, where they cause latent infections; in man, they are of widespread occurrence in tonsils, adenoids, and lymphoid tissue. The ability of these ubiquitous viruses to induce cancer in experimental animals strengthens the view that human cancers may be caused by latent viruses acquired early in the life of the individual.

**The transformation of normal cells into tumor cells**

The induction of a tumor by a virus in vivo can be reproduced in tissue cultures. When a suspension of tumor virus particles is used to infect a susceptible tissue culture, some of the infected cells are transformed into a type that exhibits unregulated growth.

The process of *transformation,* as it is called,* causes a number of striking changes in the cell. Most animal cells in tissue culture exhibit the phenomenon of *contact inhibition:* the cells move about randomly by ameboid motion and divide repeatedly until they come into contact with one another; contact between cells inhibits both movement and cell division. The result is that normal cells in tissue culture form a monolayer on the surface of the glass vessel; transformed cells, on the other hand, do not exhibit contact inhibition, and form tumorlike cell masses in tissue culture. Furthermore, cells transformed in tissue culture will initiate tumors when inoculated into host animals.

The efficiency of transformation varies greatly among tumor viruses. The Rous sarcoma virus, for example, transforms virtually every cell that it infects. Other tumor viruses may transform only one out of $10^5$ infected cells.

**The production of virus by transformed cells**

Following transformation by DNA viruses, cells of a virus-induced tumor or transformed cells of a tissue culture are completely free of detectable, infectious virus. Nevertheless, it can be demonstrated that *part or all of the viral genome persists in the transformed cell and is reproduced at each cell generation.* For example, when transformed cells are grown in mixture with susceptible, normal cells, under conditions that promote cell fusion, the normal cells may become infected and liberate infectious virus particles. In such cases, the fusion of a transformed cell with a normal cell allows the full expression of the latent viral genome.

It is known that Rous sarcoma virus will transform both chicken

*The term "transformation" is also used in biology to describe a totally different process: the formation of bacterial recombinants by the transfer of DNA extracted from donor cells (Chapter 14).

cells and mammalian cells in tissue culture. Transformed chick cells continue to produce active virus, but transformed mammalian cells do not. If transformed mammalian cells are injected into chicks, however, tumors of *chick cells* are formed and these produce active virus particles. Mammalian cells transformed by polyoma virus will similarly reveal the presence of the latent virus genome if they are allowed to fuse with susceptible uninfected mammalian cells.

In tumors induced by other types of viruses, the presence of at least part of the viral genome can be demonstrated in two ways. First, transformed cells can be shown to contain *viral antigens*—proteins that are coded for by viral genes. The nature of these antigens is not known; they are not capsid proteins, however. Second, some transformed cells can be shown to contain *viral RNA:* their extracts contain RNA that is hybridizable with viral DNA.

**The relationship of viral transformation to lysogeny**

The analogy of transformation with the lysogenization of bacteria by bacteriophages is striking, particularly in the case of the DNA viruses. In both processes, the persistence of the viral genome is accompanied by the following important changes in virus and host.

(1) Many of the viral genes are repressed, including the genes governing autonomous viral reproduction, the genes for capsid production, and the genes whose products inhibit host syntheses or damage the host cell.

(2) The cell exhibits immunity to superinfection with a similar virus.

(3) The physiology of the host cell is altered, as a result of the continued expression of some viral genes. In particular, both lysogenized bacteria and transformed animal cells often exhibit altered *surface properties;* new surface antigens appear, and transformed animal cells become resistant to contact inhibition. The latter effect may be a primary cause of the ability of a transformed cell to escape normal regulation and thus to initiate neoplasia.

**The role of viruses in human cancer**

There is little epidemiological evidence to suggest that human cancers are contagious, and no one has yet succeeded in isolating a virus from human cancer tissue and demonstrating its ability to induce neoplasia. Nevertheless, it remains a distinct possibility that some human cancers are of viral origin.

The failure to demonstrate contagion does not rule out viral agents, since our experience with tumor viruses in animals shows that a very long time may elapse between infection and the appearance of a tumor.

The failure to isolate a tumor virus from human cancer may simply reflect our inability to provide the conditions necessary for the derepression of a latent viral genome. The fact that most individuals show immunological evidence of having been infected at one time or another with adenoviruses that are capable of producing tumors in animals suggests that these or other viruses, acquired early in life, may be responsible for malignant neoplasias.

## Further reading


**Books**

Cairns, J., G. Stent, and J. Watson (editors), *Phage and the Origins of Molecular Biology*. Cold Spring Harbor, N.Y.: Cold Spring Harbor Laboratory, 1966.

Davis, B., R. Dulbecco, H. Eisen, H. Ginsberg, and W. Wood, *Microbiology*, Chaps. 41–57. New York: Harper and Row, 1967.

Fenner, F., *The Biology of Animal Viruses*. New York: Academic Press, 1967.

Luria, S., and J. Darnell, *General Virology*, 2nd ed. New York: Wiley, 1967.

Stent, A. (editor), *Papers on Bacterial Viruses*, 2nd ed. Boston: Little, Brown, 1965.

Stent, G., *The Molecular Biology of Bacterial Viruses*. San Francisco: Freeman, 1963.

**Reviews and original articles**

Horne, R. W., and P. Wildy, "Symmetry in Virus Structure." *Virology* **15**, 348 (1961).

Lwoff, A., R. Horne, and P. Tournier, "A System of Viruses." *Cold Spring Harbor Symp. Quant. Biol.* **27**, 51 (1962).

# Chapter twelve Mutation and gene function at the molecular level

*The study of genetics* was brought to the molecular level in one swift stroke when, in 1953, J. D. Watson and F. H. C. Crick published a proposal for the structure of DNA. The great significance of their model is that it provides an explanation in terms of molecular structure for both DNA replication and gene mutation. Since these processes underlie heredity and variation, respectively, it can safely be said that in the entire history of biological science only Darwin's recognition of the existence and fundamental mechanism of evolution has equaled their discovery in importance.

The Watson–Crick structure for DNA and its implications are now familiar to all students of biology; a brief summary of this topic was presented in Chapter 7. Given this background, it is now possible to define the gene, as well as mutation, in precise molecular terms. *A gene is a segment of DNA* in which the sequence of bases determines the sequence of bases in an RNA molecule, and thus—by the process of translation—the sequence of amino acids in a polypeptide chain. Mutation can be defined as any permanent alteration in the sequence of bases of DNA,* even if this alteration does not have a detectable phenotypic effect.

Although polypeptides vary considerably in chain length, the majority of those which have been accurately measured have a molecular weight of 30,000 to 40,000, corresponding to chains containing be-

*Or of RNA, in the case of the RNA viruses. The ability of RNA to serve as the genetic material of certain viruses was discussed in Chapter 11.

tween 250 and 300 amino acid residues. If we take 300 amino acid residues to represent the size of an average polypeptide chain, then—given the triplet code—the average gene must contain about 1,000 nucleotide pairs.

A given gene can exist in a variety of different forms as a result of mutational changes in its nucleotide sequence; the different mutational forms of a gene are called *alleles*. The form in which a given gene exists in a microorganism as it is first isolated from nature (the wild-type organism) is defined as the *wild-type allele* of that gene; altered forms resulting from mutations are called *mutant alleles*.

## *The chemical basis of mutation*

**Mutagenesis**

Most of our understanding of the mechanisms of mutation derives from the results of experiments in which mutations are induced by chemical agents whose mode of action on DNA is known. Such mutagens include base analogs, which are incorporated into DNA in place of the natural bases; nitrous acid, which deaminates the purine and pyrimidine residues of DNA; proflavin, which intercalates between the stacked bases of DNA, stretching the distance between them from the normal 3.4 to 6.8 A; and alkylating agents, the action of which causes depurination of DNA.

Experiments on the chemical bases of mutation have been done almost exclusively with the bacterium *Escherichia coli* and with a number of phages for which *E. coli* is the normal host. In experiments with bacteria involving mutagens that are nontoxic, such as the base analogs, the mutagen is added to the growth medium for a number of generations, after which the cells are plated on a medium selective for a particular class of mutants. The class of mutants chosen is entirely a matter of technical convenience and experimental strategy, since *mutagens tend to raise the mutation rates of all genes more or less equally.* The mutation rate in the presence of the mutagen can be determined by various methods (of which one is described in Chapter 13) and compared with the rate in a control culture without mutagen.

For toxic mutagens, the experimental procedure that is generally followed is to treat a suspension of cells with the mutagen for a defined period of time. The cells are then washed free of mutagen and plated on a medium that selects for the growth of mutants in the surviving population. The results may be expressed quantitatively as the number of mutants induced per treated cell at a given level of survival.

Experiments on the mechanism of mutation are generally performed

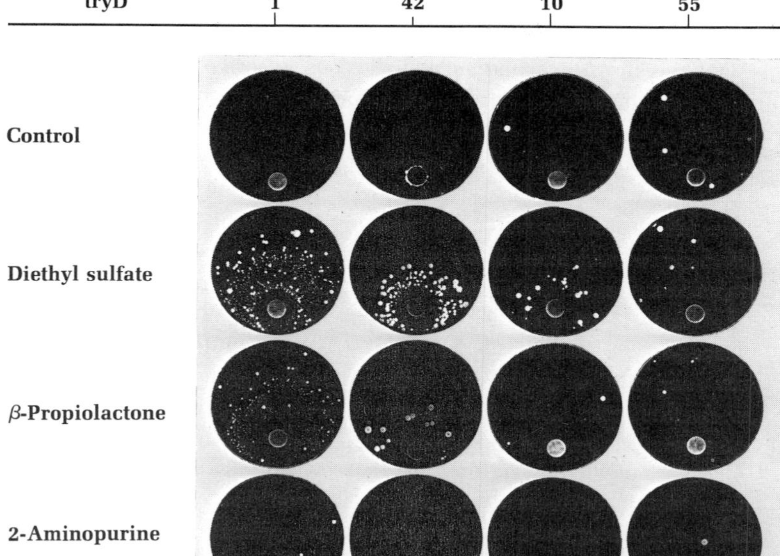


*Figure 12.1 The induction of genetic reversions by chemical mutagens. Sixteen petri plates were spread with about $2 \times 10^8$ cells each of a tryptophan-dependent strain of E. coli. The agar contained sufficient tryptophan to allow about a tenfold increase in cell number, which is not enough to appear as visible growth to the naked eye. A drop of a mutagen solution (or sterile water as control) was placed on a filter-paper disc or in a hole made in the agar. The horizontal rows were treated with the mutagenic agents shown; the vertical rows represent different mutations in the tryD locus. Note that tryD1 is revertible by all three mutagens; tryD42 is revertible only by diethyl sulfate and β-propiolactone; tryD10 is revertible only by diethyl sulfate and 2-aminopurine; and tryD55 is revertible by none of the agents tested. From E. Balbinder, "The fine structure of the loci tryC and tryD of Salmonella typhimurium. II. Studies of reversion patterns and the behavior of specific alleles during recombination." Genetics* **47**, *545 (1962).*

by inducing mutants with one mutagen and then testing their susceptibility to *reversion* by other mutagens. A common method of performing reversion tests is to spread a washed suspension of a mutant strain on agar that is selective for the revertant type—for example, auxotrophic cells may be spread on agar selective for prototrophic revertants. A drop of a solution of a mutagen is placed directly on the plate so that the mutagen diffuses outward through the agar, forming a concentration gradient. If the mutant type is susceptible to reversion by that mutagen, a ring of revertant colonies will appear around the drop (Figure 12.1).

Mutagenesis experiments with bacteriophage are carried out in much the same way. Chemical agents that are capable of modifying existing DNA structures are added directly to suspensions of phage particles. The treated particles are then plated on lawns of appropriate bacterial strains to detect those particles with mutant properties, such as altered plaque type or host range. Agents that must be present during replication, such as base analogs, are added to host cells that have been infected with phage, and the liberated phage progeny are tested by plating on appropriate bacterial strains.

## The molecular basis of mutagen-induced reversion

Simple reversion experiments of the types described above can only show that *the original phenotype is restored;* for example, if the primary mutation caused the wild-type strain to lose the activity of a particular biosynthetic enzyme, a revertant is a mutant which has regained that enzyme function.

Such a restoration of enzyme function can occur in three different ways. First, it can occur by a true reverse mutation, which restores the original base-pair sequence. For example, if the primary mutation caused the substitution of an adenine-thymine (AT) base pair for a guanine-cytosine (GC) base pair, a true reverse mutation restores the GC pair at that position. Second, it can occur by substitution of a new base pair, different from that of the wild type or of the primary mutant, at the original site; e.g., the primary mutation may have been a GC → AT change, and the reversion an AT → CG (rather than an AT → GC) change. In such a case the revertant differs from wild type in one base pair, but enzyme activity may be normal – either because the code is degenerate and the revertant has the same amino acid as the wild type at that position, or because a new amino acid at that position is functionally equivalent to the amino acid of the wild type that it replaces. Finally, a second mutation may occur elsewhere on the chromosome, which cancels the effect of the primary mutation. Such *suppressor mutations,* as they are called, usually act by altering a component of the translation system. Their mechanism will be discussed later in this chapter; for the moment, it is sufficient to note that suppressor mutations are frequent.

In most experiments on mutational mechanisms, the validity of the conclusions that are drawn rests on the certainty with which it can be inferred that *true reverse mutation* has occurred. In a true reverse mutation, the phenotype of the revertant is indistinguishable from that of the wild type in all respects, and crosses between revertant and wild type fail to segregate the primary mutation, a phenomenon that should occur if the revertant carries both the primary mutation and a suppressor mutation.

Since both of these criteria are negative ones (*failure* to observe differences, and *failure* to segregate a primary mutation), they cannot provide unequivocal proof of true reverse mutation. Nevertheless, some of the mechanisms which have been deduced from experiments on reverse mutation have now been directly confirmed by analysis of the amino acid sequence of enzymes isolated from wild-type, mutant, and revertant strains.

### Mutagen-induced transitions

Mutations that represent the substitution of one base pair for another can be divided into two classes: those in which a purine is replaced by a different purine, or a pyrimidine by a different pyrimidine *(transitions);* and those in which a purine is replaced by a pyrimidine, and vice versa *(transversions)* (Figure 12.2).

In their original paper, Watson and Crick pointed out that mutations of the type that we are now calling transitions should occur if, at the time they are acting as templates, any one of the bases underwent a *tautomeric shift* of electrons. The bases are presumed to exist in their energetically most probable forms, as shown in Figure 7.36. A shift from the keto to the enol form of thymine, for example, or a shift from the amino to the imino form of adenine, would change the hydrogen-bonding specificities of these bases so that thymine would pair with guanine and adenine with cytosine (see Figure 12.3).

The types of tautomeric shift postulated by Watson and Crick appear to account for the mutagenic activities of base analogs, notably 5-bromouracil and 2-aminopurine. When added to the culture medium, both of these compounds are readily incorporated into newly synthesized DNA, bromouracil substituting for thymine, and aminopurine substituting mainly for adenine. At subsequent rounds of replication

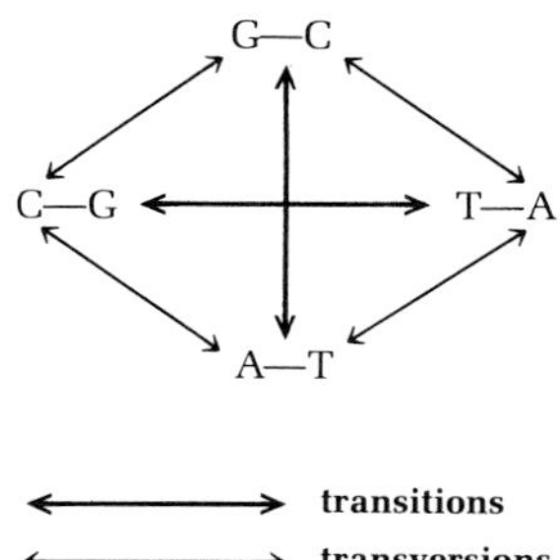


*Figure 12.2 Base-pair substitutions. Those in which a purine is replaced by a different purine and a pyrimidine are called transitions. Those in which a purine replaces a pyrimidine, or vice versa, are called transversions.*

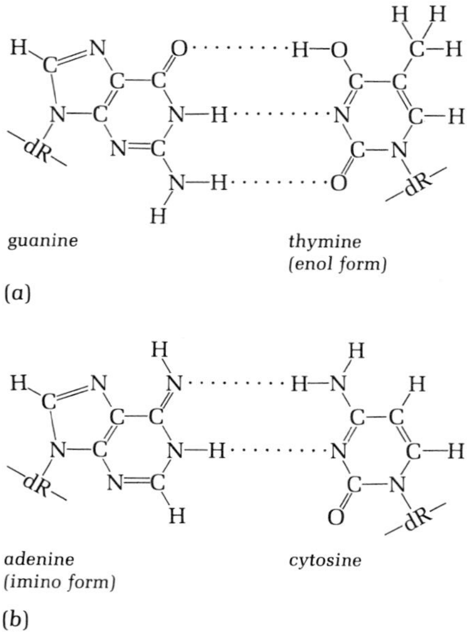

guanine

thymine (enol form)

(*a*)

adenine (imino form)

cytosine

(*b*)

*Figure 12.3 Changes in base pairing as the result of tautomeric shifts. In the enol form (a), thymine forms hydrogen bonds with guanine, instead of with adenine. In the imino form (b), adenine forms hydrogen bonds with cytosine, instead of with thymine. Similar shifts in guanine and cytosine will also cause changes in base pairing. (Compare the structures in this figure with those in Figure 7.36.)*

these analogs appear to tautomerize much more frequently than do the natural bases, with the result that both compounds promote AT→ GC transitions. On the other hand, if it happens that either one of these compounds undergoes abnormal base pairing *at the time it is being incorporated*, a GC→ AT transition must result. For example, if bromouracil were in the enol form at the time of incorporation it would pair with guanine in the template and be substituted for cytosine. At the next replication cycle, by which time bromouracil would have resumed its more probable keto form, it would cause the incorporation of thymine in the opposite strand. Thus, an AT pair would eventually appear at the former site of a GC pair (Figure 12.4).

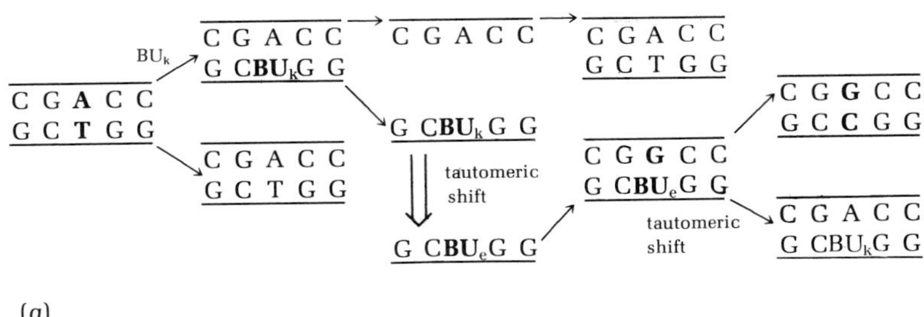


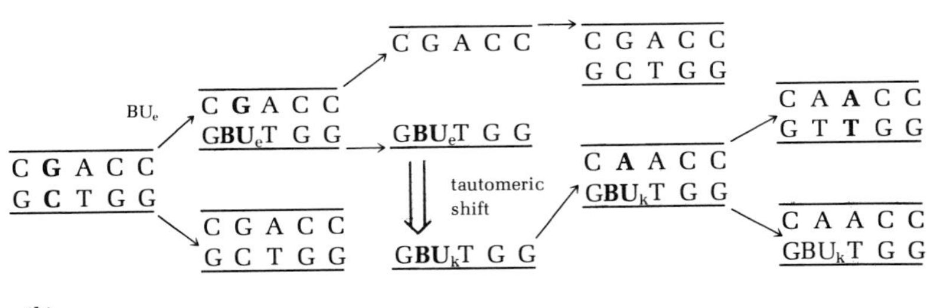


*Figure 12.4 (a) When bromouracil (BU) is incorporated in its keto form ($BU_k$), a subsequent tautomeric shift during replication causes an AT → GC transition. (b) When BU is incorporated in its enol form ($BU_e$), a subsequent tautomeric shift during replication causes a GC → AT transition.*

Nitrous acid is a third mutagen which is capable of inducing transitions in either direction. Nitrous acid deaminates the bases of DNA, a hydroxyl group replacing the amino group at the end of the reaction. As shown in Figure 12.5, deamination of the adenine of an AT pair will convert the adenine to hypoxanthine. At the next replication cycle hypoxanthine will pair with cytosine, resulting in an AT→ GC transition. Nitrous acid will also cause transitions in the opposite direction (GC→ AT) by the deamination of cytosine to uracil.

Another transition inducer that has been studied is hydroxylamine. This compound reacts almost exclusively with cytosine, the net result

Figure 12.5 The conversion of an A-T (adenine-thymine) pair to a G-C (guanine-cytosine) pair by the action of nitrous acid ($HNO_2$), followed by replication. Adenine is converted to hypoxanthine (H), which changes after pair separation into a form that can pair with cytosine instead of thymine. The time of change is marked by an asterisk (*). The structures of the bases are represented schematically; dotted lines represent the hydrogen bonds that hold the base pairs together.

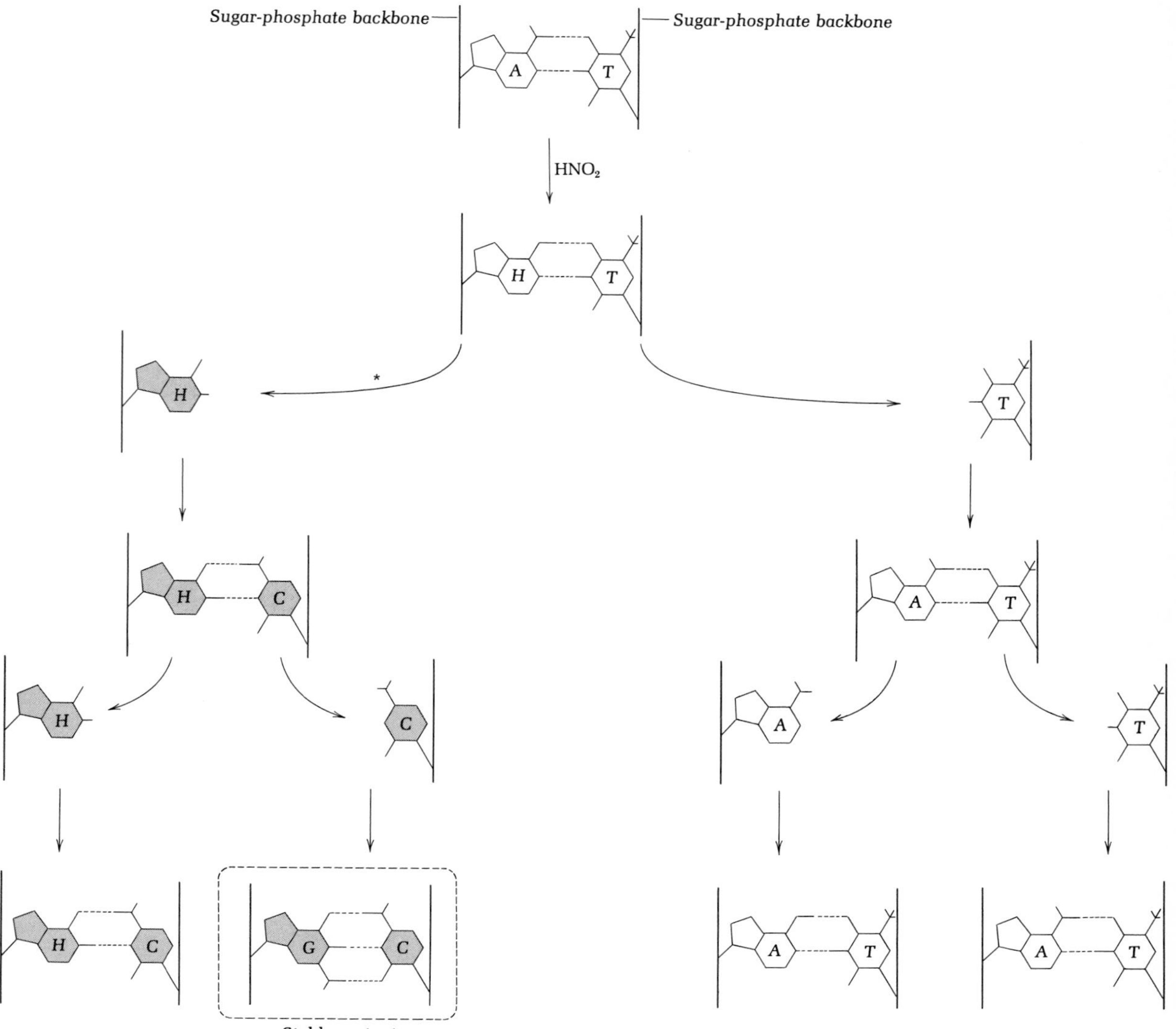

of the reaction being the replacement of the amino group by a hydroxylamino group. The hydroxylamino derivative of cytosine appears to undergo frequent tautomerization, leading exclusively to GC→AT transitions.

**Mutagen-induced insertions and deletions**

Acridine dyes, of which proflavin is the best-studied example, induce a unique class of mutations. Although they revert spontaneously and can be induced to revert at a higher rate by acridine dyes, such mutations are never reverted by base analogs, nitrous acid or hydroxylamine. When reversion does occur, it always proves to be the consequence of a second-site, or suppressor, mutation, within the same gene as the primary mutation. Furthermore, the suppressor mutation, when separated from the primary mutation by recombination, acts exactly like the primary mutation in every way: it inactivates the gene product and reverts only spontaneously or in response to acridine dyes. To put it another way, two acridine-induced mutations within the same locus may cancel each other out so that the gene product remains functional.

Acridine-induced mutations exhibit an additional unique property: they are never "leaky" (i.e., they always produce a total loss of function of the gene product). Mutations of the transition type, on the other hand, are often leaky.

These special properties of acridine-induced mutants can be accounted for by assuming that *acridines cause the insertion or deletion of one or a few base pairs in DNA.* Since the message encoded in mRNA must be read as a series of triplets, insertion or deletion causes a shift in "reading frame," with the result that all codons beyond the site of mutation are changed (Figure 12.6).

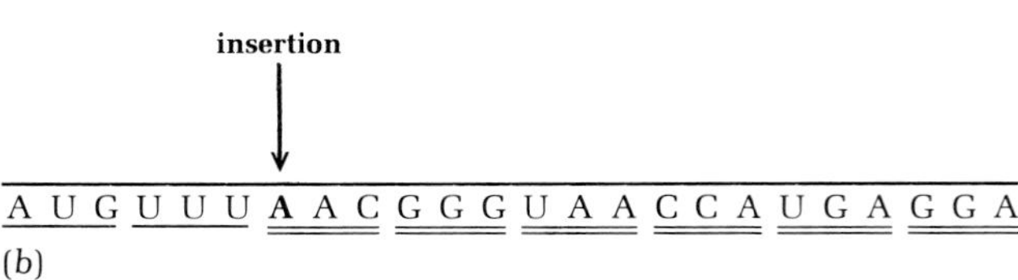


*Figure 12.6 A segment of an mRNA molecule. The message is read as a series of triplets, or codons, from left to right. The insertion of an "A" at the seventh position shifts the reading frame, changing all the codons to the right of that point. The changed codons are indicated by double underlining.*

Obviously, an insertion can be reversed by a deletion of the inserted base, and vice versa. It has been found, however, that a mutation caused by an insertion can be reverted by a deletion at a site many base pairs

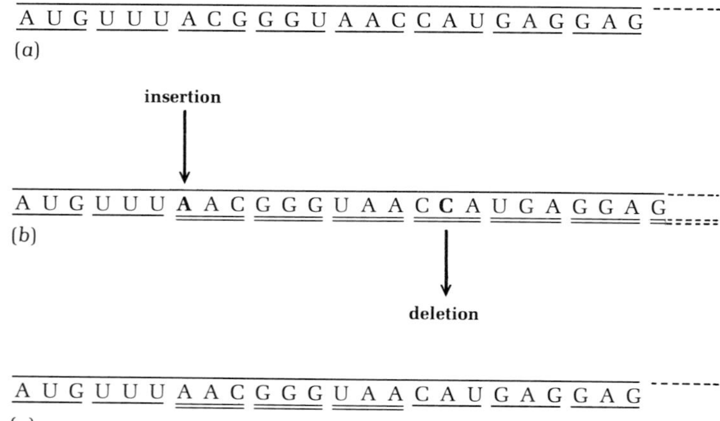


*Figure 12.7 The reversion of a single insertion by a single deletion. Note that the revertant contains a series of altered codons (doubly underlined) between the mutant sites.*

away; as shown in Figure 12.7, the revertant strain is normal except for a series of altered codons between the two mutational sites. The ability of such revertants to be selected means that many proteins can be functional even though extensive sequences of amino acids within their structures are altered.

The induction of insertions and deletions is related to the ability of proflavin to intercalate between the stacked bases of DNA, as discussed earlier. It cannot be simply a matter of altering the DNA templates at the time of replication, however, since acridines will induce mutations in phages while they are replicating inside host bacteria, without having any mutagenic effect on the replicating bacterial DNA.

As an explanation of this curious phenomenon, it was suggested that proflavin exerts its mutagenic effect only in DNA that is undergoing *recombination*, since replicating phage DNA is known to recombine frequently, whereas bacterial DNA ordinarily does not. This proposal was supported by experiments with partial diploids of *E. coli*. Partially diploid zygotes are formed in bacterial conjugation, as the result of incomplete transfer of the chromosome of one bacterial cell to another. When zygotes of *E. coli* are treated with proflavin, frameshift mutations are induced in the diploid, but not in the haploid, region of the chromosome.

The mechanism of recombination involves *breakage and reunion* of DNA molecules (see Chapter 14). Proflavin appears to act at an early stage in this process—perhaps by binding to DNA in which a single-strand break has occurred, thereby stabilizing a mispairing of strands (Figure 12.8).

Figure 12.8 *A possible mechanism of frame-shift mutation. (a) A single-strand break occurs in a DNA duplex; such breaks must occur at an early stage in recombination. (b) The duplex partially unwinds; the free end occasionally undergoes a displacement of its pairing sites, and this displacement may be stabilized by the binding of a proflavin molecule. (c) The gap is closed by resynthesis, leading to the insertion of an extra base in one strand. The gap-closing mechanism is shown in more detail in Figure 12.10.*

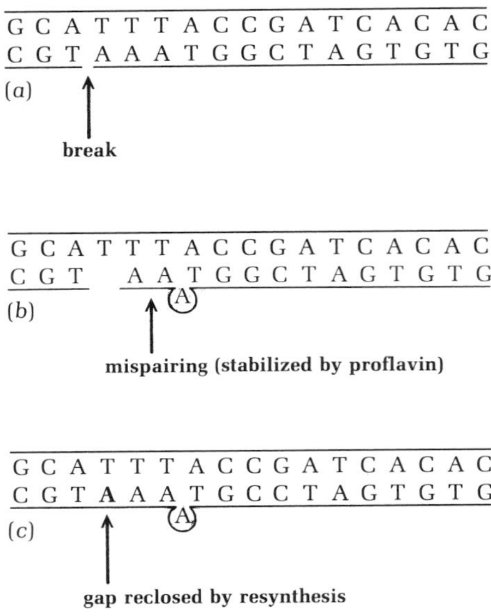


**Mutations induced by ultraviolet light**

DNA absorbs ultraviolet light strongly; the absorption maximum of DNA lies at a wavelength of 260 m$\mu$. Cells are rapidly killed by ultraviolet absorption, and a high rate of mutation occurs among the survivors.

When solutions of DNA are irradiated with ultraviolet light, two types of chemical changes take place. First, covalent bonds are formed between pyrimidine residues adjacent to each other in the same strand, forming *pyrimidine dimers.* These dimers distort the shape of the DNA molecule and interfere with normal base pairing. Second, pyrimidine residues are *hydrated* at the 4,5 double bond.

It is quite clear that much, if not all, of the lethal effect of ultraviolet light is attributable to the formation of pyrimidine dimers. Mutations, on the other hand, can probably result from either type of chemical change. Transitions might be expected to occur as a result of the hydration of pyrimidine residues, and indeed many ultraviolet light–induced mutations in phage can be reverted by base analogs. The mechanism by which pyrimidine dimers cause mutations, on the other hand, is far from clear. The importance of pyrimidine dimers as a cause of ultraviolet light–induced mutation is shown by the fact that treatments which lead to removal or cleavage of dimers also reverse most of the mutagenic effect of ultraviolet light. If ultraviolet light–treated cells are immediately irradiated with visible light in the range 300 to 400

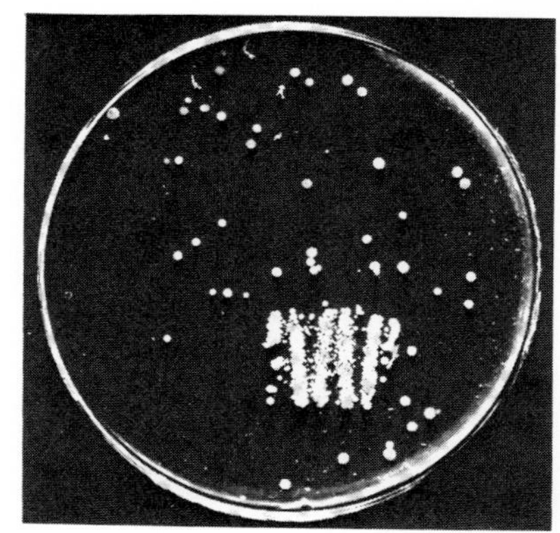

*Figure 12.9 Photoreactivation. The surface of an agar plate was evenly and heavily inoculated with E. coli, and then was exposed to a dose of ultraviolet light sufficient to kill nearly all the cells. Thereafter, a portion of the plate was illuminated for a short time with visible light by focusing on its surface the image of a tungsten lamp filament. The plate was subsequently incubated in darkness until bacterial growth was complete. On most of the plate, only a few survivors of the ultraviolet irradiation gave rise to colonies. However, in the region illuminated with visible light, dense growth occurred. In this region, the bacterial growth has the form of the tungsten lamp filament that was used for photoreactivation. From A. Kelner, "Ultraviolet irradiated E. coli." J. Bacteriol.* **58,** *512 (1949).*

nm, for example, both mutation frequency and lethality are greatly reduced, a phenomenon known as *photoreactivation* (Figure 12.9). This has been shown to result from the activation, by light of the particular wavelengths in question, of an enzyme that hydrolyzes pyrimidine dimers.

All cells also possess an elaborate set of enzymes that affect the "dark repair" of ultraviolet light–damaged DNA. The repair process, which is apparently triggered by the distortion of the double helix in the region of a dimer, occurs in a series of steps. First, an endonuclease makes a cut in the sugar-phosphate backbone on either side of the dimer, *excising the dimer* as part of an oligonucleotide. Second, an exonuclease *widens the gap* by sequentially digesting the broken chain, starting at the 3′ end.* Third, DNA polymerase *resynthesizes the missing segment*, using the opposite surviving strand as template. Fourth, polynucleotide ligase *closes the final gap* (Figure 12.10).

Postirradiation conditions that favor such repair processes, while inhibiting DNA replication, cause many potential ultraviolet light–induced mutations to be lost. Conversely, conditions favoring replication but inhibiting repair increase the number of mutations in ultraviolet light–irradiated cells. The occurrence of a dimer-induced mutation therefore depends on the outcome of a race between the repair enzyme system and the replicating enzyme system. If the replicating enzymes reach the dimer first, a mutation may result; if the repair enzymes reach the dimer first, the dimer is excised and mutation is prevented.

*Endonucleases are enzymes that catalyse the hydrolysis of phosphodiester bonds within a polynucleotide chain. Exonucleases are enzymes that hydrolyse the terminal phophodiester bond of a polynucleotide chain, splitting off the terminal nucleotide. Exonuclease action causes the *sequential digestion* of the chain.

Figure 12.10 *Repair of DNA damaged by ultraviolet irradiation. (a) A thymine dimer is produced as a result of irradiation. (b) A specific endonuclease makes a cut on either side of the dimer, excising a small oligonucleotide. (c) The exposed 3′ end is attacked by exonuclease, which widens the gap by sequentially digesting the strand. (d) DNA polymerase resynthesizes the missing segment. (e) Polynucleotide ligase seals the gap by joining the 3′-hydroxyl and 5′-phosphate.*

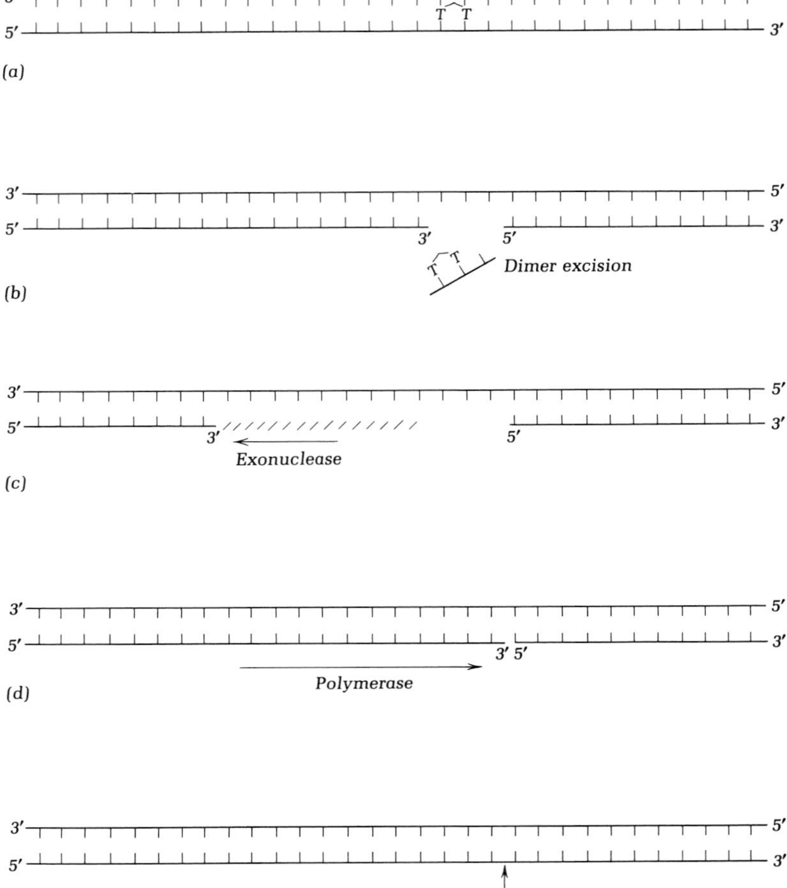


**Spontaneous mutations**

The term *spontaneous mutation* refers to those mutations which occur in the absence of known mutagenic treatment. The rate of spontaneous mutation is not a constant, however; many environmental conditions will affect it. In cells of *E. coli* grown at a very low rate in the chemostat, for example, the rate of spontaneous mutation depends on the nutrient that is limiting growth, and can vary widely. The mutation rate is lower in cells growing anaerobically than in cells growing aerobically.

There are probably many different mechanisms that can produce spontaneous mutations. Many products or intermediates of cell metabolism are demonstrably mutagenic; these include peroxides, nitrous acid, formaldehyde, and purine analogs. Some spontaneous mutations may thus in reality be induced by endogenous mutagens. These would induce a mixture of types, including transitions, transversions, and possibly insertions and deletions. Reversion studies have been carried out show that all these types do indeed occur among spontaneous mutations.

A mutation in one of the genes of *E. coli* and in a similar gene in *Salmonella typhimurium* has been shown to cause an increase in the spontaneous mutation rate for all loci by a factor of 100 to 1,000. The mutations caused by the new allele of the "mutator" locus are transitions. The product of the bacterial mutator gene has not been identified, but a similar mutator gene has been discovered in coliphage T4 and the product of this gene has been identified as DNA polymerase. The implications of this discovery are profound, since it demonstrates that the selection of nucleotides for the synthesis of a new strand of DNA depends not only on template-directed hydrogen bonding, but also on the stereospecificity of the polymerase. Since the mutationally altered polymerase causes a very high rate of "mispairings" leading to transitions, there is every reason to believe that the wild-type polymerase, of phage and of bacteria, is also responsible for spontaneous mispairings, although at a much lower rate than that of the altered polymerase.

Finally, a small fraction of spontaneous mutations has proved to reflect *deletions* of DNA segments. Their span varies from a large part of a single gene to several genes. For example, if one selects for mutants of *E. coli* strain B resistant to phage T1, a small fraction of the resistant mutants require tryptophan for growth. Genetic mapping experiments have shown that the genes for the T1 receptor and for the tryptophan biosynthetic enzymes are adjacent to each other, and in these "pleiotropic" mutants a spontaneous deletion has removed both. No known mutagen has been found to induce such large deletions; these are always spontaneous.

With the above mechanisms in mind, we can now understand two general ways in which environmental conditions could influence the rate of spontaneous mutation. First, the environment may be expected to influence appreciably the rate of formation of endogenous mutagens. Second, a number of the mechanisms described implicate the action of enzymes (DNA polymerase, recombination enzymes, repair enzymes), and both the synthesis and function of these enzymes may be subject to environmental influence.

## *The effects of mutation on the translation process*

**Nonsense mutations**

A codon is said to make "sense" if it is translated into the correct amino acid of the polypeptide for which the gene codes. A mutant codon that is translated into a different (incorrect) amino acid is said to contribute "missense" to the message, and the mutation which produced that codon is called a "missense mutation."

Some mutant codons (UAG, UAA, and UGA) have been found to cause *chain termination:* when the ribosome reaches such a codon, the process of polypeptide chain elongation is terminated, and the incomplete polypeptide is released. Such codons are called *nonsense codons,* and the mutations producing them are called *nonsense mutations.*

The nature of the nonsense mutations was not understood at the time they were discovered. They were first recognized only as mutations which inactivate a long stretch of DNA, in spite of the fact that they are revertible by base analogs and hence are neither deletions nor frameshift mutations. Their function being unknown, they were deliberately given the trivial names *amber* (mutations now known to produce the codon UAG) and *ochre* (mutations now known to produce the codon UAA). The codons so produced are commonly referred to as the *amber codon* (UAG) and the *ochre codon* (UAA). A corresponding name for the UGA codon has not come into common usage.

Following the discovery that nonsense codons cause chain termination, it was suggested that they might do so because there is no corresponding species of tRNA. This theory was confirmed by the studies of the binding of tRNA molecules to ribosomes in the presence of synthetic nucleotide triplets: of the 64 possible triplets, only UAG, UAA, and UGA do not stimulate the binding of any tRNA.

A glance at Table 7.6 will reveal the many possible base-pair substitutions that can change a "sense" codon to a nonsense codon: UGG, for example, which codes for tryptophan, changes to a nonsense codon (UGA) by a transition in the third letter.

The existence of three codons which can bring about chain termination raises the strong probability that one of them is the *natural chain terminator* in the translation process. As we shall see later, some groups of genes are transcribed into a single *polygenic messenger* RNA molecule; when a ribosome proceeds along a polygenic messenger, it must be given a signal to terminate one polypeptide chain and to start another. The termination signal is probably one of the nonsense codons.

**Suppressor mutations**

The loss of activity of a protein resulting from a mutation can be at least partially restored by a second mutation at a different site. Mutations of the latter type are called *suppressor mutations*, and the genes in which they occur are called *suppressor genes*. The term "suppressor" is generally used to refer to the mutant allele of the suppressor gene.

Suppressor mutations can occur within the same gene as the primary mutation *(intragenic suppressors)* or within different genes *(extragenic suppressors)*. Intragenic suppressors are of two types: one type causes an amino acid substitution which compensates for the primary missense mutation, restoring some function to the protein; the other type is an insertion or deletion which compensates for a primary frame-shift mutation.

Extragenic suppressors have a more complex mode of action, involving the mechanism of translation. At least two components of the translation machinery have *dual specificity:* tRNA molecules, and amino acyl synthetases (amino acid-activating enzymes; see Chapter 7). A given tRNA molecule has one site—the anticodon—which specifically recognizes a codon in mRNA, and a second site which is recognized by a specific amino acyl RNA synthetase. Alanine tRNA, for example, has the anticodon CGI,* which is complementary to the alanine codon, GCC; it also possesses a site that is recognized only by the alanine-activating enzyme, so it is charged only with alanine (Figure 12.11).

Correspondingly, the alanine-activating enzyme has two specific sites: one which recognizes alanine tRNA, and one which recognizes alanine. We can now consider how a mutation in the gene determining alanine tRNA or in the gene determining the alanine-activating enzyme may bring about the suppression of a mutation.

Let us suppose that a certain gene has the codon for alanine, GCC, at a certain position, and that a primary mutation changes this to GUC, the codon for valine. The protein formed in the mutant is inactive because it has valine in place of alanine. If a mutation now occurs in the gene governing the structure of the alanine-activating enzyme, causing the enzyme occasionally to transfer alanine to valine tRNA, a correct protein—with alanine instead of valine at the primary mutant site—will occasionally be formed. The cell will then produce a mixture of mutant and normal enzyme, and if enough of the latter is made the original mutation will be "suppressed." In this example, the suppressor locus is the gene for the alanine-activating enzyme, and the suppressor

*I stands for the base inosine (deaminated adenine), which is equivalent to G in base-pairing specificity.

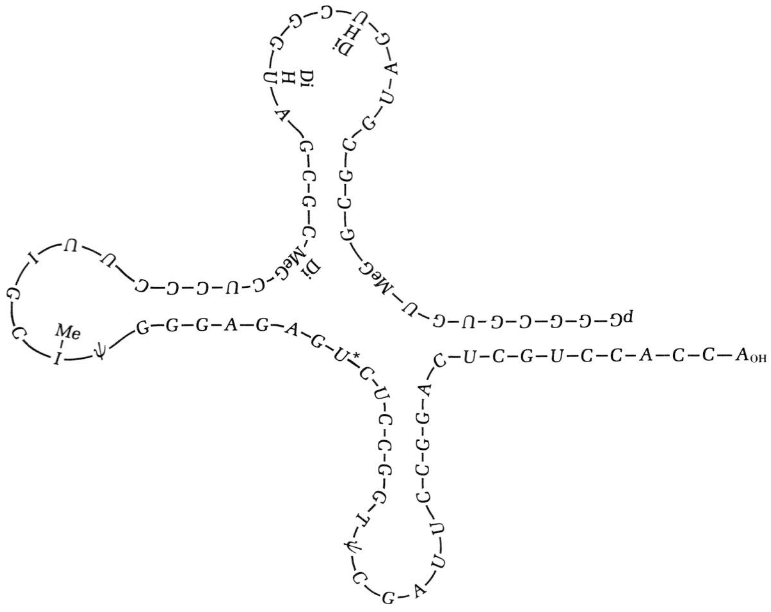


Figure 12.11 *Schematic representation of alanine tRNA in one of its possible conformations. In this conformation, there are four regions of hydrogen bonding between complementary bases. Abbreviations: p, phosphate; A, adenosine-3′-phosphate; C, cytidine-3′-phosphate; $A_{OH}$, adenosine; DiHU, 5,6-dihydrouridine-3′-phosphate; DiMeG, $N^2$-dimethyl-guanosine-3′-phosphate; I, inosine-3′-phosphate; MeG, 1-methylguanosine-3′-phosphate; MeI, 1-methylinosine-3′-phosphate; ψ, pseudouridine-3′-phosphate; T, ribothymidine-3′-phosphate; U, uridine-3′-phosphate; U*, a mixture of U and DiHU. From R. W. Holley et al., "Structure of a ribonucleic acid." Science* **147,** *1462 (1965). Copyright 1965 by The American Association for the Advancement of Science.*

allele is one which modifies the specificity of the enzyme for tRNAs.

One more example will illustrate how the genes governing the components of translation can mutate to become "suppressors." When a primary mutation has caused a substitution of valine for alanine, supression will result if a subsequent mutation occurs in the gene that determines valine tRNA, causing it to be occasionally charged with alanine by the alanine-activating enzyme. In this case the suppressor locus is the gene for valine tRNA.

The general theory of suppression outlined above has been confirmed by the finding that one suppressed mutant of *E. coli*, which incorporates serine at the site of an amber mutation, possesses an altered serine tRNA that can mimic the suppression effect in a protein-synthesizing system in vitro.

Since the binding of tRNA to a codon in mRNA requires the presence of the ribosome, structural alterations in the ribosome should also affect the recognition process—in other words, a mutationally altered ribosome might occasionally permit a mutant codon to bind a different tRNA and thus to be suppressed. Such suppression by altered ribosomes themselves has indeed been observed; furthermore it has been found that streptomycin and chemically related antibiotics will bind to ribosomes and alter the reading of the genetic code in an in vitro system. The synthetic messenger polyuridylic acid, for example, normally promotes only the incorporation of phenylalanine into polypeptide, since it contains only UUU codons. In the presence of streptomycin, however, polyuridylic acid promotes the incorporation of isoleucine (codon: AUU), serine (codon: UCU), and other amino acids, as well as phenylalanine. This effect of streptomycin leads to suppression of mutations in vivo. Some growth-factor-dependent mutants, for example, will grow if furnished either with their required growth factor or with streptomycin; the cells grown on streptomycin are found to synthesize a small amount of the normal enzyme which had been lost as a consequence of the primary mutation.

Suppression by any of the above mechanisms has to be highly *inefficient* if the cell is to survive. A given suppression mechanism will alter the reading of the suppressible codon wherever it occurs in the DNA of the cell, and every codon must occur several times in almost every gene. If suppression is too efficient, it will lead to the inactivation of every protein in the cell. If suppression occurs at low efficiency, however, all the proteins in the cell will be made correctly most of the time. When suppression operates with an efficiency of 5 percent, for example, the mutant enzyme will be synthesized in its active form 5 percent of the time, whereas the other proteins in the cell will be made correctly 95 percent of the time. For many enzymes, 5 percent of wild-type activity is sufficient to permit growth of the cell. In fact, it is generally found that suppression restores the activity of the affected enzyme to a level that is a few percent of the level in the original wild type.

## *The genetic aspects of regulation*

### The induction and repression of enzyme synthesis

The physiological aspects of the regulation of enzyme synthesis have been discussed in Chapter 8. It will be recalled that substrates and cellular metabolites of low molecular weight can specifically affect the rate of synthesis of enzymes in two ways. Some act as *inducers*, causing a marked increase in the rate of synthesis of one or more specific enzymes; this type of physiological control is particularly frequent in the regulation of catabolic pathways. Some act as *repressors*, causing a marked decrease in the rate of synthesis of one or more specific enzymes; this type of control is characteristic of biosynthetic pathways, the end-products of which repress synthesis of the specific enzymes that mediate their formation.

Much of our present knowledge about regulation of enzyme synthesis, at both the physiological and genetic levels, is derived from the intensive studies on the $\beta$-galactosidase system of *E. coli* carried out at the

Pasteur Institute by J. Monod, F. Jacob, and their collaborators. The work of this group culminated in the development of a general hypothesis to explain the control, at the genetic level, of the rate at which gene expression occurs. The following discussion will, accordingly, be centered on the studies of the Paris school with the β-galactosidase system of *E. coli*.

The first clue to the mechanism by which this regulation is achieved was provided by the discovery of mutants of *E. coli* in which the synthesis of β-galactosidase had become constitutive. Genetic experiments showed that the mutations which make β-galactosidase synthesis constitutive all map in one locus, which was named the *i* locus, for "inducibility"; by present-day convention, this locus is now designated *lacI,* as one of a set of loci concerned with lactose utilization.

The product of the wild-type allele of the *lacI* locus *(lacI$^+$)* determines the inducible state, and the product of the mutant allele *(lacI$^-$)* determines the constitutive state. To ascertain which of these two alleles is dominant, β-galactosidase formation was measured in *partial diploids* of *E. coli,* which were formed as transient zygotes in a mating between male and female cells.*

The zygotes, which were diploid and heterozygous for the *lacI* locus, were found to be inducible, establishing that *lacI$^+$* is dominant over *lacI$^-$*. Evidently, the dominant *lacI$^+$* allele produces a gene product that actively inhibits β-galactosidase formation; this product has been designated *repressor.*

The discovery of the *lac* repressor led to the general theory that inducible enzymes are ones for which the cell constantly synthesizes repressors; *the role of the inducer is to combine with and inactivate the repressor.* The theory was extended to explain end-product repression as well, by postulating the existence of genes that form repressors of biosynthetic enzymes. A repressor of this type would be nonfunctional as an inhibitor of enzyme formation, *unless activated by combination with the end-product of the biosynthetic pathway.*

This extension of the repressor theory was soon confirmed by the discovery of genes that, when mutated, *derepress* the formation of biosynthetic enzymes, even in the presence of excess amounts of end-products. The genes producing repressors have been given the general name *regulator genes* and are often symbolized by the letter R. The regulator gene for the arginine biosynthetic enzymes is thus *argR;* that for the tryptophan biosynthetic enzymes, *trpR;* and so on.

The general theory of genetic repression has been tested in cells that are diploid and heterozygous for regulator loci. Thus, an *argR$^+$/argR$^-$*

*The nature and formation of zygotes in *E. coli* will be described at length in Chapter 14.

diploid is found to be repressed for the synthesis of the arginine biosynthetic enzymes, since the *argR*$^+$ allele of the diploid continues to synthesize a repressor that is activated by arginine. Regulator loci have now been identified for a large number of catabolic and biosynthetic pathways.

The chemical nature of repressors remained obscure for some time. With the discovery that mutations in regulator genes can be suppressed by extragenic suppressors, however, it was necessary to conclude that they are proteins, since suppression acts only at the level of translation. The repressor protein produced by the *lacI* locus has indeed been isolated and identified by its ability to bind specific inducers; the mechanism by which repressors, as proteins, inhibit enzyme formation will be discussed in the following section.

**Operator genes**

The existence of an inhibitor of enzyme synthesis, the repressor, implies the existence of a target in the cell for its action: there must be a *receptor site* which binds the repressor, shutting off synthesis of the specific enzyme.

From consideration of the general mechanism of protein synthesis, it is obvious that at least two possible sites exist: the repressor could bind either to a site on DNA, blocking transcription of the specific gene, or to a site on messenger RNA, blocking the translation of the specific mRNA molecule. Experiments with purified repressor and DNA corresponding to the *lac* region* have shown that the *lacI* repressor binds to DNA. It remains possible, however, that other repressors may act at the level of mRNA, because there is no reason to assume that the mechanism of action demonstrated for the *lacI* repressor is universal.

The segment of DNA that determines the binding site of repressor is called the *operator locus*. In the specific case of regulation of $\beta$-galactosidase synthesis, the repressor binds to the operator locus itself. The operator locus (*lacO*) can be identified by its capacity to mutate to a form (*lacO*$^c$) which cannot bind repressor, thus making enzyme synthesis constitutive.

An "operator constitutive" mutant can generally be distinguished from a "regulator constitutive" mutant of the *lacI*$^-$ type in partial diploids. Whereas a *lacI*$^+$/*lacI*$^-$ diploid is inducible, by virtue of the production of repressor by the *lacI*$^+$ allele, a *lacO*$^+$/*lacO*$^c$ diploid† is *constitutive*, since the $\beta$-galactosidase gene adjacent to the *lacO*$^c$

*Such DNA can be extracted from certain transducing phages which specifically incorporate a fragment of the bacterial chromosome that includes the *lac* genes (see Chapter 14).

†The *lacO* diploids were made with the use of *F-lac* episomes, which are described in Chapter 14 (*lacO*$^+$, wild-type allele; *lacO*$^c$, constitutive allele).

allele continues to function, even though repressor is present and able to bind to the $lacO^+$ allele on the other chromosome of the diploid.

To confirm that the operator locus controls the expression only of adjacent genes on the same chromosome, diploids constitutive for the operator locus and heterozygous for the β-galactosidase locus were prepared. The latter locus, which was designated Z (*lacZ* in the new terminology), was present in the diploids in two forms: $lacZ^+$, making normal β-galactosidase; and $lacZ^-$, making an inactive form of β-galactosidase which could be detected immunochemically. Diploids of the types shown in Figure 12.12 were constructed, and both β-galactosidase and the inactive protein were measured in induced and uninduced cells. The results fully confirmed the theory: the *lacZ* allele adjacent to $lacO^+$ was always inducible, whereas the *lacZ* allele adjacent to $lacO^c$ was always constitutive, although both were present in the same cell. This result ruled out the possibility that the operator produces a diffusible gene product which is able to influence the expression of a *lacZ* allele on the opposite chromosome of the pair.

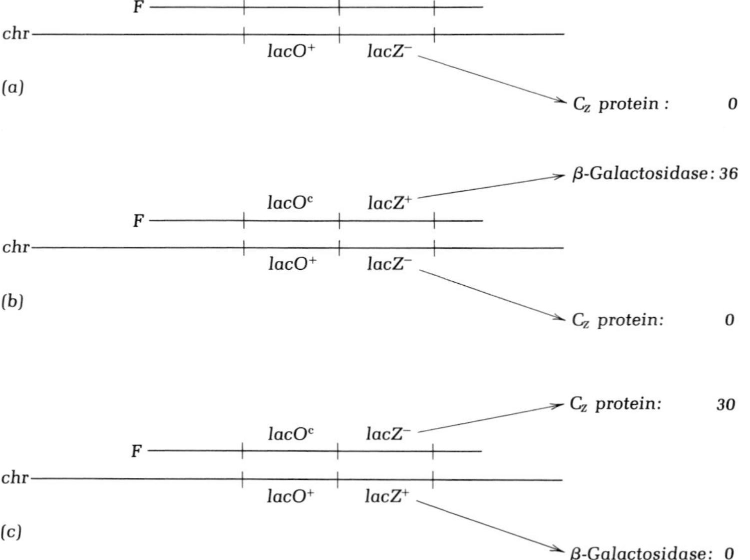


Figure 12.12 Test of the operator gene concept. (a) In a diploid in which both lacO alleles are wild type, neither β-galactosidase nor $C_Z$ protein is produced in uninduced cells. (b) When $lacZ^+$ is adjacent to the $lacO^c$ allele, β-galactosidase is made constitutively. (c) When $lacZ^-$ is adjacent to the $lacO^c$ allele, the $C_Z$ protein is made constitutively. The product of the gene adjacent to $lacO^+$ is not found in uninduced cells. Abbreviations: F, sex factor; chr, chromosome. Numbers represent amounts of gene products in arbitrary units.

## Operons

Early in the work on the genetic control of β-galactoside metabolism, it was discovered that some mutants, unable to ferment lactose, contain normal amounts of β-galactosidase and are deficient for a specific permease. This permease actively transports β-galactosides into the cell; mutants lacking it can grow on lactose only if lactose is added to the medium at very high concentrations.

The mutations that affect the β-galactoside permease were all found to map in one locus, which was designated *Y (lacY)*. Mapping experiments showed that the genes *lacO-lacZ-lacY* are tightly linked, forming a linear array in the order shown. When the experiments described above on the functions of the *lacI* and *lacO* genes were carried out, it was observed that the synthesis of β-galactosidase and permease are always affected in an identical manner: both are inducible in $lacI^+$ cells and constitutive in $lacI^-$ cells; both are inducible when their respective genes are adjacent to $lacO^+$, and both are constitutive when their respective genes are adjacent to $lacO^c$, even in diploid cells.

*LacZ and lacY thus behave as a unit of coordinated expression*, and such a unit—together with its operator—is called an *operon*. Many other operons have since been discovered, functioning in both catabolic and biosynthetic pathways.

Since repressors are freely diffusible molecules, a regulator locus, in contrast to an operator locus, need not be immediately adjacent to the genes it regulates, and many regulator loci are located on the chromosome some distance from the operons they control.

The operon provides an extremely efficient system for the regulation of metabolic pathways. The enzymes of a metabolic pathway must function as a unit: if the pathway is operative, all its enzymes must function at roughly equivalent rates; if a pathway is not operative, none of its constituent enzymes is required for cell function. The clustering of the genes for the enzymes of a pathway provides a mechanism for coordinate control.

It only remains to ask how an operator, which has bound a molecule of repressor, can inhibit the transcription of a sequence of adjacent genes. This question has been answered by the discovery that all the genes of the operon are transcribed as a unit, forming a very long molecule of *polygenic messenger RNA*. The transcribing enzyme, RNA polymerase, must attach to DNA at the operator end of the operon and transcribe sequentially all the genes within the operon. The initiation of this process appears to be sterically hindered when a repressor molecule is bound to the operator. This mechanism is shown in Figure 12.13.

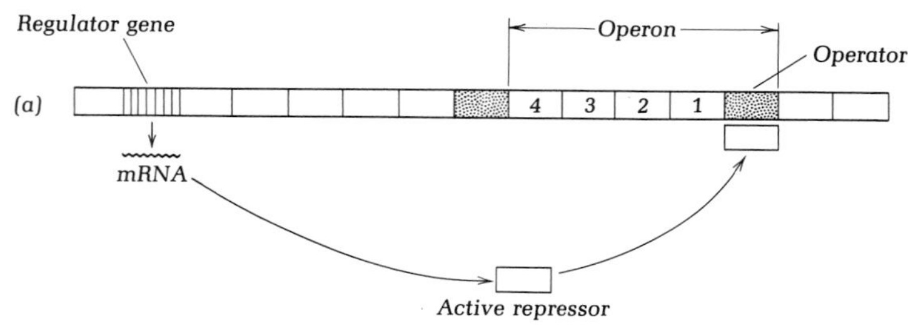


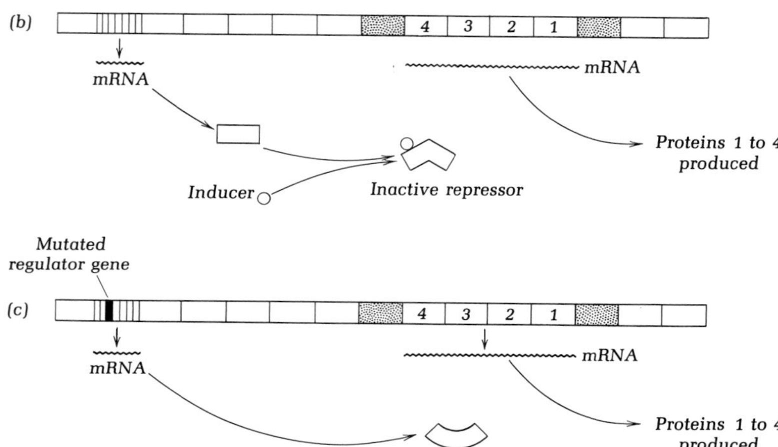


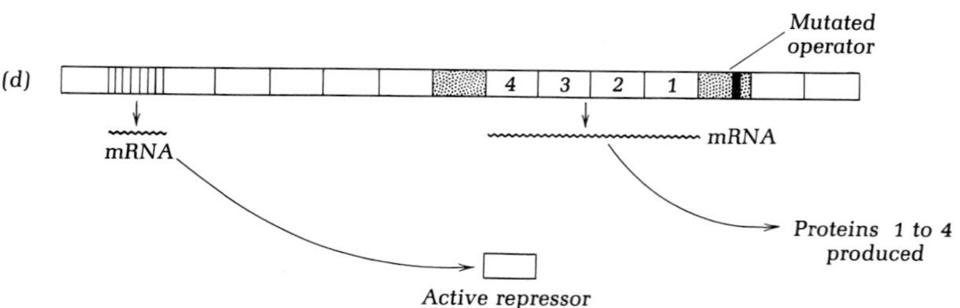


*Figure 12.13 Regulation of an operon. The segmented horizontal bar represents a section of a chromosome, with each segment representing a gene. (a) Active repressor, produced by a regulator gene, binds to its specific operator and blocks transcription of the adjacent set of four genes. (b) An inducer molecule combines with repressor and inactivates it, preventing its binding to the operator. The operon is then expressed. (This situation is typical of an operon governing the enzymes of a catabolic pathway; in the case of biosynthetic pathways, the repressor is normally inactive unless it combines with a corepressor, such as the end-product of the pathway.) (c) Inactive repressor is produced as a result of a mutation in the regulator gene; the operon is constitutively expressed. (d) A mutation in the operator prevents the binding of active repressor; the operon is constitutively expressed.*

**The translation of polygenic messenger RNA**

The transcription of an entire operon to form a single molecule of polygenic mRNA explains not only the ability of operator-bound repressor to inhibit the expression of all the genes in the operon simultaneously, but also the fact that nonsense mutations exert polarized effects on translation. The *polarity* resulting from a nonsense mutation is characterized by the greatly reduced synthesis of enzymes governed by all genes distal to the mutation (i.e., located on the opposite side of the mutation from the operator).

To illustrate the nature of a polarity effect, let us consider the case of a hypothetical operon containing five genes: operator-A-B-C-D. A nonsense mutation in A results in the total absence of enzyme A and reduced amounts of B, C, and D; a nonsense mutation in gene B results in the total absence of enzyme B and reduced amounts of enzymes C and D but normal synthesis of enzyme A. In some cases polarity is absolute, the genes distal to the nonsense mutation being completely unexpressed.

Polarity effects can be interpreted as follows. The translation of a polygenic messenger must begin at the operator end, the only point to which ribosomes can attach. Once attached, the ribosome translates each gene in turn, an individual polypeptide being released when the ribosome reaches a chain-terminating codon. If a nonsense mutation produces a chain-terminating codon in the middle of a gene, however, the ribosome is discharged of its partial polypeptide, and such *a discharged ribosome has a high probability of becoming irreversibly detached from the messenger* before it reaches the codon marking the beginning of the next gene. The fact that varying amounts of enzyme are made from the genes distal to the mutation means that occasionally a discharged ribosome remains attached and continues its passage to the next gene, where it can reinitiate translation. In some cases polarity is absolute, no detectable polypeptides being formed by the genes distal to the mutation. This suggests that detached ribosomes cannot reattach to the messenger at the starting codons of later genes.

**Diversity in the mechanisms of genetic regulation**

We have presented above only one mechanism of genetic regulation, that which is operative for the *lac* genes and their products. To recapitulate the essential features of this system, a series of genes with related functions are arranged in a sequence on the chromosome, adjacent to a special site called the *operator*. A *regulator gene* produces a product, the *lac repressor*, which binds to the operator and prevents the initiation of transcription of the *lac* genes. The *lac* genes remain unexpressed until an *inducer* enters the cell and combines with the repressor, inactivating it. The *lac* genes are then transcribed as a unit to form a polygenic messenger.

Many other operons have been detected, including sets of genes for the enzymes of biosynthetic pathways. In these cases the repressor is normally inactive and is *activated* by combination with the end-product of a specific biosynthetic pathway. This ability of a protein to be activated by a small molecule structurally unrelated to the substrate suggests that the protein is allosteric (see Chapter 9).

The genes for the enzymes of metabolic pathways are not always organized in operons. In *E. coli*, for example, many of the pathways are governed by genes that are scattered along the chromosome. In some of these systems, enzyme synthesis is coordinated by the product of a single regulator locus, indicating that one repressor can bind to a number of different receptors.

Some regulator gene products *activate*, rather than repress, the formation of enzymes. Several such cases have been discovered, including the regulation of alkaline phosphatase synthesis in *E. coli*. This enzyme allows the cell to obtain phosphate by the hydrolysis of organic phosphates in the medium. When the medium contains high levels of inorganic phosphate, the synthesis of alkaline phosphatase is repressed. This is brought about by the activation (by inorganic phosphate) of a repressor produced by a regulator gene. In the absence of inorganic phosphate, however, the repressor protein exists in a form in which it *activates* (rather than represses) the synthesis of alkaline phosphatase. Without this regulator gene product no alkaline phosphatase is made, whether inorganic phosphate is present or absent. The effect of this arrangement is to render alkaline phosphatase synthesis doubly sensitive to the presence of inorganic phosphate, which converts an activator of enzyme synthesis to a repressor.

## *Genetic complementation*

**The cis-trans test of genetic function**

At the beginning of this chapter we defined the gene as a segment of nucleic acid within which the sequence of bases determines the sequence of amino acids in a specific polypeptide chain. The gene is thus the *unit of genetic function*.

When two or more independently occurring mutations are observed to alter the same general property of a virus or cell, the question arises whether the different mutations have all occurred in the same gene or whether they have occurred in a number of different genes all of which must function for the observed property to be normal. For example, wild-type cells of *E. coli* possess an enzyme, tryptophan synthetase, which catalyses a complex reaction between indoleglycerol phosphate and serine to produce tryptophan plus glycerol phosphate (the final step in tryptophan biosynthesis: see Figure 7.16). Many mutants of *E. coli* have been isolated which lack tryptophan synthetase activity. Have all the mutations occurred in a single gene, or have they occurred in several different genes, all of which are involved in the formation of tryptophan synthetase? The latter would be true if tryptophan synthetase were composed of two or more *different* polypeptide chains.

Such a question can be tentatively answered by carrying out the *cis-trans test* of genetic function, devised by S. Benzer in the course of his studies on phage genetics. In this test two haploid genomes, each carrying an independent mutation affecting the same function, are brought together in the same diploid cell. The two mutations are said to be in the *trans* position. A second diploid, in which both mutations are present on the same genome, is also constructed; in this case the two mutations are said to be in the *cis* position. The two types of diploids are shown in Figure 12.14.

Inspection of Figure 12.14 will show that, if the two mutations are in two different genes, both the *cis* and the *trans* diploids should produce the same amount of active gene product as a wild-type haploid

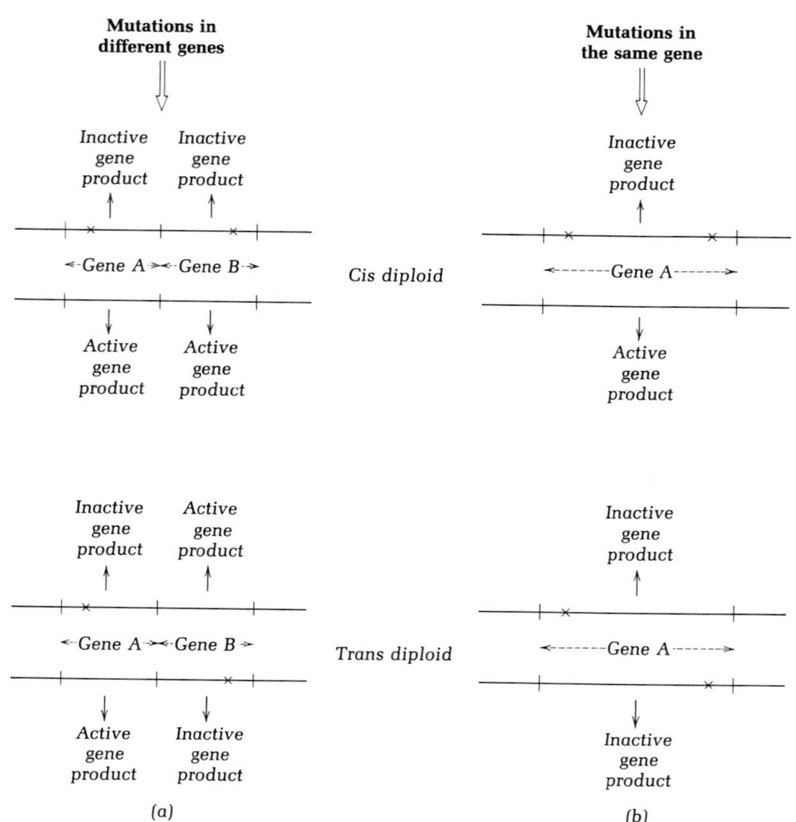


*Figure 12.14 The cis-trans test of genetic function. (a) When the two mutations are in different genes, the trans diploid produces as much active gene product as does the cis diploid. (b) When the two mutations are in the same gene, only the cis diploid produces active gene product.*

cell. On the other hand, if the two mutations lie in the *same* gene, the *cis* diploid will again produce the wild-type haploid amount of active gene product, but the *trans* diploid will produce a much lower amount or none at all.* The latter result therefore constitutes a "negative" *cis-trans* test.

Many pairs of tryptophan synthetase mutants were subjected to the *cis-trans* test. The results of the tests revealed that tryptophan synthetase is determined by two genetic regions, designated A and B. *Trans* diploids carrying either two mutations in the A region or two mutations in the B region produce no tryptophan synthetase. When, however, any A-region mutation is put into a *trans* diploid with a B-region mutation, the wild-type level of tryptophan synthetase is formed.

The A and B regions thus constitute units of genetic function. Benzer coined the term *cistron* to denote a functional unit that has been identified by *cis-trans* tests; operationally, a cistron is a genetic region within which any two mutations produce a negative *cis-trans* test. Cistrons are generally presumed to correspond to genes; final confirmation, however, requires that the products of the cistrons be identified as separate polypeptides. This has been accomplished in the case of the A and B cistrons of tryptophan synthetase in *E. coli:* the two polypeptides, designated the A protein and B protein, combine to form an active molecule of enzyme.

**Intragenic complementation**

Many enzymes are polymeric proteins, containing two or more identical subunits. A single gene governs the structure of the subunit, several of which associate to form the active enzyme. When a mutation occurs in this gene, a defective subunit is formed and the enzyme is inactivated. When two different mutations in the same gene are brought together in a *trans* diploid, however, the two different types of defective subunits they produce may associate to produce an enzyme with partial activity (Figure 12.15). This phenomenon is called *intragenic complementation.*

For example, let us suppose that a haploid wild-type cell produces 100 units of a particular enzymatic activity and that two different haploid mutants each produces less than one unit. A *cis* diploid formed from the two mutants produces 100 units of activity. If a *trans* diploid formed from the two mutants is found to yield 30 units of activity, the results constitute a *positive complementation test,* indicating that

*According to Figure 12.14, no functional gene product should appear in the *trans* diploid if both mutations lie in the same gene. Low amounts are often found, however, as a result of intragenic complementation. This phenomenon will be discussed in the next section.

*Figure 12.15 Intragenic complementation. (a) If, in a diploid cell, both alleles of the gene are wild type, the subunits that represent the gene products aggregate to produce a fully active polymeric enzyme. (b) If both alleles of the diploid carry the same mutation, the defective subunit produced will aggregate to form an inactive polymeric enzyme. (c) If the two alleles of the diploid carry two different mutations that are widely separated, the defective subunits produced may aggregate to form a polymer with partial activity.*

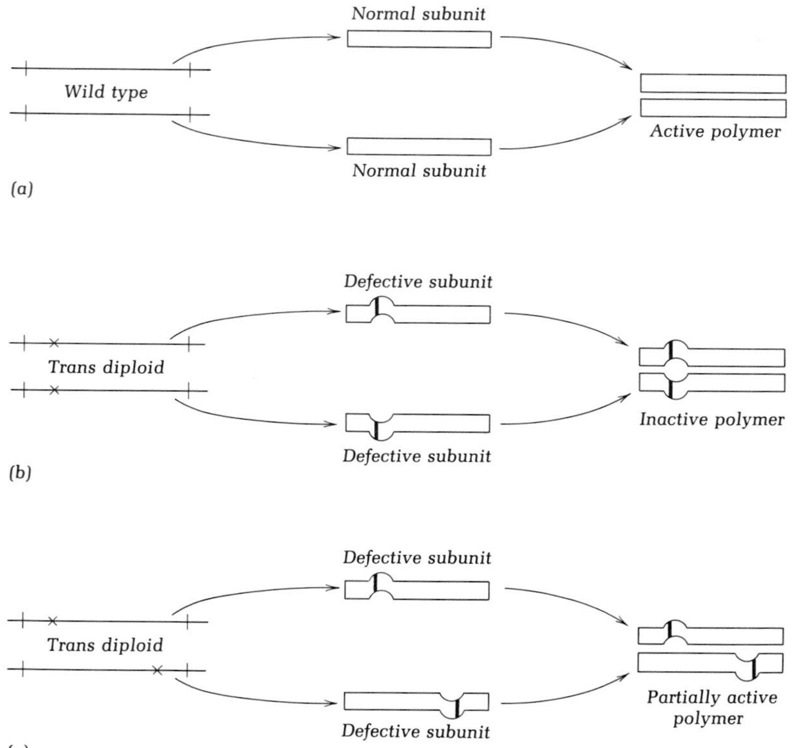


the defective subunits produced by the two mutants are capable of complementing each other in the final polymer. These results, however, constitute a negative *cis-trans* test, since the *trans* diploid produces much less enzyme activity than does the *cis* diploid. The two mutations thus lie in the same cistron (Figure 12.16).

Figure 12.16 A comparison of the cis-trans test of genetic function, and the complementation test. If the trans diploid produces significantly less active enzyme than does the cis diploid, the cis-trans test is negative (indicating that the two mutations are in the same gene). The ability of the different defective subunits to complement each other is indicated by the fact that the trans diploid produces much more active enzyme than does either haploid mutant (positive complementation test).

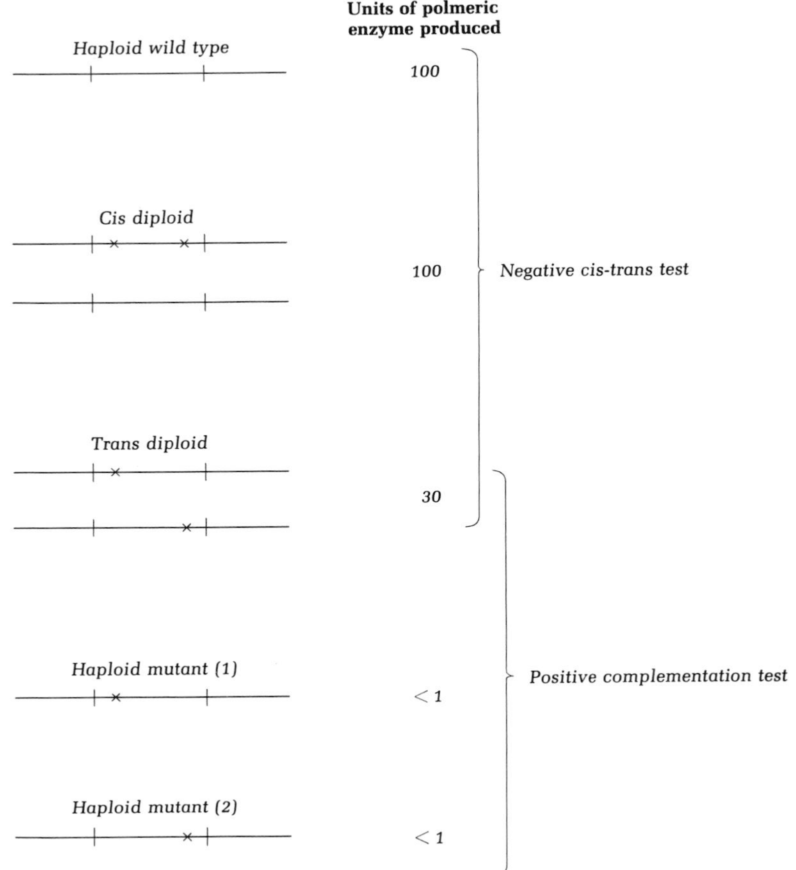


## Mutations in bacteriophage

The nucleic acid of the mature virion is extremely stable; spontaneous mutations do not occur with a detectable frequency, even when suspensions of bacteriophage particles are stored for long periods of time.

Agents that react directly with DNA, however, can be used to induce mutations in extracellular phage particles. Thus, mutations can be induced by treating phage particles with such agents as nitrous acid, alkylating agents, or ultraviolet light. In each case, the nucleic acid is so altered that base-pair substitutions or other permanent changes in base sequence will result during the next replication cycle, when the treated particles infect new host cells.

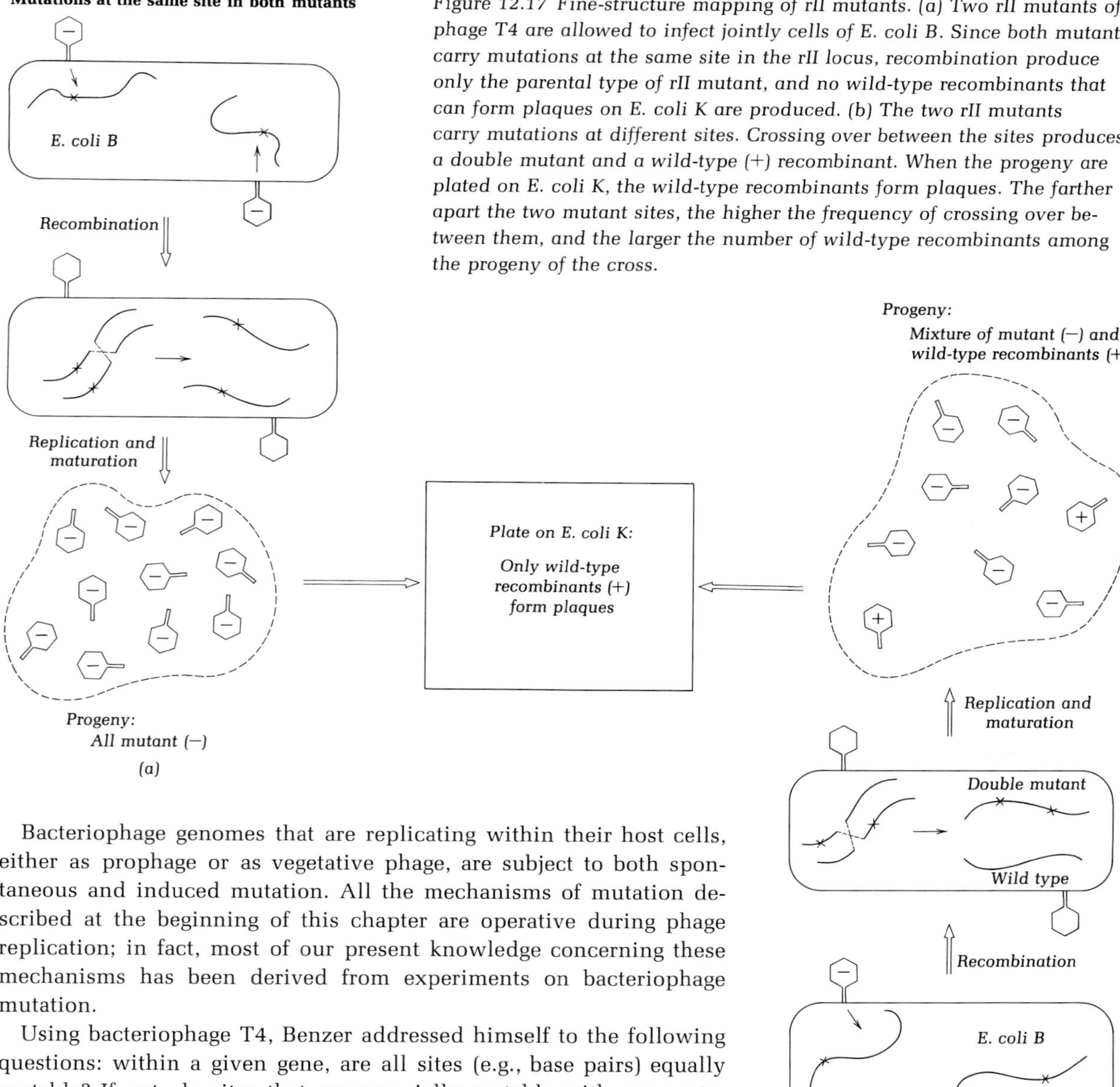


*Figure 12.17 Fine-structure mapping of rII mutants. (a) Two rII mutants of phage T4 are allowed to infect jointly cells of E. coli B. Since both mutants carry mutations at the same site in the rII locus, recombination produce only the parental type of rII mutant, and no wild-type recombinants that can form plaques on E. coli K are produced. (b) The two rII mutants carry mutations at different sites. Crossing over between the sites produces a double mutant and a wild-type (+) recombinant. When the progeny are plated on E. coli K, the wild-type recombinants form plaques. The farther apart the two mutant sites, the higher the frequency of crossing over between them, and the larger the number of wild-type recombinants among the progeny of the cross.*

Bacteriophage genomes that are replicating within their host cells, either as prophage or as vegetative phage, are subject to both spontaneous and induced mutation. All the mechanisms of mutation described at the beginning of this chapter are operative during phage replication; in fact, most of our present knowledge concerning these mechanisms has been derived from experiments on bacteriophage mutation.

Using bacteriophage T4, Benzer addressed himself to the following questions: within a given gene, are all sites (e.g., base pairs) equally mutable? If not, do sites that are especially mutable with one mutagenic agent show similar mutability with other agents?

To answer these questions, Benzer carried out *fine-structure mapping* of mutations within the *rII* gene of phage T4. Phage particles carrying the wild-type allele of this gene form small, irregular plaques when plated on a lawn of sensitive cells of *E. coli* strain B. Phage particles carrying mutant alleles of the *rII* gene form large, sharply defined plaques: *rII* mutants are thus easily detected and may be isolated without difficulty.

Benzer discovered that wild-type ($rII^+$) particles, but not *rII* mutant particles, would form plaques on a different host, *E. coli* strain K. When he carried out a mixed infection of strain B with two different *rII* mutants, most of the progeny were of the parental types and thus failed to form plaques on strain K; if two *rII* mutant genomes had *recombined*, however, to form an $rII^+$ recombinant, the resulting particle would be detected by its ability to form a plaque on strain K (Figure 12.17).

Benzer isolated a large number of *rII* mutants and crossed them two at a time. Each pair of mutants was used to mixedly infect *E. coli* strain B, and the progeny were tested for the presence of recombinants able to form plaques on strain K. The farther apart the two *rII* mutations, the higher the number of recombinants;* if two *rII* mutations occupied the *same* site, no recombinants were produced (Figure 12.17).

In this way Benzer constructed a map of the entire *rII* gene. Spontaneous mutations were observed at over 300 distinct sites, each representing a different base pair. The frequency of mutation at each site, however, was far from random: two "hot spots" were observed, one that had mutated 52 times and the other 275 times. Furthermore, the distribution of mutational sites was different for induced mutations, no two agents giving the same pattern of hot spots.

It is highly unlikely that the hot spots represent unusual base pairs, since only the four classical base pairs are known to exist in T4 DNA. The unique susceptibility of the hot spots to mutation must then represent the influence of neighboring sequences of bases; the nature of this influence is not known.

## *Further reading*

**Books**

Adelberg, E. (editor), *Papers on Bacterial Genetics,* 2nd ed. Boston: Little, Brown, 1966.

Hartman, P. E., and S. R. Suskind, *Gene Action,* 2nd ed. Englewood Cliffs, N.J.: Prentice-Hall, 1969.

Hayes, W., *The Genetics of Bacteria and Their Viruses,* 2nd ed. Oxford, England: Blackwell, 1968.

"The Genetic Code." *Cold Spring Harbor Symp. Quant. Biol.* **31** (1966).

*The mechanism of recombination will be discussed in Chapter 14.

**Reviews and original articles**

Howard-Flanders, P., "DNA Repair." *Ann. Rev. Biochem.* **37,** 175 (1968).

Jacob, F., and J. Monod, "Regulatory Mechanisms in the Synthesis of Protein." *J. Mol. Biol.* **3,** 318 (1961).

Orgel, L., "The Chemical Basis of Mutation." *Advances in Enzymol.* **27,** 289 (1965).

Speyer, J., "The Genetic Code," in *Molecular Genetics* (J. Taylor, editor), Pt. II. New York: Academic Press, 1967.

# Chapter thirteen The expression of mutation in viruses, cells, and cell populations

*All the properties* of an organism are ultimately determined by its genes, including the genes of the organelles (the mitochondria and chloroplasts of eucaryotic cells) as well as the chromosomal genes in the nucleus. Each gene can exist in a variety of structural forms, or alleles, the nature of which was discussed in Chapter 12. The allelic states of all the genes in a cell constitute its *genotype*.

The structural and physiological properties of a cell constitute its *phenotype*. In the following sections, we shall discuss the general ways in which the genotype of the cell determines its phenotype, and the manner in which changes in the genotype (mutations) bring about changes in the phenotype.

## The effects of mutation on phenotype

### The relation of phenotype to genotype

In Chapter 12 it was shown that each gene in the cell determines the structure of a single protein.* The structural and catalytic properties of the proteins determine in turn the anatomical and metabolic properties of the cell. The proteins interact with each other structurally, to form the complex organelles of the cell, as well as functionally, to form coordinated and regulated metabolic pathways. Thus, a mutation, by altering the structure of a single protein, can bring about a profound change at the *secondary level* of cell structure and function.

*With the sole exception of the genes that determine ribosomal RNA and tRNA.

### The effects of mutation on primary gene products

Protein structure is determined by the gene in two steps: the DNA of the gene is transcribed to form messenger RNA, the base sequence of which is then translated into a sequence of amino acids in a polypeptide. In many cases a third step follows: the polypeptide chains, folded into their tertiary structures, associate to form polymers. A polymeric protein may contain identical polypeptide subunits, or two different polypeptide subunits; such polymers may exhibit allosteric properties, as discussed in Chapters 7 and 8.

For the purposes of this discussion, we shall consider the proteins to be the *primary gene products* of the cell, since mutational effects on phenotype are usually the result of changes in protein structure. Such changes always consist of alterations in amino acid sequence resulting from mutational alterations in the base-pair sequence of DNA. The amino acid sequence may be altered by the substitution of one amino acid for another or by the deletion of a subset of amino acids.

The alteration in amino acid sequence of a polypeptide may produce any one of the following alterations in the properties of the final protein product.

(1) The protein may have an altered catalytic site; such alteration may cause the protein to be partially or completely inactivated as an enzyme.

(2) The protein may become unusually sensitive to any of a number of agents: it may be inactivated by high temperatures, by an allosteric effector, by a metal ion, and so on. Thermolability as a consequence of mutation is a phenomenon of major interest, since it permits the study of mutations affecting indispensable cell functions (conditional lethal mutations, page 455).

(3) The polypeptide subunits of a polymeric protein may undergo abnormal association, with a consequent loss of normal allosteric function.

(4) There may be no detectable change in the physicochemical properties or catalytic activities of the protein. This may occur if one amino acid is substituted by another with very similar properties (e.g., the substitution of an acidic amino acid, glutamate, by another acidic amino acid, aspartate).

### Phenotypic changes in dispensable cell functions

Depending upon the environmental conditions, many cell functions are dispensable; mutations conferring the loss or alteration of these functions may thus be studied under conditions that permit the mutants to grow and divide. Such studies, often referred to collectively

*Table 13.1 Some representative types of bacterial mutants*

| Type of mutant | Selection methods | Detection |
|---|---|---|
| *Able to use as carbon source a compound not utilizable by the wild type* | *Plate on agar containing the compound in question as the only available carbon source* | |
| *Resistant to inhibitory chemical agents, such as penicillin, streptomycin, sulfonamides, dyes* | *Plate on agar containing the inhibitor* | *Plating method is absolutely selective; only the desired type will form colonies* |
| *Resistant to bacteriophage* | *Plate on agar previously spread with a suspension of phage* | |
| *Auxotrophic (requirement for one or more growth factors not required by wild-type cells)* | *Incubate in growth medium lacking the growth factor in question, but containing penicillin. Wild-type cells multiply and most are killed; auxotrophic mutants are unable to multiply without the growth factor and survive (penicillin kills only actively growing cells)* | *Plate on agar lacking the growth factor. Mark the few wild-type colonies that appear, then add layer of agar containing growth factor. Auxotrophs form colonies only after addition of growth factor ("delayed enrichment method")* |
| *Unable to ferment a given sugar* | *Apply penicillin technique as above, using sugar in question as only fermentable carbon source* | *Plate on agar containing sugar in question, plus chemical indicator that changes color in the presence of fermenting cells (e.g., acid–base indicator). Mutant colonies appear a contrasting color* |
| *Temperature-sensitive DNA synthesis* | *Incubate cell suspension at 42°C for 15 minutes, add 5-bromo-uracil (5-BU), and continue incubation an additional 60 minutes; irradiate with light of 310 nm wavelength (cells that have incorporated 5-BU into their DNA are killed; cells that have failed to replicate their DNA at 42°C are spared)* | *Test colonies for ability to grow at 42°C (e.g., by replica-plating)* |
| *Sensitive to ultraviolet irradiation (unable to repair ultraviolet light damage)* | *Infect culture with ultraviolet light–inactivated bacteriophage; normal cells repair the ultraviolet light lesions of the phage and are subsequently killed by the phage; repair-deficient cells are spared* | *Make suspensions from isolated colonies, measure survival at several ultraviolet light doses* |

as "biochemical genetics," have contributed a substantial portion of our knowledge concerning metabolic pathways, regulatory functions, and other aspects of cellular physiology.

The cell functions that are dispensable under some environmental conditions include the utilization of alternative forms of carbon, nitrogen, sulfur, or phosphate as the nutritional sources of those elements; the synthesis of the precursors of macromolecules and of coenzymes; the regulation of enzyme synthesis and activity; and the synthesis of various components of the cell surface: murein, capsule, flagella, and pili. Some examples of the phenotypes that result from the loss of these functions will be briefly described in the following sections. Methods for selecting some of these mutant types have been briefly summarized in Table 13.1.

**Phenotypic changes in nutrition: sources of elements**

Many microorganisms are capable of utilizing *alternative* forms of carbon, nitrogen, sulfur, and phosphate. The loss of ability to use a particular form of an element (as a result of mutational inactivation of a protein involved in the pathway of utilization) is dispensable, provided that another utilizable form of that element is available. The activity lost by mutation may be either an enzymatic or a transport activity. The ability to use lactose as a source of carbon, for example, may result either from the loss of $\beta$-galactosidase activity or from the loss of $\beta$-galactoside transport (permease) activity.

Some of the catabolic pathways by means of which organic compounds are utilized were described in Chapter 6. Figure 6.29, for example, illustrates the pathways by means of which the pseudomonads utilize various aromatic compounds. Inspection of Figure 6.29 will reveal that the loss of any one enzyme of the convergent pathways will lead to the failure of the cell to utilize one or more aromatic compounds as sources of carbon and energy.

Many microorganisms are able to reduce sulfate and nitrate as sources of sulfhydryl and amino groups, respectively. In each case, a series of enzymatic reactions is required; the mutational loss of any of the enzymes involved will lead to the nutritional requirement for a more reduced form of sulfur or of nitrogen.

Most organisms can use organic phosphates as their source of phosphorus, hydrolyzing them to yield inorganic phosphate. In *E. coli*, for example, this is brought about by alkaline phosphatase, which is localized between the cell membrane and cell wall. The mutational loss of alkaline phosphatase activity confers on *E. coli* a nutritional requirement for inorganic phosphate.

**Phenotypic changes in nutrition: growth factors**

When mutation causes the inactivation of an enzyme of a dispensable *biosynthetic* pathway, the cell becomes dependent on the environment for one or more growth factors. This phenotypic condition is termed *auxotrophy;* the original state of growth-factor independence is termed *prototrophy.* The dispensable biosynthetic pathways are restricted to those which produce small molecules, capable of entering the cell and thus of serving as growth factors. The most common growth factors that may become required as a result of auxotrophic mutations are amino acids (precursors of proteins), purines and pyrimidines (precursors of nucleic acids), and vitamins (precursors of coenzymes).

The biosynthetic pathways for amino acids, purines, and pyrimidines, were outlined in Chapter 7. It will be noted that the purine and pyrimidine pathways involve phosphorylated intermediates and that the end-products are nucleotides; neither the intermediates nor the end-products can be supplied as growth factors to mutants that are blocked in these pathways, because the cell is impermeable to highly ionized (and thus to phosphorylated) compounds. Nevertheless, these pathways are dispensable if the medium contains the corresponding free purines or pyrimidines, since the cell is capable of taking up the free bases and converting them to nucleotides by the action of pyrophosphorylases; for example,

$$\text{free base} + \text{phosphoribosyl pyrophosphate} \rightleftharpoons \text{nucleotide} + \text{P—P}$$

The biosynthetic pathways for the amino acids are so constituted that the loss of a single enzymatic activity may lead to one of several different *patterns of auxotrophy.* One pattern, which results from the loss of an enzyme prior to a branch in the pathway, is that of *multiple requirements.* For example, the loss of any enzymatic reactions leading to chorismic acid formation in the aromatic pathway (Figure 7.16), causes the cell to require phenylalanine, tyrosine, and tryptophan for growth. All three amino acids can be replaced, however, by shikimic acid, if the pathway is blocked prior to the formation of this compound. (Later intermediates, which are phosphorylated, cannot serve as growth factors.) Multiple requirements may also arise by the loss of an enzyme activity which is common to two different pathways; the inactivation of reductoisomerase, for example, causes the cell to require both isoleucine and valine (Figure 7.19).

A second pattern produced by a single mutation is that of *alternative requirements.* This pattern arises as a consequence of the reversibility of certain reactions: the conversion of serine to glycine, for example, is freely reversible, so that a cell which is blocked in the biosynthesis of serine can use either serine or glycine as a growth factor (Figure 7.18).

**Resistance and sensitivity to antimicrobial agents**

Mutational changes in protein structure can lead to increased resistance to antimicrobial agents, both chemical and physical. Antimicrobial chemical agents used in chemotherapy are commonly called "drugs." Drug resistance may reflect any of several different changes in primary gene products: the cell membrane may become impermeable to the drug; the cell may gain an enzymatic activity that confers on it the ability to degrade the drug or to inactivate it by chemical modification; or the enzyme of the cell that is the primary target of drug action may acquire a reduced affinity for the drug.

Some structural analogs of normal metabolites are inhibitory to the growth of microorganisms. The reason for the inhibitory action of analogs varies; in some cases, the analog is incorporated into a macromolecule in place of the normal metabolite, thus producing an inactive macromolecule (e.g., *p*-fluorophenylalanine is incorporated into protein in place of phenylalanine); in other cases, the analog mimics the normal metabolite in producing end-product inhibition of the specific biosynthetic pathway, without replacing the metabolite for normal cell functions.

Resistance to analogs may thus arise in a number of different ways. Incorporation, for example, may be prevented by a mutation that alters the affinity of an amino acid–activating enzyme for the analog, whereas resistance to false end-product inhibition may result from a change in the allosteric receptor of the sensitive enzyme.

Mutations can confer increased sensitivity, as well as increased resistance, to antimicrobial agents. The sensitivity of bacteria to ultraviolet light, for example, is greatly increased by the loss of one of the enzymes that effect the *repair* of DNA damaged by ultraviolet light. The enzymes of DNA repair were discussed in Chapter 12.

**Phenotypic changes resulting from the loss of ability to synthesize components of the cell surface**

The structures of the cell which are exterior to the cell membrane are all dispensable under some environmental conditions. The cell wall, for example, is dispensable for bacterial cells growing in an isotonic medium; capsules, flagella, and pili are all dispensable under ordinary laboratory conditions. Mutations that deprive the cell of one or another of these structures are thus readily observed in the laboratory, and the changes can be traced to changes in the corresponding primary gene products.

Flagella and pili are each assembled from monomeric polypeptide subunits. Mutations in the genes for these polypeptides can prevent the synthesis of the polymerized appendages, and this in turn can produce secondary phenotypic changes. Flagellated cells, for example, may be sufficiently motile to produce spreading colonies on moist agar surfaces; nonflagellated mutants produce compact colonies on the same media.

The murein layer of the bacterial cell wall contains diaminopimelic acid (DAP), which is formed as the last intermediate in the biosynthesis of lysine by bacteria (see Figure 7.13). If a mutation inactivates one of the enzymes that acts prior to the formation of DAP in the lysine biosynthetic pathway, the cell will require both DAP and lysine for growth. The cell will grow normally if provided with DAP, since DAP also furnishes a supply of lysine. If the medium contains only lysine, however, the cell can synthesize proteins but lacks DAP for the synthesis of murein, and its wall becomes osmotically fragile. Under these conditions the cells will grow as L forms if placed in a medium of high osmotic strength; in ordinary media they will lyse. (L forms were discussed in Chapter 5.)

The production of a polysaccharide capsule by a bacterial cell involves a series of enzymatic reactions, forming a biosynthetic pathway. Cells that carry out the complete series of reactions are capsulated and form smooth, often mucoid, colonies; mutants, lacking one or another of the enzymes of the biosynthetic pathway, are noncapsulated and form "rough" colonies. This phenotypic change is accompanied, in pathogenic bacteria, by the loss of virulence, since the bacterial capsule affords the bacterium considerable resistance to phagocytosis in mammalian hosts.

**Pleiotropic mutations**

A mutation that confers on the cell changes in two or more different phenotypic properties is said to be *pleiotropic.* Pleiotropy results when any gene product, either primary or secondary, has more than one function. We have already mentioned one example: a mutation that inactivates an enzyme in a biosynthetic pathway and thus deprives the cell of its capsule leads to the two secondary phenotypic changes: rough colony formation and the loss of virulence. Similarly, a mutation that inactivates an enzyme responsible for the addition of the terminal sugar of a cell wall polysaccharide in *Salmonella* can confer on the cell altered antigenicity as well as resistance to a specific phage, because the terminal sugar of the polysaccharide is an essential component of a particular antigenic structure (see Chapter 28) that serves also as a specific phage receptor.

### Conditional lethal mutations

If the primary product of a given gene is *indispensable* to the cell under all conditions of cultivation, mutations that inactivate the product will be *lethal*. Such products include the polymerases that replicate and transcribe DNA; the various components of the apparatus for the translation of messenger RNA (ribosomal proteins, ribosomal RNA, activating enzymes, transfer RNAs, and so on); proteins that are essential for the process of cell division; and membrane proteins. Most viral gene products are indispensable for the production of normal virus particles: although some alterations in the coat proteins and viral-governed enzymes may be tolerated, most alterations prevent the normal synthesis and assembly of viral components.

The desire to apply the techniques of biochemical genetics to the analysis of indispensable gene functions led to the exploitation of the phenomenon of *conditional lethal mutation*. A conditional lethal mutation is a lethal mutation that is expressed under one set of environmental conditions but not under another. The conditions under which the lethal mutation is expressed are called *nonpermissive*, because they do not permit the cell (or virus) to develop. The conditions under which the lethal mutation is not expressed are called *permissive*, because they permit the cell (or virus) to develop normally.

The great utility of conditional lethal mutations is that the mutant microbe or virus can be cultured normally under permissive conditions. At a given moment, the cells or virions can be transferred to a nonpermissive environment, and the events that follow the expression of the lethal mutation can then be followed by physical, chemical, and microscopic analysis. The two most widely studied systems of this type are *temperature-sensitive mutants* of microorganisms and viruses and *genetically suppressible mutants* of bacteriophages.

### Temperature-sensitive mutants

Some alterations in amino acid sequence may cause the polypeptide to become nonfunctional when the temperature approaches the higher end of the normal physiological range of the organism, while remaining functional at temperatures approaching the lower end of the range. For example, many proteins of *E. coli* may be altered by mutation such that they are denatured at 42 to 44°C, while remaining intact at 30°C. Such proteins are said to be *temperature sensitive*, or *thermolabile*. When the temperature-sensitive mutation occurs in a gene governing an indispensable function, the cell will be nonviable at the higher temperature but will grow normally at a lower temperature; such a mutation is conditionally lethal.

Many temperature-sensitive, conditional lethal mutants of *E. coli* have been isolated. For example, mutants have been isolated which cease protein synthesis when transferred from 30 to 42°C; extracts of several such mutants have been found to contain thermolabile amino acid–activating enzymes. Other mutants have been found that are unable to synthesize DNA or RNA, or to undergo normal cell division, at the nonpermissive temperature.

Temperature-sensitive, conditional lethal mutants of viruses constitute an extremely useful class of mutants for the study of viral genetics and viral reproduction. For example, mutants of bacteriophage T4 have been isolated which form plaques at 28°C but not at 43°C. When host cells are infected with such a mutant at 43°C and examined in the electron microscope, the nature of the temperature-sensitive function is often revealed; for example, the cells may contain complete phage tails but no phage heads.

**Genetically suppressible mutations of bacteriophages**

For a base-pair change in DNA to produce an altered amino acid sequence in a protein, the normal functioning of the translation apparatus is required. If one of the components of the translation apparatus (e.g., an activating enzyme or a species of transfer RNA) is altered by a second mutation, the "error" in the genetic message may be corrected in a certain fraction of translational events. This process of correction, called *genetic suppression*, was discussed in Chapter 12.

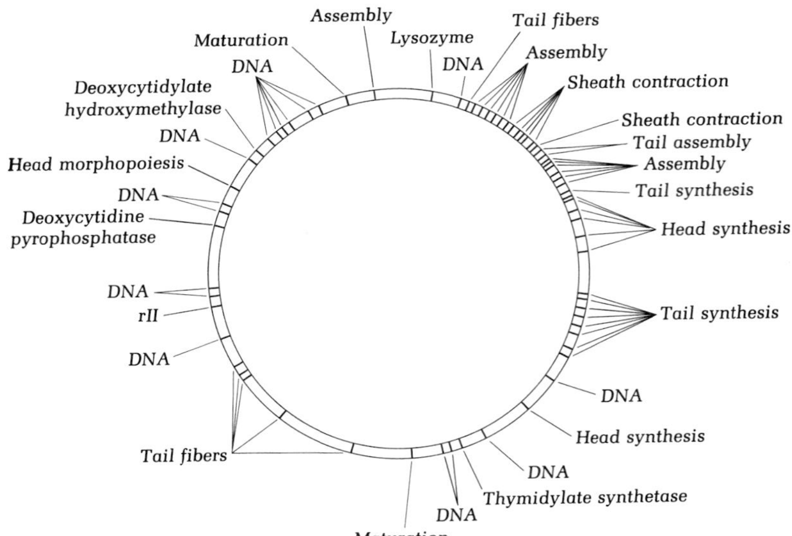


Figure 13.1 The genetic map of bacteriophage T4. The chromosome of T4 is represented by a ring; the dark lines in the ring represent genes. Note that many genes are clustered according to function.

In Chapter 12 we also described nonsense mutations, one type of which—the amber mutations—results in the formation of the nonsense codon, UAG. The UAG codon, wherever in the genome it is produced, is suppressible by a particular class of suppressors (the amber suppressor genes). When phages bearing amber mutations are grown in a host that carries an amber suppressor gene, they replicate and mature to produce a crop of normal phage progeny, since their mutations are not expressed. The host carrying the suppressor is thus a *permissive host*. If the progeny phage are now used to infect a bacterial strain that does not carry an amber suppressor (i.e., a *nonpermissive host*), the amber mutation is expressed, and an altered gene product is produced. Since the amber codon, UAG, is a chain-terminating codon, amber mutations are always lethal in the nonpermissive host.

Amber mutations have been extensively studied in bacteriophage T4. When the defects produced by these mutations were analyzed in the nonpermissive hosts, they were found to have affected all the known phage functions: DNA synthesis, head synthesis, tail synthesis, and assembly. The sites of the mutations have been located on the chromosomal map of phage T4 by genetic recombination analysis; the resulting map is shown in Figure 13.1, which combines the results of analyses of both amber and temperature-sensitive mutations.

**Other mutant types of bacteriophages**

In addition to the lethal (and conditional lethal) mutations described above, phages can undergo mutations that produce nonlethal changes in phenotype. The most readily observed nonlethal changes are those which produce *alterations in plaque morphology* and *alterations in host range*.

When a wild-type phage, such as T4, is plated on a sensitive host, such as *E. coli* strain B, under carefully standardized conditions, the plaques that appear are homogeneous and characteristic in appearance: they are small, with irregular fuzzy edges. When a large number of plaques is examined, however, a few aberrant types are always observed; when particles from such plaques are picked and replated, the aberrant plaque type is found to breed true and thus to reflect a genetic mutation. A number of such mutant phenotypes are listed in Table 13.2.

Later in this chapter we will discuss the occurrence of phage-resistant mutants in populations of phage-sensitive bacteria. These mutants owe their resistance to the production of altered surface receptors, such that they no longer adsorb wild-type phage particles; *E. coli* strain B, for example, can mutate to the state designated B/2, which does not adsorb phage T2. If $10^6$ or more particles of T2 are plated

*Table 13.2 Some mutant types of T-even bacteriophages*[a]

| Type | Phenotype | Primary effect of mutation |
|---|---|---|
| *Rapid lysis* | *Large plaques with sharp edges* | *Unknown* |
| *Minute* | *Very small plaques* | *Slow synthesis of phage, or precocious lysis of host cell* |
| *Host range* | *Adsorbs to bacteria resistant to the wild type* | *Altered polypeptides of the tail fibers* |
| *Cofactor-requiring* | *Requires a cofactor such as tryptophan for adsorption to host* | *Abnormal tail fibers bind to sheath, require cofactor to be released* |
| *Acriflavin-resistance* | *Forms plaques on agar containing concentrations of acriflavin lethal for wild type* | *Causes host cell membrane to have reduced permeability for acriflavin* |
| *Osmotic shock* | *Survives rapid dilution from 3.0 M NaCl into distilled water* | *Alteration in head protein increases permeability of head* |
| *Lysozyme* | *Does not produce halo around plaque* | *Abnormal lysozyme synthesis* |

[a]Modified from G. Stent, *Molecular Biology of Bacterial Viruses*. San Francisco: Freeman, 1963.

on a lawn of B/2 cells, however, a few plaques appear; when particles from these plaques are isolated and purified, they are found to be *host-range mutants*, which can now adsorb to cells of B/2 as well as to cells of *E. coli* strain B. The mutation in this case consists of a base-pair change in the gene governing the structure of the tail-fiber proteins, which are the adsorption organs of phage T2. The mutant phage is designated T2h.

By plating cells of B/2 with the mutant phage, one can select a new class of mutant bacteria, resistant to phage T2h. The entire cycle can now be repeated: a second-step host-range mutant of the phage can be selected, which can adsorb to the new resistant bacterium. Apparently, any altered configuration of the bacterial surface receptor can be matched by an alteration in the adsorption organ of the phage. In nature, the mutational capacities of cell and virus permit both to exist: at any given moment there are both susceptible hosts available to the virus as well as cells that can resist viral attack.

## The time course of phenotypic expression of mutation

When certain types of mutation occur in a bacterial cell, there may be a considerable delay before the mutation is phenotypically expressed.

Two phenomena account for this delay, which may last for several generations: phenotypic lag and nuclear segregation.

### Phenotypic lag

When a suspension of phage-sensitive bacteria is treated with a mutagenic agent (see Chapter 12) and the survivors are plated immediately on phage-coated agar, virtually no induced phage-resistant mutants develop. If, however, the survivors are permitted to undergo several generations of growth in nutrient broth before plating with phage, a large number of resistant mutants is obtained. The results of a typical experiment are shown in Figure 13.2, which shows that some of the induced mutations take as long as 14 generations to be expressed.

The basis of this delay in phenotypic expression became clear when the mechanism of phage resistance was elucidated. A sensitive bacterium adsorbs phage by means of specific receptors in the cell wall; resistant mutants lack these receptors. At the time that the sensitive cell undergoes the genetic change to resistance, its wall still possesses the preexisting receptor sites and the cell remains phenotypically sensitive. During subsequent growth, however, receptors are no longer synthesized and the old ones are diluted by the formation of new cell wall material. Phenotypic expression of the mutation to resistance is thus delayed until the mutated cells possess too few receptors to allow phage adsorption.

*Figure 13.2 Delay in phenotypic expression of mutation. A suspension of phage-sensitive bacteria is treated with a mutagen and the survivors are plated on phage-coated agar. Only induced phage-resistant mutants appear. In the experiment below, the survivors were allowed to produce varying numbers of generations of growth before plating.*

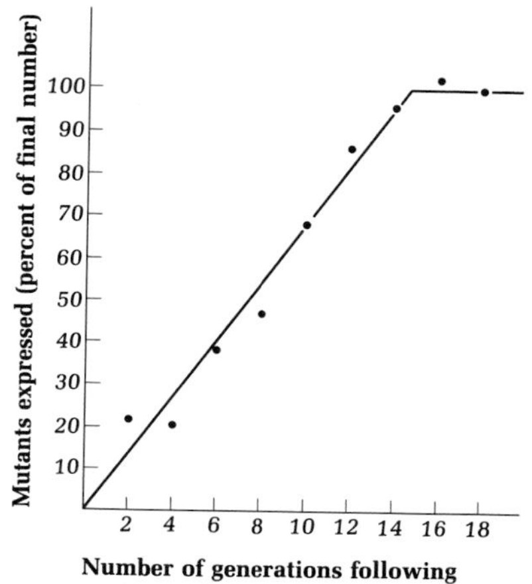


This type of delayed expression is called *phenotypic lag.* In general, a phenotypic lag occurs whenever the primary effect of a mutation is the loss of a stable gene product. Phenotypic lags will thus be observed in mutations from prototrophy to auxotrophy and in mutations that deprive the cell of the ability to use a particular form of carbon or nitrogen source. In each of these cases the primary effect of the mutation is the loss of an enzyme activity; at the time of the mutation, the cell possesses a large number of enzyme molecules which must be diluted out by cell growth and division before the effect of the mutation is observed.

The practical significance of phenotypic lag is seen in the use of the "penicillin technique" for the selection of auxotrophic or other mutants with loss of enzymatic function (Table 13.1). This technique exploits the fact that penicillin kills only those cells which are actively growing; if the survivors of mutagenic action are treated with penicillin immediately, the mutants are killed as effectively as the wild type. It is necessary to permit the survivors to undergo several generations of growth before treatment with penicillin, to allow the delayed phenotypic expression of "loss" mutations.

When a mutation results in the gain, rather than the loss, of a gene product, phenotypic expression is for all practical purposes immediate. If an auxotroph, for example, mutates to regain the ability to form a biosynthetic enzyme, active enzyme molecules begin to be synthesized immediately and to function in the biosynthetic pathway that had previously been blocked.

### Nuclear segregation

The phenomenon of phenotypic lag, which may last several generations, tends to mask another factor that delays the phenotypic expression of mutation: the *multinucleate* nature of most bacterial cells.

Most bacteria contain an average of two to four nuclei per cell during the exponential phase of growth. These nuclei are genetically identical, since they are derived from a single nucleus that existed in the cell line one or two generations earlier. When a mutation occurs in one of the nuclei, the cell becomes a *heterocaryon*. If the mutation is

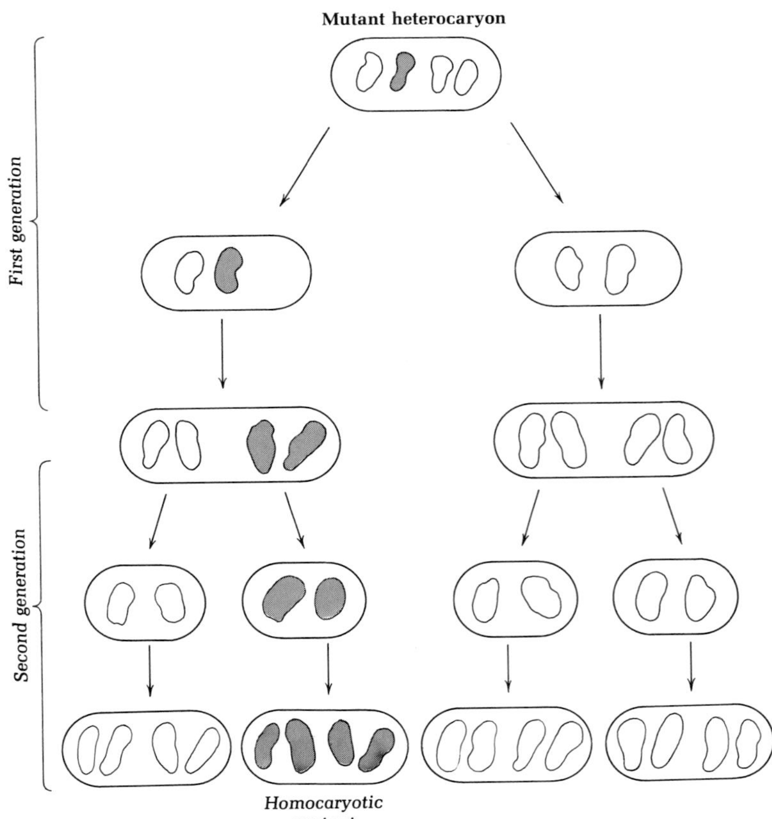


*Figure 13.3 Bacterial cells with either two or four nuclei. If a mutation first occurs in a tetranucleate cell, two generations are required before a homocaryotic mutant cell appears. If the mutation is recessive, it cannot be expressed until the mutant nucleus has completely segregated from the unmutated nuclei.*

one that causes the loss of a gene product it will be *recessive* in the heterocaryon, since the other nuclei continue to make the gene product. Thus, even in the absence of phenotypic lag, one or two cell generations are required for nuclear segregation to produce a homocaryotic mutant cell, permitting a loss mutation to be phenotypically expressed. This situation is shown in Figure 13.3.

A bacterium is a *haploid* organism, because all its nuclei are identical. Even when genetic heterogeneity is produced by mutation or—as will be described in Chapter 14—by intercellular gene transfer, the haploid condition is restored within a few generations by the process of cell division and nuclear segregation.

## *The spontaneous nature of bacterial mutation*

**Early concepts of bacterial variation**

Before the invention of pure-culture methods, bacteria were believed to have a limitless capacity for variation—indeed, all the different types observed were thought by some biologists (the pleomorphists) to represent different stages in the life cycles of a few kinds of microbe. Robert Koch's demonstration that each type of bacterium breeds true when grown in pure culture caused the pendulum to swing the other way; monomorphism became the prevalent opinion. At the beginning of the twentieth century, however, it was recognized that the characters of even the most impeccably pure cultures were able to change, and many bacteriologists began to study the mechanisms of such variations.

There were two kinds of variation to be explained. One was *adaptation*—the kind of change that confers on an organism increased fitness to its environment. Bacferial cultures placed under changed conditions were found to adjust rapidly to their new environment and to remain adapted indefinitely. For example, cultures often become drug resistant when grown in the presence of drugs, or they may adapt to use new sources of carbon or nitrogen.

The other sort of change was called *dissociation*. Many virulent pathogens, when first isolated from animal hosts and cultivated on agar, form colonies that are characteristically *smooth* and glistening, with regular edges, as a result of capsule formation by the individual cells. The animal body is an environment that strongly selects for encapsulated forms of pathogenic bacteria, since nonencapsulated forms are quickly phagocytized and destroyed. In other environments, however, the selective pressure is reversed, the nonencapsulated variants having higher growth rates. Such variants form rough colonies, which are wrinkled and dull, with irregular edges (Figure 13.4). Thus, in the course of serial transfer on laboratory media, a culture that on primary

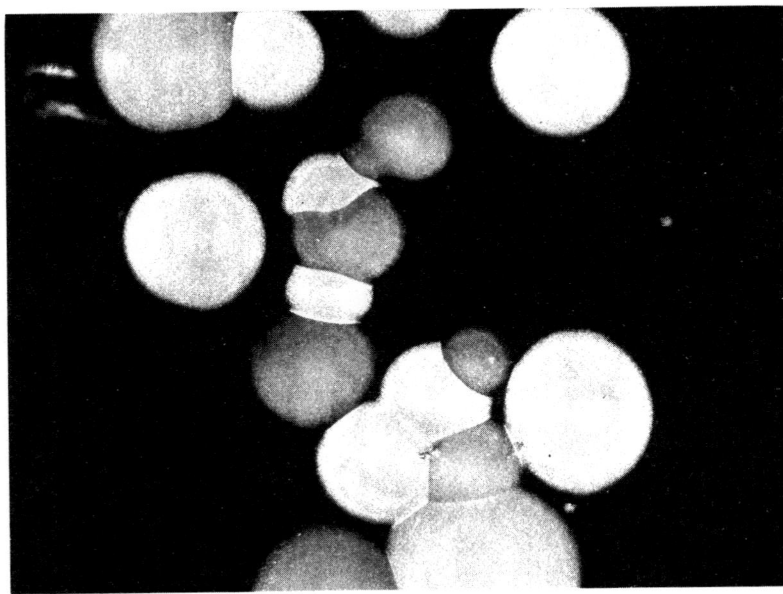

*Figure 13.4 Smooth and rough colony types of a true bacterium, Brucella abortus. Courtesy of W. Braun.*

isolation gave only smooth colonies gradually changes to produce a mixture of smooth and rough colonies and finally to a pure culture of the rough type. It was this smooth-to-rough transition to which the term "dissociation" was first applied; but eventually it was used to describe any appearance of a novel colony type in a culture previously homogeneous with respect to colony form.

We now know that both adaptations and dissociations result from the selection of rare mutant types that arise by chance in the population. Why, then, did it take so long for bacteriologists to accept this explanation, when mutation had already been recognized as the ultimate basis of variation in higher oganisms? There seem to have been two reasons. In the first place, the belief was widespread that bacteria have no nuclei and therefore no hereditary apparatus comparable to that of a higher organism. Even though workers had occasionally reported the presence of nuclear bodies in bacterial cells, the universal presence of nuclei in bacteria was not generally accepted until, in 1942, C. Robinow devised a staining method that showed their existence in all bacteria. Second, one never sees a mutant bacterial cell when it first appears. Spontaneous mutations are so rare, and bacteria so small and simple in structure that it would be a hopeless task to discover a recognizable mutant by direct microscopic examination. As a result, the occurrence of a mutation can be recognized only after the new mutant type has become a major fraction of the population. This is exactly what happens when bacteria are placed in an adverse environment: any mutant that chances to be present and that can grow faster than the parental type will be selected and will rapidly become predominant. The population change takes place so rapidly that to the early bacteriologists it appeared as though many or all of the cells of the original inoculum had changed in direct response to the environment. Such direct adaptations do occur (e.g., the induction of enzymes by their substrates), but changes of this type do not normally persist after the stimulus is removed. Mutations, in contrast, are truly heritable and do not depend directly on the environment for their maintenance. From time to time it had been suggested that bacterial variation had its basis in spontaneous gene mutation, but no critical evidence for this theory was put forth until S. Luria and M. Delbrück introduced the fluctuation test in 1943.

**The fluctuation test**

To those who looked favorably on the hypothesis of gene mutation, the great challenge was adaptive variation. In higher organisms, mutations were known to be chance events not directed by the environment. Even with deleterious agents that caused mutations, the mutation rate of all genes was raised more or less equally, and these agents did not specifically induce changes conferring resistance to their action. In bacteria, however, an adaptive mutant could be detected only by cultivating the entire population in the presence of the adverse agent itself, to permit selection to reveal the variant. For example, if a population of phage-sensitive bacteria contains one cell in 10 million that is phage-resistant, the only way to detect it is to spread at least 10 million cells on an agar plate coated with phage. All the phage-sensitive cells are then killed, and only the rare resistant variant survives to form a colony. How can it be ruled out that resistance reflects a "directed variation" brought about in some way by the phage itself? The answer to this question was provided when Luria and Delbrück devised the fluctuation test. An example of this test is outlined in Figure 13.5. A young broth culture of a bacterium is diluted with fresh medium and divided into two 10-ml aliquots. Each aliquot contains, let us say, 1,000 cells. One aliquot is not subdivided, whereas the other is distributed equally among 50 tubes. All the cultures are incubated, and after they have reached a suitable density they are plated on phage-coated agar. Fifty equal samples taken from the large culture are plated separately. The contents of each of the 50 small cultures are likewise plated separately. As shown in Figure 13.5, roughly the same number of phage-resistant colonies is found on each of the 50 plates made from the large culture. If the phage were inducing the mutation to resistance after the samples

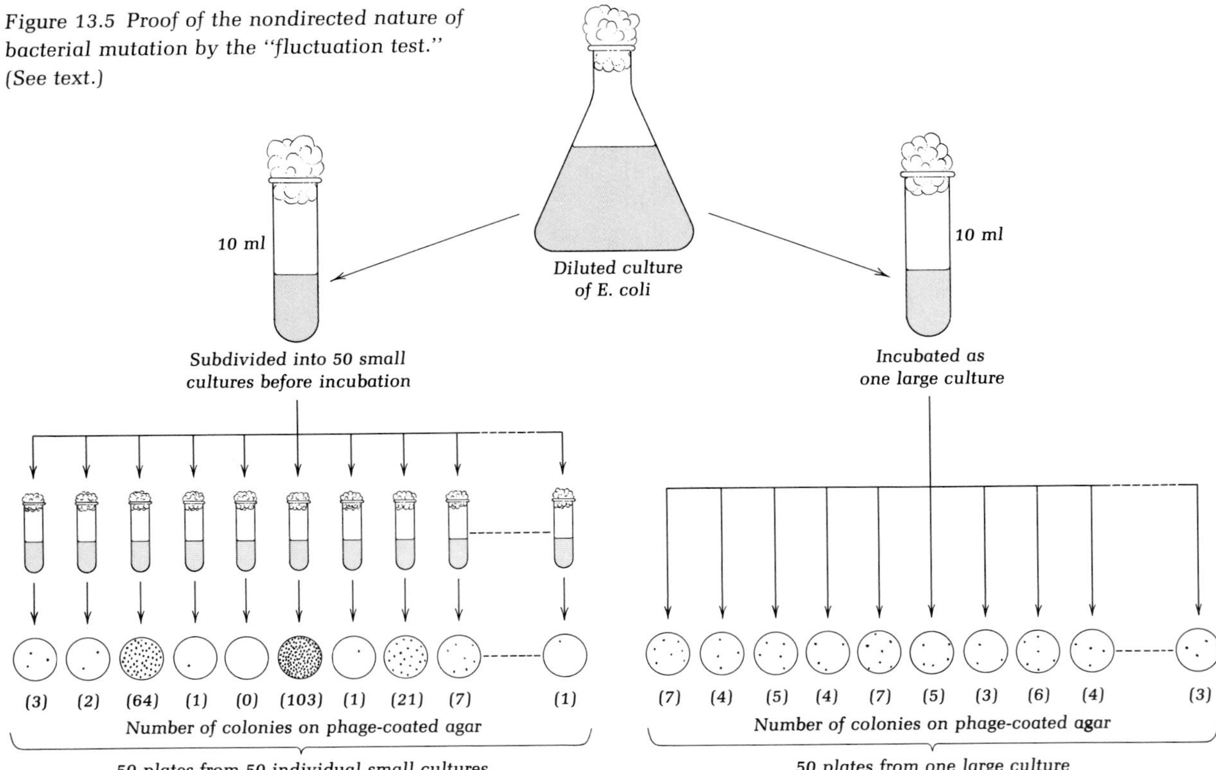


Figure 13.5 *Proof of the nondirected nature of bacterial mutation by the "fluctuation test." (See text.)*

had been plated, a similar distribution of resistant colonies should be found on the plates made from the 50 small cultures, for all the bacterial cells would be alike at the time of plating. If, however, the mutations occurred as chance events in the cultures before they were plated with the phage, some cultures might contain no mutants, whereas those in which mutations had occurred early in the incubation period would contain many mutant cells. The numbers of mutants from the 50 small cultures would then show a high degree of fluctuation compared with the numbers of mutants in the samples that all came from the one large culture.

When Luria and Delbrück carried out such a test with the bacterium *E. coli*, a high fluctuation in the numbers of mutants arising from the small cultures was observed. Statistical analysis showed that the variance among the samples from the large culture was that expected from sampling error, whereas the variance between the samples from the many small cultures was several hundred times higher. Thus, for the first time, bacterial mutations were shown to be rare chance events not directed by the environment. The fluctuation test has since been used to analyze other adaptive changes and has consistently shown that mutations are spontaneous and not induced by the change of environment.

**Replica plating**

The fluctuation test is a statistical method and only indirectly proves the spontaneous nature of mutations. In 1952, J. Lederberg with E. Lederberg devised a procedure, called *replica plating*, that permits this to be demonstrated directly. For example, about $10^8$ cells of a phage-sensitive strain of *E. coli* are spread on a plate of nutrient agar and incubated until each cell has formed a clone of a few hundred progeny. This will appear to the eye as confluent growth on the surface of the agar. If a few of the $10^8$ cells are spontaneous phage-resistant mutants, they will produce phage-resistant clones on the agar. The problem is thus to find these invisible clones within the confluent lawn of cells and to isolate a phage-resistant mutant from one of them.

This is accomplished by replica plating (Figure 13.6). A piece of sterile velvet is stretched over a cylindrical block of metal, slightly smaller in diameter than a petri dish. The block is placed with the velvet surface facing upward; the petri dish with the lawn of bacterial cells is inverted, and its surface is gently pressed against the velvet. The projecting fibers of the velvet, numbering thousands to the square inch, act as inoculating needles, sampling every clone of cells in the lawn. The petri dish is removed, and two or more fresh plates, this time containing *phage-coated agar*, are pressed against the velvet in turn. The original "master plate" is saved, and the inoculated phage-coated plates are incubated.

After incubation, a few colonies appear on the phage-coated plates. Some of these may represent mutants that arose during the few cell divisions that occurred after replica plating; if a colony is found at the identical position on every replica plate, however, they may be presumed to have arisen from an inoculum of phage-resistant cells transferred via the velvet from a phage-resistant clone on the master plate.

The investigator marks this position and with an inoculating needle carefully picks some material from that location on the master plate. Let us assume that his needle removes $10^5$ cells, of which two or three are phage-resistant. He transfers this inoculum to a tube of broth, incubates it for several hours to increase the total cell number, and then spreads a sample of this culture on a fresh agar plate. This time, however, his inoculum is *enriched* in phage-resistant mutants as a result of picking from the neighborhood containing a clone of resistant cells.

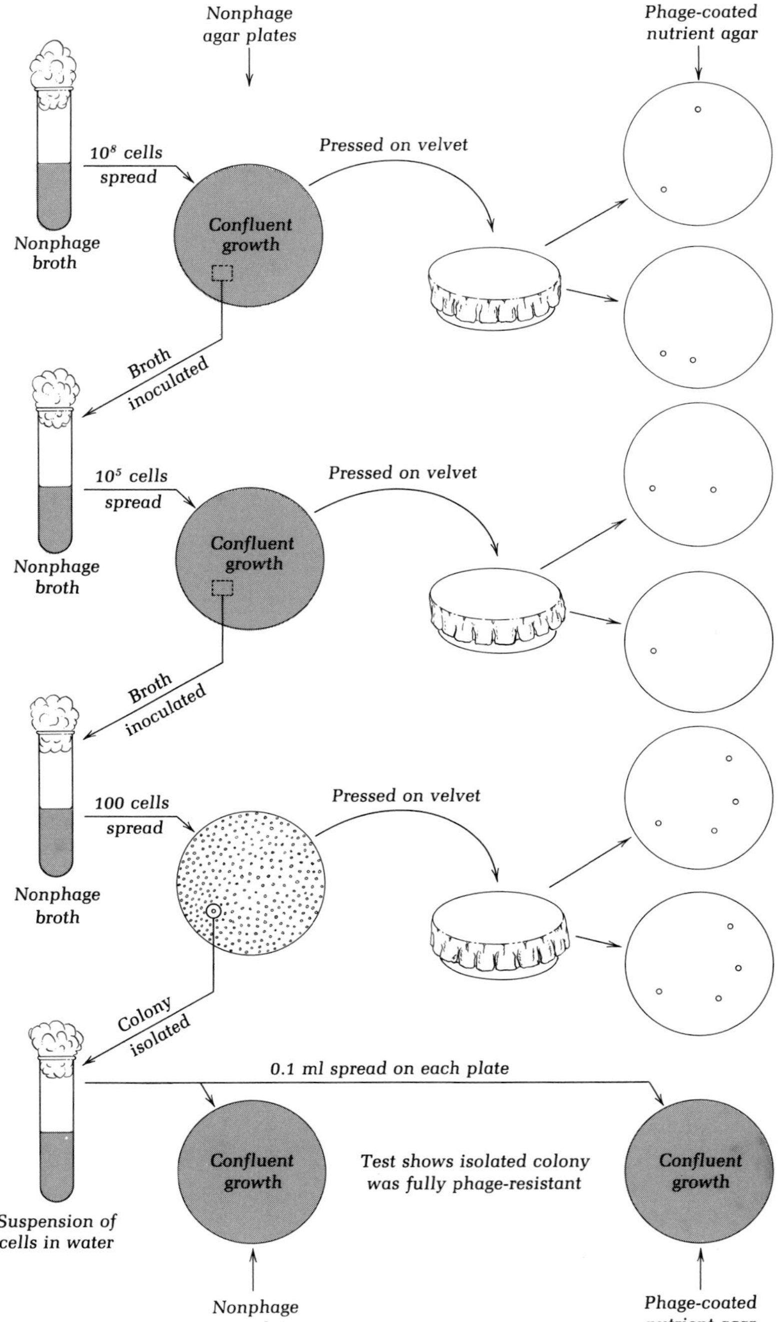


Figure 13.6 Use of the replica plate method to prove the nondirected nature of bacterial mutation (see text). The bacteria are successively transferred between nonphage broth and nonphage agar (left side of diagram). At each stage, replica plates are made on the phage agar (right side of diagram). The final suspension of bacteria is found to be completely phage resistant, as is indicated by confluent growth on the two plates at the bottom of the diagram.

Thus, he has only to spread $10^5$ cells, instead of $10^8$ cells, to be sure of having several phage-resistant clones on the plate.

When the spread plate is incubated and the resulting lawn of cells is replicated to phage-coated agar, phage-resistant clones can again be located on the master plate. Now, however, the clones are much larger, because when $10^5$ cells grow into a confluent lawn each cell can produce a clone numbering 100,000 cells or more. An inoculum taken from the suspected site is thus greatly enriched for phage-resistant cells, and one more cycle of replica plating is sufficient to obtain well-isolated colonies of the phage-resistant mutants.

The important feature of the experiment is this: the cells were spread on phage-free agar, picked and subcultured in phage-free broth, and respread on phage-free agar several times in succession. Nevertheless, by picking each time from a region on the plate shown by replica plating to contain phage-resistant mutants, it was possible to increase the proportion of mutants at each round of plating so that eventually a pure culture of a phage-resistant strain could be obtained. During the entire process the population was never exposed to phage, proving that the original mutation to phage resistance occurred in the complete absence of the selective agent.

Replica plating has become a major tool of bacterial genetics, since it permits the simultaneous transfer of hundreds of colonies from one medium to a series of different media. In the following chapters we shall discuss the analysis of recombination experiments in which many thousands of colonies of recombinant bacteria must be tested for the ability to grow on several different media. This operation can be done in a few minutes by replica plating. An example of such a test is shown in Figure 13.7.

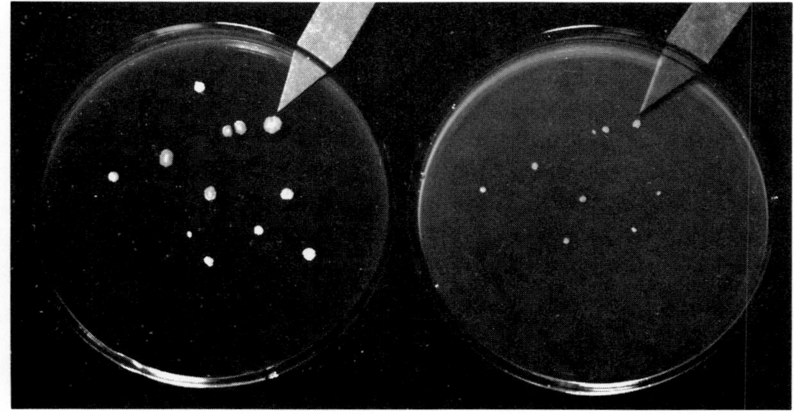

*Figure 13.7 Replica plating. The master plate of nutrient agar (not shown) bore twelve colonies. One replica was prepared on nutrient agar (left) and one on a synthetic medium lacking growth factors (right). The two plates are similarly oriented, and the arrow points to sister replicas of one colony. Note that while twelve colonies developed on the complex medium, only nine were formed on the synthetic medium. The three colonies that failed to give replicas on the synthetic medium were composed of mutants that required growth factors for their development.*

## *Population dynamics*

A growing population of microbial cells is in a dynamic state with respect to the presence of mutant types. Two parameters are involved in this phenomenon: *mutation rate,* which can be assigned a rate constant; and *mutant frequency* (or mutant proportion), which is a variable parameter determined both by the rate of mutation and by the *rate of selection* of the mutant type.

### The estimation of mutation rate

For a population of microbial cells the *mutation rate* can be defined as the *probability that any one cell will mutate during a defined interval of time.* The first measurements of bacterial mutation rates were performed by Luria and Delbrück in 1943, at a time when the nature of the genetic material and the mechanism of mutation were unknown. They chose as their time parameter the *division cycle,* or the bacterial generation; their formula for mutation rate is thus *the number of mutations per cell per generation,* averaged over many generations. As discussed in Chapter 12, most spontaneous mutations represent errors in template action which occur when the DNA double helix replicates. Since under normal conditions DNA replication and cell division are coordinated, the Luria–Delbrück formula is valid under normal physiological conditions of microbial growth.

In the Luria–Delbrück formula, mutation rate is the probability of a mutation occurring when one cell doubles in size and divides to form two cells. This series of events is called a *cell-generation.* The number of cell-generations can be simply determined for any culture, since each cell-generation increases the number of cells in the culture by one. Thus, the number of cell-generations equals the net increase in cells over the period of cultivation, and is given by the expression

$$n - n_0$$

where $n$ is the final number of cells and $n_0$ the number of cells at time zero. A small correction has to be applied to this expression, since at the moment that the culture is sampled to measure $n$, the cells are in varying stages of completion of their next division cycle. In an exponentially growing, nonsynchronized culture, the average progress toward the next generation is such that the true number of cell-generations accomplished by the culture is

$$\frac{n - n_0}{\ln 2} = \frac{n - n_0}{0.69}$$

where ln 2 is the logarithm of 2 to the base $e$. The mutation rate is equal to the average number of mutations per cell-generation; we then have the equation

$$a = \frac{m}{\text{cell-generations}} = \frac{m}{(n - n_0)/0.69} = (0.69)\frac{m}{n - n_0}$$

where $a$ stands for the mutation rate and $m$ for the average number of mutations occurring when $n_0$ cells increase in number to $n$ cells.

To determine the mutation rate of a given culture, it is thus necessary to determine $m$. A simple way to achieve this is to allow the mutations to take place in a population of cells growing on a solid medium. Under such conditions, each mutation gives rise to a mutant clone that is fixed in situ and—with appropriate manipulations—can be detected as a single colony.

In practice, this means that a population of cells must be permitted to undergo a limited number of cell divisions on an agar plate, following which the conditions must be changed so that only the mutant clones can continue growing to form visible colonies. A variety of methods has been introduced to achieve these conditions; two examples will suffice.

(1) Cells of an auxotrophic strain are spread on minimal agar containing a sufficient amount of the required growth factor to permit a limited number of divisions. Growth of the parental type then ceases; any prototrophic mutants, which no longer require the growth factor, are able to continue growth and to form visible colonies. The number of such mutants present in the inoculum must be subtracted; this number is determined by including one set of plates with no growth factor in the agar.

(2) A population of streptomycin-sensitive cells is deposited on a membrane filter, and the membrane is placed on nutrient agar for a limited time. The membrane is then transferred to the surface of nutrient-streptomycin agar; the sensitive parents are killed, while resistant mutants that arose during growth on the nutrient agar form visible colonies. The number of resistant mutants present in the inoculum is determined by plating one set of membranes on nutrient-streptomycin agar at zero time.

In each of these examples, the number of colonies per plate (corrected by subtraction of the number of mutants in the inoculum) equals the number of *mutations* per plate. It then remains only to determine the number of cell-generations per plate, which is accomplished by washing the cells off several plates in a known volume of liquid and performing a viable count. The number of cell-generations is, as stated above, equal to $0.69(n - n_0)$; here $n$ is the average number of cells per plate at the time that the parental population is killed or inhibited and $n_0$ the average number of cells per plate in the original inoculum.

As an example, $1.0 \times 10^6$ streptomycin-sensitive cells are deposited on each of a series of membrane filters. One set of filters is put on streptomycin agar at zero time, and the average number of resistant mutants in the inoculum is found to be 1.2 per filter. Another set of membranes is placed on nutrient agar and incubated for 6 hours. At the end of that time, half the membranes are placed on streptomycin agar, and half are used to determine the viable count. The viable count is found to be $1.0 \times 10^9$ cells per filter; the streptomycin plates, after incubation, show an average of 4.2 colonies per filter. The mutation rate, $a$, is calculated as follows:

$$a = (0.69)\frac{m - m_0}{n - n_0}$$

where $m$ is the final number of mutant colonies and $m_0$ the number of mutants in the inoculum. Substituting the experimentally determined figures,

$$a = (0.69)\frac{4.2 - 1.2}{(1.0 \times 10^9) - (1.0 \times 10^6)} = \frac{2.1}{1.0 \times 10^9}$$

the mutation rate to streptomycin resistance was thus $2.1 \times 10^{-9}$ per cell-generation.

**Mutational equilibrium**

There is a direct relationship between the mutation rate and the increase in the proportion of mutants in a culture at each generation, assuming that neither the mutant nor the parent type has a selective ad-

*Table 13.3 Hypothetical increase in proportion of mutants in a growing culture as the result of new mutations*[a]

| Generation | Average number of parent cells during generation | Number of mutant cells: New | Old | | | | Total | Proportion: mutants/parents |
|---|---|---|---|---|---|---|---|---|
| $n$ | $1 \times 10^8$ | 2 | | | | | 2 | $2 \times 10^{-8}$ |
| $n+1$ | $2 \times 10^8$ | 4 | 4 | | | | 8 | $4 \times 10^{-8}$ |
| $n+2$ | $4 \times 10^8$ | 8 | 8 | 8 | | | 24 | $6 \times 10^{-8}$ |
| $n+3$ | $8 \times 10^8$ | 16 | 16 | 16 | 16 | | 64 | $8 \times 10^{-8}$ |
| $n+4$ | $16 \times 10^8$ | 32 | 32 | 32 | 32 | 32 | 160 | $10 \times 10^{-8}$ |

[a]The numbers enclosed in the dotted line show how many mutants there would be if no further mutations took place after the first two. Note that the *proportion* of mutants then would have remained constant at $2 \times 10^{-8}$.

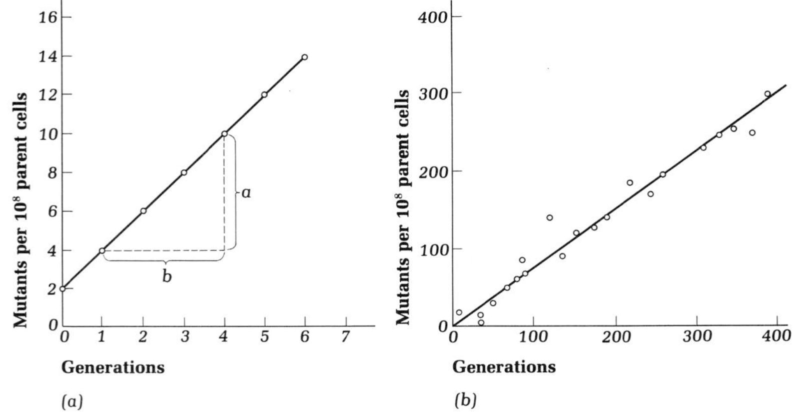


*Figure 13.8 The increasing proportion of mutants in a culture as a result of spontaneous mutation. (a) The theoretical increase that results from exactly two new mutants per $10^8$ cells appearing each generation. The mutation rate ($2 \times 10^{-8}$ per generation) is expressed by the slope of the line, which is a/b, or $(6/10^8)/3$. (b) The results of an actual experiment in which the proportion of mutants in a culture has been determined by plating at successive times. The mutation rate is found to be $0.75 \times 10^8$ per generation since this is the slope of the plotted line.*

vantage over the other. Suppose, for example, that a culture is started from a small inoculum and that the first two mutations occur when there are $1 \times 10^8$ cells in the culture. The proportion of mutants in the culture is then $2 \times 10^{-8}$. The culture continues to grow, and at the next generation there are $2 \times 10^8$ parent cells. The mutants also divide, however, and there are now four mutants in $2 \times 10^8$ cells; the proportion thus remains $2 \times 10^{-8}$. If no further mutations were to occur, and the mutant cells divided as often as the parent cells, the proportion would remain constant.

Let us assume that mutations do continue to occur, however, with a probability of $2 \times 10^{-8}$ per cell-generation. Then at each generation, for every $10^8$ parent cells, two new mutants will be added to the culture, and the proportion of mutants to parents is also increased by just that amount. This is shown in Table 13.3 and in Figure 13.8. In Figure 13.8, the slope of the line *a/b* directly represents the mutation rate; the units in which the slope is expressed are "mutants per $10^8$ cells per generation."

What happens when such a population grows indefinitely? At first thought, one might expect the proportion of mutant cells to increase until it reaches 100 percent. This is prevented, however, by the phenomenon of *reverse mutation*. Many mutations are capable of mutating back to the original state, and this reverse mutation will have its own characteristic rate. When a population of bacteria has accumulated a high enough number of mutants, reverse mutations will become significant; the proportion of mutants will ultimately level off at the point where the forward mutations and reverse mutations just balance each other. Assume, for example, that in a population of cells of type X, the mutation X → Y occurs at a certain rate. As the population grows, the

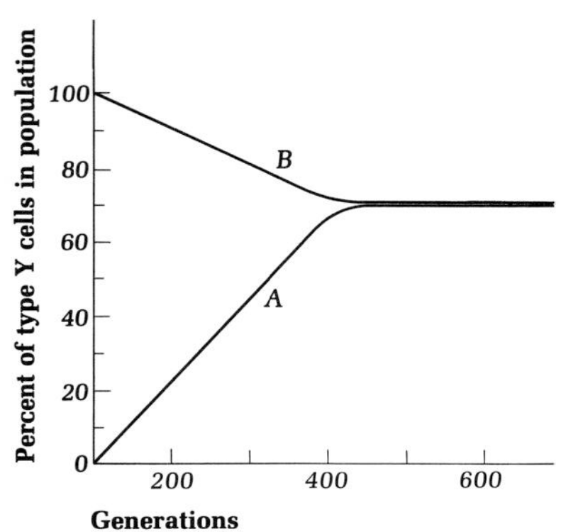


*Figure 13.9 The attainment of an equilibrium proportion of mutants in a culture. In the case of curve A, the experiment was begun with a population having no cells of type Y. As a result of forward mutation, the proportion of type Y cells rose until they constituted about 70 percent of the population. At this point, back mutation and forward mutation just balanced each other. In the case of curve B, the experiment was begun with a pure culture of type Y cells. The proportion of Y cells decreased as the result of Y → X mutations, until again an equilibrium was reached at 70 percent.*

proportion of Y mutants will increase. When there are sufficient Y cells, mutation Y → X will have a chance to occur, and eventually the population will reach a true equilibrium state in which the number of forward mutations (X → Y) just equals the number of reverse mutations (Y → X) at each generation. From then on, the proportion of mutants remains constant.

Figure 13.9 illustrates the fact that the same equilibrium proportion of Y mutants is reached whether one starts with a pure culture of cells of type X or a pure culture of cells of type Y. The actual proportion attained is a function of the relative rates of forward and reverse mutation. For example, if these rates are equal, there will be an equal number of X and Y cells at equilibrium. The relationship is simply expressed as

$$\text{equilibrium proportion of Y cells} = \frac{\text{rate of mutation (X} \rightarrow \text{Y)}}{\text{rate of mutation (Y} \rightarrow \text{X)}}$$

Thus, in the absence of selection, an equilibrium proportion of mutants should eventually be achieved if the population is allowed to multiply indefinitely.

## *Selection and adaptation*

**The genetic variability of pure cultures**

As a general rule, any one gene has only one chance in about 100 million of mutating at each cell division. At first sight, therefore, mutation might appear too rare to be of much significance. Suppose, however, that we have a "pure culture" of a bacterium in the form of 10 ml of a broth culture that has grown to the stationary phase. Such a culture will contain about 10 billion cells; for any given gene there may well be several thousand mutant cells present in the culture. Thus, a large population of bacteria is endowed with a high degree of potential variability, ready to come into play in direct response to changing environmental conditions. Because of their exceedingly short generation times and the consequent large sizes of their populations, these haploid organisms possess a store of latent variation despite the fact that they cannot accumulate recessive genes as can a population of diploid organisms. In practice, this means that no reasonably dense culture of bacteria is genetically pure; even a slight change in the medium may prove selective and bring about a complete change in the population within a few successive transfers. This explains, for example, why many "delicate" pathogenic bacteria, which prove difficult to cultivate when first isolated from their hosts, gradually become better and better adapted to the conditions of artificial media.

**Modes of selection in microbial cultures**

If a culture could be maintained indefinitely under strictly constant environmental conditions, it would reach a state of populational stability: all mutations capable of conferring fitness to that environment would have occurred and come to expression, and would have been selected. However, such constancy of conditions is never achieved in practice; no matter how carefully the environment is controlled, minor variations in the composition of the medium will occur. Such variations originate from changes in commercial sources of nutrients, as well as from variations in the sterilization process. The conditions of aeration and temperature may also vary, even though carefully regulated.

Environmental changes must necessarily be selective for any mutants in the population that are better fitted to the new conditions than is the parent type. The changes are not always absolutely selective, however; they often do not completely suppress the growth of the parent cells. Instead, parent and mutant cells may grow side by side until the mutant type, with its selective advantage, finally becomes predominant. Absolute selection is, of course, readily achieved by adding to the medium antibiotics or other chemical agents to which the parent type is highly sensitive. Absolute selection may also occur naturally, as happens when the medium becomes exhausted of a nutrient that is required by the parental type but not by a mutant.

When a culture reaches the stationary phase, in which the number of new cells formed by division is balanced by an equal number of cell deaths, selection may operate in a different way. In addition to mutants with a higher rate of growth, those with a lower rate of death will also be selected. Some of the commonly observed dissociations of smooth cultures into a mixture of rough (R) and smooth (S) cells result from this type of selection.

Earlier in this chapter we noted that some of the first bacterial variations to be studied were these so-called dissociations. Often the succession S → R → S was observed, and such changes were believed to represent life cycles of the organisms. Genetic change was not suspected. When the fundamental roles of mutation and selection in the population dynamics of bacteria became established, however, these supposed life cycles were reinvestigated by W. Braun and his colleagues. Their studies on cultures of *Brucella abortus* established that dissociation, like many other population changes, could be entirely accounted for in terms of selection operating on chance mutations.

When a culture of S cells of *B. abortus* was grown in a medium containing D-asparagine as one of the sources of nitrogen, the cells accumulated and excreted into the medium the amino acid D-alanine. This compound proved to be inhibitory to the growth of the S cells that produced it, but R cells, present as chance mutants in the population, were resistant. A gradual shift in the population thus took place, R cells replacing the parent type. On prolonged incubation, S cells again appeared in the population and once more became predominant. However, it was found that the new S cells differed from the original S cells in one very important respect: they were even more resistant to D-alanine than had been the R cells. Thus, the series of population changes did not represent a true cycle, but in fact resulted from the continuing selection of increasingly D-alanine–resistant types. One should not get the impression, however, that selection always favors the rare mutant types in the population. On the contrary, most mutations occurring in a culture are unfavorable ones, and in such cases it is the parent type that is selected by the environment.

Mutations occur, and selection operates, during the growth of a

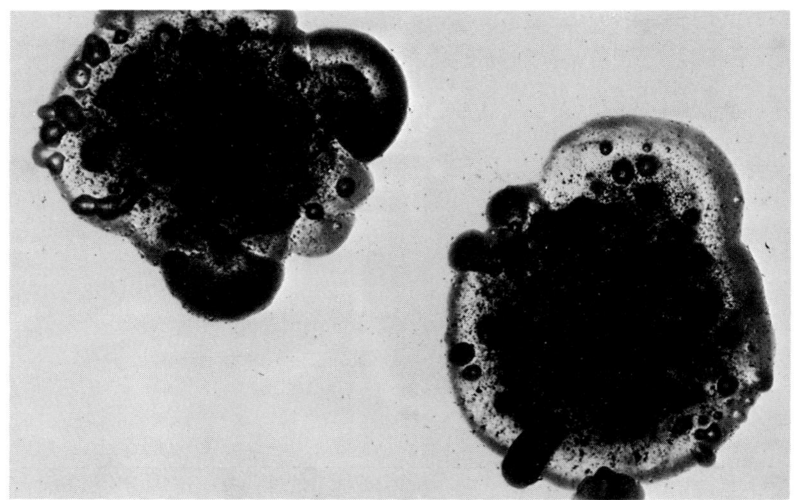

*Figure 13.10 Two bacterial colonies showing papillae, which represent secondary growth of mutants that arose during the formation of the original colonies. From V. Bryson, in W. Braun, Bacterial Genetics. Philadelphia: W. B. Saunders Co., 1953.*

*Figure 13.11 Staphylococcus colonies that have been flooded with N-phenyl-1-naphthylamineazo-O-carboxybenzene plus penicillin. The dark sectors contain penicillinase-positive cells; the light sectors contain penicillinase-negative cells. Courtesy of R. Novick.*

single colony. If the mutants in the colony begin to grow only after the parent type has ceased growing, papillae may appear. Papillae are small outgrowths consisting entirely of mutant cells, which form at the edges or on the surface of the colony (Figure 13.10). When mutants appear early in the growth of the colony, and neither mutant nor parent has a strong selective advantage, mutant sectors develop. These are often visible because the mutant grown has a distinctive texture or pigmentation. Sometimes sectors can be made visible by the use of indicators which give the parent and mutant cells contrasting colors. Figure 13.11 shows colonies of penicillinase-forming staphylococci, in which sectors arising from penicillinase-negative cells have been revealed by flooding the plates with penicillin plus an acid–base indicator. The penicillinase-positive cells hydrolyze the $\beta$-lactam ring of penicillin, producing a carboxyl group; the resulting acid turns the indicator purple. The penicillinase-negative cells do not produce acid from penicillin, and are consequently stained yellow.

**The effects of selection of one mutant type on the proportions of other mutant types in the culture**

In the section above on mutational equilibrium, we saw that the proportion of a given mutant type in a microbial population increases, in the absence of any selective advantage, in proportion to the mutation rate. Suppose, for example, that we have produced by mutagenesis a strain of *E. coli* requiring the amino acid histidine for growth. A pure culture of this strain, designated $h^-$, is put on a slant. The fully grown slant culture will probably contain one $h^+$ mutant (able to synthesize histidine and hence not requiring it for growth) for every million or so $h^-$ cells, and this proportion will increase with succeeding generations as the stock culture is transferred from slant to slant. The medium contains ample histidine, so there is no selective advantage for either type of cell. Assuming that about 10 generations are accomplished on each slant and that the culture is transferred several times a year, one should expect that in a few years the culture would contain a greatly increased proportion of $h^+$ cells.

In practice, however, this rarely happens, even when calculations based on observed forward and reverse mutation rates predict that it will. Instead, we find that an apparent equilibrium is reached long before it should be, and always in favor of the genetic type with which the culture was started. The proportion of mutant cells in the culture increases only up to a very low value—perhaps $1 \times 10^{-6}$ (one per million)—and remains there indefinitely.

This puzzling observation has been found to result from the phenomenon of *periodic selection.* At fairly regular intervals in a population of bacteria, mutants arise that are better fitted to the environment and that eventually displace the parental type as a result of selection. We cannot always define the properties of the new mutant that give it this advantage. It might be an intrinsically faster growth rate, or it might be that the new type produces metabolic products that inhibit the parent type. In any case, the better adapted mutant overgrows the culture, only to be replaced in turn by a mutant that is still better adapted. The process of replacement may be repeated many times, for new mutants that can displace the predominant type from the population continue to arise. *This periodic change in the population has a direct effect on the equilibrium proportion of all other mutants.* Let us consider the specific case of the $h^+$ mutants mentioned earlier. Suppose that a better-adapted mutant appears in the culture at the moment when the proportion of $h^+$ cells has risen to $1 \times 10^{-6}$. The better-adapted mutant could theoretically arise from either an $h^-$ cell or an $h^+$ cell, but since there are $10^6$ $h^-$ cells for every $h^+$ cell, the odds are a million to one that the new type will arise in the $h^-$ population.

The better-adapted mutant will thus have a selective advantage over all other cells in the population, which it will soon displace. Since the better-adapted type is genetically $h^-$, all $h^+$ cells in the population should disappear as the result of selection.

The total disappearance of $h^+$ cells is prevented, however, by the occurrence in the better-adapted $h^-$ population of new mutations to $h^+$. Since the new $h^+$ cells are not at a selective disadvantage, they will increase in proportion until the cycle is started over again by the appearance of an even better-adapted type.

This process can be expressed symbolically as follows. Let us call the original cells $h_0^-$ and $h_0^+$, and the first better-adapted mutant $h_1^-$. In the new $h_1^-$ population, mutations to $h_1^+$ will occur. As time goes on and the proportion of $h_0^+$ cells drops, there is a corresponding increase in the number of $\mathrm{h}_1^+$ cells. The cycle is repeated again and again. The $h_1^-$ cells give rise to a still better-adapted type, $h_2^-$, which displaces the

Figure 13.12 *Periodic selection. (a) The successive appearance and disappearance of different $h^+$ mutants. (b) The same data are replotted in terms of total $h^+$ mutants of all types. The resulting slightly fluctuating curve represents the pseudo-equilibrium level that the proportion of $h^+$ mutants reaches.*

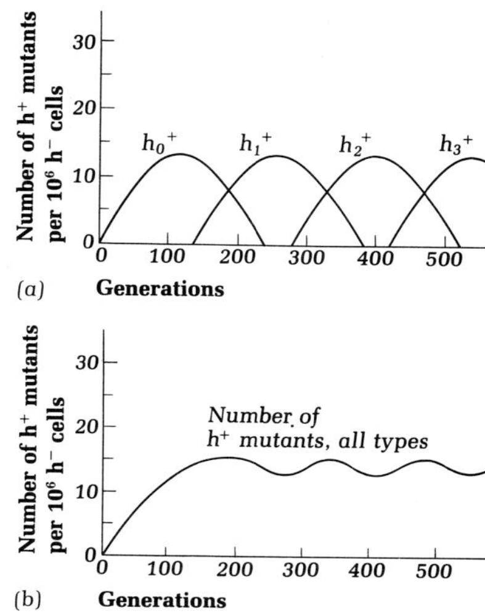


$h_1^-$ and $h_1^+$ cells. The loss of $h_1^-$ cells is compensated for by the appearance of $h_2^+$ mutants. The mutational pattern can be diagrammed as

$$\begin{array}{cccc} h_0^- \rightarrow & h_1^- \rightarrow & h_2^- \rightarrow & h_3^- \\ \downarrow & \downarrow & \downarrow & \downarrow \\ h_0^+ & h_1^+ & h_2^+ & h_3^+ \end{array}$$

The upper graph in Figure 13.12 shows the way that successive waves of $h^+$ mutants rise and fall in the population. The lower graph in Figure 13.12 shows the apparent stability of the population with respect to the characters $h^+$ and $h^-$, when $h^+$ mutants are considered as a single class.

The level that the proportion of mutants reaches can be called a *pseudoequilibrium* because it is really the composite result of a series of discrete, nonequilibrium events. With ordinary mutation rates, which are very low, the occurrence of periodic selection results in the attainment of such pseudoequilibria. It is only when both the forward and back mutation rates are extremely high that true equilibria, as illustrated in Figure 13.9, can be attained. In such cases, the proportion of mutant cells rises so rapidly that better-adapted mutants have an equal chance of appearing in the mutant or in the parent population.

Periodic selection is a subtle phenomenon, since the mutant type that is being experimentally observed (e.g., the $h^+$ mutant in the case described above) is not the one subject to selection. As the above example shows, the selection of one type of mutant in a population may prevent any other mutant type from increasing in proportion.

### Selective pressures in natural environments

So far we have considered only the selective forces that may operate in artificial cultures. In nature, however, selection acts in an even more stringent fashion. A microbe in the soil, for example, must be able not only to survive under a given set of physicochemical conditions but must also to survive in competition with the numerous other microbial forms that occupy the same niche. Any mutation that decreases, even to the slightest extent, the ability of the organism to compete, will be selected against and quickly eliminated. Nature tolerates little variation within microbial populations, for the laws of competition demand that each type retain the array of genes that confers maximum fitness.

As soon as an organism is isolated in pure culture, the selective pressures resulting from biological competition are removed. The isolated population becomes free to vary with respect to characters that are maintained stable in nature by selection. In adapting to existence in laboratory media, organisms may undergo genetic modifications that would lead to their speedy suppression in a competitive environment.

**Genetic and nongenetic adaptation**

In a population, adaptation may be either genetic or nongenetic. Genetic adaptation consists of selection acting on heritable variations. We have already considered this phenomenon in some detail. In nature, genetic adaptation is the principal mechanism of evolution. Nongenetic or physiological adaptation, in contrast, involves direct phenotypic accommodation; the cell undergoes a nonheritable, physiological change in direct response to the environment, thereby becoming better fitted.

A good illustration of nongenetic adaptation is the regulation of chlorophyll synthesis by certain photosynthetic bacteria, the purple bacteria. When grown anaerobically in the light (the conditions for bacterial photosynthesis), these organisms can modify the rate of synthesis of chlorophyll and the other photopigments in response to light intensity (Figure 13.13). At low light intensities, the cells have a high pigment content; at high light intensities, a low one. Some of these organisms can also use aerobic respiration as an alternative to photosynthesis. Growth in air, either in light or in dark, is accompanied by a progressive loss of photopigments, the synthesis of which is specifically inhibited by oxygen; consequently, the cells eventually become completely depigmented. The change is freely reversible.

*Figure 13.13 The relationship between light intensity and bacteriochlorophyll concentration in the purple bacterium* Rhodospirillum rubrum. *The bacteriochlorophyll content of the cells (expressed relatively to total cell protein) is plotted as a function of the light intensity to which the cells were exposed during anaerobic growth. Redrawn from L. P. Vernon and G. R. Seely, editors, The Chlorophylls, p. 334. New York: Academic Press, 1966.*

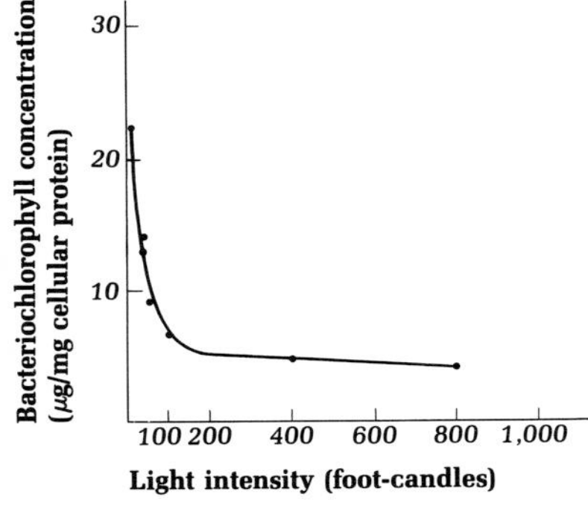


## *The consequences of mutation in cellular organelles*

Part of the genome of eucaryotic organisms is carried in the mitochondria and chloroplasts. Each kind of organelle contains and reproduces DNA that determines certain of its phenotypic properties. The properties of a chloroplast or a mitochondrion are thus controlled in part by nuclear and in part by organellar genes, both subject to change by mutation. Until recently, it has been difficult to select mutations that specifically affect organellar DNA. However, it has been found that mutations in yeast which confer resistance to certain antibiotics (those known to affect protein synthesis in bacteria) take place in the mitochondrial DNA. This is no doubt a reflection of the fact that protein synthesis in organelles is mediated by ribosomes of the 70-S, or bacterial, type (see Chapter 2). It has thus become possible to study experimentally the transmission of many different mitochondrial mutations. Each cell contains a population of mitochondria; hence the relative growth rates of normal and mutated mitochondria determine the stability of such a mutation during vegetative growth. The situation is entirely comparable to that of a growing bacterial population which contains two genetically different kinds of cells, and the outcome can also be determined by environmental factors.

A mutation in the DNA of a mitochondrion or a chloroplast can be detected by its segregation in a genetic cross. This segregation does not obey Mendelian laws, since the organellar phenotype of the progeny is determined by the number and nature of the organelles that it receives through the cytoplasm of the zygote. Examples of such non-Mendelian segregation had been known in eucaryotic microorganisms long before the discovery that mitochondria and chloroplasts contain DNA, and were described as "cytoplasmic inheritance." They can now be more precisely interpreted as reflecting the inheritance of organellar genes.

## Further Reading

**Books**

Adelberg, E. (editor), *Papers on Bacterial Genetics,* 2nd ed. Boston: Little, Brown, 1966.

Braun, W., *Bacterial Genetics,* 2nd ed. Philadelphia: Saunders, 1965.

Hayes, W., *The Genetics of Bacteria and Their Viruses,* 2nd ed. Oxford, England: Blackwell, 1968.

Stent, G., *The Molecular Biology of Bacterial Viruses.* San Francisco: Freeman, 1963.

———, *Papers on Bacterial Viruses,* 2nd ed. Boston: Little, Brown, 1965.

# *Chapter fourteen*
# *Genetic recombination*

*In eucaryotic organisms,* new combinations of genes arise not only by mutation but also by the sexual process of nuclear fusion followed by *meiosis.* The essential features of sexuality are (1) the formation of the diploid fusion cell, or *zygote,* containing the fusion nucleus; (2) the synapsis of homologous chromosomes, followed by the formation of *recombinant chromosomes* through the process of "crossing over"; and (3) the *segregation* of chromosomes through the process of nuclear division. Steps 2 and 3 normally occur at meiosis.

Among the bacteria, a number of unique processes occur that may be regarded as primitive forms of sexuality. These processes, together with recombination in bacterial viruses, will be the subject of this chapter.

## *The general features of bacterial recombination*

In molecular terms, recombination is the process by which a recombinant chromosome is formed from DNA derived from two different parental cells. Three processes that lead to the formation of recombinant chromosomes are known to occur in bacteria. In order of their discovery, these are *transformation, conjugation,* and *transduction.* They differ from the sexual process of eucaryotes in that a true fusion cell is not formed. Instead, a part of the genetic material of a donor cell is *transferred* to a recipient cell; the recipient cell thus becomes diploid for only part of its genetic complement, or genome.

The original genome of the recipient is termed the *endogenote*, and the fragment of DNA introduced into the recipient cell, forming a *partial zygote*,* is termed the *exogenote*. Both the nature and the size of the exogenote differ in the three processes. In transformation, a single-stranded piece of donor DNA replaces a short segment of one strand of the endogenote. In transduction, a small (presumably double-stranded) fragment of DNA is brought from the donor cell into the recipient cell by a bacteriophage particle, in some cases attached to a piece of phage DNA. In conjugation, the exogenote is transferred between cells which are in direct contact, and may be a major fraction of the donor genome.

Before we start discussion of these different processes, a few features common to all three will be described.

### The fate of the exogenote

If the exogenote has a sequence of base pairs which is homologous with a segment of the endogenote, *pairing* occurs rapidly and a recombinant chromosome is immediately formed. This molecular recombination event is called *integration*.

If pairing and integration are prevented for any reason, the exogenote may undergo one of several alternative fates. If the exogenote carries the genetic elements necessary for its own replication, it may *persist* and *replicate*, so that the zygote gives rise to a clone of partially diploid cells. This has been observed in special cases of both transduction and conjugation, as will be discussed later. If the exogenote lacks such elements it may persist but not replicate, so that in the clone of cells that arise from the zygote only one cell at any time is a partial diploid. This phenomenon, which occurs only in transduction, is called *abortive transduction*.

Finally, the exogenote may be enzymatically degraded; this phenomenon is called *host restriction*.

### Restriction and modification of foreign DNA

The degradation of foreign DNA that has penetrated a bacterial cell was first discovered in bacteriophage infections. It will be profitable to describe this phenomenon before going on to consider how restriction operates in bacterial recombination.

The early literature on bacteriophage contains a number of reports of what were called "host-induced modifications of bacterial viruses." In each case it was observed that the passage of a phage through one bacterial host greatly reduced its *efficiency of plating* on a second bac-

*We shall use the term "zygote" from now on to refer to the partial zygotes formed during bacterial recombination.

*Table 14.1 Efficiency of plating of bacteriophages grown in different hosts*[a]

| Hosts in which phage is grown | Efficiency of plating on: | |
|---|---|---|
| | *E. coli K12* | *E. coli B* |
| *E. coli K12* | 1.0 | $1 \times 10^{-4}$ |
| *E. coli B* | $4 \times 10^{-4}$ | 1.0 |

[a]Modified from W. Arber and D. Dussoix, "Host Specificity of DNA Produced by *Escherichia coli*. I. Host Controlled Modification of Bacteriophage λ." *J. Mol. Biol.* **5**, 18 (1962).

terial host. An example of such an experiment is given in Table 14.1; the table shows that particles of phage λ, produced during replication in *Escherichia coli* strain K12, plate with an efficiency* of 1.0 on strain K12 itself but with an efficiency of $1 \times 10^{-4}$ on *E. coli* strain B. If, however, the phage particles released from strain B are tested, their properties are reversed: their efficiency of plating is now 1.0 on strain B but only $4 \times 10^{-4}$ on strain K12. In other words, the phage particle can successfully infect only the type of host cell in which its DNA has been produced. If the DNA of the phage is "foreign" to the host cell, it fails to establish a productive infection in 99.99 percent of the cells.

The failure of "foreign" phage to infect productively a bacterial host results from the existence in the host cell of a specific endonuclease that initiates the degradation of foreign DNA; if the phage is labeled with $^{32}P$ and allowed to inject its DNA into the foreign host, the radioactive DNA is rapidly released in the form of small fragments.

This phenomenon is called *restriction*, and the endonuclease is called the "restricting enzyme." The corollary of this observation is that phage DNA which has been produced in a given type of host cell is no longer a substrate for the restricting enzyme; in other words, it has been enzymatically modified by the host so that it is henceforth protected against restriction.

Going back to Table 14.1, we can now reconsider the results in terms of restriction and modification. The phage grown in strain K12 (we will call it λK) was specifically modified so that the restricting enzyme of strain K12 could not attack it. It was still a good substrate for the restricting enzyme of strain B, however, so that the DNA of λK was degraded in strain B. Nevertheless, in one out of $10^4$ cells the phage succeeded in growing, having been modified by the *E. coli* strain B cell before degradation could begin. The DNA of the particles released from this cell carried the B type of modification; these particles, which

*A plating efficiency of 1.0 means that every particle causes a productive infection and hence a plaque. A plating efficiency of $1 \times 10^{-4}$ means that only one particle in $10^4$ can cause a productive infection of the host cell.

we can call λB, became immune to attack by the restricting enzyme of *E. coli* strain B but were quickly degraded by the restricting enzyme of strain K12.

The modification process appears to involve the *methylation* of bases at a very few, highly specific sites on the DNA, which are also the sites of attack by the restricting enzyme. The DNA of the bacterial chromosome is, of course, modified in the same way, since otherwise the chromosome would be degraded by the restricting enzyme system. Thus, restriction and modification also come into play when bacterial DNA is transferred from one cell to another by one of the recombination processes. For example, *E. coli* strain B will conjugate with *E. coli* strain K12, but the number of recombinants produced is one-thousandth of that produced in a similar cross between two strains derived from *E. coli* strain B or between two derived from *E. coli* strain K12.

The genes responsible for the production of the restricting and modifying enzymes are tightly linked on the bacterial chromosome. By recombination, it has been possible to produce a strain of K12 which carries the restriction and modification alleles of *E. coli* strain B; this K12 strain now recombines with *E. coli* strain B with high efficiency.

The existence and specificities of restricting and modifying enzymes produce a complex pattern of compatibilities between bacterial strains, with respect to the ability of the DNA of one to escape degradation in another. The pattern is further complicated by the fact that some phage genomes also possess genes for restriction and modification. Thus, a derivative of *E. coli* strain K12 that is lysogenic for phage P1 is a poor recipient in recombination with K12 donors that are nonlysogenic.

### The integration of exogenote and endogenote

Two molecular models to explain recombination have been proposed. According to one, the two parental DNA molecules undergo breakage, followed by reciprocal reunion of the broken ends to form recombinant molecules. According to the other, the recombinant molecule is formed by the replicating enzyme switching back and forth between the parental DNA templates within the region of pairing.

The second, or *copying-choice model*, became less plausible when DNA replication was shown, in 1958, to be *semiconservative*. As seen in Figure 14.1, the copying-choice model requires conservative replication, the parental duplexes remaining intact, and the recombinant molecule containing two newly synthesized polynucleotide strands. Advocates of this model, however, suggested that conservative replication might take place only in the region of a crossover and thus might not be detected in ordinary replication experiments.

Figure 14.1 *Recombination according to the "copying-choice" model. (a) Two parental DNA molecules are paired; genetic markers are indicated by the letters ABCDE. (b) Conservative replication has begun, using the lower DNA molecule as template. (c) The replicating enzyme system has switched to the upper DNA molecule as template, between markers A and B. The result is the production of a recombinant molecule carrying the markers aBCDE.*

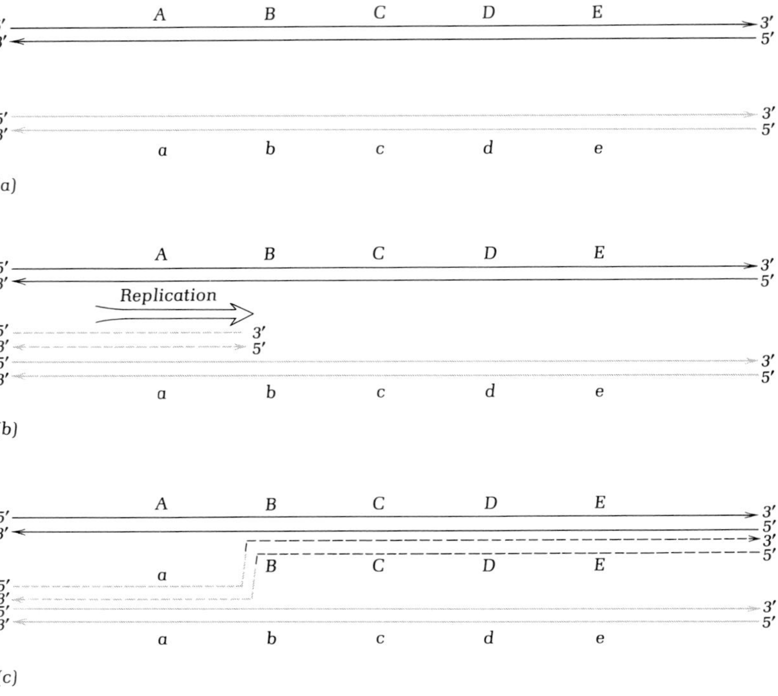


About 1960, experiments with density-labeled DNA conclusively established that recombination between phage genomes, as well as between exogenote and endogenote in transformation, occurs by *breakage* and *reunion*. It is now generally accepted that all recombination—including the integration of exogenote and endogenote in bacteria, as well as crossing over in eucaryotes—does indeed occur by this mechanism.

The most remarkable feature of the recombination process, and one that is essential to its success, is the *conservation of base-pair sequence*. With the exception of mutational differences, the recombinant DNA molecule must have the same base-pair sequence as that of the parents. For this to happen, the breaks in the two parental molecules must occur at precisely the same sites; otherwise, the process of reunion would create recombinant molecules that were either longer or shorter than the parental molecules (Figure 14.2).

The mechanism that ensures this precise alignment of the parental molecules is believed to involve the complementary pairing of bases between single-stranded regions of the parental double helices. For example, recombination between DNA fragments might occur as shown

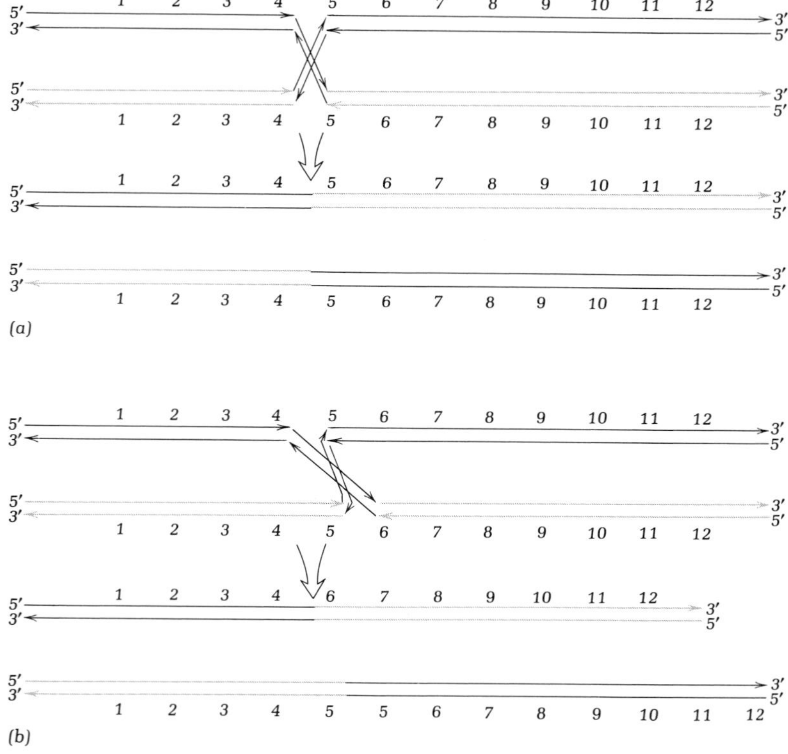


Figure 14.2 (a) Two parental DNA molecules are paired. The numbers indicate the positions of base pairs; in each molecule, a break has occurred between base pairs 4 and 5. Reunion of the broken ends as shown produces reciprocal recombinant molecules having the same base-pair sequence as the parental molecules. (b) The breaks have occurred at different positions in the two parental molecules. Reunion of the broken ends as shown produces one recombinant molecule having a deletion of base pair 5, and the other a duplication of base pair 5.

in Figure 14.3. Two fragments with overlapping base sequences undergo base pairing between complementary strands, so that the opposite strand in each case is *displaced* from the duplex and is present in the unpaired state. The displaced single strands are then *digested* by a cellular exonuclease, the digestion process continuing beyond the region of displacement. At some moment, DNA polymerase begins the *resynthesis* of the missing segment, and the process is completed by the rejoining of the strands through the action of polynucleotide ligase.

The resemblance of this process to that of *DNA repair* (p. 427) is striking. Indeed, it is likely that the enzymes of digestion, resynthesis, and reunion, which effect DNA repair, are also those involved in recombination.

When recombination occurs in the middle, rather than at the ends, of two DNA duplexes, the process must be more complicated. Again, it is believed that a single-stranded region of one parental duplex displaces the homologous region of a strand in the other duplex, but

Figure 14.3 *Hypothetical mechanism of recombination between DNA fragments with overlapping base-pair sequences. (a) A strand from one molecule has paired by hydrogen bonding with the complementary strand from the other molecule. (b and c) The displaced (unpaired) strands are sequentially digested by an exonuclease. (d) Resynthesis by DNA polymerase (dashed lines), followed by rejoining by polynucleotide ligase, closes the gaps and produces a recombinant molecule. (Note that the molecule is heterozygous for marker C. At the next round of replication, one daughter molecule will have the genotype ABCde, and the other ABcde).*

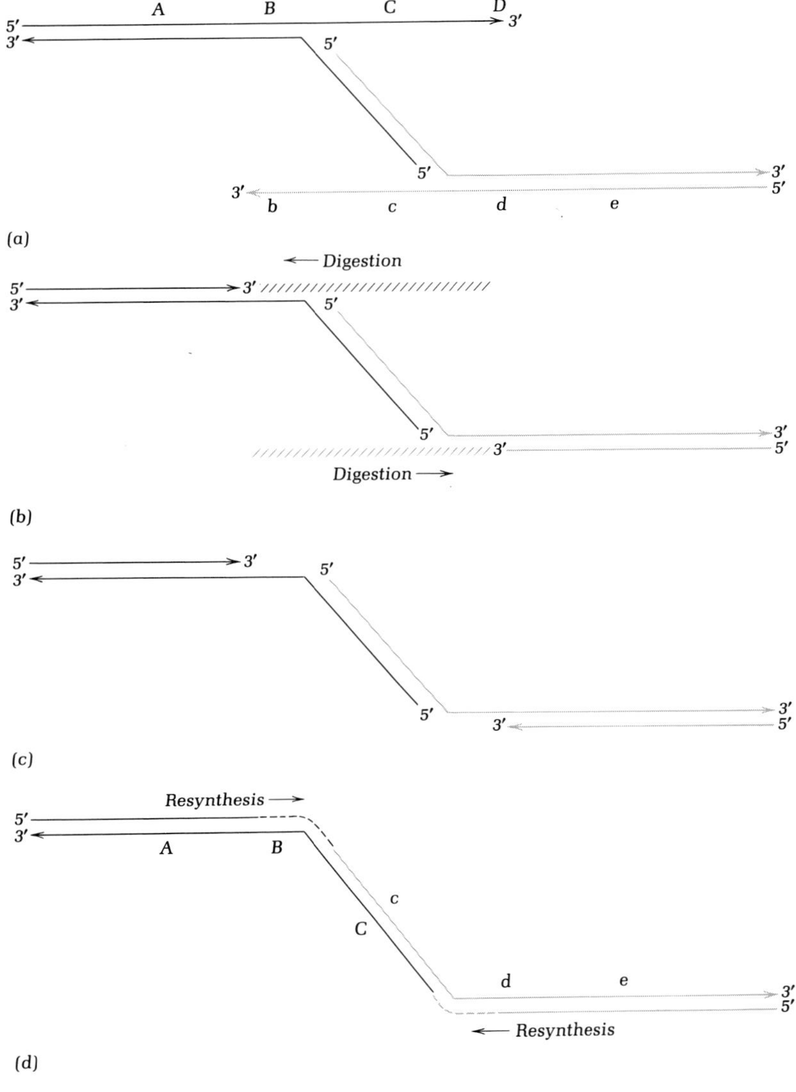


the exact sequence of events is still unknown. One possible sequence of events is shown in Figure 14.4. Whatever its mechanism, this must be the type of recombination which occurs in eucaryotic chromosomal crossing over, as well as in the integration of two different replicons* in bacteria.

*This occurs when an episome, such as the sex factor (F) or the temperate phage λ, integrates with the bacterial chromosome. The integration of episomes will be discussed later.

Figure 14.4 *Hypothetical mechanism of recombination between DNA molecules. (a) One parental molecule carries the markers AB, the other ab. Each parental molecule has a single strand cut at a different position, between markers A and B. (b) The parental duplexes have partially unwound, and a strand from one molecule has paired with a strand from the other molecule. (c) Resynthesis occurs along the single-stranded region of each parental molecule (dashed lines). (d) The newly synthesized strands pair with each other. (e) An endonuclease makes a single-strand cut in each parental duplex, and the cut 3′ ends are sequentially digested by an exonuclease. (f and g) Resynthesis and rejoining produce two recombinant molecules, Ab and aB. (After H. Whitehouse.)*

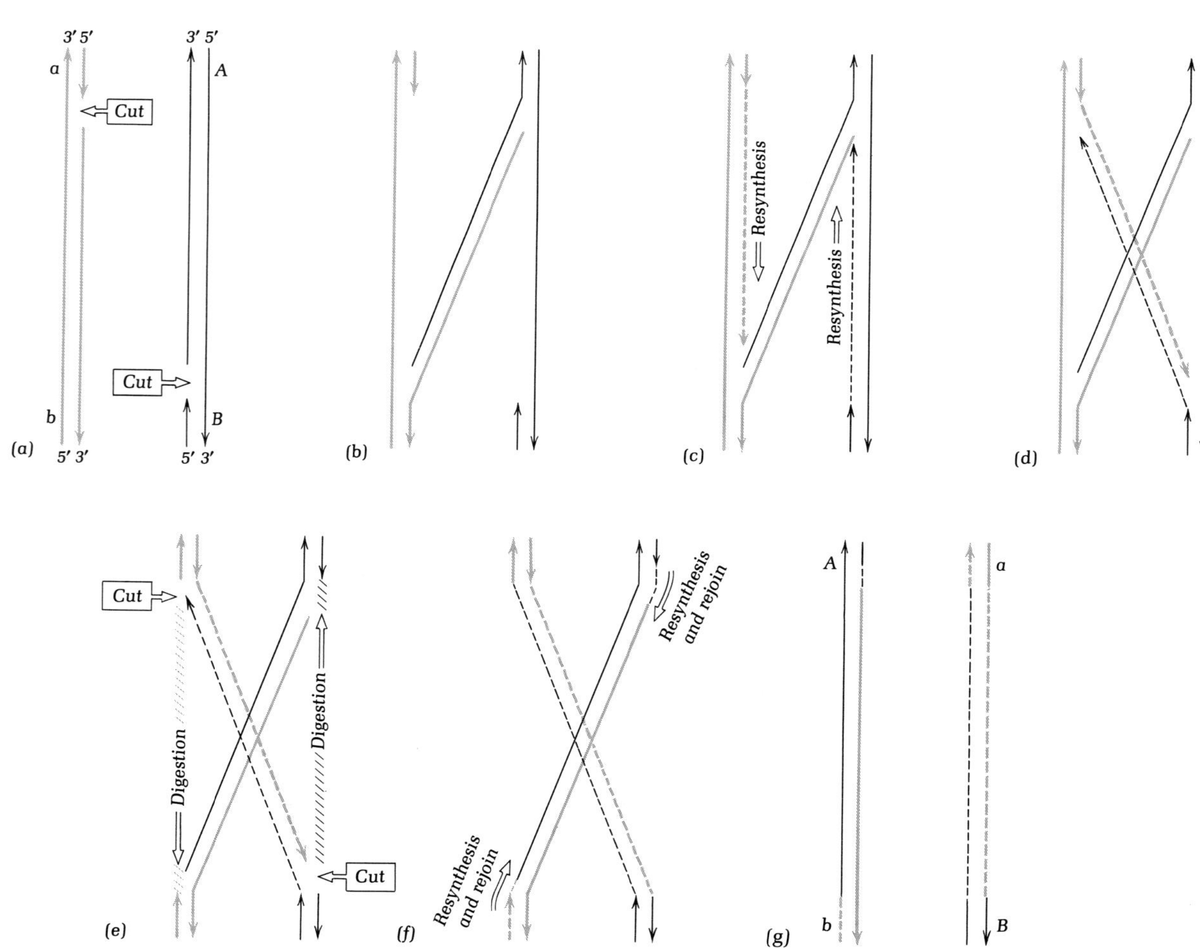

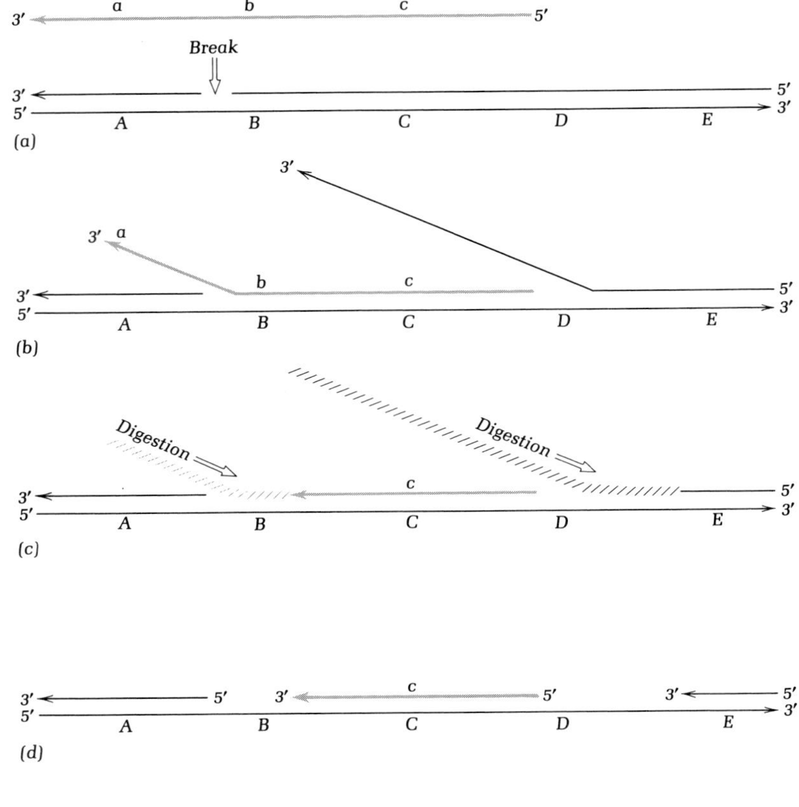


*Figure 14.5 Integration of a single-stranded exogenote with a double-stranded endogenote. (a) A single-stranded fragment, carrying the markers abc, lies near a double-stranded molecule carrying the markers ABCDE. There is a single-strand break between markers A and B. (b) Partial unwinding of the duplex permits pairing between exogenote and endogenote. (c and d) An exonuclease digests the free 3′ ends. (e) Resynthesis by polymerase and rejoining by polynucleotide ligase produce a recombinant molecule, in which one of the strands carries the markers ABcDE.*

In transformation, the situation is somewhat different. Here, isotope-labeling experiments show that only *one strand* of the exogenote is integrated into the double-stranded endogenote. A possible mechanism that could produce this result is shown in Figure 14.5.

**The segregation of the recombinant cell**

Bacteria are multinucleate, but the process of recombination involves only one nucleus in the recipient cell. Following integration of the exogenote, the multinucleate cell is thus a heterocaryon; the formation of a homocaryotic recombinant cell requires the same process of *nuclear segregation* as that described earlier for the formation of a homocaryotic mutant.

## *Bacterial transformation*

**The discovery of transformation**

Transformation was discovered in the pneumococcus *(Streptococcus pneumoniae).* The pneumococci in the sputum or tissues of a victim of pneumonia are always surrounded by large capsules composed of polysaccharide; on agar plates the encapsulated cells form smooth (S) colonies. Pneumococci can be separated into a great many types on the basis of chemical differences between their capsular polysaccharides. The different types (designated I, II, III, and so on) can be distinguished immunochemically.

When an S strain (forming smooth colonies) is serially subcultured, R cells (forming rough colonies) appear in the population. R cells have no capsules and are avirulent. In 1928 F. Griffith observed that when a very large inoculum of R cells, derived from what had originally been a type I smooth culture, was inoculated under the skin of a mouse together with heat-killed S cells of type II, the mouse died within a few days. Blood from such an animal yielded only type II smooth cells. The dead type II cells thus had liberated something which conferred on the R cells the ability to make a new type of capsular polysaccharide. The transferred property proved to be heritable.

A few years afterward, other groups of investigators succeeded in carrying out such *type transformations* by mixing R cells and heat-killed S cells in vitro. It was later found that type transformations could be brought about with cell-free extracts of S cells. In other words, when a chemical substance extracted from S cells of type II is added to a culture of R cells derived from type I, some of the cells are genetically changed (transformed) to type II. The unidentified chemical substance responsible was called *transforming principle.* Transforming principle has two properties that are generally associated only with genes: (1) it is self-duplicating, for a culture of transformed cells can be extracted and shown to contain much more transforming principle than the amount used originally to transform them; (2) it directs a specific function of the cell—the production of one type of polysaccharide.

**The nature of transforming principle**

In 1944 O. T. Avery, C. M. MacLeod, and M. McCarty succeeded in purifying pneumococcal transforming principle and identified it as DNA. Until that time it was generally believed that the specificity of the gene is determined by the protein moiety of nucleoprotein; the chemical characterization of transforming principle provided the first direct evidence that DNA is the carrier of genetic information.

Since 1944 similar transformations have been effected in other genera of bacteria, notably in *Hemophilus*, *Neisseria*, and *Bacillus*. However, all attempts to carry out transformation in certain species (for example, *E. coli*) have failed, because in these species DNA molecules cannot penetrate the recipient cells.

**The transformation of genetic markers**

All mutant loci (or "markers") of a recipient cell are capable of being transformed. The genetic fragment that is transferred is usually very small, however; although it may carry a number of genes, it rarely carries more than one marker. This is an artifact that results from the method by which transforming DNA is usually prepared. In this method, which involves stirring the preparation with phenol to remove proteins and repeated precipitations with ethanol, the DNA is usually sheared into fragments which, even in the most carefully handled preparations, rarely have a molecular weight that exceeds $1 \times 10^7$. This represents about 0.3 percent of the bacterial chromosome, corresponding to about 15 genes.* However, only a few dozen mutant loci have been used in transformation experiments with any one species, so that the chance of finding two markers on the same DNA fragment is low. Nevertheless, such "linked transformations" have occasionally been observed.

When crude preparations of transforming DNA are used, in which shearing and enzymatic degradation have been minimized, it is possible to obtain the linked transformation of many markers. As much as a third of the chromosome can be transformed with such preparations.

Recombination usually occurs with a relatively low frequency, so that selective markers must be used; that is, the donor DNA must contain markers which permit the recombinants to be selected in the presence of a large excess of viable parental recipient cells. Two convenient types of selective markers are drug resistance and nutritional independence. For example, if the recipient culture is streptomycin sensitive and the donor culture is streptomycin resistant, transformation of even a few cells of the former can be detected by plating the DNA-treated culture on streptomycin-containing agar. In this case, the mutant locus determining resistance is a *selective marker*. Similarly, if the recipient culture is auxotrophic (e.g., requiring arginine for growth) and the donor culture is arginine independent, transforma-

*The average gene size is estimated as 1,000 base pairs, coding for a polypeptide of about 330 amino acid residues. The bacterial chromosome contains about $5 \times 10^6$ base pairs of DNA, which is the equivalent of 5,000 genes of average size.

tion of the former can be detected by plating the DNA-treated population on agar lacking arginine.

*Streptococcus, Hemophilus,* and *Neisseria* require complex media for growth, so it is difficult to work with nutritional markers in these organisms. *Bacillus subtilis,* however, grows on a simple mineral medium containing a suitable carbon source, and many different auxotrophic mutants have been produced. Transformation of each auxotrophic type is readily detected using DNA taken from the wild-type strain and plating the treated auxotrophic recipient cells on *minimal medium* (medium containing the minimal number of nutrients capable of supporting growth of the wild type).

## The transformation process

The limiting factor in the yield of transformants is usually the *competence* of the recipient cell population to take up transforming DNA.

Competence is a physiological state that fluctuates greatly during the cell division cycle. Very little is known about competence, except that competent cells produce a protein that can be isolated from the medium and used to confer competence on other cells. This protein could conceivably be a component of the membrane which catalyzes the uptake of DNA; it is equally possible that it is an enzyme which degrades some component of the cell surface, unmasking the receptor for DNA. The change from noncompetence to competence appears to reflect the *synthesis* of this protein, since competence does not develop in the presence of agents, such as chloramphenicol, which block protein synthesis.

Uptake requires an expenditure of metabolic energy. The DNA must be double-stranded, with a molecular weight greater than $5 \times 10^5$. Competent cells are not specific with respect to the DNA which they take up. Pneumococci, for example, take up calf thymus DNA as well as pneumococcal DNA. Recombinants are not formed, however, unless the exogenote and endogenote have a considerable degree of homology in the base-pair sequences of their DNAs.

We have already discussed the *integration* of transforming DNA in the endogenote. Integration takes place very rapidly; if transforming DNA carrying the linked markers $A^+B^-$ is taken up by cells that are genetically $A^-B^+$, fragments of $A^+B^+$ DNA can be extracted from the cell shortly afterward.* This can be shown by testing extracts of the recipients for the presence of DNA which can "cotransform" an $A^-B^-$ recipient to $A^+B^+$; $A^+B^+$ DNA is formed at a linear rate starting with almost no lag and reaches the half-maximum level after 6 minutes.

*The designations *A* and *B* represent two linked markers; + represents the wild-type allele and − the mutant allele in each case.

This process takes place *in the absence of significant DNA replication*, a fact which itself rules out the copying-choice model of recombination.

**The occurrence of transformation in nature**

Transformation, as discovered in the laboratory, requires the artificial extraction of donor DNA, but it was recognized very early that recombination might take place by transformation in nature. To test this possibility, mixed cultures of genetically marked pneumococci were prepared under conditions in which many of the cells were lysing. As predicted, recombinants were produced as a result of the release of DNA from some cells and its uptake by others. Since recombination vastly increases the number of gene arrays upon which natural selection can act, recombination in nature—even at low frequency—must play a major role in the evolution of a species.

## *Bacterial conjugation*

**The discovery of conjugation**

The discovery of transformation revealed for the first time the existence of recombination in bacteria. Once the existence of transformations in bacteria had been established, a search was undertaken for processes of genetic recombination that might resemble eucaryotic sexual reproduction more closely. In 1946 J. Lederberg and E. L. Tatum carried out an experiment with *E. coli* that was designed to reveal the occurrence of recombination by conjugation.

*E. coli* requires no growth factors. By mutagenesis, Lederberg and Tatum produced two auxotrophic strains of *E. coli* strain K12, differing from each other with respect to four genes governing biosynthetic enzymes. Each mutation conferred a different growth-factor requirement: one strain required the compounds biotin and methionine, and the other required the compounds threonine and leucine. The loci in which the mutations occurred have now been designated *bio*, *met*, *thr*, and *leu*, respectively. The two parental genotypes can thus be partially described as follows:

| | | | | |
|---|---|---|---|---|
| parental type I: | $bio^-$ | $met^-$ | $thr^+$ | $leu^+$ |
| parental type II: | $bio^+$ | $met^+$ | $thr^-$ | $leu^-$ |

The + sign in the genotype indicates that the gene is functional, or wild type. The − sign indicates that the gene is present as a mutant allele, producing an inactive enzyme (in these cases blocking biosynthetic pathways).

About $10^8$ cells of each type were mixed together and plated on minimal medium containing none of the four growth factors. Although neither auxotroph could grow on this medium, a few hundred colonies developed. On isolation, these proved to have the genotype *bio*$^+$*met*$^+$-*thr*$^+$*leu*$^+$ (i.e., they consisted of cells having the heritable capacity to synthesize all four growth factors). The first problem was to determine whether or not some sort of transforming principle was involved. Exhaustive attempts were made to find a diffusible chemical substance that could pass from one cell to another and bring about the observed genetic changes, but such attempts were uniformly negative. It was finally established by direct microscopic observation that recombination in *E. coli* requires direct cell-to-cell contact, or *conjugation*.

**The polarity of conjugation**

It was initially assumed that bacterial conjugation involves cell fusion with the formation of true zygotes. Ultimately, however, it was discovered that there are two mating types in *E. coli* and that during conjugation *one partner acts only as genetic donor, or male, and the other only as genetic recipient, or female.* A male is recognized by the fact that it can be killed with streptomycin or other agents and still retain its fertility, while a female is recognized by the fact that its fertility is destroyed by lethal agents. In other words, since the only function of the male is to transfer some of its DNA, it need not remain viable, whereas the female cell must remain viable in order for the zygote to develop. A pair of conjugating cells is shown in Figure 14.6.

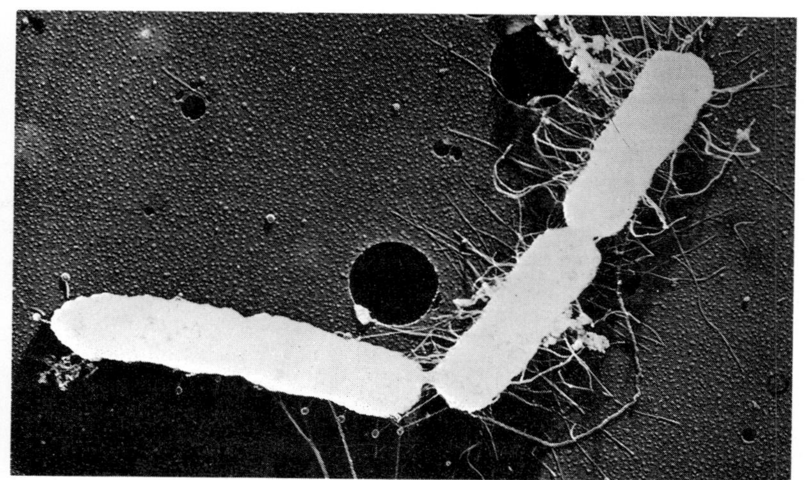

*Figure 14.6 Mating cells of E. coli (× 21,300). This electron micrograph was taken shortly after mixing together donor (Hfr) cells and receptor (F$^-$) cells. Before mixing, the Hfr cells were "marked" by causing them to absorb inactive particles of bacteriophage; the F$^-$ cells are easily recognized by the fact that in this strain they are covered heavily with pili. The micrograph clearly shows the conjugation bridge that has formed between the Hfr cell and one of the F$^-$ cells. Note the bacteriophage particles adsorbed by their tails onto the Hfr cell. Courtesy of F. Jacob, T. F. Anderson, and the Long Island Biological Association.*

When the recombinants issuing from a variety of crosses were analyzed for mating type, it was discovered that maleness in bacteria is determined by the presence in the cell of an infectious, viruslike agent. When male and female bacteria conjugate, many of the females are infected by this agent and converted to males. The infectious agent is called the *sex factor,* or the *F agent* (for "fertility"). The F agent is transferred only by cell-to-cell contact, never kills its host, and is never released into the medium. In other respects it has many properties in common with temperate bacteriophages, such as λ, as will be discussed later. Cells that harbor F and transmit it as an infectious particle are called $F^+$; females, which lack F, are called $F^-$.

Although in $F^+ \times F^-$ crosses the sex factor can be transferred with an efficiency of 100 percent, the transfer of chromosomal material is a very rare event. Under ideal conditions, when $10^7$ male cells are mated with an excess of females, only a few thousand recombinants may be formed. Thus, although every $F^+$ cell conjugates with an $F^-$ cell (as revealed by F transfer), only a small fraction of $F^+$ cells transfer chromosomal markers under the same conditions.

**The nature and replication of F**

The sex factor, F, of *E. coli* strain K12 is a circular DNA structure with a length approximately 2 percent of that of the bacterial chromosome. It has been possible to isolate F DNA, by transferring F from *E. coli* to an organism that has chromosomal DNA of a markedly different buoyant density, owing to a difference in guanine–cytosine (G + C) content. For example, $F^+$ *E. coli* will conjugate with *Proteus mirabilis. Escherichia coli* DNA has a G + C content of 50 moles percent and a buoyant density of 1.710 g/cm$^3$, whereas the DNA of *P. mirabilis* has a G + C content of 39 moles percent and a buoyant density of 1.698 g/cm$^3$. These two types of DNA can be separated from each other by ultracentrifugation in a cesium chloride gradient; when *P. mirabilis* has been made $F^+$, a new peak with a buoyant density characteristic of F appears in the sedimentation profile of its DNA (Figure 14.7).

The F DNA has been isolated from $F^+$ cells of *P. mirabilis* and characterized with respect to its buoyant density as well as to its ability to hybridize with chromosomal DNA. It was found that F is composed of DNA of two types: one type, accounting for 10 percent of the total, has a G + C content of 44 percent, unlike that of either of its hosts; the other type, accounting for 90 percent of the total, has a G + C content of 50 percent—exactly like that of *E. coli*. Half of the latter DNA is hybridizable with chromosomal DNA from *E. coli*, which indicates very strong homology of base-pair sequences in the two structures.

From a number of observations, it is clear that the sex factor exists

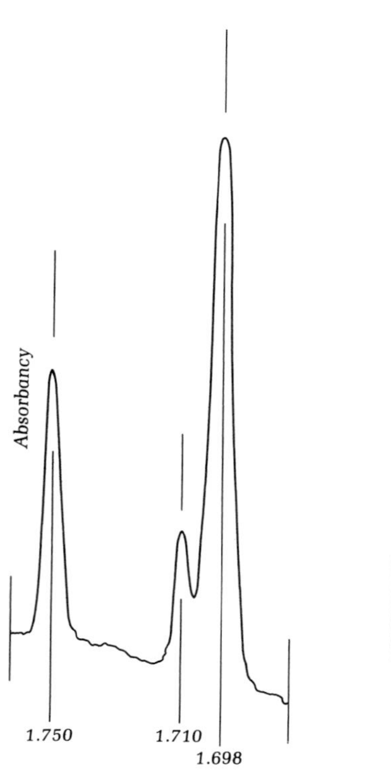

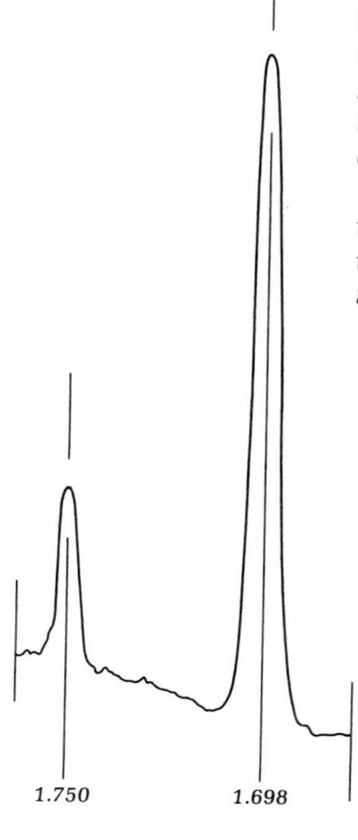


*Figure 14.7 Effects of plasmid infection on the DNA of Proteus mirabilis, strain PM-1. Microdensitometer tracings of ultraviolet absorption photographs taken after equilibrium was reached in cesium chloride density gradient centrifugation: (left) DNA extracted from strain PM-1 infected with the plasmid F13; (right) DNA extracted from PM-1 before infection. The band at density 1.750 g/cm³ is a density standard that was added to both tubes. The band at density 1.698 g/cm³ is Proteus DNA; the small "shoulder" at density 1.710 is F13 DNA. From S. Falkow et al., J. Bacteriol.* **87**, *209 (1964).*

in $F^+$ cells as an *independent replicon*, three to four copies being present in each cell. The independent replication of F is indicated not only by its independent transfer during conjugation but also by the fact that its replication can be differentially inhibited by acridine dyes. If a culture of $F^+$ cells is grown in the presence of a critical concentration of acridine orange, chromosomal replication and cell division occur normally, but F replication is strongly inhibited, with the result that at each division cycle a substantial fraction of the daughter cells are $F^-$. This phenomenon is called the *curing* of $F^+$ cells.

In Chapter 10 we discussed the attachment of the bacterial chromosome to the cell membrane and the role of this attachment in chromosomal segregation at cell division. By an ingenious experiment, it has been established that the F replicon and the chromosome are both attached to the same structure, known in the case of chromosomal attachment to be the membrane. This experiment was made possible by the isolation of strains of *E. coli* in which chromosomal genes, such as those of the *lac* operon, are attached to the autonomous sex factor. Starting with one such strain, it was possible to obtain a thermosensitive mutant of the *F-lac* replicon, which mutant replicates normally at 30°C but does not replicate at 42°C.

A culture of the strain carrying the thermosensitive *F-lac* was grown at 30°C in a medium containing $^{32}P$-labeled phosphate, so that the DNA became radioactive. The cells were then transferred to nonradioactive medium and incubated at 42°C. After a number of generations, only a few percent of the cells contained radioactive chromosomes, these being the cells which had inherited the original $^{32}P$-labeled strands of DNA. Similarly, only a few percent of the cells contained the *F-lac* replicon, since *F-lac* had not replicated further after the shift to 42°C. The radioactivity and the *F-lac* DNA were shown to be present in the *same cells:* the ancestral radioactive chromosomal strand and the nonreplicating *F-lac* had remained physically associated through many cell divisions. If F and chromosome were not attached to the same structure, the radioactivity would have been randomly distributed between *F-lac* and $F^-$ cells.

In summary, F is a circular DNA structure containing about $10^5$ base pairs. Its replication, which is autonomous, is peculiarly sensitive to inhibition by acridine dyes. It appears to be attached to the cell membrane and thus to segregate at cell division exactly in the same manner as the chromosome. F contains a small segment of DNA with a unique G + C content; much of its remaining DNA is homologous with that of the *E. coli* chromosome. Only those cells which carry F conjugate, so the genes that govern some or all of the various conjugational functions must be located on the F replicon. These genes dictate that an $F^+$ cell will conjugate with an $F^-$ cell and transfer F itself with an efficiency of 100 percent; however, only one $F^+$ cell in approximately $10^4$ transfers chromosomal DNA.

**The discovery of Hfr males and the analysis of the mating process**

The study of chromosome transfer was facilitated by the discovery of mutant male strains which transfer their chromosomal genes with very high efficiency. These mutants are designated *Hfr*, standing for "high frequency of recombination." A very large fraction (even 100 percent) of the cells in an Hfr culture may act as genetic donors in a given mating. A second important difference between $F^+$ and Hfr males concerns the transfer of the sex factor. When recombinants are isolated from an $F^+ \times F^-$ cross they are all $F^+$, the zygote having received an infectious F particle from the male parent in every case. When recombinants are isolated from an Hfr $\times$ $F^-$ cross, however, they are usually $F^-$. In other words, Hfr males do not transmit an infectious F particle.

The mutation from $F^+$ to Hfr thus produces two major changes: the efficiency of chromosome transfer increases by at least a thousandfold; and the sex factor is no longer infectious.

The first clue to the nature of the mating process was the discovery that, in most cases, a large share of the donor genome (whether $F^+$ or Hfr) simply does not appear in the recombinant progeny. In other words, when recombinants are tested for as many unselected genetic traits as possible, most of the genes are derived solely from the recipient ($F^-$) parent and only a few characters from the donor parent. This results from the fact that *the donor cell transfers only a part of its genetic complement during conjugation*.

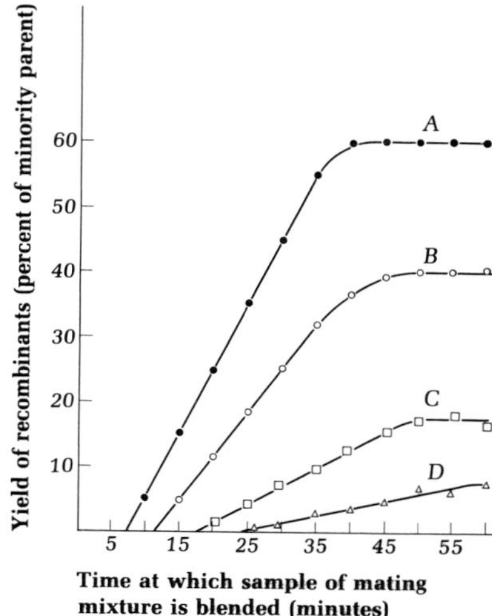


*Figure 14.8 The kinetics of recombinant formation. See text for explanation. After E. Wollman and F. Jacob.*

The nature of chromosome transfer was discovered by E. L. Wollman and F. Jacob, in the course of studies on the kinetics of recombinant formation. Such kinetic experiments were performed as follows. Samples from a mating mixture of Hfr and $F^-$ cells were withdrawn at intervals and agitated in a blender for several minutes to shear apart mating pairs. Each sample was plated on several different selective media to determine the number of different types of recombinants that had been formed by the time of sampling. The results of such an experiment are shown in Figure 14.8; the markers studied are designated simply as A, B, C, and so on. As this figure shows, the longer the mating couples are allowed to conjugate before being separated in the blender, the more genetic material is transferred from the Hfr cells to the $F^-$ cells. For example, if the mating pairs are separated after 10 minutes, only gene A will have been transferred. If conjugation is allowed to proceed for 15 minutes, genes A and B are transferred; after 20 minutes genes A, B, and C are transferred; and so on. The order in which the genes appear to be transferred is precisely the order in which the genes are arranged along the chromosome as deduced by linkage analysis.

These experiments reveal that the donor cell slowly injects a chromosomal thread into the receptor cell so that the genes along the thread enter the receptor cell in a linear order. If conjugation is not interrupted, it proceeds until the entire chromosome has been injected, a process that requires about 90 minutes at 37°C. The interpretation of this type of experiment is shown in Figure 14.9.

The slopes of the lines in Figure 14.9 show that the donor cells are not synchronized with respect to the initiation of transfer. Some cells begin to transfer almost immediately, but in others there is a delay. Within 25 minutes, however, all donor cells begin to transfer. The intercept of each curve with the abscissa indicates the time at which a given gene arrives in the zygote in the first mating couples which initiate transfer.

If a mating is allowed to proceed for 90 minutes without artificial

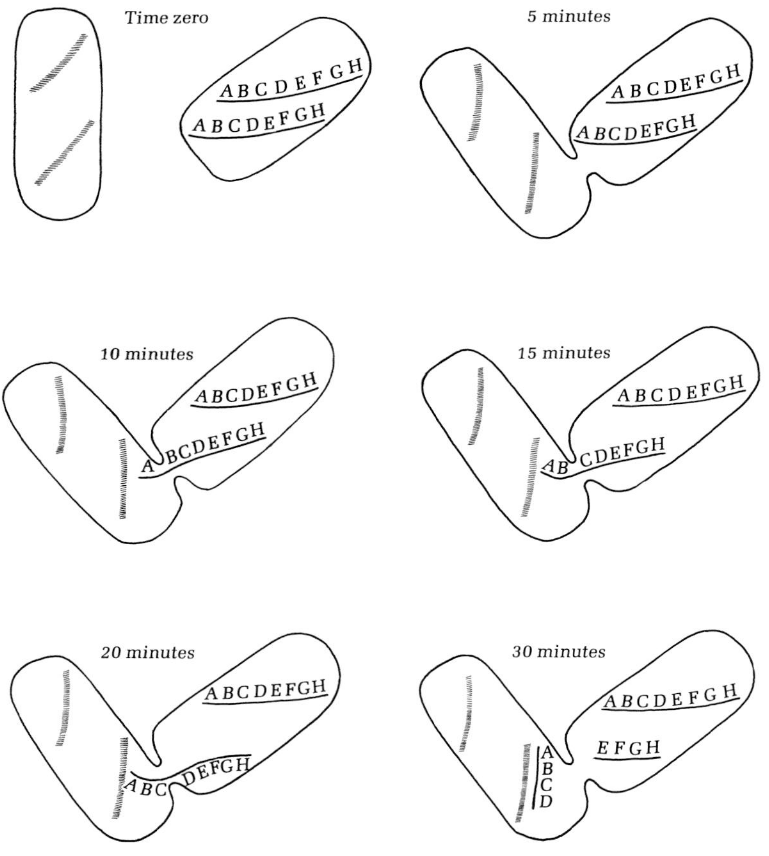


*Figure 14.9 Interpretation of the data plotted in Figure 14.8. See text for explanation. After E. Wollman and F. Jacob.*

interruption, considerable spontaneous separation of mating couples occurs. The time at which any given couple breaks apart is random; very few couples persist in conjugation for the entire 90 minutes. Thus, the farther away a given marker is from the leading point in transfer, the less chance it has of being transferred before breakage occurs. Most couples will achieve the transfer of the earliest marker, A; fewer will achieve the transfer of B; still fewer the transfer of C; and so on. This phenomenon is reflected in the *heights* of the plateaus of the different curves in Figure 14.8.

The order of marker transfer can thus be recognized in two ways: by the time of entry of each marker in interrupted-mating experiments, and by the frequency of recombination for each marker in uninterrupted matings. For example, marker A, which first enters the female cell at 10 minutes, may have a final recombination frequency of 40 percent (i.e., 40 percent of the mating couples produce recombinants for marker A). Marker B, which enters at 20 minutes, may have a recombination frequency of 10 percent; and marker C, which enters at 30 minutes, a recombination frequency of 2 percent. Such a *gradient of recombination frequencies* is characteristic of an Hfr × $F^-$ cross and is so steep that the last marker to penetrate the female may have a recombination frequency of only 0.05 percent.

The speed at which the chromosome is transferred has been found to be relatively constant throughout the mating process, so that the *times of entry of markers* provide an accurate measure of the distances between them. Since the entire chromosome, representing $5 \times 10^6$ base pairs, is transferred in 90 minutes, each minute on the time scale in Figure 14.8 is equivalent to approximately $5 \times 10^4$ base pairs.

**The use of Hfr mutants to reconstruct the genetic map of *E. coli***

*Linkage* may be defined as the tendency of two genetic markers to remain together during recombination. Markers that are linked are located on a common physical structure, the chromosome, and can only be separated if a crossover occurs between them. The closer the juxtaposition of two markers on the chromosome, the less chance there will be for a crossover to separate them (Figure 14.10). By experiments of the type described below, it is possible not only to show linkage between markers but also to ascertain their relative positions on the chromosome.

As an example of a linkage test, let us consider an $F^-$ strain that is genetically *thr*$^-$ (requiring threonine), *leu*$^-$ (requiring leucine), *pro*$^-$ (requiring proline), *lac*$^-$ (unable to ferment lactose), *gal*$^-$ (unable to ferment galactose), and *str*$^r$ (resistant to streptomycin). If this strain is crossed with an Hfr which is wild type for all these markers (*thr*$^+$, *leu*$^+$, *pro*$^+$, *gal*$^+$, *str*$^s$), recombinants can be selected which receive *str*$^r$ from the $F^-$ parent and various markers from the Hfr parent. For example, plating on minimal medium containing streptomycin, glucose as the carbon source, and proline as a growth factor will select for recombinants receiving the markers *thr*$^+$ and *leu*$^+$ from the Hfr parent. The growth of recombinants on such a medium does not depend on the presence of the *pro*$^+$, *lac*$^+$, or *gal*$^+$ markers. Similarly, plating on a medium containing streptomycin, lactose as a carbon source, and pro-

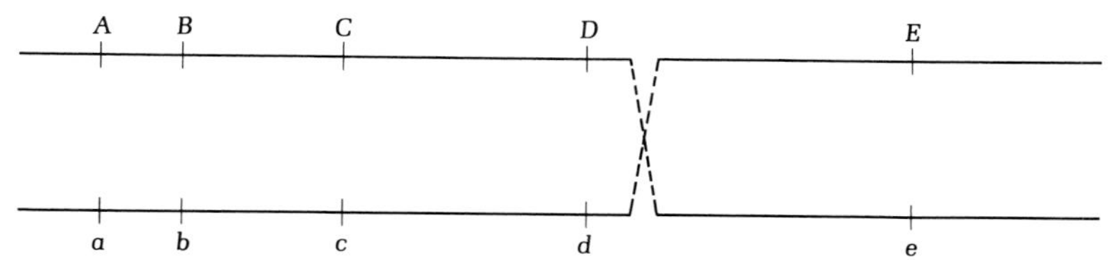


*Figure 14.10 The separation of linked markers by crossing over. In the diagram, the distance between markers A and E is such that half the zygotes will undergo one crossover between them. Thus, 50 percent of the zygotes will produce recombinants of the type Ae or aE. The distance between markers A and B, however, is only one-tenth the distance between A and E; since the probability of crossing over between two markers is directly proportional to the distance between them, crossovers between A and B will occur in only 5 percent of the zygotes.*

line, threonine, and leucine as growth factors, will select for recombinants that have received the $lac^+$ marker from the Hfr parent.

The selected recombinants can be isolated and tested for the inheritance of other, unselected markers. If two markers transferred by the male parent are linked, they will tend to be inherited together. For example, when the markers $thr^+$ and $leu^+$ are selected from the Hfr, about 40 percent of the recombinants inherit the $lac^+$ marker. When the $pro^+$ marker is selected from the Hfr, however, 80 percent of the recombinants inherit the $lac^+$ marker. Thus, the $lac^+$ and $pro^+$ markers tend to stay together, showing that they are physically linked. Using this type of test, it has been established that *E. coli* contains a single, circular linkage group.

The first Hfr mutant to be isolated was found to transfer its markers in the order $thr^+$-$leu^+$-$pro^+$-$lac^+$-$gal^+$. Other markers, such as $his^+$ (histidine synthesis), $ser^+$ (serine synthesis), $str^s$ (streptomycin sensitivity), and $met^+$ (methionine synthesis), were transferred at a much lower frequency. Linkage tests of the type described above proved that all the markers are linked and are aligned on the chromosome in the order in which they are transferred.

When a second Hfr mutant was similarly analyzed, however, a

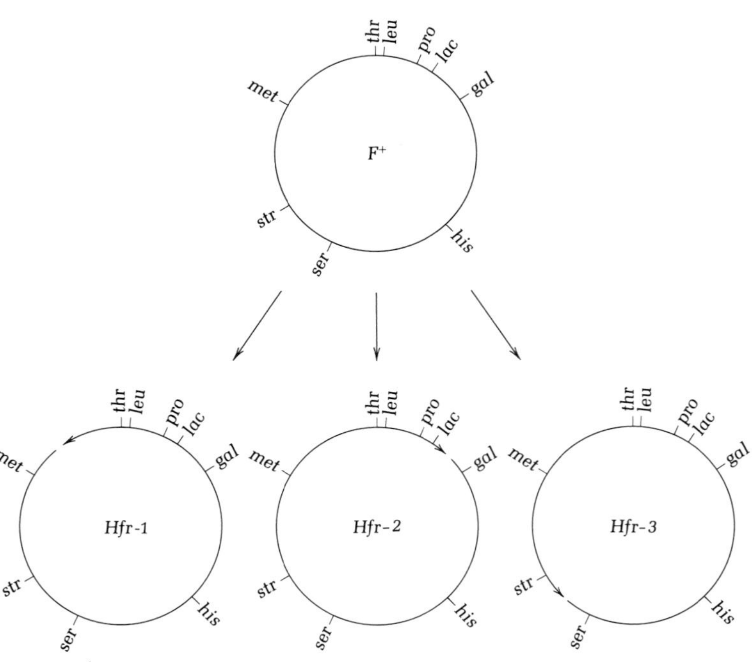


*Figure 14.11 Diagram showing the origin of three different Hfr males from a common F⁺ parent. Arrowheads on chromosomes indicate the leading point and direction of chromosome transfer in each case.*

strange fact was revealed. This Hfr transferred its markers in the order *lac*⁺-*pro*⁺-*leu*⁺-*thr*⁺-*met*⁺-*str*ˢ, and the *gal*⁺ marker was transferred only at a very low frequency. When many different Hfr mutants were analyzed, each was found to show a unique order of marker transfer. The cumulative genetic data could be interpreted by assuming that the F⁺ strain of *E. coli has a single linkage group, or chromosome, which is a closed circle.* The F⁺ → Hfr mutation is accompanied by a breakage of the circle at one point; one end of the broken chromosome becomes the leading point in chromosome transfer. In each Hfr, the break has occurred at a different point. This is shown in Figure 14.11.

The chromosome map in Figure 14.11 shows only a few of the several hundred markers that have now been mapped on the *E. coli* chromosome. Figure 14.12 shows a more complete version of this map. The discovery of the circularity of the genetic map of *E. coli* preceded by several years the demonstration by Cairns, using physical techniques, that the chromosome of *E. coli* is a circular DNA molecule (Chapter 10).

**The attachment of F to the bacterial chromosome**

In Hfr × F⁻ crosses the recombinants generally emerge in the F⁻ state; they do not receive a sex factor from the donor. The reason for this came to light when a number of crosses were analyzed more intensively. In every case it was found that those recombinants which are selected to possess the last marker to be transferred (e.g., *met*⁺ in crosses with Hfr-1, or *gal*⁺ in crosses with Hfr-2, in Figure 14.12) behave as Hfr males and transfer genes in the same order as their parent. In other words, a genetic determinant of maleness is linked to the last marker to be transferred.

Wollman and Jacob explained this observation by postulating (1) that the F⁺ → Hfr "mutation" consists of *the attachment of the sex factor to the chromosome*, and (2) that at conjugation the chromosome of the Hfr cell breaks at the site of F attachment. The break must occur in such a way that one end becomes the leading end, or *origin*, in transfer, part or all of the sex factor being attached to the other end.

We have already pointed out that F resembles certain bacteriophages, such as λ, in being a circular DNA element about $10^5$ base pairs in length, capable of independent replication in a host cell. We now see another similarity to λ—the ability to attach to the bacterial chromosome. In Chapter 11 we described the mechanism by which λ undergoes such attachment. It will be recalled that λ DNA becomes circular soon after penetration into the cell; a region of λ DNA then pairs with its homologous site on the chromosome, following which a crossover within the region of pairing integrates the two structures (Figure 11.15 illustrates this).

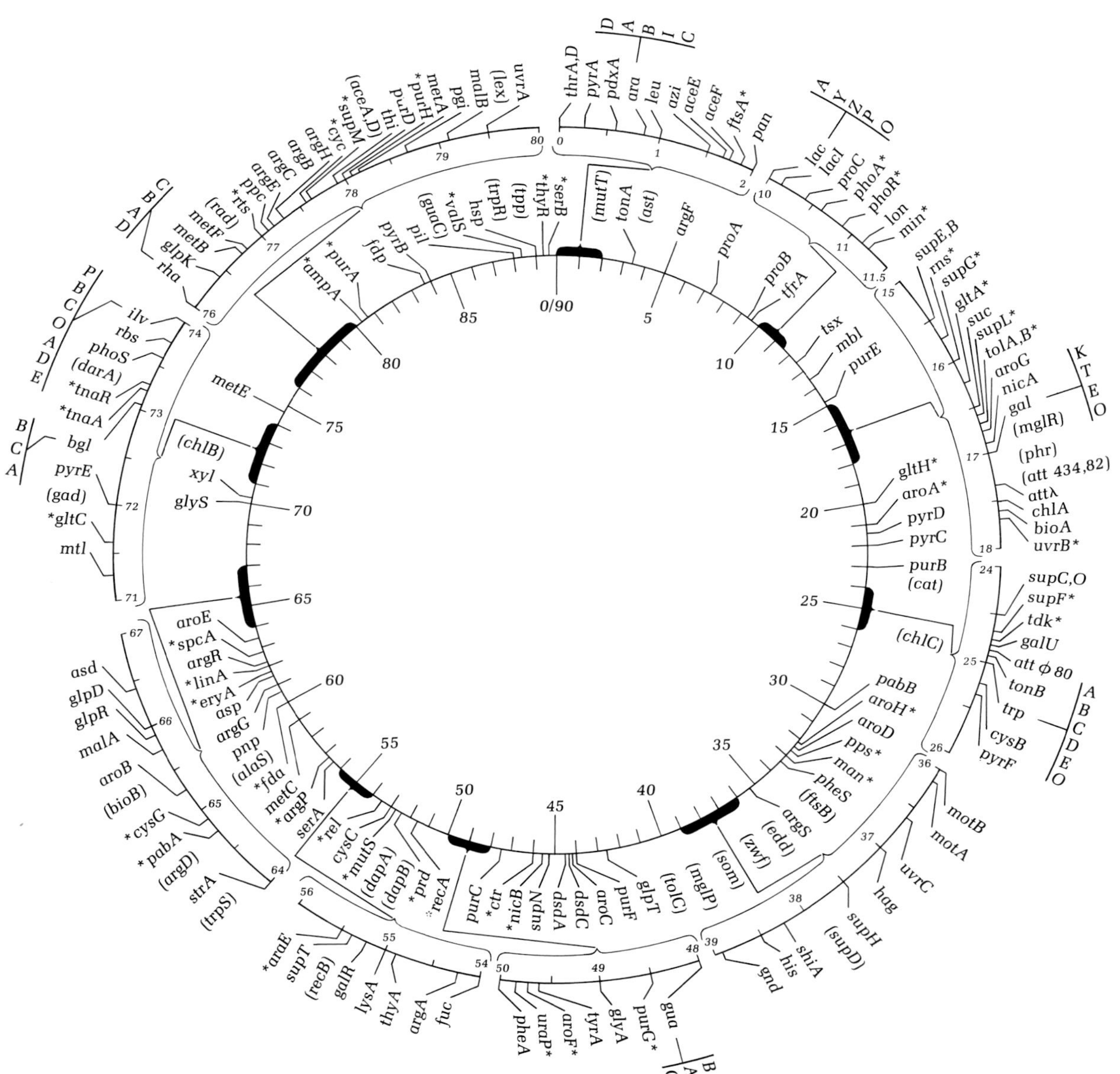

Figure 14.12 The genetic map of *E. coli* K12. The units of distance along the chromosome equal the relative times of entry of the markers during conjugal transfer at 37°C. The zero point has arbitrarily been chosen as the *thrA* locus. The map shown represents a compilation of map data available in 1967; loci whose positions are only approximate are shown in parentheses. The loci marked with an asterisk are those whose positions relative to adjacent loci are uncertain. From A. L. Taylor and C. D. Trotter, "Revised linkage map of *E. coli*." Bacteriol. Rev. **31,** 332 (1967).

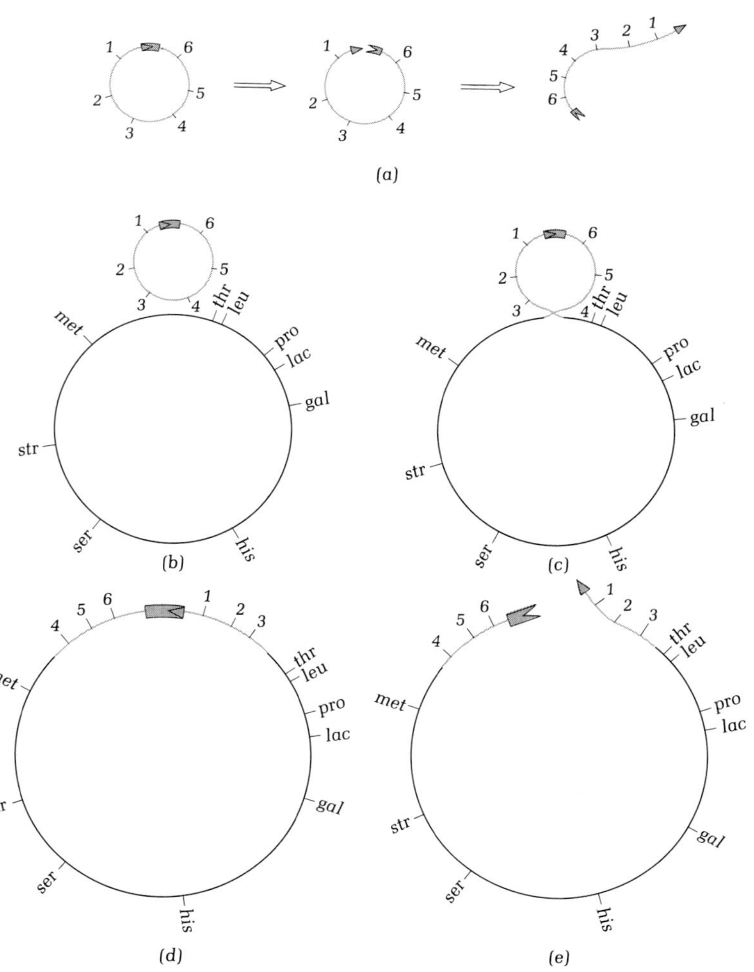


Figure 14.13 The breakage and transfer of the Hfr chromosome as a consequence of its integration with F. (a) F is shown as a circle, which has a special site of breakage between markers 1 and 6. Breakage is followed by transfer (during conjugation), such that F markers penetrate the recipient in the order 1-2-3-4-5-6. (b) F has paired with a chromosomal site between met and thr. (c) A crossover in the region of pairing integrates the two circles. (d) Same as (c), but redrawn as a single circle. (e) Breakage at the special F site leads to transfer causing F markers 1-2-3 to enter the recipient first, followed by chromosomal markers thr, leu, pro, lac, and so on; F markers 4-5-6 enter last. Recombinants will be males if they receive all six F markers; thus the terminal end of the chromosome must be transferred to produce an Hfr recombinant.

All the observations concerning F attachment can be explained by assuming that there are several regions of F homology scattered around the bacterial chromosome; the chance pairing of F with one of these regions, followed by crossing over, integrates F and chromosome and produces the different Hfrs with their specific breakage points. The postulated pairing of F and chromosome is compatible with the finding, mentioned earlier, that about half the DNA of F is hybridizable with chromosomal DNA of *E. coli*.

The sex factor, a circular replicon, must itself break open to be transferred as an autonomous element in an $F^+ \times F^-$ cross. *The breakage and transfer of the bacterial chromosome is a consequence of its integration with F,* as shown in Figure 14.13. F is postulated to have a special site within its structure at which breakage occurs; F transfer proceeds from one end of the break. *Transfer is thus uniquely a property of F;* the integration with F of any other circular DNA element (such as the bacterial chromosome) leads to the cotransfer of that element during conjugation.

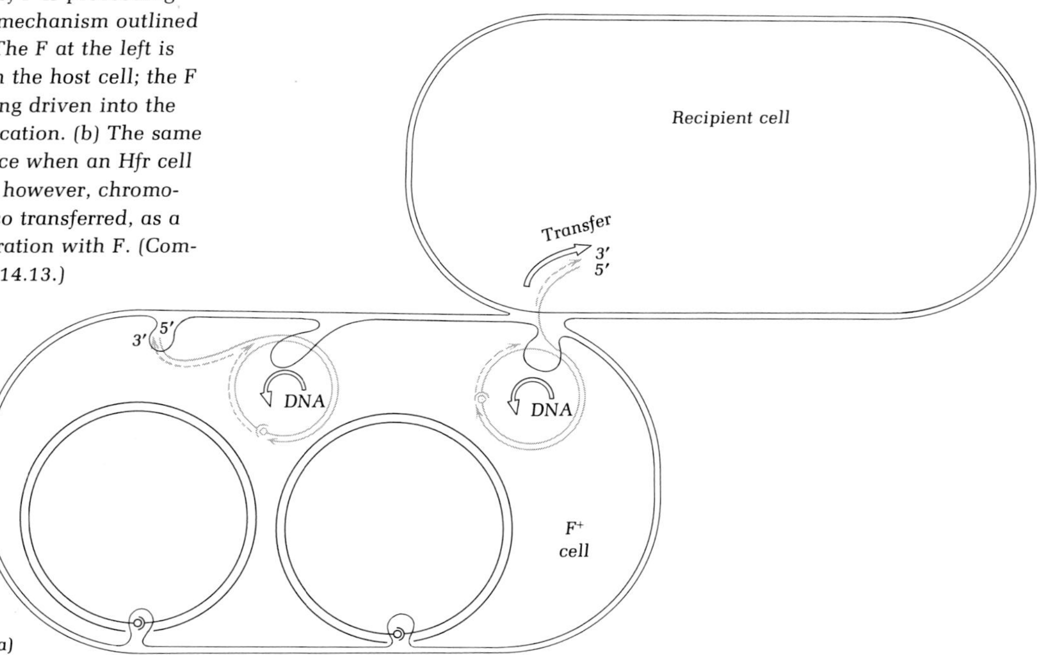


*Figure 14.14 The hypothetical mechanism of DNA transfer as a consequence of F replication. (a) An $F^+$ cell, containing two autonomous F replicons and two chromosomes, is shown conjugating with a recipient cell. Replication of F is proceeding according to the mechanism outlined in Figure 10.59. The F at the left is replicating within the host cell; the F at the right is being driven into the recipient by replication. (b) The same process takes place when an Hfr cell conjugates. Now, however, chromosomal DNA is also transferred, as a result of its integration with F. (Compare with Figure 14.13.)*

The fundamental event in bacterial conjugation is the breakage of the circular F replicon at a specific site, followed by the oriented movement of the open DNA structure from the male to the female cell. In 1962 F. Jacob and S. Brenner proposed a model that would explain both the breakage and the movement of the DNA. According to their model, the "breakage site" is the *replicator* of the F replicon, and the free end which becomes the transfer origin arises not by actual breakage of DNA but rather by the *initiation of replication*. The ensuing replication of F DNA (together with any other replicon that has become integrated with it) provides the driving force for the movement of the DNA into the recipient cell.

**The mechanism of DNA transfer in bacterial conjugation**

The Jacob–Brenner model, illustrated in Figure 14.14, makes transfer a special case of normal F replication. In a nonconjugating $F^+$ cell, F replication leads to the "transfer" of one replica to a new attachment site inside the same cell; in a conjugating $F^+$ cell, F replication leads to

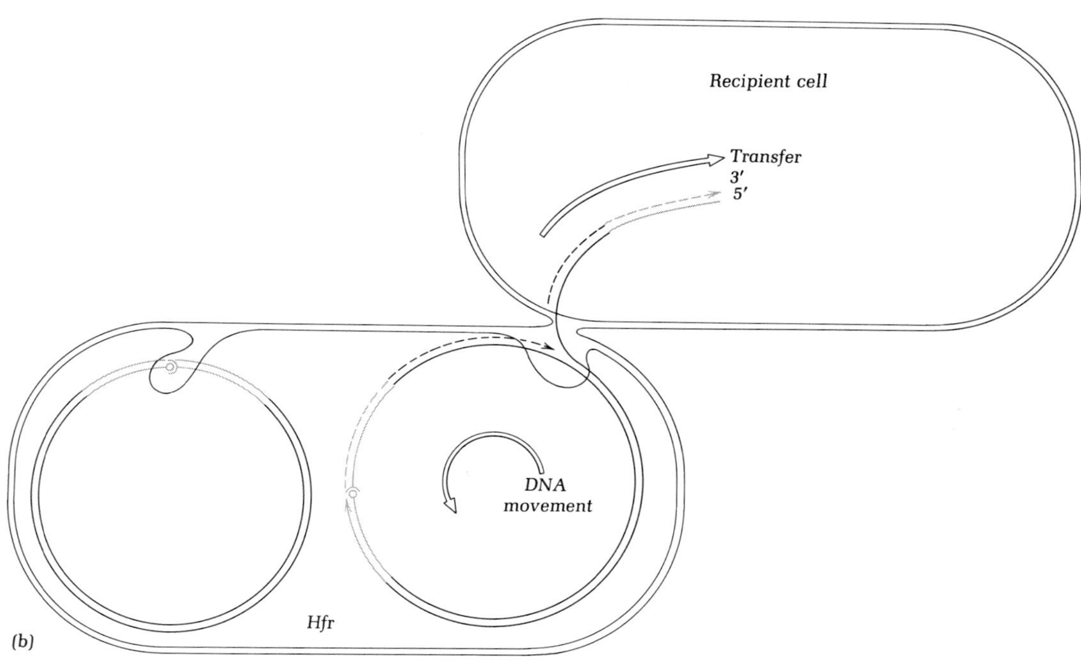

the transfer of one replica into the conjugating partner (Figure 14.14*b*).

It has been shown that any treatment of mating cells which inhibits DNA replication also blocks transfer, even after the process has been initiated. It has also been found that the DNA which is transferred contains one strand synthesized prior to conjugation and another synthesized during the period of transfer. Both observations are compatible with the model of transfer by replication.

**The role of the cell surface in bacterial conjugation**

Microscopic observation of cell suspensions reveals that cells containing F form pairs with other *E. coli* cells, male or female; cured strains, made $F^-$ by acridine orange treatment, no longer exhibit this property. F thus determines the formation of one or more surface receptors to which other cells may attach.

Immunochemical tests show the presence of a special *F antigen* on male cells. Brief exposure to periodate destroys the ability of male cells, but not of female cells, to conjugate; this suggests that a polysaccharide component of the cell wall is involved in mating; it may be a haptenic group of the F antigen. A much more dramatic difference between the surfaces of male and female cells was revealed by the discovery of phages that adsorb only to male bacteria. When cells that have adsorbed saturating amounts of these male-specific phages are examined with the electron microscope, the phage receptor sites are found to be present on special pili. There are two types of male-specific phages: small, spherical phages containing RNA, and long, filamentous phages containing DNA. The RNA phages adsorb to the sides of the "F pili" (Figure 14.15), while the DNA phages adsorb to the tips of the F pili. The F pili are produced only by cells that harbor F and are composed of a repeating protein subunit the structure of which is determined by a gene of F.

The discovery of the F pili and of the phages that adsorb to them immediately suggested to many investigators that the male-specific phages inject their nucleic acids into the host through the hollow core of the F pilus, and that the male bacterium might use the F pilus as a tube through which it injects DNA into the female partner. This hypothesis appeared to be supported by the observation that mutations which

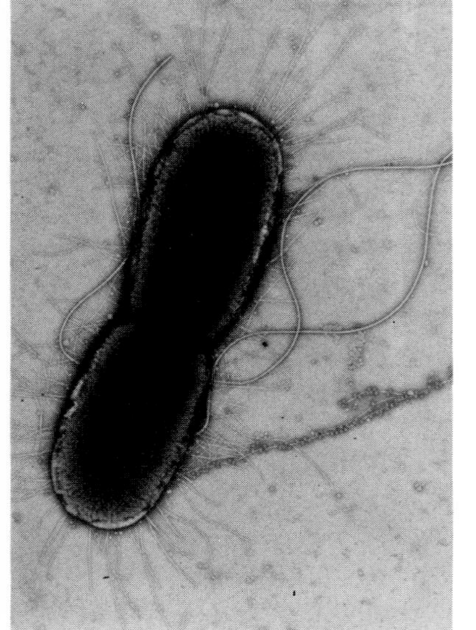

*Figure 14.15 A dividing cell of an Hfr strain of E. coli, showing three types of appendages. The long, curved, thick appendages are flagella. The short, thin, straight appendages are ordinary pili. The long, thin appendage (lower right) coated with phage particles is an F pilus. Courtesy of Judith Carnahan and Charles Brinton.*

block F-pilus formation also seem to deprive the cell of the ability to conjugate.

However, all attempts to show that nucleic acid of either phage or bacterial origin can be transferred through the core of the F pilus have so far failed. Furthermore, the filamentous DNA phage which adsorbs to the tip of the F pilus has been found to enter the cell in its entirety; it does not inject its DNA, like other DNA phages. As shown in Figure 14.15, this phage is much too large to pass intact through the core of the F pilus. Its passage from the adsorption site at the tip of the pilus to the interior of the cell must therefore take place in some other manner.

The pilus is clearly essential for the formation of an effective contact between a male and a female cell. At the present time, however, its precise role in the conjugational process remains unclear. The pilus may function in facilitating cellular attachment and in initiating a subsequent, much more intimate contact between mating cells. This is suggested by the fact that mating cells can be found coupled in two different ways: either tethered by an F pilus 10 to 15 $\mu$m long, or closely joined at the cell wall (Figure 14.16).

*Figure 14.16 (a) Mating pairs of E. coli cells. Hfr cells are elongated. (b) Electron micrograph of a thin section of a mating pair. The cell walls of the mating partners are in intimate contact in the "bridge" area. Photographs (a) and (b) from J. D. Gross and L. G. Caro, "DNA transfer in bacterial conjugation." J. Mol. Biol.* **16,** *269 (1966). (c) A male and a female cell joined by an F pilus. The F pilus has been "stained" with male-specific RNA phage particles. The male cell also possesses ordinary pili, which do not adsorb male-specific phages and which are not involved in mating. Courtesy of Judith Carnahan and Charles Brinton.*

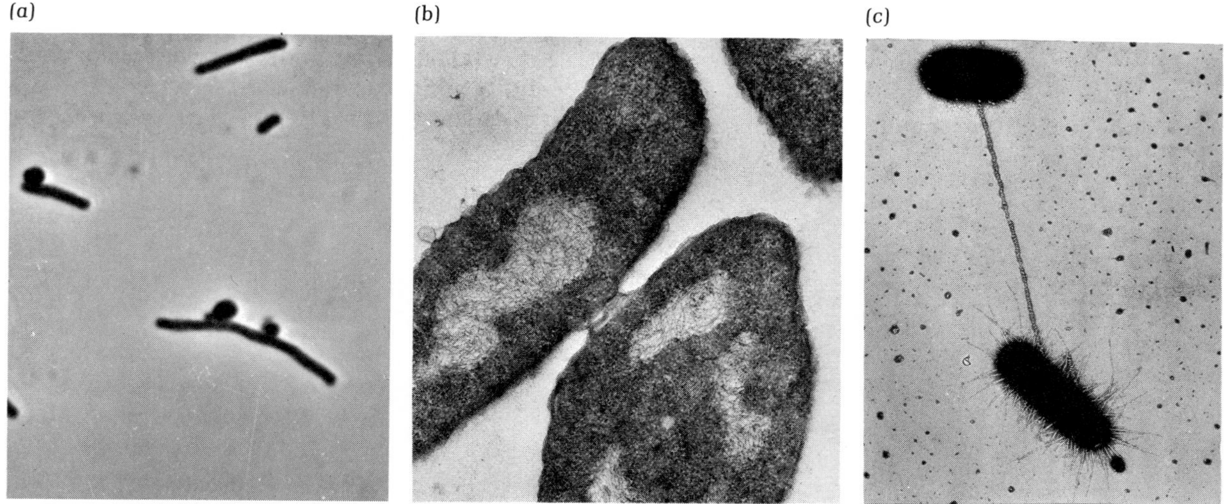


### The origin and behavior of F′ strains

The change from the $F^+$ to the Hfr state depends on the integration of F and chromosome by a recombinational event (or crossover). This process is *reversible:* in any culture of an Hfr strain, F may detach from the chromosome in an occasional cell, giving rise to a clone of $F^+$ cells.

The rate at which the Hfr → $F^+$ reversion occurs varies from strain to strain but is similar to the rate of the $F^+$ → Hfr change. The probability of either integration or detachment is usually on the order of $10^{-5}$ per cell per generation.

For a true reversion to the $F^+$ state to occur, recombination must take place in a region of pairing which is exactly the same as that which existed when integration occurred (Figure 14.17, top). Much more rarely, however, *exceptional pairing* takes place. A crossover in the

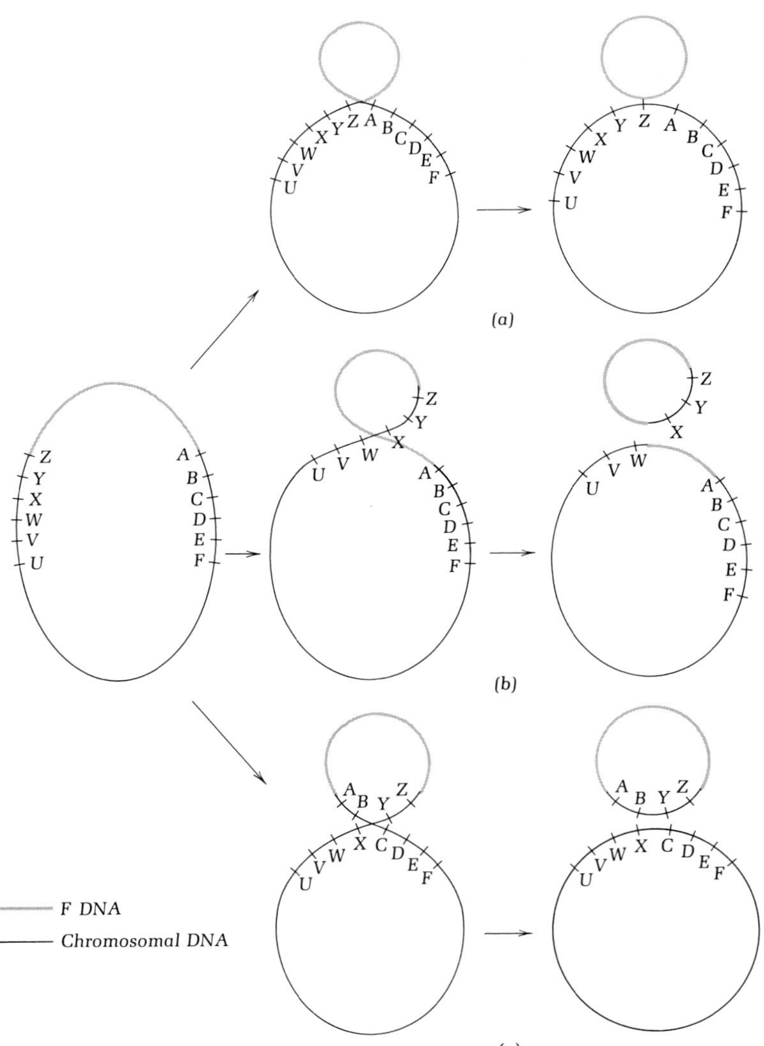


*Figure 14.17 The generation of F genotes, in primary F′ cells. At the left is an Hfr chromosome, with the integrated F DNA at the top. Letters A to F and U to Z represent chromosomal markers. (a) Crossing over within the original region of pairing between F and chromosome regenerates a normal F. (b) Pairing in an exceptional region, followed by crossing over, generates an F genote carrying the chromosomal markers XYZ. The chromosome of the primary F′ contains a segment of F DNA, and has a deletion of the XYZ segment. (c) Exceptional pairing in a different region has generated an F-genote containing a full complement of F DNA, plus chromosomal genes from both sides of the former attachment site.*

region of exceptional pairing does not regenerate an ordinary sex factor but rather *a sex factor that contains within its circular structure a segment of chromosomal DNA* (Figure 14.17, bottom two rows). As shown in Figure 14.17, the F *genote*, as it is called, may or may not lack a segment of F DNA.

The cell in which this event takes place and the clone derived from it are called *primary F′ (F-prime) cells.* Such cells carry a chromosomal deletion, corresponding to the segment that is now an integral part of the sex factor. When this segment includes genes for indispensable functions of the cell (e.g., activating enzymes, nucleic acid polymerases, ribosomal components), the sex factor becomes for all practical purposes a second chromosome of the cell; loss of the sex factor leads to death of the cell.

When primary F′ cells are mated with cells of an $F^-$ strain, F is transferred (together with the integrated chromosomal segment) at high efficiency. Chromosomal transfer, however, occurs with an efficiency of $10^{-5}$ per cell or less, the primary F′ cells having the same low probability of F integration with chromosome as do ordinary $F^+$ cells.

An F genote that is transferred to an $F^-$ cell reestablishes its circular form and becomes an autonomous replicon, producing a *secondary F′ cell.* The secondary F′ cell differs from the primary F′ cell in that the chromosomal segment of the F genote is also present in the host chromosome; the secondary F′ cell is thus a *partial diploid* (Figure 14.18), and the F genote is a dispensable element.

The region of diploidy in a secondary F′ cell makes it a moderately efficient donor of chromosomal genes, since pairing and recombination occur readily in such a region. In fact, a culture of a secondary F′ strain is always a mixture of F′ and Hfr cells, the two types being in dynamic

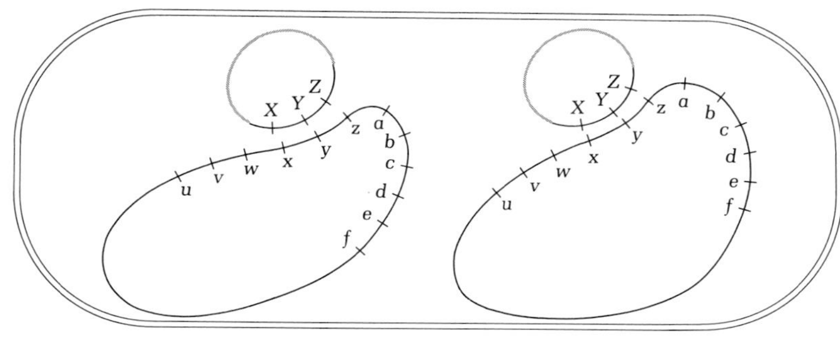


*Figure 14.18 A secondary F′ cell. The cell shown is a heterozygous diploid for genes X, Y, and Z.*

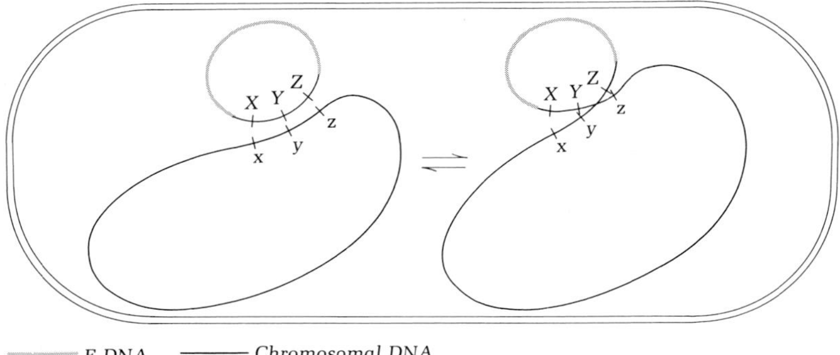


*Figure 14.19 The dynamic equilibrium between the integrated and detached states of the F genote in a secondary F′ cell.*

equilibrium, as shown in Figure 14.19. When such a culture is mated with an $F^-$ strain, those cells which at the moment of mating contain a detached F genote transfer an F genote alone; other cells, in which the F genote and chromosome are at that moment integrated, transfer the entire integrated structure.

The first F′ strain was discovered accidentally during the isolation of high-frequency donor cells from an Hfr culture which had reverted to the $F^+$ state; one of the high frequency cell types which was isolated proved to be a primary F′. A method was then devised for the selection of secondary F′ strains. Designating the markers transferred by an Hfr strain as $A^+$ through $Z^+$, it will be recalled that $Z^+$, the last marker transferred, does not normally appear in recombinants before 90 minutes of mating. If the Hfr culture contains a rare F′ cell, however, in which marker $Z^+$ is attached to the sex factor, that cell will transfer *F-Z*$^+$ within the first 10 or 20 minutes. The procedure is thus to interrupt mating after about 20 minutes and to select for the rare recipient cell that has received the $Z^+$ marker. These are usually found to be secondary F′ strains, carrying *F-Z*$^+$. It was by this procedure that the *F-lac* diploids (see the section on genetic regulation in Chapter 12) were isolated.

Wild-type F contains some DNA that is homologous with the DNA of the *E. coli* chromosome. This DNA may represent chromosomal fragments that have been picked up from the chromosome during the evolution of the *E. coli*–sex factor association. This suggests that wild-type F is itself an F genote, containing chromosomal genes that are now entirely dispensable to the cell.

The dynamic equilibrium between the integrated and attached states of the F genote in a secondary F′ cell, and the dynamic equilibrium in an $F^+$ or Hfr culture, differ only in the respective rates of attachment and detachment. In the case of the secondary F′, both integration and detachment occur about once every 10 cell-generations; in the $F^+ \rightleftharpoons$

Hfr system, both integration and detachment occur with a frequency of about $10^{-5}$ per cell-generation.

**Conjugation in the enteric group**

Male cells of *E. coli* will conjugate with any other Gram-negative enteric bacterium. The efficiency of intergeneric mating varies widely, however, ranging from $10^{-4}$ per donor cell in *Escherichia–Salmonella* mating to $10^{-8}$ per donor cell in *Salmonella–Serratia* matings (Table 14.2). Integration of exogenotic material with the recipient chromosome does not occur unless there is sufficient base-sequence homology with the endogenote to permit pairing. True recombinants are formed only in matings among *Escherichia*, *Salmonella*, and *Shigella*, where DNA homology is high.

Sex factors, on the other hand, do not need to integrate with the chromosome of the recipient cell to establish themselves in a new host. Thus, $F^+$ or F′ strains of any enteric organism can be created by the transfer of F or F genotes from *E. coli*. Since many enteric organisms are genetically *lac*$^-$, the selection of F′ strains resulting from mating with an *F-lac E. coli* donor can be readily accomplished: F′ strains of *Proteus mirabilis*, produced in this way, made possible the isolation and characterization of F DNA (see page 494).

When the chromosome of the host bacterium has one or more regions of base-sequence homology with F DNA, integration may occur to form an Hfr cell. This has been accomplished in *Salmonella*, permitting intraspecific recombination at high frequency as well as the transfer of *Salmonella* chromosomal genes to *Shigella* and *Escherichia*.

*Table 14.2 Efficiency of conjugation between different genera of Gram-negative bacteria*[a]

| *F-lac* **donor** | **Recipient** | **Frequency of** *F-lac* **transfer**[b] |
|---|---|---|
| *Escherichia coli* | *Escherichia coli* | $10^{-1}$–$10^{-3}$ |
| *Escherichia coli* | *Salmonella typhosa* | $10^{-4}$–$10^{-5}$ |
| *Escherichia coli* | *Proteus mirabilis* | $10^{-5}$–$10^{-6}$ |
| *Salmonella typhosa* | *Escherichia coli* | $10^{-4}$–$10^{-5}$ |
| *Salmonella typhosa* | *Proteus mirabilis* | $10^{-4}$–$10^{-5}$ |
| *Salmonella typhosa* | *Serratia marcescens* | $10^{-7}$–$10^{-8}$ |
| *Salmonella typhosa* | *Vibrio comma* | $10^{-5}$–$10^{-6}$ |
| *Proteus mirabilis* | *Escherichia coli* | $10^{-5}$–$10^{-6}$ |
| *Proteus mirabilis* | *Proteus mirabilis* | $< 10^{-10}$ |

[a]Data supplied by L. S. Baron.
[b]These frequencies do not reflect the fact that higher-frequency recipients can be obtained from wild-type populations.

Hfr cells have also been produced in *E. coli* strain B, following the transfer into this strain of the sex factor of strain K12.

Chromosome transfer, mediated by a sex factor with certain of the characters described for F, has been demonstrated in a second major group of Gram-negative bacteria, the pseudomonads, where it occurs in the species *Pseudomonas aeruginosa*. Conjugation has not been observed in Gram-positive bacteria.

However, F is not the only transmissible agent that possesses sex-factor activity. Similar agents, such as the *resistance transfer factors* and *colicinogeny factors*, are also capable of mobilizing the chromosome for intercellular transfer. Resistance transfer factors and colicinogeny factors will be discussed in the following section. The ability of these agents to carry chromosomal genes from one species or genus of enteric bacteria to another must have played an important role in the evolution of these Gram-negative organisms.

**Summary**

The Gram-negative enteric bacteria are hosts to a number of genetic elements, which are characterized by the following properties: (1) they are small, circular, DNA structures that replicate autonomously in the host cell; (2) they carry genes that govern the synthesis of specific surface substances, including special pili, which act as receptors for the conjugal attachment of other cells; and (3) following such attachment, one strand of the DNA structure is transferred to the attached cell, being replicated in the process. We shall refer to such structures as *autotransferable genetic elements*.

If an autotransferable element contains a sequence of base pairs that is homologous with a sequence on another replicon in the same host cell, pairing may occur. A single crossover in the region of pairing will then integrate the two structures, so that both are transferred at conjugation.

In summary, the bacterial chromosome is transferred during conjugation only if it can integrate with an autotransferable element such as an F, a resistance transfer factor, or a colicinogeny factor. Autotransferable elements that possess sufficient base-sequence homology with the host chromosome to permit transfer of the chromosome at a detectable frequency are called *sex factors*.

## *Plasmids and episomes*

Autonomous, dispensable genetic elements that are not capable of integration with the host chromosome are called *plasmids*. Elements that are capable of existing in either the autonomous or integrated state are called *episomes*. From the foregoing discussion of homology, it is obvious that a given element may exist as a plasmid in one species of bacterial host and as an episome in another. *F-lac*, for example, must be called a plasmid when harbored by *Proteus mirabilis* but an episome when harbored by *E. coli*.

Not all plasmids and episomes are autotransferable. Staphylococci, for example, are hosts to plasmids that carry a number of genes determining drug resistance, including both structural and regulatory genes for penicillinase production. These plasmids do not promote conjugation and are not autotransferable. Like all plasmids and episomes, however, they are lost spontaneously at significant rates, as a result of failure to replicate or segregate at cell division. There must, therefore, be another mechanism for their transfer from cell to cell, since without such a process to compensate for their loss, they could not be maintained in nature. This mechanism has been discovered to be *phage transduction*. Certain phages are found to transduce the entire "penicillinase plasmid" at a rate which is sufficient to compensate for its spontaneous loss.

**The known plasmids and episomes**

The distinguishing properties of plasmids and episomes are (1) their ability to exist in the autonomous, nonintegrated state and (2) their dispensability to the host cell. The former property is reflected in the capacity for independent transfer, as well as in the capacity for irreversible loss at a relatively high rate (of the order of $10^{-3}$ per cell-generation). The presence of plasmids and episomes is most easily detected if the genes they carry determine recognizable cell properties. Of the plasmids and episomes discovered to date, the best studied are the sex factor (F) of *E. coli* strain K12, the resistance transfer factors (RTFs), the colicinogeny (Col) factors, and the penicillinase plasmids of the staphylococci. The term "episome" is also applicable to the temperate bacteriophages.

**Resistance transfer factors**

The resistance transfer factors were first discovered in *Shigella dysenteriae*, during an outbreak of dysentery in Japan in 1959. Antibiotics and sulfonamides were widely used during the epidemic, and *Shigella* strains, isolated from patients undergoing chemotherapy, were found to be simultaneously resistant to four drugs: streptomycin, chloramphenicol, tetracycline, and sulfonamide. When cells of a multiply resistant strain were mixed with cells of a sensitive strain, multiple resistance was found to be transferred at high frequency. Ultimately, it was discovered that resistance to each drug is determined by a separate gene, all four genes being located on an autotransferable plasmid; another gene on the plasmid determines synthesis by the host cell of a specific kind of pilus.

Resistance transfer factors have since been discovered in many countries and in a variety of pathogenic organisms belonging to the enteric group. Two types of RTF have been recognized, on the basis of their interaction with F when both are harbored in the same cell. One type, called $fi^+$ (for "fertility inhibition"), produces a repressor that inhibits the production of pili by both F and RTF; the other type, called $fi^-$, represses its own pilus formation but not that of F. The $fi^+$ type of RTF appears to be closely related to F, since both produce the same type of pilus.

A given line of cells can harbor both an $fi^+$ and an $fi^-$ RTF but cannot harbor two $fi^+$ RTFs or two $fi^-$RTFs. Similarly, a given line of cells cannot harbor two different Fs. This strongly suggests that the cell has a single, specific attachment site in the membrane for which all RTFs of a given type must compete, and another site for which all Fs must compete. This theory can account for many observations, including the inability of an Hfr cell to harbor an autonomous F.

**Colicinogeny factors**

Colicines are substances produced by certain strains of enteric bacteria which are lethal for other strains of enteric bacteria. Their chemical nature varies widely; they range from relatively simple proteins to a lipopolysaccharide–protein complex resembling a cell wall antigen. The genes that determine the production of certain colicines are carried by autotransferable plasmids known as Col factors; some of these have properties in common with the $fi^+$ RTF-F group, whereas others are more like the $fi^-$ RTFs.

**Phylogenetic relationships**

Some RTFs have been found to act as true sex factors in *E. coli*, and some Col factors act as sex factors in *Salmonella*, indicating the existence of some base-sequence homologies with bacterial chromosomes. Different plasmids occupying the same Gram-negative cell have been observed to recombine with each other. Resistance genes carried by an RTF have become incorporated into an F factor in one such instance and into a temperate bacteriophage in another. All these recombination potentialities would seem to indicate a phylogenetic relatedness among the various plasmids and episomes, but their origins remain essentially obscure.

## *Transduction by bacteriophage*

**The discovery of transduction**

In 1952 N. Zinder and J. Lederberg began experiments to determine whether or not conjugation occurred between strains of *Salmonella*. When they mixed strains carrying appropriate genetic markers and plated them on media selective for recombinant types, colonies appeared in low yields.

In a series of experiments, Zinder and Lederberg discovered the basis of this genetic exchange. They found that one of the parent strains had released particles of a temperate bacteriophage which had infected cells of the other parent strain. Many of the receptor cells survived the infection, and some of the survivors proved to have received fragments of genetic material derived from the cells in which the phage had originally grown.

In *transduction*, as this phenomenon was named, a small piece of bacterial chromosome is incorporated into a maturing phage particle. When this particle infects a new host cell, it injects the genetic material from the former host. The recipient thus becomes a partial zygote.

The first transducing phage to be discovered, called P22, transduces all *Salmonella* markers with roughly equal efficiencies. When other temperate phages of *Salmonella* and *E. coli* were examined, however, only a few appeared to be capable of transduction. In the meantime, the discovery had been made that certain temperate phages of *E. coli* occupy fixed positions on the bacterial chromosome when in the prophage state. For example, λ always attaches close to the *gal* loci, as shown in Figure 14.12 (p. 502). When λ phage was tested for the ability to transduce markers between strains of *E. coli*, it was found that it could transduce the *gal* loci but not markers located farther from the λ attachment site.

The type of transduction performed by P22, in which any chromosomal marker has a roughly equal chance of being incorporated into a phage coat, is called *generalized transduction*. The type of transduction carried out by λ, in which only those loci immediately adjacent to the prophage attachment site are incorporated, is called *restricted transduction*. Restricted and generalized transduction differ in the mechanism by which the genetic material of the transducing phage particles is formed.

In a culture of *E. coli* lysogenic for λ, every cell carries a λ prophage inserted into the chromosome at a site adjacent to the *gal* loci. When such a culture is induced with ultraviolet light, the prophage in each cell detaches by a recombinational event within the normal region of pairing between a chromosomal site and a site on the prophage. As a rare event, however, pairing occurs between two different sites, and the recombinational event generates a circle of DNA composed of a major fraction of λ together with a segment of chromosomal DNA (Figure 14.20). This element, packaged in a phage coat, produces a particle called λ*dg*, standing for "λ defective, carrying the *gal* loci."

The induction of a λ-lysogenic $gal^+$ culture produces a lysate containing about one λ*dg* for every $10^5$ normal phage particles. When this lysate is used to infect a culture of nonlysogenic $gal^-$ bacteria, some of the cells receive λ*dg* DNA, which becomes integrated into the chromosome as a prophage. Each of these cells, however, still retains its $gal^-$ chromosomal locus and thus becomes a $gal^+/gal^-$ partial diploid. *Restricted*

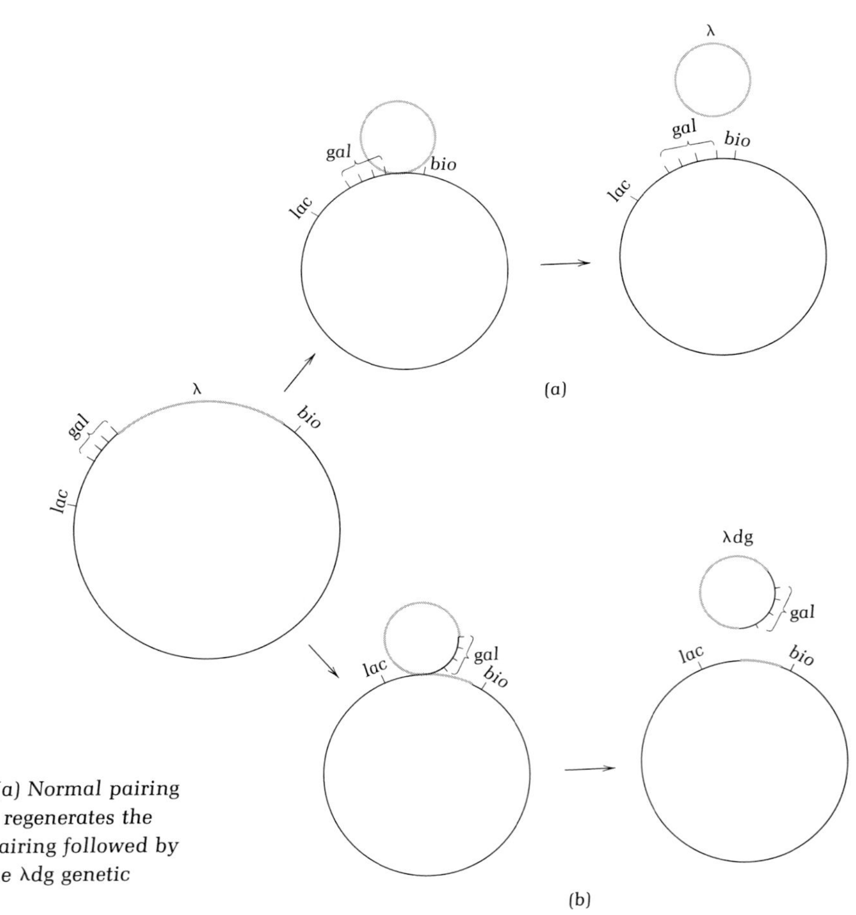


*Figure 14.20 The formation of λdg. (a) Normal pairing followed by a single crossover event regenerates the wild-type λ genome. (b) Abnormal pairing followed by a single crossover event generates the λdg genetic element.*

*transduction thus involves the addition to the chromosome of a defective prophage carrying a segment of DNA taken from its former attachment site.*

If the infection of the $gal^-$ culture is carried out at a high phage multiplicity, any cell receiving a $\lambda dg$ particle also receives a particle of normal $\lambda$. Such a cell becomes a double lysogen, carrying both a normal and a $\lambda dg$ prophage. When a culture derived from the double lysogen is induced, the normal $\lambda$ prophage provides the phage functions for which $\lambda dg$ is defective. The result is that both types of phage mature, and the lysate that is produced contains equal numbers of $\lambda$ and $\lambda dg$ phages. When this lysate is now used to transduce a fresh $gal^-$ culture, roughly half the phage particles transduce the $gal^+$ loci. This phenomenon is called *high-frequency transduction.*

A second restricted transducing phage, $\phi 80$, has been discovered. The attachment site of $\phi 80$ is close to the *trp* loci, governing the enzymes of the tryptophan biosynthetic pathway. Phage $\phi 80dt$ ($\phi 80$ defective, carrying the *trp* genes) can be produced by the procedure used to produce $\lambda dg$. Another recombinant derivative of $\phi 80$, carrying the *lac* genes, has also been produced from a strain in which *F-lac* had previously been integrated adjacent to the *trp* loci. These defective transducing phages have provided a source of purified DNA containing well-defined sets of genes.

### The mechanism of generalized transduction

In restricted transduction, the fragment of host DNA carried by a transducing particle is always covalently linked to a large segment of phage DNA. In generalized transduction, on the other hand, the transducing phage particles contain *no detectable phage DNA;* they appear to constitute a fragment of bacterial DNA packaged in a phage coat. Thus, a lysate produced by a generalized transducing phage contains two types of particles: a majority type, containing only phage DNA, and a minority type (the transducing particles), containing only host DNA. This has been shown in the following manner.

Donor cells of *E. coli* were grown in a medium containing 5-bromouracil (5-BU), with the result that all their DNA became "heavy," because the thymine of the DNA had been replaced by the bromine-containing analog. The cells were then transferred to a medium containing thymine instead of 5-BU and were infected with the generalized transducing phage, P1. The phage progeny from this infection were banded in a cesium chloride density gradient, and phage fractions of different densities were tested for their ability to transduce a recipient culture. If the transducing particles had contained donor DNA covalently linked to phage DNA, they would have banded at a position intermediate between the positions of "heavy" host DNA and "light" phage DNA, since the transducing phage particles were produced in the absence of 5-BU. Instead, however, the experiment showed that the transducing phage particles had a density corresponding to that of particles containing only "heavy" (bacterial) DNA. If 5 percent or more of the DNA of the transducing particles had been phage DNA, it would have been detected.

The difference between the nature of the DNA in a restricted transducing particle and in a generalized transducing particle is consistent with the manner in which the transducing lysate is formed. If λ is produced in its host by a lytic cycle of growth, no λ*dg* particles are formed. To obtain such particles the lysate must be prepared by inducing a λ-lysogenic culture (i.e., λ must be present in the prophage state). A generalized transducing phage such as P1, on the other hand, yields transducing lysates by a cycle of lytic growth. This difference is related to the difference in prophage state of λ and P1, as described in Chapter 11.

A generalized transducing phage is not totally incapable of forming an occasional λ*dg*-like particle. When phage P1 is grown on a $lac^+$ donor and used to transduce a recipient in which the *lac* region of the chromosome is deleted, integration of the *lac* gene cannot occur, since there is no region of homology on the chromosome. Some $lac^+$ transductants are nevertheless recovered; these prove to have received in each case a defective P1 prophage carrying *lac* genes. Such transducing particles are extremely rare in lysates of a generalized transducing phage.

**The fate of the exogenote formed by transduction**

In many instances of transduction, the exogenote immediately pairs with the endogenote, and donor genes are incorporated into the recombinant chromosome by multiple crossovers within the region of pairing. The partial zygote then segregates a haploid recombinant type. In the case of *gal* transduction by λ, in contrast, the exogenote *persists and replicates* as part of a prophage, giving rise to a heterozygous, partially diploid clone.

The exogenote may also *persist and function but not replicate*, so that in the clone arising from the partial zygote only one cell at any given moment contains an exogenote. This relatively common phenomenon is called *abortive transduction*. If, for example, a phage particle injects a $his^+$ gene into a $his^-$ bacterium and the transduction is abortive, the following events occur. The $his^+$ exogenote functions in the partial zygote, producing enzyme for the synthesis of histidine. If plated on agar lacking histidine, the zygote grows and divides, but the exogenote does not replicate. Thus, one daughter cell does not receive the *his*$^+$ gene and can continue to divide only until the initial supply of enzyme has been diluted out by growth. The other daughter cell, however, is a heterogenote like its parent and continues to form enzyme. At its next division, it again segregates one *his*$^-$ cell containing a limited supply of enzyme and one *his*$^+$/*his*$^-$ heterogenote.

The result is the production of a *slow-growing minute colony*, which becomes barely visible after a few days of growth. When a minute colony of this type is picked and restreaked, it gives rise to one minute colony again, since only one of the cells that were picked contains the *his*$^+$ gene. The events in abortive transduction are shown in Figure 14.21.

In some cases, a given suspension of phage will produce 10 times as many abortive transductions as normal ones. An abortive transduction results when recombination fails to occur between the exogenote and endogenote. This failure can be overcome by ultraviolet irradiation of the transduced cells. The ability of ultraviolet light to convert abortive transductions to normal transductions is consistent with its known ability to stimulate recombination in many diploid systems.

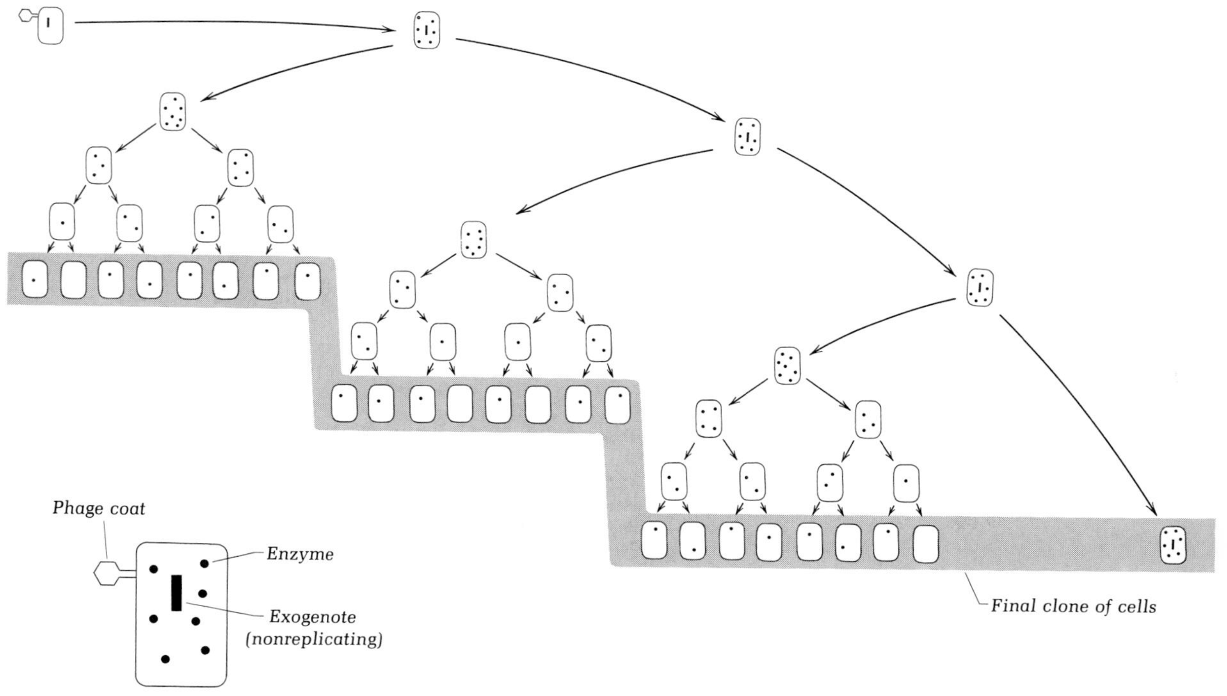


*Figure 14.21 Diagrammatic representation of the events that take place in abortive transduction. After H. Ozeki.*

Transduction can take place between various members of the enteric group of bacteria and has also been observed in *Pseudomonas*, *Vibrio*, *Staphylococcus*, and *Bacillus*. Since most species of bacteria harbor temperate phages, transduction may well be the most common mechanism of recombination in bacteria.

**Phage conversion**

Certain properties of bacterial cells are controlled solely by phage genes. Such properties are manifested only by lysogenic or phage-infected strains and never by phage-free strains. For example, *Corynebacterium diphtheriae* produces toxin only if it is infected with a particular strain of phage. Every phage-infected cell produces toxin, but phage-free cells do not and are incapable of mutating to toxigenicity. Similarly, certain antigenic components of the cell wall of *Salmonella* are produced in every cell infected with a particular phage and are always absent if the phage is absent. The acquisition of a new property solely as a result of infection by a phage is called *phage conversion*.

Conversion differs from restricted transduction, which involves the acquisition by a phage of a gene that is normally found in the bacterial chromosome. There is a second difference between conversion and restricted transduction: converting phage particles are completely normal and possess all necessary phage functions, whereas restricted transducing phages are *defective* for some phage functions, having lost some phage genes by exchange. With these exceptions, however, phage conversion and high-frequency restricted transduction are essentially similar.

It is not difficult, in the light of present knowledge, to visualize a process of evolution in the course of which both bacterial and phage chromosomes evolved from a common ancestor. The ability of bacterial genes to replace phage genes, and vice versa, points to a close phylogenetic relationship between bacterial viruses and their hosts. Thus, a gene controlling a given host property may be found only in the bacterium, only in the phage (conversion), or in both (transduction).

## Recombination in bacterial viruses

A bacterial cell may be simultaneously infected with two closely related phages which differ from each other in a number of genetic markers. When the host cell lyses, the progeny particles are found to include not only the two parental types but also a variety of genetic recombinants.

The first phage-recombination studies, carried out by A. Hershey, involved two of the mutant types described in Chapter 13: rapid lysis (r) and host range (*h*). A double mutant of phage T2 was obtained in two steps: an r mutant was isolated from a mutant plaque and was plated on *E. coli* strain B/2 to obtain the double mutant, T2*rh*. (The parental T2, carrying the wild-type alleles $r^+$ and $h^+$, can be designated T2++.)

Hershey infected *E. coli* strain B with T2*rh* and T2++ simultaneously and observed that each infected cell yielded a mixed burst containing the four expected types: the two parental types and the recombinants T2+*h* and T2r+. All four types could be scored simultaneously by plating the progeny on a "mixed indicator"—in this case a lawn of cells containing equal numbers of *E. coli* strain B and *E. coli* strain B/2. The four plaque types that result are shown in Figure 14.22.

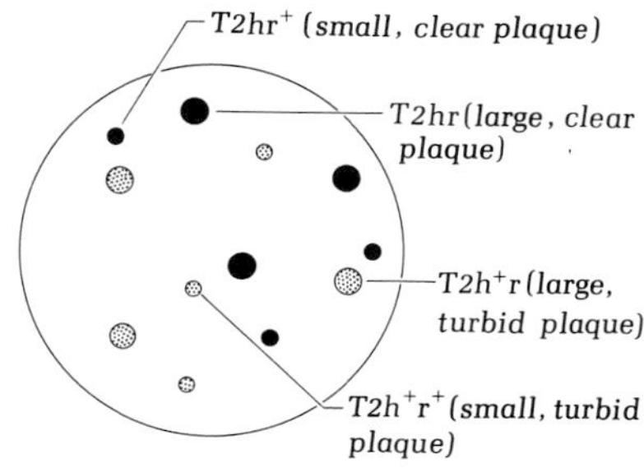


*Figure 14.22 Diagrammatic representation of the appearance of the plaques of four types of phages.*

In the course of a mixed infection, both parental genomes as well as any recombinant genomes continue to replicate until lysis of the host cell occurs. Does recombination take place only between the original two infecting genomes, or can it take place between any two genomes at any time during replication?

If the former were true, then a single infected cell should yield only the two parental types of phage or the two recombinant types, but not all four. As we have seen, a single infected cell does yield all four types; consequently, recombination cannot be restricted to the prereplication stage. Furthermore, it is possible to carry out a *triparental cross*, by infecting host cells with three phages each of which carries a different mutant marker. If the three types are designated x++, +y+, and ++z, then from the triply infected cell one can recover the recombinant, xyz, carrying all three mutant markers. The formation of such a recombinant requires at least two successive genetic exchanges between pairs of genomes; for example, the first two parental types might recombine to yield the recombinant, xy+, and this recombinant might then recombine with ++z to produce the xyz genome.

The dynamics of phage recombination have been analyzed by N. Visconti and M. Delbrück, who derived a set of equations relating three parameters: the probability of a genetic exchange between two given markers during each "mating"; the number of matings in the ancestry of the average phage particle ("rounds of mating"); and the final, observed frequency of recombination between the two markers. Their calculations indicated that the average T2 particle experiences five rounds of mating before lysis occurs, or about one mating per replication cycle. The infected cell contains a pool of replicating phage genomes, any two of which may, at any time, pair and recombine at random. Following recombination, the two recombinants reenter the replicating pool. Labeling studies have shown that even in the pool of identical phage genomes arising from a singly infected cell exchanges occur repeatedly by recombination.

The mechanism of phage recombination has been analyzed by permitting recombination to occur between genetically marked phage genomes of different density. The progeny particles were collected and centrifuged in a cesium chloride gradient, which separated the particles according to their density; they were then analyzed for their genetic markers. A particular recombinant class, which arose by a recombination event in the middle of the phage genome, was shown to have exactly the density predicted if half of its DNA was derived from one parent and half from the other. Further recombination events between phages of intermediate density yielded other classes with the properties predicted by the model of breakage and reunion. These have been the most decisive experiments in establishing the molecular basis of genetic recombination.

## Further reading

**Books**

Adelberg, E. (editor), *Papers on Bacterial Genetics*, 2nd ed. Boston: Little, Brown, 1966.

Hayes, W., *The Genetics of Bacteria and Their Viruses*, 2nd ed. Oxford, England: Blackwell, 1968.

Jacob, F., and E. Wollman, *Sexuality and the Genetics of Bacteria*. New York: Academic Press, 1961.

Stent, G., *The Molecular Biology of Bacterial Viruses*. San Francisco: Freeman, 1963.

———, *Papers on Bacterial Viruses*, 2nd ed. Boston: Little, Brown, 1965.

**Reviews and original articles**

Meselson, M., "On the Mechanism of Genetic Recombination between DNA Molecules." *J. Mol. Biol.* **9**, 734 (1964).

———, "The Molecular Basis of Recombination," in *Heritage from Mendel* (R. Brink, editor). Madison, Wisc.: University of Wisconsin Press, 1967.

Whitehouse, H., "A Theory of Crossing-over by Means of Hybrid Deoxyribonucleic Acid." *Nature* **199**, 1034 (1963).

# Chapter fifteen The classification of bacteria

*The art of biological classification* is known as *taxonomy*. It has two functions. The first is to identify and describe as completely as possible the basic taxonomic units, or *species;* the second, to devise an appropriate way of arranging and cataloging these units. These are, respectively, the operations of *data accumulation and storage* and of *data retrieval*.

## Species: the units of classification

The notion of a species is exceedingly complex, difficult to define in an operationally useful and precise way. Speaking broadly, a species consists of an assemblage of individuals (or, in microorganisms, of clonal populations) that share a high degree of phenotypic similarity, coupled with an appreciable dissimilarity from other assemblages of the same general kind. The recognition of species would not be possible if natural variation were continuous, so that an intergrading series spanned the gap between two assemblages of markedly different phenotype. However, it became evident early in the development of biology that, among most groups of plants and animals, reasonably sharp discontinuities do separate the members of a group into distinguishable assemblages. Hence, the notion of the species as the base of taxonomic operation proved workable.

Every assemblage of individuals shows some degree of internal phenotypic diversity, since genetic variation is always at work. Hence it becomes a matter of scientific tact to decide what *degree* of phenotypic dissimilarity justifies the breaking up of an assemblage into two or more species; or, to put the matter another way, how much internal diversity is permissible in an assemblage that one regards as a single species. Opinions on this question naturally vary. Taxonomists themselves can be broadly divided into two groups: "lumpers," who set wide limits to a species; and "splitters," who differentiate species on more slender grounds.

For plants and animals that reproduce sexually, a species can be defined in genetic and evolutionary terms. As long as a sexually reproducing population is free to interbreed at random, its total gene pool undergoes continuous redistribution, and new mutations, the source of phenotypic variation, get dispersed throughout the population. Such an interbreeding population may evolve, in response to environmental changes, but it will evolve with reasonable uniformity. *Divergent* evolution, eventually leading to the emergence of new species, can occur only if a segment of the population becomes reproductively isolated in an environment that is different from that occupied by the rest of the population. Reproductive isolation is probably always geographic in the first instance; a physical barrier of some sort (for example, a mountain range or a body of water) is interposed between two parts of the initially continuous population. Within each of these subpopulations, a common gene pool is maintained by interbreeding, but through chance mutation and selection, the two subpopulations are now free to evolve along different lines. They will continue to diverge, as long as the geographical barrier persists. Eventually, the cumulative differences become so great that *physiological* isolation is superimposed on geographic isolation; members of the two populations are no longer capable of interbreeding if they are brought together. Hence even if the two populations subsequently commingle once more, their gene pools remain permanently separated; a point of no return has been reached. These evolutionary considerations lead to a dynamic definition of the species, as a stage in evolution at which actually or potentially interbreeding arrays have become separated into two or more arrays physiologically incapable of interbreeding. This definition is, in fact, an *explanation* of the origin of specific discontinuities in nature. At the same time, it provides an experimental criterion for the recognition of species differences: inability to interbreed. For many taxonomists of plants and animals, this so-called *biological species* is the only meaningful one.

Unfortunately, the biological species concept is completely irrelevant in the context of microorganisms, which reproduce either exclusively or predominantly by asexual means. The two offspring resulting from a bacterial cell division are reproductively isolated from one another, barring a later rare recombinational event among their progeny. In a certain sense, accordingly, a bacterial clone is the genetic and evolutionary equivalent of the biological species in an organism that reproduces by sexual means. A workable taxonomy of microorganisms cannot, however, be based on the use of the clone as the taxonomic unit; if this were done, every microbiologist could create at will as many species as he wished.

The dynamics of the evolution of microbial populations is obviously very different from that of populations of plants or animals. We have accordingly no theoretical reason to believe that microbial evolution does lead to phenotypic discontinuities which would justify the recognition as species of assemblages above the clonal level. The experience of bacterial taxonomists shows, however, that when many strains belonging to a given bacterial group are carefully compared, they can usually be divided into a series of discontinuous clusters; and it is such *clusters of strains* that the microbiologist terms species. A microbial species is thus a strictly empirical entity. Perhaps greater knowledge of the dynamics of microbial evolution will eventually make possible a formal definition of the microbial species, analogous to the biological species concept that exists for plants and animals. If so, it will certainly be couched in terms different from these now used to define the species in groups that reproduce by sexual means.

Genetic change can occur so rapidly in bacterial populations that it is unwise to distinguish bacterial species on the basis of a small number of character differences, governed by single genes. The best working definition that can be offered is the following: a bacterial species consists of an assemblage of strains showing a high degree of overall phenotypic similarity and distinguishable from other assemblages by a large number of independent characters.

## The characterization of species

Ideally, species should be characterized by complete descriptions of their phenotypes or—even better—of their genotypes. Taxonomic practice falls far short of these ideals; in most biological groups, even the phenotypes are only fragmentarily described, and genotypic characterizations are extremely rare.

As a general rule, the phenotypic characters that can be most easily determined are structural or anatomical ones, which can be directly observed. For this reason, biological classification is still based, at most levels, almost entirely on structural properties. Virtually the only exception is the bacteria. The extreme structural simplicity of bacteria offers the taxonomist too small a range of characters upon which to base adequate characterizations. Hence, the bacterial taxonomist has always been forced to seek other kinds of characters—biochemical, physiological, ecological—with which to supplement the scanty structural data obtainable. The classification of bacteria is based, to a far greater extent than that of any other biological groups, on *functional* attributes. Most bacteria can be identified only by finding out what they can do, not simply how they look.

This confronts the bacterial taxonomist with an additional problem. To find out what a bacterium can do, he has to perform experiments with it. The number of possible experiments that can be performed is extremely large, and although all will reveal facts, the facts so revealed will not necessarily be taxonomically significant ones, in the sense of contributing to a differentiation of the organism under study from related assemblages. The bacterial taxonomist is in the position of a color-blind man, asked to classify a collection of red and green balls of different sizes. Being ignorant of the color difference, the man would be forced to classify the collection of balls in terms of size differences, and might well find that on this basis they constituted a continuum, without any sharp clusters. Similarly, the bacterial taxonomist can never be sure that he has performed the right experiments for taxonomic purposes; he may well have failed to perform certain experiments that would have shown him significant clustering in a collection of strains, and therefore erroneously conclude that he is dealing with a continuous series. There is no obvious way to get around this difficulty, except to make phenotypic characterizations as exhaustive as possible.

## The naming of species

According to a convention known as the binomial system of nomenclature, every biological species bears a latinized name that consists of two words. The first word indicates the taxonomic group of immediately higher order, or *genus* (plural, *genera*) to which the species belongs, and the second word identifies it as a particular species of that genus. The first letter of the generic (but not of the specific) name is capitalized, and the whole phrase is italicized: *Escherichia* (generic name) *coli* (specific name). In contexts where no confusion is possible, the generic name is often abbreviated to its initial letter: *E. coli.*

A rigid and complex set of rules governs biological nomenclature; the rules are designed to keep nomenclature as stable as possible. The specific name given to a newly recognized species cannot be changed, unless it can be shown that the organism has previously been described under another specific name, in which case the older name has priority. Unfortunately, the same stability does not govern the generic half of the name, since the arrangement of related species into genera is an operation that can be carried out in different ways and that often changes in the course of time as new information becomes available. For example, *E. coli* has in the past been placed in the genus *Bacterium*, as *Bacterium coli* and in the genus *Bacillus*, as *Bacillus coli*. These three double names are synonyms, since they all refer to one and the same species. This consequence of the binomial system can be very confusing, and taxonomic descriptions usually list all such synonyms to minimize the confusion. Binomial nomenclature is used for all biological groups except viruses. The virologists are at present divided over the best way to designate members of this group; some wish to extend the binomial system to the viruses, whereas others would prefer another system, which gives in coded form information about the properties of the organism.

In bacterial taxonomy, when a new species is named, a particular strain is designated as the *type strain*. Such strains are preserved in culture collections; if one is lost, a *neotype strain*, which resembles as closely as possible the description of the type strain, is chosen. The type strain is important for nomenclatorial purposes, since the specific name is attached to it. If other strains, originally included in the same species, prove on subsequent study to deserve recognition as separate species, they must receive new names, the old specific name resting with the type strain and related strains.

In the taxonomic treatment of a biological group, the individual species are usually grouped in a series of categories of successively higher order: genus, family, order, class, and division (or phylum). Such an arrangement is known as a hierarchical one, because each category in the ascending series unites a progressively larger number of taxonomic units in terms of a progressively smaller number of shared properties. It should be noted that the genus has a position of special importance, since according to the rules of nomenclature a species cannot be named unless it is assigned to a genus. The allocation of a species to a taxonomic category higher than the genus does not carry any essential *nomenclatorial* information; it is merely indicative of the position of an organism, relative to other organisms, in the system of arrangement adopted.

## *The problems of taxonomic arrangement*

In dealing with a large number of different objects, some system of orderly arrangement is essential for purposes of data storage and retrieval. It does not matter what criteria for making the arrangement are adopted, provided that they are unambiguous and convenient. Books can be arranged in different ways: for example, by subject, by author, or by title. Different individuals tend to adopt different systems, depending on their particular needs and tastes. Such a system of classification, based on arbitrarily chosen criteria, is termed an *artificial* one.

The earliest systems of biological classification were largely artificial in design. However, as knowledge about the anatomy of plants and animals increased, it became evident that these organisms can be arranged into a number of major patterns or types, each of which shares many common properties, including ones that are not necessarily obvious upon superficial examination. Examples of such types are the mammalian, avian, and reptilian types among vertebrate animals. The first system of biological classification that attempted to group organisms in terms of such typological resemblances and differences was developed in the middle of the eighteenth century by Linnaeus (Carl von Linné). The Linnaean arrangement was clearly more useful than previous artificial arrangements, since the taxonomic position of an organism immediately furnished a large body of information about its properties: to say that an animal belongs to the vertebrate class Mammalia immediately tells one that it possesses all those properties which distinguish mammals collectively from other vertebrates. Because Linnaean classification expressed the *biological nature* of the objects that it classified, it became known as a *natural* system of classification, in contrast to preceding artificial systems.

**The phylogenetic approach to taxonomy**

When the fact of biological evolution was recognized, another dimension was immediately added to the concept of a natural classification. For biologists of the eighteenth century, the typological groupings merely expressed *resemblances;* but for post-Darwinian biologists, they revealed *relationships.* In the nineteenth century, the concept of a "natural" system accordingly changed: it became on that grouped organisms in terms of their *evolutionary affinities.* The taxonomic hierarchy became in a certain sense the reflection of a family tree. The analogy between a family tree and a hierarchy cannot be pushed very far, however, since the dimension of time, implicit in a family tree, is completely absent from a taxonomic hierarchy. This point was not very clearly grasped by many evolutionary biologists, for whom taxonomy suddenly acquired a new goal: the restructuring of hierarchies to mirror evolutionary relationships. A taxonomic system in which this is an avowed goal is known as a *phylogenetic system.*

Reflection and experience have shown, however, that the goal of a phylogenetic classification can seldom be realized. The course that evolution has actually followed can be ascertained only from direct historical evidence, contained in the fossil record. This record is at best fragmentary and becomes almost completely illegible in Precambrian rocks more than 400 million years old. By the beginning of the Cambrian period, most of the major biological groups that exist had already made their appearance; vertebrates and plants are the principal evolutionary newcomers in Postcambrian time. For these two groups, the fossil record is, accordingly, reasonably complete, and the main lines of plant and vertebrate evolution can be retraced with some assurance. For all other major biological groups, the general course of evolution will probably never be known, and there is simply not enough objective evidence to base their classification on phylogenetic grounds.

**Numerical taxonomy**

For these and other reasons, most modern taxonomists have explicitly abandoned the phylogenetic approach, in favor of a more empirical one: the attempt to base taxonomic arrangement upon quantification of the similarities and differences among organisms. This was first suggested by Michel Adanson, a contemporary of Linnaeus, and is known as *Adansonian* (or *numerical*) *taxonomy*. The underlying assumption is that, provided each phenotypic character is given equal weighting, it should be possible to express numerically the taxonomic distances between organisms, in terms of the number of characters they share, relative to the total number of characters examined. The significance of the numerical relationships so determined is greatly influenced by the number of characters examined; these should be as numerous and as varied as possible, to obtain a representative sampling of phenotype.

*Table 15.1 The determination of similarity coefficient and matching coefficient for two bacterial strains, both characterized with respect to many (a+b+c+d) different characters*

| | |
|---|---|
| *Number of characters positive in both strains:* | a |
| *Number of characters positive in strain 1 and negative in strain 2:* | b |
| *Number of characters negative in strain 1 and positive in strain 2:* | c |
| *Number of characters negative in both strains:* | d |

*Similarity coefficient* $(S_J) = \dfrac{a}{a+b+c}$

*Matching coefficient* $(S_S) = \dfrac{a+d}{a+b+c+d}$

Until recently, the Adansonian approach appeared impractical on account of the magnitude of the numerical operations involved. This difficulty has been obviated by the advent of computers, which can be programmed to compare data for a large number of characters and organisms and to compute the degrees of similarity. For any pair of organisms, the calculation of similarity can be made in two slightly different ways (Table 15.1). The similarity coefficient $S_J$ does not take into account characters negative for both organisms, being based only on positive matches; the matching coefficient $S_S$ includes both positive and negative matches in the calculation.

After similarity (or matching) coefficients have been calculated pairwise for all the organisms under study, the data are arranged in a similarity matrix, an example of which is shown in Figure 15.1(*a*) for a series of 10 bacterial strains. By inspection, such a matrix can be reordered so as to bring into juxtaposition similar strains [Figure 15.1(*b*)]. The data can then be transposed into a dendrogram (Figure 15.2), as a basis for determining taxonomic arrangement in terms of numerical relationships. The two dotted vertical lines in Figure 15.2 indicate similarity levels which might be considered appropriate for recognizing two different taxonomic ranks (e.g., genus and species).

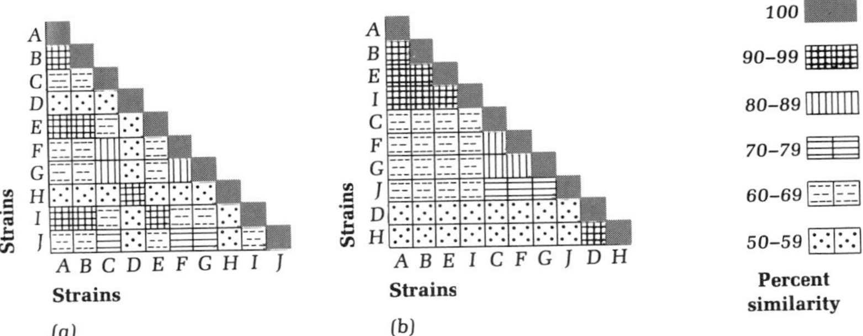


*Figure 15.1 Similarity matrices for a collection of ten bacterial strains, designated as A through J: (a) similarity matrix before rearrangement; (b) similarity matrix after rearrangement. The strains have been so ordered as to bring into juxtaposition strains that most closely resemble one another in overall phenotype. After P. H. A. Sneath, "The construction of taxonomic groups," in Microbial Classification (G. C. Ainsworth and P. H. A. Sneath, editors). New York: Cambridge University Press, 1962.*

Numerical taxonomy does not have the evolutionary connotations

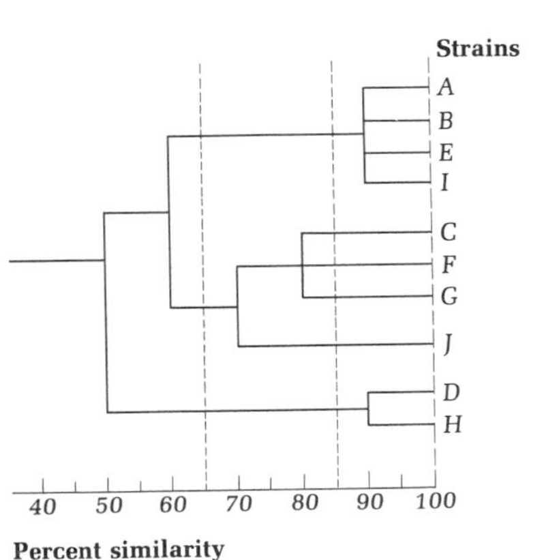


*Figure 15.2 A dendrogram showing similarity relationships among the 10 bacterial strains, using the data from Figure 15.1. The two dotted vertical lines indicate possible similarity levels at which successive ranks (e.g., genus and species) in the taxonomic hierarchy might be established. After P. H. A. Sneath, "The construction of taxonomic groups," in Microbial Classification (G. C. Ainsworth and P. H. A. Sneath, editors). New York: Cambridge University Press, 1962.*

of phylogenetic taxonomy, but it provides a more objective and stable basis for the construction of taxonomic groupings. Perhaps its greatest advantage is that it cannot be applied at all until a relatively large number of characters have been determined, so that its use encourages a fairly thorough examination of phenotypes. Furthermore, the analyses are open to continuous revision and refinement as more characters in a given group are determined.

## *New approaches to bacterial taxonomy*

In recent years, the growth of molecular biology has opened up a number of new approaches to the characterization of organisms, and these approaches have already had a profound impact on the taxonomy of bacteria. Of particular value are certain techniques that give insights into the genotypic properties of bacteria and thus complement the hitherto exclusively phenotypic characterizations of these organisms. Two different kinds of analysis performed upon isolated nucleic acids furnish information about genotype: the analysis of the base composition of DNA, and the study of chemical hybridization between DNA and DNA or between DNA and RNA.

### The base composition of DNA: its determination and significance

DNA contains four bases: adenine (A), thymine (T), guanine (G), and cytosine (C). For double-stranded DNA, the base-pairing rules (see Chapter 7) require that A = T and G = C. However, there is no chemical restriction on the molar ratio (G + C) : (A + T). Early in the chemical study of DNA, analyses showed that this ratio in fact varies over a rather wide range in DNA preparations from different groups of organisms, and subsequent work has revealed that the base composition of DNA is a character of profound taxonomic importance, particularly among microorganisms.

Although DNA base composition may be determined chemically, after hydrolysis of a DNA sample and separation of the free bases, it can be determined more easily by physical methods, and these are now the ones principally used. The "melting temperature" of DNA (i.e., the temperature at which it becomes denatured, by breakage of the hydrogen bonds that hold together the two strands) is directly related to G + C content, because hydrogen bonding between G–C pairs is stronger than that between A–T pairs. Strand separation is accompanied by a marked increase in absorbance at 260 nm, the absorption maximum of DNA, and this can be easily measured in a spectrophotometer. When a DNA sample is gradually heated, the absorbance increases as the hydro-

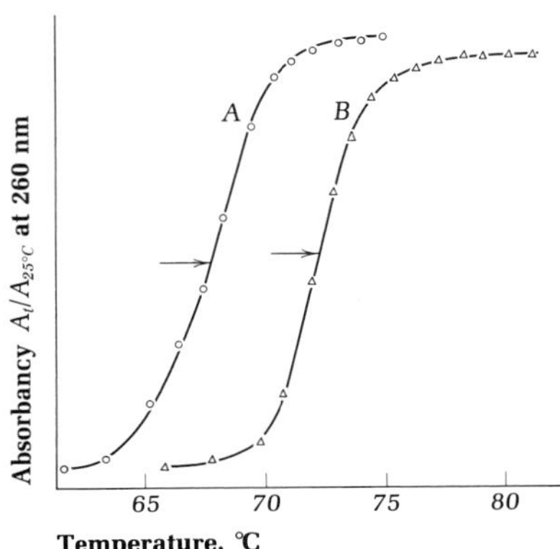


*Figure 15.3* Melting curves determined optically for two samples of bacterial DNA. Curve A: DNA of *Lactobacillus acidophilus* ($T_m$ 67.7°C). Curve B: DNA of *Leptospira* sp. ($T_m$ 72.1°C). The ordinate expresses the absorbancy at 260 nm of the DNA sample at each temperature on the abscissa, relative to its absorbancy at 25°C. The midpoint of the absorbancy increase (arrows) is the temperature ($T_m$) at which approximately half of the hydrogen bonds holding together the DNA double helices have been broken. This temperature is directly related to the GC content of the DNA sample; the higher the GC content, the higher the $T_m$. Data courtesy of M. Mandel.

*Figure 15.4* The positions in a CsCl density gradient assumed after centrifugation by three different DNAs of differing GC content. (A) DNA of a bacteriophage of *Bacillus subtilis*. (B) DNA of *Thiobacillus novellus*. (C) DNA of *Leptospira* sp. Note that centrifugation sharply separates the three DNAs, each of which bands at a position in the CsCl density gradient which corresponds to its GC content: the lower the GC content of a given DNA, the lower the density at which it forms a band. The order of GC content for the three DNAs is: *B. subtilis* phage > *T. novellus* > *Leptospira* sp. Data courtesy of M. Mandel.

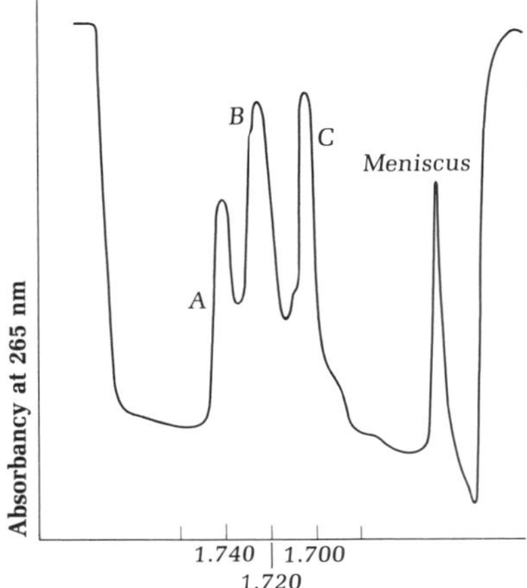


gen bonds are broken and reaches a plateau at a temperature at which the DNA has all become single-stranded (Figure 15.3). The midpoint of this rise, the melting temperature ($T_m$), is a measure of the G+C content. The G + C content of DNA may also be determined by subjecting a DNA sample to centrifugation in a CsCl gradient, and determining optically the position at which the DNA bands in the gradient, which affords a precise measure of its density (Figure 15.4). This method can be used because the density of DNA is also a function of the (G+C):(A+T) ratio.

Physical methods of analysis also provide an indication of the *molecular heterogeneity* of a DNA sample. If every molecule of DNA had the same G + C content, both the thermal transition in a melting curve and the band position in a CsCl gradient would be extremely sharp. The steepness of the curve for thermal transition and the narrowness of the band in a gradient are therefore directly related to the homogeneity of G + C content in a population of DNA molecules. Even when DNA has been considerably fragmented by shearing, preparations from most organisms remain relatively homogeneous by these criteria, which indicates that the mean G + C content varies little in different parts of the genome. The only major exceptions are preparations from organisms that contain two genetic elements of different G + C content. Thus, DNA of mitochondrial or chloroplast origin may differ appreciably in G + C content from the nuclear DNA, in preparations from certain eucaryotic organisms; and there is sometimes a marked molecular heterogeneity in the DNA of a bacterium that harbors an episome. In such cases, the minor constituent may form a distinct *satellite band* in a CsCl gradient; this phenomenon provided one of the clues that led to the discovery of DNA in mitochondria and chloroplasts.

Since no DNA preparation shows *absolute* molecular homogeneity, the G + C content is always a *mean* value and represents the peak in a normal distribution curve.

The mean DNA base compositions characteristic of the nuclear DNA in major groups of organisms are shown in Figure 15.5. In both plants and animals, the ranges are relatively narrow and quite similar, centering about a value of 35 to 40 mole percent G + C. Among the protists, the ranges are much wider. The widest range of all occurs among the bacteria, where mean G + C contents extend from about 30 to 75 mole percent G + C. If, however, one examines the mean G + C content of many different strains that belong to a *single microbial species*, the values are closely similar or identical, as shown by the data for several

**The taxonomic implications of DNA base composition**

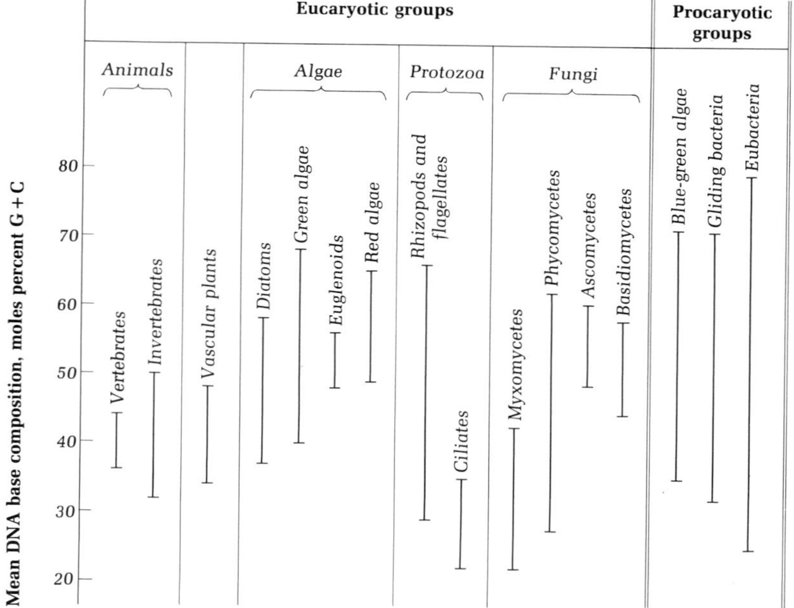


*Figure 15.5 The ranges of mean DNA base composition (moles percent G + C) characteristic of major biological groups.*

*Pseudomonas* species assembled in Table 15.2. Each bacterial species accordingly has DNA with a characteristic mean G + C content; this can be considered one of its important specific characters.

Why do organisms differ so widely in the mean base composition of their DNA? This cannot reasonably be ascribed to coding differences, since there is considerable evidence that the genetic code is universal. It could reflect differences between organisms in the amino acid composition of the cellular proteins, some having a preponderance of amino acids coded for by G + C-rich triplets, others a preponderance coded

*Table 15.2 Constancy of G+C content in strains of bacteria belonging to a given species*[a]

| *Pseudomonas* spp. | Number of strains examined | G + C content of DNA, mole percent (mean value ± standard deviation) |
|---|---|---|
| *P. aeruginosa* | 11 | 67.2 ± 1.1 |
| *P. acidovorans* | 15 | 66.8 ± 1.0 |
| *P. testosteroni* | 9 | 61.8 ± 1.0 |
| *P. multivorans* | 12 | 67.6 ± 0.8 |
| *P. pseudomallei* | 6 | 69.5 ± 0.7 |
| *P. putida* | 6 | 62.5 ± 0.9 |

[a]From M. Mandel, *J. Gen. Microbiol.* **43**, 273 (1966).

for by A + T-rich triplets. There is a certain amount of evidence to suggest that this factor does play a role. In addition, as a result of the degeneracy of the genetic code, substantial differences in mean G + C content could occur, even between two organisms of identical amino acid composition, if there were systematic selection for the use of specific triplets. Thus, one organism might systematically employ a preponderance of G + C-rich triplets to code for amino acids represented by several triplets, and another might systematically employ a preponderance of A + T-rich triplets to code for the same amino acids. It is possible that both factors have played a role in producing the existing biological divergences with respect to mean G+C content.

One point is clear: *a substantial divergence between two organisms with respect to mean DNA base composition reflects a large number of individual differences between the specific base sequences of their respective DNAs.* It is *prima facie* evidence for a major genetic divergence and hence for a wide evolutionary separation. The very broad span of values characteristic of the bacteria as a whole accordingly reveals the great evolutionary diversity of this particular biological group and also suggests their evolutionary antiquity.

The converse is not necessarily true, of course: two organisms with identical mean DNA base compositions may differ greatly in genetic constitution. This is evident from the very similar base ratio values for DNA from all plants and animals. Hence, major evolutionary divergence is not *necessarily* expressed by a divergence of mean base composition. When two organisms are closely similar in their DNA base composition, this fact can only be construed as indicative of genetic and evolutionary relatedness if the organisms *also* share a large number of phenotypic properties in common, or are known to resemble one another in genetic constitution (e.g., different strains that belong to a single bacterial species). In such a case, near-identity of DNA base composition provides supporting evidence for their genetic and evolutionary relatedness.

Mean DNA base compositions are of particular taxonomic value for bacteria, since the range for the bacteria as a whole is so wide. They have now been determined for many bacteria, belonging to every major bacterial group. The ranges for some bacterial genera are presented in Figure 15.6. Although the values for the species in a genus differ (see Table 15.2) the total range for most bacterial genera is fairly narrow and can be considered a useful character for the definition of a bacterial genus. In some cases, base compositions substantiate multigeneric clusters that had been recognized as orders or families by bacteriologists on the basis of common phenotypic traits. Thus, the fruiting myxobacteria, and the actinomycetes—both multigeneric clusters—each fall in a relatively narrow span at the high end of the G + C scale. Similarly, all spore-forming eubacteria (genera *Bacillus, Clostridium,* and *Sporosarcina*) cover a fairly narrow span, near the low end of the G + C scale.

In a few cases, data on DNA base composition have shown that earlier taxonomic groupings based on phenotypic resemblances cannot reflect close genetic or evolutionary relationships. Among gliding organisms, the cytophagas lie almost at the opposite extreme of the G + C

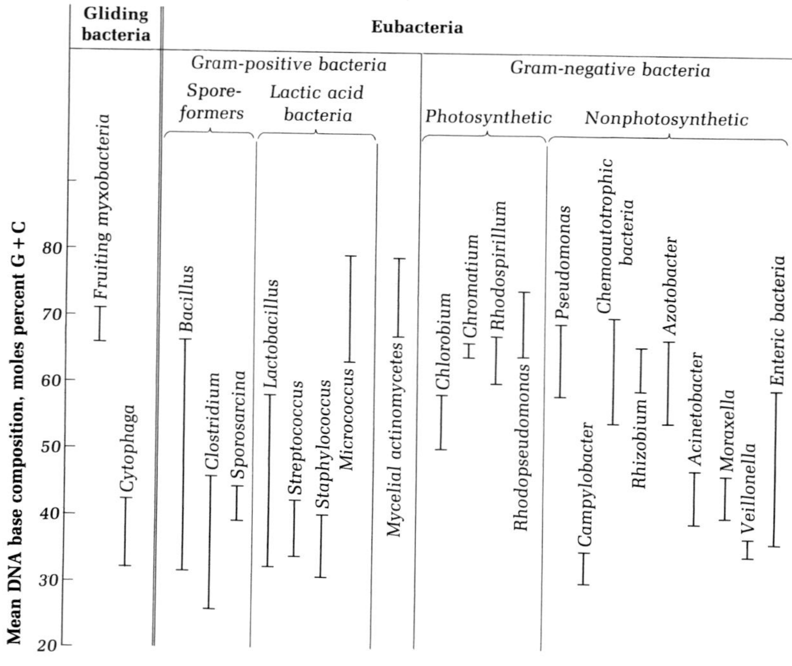


*Figure 15.6 The ranges of mean DNA base composition (moles percent G + C) characteristic of certain bacterial genera and generic clusters.*

scale from the fruiting myxobacteria. Accordingly, these two groups cannot be at all closely related, despite their shared mechanism of movement and certain other phenotypic resemblances. In such cases, the data strongly suggest that previous phenotypic characterizations have been inadequate and that a closer scrutiny of phenotypes is desirable.

**The techniques of nucleic acid hybridization**

If native DNA is subjected to thermal denaturation and then slowly brought to a lower temperature, the single strands anneal with one another, to form again double-stranded molecules. The rate and efficiency of this process depend on many factors, including the mean chain length of the DNA, the annealing temperature, and the ionic strength of the medium. Some random pairing always occurs, but since it is the complementary regions of the two strands that form the most stable duplexes (as a result of the high efficiency of their pairing), their reassociation is favored. Shortly after this phenomenon had been discovered, it was shown that when DNA preparations from two related strains of bacteria are mixed and treated in this manner, *hybrid DNA molecules are formed.* One bacterial strain was grown in a medium containing $D_2O$, so that its DNA was "heavy" as a result of deuterium incorporation. After the two DNA samples had been mixed, denatured, and annealed, hybrid molecules could be detected by centrifugation in a CsCl gradient, where they formed a band intermediate in position between those of the "light" and "heavy" duplexes (Figure 15.7).

When similar experiments were conducted with DNA preparations from two unrelated bacteria, no hybridization could be detected; upon annealing, duplexes were formed only by specific pairing between single strands originally derived from the same DNA.

To be taxonomically useful, the data from experiments on nucleic acid hybridization must be expressible in quantitative terms. It is therefore necessary always to start from a particular reference point, represented by the DNA derived from a *reference strain.* One then determines the amount of hybridization obtainable when single-stranded DNA from the reference strain is paired with single-stranded DNA from other (heterologous) strains, relative to the amount obtainable in a control experiment, when it is paired with homologous single-stranded DNA from the reference strain itself. For example, single-stranded reference DNA may be immobilized in a gel and then mixed with isotopically labeled homologous and heterologous single-stranded DNA preparations under conditions that permit annealing. The amount of radioactivity bound to the reference DNA is an index of the magnitude of hybridization.

Figure 15.7 *An experiment demonstrating the formation of hybrid DNA molecules through reassociation of single-stranded (denatured) DNA molecules to form double helical DNA. DNA was prepared from two different strains of Pseudomonas aeruginosa: strain A was grown in a normal medium, and strain B was grown in a medium containing "heavy water" ($D_2O$) and $N^{15}H_4Cl$. Consequently, the DNA of strain B, although identical in base composition to the DNA of strain A, had a higher density as a result of its content of heavy atoms. The two DNAs were isolated, denatured by heating, mixed, and then annealed, after which residual single-stranded molecules were eliminated by treatment with a specific hydrolase. The preparation was then centrifuged in a CsCl gradient. Three peaks of double-stranded DNA are apparent in the density gradient. Peak A corresponds to "light" double-helical DNA, formed by reassociation of single-stranded DNA from strain A; peak B corresponds to "heavy" double-stranded DNA, formed by reassociation of single-stranded DNA from strain B. Between these two peaks, a third peak of intermediate density occurred; this corresponds to hybrid double-stranded DNA, formed by specific reassociation of single strands, one of which was derived from strain A, and one from strain B. Data courtesy of M. Mandel.*

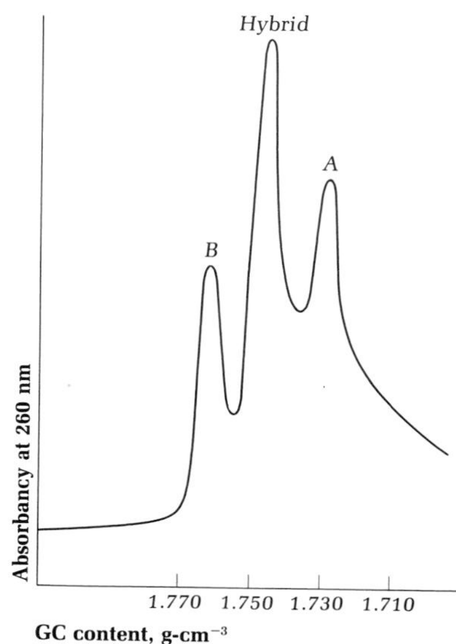


A DNA preparation of high molecular weight is made from the reference strain, thermally denatured, and mixed with molten agar, which is then allowed to form a gel. In the gel, the long single strands are immobilized and hence cannot pair with each other. The DNA agar is then broken into small fragments of uniform size.

Isotopically labeled DNA from the heterologous strains and the reference strain is derived from cultures grown with an isotopically labeled nucleic acid precursor (e.g., adenine-$^{14}C$). After extraction from the cells, the $^{14}C$-labeled DNA is sheared into fragments of short chain length, able to diffuse readily into the DNA agar. These DNA samples are then thermally denatured and added separately to portions of DNA agar, after which the mixtures are placed at a temperature that permits annealing. If the labeled DNA that has diffused into the gel can form duplexes with the immobilized DNA in the agar, it will be trapped there. After an appropriate time, the DNA agar is washed to remove unbound radioactive DNA, and the amount of radioactivity in the agar is determined.

An analogous type of binding experiment can be performed, to determine the binding of labeled messenger RNA to the reference DNA. Labeled messenger RNA can be prepared by brief labeling (pulse labeling) of growing bacteria with an RNA precursor. Since messenger RNA has a base sequence complementary to that of one strand of DNA, in principle such experiments to test DNA–RNA hybridization should give data entirely comparable to those obtained in experiments which test DNA–DNA hybridization. They thus afford a valuable control of the validity of the results of DNA–DNA hybridizations.

The results of a series of hybridization experiments, where both DNA–DNA and DNA–mRNA binding were measured, are shown in Table 15.3. The reference strain for this series of experiments was a strain of *Aerobacter aerogenes,* belonging to the enteric group, and the amounts of radioactivity bound with homologous DNA and mRNA derived from the same strain are assigned arbitrary values of 100. It can be seen that in each experiment with heterologous nucleic acid preparations, the relative amounts of DNA and of mRNA bound were practically identical, as would be expected on theoretical grounds. One of the test organisms employed was a second strain belonging to the same species as the reference strain, and the nucleic acids from it gave essentially the same amounts of binding. Two additional test organisms were strains representative of other genera of enteric bacteria, which show phenotypic relationships to *Aerobacter: E. coli* and *Salmonella typhimurium.* In each case, nucleic acid binding was about half that obtained with homologous preparations. The genetic homology of these bacteria with *A. aerogenes* is accordingly substantial, but far from complete. The last organism examined was a strain of *Pseudomonas aeruginosa,* a bacterium that belongs to a genus showing few phenotypic similarities to the enteric group; virtually no binding occurred, demonstrating the almost complete absence of genetic homology between *P. aeruginosa* and *A. aerogenes.*

*Table 15.3 Results of a series of experiments on the chemical hybridization of DNA with DNA and with messenger RNA*[a]

| | Relative binding of: | |
|---|---|---|
| **Bacterial strain** | **mRNA** | **DNA** |
| *A. aerogenes (homologous)* | 100 | 100 |
| *A. aerogenes, strain 2* | 97 | 105 |
| *S. typhimurium* | 56 | 60 |
| *E. coli* | 52 | 49 |
| *P. aeruginosa* | 1 | 2 |

[a]The reference DNA and mRNA for these experiments were derived from *A. aerogenes,* strain 1. The amounts of binding observed with homologous nucleic acid preparations, derived from this strain, are assigned arbitrary values of 100, and all other data are normalized to them. Data have been taken from B. J. McCarthy and E. T. Bolton, *Proc. Natl. Acad. Sci. U.S.* **50,** 156 (1963).

The total amount of information concerning bacterial genetic homologies is still relatively small. Nevertheless, as the example analyzed above shows, the technique is a powerful one. Its great potential advantage is that it can be used to explore gross genetic homology in the many bacterial groups where no mechanisms of genetic transfer are known, and where, in consequence, biological hybridization experiments are precluded.

## *The comparison of bacterial genotypes by genetic analysis*

Within those bacterial groups where genetic transfer by transduction, transformation, or conjugation has been shown to occur, it is possible to explore interspecific (and occasionally intergeneric) relationships at the genetic level by the appropriate technique of biological hybridization. Interspecific hybridization may be prevented by extrinsic barriers: for example, by a difference in the surface structure of the cell, which prevents conjugation, or the phage attachment necessary for transduction, or by the enzymatic destruction of "foreign" DNA after entry into the cell, as a result of host restriction (see Chapter 14). Hence, a failure to demonstrate hybridization does not necessarily indicate a lack of genetic homology, unless the absence of extrinsic barriers can be shown.

Biological hybridization involves the integration of a piece of DNA derived from the donor with the chromosome of the recipient, by a process of recombination. It tests for a far finer level of genetic homology than does in vitro hybridization, because the integration of any particular segment of DNA from the donor depends on the extent of its homology with the DNA of the recipient in that small, specific region of the chromosome where recombination must occur. The rates of evolution of the various genes in an organism may differ considerably. Since hybridization is detected by examining the incorporation of a specific marker into the recipient genome, the results obtained with a particular pair of donor and recipient strains may differ markedly, depending on the particular gene selected as a marker. This has been shown very clearly in experiments on interstrain hybridization in *Bacillus*, where the integration of a donor marker for streptomycin resistance consistently occurred with much higher frequencies than the integration of donor nutritional markers (Table 15.4).

Particularly extensive hybridization studies by transformation, using streptomycin resistance as a marker, have been carried out in the *Neisseria–Moraxella* group of Gram-negative bacteria; some of the data

*Table 15.4 Effect of the choice of genetic marker on the frequencies of genetic recombination observed in interstrain and interspecies hybridization by transformation*[a]

| Donor strain | Relative frequency of transformation (normalized to values obtained with homologous donor) | | |
|---|---|---|---|
| | **Streptomycin** | **Tryptophan** | **Leucine** |
| *B. subtilis 168-2* (homologous) | 1.0 | 1.0 | 1.0 |
| *B. subtilis 6633* | 0.6 | $<5 \times 10^{-5}$ | $<5 \times 10^{-5}$ |
| *B. subtilis* var. *niger* | 0.54 | $<5 \times 10^{-5}$ | $1.4 \times 10^{-3}$ |
| *B. pumilus* | $2.3 \times 10^{-2}$ | $<5 \times 10^{-5}$ | $<5 \times 10^{-5}$ |
| *B. licheniformis* | $4 \times 10^{-3}$ | $<5 \times 10^{-5}$ | $<5 \times 10^{-5}$ |
| *B. megaterium* | $<2 \times 10^{-5}$ | $<5 \times 10^{-5}$ | $<5 \times 10^{-5}$ |
| *B. polymyxa* | $<2 \times 10^{-5}$ | $<5 \times 10^{-5}$ | $<5 \times 10^{-5}$ |

[a]A strain of *B. subtilis* 168-2, which was streptomycin sensitive and required tryptophan and leucine as growth factors, was transformed for streptomycin resistance or nutritional independence with DNA from several different donors (all streptomycin resistant and prototrophic for the two amino acids). Data of D. Dubnau, I. Smith, P. Morell, and J. Marmur, *Proc. Natl. Acad. Sci. U.S.* **54,** 491 (1965).

are summarized in Figure 15.8. These experiments have thrown considerable light on genetic relationships within the group in question. Although most of the coccoid *Neisseria* show close genetic interrelationships, one species, *N. catarrhalis*, is genetically isolated from the rest and actually appears to be more closely related on the genetic level to some of the rod-shaped organisms of the genus *Moraxella*. Within the *Moraxella* group, the species established in terms of phenotypic properties appear to correspond to genetically well-separated entities, since the intraspecific frequencies of transformation are far higher than the interspecific ones.

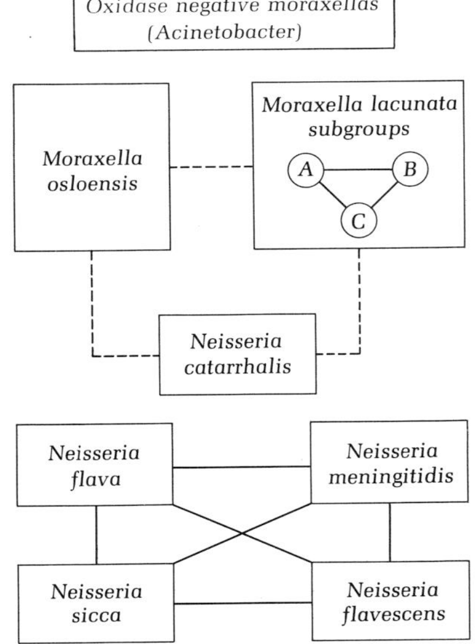


*Figure 15.8 A simplified diagram illustrating genetic relationships among the Moraxella and Neisseria groups, as inferred from studies on transformation frequencies for a particular genetic marker, streptomycin resistance. Solid lines connecting boxes or circles indicate intergroup transformation frequencies ranging from $10^{-2}$ to $10^{-3}$ times those observed within each group; dashed lines, transformation frequencies from $10^{-3}$ to $10^{-5}$ times those observed within each group. The absence of connecting lines indicates that intergroup transformations have not been detected. Note that in terms of this criterion, "Neisseria" catarrhalis appears to be more closely related to Moraxella spp. than to other Neisseria spp., with which it has been classified on the basis of morphological resemblances (coccoid cell form).*

Genetic transfer by conjugation has so far been studied principally among the enteric bacteria, where considerable information on interspecific and intergeneric hybridization has been accumulated. True hybrids can be obtained between members of the three genera *Escherichia, Salmonella,* and *Shigella.* On the other hand, attempts to obtain hybrids between *E. coli* and members of other genera in the group (e.g., *Aerobacter, Proteus*) have failed. These results confirm the inferences concerning genetic relatedness that can be drawn from phenotypic comparisons. In phenotypic terms, the species of the genera *Escherichia, Salmonella,* and *Shigella* are much more similar to one another than to other members of the enteric group; indeed, it is questionable whether generic separations between them are justified (see Chapter 18).

**The taxonomic significance of episomal transfer**

Conjugation may also permit the transfer between species of such episomal elements as sex factors (F) or drug-resistance factors (RTF). When episomal transfer between bacteria occurs, no act of recombination is required for the episome to become established in the recipient cell (see Chapter 14). Therefore, the successful transfer of an episome between two different bacterial species does not constitute evidence that they are *genetically* homologous. However, the ability of two bacteria to conjugate and exchange episomes does imply that they share a series of important *phenotypic* properties, such as the existence of appropriate attachment sites for episomal DNA. The biological range of possible episomal transfer can therefore be construed as *indirect* evidence of a certain degree of genetic relatedness. Experiments have shown that the F-*lac* episome of *E. coli* can be successfully transferred to bacteria with which *E. coli* cannot form hybrids; these include members of the genera *Serratia, Proteus, Yersinia,* and *Vibrio.* On the other hand, attempts to transfer this episome to *Pseudomonas* spp. have failed. Episomal transfer may therefore prove of value in revealing relatively remote relationships among bacteria, which cannot be detected by biological hybridization.

## *The problem of higher taxa in the classification of the bacteria*

For nomenclatorial reasons, the assignment of species to genera is an essential operation. Furthermore, a number of different criteria for the recognition of a bacterial genus readily suggest themselves. Ideally, the genus should group species which share the same gross structural characters (e.g., cell shape, flagellation, Gram reaction) and which also share major functional characters (e.g., mode of energy-yielding metabolism, relations to oxygen). Recent work on the DNA base composition of bacteria suggests another important generic criterion: the range of DNA base composition of the included species should be relatively narrow. In terms of such criteria, it is possible to arrange the bacteria in a fairly large number of reasonably well-defined genera.

The principal utility of ordering genera into higher taxa—families, orders, and classes—is to provide a device for indicating typological similarities that transcend the generic level. Many attempts have been made to develop a complete hierarchical framework for the classification of the bacteria; a typical example can be found in the seventh edition of *Bergey's Manual of Determinative Bacteriology* (see the bibliography). In a few large bacterial subgroups (e.g., the fruiting myxobacteria, the spirochetes, and the actinomycetes) there are a sufficient number of shared properties to justify the recognition of suprageneric clusters. However, in the largest and most varied bacterial group, the unicellular true bacteria, this operation becomes extremely difficult, *primarily because the correlations between major structural and functional characters are scanty and irregular.* If one attempts to define suprageneric groups in structural terms, organisms that resemble one another in major functional respects become widely separated; and, conversely, a hierarchical classification based on major functional characters leads to the separation of organisms that share important structural properties. One illustration will serve to make clear the nature of the difficulty. The members of the largest group of photosynthetic bacteria, the purple bacteria, all share a number of major functional attributes: they possess a special type of photosynthetic pigment system, readily distinguishable from that of all other photosynthetic organisms, and their photosynthetic metabolism has many common features. In physiological terms, accordingly, it is very tempting to treat these organisms as a major eubacterial subgroup. However, in structural terms, the purple bacteria are extremely diverse, apart from the fact that they are all Gram negative; and it is quite easy to find structural counterparts among the nonphotosynthetic unicellular true bacteria for many of the structural types that occur among them (Table 15.5). Consequently, in attempting to group the genera of purple bacteria into higher taxa, the bacterial taxonomist is faced with two alternatives, both equally arbitrary: to maintain them as a single large group, defined in physiological terms, or to split them up, and assign the constituent genera to suprageneric taxa defined in structural terms, but heterogeneous with respect to major physiological properties. The necessity of making such choices explains why bacterial taxonomists have dif-

*Table 15.5 Purple bacteria and nonphotosynthetic bacteria that possess similar cell shapes and modes of cell division*

| Cell shape | Mode of cell division | Purple bacteria | Nonphotosynthetic bacteria |
|---|---|---|---|
| *Spiral* | *Binary fission* | *Rhodospirillum, Thiospirillum* | *Spirillum* |
| *Curved rod* | *Binary fission* | *Ectothiorhodospira* | *Vibrio* |
| *Straight rod* | *Binary fission* | *Chromatium, Rhodopseudomonas* | *Pseudomonas* |
| *Rod* | *Budding* | *Rhodomicrobium* | *Hyphomicrobium* |
| *Coccus* | *Binary fission* | *Thiocapsa* | *Gram-negative cocci* |

ficulty in agreeing on a general hierarchical framework for the classification of the bacteria.

In the ensuing chapters of this book, we shall not attempt to describe the various groups of bacteria in terms of a hierarchical system. In many cases, the bacterial genus is the largest taxonomic entity that can be defined in a meaningful way, and when a number of genera do share common properties that make it convenient to discuss them collectively, we shall designate the group by a common name (e.g., "purple bacteria") without attempting to define it as a family, an order, or a class.

It should be specifically noted that for *determinative purposes*, limitation of bacterial classification to the generic level is entirely satisfactory, provided that the genera are unambiguously defined and that their descriptions are accompanied by an adequate system of keys, which permits the ready assignment of a newly isolated organism to the correct genus. There is an excellent *Guide to the Identification of the Genera of Bacteria* (see the bibliography) which meets this determinative need.

The characters that can be used to make a primary subdivision of the bacteria into the three groups of myxobacteria, true bacteria, and spirochetes have been discussed in Chapter 5. The myxobacteria and the spirochetes are relatively small and taxonomically simple groups. However, the true bacteria are a very large and complex assemblage. Experience suggests that the most useful single character for the primary subdivision of the true bacteria is the Gram reaction. As pointed out in Chapter 10, this simple staining method provides a good, although not infallible, guide to the structure and composition of the cell wall. A primary division on the basis of the Gram reaction therefore divides the true bacteria into two large subgroups with different types of walls.

The Gram-positive true bacteria include the mycelial actinomycetes, as well as a considerable number of groups of unicellular rod-shaped or spherical organisms, including the endosporeformers (see Chapters 20 and 21). The Gram-negative forms are far more diverse. They include all photosynthetic bacteria (Chapter 17); all prosthecate groups; and a wide variety of rod-shaped, helical, and spherical nonphotosynthetic groups (Chapters 18 and 19).

## Further reading


**Books**

Ainsworth, G. C., and P. H. A. Sneath (editors), *Microbial Classification*. New York: Cambridge University Press, 1962.

*Bergey's Manual of Determinative Bacteriology*, 7th ed. Baltimore: Williams & Wilkins, 1957.

Skerman, V. B. D., *A Guide to the Identification of the Genera of Bacteria*, 2nd ed. Baltimore: Williams & Wilkins, 1967.

Sokal, R. R., and P. H. A. Sneath, *Principles of Numerical Taxonomy*. San Francisco: Freeman, 1963.

**Reviews**

Hill, L. R., "An Index to Deoxyribonucleic Acid Base Compositions of Bacterial Species." *J. Gen. Microbiol.* **44,** 419 (1966).

McCarthy, B. J., "Arrangement of Base Sequences in Deoxyribonucleic Acid." *Bacteriol. Rev.* **31,** 215 (1967).

Mandel, M., "New Approaches to Bacterial Taxonomy." *Ann. Rev. Microbiol.* **23,** 239 (1969).

Marmur, J., S. Falkow, and M. Mandel, "New Approaches to Bacterial Taxonomy." *Ann. Rev. Microbiol.* **17,** 329 (1963).

# Chapter sixteen
# The gliding bacteria

*Gliding movement,* characteristic of most blue-green algae, is also the mode of movement in several groups of nonphotosynthetic procaryotic protists. Gliding movement can be recognized by its slowness relative to flagellar movement, as well as by the fact that it is shown only by cells in contact with a solid substrate, never by cells completely suspended in a liquid. No organelles associated with this kind of movement have been demonstrated, and its mechanism is obscure.

The gliding bacteria fall into three main groups, distinguishable from one another by structural and developmental properties (Table 16.1). The constituent groups can also be more or less sharply distinguished by the base composition of their DNA. The filamentous gliding organisms have DNA with a G+C content in the range 35 to 49 mole percent, closely similar to the range for filamentous blue-green algae (39 to 51 mole percent G+C). Since there are also structural parallelisms

*Table 16.1 Primary subdivision of the gliding bacteria*

| Group | Vegetative structure | Aggregation to form fruiting bodies | G+C content of DNA, mole % |
|---|---|---|---|
| *Filamentous gliding bacteria* | *Multicellular filaments* | — | 35–49 |
| *Fruiting myxobacteria* | *Unicellular rods* | + | 67–71 |
| *Cytophagas* | *Unicellular rods* | — | 33–42 |

between the two groups (as discussed below), it seems plausible to assume that the various filamentous nonphotosynthetic gliders have been derived evolutionarily from structural homologs among the blue-green algae. The unicellular gliding bacteria fall into nonoverlapping groups with respect to their DNA base composition. The fruiting myxobacteria lie near the high end of the span of bacterial DNA base composition; those species analyzed fall in the narrow range 67 to 71 mole percent G+C. The recorded values for the cytophaga group range from 33 to 42 mole percent G+C, overlapping with the low end of the range characteristic for filamentous gliders. Hence, despite their apparent resemblances in cell structure, the two unicellular gliding groups must be widely separated in evolutionary terms.

## *The filamentous gliders*

The characters that serve to distinguish the principal genera among the filamentous, nonphotosynthetic gliders are shown in Table 16.2, which also lists the structural counterparts for each genus among filamentous blue-green algae. It should be emphasized that the differences between the photosynthetic forms and their nonphotosynthetic structural counterparts in physiological respects are considerable. The filamentous blue-green algae are almost all obligate

*Table 16.2 Distinguishing characters of the principal genera of filamentous nonphotosynthetic gliding organis[ms]*

| **Structural properties** | | | | **Energy-yielding metabolism** | | | |
|---|---|---|---|---|---|---|---|
| **Shape of filament** | **Cross section of filament** | **Motility of filament** | **Mode of reproduction** | **Chemoautotrophic (oxidation of $H_2S$)** | **Chemoheterotrophic** | **Genus** | **Structural counterpart among filament[ous] blue-green alga[e]** |
| *Straight* | *Cylindrical* | + | *Short chains of cells (hormogonia)* | + | + (?) | *Beggiatoa* | *Oscillatoria* |
| *Straight* | *Cylindrical* | + | *Short chains of cells (hormogonia)* | − | + | *Vitreoscilla* | *Oscillatoria* |
| *Straight* | *Flattened, ribbon-shaped* | + | *Short chains of cells (hormogonia)* | − | + | *Simonsiella* | *Crinalium* |
| *Helical* | *Cylindrical* | + | *Short chains of cells (hormogonia)* | − | + | *Saprospira* | *Spirulina* |
| *Straight* | *Cylindrical* | − | *Single gliding cells* | + | ? | *Thiothrix* | *Calothrix* |
| *Straight* | *Cylindrical* | − | *Single gliding cells* | − | + | *Leucothrix* | *Calothrix* |

photoautotrophs, incapable of growing in the dark at the expense of any organic compound. The metabolic basis of obligate photoautotrophy among the blue-green algae has been traced to the lack of key enzymes involved in respiratory metabolism. Consequently, although these organisms can assimilate certain organic compounds and use them in the light for the synthesis of cell material, they cannot use them as oxidizable energy sources. The nonphotosynthetic filamentous gliders are strict aerobes, dependent on reduced inorganic or organic compounds as energy sources. *Beggiatoa* and *Thiothrix* are chemoautotrophs, which can use $H_2S$ as an energy source, but most members of the group are chemoheterotrophs.

The chemoautotrophic members of the group are aquatic organisms, which occur in ponds, streams, and seawater containing dissolved $H_2S$ formed either by geochemical or microbial means; they grow at or near the air-water interface. A habitat where they often occur in great abundance, forming white mats or tufts on the substrate, is the effluent from sulfur springs. Natural growths of *Beggiatoa* and *Thiothrix* have a very characteristic microscopic appearance (Figure 16.1; see also Figure 5.6), since the filaments are normally packed with highly refractile inclusions of elementary sulfur. The striking appearance of the sulfur-filled filaments of *Beggiatoa* led to its description early in the nineteenth century. The correct interpretation of the physiological role of these sulfur deposits was first made by S. Winogradsky about 1885, as a result of experiments with natural populations of *Beggiatoa* and *Thiothrix*. Winogradsky showed that these organisms are aerobes. Maintained in the absence of $H_2S$ they gradually lose their stored sulfur, as a result of its oxidation to $H_2SO_4$. When filaments have been so depleted, the addition to the filaments of a small amount of dissolved $H_2S$ causes a rapid re-formation of sulfur deposits. Winogradsky also observed that the maintenance and growth of *Beggiatoa* in crude cultures are dependent on the provision of $H_2S$; and it was this observation which first led him to propose the concept that the oxidation of *inorganic* compounds can provide a source of energy for certain microorganisms. The concept of chemoautotrophy was subsequently confirmed, in much greater experimental detail, through Winogradsky's work on the nitrifying bacteria.

*Figure 16.1 A characteristic rosette of Thiothrix, the filaments of which are filled with sulfur droplets (× 670). Courtesy of E. J. Ordal.*

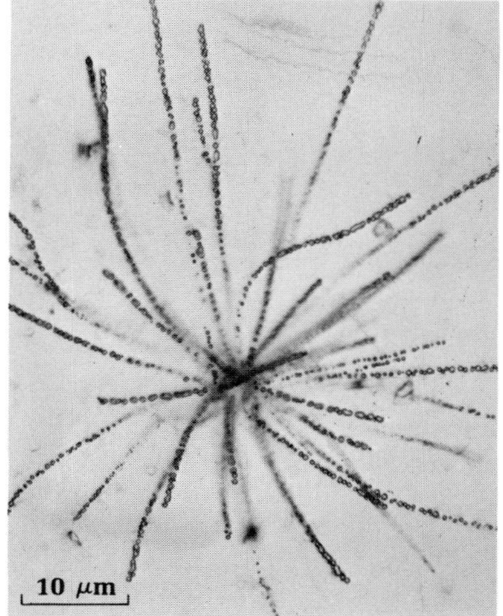


In view of this history, it is ironical that our present knowledge about *Beggiatoa* and *Thiothrix* is still extremely limited. This is largely a result of the difficulties encountered in isolating and cultivating these organisms under conditions of chemoautotrophic growth. Unlike the sulfur-oxidizing chemoautotrophic true bacteria (*Thiobacillus* spp.), *Beggiatoa* and *Thiothrix* appear to be unable to oxidize any externally supplied reduced sulfur compounds except $H_2S$. Since

$H_2S$ readily undergoes autoxidation in contact with air, a chemically stable inorganic medium suitable for their growth cannot be prepared. Certain strains of *Beggiatoa* have been isolated by the use of organic media, containing acetate as a carbon and energy source; these particular strains are accordingly facultative chemoautotrophs. It remains possible (as indicated by queries in Table 16.2) that other strains of *Beggiatoa* and *Thiothrix* are obligate chemoautotrophs; such forms could not be isolated by the procedures so far used to obtain pure cultures.

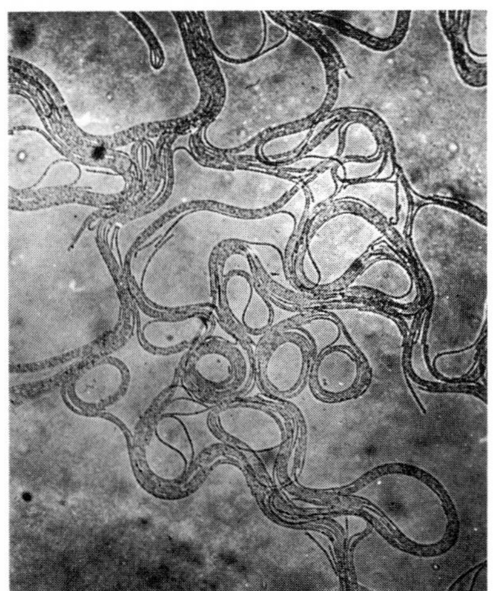

*Figure 16.2 Vitreoscilla filaments growing on the surface of an agar plate. Courtesy of E. G. Pringsheim.*

The aerobic chemoheterotrophs of the genera *Vitreoscilla* and *Leucothrix* are similar in structure to *Beggiatoa* and to *Thiothrix*, respectively. *Leucothrix* is a marine organism that grows abundantly in decomposing algal material; *Vitreoscilla* (Figure 16.2) occurs in soil and fresh water.

The filamentous organisms of the genus *Simonsiella* occur in a specialized habitat: they are parasites in the oral cavity of man and the breaking off of short filaments of cells (hormogonia). In *Leucothrix* and *Thiothrix* the filaments are immotile and have a secreted holdfast at their base, by means of which they attach to substrates. Frequently, rosettes of filaments are united at the base by their holdfasts, as illustrated in Figure 16.1 for *Thiothrix*. This habit of growth is also characteristic of filamentous blue-green algae such as *Calothrix* and *Rivularia*. The mechanisms of reproduction and rosette formation have been studied in *Leucothrix*. Gliding cells, single or in short chains, break off from the tips of the filaments (Figure 16.3). When they are liberated

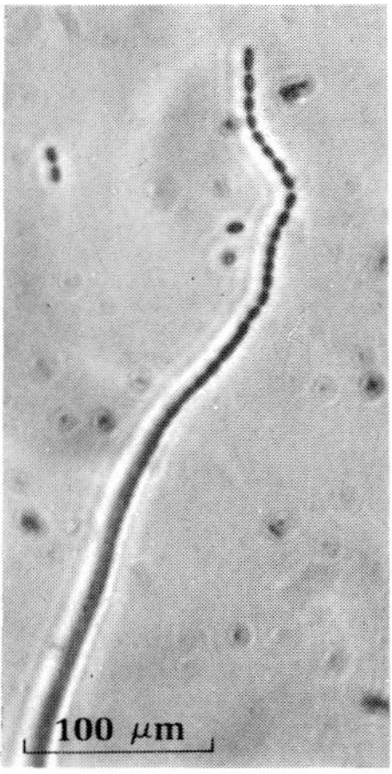


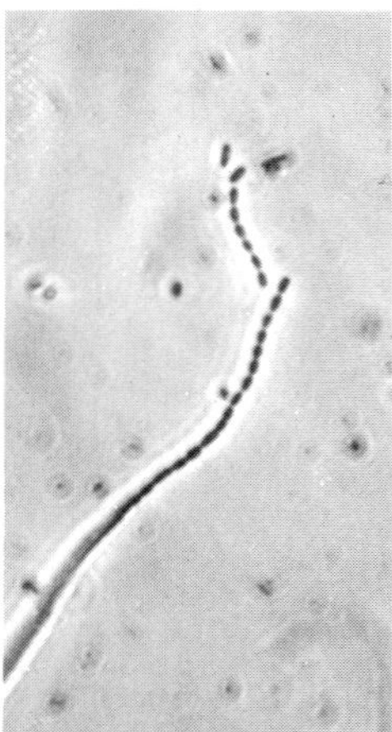

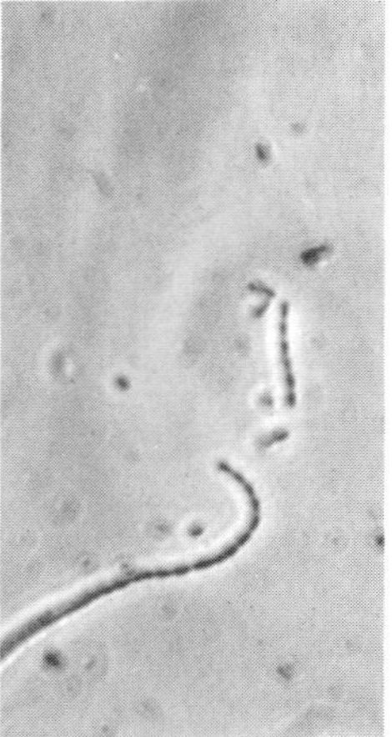

*Figure 16.3 Successive pictures of a Leucothrix filament, showing liberation of gonidia (phase contrast, × 309). From Ruth Harold and R. Y. Stanier, "The Genera Leucothrix and Thiothrix." Bacteriol. Rev.* **19**, *49 (1955).*

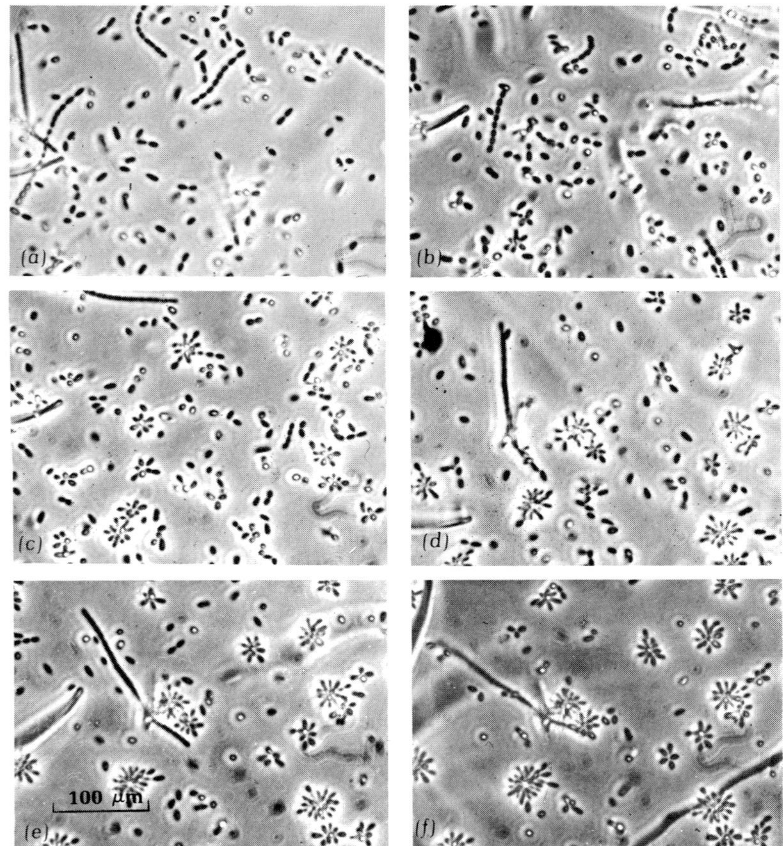


*Figure 16.4* Successive stages in the aggregation of Leucothrix gonidia to form rosettes (phase contrast photographs of a single field, taken over a period of 55 minutes, × 279). From Ruth Harold and R. Y. Stanier, "The Genera Leucothrix and Thiothrix." Bacteriol. Rev. **19,** 49 (1955).

in large numbers, these motile cells aggregate to form tightly packed rosettes, held together by their holdfasts (Figure 16.4). Each cell in the rosette subsequently develops into a filament, giving rise to the radial arrangement of filaments shown in Figure 16.1.

The members of the genus *Saprospira* are chemoheterotrophs that occur both in fresh water and in the sea. They are distinguished from *Vitreoscilla* by the helical form of the filament (Figure 16.5), a structural feature also characteristic of blue-green algae belonging to the genus *Spirulina*.

The filamentous organisms of the genus *Simonsiella* occur in a specialized habitat: they are parasites in the oral cavity of man and other warm-blooded animals. The cells that compose the filaments are

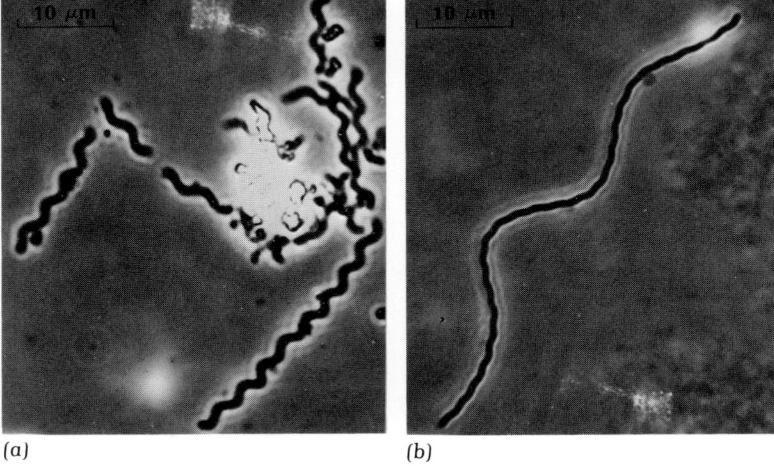


Figure 16.5 *Phase contrast photomicrographs of Saprospira (× 525): (a) S. albida; (b) S. grandis. Compare with the photomicrographs of the photosynthetic counterpart, Spirulina (Figure 5.5). Courtesy of Ralph Lewin.*

markedly flattened, so that each filament is ribbon-shaped (Figure 16.6). Gliding movement occurs only when the broad face of the filament is in contact with the substrate. Reproduction occurs by formation of hormogonia. The nutrition appears to be complex; good growth requires complex media supplemented with serum. Comparable ribbon-shaped filaments are characteristic of blue-green algae belonging to the genus *Crinalium*.

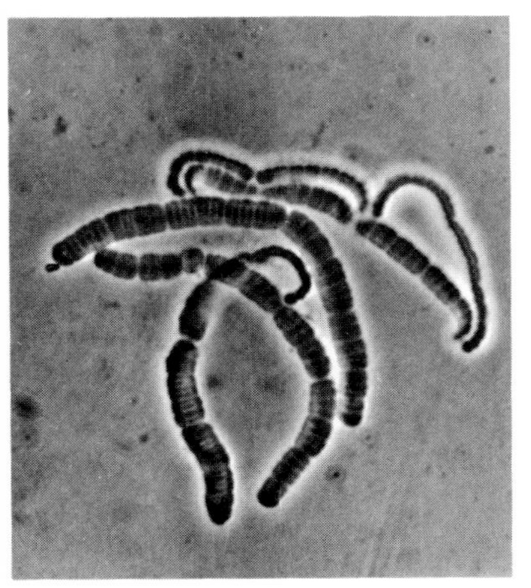

Figure 16.6 *Simonsiella, a gliding bacterium which forms ribbon-shaped filaments of flattened cells (phase contrast, × 620). Some of the filaments are viewed on edge, and appear much thinner than the filaments that lie flat. Courtesy of Mrs. P. D. M. Glaister.*

## *The fruiting myxobacteria*

The fruiting myxobacteria have rod-shaped vegetative cells, flexible and weakly refractile, which multiply by binary transverse fission. Under appropriate conditions, the vegetative cells aggregate to form macroscopically visible fruiting bodies, often of considerable complexity, which bear resting structures. This remarkable bacterial group was characterized and explored in considerable detail about 1895 by a mycologist, R. Thaxter. Myxobacterial fruiting bodies occur in nature on decayed wood and other plant materials, as well as on the dung of animals. They had been observed as early as 1809 by mycologists, but their bacterial nature was not suspected until Thaxter succeeded in cultivating them and discovered the unicellular vegetative stage in the life cycle.

Fruiting myxobacteria are soil inhabitants. They can be isolated by placing moistened, sterile dung pellets on a dish of soil; after 1 week or so, the characteristic fruiting bodies form on the surface of the dung. Purification can be effected by transfer of material from the fruiting bodies to appropriate agar media, followed by repeated transfers of material taken from the edge of the spreading myxobacterial colony.

**The developmental cycle**

The characteristic developmental cycle of a fruiting myxobacterium is best observed on the surface of an agar plate. The growth of the vegetative cells produces a rather flat colony, which spreads radially from the point of inoculation as a result of gliding movement. At a magnification of 20 to 50 diameters, the edge of the colony can be seen to have a very irregular structure, as a result of the fact that single cells or small groups are continuously breaking away from the mass of growth and gliding outward (Figure 16.7). As it moves, each cell leaves a thin trail of slime behind it; consequently, the entire colony rests on a more or less continuous thin layer of secreted slime. In some groups the cells are relatively thick (~1.5 μm) with blunt ends; in others, they are thinner (~0.5 μm) with slightly tapered ends. Although each cell is enclosed by a wall (Figure 16.8), the cells are highly flexible, and frequently bend into arcs or semicircles in the course of movement. As a result of slime formation, growth in liquid media usually takes the form of a sheet of cells, attached to the wall of the culture vessel or spreading out over the air–liquid interface; relatively few of these organisms grow dispersed in liquid cultures, even when subjected to mechanical agitation.

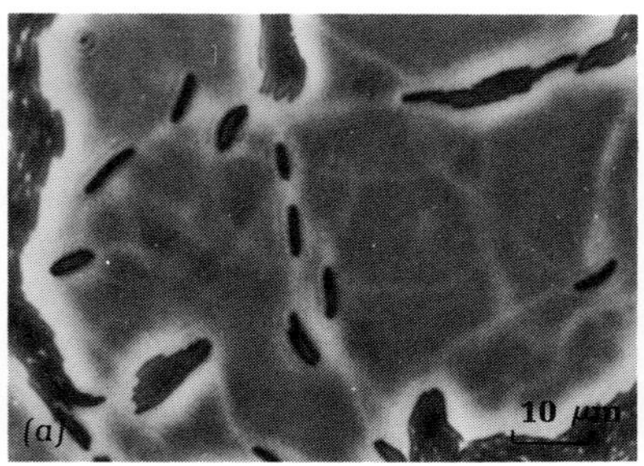

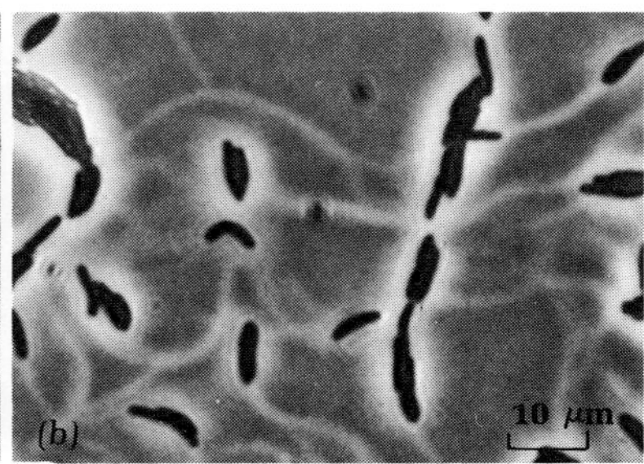


*Figure 16.7 Edge of the growth of Myxococcus fulvus on agar. Note slime tracks behind gliding cells. (a) Phase contrast, ×630. (b) Phase contrast, ×700. Courtesy of M. Dworkin and H. Reichenbach.*

On a solid medium, the fruiting process usually begins in the oldest, central portion of the colony and later extends outward toward the

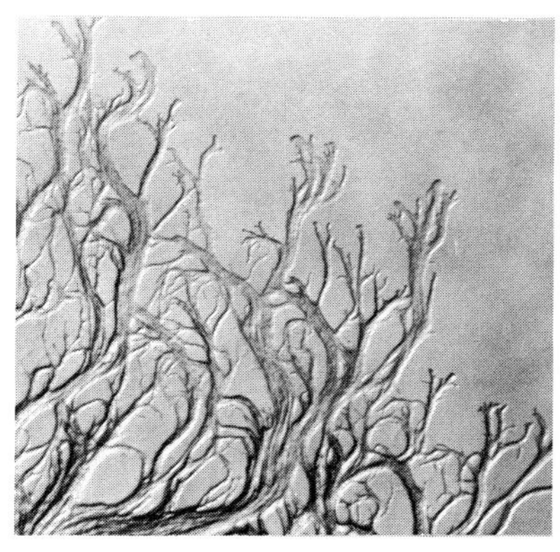

*Figure 16.8 Edge of the growth of a Sorangium species on agar (× 20). Courtesy of M. Dworkin and H. Reichenbach.*

Figure 16.9 *Fruiting bodies of several species of myxobacteria: (a) Podangium lichenicolum (× 232); (b) Stigmatella aurantiaca (× 318); (c) Polyangium fuscum (× 106); (d) Sorangium sp. (× 573). Courtesy of M. Dworkin and H. Reichenbach.*

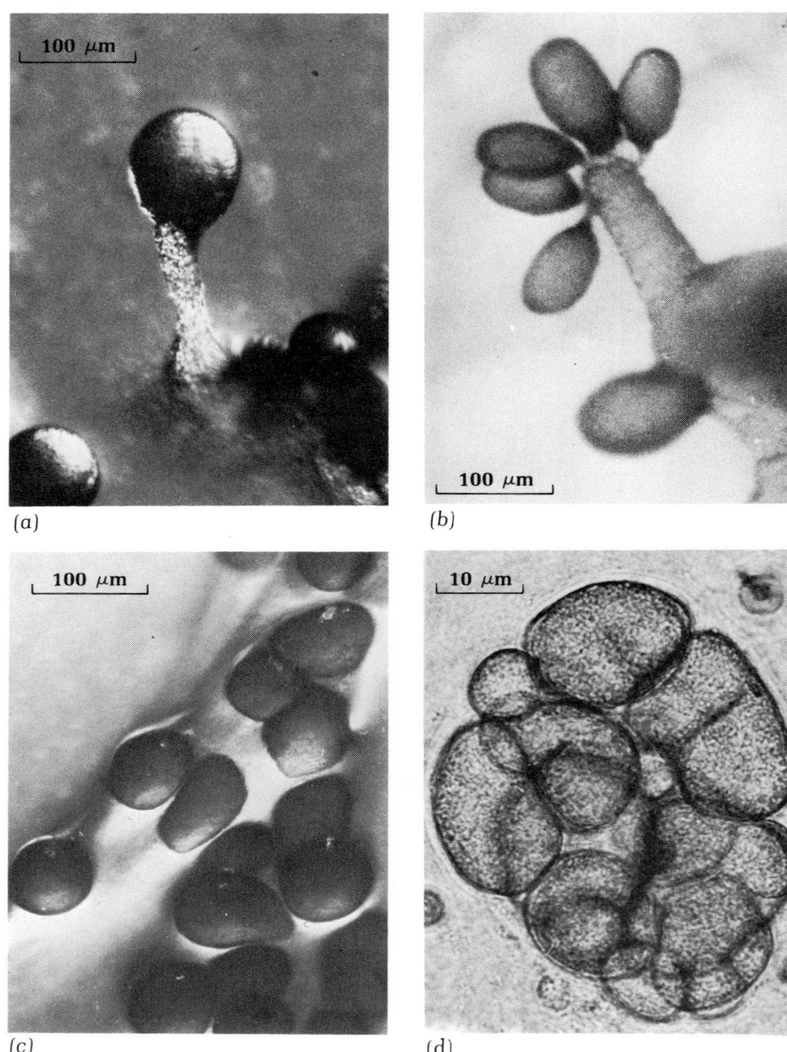


periphery. The first indication of this phenomenon is the localized migration of the rods into aggregation centers, where they pile up in raised masses. The aggregation process is often so efficient that the adjacent areas of the colony are swept almost clean of cells. Each heap of cells then gradually differentiates into a mature fruiting body with a definite size, structure, and color, characteristic for each species (Figure 16.9). The taxonomic subdivision of the fruiting myxobacteria is based largely on the properties of the fruiting body (Table 16.3).

*Table 16.3 Distinguishing characteristics of the principal genera of fruiting myxobacteria*

| Genus | Structure of vegetative rods | Nature of resting cells | Association of resting cells in cysts | Primary cysts, containing smaller cysts | Fruiting body carried on stalk |
|---|---|---|---|---|---|
| *Archangium* | *Thin (~0.5 μm), tapering* | *Shortened rods* | − | − | − |
| *Myxococcus* | *Thin (~0.5 μm), tapering* | *Microcysts* | − | − | − |
| *Polyangium* | *Thin (~0.5 μm), tapering* | *Shortened rods* | + | − | − |
| *Podangium* | *Thin (~0.5 μm), tapering* | *Shortened rods* | + | − | +[a] |
| *Chondromyces* | *Thin (~0.5μm), tapering* | *Shortened rods* | + | − | +[b] |
| *Sorangium* | *Thick (~1.5 μm), blunt ends* | *Shortened rods* | + | + | − |
| *Stigmatella* | *Thick (~1.5 μm), blunt ends* | *Shortened rods* | + | − | + |

[a] Stalk simple. [b] Stalk branched.

In the genus *Archangium*, the fruiting body shows little differentiation. As it matures, the component cells become immotile, slightly shortened, and somewhat more refractile but still retain their rod shape. They also synthesize pigment, which imparts a characteristic color to the fruiting body.

In *Myxococcus* and related genera, the maturation of the relatively simple fruiting body is accompanied by a conversion of its constituent cells into spherical, highly refractile *microcysts* (Figure 16.10). This involves a progressive shortening and thickening of the rods, which become enclosed by a heavy wall.

Morphogenesis of the resting structures is much more complex in the genera *Polyangium*, *Podangium*, and *Chondromyces*. The mature fruiting bodies of these organisms contain a relatively small number of large cysts, as much as 50 by 75 μm in size, with brightly colored walls. Each cyst encloses thousands of shortened, rod-shaped resting cells. The cysts of *Polyangium* are held together in a common mass of slime, formed by the autolysis of those cells in the aggregate which do not enter cysts. In *Podangium*, single cysts are borne at the tips of simple stalks; in *Chondromyces* and *Stigmatella*, a large number of cysts are borne at the tips of a branched, treelike stalk. Although they contain

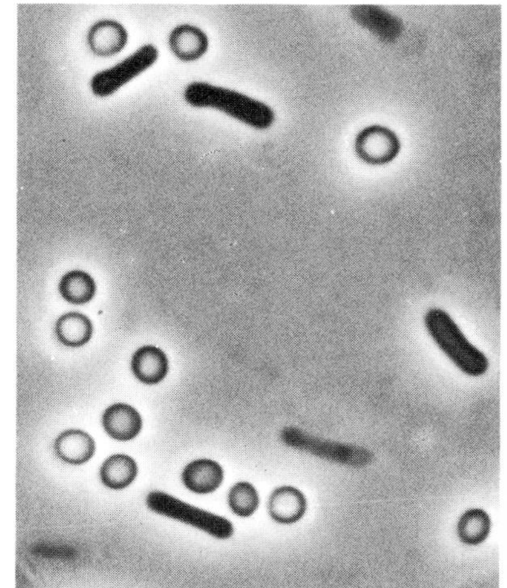

*Figure 16.10 Microcysts and shortened rods from a fruiting body of Myxococcus fulvus (phase contrast, × 2,540). Courtesy of M. Dworkin and H. Reichenbach.*

Figure 16.11 Successive stages in the development of the complex, branched fruiting body of *Chondromyces*. From J. T. Bonner, *Morphogenesis*, p. 104. Princeton, N.J.: Princeton University Press, 1952.

few detectable cells at maturity, these stalks have a cellular origin: the cysts as they develop are borne upward by other cells that have entered the aggregate and which eventually autolyse. Successive stages in the formation of a *Chondromyces* fruiting body are shown in Figure 16.11.

In the genus *Sorangium* many-celled cysts are also formed, but they have a slightly more complex structure. The resting shortened rods are contained in thin-walled cysts with polygonal structure, caused by their compression within larger primary cysts.

It has been shown that the microcysts of myxococci are somewhat more resistant to heat and to radiation than the vegetative cells, although the levels of resistance are far below those characteristic of eubacterial endospores. The principal selective value of myxobacterial resting structures appears to be their ability to survive in the dry state for long periods. Survival of dried fruiting bodies for as much as 10 years has been recorded.

When germination of microcysts occurs, the thick wall ruptures and a vegetative rod emerges. Germination of large, many-celled cysts occurs by rupture of the cyst wall and simultaneous emergence of many vegetative cells.

The fascinating developmental problems presented by the fruiting myxobacteria have so far been little explored. The mechanism of aggregation, basic to the fruiting process, is not known, although from the behavior of the participating cells it seems certain that a specific chemotactic response is involved. Most experimental work has been done with strains of *Myxococcus*, some of which will grow dispersed in agitated liquid cultures and all of which fruit readily and abundantly in culture. Nutritional studies have shown that these forms can grow on a medium containing a mixture of amino acids and that fructification takes place when tyrosine and phenylalanine become limiting. The effect is specific for these two aromatic amino acids; depletion of other amino acids does not induce fruiting. The conversion of rods to microcysts normally occurs in the fruiting body, after aggregation has taken place. However, microcyst formation can be induced experimentally in suspensions of vegetative cells, by the addition of 0.5 *M* glycerol. The conversion is massive and relatively rapid; mature microcysts are formed 2 hours after glycerol addition.

## Nutrition and ecology

All fruiting myxobacteria are strictly aerobic chemoheterotrophs. Some members of the genera *Sorangium* and *Archangium* are cellulose decomposers. These organisms can be grown in a defined medium consisting of a mineral base furnished with filter paper or some other source of cellulose; soluble sugars can also be utilized. However, most fruiting myxobacteria do not attack cellulose and will not grow on simple defined media. Detailed nutritional studies with *Myxococcus* strains have shown that the simplest medium able to support growth consists of a mixture of amino acids; growth on such a medium is slow, however. Considerably more rapid growth is obtainable with an enzymatic protein hydrolyzate, suggesting that peptides are more favorable carbon sources than free amino acids. Vitamins are not required.

During the isolation of myxobacteria on solid complex media, it has been observed that both growth and fructification are often markedly enhanced by the presence of other kinds of bacteria. Systematic studies of this phenomenon have revealed that many myxobacteria can kill and lyse microorganisms belonging to other groups (true bacteria, both

Gram positive and Gram negative; algae; yeasts; and fungi). Many myxobacteria thus appear to be predators, which kill their microbial prey by the secretion of antibacterial substances, and subsequently digest the host proteins (and possibly other host cell constituents) by means of extracellular hydrolases. This ecological interpretation is fully consonant with the peculiar nutritional requirements established for *Myxococcus*. The predacious character of the fruiting myxobacteria probably accounts for their abundant development on dung, which has a very high bacterial content.

## The Cytophaga group

In 1930, toward the end of his long scientific career, Winogradsky made a systematic study of the microorganisms responsible for the aerobic decomposition of cellulose in soil. Prominent among the bacteria that possess this ability was a group of organisms with unusual slender, flexible, pointed cells for which he created a new genus, *Cytophaga*. These bacteria can completely destroy the structure of the cellulose fibers in a sheet of filter paper laid on the surface of a plate of mineral agar; the area attacked is converted to a pigmented, slimy patch, filled with bacterial cells (Figure 16.12). Another characteristic feature of the growth of these bacteria on cellulose is the regularity with which the rod-shaped cells are aligned on the surface of the fibers; they are oriented parallel to the fibrillar substructure of the polysaccharide (Figure 16.13). Subsequent work revealed that the cytophaga group are gliders. Furthermore, certain of these organisms (genus *Sporocytophaga*) form microcysts, similar in structure, mode of formation, and germination to the microcysts produced by fruiting myxobacteria of the genus *Myxococcus*. In *Sporocytophaga*, however, aggregation of the vegetative cells never precedes microcyst formation.

The cytophagas accordingly constitute a second group of unicellular rod-shaped gliding bacteria, distinguishable from the fruiting myxobacteria by their inability to undergo aggregation and fructification. The members of the cytophaga group are widely distributed in soil, fresh water, and the sea.

*Figure 16.12 Agar plate covered with a layer of filter paper and streaked with a culture of Cytophaga. Note that the filter paper has been completely dissolved where growth has occurred.*

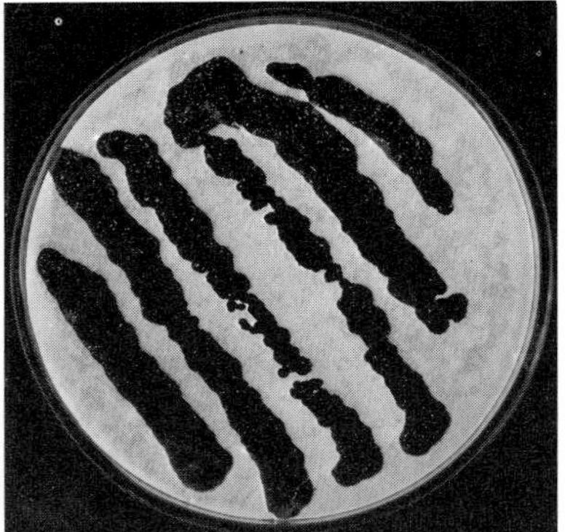

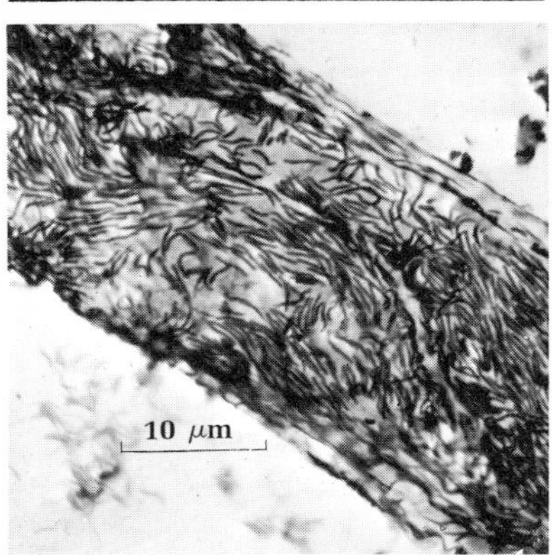


*Figure 16.13 Cellulose fiber heavily attacked by Cytophaga (stained preparation). Note the characteristic regular arrangement of the cells. From S. Winogradsky, Microbiologie du Sol. Paris: Masson, 1949. Reprinted with permission of M. Manigault and the publisher.*

**Physiological properties**

The strictly aerobic cellulose-decomposing soil cytophagas are physiologically very highly specialized: the only substrate they can use as a carbon and energy source is cellulose. It is possible to obtain mutants that can grow on the soluble sugars derived from cellulose by hydrolysis, cellobiose, and glucose; but other classes of organic compounds commonly attacked by aerobic chemoheterotrophs, such as organic acids, are never used. Another very peculiar feature of these bacteria is their failure to secrete an extracellular hydrolase for the initial attack on cellulose: the cells must be in direct physical contact with cellulose in order to decompose it. Since the substrate is completely insoluble in water, this behavior suggests that the cellulase formed by the cytophagas is an *exocellular* enzyme, which is associated with the cell surface but cannot diffuse away from it, as does a truly *extracellular* hydrolytic enzyme.

The ability to use complex polysaccharides as sources of carbon and energy is characteristic of many members of the cytophaga group. Other soil species can utilize chitin, a polymer of N-acetylglucosamine that is the principal constituent of the cuticle of insects and the cell wall of many fungi. Many marine cytophagas can grow at the expense of agar, a polysaccharide constituent of the red algae. These organisms are, however, not as specialized physiologically as the soil types that attack cellulose and can use a wide range of other organic compounds as sources of carbon and energy. Some cytophagas are facultative anaerobes: certain marine species can ferment carbohydrates, and denitrification may be used as a means of anaerobic energy-yielding metabolism by chitin decomposers.

**The pathogenic members of the cytophaga group**

The aquatic cytophagas include several species that are pathogens of fish. The most important of these organisms is *Cytophaga columnaris*, which is associated with massive epidemics in salmonid fishes. Highly virulent strains are infective by contact and give rise to systemic infections which can decimate the populations of salmon fingerlings in hatcheries within a few days.

## Further reading

**Reviews**

Dworkin, M., "Biology of the Myxobacteria." *Ann. Rev. Microbiol.* **20,** 75 (1966).

Harold, R., and R. Y. Stanier, "The Genera *Leucothrix* and *Thiothrix*." *Bacteriol. Rev.* **19,** 49 (1955).

Pringsheim, E. G., "The Vitreoscillaceae: a Family of Colourless, Gliding, Filamentous Organisms." *J. Gen. Microbiol.* **5,** 124 (1951).

Stanier, R. Y., "The Cytophaga Group." *Bacteriol. Rev.* **6,** 143 (1942).

Steed, P. D. M., "Simonsiellaceae fam. nov. with Characterization of *Simonsiella crassa* and *Alysella filiformis.*" *J. Gen. Microbiol.* **29,** 615 (1963).

Thaxter, R., "On the Myxobacteriaceae, a New Order of Schizomycetes." *Botan. Gaz.* **17,** 389 (1892).

# Chapter seventeen
# Photosynthetic bacteria

*Among the bacteria,* photosynthetic ability is confined to a small number of unicellular eubacteria, belonging to two biological groups, the *purple bacteria* and the *green bacteria.* The photosynthetic nature of these organisms was definitively established in 1930 by C. B. van Niel. Aside from its intrinsic interest, this discovery led to a profound reevaluation of the nature of photosynthetic processes.

In terms of gross cellular characters, there is nothing that distinguishes the purple and green bacteria from other unicellular eubacteria, except the color that is conferred on their cells by the photosynthetic pigment system. All photosynthetic bacteria are Gram negative and, if motile, polarly flagellated. Binary fission is the most common mode of multiplication, but a few purple bacteria reproduce by budding. The cells may be spherical, rod-shaped, or spiral in the case of purple bacteria, and are always rod-shaped in the case of green bacteria. The base content of the DNA spans a fairly broad range: it extends from 46 to 73 mole percent G+C in purple bacteria, and from 48 to 58 mole percent G+C in green bacteria.

Purple and green bacteria can be readily distinguished from the other photosynthetic procaryotes, the blue-green algae, on biochemical and physiological, as well as structural, grounds (Table 17.1). The bacteria are unable to produce oxygen photosynthetically, as a result of their inability to use water as a photosynthetic electron donor, and are instead dependent either on reduced inorganic compounds such as $H_2S$, thiosulfate, and $H_2$, or on organic compounds for photosynthetic

*Table 17.1 Distinctions among purple bacteria, green bacteria, and blue-green algae*

| Group | Principal photosynthetic electron donors | Relations to oxygen | Major chlorophylls | Region of long-wave-length absorption band, nm |
|---|---|---|---|---|
| *Blue-green algae* | $H_2O$ | *Aerobic* | *Chlorophyll a* | *670–685* |
| *Purple bacteria* | $H_2S$ *and/or organic compounds* | *Mostly anaerobic* | *Bacteriochlorophyll a or d* | *850–1,000* |
| *Green bacteria* | $H_2S$ *and/or organic compounds* | *Anaerobic* | *Bacteriochlorophyll b or c*[a] | *735–755* |

[a]Together with a smaller amount of bacteriochlorophyll a.

growth. In fact, *bacterial photosynthesis is an anerobic metabolic process*, and most photosynthetic bacteria are strict anaerobes.

The pigment systems of the photosynthetic bacteria are distinctive. They never contain chlorophyll a, universal in oxygen-evolving photosynthetic organisms; their chlorophyllous pigments have slightly different chemical structures and are termed *bacteriochlorophylls*. The principal carotenoid pigments of photosynthetic bacteria also differ chemically from those of the algae. The special photosynthetic pigment systems of the purple and green bacteria enable them to use for photosynthesis wavelengths lying in the far-red and infrared regions of the spectrum, which are not utilizable by algae and higher plants.

## *The differences between purple and green bacteria*

The two major groups of photosynthetic bacteria are distinguished by their special pigment systems and by the fine structure of the photosynthetic apparatus.

**The pigments and absorption spectra of purple and green bacteria**

Four different molecular kinds of bacteriochlorophyll occur in photosynthetic bacteria. Their chemical differences from chlorophyll a, common to all algae and plants, are summarized in Figure 17.1. Purple bacteria contain a single kind of chlorophyll, which is either bacteriochlorophyll a or d, depending on the species. Green bacteria contain as a major chlorophyll either bacteriochlorophyll b or c, depending on the strain, but all green bacteria contain in addition a small amount of bacteriochlorophyll a. In contrast to purple bacteria, they thus have two kinds of chlorophyll in the cell.

The principal carotenoids in most purple bacteria are open-chain

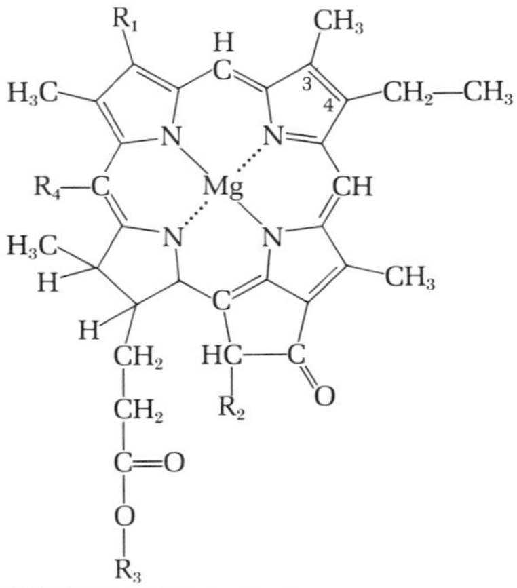

| | $R_1$ | 3,4 | $R_2$ | $R_3$ | $R_4$ |
|---|---|---|---|---|---|
| *Chlorophyll a* | $—HC{=}CH_2$ | — | $—C(=O)—O—CH_3$ | *Phytyl ester* ($C_{20}H_{30}O—$) | —H |
| *Bacteriochlorophyll a* | $—C(=O)—CH_3$ | *Dihydro* | $—C(=O)—O—CH_3$ | *Phytyl ester* | —H |
| *Bacteriochlorophyll c* | $—C(OH)(H)—CH_3$ | — | —H | *Farnesyl ester* ($C_{15}H_{25}O—$) | $—CH_3$ |
| *Bacteriochlorophyll d* | $—C(OH)(H)—CH_3$ | — | —H | *Farnesyl ester* | —H |
| *Bacteriochlorophyll b*[a] | — | — | — | — | — |

[a]Structure unknown.

*Figure 17.1 Structural relationships of the chlorophylls of purple and green bacteria to chlorophyll a; general formula is at left.*

(aliphatic) compounds, which frequently bear one or two terminal methoxyl ($—OCH_3$) groups. A few purple bacteria contain aryl carotenoids, which have an aromatic ring at one end of the chain. The green bacteria always contain aryl carotenoids, different from those in purple bacteria. The structures of some major carotenoids of purple and green bacteria are compared in Figure 17.2 with the structure of β-carotene, a typical carotenoid of the alicyclic type common to oxygen-evolving photosynthetic organisms.

Since the composition of the pigment system varies somewhat within each group of photosynthetic bacteria, the absorption spectrum of the cells is not identical for all purple bacteria or for all green bacteria. Nevertheless, the cellular absorption spectra of purple and green bacteria have certain distinctive features, which permit their ready differentiation from one another and from algae (Figure 17.3). The most important distinguishing feature is the position of the major bands at the long-wavelength end of the absorption spectrum, in the red or infrared region. These bands are caused by the specific major chlorophylls of the cell. In algae, the peak lies at 675 to 685 nm, and in green bacteria at 735 to 755 nm. In purple bacteria, it lies completely in the infrared region of the spectrum and often shows several minor peaks. The peaks are located between 850 and 910 nm for purple

*Figure 17.2 The structures of some major carotenoids of purple and green bacteria, compared with the structure of β-carotene, a major carotenoid in most algae and plants.*

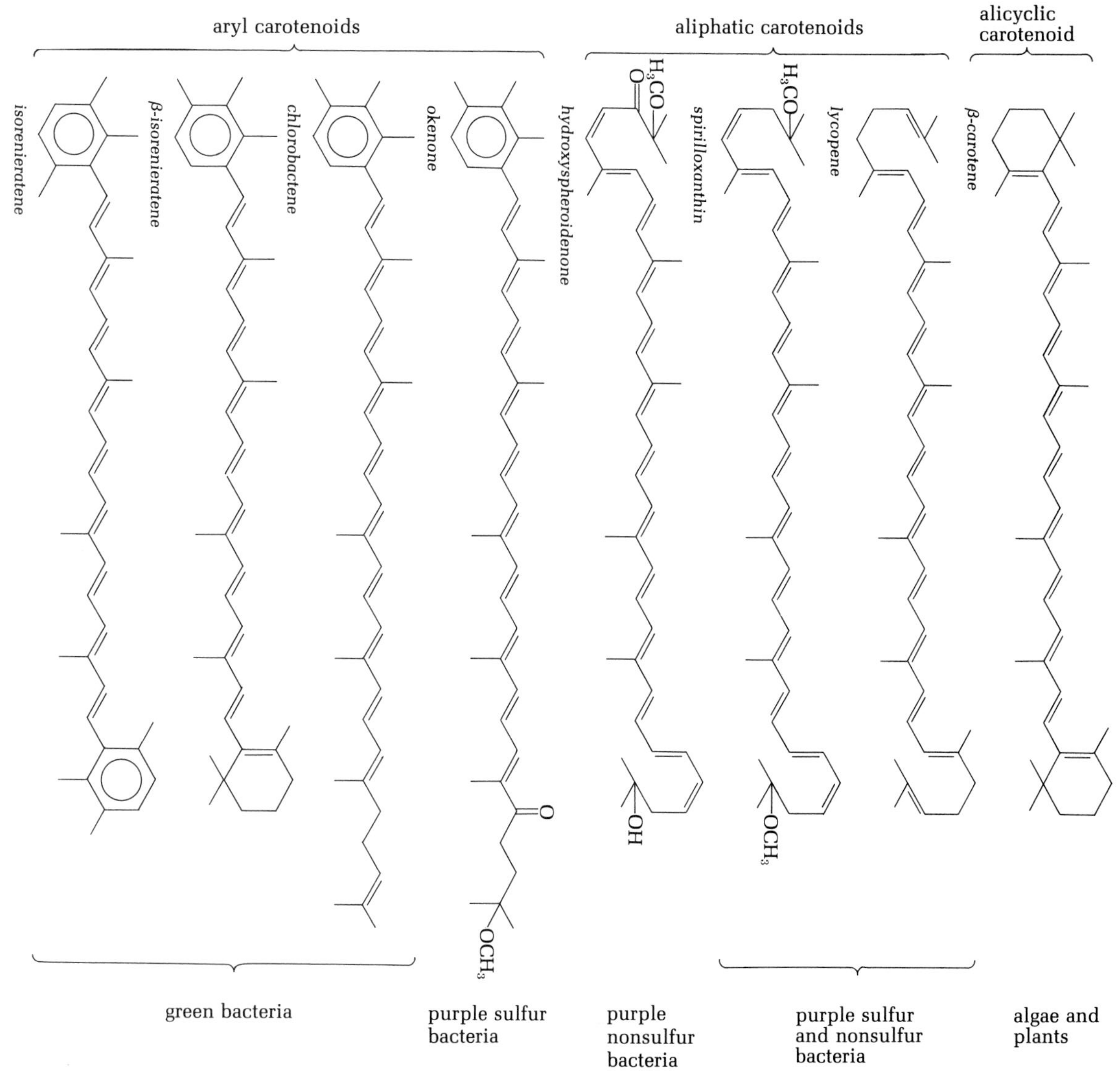

aryl carotenoids

aliphatic carotenoids

alicyclic carotenoid

isorenieratene

β-isorenieratene

chlorobactene

okenone

hydroxyspheroidenone

spirilloxanthin

lycopene

β-carotene

$H_3CO$

O

$OCH_3$

OH

green bacteria

purple sulfur bacteria

purple nonsulfur bacteria

purple sulfur and nonsulfur bacteria

algae and plants

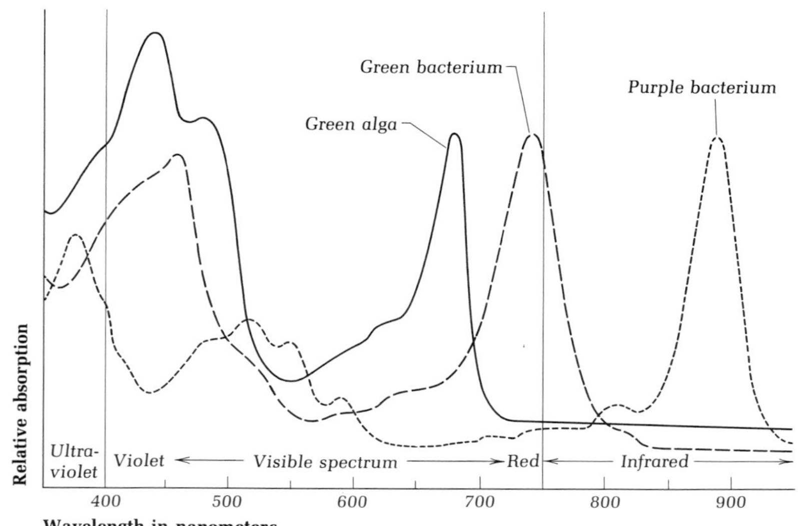


*Figure 17.3 The patterns of light absorption by a green alga, a green bacterium, and a purple bacterium. Note the complementarity of light absorption by these three types of photosynthetic organisms in the red and infrared regions of the spectrum.*

bacteria that contain bacteriochlorophyll a, and at about 1,000 nm for purple bacteria that contain bacteriochlorophyll b.

Since the major absorption bands of all bacteriochlorophylls in the cell lie in the far-red or infrared region, to which the eye is almost completely insensitive, they contribute only a very pale blue color to the cell. This color is largely masked by that of the carotenoids. Depending on their specific carotenoids, the "purple" bacteria may appear brown, red, or purple; the "green" bacteria, either yellow-green or brown. The common names of these two groups do not, accordingly, always correspond to their color as perceived by the human eye.

**The structure of the photosynthetic apparatus**

The photosynthetic apparatus of bacteria is too small to be visible with the light microscope; its structure can be ascertained only by electron microscopy. In purple bacteria, the photosynthetic pigment system is located in intracytoplasmic unit membranes, which typically occupy a substantial portion of the cytoplasmic region and arise from infoldings of the cell membrane, to which they probably always remain attached. These membrane systems can assume a number of different forms and arrangements, characteristic for each species. The principal variations of structure are illustrated in Figure 17.4. In many species, the membrane system has the form of interconnected hollow vesicles.

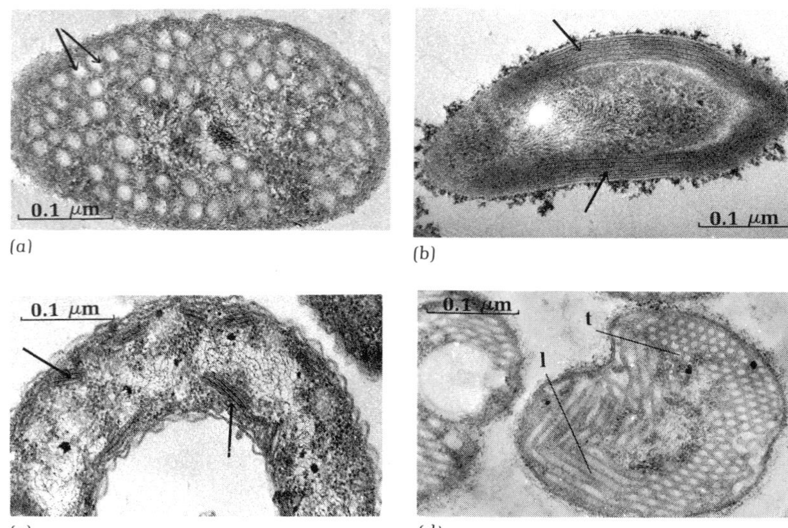


Figure 17.4 *Electron micrographs of thin sections of several purple bacteria, illustrating variations in the structure of the internal membranes. (a) Rhodopseudomonas spheroides, in which the membranes occur as hollow vesicles (arrows) (× 5,640). (b) Rhodopseudomonas palustris, in which the membranes occur in regular parallel layers in the cortical region of the cell (arrows) (× 5,640). (c) Rhodospirillum fulvum, in which the membranes occur in small, regular stacks (arrows) (× 42,300). (d) Thiocapsa sp., in which the membranes are tubular; some of these tubes are sectioned longitudinally (l), others transversely (t) (× 40,900). Micrographs (a), (b), and (c) courtesy of Germaine Cohen-Bazire; (d) courtesy of K. Eimhjellen.*

In some it consists of parallel, straight tubes; in others of short stacks of lamellae, superficially similar to the grana of chloroplasts; and in still others, rows of lamellae, arrayed parallel to and adjoining the cell membrane.

In green bacteria, the photosynthetic pigments are located in a completely different kind of structure, the chlorobium vesicle. These cigar-shaped objects, each some 400 A wide and 1,500 A long, form a cortical layer that immediately underlies the cell membrane. They are not, however, connected directly with the cell membrane, each vesicle being enclosed completely by a nonunit membrane about 30 A thick (Figure 17.5).

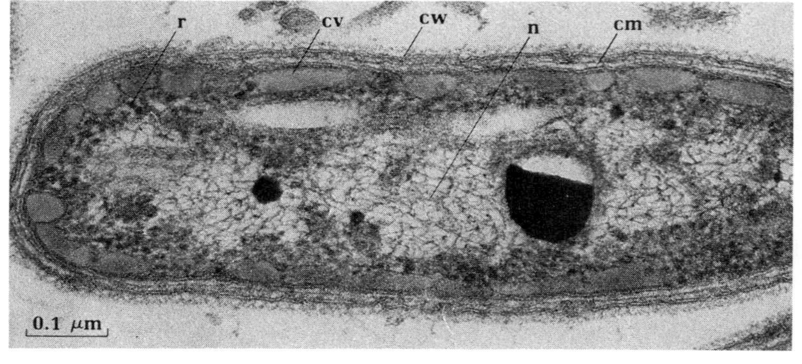


Figure 17.5 *Electron micrograph of a thin section of the green bacterium Pelodictyon, showing the relationship of the chlorobium vesicles (cv) to other parts of the cell (× 74,400): cw, cell wall; cm, cell membrane; r, ribosomes; n, nucleoplasm. Courtesy of Germaine Cohen-Bazire.*

## The biological properties of green bacteria

Three genera of green bacteria can be recognized on the basis of structural differences. The most common representatives are members of the genus *Chlorobium,* with small, immotile, rod-shaped or curved cells that occur singly or in short chains (Figure 17.6). In the genus *Chloropseudomonas,* the rod-shaped cells are motile by means of polar flagella. In the genus *Pelodictyon,* the relatively large, immotile rods are united in chains that take the form of loose, irregular nets (Figure 17.7). The meshes of these nets arise as a result of the fact that two adjacent cells in a chain sometimes undergo ternary, rather than the normal binary, fission, to form two apposed, Y-shaped structures (Figure 17.8). Provided that the cells produced by such a double ternary fission remain apposed, subsequent growth and binary fission leads to the formation of a closed chain of cells. Another distinctive property of the genus *Pelodictyon* is the ability to form gas vacuoles (Figure 17.9).

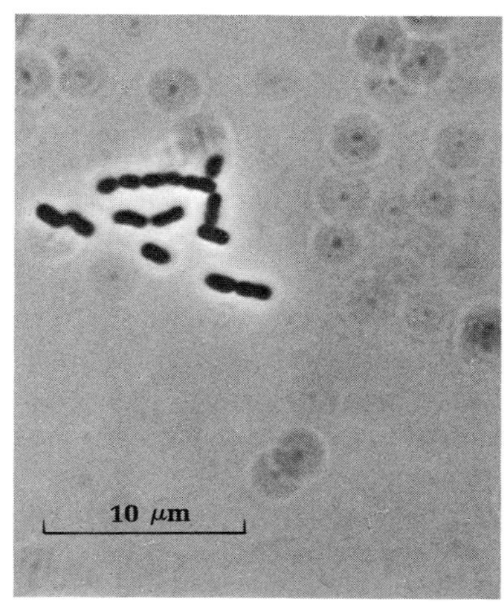


*Figure 17.6 The green bacterium, Chlorobium (phase contrast, × 2,000). Courtesy of Riyo Kunisawa and Rosmarie Rippka.*

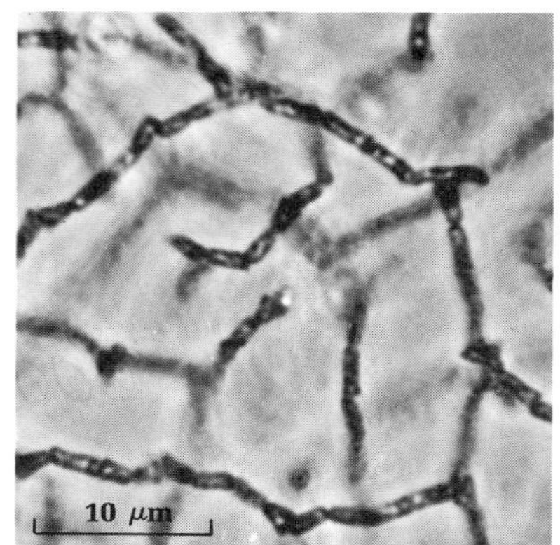


*Figure 17.7 The green bacterium, Pelodictyon, showing the characteristic growth in the form of loose, irregular nets (phase contrast, × 2,000). The light areas within the cells are gas vacuoles. Courtesy of Norbert Pfennig.*

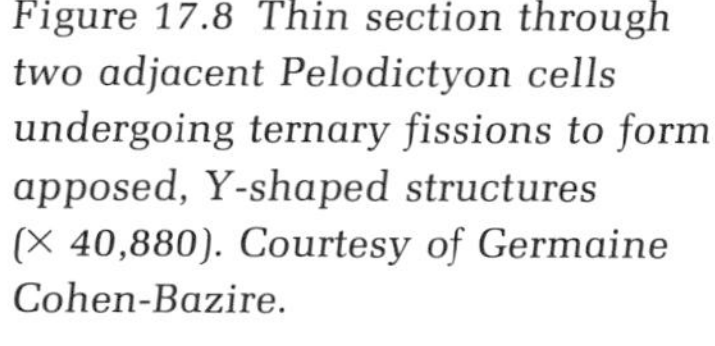

*Figure 17.8 Thin section through two adjacent Pelodictyon cells undergoing ternary fissions to form apposed, Y-shaped structures (× 40,880). Courtesy of Germaine Cohen-Bazire.*

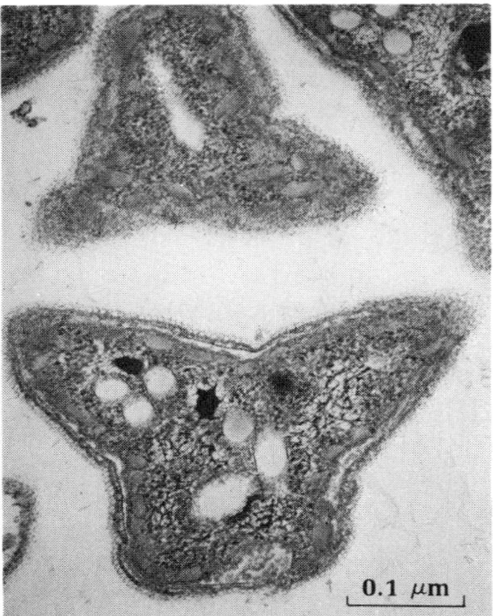


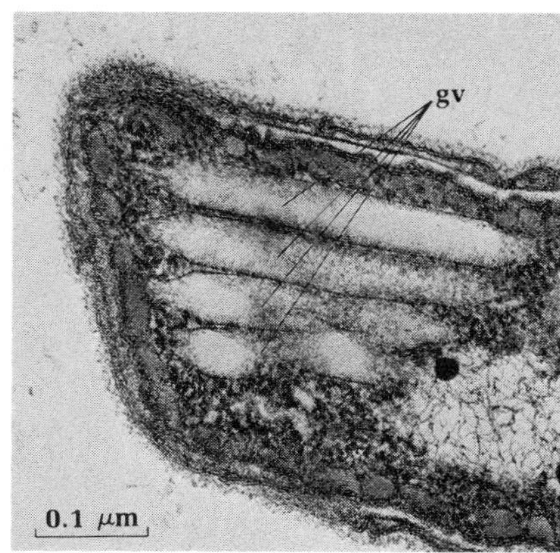


*Figure 17.9 Thin section of part of a Pelodictyon cell, showing profiles of four adjacent gas vesicles (gv), which compose a gas vacuole (× 68,250). Courtesy of Germaine Cohen-Bazire.*

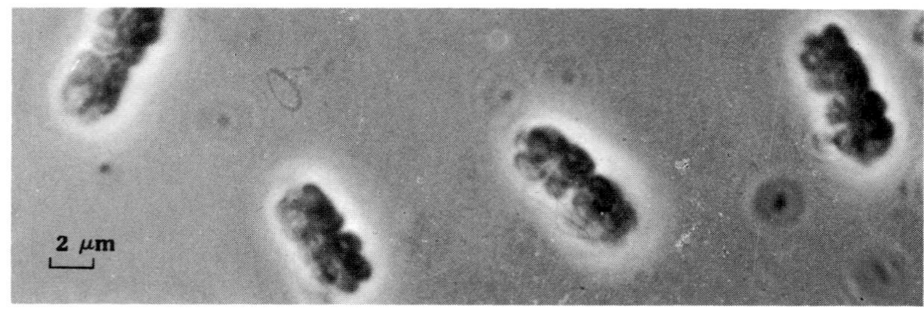


*Figure 17.10 The ectosymbiosis of green bacteria with a larger, colorless bacterium (the so-called "chloro-chromatium" association). Each object is a rod-shaped colorless bacterium coated with the regularly arranged smaller cells of a Chlorobium (phase contrast, × 2,500). Courtesy of Norbert Pfennig.*

Certain green bacteria of the *Chlorobium* type may participate in a curious ectosymbiosis (see Chapter 24). They adhere in regular, parallel, longitudinal rows to the surface of a much larger, motile nonphotosynthetic bacterium (Figure 17.10).

**The metabolism of green bacteria**

The green bacteria are all strictly anaerobic and obligately photosynthetic. They can use $H_2S$ as an electron donor for the reduction of $CO_2$; some can also use thiosulfate, and probably all can use hydrogen. The sulfur formed as an intermediate in the oxidation of sulfide to sulfate is always deposited outside the cells; unlike certain purple bacteria, the green bacteria never contain intracellular sulfur inclusions.

The photometabolism of green bacteria is, accordingly, autotrophic. Most green bacteria are unable to use organic compounds as photosynthetic electron donors; however, if sulfide and $CO_2$ are present, they can assimilate acetate and use it for the synthesis of cell material. This assimilation occurs through a ferredoxin-mediated reductive synthesis of pyruvate,

$$CH_3COOH + CO_2 + 2H \rightarrow CH_3COCOOH + H_2O$$

a reaction that provides a means of $CO_2$ assimilation in addition to the reactions of the Calvin cycle (see Chapter 7). In contrast to other members of the group, *Chloropseudomonas* can use organic compounds alone for photosynthesis, growing at the expense of ethanol, acetate, or glucose.

Although many strains of green bacteria have no growth-factor requirements, some require specifically vitamin $B_{12}$.

## *The biological properties of purple bacteria*

It is customary to recognize two major groups among the purple bacteria: purple sulfur bacteria (Thiorhodaceae) and purple nonsulfur bacteria (Athiorhodaceae). The former can use hydrogen sulfide as a photosynthetic electron donor and, like green bacteria, characteristically develop in natural environments where sulfide is available. The latter are unable to use hydrogen sulfide and depend on the availability of organic compounds for their photosynthetic metabolism; they are sensitive to sulfide and develop only in natural environments where its concentration is very low. The differences between these two groups of purple bacteria are accordingly nutritional and ecological ones; they cannot be distinguished from one another in terms of their cellular structures or pigment systems.

**The purple sulfur bacteria**

Within the purple sulfur bacteria a further subdivision can be made in terms of the site at which sulfur, formed during the oxidation of hydrogen sulfide, is deposited. In the genus *Ectothiorhodospira,* sulfur deposition is extracellular, as in the green bacteria; in all other purple sulfur bacteria, sulfur is deposited within the cells as oily, refractile inclusions. Further generic and specific subdivisions are based on structural properties: cell size and shape, and the presence or absence of flagella and gas vacuoles (Table 17.2). Several different purple sulfur bacteria are illustrated in Figure 17.11.

The metabolism of purple sulfur bacteria is similar to that of green bacteria: they use sulfide, and in some cases thiosulfate and molecular hydrogen, as inorganic electron donors for $CO_2$ assimilation. Many purple sulfur bacteria can also use a few organic compounds – notably, fatty acids, pyruvate, and dicarboxylic acids – in place of inorganic compounds for their photosynthetic metabolism. Some require a small amount of sulfide for growth even in the presence of a utilizable organic

*Table 17.2 Generic subdivision of the purple sulfur bacteria*

| Characteristics | Genera |
|---|---|
| I. *Sulfur deposited externally* | |
| A. *Straight or curved rods, motile* | *Ectothiorhodospira* |
| II. *Sulfur deposited internally* | |
| A. *Do not contain gas vacuoles* | |
| 1. *Cells spiral, motile* | *Thiospirillum* |
| 2. *Cells spherical, motile* | *Thiocystis* |
| 3. *Cells spherical, immotile* | *Thiocapsa* |
| 4. *Cells kidney-shaped or short rods, motile* | *Chromatium* |
| B. *Contain gas vacuoles* | |
| 1. *Cells spherical, motile* | *Lamprocystis* |
| 2. *Cells spherical, immotile* | *Rhodothece* |
| 3. *Cells rod-shaped, immotile* | *Amoebobacter* |

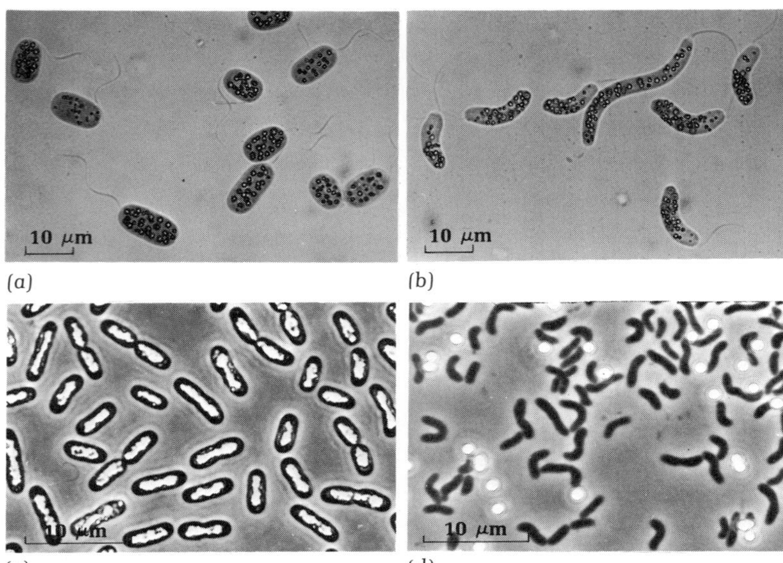


Figure 17.11 Purple sulfur bacteria grown in sulfide-containing medium. (a) *Chromatium okenii* (× 653). (b) *Thiospirillum jenense* (× 653). Note sulfur inclusions and flagellar tufts. (c) *Thiodictyon elegans* (× 1,340). The large irregular light area in the center of each cell is a gas vacuole; smaller peripheral spherical sulfur inclusions can be seen in some cells. (d) *Ectothiorhodospira mobilis* (× 1,340). Note that the sulfur granules (light bodies) are extracellular. Micrographs (a) and (b) from H. G. Schlegel and N. Pfennig, "Die Anreicherungskultur einiger Schwefelpurpurbakterien." Arch. Mikrobiol. **38,** 1 (1961); (c) courtesy of N. Pfennig; (d), courtesy of H. G. Truper.

substrate, because they are unable to use sulfate as a sulfur source for the synthesis of cell material. Most purple sulfur bacteria require no growth factors, but certain of the large *Chromatium* species and *Thiospirillum* require vitamin $B_{12}$, also a growth factor for some green bacteria.

**The purple nonsulfur bacteria**

Generic subdivisions of this group are based on structural properties (Table 17.3). Most species are rod-shaped or spiral organisms with polar flagella and multiply by binary fission (Figure 17.12). In *Rhodo-*

*Table 17.3 Generic subdivision of the purple non-sulfur bacteria*[a]

| Characteristics | Genera |
|---|---|
| I. Cells spiral | *Rhodospirillum* |
| II. Cells rod-shaped | *Rhodopseudomonas* |
| A. Division by binary fission | *R. spheroides, R. capsulata, R. gelatinosa* |
| B. Division by unequal fission (budding?) | *R. palustris, R. viridis* |
| III. Cells oval, equipped with filiform appendages, at the tips of which daughter cells are formed | *Rhodomicrobium* |

[a]These bacteria have two common structural properties: they are motile by polar flagella and they never contain gas vacuoles.

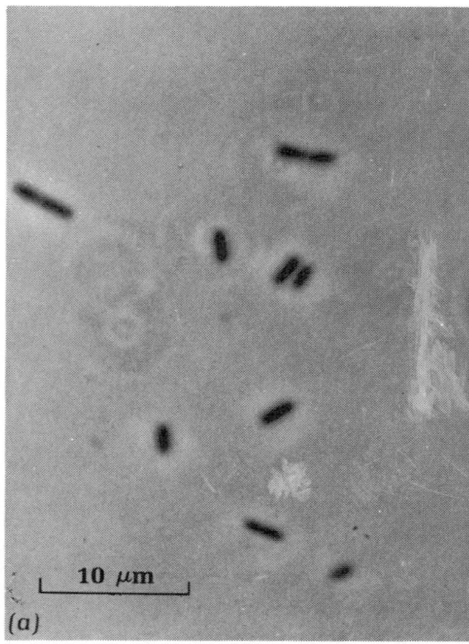


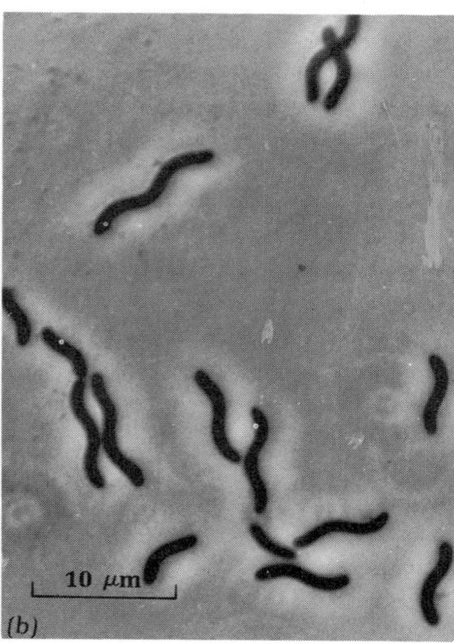


*Figure 17.12 Nonsulfur purple bacteria (× 1,500). (a) Rhodopseudomonas spheroides. (b) Rhodospirillum rubrum.*

*microbium*, multiplication occurs by budding, in a manner very similar to that found in the nonphotosynthetic *Hyphomicrobium*. Slender filiform extensions are formed from the oval mother cell, at the tip of which the daughter cell forms and enlarges (Figure 17.13). In two species of the genus *Rhodopseudomonas*, cell division is notably

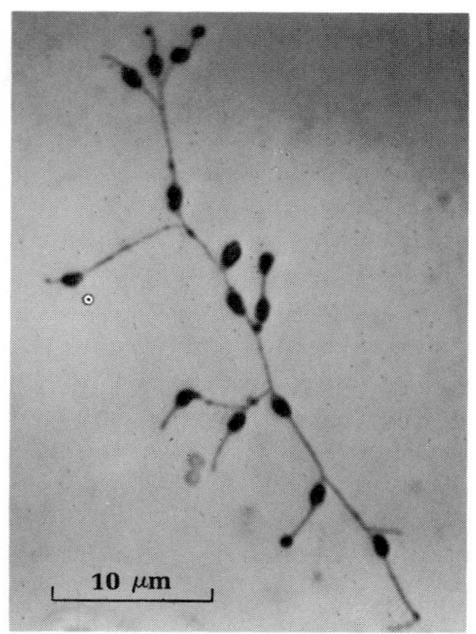


*Figure 17.13 A small colony of the budding nonsulfur purple bacterium Rhodomicrobium vannielii, showing the formation of buds at the tips of filaments (× 1,800). From Esther Duchow and H. C. Douglas, "Rhodomicrobium vannielii, a new photoheterotrophic bacterium." J. Bacteriol.* **58,** *411 (1949).*

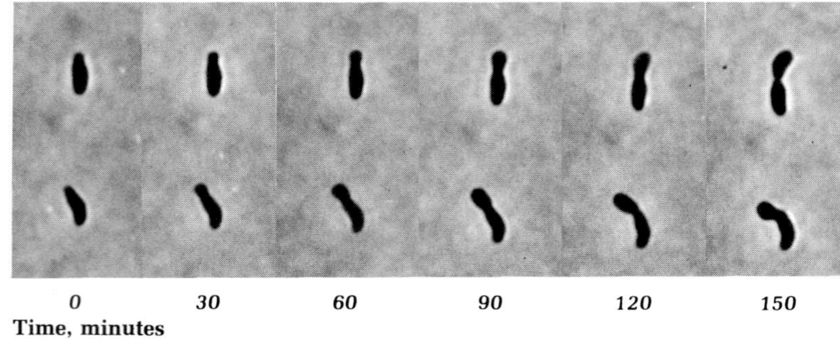


*Figure 17.14 Photomicrographs of budding growth of two cells in a slide culture of a Rhodopseudomonas species, R. acidophila, taken at intervals of 30 minutes to show the outgrowth of buds at the poles of the rod-shaped cells (× 1,500). Courtesy of Norbert Pfennig and Heather Johnston. From N. Pfennig, "Rhodopseudomonas acidophila, a new species of the budding nonsulfur purple bacteria." J. Bacteriol.* **99,** *597 (1969).*

asymmetric; although the daughter cell does not arise at the end of a filiform extension as in *Rhodomicrobium,* the mode of division of these species may also be regarded as "budding" (Figure 17.14).

For purple nonsulfur bacteria, organic compounds serve as the principal photosynthetic electron donors and source of carbon. The range of substrates that can be used by the various species of the group is wide. It includes many fatty acids, primary and secondary alcohols, dicarboxylic and other organic acids, carbohydrates, and aromatic compounds. One species can utilize thiosulfate, and probably all can use hydrogen; with these inorganic electron donors, $CO_2$ provides the principal carbon source, as in purple sulfur and green bacteria. With the exception of *Rhodomicrobium,* purple nonsulfur bacteria all require growth factors; in some species the requirements are complex, and include the vitamins biotin, thiamine, and niacin in various combinations. Vitamin $B_{12}$, the only growth factor required by purple sulfur and green bacteria, is not known to be required by any of the purple nonsulfur bacteria.

**Relations to oxygen**

The purple nonsulfur bacteria are the only photosynthetic bacteria that are not all strict anaerobes. Some species are oxygen tolerant, and can grow aerobically in the dark, obtaining energy from a respiratory metabolism of organic compounds. In contrast to other photosynthetic bacteria, these species are accordingly not obligately photosynthetic.

The synthesis of photosynthetic pigments by the purple nonsulfur bacteria is specifically inhibited by oxygen at relatively low concentrations irrespective of whether the cells are kept in the dark or in the light. Consequently, the introduction of air into a growing culture leads to a progressive decline in the cellular content of bacteriochlorophyll and carotenoids (Figure 17.15), and the cells in such a population

Figure 17.15 *The effect of introducing oxygen into a culture of the nonsulfur purple bacterium, Rhodopseudomonas spheroides, growing exponentially under anaerobic conditions ($N_2$ atmosphere) in the light. Note that the change from $N_2$ to air does not diminish the growth rate but almost instantly abolishes chlorophyll and carotenoid synthesis. When anaerobic conditions are reestablished about 3 hours later, photosynthetic pigment synthesis resumes at a rapid rate. After G. Cohen-Bazire, W. R. Sistrom, and R. Y. Stanier, "Kinetic studies of pigment synthesis by non-sulfur purple bacteria." J. Cellular Comp. Physiol.* **49,** *25 (1957).*

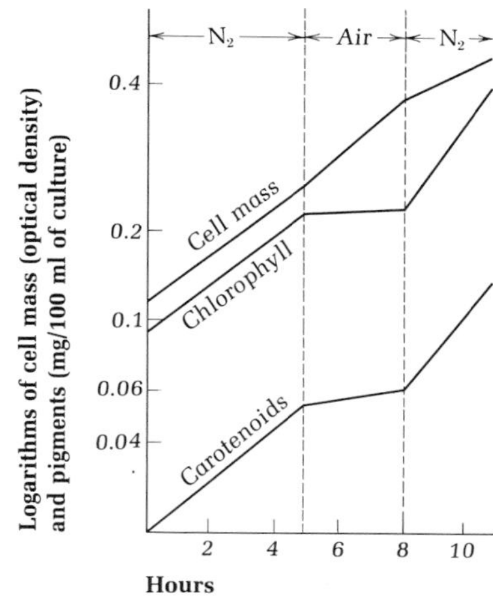


eventually become almost completely decolorized. The decline in the concentration of the photosynthetic pigment system is accompanied by a progressive reduction in the extent of the intracellular membrane system with which these pigments are associated, so that cells grown under strictly aerobic conditions can no longer be distinguished structurally from those of nonphotosynthetic bacteria (pseudomonads or spirilla) having cells of similar form. These changes induced by oxygen in purple nonsulfur bacteria are, however, reversible. If bleached cells are once more placed in an appropriate growth medium under anaerobic conditions in the light, the photosynthetic pigments and intracellular membrane system are resynthesized and photosynthetic growth can then resume, typically after a lag of many hours.

The cellular content of photosynthetic pigments and the extent of the

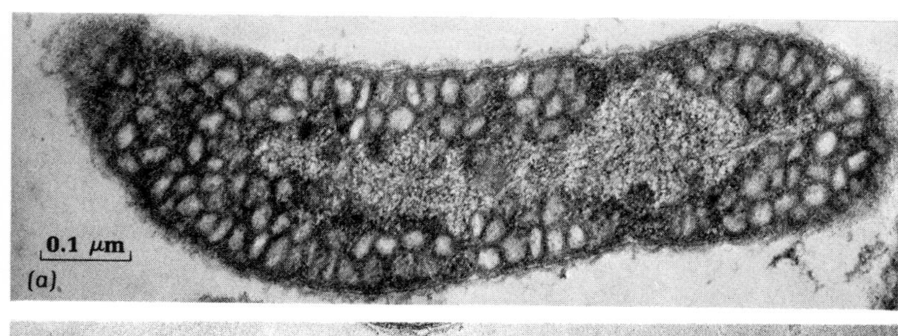


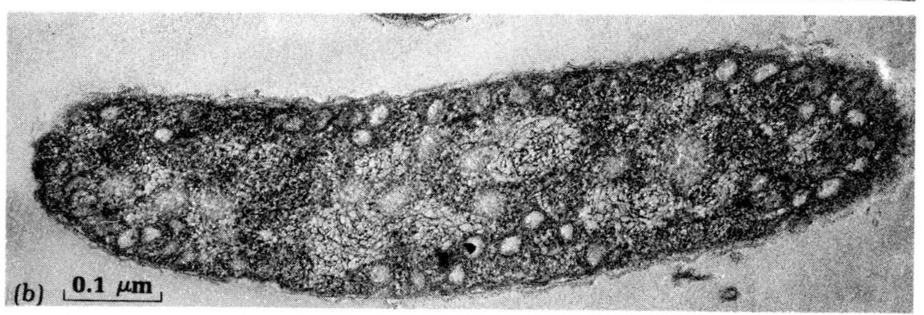


Figure 17.16 *Electron micrographs of thin sections of the purple nonsulfur bacterium, Rhodospirillum rubrum, illustrating changes in the quantity of internal membranes as a function of growth conditions (× 46,200). (a) Cells grown anaerobically in dim light. (b) Cells grown anaerobically in bright light. Courtesy of Germaine Cohen-Bazire.*

internal membrane system are also affected, in cells growing anaerobically, by light intensity. In dim light, cells have a high pigment and membrane content; in bright light, the cellular content of pigments and membrane material is greatly diminished (Figure 17.16).

**The metabolism of organic compounds in the light and in the dark**

With rare exceptions, any organic compound that supports photosynthetic growth of a purple nonsulfur bacterium under anaerobic conditions will also support respiratory growth of the same organism under aerobic conditions. However, the pathways of metabolism under photosynthetic and respiratory conditions are usually entirely different. Under respiratory conditions, much of the carbon of the organic substrate is completely oxidized to $CO_2$ through the tricarboxylic acid cycle, providing ATP through the reactions of oxidative phosphorylation. As in the respiratory metabolism of nonphotosynthetic bacteria, the fraction of substrate carbon assimilated rarely exceeds 35 percent. However, under anaerobic conditions in the light, the reactions of cyclic photophosphorylation provide a potentially unlimited supply of ATP for the purposes of biosynthesis and thus permit an almost total assimilation of the carbon contained in organic substrates. Little, if any, of the organic substrate is therefore oxidized through the tricarboxylic acid cycle, which plays a minor role in the anaerobic photometabolism of most organic substrates.

During the anaerobic metabolism of organic substrates in the light, the balance between oxidation and reduction is maintained either by conversion of part of the substrate to $CO_2$, if it is more oxidized than cell material, or by a concomitant assimilation and reduction of $CO_2$, if the organic substrate is more reduced than cell material. Grossly speaking, the photosynthetic metabolism of organic compounds by purple bacteria can be schematized as

$$\text{organic substrate} \pm CO_2 \xrightarrow{\text{light}} \text{organic cell material}$$

The biochemical mechanisms involved can be illustrated by considering the photometabolism of two fatty acids: acetate, which is slightly more oxidized than cell material, and butyrate, which is slightly more reduced. The assimilation of both these fatty acids leads in the first instance to the formation of a cellular reserve material, poly-$\beta$-hydroxybutyric acid, through the sequences of reactions shown in Figure 17.17.

The overall conversion of acetate to poly-$\beta$-hydroxybutyrate is a reductive process:

$$2nCH_3COOH + 2nH \rightarrow (C_4H_6O_2)_n + 2nH_2O$$

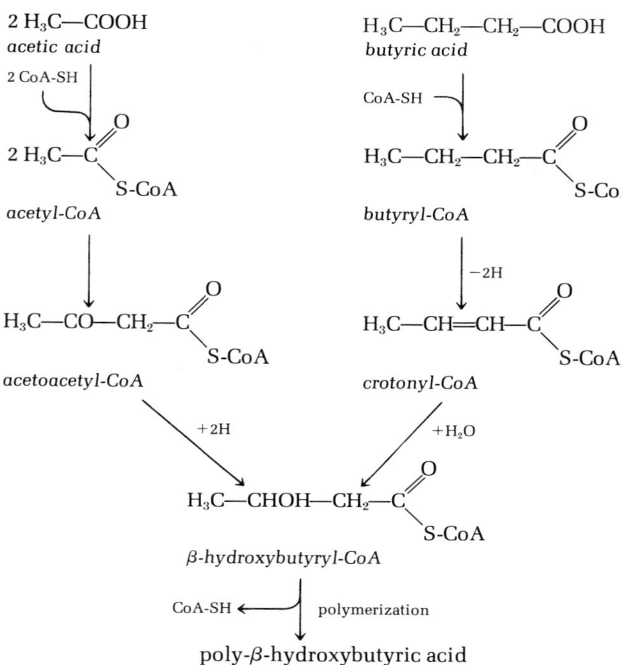


*Figure 17.17 The conversions of acetic and butyric acids to poly-β-hydroxybutyric acid by nonsulfur purple bacteria.*

The necessary reducing power is produced by the concomitant oxidation of some acetate anaerobically through the reactions of the tricarboxylic acid cycle, according to the overall equation

$$CH_3COOH + 2H_2O \rightarrow 2CO_2 + 8H$$

The balanced equation for these two paths of acetate metabolism is

$$9nCH_3COOH \rightarrow 4(C_4H_6O_2)_n + 2nCO_2 + 2nH_2O$$

Assimilation is thus extremely efficient, almost 90 percent of the acetate carbon being assimilated, thanks to the availability from cyclic photophosphorylation of the ATP necessary to drive the assimilatory process. Assimilation of acetate carbon becomes almost total if molecular hydrogen is provided as an external source of reducing power:

$$2nCH_3COOH + nH_2 \rightarrow (C_4H_6O_2)_n + 2nH_2O$$

The conversion of butyrate to poly-β-hydroxybutyrate corresponds to an oxidation:

$$nCH_3CH_2CH_2COOH \rightarrow (C_4H_6O_2)_n + 2nH$$

During the photosynthetic assimilation of this fatty acid, the excess reducing power generated is used for a coupled assimilation of $CO_2$ through the reactions of the Calvin cycle, which yields polysaccharide, schematized as $(CH_2O)_n$, as the primary assimilatory product:

$$nCO_2 + 4nH \rightarrow (CH_2O)_n + nH_2O$$

During the photometabolism of butyrate there is a total assimilation of organic substrate carbon, obligatorily coupled with the assimilation of some $CO_2$:

$$nC_4H_8O_2 + nCO_2 \rightarrow (C_4H_6O_2)_n + (CH_2O)_n$$

The reactions of the tricarboxylic acid cycle, necessary for the respiratory metabolism of butyrate, accordingly play no role in its photometabolism.

**Photosynthetic nitrogen fixation and hydrogen evolution**

All purple and green bacteria so far examined are highly effective fixers of nitrogen under anaerobic conditions in the light. They can also be induced to form molecular hydrogen in the light from appropriate inorganic or organic electron donors. When an organic electron donor serves for hydrogen production in the light it is largely oxidized through the tricarboxylic acid cycle, and the electrons so removed are discharged as molecular hydrogen. *This hydrogen production is specifically inhibited by* $N_2$. Hydrogen production in the light is probably therefore mediated by the enzyme system responsible for nitrogen fixation: when $N_2$ is absent, the reducing power derived from the electron donor and the ATP produced by photophosphorylation are employed for the formation of $H_2$, instead of for the reduction of $N_2$ to ammonia. The ecological significance of light-mediated hydrogen production is not known.

## *Ecological considerations*

Since all photosynthetic bacteria require anaerobiosis and an electron donor other than water for the performance of photosynthesis, the environmental conditions necessary for their development in nature are far more restrictive than those for oxygen-evolving photosynthetic organisms. Consequently the overall role of the bacteria in photosynthetic activity on the earth today is negligible compared to that of algae; the purple and green bacteria flourish only in certain specialized habitats. They are aquatic organisms and may occur either in freshwater or marine environments which provide light, anaerobiosis, and a suitable electron donor.

The purple sulfur and green bacteria often develop abundantly in natural sulfur springs, the waters of which contain dissolved hydrogen sulfide together with a low content of organic matter. A favorable environment for all types of photosynthetic bacteria is provided by shallow, muddy ponds, particularly those subject to considerable organic pollution. The fermentation of organic matter and the activities of sulfate-reducing bacteria in the sediment provide an anaerobic environment, together with the electron donors, both organic and inorganic, necessary for the development of the photosynthetic bacteria. In this type of habitat, only a thin surface layer, typically densely populated with algae, is aerobic; and almost immediately below this layer, the photosynthetic bacteria may be present in great abundance. Their photosynthetic development despite the overlying light filter provided by the algae is possible as a result of their ability to use the far-red and infrared portions of the solar spectrum, transmitted by algae (see Figure 17.3).

Photosynthetic bacteria also find favorable conditions for growth in the deeper layers of freshwater lakes, particularly in narrow, deep lakes where the water is permanently stratified, called *meromictic lakes* (Figure 17.18). In lakes of this type, the circulation of water is confined to the upper layers, and below a depth of some 15 to 20 m, the water is permanently cold and oxygen-free. The photosynthetic bacteria develop in a narrow horizontal band, just within the cold anaerobic layer, where they receive from below the necessary nutrients, formed in the sediment, and from above a weak intensity of light. Water samples from this layer are often brightly colored as a result of their large population of photosynthetic bacteria. At these depths, practically the only region of the solar spectrum that is not absorbed by the

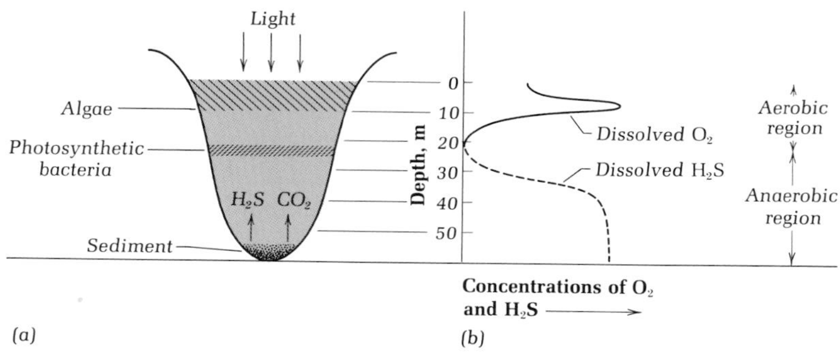


*Figure 17.18 Diagram of the structure of a meromictic lake, showing (a) the vertical distributions of algae and photosynthetic bacteria, and (b) the relative concentrations in the water profile of dissolved oxygen and $H_2S$.*

overlying water is the blue and blue-green light of wavelengths between 450 and 550 nm. Photosynthesis is therefore mediated entirely by light energy that is absorbed by carotenoid pigments and subsequently transferred to chlorophyll. The photosynthetic bacteria from these deep habitats are typically very rich in carotenoids.

## *The evolutionary significance of bacterial photosynthesis*

Geochemical evidence suggests that the atmosphere of the early earth was oxygen-free and that the oxygen which now comprises some 20 percent of the atmosphere is of biological origin, the result of the activity of photosynthetic organisms. Photosynthesis must therefore have evolved as a biochemical mechanism prior to the shift from an anaerobic to an aerobic atmosphere. It seems probable that the purple and green bacteria represent very ancient surviving lines, derived from photosynthetic groups that antedated the emergence of oxygen-producing photosynthetic organisms and were at one time the only forms able to use radiant energy on the primitive, oxygen-free earth.

The evolution of the ability to use water as a photosynthetic electron donor involved the addition to the photosynthetic apparatus of a second type of photochemical reaction center, as has been discussed in Chapter 6. With the emergence of this type of photosynthesis, today characteristic of the overwhelming majority of photosynthetic organisms, a major turning point in biological evolution had been reached; the way lay open for the massive production of molecular oxygen on earth and hence for the eventual development of aerobic respiratory metabolism, destined to become, after photosynthesis, the most widespread biological energy-yielding mechanism.

How is the continued survival of the photosynthetic bacteria, embodying the older, more restricted anaerobic modes of photosynthetic metabolism, to be explained? The most important factor in their survival was probably that they possessed (or evolved) pigment systems with absorption spectra significantly different from those of oxygen-evolving photosynthetic organisms. This enabled them to avoid direct competition for solar radiant energy and thus to continue to develop in the deeper layers of aqueous environments, while their more advanced photosynthetic competitors took over the surface layers, in direct contact with the increasingly oxygenated atmosphere.

## *Further reading*

**Books**

Gest, H., A. San Pietro, and I. Vernon (editors), *Bacterial Photosynthesis*. Yellow Springs, Ohio: Antioch Press, 1963.

Kondratieva, E. N., *Photosynthetic Bacteria*. Jerusalem: Israel Program for Scientific Translations, 1965.

**Review**

Pfennig, N., "Photosynthetic Bacteria." *Ann. Rev. Microbiol.* **21**, 286 (1967).

# *Chapter eighteen*
# *The enteric group and related organisms*

*In this chapter* we shall describe one of the largest distinctive subgroups among the Gram-negative nonphotosynthetic true bacteria. The members of this group have small, rod-shaped cells, either straight or curved, not exceeding 0.5 $\mu$m in width. Some are permanently immotile; others are motile by means of either peritrichous or polar flagella. What distinguishes them from all other Gram-negative bacteria with similar structural properties is their *facultatively anaerobic* nature; they are capable of using the fermentation of carbohydrates as a means of anaerobic growth, but possess a respiratory electron transport system which enables them to grow aerobically at the expense of a wide range of oxidizable organic substrates.

The classical representative of this group is *Escherichia coli*, an inhabitant of the intestinal tract of vertebrates and one of the most characteristic members of the normal intestinal flora. Closely related to *E. coli* are the other *enteric* or *coliform bacteria* of the genera *Salmonella* and *Shigella*. In contrast to *E. coli*, they are pathogens, agents of enteric diseases. Motile members of the enteric group are peritrichously flagellated.

Many other peritrichously flagellated bacteria share the characteristic physiological properties of the enteric group but have a different ecology, not being inhabitants of the intestinal tract. Members of the genera *Aerobacter*, *Serratia*, and *Proteus* occur primarily in soil and water, and the members of the genus *Erwinia* are plant pathogens or parasites. The resemblances between these organisms and the enteric group have long been recognized by their inclusion in a single family, the Enterobacteriaceae.

In recent years, it has become evident that certain other Gram-negative bacteria, not included in the Enterobacteriaceae, also have many phenotypic resemblances to the enteric group. This applies to the facultatively anaerobic, polarly flagellated aquatic organisms of the genera *Aeromonas*, *Photobacterium*, and *Vibrio*. Some animal pathogens that do not cause enteric diseases (genus *Yersinia*) share the physiological properties of the enteric group.

There is as yet no appropriate collective name for this very large assemblage among the Gram-negative bacteria, which extends beyond the traditional confines of the Enterobacteriaceae. For this reason, we shall refer to it as "the enteric group and related organisms."

## *The common properties of the group*

The enteric group and related organisms can use a considerable variety of simple organic compounds as substrates for aerobic respiratory metabolism; organic acids, amino acids, and carbohydrates are universally utilized, and some can use aromatic compounds. Under aerobic conditions, these bacteria grow well at the expense of peptone or yeast extract, the nitrogenous constituents of which (amino acids and peptides) serve as energy-yielding respiratory substrates. Under anaerobic conditions, on the other hand, growth becomes strictly dependent on the provision of a fermentable carbohydrate. Monosaccharides, disaccharides, and polyalcohols are fermented by all members of the group. The fermentation of polysaccharides is less common, but some species can attack starch or pectin.

Although these organisms are customarily grown on complex media, nutritional studies have shown that the minimal requirements are often rather simple. In the Enterobacteriaceae, growth factors are not required by *Escherichia*, *Aerobacter*, *Serratia*, and most *Salmonella* and *Erwinia* species; all these organisms grow well in a synthetic medium with a single organic compound (e.g., glucose) as carbon and energy source and ammonia as nitrogen source. Other members of the Enterobacteriaceae have specific growth-factor requirements. Thus, *Salmonella typhosa* needs one amino acid, tryptophan; *Proteus vulgaris* and *Erwinia amylovora* require one vitamin, nicotinic acid. All *Shigella* species require nicotinic acid. There are usually additional requirements for other vitamins or amino acids in the *Shigella* group.

The characteristic cellular reserve material is glycogen. None of these bacteria can synthesize poly-$\beta$-hydroxybutyrate.

**Fermentative metabolism**

Sugar fermentation by the enteric bacteria and related groups occurs through the Embden–Meyerhof pathway. The products vary, both qualitatively and quantitatively. These fermentations have one characteristic biochemical feature, however, which is not encountered in any other bacterial fermentations. This is a special mode of cleavage of the intermediate, pyruvic acid, to yield formic acid:

$$CH_3COCOOH + CoASH \rightarrow CH_3COSCoA + HCOOH$$

Formic acid is, therefore, frequently a major fermentative end-product. It does not always accumulate, however, since some of these bacteria possess the enzyme formic hydrogenlyase, which splits formic acid to $CO_2$ and $H_2$:

$$HCOOH \rightarrow CO_2 + H_2$$

In such organisms, formic acid is largely replaced as a fermentative end-product by equimolar quantities of $H_2$ and $CO_2$.*

The most frequent mode of fermentative sugar breakdown in the

*Formation of molecular hydrogen as an end-product of sugar fermentation is also characteristic of many sporeformers of the genera *Clostridium* and *Bacillus* (see Chapter 20). The biochemical mechanism responsible for its production is, however, different. In sporeformers, hydrogen is formed as a *direct* product of pyruvic acid cleavage:

$$CH_3COCOOH + CoASH \rightarrow CH_3COSCoA + H_2 + CO_2$$

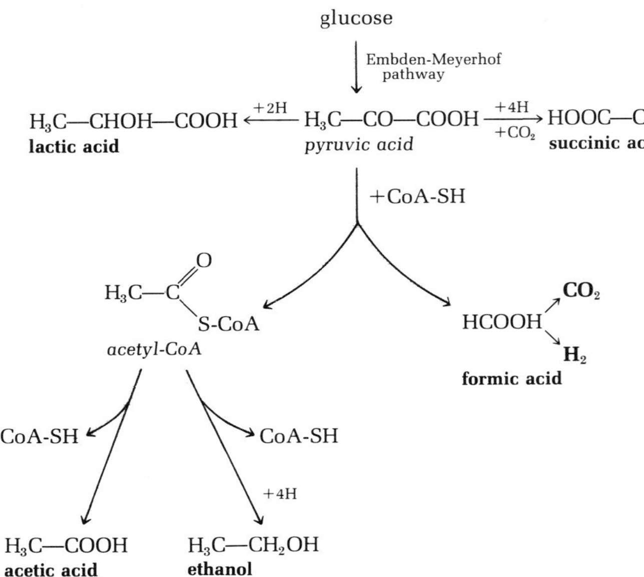


*Figure 18.1 Mechanisms of formation of the characteristic end-products (boldface) of a mixed acid fermentation from pyruvic acid.*

enteric group and related organisms is the *mixed-acid fermentation*, which yields lactic, acetic, and succinic acids; formic acid (or $CO_2$ and $H_2$); and ethanol. This fermentation is characteristic of the genera *Escherichia, Salmonella, Shigella, Proteus, Yersinia, Vibrio,* and *Photobacterium,* and occurs in some *Aeromonas* species. The ratios of the end-products may vary considerably, both from strain to strain and in a single strain grown under different environmental conditions (e.g., at different pH values). This variability reflects the fact that the end-products arise from pyruvic acid through three independent pathways (Figure 18.1).

In some of these organisms, sugar fermentation gives rise to an additional major end-product, 2,3-butanediol, which is formed from pyruvic acid by a fourth independent pathway (Figure 18.2). This *butanediol fermentation* is characteristic of *Aerobacter* and *Serratia,* most species of *Erwinia,* and some *Aeromonas* species. The formation of butanediol is accompanied by increased formation of the reduced end-product ethanol, since the formation of butanediol from glucose results in a net generation of reducing power:

$$C_6H_{12}O_6 \rightarrow CH_3CHOHCHOHCH_3 + 2CO_2 + 2H$$

Hence neutral end-products (butanediol and, to a lesser extent, ethanol) predominate over the acidic end-products; as a result, the total amount of acid formed per mole of glucose fermented is considerably less in a

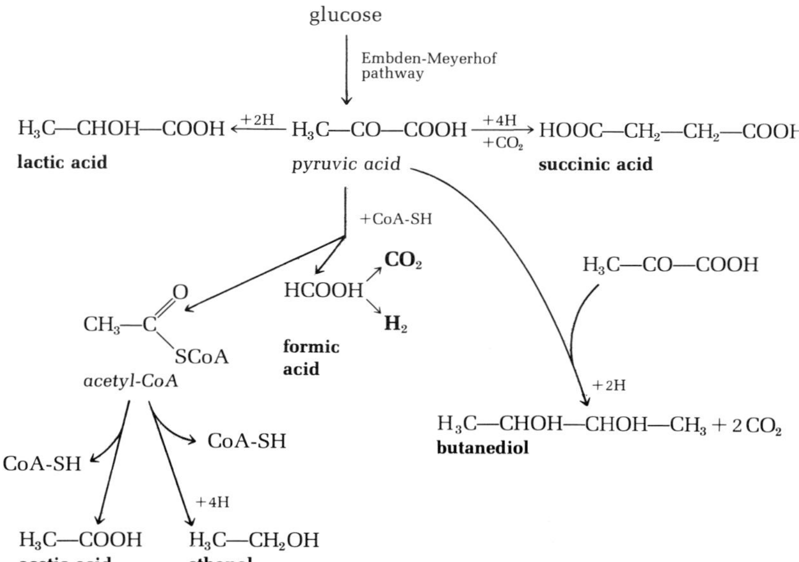


*Figure 18.2 Mechanisms of formation of the characteristic end-products (boldface) of a butanediol fermentation from pyruvic acid.*

*Table 18.1 Products of glucose fermentation by Escherichia coli and related organisms*

| | **Products, moles per 100 moles of glucose fermented** | | | | |
|---|---|---|---|---|---|
| | **Mixed-acid fermentations** | | | **Butanediol fermentations** | |
| | *Escherichia coli* | *Aeromonas formicans* | *Vibrio cholerae* | *Aerobacter aerogenes* | *Serratia marcescens* |
| *Ethanol* | 50 | 64 | 51 | 70 | 46 |
| *2,3-Butanediol* | — | — | — | 66 | 64 |
| *Acetic acid* | 36 | 62 | 39 | 0.5 | 4 |
| *Lactic acid* | 79 | 43 | 53 | 3 | 10 |
| *Succinic acid* | 11 | 22 | 28 | — | 8 |
| *Formic acid* | 2.5 | 105 | 73 | 17 | 48 |
| $H_2$ | 75 | — | — | 35 | — |
| $CO_2$ | 88 | — | — | 172 | 116 |
| *Total acid formed* | 129 | 232 | 193 | 20 | 70 |

butanediol fermentation than in a simple mixed-acid fermentation. This is shown by the fermentation balances for a number of different species assembled in Table 18.1.

The formation of gas as a result of sugar fermentation is a character of considerable differential value in the enteric group, since it distinguishes the gasformers of the genus *Escherichia* from the pathogens of the *Shigella* group and *Salmonella typhosa,* which ferment sugar without gas production. In a simple mixed-acid fermentation, gas can be formed only by the cleavage of formate; gas production therefore reflects the possession of formic hydrogenlyase. This enzyme is not, of course, essential for fermentative metabolism and can be lost by mutation without effect on fermentative capacity. In fact, experience has shown that "anaerogenic" (i.e., non-gas-producing) strains of such a typical gas producer as *Escherichia coli* exist in nature. Hence, although gas production is a useful differential character in the enteric group, it is by no means an infallible one.

The bacteria that perform a butanediol fermentation also differ with respect to the possession of formic hydrogenlyase. Members of the genus *Aerobacter* almost always contain formic hydrogenlyase and are vigorous gas producers; members of the genus *Serratia* do not contain the enzyme and produce little or no visible gas, as judged by the customary criterion (formation of a bubble in an inverted vial placed in the fermentation tube). This may appear paradoxical, since the formation of butanediol from sugars is accompanied by a considerable net production of $CO_2$. However, this gas is very soluble in water, so that most (or all) of the $CO_2$ produced tends to remain dissolved in the medium. When $CO_2$ is the sole gaseous product of a bacterial fermentation, special cultural methods may be required to demonstrate its formation, a point discussed in connection with the lactic acid bacteria (Chapter 21).

Another character of considerable diagnostic importance in the enteric group is the ability to ferment the disaccharide, lactose, which depends on possession of $\beta$-galactosidase. Effective lactose utilization also depends on possession of a specific galactoside permease which facilitates entry of lactose into the cell. Strains lacking permease but containing $\beta$-galactosidase cannot take up lactose at a sufficient rate to produce a prompt and vigorous fermentation and will normally be classified as nonfermenters of this sugar. Lactose fermentation is characteristic of *Escherichia* and *Aerobacter* and is absent from *Shigella*, *Salmonella*, and *Proteus*. It should be noted that some *Shigella* strains produce $\beta$-galactosidase but cannot ferment lactose because they lack permease.

The fermentative characteristics of the various genera are summarized in Table 18.2.

*Table 18.2 Summary of fermentative patterns in the enteric group and related organisms*

- I. *Mixed-acid fermentation*
  - A. Produce $CO_2 + H_2$ *(contain formic hydrogenlyase)*
    - *Escherichia*
    - *Proteus*
    - *Salmonella (most spp.)*
    - *Photobacterium (some spp.)*
  - B. *No gas produced (formic hydrogenlyase absent)*
    - *Shigella*
    - *Salmonella typhi*
    - *Yersinia*
    - *Vibrio*
    - *Aeromonas formicans and A. shigelloides*
    - *Photobacterium (some spp.)*
- II. *Butanediol modification of mixed-acid fermentation*
  - A. Produce $CO_2 + H_2$ *(contain formic hydrogenlyase)*
    - *Aerobacter*
    - *Aeromonas hydrophila*
  - B. *Produce only* $CO_2$ *(formic hydrogenlyase absent); visible gas formation slight or undetectable*
    - *Serratia*
    - *Erwinia herbicola and E. carotovora*

**The oxidase and arginine dihydrolase reactions as differential characters**

The polarly flagellated organisms of the genera *Aeromonas*, *Photobacterium*, and *Vibrio* can be distinguished from the other genera under consideration by means of the *oxidase reaction*. This is a test for the ability to oxidize aromatic amines to colored products. The usual substrate, dimethyl-*p*-phenylenediamine, is almost immediately converted to a red oxidation product by oxidase positive bacteria, whereas oxidase negative bacteria do not produce a color change. The oxidase test is an indirect test for the presence of a cytochrome of the c type in the respiratory electron transport chain; it therefore reveals differences in the composition of the electron transport system. *Aeromonas*, *Photobacterium*, and *Vibrio* are oxidase positive; all the other genera in this group are oxidase negative.

One additional biochemical character which distinguishes *Aeromonas*, *Photobacterium*, and *Vibrio* from the members of the Enterobacteriaceae and from *Yersinia* is the ability to decompose arginine anaerobically to ornithine, $CO_2$, and ammonia by means of the arginine dihydrolase enzyme system. Both the production of arginine dihydrolase and the positive oxidase reaction are characters these three genera share with the strictly aerobic polarly flagellated bacteria of the genus *Pseudomonas*, nearly all of which are oxidase positive and many of which are producers of arginine dihydrolase.

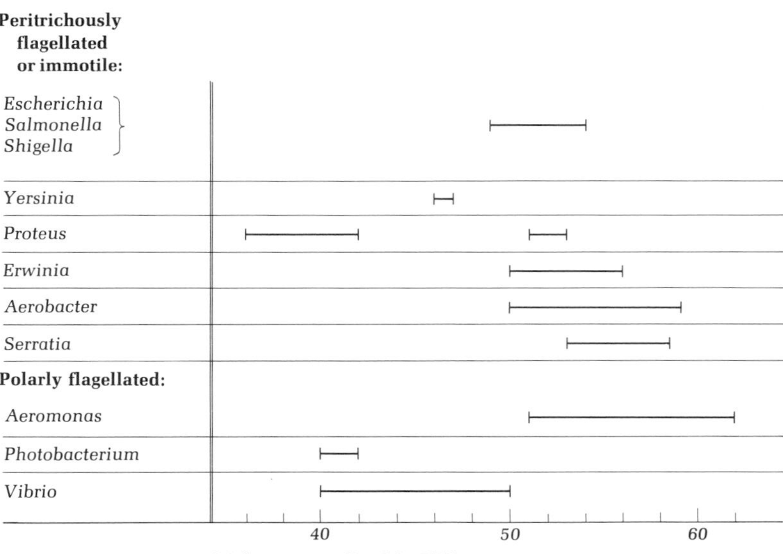


*Figure 18.3 The ranges of DNA base composition (moles percent G + C) characteristic of different genera in the enteric group.*

The DNA base composition of the enteric group and related organisms ranges from a low value of 37 mole percent in the genus *Proteus* to a high value of almost 60 mole percent in *Aeromonas*. The characteristic ranges for each genus are summarized in Figure 18.3.

**The genetic properties of the enteric group and related organisms**

The discovery of conjugation and transduction as means of genetic transfer in the enteric group has permitted a detailed analysis of

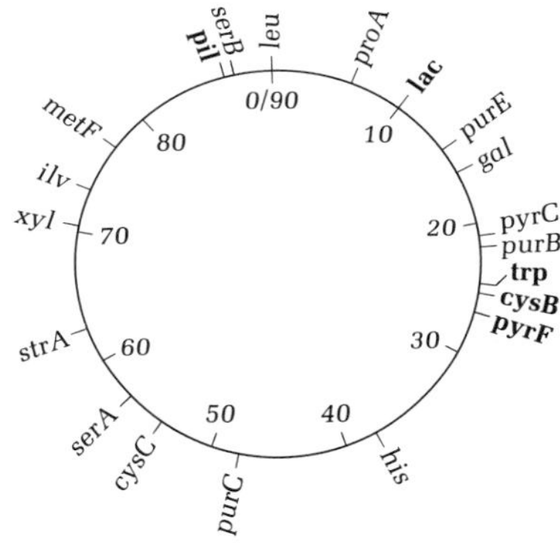


*Figure 18.4 Partial genetic map of E. coli strain K12. Compare with the partial genetic map of Salmonella typhimurium shown in Figure 18.5, and with the more complete genetic map of E. coli shown in Figure 14.12. The numbers along the inside of the circle are units of distance and equal the number of minutes required to transfer that length of the chromosome during conjugal transfer at 37°C (see Chapter 14). Different genetic loci which have been identified and mapped are indicated by three-letter or four-letter symbols. Over 200 loci have been mapped in E. coli strain K12 and almost as many in S. typhimurium, of which approximately 100 have the same functions in the two species. Of these 100, only a few have different map locations in the two species. The locus for pilus formation (pil) for example, is at minute 88 in E. coli but at minute 23 in S. typhimurium; E. coli contains the genes for lactose utilization (lac) at minute 10, whereas the lac genes are totally absent from S. typhimurium. Finally, the chromosomal segment trp–cysB–pyrF, located at minute 25 on the E. coli chromosome, is inverted in S. typhimurium. Modified from A. Taylor and C. Trotter, "Revised Linkage Map of Escherichia coli." Bacteriol. Rev.* **31,** *332 (1967).*

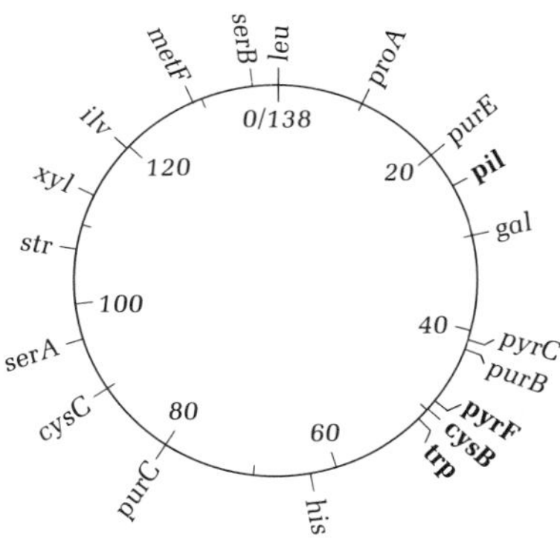


*Figure 18.5 Partial genetic map of S. typhimurium. Compare with the partial genetic map of E. coli strain K12 shown in Figure 18.4; the similarities and differences are described in the legend to Figure 18.4. In both cases, the maps have been assigned units of distance corresponding to the number of minutes required to transfer corresponding map segments during conjugal transfer at 37°C. The difference in total "length" (90 minutes versus 134 minutes) probably does not represent a difference in physical length but rather a difference in speed of transfer or in measurement techniques. Modified from K. Sanderson, "Revised Linkage Map of Salmonella typhimurium." Bacteriol. Rev.* **31,** *354 (1967).*

genetic relations among these bacteria. It has proved possible to obtain true hybrids of *E. coli* with members of both the genera *Salmonella* and *Shigella*, which indicates that these bacteria all share a relatively high degree of genetic homology. This is confirmed by the many resemblances between the chromosomal maps of the two species that have been subjected to detailed genetic analysis, *E. coli* and *Salmonella typhimurium* (Figures 18.4 and 18.5). It should also be noted that *Escherichia*, *Salmonella*, and *Shigella* are indistinguishable in DNA base composition.

The attempts to obtain intergeneric hybrids between members of the *Escherichia–Salmonella–Shigella* complex and representatives of other genera of the Enterobacteriaceae (*Aerobacter*, *Serratia*, *Proteus*) have failed. Moreover, experiments on nucleic acid hybridization in vitro, of the type described in Chapter 15, indicate that the degree of genetic homology between the enteric complex and the other genera in question is relatively low. However, it has been found that bacteria belonging to the genera *Serratia*, *Proteus*, *Yersinia*, and *Vibrio* can acquire episomes by conjugational transfer from donor strains of *E. coli*, and subsequently maintain them as extrachromosomal genetic elements. Such experiments have been performed largely with F-*lac* episomes, since transfer to lactose-negative recipients can be detected

*Table 18.3 Genetic evidence concerning relationships among the enteric bacteria and related groups*[a]

I. *Organisms between which hybridization of chromosomal genes can occur*

| *Genetic donors:* | *Genetic recipients:* |
|---|---|
| *Escherichia coli* | *Many E. coli strains* |
| *Escherichia coli* | *Shigella flexneri and other Shigella spp.* |
| *Escherichia coli* | *Salmonella typhimurium, S. typhi* |
| *Salmonella typhimurium* | *Salmonella typhi, other Salmonella spp.* |
| *Shigella flexneri* | *Escherichia coli, Salmonella typhi* |

II. *Organisms between which episomes can be transferred*

A. *Substituted F factors of E. coli (e.g., F-lac):*
*Salmonella spp.; Shigella spp.; Serratia marcescens; Proteus spp.; Aerobacter–Klebsiella group; Vibrio cholerae; Yersinia pestis*

B. *Drug resistance (R) factors:*
*Escherichia coli; Shigella spp.; Salmonella spp.; Serratia marcescens; Aerobacter–Klebsiella group; Proteus spp.; Vibrio cholerae; Yersinia pestis; Aeromonas hydrophila*

[a]Data courtesy of D. J. Brenner, R. V. Citarella, and Stanley Falkow.

easily; the recipients become lactose positive. More recently episomes bearing drug-resistance markers (R factors) have also been used. There appear to be definite limits to the range of transfer of such episomes; only bacteria that share the general physiological properties of the enteric group can serve successfully as recipients (Table 18.3). Hence, this constitutes a significant criterion of evolutionary affinity even though it provides no evidence of true genetic homology, since the episome is maintained and reproduced separately from the chromosome of the recipient.

## *Taxonomic problems*

A simplified scheme for the subdivision of the enteric group and related organisms is presented in Table 18.4. On the basis of three characters, the *Aeromonas–Photobacterium–Vibrio* group can be separated from the remaining genera. It is not at all easy to describe the taxonomy of the organisms that comprise the Enterobacteriaceae, since the specialists who have studied these organisms have created a very large number of genera, for the most part distinguished from one another by phenotypic differences that are very minor and which would certainly be used only for the differentiation of species in the classification of other bacterial groups. A few examples will illustrate the problem. The maintenance of three genera, *Escherichia*, *Salmonella*, and *Shigella*, is not really justifiable. Particularly in the light of their close genetic relationships, the members of these three genera should certainly all be placed in a single genus. Yet the specialists have carved out two additional genera (*Arizona* and *Citrobacter*) for certain organisms that fall into the confines of this group. Similarly, the differences between *Aerobacter* and *Serratia* scarcely justify a generic separation, and most *Erwinia* species could also be accommodated in a single genus centered around the *Aerobacter–Serratia* complex. The scheme for the classification of these organisms outlined in Table 18.4 attempts to simplify somewhat the very complex subdivision of the Enterobacteriaceae that is now current, although even greater simplification would be desirable.

Broadly speaking, the members of the Enterobacteriaceae can be split into three major groups on the basis of DNA base composition and fermentative patterns; these are designated as groups I to III in Table 18.4. The organisms of the genus *Yersinia* probably should constitute a fourth group. Each of these four groups might appropriately constitute a single genus.

*Table 18.4 Simplified taxonomic subdivision of the enteric group and related organisms*

**A. Straight rods, immotile or motile by peritrichous flagella; oxidase negative; arginine dihydrolase negative (family Enterobacteriaceae and genus *Yersinia*)**

| | | | Fermentative characters | | Production of: | | | |
|---|---|---|---|---|---|---|---|---|
| Major subgroup | DNA base composition, mole % G + C | Motility | Butanediol production | Production of $H_2 + CO_2$ | Urease | Tryptophan deaminase | Constituent genera | Other generic names frequently applied to members of group |
| I | 50–52 | V[a] | – | V | – | – | *Escherichia, Salmonella, Shigella* | *Arizona* and *Citrobacter* for intermediate types |
| II | 50–58 | V | + | V | – | – | *Aerobacter, Serratia, Erwinia* | *Klebsiella* and *Enterobacter* for forms here treated as *Aerobacter* |
| III | 37–53 | + | – | + | + | + | *Proteus* | *Providencia* |
| IV | 46–47 | V | – | – | + | – | *Yersinia* | Formerly placed in *Pasteurella* |

**B. Straight or curved rods, motile by polar flagella (rarely immotile); oxidase positive; arginine dehydrolase positive**

| | | Fermentative characters | | | |
|---|---|---|---|---|---|
| DNA base composition, mole % G + C | Shape of cell | Butanediol production | Production of $H_2 + CO_2$ | Bioluminescence | Genus |
| 52–59 | Straight | V | V | – | *Aeromonas* |
| 40–42 | Straight or curved | – | V | + | *Photobacterium* |
| 40–49 | Curved | – | – | – | *Vibrio* |

[a]V denotes a variable reaction within the group.

**Group I:** *Escherichia-Salmonella-Shigella*

The members of this group are all inhabitants of the intestinal tract of man and other vertebrates. The principal generic distinctions are shown in Table 18.5. It should be noted that certain strains of intestinal origin—the so-called *paracolon group*—have characters intermediate between those shown for the genera in Table 18.5, so that distinctions are not always as clear-cut as the table suggests. The additional genera, *Arizona* and *Citrobacter*, were created for the intermediate forms of the paracolon group, but this generic hypertrophy really does nothing to help the problem of differentiation. In addition to cultural characters, the analysis of the surface antigens of these bacteria (particularly their wall polysaccharides) has played an important part in their detailed classification. Space does not permit a discussion of this very complex topic (see page 342).

*Escherichia coli* and some members of the paracolon group are components of the normal intestinal flora and give rise to disease only under exceptional conditions. The genera *Salmonella* and *Shigella* comprise pathogens that cause a wide variety of enteric diseases in man and other animals. In all cases, entry occurs through the mouth; and the small intestine is the primary locus of infection, although some of these pathogens may subsequently invade other body tissues and cause more generalized damage in the infected host. The members of the genus *Shigella* are the agents of a specifically human enteric disease, bacterial dysentery. In the genus *Salmonella*, both the host range and the variety of diseases produced are much broader. *Salmonella typhi* and *S. paratyphi*, the agents of typhoid fever, are specific

*Table 18.5 Internal differentiation of the major genera of group I*

| **Characteristics** | *Escherichia* | *Salmonella* | *Shigella* |
|---|---|---|---|
| *Pathogenicity for man or animals* | − | V[a] | + |
| *Motility* | V | + | − |
| *Gas* ($CO_2 + H_2$) *from fermentation of glucose* | + | +[b] | − |
| *Fermentation of lactose* | + | − | − |
| *β-Galactosidase* | + | − | V |
| *Utilization of citrate as carbon source* | − | + | − |
| *Production of indole from tryptophan* | + | − | V |

[a]V denotes a variable reaction within the group.
[b]Except for S. *typhosa*.

pathogens of man, whereas certain other species (e.g., *S. typhimurium*) are specific pathogens of other mammals or of birds. The great majority of the *Salmonella* group, however, have a low host specificity. They exist, often without causing disease symptoms, in the intestine and in certain tissues of animals or birds. If these forms gain access to and develop in foods, their subsequent ingestion by man can give rise to food poisoning. This can be defined as a gastrointestinal infection which is usually acute but transient, although in some cases it may assume a graver form. Outbreaks of food poisoning often have an epidemic character, because food preparation on a large scale provides many favorable opportunities for the growth of these organisms.

**Group II:** *Aerobacter–Serratia–Erwinia*

*Aerobacter aerogenes*, the principal *Aerobacter* species, sometimes occurs in the intestinal tract. It is not, however, a characteristic intestinal inhabitant, being widely present in water and soil. Very similar forms, distinguishable largely by their possession of capsules and their permanent immotility, occur in the human respiratory tract and bear the name of *Klebsiella pneumoniae;* this name may be considered a synonym for *Aerobacter aerogenes*. Many strains of this group are active nitrogen fixers; nitrogen fixation occurs only in the absence of oxygen.

*Serratia marcescens*, also a common soil and water organism, differs from *Aerobacter* principally by its failure to produce formic hydrogenlyase (little or no visible gas formed during sugar fermentation) and by its inability or weak ability to ferment lactose. Many (but by no

*Table 18.6 Internal differentiation of the major genera of group II*

| **Characteristics** | *Aerobacter* | *Serratia* | *Erwinia*[a] |
|---|---|---|---|
| *Motility* | V[b] | + | + |
| $CO_2 + H_2$ *formed by glucose fermentation* | + | – | – |
| *Lactose fermentation* | + | –(+)[c] | V |
| *β-Galactosidase* | + | + | + |
| *Gelatin liquefied* | – | + | +(–)[d] |
| *Pectinolytic enzymes produced* | – | – | V |
| *Yellow cellular pigments* | – | – | V |
| *Red cellular pigments* | – | +(–) | – |
| *Plant pathogens or parasites* | – | – | + |

[a]Excluding *E. amylovora*.
[b]V denotes a variable reaction within the group.
[c]–(+) denotes predominantly negative, with rare positive strains.
[d]+(–) denotes predominantly positive, with rare negative strains.

means all) strains of *Serratia* produce a characteristic red cellular pigment, prodigiosin, a tripyrrole derivative (Figure 18.6).

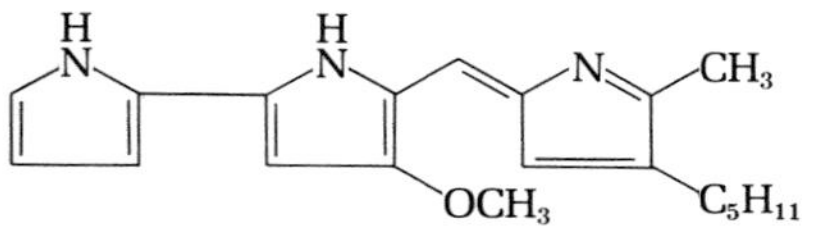

*Figure 18.6 The structure of prodigiosin, the red pigment formed by Serratia.*

The members of the genus *Erwinia* are probably a heterogeneous collection of organisms, placed in a single genus entirely because of their association with plants. Of the three principal species, two—*E. herbicola* and *E. carotovora*—belong, in terms of DNA base composition and physiological properties, near the *Aerobacter–Serratia* complex. Like *Serratia* they lack formic hydrogenlyase and therefore carry out a butanediol fermentation with little or no visible production of gas. Lactose fermentation is variable. *Erwinia herbicola*, although it has occasionally been reported to cause plant disease, is commonly found as a harmless epiphyte on the surface of plants; strains have also been isolated from the intestinal tract of man and animals. It is characterized by the production of a yellow cellular pigment. *Erwinia carotovora* causes a *soft rot* of the fleshy parts of many plants and can be distinguished by its ability to produce pectinolytic enzymes, a trait that accounts for the nature of its attack on plants.

The third major *Erwinia* species, *E. amylovora*, causes a different type of plant disease: a dry necrosis or wilt. This organism is probably not related to the other *Erwinia* species or the *Aerobacter–Serratia* group. This is suggested by its type of glucose fermentation, unlike that of any other members of the Enterobacteriaceae; glucose is converted almost entirely to lactic acid, ethanol, and $CO_2$ (Figure 18.7). A decomposition of pyruvic acid with production of formic acid, otherwise universal in the enteric group and related organisms, clearly does

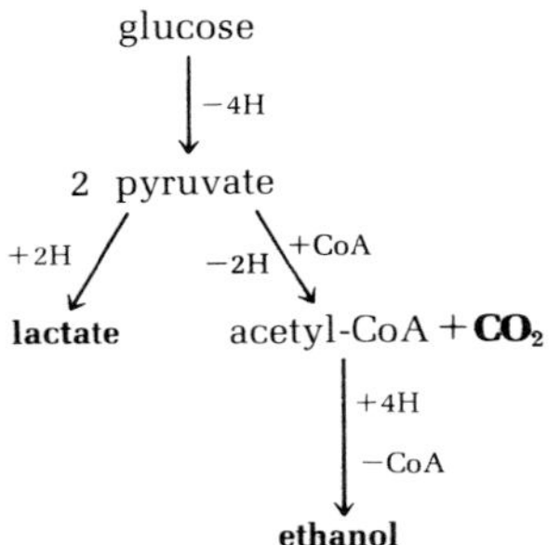


*Figure 18.7 Probable pathways for the formation of the end-products (boldface) of glucose fermentation by Erwinia amylovora.*

not occur, since neither formic acid nor $H_2$ is produced in significant amounts.

**Group III:** *Proteus*

The members of the *Proteus* group are probably soil inhabitants, although they are found in particular abundance in decomposing animal materials. The relatively low G + C content of their DNA distinguishes most species from the groups so far discussed, from which they are also distinguishable by certain physiological properties. These include strong proteolytic activity (gelatin is rapidly liquefied) and ability to hydrolyze urea. Most members of the *Proteus* group are very actively motile and can spread rapidly over the surface of a moist agar plate, a phenomenon known as *swarming*. A curious feature of the swarming phenomenon is its periodicity: it occurs in successive waves, separated by periods of quiescence and growth. This produces a characteristic zonate pattern of development on an agar plate (Figure 18.8).

*Figure 18.8* *Swarming of Proteus on the surface of nutrient agar plate. The plate was inoculated in the center with a drop of a bacterial suspension, and was photographed after incubation at 37°C for 20 hours. From H. E. Jones and R. W. A. Park, "The influence of medium composition on the growth and swarming of Proteus." J. Gen. Microbiol.* ***47****, 369 (1967).*

**Group IV:** *Yersinia*

The genus *Yersinia* contains two or three species which are agents of disease in rodents. *Yersinia pestis* can be transmitted by fleas from its normal rodent hosts to man; it is the cause of human bubonic plague, a disease that has been responsible throughout human history for massive epidemics with a very high mortality. In man, the disease can also be transmitted through the respiratory route. Both in their mode of transmission and in their symptoms, the diseases caused by *Yersinia* species are entirely different from the major enteric diseases, such as dysentery and typhoid fever.

Until recently the plague organism and related species have been classified in the genus *Pasteurella*, a poorly defined group of Gram-negative rods responsible for an assortment of animal diseases. Their segregation into a new genus *Yersinia* followed the realization that biochemically and physiologically (although not in pathological respects) they resemble the enteric group far more closely than they resemble the other members of *Pasteurella*.

The members of the genus *Yersinia* carry out a mixed-acid fermentation without production of $H_2$ and $CO_2$; in this respect, they resemble the *Shigella* group. They produce $\beta$-galactosidase and have a powerful urease. Motility is a variable character; *Y. pestis* is permanently immotile. The G+C content of the DNA is significantly lower than that of the *Escherichia–Salmonella–Shigella* group. A cultural character that distinguishes them from the Enterobacteriaceae is their relatively slow growth on complex media.

**The polar flagellates:** *Aeromonas–Photobacterium–Vibrio*

The polarly flagellated facultative anaerobes have been studied far less intensively and systematically than the Enterobacteriaceae, and their classification is still uncertain. Provisionally, three groups may be recognized (Table 18.4).

The genus *Aeromonas* contains organisms which differ with respect to the nature of sugar fermentation. *Aeromonas hydrophila* performs a butanediol fermentation, accompanied by $H_2$ and $CO_2$ production, similar to that of *Aerobacter*. *Aeromonas formicans* and *A. shigelloides* perform a mixed-acid fermentation without gas production, similar to that of *Shigella*. *Aeromonas shigelloides* also closely resembles the *Shigella* group in DNA base composition and has been shown to share certain somatic antigens with this enteric group.

In this genus *A. formicans* has been most closely studied with respect to its biochemical properties, and the biochemical data suggest that it resembles the enteric group in many ways that are not revealed by simple physiological and cultural studies. For example, the modes of regulation of animo acid biosynthesis are closely similar to those operative in *E. coli* and not at all like those operative in the aerobic pseudomonads. This species produces a $\beta$-galactosidase, which shows antigenic affinities to the $\beta$-galactosidase of *E. coli*. These facts suggests that the great emphasis long placed on the mode of flagellar insertion as a primary taxonomic character among the true bacteria, which has resulted in a wide taxonomic separation between polarly and peritrichously flagellated groups, is not really justified. In terms of their other phenotypic characters, the polarly flagellated facultative anaerobes show many more similarities to the enteric group than to the aerobic pseudomonads.

The members of the genus *Aeromonas* are freshwater organisms. *Aeromonas hydrophila* is a pathogen for cold-blooded animals, causing a systemic infection in frogs.

The members of the genus *Photobacterium* occur exclusively in the sea. Like many other indigenous marine bacteria, they are osmotically sensitive, being rapidly lysed by suspension in fresh water, and requiring at least one percent NaCl for growth. Their other distinguishing group character is bioluminescence. The manifestation of this property is dependent on the presence of free oxygen, as discussed in Chapter 6 (p. 203). It is also to some degree dependent on the nutritional environment. Although many of these organisms can be grown aerobically in a synthetic medium containing only glucose as an organic nutrient, they do not luminesce under these conditions; amino acids are required for light production. Under anaerobic growth conditions, these organisms perform a mixed-acid fermentation of sugars. Some produce $CO_2$ and $H_2$; others do not. Some of the luminous bacteria are straight rods, while others are curved. It has been suggested that the recognition of a special genus on the basis of bioluminescence is not justified and that these organisms might be more appropriately classified, depending on their cell shape, in the genera *Aeromonas* and *Vibrio*.

Luminous bacteria are present in all marine environments and develop abundantly on the surface of dead fish. Many of them are symbiotic, populations being harbored in the special luminous organs of certain fishes and invertebrates (see Chapter 24).

The facultative anaerobes of the genus *Vibrio* have curved cells. They perform a mixed-acid fermentation without gas production. The best known species is *Vibrio cholerae*, which causes a typical enteric disease, cholera. Nonpathogenic organisms similar to *V. cholerae* in structural and physiological respects are not uncommon in fresh water; the group is also represented in the sea.

### *Zymomonas mobilis*

As already mentioned, one peritrichously flagellated facultative anaerobe, the plant pathogen *Erwinia amylovora*, has a pattern of carbohydrate fermentation significantly different from that otherwise characteristic of the Enterobacteriaceae. One facultatively anaerobic polar flagellate, placed in a special genus *Zymomonas*, ferments glucose in a special manner. This organism, which has been found in fermenting plant juices, carries out a virtually pure alcoholic fermentation, accompanied by production of a small amount of lactic acid. It is a powerful fermenter, and can grow with high sugar concentrations, producing as much as 10 percent ethanol. Biochemical studies have shown that the pathway of this fermentation is different from that of the alcoholic fermentation performed by yeast; glucose is dissimilated through the hexose monophosphate shunt, not through the Embden–Meyerhof pathway.

**Coliform bacteria in sanitary analysis**

The enteric diseases caused by the coliform bacteria and by *V. cholerae* are transmitted almost exclusively by the fecal contamination of water and food materials. Transmission through contaminated water supplies is by far the most serious source of infection and was responsible for the massive epidemics of the more serious enteric diseases (particularly typhoid fever and cholera) that periodically scourged all countries until the beginning of the present century. Today these diseases are almost unknown in most parts of the Western world. Their eradication was achieved primarily by appropriate sanitary controls. An essential part of this operation was the *development of bacteriological methods for ascertaining the occurrence of fecal contamination in water and foodstuffs.*

It is seldom possible to isolate enteric pathogens directly from contaminated water, since they are usually present in small numbers unless contamination from an infected individual has been recent and massive. To demonstrate the fact of fecal contamination, it is sufficient to show that the sample under examination contains bacteria known to be specific inhabitants of the intestinal tract, even though they may not be agents of disease. The bacteria that have principally served as indices of such contamination are the fecal streptococci (discussed in Chapter 21) and *E. coli*. The sanitation methods of analysis developed by bacteriologists differ somewhat from country to country.

One method for detecting *E. coli* is to inoculate dilutions of the sample under test into tubes of lactose broth, which are then incubated at 37°C, and examined after 1 and 2 days for acid and gas production. Cultures showing acid and gas formation are then streaked on a special medium, with a composition that facilitates recognition of *E. coli* colonies. One of the media most commonly used is a lactose–peptone agar containing two dyes, eosin and methylene blue (EMB agar). On this medium, *E. coli* produces blue-black colonies with a metallic sheen, whereas the other principal member of the group capable of fermenting lactose with acid and gas production, *Aerobacter aerogenes* (not necessarily indicative of fecal contamination) produces pale pink mucoid colonies without a sheen (Figure 18.9). For a final distinction between these two organisms, a series of physiological tests, known as the IMViC tests, can be performed on material from an isolated colony.

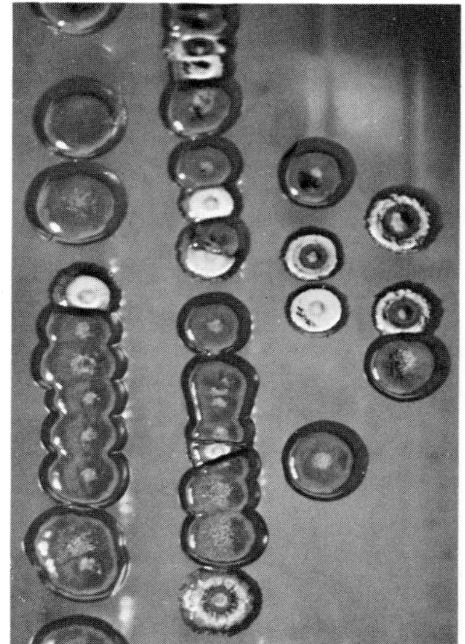

*Figure 18.9 A plate of EMB agar streaked with a mixture of Escherichia coli and Aerobacter aerogenes. The colonies of E. coli are relatively small and appear light as a result of their metallic sheen. Courtesy of N. J. Palleroni.*

*Table 18.7 IMViC tests for the differentiation between Escherichia coli and Aerobacter aerogenes*

| | **Typical reactions** | | | |
|---|---|---|---|---|
| | **Indole** | **Methyl Red** | **Voges–Proskauer** | **Citrate** |
| *Escherichia coli* | + | + | − | − |
| *Aerobacter aerogenes* | − | − | + | + |

The typical results obtained with the two species are shown in Table 18.7. Of these four tests, the Methyl Red and Voges–Proskauer tests are the most significant, since they indirectly reveal the mode of fermentative sugar breakdown. Both are performed on cultures grown in a glucose–peptone medium. The Methyl Red test affords a measure of the final pH: this indicator is yellow at a pH of 4.5 or higher and red at lower pH values. A positive test (red color) is therefore indicative of substantial acid production, characteristic of a mixed-acid fermentation. The Voges–Proskauer test is a color test for acetoin, an intermediate in the formation of butanediol from pyruvic acid; a positive reaction is therefore indicative of a butanediol fermentation. The test for indole production from tryptophan, performed on a culture grown in a peptone medium rich in tryptophan, is a test for the presence of the enzyme tryptophanase, which splits tryptophan to indole, pyruvate, and ammonia. This enzyme is present in many bacteria of the enteric group and related forms (including *E. coli*) but is not found in *A. aerogenes*. The citrate utilization test determines ability to grow in a synthetic medium containing citrate as the sole carbon source. This ability is lacking in most strains of *E. coli*, as a result of the absence of a citrate permease.

The first step of the analytical procedure described above is relatively nonspecific, since many bacteria, not even necessarily members of the Enterobacteriaceae, can grow at 37°C in lactose broth with acid and gas production. A much more specific primary enrichment of *E. coli* can be achieved by the Eijkman method: use of a glucose broth, incubated at 44°C. This slight elevation of incubation temperature eliminates *Aerobacter aerogenes* and most other organisms that ferment lactose with gas production but permits growth of *E. coli*.

## *Escherichia coli* and the rise of molecular biology

The revolution in biological thought which has occurred in the past two decades reflects the acquisition of a new understanding of the molecular basis of many central biological phenomena which previously could not be interpreted at a molecular level. The discoveries in this area were largely made through the study of the simplest kinds of biological systems: bacteria and bacterial viruses. Primarily because it could be easily and rapidly cultivated, and was amenable to genetic analysis, *E. coli* has served as the principal cellular model system for this research. As a result, its genetics, physiology, and biochemistry are known today in far greater detail than the genetics, physiology, and biochemistry of any other organism, plant, animal, or microbe. A large treatise could (and probably will) be written on the properties of this one bacterial species.

## Further reading

Surprisingly, there are no books or comprehensive reviews that give a good account of the general properties of the enteric group, or of the related polarly flagellated Gram-negative bacteria. Perhaps the best general source of detailed information on these organisms can be found in G. S. Wilson and A. A. Miles, *Topley and Wilson's Principles of Bacteriology and Immunology*, 4th ed., Vol. I. Baltimore: Williams and Wilkins, 1955; the relevant chapters are 22, 27, 28, 29, 30 and (in part) 32. This work is, however, considerably out of date. The following references provide more recent information on certain aspects of the groups in question.

**Reviews and original articles**


Anderson, E. S., "The Ecology of Transferable Drug Resistance in the Enterobacteria." *Ann. Rev. Microbiol.* **22,** 131 (1968).

Brenner, D. J., G. R. Fanning, K. E. Johnson, R. V. Citarella, and S. Falkow, "Polynucleotide Sequence Relationships among Members of Enterobacteriaceae." *J. Bacteriol.* **98,** 637 (1969).

Eddy, B. P., "Cephalotrichous, Fermentative Gram-Negative Bacteria: The Genus Aeromonas." *J. Appl. Bacteriol.* **23,** 216 (1960).

Ewing, W. H., and P. R. Edwards, "The Principal Divisions of Enterobacteriaceae and their Differentiation." *Intern. Bull. Bacteriol. Nomencl. Taxon.* **10,** 1 (1960).

Lüderitz, O., A. M. Staub, and O. Westphal, "Immunochemistry of O and R Antigens of *Salmonella* and Related Enterobacteriaceae." *Bacteriol. Rev.* **30,** 192 (1966).

Schubert, R. H. W., "The Taxonomy and Nomenclature of the Genus *Aeromonas*." *Intern. J. Syst. Bacteriol.* **17,** 23; **17,** 273 (1967).

Graham, D. C., "Taxonomy of the Soft Rot Coliform Bacteria." *Ann. Rev. Phytopathol.* **2,** 13 (1964).

# *Chapter nineteen Other groups of Gram-negative nonphotosynthetic true bacteria*

*Chapter 18 was devoted* to an account of one large group of Gram-negative nonphotosynthetic true bacteria: the enteric organisms and other facultatively anaerobic carbohydrate-fermenting forms. The present chapter will be devoted to a survey, necessarily more cursory, of other groups among the Gram-negative nonphotosynthetic true bacteria. The biological diversity of these organisms is very great. In terms of DNA base composition, perhaps the most reliable indication of evolutionary diversity, they cover a wide span. They can be divided into a large number of subgroups, some distinguishable by special structural characters, some by special physiological characters, and some by a combination of both. The sequence of discussion will be an arbitrary one, starting with the aerobic chemoautotrophs, continuing with aerobic chemoheterotrophs, and concluding with strict anaerobes.

## *The aerobic chemoautotrophs*

With the exception of the filamentous gliders (Chapter 16), most chemoautotrophs belong to the general category of Gram-negative unicellular true bacteria. These organisms have long been classified as a special subgroup among the true bacteria, because the ability to grow in the dark on a completely inorganic medium is such an unusual and distinctive physiological property. However, in terms of their total phenotypes, these bacteria appear to be a very heterogeneous group.

By definition, a chemoautotroph must possess two special biochemical capacities: the ability to derive ATP and reducing power from the oxidation of a reduced inorganic compound, and the ability to use $CO_2$ as its principal or sole source of carbon, an attribute that implies possession of the enzymatic machinery of the Calvin cycle (see Chapter 7). Any given chemoautotroph can use only a limited range of inorganic energy sources, and a primary subdivision of the aerobic chemoautotrophs can be made on this basis (Table 19.1). Within each category a further subdivision can be made with respect to the degree of physiological specialization: some of these organisms are *obligate* autotrophs, unable to use any organic compound as sole source of carbon and energy; others are *facultative,* and can use organic compounds as energy sources in place of the inorganic oxidizable substrate, at the same time deriving most of their cellular carbon from the organic source. It should be emphasized that obligate autotrophs are not necessarily unable to metabolize exogenously furnished organic substrates; many (possibly all) can take up certain organic compounds

*Table 19.1 Principal groups of aerobic Gram-negative chemoautotrophic true bacteria*

| Physiological group | Inorganic energy source | Chemoautotrophy | | Range of DNA base composition, mole % G+C | Structural properties | Taxonomic assignment |
|---|---|---|---|---|---|---|
| | | Facultative | Obligate | | | |
| *Hydrogen bacteria* | $H_2$; *some can also use CO* | + | − | *60–70* | *Rods, with polar or peritrichous flagella, some are immotile cocci; division by binary fission* | *Sometimes placed in a special genus, Hydrogenomonas; probably better assigned to several different genera among the chemoheterotrophs* |
| *Nitrifying bacteria* | $NH_3$ | − | + | 50–58 | *Various cell forms, polarly flagellated when motile; some reproduce by budding, others by binary fission* | *Several genera, distinguished on structural grounds* |
| | $NO_2$ | + | + | *60–61* | *Various cell forms, polarly flagellated if motile; some reproduce by budding, others by binary fission* | *Several genera, distinguished on structural grounds* |
| *Sulfur-oxidizers* | $H_2S$, S, $S_2O_3^{2-}$; *some can also use* $Fe^{2+}$ | + | + | 52–65 | *Small, polarly flagellated rods; division by binary fission* | *Thiobacillus* |
| *Iron bacteria* | | ? | + | Not determined | *Many different structural types* | *Gallionella and other genera* |

(e.g., acetate, amino acids) and use them in part for the synthesis of cellular material. *It is the inability of these organisms to use organic compounds as energy sources that confers on them the status of obligate autotrophs.*

The ability to oxidize reduced inorganic compounds is by no means confined to autotrophs. Many chemoheterotrophic true bacteria possess the enzymes necessary for oxidation of molecular hydrogen but are not classified as chemoautotrophs because of their failure to grow in a completely inorganic medium. Such failure presumably reflects the absence of key enzymes of the Calvin cycle, thus precluding the use of $CO_2$ as sole carbon source. Even the ability to oxidize reduced inorganic nitrogen compounds (ammonia and nitrite) is not an exclusive attribute of chemoautotrophs; some chemoheterotrophic actinomycetes can perform these oxidations, although far less efficiently than the physiologically specialized nitrifying bacteria.

## The hydrogen bacteria

In terms of number of species, the largest group of chemoautotrophs among true bacteria consists of the hydrogen bacteria, which can use $H_2$ as an energy source. They are without exception facultative chemoautotrophs, which can also grow well on a very wide range of simple organic compounds. Adaptation to the use of this particular inorganic energy source seems never to have been accompanied by a high degree of physiological specialization. Although sometimes placed in a special genus *Hydrogenomonas*, the hydrogen bacteria can be assigned to a variety of genera of chemoheterotrophic bacteria with which they share phenotypic properties; some, for example, can be placed in the genus *Pseudomonas*. This taxonomic treatment seems preferable for two reasons. First, certain hydrogen bacteria lose rather readily the capacity for chemoautotrophy when maintained on organic media; second, some hydrogen bacteria are almost identical phenotypically with chemoheterotrophic forms, isolable from nature by enrichment with organic substrates and lacking the capacity to grow as autotrophs. The hydrogen bacteria include polarly flagellated rods, peritrichously flagellated rods, and immotile cocci. A few of these bacteria can also grow chemoautotrophically with CO, an inorganic substrate not used by other chemoautotrophs.

## The nitrifying bacteria

The nitrifying organisms fall into two separate physiological groups: bacteria that oxidize ammonia to nitrite, and bacteria that oxidize nitrite to nitrate. Representatives of both groups were first isolated and described about 1890 by S. Winogradsky, who thus established the bio-

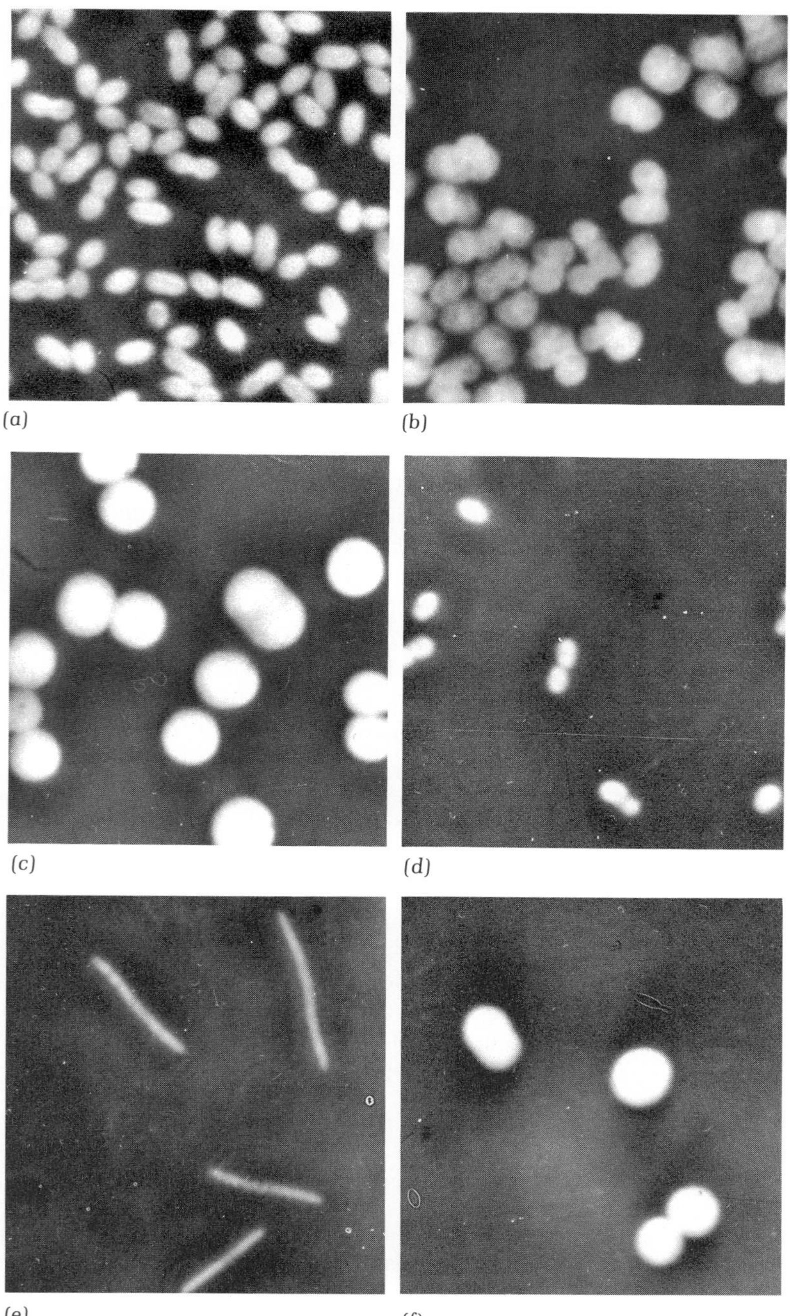


*Figure 19.1* Bright phase contrast photomicrographs of several species of nitrifying bacteria, to illustrate the diversity of cell form in this group (× 4,000). Ammonia-oxidizers: (a) *Nitrosomonas europea*; (b) *Nitrosolobus multiformans*; (c) *Nitrosocystis oceanus*. Nitrite-oxidizers: (d) *Nitrobacter* sp.; (e) *Nitrospina gracilis*; (f) *Nitrococcus mobilis*. Courtesy of S. W. Watson, Woods Hole Oceanographic Institution, Woods Hole, Mass.

logical nature of nitrification. This work at the same time provided a firm experimental foundation for Winogradsky's concept of chemoautotrophy as a mode of biological existence. Winogradsky was unable to demonstrate growth at the expense of organic compounds by any of the nitrifying strains that he isolated; and since that time, it has been generally believed that nitrifying bacteria are all obligate chemoautotrophs. This has been challenged recently for the nitrite oxidizer, *Nitrobacter*, which has been shown to grow with acetate in the absence of nitrite. Such growth is slow, relative to growth with nitrite; and there can be no doubt that *Nitrobacter*, like all other autotrophic nitrifying bacteria so far isolated, is very highly adapted to the use of an inorganic substrate as an energy source.

Even under optimal conditions, the growth of nitrifying bacteria is very slow, and the isolation of pure cultures is consequently a long and tedious process. For this reason, relatively few pure strains have been available, and the group is not well known.

The commonest ammonia oxidizer in soil is *Nitrosomonas*, first described by Winogradsky. The small, rod-shaped cells may be immotile or motile by a single polar flagellum; cell division occurs by binary fission. In liquid cultures, the cells tend to grow in zoogleal masses, held together by copious slime formation. Other ammonia oxidizers have been isolated, either from soil or from water. *Nitrosocystis* has spherical cells; *Nitrosouva* has peculiar, irregular cells, which reproduce by budding. In physiological respects, all these organisms are similar.

Among the nitrite oxidizers, the commonest soil form is *Nitrobacter*, also first described by Winogradsky. The pear-shaped cells reproduce by budding. *Nitrococcus*, a marine nitrite oxidizer, has spherical cells with a single flagellum and reproduces by binary fission. The various structural types are illustrated in Figure 19.1.

Nitrifying bacteria have a very high content of cytochrome c and other respiratory pigments, and electron microscopic studies reveal that most of these bacteria also possess elaborate internal membrane systems bearing the machinery of electron transport. The internal membrane structure is characteristic for each genus (Figure 19.2).

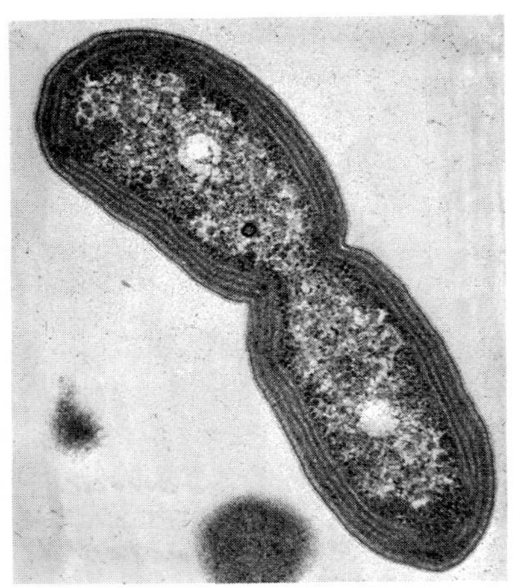

(a)

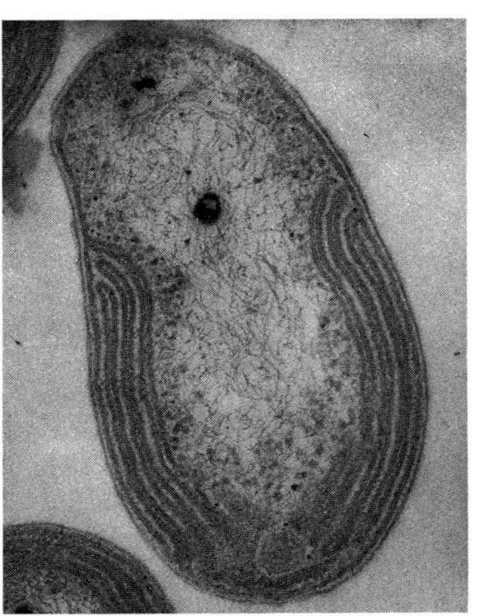

(b)

*Figure 19.2 Electron micrographs of thin sections of two nitrifying bacteria, showing the complex internal membranes characteristic of most of these organisms. (a) Nitrosomonas marina (× 30,000); (b) Nitrobacter agilis (× 51,250). Courtesy of S. W. Watson, Woods Hole Oceanographic Institution, Woods Hole, Mass.*

## The thiobacilli

The chemoautotrophic sulfur-oxidizing true bacteria are all small, polarly flagellated rods, which multiply by binary fission. They are placed in a single genus, *Thiobacillus*. All can grow at the expense of elemental sulfur, and many can use thiosulfate as well (Figure 19.3). Their growth in mineral media is rapid relative to that of nitrifiers; generation times as short as 4 to 5 hours have been recorded. For these reasons, the thiobacilli have been studied far more extensively than any other group of chemoautotrophs.

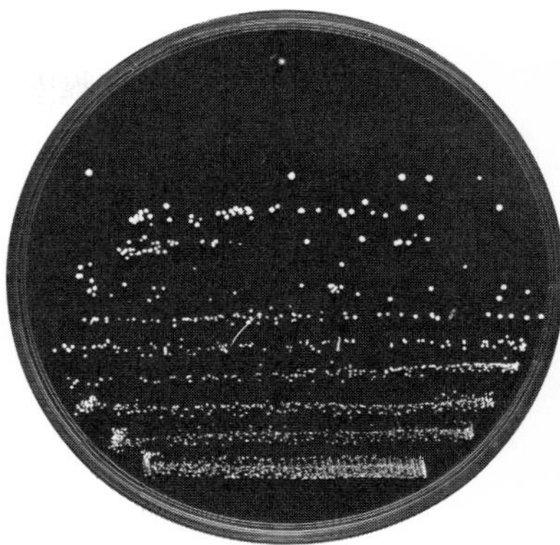

*Figure 19.3 Thiobacillus thioparus colonies growing on a mineral agar plate with thiosulfate as the energy source. The colonies are pale yellow and very refractile, as a consequence of the deposition of elemental sulfur among the cells.*

Some species are obligate chemoautotrophs, whereas others can be grown, more or less readily, with organic substrates. In fact, this group contains an interesting spectrum of physiological types, covering the range from highly specialized chemoautotrophs to relatively unspecialized ones. The three principal species of obligate chemoautotrophs are *Thiobacillus thiooxidans*, *T. thioparus*, and *T. denitrificans*. *Thiobacillus thiooxidans* is notable for its extreme tolerance of acidity. The oxidation of either sulfur or thiosulfate to sulfate is accompanied by a considerable generation of hydrogen ions; in cultures with either substrate, *T. thiooxidans* will continue to grow until the pH has fallen to approximately 1.0, and at least some strains remain active at a pH below zero. *Thiobacillus thioparus*, on the other hand, grows best under neutral conditions, and its development ceases at a pH of approximately 5. Both these species are dependent on $O_2$ as an electron acceptor. *Thiobacillus denitrificans* can grow either aerobically or under anaerobic conditions, using denitrification as a mechanism for the anaerobic oxidation of sulfur or thiosulfate. This species is unable to perform assimilatory nitrate reduction and therefore must be furnished with ammonia as a nitrogen source, even when grown with nitrate as terminal electron acceptor.

Another special physiological type, *T. ferrooxidans*, has been isolated from acid mine waters having a high content of ferrous salts. This bacterium can oxidize ferrous ions in addition to reduced inorganic sulfur compounds. Because it is acid tolerant, capable of supporting a pH of 3 to 3.5, it can be grown aerobically under conditions where dissolved ferrous ions are not subject to rapid spontaneous oxidation, as is the case in neutral or alkaline solutions. Hence, it has been possible to show unambiguously that *T. ferrooxidans* can obtain the energy necessary for its growth from the reaction

$$4Fe^{2+} + O_2 + 4H^+ \rightarrow 4Fe^{3+} + 2H_2O$$

It is therefore a chemoautotroph which can use two quite different inorganic energy sources, reduced sulfur compounds and $Fe^{2+}$.

## The iron bacteria

Certain freshwater ponds and springs have a high content of reduced iron salts. It has long been known that a distinctive bacterial flora is associated with such habitats. These *iron bacteria* form natural colonies that are heavily encrusted with ferric oxide. However, since most iron springs are neutral or alkaline, the ferrous ions contained in them undergo rapid spontaneous oxidation in contact with the atmosphere, so it has proved very difficult to ascertain the role which irons plays in the metabolism of these bacteria. The most conspicuous iron bacteria are filamentous, ensheathed bacteria of the *Sphaerotilus* group, in many of which the sheaths are encrusted with iron oxide. They can be readily grown as chemoheterotrophs and so isolated in pure culture. Although it is possible to show that such pure cultures will accumulate iron oxide on the sheaths, there is no evidence that such deposition is a

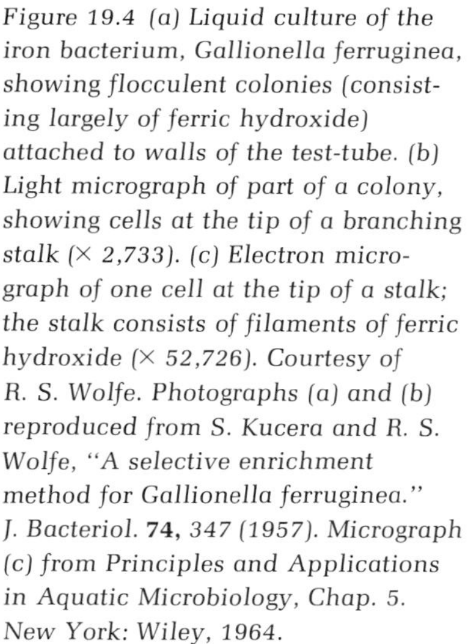

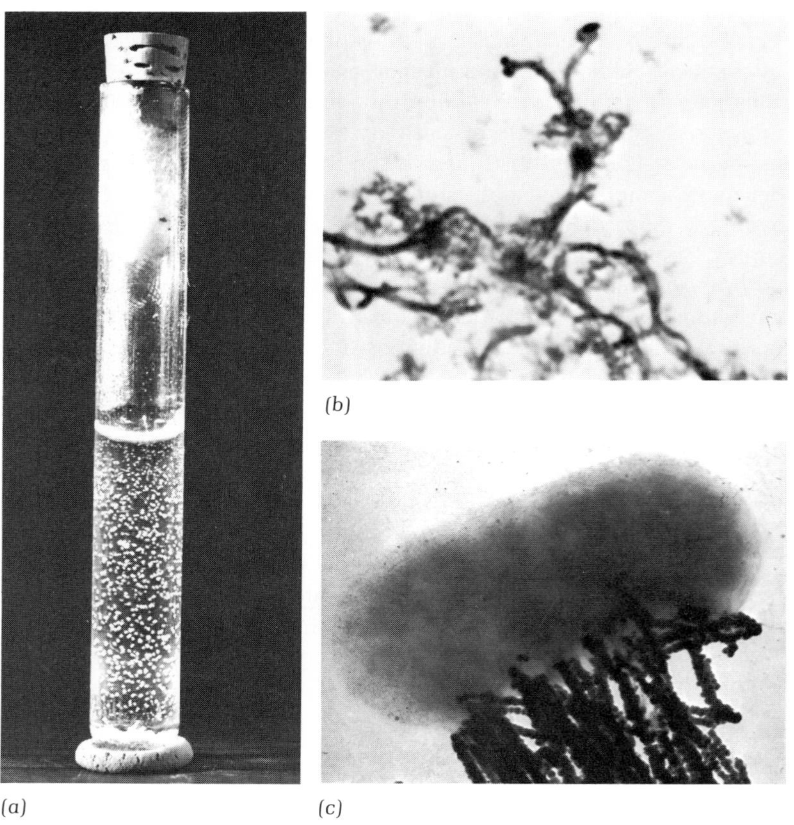

(a) (b) (c)

*Figure 19.4 (a) Liquid culture of the iron bacterium, Gallionella ferruginea, showing flocculent colonies (consisting largely of ferric hydroxide) attached to walls of the test-tube. (b) Light micrograph of part of a colony, showing cells at the tip of a branching stalk (× 2,733). (c) Electron micrograph of one cell at the tip of a stalk; the stalk consists of filaments of ferric hydroxide (× 52,726). Courtesy of R. S. Wolfe. Photographs (a) and (b) reproduced from S. Kucera and R. S. Wolfe, "A selective enrichment method for Gallionella ferruginea." J. Bacteriol.* **74,** *347 (1957). Micrograph (c) from Principles and Applications in Aquatic Microbiology, Chap. 5. New York: Wiley, 1964.*

physiologically significant process or that these bacteria can develop as chemoautotrophs. Chemoautotrophic growth at the expense of iron oxidation seems more probable in the case of another structurally distinctive iron bacterium, *Gallionella*. All attempts to obtain cultures of *Gallionella* in organic media have failed, but it has been grown artificially (although not in the pure state) in a mineral medium containing a deposit of ferrous sulfide as a source of reduced iron. The use of this virtually insoluble ferrous salt minimizes the problem of iron autoxidation under neutral conditions. In these cultures, *Gallionella* will form cottony colonies attached to the wall of the vessel. Much of the "colony" is inorganic: the small, bean-shaped bacterial cells are located at the branched tips of the excreted stalks, which are heavily impregnated with ferric oxide (Figure 19.4).

**Aerobic bacteria that utilize organic one-carbon compounds**

Although they are, technically speaking, chemoheterotrophs, the aerobic Gram-negative bacteria that use organic one-carbon compounds (methane, methanol, methylamine, formate) constitute a very highly specialized physiological group unlike other chemoheterotrophs and having some resemblances to the chemoautotrophs (Table 19.2).

The ability to grow at the expense of the gas methane ($CH_4$) is confined to a few Gram-negative rods and cocci, of which the best known is a polarly flagellated rod, *Methanomonas methanica*. The only organic compounds that support the growth of this organism are methane and its oxidation product, methanol ($CH_3OH$). Some higher homologs of methane, such as ethane ($C_2H_6$) and propane ($C_3H_8$), can be oxidized but do not support growth. Biochemical studies have shown that the methane oxidizers derive most of their cell carbon from methane, through a unique biosynthetic pathway which involves the formation of a special hexose phosphate, allulose phosphate, from formaldehyde units.

*Table 19.2 Principal types of aerobic bacteria able to use organic one-carbon compounds for growth*

| **Physiological type** | **Growth with:** $CH_4$ | $CH_3OH$ | $CH_3NH_2$ | $HCOOH$ | **Growth with multi-carbon organic compounds** | **Examples** |
|---|---|---|---|---|---|---|
| *Methane-utilizing bacteria* | + | + | − | − | − | *Methanomonas methanica* |
| *Nonmethane utilizers* | − | + | + | + | + or − | *Hyphomicrobium* spp., *Pseudomonas* spp. |

The ability to use methanol for growth is somewhat more widespread, being possessed by a number of Gram-negative bacteria, including polarly flagellated rods (*Pseudomonas* spp.) as well as the prosthecate bacteria of the genus *Hyphomicrobium* (see p. 625). These bacteria cannot grow with methane, but use other $C_1$ compounds (methylamine, formate) and, in some cases, multicarbon organic compounds such as acetate and succinate. When growing with methanol or methylamine, they derive cell carbon from the organic substrate through another special biosynthetic pathway, distinct from that characteristic of *Methanomonas:* formaldehyde units are added to the amino acid glycine to form serine, which is in turn converted to pyruvic acid. However, when such bacteria grow with formate, cell carbon is synthesized from $CO_2$ through the Calvin pathway, as in autotrophs; formic acid serves exclusively as an energy source. Growth with formate is, therefore, semiautotrophic; it is formally comparable to the growth of hydrogen bacteria with hydrogen gas ($HCOOH = H_2 + CO_2$).

## *The aerobic Gram-negative chemoheterotrophs*

The biological spectrum of the aerobic, chemoheterotrophic, Gram-negative true bacteria is particularly wide, and the possible relationships between the constituent subgroups are far from clear. The taxonomic treatment adopted here is a provisional one, which might well be considerably modified by future work. For example, it is becoming evident that the mode of flagellar insertion (polar or peritrichous), long used as a primary character for the subdivision of these organisms, is not necessarily a good indication of internal relationships. Analyses of DNA base composition, which is a recently introduced character in bacterial taxonomy, have led in some cases to the recognition that groups previously united on the basis of common phenotypic characters have little in common at the genetic level.

*Figure 19.5 Phase contrast photomicrograph of cells of Pseudomonas aeruginosa (× 1,102). Inset (upper right): flagella stain of Pseudomonas stutzeri, showing the polar monotrichous flagellation characteristic of many Pseudomonas species (× 1,102). Courtesy of N. J. Palleroni.*

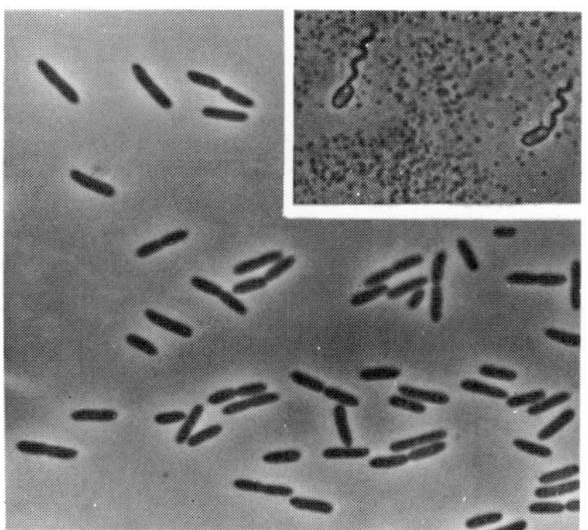

## *The aerobic pseudomonads*

One of the largest groups of strictly aerobic, chemoheterotrophic, Gram-negative bacteria consists of the aerobic pseudomonads (*Pseudomonas* and related genera). The cells are small straight or curved rods, not exceeding 0.8 $\mu$m in width, and motile by one or several polar flagella (Figure 19.5). Reproduction occurs by binary fission. The G + C content of the DNA is relatively high, ranging from 57 to 70 mole percent. The aerobic pseudomonads include bacteria that are among the most metabolically versatile organisms known, capable of using as many as 100 different organic compounds as sole sources of carbon and energy. In addition to the organic compounds commonly used by most aerobic bacteria (simple carbohydrates, amino acids, organic acids), members of this group can attack many more exotic organic compounds: terpenes, such as camphor; steroids, such as testosterone; and polycyclic aromatic compounds, such as naphthalene, to mention only a few. The nutrient requirements are simple, most species not requiring growth factors. Some aerobic pseudomonads are denitrifiers and can thus grow anaerobically by the use of nitrate as a terminal electron acceptor. Unless they possess this ability, they are strict aerobes.

The aerobic pseudomonads can be divided into a number of physiologically different subgroups, some of which are shown in Table 19.3. The forms most commonly encountered are members of the fluorescent group, which produce characteristic, water-soluble, yellow-green fluorescent pigments, particularly when grown in media with a low iron content. There are four principal species (Table 19.4). *Pseudomonas aeruginosa*, *P. putida*, and *P. fluorescens* are metabolically versatile free-living organisms which occur in soil and water. *Pseudomonas aeruginosa*, which has a markedly higher temperature range than the other two species, is occasionally pathogenic for man, causing hospital infections that are very difficult to treat, as a result of the resistance of this bacterium to antibiotic therapy. However, *P. aeruginosa* does not have the character of a true parasite; it is an "accidental pathogen," able to establish infection only in individuals with a severely lowered natural resistance, such as victims of severe burns or cancer patients who have been treated with highly toxic drugs. Since infections are not transmissible, development in animal hosts is a rare and discontinuous event in the life of the species.

The only truly parasitic members of the fluorescent group are the forms pathogenic for plants. Fluorescent pseudomonads are responsible for many plant diseases, and a large number of species have been named, generally on the basis of the specific plant hosts in which they have been found. Most of them are very similar, and should probably be regarded as varieties of one species, *P. syringae*. This organism can be readily distinguished from the three free-living fluorescent species: it is oxidase negative, does not produce arginine dihydrolase, attacks a much smaller range of organic compounds, and grows more slowly, even on complex media.

The two species of the pseudomallei group are of considerable ecological and evolutionary interest. *Pseudomonas pseudomallei* was discovered as the agent of a highly fatal tropical disease of man and

Table 19.3 *Major species and species groups among the aerobic pseudomonads*

| Group | DNA base composition, mole % G+C | Poly-β-hydroxybutyrate as reserve material | Pigments: Soluble fluorescent | Pigments: Phenazines | Pigments: Carotenoids | Number of polar flagella | Sensitivity to lysis by EDTA[a] | Growth-factor requirements | Chemoautotrophic growth with $H_2$ | Utilization of: Glucose | Utilization of: Starch | Utilization of: Aromatic compounds | Typical species |
|---|---|---|---|---|---|---|---|---|---|---|---|---|---|
| Fluorescent group | 60–67 | – | + | V[b] | – | 1 or > 1 | + | – | – | + | – | + | *P. aeruginosa* |
| Pseudomallei group | 67–69 | + | – | V | – | >1 or immotile | + | – | – | + | + | + | *P. pseudomallei* |
| Acidovorans group | 63–66 | + | – | – | – | >1 | + | – | – | – | – | + | *P. acidovorans* |
| Hydrogen-oxidizing pseudomonads | 61–69 | + | – | – | V | 1 | n.t.[c] | – | + | + | Rare | Rare | *P. saccharophila* |
| Diminuta group | 62–67 | + | – | – | V | 1 | – | +[d] | – | V | – | – | *P. diminuta* |
| *P. maltophilia* | 67 | – | – | – | – | >1 | – | +[e] | – | + | – | – | *P. maltophilia* (only species) |

[a]EDTA denotes ethylenediaminetetraacetic acid.
[b]V denotes a character that is variable for the group in question.
[c]n.t. denotes that the character has not been determined.
[d]Complex. [e]Methionine only.

Table 19.4 *Distinguishing characters of the major species of fluorescent pseudomonads*

| Species | G+C content of DNA, mole % | Production of pyocyanin | Maximal growth temp., °C | Oxidation of aliphatic hydrocarbons | Gelatin liquefaction | Denitrification | Oxidase reaction | Arginine dihydrolase |
|---|---|---|---|---|---|---|---|---|
| *P. aeruginosa* | 67 | + | 44 | + | + | + | + | + |
| *P. putida* | 62 | – | 35–37 | – | – | – | + | + |
| *P. fluorescens* | 60–63 | – | 35–37 | – | + | V[a] | + | + |
| *P. syringae* | 58–60 | – | 35–37 | – | + | – | – | – |

[a]V denotes variable character.

other animals, melioidosis. Even in areas where melioidosis is endemic (e.g., southeast Asia), it is a rare infection, characteristically acquired traumatically, through contamination of wounds with soil or mud. The bacterial agent is, in fact, a common soil and water inhabitant in tropical regions, and has the metabolic versatility characteristic of so many aerobic pseudomonads; it is also a powerful denitrifier. There can be little doubt, accordingly, that *P. pseudomallei* is a natural member of the soil and water population of the tropics, and, like *P. aeruginosa*, is an accidental pathogen. Human infections occur only in individuals with a lowered natural resistance, and are nontransmissible; the organism is therefore not a true parasite.

At the time of its isolation, *P. pseudomallei* was recognized to share many of the characters of a *nonmotile*, Gram-negative bacterium which is a true parasite: the agent of the disease of horses known as glanders. This bacterium, *P. mallei*, is now placed in the genus *Pseudomonas* despite its permanent immotility, because more detailed studies have confirmed its very close similarity in both phenotypic and genetic respects to *P. pseudomallei*. The two species are indistinguishable in DNA base composition; and their DNA homology is high, as shown by in vitro hybridization. In ecological respects, however, they differ greatly. *Pseudomonas mallei* exists only in the specific animal hosts; its geographic range is therefore a much wider one than that of *P. pseudomallei*, being coterminous with that of the horse. Apart from its permanent immotility, *P. mallei* differs from *P. pseudomallei* by its lesser degree of metabolic versatility and also by its much slower growth, both on synthetic and on complex media. There can be little doubt, however, that it is derived evolutionarily from a free-living ancestor of the *P. pseudomallei* type. *Pseudomonas mallei* provides a good illustration of a phenomenon that appears to be generally characteristic of evolution toward a parasitic mode of existence; such evolution is accompanied by *losses of functions* possessed by the free-living ancestral type.

### *Halobacterium*

The extreme halophiles of the genus *Halobacterium* are physiologically highly specialized organisms that resemble the aerobic pseudomonads in structure and DNA base composition. The cells of these bacteria are colored pink or red as a result of the presence of carotenoid pigments. The group was discovered as a result of the ability of its members to grow on the surfaces of salted fish and salted animal hides, producing red discolorations. These bacteria occur in brine ponds, used for the manufacture of salt from seawater by solar evaporation, and can be isolated from salt produced by this method, in which they remain

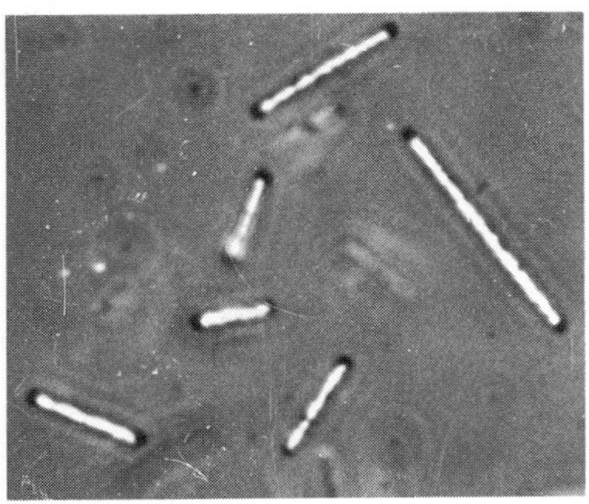

*Figure 19.6 Phase contrast photomicrograph of an extreme halophile, Halobacterium sp. (× 1,956). The bright, refractile areas in each cell are caused by the presence of numerous gas vesicles. Courtesy of Germaine Cohen-Bazire.*

viable for some time. They are also present in natural bodies of water with an extremely high salt content, such as the Dead Sea and the Great Salt Lake.

*Halobacterium* has a high and absolute requirement for NaCl, which is not replacable by any other salt. The minimal concentration that supports growth is about 15 percent NaCl, and growth is optimal at about 25 percent (4 *M*). A relatively high concentration of $Mg^{2+}$ (about 0.1 *M*) is also essential. The nutritional requirements are complex; these bacteria are best grown on peptone or yeast extract media supplemented with the necessary salts. Growth is slow, and colonies become visible on plates only after incubation for 5 to 7 days.

Grown under optimal conditions, *Halobacterium* has rod-shaped cells, motile by polar flagella; many strains contain gas vacuoles (Figure 19.6). At lower salt concentrations, the cells become spherical, lysing when suspended in solutions containing 10 percent or less NaCl. The cell walls contain no detectable murein, consisting only of protein and lipid. The maintenance of the integrity of the wall requires a high concentration of salt, and it is probably this fact which accounts for the conversion of the rods to spheres as the salt concentration is lowered.

The extreme physiological specialization of these bacteria is revealed by the properties of their enzymes in cell-free extracts. Enzyme activity is dependent on a high NaCl concentration (optimal 2 *M*), and most isolated enzymes from *Halobacterium* are rapidly and irreversibly denatured in the absence of salt. Adaptation to growth with very high levels of salt has, consequently, led to a modification of the properties of all or most of the cell proteins.

## *The acetic acid bacteria*

The acetic acid bacteria comprise an ecological group, consisting of Gram-negative, rod-shaped bacteria which grow characteristically in plant materials that have undergone an alcoholic fermentation; these bacteria convert the alcohol to acetic acid. In addition to their ability to form relatively large amounts of acetic acid from ethanol, these organisms show a degree of acid tolerance that is rare among aerobic chemoheterotrophs; they can grow in media with an initial pH of 4.5. Their cultivation is most easily effected in complex media supplemented with one of the readily oxidizable substrates used by the group: ethanol or a soluble sugar, such as glucose.

The acetic acid bacteria are subdivided into two genera: the polarly flagellated *Gluconobacter* and the peritrichously flagellated *Acetobacter* (Table 19.5).

*Table 19.5 Distinguishing properties of the two groups of acetic acid bacteria*

| Genus | Range of DNA base composition, mole % G + C | Mode of flagellar insertion | Ability to oxidize acetate |
|---|---|---|---|
| *Gluconobacter* | 60–62 | *Polar* | – |
| *Acetobacter* | 56–62 | *Peritrichous* | + |

The genus *Gluconobacter* contains only one species, *G. suboxydans*. This organism has a remarkable metabolic defect for a strict aerobe: it lacks a functional tricarboxylic acid cycle. As a result, it oxidizes ethanol stoichiometrically to acetic acid; other substrates, such as lactate, which are oxidized through pyruvate, are also converted stoichiometrically to acetic acid. Sugars can be oxidized more extensively, through the reactions of the pentose cycle. Glucose is also degraded by an alternative pathway, involving oxidation through gluconic acid to ketogluconic acids. Certain strains (sometimes placed in a separate species, *G. melanogenum*) produce a dark-brown discoloration of the medium when grown with glucose as a result of the production of a highly unstable diketo derivative, 2,5-diketogluconic acid, which decomposes spontaneously to colored products.

The nutrition of *Gluconobacter* is complex, several vitamins of the B group being required. Growth in a defined medium with ethanol or lactate as the oxidizable substrate also requires the presence of a small amount of a utilizable sugar, to provide carbon for biosynthesis, which cannot be obtained from the oxidation of ethanol or lactate to acetic acid.

The members of the genus *Acetobacter*, which contains several species, possess the machinery of the tricarboxylic acid cycle and can therefore oxidize acetate. Although acetic acid accumulates in cultures growing with ethanol, this accumulation is transient; eventually, the acetic acid is further oxidized to $CO_2$. For this reason, the *Acetobacter* group are sometimes termed "overoxidizers," in contrast to the "underoxidizers" of the genus *Gluconobacter*, which cannot further oxidize the acetic acid that they form from ethanol.

The species *Acetobacter xylinum* is distinguished by its ability to form an extracellular slime layer of cellulose, when grown with glucose or other sugars. The fibrils of cellulose form a loose mesh enclosing the cells (Figure 19.7). In stationary liquid cultures, this bacterium forms a very thick, tough pellicle, often 1 cm or more in depth, composed of cellulose and bacterial cells.

*Figure 19.7 Extracellular formation of cellulose by Acetobacter xylinum (× 1,213). The bacterial cells are entangled in a mesh of cellulose fibrils. From J. Frateur, "Essai sur la systématique des acétobacters." La Cellule* **53,** *3 (1950).*

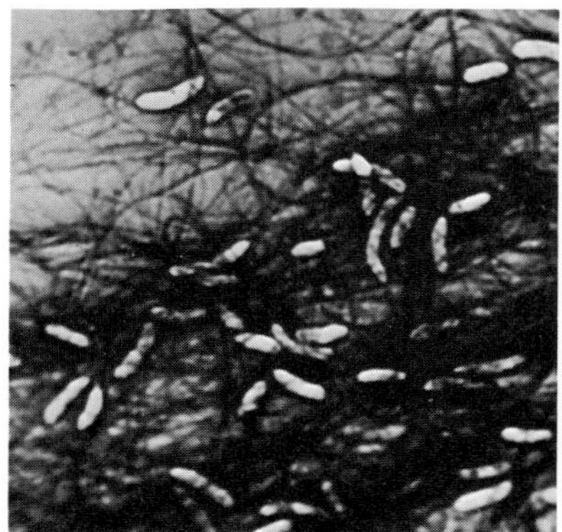

Characteristic to some degree of all acetic acid bacteria, both *Glu-*

*conobacter* and *Acetobacter*, is the oxidation of a number of organic substrates with the accumulation of partially oxidized metabolic products. Such reactions include the oxidation of glucose to sugar acids and the oxidation of polyalcohols to ketosugars (e.g., mannitol to fructose, glycerol to dihydroxyacetone).

## *The Azotobacter group*

The primary defining character of the aerobic chemoheterotrophs of the *Azotobacter* group is their ability to fix large amounts of $N_2$ (as much as 10 mg/g of organic substrate respired). The capacity for nitrogen fixation under aerobic conditions distinguishes them from other free-living, nonphotosynthetic nitrogen-fixing bacteria (*Bacillus* and *Clostridium* spp., *Aerobacter*) which fix nitrogen effectively only under anaerobic growth conditions or at very low oxygen tensions. Together with certain filamentous blue-green algae, members of the *Azotobacter* group are the principal agents of aerobic nitrogen fixation in soil and water (Table 19.6).

The cells of the *Azotobacter* group are plump rods or cocci, characteristically paired, and often containing conspicuous granules of poly-$\beta$-hydroxybutyrate, the characteristic reserve material. Motility is a variable character; motile species are either polarly or peritrichously flagellated. Most of these organisms produce gummy extracellular polysaccharides, giving their colonies a mucoid appearance (Figure 19.8). Two principal genera are recognized, primarily on physio-

*Table 19.6 Characteristics of the principal species of the Azotobacter group*

| Species | Flagellation | Micro-cyst forma-tion | DNA base compo-sition, mole % G+C | Growth at pH 4.5 | Pigments | | Utilization of: | | Typical habitat | |
|---|---|---|---|---|---|---|---|---|---|---|
| | | | | | Cellular | Soluble | Rhamnose | Aromatic acids[a] | Soil | Water |
| *Azotobacter chroococcum* | Peritrichous | + | 66 | − | Brown | − | − | + | + | − |
| *Azotobacter beijerinckii* | Nonmotile | + | 66 | − | Yellow or − | − | − | + | + | − |
| *Azotobacter vinelandii* | Peritrichous | + | 66 | − | − | Green | + | + | − | + |
| *Azotobacter agilis* | Peritrichous | − | 53 | − | − | Green | − | − | − | + |
| *Azotobacter macrocytogenes* | Polar | − | 58–59 | − | − | Pink | − | − | + | − |
| *Beijerinckia indica* | Peritrichous | − | 54–59 | + | − | − | − | − | +[b] | − |

[a]Benzoate or p-hydroxybenzoate. [b]Largely tropical.

logical and ecological grounds (Table 19.6). Members of the genus *Azotobacter* grow under neutral or alkaline conditions, the minimal pH for their development being about 6.5. They are found throughout the world in neutral or alkaline soils and water. Members of the genus *Beijerinckia* have a much wider pH range, developing at pH values as low as 3. *Beijerinckia* spp. occur principally in the tropics, as a result of their confinement to a specific kind of soil: the acidic reddish lateritic soils that are characteristic for many tropical regions of the earth and of rare occurrence in temperate areas.

The cells of *Azotobacter* are relatively large, plump rods, attaining a width of as much as 4 μm in *Azotobacter agilis.* Some species are immotile; others are polarly or peritrichously flagellated. Three species produce distinctive resting cells termed *microcysts.* These are formed by the synthesis of a thick, multilayered wall around the vegetative cell (Figure 19.9); they are resistant to desiccation but not notably heat resistant. Microcyst formation is not accompanied by the extensive changes in enzymatic constitution that occur during the formation of endospores in the spore-forming bacteria; the microcysts appear to retain the enzymatic constitution of the vegetative cells from which they develop, although enzyme levels are somewhat lower. In both structural and physiological respects, the microcysts of *Azotobacter* resemble the microcysts of gliding organisms (*Myxococcus* and *Sporocytophaga*) rather than the endospores that are formed by other true bacteria.

*Azotobacter* spp. can utilize a limited range of simple organic compounds as carbon and energy sources. The principal substrates are

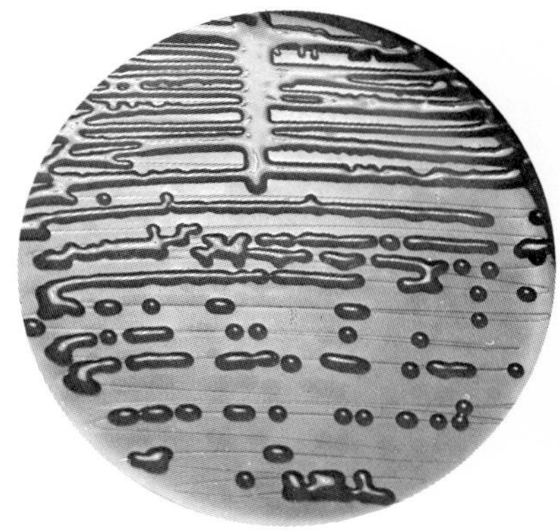

*Figure 19.8 A streaked plate of Azotobacter vinelandii, showing the smooth, glistening colonies typical of Azotobacter. Courtesy of O. Wyss; reproduced from his Elementary Microbiology. New York: Wiley, 1963.*

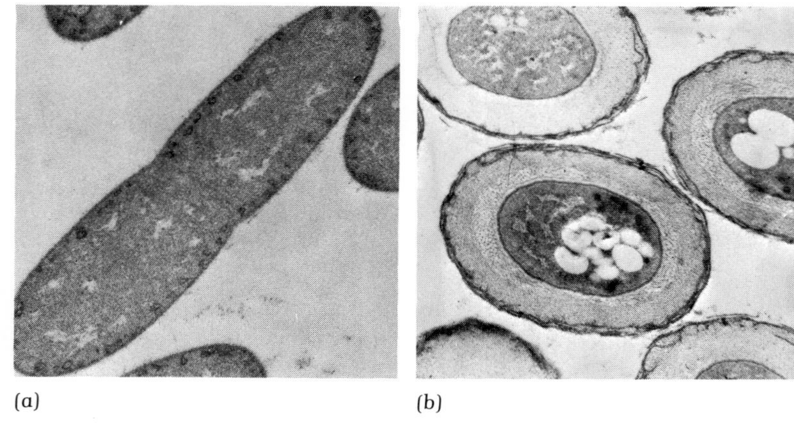


*Figure 19.9 Electron micrographs of thin sections of vegetative cells and cysts of Azotobacter vinelandii. (a) Vegetative cells (× 16,450). (b) Cysts (× 9,870). Both from O. Wyss, M. G. Neumann, and M. D. Socolofsky, "Development and germination of the Azotobacter cyst." J. Biophys. Biochem. Cytol.* **10,** *555 (1961).*

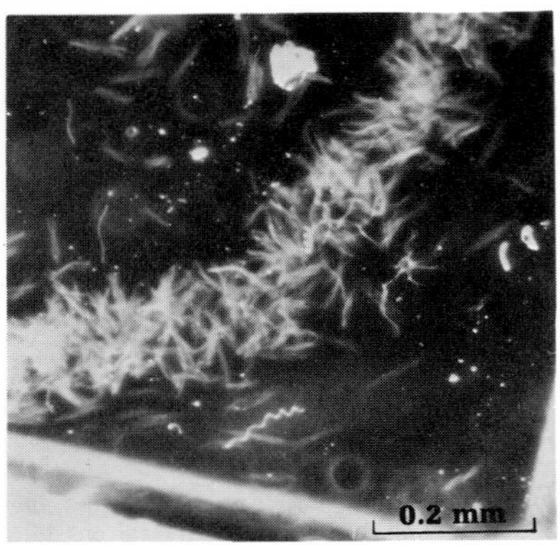


*Figure 19.10 Typical aerotactic pattern of Spirillum in a wet mount, showing the accumation of the cells in a very narrow band some distance from the air-water interface. From N. R. Krieg, "Cultivation of Spirillum volutans in a bacteria-free environment." J. Bacteriol.* **90,** *817 (1965).*

*Figure 19.11 A single cell of the largest spirillum, S. volutans (phase contrast). Note the flagellar tufts at one pole of the cell. From N. R. Krieg, "Cultivation of Spirillum volutans in a bacteria-free environment." J. Bacteriol.* **90,** *817 (1965).*

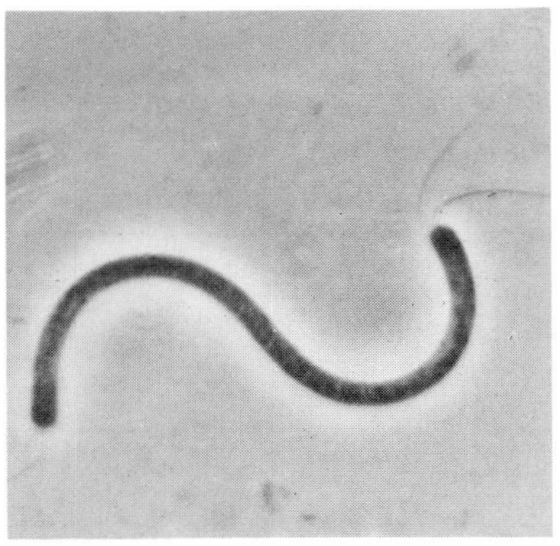

sugars, organic acids, and primary alcohols; some species also use a few aromatic compounds, such as benzoate. Growth factors are not required, and growth is poor in complex media such as peptone or yeast extract. Inorganic forms of combined nitrogen (ammonia and nitrate) support growth at rates equal to, or more rapid than, the rate of growth with $N_2$ as a nitrogen source. However, enrichment experiments show that when a source of combined nitrogen is present, *Azotobacter* cannot compete successfully with non-nitrogen-fixing chemoheterotrophic aerobes that utilize the same types of organic substrates (e.g., aerobic pseudomonads). Successful enrichment of *Azotobacter* spp. from soil or water is strictly dependent on the use of a medium free of combined nitrogen. The development of these organisms in nature is therefore confined to nitrogen-poor environments, where the ability to fix nitrogen confers a decisive advantage.

## The Spirillum group

Chemoheterotrophic bacteria with helical cells, bearing a tuft of 10 to 30 flagella at one or both poles, are placed in the genus *Spirillum*. The spirilla are aquatic organisms, which occur in both freshwater and marine environments where plant materials are undergoing decomposition. The freshwater forms can be enriched by adding sterilized hay to a fairly large volume of pond water (~200 ml), and allowing the infusion to incubate for several days at 25 to 30°C. The marine forms can be enriched in infusions of algal material.

The spirilla use a limited range of organic compounds, particularly amino acids and organic acids, as respiratory substrates. With the exception of *S. itersonii*, which is capable of denitrification, they grow only under aerobic conditions. Some species are microaerophilic. This is evident from the accumulation pattern of the highly motile cells in wet mounts; they accumulate in a narrow band at some distance from the air–water interface at the edge of the cover slip (Figure 19.10). Although most of these bacteria can be readily isolated by streaking on nutrient agar with a low peptone concentration (0.2 percent), *S. volutans*, the largest member of the group and one of the largest known bacteria, long resisted attempts at isolation. This organism, the cells of which are 1.5 $\mu$m wide and as much as 40 $\mu$m long (Figure 19.11), is considerably more sensitive to oxygen than the other species and can initiate growth only in cultures exposed to an oxygen-poor atmosphere, containing from 3 to 9 percent $O_2$. Hence, despite the fact that it is an obligate aerobe, it cannot form colonies on plates that have been exposed to air.

*Bdellovibrio* **and** *Campylobacter*

Two genera of parasitic bacteria, *Bdellovibrio* and *Campylobacter*, although not necessarily related, share a number of properties which set them apart from other groups of polar flagellates. The very small curved or helical cells (0.3 to 0.4 $\mu$m wide) are motile by means of a single, rather thick polar flagellum (Figure 19.12). The *Campylobacter* spp. (and possibly some *Bdellovibrio* strains) are microaerophiles, growing best at low oxygen tensions. The nutritional requirements are complex, and the physiological properties are poorly known. The G+C content of the DNA is low (30 to 35 mole percent).

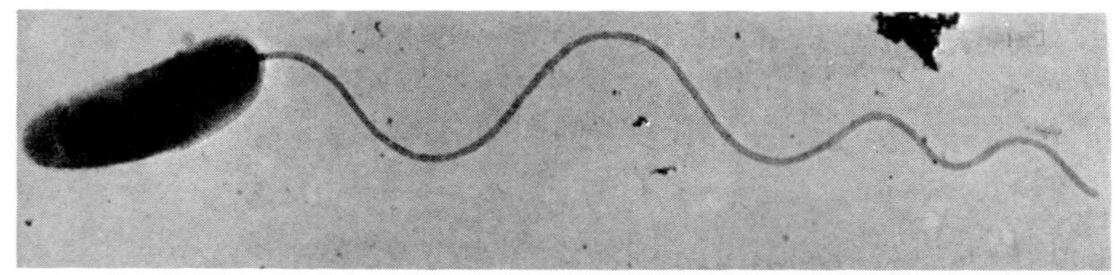

*Figure 19.12 Electron micrograph of a cell of Bdellovibrio, showing the unusually thick single polar flagellum (uranyl acetate stain;* × *24,338). From R. J. Seidler and M. P. Starr, "Structure of the flagellum of Bdellovibrio bacteriovorus." J. Bacteriol.* **95,** *1952 (1968).*

*Bdellovibrio bacteriovorus* is parasitic on other Gram-negative bacteria. The name (derived from the Latin *bdellus*, a leech) refers to the characteristic mode of attack on other bacteria. When a moving *Bdellovibrio* cell impinges on a susceptible host bacterium, it usually adheres tenaciously. After some time, the host cell rounds up, eventually lysing (Figure 19.13). Electron microscopic studies have shown that the *Bdellovibrio* penetrates the host cell wall, and subsequently grows inside the host cell, although without penetrating into the cyto-

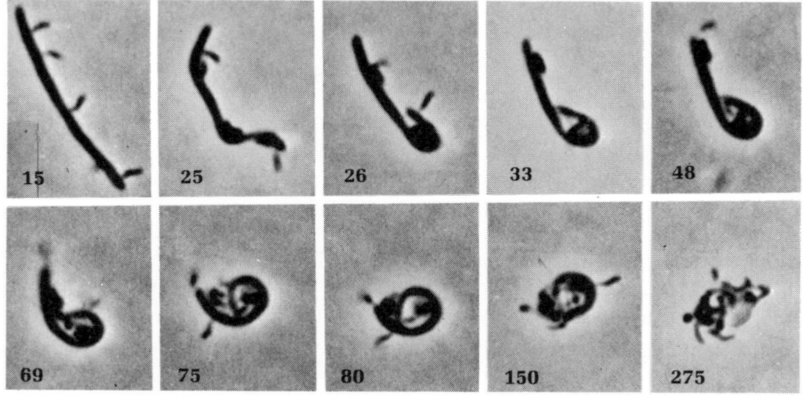


*Figure 19.13 Sequence of phase contrast photomicrographs illustrating the attack of Bdellovibrio on Pseudomonas tabaci. The time in minutes after mixing suspensions of host and predator is indicated on each photograph. After 15 minutes, several small Bdellovibrio cells are attached to the host cell, which still has its normal shape. Thereafter, the host cell undergoes progressive distortion, culminating in complete lysis after 275 minutes. From M. P. Starr and N. L. Baigent, "Parasitic interaction of Bdellovibrio bacteriovorus with other bacteria." J. Bacteriol.* **91,** *2006 (1966).*

Figure 19.14 *Sequence of electron micrographs of thin sections illustrating the interaction of* Bdellovibrio *with a susceptible host bacterium,* Erwinia amylovora *(× 37,600). (a) Uninfected host cell. (b) Attachment of* Bdellovibrio. *(c) Penetration of* Bdellovibrio *through cell wall of host. (d) A late stage of infection. From M. P. Starr and N. L. Baigent, "Parasitic interaction of* Bdellovibrio bacteriovorus *with other bacteria."* J. Bacteriol. **91,** *2006 (1966).*

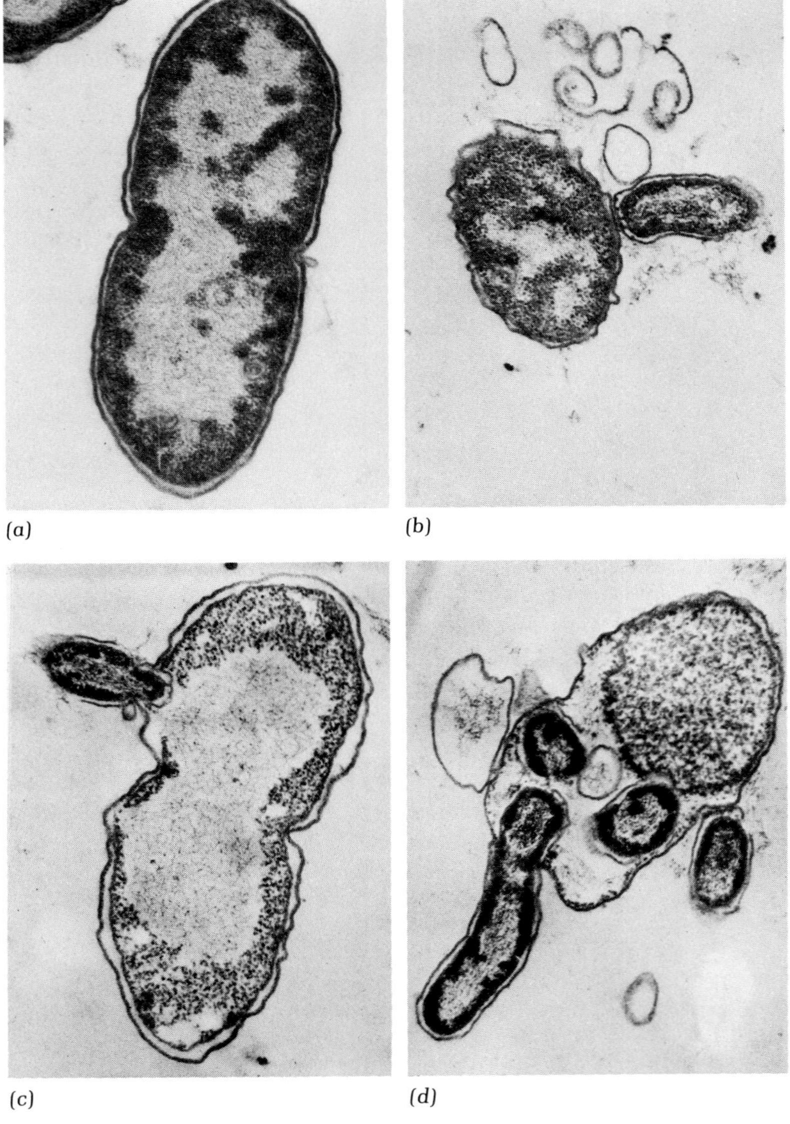


plasm (Figure 19.14). On the surface of a plate covered with a growth of host cells, *Bdellovibrio* will form lytic plaques, superficially similar to those produced by a phage infection (Figure 19.15). In fact, this remarkable bacterium, discovered only in 1962, was isolated during a search for phages active against pseudomonads. The host range is wide,

Figure 19.15 *Macroscopically visible lysis of susceptible bacteria by Bdellovibrio. (a) Plaque formation in a lawn of Pseudomonas putida (poured plate). (b) Partial lysis of surface colonies of Escherichia coli on a nutrient agar plate streaked with a nixture of host and parasite. From H. Stolp and M. P. Starr, "Bdellovibrio bacteriovorus gen. et. sp. n., a predatory, ectoparasitic, and bacteriolytic microorganism." Antonie van Leeuwenhoek* **29**, *217 (1963).*

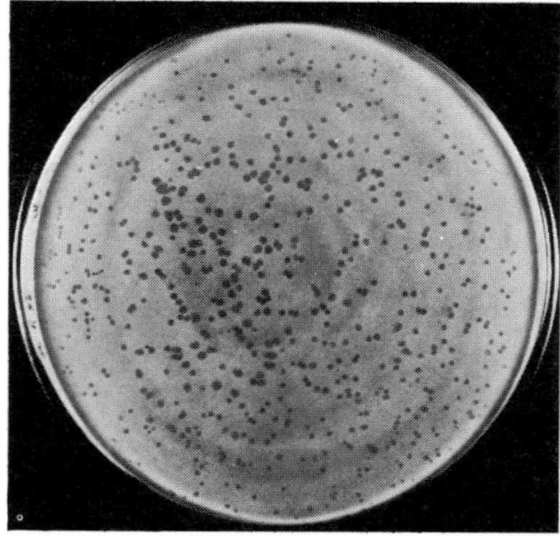

(a)

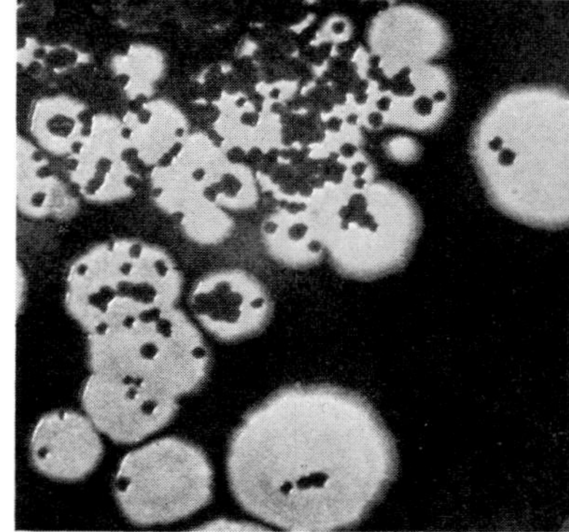

(b)

including pseudomonads and bacteria of the enteric type. It has proved possible to obtain a few strains in pure culture, on complex media; most isolates, however, have so far been grown only in association with bacterial hosts. The group is common in soil, where *Bdellovibrio* populations as large as $10^5$ per gram have been recorded.

The *Campylobacter* group have a completely different ecology. They are parasites of man and animals, occurring in either the mouth or the genital tract. *Campylobacter sputorum* is a common inhabitant of the human mouth, while *C. fetus* is a pathogen for cattle. These bacteria have complex nutritional requirements and are usually cultivated on complex media. They are markedly microaerophilic, many strains growing only at reduced oxygen tensions. All members of the group are denitrifiers, and even in the presence of oxygen, growth is considerably increased by the inclusion of $KNO_3$ in the medium. *Campylobacter sputorum* is catalase negative, but contains cytochromes, a unique combination among aerobic bacteria.

## The Moraxella–Neisseria group

The organisms in the *Moraxella–Neisseria* group are nonflagellated short rods or cocci, with a DNA base composition in the range of 40 to 50 mole percent G + C. There are three genera, the distinguishing characters of which are shown in Table 19.7. In *Neisseria*, the cells are

*Table 19.7 Properties of the Moraxella–Neisseria group*

| | Cell shape | | | | | | |
|---|---|---|---|---|---|---|---|
| **Genus** | **Exponential phase** | **Stationary phase** | **Range of DNA base composition, mole % G + C** | **Oxidase reaction** | **Sensitivity to penicillin (10 units/ml)** | **Growth factors required** | **Ability to utilize carbohydrates** |
| *Neisseria* | Cocci | Cocci | 48–52 | + | + | Yes | + |
| *Moraxella* | Rods | Cocci or shortened rods | 40–46 | + | + | Variable | − |
| *Acinetobacter* | Rods | Cocci or shortened rods | 40–46 | − | − | No | − |

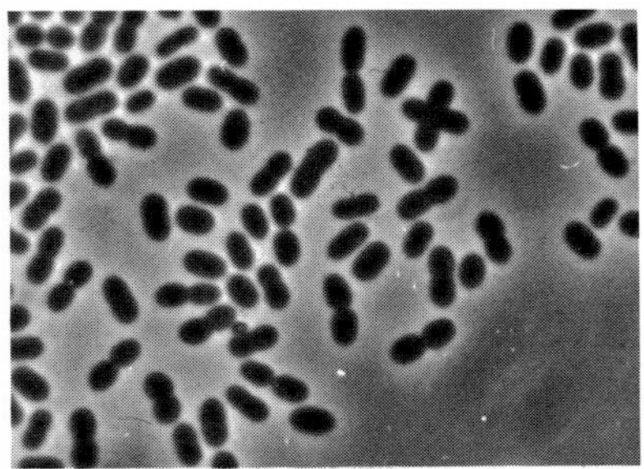

(a)

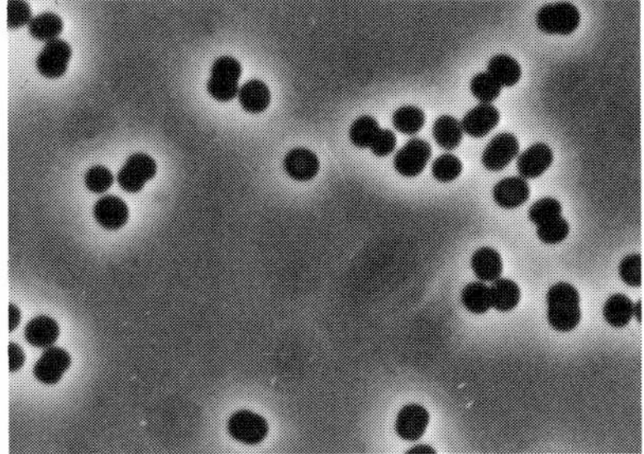

(b)

*Figure 19.16 The Moraxella–Neisseria group (phase contrast, × 2,000). (a) Moraxella osloensis. (b) Neisseria catarrhalis.*

always spherical, wheras in *Moraxella* and *Acinetobacter* they are plump, paired rods during exponential growth (Figure 19.16). After entry into the stationary phase, the cells of the two latter genera often become spherical, so that these organisms can easily be mistaken for cocci. The members of *Neisseria* and *Moraxella* are extremely sensitive to penicillin, growth of most strains being inhibited by 1 unit per milliliter, whereas for other Gram-negative bacteria the inhibitory concentration generally lies between 100 and 1,000 units per milliliter.

The genera *Neisseria* and *Moraxella* consist exclusively of animal parasites, commonly present in the nasopharynx. Several species are pathogenic: *N. meningitides* and *N. gonorrheae* are agents of meningitis and gonorrhea respectively, while *M. lacunata* is the agent of conjunctivitis. Most of these bacteria are nutritionally exacting and use a rather limited range of organic compounds as carbon and energy sources.

Organisms of the genus *Acinetobacter* are common members of the microflora of soil and fresh water. They do not require growth factors and are nutritionally highly versatile, able to utilize a range of simple organic compounds almost as extensive as the range used by the aerobic pseudomonads. They appear to play a role analogous to that of the aerobic pseudomonads in the mineralization of organic matter and can be enriched by similar techniques; however, they can tolerate somewhat more acidic conditions than the aerobic pseudomonads and can be selectively isolated by the use of enrichment media containing acetate as sole carbon source at an initial pH of 5.5 to 6.

## *Gram-negative bacteria that form distinctive cell aggregates*

Among the aerobic Gram-negative chemoheterotrophs, three groups are distinguished by the formation of distinctive cell aggregates. *Sphaerotilus* is characterized by the formation of tubular sheaths, enclosing chains of cells. *Zoogloea* forms large flocs, consisting of many cells held together by a gelatinous slime layer. In *Lampropedia*, the spherical cells are arranged in very regular rectangular sheets.

### The *Sphaerotilus* group

The members of the *Sphaerotilus* group are motile rods, which bear a polar tuft of 10 to 20 flagella. Their distinguishing structural feature, which differentiates them from all other Gram-negative polar flagellates, is the formation of a tubular sheath, which encloses the rods in loose chains (Figure 19.17). Reproduction occurs by the liberation of motile cells from the open ends of the sheath. The sheaths are chemically complex, containing the characteristic major components also found in Gram-negative cell walls: proteins, polysaccharides, and lipids. They can, therefore, be interpreted as modified wall structures, synthesized by and common to a chain of cells rather than associated specifically with individual cells. Labeling experiments with fluorescent antibodies have shown that synthesis of new sheath material takes place exclusively at the growing tip of the sheath; presumably, therefore, it is produced and extended by the terminal cell in the chain.

The typical habitat is fresh water. *Sphaerotilus natans* develops abundantly in streams that are polluted with organic wastes, where it grows as long, slimy tassels attached to submerged stones and plants. It is also a conspicuous component of the microflora in aerobic sewage treatment plants using the activated-sludge process. Indeed, this bacterium can be regarded as one of the best biological indicators of the pollution of water with organic wastes.

Other members of the group, such as *S. discophorus*, belong to the ecological group of iron bacteria; the sheaths are usually thick and golden brown in color, as a result of heavy incrustation with iron and manganese oxides. Some authors place these forms in a separate genus, *Leptothrix*.

Although all members of the group, including *S. natans*, can deposit ferric oxide in the sheath, *S. discophorus* and related species can also oxidize manganous compounds to the dark-brown manganic oxide. This character is readily evident from the color of colonies grown on plates of a medium supplemented with $MnCO_3$. *Sphaerotilus natans* lacks the ability to perform this oxidation (Figure 19.18). Experiments

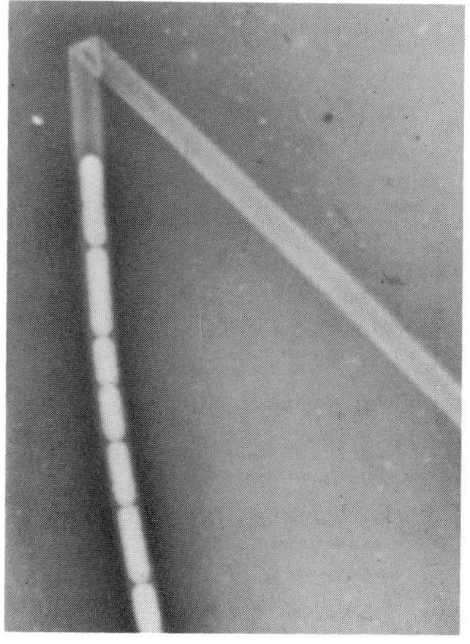

*Figure 19.17 Sphaerotilus, showing a chain of cells enclosed within the sheath. From J. L. Stokes, "Studies on the filamentous sheathed iron bacterium Sphaerotilus natans." J. Bacteriol.* **67**, *281 (1954).*

(a)

(b)

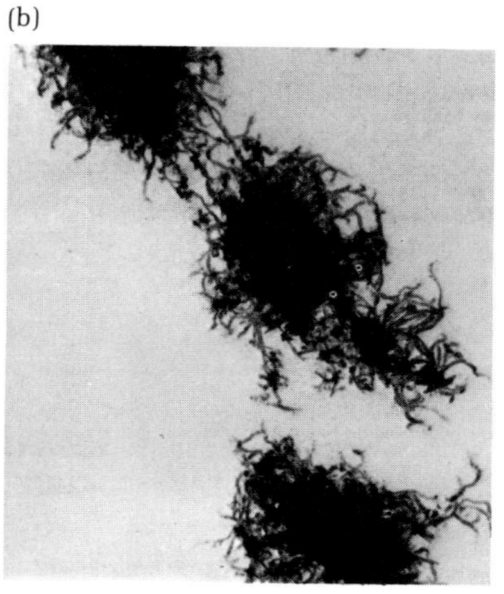

*Figure 19.18 Colonies of two Sphaerotilus species. (a) Sphaerotilus natans (× 20). (b) Sphaerotilus discophorus, growing on agar containing $MnCO_3$ (× 135); the dark color of the colonies is caused by heavy incrustation of the sheaths with manganic oxide. Courtesy of J. L. Stokes and M. A. Rouf.*

suggest that the oxidation of $Mn^{2+}$ to $Mn^{4+}$ by *S. discophorus* is an enzymatic process. There is no evidence, however, that the oxidation of either iron or manganese by bacteria of the sphaerotilus group is of significance as a mechanism of energy generation; all these bacteria appear to be dependent on organic compounds for the provision of carbon and energy.

### *Zoogloea ramigera*

Like *Sphaerotilus natans*, *Zoogloea ramigera* occurs characteristically in sewage. This bacterium has a very peculiar growth habit: in liquid media, it develops as large, macroscopically visible flocs (Figure 19.19), consisting of a coherent gelatinous slime layer containing many cells (Figure 19.20). The rod-shaped cells bear a polar flagellum and usually contain conspicuous deposits of poly-β-hydroxybutyrate. *Zoogloea* is nutritionally versatile, using many carbohydrates, amino acids, and organic acids as carbon and energy sources. Vitamin $B_{12}$ and biotin are required growth factors.

*Figure 19.19 Star-shaped flocs of Zoogloea ramigera in a broth culture (× 23). From K. Crabtree and E. McCoy, "Zoogloea ramigera Itzigsohn, identification and description." Intern. J. Syst. Bacteriol.* **17**, *1 (1967).*

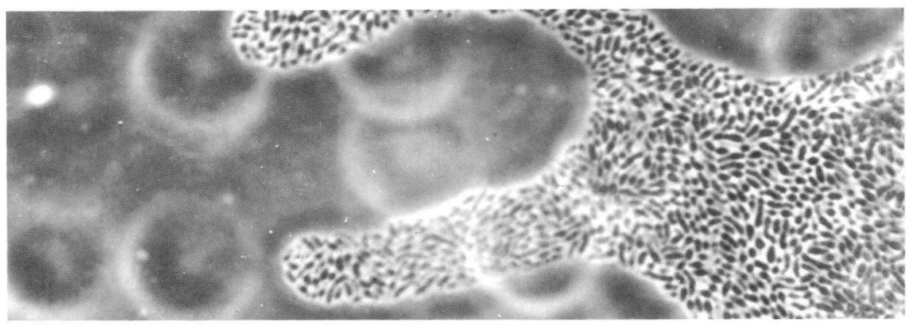

*Figure 19.20 Phase contrast photomicrograph of projections at the periphery of a floc of Zoogloea ramigera, showing cells contained in slime layer (× 1,804). From K. Crabtree and E. McCoy, "Zoogloea ramigera Itzigsohn, identification and description." Intern. J. Syst. Bacteriol.* **17**, *1 (1967).*

*Lampropedia hyalina* (Figure 19.21) was first described almost 100 years ago but has only recently been isolated and cultivated. It is a freshwater bacterium and has also been observed in the gut contents of

*Lampropedia hyalina*

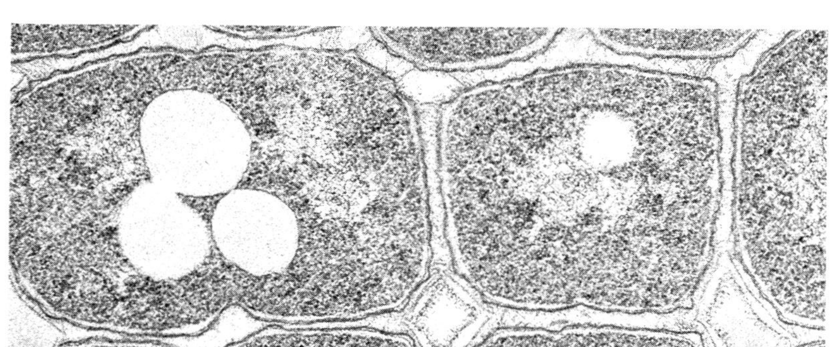

*Figure 19.21 Electron micrograph of a transverse section through a sheet of Lampropedia cells (× 35,235). Note the complexity of the wall investment between adjoining cells in this sheet. The large clear intracellular areas are sites of poly-β-hydroxybutyrate deposits. From J. Pangborn and M. P. Starr, "Ultrastructure of Lampropedia hyalina." J. Bacteriol.* **91,** *2025 (1966).*

reptiles. The spherical cells grow in exceedingly regular rectangular sheets, one cell thick. Cell division occurs in two successive planes at right angles to one another; and the very orderly array of the cells is maintained by adherence through the outer layers of the cell wall. The wall structure is extremely elaborate; electron microscopy reveals the presence of three quite distinct layers (Figure 19.22). *Lampropedia hyalina* grows well in complex media. It requires vitamins (biotin and thiamine), and can utilize a small number of organic acids (e.g., lactate and succinate) as carbon and energy sources.

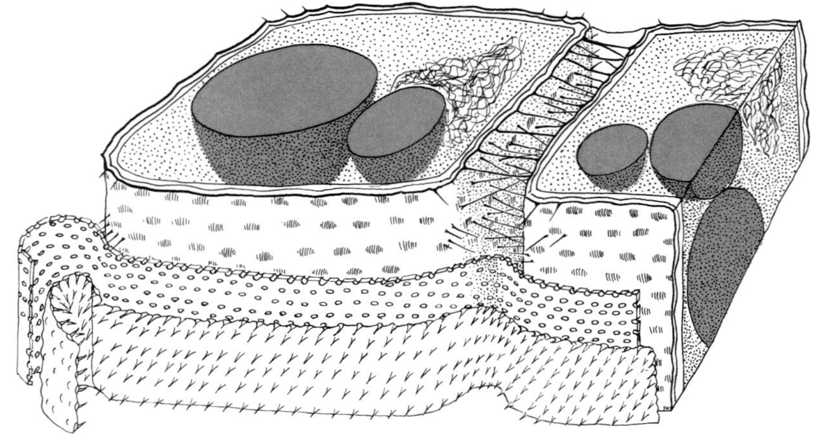

*Figure 19.22 A schematic reconstruction of parts of two adjoining cells in a sheet of Lampropedia, showing the arrangement of the sucessive wall layers. From J. Pangborn and M. P. Starr, "Ultrastructure of Lampropedia hyalina." J. Bacteriol.* **91,** *2025 (1966).*

Table 19.8 *Distinctions among the genera of prosthecate bacteria*

| Genus | Range of DNA base composition, mole % G+C | Mode of cell division | Prosthecae have reproductive function | Number of prosthecae per cell | Site of insertion of prosthecae | Holdfasts | Flagella | Gas vacuoles |
|---|---|---|---|---|---|---|---|---|
| *Caulobacter* | 62–67 | Binary fission | No | 1 | Polar | + | + | − |
| *Asticcacaulis* | 55 | Binary fission | No | 1 or 2 | Subpolar | + | + | − |
| *Hyphomicrobium* | 60–67 | Budding | Yes | 1 | Polar or subpolar | V[a] | + | − |
| *Ancalomicrobium* | − | Budding | No | Several | Various sites | − | − | + |

[a]Variable.

## The prosthecate bacteria

A distinctive feature of many aerobic, Gram-negative true bacteria is the formation of *prosthecae:* filiform or blunt extensions of the cell surface (wall and membrane), which confer a complex geometric form on the cell. One prosthecate organism, *Rhodomicrobium,* is photosynthetic, a typical purple bacterium in physiological respects; it has been described in Chapter 17. The other prosthecate organisms so far described are all aerobic chemoheterotrophs. They can be divided into several subgroups (Table 19.8).

### The *Caulobacter* Group

The organisms of the *Caulobacter* group are small straight or curved rods with a characteristic dimorphism. A growing population always consists of two kinds of cells: immotile cells bearing a very fine filiform prostheca, approximately 0.2 $\mu$m thick, known as a stalk; and nonprosthecate, motile cells, bearing a single flagellum, either polar or subpolar. These two kinds of cells arise by a process of unequal binary fission (Figure 19.23). A dividing cell bears a prostheca at one pole and a flagellum at the other pole; division therefore produces an immotile,

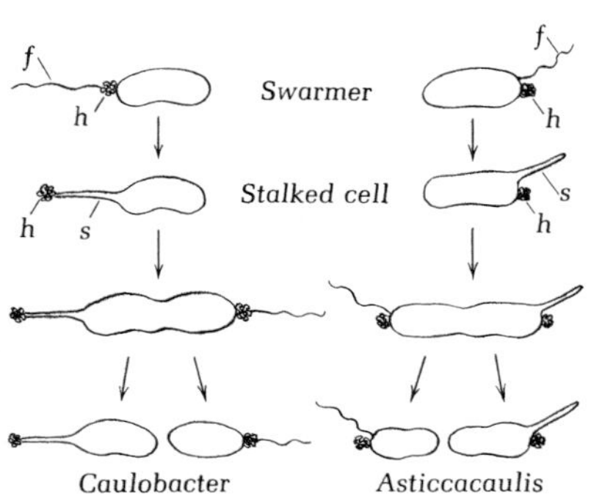


Figure 19.23 *A diagrammatic representation of cellular differentiation and division in two genera of stalked bacteria, Caulobacter and Asticcacaulis: h, holdfast; s, stalk; f, flagellum. From J. M. Schmidt and R. Y. Stanier, "The development of cellular stalks in bacteria." J. Cell Biol.* **28,** *423 (1966).*

stalked basal cell, and a motile, apical cell, or swarmer. The swarmer soon loses its flagellum and develops in its place a stalk; prior to its next division, a flagellum is synthesized at the nonstalked pole. The basal stalked cell, on the other hand, simply enlarges and divides with the liberation of a new apical swarmer cell. Since every swarmer cell must synthesize a stalk prior to division, the mean generation time of swarmer cells is significantly longer than that of stalked cells, as shown in Figure 19.24.

At the time of its liberation, each swarmer bears an adhesive disc of secreted material, or holdfast, at the apical pole, through which it can attach to solid objects. In cultures, many swarmers become attached

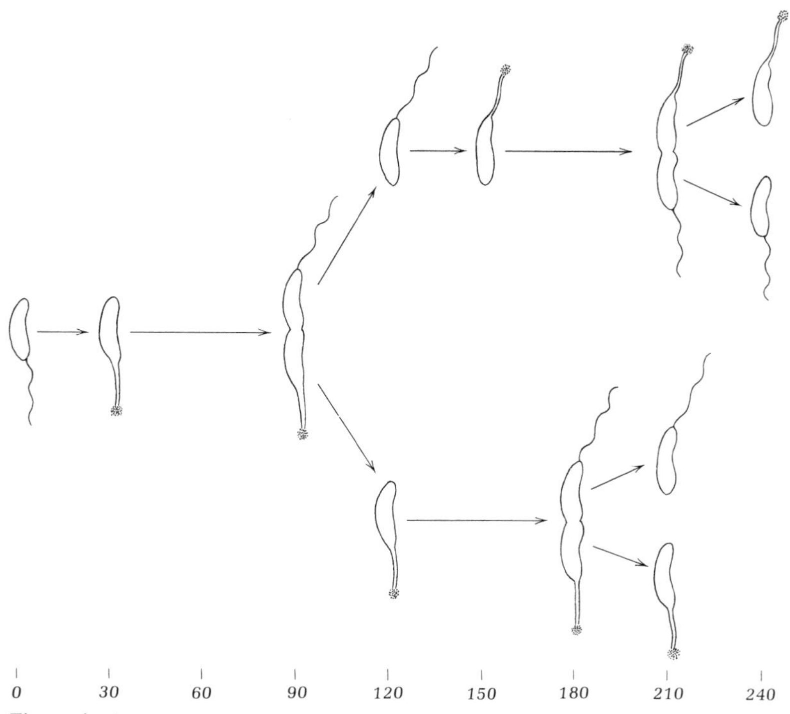


*Figure 19.24 A diagrammatic representation of clonal growth in Caulobacter, based on continuous microscopic observations. Note that the time required for the division of a swarmer cell is considerably longer than that required for the division of its stalked sibling. After J. L. S. Poindexter, "Biological properties and classification of the Caulobacter group." Bacteriol. Rev.* **28**, *231 (1962).*

to one another through their holdfasts; and subsequent stalk growth gives rise to characteristic rosettes (see Figure 5.4). In nature, rosettes are rarely formed; instead, the caulobacters attach to cells of other microorganisms. This type of attachment can be obtained experimentally by mixing a suspension of swarmers with cells of other kinds of bacteria (Figure 19.25).

Figure 19.25 The attachment of *Caulobacter* to (a) *Bacillus* and (b) *Azotobacter* (scale marker, 1 μm). From J. L. S. Poindexter, "Biological properties and classification of the Caulobacter group." Bacteriol. Rev. **28,** 231 (1962).

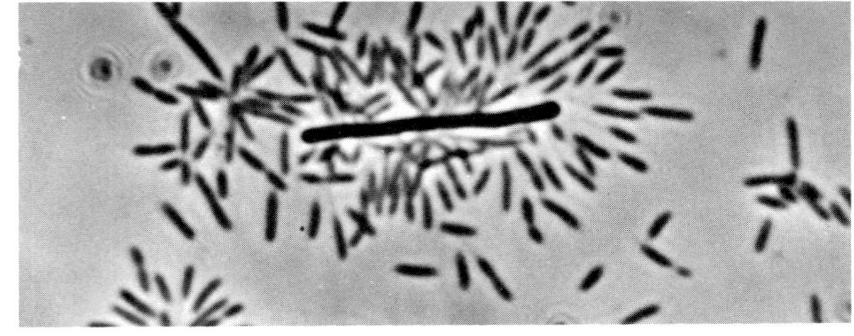

(a)

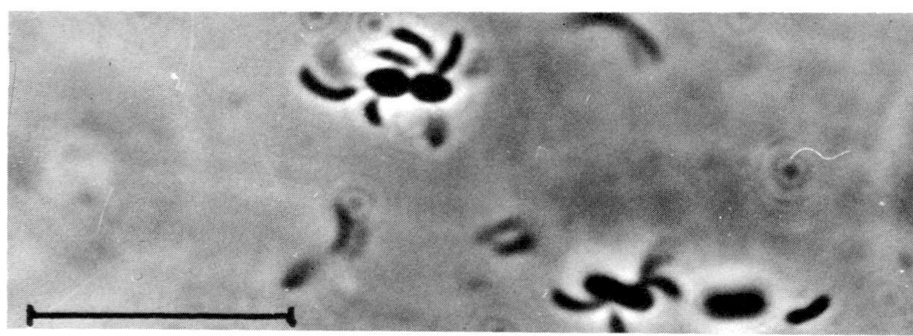

(b)

There are two genera, distinguished on structural grounds (Table 19.8). In *Caulobacter*, the flagellum, the stalk, and the holdfast are all located at the pole of the cell; in *Asticcacaulis*, the stalk and flagellum are subpolar, while the holdfast is polar. Cell division in *Asticcacaulis* is unequal, the apical swarmer being notably smaller than the stalked basal cell; in *Caulobacter* the two cells are of equal size.

Caulobacters are in the main aquatic organisms, which can survive and grow in very dilute organic media. This character serves in their isolation, achieved by adding a very low peptone concentration (0.01 percent) to a water sample, which is kept in a beaker or flask at room temperature. After several days, the caulobacters accumulate in the surface film, from which they can be isolated by streaking on appropriate media (dilute peptone agar). The caulobacters can use soluble carbohydrates, amino acids, and organic acids as carbon and energy sources. Although a few species do not require growth factors, most require at least one B vitamin, and some have more complex requirements. Relative to other aerobic chemoheterotrophs, such as most *Pseudomonas* spp., their growth is slow, even in complex media, generation times being as much as 5 hours under optimal conditions. It is probable that in nature the caulobacters develop primarily as ectosymbionts (see Chapter 23), attached to the surface of algae and other larger aquatic microorganisms. In this situation, they can derive a nutritional advantage from any organic materials secreted by the host cell. There is no evidence that the relationship is a parasitic one; even when heavily invested by caulobacter cells, other kinds of bacteria do not suffer demonstrable ill effects.

### The *Hyphomicrobium* group

These prosthecate bacteria are small plump rods that bear long filiform prosthecae, like those of the caulobacters. The developmental cycle is, however, radically different, since the prosthecae of the hyphomicrobia have a reproductive function. Cell division occurs by budding: the formation and progressive enlargement of a daughter cell at the tip of the prostheca. This cell is eventually liberated as a motile, polarly flagellated swarmer (Figure 19.26). In some strains the prosthecae are branched, and the cells adhere in small colonies, a mode of development also characteristic of the purple bacterium *Rhodomicrobium*. Some strains produce holdfasts and form rosettes in which the cells adhere through their nonprosthecate poles. There is one genus, *Hyphomicrobium*.

The hyphomicrobia are also a distinctive group in physiological

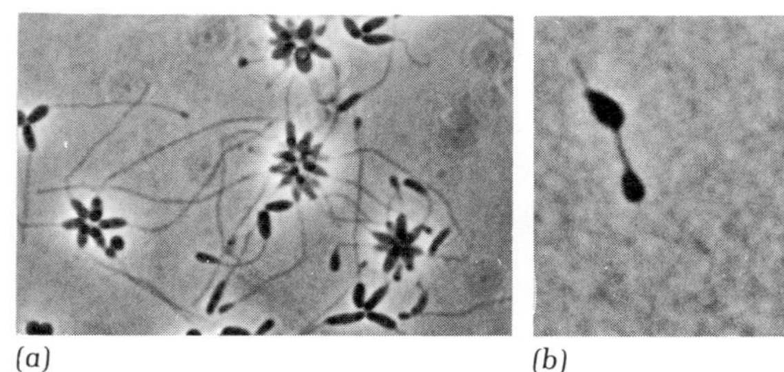

(a)

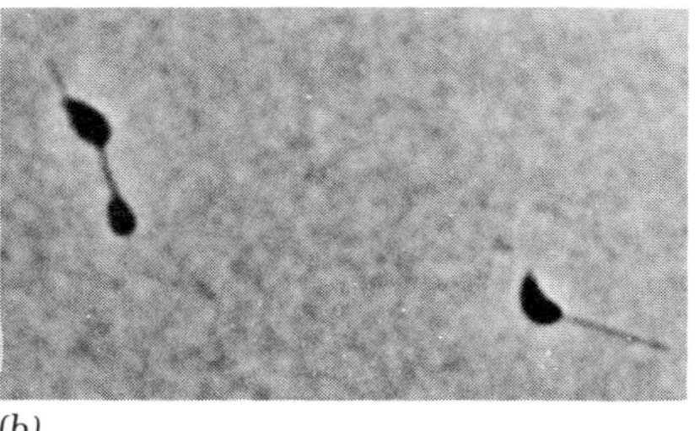

(b)

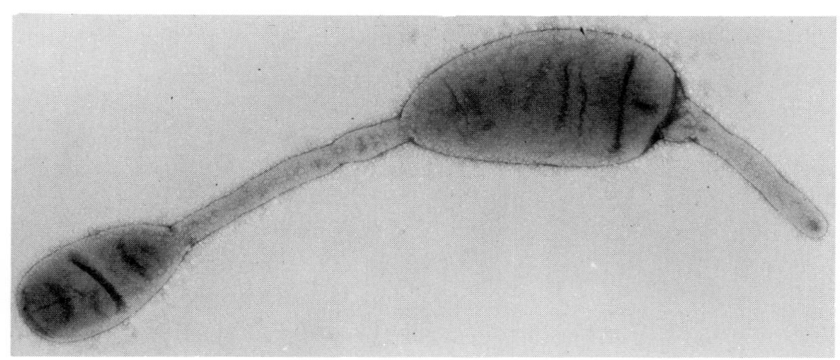

(c)

*Figure 19.26 Hyphomicrobium vulgare. (a) Phase contrast micrograph of a strain that forms rosettes and has very long prosthecae (× 1,440). Courtesy of Rosmarie Rippka. (b) Phase contrast micrograph of two budding cells (× 2,272). (c) Electron micrograph of a whole mount of a budding cell, showing continuity between cell wall and prosthecae. Micrographs (b) and (c) courtesy of Peter Hirsch.*

respects. Most strains can use a very limited range of organic compounds, and the preferred sources of carbon and energy are one-carbon compounds: methanol, methylamine, and formate, rarely utilizable by other aerobic chemoheterotrophs. The growth of these bacteria is always very slow, generation times of as much as 10 hours under optimal conditions being common.

Thanks to their very simple nutritional requirements, the hyphomicrobia can grow in mineral solutions to which no organic substrates have been added, utilizing traces of volatile organic compounds present in air. As a result of this property, they were first discovered as contaminants in enrichment cultures for nitrifying bacteria. However, they are not chemoautotrophs; growth does not occur in mineral solutions supplied with air that has been treated to destroy organic materials.

**Other prosthecate bacteria**

By direct observation with the electron microscope, a variety of other prosthecate bacteria having bizarre cell shapes have been detected in soil and water (Figure 19.27); relatively few of these organisms have been cultivated, however. The most remarkable form so far obtained in

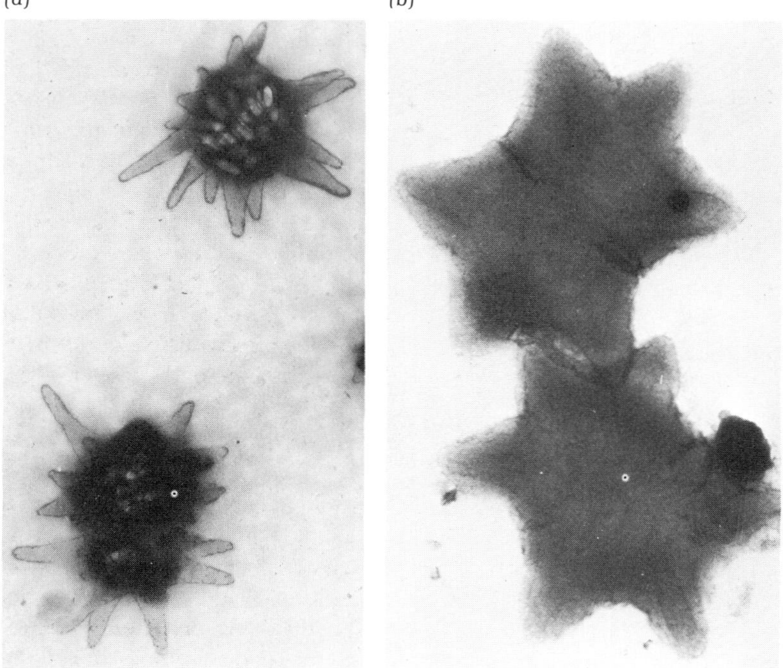

*Figure 19.27 Electron micrographs of two prosthecate freshwater bacteria. (a) Prosthecomicrobium pneumaticum (× 15,666). Note gas vesicles (light areas) within the cells. (b) Unidentified dividing, star-shaped bacterium. From J. T. Staley, "Prosthecomicrobium and Ancalomicrobium: new prosthecate freshwater bacteria." J. Bacteriol.* **95**, *1921 (1968).*

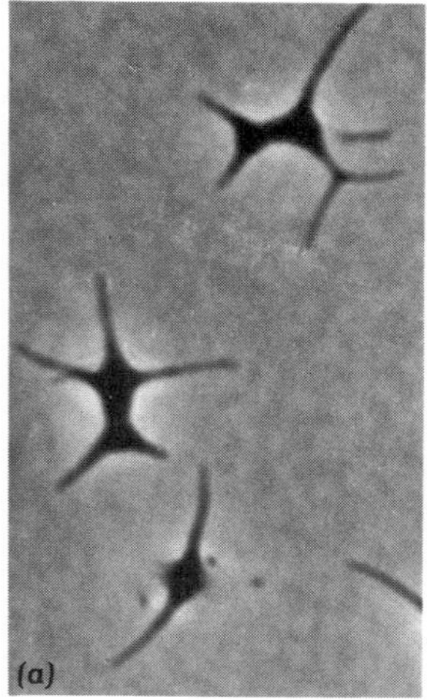

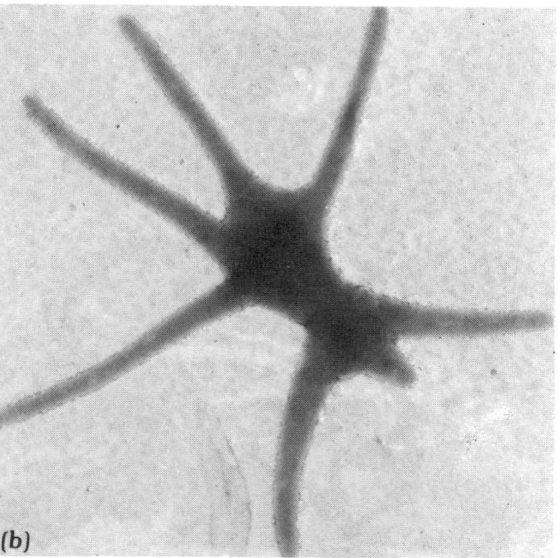


*Figure 19.28 A prosthecate, freshwater bacterium, Ancalomicrobium adetum. (a) Phase contrast micrograph of a group of cells (× 2,800). (b) Electron micrograph of a single cell, showing continuity of prosthecae with the body of the cell (× 10,938). From J. T. Staley, "Prosthecomicrobium and Ancalomicrobium: new prosthecate freshwater bacteria." J. Bacteriol.* **95,** *1921 (1968).*

pure culture is a freshwater organism, *Ancalomicrobium* (Figure 19.28). The immotile cells bear several long spine-shaped prosthecae and contain gas vacuoles. Division occurs by budding from the body of the cell; as the bud grows, it forms new prosthecae, and at the time of division the mother and daughter cells are equivalent in size and form.

**The functions of prosthecae**

When the caulobacters were first discovered, their filiform prosthecae were interpreted as stalks, which served to anchor the cell to substrates. However, the anchoring of sessile prosthecate bacteria is, in fact, always mediated by the secreted holdfast. Even in the caulobacter group, the topological separation between holdfast and stalk characteristic of *Asticcacaulis* shows that the so-called "stalk" is not necessarily involved in the attachment process. In other prosthecate groups, the prosthecae clearly never play a role in cellular attachment. It is necessary, accordingly, to seek other explanations for these peculiar bacterial organelles. One obvious consequence of their presence on the cell is to decrease the rate of its sedimentation. This can be very easily shown for caulobacters, where simple centrifugation of the dimorphic population results in a remarkably efficient physical separation of swarmers and stalked cells: the swarmers form a compact pellet at a gravitational force which leaves most stalked cells in suspension. Since most prosthecate bacteria are strictly aerobic aquatic organisms, immotile in the prosthecate state, the primary function of the prosthecae may be to maintain cells near the air–water interface, by slowing their sedi-

*Table 19.9 Characteristics of major genera among the Gram-negative strictly anaerobic bacteria*

| **Genus** | **Cell shape** | **Flagellation** | **DNA base composition, mole % G+C** | **Energy-yielding substrate(s)** | **Products of energy-yielding metabolism** | **Other characters** |
|---|---|---|---|---|---|---|
| *Bacteroides* | *Straight rods* | *Immotile* | 41–43 | *Carbohydrates* | *Succinic and acetic acids* | — |
| *Fusobacterium* | *Straight rods with tapered ends* | *Immotile* | 32–38 | *Carbohydrates* | *Butyric or acetic acids,* $CO_2$, $H_2$ | — |
| *Veillonella* | *Cocci* | *Immotile* | 36 | *Lactic acid, succinic acid* | *Propionic and acetic acids,* $CO_2$, $H_2$ | *Do not attack carbohydrates* |
| *Peptococcus* | *Cocci* | *Immotile* | — | *Glutamate, possibly other amino acids* | — | *Do not attack carbohydrates* |
| *Butyrivibrio* | *Curved rods* | *Single polar* | — | *Carbohydrates* | *Butyric and acetic acids,* $CO_2$, $H_2$ | — |
| *Succinivibrio* | *Curved rods* | *Single polar* | 49 | *Carbohydrates* | *Lactic, succinic, and acetic acids* | — |
| *Desulfovibrio* | *Curved rods* | *One or more polar flagella* | 46–62 | *Lactic acid, malic acid (sulfate reduction)* | *Acetic acid,* $H_2S$ | *Contain heme pigments* |
| *Selenomonas* | *Curved rods* | *Tuft of flagella inserted at center of cell* | — | *Carbohydrates* | *Lactic, propionic and acetic acids,* $CO_2$ | — |

mentation. Comparable structural adaptations are known in several groups of marine eucaryotic protists, such as diatoms and dinoflagellates, which comprise part of the floating (planktonic) population of the oceans.

## *The strictly anaerobic Gram-negative bacteria*

The Gram-negative true bacteria include a large number of strictly anaerobic groups. With a few exceptions, these groups are still rather poorly known, since their isolation and purification is difficult. They are classified in terms of structural characters and modes of anaerobic energy-yielding metabolism; some of the principal genera are described in Table 19.9.

The members of the *Desulfovibrio* group are free-living organisms, with a very wide natural distribution. They are the principal biological agents of sulfide formation and can be found in all natural anaerobic environments where sulfate is available. These organisms have simple nutritional requirements and do not need organic growth factors.

The other groups of bacteria listed in Table 19.9 have almost all been isolated from anaerobic environments in the animal body, particularly the mouth and the intestinal tract. A frequent source has been the rumen of herbivorous animals; however, this may simply reflect the fact that the microbiology of the rumen has been studied very intensively. These bacteria can be regarded as animal symbionts; only very few of them are known to cause disease. Their nutritional requirements are in all cases complex, and they are customarily grown in media containing peptone, yeast extract, or (for rumen bacteria) rumen fluid, and supplemented with a suitable energy source (a carbohydrate or organic acid). Many of these organisms perform fermentations in which succinic acid is a major end-product; typically, these forms require an addition of $CO_2$ (or bicarbonate) to the medium, since the formation of succinate involves a carboxylation. Other common fermentative end-products include lactic, propionic, and acetic acids.

### **The sulfate-reducing bacteria**

The sulfate-reducing bacteria constitute a distinctive physiological group. They are dependent on the reduction of sulfate as a mode of anaerobic energy-yielding metabolism. Sulfate is therefore required as a nutrient at high levels, since it serves as the terminal electron acceptor. The range of utilizable organic substrates is narrow, being confined for most strains to a few organic acids, notably lactate, malate, and pyruvate. The end-product of the anaerobic oxidation of organic substrates is in all cases acetic acid, since these bacteria do not possess a functional tricarboxylic acid cycle. Under favorable conditions, *Desulfovibrio* can form very large amounts of sulfide during growth: the concentration may attain a level of as much as 10 g/liter. Unlike nearly all other strictly anaerobic chemoheterotrophs, *Desulfovibrio* spp. contain heme pigments: specifically, a cytochrome of the c type, which participates in the anaerobic electron transport system. The cells also contain a colored protein, desulfoviridin, the function of which is not known.

Most species of *Desulfovibrio* have small, vibrioid cells, 0.5 to 1 by 2 to 5 $\mu$m, very actively motile by means of a single polar flagellum. Some species have helical cells, bearing a tuft of polar flagella. Although the span of DNA base composition characteristic of each species is narrow, the span for the genus is unusually wide, extending from 46 to 62 mole percent G+C.

**Other vibrioid genera**

Small vibrioid rods (genera *Butyrivibrio* and *Succinivibrio*) are common anaerobic forms in the rumen. In contrast to *Desulfovibrio*, these organisms are not sulfate reducers, depending on the fermentation of carbohydrates for provision of energy. They appear to play a role in the decomposition of plant polysaccharides in the rumen, being able to attack such compounds as pectin and xylan. A structurally distinctive group, which occurs both in the rumen and in the human mouth, is *Selenomonas*. The relatively large, crescent-shaped cells, 1 $\mu$m wide and up to 7 $\mu$m long, show a unique mode of flagellation: they bear a tuft of flagella, inserted centrally on the concave side of the cell (Figure 19.29).

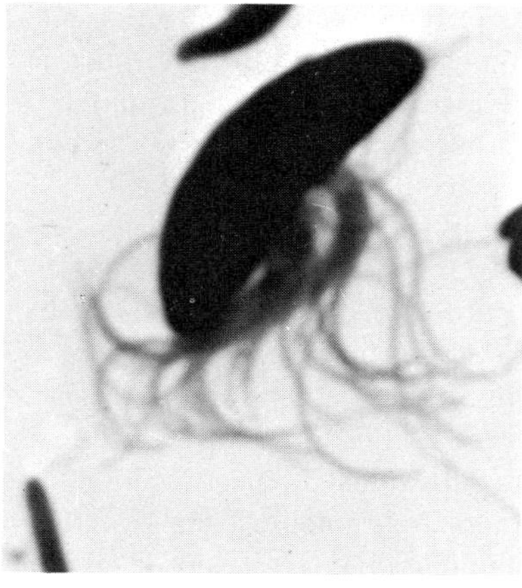

*Figure 19.29 Selenomonas: flagella stain, showing the crescent-shaped cell and the tuft of flagella inserted laterally on the concave surface of the cell (× 3,600). Courtesy of C. F. Robinow.*

**The anaerobic cocci**

The best studied of the Gram-negative anaerobic cocci are the members of the genus *Veillonella*, which are abundant in the saliva of man and other mammals. The cells are small (0.3 to 0.4 $\mu$m in diameter) and arranged in pairs or masses. These bacteria cannot ferment carbohydrates, the fermentable substrates being organic acids, particularly lactic acid, which is converted to acetic and propionic acids, $CO_2$, and $H_2$. Other anaerobic cocci *(Peptococcus)* utilize amino acids (particularly glutamate) as fermentable substrates.

**The rod-shaped anaerobes**

Many rod-shaped Gram-negative anaerobic bacteria have been described, for the most part fragmentarily; and the classification of these forms is still unclear, since their mechanisms of energy-yielding metabolism are often unknown. The immotile rods of the genus *Bacteroides* are abundant in the animal intestine, and members of this group also occur in the rumen. Some ferment carbohydrates with formation of succinic and acetic acids as major end-products, but additional mechanisms of anaerobic energy-yielding metabolism may well occur in the bacteria assigned to this group. The members of the genus *Fusobacterium* have distinctive, fusiform cells with pointed ends. They are common in saliva and perform a butyric acid fermentation of carbohydrates.

## *Further reading*

There are no books that present a detailed account of the bacterial groups described in this chapter; for many of these groups, even reviews are lacking. The following reading list contains both reviews and important original articles devoted to certain of the groups which have been discussed.

**Reviews and original articles**

Baumann, P., M. Doudoroff, and R. Y. Stanier, "Study of the *Moraxella* Group. I. Genus *Moraxella* and the *Neisseria catarrhalis* Group." *J. Bacteriol.* **95,** 58 (1968).

———, "A Study of the *Moraxella* Group. II. Oxidase-Negative Species (Genus *Acinetobacter*)." *J. Bacteriol.* **95,** 1520 (1968).

Bryant, M. P., "Bacterial Species of the Rumen." *Bacteriol. Rev.* **23,** 125 (1959).

Crabtree, K., and E. McCoy, "*Zoogloea ramigera.*" *Intern. J. Syst. Bacteriol.* **17,** 1 (1967).

DeLey, J., "*Pseudomonas* and Related Genera." *Ann. Rev. Microbiol.* **18,** 17 (1964).

Jensen, H. L., "The Azotobacteriaceae." *Bacteriol. Rev.* **18,** 195 (1954).

Kushner, D. J., "Halophilic Bacteria." *Adv. Appl. Microbiol.* **10,** 73 (1968).

Mulder, E. G., "Iron Bacteria, Particularly Those of the *Sphaerotilus–Leptothrix* group." *J. Appl. Bacteriol.* **27,** 151 (1964).

Poindexter, J. S., "Biological Properties and Classification of the *Caulobacter* group." *Bacteriol. Rev.* **28,** 231 (1964).

Postgate, J. R., and L. L. Campbell, "Classification of *Desulfovibrio* Species, the Nonsporulating Sulfate-Reducing Bacteria." *Bacteriol. Rev.* **30,** 732 (1966).

Staley, J. T., "*Prosthecomicrobium* and *Ancalomicrobium,* New Prosthecate Freshwater Bacteria." *J. Bacteriol.* **95,** 1921 (1968).

Starr, M. P., and V. B. D. Skerman, "Bacterial Diversity: the Natural History of Selected Morphologically Unusual Bacteria." *Ann. Rev. Microbiol.* **19,** 407 (1965).

Stolp, H., and M. P. Starr, "*Bdellovibrio bacteriovirus,* a Predatory, Ectoparasitic and Bacteriolytic Microorganism." *Antonie van Leeuwenhoek* **29,** 217 (1963).

Vishniac, W., and M. Santer, "The Thiobacilli." *Bacteriol. Rev.* **21,** 195 (1957).

Williams, M. A., and S. C. Rittenberg, "A Taxonomic Study of the Genus *Spirillum.*" *Intern. Bull. Bacteriol. Nomencl. Taxon.* **7,** 49 (1957).

# Chapter twenty
# The endospore-forming bacteria

*Many unicellular true bacteria* share a distinctive developmental character: the ability to form *endospores*. Endospores (Figure 20.1) can be readily recognized microscopically by their characteristic intracellular location, by their extreme refractility, and by their resistance to staining by basic aniline dyes that readily stain the vegetative cells. They also possess remarkable physiological properties. Endospores can remain dormant for long periods (in some cases, as much as half a century), and are highly resistant to heat, shortwave radiation, and toxic chemicals. The heat resistance of spores provides a convenient general technique for the selective isolation of spore-forming bacteria from nature: pasteurization of aqueous suspensions of the source material.

Although diverse in physiological and biochemical respects, the endospore-forming true bacteria share certain general properties, in addition to spore formation, which suggests that they constitute a large natural group. With one exception, the vegetative cells are rod-shaped. The Gram reaction is typically positive. It must be determined on exponentially growing cells, however, since many sporeformers tend to become Gram negative very rapidly after populations enter the stationary phase of growth. Most sporeformers are motile, and the mode of insertion of flagella is always peritrichous. The G + C content of the DNA ranges from 25 to 50 mole percent, and for most species lies between 30 and 40 mole percent. All sporeformers (with the exception of one strict anaerobe) are chemoheterotrophs; the modes of energy-

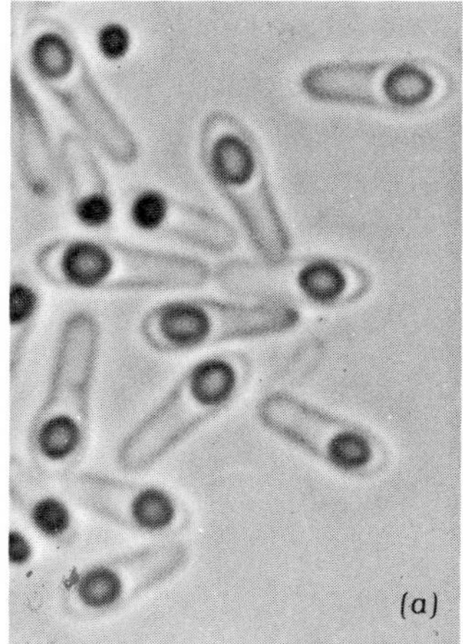

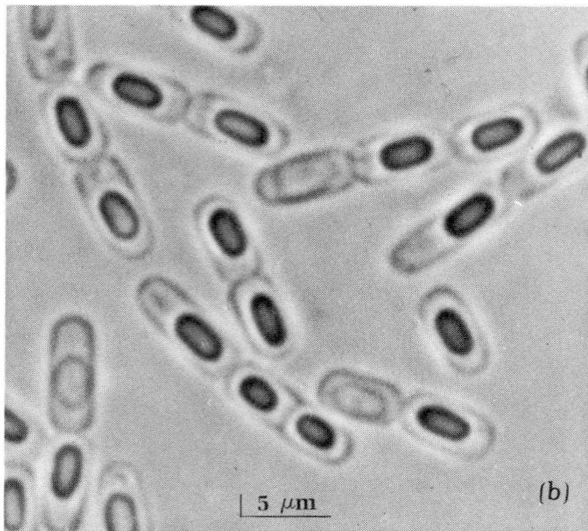

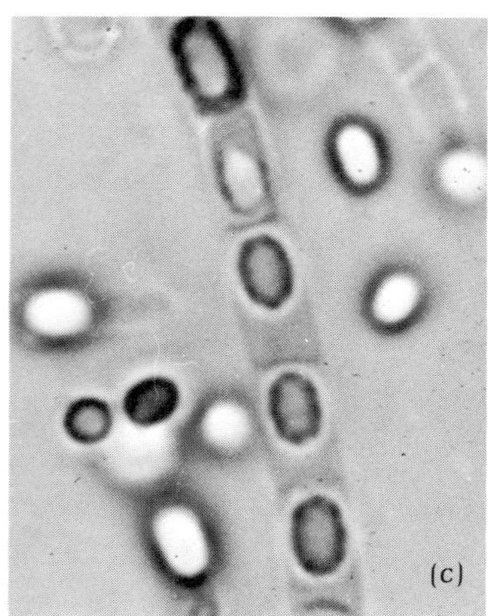


*Figure 20.1 Sporulating cells of Bacillus species: (a) unidentified bacillus from soil; (b) B. cereus; (c) B. megaterium. From C. F. Robinow, in The Bacteria (I. C. Gunsalus and R. Y. Stanier, editors), Vol. 1, p. 208. New York: Academic Press, 1960.*

yielding metabolism found in the group include aerobic and anaerobic respiration and fermentation. The typical habitat for most sporeformers is soil. A few species are pathogenic for animals, but pathogenicity is typically the consequence of the elaboration of specific toxins and only rarely reflects the ability to develop extensively in the animal host. Paradoxically, therefore, most pathogenic sporeformers are not animal parasites.

The primary taxonomic subdivision of the sporeformers is based on a physiological property: *the relation to free oxygen*. The anaerobic sporeformers (genus *Clostridium*) are strict anaerobes, rapidly killed by exposure to $O_2$. Their metabolism is fermentative, with the exception of a few species that obtain energy by anaerobic respiration. The aerobic sporeformers (genera *Bacillus* and *Sporosarcina*) are either strict aerobes or facultative anaerobes. The one spore-forming species that possesses spherical vegetative cells, a strict aerobe, is placed in a special genus, *Sporosarcina* (Table 20.1).

*Table 20.1 Properties of the principal genera of endospore-forming bacteria*

| Genus | Cell shape | Relations to oxygen | Range of G + C content of DNA, mole % |
|---|---|---|---|
| *Bacillus* | Rods | Aerobes, facultative anaerobes | 32–52 |
| *Sporosarcina* | Spheres | Aerobes | 40 |
| *Clostridium* | Rods | Strict anaerobes | 25–45 |

## *The aerobic sporeformers*

Most aerobic sporeformers are versatile chemoheterotrophs, able to use a considerable range of simple organic compounds (sugars, amino acids, organic acids) as carbon and energy sources. Some species require no organic growth factors; others may require amino acids, vitamins of the B group, or both. However, the growth-factor requirements are as a rule relatively simple. Although most aerobic sporeformers are mesophilic, with a temperature optimum in the range 30 to 45°C, the group also includes a number of thermophilic species, characterized by temperature optima as high as 65°C, and incapable of growth at temperatures below 45°C.

**The genus *Bacillus***

The genus *Bacillus* can be subdivided into three major groups on the basis of the structure and intrasporangial location of the endospore, a constant and distinctive character (Table 20.2). In group I, spores are

*Table 20.2 Primary subdivision of the genus Bacillus*

| Group | Characteristics | Major species |
|---|---|---|
| I | *Spores oval, thin-walled, located centrally or subterminally in sporangium. Sporangium not swollen by spore.* | *B. subtilis, B. licheniformis, B. megaterium, B. cereus, B. anthracis, B. thuringensis* |
| II | *Spores oval, thick-walled, located centrally or subterminally in sporangium. Sporangium swollen by spore.* | *B. polymyxa, B. macerans, B. circulans* |
| III | *Spores spherical, thick-walled, located terminally in sporangium. Sporangium swollen by spore.* | *B. pasteurii* |

oval, thin-walled, and never of greater diameter than the sporangium, which is consequently not distended at the time of sporulation. In group II, the spores are oval, thick-walled, and of greater diameter than the sporangium, which is swollen at the time of sporulation. In group III, the spores are spherical, always terminally located in the sporangium, and of greater diameter than the sporangium, which is consequently swollen at one pole.

**Species of group I**

Within group I of the genus *Bacillus* (Table 20.3) a further subdivision can be made between species with relatively small cells (0.8 μm or less in diameter) and those with relatively large cells (1 to 1.5 μm in diameter). The former (*B. subtilis* group) do not form poly-β-hydroxybutyrate as a reserve material, whereas the latter (*B. megaterium–B. cereus* group) always do so.

*Bacillus subtilis* is a strict aerobe that does not require growth factors. This organism is a powerful producer of extracellular amylases (starch-splitting enzymes) and proteases, and serves as a source for the commercial production of these enzymes. The other principal small-celled species, *B. licheniformis*, differs from *B. subtilis* by virtue of its ability to grow under anaerobic conditions, either by the fermentation of sugars or by denitrification. The sugar fermentation of *B. licheniformis* is distinctive: the major end-products are 2,3-butanediol, $CO_2$, and glycerol. Unlike most other bacteria that perform a butanediol fermentation, this species does not produce hydrogen; the surplus reducing power available from the formation of butanediol is used instead to reduce triose phosphate to glycerol, and the balanced equation for the fermentation of glucose is

$$3 \text{ glucose} \rightarrow 2CH_3CHOHCHOHCH_3 + 4CO_2 + 2CH_2OHCHOHCH_2OH$$

*Bacillus licheniformis* is also a denitrifier; it is the only species of the genus to possess this metabolic character. Since *B. licheniformis* can tolerate far higher nitrate concentrations than most other denitrifying bacteria, it can be selectively enriched by anaerobic growth in a peptone medium containing 8 percent $KNO_3$.

*Table 20.3 Some properties that distinguish Bacillus spp. of Group I*

| Species | Width of vegetative cells, μm | Formation of poly-β-hydroxybutyrate | Motility | Requirement for growth factors | Anaerobic growth: Sugar fermentation | Anaerobic growth: Denitrification | Other properties |
|---|---|---|---|---|---|---|---|
| *B. subtilis* | <0.8 | − | + | − | − | − | − |
| *B. licheniformis* | <0.8 | − | + | − | + | + | − |
| *B. megaterium* | >1.0 | + | + | − | − | − | − |
| *B. cereus* group | | | | | | | |
| *B. cereus* | >1.0 | + | + | + | + | − | − |
| *B. anthracis* | >1.0 | + | − | + | + | − | *Pathogenic for mammals* |
| *B. thuringensis* | >1.0 | + | + | + | + | − | *Pathogenic for insects* |

*Figure 20.2 Mature endospores of Bacillus cereus (× 3,600). Each stained spore is surrounded by a less deeply stained exosporium. Courtesy of C. F. Robinow.*

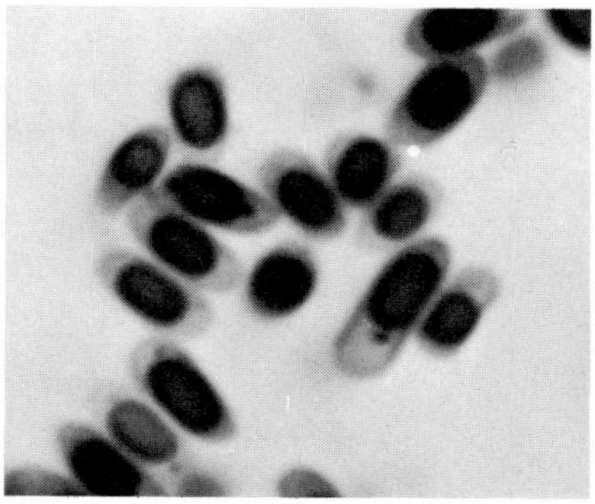

The large-celled species of group I fall into two clusters: *B. megaterium* and the *B. cereus* species group. The former is a strict aerobe that requires no growth factors: the latter are able to grow anaerobically when provided with fermentable sugars and require amino acids as growth factors. In the *B. cereus* group the spore is surrounded by a loose outer coat known as the *exosporium* (Figure 20.2), which is lacking in *B. megaterium*.

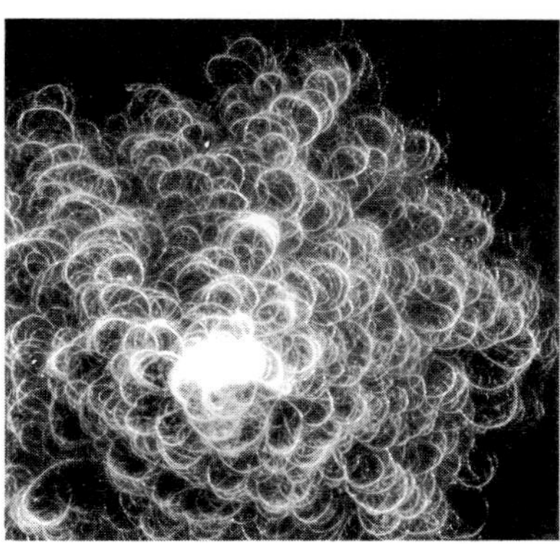

*Figure 20.3 A colony of Bacillus cereus var. mycoides (× 1.1). Courtesy of David Cornelius and C. F. Robinow.*

*Bacillus cereus* is one of the most abundant aerobic sporeformers in soil. Certain strains *(B. cereus* var. *mycoides)* produce distinctive colonies (Figure 20.3), somewhat resembling those of a fungus. This reflects the fact that the cells tend to remain adherent to one another after division, producing very long chains.

Closely related to *B. cereus* with respect to most properties are two other species, *B. anthracis* and *B. thuringensis,* which are pathogenic, respectively, for mammals and for certain insects. *Bacillus anthracis* is the causative agent of anthrax, a disease of cattle and sheep that is also transmissible to man. It is the only *Bacillus* that can be considered a true parasite of higher animals, in the sense that it is capable of extensive growth within the body of the host. Apart from its pathogenicity, the mechanism of which has been studied in considerable detail (see Chapter 27), it can be distinguished from *B. cereus* by its permanent immotility.

*Bacillus thuringensis* causes a paralytic disease in the caterpillars of many lepidopterous insects. Paralysis results when the caterpillars ingest plant material which carries spores or sporulating cells of *B. thuringensis,* and disease is produced by a toxic protein that is formed by the bacteria at the time of sporulation. This toxic protein can, in fact, be visualized microscopically in the sporangia of *B. thuringensis* (Figure 20.4); each sporangium contains, adjacent to the spore, an extremely regular bipyramidal crystal. After autolysis of the sporangia,

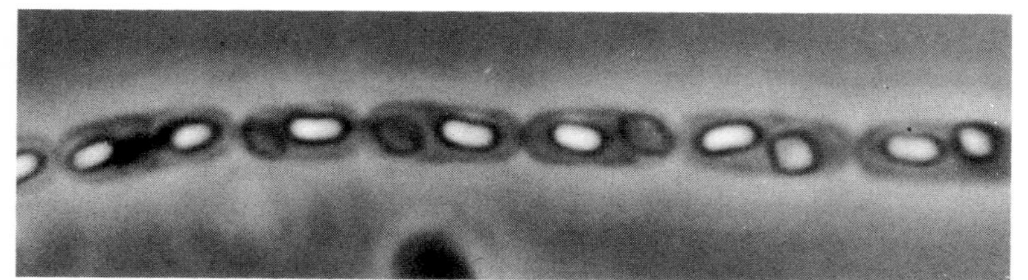

*Figure 20.4 A chain of sporulating cells of Bacillus thuringensis (phase contrast, × 3,900). Each cell contains, in addition to the bright, refractile spore, a less refractile bipyramidal crystalline inclusion. Courtesy of P. FitzJames.*

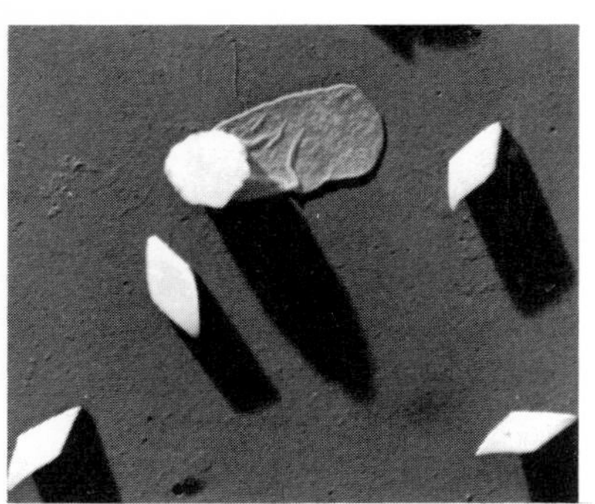

*Figure 20.5 Electron micrograph of free crystals and a free spore surrounded by an exosporium of a crystal-forming Bacillus (metal-shadowed preparation, × 7,500). From C. L. Hannay and P. FitzJames, "The protein crystals of B. thuringensis." Can. J. Microbiol.* **1**, *694 (1955).*

the crystals can be separated from the spores (Figure 20.5), and feeding experiments have shown that they are capable of reproducing all the symptoms of the disease, which is consequently an intoxication, not an infection. The crystal protein is soluble only under alkaline conditions and dissolves after ingestion in the alkaline contents of the caterpillar gut. The soluble protein produces a loosening of the epithelial tissue lining the gut, which permits diffusion of the alkaline gut contents into the blood of the insect. The resulting change in the pH of the blood leads to rapid general paralysis.

It is possible to isolate mutants of *B. thuringensis* which still form spores but have lost the property of crystal formation. Such mutants have become completely nonpathogenic for insects and cannot be distinguished from *B. cereus. Bacillus thuringensis* could, accordingly, be regarded as a special variety of *B. cereus.* The significance of crystal formation for the bacterium is unknown, but the fact that mutants which have lost this property are unaffected in other respects shows that crystal formation does not play an essential role in either normal growth or sporulation.

The *Bacillus* spp. of group I also include one representative, *B. fastidiosus,* which shows an extreme degree of nutritional specialization. In contrast to the other members of group I, which can use a wide variety of organic compounds as carbon and energy sources, *B. fastidiosus* is strictly confined to the use of a single compound as its carbon and energy source: the purine, uric acid.

**Species of group II**

The *Bacillus* spp. of group II, in which the relatively thick-walled spores cause a marked distension of the sporangium, are all facultative anaerobes which ferment sugars. In fact, some of them (e.g., *B. polymyxa*) appear to be restricted to the use of sugars as carbon and energy sources, growing poorly if at all on complex media unsupplemented with a fermentable carbohydrate. All species of this group require organic growth factors; thiamin and biotin are commonly required, and in some species amino acids.

The most distinctive species of group II is *Bacillus polymyxa.* Its spores have a characteristic ridged, barrel-shaped structure (Figure 20.6); *B. polymyxa* performs a butanediol fermentation of sugars, with the formation of products different from those characteristic of *B. licheniformis.* The products are 2,3-butanediol, ethanol, acetic acid, $CO_2$, and $H_2$.

Another unusual physiological property of *B. polymyxa* is its ability to fix atmospheric nitrogen; it is the only *Bacillus* species known to

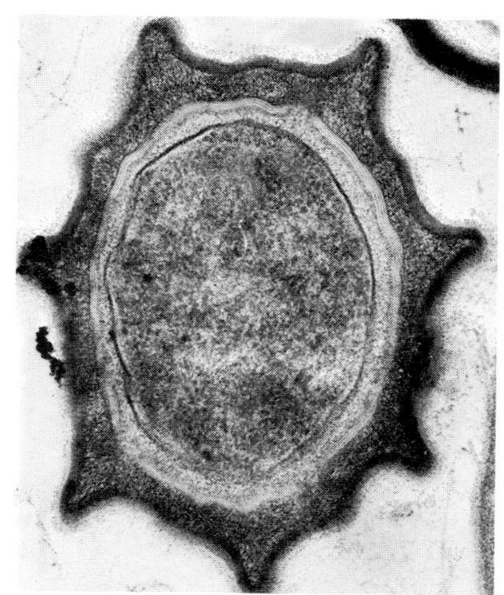

*Figure 20.6 Electron micrograph of a transverse section of a spore of Bacillus polymyxa, showing the elaborate stellate contour of the spore coat (× 45,600). Courtesy of R. G. E. Murray.*

possess this character. Although *B. polymyxa* is a powerful nitrogen fixer, nitrogen fixation takes place only when it is grown under anaerobic conditions, because the enzyme system responsible is inhibited (or possibly repressed) by relatively low concentrations of free oxygen. This character is shared by one other facultatively anaerobic nitrogen fixer, *Aerobacter aerogenes* (see Chapter 18).

*Bacillus macerans* performs a fermentation of sugars in which acetone and ethanol are the main organic end-products.

*Bacillus circulans* has a distinctive property, also found in some species of group III: the ability to produce *motile colonies* on well-dried agar plates. The course followed by these colonies during their movement over the agar surface is readily evident, since some cells get left behind and subsequently give rise to growth (Figure 20.7).

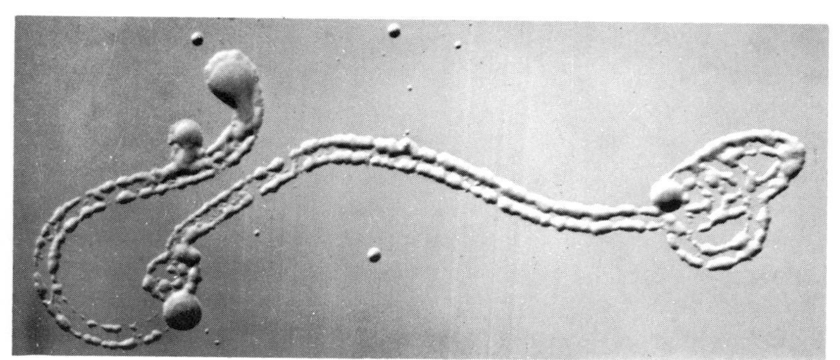

*Figure 20.7 Motile colony of Bacillus on the surface of an agar plate. Some cells get left behind as the colony moves, and the path of movement is marked by the subsequent growth of these stragglers. Nonmotile colonies of several other bacteria are also present on the plate. Note the difference.*

**Species of group III**

The species of group III, which form spherical spores producing a terminal distension of the sporangium, are all strict aerobes which require organic growth factors. Unlike members of groups I and II (with the exception of *B. fastidiosus*) they are unable to use carbohydrates, the characteristic carbon and energy sources being amino acids or organic acids. Group III includes a distinctive physiological class, known as the *ureolytic bacilli*. These organisms produce large amounts of the enzyme *urease*, which hydrolyzes urea to ammonia and $CO_2$:

$$CO(NH_2)_2 + H_2O \rightarrow CO_2 + 2NH_3$$

The decomposition of urea, being a hydrolytic reaction, cannot yield energy. It is nevertheless a reaction of physiological and ecological importance for the ureolytic bacilli. This is shown by the fact that the principal ureolytic species, *B. pasteurii*, is unable to grow on complex media at neutral pH in the absence of urea. Urea is a neutral compound, but its hydrolysis results in a considerable net production of alkali, because 2 moles of ammonia are formed from each mole of urea decomposed. A urea-containing medium inoculated with *B. pasteurii* therefore rapidly becomes highly alkaline as a result of urea hydrolysis. The requirement for urea can be replaced by ammonia, in conjunction with a high pH (8.5). Ammonia cannot be substituted by any other monovalent cation and is essential for respiration, as well as for growth. *Bacillus pasteurii* can therefore grow only under highly alkaline conditions, and requires ammonia not merely as a nitrogen source but also as a specific cation.

The ecological advantage that accrues to these organisms from their ureolytic ability is readily demonstrable by enrichment experiments. If a peptone medium containing 2 to 5 percent urea is inoculated with soil, the enriched population consists exclusively of ureolytic bacilli, even though a large variety of soil bacteria is capable of growth in pure culture on this medium. Under enrichment conditions, other bacteria are killed by the free ammonia produced as a result the hydrolysis of urea by the ureolytic bacilli.

*Sporosarcina ureae*

The only sporeformer that does not possess rod-shaped vegetative cells is *Sporosarcina ureae*. The motile, spherical cells are arranged in cubical packets. Physiologically, this organism closely resembles the ureolytic bacilli.

**Thermophilic bacilli**

A special physiological group among the aerobic sporeformers consists of the extreme thermophiles, which are capable of growing at temperatures as high as 65°C, and usually fail to grow at temperatures below 45°C. There are probably a number of species with this attribute, but their taxonomy has not been studied in detail. The characteristic environment for these organisms is decomposing plant material, in which the heat generated by microbial metabolic activity cannot be readily dissipated. The classical example is a moist haystack, in which the rise of temperature is so marked that it sometimes leads to spontaneous combustion. As the temperature rises, in the first place through the metabolic activities of mesophilic microorganisms, the primary microbial population is displaced by extreme thermophiles, principally bacilli and actinomycetes.

The ability to grow at very high temperatures clearly involves a considerable degree of physiological specialization. Its basis has been investigated in some detail in the thermophilic bacilli, particularly the species *B. stearothermophilus.* It reflects the synthesis by these organisms of proteins that are much more thermostable than the corresponding proteins of mesophilic organisms; both the protein-synthesizing machinery (ribosomes) and the enzymes of a thermophile can therefore operate at temperatures where the corresponding systems of a mesophile undergo rapid denaturation. The thermostability of a protein is determined by its primary structure: its specific amino acid sequence. The extreme thermophiles must therefore represent a highly specialized evolutionary subgroup among the aerobic sporeformers, since this physiological character could be gained only by mutational changes affecting virtually every protein made by the cell.

## *The anaerobic sporeformers (genus Clostridium)*

The sensitivity to free oxygen which is the distinguishing character of sporeformers belonging to the genus *Clostridium* was first demonstrated by Pasteur. For most clostridia, oxygen is not merely growth inhibitory, but lethal: exposure of vegetative cells to air for 5 to 10 minutes may cause massive death of the population. Spores, however, are not killed by oxygen. The lethal effect of free oxygen on the clostridia (as well as on many other strictly anaerobic bacteria) may have several mechanisms. In many cases, it results from the formation, by the bacteria themselves, of hydrogen peroxide, which is highly toxic. Most clostridia completely lack heme proteins, including catalase, and are incapable of decomposing hydrogen peroxide. However, they have a high content of flavoproteins, some of which can act as oxidases in the presence of oxygen:

$$\text{substrate} + O_2 \xrightarrow{\text{flavoproteins}} \text{oxidized substrate} + H_2O_2$$

Consequently, exposure of cells to air is followed by a rapid formation and intracellular accumulation of hydrogen peroxide.

The development of satisfactory methods for the isolation and cultivation of strict anaerobes was slow, and consequently the study of the clostridia lagged behind that of the aerobic sporeformers. It had been shown by Pasteur that some of these organisms are agents of butyric acid fermentation, and at about the same period, the role of clostridia in the anaerobic decomposition of proteins was established. Late in the nineteenth century, some clostridia were found to be pathogenic for man and other vertebrates. The species in question are not true parasites; their primary habitat, like that of the nonpathogenic clostridia, is soil. Their pathogenic abilities reflect the synthesis of extracellular proteins of very high toxicity for animals. One of these species, *C. botulinum*, never develops in the animal host; the disease that it causes, botulism, is a pure intoxication, resulting from growth

*Table 20.4 Subdivisions among the clostridia, primarily in terms of energy-yielding substrates and modes of energy-yielding metabolism*

| **Substrates used** | **Mode of energy-yielding metabolism** | **Other distinctive properties** | **Typical species** |
|---|---|---|---|
| *Sugars, starch, pectin* | *Butyric acid (or butanol–acetone) fermentation* | *Many can fix nitrogen* | *C. butyricum, C. pastorianum, C. welchii, C. acetobutylicum* |
| *Pairs of amino acids* | *Stickland reaction* | *Proteolytic; many can also perform butyric fermentations of sugars* | *C. sporogenes, C. tetani, C. botulinum* |
| *Single amino acids (e.g., glutamate)* | *Butyric or propionic acid fermentation* | *Nonproteolytic; often unable to attack sugars* | *C. tetanomorphum, C. cochlearum, C. propionici* |
| *Purines (e.g., uric acid)* | *Fermentation to* $NH_3$, $CO_2$, *acetic acid* | *Restricted to purines as energy sources* | *C. acidi-urici* |
| *Cellulose* | *Fermentation to acetic and succinic acids, ethanol,* $CO_2$, $H_2$ | *Attack few or no carbohydrates other than cellulose* | *C. dissolvens* |
| *Sugars* | *Fermentation to acetic acid* | – | *C. thermoaceticum* |
| *Sugars* | *Fermentation to ethanol and* $CO_2$ | *Also able to ferment a pyrimidine, orotic acid* | *C. oroticum* |
| *Ethanol + acetic acid* | *Fermentation to higher fatty acids* | *No other substrates fermentable* | *C. kluyveri* |
| *Lactic acid, some other organic acids* | *Anaerobic respiration, with sulfate as electron acceptor* | *Contain cytochromes* | *C. nigrificans* |
| $H_2$ | *Anaerobic respiration with* $CO_2$ *as electron acceptor;* $CO_2$ *converted to acetic acid* | *No other substrates attacked* | *C. aceticum* |

of the bacterium and toxin formation in foodstuffs subsequently ingested by the affected animal. The other pathogenic clostridia cause disease only after their introduction into the tissues of an animal, typically through the contamination of wounds. *Clostridium tetani*, the causative agent of tetanus, develops locally in wounded tissues, where it produces a diffusible neurotoxin which eventually affects all parts of the host's nervous system. Several other species, of which *C. welchii* is representative, cause a disease known as gas gangrene. They grow somewhat more extensively than *C. tetani* in wounds and produce a number of different toxic proteins, some of which cause lysis of red blood cells and others tissue necrosis. The detailed study of these pathogenic clostridia was undertaken during World War I, when tetanus and gas gangrene as a result of wound infections were major medical problems.

At about the same time, the clostridia that bring about an acetone–butanol fermentation were also subjected to investigation, with the aim of establishing this fermentation on an industrial basis (see Chapter 29).

Although many species of clostridia have been named, the descriptions are for the most part fragmentary, and the taxonomy of this genus is still in a very unsatisfactory state. It is therefore not possible to give a systematic account of the principal species, as was done for the genus *Bacillus*. It is more useful to discuss the group primarily in terms of its modes of anaerobic energy-yielding metabolism, which are extremely varied and probably afford the best basis for the taxonomic subdivision of the genus (Table 20.4).

**The butyric acid bacteria**

Many *Clostridium* species, exemplified by the species *C. butyricum*, ferment soluble sugars, and sometimes starch and pectin, with the formation of butyric and acetic acids, $CO_2$, and $H_2$ as the principal end-products. This character defines a large number of species, sometimes known collectively as the *butyric acid bacteria*. In some species (e.g., *C. acetobutylicum*), the fatty acids are further converted to butanol, ethanol, and acetone; this represents a minor biochemical variant of the butyric acid fermentation. One member of this group, *C. pastorianum*, was shown by Winogradsky to be an active nitrogen fixer, and more recent work indicates that the capacity for nitrogen fixation is widespread among butyric acid bacteria. A starchlike polysaccharide, which stains blue with iodine, is commonly formed by these organisms as a reserve material. Many produce capsules (Figure 20.8).

Some clostridia that perform a butyric fermentation of sugars are also proteolytic and can use amino acids as alternative fermentable

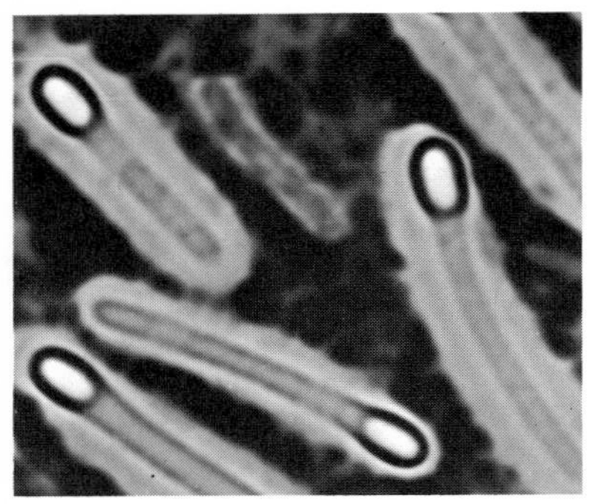

*Figure 20.8 Sporulating cells of Clostridium (× 2,880). The unstained cells are surrounded by capsules, demonstrated by preparing the smear in nigrosin, which provides the dark background. Courtesy of C. F. Robinow.*

substrates. In fact, it is possible to find among a large number of strains a virtually complete spectrum of physiological types, ranging from those which are completely nonproteolytic and depend exclusively upon fermentable carbohydrates as energy sources, to those which are very strongly proteolytic and ferment sugars weakly, if at all.

**Clostridia that ferment amino acids**

Among the proteolytic clostridia, the characteristic mechanism of energy-yielding metabolism at the expense of amino acids is the Stickland reaction. This involves a coupled oxidation–reduction between pairs of amino acids, one amino acid serving as a hydrogen donor and the other as a hydrogen acceptor (see p. 192).

Certain other clostridia, exemplified by *C. tetanomorphum*, have no proteolytic ability but can obtain energy by the fermentation of single amino acids; *C. tetanomorphum* ferments histidine and glutamic acid. Histidine is decomposed via glutamate, which is fermented with the formation of the end-products characteristic of a butyric acid fermenta-

$^{1}COOH-{}^{2}CHNH_2-{}^{3}CH_2-{}^{4}CH_2-{}^{5}COOH$ *glutamate* → ($NH_3$) → $^{1}COOH-{}^{2}CH={}^{4}CH({}^{5}COOH)-{}^{3}CH_3$ *mesaconate* → ($H_2O$) → $^{1}COOH-{}^{2}CH_2-HO{}^{4}C({}^{5}COOH)-{}^{3}CH_3$ *citramalate*

*citramalate* → $^{5}COOH-{}^{4}CO-{}^{3}CH_3$ *pyruvate* and $^{4,1}COOH-{}^{3,2}CH_3$ **acetate**

*pyruvate* → $^{4,1}COOH-{}^{3,2}CH_3$ **acetate** + $^{5}CO_2$ + $H_2$ + $^{3}CH_3-{}^{4}CH_2-{}^{3}CH_2-{}^{4}COOH$ **butyrate**

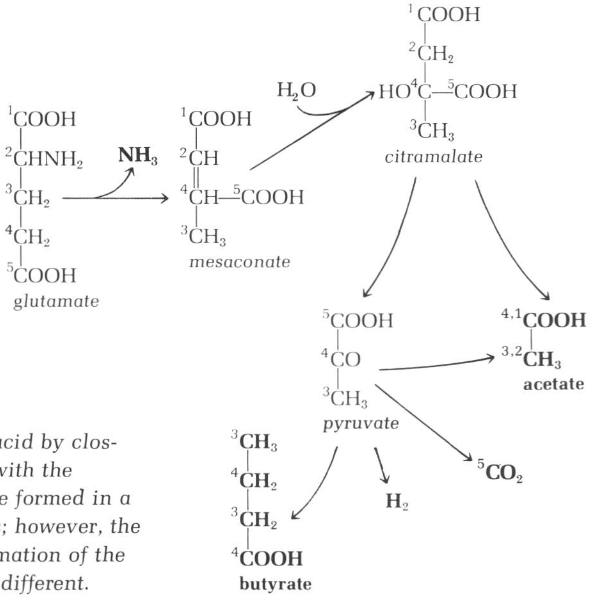

*Figure 20.9 The fermentation of glutamic acid by clostridia. The products (set in boldface) are, with the exception of ammonia, identical with those formed in a butyric acid fermentation of carbohydrates; however, the biochemical pathway that leads to the formation of the key intermediate, pyruvic acid, is entirely different.*

tion, together with ammonia. Although the products of glutamate fermentation are identical with those arising from a butyric fermentation of sugars, the pathway is entirely different, as shown in Figure 20.9. *Clostridium tetanomorphum* can likewise ferment sugars. Another glutamate-fermenting species, *C. cochlearum*, is completely unable to attack carbohydrates.

*Clostridium propionicum* can ferment alanine, serine, or threonine with formation of propionic acid as a major end-product, e.g.,

$$3 \text{ alanine} + 2H_2O \rightarrow 3NH_3 + 2CH_3CH_2COOH + CH_3COOH + CO_2$$

It is also unable to attack carbohydrates.

**Clostridia that ferment ring compounds**

Several nutritionally specialized clostridia can ferment nitrogen-containing ring compounds: purines, pyrimidines, or pyridines. The best studied of these processes is the fermentation of purines (e.g., uric acid) by *C. acidi-urici*, a species that is unable to use any other type of energy source. The products are acetic acid, ammonia, and $CO_2$.

**Other clostridial fermentations of carbohydrates**

Although butyric acid fermentation is the most common mode of carbohydrate breakdown among the clostridia, it is by no means the only one. Several species of clostridia (e.g., *C. dissolvens*) ferment the polysaccharide cellulose, with formation of acetic and succinic acids, ethanol, $CO_2$, and $H_2$ as the major end-products. *Clostridium thermoaceticum* ferments soluble sugars with the formation of acetic acid as the principal end-product. The mechanism of this unique fermentation has been studied in some detail. By classical fermentative mechanisms, only 2 moles of acetic acid can be formed from 1 mole of glucose. The very high yield of acetic acid in this fermentation is a result of the reduction of $CO_2$:

$$\text{glucose} + 2H_2O \rightarrow 2CH_3COOH + 2CO_2 + 8H$$
$$2CO_2 + 8H \rightarrow CH_3COOH + 2H_2O$$

*Clostridium oroticum* performs a fermentation of sugars in which ethanol and $CO_2$ are the major end-products. It can also ferment a pyrimidine derivative, orotic acid.

**The fermentation of ethanol and acetate**

The unusual range of substrates that can be fermented by clostridia is strikingly exemplified by *C. kluyveri*, which ferments mixtures of ethanol and acetic acid with the production of higher fatty acids (primarily butyric and caproic acids) as end-products.

**The sulfate-reducing clostridia and** *C. aceticum*

Several species, exemplified by *C. nigrificans*, employ the anaerobic respiration of organic compounds as a means of obtaining energy. Like the members of the genus *Desulfovibrio*, these organisms use sulfate (reduced to $H_2S$) as a terminal inorganic electron acceptor. The sulfate-reducing clostridia are consequently the only species of the genus which contain cytochromes, a property that is unusual among strictly anaerobic bacteria. These heme pigments comprise part of the electron transport chain involved in sulfate reduction.

Finally, it should be noted that one species of the genus, *C. aceticum*, uses $CO_2$ as a terminal electron acceptor. The unique electron donor is molecular hydrogen, and the energy-yielding reaction is

$$2CO_2 + 4H_2 \rightarrow CH_3COOH + 2H_2O$$

This species is, strictly speaking, a chemoautotroph, since its energy source is inorganic, even though the product of the energy-yielding reaction is an organic compound.

## Other endospore-forming bacteria

A number of very large bacteria which appear to form endospores occur in the intestinal contents of animals. However, none of them has been cultivated, so their nature and systematic position are not known. The most remarkable of these organisms is *Metabacterium*, which occurs in the intestinal tract of guinea pigs. Each sporulating cell contains several endospores (Figure 20.10).

*Figure 20.10 Metabacterium polyspora: smear from intestinal tract of guinea pig, showing three sporulating cells, each containing two or more rod-shaped endospores (× 2,400). Courtesy of C. F. Robinow.*

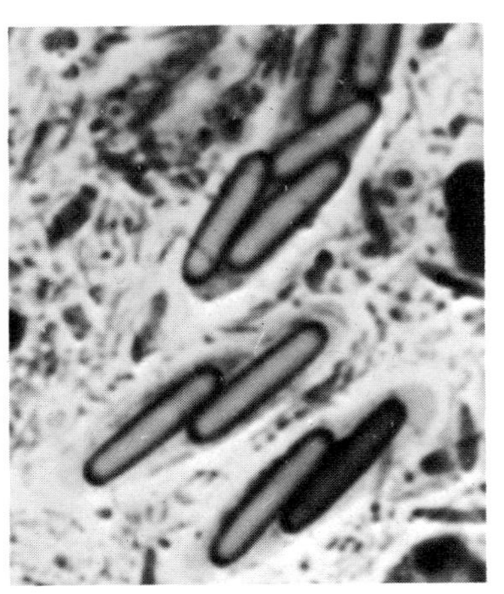

## The sporulation process

The development of an endospore involves the formation, within a vegetative cell, of a *new kind of cell*, with a completely different fine structure, chemical composition, and enzymatic constitution. Upon germination, an endospore gives rise once more to a typical vegetative cell. The fascinating problems posed by endospore formation and germination have led to intensive experimental studies on these processes. Although most of the work has been performed on aerobic sporeformers, the less detailed information about sporulation in clostridia indicates that the endospores of this group are fundamentally similar in their structure, their mode of formation, and their mode of germination.

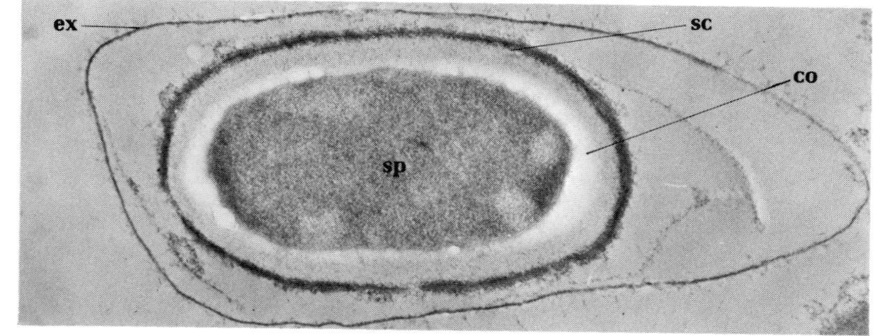


*Figure 20.11 Electron micrograph (× 37,500) of a thin section of an endospore of Bacillus cereus (compare with light micrograph of same object, Figure 20.2): ex, exosporium; sc, spore coat; co, cortex; sp, spore cell. Courtesy of Dr. C. F. Robinow.*

**The formation of endospores: cytological events**

At the level of resolution of the light microscope, little can be learned about the sporulation process. It normally begins some hours after a population of vegetative cells enters the stationary phase and is first manifested by the appearance in these cells of a relatively transparent *forespore*, which then gradually changes into the characteristically refractile, thick-walled mature spore. The development of the forespore is preceded by unusual nuclear changes: the pair of nuclear bodies in the vegetative cell fuse into a rod-shaped structure, which then divides; part of the DNA enters the developing spore and part remains in the sporangium. Electron microscopy of thin sections of sporulating bacilli has added much further information. Division of the rod-shaped nucleus is accompanied by *unequal cell division:* a new septum grows across the cell near one pole, segregating in the smaller daughter cell the DNA destined to enter the spore. The formation of the septum is not followed, as in normal cell division, by the development of a transverse cell wall which separates the daughter cells; instead, the newly formed septum grows back around the pole of the smaller cell, which thus becomes cut off from, and finally completely enclosed by, the larger cell. It is only at this stage that the forespore becomes visible with the light microscope. Once completely enclosed by the larger cell, or sporangium, the forespore develops rapidly into a mature endospore. The maturation process involves the elaboration around the forespore of additional outer layers. These include a layer known as the *cortex*, which is enclosed by a thick (usually multilayered) *spore coat*, which may have a sculptured surface. In certain bacilli, the spore coat is enclosed by a loose outer investment, the exosporium. The process of spore maturation may be accompanied by the development, in the sporangium, of special *parasporal bodies*, of which the bipyramidal crystals characteristic of *B. thuringensis* (see Figure 20.5) constitute one of the most striking examples. The fine structure of a mature spore is shown in Figure 20.11 and its development in Figure 20.12.

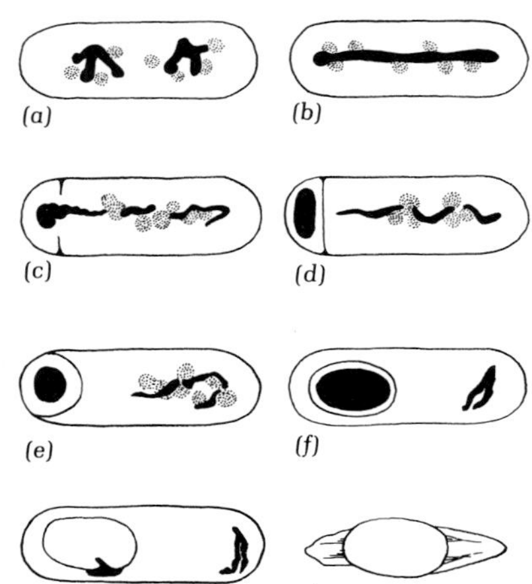


*Figure 20.12* *A diagrammatic representation of the cytological changes accompanying endospore formation in Bacillus cereus. (a) Vegetative cell, containing two nuclear bodies (black areas) and granules of reserve material (stippled areas). (b) Condensation of the nuclear material into a rod-shaped element. (c) Beginning of transverse wall formation. (d) Completion of the transverse wall; the forespore with its nuclear material is now cut off from the vegatative cell. (e) Growth of the new wall around the forespore. (f) Completion of the enclosing wall of the spore. (g) Maturation of the spore. (h) Liberated spore, surrounded by a loose outer coat, the exosporium. After I. E. Young and P. FitzJames, "Chemical and morphological studies of bacteria spore formation. I." J. Biophys. Biochem. Cytol.* **6,** *467 (1959).*

## The formation of endospores: physiological and biochemical events

In endospore formation, the cytological transformations are accompanied by profound changes in both the chemical and enzymatic composition of the sporulating cell. Many enzyme systems characteristic of the vegetative cell are reduced to very low levels, or completely absent, in endospores. In contrast, other enzymes are synthesized preferentially during the maturation of the spore; for example, alanine racemase is present at a far higher level in spores than in vegetative cells.

There is also a massive synthesis of new structural constituents, absent from the vegetative cell. The spore coat has a completely different composition from the wall of the vegetative cell, as shown by the fact that it is immunologically different and contains amino acids not present in the peptidoglycan of the vegetative cell walls. The most remarkable chemical change associated with spore formation is the synthesis, in large amounts, of a substance of low molecular weight, *dipicolinic acid* (Figure 20.13). This substance, present in the mature spore as the calcium salt, accounts for some 10 to 15 percent of its dry weight but is undetectable in vegetative cells. The calcium dipicolinate of the spore appears to be localized in the cortex, between the spore cell and the outer spore coat.

Many years before the discovery of dipicolinic acid as a spore con-

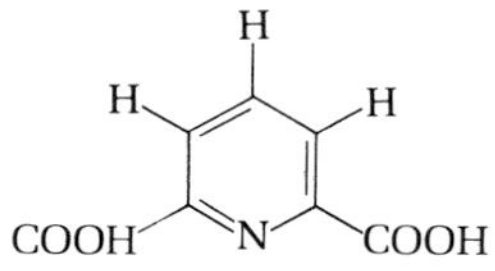

H
H H
COOH N COOH

*Figure 20.13* *The structure of dipicolinic acid. In endospores, this substance is principally present as the calcium salt.*

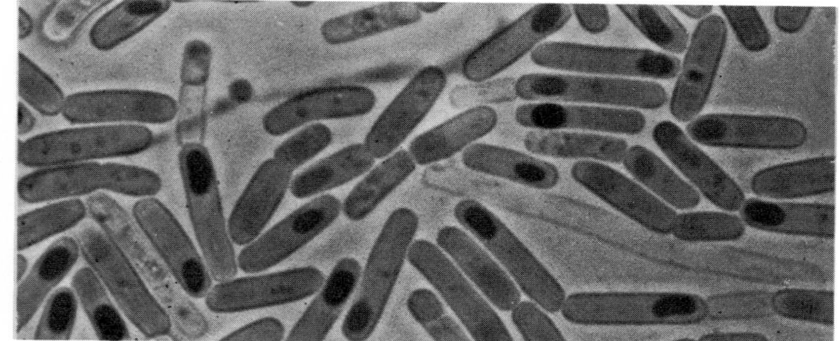

*Figure 20.14 Developing endospores in Bacillus. Photograph of an unstained preparation with ultraviolet light (× 2,000, approximately). Note the intense absorption of ultraviolet light by the developing endospores, caused by their content of dipicolinic acid, a substance that does not occur in the vegetative cells. From R. G. Wyckoff and A. L. ter Louw, "Some ultraviolet photomicrographs of B. subtilis." J. Exptl. Med.* **54,** *No. 3, p. 451 (Sept. 1931).*

stituent, it had been observed that spores absorb ultraviolet light much more intensely than vegetative cells (Figure 20.14). This is attributable to their high content of dipicolinic acid, which has a strong absorption band in the ultraviolet region.

**The problem of resistance to heat**

The resistance of endospores to toxic chemicals can be easily accounted for by assuming that the outer layers of the spore are highly impermeable, an assumption that is reasonable in view of the resistance of spores to the penetration of stains. However, the extreme thermal resistance of spores, which can survive heat exposures sufficient to denature solutions of both proteins and nucleic acids, is less easy to account for. The probable explanation has become evident from careful measurements of the density and the refractive index of spores, which have shown that *their water content is extremely low*, not much greater than that of dry wool or casein (Table 20.5). In the dehydrated state,

*Table 20.5 Water content of bacilli (spores and vegetative cells) and of dry protein, determined from measurements of refractive index*[a]

| | **Water content, g/100 ml** | |
|---|---|---|
| *Proteins* | | |
| *Wool* | 15 | |
| *Dry casein* | 14 | |
| *Bacillus spp.* | | |
| *B. megaterium* | 77.5[b] | 15[c] |
| *B. cereus* | 78[b] | 23[c] |

[a]From K. Ross and E. Billing, *J. Gen. Microbiol.* **16,** 418 (1957).
[b]Vegetative cells.
[c]Spores.

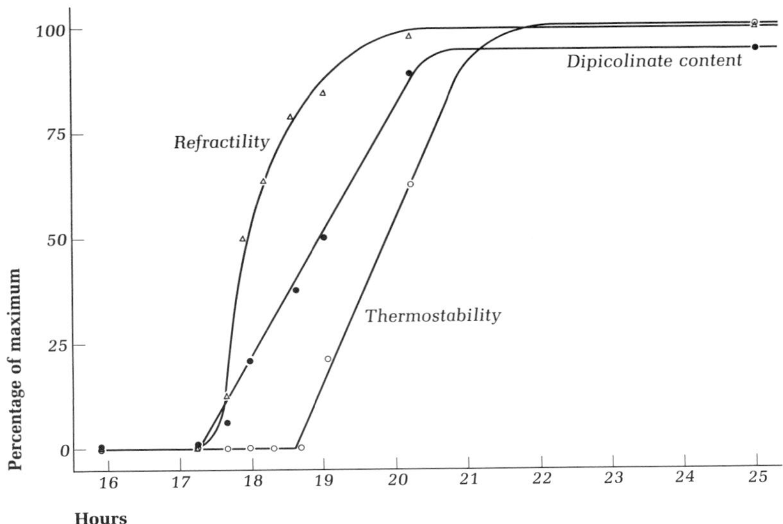


*Figure 20.15 The increases in the refractility, thermostability of the cells, and dipicolinate content of the population that occur during sporulation in a culture of Bacillus cereus. All values are plotted against the age of the culture in hours. After T. Hasimoto, S. H. Black, and P. Gerhardt, "Development of fine structure, thermostability, and dipicolinate during sporogenesis in a bacillus." Can. J. Microbiol.* **6,** *203 (1960).*

proteins and nucleic acids become far more resistant to thermal denaturation; and even vegetative cells of bacteria, if dry, are as thermoresistant as spores. Hence, the thermal resistance of endospores can probably be attributed to the fact that the enclosed cell is essentially water-free; progressive dehydration of the spore cell is accordingly one of the major physiological events accompanying the maturation of the endospore.

Although the mechanism of dehydration is not known, there is considerable circumstantial evidence to suggest that the development of the cortex and the accompanying massive synthesis of dipicolinic acid play important roles. The acquisition by the spore of refractility and heat resistance follows very closely the formation of dipicolinic acid during the maturation process, as shown in Figure 20.15. If sporulation occurs in a medium that does not contain calcium ions, the spores formed are weakly refractile, of low thermal resistance, and have a very low content of dipicolinic acid. Finally, certain mutants of bacilli make spores that are devoid of a cortex and do not contain dipicolinic acid; such spores have low refractility and heat resistance.

## The genetic and regulatory aspects of sporulation

As shown by the preceding account, the formation of endospores involves an elaborate sequence of special synthetic events in the sporulating cell, entirely different from the synthetic events associated with vegetative growth. This shows that there must reside in the genome of every spore-forming bacterium a considerable amount of genetic information specifically concerned with the sporulation process and which is not expressed phenotypically during vegetative growth. The genetic complexity of the system has been shown by the properties of mutants with derangements of the sporulation process. From any spore-forming bacterium, it is possible to obtain mutants that are completely incapable of initiating sporulation, as well as mutants in which the events leading to spore formation are blocked at later stages, resulting in the formation of abnormal or immature spores. At a conservative estimate, some 50 genes must be involved in the sporulation process. The genetic analysis of sporulation mutants in *B. subtilis* by the techniques of transformation and transduction has shown that these genes are widely distributed over the bacterial chromosome.

Since the genes governing the sporulation process are not normally expressed during the vegetative growth of spore-forming bacteria, it must be assumed that they are subject to a system of repression, which is relieved when vegetative growth ceases. Massive sporulation can be induced experimentally by removing cells from a growing culture and transferring them to a nitrogen-free medium. For 2 or 3 hours after such a transfer, the provision of a nitrogen source leads to a resumption of vegetative growth. After a certain point, however, the cells become *committed to sporulation*, in the sense that the addition of a nitrogen source no longer causes a resumption of vegetative growth, the population instead proceeding to sporulate. This fact suggests that the genes which govern sporulation are expressed sequentially, the initiation of expression being prevented by the repression of the first gene of the sequence. Commitment to sporulation occurs when this repression has been relieved, and thereafter the expression of genes operative at later steps in the sequence can no longer be prevented.

Although repression of sporulation is normally extremely effective during vegetative growth, it can be overcome under certain circumstances. In synthetic media where the growth rate is relatively low, aerobic sporeformers form a small number of spores in the course of growth; these spores are formed at a constant rate, so that their increase in the population parallels the increase in the number of vegetative cells. This behavior suggests that there is a certain probability under such conditions for a vegetative cell to undergo commitment. An even more striking demonstration of the fact that complete arrest of growth is not essential to initiate sporulation has been obtained from experiments on the growth of *B. megaterium* in a chemostat. When glucose is supplied as the rate-limiting nutrient, it is impossible to maintain a vegetative population at a growth rate of less than 0.5 division per hour; instead, mass sporulation occurs, and the population is washed out of the growth chamber.

**The formation of extracellular products during sporulation**

Many aerobic sporeformers produce antibiotic substances. These antibiotics are peptides of peculiar structure: they are cyclic polypeptides, and contain some amino acids of the unnatural, D configuration. Examples are bacitracin, produced by *B. licheniformis;* polymyxin, produced by *B. polymyxa;* and gramicidin, produced by *B. subtilis.* Furthermore, the aerobic sporeformers frequently produce extracellular proteases and ribonucleases. Kinetic studies have shown that the synthesis both of peptide antibiotics and of these hydrolases does not occur during exponential growth but is characteristically associated with entry of the population into the stationary phase. It has recently been found that the formation of these extracellular substances is intimately associated with the sporulation process. Certain mutants that are blocked at an early stage of sporulation simultaneously lose the ability to produce antibiotics and hydrolytic enzymes. These extracellular products are still synthesized by mutants blocked at late stages in the sporulation process. Furthermore, mutants selected for inability to synthesize protease or ribonuclease are asporogenous. Hence it seems likely that the production of antibiotics, protease, and ribonuclease is controlled by genes concerned with the sporulation process and which are subject to the same mechanism of repression during vegetative growth. It is possible that the formation of the exotoxins characteristic of pathogenic clostridia may be associated with the sporulation process in these species.

**Endospore germination and outgrowth**

The germination of endospores is manifested by simultaneous structural and physiological changes. The spores lose refractility and become readily stainable; at the same time, heat stability is lost. These changes occur very rapidly. They are accompanied by a disappearance of the cortex, and the liberation of soluble organic materials, representing about 30 percent of the dry weight of the spore; approximately half the material released upon germination consists of calcium dipicolinate, the rest being represented by protein, peptides, and peptidoglycan material of relatively low molecular weight. If the medium in which germination takes place is not adequate to support vegetative growth, no further changes occur. However, if the nutrients necessary for growth are present, germination is followed by the conversion of the spore cell into a vegetative cell: after 30 to 60 minutes, the outer spore coat ruptures and the vegetative cell emerges (Figure 20.16).

The factors required to trigger germination vary from species to species. In some sporeformers, rapid and massive germination can be

*Figure 20.16 Spore germination in (a) Bacillus polymyxa and (b) B. circulans (stained preparations, × 4,500). Courtesy of C. F. Robinow and C. L. Hannay.*

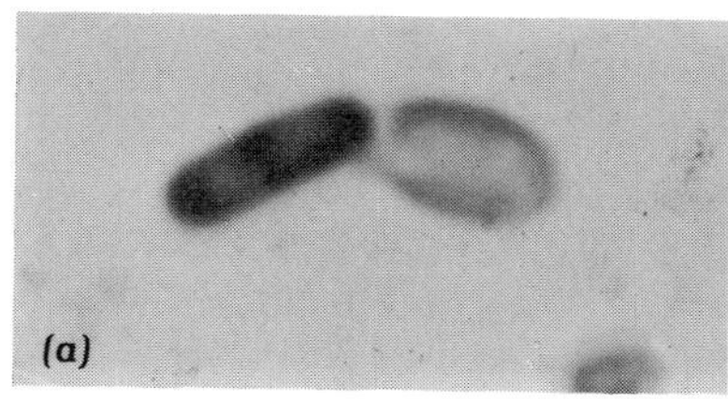


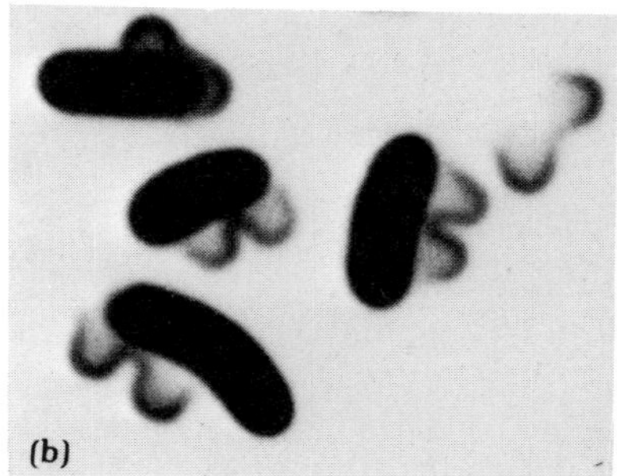


elicited by *heat activation:* after exposure for some minutes to a temperature of 60 to 80°C, spores will germinate in water. In other species, specific chemicals are required for germination. In the *B. cereus* group, adenosine triggers the process; in *B. subtilis, B. polymyxa*, and *B. sphaericus*, L-alanine is required. The mechanism by which heat shock and specific chemicals break the dormancy of spores is not understood. Spore germination can also be induced by a mechanical treatment, rapid shaking of a suspension of spores for a few minutes with very fine glass beads. After this treatment, most of the spores have germinated, as judged by the criteria listed above, and are capable of outgrowth when placed in a suitable medium. In this case, mechanical abrasion of the spore coat appears to be responsible for breaking dormancy.

## Further reading

**Books**


Smith, N. R., R. E. Gordon, and F. E. Clark, *Aerobic Sporeforming Bacteria*, Agricultural Monograph 16, U.S. Department of Agriculture. Washington, D.C.: U.S. Government Printing Office, 1952.

Sussman, A. S. and H. O. Halvorson, *Spores, Their Dormancy and Germination*. New York: Harper & Row, 1966.


**Symposia, reviews, and original articles**


Gibson, T., "An Investigation of the *Bacillus pasteurii* group." *J. Bacteriol.* **28,** 295 (1934); **28,** 313 (1934); **29,** 491 (1935).

Heimpel, A. M., "A Critical Review of *Bacillus thuringensis*." *Ann. Rev. Entomol.* **12,** 287 (1967).

Knight, B. C. J. G., and H. Proom, "A Comparative Survey of the Nutrition and Physiology of Mesophilic Species in the Genus *Bacillus*." *J. Gen. Microbiol.* **4,** 508 (1950).

Murrell, W. G., "The Biochemistry of the Bacterial Endospore." *Adv. Microb. Physiol.* **1,** 133 (1967).

Perkins, W. E., and other authors, "Symposium on Clostridia." Published in *J. Appl. Bacteriol.* **28,** 1–152 (1965).

Schaeffer, P., "Sporulation and the Production of Antibiotics, Exoenzymes and Exotoxins." *Bacteriol. Rev.* **33,** 48 (1969).

# *Chapter twenty-one Gram-positive eubacteria not forming endospores*

*The Gram-positive eubacteria* that do not form endospores include a number of unicellular groups, as well as the mycelial bacteria, or actinomycetes. Since these unicellular and mycelial bacteria are linked through a long series of transitional forms, it is convenient to discuss them all in one chapter.

## *Structure and reproduction: major groups*

The eubacteria to be described here can be divided into three principal groups, in terms of their structure and modes of reproduction (Table 21.1). Group I consists of rod-shaped or spherical organisms that are unicellular and reproduce exclusively by binary fission. Group III consists of what some authors call the euactinomycetes: organisms that grow in the form of a largely coenocytic, much-branched mycelium, and form specialized reproductive cells, late in the development of the mycelium. These reproductive cells are formed at the tips of hyphae. In some euactinomycetes, they are conidiospores, formed by transverse divisions at the hyphal tips. In others, they are sporangiospores, formed within terminal swellings of the hyphae. After liberation, each spore can germinate and give rise to a new mycelial colony. The euactinomycetes are strikingly analogous to certain eucaryotic fungi in structure and modes of reproduction.

Between the two easily distinguishable structural extremes repre-

*Table 21.1 Grouping of Gram-positive eubacteria (exclusive of sporeformers) in terms of structural properties*

| Cell form | Genus | Suprageneric group |
|---|---|---|
| *Group I. Unicellular rod-shaped or spherical organisms of regular cell form, reproducing by binary transverse fission* | | |
| Spheres | *Streptococcus* | *Lactic acid bacteria* |
| Spheres | *Leuconostoc* | |
| Spheres | *Pediococcus* | |
| Rods | *Lactobacillus* | |
| Rods | *Methanobacterium* | *Methane bacteria* |
| Spheres | *Methanococcus* | |
| Spheres | *Methanosarcina* | |
| Spheres | *Staphylococcus* | |
| Spheres | *Micrococcus* | |
| Spheres | *Sarcina* | |
| Rods | *Butyribacterium* | |
| *Group II. Unicellular, with tendency to irregular cell form and nonbinary fission; or mycelial, reproducing by massive fragmentation of the mycelium* | | |
| Unicellular | *Corynebacterium* | *Coryneform bacteria* |
| Unicellular | *Propionibacterium* | |
| Unicellular | *Arthrobacter* | |
| Unicellular | *Cellulomonas* | |
| Unicellular | *Microbacterium* | |
| Unicellular | *Brevibacterium* | |
| Unicellular | *Bifidobacterium* | |
| Mycelial | *Actinomyces* | *Proactinomycetes* |
| Unicellular–mycelial | *Mycobacterium* | |
| Mycelial | *Nocardia* | |
| *Group III. True actinomycetes; vegetative development exclusively mycelial; reproduction by formation of specialized conidiospores or sporangiospores* | | |
| | *Genera that produce conidiospores:* | |
| | *Streptomyces* | |
| | *Micromonospora* | |
| | *Thermoactinomyces* | |
| | *Genera that produce motile sporangiospores:* | |
| | *Actinoplanes* | |
| | *Spirillospora* | |
| | *Genera that produce immotile sporangiospores:* | |
| | *Streptosporangium* | |

sented by the eubacteria of groups I and III there lie the organisms of group II, which show all possible transitions between the unicellular and mycelial habits of growth. Near the unicellular end of the spectrum are the organisms known collectively as *coryneform bacteria*. Although these bacteria are technically unicellular, they show irregularities of cell shape and cell division that set them apart from the unicellular organisms of group I. At the mycelial end of the spectrum are the organisms known collectively as *proactinomycetes*. These organisms all have some capacity for mycelial growth; however, the mycelium is unstable and undergoes a more or less complete fragmentation into short, rod-shaped cells. Some proactinomycetes form a very rudimentary mycelium, which fragments early in growth; in others, the mycelium is more extensive, and never fragments completely. From proactinomycetes of the latter type, it is only a short step to the euactinomycetes.

The structural continuum represented by the organisms of group II poses a number of very difficult taxonomic problems, still unsettled. There is still no general agreement about where (if anywhere) a formal taxonomic line should be drawn between the Gram-positive unicellular eubacteria and the actinomycetes. Most bacteriologists include the coryneform group with the unicellular eubacteria, a taxonomic placement that can be justified in purely structural terms. In other respects, however, there are many resemblances between the coryneform group and the frankly mycelial proactinomycetes. If, on the other hand, all the organisms of group II are included among the actinomycetes, these eubacteria can no longer be defined in terms of the mycelial habit of growth, and it is not easy to find any other property that is shared by all the members of groups II and III upon which a group definition can be based.

To make clear the subtlety of the structural transitions evidenced by the organisms of group II, it may be useful to describe in detail the growth patterns of *Arthrobacter atrocyaneus*, a typical member of the coryneform group. In the stationary phase (Figure 21.1) the cells of *A. atrocyaneus* are oval or spherical; at this stage of its development, the organism could easily be mistaken for a somewhat irregular *Micrococcus* of group I. Upon transfer to a fresh medium, these coccoid cells

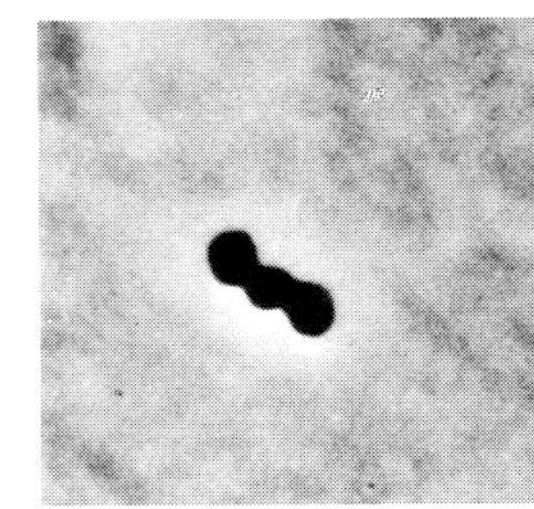
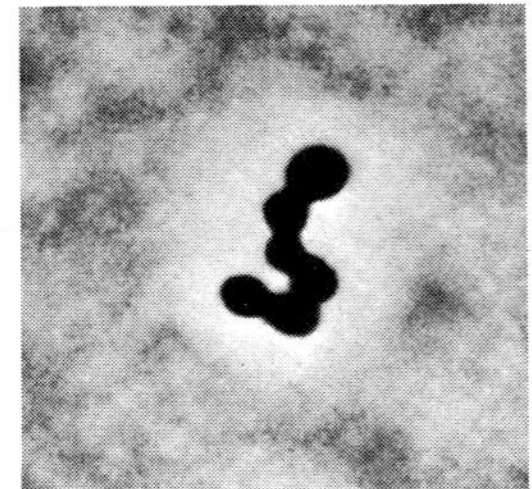
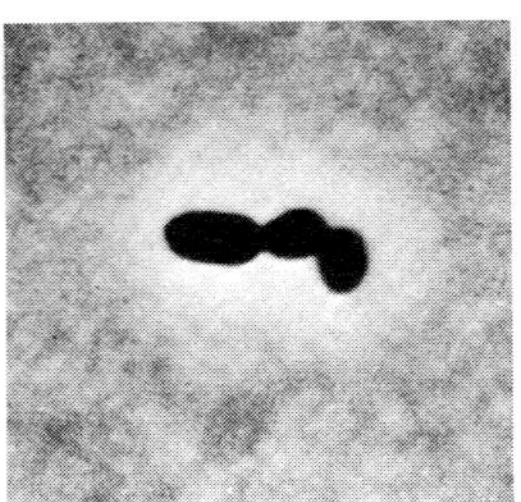

*Figure 21.1 Coccoid cells of Arthrobacter from a broth culture in the stationary phase (phase contrast, × 1,771). From M. P. Starr and D. A. Kuhn, "On the origin of V-forms in Arthrobacter atrocyaneus." Arch. Mikrobiol.* **42,** *289 (1962).*

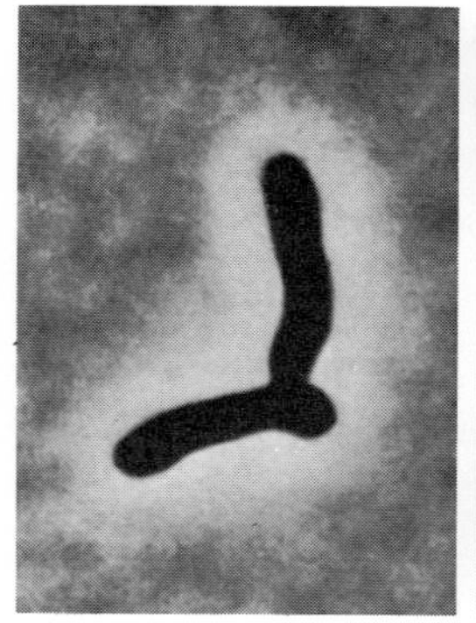
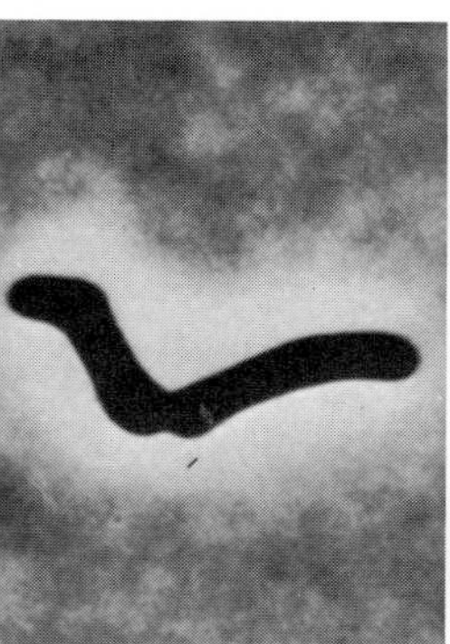
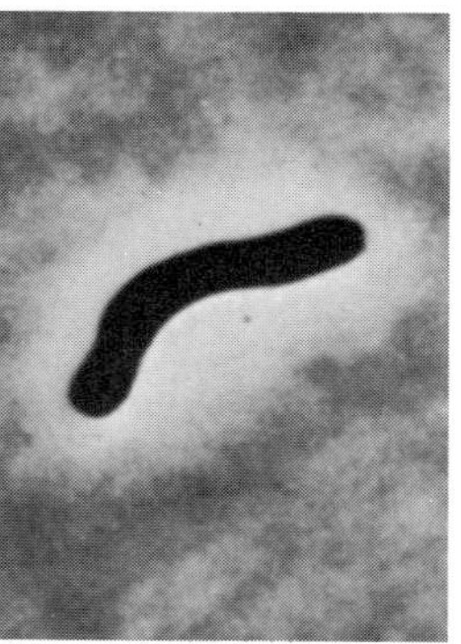

*Figure 21.2* Rod-shaped cells of *Arthrobacter* from a broth culture in the early exponential phase of growth (phase contrast, × 2,506). From M. P. Starr and D. A. Kuhn, "On the origin of V-forms in *Arthrobacter atrocyaneus*." Arch. Mikrobiol. **42**, 289 (1962).

elongate to rods of irregular shape (Figure 21.2), which reproduce in the main by binary transverse fission. Cell division is usually followed by a so-called "snapping movement," which brings the two daughter cells into V-shaped apposition: this is an almost universal character of coryneform bacteria. Snapping postfission movement in *A. atrocyaneus* is illustrated by a series of three photomicrographs of a growing group of cells, taken over a period of 45 minutes (Figure 21.3). The movement is very rapid; in the first photomicrograph, the cell at the upper left has snapped during the photographic exposure time of 15 seconds, so that its positions before and after movement are both recorded on the film. A second snapping movement, of the dividing cell at the lower right, occurred between the second and third photomicrographs. During the growth of the rod forms, not all cell divisions occur in a regular fashion. Daughter cells sometimes arise by so-called "angular growth": the development of thinner rods, which seem to arise through outgrowth from a subpolar position on the mother cell,

**Time, minutes**

*0* *15* *45*

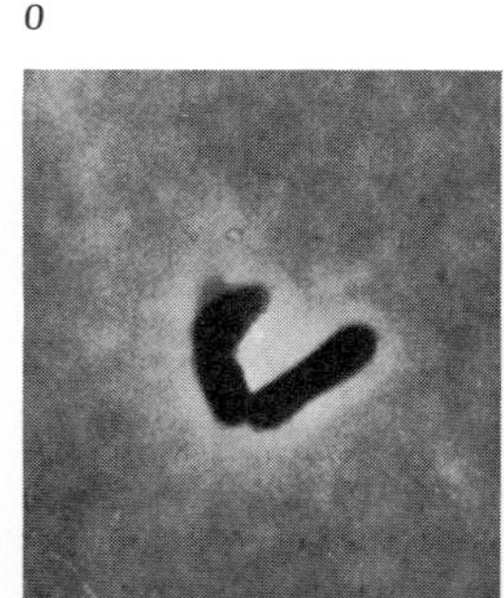
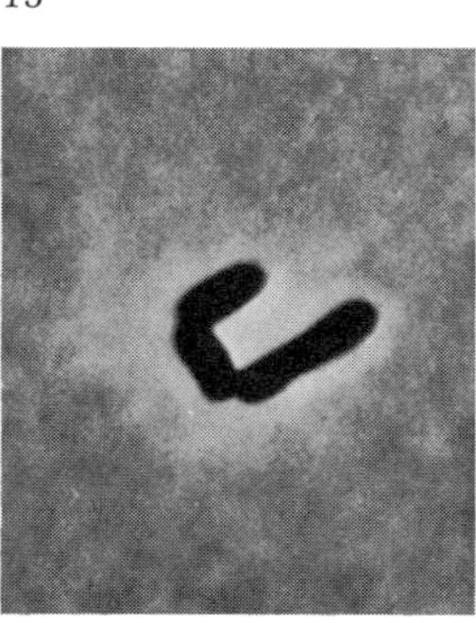
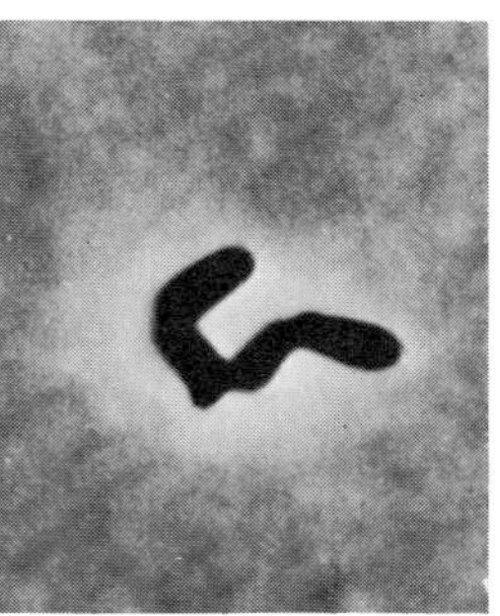

*Figure 21.3* Postfission movement in *Arthrobacter* illustrated by time-lapse pictures of a group of cells growing on agar (phase contrast, × 1,620). From M. P. Starr and D. A. Kuhn, "On the origin of V-forms in *Arthrobacter atrocyaneus*." Arch. Mikrobiol. **42**, 289 (1962).

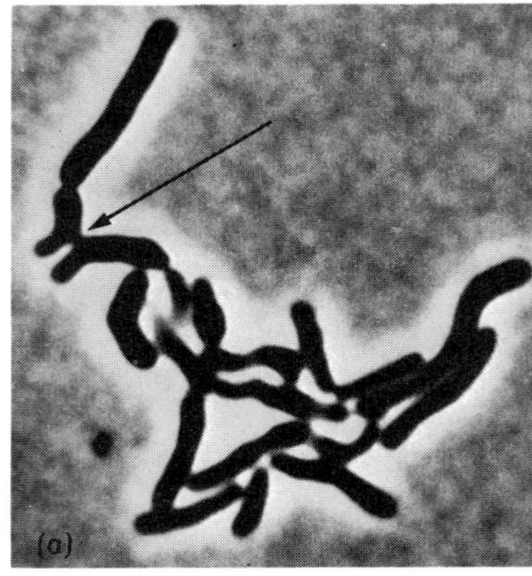

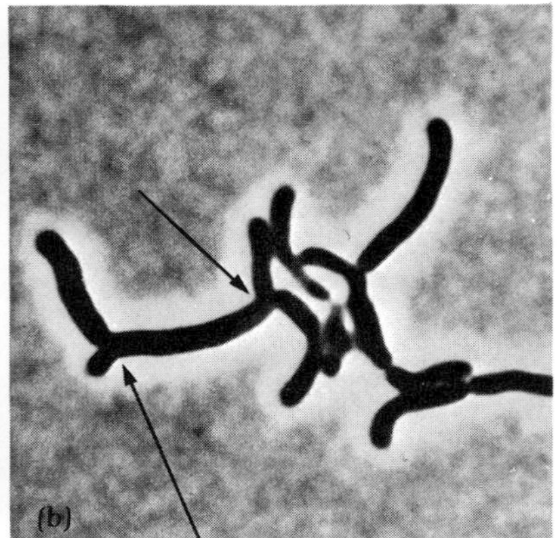


*Figure 21.4 Two microcolonies of Arthrobacter on agar, showing angular growth (arrows) (phase contrast, × 1,344). From M. P. Starr and D. A. Kuhn, "On the origin of V-forms in Arthrobacter atrocyaneus." Arch. Mikrobiol.* **42,** *289 (1962).*

to produce Y-shaped configurations (Figure 21.4). It can be readily appreciated that if such angular growth were not immediately followed by cell division, a mycelial mode of development would result. Coryneform bacteria such as *A. atrocyaneus* can, accordingly, be construed as bacteria in which there is a limited potential for mycelial growth.

**The Gram reaction and the composition of the cell wall**

In terms of the fine structure and chemical compositions of their walls (insofar as these are known) all the organisms in groups I, II, and III are of the Gram-positive type. Most of them are also Gram positive in the narrower technical sense of retaining the gentian violet–iodine complex upon treatment with ethanol. In a few of the constituent groups, however, Gram staining gives irregular results; at certain stages of growth, the cells are partly or entirely decolorized. This "Gram variability" is particularly characteristic of many coryneform bacteria.

The chemical composition of the cell wall polymers is a character of considerable taxonomic significance among these eubacteria. The constituent groups differ with respect to the diamino acid incorporated in their mureins. In some groups it is *meso*-diaminopimelic acid; in others, it is either LL-diaminopimelic acid or lysine. The walls of some groups also contain polysaccharides with characteristic sugar constituents, which accordingly serve as group markers. The more significant features of wall composition in groups I, II, and III are summarized in Table 21.2. In particular, it should be noted that the possession of a wall polysaccharide containing galactose and arabinose is a chemical character that unites certain structurally dissimilar members of group II: specifically, many coryneform bacteria, mycobacteria, and nocardias.

*Table 21.2 Taxonomically significant features of cell wall composition in unicellular Gram-positive eubacteria and actinomycetes*

| Genus | Diamino acid in murein | Specific wall sugars |
|---|---|---|
| *Group I* | | |
| *Streptococcus* | L-*lysine* | *Rhamnose*[a] |
| *Leuconostoc* | L-*lysine* | – |
| *Lactobacillus* | L-*lysine*[b] | – |
| *Staphylococcus* | L-*lysine* | – |
| *Micrococcus* | L-*lysine* | – |
| *Group II* | | |
| *Propionibacterium* | *meso-DAP* | – |
| *Actinomyces* | L-*lysine* | – |
| *Bifidobacterium* | L-*lysine* | – |
| *Coryneform bacteria (Corynebacterium, Arthrobacter)* | *meso-DAP* | *Arabinose, galactose* |
| *Mycobacterium* | *meso-DAP* | *Arabinose, galactose* |
| *Nocardia* | *meso-DAP* | *Arabinose, galactose* |
| *Group III* | | |
| *Streptomyces* | LL-*DAP* | – |
| *All other genera* | *meso-DAP* | – |

[a]In many (but not all) species.
[b]Except *L. plantarum*, which contains *meso*-DAP.

**Acid fastness**

The walls of some members of group II contain large amounts of waxy materials. These waxes, which are soluble in chloroform and have an extremely complex and unusual chemical composition, confer on the cell a special staining property, termed *acid fastness*. Acid fastness is determined by a procedure known as the Ziehl–Neelsen stain. The fixed cells are treated for 10 minutes with a hot, dilute phenolic solution of the red basic dye, fuchsin. They are then washed and treated for 10 minutes at room temperature with a 20 percent solution of a strong mineral acid (HCl or $H_2SO_4$). The cells of acid-fast bacteria retain the red color of fuchsin through this treatment, whereas other bacteria are rapidly decolorized.

Acid fastness is characteristic of the members of the genus *Mycobacterium* and is also displayed by a few members of the genus *Nocardia*. When present, it is a very valuable diagnostic character; however, in many mycobacteria acid fastness tends to be a somewhat variable property, dependent to a considerable extent on the nature of the medium in which the organism is grown. Consequently, it is not an infallible diagnostic test for these members of group II.

**Motility** The possession of flagella, and hence the capacity for active movement, is relatively rare among these Gram-positive bacteria; many of the constituent genera consist of permanently immotile organisms. In group I, the lactic acid bacteria and most cocci are immotile; however, in some species of lactic acid bacteria (e.g., *Streptococcus fecalis*), very rare motile strains are occasionally isolated. In group II, *Bifidobacterium, Propionibacterium, Actinomyces, Mycobacterium,* and *Nocardia* are generally immotile, although rare motile *Nocardia* strains have been reported. In the coryneform group, motility is a variable character, shown by some species but not by others.

In group III (the euactinomycetes), permanent immotility was long believed to be universal. However, when actinomycetes that reproduce by sporangiospores were discovered in 1950, it was found that some of these forms regularly produce motile sporangiospores, equipped with a polar tuft of flagella.

The rare and sporadic presence of flagella in some strains, species or genera of groups I, II, and III, suggests that the permanent immotility characteristic of so many of these eubacteria reflects an evolutionary loss of the power of active movement. Flagellar locomotion can occur only in an aqueous environment, and very few members of these groups are aquatic. The primary habitats of most are either soil or the animal body.

## *DNA base composition*

The evolutionary diversity of these Gram-positive eubacteria, already indicated by their wide range of structure, is strongly supported by the available data on mean DNA base composition, summarized in Table 21.3; the values cover almost the entire range found in procaryotic organisms as a whole. Among the unicellular forms of group I, most of the lactic acid bacteria have DNA of low G+C content, in the range 33 to 45 mole percent; the range is slightly wider for *Lactobacillus* spp. The spherical unicellular organisms of the *Staphylococcus* group are also near the low end of the G+C scale, whereas the organisms of the genus *Micrococcus*, with similar cell structure, have DNA with an exceptionally high G + C content, in the range of 66 to 72 mole percent G+C. The mycelial organisms of group III are also characterized by DNA of an extremely high G+C content, over 70 mole percent. The organisms of group II, which are transitional in structural respects, also occupy the middle portion of the scale of DNA base composition,

Table 21.3 *Ranges of mean base composition of DNA in unicellular Gram-positive eubacteria and actinomycetes*

| Organisms | Mole % G + C |
|---|---|
| *Group I* | |
| *Lactic acid bacteria* | |
| *Streptococcus* | 33–41 |
| *Leuconostoc* | 39–41 |
| *Pediococcus* | 38–44 |
| *Lactobacillus* | 35–53 |
| *Gram-positive cocci* | |
| *Staphylcococcus* | 30–35 |
| *Micrococcus* | 66–72 |
| *Sarcina* | 29–31 |
| *Group II* | |
| *Bifidobacterium* | 58 |
| *Propionibacterium* | 67–70 |
| *Corynebacterium, Arthrobacter* | 51–64 |
| *Mycobacterium* | 63–70 |
| *Nocardia* | 64–72 |
| *Group III* | |
| *Streptomyces* | 70–74 |
| *Micromonospora* | 72 |
| *Actinoplanes* | 73 |

with values ranging from 51 to 72 mole percent G+C. The lowest values are found in some coryneform bacteria; the highest ones in the mycelial organisms of the genus *Nocardia*.

## Energy-yielding metabolism and relations to oxygen

The methane bacteria are strictly anaerobic chemoautotrophs, which obtain energy by a unique mode of anaerobic respiration. Apart from these organisms, all the members of groups I, II, and III are chemoheterotrophs, which obtain energy either by aerobic respiration or by fermentation (Table 21.4).

Nearly all the members of group III have a purely respiratory metabolism and are strict aerobes. This is also true of the proactinomycetes and many coryneform bacteria in group II and *Micrococcus* in group I. The capacity for fermentative metabolism occurs in many members of

*Table 21.4 Modes of energy-yielding metabolism and relations to oxygen among the Gram-positive eubacteria*

| Group | Organisms | Modes of energy-yielding metabolism | Relations to oxygen |
|---|---|---|---|
| I | *Lactic acid bacteria* | *Fermentation* | *Facultative anaerobes* |
| | *Methane bacteria* | *Anaerobic respiration* | *Strict anaerobes* |
| | *Staphylococcus* | *Aerobic respiration or fermentation* | *Facultative anaerobes* |
| | *Micrococcus* | *Aerobic respiration* | *Strict aerobes* |
| | *Sarcina*<br>*Butyribacterium* | *Fermentation* | *Microaerotolerant anaerobes* |
| II | *Propionibacterium*<br>*Bifidobacterium*<br>*Actinomyces* | *Fermentation* | *Microaerotolerant anaerobes* |
| | *Corynebacterium*<br>*Arthrobacter, etc.* | *Aerobic respiration; varying degrees of fermentative ability* | *Strict aerobes or facultative anaerobes* |
| | *Mycobacterium*<br>*Nocardia* | *Aerobic respiration* | *Strict aerobes* |
| III | *Euactinomycetes*[a] | *Aerobic respiration* | *Strict aerobes* |

[a]With the exception of one *Micromonospora* species, an anaerobic, fermentative organism.

groups I and II. In nearly every case, the fermentable substrates are carbohydrates; the principal exceptions are the propionic acid bacteria, which can ferment lactic acid as well as carbohydrates. The most common modes of fermentation are the lactic and propionic fermentations, although a few members of group I perform either a butyric or an alcoholic fermentation.

Among the members of groups I and II that are capable of fermentation, there are wide differences with respect to the relations to oxygen. Three distinct physiological categories can be distinguished. Some of these organisms (for example, the staphylococci and many coryneform bacteria) are true facultative anaerobes. They possess a respiratory electron transport chain and can grow by means of respiration in the presence of oxygen or by fermentation in its absence. The lactic acid bacteria, although likewise able to grow readily in the presence of air, do not have a normal respiratory electron transport chain and are dependent for energy on the fermentation of carbohydrates under all environmental conditions. Finally, some of the fermentative organisms belonging to groups I and II are more or less sensitive to free oxygen;

they are dependent on fermentation for the provision of energy and can grow only under anaerobic conditions or in the presence of a very low partial pressure of oxygen. This behavior is characteristic of *Sarcina*, *Bifidobacterium*, *Propionibacterium*, and *Actinomyces*. These organisms are not microaerophiles, like *Spirillum* species, which have a respiratory metabolism, but are sensitive to normal partial pressures of oxygen; nor are they really strict anaerobes, like *Clostridium* and *Desulfovibrio*, which are rapidly killed by exposure to traces of free oxygen. The best term to describe their relations to oxygen is the clumsy phrase, *microaerotolerant anaerobes*.

## The lactic acid bacteria

Among the unicellular bacteria of group I, the largest constituent group consists of the rod-shaped or spherical organisms known collectively as *lactic acid bacteria* (Figure 21.5). These bacteria share an unusual constellation of physiological and biochemical properties. Their energy-yielding metabolism is fermentative. The fermentable substrates are carbohydrates and include a variety of monosaccharides, disaccharides, polyalcohols, and (in some species) starch. Lactic acid is always a major fermentative end-product, and in some cases virtually the only end-product.

The lactic acid bacteria grow readily on the surface of solid media exposed to air, even though they cannot derive energy from respiration. The absence of respiratory energy-yielding metabolism is a reflection of the fact that these organisms do not synthesize heme and therefore do not contain cytochromes or other heme-containing enzymes. They can perform limited oxidations of a few organic compounds; these oxidations are mediated by flavoprotein enzymes and are not accompanied by ATP synthesis. Consequently, the growth yields of lactic acid bacteria are the same in air and in anaerobiosis, the fermentation of sugars being the source of energy under both conditions.

Many strict anaerobes (e.g., the clostridia) which lack heme pigments die rapidly upon exposure to oxygen. Death results from the formation of hydrogen peroxide through flavoprotein-mediated oxidations, because in the absence of the heme protein catalase, this highly toxic substance cannot be destroyed by the cells. Lactic acid bacteria are also

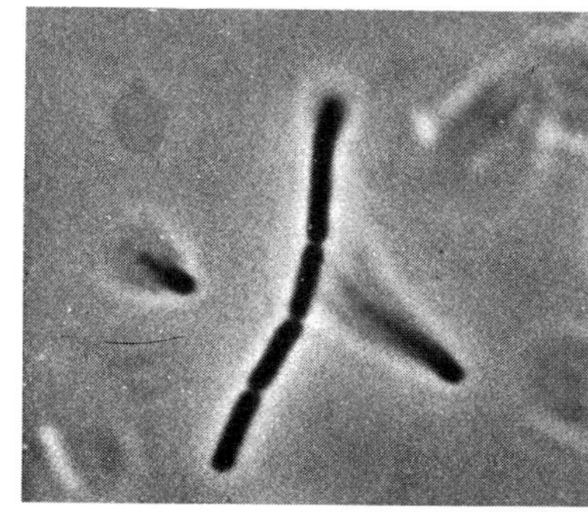

(a)

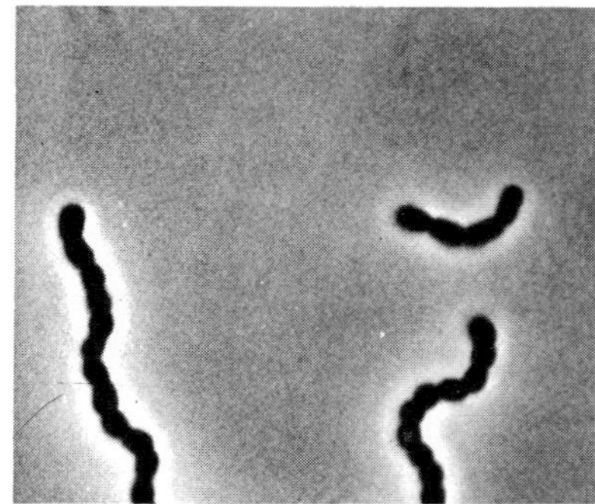

(b)

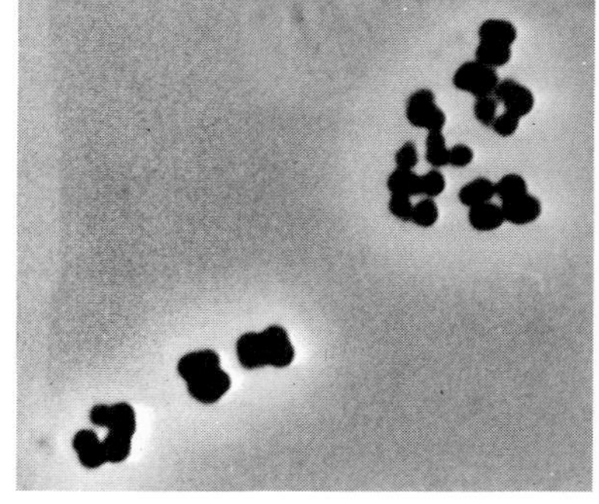

(c)

*Figure 21.5 The form and arrangement of cells in three genera of lactic acid bacteria: (a) Lactobacillus; (b) Streptococcus; (c) Pediococcus (phase contrast, × 2,000).*

catalase negative. Indeed, the absence of this enzyme, readily demonstrable by the absence of oxygen formation when cells are mixed with a drop of dilute hydrogen peroxide, provides one of the most useful diagnostic characters for the recognition of these organisms, since they are virtually the only catalase-negative bacteria able to grow in the presence of oxygen. How is their immunity to the lethal effects of hydrogen peroxide to be explained? It has been found that these bacteria contain another class of peroxide-destroying enzymes, *peroxidases*, which are flavoproteins. They mediate oxidations of organic substrates by means of $H_2O_2$. Consequently, the $H_2O_2$ formed by lactic acid bacteria under aerobic conditions can be destroyed by peroxidase action.

*Figure 21.6 The colony structure of (a) Lactobacillus plantarum and (b) Streptococcus lactis (× 8.9).*

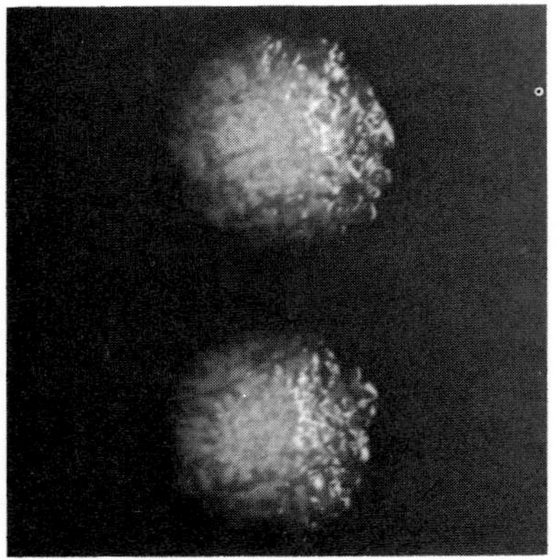

(a)

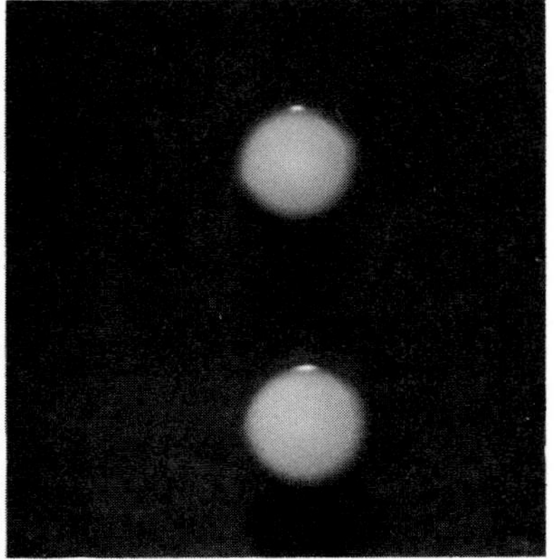

(b)

Certain lactic acid bacteria that are normally catalase negative become strongly catalase positive if they are grown on media that contain hemin (e.g., blood-containing media). Apparently, these species synthesize a protein that can combine with exogenously supplied hemin to form a conjugated protein with the properties of catalase.

The inability to synthesize hemin is only one manifestation of the *extremely limited synthetic abilities* characteristic of the lactic acid bacteria. All these organisms have complex growth-factor requirements; they invariably require B vitamins and, with one exception (*Streptococcus bovis*), a considerable number of amino acids. The amino acid requirements of many lactic acid bacteria are considerably more extensive than those of higher animals. As a result of their complex nutritional requirements, lactic acid bacteria are usually cultivated on media containing peptone, yeast extract, or other digests of plant or animal material. These must be supplemented with a fermentable carbohydrate to provide an energy source.

Even when growing on very rich media, the colonies of lactic acid bacteria (Figure 21.6) always remain relatively small (at most, a few millimeters in diameter). They are never pigmented; as a result of the absence of cytochromes, the growth has a chalky white appearance which is very characteristic. The small colony size of these bacteria is attributable primarily to low growth yields, a consequence of their exclusively fermentative energy-yielding metabolism. Some species can produce unusually large colonies when grown on sucrose-containing media, as a result of the massive synthesis of extracellular polysaccharides (either dextran or levan) at the expense of this disaccharide; in this special case, much of the volume of the colony consists of polysaccharide. Since dextrans and levans are synthesized only from sucrose, the species in question form typical small colonies on media containing any other utilizable sugar. In the isolation of the spherical lactic acid bacteria, which can grow in media that have an initial pH of 7 or above, the incorporation of finely divided $CaCO_3$ in the plating medium is useful, since the colonies can be readily recognized by the surrounding zones of clearing, caused by acid production (Figure 21.7).

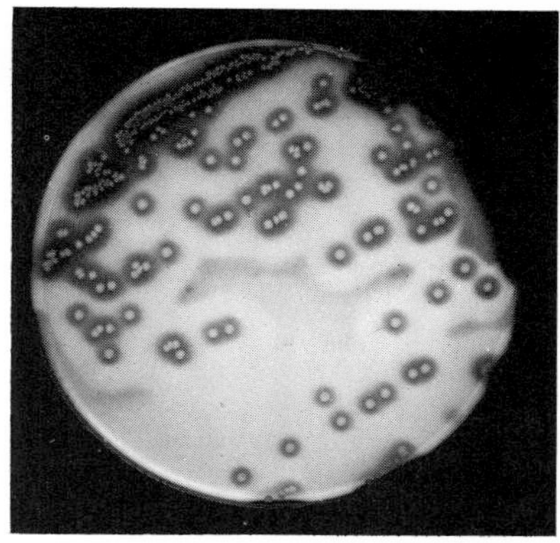

*Figure 21.7 Colonies of Streptococcus growing on an agar medium containing a suspension of calcium carbonate. Lactic acid production has dissolved the calcium carbonate, producing clear zones around each colony.*

Another distinctive physiological feature of lactic acid bacteria is *their high tolerance of acid,* a necessary consequence of their mode of energy-yielding metabolism. Although the spherical lactic acid bacteria can initiate growth in neutral or alkaline media, most of the rod-shaped forms cannot grow in media with an initial pH greater than 6. Growth of all lactic acid bacteria continues until the pH has fallen, through fermentation, to a value of 5 or less.

The capacity of lactic acid bacteria to produce and tolerate a relatively high concentration of lactic acid is of great selective value, since it enables them to eliminate competition from most other bacteria in environments that are rich in nutrients. This is shown by the fact that lactic acid bacteria can be readily enriched from natural sources through the use of complex media with a high sugar content. Such media can, of course, support the growth of many other chemoorganotrophic bacteria, but the competing organisms are almost completely eliminated as growth proceeds by the accumulation of lactic acid, formed through the metabolic activity of the lactic acid bacteria.

As a result of their extreme physiological specialization, the lactic acid bacteria are confined to a few characteristic and restricted natural environments. Some live in association with plants and grow at the expense of the nutrients liberated through the death and decomposition of plant tissues. They occur in foods and beverages prepared from plant materials: pickles, sauerkraut, ensilaged cattle fodder, wine, and beer. A lactic fermentation of the sugar initially present occurs during the preparation of pickles, sauerkraut, and ensilage. In fermented beverages, the lactic acid bacteria are potential spoilage agents, which sometimes grow and produce an undesirable acidity.

Other lactic acid bacteria constitute part of the normal flora of the animal body and occur in considerable numbers in the nasopharynx, the intestinal tract, and the vagina. These forms include a number of important pathogens of man and other mammals, all belonging to the genus *Streptococcus.*

A third characteristic habitat of the lactic acid bacteria is milk, to which they gain access when the milk is drawn, either from the body of the cow or from plant materials. The normal souring of milk is caused by certain streptococci, and both rod-shaped and spherical lactic acid bacteria play important roles in the preparation of fermented milk products (butter, cheeses, buttermilk, yogurt).

As a result of their activities in the preparation of foods and as agents of human and animal disease, the lactic acid bacteria are a group of major economic importance.

**The fermentative pattern of lactic acid bacteria**

The first systematic study of the lactic acid bacteria, carried out about 1920 by S. Orla Jensen, showed that these organisms can be divided into two physiological subgroups, defined by the products of glucose fermentation. The *homofermentative lactic acid bacteria* convert glucose almost quantitatively to lactic acid. The *heterofermentative lactic acid bacteria* convert it to a mixture of lactic acid, ethanol, and $CO_2$, in a molar ratio 1:1:1. The explanation in biochemical terms of these two fermentative patterns was obtained only much later. It is now known that homofermentative species dissimilate glucose through the Embden–Meyerhof pathway, whereas heterofermentative species dissimilate this sugar through the hexose monophosphate shunt pathway (see p. 189). The differences with respect to end-products accordingly express a basic difference with respect to the biochemical mechanism of sugar breakdown.

It should be specifically noted that the characterization of a lactic acid bacterium as homofermentative or heterofermentative must be made on the basis of the products of *glucose* fermentation. Some homofermentative and heterofermentative species can ferment pentoses such as arabinose and xylose, with formation of identical end-products: equimolar quantities of lactic and acetic acids.

In practice, the mode of glucose fermentation is usually ascertained by determining whether $CO_2$ is produced; this gas is an end-product only of heterofermentative glucose breakdown. Because $CO_2$ is very soluble in water and only 1 mole is formed per mole of glucose fermented, special cultural procedures must be used to demonstrate its production. The bacteria are grown in a sugar-rich, well-buffered medium provided with an agar seal, which traps the $CO_2$ produced (Figure 21.8).

Homofermentative and heterofermentative lactic acid bacteria also differ with respect to the fermentation of fructose. Heterofermentative species can reduce this ketohexose to the corresponding polyalcohol, mannitol:

$$\text{fructose} + \text{NADH}_2 \rightarrow \text{mannitol} + \text{NAD}$$

Consequently, fructose fermentation leads to the formation of considerable amounts of mannitol, coupled with the formation of acetic acid instead of ethanol as an end-product. The ideal overall equations

*Figure 21.8 The demonstration of $CO_2$ production by lactic acid bacteria in tubes of a sugar-rich medium with agar seals: (a) Streptococcus lactis; (b) Leuconostoc mesenteroides.*

for glucose and for fructose fermentation by a heterofermentative lactic acid bacterium are, accordingly,

glucose → lactic acid + ethanol + $CO_2$
3 fructose → lactic acid + acetic acid + $CO_2$ + 2 mannitol

The homofermentative lactic acid bacteria do not reduce fructose to mannitol, and the products of glucose and fructose fermentation are identical.

Another taxonomically useful character among lactic acid bacteria, also discovered by Orla Jensen, is the *optical configuration of the lactic acid formed.* This reflects the stereospecificity of the NAD-linked lactic dehydrogenases operative in the ultimate step reaction of the fermentation, reduction of pyruvic acid:

$CH_3COCOOH + NADH_2 \rightarrow CH_3CHOHCOOH + NAD$

In some lactic acid bacteria, this reduction is mediated by a single, stereospecific enzyme, yielding either the L (+) or the D (−) isomer of lactic acid. Other species possess two lactic dehydrogenases of differing sterospecificity, and the fermentative end-product is a racemic mixture of the D and L isomers.

**Taxonomic subdivision of the lactic acid bacteria**

The lactic acid bacteria are commonly placed in four genera (Table 21.5*a*,*b*). The genera *Streptococcus* and *Leuconostoc* contain organisms with spherical cells, dividing in a single plane; the genus *Pediococcus*

*Table 21.5a Taxonomic subdivision of the lactic acid bacteria*

| Cell shape and arrangement | Mode of glucose fermentation | Configuration of lactic acid | Genus |
|---|---|---|---|
| *Spheres in chains* | *Homofermentative* | L- | *Streptococcus* |
| *Spheres in chains* | *Heterofermentative* | D- | *Leuconostoc* |
| *Spheres in tetrads* | *Homofermentative* | DL- | *Pediococcus* |
| *Rods* | *Varied* | *Varied* | *Lactobacillus* |

*Table 21.5b Subgenera of Lactobacillus*

| Mode of glucose fermentation | Configuration of lactic acid | Growth rate at: 15°C | 45°C | Subgenus |
|---|---|---|---|---|
| *Homofermentative* | L-, L-, *or* DL- | − | + | *Thermobacterium* |
| *Homofermentative* | L-, *rarely* DL- | + | − | *Streptobacterium* |
| *Heterofermentative* | DL- | V[a] | V | *Betabacterium* |

[a]V denotes a variable reaction within the group.

also contains spherical forms, dividing in two planes at right angles to one another. Fermentation is heterofermentative in *Leuconostoc*, homofermentative in the other two genera. Somewhat illogically, all rod-shaped lactic acid bacteria, both homofermentative and heterofermentative, are placed in one genus, *Lactobacillus*. This genus is divided by some authors into three subgenera, distinguished partially by fermentative properties and partially by temperature relationships, as indicated in Table 21.5.

**The genus *Streptococcus***

The members of the genus *Streptococcus* can be subdivided into a number of groups, most of which contain several species. These groups are distinguishable from one another in both ecological and physiological terms (Table 21.6).

The *pyogenic group*, represented by *S. pyogenes*, consists of animal pathogens which characteristically produce an extracellular enzyme that lyses red blood cells. Hence, their colonies on agar media containing whole blood are surrounded by clear zones in which the erythrocytes have been completely lysed; as a result of the diffusion of the liberated hemoglobin, these cleared areas appear colorless or pale pink. This mode of action on blood is sometimes termed *β-hemolysis*. The pyogenic streptococci are delicate organisms. They have a narrow temperature range, being unable to grow at 10 or 45°C, and they are rapidly killed by exposure to 60°C. *Streptococcus pyogenes* is the agent of many major human streptococcal infections, including scarlet fever. It produces several toxic substances, in addition to hemolysin. These include fibrinolysin, an enzyme that dissolves clots of fibrin; hyaluronidase, which attacks hyaluronic acid, a constituent of connective tissue; an erythrogenic toxin, responsible for the rash characteristic of scarlet fever; and leucocidin, a substance that destroys white blood cells.

*Table 21.6 Principal groups among the streptococci*

| Group | Representative species | Hemolysis | Greening of blood | Growth at: 10°C | Growth at: 45°C | Arginine dihydrolase activity | Growth at pH 9.6 or with 6.5% NaCl | Bile solubility |
|---|---|---|---|---|---|---|---|---|
| *Pyogenic* | *S. pyogenes* | + | − | − | − | + | − | − |
| *Viridans* | *S. salivarius*<br>*S. bovis* | − | + | − | + | − | − | − |
| *Pneumococcus* | *S. pneumoniae* | − | + | − | − | | − | + |
| *Lactic* | *S. lactis* | − | − | + | − | V | − | − |
| *Enterococcus* | *S. fecalis* | V[a] | − | + | + | + | + | − |

[a]V denotes a variable reaction within the group.

The streptococci of the *viridans group* are so named because their colonies on blood agar plates are surrounded by a zone of greenish discoloration, in some cases accompanied by limited hemolysis. The greening occurs as a result of the oxidation of hemoglobin to methemoglobin, a change sometimes termed *α-hemolysis.* This is an incorrect name, since the color change is not necessarily accompanied by an actual lysis of red blood cells. In contrast to the pyogenic group, these streptococci grow well at 45°C and survive exposure to 60°C for 30 minutes. The members of the viridans group are animal parasites, which occur in the human mouth and throat and the intestinal tract. Most of them have little or no ability to cause disease. One distinctive member of this group is *S. salivarius*, which can be readily isolated from saliva by streaking it on a yeast extract–sucrose plate. On this medium, *S. salivarius* forms large, tough, gelatinous, translucent colonies, as a result of its ability to produce levan from sucrose.

The intestinal inhabitants among the viridans group include *S. equinus* and *S. bovis*, the predominant streptococci in the intestinal contents of horses and cattle, respectively. *S. bovis* is unique among the lactic acid bacteria by virtue of its relatively simply nutritional requirements; it does not require amino acids, although B vitamins are required.

The principal causative agent of bacterial pneumonia, *S. pneumoniae*, often referred to as the pneumococcus, also produces a greening of blood agar but possesses a number of special properties that set it apart from the members of the viridans group. The cells are characteristically paired and surrounded by a sharply circumscribed capsule, which confers on the colonies a smooth, glistening structure. The capsular substances are polysaccharides, which are strongly antigenic; and the species can be subdivided into a large number of types, of differing antigenic specificities, determined by the specific chemical nature of their capsular polysaccharides. Under conditions of laboratory cultivation, so-called "rough" mutants, devoid of capsules, readily arise. As described in Chapter 14, the study of the heritable transfer of the ability to form capsules in pneumococci led to the discovery of transformation, one of the three major modes of genetic transfer among bacteria. This discovery was a landmark in the development of twentieth-century biology, since it provided the first decisive experimental evidence that DNA is the carrier of genetic information.

The streptococci of the *lactic group*, which have no action on blood, are the lactic acid bacteria primarily responsible for the normal souring of raw milk. There are two closely similar species, *S. lactis* and *S. cremoris*. These species are characterized by their low temperature range, unusual among streptococci; they grow well at 10°C but cannot grow at temperatures in excess of 40°C. They are not found on the surface of the animal body or in the udder, and probably grow normally on plant materials, from which they gain access to milk at the time it is drawn.

The last major group among the streptococci consists of the *enterococci,* so named because they occur exclusively in the intestinal tract of man and other animals. The principal species is *S. fecalis.* The enterococci have distinctive physiological characters; their temperature range is wide (growth at both 10 and 45°C), and they are the only streptococci that can grow at pH 9.6 and in the presence of 6.5 percent NaCl. They are also the only streptococci that produce proteolytic enzymes; many strains can liquefy gelatin.

**The genus *Lactobacillus***

The rod-shaped lactic acid bacteria (genus *Lactobacillus*) provide a striking contrast in ecological respects to the streptococci. Although members of this group occur on or in the animal body—the mouth, the intestinal tract, the vagina—they are completely devoid of pathogenic ability. The commonest habitats are decomposing plant materials and milk or milk products. The lactobacilli are much stronger acid producers than the streptococci. They have a high tolerance for acetic acid as well as lactic acid and can therefore be isolated selectively by the use of a complex, carbohydrate-containing medium with a low initial pH (4.5) and containing acetic acid, which prevents the growth of nearly all other bacteria, including most streptococci.

Although the individual cells of the lactobacilli are short rods, they often have a tendency to remain adherent after division, forming very long chains. This feature, particularly pronounced in the thermophilic species, is grossly reflected in the formation of rough colonies with irregular edges, different from the smooth colonies with entire edges formed by streptococci (see Figure 21.6).

As shown in Table 21.3, the span of DNA base composition found in true lactobacilli is much wider than that in other genera of lactic acid bacteria. Within the genus, this character permits a division into several species groups, each defined by a characteristic base ratio. These groups do not correspond to the traditional division into the subgenera *Thermobacterium, Streptobacterium,* and *Betabacterium* (Table 21.7). In particular, it can be seen that the thermophilic species, traditionally assigned to the subgenus *Thermobacterium,* fall into two entirely separate groups in terms of DNA base composition: *L. jugurti–L. acidophilus–L. salivarius,* with DNA containing 35 to 39 mole percent G+C; and *L. bulgaricus–L. leichmannii,* with DNA containing 50 to

*Table 21.7 Subgroups among the lactobacilli, as indicated by DNA base composition and physiological characters*

| Subgenus | Species | DNA base composition, mole % G+C | Mode of glucose fermentation[a] | Optical configuration of lactic acid formed | Growth at: 15°C | Growth at: 45°C |
|---|---|---|---|---|---|---|
| Thermobacterium | *L. jugurti* | 39 | Ho | DL- | − | + |
| | *L. acidophilus* | 36 | Ho | DL- | − | + |
| | *L. salivarius* | 35 | Ho | L- | − | + |
| Streptobacterium | *L. casei* | 46 | Ho | L- | + | ± |
| | *L. plantarium* | 45 | Ho | DL- | + | − |
| Betabacterium | *L. buchneri* | 45 | He | DL- | + | − |
| | *L. brevis* | 43 | He | DL- | + | − |
| Thermobacterium | *L. bulgaricus* | 50 | Ho | D- | − | + |
| | *L. leichmannii* | 51 | Ho | D- | − | + |
| Betabacterium | *L. fermenti* | 53 | He | DL- | − | + |
| | *L. cellobiosus* | 53 | He | DL- | ± | − |

[a]Ho, homofermentative; He, heterofermentative.

51 mole percent G+C. This example provides a good illustration of the value of DNA base composition in the reassessment of traditional taxonomic arrangements, which were based entirely on phenotypic characters.

## The methane bacteria

The ability to use $CO_2$ as an electron acceptor for anaerobic respiration, resulting in its reduction to the gas methane ($CH_4$), is a characteristic feature of the energy-yielding metabolism of a small group of non-spore-forming unicellular eubacteria. Reports on the Gram reaction of the methane bacteria vary, but since some species are definitely Gram positive, they will be described in this chapter. Three genera—*Methanobacterium, Methanococcus,* and *Methanosarcina*—are recognized (Figure 21.9). The first genus has rod-shaped cells; the latter two, spherical cells. In *Methanosarcina*, the cells are arranged in regular cubical packets, whereas in *Methanococcus* they occur singly.

Although methane-producing bacteria were discovered in 1905, the group is still poorly known, largely because they are difficult to cultivate. All are strict anaerobes and require a very low oxidation–reduction potential for growth. Even under optimal conditions, their development is exceedingly slow. Consequently, few of these bacteria have

*Figure 21.9 Methane-producing bacteria: (a) Methanobacterium (phase contrast, × 1,800). Courtesy of R. E. Hungate. (b) Methanosarcina (phase contrast, × 1,206). Courtesy of R. L. Uffen.*

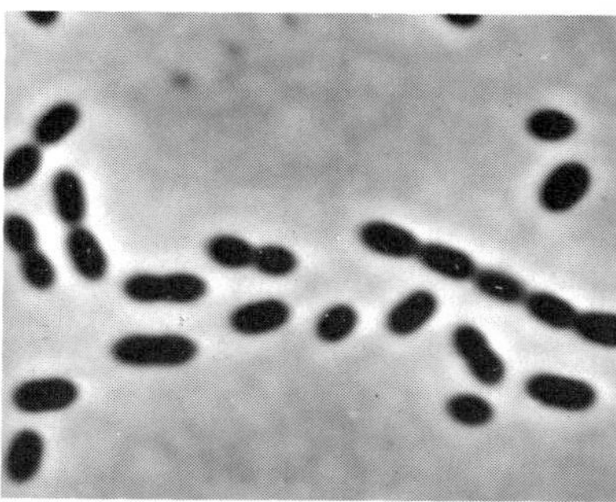

(a)

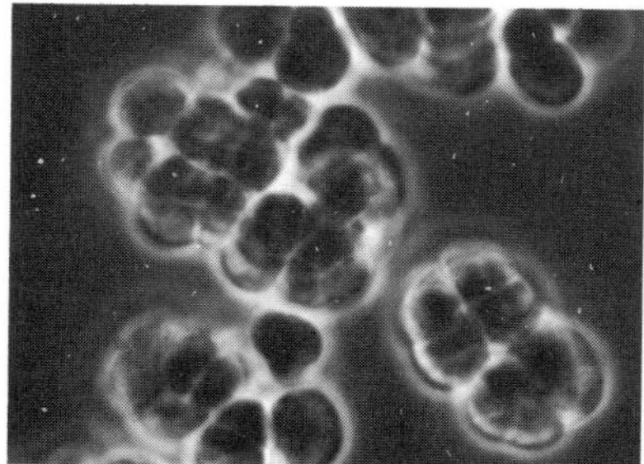

(b)

been obtained in pure culture, and much of the information about their properties has been derived from studies with mixed cultures.

All methane bacteria can obtain the energy required for growth by the anaerobic oxidation of molecular hydrogen, as shown in the following reaction:

$$4H_2 + CO_2 \rightarrow CH_4 + 2H_2O$$

Some can also utilize the one-carbon organic compounds, methanol and formic acid, as substrates; and some can utilize acetic acid. There have been many reports of the metabolism of other organic compounds containing more than one carbon atom by these bacteria. For example, the organism formerly known as *Methanobacillus omelianskii* was presumed for many years to perform the reaction

$$2CH_3CH_2OH + CO_2 \rightarrow 2CH_3COOH + CH_4$$

However, the presumed pure culture of "*M. omelianskii*" has recently been found to consist of an association of two strictly anaerobic bacteria, each of which performs one of the following reactions:

$$CH_3CH_2OH + H_2O \rightarrow CH_3COOH + 2H_2$$
$$CO_2 + 3H_2 \rightarrow CH_4 + 2H_2O$$

The methane bacterium in this association cannot, therefore, attack ethanol. In the light of this discovery, it has become an open question as to whether any methane-producing bacterium can metabolize organic compounds that contain more than one carbon atom, other than acetic acid. The group may consist of organisms that are primarily (or exclusively) anaerobic chemoautotrophs which use $H_2$ as their energy source.

The most general habitat for the methane bacteria is the anaerobic sediment at the bottom of lakes, ponds, and marshes; the common name of methane ("marsh gas") is a reflection of the abundance with which it is often produced by these bacteria in such environments. Members of the group are also present in great abundance as components of the very complex anaerobic microflora in the rumen of herbivorous mammals (see Chapter 24). Under normal circumstances, methane is the major gaseous end-product of the rumen fermentation. Its formation in the rumen can be completely suppressed by the introduction of a small amount of chloroform ($CHCl_3$), a specific inhibitor of the enzyme system involved in methane formation, and under such circumstances, the gases produced consist of $H_2$ and $CO_2$. This fact indicates that the methane bacteria of the rumen grow largely, if not entirely, at the expense of $H_2$, formed through the fermentative activities of other members of the rumen microflora.

## *Micrococcus, Staphylococcus,* and *Sarcina*

In addition to the Gram-positive cocci that belong to the physiologically defined groups of lactic acid bacteria and methane bacteria, unicellular bacteria with this cell structure fall into three other genera, distinguishable by a combination of physiological characters (primarily, the relation to oxygen) and structural characters (size and grouping of cells). The cells of staphylococci and micrococci are relatively small (generally, about 1 to 2 μm in diameter); those of sarcinae are considerably larger.

The members of the genus *Staphylococcus* are facultative anaerobes, which ferment sugars with the formation of considerable amounts of lactic acid. Although they have DNA of low G+C content (30 to 35 mole percent), in the same range as the spherical lactic acid bacteria, they are excluded from this group by their possession of heme pigments and their capacity for normal respiratory metabolism, as well as by their much less restrictive nutritional requirements (growth can occur in the absence of carbohydrates). Many of them produce yellow cellular carotenoid pigments, a character absent from lactic acid bacteria. These bacteria are typical members of the normal microflora of skin; some are also pathogenic.

In contrast to the staphylococci, the members of the genus *Micrococcus* are obligate aerobes, with a strictly respiratory energy-yielding metabolism. They also have a completely different DNA base composition (66 to 72 mole percent G+C). The normal habitat of these cocci is obscure; they are common in air and also occur in milk and dairy equipment.

The members of the genus *Sarcina* are microaerotolerant anaerobes, distinctive in structural and physiological respects. The extremely large spherical cells (approximately 4 μm in diameter) are arrayed in regular cubical packets (Figure 21.10). The complex cell wall of *S. ventriculi* (Figure 21.11) contains a cellulosic layer; this bacterium and *Acetobacter xylinum* (Chapter 19) are the only procaryotic organisms known to be capable of cellulose synthesis. *Sarcina* species are vigorous fermenters of soluble sugars. *Sarcina ventriculi* carries out an alcoholic fermentation, accompanied by the production of small amounts of $H_2$, acetic acid, and formic acid; *S. maxima* carries out a butyric acid fermentation.

*Sarcina ventriculi* is one of the oldest described bacteria that can still be recognized. It was discovered in 1844 by microscopic examination of the stomach contents of a patient with gastric disease. Subse-

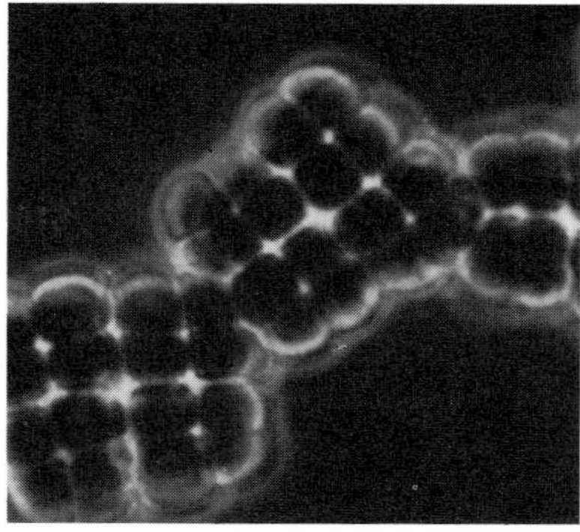

*Figure 21.10 Sarcina maxima (phase contrast, × 2,295). From S. Holt and E. Canale-Parola, "Fine structure of Sarcina maxima and Sarcina ventriculi." J. Bacteriol.* **93**, *399 (1967).*

*Figure 21.11 Electron micrograph of a thin section of a group of cells of Sarcina ventriculi, showing the heavy cellulose layer (c) that encloses each cell (× 23,864). Courtesy of S. Holt and E. Canale-Parola.*

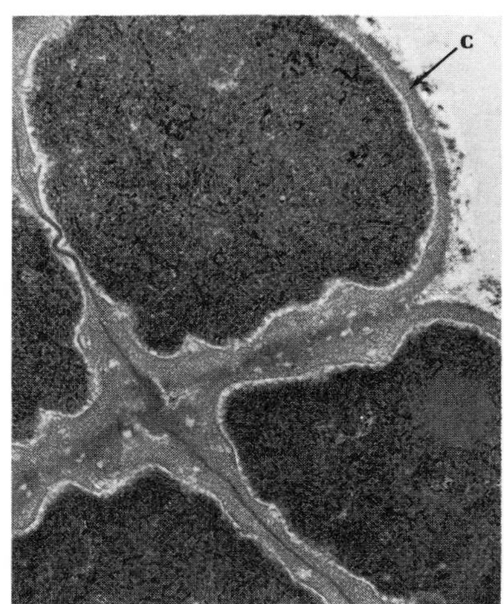

quent work has shown that it never occurs in the stomach contents of healthy individuals, developing only in various pathological conditions, when food remains in the stomach for an unusually long time. The organism is not a pathogen, however; in 1905 M. W. Beijerinck showed that it can be selectively isolated from soil by anaerobic enrichment in complex media containing a fermentable carbohydrate, provided that the medium is acidified with a strong mineral acid (phosphoric or hydrochloric acid) to a final concentration of approximately 0.06 to 0.08 *N*. Experiments with pure cultures show that the organism can grow at pH values as low as 1.5 (the limit varies with the acid used, being higher for organic acids than for mineral acids). *Sarcina ventriculi* is, accordingly, one of the most acid-tolerant bacteria known. Its occasional growth in the stomach is attributable to this property, as is also its highly selective enrichment by the technique of Beijerinck. Surprisingly, growth is not confined to acid conditions and can be initiated in appropriate media with a pH as high as 9.5.

*Sarcina maxima* is also an extremely acid-tolerant organism. It appears to occur less frequently in soil than *S. ventriculi;* the most favorable source materials are grains.

It should be noted that many authors incorrectly use the generic name *Sarcina* for certain small, strictly aerobic Gram-positive cocci that belong to the genus *Micrococcus.*

## *The coryneform bacteria*

There is perhaps no other large and important group of bacteria whose taxonomic limits and internal subdivision are as uncertain and controversial as those of the coryneform bacteria. To appreciate the reasons for this situation, it is necessary at this point to recapitulate the history of the group.

The first bacterium of this type to be discovered, and the type species of the genus *Corynebacterium,* is *Corynebacterium diphtheriae.* This organism is a parasite of the human mouth and throat; some strains are pathogenic and give rise to the disease known as diphtheria. Subsequently, several other animal parasites that resemble *C. diphtheriae* more or less closely were described. These "animal diphtheroids" share many characters with *C. diphtheriae.* All are permanently immotile. All are capable of performing both respiration and fermentation and are facultative anaerobes, although growth under anaerobic conditions is often poorer than under aerobic ones. In the case of *C. diphtheriae,* it has been shown that sugar fermentation yields propionic acid as a major end-product; the nature of the fermentative products formed by other members of the group is not known. The nutritional requirements are in all cases complex.

One organism originally assigned to the animal diphtheroid group differs markedly from the other members in physiological respects. This is the acne bacillus, an organism that grows in a special habitat on human skin: in the deeper regions of hair follicles. This species is a strictly fermentative, microaerotolerant anaerobe, and performs a typical propionic acid fermentation of sugars. Because of its physiological properties, it is now placed in the genus *Propionibacterium*, described below.

The immotile coryneform bacteria of the genus *Propionibacterium* were first isolated from Swiss cheese, in the ripening of which they play an important role. These microaerotolerant anaerobes perform a propionic acid fermentation of both carbohydrates and lactic acid. Paradoxically, they are catalase positive and contain cytochrome pigments, despite their low oxygen tolerance and fermentative metabolism. During the ripening of Swiss cheese, they develop as a secondary microflora, and convert lactic acid, originally formed in the curd through the activities of the lactic acid bacteria, to propionic acid, acetic acid, and $CO_2$. Both the characteristic sharp flavor and the "eyes" of Swiss cheese result from this fermentation. The source from which these bacteria enter Swiss cheese remained unclear for some time after their discovery but was eventually traced to the rennet used for the initial curdling of the milk. Rennet is an aqueous extract of the stomachs of calves, and the primary natural habitat of the propionic acid bacteria is the rumen of cattle. In this environment, they occur in enormous numbers and ferment the lactic acid produced through the fermentative activities of other members of the rumen microflora. Like the human diphtheroids, propionic acid bacteria have complex nutritional requirements.

The next major enlargement of the coryneform group came with the recognition that bacteria with this cell structure are very common in soil, and often constitute a major fraction of the normal bacterial microflora of soil. Unlike the coryneform bacteria so far described, many of these soil forms are motile; and they also differ markedly in physiological and nutritional respects from the animal diphtheroids and propionic acid bacteria. Some (but not all) do not require organic growth factors. Many are strict aerobes, capable of oxidizing a wide range of organic substrates, including some rarely attacked by other kinds of bacteria, such as halogenated aromatic compounds, purines, and pyrimidines. Some authors place these bacteria in a special genus, *Arthrobacter*. The soil coryneforms also include a number of cellulose decomposers, sometimes placed in a separate genus, *Cellulomonas*.

Closely related in physiology to the soil coryneforms (and also frequently motile) are certain plant pathogens, such as *Corynebacterium poinsettiae.*

Finally, it should be noted that a distinctive subgroup of coryneform bacteria is common in milk and milk products. These bacteria have been placed in two special genera, *Microbacterium* and *Brevibacterium.* They are strict aerobes, with complex nutritional requirements. They possess one very unusual physiological property. Although they do not grow at high temperatures (the maxima are 35 to 40°C) they are *thermoduric,* having an exceptional ability to survive exposure to high temperatures. Many strains can survive a temperature of 80°C for as long as 15 minutes, a degree of heat tolerance otherwise found only in bacteria that form endospores.

As this summary shows, the coryneform group, defined in structural terms, encompasses an exceptionally wide range of bacteria, differing from one another in ecological, nutritional, and metabolic respects. It is generally agreed that the propionic acid bacteria (including the acne bacillus, *P. acnes*) are sufficiently distinctive to deserve recognition as a special genus, *Propionibacterium.* The main taxonomic arguments center around the treatment of the facultatively anaerobic and strictly aerobic types. Some authors prefer to place all these organisms in the genus *Corynebacterium,* which thus becomes very broad. Others would restrict *Corynebacterium* to the animal diphtheroids and place the other species in different genera (e.g., *Arthrobacter, Cellulomonas, Microbacterium, Brevibacterium).* Although such a subdivision seems desirable, the real problem, so far unsolved, is to define these additional genera in such a manner that they can be unambiguously distinguished from one another, as well as from *Corynebacterium,* narrowly defined.

## *The catalase-negative members of group II: Bifidobacterium and Actinomyces*

Two genera assignable on structural grounds to group II—*Bifidobacterium* and *Actinomyces*—show physiological and biochemical resemblances to the lactic acid bacteria. Their members ferment sugars with the formation of lactic acid as a major end-product and are catalase negative. Furthermore, like lactic acid bacteria and unlike other members of group II, they contain lysine as a murein constituent. These organisms do not share the oxygen relations of the lactic acid bacteria and can be characterized as microaerotolerant anaerobes.

*Bifidobacterium* has a very striking and distinctive ecology. It is the predominant member of the intestinal flora of breast-fed infants but is rapidly displaced by other organisms after infants are weaned. It has a narrow temperature range and restrictive nutritional requirements, growing best in milk. The products of hexose fermentation are unlike those of any rod-shaped lactic acid bacteria; they consist of lactic and acetic acids. As was recently discovered (see Chapter 6), this bacterium uses a unique pathway of sugar breakdown, involving an initial $C_4$–$C_2$ cleavage of hexose phosphate to yield acetylphosphate and erythrose phosphate. The cells of *Bifidobacterium* are characteristically swollen, irregular, and branched; its cell structure is thus typically coryneform. This accords with the base content of the DNA (58 mole percent), which lies in the range characteristic of other coryneform organisms (51 to 64 mole percent), and is considerably higher than the highest values found in rod-shaped lactic acid bacteria.

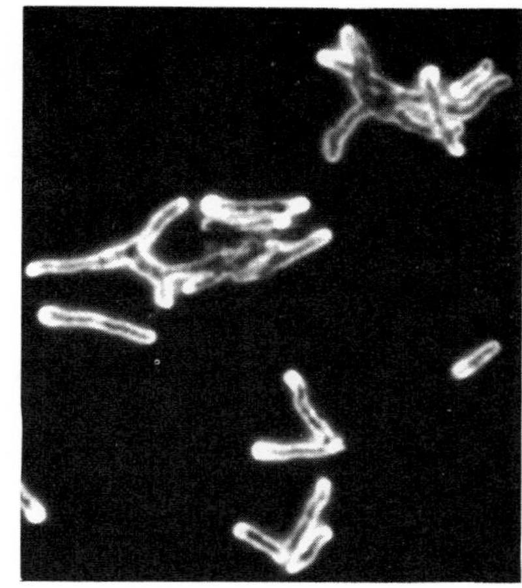

*Figure 21.12 Actinomyces israelii from a broth culture, showing branched cells and short mycelial fragments (dark field illumination, × 1,577). From J. M. Slack, S. Landfried, and M. A. Gerencer, "Morphological, biochemical and serological studies on 64 strains of Actinomyces israelii." J. Bacteriol.* **97**, *873 (1969).*

Work on the complex nutritional requirements of *Bifidobacterium* has shown that they include an unusual organic growth factor: N-acetylglucosamine, or a $\beta$-substituted disaccharide containing this residue (e.g., N-acetyllactosamine). The presence of such compounds in milk explains the marked preference of bifidobacteria for this medium. N-acetylglucosamine is a biosynthetic precursor of N-acetylmuramic acid; and both compounds are components of mureins. When *Bifidobacterium* is grown in a medium that contains an excess of this specific growth factor, the cell structure becomes much more regular; the branched, swollen rods considered characteristic of this organism may, therefore, reflect a deficiency of an essential murein precursor in the media normally used for its cultivation.

The members of the genus *Actinomyces* (Figure 21.12) are typical proactinomycetes, which can form a mycelium. Upon original isolation, most of these organisms show little tendency to mycelial fragmentation, producing rough colonies on solid media and coherent clumps of growth in liquid media. Many strains can, however, change to a smooth form, in which the mycelium fragments more readily; these variants form smoother colonies and produce turbid growth in liquid media. Some members of the genus are pathogens, producing the disease known as actinomycosis in man and other animals. Others are present as members of the normal microflora of the mouth and throat.

## *Mycobacterium* and *Nocardia*

These members of group II are all strict aerobes and can be distinguished collectively from the strictly aerobic coryneform bacteria by their capacity to undergo some degree of mycelial growth (Figure 21.13). The line of demarcation with respect to the coryneform group is,

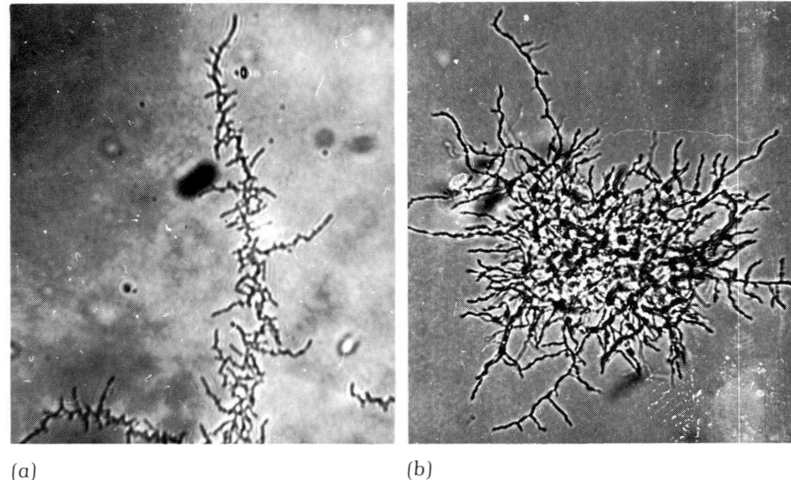


*Figure 21.13 The surface growth on agar plates of (a) Mycobacterium fortuitum and (b) Nocardia asteroides (× 600). Courtesy of Ruth Gordon and H. Lechevalier.*

however, not a very sharp one; aerobic coryneform bacteria and members of the *Mycobacterium–Nocardia* group share the same characteristic wall constituents (Table 21.2) and are broadly similar in nutritional and metabolic respects. The ranges of DNA base composition for the two groups overlap, although the values for *Mycobacterium* and *Nocardia* are characteristically somewhat higher. The property of acid fastness, although confined to the *Mycobacterium–Nocardia* group, is not shown by all members and therefore cannot serve as a general differential character.

The earliest described species of both *Mycobacterium* and *Nocardia* (and hence the type species of these genera) were pathogenic organisms: *M. tuberculosis*, the causative agent of human tuberculosis, and *N. asteroides*, the agent of human and animal infections known as nocardioses. Other pathogenic *Mycobacterium* species include the agents of tuberculosis in birds and cattle, as well as the agents of leprosy. More or less pronounced acid fastness is common to all these organisms. The distinction between the two genera rests essentially on the extent of mycelial growth: in *Mycobacterium*, mycelial development is transitory, and growth occurs in large part by binary fission; in *Nocardia*, mycelial growth continues through most of the growth period, and fragmentation occurs as cultures enter the stationary phase. It must be admitted, however, that the distinction is a tenuous one.

The pathogenic members of the *Mycobacterium–Nocardia* group are extremely slow growing bacteria. As a result of their waxy coating, the cells do not form an even suspension in a liquid medium and tend to grow in the form of a wrinkled surface pellicle. The colonies on solid media are also characteristically wrinkled and dry. Nonpathogenic members of the *Mycobacterium–Nocardia* group are common soil inhabitants. These forms grow much more rapidly, and frequently are not acid fast. They can be selectively isolated from soil by taking advantage of their ability to use as energy sources certain organic compounds that are not readily attacked by other aerobic bacteria; notably, cholesterol, amyl alcohol, and paraffin. With the exception of some pathogenic species, *Mycobacterium* and *Nocardia* have simple nutritional requirements and do not need organic growth factors.

## *The euactinomycetes: Streptomyces*

By far the largest genus of group III is *Streptomyces*. The streptomycetes are one of the most abundant bacterial groups in soil, and the characteristic smell of damp soil is attributable to the presence of a volatile substance formed by many of these organisms. In the past 30 years, the streptomycetes have acquired great economic importance by virtue of their ability to produce commercially valuable antibiotics belonging to the classes streptomycins, actinomycins, tetracyclines, and macrolides (see Chapter 29 for further discussion of this aspect of their biology). As a result, the group has been intensively explored in terms of antibiotic-producing ability, and literally hundreds of species have been named, frequently on very questionable grounds.

**The developmental cycle**

Germination of the small rod-shaped to spherical reproductive cells (conidiospores) occurs by hyphal outgrowth; generally three or four hyphae develop from a spore (Figure 21.14). Through repeated branching, a compact, dense, and coherent *substrate mycelium* is formed. At this stage of development, the colonies on solid media are cartilaginous in consistency and have a smooth surface. At numerous points in the surface of the substrate mycelium, a looser layer of slightly thicker hyphae develops, forming a powdery *aerial mycelium*, often different in color from the underlying substrate mycelium. Conidiospores are produced in chains, by a regular fragmentation of the

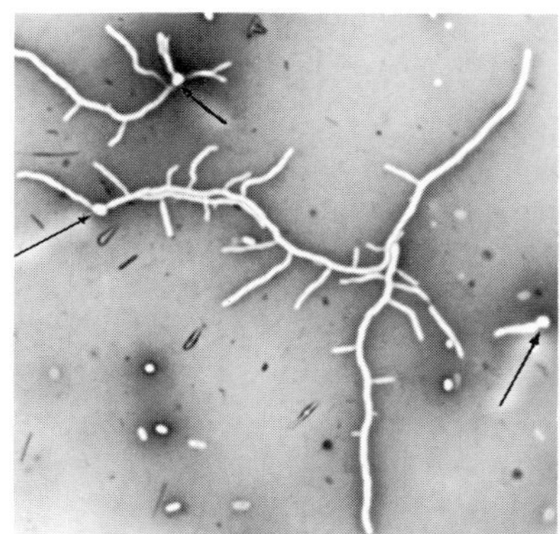

*Figure 21.14 Germinating conidia of a Streptomyces, showing early stages in the development of the substrate mycelium nigrosin mount, × 1,350). Arrows point to three conidia from which mycelial outgrowth has begun. The field also contains several conidia that have not germinated. Courtesy of C. F. Robinow.*

*Figure 21.15 Portions of the aerial mycelium of two streptomycetes, illustrating two different kinds of arrangement of the conidiating hyphae. Photograph (a) × 2,673, courtesy of H. Lechevalier; (b) × 2,322, courtesy of Peter Hirsch.*

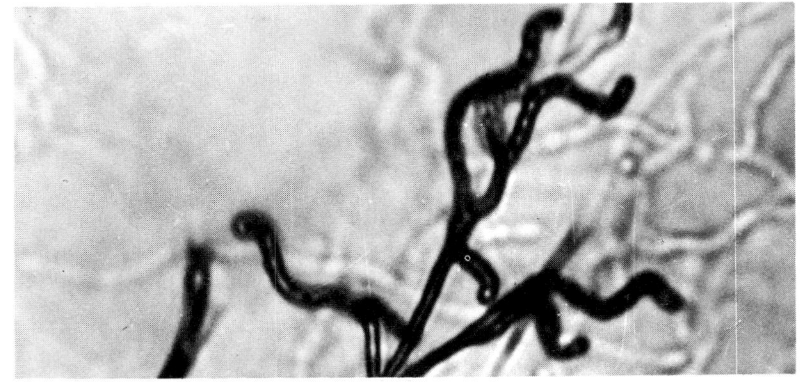

(a)

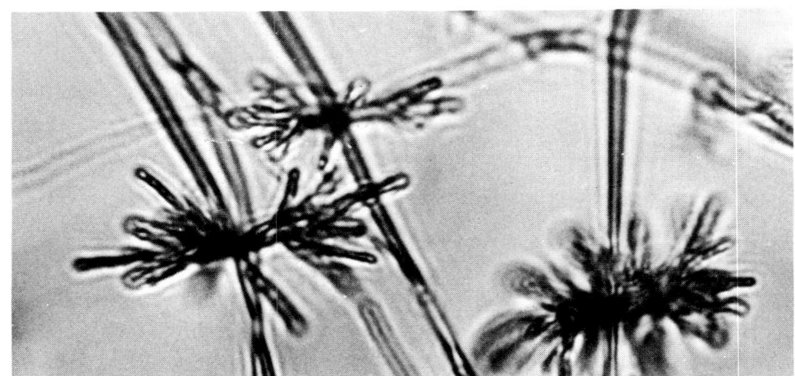

(b)

terminal hyphae in the aerial mycelium. At maturation, they detach very readily from the sporogenous hyphae. The conidiospores are immotile and are dispersed through the atmosphere by air currents.

The pattern of organization of the spore-bearing hyphae of the aerial mycelium is a constant and distinctive character, which allows the genus *Streptomyces* to be divided into a number of generic subgroups; two different hyphal arrangements are illustrated in Figure 21.15. Another very useful determinative character, which can be ascertained only by electron microscopic examination, is the *shape and surface structure of the conidiospores.* In some species, they have a smooth surface; in others, they are covered by a coat with hairy, spiny, or warty protuberances (Figure 21.16). The regular patterns of organization of the spore-bearing hyphae, together with those distinctive features of conidiospore structure, both testify to the *high degree of differentiation*

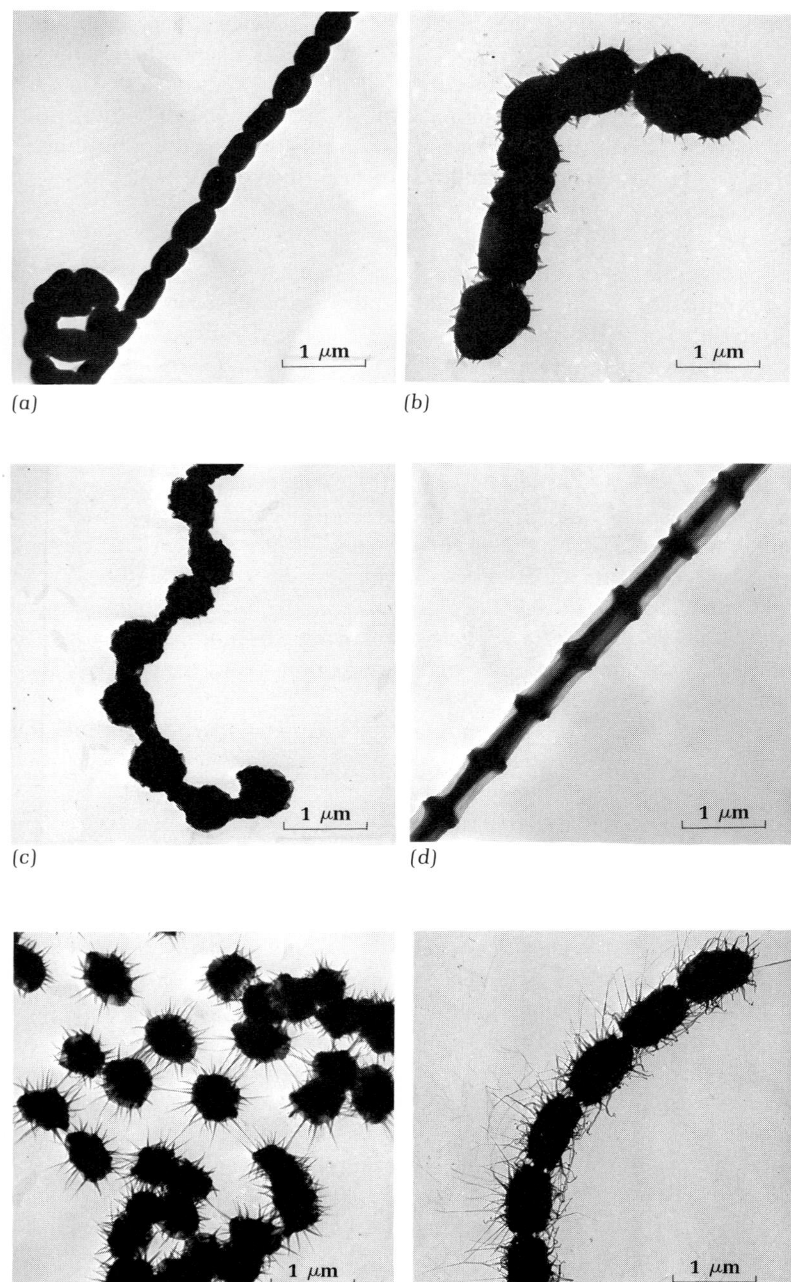


Figure 21.16 *Electron micrographs of the conidiospores of six different* Streptomyces *species, which illustrate various types of surface structure and ornamentation: (a)* S. cacaoi, *showing smooth spores; (b)* S. hirsutus, *showing spiny spores with obtuse spines; (c)* S. griseoplanus, *which has warty spores; (d)* S. aureofaciens, *of a smooth but special "phalangiform" type; (e)* S. fasciculatus, *showing spiny spores wth long acute spines; (f)* S. flavoviridis, *showing hairy spores.* Courtesy of H. D. Tresner, Lederle Laboratories; reproduced in part from Intern. J. Syst. Bacteriol. **18,** 69–189 (1968).

characteristic of these euactinomycetes, relative to the proactinomycetes of group II, where mycelial fragmentation is a generalized process and does not give rise to cells with a distinctive structure.

One chemical property distinguishes the *Streptomyces* group from all other bacteria: the presence of the LL isomer of diaminopimelic acid as a constituent of the murein layer in the cell wall.

**Physiological properties**

The streptomycetes are all strict aerobes and have simple nutritional requirements; none is known to require organic growth factors. Although the range of utilizable carbon and energy sources has not been systematically examined, it appears to be wide for most species. One natural substrate that is uniquely attacked by streptomycetes is rubber; enrichment cultures containing a suspension of latex particles as the only organic constituent yield only organisms of this group. Most strains secrete proteolytic enzymes, as evidenced by their ability to liquefy gelatin. Another class of extracellular enzymes produced by many of these bacteria consists of hydrolases which specifically attack certain bonds in mureins; these enzymes are, consequently, able to lyse other bacteria, the walls of which contain linkages susceptible to attack. The bacteriolytic properties of many *Streptomyces* strains can be readily demonstrated by streaking them on plates containing a suspension of bacterial cells.

The streptomycetes are mostly mesophiles, growing well in the temperature range 25 to 40°C.

**The genetics of *Streptomyces***

Genetic transfer between *Streptomyces* strains was first demonstrated about 1956; since then, extensive genetic studies have been conducted with one species, *S. coelicolor*. Since streptomycetes are among the structurally most complex procaryotic organisms, it has been of considerable interest to determine whether they share the distinctive genetic properties first shown for a structurally very simple procaryote, *Escherichia coli*.

Matings in *S. coelicolor* are performed by growing conidiospores or hyphal fragments of mutant strains together; during the development of such genetically mixed cultures, zygotes are produced, and subsequently recombinations occur; the properties of the recombinants can be studied by isolating conidia and subjecting them to analysis. There seems to be no mating-type specificity of the kind characteristic of *E. coli;* all strains are interfertile. In other respects, the genetic systems of the two organisms appear broadly similar. In both cases, the zygotes are partial diploids or merozygotes, and the normal haploid state is reestablished by eventual elimination of one of the alleles. *Streptomyces coelicolor* has a single linkage group, and hence a single chromosome, most simply interpreted as being circular, like that of *E. coli*.

## *Micromonospora*

The euactinomycetes of the genus *Micromonospora* form compact, smooth colonies devoid of an aerial mycelium. Spherical or oval conidiospores are formed singly at the tips of short hyphal branches throughout the colony (Figure 21.17).

*Micromonospora* species occur in soil, although they are less abundant than the streptomycetes; they have also been isolated from aquatic environments (e.g., lake mud). Most are strict aerobes; however, it should be noted that this is the only group of euactinomycetes that also includes a few anaerobic members. The anaerobic *Micromonospora* species have been isolated from the intestinal tract of termites; one is a cellulose-fermenter.

Although the *Micromonospora* group are typical mesophiles, with a maximal growth temperature of 40°C, their conidiospores are extremely heat-resistant, surviving exposure to 80°C for as long as 5 minutes.

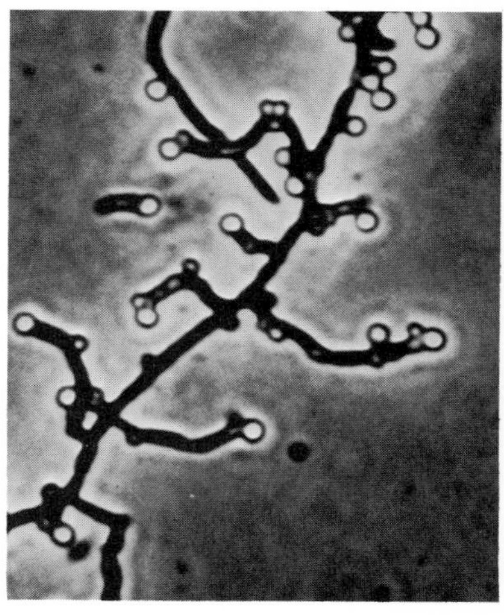

*Figure 21.17 Micromonospora chalcea, showing spherical conidiospores borne singly at the tips of hyphae (phase contrast, × 1,965). Courtesy of G. M. Luedemann and the Schering Corporation.*

## *The thermophilic actinomycetes*

Composts of plant materials that undergo spontaneous heating (e.g., damp haystacks) always contain thermophilic actinomycetes as a characteristic component of the thermophilic microflora. The most characteristic of these thermophilic forms is *Thermoactinomyces*, which can grow at temperatures as high as 68°C and fails to grow at temperatures of 45°C or below. The colonies on solid media are more spreading than those of most euactinomycetes and are covered at maturity with a white, powdery aerial mycelium. Conidiospores are borne singly on the tips of the aerial hyphae. Their location distinguishes *Thermoactinomyces* from *Micromonospora*.

## *The sporangial actinomycetes*

The members of this group of euactinomycetes were discovered only in 1950. Most members of the group do not produce an aerial mycelium. Swellings form at the tips of hyphae in the compact substrate myce-

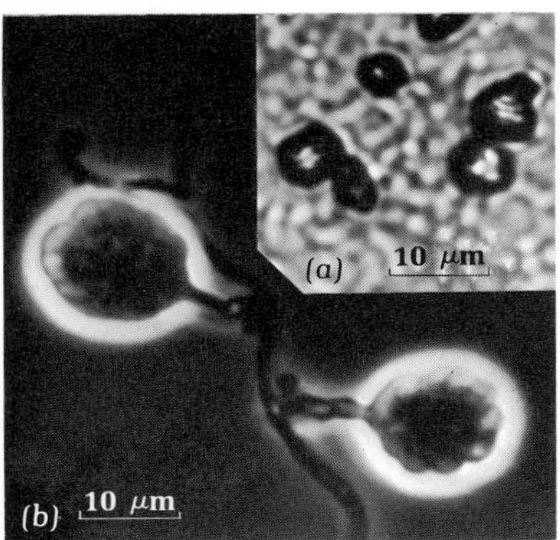


*Figure 21.18 Sporangia of Actinoplanes: (a) (inset) A group of mature sporangia viewed on the surface of a colony (bright field illumination). (b) Two mature sporangia attached to a hypha, mounted in water (phase contrast illumination). From H. Lechevalier and P. E. Holbert, "Electron microscopic observation of the sporangial structure of a strain of Actinoplanes." J. Bacteriol.* **89,** *217 (1965).*

lium; the swollen portions are then cut off by cross walls and become a sporangium, within which spores are formed. Sporangial structure is illustrated for an *Actinoplanes* species in Figure 21.18. The sporangiospores are eventually liberated by rupture of the sporangial wall. At least in some members of this group, rupture of the mature sporangia and liberation of the enclosed spores is triggered by their immersion in water, a feature also characteristic of the aquatic phycomycetes among the fungi. Although members of this group can be isolated from soil, they were originally discovered in fresh water and appear to be a characteristically aquatic group. This ecological interpretation is consonant with the mode of sporangial rupture and is further supported by the fact that some members of the group produce motile, flagellated sporangiospores—the only instance of flagellation among euactinomycetes.

A number of genera have been described; they are distinguished primarily by the shape of the sporangium and the structure of the sporangiospores. In *Actinoplanes* and *Spirillospora,* the sporangia are spherical, and the spores are motile by means of a tuft of inserted flagella. The spores of *Actinoplanes* are oval, whereas those of *Spirillospora* are curved or helical in form. In *Streptosporangium,* the spores are immotile, short rods.

In physiological respects, the sporangial actinomycetes appear similar to the streptomycetes or the aerobic micromonosporas. They are strictly aerobic mesophilic organisms, with simple nutritional requirements.

## Further reading

**Books**

Waksman, S. A., *The Actinomycetes: A Summary of Current Knowledge.* New York: Ronald Press, 1967.

**Symposia, reviews, and original articles**

Canale-Perola, E., "Biology of the Sugar-Fermenting Sarcinae." *Bacteriol. Rev.* **34,** 82 (1970).

Cowan, S. T., and other authors, "Symposium on Staphylococci and Micrococci." Published in *J. Appl. Bacteriol.* **25,** 324–440 (1962).

Cross, T., "Thermophilic Actinomycetes." *J. Appl. Bacteriol.* **32,** 36 (1968).

Garvie, E. J., "The Genus *Leuconostoc* and its Nomenclature." *J. Dairy Res.* **27,** 383 (1960).

Gunther, H. I., and H. R. White, "The Cultural and Physiological Characters of the Pediococci." *J. Gen. Microbiol.* **26,** 185 (1961).

Jensen, H. L., "The Coryneform Bacteria." *Ann. Rev. Microbiol.* **6,** 77 (1952).

———, and other authors, "Symposium on Coryneform Bacteria." Published in *J. Appl. Bacteriol.* **29,** 13–184 (1966).

Lechevalier, H. A., and M. P. Lechevalier, "Biology of Actinomycetes." *Ann. Rev. Microbiol.* **21,** 71 (1967).

Rogosa, M., and M. E. Sharpe, and other authors, "Symposium on the Lactobacilli." Published in *J. Appl. Bacteriol.* **22,** 329–416 (1959).

Sherman, J. M., "The Streptococci." *Bacteriol. Rev.* **1,** 1 (1937).

Stadtman, T. C., "Methane Fermentation." *Ann. Rev. Microbiol.* **21,** 121 (1967).

# *Chapter twenty-two Microorganisms as geochemical agents*

*All regions of the earth* that contain living organisms are known collectively as the *biosphere.* They include the oceans; the land surface of continents, to a depth of a few feet; lakes and rivers; and the lower part of the atmosphere. The thin film of life distributed through the biosphere performs many chemical transformations which would not occur at measurable rates in its absence. The activities of living organisms have profoundly affected the composition and structure of the surface layers of our planet. Living organisms are accordingly one of the most powerful agents of *geochemical change.*

## *The biosphere*

The maintenance of the biosphere in a more or less steady state depends on two factors: the cyclic turnover of elements essential for life, particularly carbon, nitrogen, sulfur, oxygen, and phosphorus; and a steady input of energy from an external source, the sun.

Through solar-energy conversion by photosynthesis many elements are withdrawn as inorganic compounds from the environment and are accumulated in the organic constituents of the cells and tissues of living organisms. The major producers of organic matter are unicellular algae (principally diatoms and dinoflagellates) in the oceans, and seed plants on land. The organic materials they accumulate provide, either directly or indirectly, the energy sources for other forms of life.

Insofar as photosynthetic organisms serve as food sources for animals or nonphotosynthetic microorganisms, the biologically important elements that they contain remain, at least in part, in the organic state, being incorporated in the cells and tissues of the primary consumers. The primary consumers may themselves provide food sources for other animals, so that these elements may persist in the organic state during their passage through more or less long and complex *food chains*, composed of many different kinds of nonphotosynthetic organisms. Before these elements again become available to support the growth of photosynthetic organisms, they must be converted once more to the inorganic form. This conversion, known as *mineralization*, is brought about largely through the decomposition of plant and animal remains and of the organic excretory products of animals. The principal agents of mineralization are nonphotosynthetic bacteria and fungi. In the turnover of the element carbon, it has been estimated that at least 90 percent of the $CO_2$ produced in the biosphere arises through the metabolic activities of these two microbial groups. The small remaining fraction of the $CO_2$ produced results from the respiratory activities of other biological groups and from combustion (forest fires, burning of organic fuels by man). The overwhelming contribution made by nonphotosynthetic microorganisms to the mineralization process reflects their ubiquity in the biosphere, their high rates of growth and metabolism, and their collective ability to degrade all naturally occurring organic materials.

## *The fitness of microorganisms as agents of geochemical change*

**The distribution of microorganisms in space and time**

The omnipresence of microorganisms throughout the biosphere is a consequence of their ready dissemination by wind and water. Surface waters, the floors of oceans, and the top few inches of soil are teeming with microorganisms that are ready to decompose any organic matter that may become available to them. It has been estimated that the top 6 inches of fertile soil may contain more than 2 tons of fungi and bacteria per acre. Any handful of soil is a small world in itself, seeded with many different kinds of microbes and presenting at different times microscopic ecological niches for different types to develop. Even on a single soil particle, the conditions may change from hour to hour and from facet to facet.

Let us consider what happens upon the death of a microscopic root hair or a worm in the soil. The complex organic compounds of the dead tissue are attacked by microorganisms that are capable of digesting and oxidizing these compounds. As oxygen is consumed, conditions become anaerobic in the immediate proximity of the dead tissue and fermentative organisms develop. The products of fermentation then may diffuse to regions in which oxygen is still present or may be oxidized anaerobically by organisms capable of reducing nitrates, sulfates, or carbonates. Ultimately, the organic compounds will be completely converted to $CO_2$, the conditions will again become fully aerobic, and the autotrophs will develop at the expense of such reduced inorganic products as ammonia, sulfide, and hydrogen. Thus, the inorganic products of the decomposition of the plant or animal are eventually completely oxidized. This sequence of events, which occurs on a microscopic scale on a particle of soil, can be observed on a macroscopic scale in nature. When a tree falls into a swamp or a whale decomposes on a beach, the eventual chemical results are essentially the same. Seasonal and climatic conditions may retard or accelerate the cyclic turnover of matter. In cold climates, decomposition is most rapid in the early spring; in semiarid areas, it is largely restricted to the rainy season.

In nature, only those microorganisms that are favored by the local and temporary environment reproduce, and their growth ceases when they have changed their environment. It is improbable that many bacteria die a "natural death." Most of them are consumed by such ever-present predators as the protozoa. A few cells of each type of microorganism persist to initiate a new burst of growth when conditions again become favorable for their development.

## The metabolic potentials of microorganisms

The relatively enormous catalytic power of microorganisms contributes to the major role they play in the chemical transformations occurring on the earth's surface. Because of their small size, bacteria and fungi possess a large surface–volume ratio compared with higher animals and plants. This permits a rapid exchange of substrates and waste products between the cells and their environment.

Per gram of body weight, the respiratory rates of some aerobic bacteria are hundreds of times greater than that of man. On the basis of the known metabolic rates of microorganisms, one can estimate that the metabolic potential of the microorganisms in the top 6 inches of an acre of well-fertilized soil at any given instant is equivalent to the metabolic potential of some tens of thousands of human beings.

An even more important factor influencing the chemical role that microorganisms play in nature is their high rate of reproduction in favorable environments. Thus, the metabolic potential of the progeny of a single bacterium with a division time of 20 minutes can increase a thousandfold in a little over 3 hours.

**The metabolic versatility of microorganisms**

Every organic compound that occurs naturally is decomposed by some microorganism. This fact explains the rarity of undecomposed organic matter on the earth's surface. Except under special conditions, any organic compound that is no longer part of a living organism is consumed and rapidly mineralized by microorganisms in the biosphere.

Although some nonphotosynthetic microorganisms (e.g., the *Pseudomonas* group) can attack many different organic compounds, the metabolic versatility of the microbial world *as a whole* is not primarily a reflection of the metabolic versatility of its individual members. Any single bacterial species is only a limited agent of mineralization. Highly specialized physiological groups of microorganisms play important roles in the mineralization of specific classes of organic compounds. For example, the decomposition of cellulose, which is one of the most abundant constituents of plant tissues, is mainly brought about by organisms that are highly specialized nutritionally. Among the aerobic bacteria capable of decomposing cellulose, the gliding bacteria that belong to the *Cytophaga* group (p. 556) are perhaps the most important. The cytophagas can rapidly dissolve and oxidize this insoluble compound, but cellulose is the only substance they can use as carbon source.

It will be recalled that the autotrophic bacteria, responsible for the oxidation of reduced inorganic compounds in nature, are also highly specific. Each type of autotroph is capable of oxidizing only one class of inorganic compounds and, in some cases (the nitrifying bacteria), only one compound.

## *The cycle of matter*

The turnover of all the elements that enter into the composition of living organisms constitutes the *cycle of matter*. In many cases, the cyclic transformations of biologically important elements are simple ones considered from the chemical standpoint. Phosphorus, for example, is assimilated by living organisms in inorganic form as phosphate. In the cell, it is incorporated into organic compounds by the esterification of the phosphate ion. Upon the death of the cell, it is again liberated as inorganic phosphate by hydrolysis. Throughout this cycle, the phosphorus atom does not change in valence and remains as part of a phosphate group. Phosphorus is an element that, in many environments, is a limiting factor in the growth of living organisms. It is a limiting factor for a purely *physical* reason—its relative scarcity in the biosphere.

Some of the biologically important elements undergo transformations that involve not only their incorporation into living matter and their return to inorganic form but also *cyclic changes in their state of oxidation (valence)*. Such cyclic changes in the oxidation state are characteristic of the elements *carbon, oxygen, nitrogen*, and *sulfur*. On a minor scale, other elements, such as hydrogen and iron, also undergo cyclic changes in oxidation state as a result of the activities of living organisms.

The oxidation state of the elements C, O, N, and S is a factor of great importance in their utilizability as nutrients for the growth of living organisms. Hence, in the case of these four elements, unavailability may be caused not by scarcity in the *physical* sense, but rather by the scarcity of the element in question in a *suitable chemical form*. As an illustration of the paramount importance of the chemical state of an element in determining its availability as a nutrient, we may consider the case of nitrogen. Molecular nitrogen ($N_2$) is the major gas of our atmosphere, constituting some 80 percent of the atmosphere. The air that overlies each acre of soil contains about 36,000 tons of gaseous nitrogen. Yet nitrogen is the element that is most frequently limiting for plant growth because most plants cannot use molecular nitrogen but require it in combined form as nitrate or ammonia. The total *combined nitrogen* content of soil ranges from 10 to 500 lb/acre. Thus, the fraction of the nitrogen content of the biosphere that is immediately available for plant growth is some 0.00001 to 0.001 percent of the total.

## *The cycles of carbon and oxygen*

**Carbon dioxide reduction and oxygen formation**

The prime moving force in the cyclic turnover of biologically important elements in nature is oxygen-evolving photosynthesis. It is mainly through this process that the oxidized form of carbon ($CO_2$, bicarbonate, carbonate) is converted to the reduced state in which it occurs in organic compounds, and that molecular oxygen is produced by the oxidation of water.

The subsequent oxidation of the organic compounds produced in photosynthesis is coupled, either directly or indirectly, with the reduction of molecular oxygen back to water. The most direct means by which this is achieved are *combustion* and *respiration*. Algae and plants, as well as the animals that feed on them, contribute their share of respiratory activity. The bacteria and fungi, however, oxidize the bulk of organic matter, for these organisms decompose the bodies of dead plants and animals as well as the various organic products excreted by animals. Through photosynthesis on the one hand, and res-

Figure 22.1 *The carbon and oxygen cycles: the oxidations of the carbon and oxygen atoms are shown as solid arrows, reductions as broken arrows, and reactions involving no valence change as dotted arrows.*

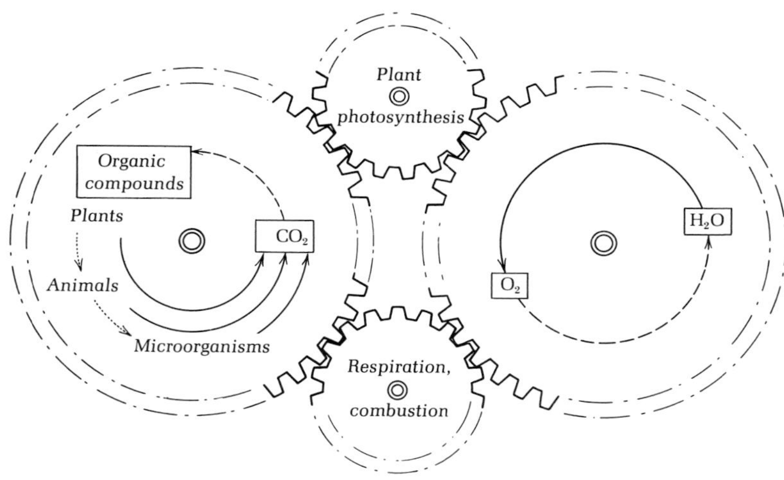


piration on the other, the cyclic transformations of carbon and oxygen are obligately linked to each other. This is shown in Figure 22.1.

The oxidized carbon that is available for photosynthesis in the biosphere is the carbon dioxide contained in the atmosphere and the soluble bicarbonate and carbonate dissolved in the waters of the earth's surface. Air contains approximately 0.03 percent $CO_2$ by volume, and this concentration is maintained at a constant level as a result of the dynamic balance between photosynthesis and mineralization. Ocean waters, being slightly alkaline, contain bicarbonate ion in addition to dissolved $CO_2$. The concentration of bicarbonate in the oceans is approximately 0.002 $M$ and remains constant because it is determined by a chemical equilibrium between bicarbonate ion in solution and $CO_2$ in the atmosphere. The oceans act as a reservoir of $CO_2$ for the atmosphere. At the same time they trap a large fraction of the $CO_2$ produced on land and keep its concentration in the atmosphere at a relatively low level.

The importance of the carbon cycle can best be emphasized by the estimate that the total $CO_2$ contained in the atmosphere, if it were not replenished, would be completely exhausted in less than 20 years at the present rate of photosynthesis. This estimate does not appear too radical when it is realized that the carbon contained in a single giant redwood tree is equivalent to that present in the atmosphere over an area of approximately forty acres. On land, seed plants are the principal agents of photosynthetic activity. A minor contribution is made by the algae. In the oceans, however, it is the unicellular photosynthetic organisms that play the most important role. The large plantlike algae (seaweeds) are confined in their development to a relatively narrow coastal strip. Since light of photosynthetically effective wavelengths is largely filtered out at a depth of about 50 feet, these sessile algae cannot grow in deeper waters. Because they are free-floating, the microscopic algae of the ocean (known as the phytoplankton) are capable of developing in the surface layers wherever the chemical environment is favorable. Their growth is largely limited by the relative scarcity of two elements: phosphorus and nitrogen. Where these elements are made available as phosphates and nitrates by the runoff of rain water from continents and subsequent distribution by ocean currents, profuse development of phytoplankton occurs. According to one estimate, the total annual fixation of carbon in the oceans amounts to approximately $1.2 \times 10^{10}$ tons, whereas that on the land is about $1.6 \times 10^{10}$ tons.

Although oxygen-evolving photosynthesis is by far the most important means of reducing $CO_2$ to organic matter, other processes contribute to a small extent. These include photosynthesis by the purple and green bacteria, and $CO_2$ reduction by the chemoautotrophs.

**The mineralization process: carbon dioxide formation and the reduction of oxygen**

The biological conversion of organic carbon to $CO_2$ with the concomitant reduction of molecular oxygen involves the combined metabolic activity of many different kinds of microorganisms. The complex constituents of dead cells must be digested, and the products of digestion must be oxidized by specialized organisms that can use them as nutrients. Many aerobic bacteria (pseudomonads, bacilli, actinomycetes), as well as fungi, carry out complete oxidations of organic substances derived from dead cells. However, it should be remembered that even those organisms that produce $CO_2$ as the only waste product of the respiratory decomposition of organic compounds usually use a large fraction of the substrate for the synthesis of their own cell material. In anaerobic environments, organic compounds are decomposed initially by fermentation, and the organic end-products of fermentation are then further oxidized by anaerobic respiration, provided that suitable inorganic hydrogen acceptors (nitrate, sulfate, or $CO_2$) are present.

**The sequestration of carbon: inorganic deposits**

The carbonate in the oceans combines with dissolved calcium ions and, under slightly alkaline conditions, becomes precipitated as calcium carbonate. Calcium carbonate is also deposited biologically in the shells of protozoa, corals, and mollusks. This is the geological origin of the calcareous rock (limestone) that is an important constituent of the surface of continents. Calcareous rock is not directly available as a source of carbon for photosynthetic organisms, and hence its formation causes a depletion of the total carbon supply available for life. Nevertheless, much of this carbon eventually reenters the cycle through weathering. The formation and solubilization of calcium carbonate are brought about primarily by changes in hydrogen ion concentration, and microorganisms contribute indirectly to both processes as a consequence of pH changes that they produce in natural environments. For example, such microbial processes as sulfate reduction and denitrification cause an increase in the alkalinity of the environment, which favors the deposition of calcium carbonate in the ocean and other bodies of water. Microorganisms also play an important role in solubilizing calcareous deposits on land by production of carbonic acid ($CO_2$) and by acid formation during nitrification, sulfur oxidation, and fermentation.

### The sequestration of carbon: organic deposits

Certain organic constituents of plants are highly resistant to microbial attack and hence tend to accumulate in nature after the more readily decomposable compounds have been mineralized. Most of the organic matter of soil consists of such resistant residues, which are collectively known as *humus*. The degree to which humus accumulates in soil is highly variable and depends on a large number of physical and chemical factors, such as temperature, moisture, pH, and the availability of oxygen. Humus is important in the maintenance of soil fertility because the physical properties of soil are modified by its presence. In particular, it causes an increased porosity of soil, which favors the diffusion of oxygen to plant roots and to the microbial soil population.

A high moisture content, causing oxygen depletion and the accumulation of acidic substances, is sometimes particularly favorable for the accumulation of humic materials. This phenomenon is most pronounced in *peat bogs*, where, in the course of time, deposits of undecomposed organic matter known as *peat* accumulate. These deposits may extend for hundreds of feet below the surface of the bog. In the course of geological time, the compression of peat deposits, probably aided by other physical and chemical factors, has resulted in the formation of coal. Much carbon has thus been sequestered from the biosphere in the form of peat and coal deposits. A second kind of sequestration of carbon in organic form has occurred in deposits of natural oil and gas (methane).

Since the Industrial Revolution, the exploitation by man of the stored deposits of organic carbon in the earth's crust has resulted in their very rapid mineralization. Although substantial deposits still remain to be exploited, it is estimated that at present rates of consumption most of the petroleum and natural gas will be used up within a hundred years and much of the coal within a few centuries. Thus, within the short span of a thousand years, the activities of man will have reversed a geological process that required millions of years.

## The nitrogen cycle

Although molecular nitrogen is a major constituent of the earth's atmosphere, the nitrogen molecule is chemically extremely inert and is not a suitable source of the element for most living forms. Almost all higher plants, animals, and microorganisms depend on combined nitrogen in their nutrition. Combined nitrogen, in the form of ammonia, nitrates, and organic compounds, is relatively scarce in soil and water, and its concentration often becomes the limiting factor in the development of living organisms. For this reason, the cyclic transformation of nitrogenous compounds, including the mineralization of nitrogenous organic matter, is of paramount importance in the total turnover of this element in the biosphere. The main features of the *nitrogen cycle* are illustrated schematically in Figure 22.2.

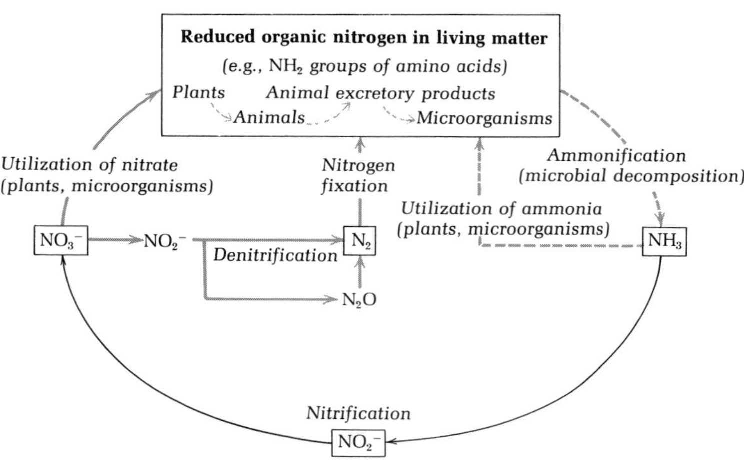


*Figure 22.2 The nitrogen cycle: the oxidations of the nitrogen atom are shown as solid black arrows, reductions as solid gray arrows, and reactions involving no valence change as broken gray arrows.*

**The assimilation of combined nitrogen by photosynthetic organisms**

Algae and plants assimilate nitrogen as either nitrate or ammonia. If nitrate is the form in which nitrogen is assimilated, it must be reduced in the cell to ammonia. Although nitrate is reduced when plants assimilate it, there is no significant excretion of reduced nitrogenous compounds by the living plant; the processes of nitrate assimilation and reduction proceed only to the extent to which nitrogen is required for growth. It is this feature in particular that distinguishes *nitrate assimilation* from *nitrate reduction*, a process of anaerobic respiration.

**The transformations of organic nitrogen and the formation of ammonia**

The organic nitrogenous compounds synthesized by algae and plants serve as the nitrogen source for the animal kingdom. During their assimilation by animals, complex nitrogenous compounds are hydrolyzed to a greater or lesser extent, but the nitrogen remains largely in reduced organic form. Unlike plants, however, animals do excrete a significant quantity of nitrogenous compounds in the course of their metabolism. The form in which this nitrogen is excreted varies from one group of animals to another. Invertebrates predominantly excrete ammonia; but among vertebrates, organic nitrogenous excretion products make their appearance as well. In reptiles and birds, uric acid is the major form in which nitrogen is excreted; in mammals, urea is the principal form. In nature, the urea and uric acid excreted by animals are rapidly mineralized by special groups of microorganisms, with the formation of $CO_2$ and ammonia.

Only part of the nitrogen stored in organic compounds through plant growth is converted to ammonia by animal metabolism and the microbial decomposition of urea and uric acid. Much of it remains in plant and animal tissues and is liberated only on the death of these organisms. Whenever a plant or animal dies, its body constituents are immediately attacked by microorganisms, and the nitrogenous compounds are decomposed with the liberation of ammonia. Part of the nitrogen is assimilated by the microorganisms themselves and thus converted into microbial cell constituents. Ultimately, these constituents are converted to ammonia following the death of the microbes.

The first step in this process of *ammonification* is the hydrolysis of the proteins and nucleic acids with the liberation of amino acids and organic nitrogenous bases, respectively. These simpler compounds are then attacked by respiration or fermentation.

Protein decomposition under anaerobic conditions (putrefaction) usually does not lead to an immediate liberation of all the amino nitrogen as ammonia. Instead, some of the amino acids are converted to amines. The putrefactive decomposition is characteristically brought about by anaerobic spore-forming bacteria (genus *Clostridium*). In the presence of air, the amines are oxidized with the liberation of ammonia.

**Nitrification**

Through all the transformations that nitrogen undergoes from the time of its reductive assimilation by plants until its liberation as ammonia, the nitrogen atom remains in the reduced form. The conversion of ammonia to nitrate *(nitrification)* is brought about in nature by two highly specialized groups of obligately aerobic chemoautotrophic bacteria. Nitrification occurs in two steps: In the first step, ammonia is oxidized to nitrite; in the second, nitrite is oxidized to nitrate. As a result of the combined activities of these bacteria, the ammonia liberated during the mineralization of organic matter is rapidly oxidized to nitrate. Thus, nitrate is the principal nitrogenous material available in soil for the growth of plants. The practice of soil fertilization with manure depends on the microbial mineralization of organic matter and results in the conversion of organic nitrogen to nitrate through ammonification and nitrification. Irrigation with dilute solutions of ammonia, which is one of the modern methods used for fertilization, is an even more direct means by which the nitrate content of soil is increased. Ammonia, which can be synthesized chemically from molecular nitrogen, is the most concentrated form of combined nitrogen available because it contains about 82 percent nitrogen by weight. Nitrates are very soluble compounds and are therefore easily leached from the soil and transported by water; hence, a certain amount of combined nitrogen is constantly removed from the continents and carried down to the oceans. In some special localities, notably in the semiarid regions of Chile, deposits of nitrate have accumulated in the soil as a result of the runoff and evaporation of surface water. Such deposits are a valuable source of fertilizer, although their importance has diminished greatly in the course of the last 50 years as a result of the development of chemical methods for making nitrogen compounds from atmospheric nitrogen.

Nitrates have played an important role not only in the development of agriculture, but also in the destructive activities of man. Gunpowder, which was the only explosive used for war before the invention of dynamite, is a mixture of sulfur, carbon, and saltpeter ($KNO_3$). During the Napoleonic wars, largely as a result of the British blockade, a shortage of nitrate for gunpowder production occurred in France. This led to the development of "nitrate gardens," in which nitrate was obtained by the mineralization of organic matter. A mixture of manure and soil was spread on the surface of the ground and frequently turned to permit aeration. After the manure had decomposed, nitrate was extracted from the residue.

## Denitrification and nitrogen fixation

Many different types of bacteria can use nitrate in place of oxygen as a final hydrogen acceptor. Whenever organic matter is decomposed in soil and conditions become anaerobic as a result of microbial respiration, any nitrate that may be present tends to become reduced. In the process of denitrification, molecular nitrogen is the principal end-product. Hence, through denitrification, combined nitrogen is removed from the soil by being converted to an inert gas which escapes into the atmosphere.

If molecular nitrogen were completely inert biologically, the activities of the denitrifying bacteria would very rapidly deplete the biosphere of all nitrogen available for growth, and life would cease on the earth. Although atmospheric nitrogen is not a suitable nutrient for most organisms, it can be used by a few specialized types as a source of nitrogen for growth. It is these *nitrogen-fixing organisms* that, by their own growth, compensate for the losses of combined nitrogen due to denitrification, and maintain a more or less constant amount of it in the biosphere.

Two principal types of nitrogen fixation are encountered in nature, both of which are dependent on the activities of microorganisms. These are *symbiotic nitrogen fixation* and *nonsymbiotic nitrogen fixation*. Symbiotic nitrogen fixation results from a mutualistic partnership between a seed plant and a bacterium, neither of which can fix nitrogen alone. The plants principally concerned are members of the group known as the legumes (e.g., peas, beans, alfalfa, clover, lupine), and the bacteria associated with them belong to a special genus, *Rhizobium*. The symbiotic relationship between these organisms will be discussed in Chapter 25.

The most important agents of nonsymbiotic nitrogen fixation are heterocyst-forming blue-green algae of the type of *Anabaena* and *Nostoc*, and aerobic bacteria of the genera *Azotobacter* and *Beijerinckia*. Although many other bacteria (e.g., photosynthetic bacteria, *Clostridium* spp., *Bacillus polymyxa*) can fix nitrogen, their quantitative contributions to this process in nature are probably small.

No satisfactory estimates have been made of the total annual fixation of atmospheric nitrogen in the entire biosphere. Such estimates are made difficult by the fact that nitrogen fixation is partially counterbalanced by denitrification. The total turnover of atmospheric nitrogen through fixation and denitrification may, however, be comparable in scale to the turnover of the combined nitrogen in the soil.

Nonsymbiotic nitrogen fixation is believed to contribute relatively little to nitrogen turnover in temperate zones. According to a widely accepted estimate, only about six pounds of nitrogen are fixed by free-living organisms in an acre of soil annually; some recent work suggests, however, that the actual amount may be several times greater. In the tropics nonsymbiotic fixation, particularly by blue-green algae, is a major factor in the maintenance of soil fertility; as much as 70 lb of nitrogen may be fixed per acre annually.

Symbiotic nitrogen fixation is, under certain conditions, a far more effective process than nonsymbiotic fixation. An acre of alfalfa may, for instance, fix as much as 400 lb of nitrogen in a single season. The growth of legumes, however, is limited by many factors, such as climatic conditions, competition with nonleguminous plants, and the mineral composition of the soil.

## *The sulfur cycle*

The element sulfur, an essential constituent of living matter, is abundant in the earth's crust. It is available to living organisms principally in the form of soluble sulfates. Reduced sulfur, in the form of $H_2S$, also occurs in the biosphere as a result of volcanic activity (for example, in sulfur springs) and of microbial metabolism. Except under anaerobic conditions, however, $H_2S$ does not accumulate in nature, because it is rapidly and spontaneously oxidized in the presence of oxygen.

The turnover of sulfur compounds in the cycle of matter, referred to as the sulfur cycle, is shown in Figure 22.3. In many respects, it shows a striking resemblance to the nitrogen cycle already described.

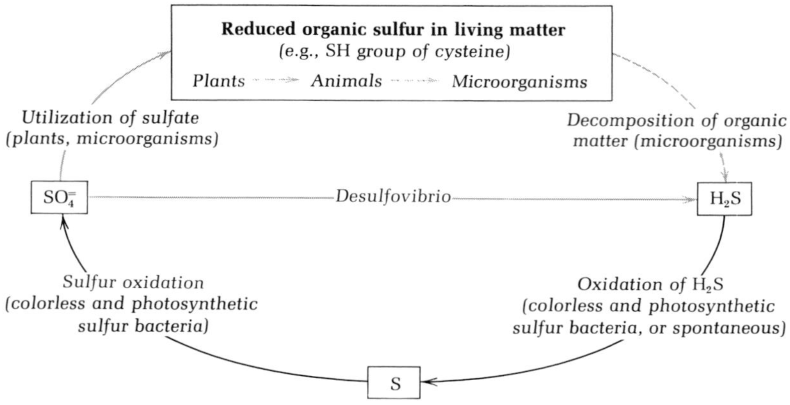


*Figure 22.3 The sulfur cycle: the oxidations of the sulfur atom are shown as solid black arrows, reductions as solid gray arrows, and reactions involving no valence change as broken gray arrows.*

**The assimilation of sulfate**

Sulfate is almost universally used as a nutrient by plants and microorganisms. The assimilation of sulfate resembles the assimilation of nitrate in two respects. First, like the nitrogen atom of nitrate, the sulfur atom of sulfate must become reduced to be incorporated into organic compounds, because in living organisms, sulfur occurs almost exclusively in reduced form as —SH or —S—S— groups. Second, in both cases, only enough of the nutrient is assimilated to provide for the growth of the organism, no reduced products being excreted into the environment.

### The transformation of organic sulfur compounds and formation of $H_2S$

When sulfur-containing organic compounds are mineralized, sulfur is liberated in the reduced inorganic form as $H_2S$. This process resembles ammonification, in which nitrogen is liberated from organic matter in its reduced inorganic form as ammonia.

### The direct formation of $H_2S$ from sulfate

The utilization of sulfate for the synthesis of sulfur-containing cell constituents and the subsequent decomposition of these compounds results in an overall reduction of sulfate to $H_2S$. $H_2S$ is also formed more directly from sulfate through the activity of the sulfate-reducing bacteria. These obligately anaerobic bacteria oxidize organic compounds and molecular hydrogen by using sulfate as an oxidizing agent. Their role in the sulfur cycle may therefore be compared to the role of the nitrate-reducing bacteria in the nitrogen cycle. The activity of the sulfate-reducing bacteria is particularly apparent in the mud at the bottom of ponds and streams, in bogs, and along the seashore. Since seawater contains a relatively high concentration of sulfate, sulfate reduction is an important factor in the mineralization of organic matter on the ocean floors. Signs of the process are the odor of $H_2S$ and the pitch-black color of the mud in which it occurs. The color of black mud is caused by the accumulation of ferrous sulfide. Some coastal areas, where an accumulation of organic matter leads to a particularly massive reduction of sulfate, are practically uninhabitable because of the odor and the toxic effects of $H_2S$.

### The oxidation of $H_2S$ and sulfur

The $H_2S$ that is produced in the biosphere as a result of the decomposition of sulfur-containing compounds, of sulfate reduction, and of volcanic activity is largely converted to sulfate. Only a small part of it becomes sequestered in the form of insoluble sulfides or, after spontaneous oxidation with oxygen, as elemental sulfur.

The biological oxidation of $H_2S$ and of elemental sulfur is brought about by photosynthetic and chemoautotrophic bacteria. It can be effected either aerobically by the colorless sulfur bacteria or anaerobically by the photosynthetic purple and green sulfur bacteria. Since these oxidations result in the production of hydrogen ions, they result in the local acidification of soils. Sulfur is commonly added to alkaline soil to increase its acidity.

## *The cycle of matter in anaerobic environments*

Regions of the biosphere that are not in direct contact with the atmosphere can remain oxygen-free for long periods of time. Such anaerobic environments harbor distinctive microorganisms; provided that light can penetrate, these microorganisms are able to bring about an almost completely closed anaerobic cycle of matter.

The primary synthesis of organic material under these conditions is mediated by photosynthetic bacteria. The purple and green sulfur bacteria convert $CO_2$ to cell material, using $H_2S$ as a reductant; the purple nonsulfur bacteria perform an almost complete assimilation of acetate and other simple organic compounds. Upon the death of the photosynthetic bacteria, their organic cell constituents are decomposed by clostridia and other fermentative anaerobes, with the formation of $CO_2$, $H_2$, $NH_3$, organic acids, and alcohols. The hydrogen and some of the organic fermentation products are anaerobically oxidized by sulfate-reducing and methane-producing bacteria. The anaerobic oxidations performed by the sulfate reducers result in the formation of $H_2S$ and acetate, both utilizable in turn by photosynthetic bacteria. The metabolism of the methane bacteria results in the conversion of $CO_2$ and of some organic carbon (e.g., the methyl group of acetic acid) to methane. Although part of this methane is probably used in photosynthesis by purple nonsulfur bacteria, much of it escapes to aerobic regions, where it is oxidized by *Methanomonas* and other aerobic methane oxidizers. The loss of methane constitutes the only significant leak from this anaerobic cycle of matter. The anaerobic sulfur cycle is completely closed, since sulfate and $H_2S$ are interconverted by the combined activities of sulfate-reducing and photosynthetic bacteria. The anaerobic nitrogen cycle is also closed and is chemically very simple, relative to the cycle in aerobic environments; the nitrogen atom does not undergo valence changes, alternating between ammonia and the amino groups ($R—NH_2$) in nitrogenous cell materials.

Since the participants in this anaerobic cycle of matter are all microorganisms, the cycle does not require much space. In fact, such a cycle can be established in the laboratory, in a closed bottle that contains the appropriate nutrients and that has been inoculated with water and mud from an anaerobic pool. Provided that the bottle is sufficiently illuminated, microbial development will continue in it over a period of years.

## The cycle of matter through geological time

The integration of the various reactions that constitute the cycle of matter results in a balanced production and consumption of the biologically important elements in the biosphere. It is probable that the cycle of matter as we know it today has operated without significant change for at least a billion years. However, there are good reasons to believe that the cycle of matter was considerably different at an early period in the history of the earth, when biological systems first developed on the planet (Figure 22.4).

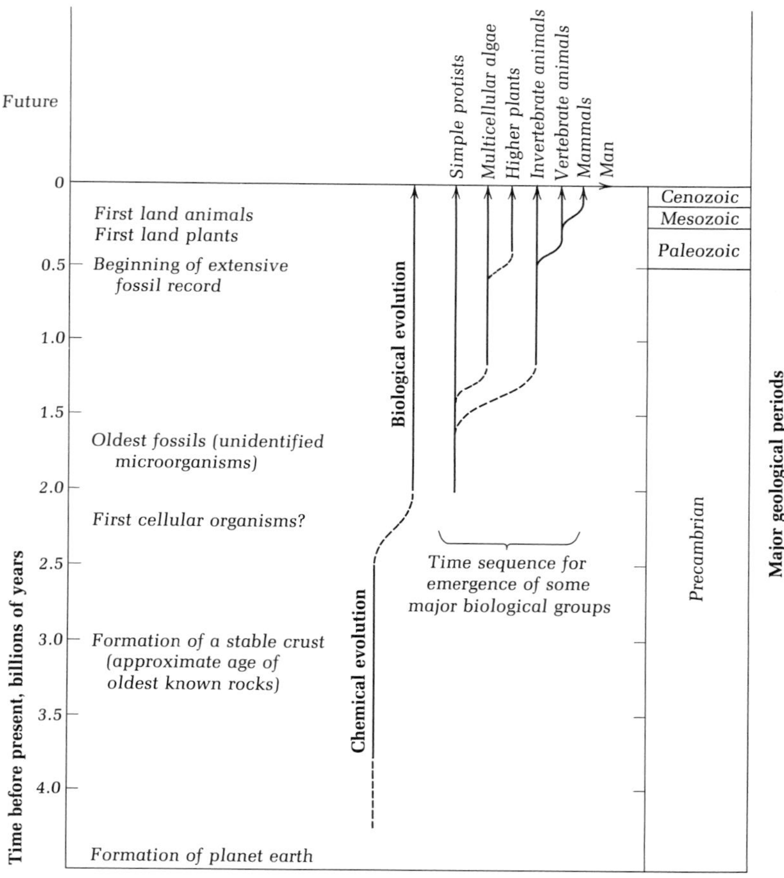


Figure 22.4 The time scale of terrestrial evolution.

On the primitive earth, prior to the emergence of life, the elements which are the principal constituents of living organisms were probably present in their reduced forms: hydrogen as $H_2$, carbon as methane, oxygen as water, nitrogen as ammonia, and sulfur as $H_2S$. Molecular oxygen was absent from the earth's original atmosphere. It is now generally believed that the emergence of living systems was preceded by a long period of chemical organic synthesis, during which organic compounds, formed as a result of reactions between atmospheric constituents, accumulated and interacted with one another in the primitive oceans. The relatively rapid formation of a great variety of organic compounds has been observed in the laboratory when mixtures of hydrogen, methane, ammonia, and water vapor are subjected to the ionizing action of an electric discharge. Among the products thus obtained artificially, some compounds characteristic of living organisms, such as the amino acids glycine and alanine, have been detected. In nature, such agencies as the ionizing rays of the sun, radioactivity, and electrical phenomena may have been responsible for the formation of a wide variety of organic compounds containing both nitrogen and sulfur. Under the reducing conditions prevailing in the absence of molecular oxygen, such compounds would not have become oxidized and would have tended to accumulate.

The formation of ever more complex organic compounds and molecular aggregates as a result of chemical interactions eventually resulted in the development of self-duplicating systems, and biological evolution began its course. The first living systems probably had very limited synthetic abilities and depended upon fermentative reactions for the generation of energy. With their growth and expansion, the preexisting store of organic raw materials gradually became depleted, favoring the emergence of organisms with an increasing degree of synthetic ability. At a relatively early stage in this primary biochemical evolution, the supply of energy-rich organic compounds must have become limiting, and the further course of biological evolution therefore depended on the acquisition by some organisms of the ability to use light as an energy source. The development of a mechanism for the performance of photosynthesis was therefore one of the earliest and most important steps in biochemical evolution. The first photosynthetic organisms must have been anaerobes, with modes of photosynthetic metabolism analogous to those of the contemporary purple and green bacteria.

The early evolution of photosynthetic organisms culminated in the emergence of forms that were able to use water as a reductant for the photosynthetic assimilation of $CO_2$. Once this point had been attained, the oxidized product of water, molecular oxygen, began to accumulate in the atmosphere, creating the conditions necessary for the evolution of organisms that obtain energy by aerobic respiration. As a consequence of the presence of molecular oxygen, the oxidized forms of nitrogen and sulfur (nitrate and sulfate) became predominant in the biosphere, and the stage was at last set for the establishment of the cycle of matter as we know it today.

## *The influence of man on the cycle of matter*

The emergence of man as a member of the biological community did not at first significantly affect the cycle of matter on earth. However, the rapid increases in the total size and local density of human populations that have occurred since the Industrial Revolution, coupled with the ever-increasing power of the human species to modify its environment, have begun to change the picture. Within the past century, these factors have led to local environmental changes comparable in scale to those produced by major geological upheavals in the past history of the earth. The spread of agriculture, the denudation of forests, the mining and burning of fossil fuels, and the pollution of the environment with human and industrial wastes have profoundly affected the distribution and growth of other forms of life.

### Sewage treatment

As a result of the concentration of the human population in large cities, which has proceeded at an ever-increasing pace during the past 150 years, the disposal of organic wastes, both domestic and industrial, has become a major ecological problem. The discharge of untreated urban wastes into adjoining rivers and lakes presents two hazards: the contamination of drinking water by microbial agents of enteric diseases, and the depletion of the dissolved oxygen supply as a result of the microbial decomposition of organic matter, leading to the destruction of animal life. For reasons of public health and of conservation, man has been forced to develop methods of *sewage treatment*, which result in the mineralization of the organic components of sewage prior to its discharge into the natural environment.

In a typical sewage treatment plant, the sewage is first allowed to settle. The precipitate, or *sludge*, undergoes a slow anaerobic decomposition, in which methane bacteria play an important part. The soluble organic compounds in the supernatant liquid are mineralized under aerobic conditions. This is sometimes achieved by spraying the liquid on a bed of loosely packed rocks, over which it trickles by gravity. In the *activated sludge process*, air is forced through the sewage, and a floc or precipitate is formed, the particles of which contain actively oxidizing microorganisms. After the organic matter has been largely oxidized, the sludge is allowed to settle. The sludges produced both by anaerobic decomposition and in the activated sludge process consist largely of bacteria, which have grown at the expense of nutrients in the sewage. These residues are eventually dried and used as fertilizer, either directly or after being ashed.

In all processes of sewage treatment, the goal is to produce a final liquid effluent in which the biologically important elements have been restored to the inorganic state. Sewage treatment accordingly involves intensive operation of a substantial segment of the cycles of matter under more or less controlled conditions. Even an effluent in which mineralization is complete may, however, produce undesirable ecological effects. If it is discharged into a lake, the consequent enrichment of lake water with nitrates and phosphates can cause an enormous increase in algal productivity, so that the water becomes colored and turbid at certain times of the year. If algal growth is sufficiently massive, the subsequent decomposition of algal organic matter may deplete the dissolved oxygen supply in the lake, with catastrophic effects on its animal life. Such progressive biological degradation of freshwater environments, first encountered in relatively small lakes near urban communities, has now become serious in some of the Great Lakes, particularly Lake Erie.

**The dissemination of synthetic organic chemicals**

In recent decades, chemical industry has produced an ever-increasing variety of synthetic organic chemicals, which are being used on an ever-increasing scale, as textiles, plastics, detergents, insecticides, and herbicides. Many of these synthetic organic compounds are more or less completely resistant to microbial decomposition, and accumulate in nature. One manifestation of this problem has occurred in areas where water is limited and subject to intensive reuse. The introduction of synthetic detergents which are not mineralized during sewage treatment has led to their reappearance in water supplies. The indiscriminate use of insecticides and herbicides that are highly resistant to microbial decomposition presents a far greater, although less obvious, danger. Many of these compounds are toxic for forms of life other than those which they are designed to control, and the long-range ecological effects of their dissemination are difficult (if not impossible) to predict.

## *Further reading*

**Books**

Alexander, M., *Introduction to Soil Microbiology*. New York: Wiley, 1961.

Bernal, J. D., *The Origin of Life*. New York: World, 1967.

Brock, T. D., *Principles of Microbial Ecology*. Englewood Cliffs, N.J.: Prentice-Hall, 1966.

Starr, M. P. (editor), *Global Impacts of Applied Microbiology*. New York: Wiley, 1964.

Wood, E. J. F., *Microbiology of Oceans and Estuaries*. Amsterdam: Elsevier, 1967.

**Reviews**

Alexander, M., "Biodegradation Problems of Molecular Recalcitrance and Microbial Fallibility." *Adv. Appl. Microbiol.* **7,** 35 (1965).

Porges, N., "Newer Aspects of Waste Treatment." *Adv. Appl. Microbiol.* **2,** 1 (1960).

Wuhrmann, K., "Microbial Aspects of Water Pollution Control." *Adv. Appl. Microbiol.* **6,** 119 (1964).

# Chapter twenty-three Symbiosis

*Each group of organisms* has had to adapt itself during its evolution not only to the nonliving environment but also to the other organisms by which it is surrounded. Adaptation to the environment sometimes involves the acquisition of special metabolic capacities which endow their possessor with the unique ability to occupy a particular physicochemical niche. The nitrifying bacteria, for example, can grow in a strictly inorganic environment with ammonia or nitrite as the oxidizable energy source; in the absence of light, no other living organisms are capable of developing in this particular environment, and the nitrifying bacteria are thus freed from biological competition. Withdrawal into a unique physicochemical niche is one means, and a highly effective one, of meeting the challenge of biological competition. A second method, however, which has been adopted by large numbers of microorganisms, has been to meet the challenge *by adapting to existence in continued close association with some other form of life.* This is the biological phenomenon known as *symbiosis.*

## Types of symbioses

The symbiotic associations which microorganisms form with plants and animals, as well as with other microorganisms, vary widely in their degree of intimacy. In terms of the closeness of the association, symbioses can be divided into two categories: *ectosymbioses* and *endo-*

*symbioses.* In an ectosymbiosis, the microorganism remains external to the cells and tissues of its host, as exemplified by the bacteria that live on the surface of leaves and in the body cavities of animals. In an endosymbiosis, the microorganism grows in the cells and tissues of its host,* as exemplified by the bacteria that inhabit the nodules of leguminous plants.

Symbioses also differ with respect to the relative advantage accruing to each partner. In *mutualistic symbioses,* both partners benefit from the association; in *parasitic symbioses,* one partner benefits while the second gains nothing and often suffers more or less severe damage. It is sometimes difficult to determine whether a given symbiosis is mutualistic or parasitic. The degree to which each partner is benefited or harmed can only be evaluated by comparing the fitness of the two members when living independently with their fitness when living in association. Furthermore, the nature of a particular symbiosis can shift under changing environmental conditions, so that a relationship that starts out as mutualistic may become parasitic, or vice versa.

The fact that two organisms have evolved a symbiosis implies that at least one partner derives some advantage from the relationship. The extent to which this partner depends on symbiosis for its existence, however, varies considerably. At one extreme are the microorganisms which populate the *rhizosphere*—the region that includes the surface of the root together with the soil immediately surrounding the root hairs of higher plants. These microorganisms live successfully in other regions of the soil but attain higher cell densities in the rhizosphere, where they derive advantages from their proximity to the root hairs. At the other extreme are the obligate parasites, which have never been successfully cultivated outside their hosts.

Thus, symbioses vary with respect to the degree of intimacy (ectosymbiosis versus endosymbiosis), the balance of advantage (mutualism versus parasitism), and extent of dependence (facultative versus obligate symbiosis).

## Mutualistic symbioses

In the succeeding chapters we will describe a variety of symbioses in which microorganisms have established mutually advantageous associations with other microorganisms, with plants, and with animals. Chapter 24 will deal with mutualistic *ectosymbioses* and Chapter 25 with mutualistic *endosymbioses.* Here we will confine our discussion to a few examples, to illustrate the varying types of relationships that may properly be regarded as mutualistic.

*The term *host* refers to the larger of two symbionts.

In the endosymbioses, the microbial symbionts live in the tissues of their hosts. In many such associations the microorganism leads a permanently intracellular existence and is passed from one generation of the host to the next in the cytoplasm of the egg. In others, the microorganism remains intracellular through part of the life cycle of the host; at some stage it is liberated into the extracellular environment, from which the next generation of host becomes infected.

Among the ectosymbioses we can discern three broad types of associations. First, there are the so-called "loose associations" between two microorganisms occupying the same natural habitat. In milk, for example, a yeast is often found growing in mixed culture with one or more types of bacteria. The bacteria hydrolyze lactose to glucose and galactose, which are then fermented by the yeast, the by-products of the fermentation being utilized by the bacteria. Second, there are the ectosymbioses in which the microbial symbiont lives directly on the external surfaces of its host. Some blue-green algae, for example, attach themselves to the gelatinous capsules of other microorganisms (Figure 23.1*a*), while certain flagellated protozoa bear on their surface a mantle of spirochetes (Figure 23.1*e*). Finally, many microbial symbionts inhabit the body cavities of their hosts. These associations are still considered as ectosymbioses, because the body cavities, though internal to the whole organism, are external to the tissues and are continuous with the external surfaces of the host.* The most familiar examples are the microorganisms that inhabit the digestive tract of mammals; to the same category belong the luminous bacteria that populate the light-emitting organs of some fishes and mollusks (Figure 23.4) and the bacteria that live in the cavities formed by the nodulation of plant leaves.

*In many texts the term *ectosymbiosis* is restricted to those associations in which the microorganism remains external to the body of the host.

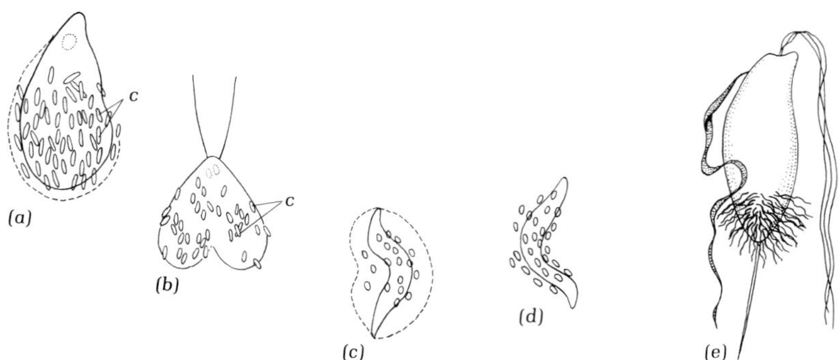


*Figure 23.1 Ectosymbioses between blue-green algae and other microorganisms. The symbol "c" stands for "cyanellae," a general name for blue-green algae. (a) Cyanellae attached to the gelatinous capsule of the eucaryotic flagellate, Oicomonas syncyanotica. (b) Same, at division stage. (c) and (d) Cyanellae attached to large spirilla. The dotted lines represent the boundary of the capsule layer. (After A. Pascher.) (e) Spirochetes attached to posterior end of termite flagellate, Glyptotermes sp. From H. Kirby, Jr., Univ. Calif. Publ. Zool.* **45,** *247 (1945).*

Of all the ectosymbioses that we shall describe later on, none is more remarkable than the cultivation of fungi by certain insects, notably the higher termite, *Termes*, and the wood-boring "ambrosia beetles." The termites have special chambers in their nests which contain "fungus gardens," on which the young nymphs and larvae browse. The ambrosia beetles have evolved elaborate methods of infecting their tunnels with fungal spores so that the tunnels become lined with mycelium, on which the beetles feed (Figure 23.2). These practices closely parallel the cultivation by man of food plants, and in fact all these activities are symbiotic.

The picture of one symbiont literally devouring its partner may at first glance appear to contradict our concept of symbiosis as "existence in continued close association." When one of the partners is a microorganism, however, we are dealing with a *population*, not with an individual. Thus, the cultivated fungus garden of the termite benefits as a population, although a fraction of the population at any given moment is being devoured. In such ectosymbioses, as well as in all endosymbioses, the relationship is one of *reciprocal exploitation*, involving a

*Figure 23.2 Tunnel of the ambrosia beetle, Platypus cylindrus, in oak. Transverse section showing palisade layer with sterile hyphae, conidia and chlamydospores of the ambrosial fungus, Endomycopsis platypodis (an ascomycete). From J. M. Baker, "Ambrosia beetles and their fungi, with particular reference to Platypus cylindrus Fab." in Symbiotic Associations: Thirteenth Symposium of the Society for General Microbiology. New York: Cambridge University Press, 1963. (Crown copyright: reproduced by permission of the Director, Forest Products Research Laboratory, United Kingdom.)*

dynamic balance between offensive and defensive activities of the two members. In the root nodule endosymbiosis, for example, the nodule bacteria infect the plant root and exploit it, but the plant ultimately digests most (but not all) the bacteria. The net result is a superficially peaceful, mutualistic symbiosis.

**Parasitic symbioses**

We have defined the parasitic symbioses as those in which one of the partners does not profit from the association and may suffer more or less severe damage. In the case of microbial symbioses with plants and animals, it is usually a simple matter to demonstrate the advantage accruing to the microorganism, but it is often difficult or impossible to evaluate the effect on the host. *Infectious disease*, in which the host is progressively weakened and may eventually die, is obviously a parasitic symbiosis. When however, there is no overt damage to the host, it is difficult to tell whether the relationship is a parasitic or a mutualistic one. For example, the normal flora of the mammalian intestinal tract was long assumed to be parasitic, only the microorganisms benefiting from the association. Work with germ-free animals, however, has revealed a number of subtle but important benefits which the intestinal flora confers on the host (see p. 738).

Our chapters on parasitism (Chapters 26 to 28) will deal almost exclusively with infectious disease, since only when damage to the host occurs is it clear that the symbiosis is not a mutualistic one. The infectious diseases provide the most vivid demonstration of a principle mentioned earlier: that symbiosis is a dynamic state, reflecting a balance between offensive and defensive activities. Chapter 26 will deal with those vertebrate host defenses which are *constitutive*, being present in normal organisms even in the absence of any inducing stimulus. Chapter 27 will describe the offensive activities of microbial parasites, by means of which they overcome the constitutive host defenses and produce disease. Chapter 28 will describe the host's final line of defense—the inducible mechanisms, such as antibody formation, by means of which balance is restored or the parasite eliminated.

**Parasitism as an aspect of ecology**

The central problem of ecology is to discover the factors that ultimately determine the survival of a species. Given the reproductive potential of a species, that is, the number of viable offspring produced per parent per unit of time, the factors that affect survival all fall into one of two categories: those that affect the available food supply and those that affect the rate of destruction of individuals. For any species other than man, who has invented extraordinary ways of destroying himself, the possible fates of an individual are restricted. Animals, other than a few domestic ones, rarely die of old age or from accidental mishaps. Most are either eaten by their natural predators or destroyed by their pathogenic parasites. The distinction between a predator and a pathogenic parasite is a fine one, however, for both satisfy their nutritional needs at the expense of their victim. The only important difference, other than relative size, is that the parasite is a member of a symbiosis, having adapted to a continued close association with an individual of a different species. A predator could also be said to exist in continued close association with its prey, but hardly with any single individual. In the final analysis, then, the difference is that the victim of a predator is killed immediately, whereas the victim of a parasite remains alive for an extended period of time.

Since the normal fate of any individual is to be killed either by predator or by parasite, each species constitutes a link in a biological *food chain*. Microbial cells, for example, are food for many species of plankton, the minute plants and animals that drift in the oceans in great abundance. The plankton serve as the major food source for many marine invertebrates and fishes, which are fed upon in turn by larger fishes and some marine mammals. The largest animals, which seemingly stand at the ends of the food chains, must eventually die and be devoured by microorganisms, so that the food chains are actually *cyclic* in nature.

The fact that every organism is part of a food chain means that—for any nonphotosynthetic type—the best source of nutrients is another organism. The factors that affect the available food supply of a species are numerous and complex, whether the organism feeds as a predator or as a parasite. For example, an excessively large population may exhaust the food supply to such an extent that the next generation will find an extreme shortage and hence will survive in much smaller numbers. The food supply may also be affected by a change in climate, by long-term geochemical evolution, or by changing competition with other predators or parasites.

The ability of a species to resist being eaten by others depends both on its own defense mechanisms (e.g., protective coloration, armor, ability to fly or burrow, immunity to disease) and on the properties of its predators or parasites. A new predator or parasite may appear on the scene, for example, or an existing one may evolve more efficient feeding habits.

The living world is thus organized into a large number of intersecting food chains. Each chain consists of a number of species, the populations of which have reached equilibrium in terms of rate of reproduction and destruction. The equilibrium size of a population may shift abruptly if any one of its complex determinants changes. In modern times, the most significant factors affecting this ecological equilibrium have been the activities of man. By building dams, destroying forests, polluting streams, slaughtering game, spraying poisons, and transporting parasites, he has exterminated many species and changed the ecology of many others.

## *The significance of symbiosis*

A symbiont substitutes for part or all of the nonliving environment in which free-living organisms must survive; among the myriads of symbioses that have evolved we can find examples in which almost every known environmental function is furnished by one or another symbiont for its partner. For convenience, we shall discuss these functions under four headings: protection, provision of a favorable position, provision of recognition devices, and nutrition.

To determine the functions fulfilled by the partners in a symbiosis it is necessary—in all but the most obvious cases—to separate the partners and study their requirements in isolation. In many cases this has not yet been achieved, either because the symbionts cannot be separated without damaging them, or because the isolated partners cannot be cultivated. Symbionts that have defied attempts to cultivate them in isolation are said to be *obligate symbionts.* The classification of a symbiont as "obligate" is always provisional, since it is always possible that identification of the function performed by its partner will permit its eventual cultivation.

### Protection

Endosymbionts, as well as those ectosymbionts which live in the body cavities of animals, are protected from adverse environmental conditions. These habitats protect the symbionts from dessication and—in the case of warm-blooded hosts—from extremes of temperature.

The microbial symbionts of plants and animals also perform functions which protect their hosts. Most notable is the protection which the normal flora of vertebrates offers against invasion by pathogenic (disease-producing) microorganisms; germ-free animals are much more susceptible to infection than their normal counterparts, as we shall describe in Chapter 24. The removal of toxic substances is another function which many microbes perform for their symbiotic partners; in some insects bacteria harbored in the excretory organs break down uric acid and urea to ammonia, which the bacteria themselves assimilate. In many natural environments strict anaerobes are protected from damage by molecular oxygen as a result of the respiratory activity of associated aerobic microorganisms.

**Provision of a favorable position**

A symbiotic association may provide one partner with a position that is favorable with respect to the supply of nutrients. Many of the marine ciliated protozoa, for example, are found only on the body surfaces of crustacea, where the host's respiratory and feeding currents assure the microbe of a constant supply of food (Figure 23.3). No less spectacular is the favorable position provided by many marine invertebrates for their photosynthetic algal symbionts. Some of these hosts, such as the tridacnid clams, house the algae in special organs which act as lenses to gather light; other hosts are phototactic, carrying their photosynthetic partners toward the light.

**Provision of recognition devices**

In many groups of invertebrates, as well as in some groups of fishes, certain members have evolved systems of bioluminescence. The emission of light by animals very often appears to be a recognition device, promoting schooling, mating, or the attraction of prey. Some of these animals are themselves unable to emit light and rely upon luminous bacterial symbionts which they harbor in special organs. The fish *Photoblepharon* carries its luminous bacteria in special pouches under the eyes; the pouches can be uncovered or covered by a shutter mechanism, permitting the fish to flash its light on and off (Figure 23.4). Some groups of cephalopods have also evolved highly specialized organs to house their luminous bacterial symbionts, even to the point of developing reflectors and lenses.

**Nutrition**

By far the most common function of symbionts is to provide nutrients for their partners. The provision of nutrients may be indirect, as in the case of fungi that infect plant roots and thereby increase the water-absorbing capacity of the root system. Usually, however, the nutritional support is direct, the symbiont furnishing one or more essential nutrients to its partner.

The most dramatic and extensively studied example is *nitrogen fixation*. Nitrogen fixation by root-nodule bacteria will be discussed in some detail in Chapter 25; originally established only for the root nodules of legumes, nitrogen fixation by the nodules of several nonleguminous plants has more recently been established using $^{15}N$ as an isotopic tracer. Many claims have also been made for $N_2$ fixation by the symbionts of insects, particularly the symbionts of termites that appear

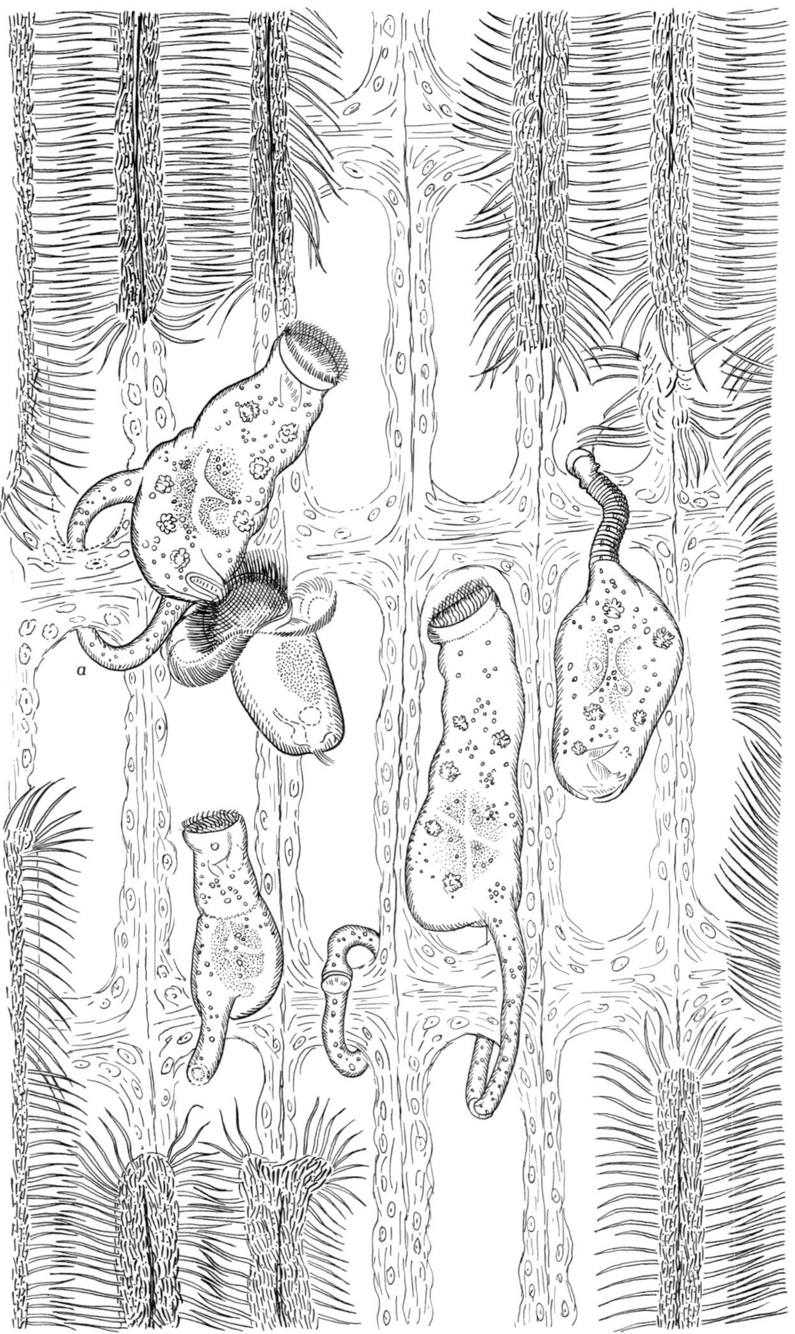


Figure 23.3 The ciliate protozoon, *Ellobiophyra donacis*, "padlocked" to the gills of the bivalve, *Donax vittatus*. In (a) *Ellobiophyra* is seen in the process of reproduction by budding. From E. Chatton and A. Lwoff, "Ellobiophyra donacis Ch. et Lw., péritriche vivant sur les branchies de l'acephale Donax vittatus da Costa." Bull. Biol. Belg. **63**, 321 (1929).

to be able to live on a diet of pure cellulose. These claims, however, have not yet been substantiated by experiments with $^{15}N$.

Cellulose, as a major plant constituent, provides the principal carbon and energy source for grazing animals as well as for wood-boring insects. Some of these animals are incapable of digesting cellulose; in the ruminants, and in at least one group of insects, the termites, cellulose digestion is performed on behalf of the host by symbiotic bacteria and protozoa. The digestion of other complex carbohydrates is often carried out by microbial symbionts living in the digestive tracts of animals.

One of the most intriguing detective stories in biology has been the discovery and elucidation of the endosymbioses between microorganisms and their insect hosts. In 1888 F. Blochmann recognized that certain special cells of cockroaches contained symbiotic bacteria, and soon entomologists discovered bacterial, fungal, and protozoan endosymbionts in a variety of other insects.

The significance of these symbioses became clear when techniques were developed for ridding the insects of their symbionts; such animals require one or more B vitamins to develop normally. The importance of the symbionts to the well-being of their hosts is emphasized by the elaborate mechanisms which insects have evolved for transmitting the symbionts to their young. These mechanisms will be described later in this chapter.

Many of the symbioses which we will consider in the succeeding chapters involve associations between a photosynthetic and a nonphotosynthetic partner. Lichens, mycorrhizas, and the algal endosymbioses with invertebrates are all examples of such associations. In these particular cases, the metabolic functions of the two partners are complementary with respect to the metabolism of carbon and oxygen; in effect, therefore, the symbiotic association carries out a complete cycle of these two elements, according to the scheme described in Chapter 22.

*Figure 23.4 Photoblepharon, a fish with organs in which luminous bacteria are harbored. The luminous organ, below the eye, can be either concealed (a) or exposed (b) by a shutter mechanism. From E. N. Harvey, Living Light, p. 266. Princeton, N.J.: Princeton University Press, 1940; and E. N. Harvey, Bioluminescence, p. 521. New York: Academic Press, 1952.*

**The movement of carbohydrate from photosynthetic to nonphotosynthetic symbiont**

One of the chief functions provided by a photosynthetic symbiont for its partner is the supply of carbohydrate. The movement of carbohydrate from one symbiont to the other has been studied by allowing $^{14}C$-labeled $CO_2$ to be assimilated in the light, following with time the incorporation of label into metabolites of each partner. In symbiotic associations of algae with invertebrate animals, as well as in associations of algae with fungi (lichens), isotope studies have revealed a number of important adaptations that facilitate the *unidirectional transport of carbohydrate* from the photosynthetic to the nonphotosynthetic partner. For example, the symbiotic algae excrete a much greater proportion of their fixed carbon than do related free-living algae; in many cases, this excretion ceases soon after the symbiotic alga is isolated, indicating a specific stimulatory effect by the nonphotosynthetic partner.

The excreted carbohydrate is usually different from the major intracellular carbohydrates of the alga: in most cases, it is a carbohydrate which the nonphotosynthetic partner, but not the alga itself, can utilize. For example, the green algae of lichens excrete polyols, such as ribitol, which are not metabolizable by the algae themselves but which are rapidly utilized by the fungal components of the lichens (Table 23.1). This phenomenon explains the unidirectional flow of excreted carbohydrate: the utilization of the excreted material by the nonphotosynthetic partner creates a *concentration gradient*, such that carbohydrate must flow steadily from the alga to its partner. In some cases, the excreted carbohydrate is one which may be metabolized by the alga also; in such cases the unidirectional flow is maintained by the rapid conversion of the carbohydrate in the fungus to a form that only the latter can utilize.

In mycorrhizas (associations between fungi and the roots of higher plants), carbohydrate is again found to move from the photosynthetic to the nonphotosynthetic partner. Here the transported carbohydrate appears to be sucrose, which is the form in which carbohydrate is also translocated within the plant. Movement to the fungus thus represents a *diversion of the translocation stream;* in part, this can be accounted

*Table 23.1 Carbohydrate movement from photosynthetic to nonphotosynthetic symbiont*[a]

| **Photosynthetic donor** | | | **Nonphotosynthetic recipient** | |
|---|---|---|---|---|
| **Organism** | **Carbohydrate released** | | **Immediate fate of carbohydrate** | **Organism** |
| *Zoochlorellae* | *Maltose, glucose* | → | *Glycogen, pentoses* | *Marine invertebrates* |
| *Zooxanthellae* | *Glycerol* | → | *Lipids, proteins* | *Marine invertebrates* |
| *Lichen algae* | | | | |
| *Chlorophyceae* | *Polyols* | → | *Polyols* | *Lichen fungi* |
| *Cyanophyceae* | *Glucose* | → | *Mannitol* | *Lichen fungi* |
| *Higher plants* | *Sucrose* | → | *Trehalose, glycogen, polyols* | *Mycorrhizal fungi* |

[a]Modified, with permission, from Table 8 in D. Smith, L. Muscatine, and D. Lewis, "Carbohydrate Movement from Autotrophs to Heterotrophs in Parasitic and Mutualistic Symbioses." *Biol. Rev.* **44,** 17 (1969).

for by the rapid conversion of sucrose to fungal carbohydrates such as trehalose and polyols, which the plant cannot utilize. It is possible, however, that the diversion is brought about through the release of plant hormones, many of which are known to be produced by fungi.

## *The establishment of symbioses*

As we shall see later on, the evolution of a symbiosis is usually characterized by a greater and greater interdependence of the two partners. This in turn places a premium on the development of mechanisms to ensure the continuity of the symbiosis from generation to generation. Such mechanisms are of two kinds: those in which the host transmits its symbionts directly to its progeny at each generation, and those in which each new generation is freshly reinfected.

**Direct transmission**

The simplest type of direct transmission is found in the endosymbioses of protozoa and algae. The protozoan and its intracellular algal symbiont divide at more or less the same rate, so that each daughter cell of the host receives a proportionate share of algal cells. In some instances cell division is beautifully regulated: the host cell, containing two algal symbionts, divides to yield two daughter cells, each containing one symbiont. The symbiont then divides, restoring the number per cell to two.

In the sexually reproducing animals direct transmission may be accomplished by infection of the egg cytoplasm. This may require an extremely elaborate sequence of host cell movements, morphological changes, and interactions. In certain insects, for example, the microbial symbionts may be contained within specialized cells, the *mycetocytes*. Mycetocytes become associated with the tissues of the ovary, often as a result of oriented movement from other parts of the body. The symbionts are liberated and coat the surface of the developing oocytes, which they penetrate. Since none of the insect endosymbionts is motile, all their movements from other parts of the insect body to the egg cells are host-mediated.

**Reinfection**

Again it is among the insects that we find the most elegant examples of mechanisms designed to ensure infection of the progeny with symbionts from the mother. Each group has evolved its own set of specialized devices: in some insects a direct anatomical connection between the intestine and vagina guarantees that intestinal symbionts will be copiously smeared on the surface of the eggs as they pass through the ovipositor. When the young larvae hatch, they infect themselves immediately by eating part or all of the egg shell.

Two groups of flies are viviparous: *Glossina* (the tsetse flies), and a large group called *Pupipara*, which are themselves ectosymbionts of mammals and birds. The larvae of these flies are retained in the uterus, where they are nourished by the secretions of greatly developed accessory glands, the "milk glands." The symbiotic microorganisms are localized in these glands and are delivered to the larvae during feeding.

Two mechanisms of transmission are particularly intriguing because they involve a stereotyped, genetically determined behavior pattern of the newly hatched larvae. In one, the hatchlings suck up drops of bacterial suspension which exude from the mother's anus during the period of brood care. In the other, discovered in the plant-juice sucking insect *Coptosoma*, the female deposits a bacterium-filled "cocoon" or capsule between each pair of eggs. When the eggs hatch each larva sinks its proboscis into a cocoon and sucks up a supply of symbionts (Figure 23.5).

Equally complex are the mechanisms that have evolved in the plant kingdom for the initiation of root nodules when the bacteria *Rhizobium* infects its leguminous host. The plant root excretes a number of substances, among which is tryptophan. The bacterial cells in the soil convert tryptophan to the plant growth hormone, indoleacetic acid, and also produce an extracellular polysaccharide capsular material which induces the plant root to excrete the enzyme polygalacturonase. The bacteria then commence penetration of those root hairs which have been induced to grow abnormally by the indoleacetic acid; it is possible that polygalacturonase plays some role in mediating the penetration. The process of nodulation will be discussed in greater detail in Chapter 25.

(a)

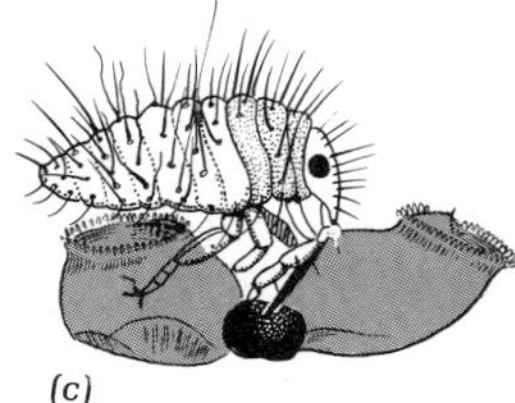


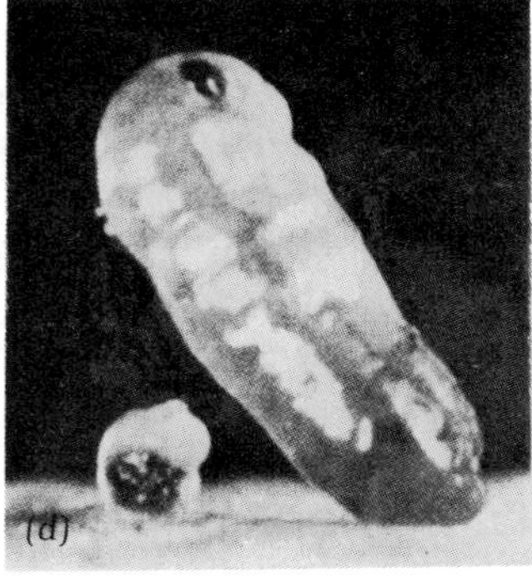


*Figure 23.5 Coptosoma scutellatum. (a) Eggs deposited on a vetch leaf. (b) The eggs seen from below, showing the symbiont-filled cocoons lying between each pair of eggs. (c) A newly hatched larva sucking the symbiont suspension from a cocoon. (d) Enlarged view of egg and cocoon. (After H. J. Müller.)*

In general, the more interdependent the symbiotic partners, the more we can expect to find that evolution has produced means for ensuring their continued association. In contrast, the formation of loose associations often seems to depend entirely on chance, and both partners may also be free-living.

## *The evolution of symbioses*

Natural selection acts on symbiotic associations as well as on individual organisms; symbioses thus have their own phylogenies. In the absence of fossil evidence, such phylogenies are necessarily speculative, but certain trends can nevertheless be deduced from the nature of contemporary symbioses.

### Evolutionary trends

It seems inescapable that symbioses evolve in the direction of increasing intimacy. Starting with a loose association, in which one or both organisms finds an optimal environment in the vicinity of the other, an ectosymbiosis may gradually develop. At a later stage in its evolution the relationship may become endosymbiotic, the small organism penetrating the host tissues and ultimately the host cells.

A lichen, for example, represents an extremely tight association between a specific fungus and a specific alga, both of which are closely related to free-living forms. The very regular arrangement of the mycelium and algal cells in the lichen has presumably evolved from loose, random aggregations; penetration of the algal cells by specialized fungal hyphae (haustoria) is probably a later development. Similarly, loose associations between plant roots and fungi in the rhizosphere have evolved into *mycorrhizas:* root systems that have been morphologically altered as a result of extensive penetration by the fungal hyphae. Both intercellular and intracellular penetrations are found, the latter presumably being the more advanced type.

Once a symbiosis has been established, selection operates to increase its efficiency. An increased degree of adaptation to one highly specialized environment necessarily implies, however, a decreased degree of adaptation to all other environments. The result is a high degree of specialization; the symbiont not only loses its ability to live freely but also becomes increasingly *specific* with respect to its choice of partner. Today we find many extreme cases, particularly in the endosymbioses, where neither partner can grow without the other.

At this point we must ask ourselves whether a distinction can be drawn between a totally interdependent pair of symbionts, such as a protozoon that carries an intracellular algal symbiont, and a eucaryotic cell with its contained organelles. Some years ago a distinction might have been proposed in genetic terms, since the coexistence of protozoon and alga involves two distinct genomes, whereas it was assumed that all parts of the eucaryotic cell are controlled by a single genome. Today this distinction can no longer be made, because it is now known that such organelles as mitochondria and chloroplasts contain DNA.

Indeed, it is entirely conceivable that the eucaryotic cell originally arose as an endosymbiosis between two (or more) primitive cell types. The general similarities between chloroplasts and endosymbiotic blue-green algae, for example, are striking (Figure 23.6).

If cell organelles that possess DNA are to be considered as symbionts, for the purposes of this discussion, then what can be said of the *plasmids* and *episomes* which were described in Chapter 14? For that matter, what of the *viruses*, which also behave as intracellular foreign genetic elements? All these elements could be classed as obligate symbionts, usually parasitic but under some circumstances entering into

*Figure 23.6 (a) Electron micrograph of a thin section of the flagellated protozoon, Cyanophora paradoxa, containing several endosymbiotic blue-green algae (× 9,100). Courtesy of William T. Hall. (b) Electron micrograph of thin section of Euglena gracilis, an algal cell containing several chloroplasts (× 13,138): g, Golgi apparatus; l, lamellae of chloroplast; m, mitochondrion; n, nucleus; p, pyrenoid; pm, paramylum; v, vacuole. Courtesy of Jerome A. Schiff.*

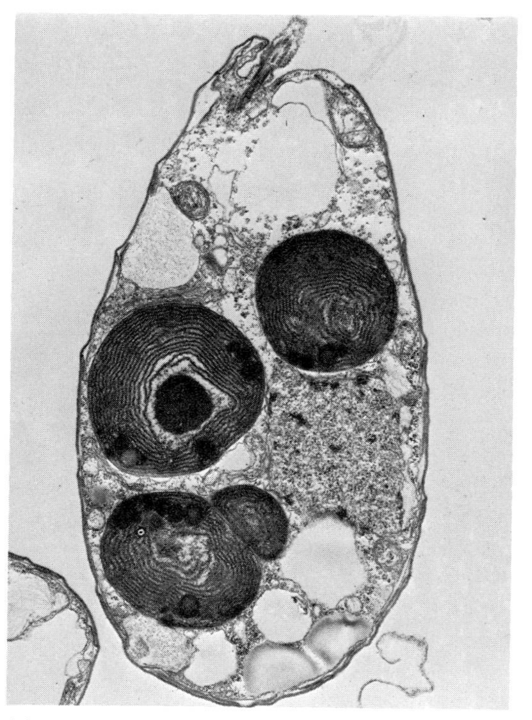

(a)

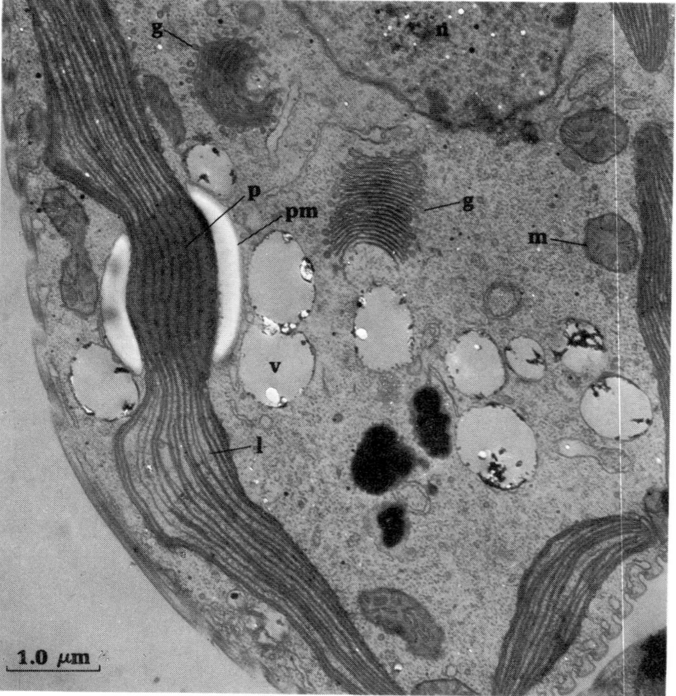


(b)

mutualistic associations with their host cells. Whether we so classify them depends entirely on whether or not we wish to restrict our definition of symbiosis to relationships between cells (or celllike organelles). Plasmids, episomes, and viruses in the intracellular state are neither cellular nor celllike; they consist of genetic material, either DNA or RNA, and are analogous to the procaryotic chromosome. Although it is obviously true that living systems involve interdependencies at all levels—molecular, organellar, and cellular—it is probably more useful to restrict the term *symbiosis* to interdependent systems in which the two partners are at least theoretically capable of autonomous existence.

**The creative manifestations of symbiosis**

In an earlier section of this chapter we enumerated some functions that symbionts provide for each other: protection, nutrition, and so on. In the early stages of the evolution of a symbiosis these functions do not differ from those performed by the free-living forms; the symbionts simply exploit their partner's normal activities and structures.

As natural selection operates to make the symbiotic association more and more effective, however, new structures and functions develop which neither partner possesses in isolation. There are a great many examples: the production of unique organic acids and pigments by lichens, the histological changes in the intestinal tract of mammals brought about by the intestinal flora, and the complex processes of root nodule formation and nitrogen-fixing activity are all creative manifestations of symbiosis. Perhaps even more remarkable is the positive phototaxis shown by marine invertebrates harboring photosynthetic algae; symbiont-free individuals, in contrast, do not move preferentially toward the light.

These new structures and activities reflect, in each situation, the activity of one or more catalytic or structural proteins. Obviously, the structures of these proteins are predetermined by the genes of the symbiotic partners, but very little is known about the interactions that permit these genes to be manifested. In some cases we may expect to find that symbionts produce substances which induce or repress enzyme formation by their partners; in other cases they may directly or indirectly regulate the partner's enzymatic activities; and in still other cases the constitutive enzymes of the two partners may act in concert to create new metabolic pathways, leading to the production of new substances. In fact, we can expect to find types of interactions similar to those which regulate and integrate the activities of different tissues of a multicellular organism.

What at first may have been an *inducible* response by one symbiont to its partner has often become *constitutive,* so that the new structure or activity persists even when the symbionts are separated. The formation of mycetocytes by insects, for example, continues after the insect has been freed of its endosymbiotic microorganisms. A symbiont constitutes a determinative environmental factor for its partner, and thereby plays a major role in shaping its evolution.

## *Further reading*


**Books**

Burnet, F., "The Natural History of Infectious Disease," 3rd ed. New York: Cambridge University Press, 1962.

Henry, S. M. (editor), *Symbiosis,* Vols. I and II. New York: Academic Press, 1966 and 1967.

*Symbiotic Associations: Thirteenth Symposium of the Society for General Microbiology*. New York: Cambridge University Press, 1963.

**Reviews and original articles**

Smith, D., L. Muscatine, and D. Lewis, "Carbohydrate Movement from Autotrophs to Heterotrophs in Parasitic and Mutualistic Symbiosis." *Biol. Rev.* **44,** 17 (1969).

# Chapter twenty-four
# Mutualistic ectosymbioses

*In an ectosymbiosis,* neither partner penetrates the cells or tissues of the other. Many microorganisms are found in nature living ectosymbiotically with other microorganisms, with plants, or with animals. These symbioses range from quite loose associations of two or more different microorganisms, to extremely tight association of a particular microorganism with a particular plant or animal host. In this chapter we will discuss ectosymbioses in which both the microorganism and its partner benefit: the mutualistic ectosymbioses.

## *Mutualistic ectosymbioses of microorganisms with other microorganisms*

Many microorganisms are capable of growth in a particular environmental niche only if a second type of microorganism is present. Such relationships may be highly specific and are often mutualistic rather than parasitic. The symbiotic nature of the relationship becomes apparent when attempts to isolate the organisms in pure culture fail, in spite of their abundant growth in mixed culture. Further experimentation is then required to establish the function that each performs for the other.

These functions can be conveniently discussed under two categories: the provision of nutrients and the removal of toxic substances.

**The provision of nutrients**

In many environments, one microorganism may depend upon another for its source of an important nutrient. This situation is particularly common in niches where carbon is present in a polymer such as a polysaccharide. Here a microorganism lacking the specific hydrolytic exoenzyme can grow only if the polymer is digested by a different organism.

In anaerobic, illuminated environments, a loose mutualistic symbiosis often develops between sulfate-reducing bacteria of the genus *Desulfovibrio* and the photosynthetic purple and green sulfur bacteria. *Desulfovibrio* decomposes organic compounds by anaerobic respiration. Such compounds as lactate are oxidized with sulfate as electron acceptor, forming as metabolic end-products acetate, $CO_2$, and $H_2S$. The $H_2S$ provides a photosynthetic electron donor, and the acetate and $CO_2$ provide carbon sources for the photosynthetic bacteria. The metabolism of the photosynthetic bacteria results in the rapid reoxidation of $H_2S$ to sulfate; in this way, the quantitatively important requirement of sulfate by *Desulfovibrio* can be met, even in environments where the absolute sulfate concentration is very low.

Photosynthetic algae are frequently found in nature in association with aerobic bacteria. Here both the oxygen and carbon cycles are continuously driven by the absorption of light in algal photosynthesis. The algae produce $O_2$ and synthesize carbohydrates from $CO_2$; the bacteria use the oxygen for the respiration of organic substrates, including those released from dead algal cells, and produce $CO_2$. Much of this $CO_2$ is then reutilized in algal photosynthesis.

In the laboratory, it is common to find pairs of microorganisms which can grow in mixed culture in a synthetic medium that will not support the growth of either organism alone. This is frequently found to result from the mutual provision of growth factors. Thus, an organism that excretes biotin and requires methionine for growth can grow in mixed culture with an organism that excretes methionine and requires biotin for growth, in a synthetic medium lacking both growth factors. It is easy to imagine that such complementary provision of growth factors takes place in nature, because many free-living microorganisms exhibit growth factor requirements.

In summary, one microorganism may depend upon a second one for its source of carbon, nitrogen, or sulfur; for one or more growth factors; or for a special type of hydrogen donor or hydrogen acceptor. When, by the liberation of metabolic products, two microorganisms provide each other with required nutrients, a mutualistic ectosymbiosis develops. This situation may be contrasted with the unilateral dependence of one microorganism upon the metabolic products of another: here the two organisms need not be living simultaneously in the same niche, but can grow in succession. It is the *mutualistic* nutritional relationship that demands simultaneous occupation of the same niche.

**The removal of toxic substances**

Among the functions performed by one ectosymbiotic partner for another is the removal of toxic substances. Obligate anaerobes, for example, can grow in environments freely exposed to the air, when aerobic organisms are present to remove molecular oxygen. If, in turn, the anaerobe performs some function useful to the aerobe, a mutualistic ectosymbiosis is established.

Every microorganism has an optimum hydrogen ion concentration for growth, and in most cases growth is severely inhibited when this concentration is either raised or lowered substantially. Since some organisms produce hydrogen ions through their metabolism whereas others remove them, two microorganisms may—through complementary metabolic activities—maintain a pH favorable to both.

Hydrogen ions are not the only metabolic products which can accumulate and inhibit the organisms that produce them. Depending on the nature of the carbon source, the presence or absence of oxygen, and other environmental conditions, a microorganism may excrete toxic metabolites that must be degraded by other microorganisms if growth is to continue. For example, a mutualistic relationship has been observed, under laboratory conditions, between a diphtheroid bacterium and a strain of *Hemophilus pertussis*. The basis for the ectosymbiosis proved to be the removal by the diphtheroid of an autotoxic long-chain fatty acid produced by *H. pertussis*, the fatty acid serving as a growth factor for the diphtheroid.

**Types of microorganisms participating in ectosymbiotic associations**

In the types of associations described above, the two partners may belong to totally different groups (e.g., one may be a bacterium and the other a fungus or a protozoon). Loose associations are generally recognized, as we have said earlier, by the failure to cultivate separately two microorganisms that grow well in mixed culture. On the other hand, tight associations—those in which the partners remain in intimate physical contact—may be detected by direct microscopic observation. Certain flagellated protozoa, for example, can be seen to bear on their surface a mantle of spirochetes (Figure 23.1e); and blue-green algae are frequently found living in the gelatinous capsules of eucaryotic algae. The common occurrence of blue-green algae in such associations may be related to their ability to fix atmospheric nitrogen, a property that would make them valuable partners in a symbiosis. Symbiotic associations involving blue-green algae were first observed by A. Pascher in 1914. Some drawings from one of Pascher's early papers are reproduced in Figure 23.1*a*.

## *Mutualistic ectosymbioses of microorganisms with plants*

**The rhizosphere**

The regions of the soil immediately surrounding the roots of a plant, together with the root surfaces, constitute that plant's *rhizosphere*. Operationally, it can be defined as the region, extending a few millimeters from the surface of each root, in which the microbial population of the soil is influenced by the chemical activities of the plant. Most studies of the rhizosphere have concentrated on the effect of the plant on bacterial populations. The major effect observed is a quantitative one: the numbers of bacteria in the rhizosphere usually exceed the numbers in the neighboring soil by a factor of 10 and often by a factor of several hundred.

There is also a qualitative effect. Short Gram-negative rods predominate in the rhizosphere, while Gram-positive rods and coccoid forms are less numerous in the rhizosphere than elsewhere in the soil. No specific association of a particular bacterial species with a particular plant has, however, been established.

The reason for the relative abundance of bacteria in the rhizosphere must certainly be the excretion by plant roots of organic nutrients, which selectively favor certain nutritional types of bacteria. However, no clear-cut nutritional relationships have been discovered, although many organic products excreted by plant roots have been identified. Our state of knowledge concerning the effects of the microbial population of the rhizosphere on the plant is even less satisfactory; despite numerous claims, it remains to be established that the plant benefits from the association. Many free-living soil bacteria, however, perform functions essential for plants, such as nitrogen fixation and the mineralization of organic compounds, so it seems reasonable to assume that some plants do profit from the proximity of some microorganisms. For this reason we have included these associations in the discussion of mutualistic ectosymbioses.

**The phyllosphere**

The surfaces of plant leaves constitute a special environment for microorganisms, called the *phyllosphere*. The most extensive studies of the phyllosphere have been conducted in tropical areas of high rainfall. After leaves have become inoculated from dust and rain spatter,

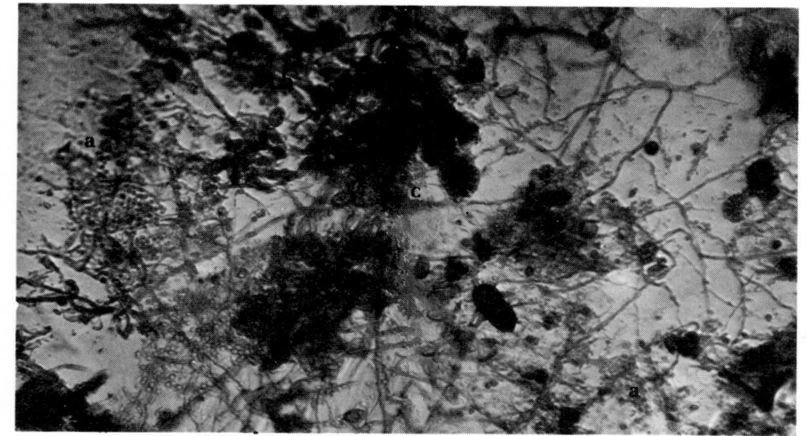


*Figure 24.1 Collodion film with embedded elements of the phyllosphere population from a mature leaf of Citrus sp. from Surinam. Only part of the population has been removed in the film. The part shown includes bacteria, algae, and fungal hyphae: (a) Azotobacter; (b) Trentepohlia sp.; (c) Phycopeltis. From J. Ruinen, "The phyllosphere. I. An ecologically neglected milieu." Plant and Soil* **15,** *81 (1961).*

they show a characteristic succession of microbial populations (Figure 24.1). At the beginning, there is a predominance of nitrogen-fixing forms belonging to the genera *Beijerinckia* and *Azotobacter;* their density reaches very high levels, often exceeding $10^7$ cells per square centimeter of leaf surface. Later on, perhaps favored by the exudation of nutrients from the leaf, a mixed population of algae, fungi, protozoa, and slime molds appears.

The mutualistic nature of the plant–microbe association is readily inferred, at least in the case of the nitrogen-fixing bacteria. The nitrogen which they fix appears first in bacterial proteins but ultimately becomes available to the plant when the bacteria die, either in the phyllosphere itself or in the adjacent soil following the washing of the leaves by rainwater.

The nitrogen-fixing bacteria are, in turn, dependent in part upon the leaf exudates for their source of carbon and other nutrients, although it is conceivable that some of these substances can be obtained from dust. The ability of airborne dust to provide at least minerals is evident from the luxuriant growth of epiphytes in the tropics on telegraph wires and other nonliving supports.

## *Mutualistic ectosymbioses of microorganisms with animals*

In this section we will describe microorganisms that live symbiotically on the surfaces of animal hosts, both the external surfaces and those which line the body cavities. We shall also include here the microorganisms that populate the liquid contents of a body cavity, such as the intestinal tract and the rumen, since such regions are topologically continuous with the external environment of the animal.

**Ectosymbioses of protozoa with insects: the intestinal flagellates of wood-eating termites and roaches**

The woody tissue of trees, consisting mainly of cellulose and lignin, is unavailable as a source of food for most animals; in general, animals do not possess the enzymes necessary to degrade these polymers. Nevertheless, many species of insects obtain the bulk of their food from wood by virtue of an ectosymbiotic relationship with cellulose- and lignin-digesting microorganisms.

Both the termites and cockroaches, which have evolved from a common ancestral group, include some species that eat wood. All the wood-eating species of both groups harbor in their gut immense numbers of flagellated protozoa belonging to the polymastigotes and hypermastigotes. The flagellates are packed in a solid mass within a saclike dilation of the hindgut; it has been reported that they constitute over one-third of the body weight of the insect in some cases. The flagellates are responsible for cellulose digestion, of which the insects themselves are incapable. The flagellates, in turn, are themselves hosts to intracellular bacteria, and it is possible that some—if not all—of the cellulases produced by the flagellates derive from their intracellular symbionts.

The mode of transmission of the flagellated symbionts from one insect generation to the next differs in the two groups. The newly hatched nymphs of termites feed on fecal droplets that exude from the adults; the droplets are laden with symbionts, which infect the young insects. The newly hatched nymphs of cockroaches eat dry fecal pellets that are excreted by the adults; the pellets are laden with flagellate *cysts*, which are able to withstand dessication. The cysts germinate in the gut of the nymphs, reestablishing the symbiosis.

One remarkable feature of the transmission cycle in cockroaches is that *the encystment of the flagellates is regulated by hormones of the insect*. The hatching of eggs in this insect coincides with the peak of the molting season, and protozoan cyst formation is induced by the molting hormone, *ecdysone*. This mechanism ensures that the flagellates will survive dessication in the fecal pellets and be available for infection of the hatching nymphs.

The flagellates enter a sexual cycle following encystment, nuclear and cytoplasmic divisions giving rise to one male and one female gamete from each cyst. Ultimately, these fuse to form a zygote. In an extensive series of studies, L. Cleveland established that sexuality in flagellates is induced by ecdysone, at concentrations of the hormone well below those required to induce molting of the insect. The adaptive significance of this regulation is not clear. It may reflect an obligatory coupling of gametogenesis with encystment.

**Ectosymbioses of fungi with insects: the ambrosia beetles**

Some species of insects, prominently the *ambrosia beetles,* have evolved symbiotic relationships with specific fungi, which they carry with them and inoculate onto the walls of tunnels which they bore in wood. Ambrosial species belong principally to three families of beetles: Scolytidae, Platypodidae, and Lymexilidae. Many, but by no means all, of the species in these groups are ambrosial; the pattern observed suggests that the ambrosial habit is polyphyletic in origin. The relationship between a fungus and an ambrosia beetle shows the following characteristics: (1) a specific fungus is always found associated with a particular insect; (2) the fungus grows primarily on the walls of tunnels made by that insect, penetrating the wood to a very limited extent; (3) the insect feeds primarily or exclusively on the fungus, rather than on wood softened by fungal digestion; and (4) the insect has a specific mechanism for transmitting the fungus to its new tunnels.

There are hundreds of species of ambrosia beetles, for several dozens of which detailed descriptions of the symbioses have been published. Although all possess the general attributes presented above, each beetle differs from the others with respect to the species and mode of transmission of the fungus with which it is associated.

The ambrosia fungi, with very few exceptions, belong to the Fungi Imperfecti. Some ambrosia fungi have been assigned to well-known fungal genera, such as *Cladosporium, Penicillium,* and *Cephalosporium.* Other ambrosia fungi are highly specialized, however, and some new genera (e.g., *Ambrosiella*) have been established for them.

The mechanisms for transmission of these fungi differ widely. Some species of ambrosia beetles have no special organs of transmission, but in most cases the insects have evolved special anatomical structures for carrying fungal spores. These structures vary from special hairs to deep, flasklike pockets of the body surface into which oil is secreted by special glandular cells (Figure 24.2). Oil secretion is greatly accelerated during periods of active tunnel boring, with the result that fungal spores are washed out of the pockets and onto the tunnel walls. In some species the spores are carried in grooves of the wings, in pouches opening into the mouth cavity, or in pouches opening into the ovipositor. In all cases, the location of the organ ensures the inoculation of the tunnel with fungal spores.

*Figure 24.2 Section through the head of the ambrosia beetle, Xyleborus monographus, showing the special pockets for the storage of fungal spores. After Schedl.*

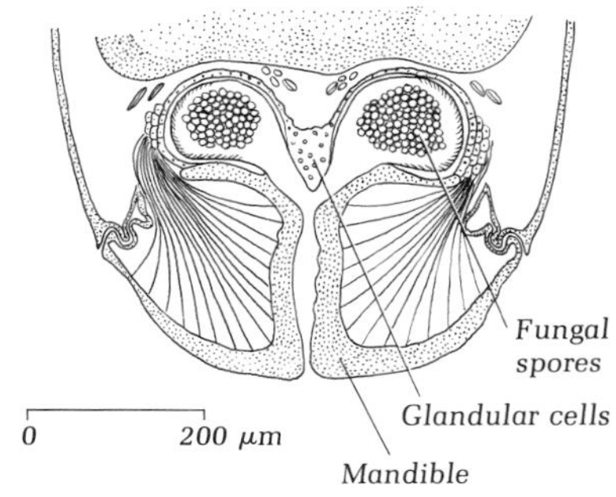


The ambrosia beetles lay their eggs in the galleries of their tunnels; the larvae that emerge, as well as the adults, feed on the mats of the ambrosia fungus (Figure 23.3). The mutualistic relationship is readily deduced from the fact that the beetle obtains all its food from the fungus, while the fungus is provided with a suitable environment as well as a means of transmission. In some cases, the symbiosis has evolved to such a high degree that the fungus grows only in the tunnels of its "host," and the beetle is unable to complete its life cycle in the absence of its fungal partner.

**Ectosymbioses of luminous bacteria with mollusks and fish**

Bioluminescence is widespread in the animal kingdom, occurring in such diverse groups as jellyfish, earthworms, fireflies, squid, and fish. In most cases the luminescence is produced by the tissues of the animal itself, but in some species of squid, and in certain fishes, it is produced by luminous bacteria living ectosymbiotically in special glands of the host.

*Figure 24.3 The cuttlefish, Euprymna morsei. (a) Male with opened mantle, showing the luminous organs embedded in the ink sac. (b) Cross section through the luminous organs, showing the reflectors, lens, and the open chambers containing the luminous bacteria. After T. Kishitani, "Studien über Leuchtsymbiose von Japanischen Sepien." Folia Anat. Japan.* **10,** *315 (1932).*

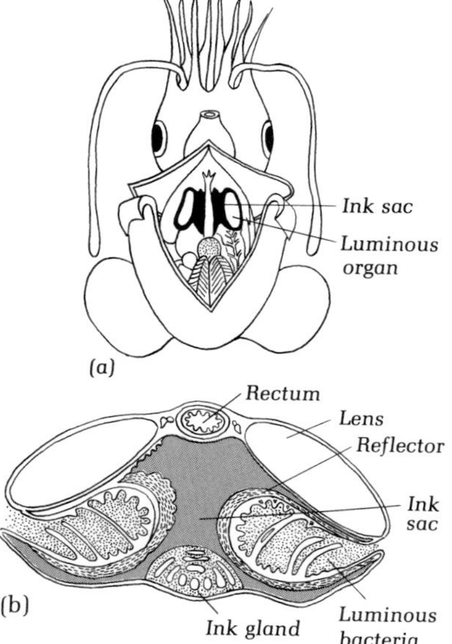


Among the squid (mollusks belonging to the class Cephalopoda), symbiotic luminous bacteria have been identified in a number of species of one suborder, Myopsida. The myopsid squids, also called cuttlefishes, are characterized by their strongly calcified shell. Figure 24.3 shows a male of the genus *Euprymna,* the light organs of which are quite typical of the myopsid squids. The luminous glands are embedded in the ink sac and are partially enclosed in a layer of reflective tissue; just above the glands are lenses composed of hyaline cells which transmit light. In some squids, the animal can control the emission of light by a muscular contraction which squeezes the ink sac, pushing it between the light source and the lens. Note in Figure 24.3 that the luminous organs are open to the exterior; all evidence suggests that young animals must be infected externally and that intracellular transmission via the egg does not occur.

A large number of unrelated species of fish also possess light organs which consist of open glands containing luminous bacteria. In a few species, the organ is provided with a reflecting layer of tissue. The most complex organs are found in the two closely related genera, *Photoblepharon* and *Anomalops,* both of which harbor luminous bacteria in special pouches under the eyes (Figure 23.2). What makes these forms particularly spectacular is their ability to control the emission of light; in *Photoblepharon,* this is accomplished by drawing up a fold of black tissue over the pouch like an eyelid, while in *Anomalops* the light organ itself can be rotated downward against a pocket of black tissue.

The complexity of the organs that have been evolved to control the symbiotic light emission implies that luminescence has great adaptive value for the host; in most cases it is believed to serve as a recognition device. The functions performed by the host on behalf of the luminous bacteria are undoubtedly those of providing nutrients and protection.

**The ruminant symbiosis**

The ruminants are a group of herbivorous mammals that includes cattle, sheep, goats, camels, and giraffes. Ruminants, like other mammals, cannot make cellulases. They have evolved an ectosymbiosis with microorganisms, however, which enables them to live on a diet in which the major source of carbon is cellulose.

The digestive tract of a ruminant contains no less than four successive stomachs. The first two, known as the rumen (Figure 24.4), are essentially vast incubation chambers teeming with bacteria and protozoa. In the cow, the rumen is a bag with a capacity of about 100 liters. The plant materials ingested by the cow are mixed with a copious amount of saliva and then passed into the rumen, where they are rapidly attacked by bacteria and protozoa. The total microbial population is enormous, and the population density is of the same order as that of a heavy laboratory culture of bacteria ($10^{10}$ cells per milliliter). Many different microorganisms are present (Figure 24.5), and the full details of their biochemical activities are not yet understood. However, the net effect is clear: the cellulose and other complex carbohydrates present in the ingested fodder are broken down with the eventual formation of simple fatty acids (acetic, propionic, and butyric) and gases (carbon dioxide and methane). The fatty acids are absorbed through

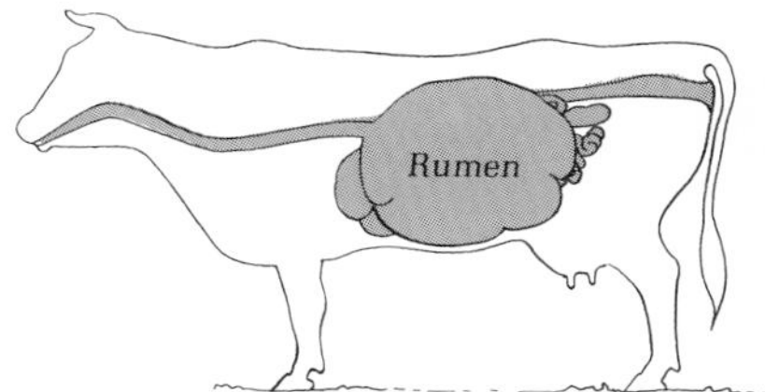


*Figure 24.4 Diagram of the intestinal tract of the cow to show the rumen.*

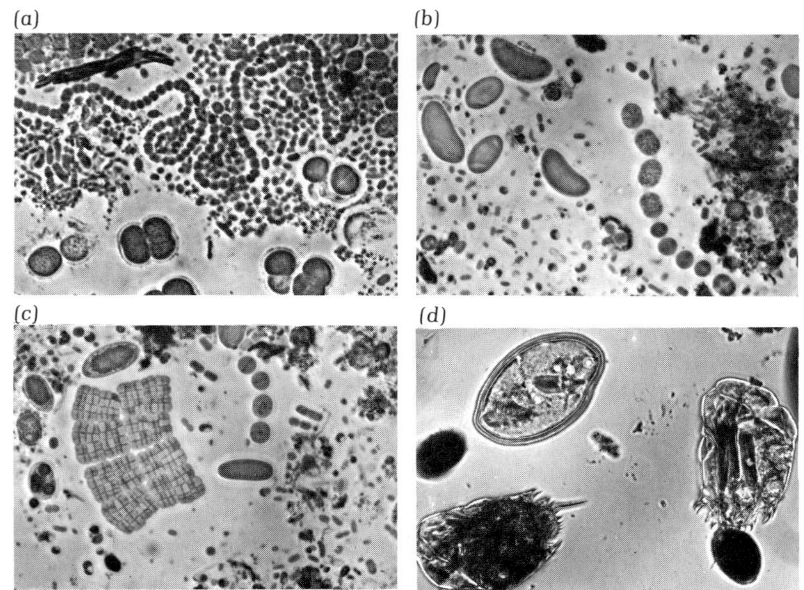


*Figure 24.5 Some microorganisms from the rumen of the sheep. (a), (b), and (c), bacteria (ultraviolet photomicrographs, × 678). From J. Smiles and M. J. Dobson, "Direct ultraviolet and ultraviolet negative phase-contrast micrography of bacteria from the stomachs of the sheep." J. Roy. Micro. Soc.* **75,** *pt. 4. (d) Ciliate protozoa (× 9). Courtesy of J. M. Eadie and A. E. Oxford.*

the wall of the rumen into the blood stream, circulating in the blood to the various tissues of the body where they are respired. The cow gets rid of the gases formed in the rumen by belching at frequent intervals. The microbial population of the rumen grows rapidly, and the microbial cells pass out of the rumen with undigested plant material into the lower regions of the cow's digestive tract. The rumen itself produces no digestive enzymes, but the lower stomachs secrete proteases, and as the microbial cells from the rumen reach this region they are destroyed and digested. The resulting nitrogenous compounds and vitamins are absorbed by the cow. For this reason, the nitrogen requirements of the cow and other ruminants are much simpler than those of other groups of mammals. Whereas man or the rat requires many amino acids (the so-called essential amino acids) preformed in the diet, the ruminant can grow on ammonia or urea, which are excretion products in most mammals. These simple nitrogenous compounds are built up into microbial proteins by the rumen population.

The evolution of the rumen has involved both structural and functional modifications of the gastrointestinal tract. The principal structural modification is the development of a complex stomach, of which the largest compartments are essentially fermentation vats. The functional modifications that ruminants have undergone are even more profound. In the first place, the salivary glands do not secrete enzymes, the saliva being essentially a dilute salt solution (principally sodium bicarbonate and sodium phosphate) that provides a suitable nutrient base for the microbes of the rumen. In the second place, the lower fatty acids have very largely replaced sugar as the primary energy-yielding substrate. This, in turn, has led to changes in the enzymatic makeup of nearly all the tissues in the body, which respire fatty acids far more rapidly than do the tissues of nonruminants. Finally, the source of amino acids and vitamins has become very largely internalized (microorganisms instead of ingested food materials).

For the microorganisms that have taken up residence in the rumen, the situation also offers advantages; they are provided with an environment always rich in fermentable carbohydrates, well buffered by the saliva, and maintained at a constant favorable temperature, the body temperature of the cow. Individually, their ultimate fate is to fall prey to the proteolytic enzymes in the lower regions of the digestive tract; for the species, however, the rumen provides a safe and constant ecological niche.

The rumen association is in a delicately balanced equilibrium, easily disturbed by slight changes of the environment. The principal failure to which this symbiosis is liable is a mechanical one. The gas production in the rumen of a cow is some 60 to 80 liters/day and, since the total volume of the rumen is only 100 liters, steady belching is necessary to get rid of the accumulating gases. For reasons that are not fully understood, certain diets lead to foaming of the rumen contents, and when this happens the belching mechanism of the cow fails to function properly. This causes a painful and, if untreated, eventually fatal affliction known as "bloat," (i.e., distention of the rumen by the trapped gases).

**Metabolic activities of rumen bacteria**

Since the $E_0'$ of the rumen contents is steadily maintained at −0.35 V, all the microbial processes that occur in the rumen are anaerobic ones. As the ruminant grazes, the rumen receives a steady flow of finely ground plant materials mixed with saliva. The plant materials consist chiefly of cellulose, pectin, and starch, together with some protein and lipid. The first stage in the process is the digestion of these polymeric macromolecules. A great deal of attention has been given to the identification of the microorganisms responsible for the digestion of cellulose, since this is the major digestive process in the rumen. Between 1 and 5 percent of the bacterial cells in the rumen have been found to be cellulolytic; they produce an extracellular cellulase that hydrolyzes cellulose, glucose appearing as the final product of digestion.

The cellulose-digesting bacteria of the rumen, like the other principal ruminant microorganisms, are all strict anaerobes. Several different species of cellulose-digesting bacteria have been isolated and described: *Bacteroides succinogenes, Ruminococcus flavofaciens, R. albus,* and *Butyrovibrio fibrisolvens.* Of these, *B. succinogenes* is a Gram-negative organism which should probably be classified as a myxobacterium, since it shows gliding movement. It forms succinic acid and lesser amounts of acetic acid. *Ruminococcus* produces principally succinic acid, and *Butyrovibrio* produces principally butyric acid.

The great bulk of the bacterial population, however, is noncellulolytic. These organisms rapidly utilize the glucose and cellobiose produced by the cellulolytic species; their great efficiency in scavenging these molecules presumably accounts for their predominance over the cellulolytic forms. Furthermore, many of the rumen bacteria (including some of the cellulolytic species) are capable of digesting starch, pectin, proteins, and lipids. Indeed, only the lignin of the ingested plant material escapes digestion by the rumen flora.

The products of digestion of polysaccharides, proteins, and lipids are fermented by the rumen bacteria. In the rumen, these metabolic activities lead to the accumulation of the gases $CO_2$ and methane and the fatty acids acetic, propionic, and butyric acids. When the predominant rumen bacteria are isolated and studied as pure cultures, however, not only are the above-listed products formed but also hydrogen gas, together with large amounts of formic, lactic, and succinic acids. By adding radioactive isotopes of these compounds to rumen contents, the reasons for their failure to accumulate in the rumen have been discovered. Thus, hydrogen gas is quantitatively combined with carbon dioxide to form methane, by the organism *Methanobacterium ruminantium;* the lactate is fermented to acetate plus smaller amounts of propionate and butyrate, by such organisms as *Peptostreptococcus elsdenii;* and the formate is converted first to carbon dioxide and hydrogen, by a variety of bacteria, and then to methane by *M. ruminantium.* Finally, the succinate is rapidly decarboxylated to propionate by *Veillonella alcalescens* and other bacteria.

The net result is the formation of carbon dioxide and methane, and acetic, propionic, and butyric acids, in remarkably constant proportions. Carbon dioxide accounts for 60 to 70 percent of the gases, methane accounting for the remainder; acetic acid represents 47 to 60 percent, propionic acid represents 18 to 23 percent, and butyric acid represents 19 to 29 percent of the fatty acids, respectively.

**The rumen protozoa**

Protozoa were seen microscopically in rumen contents as early as 1843, but almost 100 years elapsed before they were successfully isolated and cultivated in vitro. Several species have now been cultivated, notably the oligotrichous ciliates *Diplodinium, Entodinium, Epidinium, Metadinium,* and *Ophryoscolex,* and the holotrichous ciliates *Isotricha* and *Dasytricha.* Unfortunately, attempts to grow these protozoa in axenic (bacteria-free) culture have not yet been successful.

All such cultures to date have been contaminated with bacteria, intracellular as well as extracellular, so a final conclusion cannot be drawn from the enzymological and metabolic experiments that have been reported. Nevertheless, some tentative conclusions have been reached, on the basis of experiments in which the protozoa were maintained alive for extended periods of time in the presence of high levels of bactericidal antibiotics. These experiments have implicated species of *Diplodinium* and *Metadinium* in the digestion of cellulose and species of *Entodinium* and *Epidinium* in the digestion of starch. Many of these protozoa are active predators on the rumen bacteria.

**Ectosymbioses of microorganisms with birds: the honey guides**

The honey guides, a group of birds belonging to the genus *Indicator,* are found in Africa and India. Their name accurately describes their behavior: they literally guide honey badgers, as well as humans, to the nests of wild bees, where they wait for their follower to break open the hive. When the badger (or human) has departed, the honey guide proceeds to feed on the remnants of honeycomb that have been left exposed.

This behavior became all the more remarkable when it was discovered that these birds do not possess enzymes for digesting beeswax. Instead, they harbor in their intestines two microorganisms which carry out the digestion for them: a bacterium, *Micrococcus cerolyticus,* and a yeast, *Candida albicans.* The micrococcus is a highly specialized symbiont, depending on a growth factor that is produced in the small intestine of the honey guide.

**Ectosymbioses of microorganisms with mammals: the normal flora of the human body**

Both the skin and the mucous membranes of the body are directly accessible to the external environment; soon after an infant is born these surfaces become populated by a characteristic flora. The skin becomes contaminated during passage through the birth canal; the mucous membranes may be sterile at birth but become contaminated within hours. Of the many microorganisms that reach these surfaces, only those that are particularly suited to growth in such environments become established; these constitute the *normal flora,* which remains remarkably constant. The bacteria of the normal flora do not cause disease unless accidentally introduced into normally protected regions of the human body or as a result of physiological changes within the host. For example, a radical change in diet, or infection with a virus, may alter the conditions within the body in such a fashion that a member of the normal bacterial flora can become pathogenic.

Each region of the body provides a distinctive ecological niche which selects for the establishment of a characteristic flora. The bacteria that populate the skin, for example, include mainly corynebacteria, micrococci, nonhemolytic streptococci, and mycobacteria. Moist regions of the skin harbor yeasts and other fungi. The skin flora is selected both by nutritional conditions and by antibacterial agents, such as the fatty acids secreted by the skin. The microorganisms of the skin flora are so firmly entrenched in the sweat glands, sebaceous glands, and hair follicles that no amount of bathing or scrubbing can totally remove them.

The throat and mouth support a variety of microorganisms, including representatives of most of the common eubacterial groups. Gram-positive cocci, including micrococci, pneumococci, and *Streptococcus salivarius,* are common throat inhabitants. The Gram-negative cocci are represented by members of both the aerobic genus *Neisseria* and the anaerobic genus *Veillonella.* The mouth and throat also harbor large numbers of both Gram-positive and Gram-negative rods. The former include mainly lactobacilli and corynebacteria, while the latter include members of the genera *Bacteroides* and *Spirillum.* Spirochetes *(Tre-*

*ponema dentium*), yeasts (*Candida albicans*), and actinomycetes (*Actinomyces israelii*) are also common mouth inhabitants. All these organisms are normally harmless, but if the mucous membranes are injured, many can invade the body tissues and produce disease.

The same types of organisms inhabit the nasopharynx and are at least potentially able to reach the lungs. However, a series of protective mechanisms keeps the trachea and bronchi relatively free of live bacteria. First, most of the bacteria adhere to the mucous lining of the nasopharynx and cannot readily move to the lungs because the ciliated epithelial cells that line the trachea constantly sweep the mucus upward. Second, the lungs are the site of extremely active phagocytosis, a mechanism whereby foreign particles are engulfed and destroyed by special amebalike cells.

The stomach is too acid to permit much bacterial development, but the intestines harbor an extremely large resident flora. In the intestines, the nature of the flora changes with diet and with age. Breast-fed infants, for example, have always a predominant intestinal population of *Bifidobacterium*, an organism not commonly found elsewhere. This organism disappears soon after weaning.

In adults, the predominant bacteria of the intestinal tract are *Bacteroides* spp., *Escherichia coli*, and *Streptococcus faecalis*. Organisms of the genus *Bacteroides* are strictly anaerobic, nonmotile, Gram-negative rods; they are the most characteristic organisms of the mammalian intestinal tract and are seldom found elsewhere. Their numbers in human feces exceed $10^9$ cells per gram, whereas the density of *E. coli* rarely exceeds $10^8$ cells per gram. Other genera, such as viridans streptococci and micrococci, are present in lower numbers. The yeasts are represented by *Candida* and *Torulopsis*, and protozoa by several genera: *Balantidium*, a ciliate found only in man; *Entamoeba*; and flagellates of the genus *Trichomonas*. The proportions of the different microbial types are largely dependent on diet. Oral chemotherapy with such agents as antibiotics or sulfonamides can cause striking changes in the intestinal flora.

The principal bacteria of the vagina are the lactobacilli, which maintain a low pH as a result of their fermentative activity. This acidity is responsible for preventing the establishment of other forms. When the lactobacilli of the vagina are reduced in numbers during chemotherapy with antibacterial drugs, vaginal infections by bacteria and by yeasts commonly occur.

In hospital practice today many serious infections are seen with bacteria that normally reside harmlessly in the host. Probably the most common source of such "endogenous infections" are the intestinal contents, which normally contain $10^{10}$ to $10^{11}$ bacteria per gram of stool.

In persons with diseases that interfere with host defense mechanisms or in circumstances when drug treatment has the same effect, it is common for bacteria that reside normally in the bowel to invade the blood stream. Examples of diseases that may interfere with host defense mechanisms include leukemia, which may produce severe lowering of circulating leukocytes, and multiple myeloma, which is a disease of antibody-producing cells. Drugs like cortisone may also increase susceptibility to invasion by organisms of the normal intestinal flora.

The normal flora of the mammalian body provides an example of a symbiotic relationship which can shift from mutualism to parasitism and back again. In the absence of circumstances which permit them to invade the tissues, organisms of the normal flora benefit the host by preventing the establishment of virulent pathogens to which the host is often exposed.

**Germ-free animals**

By the use of complex equipment and elaborate techniques, it has been possible to deliver grem-free animals by caesarean section and to rear them in a germ-free environment. Adult animals in such an environment will mate and produce germ-free litters, so that colonies of germ-free animals can be maintained.

The chambers in which the animals are delivered and reared are equipped to prevent the entry of organisms such as bacteria and fungi. The equipment in common use does not exclude viruses. Except for viruses, accordingly, the germ-free animals are devoid of their normal microbial flora. Such animals permit experimentation on the role of the normal flora in the growth and development of the host, as well as on resistance to infection by virulent pathogens.

The development of the germ-free animal is abnormal in several respects. The cecum of the germ-free animal is greatly enlarged (Figure 24.6), the lymphatic system is poorly developed, and the germ-free animal makes much less immunoglobulin than does the normal animal.

Comparisons of normal and germ-free animals have not revealed any effects of the normal flora on the nutritional requirements of the host. On the other hand, striking effects have been observed on the *resistance or susceptibility of the host to infectious diseases.* For example, germ-free rats do not develop dental caries, whereas control animals do. Cavities in the teeth appear, however, when the germ-free rats are infected with streptococci. A more complicated role of the normal flora in producing disease is revealed in the case of infection by the protozoan pathogen, *Entamoeba histolytica,* the agent of amebic dysentery. This organism cannot produce disease in germ-free guinea pigs, but does so if the animals are first infected with *Escherichia coli* or *Aero-*

Figure 24.6 *The effect of normal flora on the development of the caecum in the guinea pig. Courtesy of the Walter Reed Army Institute of Research.*

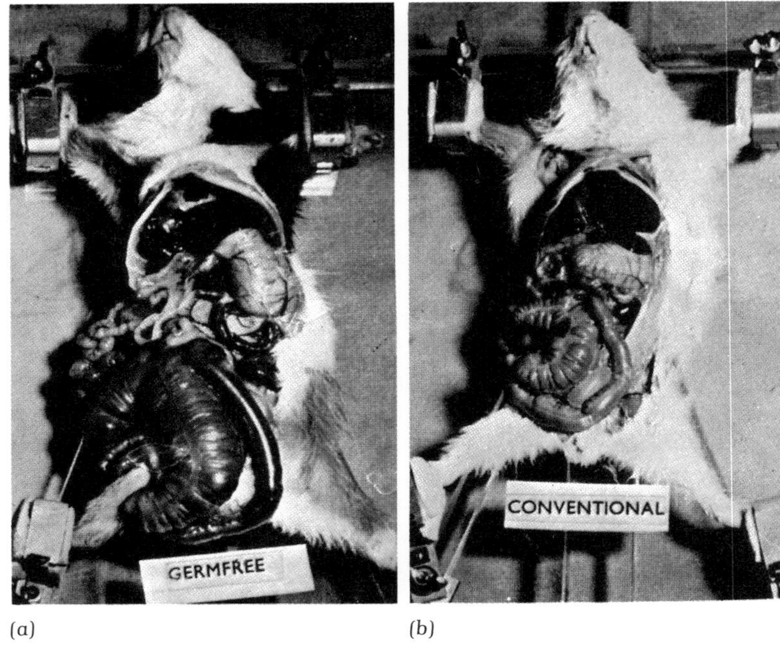


(a) (b)

*bacter aerogenes.* Thus, organisms typical of the normal flora potentiate the virulence of a pathogen which, by itself, cannot survive in the intestine.

In contrast, the presence of the normal flora protects the host against some infectious diseases. For example, normal rats develop resistance to *Bacillus anthracis* spores at an early age, the number of spores in a dose required to kill 50 percent of the animals in a given experiment rising from $10^4$ to $10^9$ very soon after birth. Germ-free rats never develop this resistance, which can thus be attributed to the presence of the normal flora. Similarly, germ-free guinea pigs are rapidly and fatally infected by *Shigella flexneri* administered orally, whereas normal animals are totally resistant. It is not yet known whether such resistance is the result of the cross reactivity of antibodies induced by the antigens of the normal flora, or whether some other mechanism is involved.

## Further reading

**Books**

Henry, S. M. (editor), *Symbiosis*, Vols. I and II. New York: Academic Press, 1966 and 1967.

*Symbiotic Associations: Thirteenth Symposium of the Society for General Microbiology.* New York: Cambridge University Press, 1963.

# Chapter twenty-five
# Mutualistic endosymbioses

*Many microorganisms are endosymbionts:* they may be totally intracellular, carrying out their complete life cycle within the cytoplasm or nucleus of a host cell, or – in the case of certain fungi – they may penetrate the cell walls of their host by rootlike extensions called *haustoria*. Some associations of this type characteristically lead to the death of the host and are thus parasitic symbioses. Others characteristically lead to the complete digestion of the intracellular microorganism and might therefore be classified as predations. In many cases, however, a balance is achieved and both partners benefit from the association.

It seems likely that the mutualistic state represents a late stage in the evolution of the association, which began either as a predation (e.g., the phagocytosis of bacteria and algae by protozoa) or as a parasitism (e.g., the invasion of plant roots by bacteria and fungi). Even in those associations which are clearly mutualistic, a change in external conditions or in the life cycle of one of the partners can cause a shift in the balance, with fatal consequences to one of the symbionts. As we pointed out in Chapter 23, a mutualistic symbiosis can best be regarded as a dynamic process of mutual exploitation.

## Endosymbioses of microorganisms with protozoan hosts

Representatives of every major microbial group – bacteria, fungi, algae, and protozoa themselves – are found living as endosymbionts in various species of protozoa.

### Bacterial endosymbionts

Bacterial endosymbionts are extremely widespread in protozoa; they have been described in amebas, flagellates, ciliates, and sporozoa. None of them has been cultivated outside its host, but their bacterial nature has been clearly established on the basis of their morphology, staining properties, and mode of cell division. Some multiply in the nucleus of the host and others in the cytoplasm.

In most cases, the contribution which the bacterium makes to the symbiosis is unknown. In one case, however, its contribution is clear: the bacterial endosymbiont provides its host with amino acids and other growth factors that most protozoa require as exogenous nutrients. The host, a trypanosomatid flagellate named *Crithidia oncopelti*, can grow in a simple synthetic medium containing glucose as carbon source together with adenine, methionine, and several vitamins as growth factors. In contrast, another species of *Crithidia* requires not only the above nutrients but also 10 other amino acids (including lysine), hemin, and several additional vitamins. Radioisotope studies showed that in *C. oncopelti* lysine is synthesized via the diaminopimelic acid pathway, characteristic of bacteria. Final proof of the role of the endosymbiotic bacterium found in this protozoon was obtained by fractionating the *Crithidia* cells and showing that diaminopimelic acid decarboxylase, the last enzyme of the biosynthetic pathway leading to lysine, is located in the fraction consisting of the cells of the endosymbiont.

Perhaps the most fascinating, and certainly the most extensively studied, protozoan symbiosis is that of *Paramecium aurelia* and its endosymbiont, *kappa*. In the first of a series of investigations extending over 20 years, T. M. Sonneborn and collaborators showed that most strains of *P. aurelia* fall into two general classes: killers and sensitives. The former liberate a particulate toxic principle, called P (or *paramecin*), to which killers are immune but which is lethal for sensitive strains. The ability to liberate P is genetically controlled by the cytoplasm of the host, rather than by its nucleus; at conjugation, when cytoplasm is exchanged, a sensitive cell mated with a killer is itself converted to a killer.

In attempts to identify the genetic material in the cytoplasm, X rays were used to inactivate it. Surprisingly, the data yielded a calculated target size for the genetic element so large that it should be visible with the light microscope. Staining experiments were then performed, and the feulgen stain – which is specific for DNA – revealed that the genetic element responsible for liberation of P is a bacteriumlike endosymbiont which divides by binary fission in the cytoplasm of the paramecium.

Kappa, as the endosymbiont was designated, has the same dimensions as a small bacterium and can be eliminated from its host by any one of a variety of antibiotics or other agents, both chemical and physical. Its loss is irreversible, and the host continues to propagate normally without it. Kappa can be transmitted to sensitive paramecia through extracts prepared from killers but so far has not been cultivated outside its host.

Kappa contains DNA, and can undergo mutations, including mutation to resistance to the antibiotic aureomycin. Its reproduction is dependent on the product of a particular nuclear gene of its host, called *K. Paramecium aurelia* is a diploid organism; the *K* gene can mutate to the recessive allele, *k,* and hence a cell may have the genotype *KK, Kk,* or *kk.* When a cross between two *Kk* killers produces a *kk* segregant, kappa can no longer reproduce and is diluted out during ensuing divisions of the *kk* host cell. Ultimately, the *kk* cell gives rise to a clone of sensitive paramecia.

The relationship between kappa and the toxic particle, P, is uncertain. It has been proposed, on the basis of much indirect evidence, that a kappa organism is transformed directly into a P particle. Actually, a complex series of developmental changes takes place, in which a kappa body develops a refractile body within its own cytoplasm, loses the capacity to divide further, and then (presumably) is converted into a P particle. The nature of the toxic substance associated with the P particle, and the mechanism by which kappa confers immunity to it, are not known.

**Algal endosymbionts: zoochlorellae and zooxanthellae**

Many protozoa of the Ciliophora and of the Rhizopoda are hosts to endosymbiotic algae. In freshwater forms, the algae are generally green types belonging to the Chlorophyta; in the marine forms, the algae are generally yellow or brown types belonging to the dinoflagellates. They are called *zoochlorellae* and *zooxanthellae,* respectively.

Figure 25.1 shows zooxanthellae liberated by crushing a foraminiferan protozoon. As found in their hosts, both zoochlorellae and zooxanthellae are invariably coccoid. When cultured free of their hosts, however, zooxanthellae are sometimes observed to form swarming zoospores, which are typical dinoflagellates. Each protozoan cell harbors from 50 to several hundred algae; maintenance of the symbiosis is ensured by similar growth rates of the two partners. Endosymbionts resist digestion by the host. This resistance is undoubtedly related to their location in the host cytoplasm. It should be recalled that microorganisms are taken into the cells of phagotrophic protozoa by phagocytosis and localized inside food vacuoles formed by invaginations of the

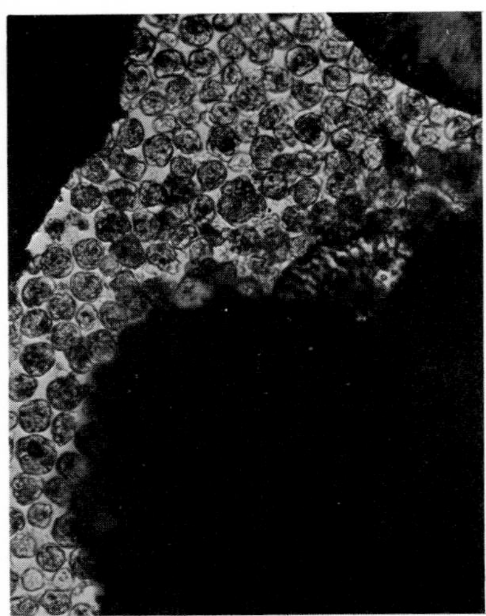

*Figure 25.1 Zooxanthellae escaping from a crushed foraminiferan. From J. McLaughlin and P. Zahl, "Endozoic Algae," in Symbiosis, (S. M. Henry, editor), Vol. I, New York: Academic Press, 1966. Photograph made by J. J. Lee and H. D. Freudenthal.*

cell membrane. Their digestion is effected by hydrolytic enzymes liberated into these food vacuoles from lysosomes. The endosymbionts in the cytoplasm are not contained in food vacuoles and are thus isolated from the digestive enzymes of the lysosomes.

As we discussed at the beginning of Chapter 24, symbioses between photosynthetic and nonphotosynthetic partners are particularly successful because together the two organisms can carry out a full carbon cycle and a full oxygen cycle. The photosynthetic partner uses light energy to fix carbon dioxide, while liberating $O_2$ from water; the nonphotosynthetic partner uses the $O_2$ to respire organic products, producing carbon dioxide as a by-product. This is presumably the basis for the extremely common occurrence of algal–protozoan endosymbioses.

The intimate nature of the relationship is dramatically demonstrated by the fact that many protozoan hosts exhibit *phototaxis* when they harbor a photosynthetic endosymbiont. In paramecia it has been shown that the alga is the photoreceptor; the movements of the protozoon seem to be controlled by the intracellular concentration of oxygen produced by algal photosynthesis, since phototaxis is exhibited only when the external supply of oxygen is limiting.

**Algal endosymbionts: cyanellae**

Blue-green algal symbionts are called *cyanellae.* They are found in a few genera of freshwater protozoa (e.g., the flagellates *Cyanophora* and *Peliaina,* and the ameboid rhizopod *Paulinella*).

*Peliaina* contains from one to six cyanellae. The symbiosis is maintained by balanced cell division, but this mechanism sometimes fails and an alga-free protozoan cell is formed. In *Cyanophora* and *Paulinella,* on the other hand, the symbiotic association is perfectly regulated: the protozoan host usually contains two cyanellae, each daughter protozoan receiving one cyanella at cell division. The algal symbiont then divides, restoring the number of two per host.

The cyanellae were discovered as symbionts of protozoa by A. Pascher in 1929. One of Pascher's drawings of *Cyanophora* is shown in Figure 25.2. In recent years, *Cyanophora* has been reinvestigated using the techniques of electron microscopy and nucleic acid analysis. Figure 23.8 shows a section of *C. paradoxa* as seen in the electron microscope. The endosymbiont has the typical fine structure of a blue-

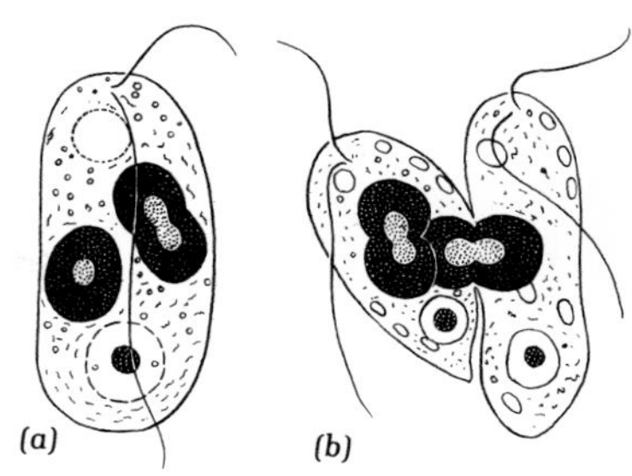


*Figure 25.2 The protozoon, Cyanophora, with its endosymbiotic blue-green algae. (a) One of the algal cells has begun to divide. (b) At the time of host cell division, each daughter cell receives one symbiont. Compare with Figure 23.6(a). (After A. Pascher.)*

green alga, except that it lacks a cell wall, having only a cell membrane as its outer layer. The nucleic acids of the host and symbiont can be separated from each other and can be shown to be typical of a eucaryote and of a procaryote, respectively.

It would be very easy to mistake the symbiotic association seen in Figure 23.8(a) for a typical flagellated green alga such as is shown in Figure 23.8(b). In fact, the chloroplasts of eucaryotic algae and of higher plants have been found to contain DNA which resembles the DNA of a procaryote in several respects. The existence of such DNA in the chloroplasts of eucaryotic organisms strongly suggests an evolutionary origin of chloroplasts from procaryotic endosymbionts.

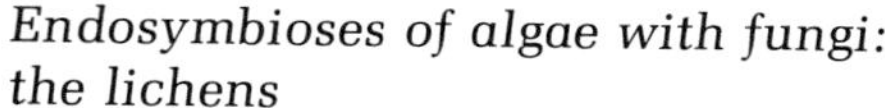

## *Endosymbioses of algae with fungi: the lichens*

**The nature of lichens**

A lichen is a composite organism, consisting of a specific fungus, usually an ascomycete, living in association with one—or sometimes two—species of algae. The symbionts form a vegetative body, or *thallus*, of which both the gross structure and the fine structure are characteristic for each lichen "species."

In terms of gross structure, the lichen thalli are divided into three types. The *crustose* lichens adhere closely to their substrate (either rocks or the bark of trees). The *foliose* lichens are leaflike, and more loosely attached to the substrate. The *fruticose* lichens form pendulous strands or upright stalks. Figure 25.3 shows a representative of each type, together with a cross section showing the internal organization of the thallus. The bulk of the thallus is made up of fungal hyphae. In most species these are differentiated into distinct tissues: a closely packed *cortex*, a loosely packed *medulla*, and (in the foliose lichens) attachment regions or *rhizinae*. The algal cells are usually found in a thin layer just below the cortex; in a few species of lichens, however, the fungal hyphae and algal cells are distributed at random throughout the thallus.

Electron micrographs of thin sections show that in most lichens each algal cell is penetrated by one or more fungal haustoria. In some lichens the haustoria penetrate deeply into the algal cells, the membrane of the

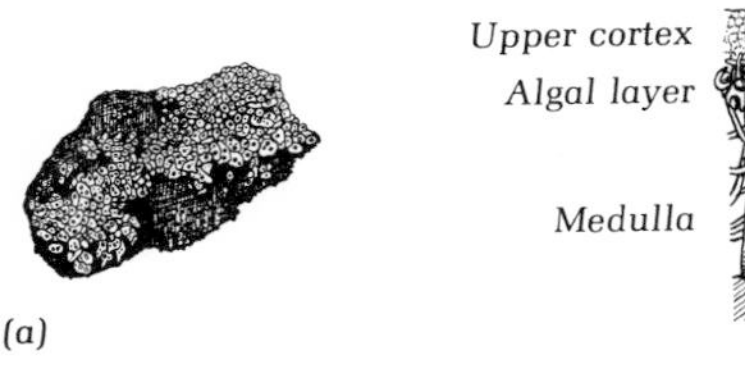


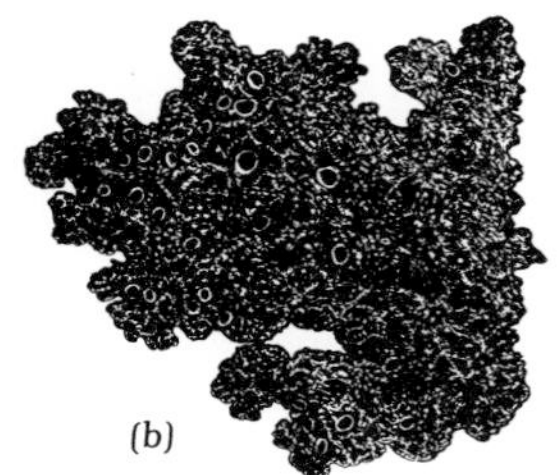


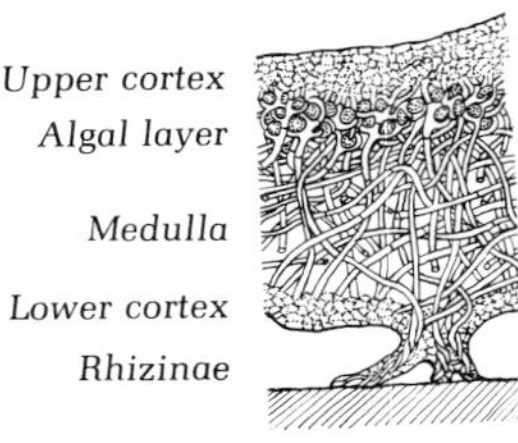


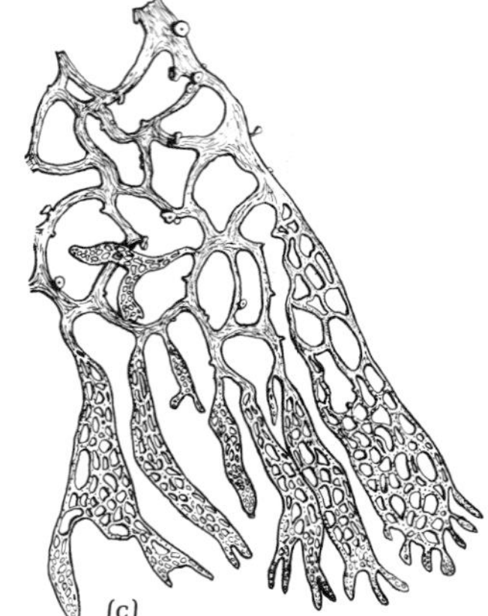


*Figure 25.3 Lichens of three major types. (a) Crustose lichens, which adhere closely to their substrate. (b) Foliose lichens, which are leafy in form and are attached to their substrates more loosely. (c) Fruticose lichens, which are either pendulous strands or hollow, upright stalks. The diagrams at the right show vertical sections of crustose and foliose lichens, and a horizontal section of a fruticose lichen. From V. Ahmadjian, The Lichen Symbiosis, Waltham, Mass.: Blaisdell, 1967.*

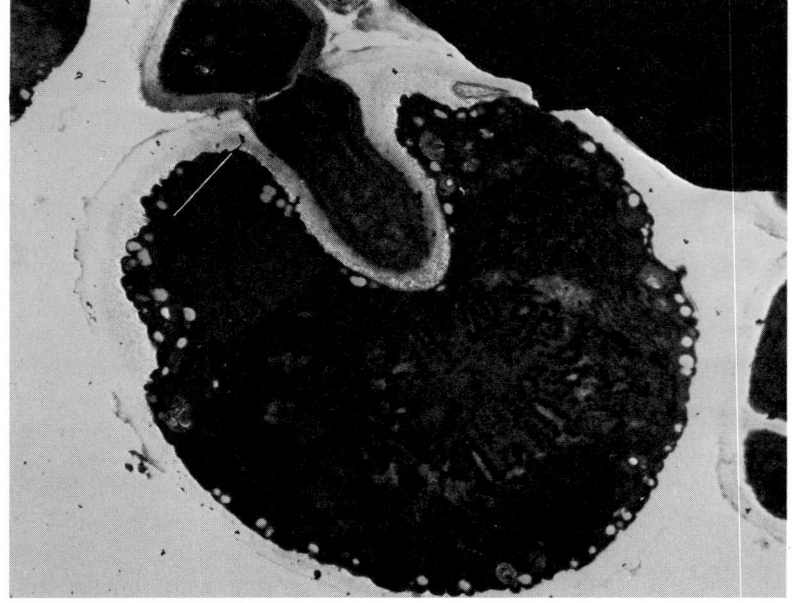

*Figure 25.4 Electron micrograph of a section through the lichen, Lecanora rubina, showing the penetration of an algal cell (Trebouxia) by a fungal haustorium (arrow). The haustorium has penetrated the outer layer of the cell wall but not the inner layer or the membrane of the algal cell. From J. B. Jacobs and V. Ahmadjian, "The Ultrastructure of Lichens. I. A General Survey." J. Phycol.* **5**, *227 (1969).*

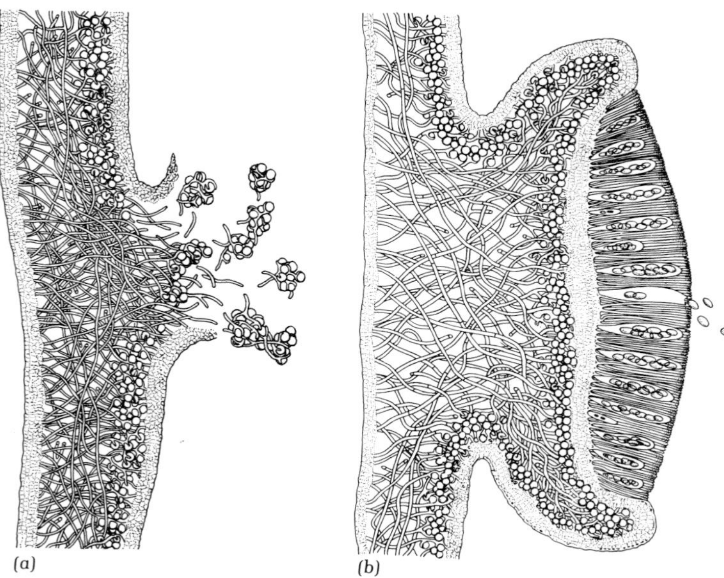


*Figure 25.5* Lichen reproduction: (a) by the liberation of soredia, composed of fungal threads and hyphae; (b) by the liberation of fungal spores (in this case, ascospores). On germination, a fungal mycelium will develop, which may form a lichen if it comes into contact with an algal cell. From V. Ahmadjian, "The fungi of lichens." Sci. Am. **208,** 122 (February, 1963).

algal cell invaginating to form a sheath around the haustorium (Figure 25.4). In other species, the haustoria penetrate only the algal cell wall. The fungi in a few lichens do not have haustoria; instead, there is an intimate contact between the algal and fungal cell walls, which in these species are very thin.

Most lichens propagate by the liberation of *soredia:* small fragments, composed of algal cells and fungal hyphae [Figure 25.5(*a*)]. In addition, lichens liberate fungal spores [Figure 25.5(*b*)]. There is considerable evidence, as will be discussed later, that the hyphae produced when these spores germinate make contacts with free-living algal cells and initiate the formation of new lichen thalli.

### The morphological consequences of symbiotic existence

It is relatively easy to separate the symbiotic partners of a lichen and to grow them in pure culture, making possible a comparison of the morphology of each partner as a free-living organism and as a symbiont.

The algal partner may be severely modified by "lichenization." Certain filamentous blue-green algae, for example, fail to form normal filaments when in the thallus; instead, each cell is separated and surrounded by fungal tissue. When isolated from the thallus, the alga regains its filamentous growth habit. The green algae found in lichens are so modified that they never produce their characteristic zoospores while part of the lichen thallus.

The fungal partner forms fruiting structures (ascospores and asexual spores) when it is lichenized, but with rare exceptions does not do so when it is isolated and cultivated in the free-living state. In the free-living state it is also incapable of forming a thallus with cortex, medulla, or other tissues. Thus, each partner affects the morphology of the other in a highly specific way.

**Species and specificity of the symbiotic partners**

The association of fungus with alga in a lichen is not specific. Thus, a given algal species may be found associated with any one of a variety of lichen fungi, and—conversely—a given fungus may be found associated with any one of a variety of algae. Altogether, algae belonging to 26 genera have been found in lichens: 17 genera of green algae, 8 of blue-green algae, and 1 of yellow-green algae. One green alga, *Trebouxia*, is found in more than half of the described lichens. Some 5 to 10 percent of all lichens contain blue-green algae, of which *Nostoc* is the most common representative.

The lichen fungi have been classified into several hundred genera; most of the fungi in lichens are ascomycetes, but a few imperfect fungi and a few basidiomycetes have also been found.

It is difficult to make any statement about the taxonomic relationships of lichen fungi to free-living ascomycetes. Historically, the lichens have been studied and classified by specialists, who have given them a unique set of names based on the morphology of the composite plant. When a lichen is experimentally separated into its two components, the fungus retains the name of the lichen. Thus, the lichen *Cladonia cristatella* is said to be composed of the fungus, *Cladonia cristatella*, and its algal symbiont, *Trebouxia erici*.

No attempt has been made to integrate the taxonomies of lichen fungi and free-living fungi. Although the morphology of the lichen is dominated by the fungus, and the description of the lichen includes many fungal characters (e.g., shape and number of ascospores), some descriptive features of the lichen may well be expected to vary according to which alga is present. It thus seems highly possible that several different lichen "species" contain the same fungus; if so, the practice of assigning the name of the lichen to the isolated fungal component is bound to prove misleading in many instances.

**The formation and maintenance of the symbiosis**

Many lichen fungi have been isolated and grown in pure culture. Under conditions of low nutrient supply, the hyphae encircle almost any rounded object of appropriate size which they encounter. If the object is an algal cell, it is penetrated by haustoria. This type of experimentally observable association is thought to mimic the first stages of true lichenization in nature. Further stages in lichenization have been achieved experimentally by allowing mixed cultures of a fungus and an alga, isolated from the same lichen, to grow together under conditions of progressive dessication. After several months, structures typical of the parental lichen thallus are formed, including fungal fruiting bodies and asexual spores.

When a lichen is cultivated, it will continue its normal development only as long as the growth conditions are unfavorable for the independent growth of the individual components. A low supply of nutrients and alternate periods of wetting and drying are thus favorable for the maintenance of the symbiosis. If the lichen is placed on a rich nutrient medium with adequate moisture over a prolonged period, the union breaks down, and the algae grow out in their characteristic free-living form.

**The physiology of the composite organism**

The lichen association seems to have evolved by selection for the ability to withstand extreme drought as well as for the ability to scavenge essential minerals. These deductions follow from the ecology of lichens, as well as from the experimental observations discussed above. Lichens are found in nature colonizing exposed rock surfaces and tree trunks where other forms of life are unable to gain a foothold.

A lichen can remain viable in the dry state for months; when submerged, its water content can change from 2 percent of its dry weight to 300 percent of its dry weight in 30 seconds. Its ability to scavenge minerals is probably related to its production and excretion of *lichen acids,* organic compounds that have the ability both to dissolve minerals and to chelate them. Chelation, the process of binding metal atoms to organic ligands, undoubtedly plays an important role in the solubilization and uptake of minerals by lichens.

The ability of lichens to scavenge nutrients at very low concentrations, normally advantageous, becomes injurious to these organisms in regions of industrial air pollution. In such regions the lichen population is greatly reduced or even totally eliminated.

More than 100 different lichen acids have been described. Most of them contain two or more phenylcarboxylic acid substituents, with aliphatic side chains. They are not produced by isolated lichen algae, nor—with a very few exceptions—by isolated lichen fungi. A few nonlichen fungi produce similar compounds, however, and it seems likely that the biosynthesis of these compounds in the composite organism is mediated by the fungus.

The production of the lichen acid is thus a creative manifestation of symbiosis, a phenomenon discussed in Chapter 23. The acids are often excreted in large quantities, crystallizing on the surface of the lichen. In addition to their role as chelating agents, described above, it has been suggested that they inhibit the growth of other microorganisms. Many of the lichen acids do possess strong antibiotic activity, and one of them—usnic acid—is widely used in some European countries as a chemotherapeutic drug for external application.

Lichens grow extremely slowly; an annual increment in radius of 1 mm or less is typical. The range of growth rates is wide, however, and in some species may average 2 to 3 cm/year. Despite their slow growth, lichens form a significant part of the vegetation in some areas; in fact, they are the primary source of fodder for reindeer and caribou in arctic regions.

**The significance of the lichen symbiosis**

Both partners of the lichen symbiosis are capable of free-living existence, as shown by the fact that, under conditions favorable for the growth of the free-living forms, the symbiosis breaks down. The association is thus of mutual benefit *only in very special ecological situations* (i.e., in environments where nutrients are extremely scarce and where extremes of wetting and drying occur).

The benefit to the fungus of the symbiosis under such conditions is clear: it depends on the alga for its source of organic nutrients. Tracer experiments have confirmed that carbon dioxide fixed by the alga passes rapidly into the fungal mycelium. In those lichens that contain blue-green algae as symbionts, the fungus also benefits directly from the atmospheric nitrogen fixed by the alga.

The contribution of the fungus to the association is less clear, but there is good reason to believe that it facilitates the uptake of both water and minerals and may also protect the alga from dessication as well as from excess light intensities. Free-living algae, however, are able to grow to a limited extent in the ecological niche inhabited by lichens, and the algal partner may thus be thought to benefit less than does the fungus.

## *Symbioses of fungi with higher plants: mycorrhizas*

The roots of most higher plants are infected by fungi. As in so many symbioses, a dynamic condition of mutual exploitation results, both partners benefiting as long as a balance between invasive and defensive forces is maintained. As a result of the infection, the plant root is structurally modified in a characteristic way. The composite root-fungus structure is called a *mycorrhiza*.

The formation of a mycorrhiza begins with the invasion of the plant root by a soil fungus; growth of the fungus toward the root is stimulated by the excretion into the soil of certain organic compounds by the plant. The fungal mycelium penetrates the root cells by means of haustoria and develops intracellularly. In some mycorrhizas the fungus forms intracellular branching structures called *arbuscules* (Figure 25.6); in others it forms characteristic coils.

Depending on the host, the fungus either maintains its intracellular state or undergoes digestion. In the latter case the fungal mycelium persists mainly in the form of *intercellular* hyphae. In all mycorrhizas, however, a large fraction of the mycelium remains in the soil, the intercellular forms tending to produce a compact sheath around the root.

With few exceptions, mycorrhizas are not species specific. A given fungus may be associated with any of several plant hosts, and in most cases a given plant may form mycorrhizas with any of a number of soil fungi. One species of pine tree, for example, has been found to associate with any of 40 different fungi. A great many free-living soil fungi are capable of forming mycorrhizas. In an experiment performed with pure cultures of free-living fungi and sterile plant roots, over 70 fungal species were found to form mycorrhizas, and many times that number are undoubtedly capable of doing so in nature.

A typical mycorrhiza is shown in Figure 25.7. The stocky, club-shaped appearance results from several effects of the fungus on the root: cell volumes increase but root elongation is inhibited, and lateral root formation is stimulated by *auxins* (plant growth hormones) produced by the fungi.

The mutualistic nature of the mycorrhiza symbiosis can be readily demonstrated in many cases. The fungi that participate are characteristically those which are unable to use the complex polysaccharides which are the principal carbon sources for microorganisms in forest soils and humus. By invasion of plant roots, these fungi avail themselves of simple carbohydrates such as glucose. In fact, the auxins excreted by the fungi induce a dramatic flow of carbohydrate from the leaves to the roots of the host plant.

The plant also benefits from the association. Many forest trees become stunted and die when deprived of their mycorrhiza. Stunted trees

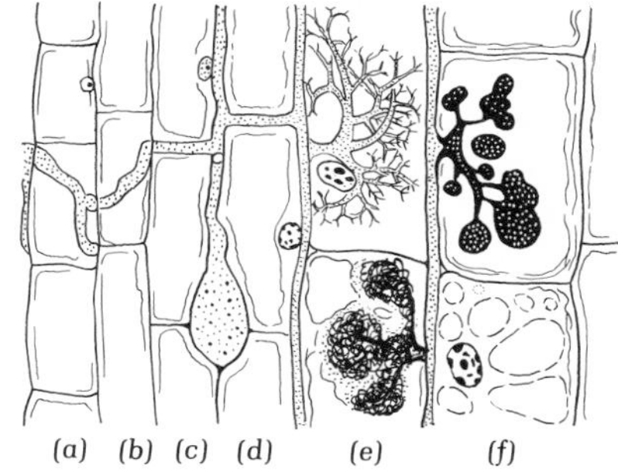


*Figure 25.6 Drawing showing the penetration of the root of Allium by a mycorrhizal fungus. In the first two cell layers (a and b), the fungal mycelium is intracellular. In the third and fourth layers (c and d) it is intercellular; a vesicular storage organ is shown between these layers. In the fifth and sixth layers (e and f), the fungus has formed intracellular branching structures (arbuscules). In (f) the arbuscules are undergoing digestion by the host cells. From F. H. Meyer, "Mycorrhiza and other plant symbioses," in Symbiosis, Vol. I (S. M. Henry, editor). New York: Academic Press, 1966.*

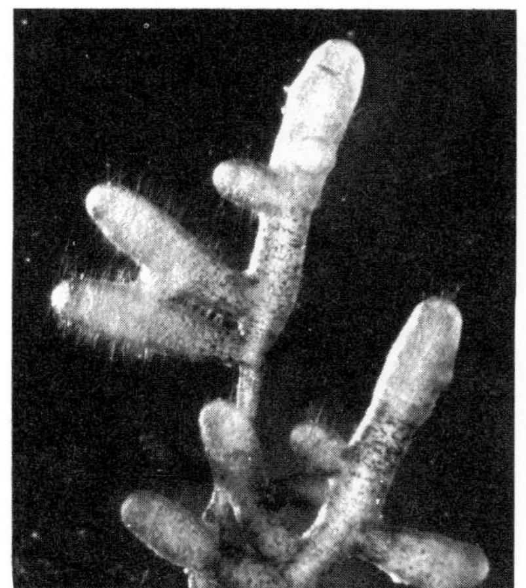

*Figure 25.7 Mycorrhiza of Fagus sylvatica, showing the club-shaped apices of roots and hyphae radiating from the surface. From F. N. Meyer, "Mycorrhiza and other plant symbioses," in Symbiosis (S. M. Henry, editor), Vol. I. New York: Academic Press, 1966.*

can be restored to health by the introduction of suitable mycorrhizal fungi into the soil. The fungus seems to facilitate the absorption of water and minerals from the soil; the absorbing surface of the plant's root system is increased manyfold by the fungal hyphae. The function of a mycorrhiza as an absorbing organ has been confirmed by comparing the uptake of minerals from the soil by plants with and without mycorrhiza. Pines, for example, absorb two to three times more phosphorus, nitrogen, and potassium when mycorrhiza are present than when they are absent.

## *Root nodule bacteria and leguminous plants*

It has long been known that the fertility of agricultural land can be maintained by a "rotation of crops." If a given plot of soil is sown year after year with a grass, such as wheat or barley, its productivity begins to decline but can be restored by interrupting this annual cycle with a crop of some leguminous plant such as clover or alfalfa. Roman writers on agriculture recognized that leguminous plants possess this ability to restore or maintain soil fertility which is not shown by other types of plants. It was also known that the leguminous plants have peculiar nodular structures on their roots (Figure 25.8). The plant anatomists of the seventeenth and eighteenth centuries, who examined these nodules in some detail, interpreted them as pathological structures analogous to the galls formed on the shoots of some plants as a result of infestation by insects.

*Figure 25.8 A seventeenth-century drawing by Malpighi of the root of a leguminous plant, showing the root nodules (m). The large dark object (n) is the coat of the seed from which the plant has developed.*

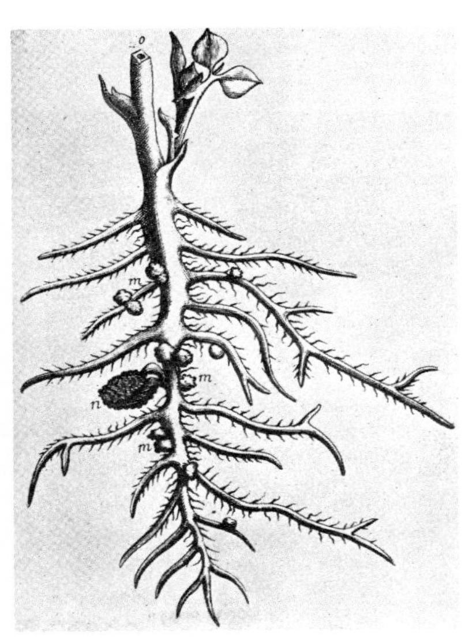


About the middle of the nineteenth century, a new interpretation of the nature of root nodules was offered. At this time, the development of chemical methods enabled scientists to start analyzing the problems of soil fertility and plant growth in chemical terms, and one of the early results of these studies was the elucidation of the role that leguminous plants play in the maintenance of soil fertility. It was found that most plants are limited in their growth by the amount of combined nitrogen in the soil but that leguminous plants are not. Furthermore, by total nitrogen analyses it could be shown that when leguminous plants are grown on nitrogen-poor soil, there is a net increase in the amount of fixed nitrogen in the soil. Since the only possible source of this extra nitrogen is the atmosphere, such experiments demonstrated that leguminous plants, unlike other higher plants, can fix atmospheric nitrogen. Hence, the growth of a crop of legumes on a nitrogen-poor soil results in an increase in the total fixed nitrogen content of the soil, particularly if the crop is plowed under. This is the chemical basis for the long-established practice of crop rotation.

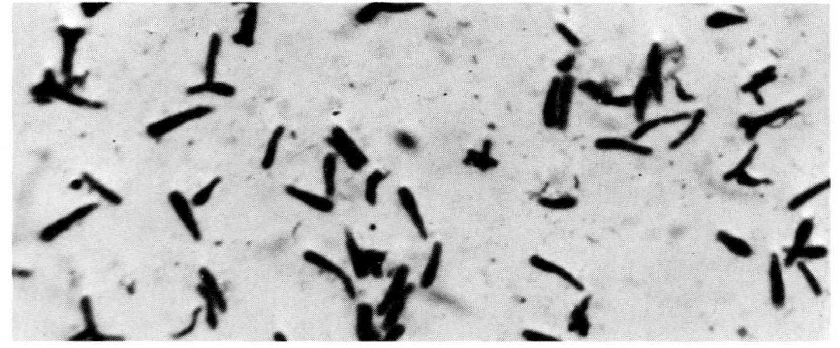

*Figure 25.9 A stained smear of the contents of a root nodule, showing bacteroids (× 1,360). Courtesy of H. G. Thornton and the Rothamsted Experimental Station, United Kingdom.*

Once these facts had been established, the question naturally arose as to whether the peculiar nodulations on the roots of leguminous plants had any connection with their ability to fix nitrogen. Occasionally, leguminous plants fail to form nodules, and analyses showed that such plants do not fix nitrogen. When the contents of nodules were examined microscopically, they were found to contain larger numbers of "bacteroids": small, rod-shaped, or branched bodies similar in size and shape to bacteria (Figure 25.9). These facts suggested that the nitrogen-fixing ability of leguminous plants is not a property of the plants as such but results from infection of their roots by bacteria in the soil, such infection leading to the formation of nodules. About 1885 the correctness of this hypothesis was established by showing that if seeds are treated with chemical disinfectants so as to sterilize their surface without impairing their germinability, and then grown in pots of sterile soil, they will never form nodules. The growth of such plants is strictly limited by the supply of combined nitrogen in the soil. Nodulation can be induced by adding crushed nodules from plants of the same species to the soil. Once nodulation has occurred, the growth of the plants becomes independent of the supply of combined nitrogen (Figure 25.10). The final proof came in 1888, when M. W. Beijerinck succeeded in isolating and cultivating the bacteria present in the nodules and demonstrated that sterile seeds produced the characteristic nodules once more when treated with pure cultures of the isolated bacteria.

*Figure 25.10 The effect of nodulation on plant growth. Two red clover plants grown in a medium deficient in combined nitrogen. The one at left, without nodules, shows very poor growth as a result of nitrogen deficiency. The plant at right, with nodules, shows normal growth. Courtesy of H. G. Thornton and the Rothamsted Experimental Station, United Kingdom.*

## The nodule bacteria

The agricultural importance of nitrogen fixation led to extensive work on the nodule bacteria. These organisms are Gram-negative motile rods that are classified in the genus *Rhizobium*. It was soon found that the nodule bacteria isolated from the roots of the various kinds of leguminous plants resemble one other closely in their morphological and cultural properties. When inoculated back into plants, however, they show a considerable degree of host specificity. The nodule bacteria isolated from the roots of lupines cannot evoke nodule formation on peas, and vice versa. In contrast, the nodule bacteria from peas, lentils, and broad beans can evoke nodulation in every member of this group of legumes. There are thus differences between the nodule bacteria of peas and those of lupines, which can be detected in terms of their host specificity. The nodule bacteria can be classified into a series of cross-inoculation groups. Strains of any one group have the same host range, which differs from those of the other groups. All attempts to induce infection of nonleguminous plants with nodule bacteria have failed.

The nodule bacteria are normally present in soil. Their numbers are very variable, depending on the nature of the soil and on its previous agricultural treatment. Hence, it not infrequently happens that a leguminous crop will develop poorly in a given plot of soil as a consequence of the fact that the nodule bacteria specific for it are either absent or present in such small numbers that effective nodulation does not occur. Nodulation can be ensured by inoculating the seed with a pure strain of nodule bacteria belonging to the correct cross-inoculation group. Bacterial cultures of proved effectiveness were first made commercially available at the beginning of the twentieth century, and seed inoculation is now a routine agricultural operation. This is by far the most important contribution that the science of soil bacteriology has made to agricultural practice.

Whereas the soil under a nonleguminous crop, such as wheat, may have fewer than 10 *Rhizobium* cells per gram, the same soil will contain between $10^5$ and $10^7$ *Rhizobium* cells per gram following the development of a flourishing legume crop. The ability of legume plants to stimulate the growth of *Rhizobium* in the soil extends as far as 10 to 20 mm from the roots. The effect is highly specific: bacteria other than *Rhizobium* show little or no stimulation, and growth of the species of *Rhizobium* able to infect that particular leguminous plant is stimulated more than the growth of other species of *Rhizobium*. The substances responsible for this stimulation have not been identified. It has been experimentally established that the high number of Rhizobium cells in the rhizosphere of legumes represents stimulation of free-living cells rather than their liberation from nodules, by showing that the increase occurs in the absence of active nodulation.

The number of nodules formed on the roots of the legumes is directly proportional to the density of *Rhizobium* in the soil, up to about $10^4$ cells per gram. Above this number, no further increase takes place, and nodule formation may even decline. When the number of *Rhizobium* cells is limited, so that fewer nodules are formed, the size of the nodules is proportionately larger. The result is that the total *volume* of nitrogen-fixing tissue remains fairly constant per acre of leguminous plants.

**The process of nodule formation**

In Chapter 23, we summarized what is known about the interactions occurring between the legume root and the free-living bacterium in the soil, interactions which serve to initiate infection. The infection itself begins with the penetration of a root hair by a group of rhizobial cells, and involves the invagination of the root hair membrane. A tube is formed containing bacteria and lined with cellulose produced by the host cell. This tube is called the *infection thread* (Figure 25.11). The infection thread penetrates the cortex of the root, passing through the cortical cells rather than between them.

As the thread passes through a cell, it may branch to produce vesicles that contain bacteria; the walls of the thread and vesicles are continuous with the host cell membrane. The bacteria are finally liberated into the

(a)

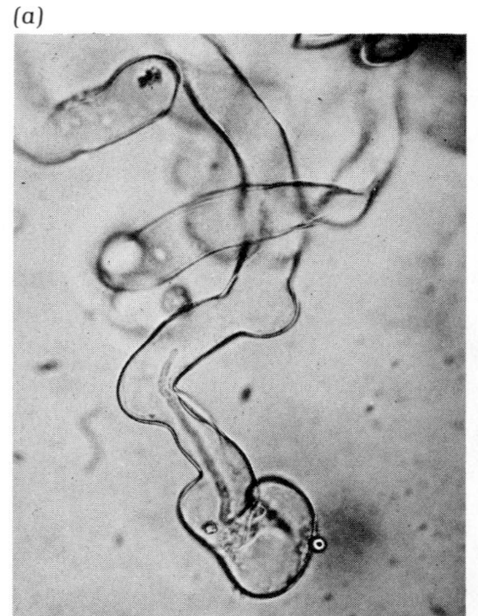

(b)

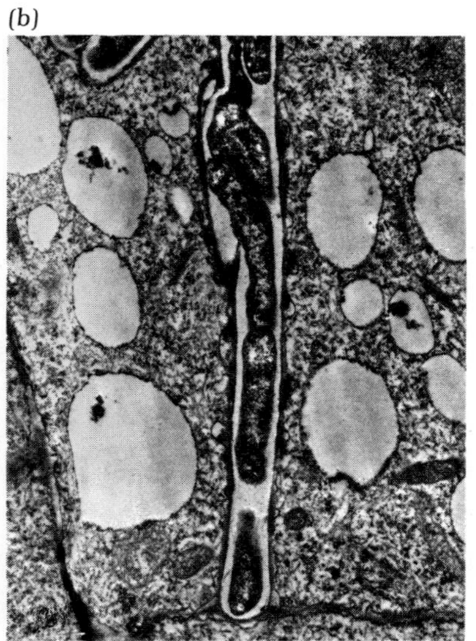

*Figure 25.11 (a) A newly infected root hair. The bacterial infection thread can be seen passing up a root hair, which has curled at the tip as a result of infection. Courtesy of H. G. Thornton and the Rothamsted Experimental Station, United Kingdom. (b) Infection thread crossing a central tissue cell of a nodule aged 1 to 2 days. From D. J. Goodchild and F. J. Bergersen, "Electron microscopy of the infection and subsequent development of soybean nodule cells." J. Bacteriol.* **92,** *204 (1966).*

Figure 25.12 Mature nodule cell with large membrane envelopes containing four to six bacteroids. No further bacterial growth occurs. From D. J. Goodchild and F. J. Bergersen, "Electron microscopy of the infection and subsequent development of soybean nodule cells." J. Bacteriol. **92,** 204 (1966).

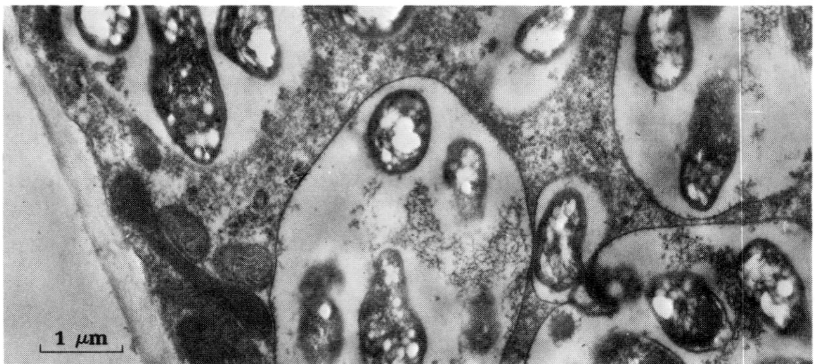


cytoplasm of the host cell; electron micrographs of thin sections of legume roots show that the bacteria are enclosed, either singly or in small groups, in a membranous envelope (Figure 25.12).

Development of the nodule itself is initiated when the infection thread reaches a tetraploid cell of the cortex. This cell, along with neighboring diploid cells, is stimulated to divide repeatedly, forming the young nodule. The rhizobial cells invade only tetraploid plant cells, the uninfected diploid tissue becoming the cortex of the nodules (Figure 25.13). In young nodules, the bacteria occur mostly as rods but subsequently acquire irregular shapes, becoming branched, club-shaped, or spherical (the typical bacteroids). At the end of the period of plant growth, the bacteria have often disappeared completely from the nodules; they die, and their cell materials are absorbed by the host plant.

Figure 25.13 (a) Section of a root nodule. The dark cells are filled with bacteria. (b) Section of a nodule at high magnification, showing the individual bacteria in the infected cells. Courtesy of H. G. Thornton and the Rothamsted Experimental Station, United Kingdom.

(a)

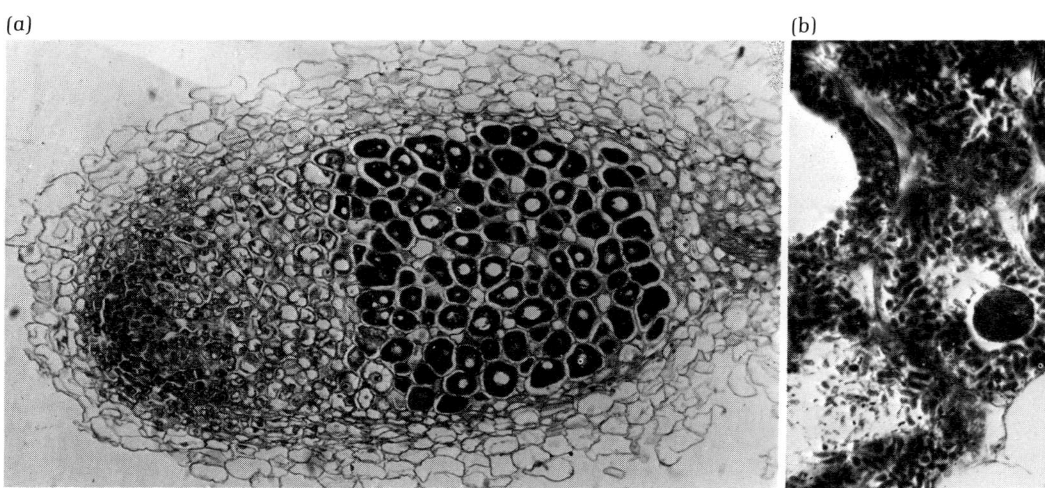

(b)

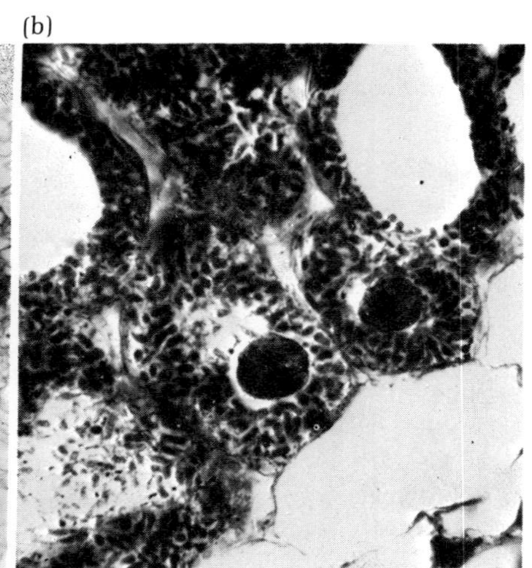

## Symbiotic nitrogen fixation

It has already been mentioned that uninfected leguminous plants are unable to fix atmospheric nitrogen, and all attempts to show significant nitrogen fixation by leguminous bacteria isolated and grown in pure culture have failed. Nitrogen fixation occurs only when a symbiosis has been established.

Excised nodules can fix nitrogen for a limited period. All attempts to obtain fixation with crushed nodules were unsuccessful, until it was discovered that oxgyen rapidly inactivates such preparations; macerated nodular tissue prepared under totally anaerobic conditions were found to be active. The activity of such preparations can be greatly increased by removal of phenolic compounds which are liberated during the maceration of the nodules. When nodule preparations are fractionated into bacteroids, membrane material, and soluble material, the activity is found to be associated with the bacteroids; in fact, *cell-free extracts of the bacteroids fix nitrogen* in the presence of an appropriate electron donor and an ATP-generating system. The fixation product is ammonia.

What, then, could be the role of the plant host in symbiotic nitrogen fixation? Two possibilities suggest themselves: either the *Rhizobium* cells grown in vitro lack one or more enzymes necessary for nitrogen fixation, formation of these enzymes being induced by growth in the nodule, or the nodule provides environmental conditions necessary for the activity of the bacterial enzymes.

It is a curious fact that healthy nodules contain hemoglobin, which neither the plant nor the bacteria can synthesize when grown in isolation. Hemoglobin has two well-known properties: it can be reversibly oxidized and it can bind oxygen avidly. In view of the sensitivity of the nitrogen-fixing activity to inhibition by oxygen, it is possible that the interior of the plant nodule provides the necessary regulation of redox potential for bacteroid activity and that hemoglobin plays some role in this regulation.

## The relation between mutualism and parasitism in the legume symbiosis

In discussing the formation of nodules, we spoke of the "infection" of the root hair by bacteria present in the soil. In many respects, the establishment of the root nodule symbiosis greatly resembles the infection of plants by pathogenic bacteria; there is an initial destruction of the surface of the root hairs, followed by penetration and proliferation of the bacteria, which then provoke abnormal growth of the surrounding plant tissues. Nevertheless, one does not normally think of the relationship between leguminous plants and root nodule bacteria as a parasitic one, for the obvious reason that the plant gains much from the association. In recent years, however, it has been found that the association does not necessarily benefit the plant. If inoculation studies are conducted with many different strains of leguminous bacteria belonging to one cross-inoculation group, all will give rise to nodulation upon a susceptible plant. With some strains, however, the symbiosis so established is "ineffective" (i.e., it does not permit active nitrogen fixation). As a result, the plant receives little or no benefit from the association, although the bacteria that have established themselves in its roots still profit because they are able to develop at the expense of materials produced by their host plant. In other words, a symbiosis that is normally mutualistic becomes, in this special case, parasitic. It is true that the parasitism is a mild one, since damage to the plant is highly localized in its root system and does not lead to its death. Nevertheless, the situation differs only in degree, not in kind, from the one that occurs when a frankly pathogenic bacterium invades a plant and establishes itself within the plant tissues. This is a good example of the fact, mentioned in Chapter 23, that the dividing line between mutualism and parasitism is a shifting one which can be crossed in either direction.

## *Root nodule symbionts and nonleguminous plants*

The formation of root nodules is not confined to the legumes. There are nine genera of nonleguminous plants, in each of which one or more species is characterized by the presence of root nodules. These genera, of which *Alnus* (the alder) is typical, are listed in Table 25.1. They are

*Table 25.1 Nitrogen-fixing genera of nonleguminous plants*[a]

| | |
|---|---|
| Order: *Coriariales*<br>Family: *Coriariaceae*<br>Genus: *Coriaria* | Order: *Casuarinales*<br>Family: *Casuarinaceae*<br>Genus: *Casuarina* |
| Order: *Myricales*<br>Family: *Myricaceae*<br>Genus: *Myrica* | Order: *Rhamnales*<br>Family: *Elaeagnaceae*<br>Genera: *Hippophaë*, *Shepherdia* |
| Order: *Fagales*<br>Family: *Betulaceae*<br>Genus: *Alnus* | Family: *Rhamnaceae*<br>Genera: *Ceanothus*, *Discaria* |

[a]Taxonomy according to J. Hutchinson, *The Families of Flowering Plants*, Vol. I. New York: Oxford University Press (Clarendon), 1959.

not particularly closely related to each other, being distributed among six families; all but one of these includes nonnodulated genera as well.

As early as 1896, it was shown that alder plants can grow well in a medium free of combined nitrogen only if nodulated; by 1962 similar experiments had been done with members of eight of the nine genera. By analogy with the legumes, it was supposed that the nodules were fixing atmospheric nitrogen, and this was eventually confirmed in all cases by the use of $^{15}N$. Other parts of the plant do not fix nitrogen.

The fixation process resembles that of legumes in many respects. In both legumes and nonlegumes fixation is inhibited by carbon monoxide and by high levels of oxygen or hydrogen and is poor when the plants are deficient in cobalt or molybdenum. Finally, hemoglobin has been detected spectroscopically in the nodules of three groups of nonlegumes, *Alnus*, *Myrica*, and *Casuarina*.

Microscopic examination of stained sections of nonleguminous nodules always reveals structures that suggest symbiotic microorganisms (Figure 25.14). Most workers have interpreted them as actinomycetes, on morphological grounds. Actinomycetes have been isolated from such material, but it is not yet clear whether these isolates represent the nitrogen-fixing symbionts.

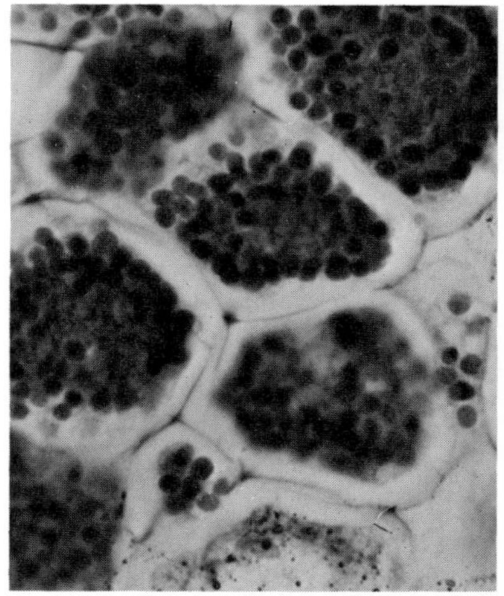

*Figure 25.14 Cortical region of a transverse section of a root nodule of Alnus glutinosa, showing the dark-stained "vesicles," which are presumed to be the nitrogen-fixing endosymbionts. From G. Bond, "The root nodules of non-leguminous angiosperms," in Symbiotic Associations: Thirteenth Symposium of the Society for General Microbiology. New York: Cambridge University Press, 1963. Section prepared by E. Boyd; micrograph by W. Anderson.*

Nodulation can be readily induced in most genera by applying suspensions of crushed nodules to the roots; untreated roots show little or no nodulation in control experiments. Whatever the nature of the microbial symbionts, cross-inoculation experiments using crushed nodules show that they possess group specificity. Cross inoculation is often possible between species of a given genus of host plant, but usually not between species of different genera, with the exception of the three genera *Elaeagnus*, *Hippophaë*, and *Shepherdia*, which belong to one family. Such specificities are reminiscent of the cross-inoculation groups of rhizobia.

## *Bacteria in leaf nodules and other plant tissues*

Certain plant species are characterized by the presence on the leaf of bacterium-filled nodules. Little is known about the symbiotic relationships of these plant–bacterium associations, but in one case—that of the flowering plant, *Ardisia crispa*—it is clear that the bacteria are necessary for normal host development.

The bacterial symbionts of *A. crispa* are present in the leaf bud, and the nodules begin to form while the leaves are still rolled within the foliar bud. The mature nodule is macroscopically visible and is densely packed with bacteria. As the leaf and nodule mature, the bacteria assume the forms of bacteroids.

The bacteria are transmitted from one plant generation to the next through the seed. Bacteria are present in the floral bud, and when the flower forms they surround the ovules and become entrapped by the integuments of the developing seed. They can be demonstrated in the mature seed and at germination infect the axillary and terminal buds (Figure 25.15).

*Figure 25.15 Schematic representation of the symbiotic cycle between Ardisia and Bacillus foliicola. The regions inhabited by the bacterial symbionts are shown stippled or black: (A) terminal bud; (B) foliar nodules; (C) dormant bud; (D) axillary bud; (E) inflorescence bud; (F to K) stages in the development of flower and fruit; (L) seedling. From R. T. Lange, "Bacterial symbiosis with plants," in Symbiosis (S. M. Henry, editor), Vol. I. New York: Academic Press, 1966.*

The host plant can be freed of its symbiotic bacteria by heating the seeds. The heated seeds germinate normally, but development of the plant stops after the first few leaves have been formed. The dwarfed plant may survive for several years. If the foliated shoots of normal plants are similarly heated, their development ceases.

Such "cripples," as they are called, are always symbiont-free. In some experiments they have been induced to resume normal growth after reinfection by isolated symbionts or with macerated nodules.

The symbiont isolated from another *Ardisia* species, *A. hortoricola*, fixes nitrogen in pure culture at a rate comparable to that of *Azotobacter*. Bacterium-free cripples, however, cannot be induced to grow normally by the application of combined nitrogen, so the symbiont must play an additional role, essential to the normal development of the host plant.

## *Endosymbioses of algae with aquatic invertebrates*

**Occurrence and significance**

Endosymbiotic algae have been recorded in over 100 genera of aquatic invertebrates, particularly in the coelenterates (jellyfish, corals, sea anemones, hydra), the platyhelminths (flatworms, principally the planarians), the Porifera (sponges), and the mollusks (clams, squid). They are usually found in the cytoplasm of cells concerned with digestion or with food transport (e.g., in the amebocytes of sponges or the phagocytic blood cells of certain clams).

Some of these animal hosts acquire their symbionts with their food, either directly—as in the case of sponges which feed on algae—or indirectly, as in the case of carnivorous animals which find the algae in the tissues of their prey. Once infection has taken place, however, a permanent symbiosis is usually established in which intracellular growth of the algae is restricted either by their digestion by the host or by regulatory mechanisms.

In most coelenterates, and in certain other invertebrates, the algae are transmitted to the next generation through the cytoplasm of the egg. In such cases it has not been possible to obtain symbiont-free animals, so the significance of the symbiosis is not known. Nevertheless, experiments with isotopically labeled $CO_2$ show that organic compounds photosynthesized by the algae are utilized by the tissues of the host, and it can be shown that the molecular oxygen generated by algal photosynthesis is several times more than that which would be necessary to provide for the respiratory needs of the host–alga complex. Although the environment of the animal provides dissolved oxygen as well as organic material (principally as plankton), the mechanisms that have evolved for ensuring symbiosis suggest that it is of great ecological significance.

In certain mollusks, the photosynthetic symbionts are not algal cells, but *intact, surviving chloroplasts* which are liberated when the algal cells undergo digestion by the host. The animal cell thus becomes photosynthetic, by acquiring a plant organelle. In a sense, a symbiotic relationship can be said to exist between the animal cell and the chloroplast, although it seems unlikely that the latter can grow and divide in its new "host."

The algae found in aquatic invertebrates are of very few types. They are either green algae or dinoflagellates: *zoochlorellae* and *zooxanthellae,* respectively. The algae are presumed to benefit by their intracellular habitat, which supplies a rich supply of essential nutrients. As an example of this type of symbiosis, we will describe briefly the case of the tridacnid clams.

**The algal endosymbiosis of the tridacnid clams**

M. J. Yonge has described in detail the algal symbioses of three members of the family Tridacnidae, all ubiquitous inhabitants of the Great Barrier Reef of Australia.

The tridacnid clams have several unique anatomical features, the most prominent of which is the location and thickening of the mantle, the epithelial tissue that lines the shell. Unlike the mantle of all other clams, the mantle of the Tridacnidae is greatly extended along the dorsal, or open, part of the shell, the visceral mass being moved to a more ventral position. The mantle, olive-green in color due to its dense population of algal symbionts, is so thick that it prevents the shell from closing, and its surface is covered with conical projections.

Sections through the mantle tissue reveal the nature and function of its conical protuberances. Each protuberance contains one or more lenslike structures, the *hyaline organs,* made up of transparent cells. Each hyaline organ is surrounded by a dense mass of zooxanthellae, living intracellularly in blood phagocytes (Figure 25.16).

Yonge deduced that the function of the lenslike hyaline organ is to permit light to penetrate deeply into the mass of zooxanthellae. These clams have thus evolved a highly specialized system for cultivating algae within their own tissues. The algae, contained within phagocytic blood cells, are eventually transported from the mantle to the visceral mass, where they are digested by the phagocytic cells that transported them.

There is much anatomical evidence to suggest that the tridacnid clams rely heavily on their algae as a source of food. The digestive sys-

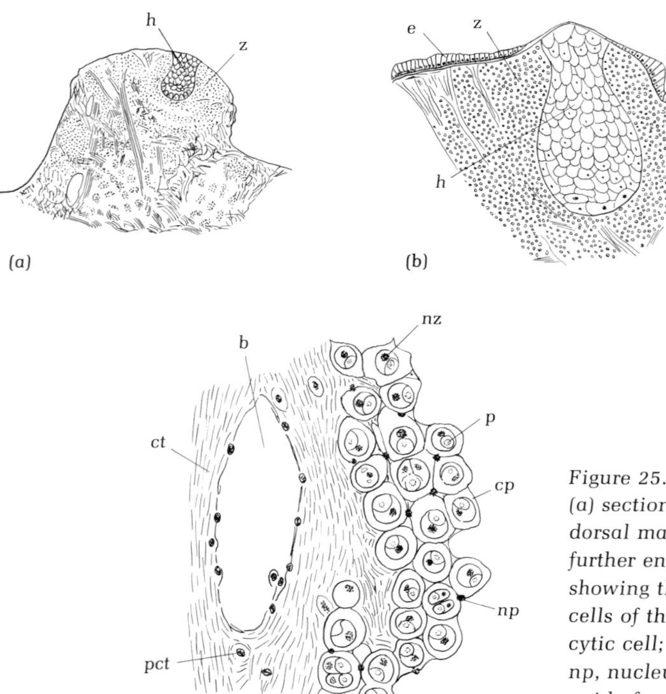


*Figure 25.16 Endosymbiotic algae of the clam, Tridacna crocca: (a) section through a protuberance on the inner fold of the dorsal mantle edge; (b) enlarged view of a hyaline organ; (c) further enlargement of the region adjacent to a hyaline organ, showing that the algal cells are harbored within phagocytic cells of the host. Key: b, blood vessel; cp, cell wall of phagocytic cell; ct, connective tissue; e, epithelium; h, hyaline organ; np, nucleus of phagocyte; nz, nucleus of zooxanthella; p, pyrenoid of zooxanthella; pct, phagocyte in connective tissue; z, zooxanthellae around hyaline organ. After M. J. Yonge.*

tem is reduced, for example, and the feeding organs are so altered that they screen out all but the most minute particles. Finally, the kidneys are vastly increased in size, presumably to handle the excretion of products formed in the phagocytes by digestion of the algae. The tridacnids thus represent an extreme example of an evolutionary response to symbiosis.

## *Endosymbioses of fungi and bacteria with insects*

**The insect hosts**

Microbial endosymbioses are extremely widespread among insects. P. Buchner, the German biologist whose pioneering work on symbiosis has spanned more than half a century, discovered a striking correlation between the diet of insects and the presence of symbionts: symbionts are never found in insects that have a nutritionally complete diet, but are present in all insects that have a nutritionally deficient diet during their developmental stages. Thus, no carnivorous insect has symbionts, whereas insects that live on blood or on plant sap all contain symbionts. The main function of the symbiont is thus to provide the host with one or more growth factors that are lacking in the insect's diet.

Certain apparent exceptions prove this rule. Mosquitoes, for example, contain no symbionts, although they suck blood. It is only the adult female, however, which takes a blood meal; the larvae and pupae have a nutritionally complete diet consisting of microorganisms and organic debris. Conversely, the granary weevil, *Sitophilus granarius*, contains symbionts although it feeds on nutritionally rich grains. This genus, however, inherits its symbionts from its wood-eating ancestors; it is able to survive and reproduce if freed of its symbionts, *provided that it is fed a nutritionally rich diet*. Without symbionts its choice of food is severely restricted.

**The microbial endosymbionts**

The microbial endosymbionts of insects include both bacteria and yeasts. Most of these have been identified as such solely on the basis of their appearance and mode of reproduction in the host. A few, however, have been successfully isolated and grown in pure culture. For example, one of the symbionts of *Rhodnius*, a kissing bug, has been isolated and identified as an actinomycete of the genus *Nocardia*. Other isolated insect symbionts have proved to be coryneform bacteria or Gram-negative rods. Some yeasts have also been successfully isolated, notably from the long-horned beetles *(Cerambycidae)* and the death-watch beetles *(Anobiidae)*. Although most insects are monosymbiotic, it is not uncommon for a particular species to harbor two or more different microorganisms. The relationship between insects and their endosymbionts appear to be highly specific; an insect species can often be identified reliably by observing the nature of its symbionts.

**The localization of the endosymbionts**

The microbial endosymbionts are housed within specialized cells of the insect. These are called *mycetocytes* when they harbor yeasts and *bacteriocytes* when they harbor bacteria. Some authors refer to both as mycetocytes, and we will use this terminology here.

In some insects, the mycetocytes are scattered randomly throughout a normal tissue, such as the wall of the midgut or the *fat body*, a loose, discontinuous tissue lining the body cavity. In many insects, however, the mycetocytes are restricted to special organs called *mycetomes*, the only function of which is to house the endosymbionts. It is possible to trace an evolutionary series of steps between ectosymbiosis, in which the symbionts develop in the lumen of the insect gut, and endosym-

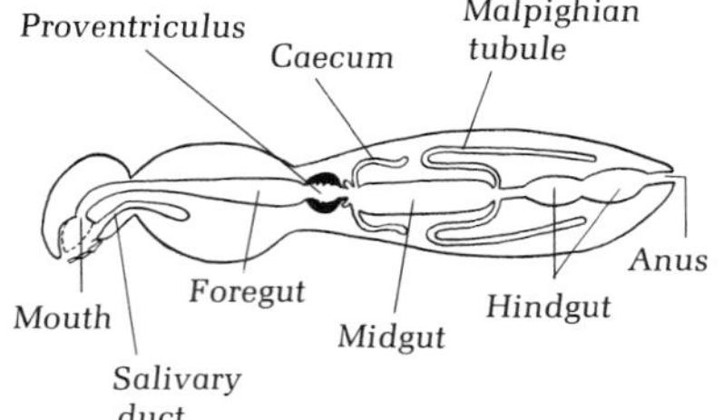


*Figure 25.17 Schematic diagram of the digestive tract of the insect.*

biosis in mycetomes. Figure 25.17 shows schematically the principal parts of the insect digestive tract. Figure 25.18 illustrates the localization of endosymbionts in out-pocketings or *blind sacs* of the insect midgut. Figure 25.19 shows how, in a series of species of anobiid beetles, the blind sacs have evolved to become more and more independent of the midgut. In the most primitive endosymbioses, the symbionts are found both extracellularly in the gut lumen and intracellularly in the blind sacs. In the most advanced forms, the symbionts are completely isolated, and the blind sacs have evolved into independent organs, or mycetomes.

In some insects the mycetocytes are localized in the Malpighian vessels, the excretory organs of the insect. In certain genera of the *Curculionidae* (the family that includes weevils, snout beetles, and curculios), two of the six Malpighian vessels have become anatomically specialized for this purpose and have evolved into club-shaped myce-

*Figure 25.18 Blind sacs of the midgut of Sitodrepa panicea, an anobiid beetle: (a) larva; (b) adult; (c) epithelium of the blind sac of the larval midgut, showing yeast-filled mycetocytes separated by sterile cells with brush borders. After A. Koch.*

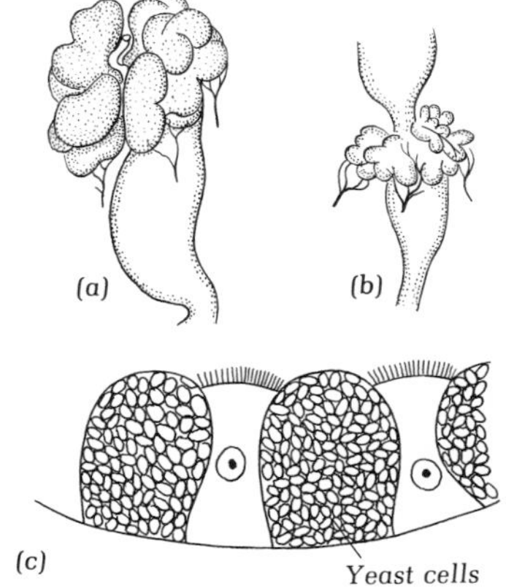


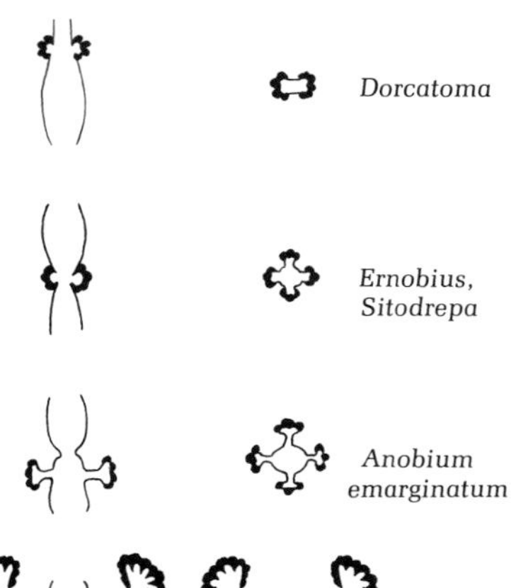


*Figure 25.19 Blind sacs of the midgut of a series of anobiid beetles, showing evolutionary development of the blind sacs as independent organs. After A. Koch.*

tomes (Figure 25.20). In a number of other insects, the mycetomes are detached from the gut, forming essentially independent structures in the body cavity.

**The transmission of endosymbionts**

*Figure 25.20 The adult gut of Apion pisi, showing the transformation of two of the six Malpighian tubules into mycetomes. After A. Koch.*

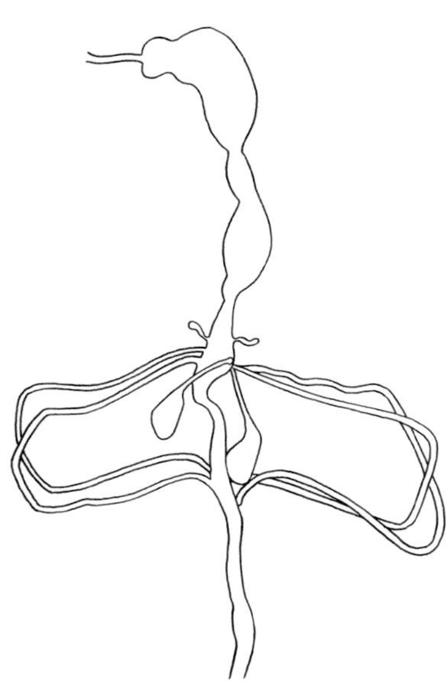

Most insects that harbor endosymbionts are totally dependent on them for certain growth factors. Accordingly, the host species cannot rely on chance reinfection at each new generation; instead, they have evolved elaborate mechanisms for the *transmission* of the symbionts from mother to young. The mechanisms of transmission fall into three categories.

First, some insects whose mycetocytes are in direct contact with the gut liberate symbionts into the alimentary tract at the time of egg laying. The egg-laying mother exudes fecal droplets which are rich in symbionts, and these are either sucked from the mother by the newly hatched young or else deposited near the eggs where the young will find them and feed on them. The evolution of these symbioses has resulted in the acquisition of a *behavioral trait:* in the former case, the mother remains in the vicinity of the brood so as to "nurse" them, and in both cases the larvae specifically feed on fecal drops. The extreme development of this form of transmission is found in the bug *Coptosoma scutellatum* (Figure 23.7). In this insect, the female forms symbiont-filled capsules by a process of secretion in the gut. One capsule is deposited between each pair of eggs as they are laid. When the larvae hatch, their first act is to insert their beaks into the capsules and suck up a supply of bacterial symbionts.

Second, some insects have special mechanisms for smearing their eggs with symbionts at the time of egg laying. The stink bugs (family Pentatomidae) smear their eggs with symbionts from the anus. In other families, special symbiont-filled organs are formed which communicate directly with the egg-laying apparatus.

Third, many insects ensure transmission to the next generation by intracellular infection of the eggs. In some cases, the germ cells become infected early in the embryonic development of the insect. If the insect matures as a male, the germ cells become testes and the symbionts disintegrate. If the insect matures as a female, however, the germ cells become ovaries and the symbionts multiply, so that each egg contains large numbers of them. In other cases, infection of the germ cells of the female occurs in the adult stage. The mechanism for transferring the symbionts from the mycetome to the egg varies from family to family and may be very complex. A common sequence of events is the following. The symbionts are liberated from the mycetocytes and—being nonmotile in all cases—are passively transported to the ovary by way of the lymph. At the ovary, the symbionts are taken up by special epithelial cells and from these are then ultimately transferred to the egg cells.

The cycle is completed during embryogenesis of the progeny, when events occur which lead to the formation of the symbiont-filled mycetome. The way in which mycetomes are formed varies from family to family, being particularly complicated in those insects which carry several different symbionts, each of which must eventually be housed in its own special type of mycetocyte or mycetome. In every case, the development of the mycetome involves a process of differentiation comparable to that which leads to the formation of any animal organ. Differentiation is initiated by a series of regulated nuclear divisions in the region of the egg that contains the symbiont mass and culminates in the formation of the mycetome of the adult insect.

**The significance of the insect endosymbioses**

The essential role played by the endosymbionts in the nutrition of the host can be demonstrated by artificial elimination of the symbionts and study of the behavior of the symbiont-free insects. Elimination has been accomplished by a variety of ingenious methods. In insects that smear their eggs, the egg surface can be sterilized. In insects with well-defined and isolated mycetomes, such as the stomach disc of *Pediculus*, the louse, the mycetome can be surgically removed. Some insects can be freed of their symbionts by the use of high temperatures or of antibiotics. In some cases, growth of symbiont-free insects is severely retarded, and the adult stage may not be reached (Figure 25.21). In other cases, the principal effect is to disturb the reproductive system: the female organs may be damaged, or their formation may be completely blocked.

In many such experiments, the loss of the symbionts can be totally compensated for by the provision of vitamins, particularly the B vitamins. In cockroaches (family Blattidae) it has also been shown that the symbionts provide the host with some essential amino acids. Feeding the young insects $^{14}$C-labeled glucose led to the appearance of labeled tyrosine, phenylalanine, isoleucine, valine, and arginine in symbiotic, but not in symbiont-free, individuals. The injection of $^{35}$S-labeled sulfate similarly showed that the methionine and cysteine of the cockroach are synthesized by the symbiotic bacteria.

The bacterial symbionts of some insects also appear to aid the host in the breakdown of nitrogenous waste products (uric acid, urea, and xanthine).

*Figure 25.21 The effect of symbiont loss on the growth of larvae of Sitodrepa panicea: (a) symbiont-free larva on normal diet; (b) symbiont-free larva on normal diet plus 25 percent dried yeast; (c) normally infected larva, on normal diet without supplementation. After A. Koch.*

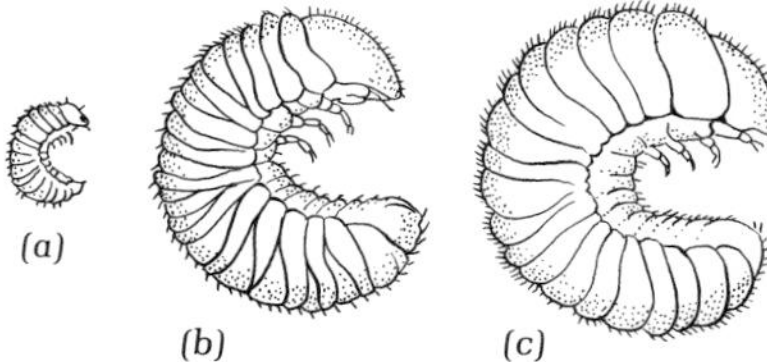

**The evolution of the insect–microbe symbioses**

The evolutionary relationship between a specific symbiosis and the diet of the host can be clearly traced in the termites. The fossil record shows that the termites and the cockroaches split off from a common ancestor about 300 million years ago. The most primitive group of termites, *Mastotermes*, harbors endosymbiotic bacteria which are identical in type and location with those harbored by all genera of cockroaches. *Mastotermes* also harbors the intestinal flagellated protozoa, which, by digesting cellulose, allow their hosts to feed on wood. All higher termites, however, have lost the endosymbiotic bacteria and rely exclusively on intestinal flagellate protozoa for all symbiotic functions.

The enzymatic activities of the symbiotic microorganisms (synthesis of growth factors and digestion of cellulose) have allowed the insects to enter new ecological niches. The ability of insects to live by sucking blood, sucking plant sap, or boring in wood is entirely dependent on their development of organs to house symbionts and to their development of mechanisms for transmitting these symbionts from generation to generation.

The microbial symbionts have also undergone adaptive evolutionary changes. Such adaptation has frequently been accompanied by the loss of ability to grow in the free-living state. The evolutionary changes that have occurred are also indicated by the specificity exhibited when symbiont-free insects are infected with "foreign" symbionts. The anobiid beetle *Sitodrepa*, for example, can be reinfected either with its normal symbiotic yeast or with a foreign yeast. The former infects only the normal mycetocytes, whereas the latter infects all the epithelial cells of the blind sacs and midgut. Furthermore, the foreign yeast is not transmitted to the adult stage during morphogenesis. It is of particular interest that a beetle which harbors its normal symbiont *cannot* be infected with the foreign yeast that easily infects symbiont-free individuals; the normal symbiont appears to confer on its host an immunity to infection by related microorganisms, a phenomenon frequently observed in symbiotic relationships of vertebrates as well as of invertebrates.

## *Further reading*

**Books**

Ahmadjian, V., *The Lichen Symbiosis*. Waltham, Mass.: Blaisdell, 1967.

Buchner, P., *Endosymbiosis of Animals with Plant Microorganisms*. New York: Wiley, 1965.

Henry, S. M. (editor), *Symbiosis*, Vols. I and II. New York: Academic Press, 1966 and 1967.

*Symbiotic Associations: Thirteenth Symposium of the Society for General Microbiology*. New York: Cambridge University Press, 1963.

**Reviews and original articles**

Burris, R., "Biological Nitrogen Fixation." *Ann. Rev. Plant Physiol.* **17,** 155 (1966).

Sonneborn, T., "Kappa and Related Particles in *Paramecium*," *Adv. Virus Res.* **6,** 229 (1959).

Stewart, W., "Nitrogen-Fixing Plants," *Science* **158,** 1426 (1967).

# *Chapter twenty-six Parasitism in vertebrates: constitutive host defenses*

*Plants and animals* have evolved numerous mechanisms that keep microbial parasites in check. Some of these mechanisms are *constitutive:* they are normal properties of the host and are present under a wide range of environmental conditions. Other defense mechanisms, most of which occur only in the animal kingdom, are *inducible;* they appear in an individual host as a response to specific stimuli, notably infection.

When a parasite comes into contact with a normal (nonimmunized) host organism, the initial barrier to its establishment in the host is provided by the mechanisms of constitutive resistance. In animals, these include the skin and mucous membranes, which form protective surface barriers; antimicrobial substances in the tissues and circulating fluids; the phagocytic cells which engulf and digest foreign materials; and a complex of reactions known as *inflammation.* In plants, the principle constitutive defense mechanisms consist of antimicrobial substances, as well as the mechanical barriers to penetration represented by the surface tissues.

In this chapter we will survey the mechanisms of constitutive host resistance. In Chapter 27, we shall consider the mechanisms by means of which the pathogenic microorganisms overcome the constitutive defenses and invade the host, as well as the ways in which the invasion leads to host damage. Finally, in Chapter 28, we shall describe the mechanisms of inducible host resistance, which, coming into play as a result of the infectious process, often turn the tide of battle in favor of the host. In all three chapters we shall restrict our discussion to vertebrate animal hosts and their microbial and viral parasites.

## *Surface barriers*

The skin and the mucous membranes constitute the surfaces of the animal host and are directly accessible to contamination from the environment. Each surface has some means for ridding itself of microorganisms, with the result that only specially adapted types of microorganisms can survive on it. The adapted types constitute the *normal flora* of the body, which has been described in Chapter 24. The metabolic activities of the normal flora probably aid in preventing the establishment of pathogens; in the vagina, for example, acid production by lactobacilli tends to prevent establishment of other bacteria.

The skin and the mucous membranes secrete antimicrobial substances, including long-chain fatty acids and the enzyme lysozyme. Lysozyme catalyzes the destruction of the cell wall of many bacteria, with the result that sensitive bacteria are lysed by its action. It is quite probable that other antimicrobial substances, as yet unidentified, are secreted by the surface tissues.

In the respiratory passages, another defense mechanism is provided by the sweeping action of the ciliated epithelial cells. The cilia of these cells beat constantly and rhythmically, sweeping the film of secreted mucous and adhering microorganisms toward the outer portals of the body.

The skin epithelium, if unbroken, appears to be completely impregnable to penetration by microorganisms, infections being limited to growth of pathogens in the hair follicles and sweat glands. The mucous membranes, on the other hand, are readily penetrated by many pathogens; the mechanism of this penetration is not known.

For parasites that have penetrated the surface barriers of the body, the circulatory system serves both as a major route for their further spread and as a transporting system for the two most important constitutive defenses: antimicrobial substances and phagocytes. Phagocytes, it will be recalled, are ameboid cells that engulf particles of foreign matter, including invading microorganisms.

## *The mammalian circulatory system*

**The circulation of the blood**

Figure 26.1 is a highly simplified diagram of the circulatory system. Oxygenated blood is pumped by the heart through the arteries to the capillaries, which ramify through the tissues of the body and to the spleen. The spleen, like many other organs, is rich in phagocytes. It is

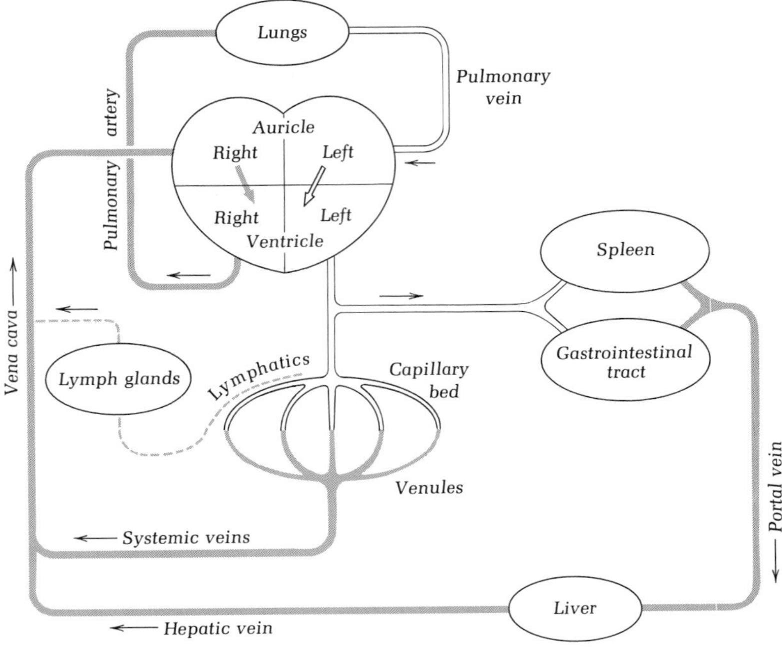


*Figure 26.1 Schematic diagram of the mammalian circulatory system. Vessels carrying oxygenated blood are shown as open lines; vessels carrying deoxygenated blood are shown as gray lines. Arrows indicate the direction of the flow. Courtesy of A. P. Krueger.*

also the site at which old red blood cells are broken down and their components made available for reuse. From the spleen, and from the capillaries surrounding the gastrointestinal tract, the deoxygenated blood flows to the liver, another center of phagocytic activity. From the liver and from the venules that drain the capillary bed, the blood is returned to the heart, which pumps it to the lungs for reoxygenation.

Blood is an extremely complex mixture, consisting of fluid, *plasma*, in which are suspended the oxygen-carrying red blood cells *(erythrocytes)*, white blood cells *(leucocytes)*, and cellular fragments called *platelets*. Plasma contains in solution the salts, proteins, and soluble metabolites that must be transported to and from the tissues. The plasma proteins include *fibrinogen*, which is converted to a network of insoluble fibrin when blood clots; a group of proteins called *globulins*; a second group of proteins called *albumins*; and *complement*.

The globulin fractions include the *antibodies* of the blood. The antibodies are protein molecules that are formed in the body in response to the presence of foreign substances called *antigens*. Antibodies have the property of being able to combine with the specific antigen molecules that induced their formation; the significance of such antigen–antibody reactions for host defense will be discussed in Chapter 28.

Complement was first recognized as a heat-labile substance in blood which can kill or lyse antibody-coated cells. Later investigations showed that complement is a group of 11 different proteins, all of which must be present for lysis to take place.

When whole blood is removed from an animal and allowed to stand in contact with air, a clot is formed. The clot is composed of a fibrin network, in which are trapped the blood cells and platelets. After a time the clot shrinks, leaving a straw-colored, supernatant fluid. This supernatant fluid is called *serum;* it differs from plasma in being free of fibrinogen and other components of the clotting system.

### The lymphatic system

Although the blood vessels shown in Figure 26.1 form a closed system, the fluid in the arteries is under sufficient hydrostatic pressure to force it into the tissues through the thin walls of the capillaries, against the osmotic gradient. Some of this fluid is reabsorbed by the venous ends of the capillaries, in which the hydrostatic pressure is lower than the opposing osmotic pressure; much of it, however, diffuses into the capillaries of the lymphatic system. Once it has gained entrance to these capillaries, the fluid is called *lymph.* The lymph passes along the lymphatic vessels, through the lymph glands, eventually reaching the venous blood via the main lymph channels. This part of the circulatory system is schematically shown in Figure 26.2. Many types of white

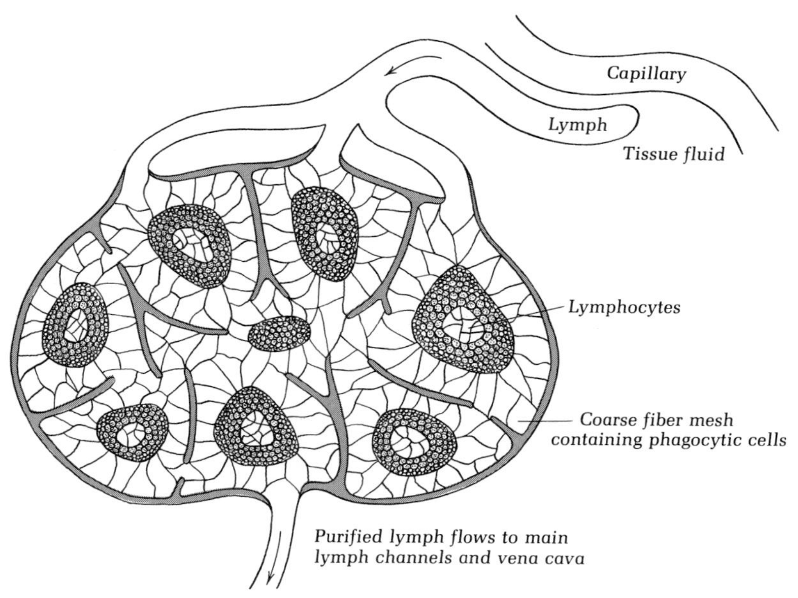


*Figure 26.2 Schematic cross section through a lymph gland. In the gland, the lymph filters through a coarse fiber mesh, in which are suspended the phagocytic cells, and past collections of lymphocytes. The lymph eventually returns to the bloodstream via the main lymph channels. Courtesy of A. P. Krueger.*

blood cells are also able to leave the capillaries and reach the lymphatic system. Thus, the blood and lymph vessels make up a continuous system of circulating fluid by means of which oxygen and metabolites are brought to all tissues of the body and waste materials are removed from them. The circulatory system also serves, however, as a potential route of invasion for many microorganisms.

**The blood cells**

Normal human blood contains about 5,000,000 erythrocytes per milliliter. They are formed continuously in the bone marrow; when erythrocytes mature, their nucleus disappears, and their cytoplasm becomes filled with hemoglobin. Hemoglobin consists of the iron-containing porphyrin, heme, bound to protein; it combines with molecular oxygen during passage of the red blood cells through the lungs and transports it to all tissues of the body for use in respiration. The erythrocytes are not known to play a direct role in the defense systems of hosts.

The leucocytes, which number 5,000 to 10,000 per milliliter of blood, play a major role in defense of the host against invasion by foreign organisms. Five distinct types of white blood cells can be recognized, on the basis of their staining characteristics and internal structures. The different types of blood cells are shown in Figure 26.3.

Three types of leucocytes have in common an irregularly shaped nucleus and a granular appearance of the cytoplasm. These types are known collectively as granular leucocytes or *granulocytes;* they are further differentiated, on the basis of the properties of their cytoplasmic granules, into *eosinophils* (granules stainable by the acid dye eosin), *basophils* (granules stainable by the basic dye methylene blue), and *neutrophils* (granules stainable by a mixture of acidic and basic dyes).

The granular leucocytes are also called *polymorphonuclear leucocytes,* because their nuclei exhibit many different shapes. The irregularity is most pronounced in the neutrophils and least in the basophils; in recent years, the use of the term "polymorphonuclear leucocyte" has been increasingly restricted to the neutrophils.

*Figure 26.3 Some types of blood cells: (a) an erythrocyte; (b) a small lymphocyte; (c) a polymorphonuclear leucocyte; (d) a monocyte; (e) a macrophage.*

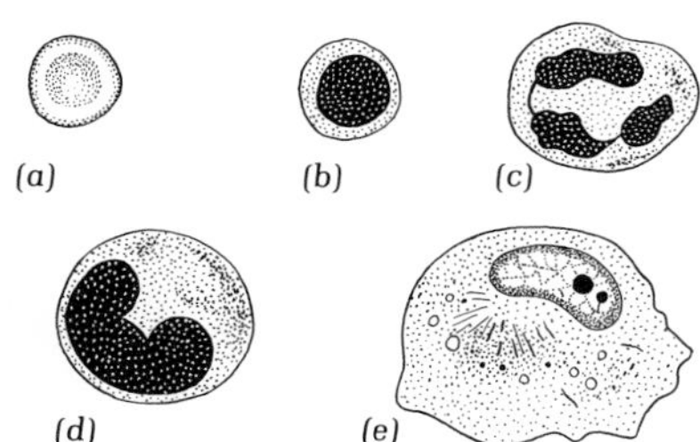

All the granular leucocytes are capable of phagocytosis, but only in the neutrophils is phagocytosis sufficiently active to be an important factor in host defense. The basophils and eosinophils, on the other hand, play special roles in the inflammatory response. Inflammation will be discussed at the end of this chapter.

The nongranular leucocytes of the blood are of two types: the *monocytes* and the *lymphocytes.* The lymphocytes are responsible for antibody formation, and we will return to them in Chapter 28. The monocytes are actively phagocytic. The importance of the leucocytes in host defense is indicated by the fact that they leave the circulatory system and congregate in any tissue which has been the site of invasion or injury.

## *Antimicrobial substances in body tissues and fluids*

**Specific antibodies, natural antibodies, and properdin**

It will be recalled that antibodies are globulins, which are capable of combination with specific large molecules called antigens. The formation of antibodies in the animal is induced by immunization with specific antigen; however, very low levels of specific antibody can also be found in the serum of normal, nonimmunized animals.

When such specific antibodies are found, there is good reason to believe that the animal has had prior contact with the inducing antigen, even if infection has not been apparent. It is probably significant that the specific antibodies found in normal animals are directed toward members of the normal enteric flora, whose presence in the digestive tract could well serve as the inducing stimulus.

The existence in normal serum of a mixture of specific antibodies present in very small amounts has led to considerable confusion. Early workers attributed the broad antibacterial activity of normal serum to the presence of a single type of nonspecific, constitutive antibody which they called "natural antibody." More recently, it has been claimed that part of the antibacterial activity of normal serum is caused by the presence of a nonspecific protein that differs in certain physical properties from true antibody. This protein has been named "properdin," from the Latin meaning "to prepare for destruction." It remains possible, however, that both "natural antibody" and "properdin" represent mixtures of specific antibodies which have been induced by contact with the corresponding specific antigens.

**Beta-lysins and lysozyme**

The serum of several species of mammals, including man, contains bactericidal substances that can be differentiated from properdin on the basis of their antibacterial spectrum: they are most effective against Gram-positive bacteria. One such group of substances is called *beta-lysins;* they are heat-stable materials of unknown chemical composition. Beta-lysins are found only in serum that has separated from clotted blood; they can be produced by the clotting of platelets in plasma, but their exact origin is not known.

Lysozyme is an enzyme that catalyzes the hydrolysis of the murein layer in procaryotic cell walls. It is found in tears, nasal secretions, saliva, mucus, and extracts of various organs, including skin.

**Antimicrobial substances in cells**

Many antimicrobial substances can be extracted from mammalian cells. One such substance is *phagocytin*, a globulin found in the phagocytic leucocytes; it is active principally against Gram-negative bacteria. The other antimicrobial substances of tissues are basic proteins, basic polypeptides, and polyamines; they are active principally against Gram-positive bacteria. They include the small proteins called histones and protamines, which are rich in the basic amino acids, arginine and lysine. Basic polypeptides of shorter chain lengths are found associated with histones and protamines; they have similar amino acid compositions and may be degradation products of the basic proteins. One such group of basic polypeptides, *leukins*, are found in leucocytes. It is noteworthy that these highly phagocytic cells contain substances active against both Gram-positive and Gram-negative bacteria (leukins and phagocytins, respectively).

Two naturally occurring polyamines, spermine and spermidine (see Figure 10.56), have also been found to have antibacterial activity, particularly against tubercle bacilli. It is believed that the antimicrobial activity of these polycationic compounds (as well as of histones and protamines) results from their ability to combine with polyanionic components of the bacterial cell surface. This hypothesis is supported by two types of experimental evidence. First, synthetic polypeptides made up entirely of lysine subunits are highly active against Gram-positive bacteria; second, polyanionic substances such as nucleic acids and acidic polysaccharides can reverse the antibacterial action of basic polypeptides.

## *Phagocytosis*

The capacity for phagocytosis exists in many unicellular protists, and phagocytic cells occur in the tissues of all animal groups. The phagocytic activity of certain cells in multicellular animals, and their role in preventing infection, were discovered by E. Metchnikoff as a result of his studies on *Daphnia*, the water flea, about 1880. *Daphnia* is a microscopic, transparent crustacean. Metchnikoff watched it ingest pointed fungal spores that penetrated the body cavity through the walls of the digestive tract. Through the transparent tissues, he could see ameboid cells approach and engulf the spores, which were soon digested. Sometimes, however, the phagocytes (as he named them) failed to dispose of the spores, which then germinated and initiated a fatal fungal infection. This work led to the discovery that phagocytosis is a general mechanism of host defense against invading microorganisms in all animals.

Metchnikoff proposed that both constitutive and inducible resistance are reflections of phagocytic activity. A lively controversy developed between Mechnikoff's followers and others who believed that antibody formation could account completely for the phenomena of immunity. Today, it is recognized that both phagocytes and antibodies play important roles in host defense. There is an interplay between these two kinds of defense systems, in that one of the functions of antibodies is to increase the efficiency of phagocytic action.

**The different types of phagocytic cells**

We have seen that two types of white blood cells, the neutrophils and the monocytes, are actively phagocytic. These cells will leave the bloodstream and enter the tissues, where they phagocytize invading microorganisms. There are, however, phagocytic cells that normally occur only in the tissues. These tissue phagocytes, which are much larger than the blood phagocytes, are called *macrophages.* One kind, the *fixed macrophages,* line the channels of certain organs of the circulatory system, notably the liver, spleen, bone marrow, and lymph nodes. The other kind, the *wandering macrophages* (now commonly called *histiocytes*), are scattered through the connective tissues.

The location of macrophages in the animal body can be revealed following intravenous injection of certain dyes, such as trypan blue. These dyes are taken up selectively by phagocytic cells, with the result that both fixed and wandering macrophages are deeply stained, having concentrated the dye from the body fluids. Since most of the macrophages which are stained by this procedure are endothelial (lining) or reticular (forming a supporting network) in organs of the circulatory system, the macrophages are collectively referred to as the reticuloendothelial system, or RE system. This system constitutes a major line of defense in the animal body.

**The operation of the reticuloendothelial system**

Referring once again to Figure 26.1, it is easy to see why the RE system is so effective. All the circulating fluid of the body must pass repeatedly through organs (liver, spleen, bone marrow, lymph nodes), the channels of which are lined with fixed macrophages. Such organs thus behave as filters, and few invading microorganisms escape phagocytosis in them. The surviving microorganisms, if any, fall prey to the phagocytic neutrophils and monocytes of the blood.

The RE system also plays an active role in clearing the body of microorganisms that invade the tissues: wandering macrophages (together with neutrophils from the blood) are attracted to the area of infection. Neutrophils are relatively short-lived; following their death they are phagocytized by the RE cells.

**The process of phagocytosis**

The ability of phagocytes to ingest microorganisms depends upon the surface properties of the microorganisms. Certain bacteria resist phagocytosis when they are enclosed in a polysaccharide or polypeptide capsule but can be rendered susceptible if their capsular layer is coated with a protein (e.g., antibody). It has also been observed that encapsulated bacteria, which cannot be phagocytized when suspended in fluid, can be ingested if the phagocytes can trap them against a solid surface, even in the absence of antibody.

A particle undergoing ingestion appears to pass rapidly through the membrane of the phagocyte (Figure 26.4); if the particle is very large,

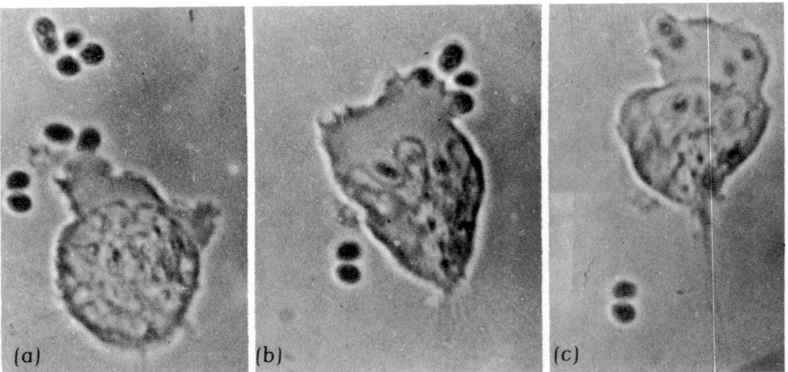


*Figure 26.4 The ingestion of pneumococci by a phagocyte in the presence of opsonizing antibody. (a) Two pneumococci are in contact with a pseudopodium of the phagocyte. (b) The same two pneumococci are inside the phagocyte, one on each side of a lobe of the nucleus; a group of four pneumococci are in the process of being ingested. (c) Six of the eight pneumococci have been engulfed. From W. B. Wood, Jr., M. R. Smith, and B. Watson, "Studies on the Mechanism of Recovery in Pneumococcal Pneumonia: IV. The mechanism of phagocytosis in the absence of antibody." J. Exptl. Med.* **84,** *402 (1946).*

the phagocyte may first extend pseudopodia around it. In fact, the particle does not actually penetrate the membrane of the phagocyte. Instead, an invagination of the membrane occurs, to form a saclike structure which encloses the particle. The sac ultimately pinches off inside the cell, becoming a *vacuole*. The membrane of the vacuole, which now contains the particle, is thus derived directly from the cell membrane of the phagocyte. The membrane has been inverted, however, the inner surface of the vacuole membrane being homologous with the outer surface of the cell membrane (Figure 26.5).

The process by which bacteria, enclosed in phagocytic vacuoles, are killed and digested was brought to light by the discovery that the characteristic cytoplasmic granules of the phagocytes are membrane-bounded sacs, containing hydrolytic enzymes and phagocytin. Phago-

Figure 26.5 *Three stages in the ingestion of a pair of cocci by a phagocyte, showing the formation of the phagocytic vacuole by an inversion of the cell membrane.*

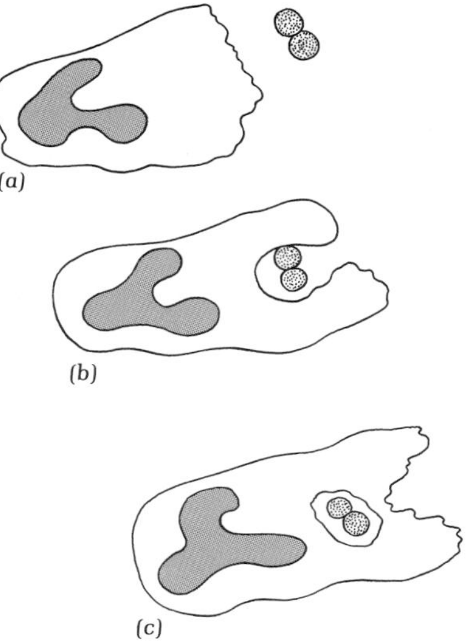


cytosis is followed by *degranulation* of the cell, the granules fusing with the phagocytic vacuoles and discharging their contents into them.

The metabolic changes that accompany the phagocytosis of bacteria and other particles by polymorphonuclear leucocytes have been studied in vitro. By the use of specific inhibitors, it has been shown that energy is required for phagocytosis and that the energy is supplied exclusively by glycolysis. Lactic acid accumulates within the phagocyte, lowering the internal pH and thereby activating a number of enzymes concerned with aerobic respiration. As a result, a marked increase in oxygen uptake is exhibited by phagocytizing cells.

Phagocytosis is not always followed by the destruction of the ingested microorganism. Sometimes, microorganisms survive and even multiply within the phagocyte, eventually killing it. This behavior is characteristic of certain bacteria, notably tubercle bacilli and brucellae. In such cases, the phagocytes may actually serve to disseminate the pathogenic organisms throughout the body, rather than protect the host.

## *Inflammation*

When a tissue of a higher animal is subjected to any of a variety of irritations, it becomes "inflamed." The characteristics of inflammation—reddening, swelling, heat, pain—are familiar. Although the causes of the heat and pain are not well understood, the reddening and swelling are readily explained. In the area of inflammation, the blood capillaries are dilated, so that the flow of blood through the area is increased; hence the reddening. The walls of the capillaries also become more porous, so that soluble proteins escape from the vessels and cause an osmotic movement of fluid into the tissues; hence the swelling.

The fact that inflammation is produced by such widely different irritants as heat, mechanical injury, and microbial infection suggests that the symptoms are caused by a substance released from the damaged cells or activated in the body fluids. Among the many different compounds that have been extracted from cells or serum and shown experimentally to produce inflammation, the most thoroughly studied are *histamine* and *serotonin* (5-hydroxytryptamine) (Figure 26.6). Both histamine and serotonin are present in a loosely held form in blood

Figure 26.6 *The structures of histamine and serotonin. Histamine is the decarboxylation product of histidine; serotonin is the decarboxylation product of 5-hydroxytryptophan.*

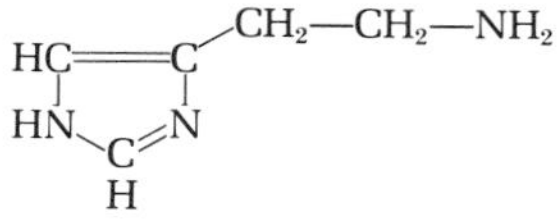

$$\begin{array}{l} \mathrm{HC{=}C{-}CH_2{-}CH_2{-}NH_2} \\ \mathrm{HN{-}CH{=}N} \end{array}$$

*histamine*

$$\begin{array}{l} \mathrm{HO{-}C{=}CH{-}C{-}C{-}CH_2{-}CH_2{-}NH_2} \\ \mathrm{HC{=}CH{-}C{-}NH{-}CH} \end{array}$$

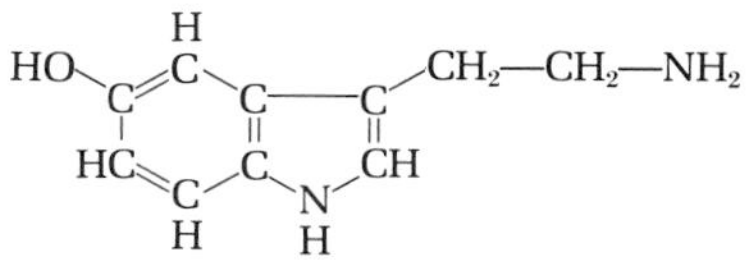

*serotonin*

Figure 26.7 *Chemotaxis of monocytes. The paths of the monocytes have been photographically recorded as light areas. Photograph (a) shows random movement of the monocytes; photograph (b) shows the directed movement of monocytes toward a clump of bacteria at the top. From H. Harris, "Chemotaxis of Monocytes." Brit. J. Exptl. Pathol.* **34,** *278 (1953).*

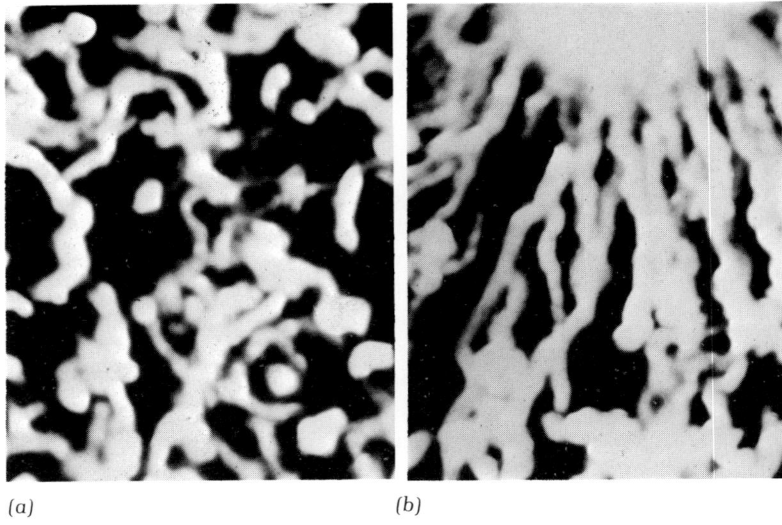

(a) (b)

platelets as well as in the cells of many tissues, and are released in response to a variety of stimuli.

As the inflammatory response develops, a striking change in the behavior of the polymorphonuclear leucocytes takes place. At first, they adhere to the inner walls of the capillaries. Next, they can be seen to

Figure 26.8 *(a to d) Stages in the migration of a small lymphocyte through an endothelial cell of a venule: (a) the lymphocyte passes into the endothelial cell; (b) the lymphocyte lies within the cytoplasm of the endothelial cell; (c) the lymphocyte has passed through the endothelial cell, but is still separated from the tissue of the lymph node by the basement membrane; (d) lymphocyte passes through basement membrane. (e to h) Stages in the migration of a polymorphonuclear leucocyte through the venule wall. The cell penetrates the intercellular junction and remains extracellular at all times. (i) Part of a venule from a normal lymph node: a lymphocyte (l) is completely enclosed by the cytoplasm of an endothelial cell; n, nucleus of the endothelial cell. (j) Part of an inflamed venule. Cell m is a monocyte which is penetrating an intercellular junction of the endothelium (e); n, nucleus of an endothelial cell; pe, perienendothelial sheath. From. V. T. Marchesi and J. L. Gowans, "The Migration of Lymphocytes through the Endothelium of Venules in Lymph Nodes: an Electron Microscope Study." Proc. Roy. Soc.* **B159,** *283 (1964).*

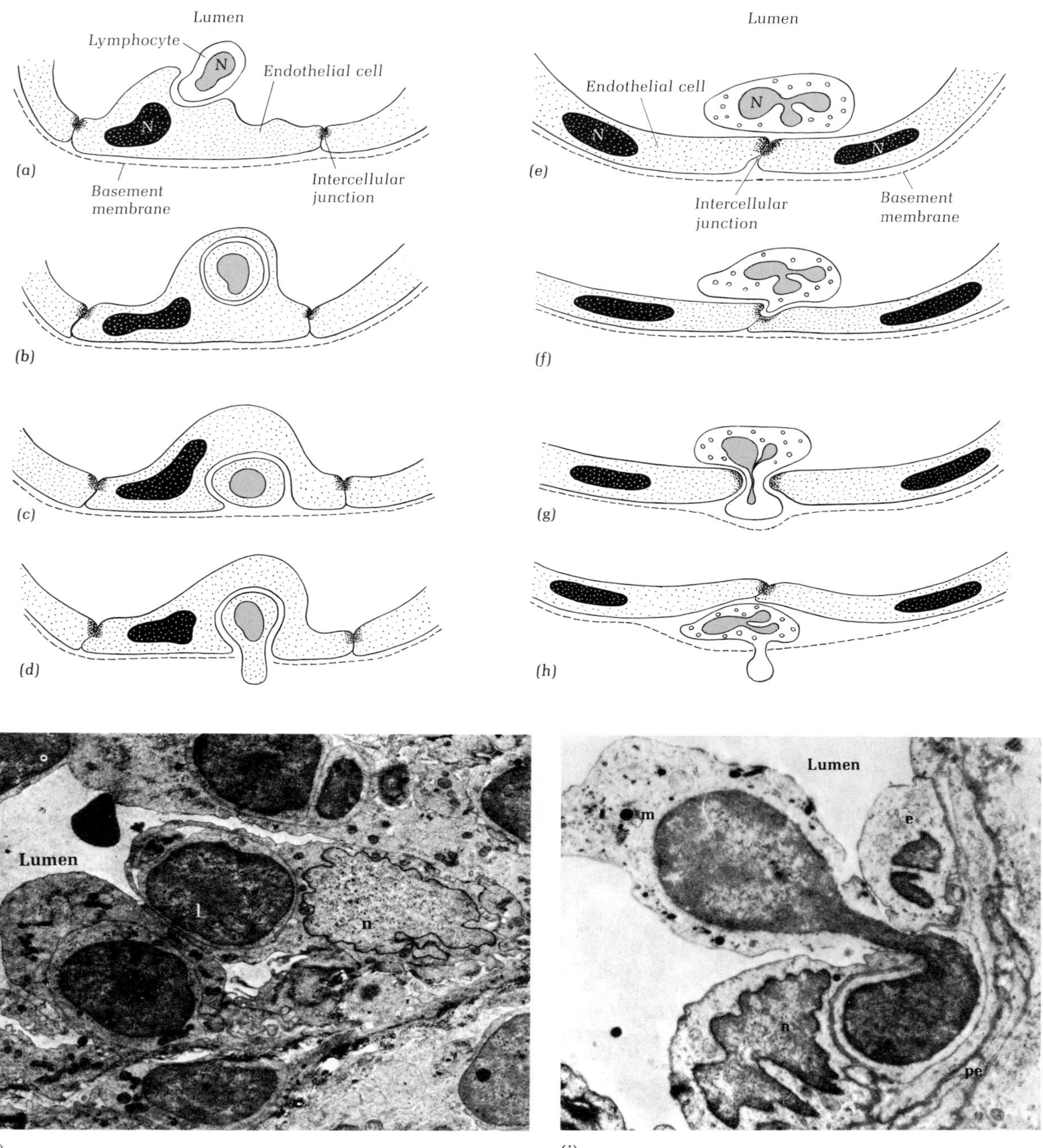
Lumen
Lymphocyte
N
Endothelial cell
N
(a)
Basement
membrane
Intercellular
junction
(b)
(c)
(d)
Lumen
Endothelial cell
N
N
N
(e)
Intercellular
junction
Basement
membrane
(f)
(g)
(h)
Lumen
l
n
(i)
Lumen
m
e
n
pe
(j)

push their way between the cells of the capillary walls and to enter the tissues, a process that can take as little as 2 minutes. If the inflammation has been initiated by a bacterial infection, the polymorphonuclear leucocytes move toward the focus of infection in response to chemotactic substances liberated by the bacteria.

In the later stages of inflammation, the polymorphonuclear leucocytes that have accumulated at the inflammatory site are replaced by monocytes. The reason for this sequential accumulation of the two types of phagocytes is not known; it has been well established, however, that monocytes respond chemotactically to the same substances as do polymorphonuclear leucocytes (Figure 26.7).

Both eosinophils and lymphocytes also accumulate in certain types of inflammatory lesions, particularly those associated with hypersensitivity reactions. Neither of these cell types exhibits any chemotactic movements, and the mechanism by which they accumulate is unknown. (It should be realized that the accumulation of mobile cells can theoretically occur either by chemotaxis or by the trapping of cells that have arrived at the inflammatory site through random movement.) The eosinophils, like the monocytes and polymorphonuclear leucocytes, leave the capillaries by squeezing between the endothelial cells of the capillary wall. The lymphocytes, on the other hand, leave in an entirely different manner; they actually *pass through* the endothelial cells, enclosed in vacuoles (Figure 26.8).

The role of the eosinophils is only now becoming clear. It appears that their granules contain substances that block the action of histamine, and that degranulation occurs when the eosinophils ingest antigen–antibody complexes. These cells thus appear to play a regulatory role in inflammation, damping the action of histamine.

Let us now consider how inflammation may act as a mechanism of defense, even though it is itself a pathological condition and in some instances may even be severe enough to cause death. In the first place, the tissues at the site of infection become richly supplied with leucocytes whose antibacterial activities include both phagocytosis and antibody formation. In the second place, the supply of plasma to the tissues is increased, raising the local concentration of antibacterial serum factors and (in immune animals) of antibodies. In the third place, progressive inflammation leads to the accumulation of dead host cells, from which antimicrobial tissue substances are released. In the center of the necrotic area, oxygen tension is diminished and lactic acid accumulates; these conditions are also inimical to the growth of many types of pathogenic bacteria.

Thus, inflammation brings into play at the site of infection all the mechanisms of constitutive host resistance. The usual result is the destruction of the invading microorganisms, unless the invaders are so virulent that they can overcome this series of defenses. The mechanisms of virulence will be discussed in Chapter 27.

## *Further reading*

**Reviews and original articles**

Cohn, Z., "The Structure and Function of Monocytes and Macrophages." *Adv. Immunol.* **9,** 164 (1968).

Housewright, R. (editor), "The 1959 Fort Detrick Symposium on Nonspecific Resistance to Infection." *Bacteriol. Rev.* **24,** 1–200 (1960).

Skarnes, R., and D. Watson., "Antimicrobial Factors of Normal Tissues and Fluids." *Bacteriol. Rev.* **21,** 273 (1957).

Spector, W., and D. Willoughby, "The Inflammatory Response." *Bacteriol. Rev.* **27,** 117 (1963).

# *Chapter twenty-seven Mechanisms of microbial pathogenicity in vertebrates*

*The term pathogenicity* denotes the ability of a parasite to cause disease. Pathogenicity is a taxonomically significant attribute, being the property of a species; thus, the bacterial species *Corynebacterium diphtheriae* is said to be pathogenic for man. The individual strains of a bacterial species may, however, vary widely in their ability to harm the host species, and this relative pathogenicity is termed *virulence.* Virulence is accordingly an attribute of a strain, not a species; one can talk of a highly virulent, a weakly virulent, or even an avirulent strain of *C. diphtheriae.*

In general, the virulence of a strain of a pathogenic species is determined by two factors: its *invasiveness,* or ability to proliferate in the body of the host, and its *toxigenicity,* or ability to produce chemical substances—*toxins*—that damage the tissues of the host. It is characteristic of bacterial toxins that they are capable of damaging or killing *normal* host cells (i.e., the cells of a host that has not previously been exposed to the infectious agent in question). Certain pathogenic bacteria, however, cause damage to the host by a mechanism that is more indirect and does not come into play unless or until the host has previously experienced specific infection. This mechanism is known as *hypersensitivity,* and involves an immune response by the already sensitive host to a cell component of the parasite which is *nontoxic* for a normal host. We shall discuss further the role of hypersensitivity in disease in Chapter 28.

The role played by invasiveness in damaging the host varies widely. Some pathogens are so toxigenic that an extremely localized infection may result in the production and diffusion through the host of sufficient toxin to cause death. The classical example of such a disease is diphtheria, in which the pathogen, *Corynebacterium diphtheriae*, multiplies in the throat, and produces a diffusible toxin that affects virtually all the tissues of the animal body. At the other extreme are pathogens that must invade and multiply extensively within the body of the host, in order to produce enough toxin to cause damage or to enable the toxin to reach specifically susceptible tissues. The classical example of such a disease is anthrax, in which the pathogen, *Bacillus anthracis*, is present in enormous numbers in the bloodstream in the terminal stages of infection.

In this chapter we shall first discuss the various properties of pathogenic bacteria that contribute to the complex character of invasiveness and then describe some of the specific toxic substances responsible for bacterial toxigenicity, as well as the ways in which they cause damage to the host.

## *Invasion*

In Chapter 26 it was pointed out that avirulent microorganisms are kept in check by surface barriers, by antimicrobial substances, and by phagocytosis. Many virulent organisms are able to counter these host defenses. To do so they must produce specific chemical substances that are not present in the noninvasive forms. These substances, which promote the invasive spread of pathogenic microorganisms, are called *aggressins*.

The experimental demonstration of an aggressin requires a specific assay of aggressin activity. This can be measured in terms of the ability of a preparation that contains an aggressin to augment the fraction of experimental animals killed upon injection of a fixed number of virulent cells. The number of virulent cells must be so chosen that in the absence of the aggressin, few or no members of a control group of animals are killed. Alternatively, the suspected material can be assayed for its ability to protect avirulent microbial strains from antimicrobial substances or from phagocytosis in vitro.

### The establishment of pathogens

In nature, the pathogen normally reaches its host in small numbers. Contact is usually made at the surface tissues of the host; in mammals, these are the skin and the mucous membranes. In such cases, an important first stage of the invasion process is the establishment of the pathogen on the surfaces, where it forms a primary focus of infection.

Experiments have shown that the mammalian host has special defense mechanisms for dealing with invading microorganisms during the first few hours following their lodgment; probably antimicrobial substances of the blood, such as those described in Chapter 26, play the major role in this defense. Many strains of a pathogenic microorganism may be completely wiped out by this process; for example, if 1 million bacterial cells are injected into the skin, fewer than 10 may survive at the end of 5 hours. More virulent strains, however, resist this host action and survive in large numbers. Such resistance contributes greatly to the overall invasiveness of the pathogen.

The mechanisms by which virulent organisms resist antimicrobial substances of the host are in general unknown. In a few instances, however, specific aggressins have been implicated. Both the capsular material of *Bacillus anthracis* (poly-D-glutamic acid) and a cell wall complex from *Brucella abortus* will protect avirulent strains of homologous species from the bactericidal action of bovine serum.

**Aggressins active against phagocytosis**

Aggressins have been found to play a role in the resistance of invasive organisms to phagocytosis. Different aggressins act at each stage of the process: during the attraction of phagocytes, during the ingestion, and during the intracellular destruction of microorganisms.

Both polymorphonuclear leucocytes and monocytes accumulate at the site of infection, as a result of positive chemotaxis; unidentified aggressins liberated by tubercle bacilli inhibit such leucocyte migration.

The *ingestion* of bacteria by phagocytes is inhibited by many, but by no means all, types of bacterial capsules; these components of the bacterial cell thus often serve as aggressins.

Some bacteria, notably the tubercle bacilli and the brucellae, owe their invasiveness not to resistance to phagocytosis but to resistance to intracellular destruction by the phagocytes. In the case of *Brucella abortus*, the bacterial cell wall contains a substance that will inhibit the intracellular destruction of avirulent cells. This material thus constitutes a third class of aggressins. Some bacteria owe their virulence not only to their resistance to phagocytosis but also to their excretion of substances that kill phagocytes. *Leucocidins*, as they are called, are produced by virulent staphylococci and streptococci. These bacteria are pyogenic, since the leucocytes they attract and kill accumulate at the site of infection. The mass of autolyzed leucocytes forms the pus which characterizes pyogenic infections.

It is important to recall that different types of phagocytes contain different antibacterial substances; macrophages, for example, lack phagocytin, which is present in neutrophils. Thus, to avoid phagocytosis, a pathogen must produce a series of different aggressins capable of countering all types of phagocytes.

## Toxins

**The nature of toxins**

The search for bacterial toxins began shortly after the discovery of the role of bacteria as etiological agents of human disease. By 1890, the toxins of two important human pathogens, *Corynebacterium diphtheriae* and *Clostridium tetani,* had been discovered. In each case, the discovery was made in the same manner: the bacterium was grown in vitro in a culture medium, and a sterile filtrate prepared from the fully grown culture was observed to cause death when injected into experimental animals. Furthermore, autopsies revealed that these animals showed the characteristic lesions associated with the specific natural infection. The toxic substances proved to be heat-labile and are now known to be proteins. Because they were present in the medium, not associated with the bacterial cells, they were termed *exotoxins.*

A number of other pathogenic bacteria have been subsequently shown by comparable methods to produce exotoxins that have specific effects, clearly significant in the causation of the specific disease. However, filtrates prepared from cultures of many important pathogens failed to show toxicity. This led to the examination of the bacterial cells themselves, killed by heat, as possible toxic agents. Such experiments showed that the cells of nearly *all Gram-negative pathogenic bacteria are intrinsically toxic;* furthermore, heat-killed cells of many *nonpathogenic* Gram-negative bacteria show similar toxic effects. The heat-stable toxins, associated with the cells of Gram-negative bacteria, came to be known as *endotoxins.* As we shall describe later, the endotoxins are relatively nonspecific, all producing much the same clinical and pathological symptoms when injected into experimental animals. Many years of intensive study were required to reveal their nature and cellular origin; it is now known that *endotoxins are lipopolysaccharide–protein complexes, derived from the outer layers of the cell walls characteristic of Gram-negative bacteria.*

The names "exotoxin" and "endotoxin" to designate these two classes of toxic substances are misleading, since there is now good evidence to show that all these toxic substances are associated with the bacterial cells during growth and are liberated only after death and lysis of the bacteria. They can be more correctly differentiated in terms of their chemical nature: "exotoxins" are proteins, whereas "endotoxins" are complexes containing protein, lipid, and polysaccharide. Nevertheless, these names are now so firmly entrenched that they are not likely to be abandoned, and we shall use them in the subsequent discussion.

The examination of the cells and culture filtrates of pathogenic bacteria grown in vitro led to the recognition of a number of microbial products that damage the host. There remained, however, many important bacterial pathogens, including the causative agents of anthrax and plague, for which this approach failed to reveal any significant toxic product. The environmental conditions in a laboratory culture are always different from those provided in the body of an infected animal, and the recognition of this obvious but previously overlooked fact has led in recent years to *the search for bacterial toxins that are produced by pathogens in the animal body*. Such work, conducted mainly by H. Smith and his collaborators, was centered originally on anthrax and led to the discovery of the specific exotoxin produced by *Bacillus anthracis*. More recently, the toxin of *Yersinia pestis*, the agent of plague, has been demonstrated for the first time by comparable methods.

The toxins of both the anthrax and plague organisms were found to be complexes of two or more substances, each of which was nontoxic by itself but which together acted synergistically to produce a toxic effect. Such knowledge permitted the refinement of the assay systems for the toxins to such an extent that—in each case—it became possible to establish the production of toxin by cultures of bacteria in vitro.

The failure to discover the toxin of a virulent organism may often result from the lack of a suitable assay system, particularly in the case of organisms specifically pathogenic for man. For example, the toxin of *Vibrio cholera*, the agent of cholera, escaped detection for many years; neither culture filtrates nor cell extracts exhibited toxicity when injected into experimental animals. The toxin was discovered, however, when filtrates of cultures of *V. cholerae* were injected into ligated intestinal loops of rabbits; such filtrates produced the gross fluid loss and mucosal damage characteristic of the natural disease. Using this assay system, it was possible to purify and characterize the cholera toxin as a heat-labile protein. The same assay system has permitted the detection and isolation of the toxins of virulent strains of *coli*.

**The ecological significance of toxins**

For an understanding of pathogenesis, it is not sufficient to isolate a toxic substance from a pathogenic bacterium. If such a substance is to be implicated as a determinant of virulence, it must be demonstrated to produce one or more of the specific symptoms of the disease. Furthermore, the site of action and the effective concentration must be ones that could plausibly obtain in the course of a natural infection. These criteria of ecological significance are extremely difficult to apply and have been fully satisfied in relatively few cases. Two other criteria are often applied: a correlation between toxin production and virulence in different strains of a given pathogenic species, and the ability of antiserum directed against the toxin to protect animals from infection.

All the abovementioned criteria have been satisfied for the toxins of botulism, tetanus, and diphtheria. These exotoxins cause all the symptoms of the respective diseases and are produced only by virulent strains. Antisera against these toxins fully protect individuals from natural disease. Many other microbial products identified as toxins are, without doubt, also ecologically significant, but in each case one or more of the criteria mentioned above is lacking.

## Bacterial exotoxins

Many virulent bacteria liberate toxic substances into the medium when grown in vitro. Some of these appear to be ecologically significant toxins; they are listed in Table 27.1. Others do not seem to be related

*Table 27.1 Exotoxins*

| | |
|---|---|
| *I. Bacteria from which have been isolated exotoxins accounting for all the pathological effects of infection* | |
| *Corynebacterium diphtheriae* | |
| *Clostridium tetani* | |
| *Clostridium botulinum* | |
| *II. Bacteria from which have been isolated exotoxins accounting for at least some of the pathological effects of infection* | |
| *Clostridium perfringens (welchii)* | |
| *Clostridium novyi* | |
| *Bacillus anthracis* | |
| *Yersinia pestis (guinea pig toxin)* | |
| *Vibrio cholerae* | |
| *Pseudomonas aeruginosa* | |
| *III. Bacteria from which have been isolated exotoxins of unknown significance in infection* | |
| *Clostridium perfringens*<br>*Clostridium septicum*<br>*Clostridium novyi* | *many necrotizing and hemolytic toxins* |
| *Staphylococcus aureus* | *many necrotizing and hemolytic toxins; leucocidin; enterotoxin* |
| *Streptococcus pyogenes* | *Streptolysin S, Streptolysin O (hemolysins); erythrogenic toxin; DPNase* |
| *Yersinia pestis* | *murine toxin* |
| *Shigella dysenteriae* | *neurotoxin* |

to the disease produced by the pathogens in vivo. *Shigella dysenteriae*, for example, produces a substance (neurotoxin) that can be shown experimentally to damage nerve cells, but such damage is never observed in natural infections with *S. dysenteriae*. *Yersinia pestis* produces a substance (murine toxin) which is toxic only to mice, whereas *Y. pestis* is pathogenic for both mice and guinea pigs. Thus, the neurotoxin of *S. dysenteriae* and the murine toxin of *Y. pestis* are of uncertain ecological significance.

**The production of exotoxins**

For many years it was assumed that exotoxins are excreted by growing cells; however, more careful studies have now shown that in a number of cases exotoxin liberation accompanies the death and lysis of the bacteria. This is true for the toxins of *C. botulinum* and *C. diphtheriae*.

In the case of *C. diphtheriae*, only cells lysogenic for a particular phage can produce exotoxin. Nontoxigenic cells can be converted to toxigenicity by infecting them with this specific bacteriophage. It will be recalled that prophages are induced by certain environmental conditions, as a result of which they proceed to multiply and lyse the cells. It has been found that diphtheria toxin is produced only during the development and liberation of phage. This finding partially explains the earlier observation that diphtheria toxin is produced only in media with a low iron content, because the prophage is induced at low iron concentrations. Furthermore, high concentrations of iron block the liberation of toxin by cells in which the prophage has been induced by ultraviolet light. Artificial lysis of the cells never leads to toxin liberation.

The role of intracellular phage maturation in toxin formation can be interpreted in two possible ways: either phage maturation causes the degradation of a preexisting cell component to toxin, or the toxin is specifically synthesized under the control of a phage gene.

It is possible that the production of other bacterial exotoxins is also determined by lysogeny with a specific phage. So far, only one other example of this phenomenon has come to light: the production of the erythrogenic toxin by *Streptococcus pyogenes*, the causal agent of scarlet fever.

**The modes of action of exotoxins**

Toxins exert their effects either by destroying specific cellular components or by inhibiting specific cellular functions; in a few cases, it has been possible to demonstrate such effects using purified toxins and preparations of sensitive cells. It is difficult, however, to be certain that the effect observed in vitro occurs during the natural infection, or

—if it does occur—that it represents the *primary* effect of the toxin. A number of exotoxins, for example, are strongly hemolytic, but the lysis of red blood cells is not of major significance in any of the diseases in which these toxins have been implicated.

Some exotoxins have been shown to act as *hydrolytic enzymes*, degrading essential components of host cells or tissues. The $\alpha$ toxin of *Clostridium perfringens*, for example, is a lecithinase. Lecithin is an important lipid constituent of cell membranes and mitochondrial membranes; its hydrolysis by the $\alpha$ toxin results in the destruction of the membrane of many types of cells and could be a primary cause of the tissue damage that occurs in gas gangrene. *Clostridium perfringens* produces a number of other exotoxins, including the *k* toxin, which is a collagenase.

The diphtheria toxin, produced by *Corynebacterium diphtheriae*, is an example of a toxin which acts by inhibiting a normal cell function rather than by destroying a cell component. When diphtheria toxin is added to mammalian cells in culture, the syntheses of DNA, RNA, and protein are almost instantly inhibited. The *primary* effect of the toxin, however, appears to be the inhibition of protein synthesis, since diphtheria toxin also inhibits the incorporation of amino acids into polypeptides in cell-free systems. Using such systems, it has been found that the primary site of toxin action is the enzyme, transferase II, which catalyses one step in the transfer of the growing polypeptide chain from one tRNA molecule to another on the surface of the ribosome (see p. 273). The toxin specifically catalyses the transfer of the ADP–ribose moiety of NAD to transferase II, which is consequently inactivated.

The work on the diphtheria toxin represents the most refined analysis of the site of action of a toxin that has been achieved to the present time. No matter how refined the analysis, however, the evidence for the *primary* site of action of an exotoxin (or of any other toxic substance) on a susceptible cell is always circumstantial.

In some cases, the pathology of the natural infection shows that the primary target of a bacterial toxin is a *specific type of cell or tissue*. For example, the exotoxins produced by *Clostridium botulinum*, the agent of botulism, and by *C. tetani*, the agent of tetanus, are both *neurotoxins*, which specifically interfere with the mechanisms by which the nerves transmit stimuli to muscles. In botulism, the toxin binds to cells of the central nervous system, blocking the release of acetylcholine from the ends of cholinergic motor nerves. In tetanus, the toxin is transported to the spinal cord via the peripheral nerves and exerts its primary effect on the anterior horn cells. In neither case, however, is the molecular basis of toxin action known.

## *Bacterial endotoxins*

**The chemistry of endotoxins**

Endotoxins are lipopolysaccharide–protein complexes derived from the cell walls of Gram-negative bacteria. They are antigenically active and are identical with the somatic, or O, antigens of the whole cell. When a culture of Gram-negative bacteria is allowed to age and autolyze, the endotoxins are liberated in soluble form. They can also be extracted from cell suspensions by appropriate procedures.

Endotoxins have been isolated from all pathogenic Gram-negative bacteria; the best known are those of the coliform bacteria belonging to the genera *Salmonella*, *Shigella*, and *Escherichia*. The endotoxins of these organisms exhibit two distinct activities: pyrogenicity (fever production) and toxicity.

An endotoxin complex can be separated into a lipopolysaccharide fraction, which is pyrogenic and toxic, and a protein fraction. The protein fraction is neither pyrogenic nor toxic but confers antigenicity on the whole complex. The role of proteins in antigenicity will be discussed in more detail in Chapter 28.

The isolated lipopolysaccharides are highly active: $10^{-6}$ g will produce fever in a horse weighing 700 kg.

**The modes of action of endotoxins**

The purified endotoxins of virulent, as well as of avirulent, enteric bacteria are capable of causing many of the symptoms of disease when injected into animals. They are inflammatory agents, causing increased capillary permeability and cellular injury. The injured host cells release additional inflammatory agents, which probably contribute significantly to the pathological picture.

The body temperature of the mammal is regulated by mechanisms that are controlled by certain centers in the brain; a number of different substances can, upon introduction into the bloodstream, interfere with the regulatory mechanisms and cause fever. As fever accompanies many different types of inflammation, a search was made for pyrogenic products arising from injured tissues. Positive results were readily obtained, but eventually these were traced to the presence in all such active preparations of contaminating bacterial endotoxins.

The analysis of pyrogenicity then focused on the endotoxins, which were found to have extremely high activities. Endotoxins, however, do not act directly on the thermoregulatory brain centers. Instead, they cause the release of an *endogenous pyrogen* from polymorphonuclear leucocytes. This substance, the chemistry of which is still unknown, is directly responsible for the fever.

The inflammatory and pyrogenic effects of the endotoxins are *non-specific.* Although they undoubtedly contribute to the general pathology of the infection, they are not responsible for the specific symptoms of disease caused by the Gram-negative pathogens. Cholera, for example, is characterized by localized damage to the mucosa of the lower bowel, resulting in severe fluid loss and consequent shock. The toxin responsible for this clinical picture has only recently been identified; it is not an endotoxin. It is also important to emphasize that fever and toxicity accompany severe infections with Gram-positive organisms that do not contain endotoxins.

## *Some important bacterial diseases of man*

The bacterial parasites of man depend for their survival on their transmission from one individual to another. They have thus evolved, in each case, a characteristic *portal of exit, mode of transmission,* and *portal of entry.* In the following sections we shall group the bacterial diseases according to their modes of transmission, because such a classification tends to link ecologically related pathogens.

### **Diseases transmitted by fecal contamination of food and drink**

The intestinal tract is the natural habitat of many kinds of bacteria, most of them harmless under ordinary conditions. A number of intestinal inhabitants are serious pathogens, however; these include the causative agents of typhoid and paratyphoid fevers, dysentery, cholera, and the *Salmonella* infections incorrectly referred to as "bacterial food poisoning." Some do their damage locally, whereas others spread from the intestinal tissues to other parts of the body. All, however, have two important attributes in common: they leave the host in excreted fecal matter and must enter the next host via the mouth to reach the intestines once again.

This chain of events forces us to accept the unpleasant fact that the enteric diseases, as they are called, are acquired principally by swallowing food or drink contaminated with feces. Before the introduction of modern sanitation, water supplies were constantly subject to direct contamination from privies and faulty sewers. Today, however, contamination by these means has become rare, and other modes of transmission have become relatively more important. The common housefly is an effective agent of transmission because it visits both feces and food indiscriminately. Furthermore, there are far more healthy carriers of enteric pathogens than frank clinical cases, so that anyone who handles

Table 27.2 *Some human diseases transmitted by fecal contamination*

| Disease | Etiologic agent | Pathogenesis |
|---|---|---|
| *Typhoid fever* | *Salmonella typhi (S. typhosa), a Gram-negative, peritrichously flagellated rod. Facultatively anaerobic; mixed-acid fermentation. (See Chapter 18.)* | *The organisms first multiply in the gastrointestinal tract. Invasion of the bloodstream via the intestinal lymphatics and thoracic duct leads to dissemination throughout the body. Heavy growth occurs in the biliary tract, from which the bowel is infected. Foci in the lungs, bone marrow, and spleen.* |
| *Enteric fevers, gastroenteritis, Salmonella septicemias ("bacterial food poisoning")* | *Salmonella typhimurium, S. schottmülleri, S. choleraesuis; Gram-negative, peritrichously flagellated rods. Facultatively anaerobic; mixed-acid fermentation. (See Chapter 18.)* | *Enteric fevers are diseases characterized by wide dissemination of the organism throughout the body. Enteric fever caused by salmonellae other than S. typhi are milder than typhoid fever and are called "paratyphoid fevers"; S. schottmülleri is the most common cause of enteric fevers in the U.S. Gastroenteritis is a salmonellosis in which the organism remains localized in the gastrointestinal tract; S. typhimurium is the most common cause of salmonella gastroenteritis in the U.S. Salmonella septicemias are most commonly caused by S. choleraesuis.* |
| *Cholera* | *Vibrio cholerae, a Gram-negative, polarly flagellated, curved rod. Facultatively anaerobic; mixed-acid fermentation. (See Chapter 18.)* | *The organism multiplies extensively in the small intestine. An exotoxin acts on the mucosal cells; water and electrolyte loss leads to shock.* |
| *Bacterial dysentery* | *Shigella dysenteriae, a Gram-negative, immotile rod. Facultatively anaerobic; mixed-acid fermentation. (See Chapter 18.)* | *Lesions are formed in the terminal ileum and colon. Abdominal cramps, diarrhea, and fever are produced.* |

food is a potential source of contamination. Hence, only the strictest attention to personal hygiene on the part of food handlers can prevent the spread of enteric disease.

Many animals, including cattle and fowl, may be naturally infected with members of the genus *Salmonella*, which cause the enteric infections known as bacterial food poisoning. It is, consequently, possible to become infected by eating contaminated meat or eggs.

Some important bacterial diseases transmitted by the fecal contamination of food and drink are described in Table 27.2.

**Diseases transmitted by droplet infection**

The transmission of disease by the respiratory route is called *droplet infection* because in such cases the pathogenic organisms are carried from person to person in microscopic droplets of saliva. In countries that practice modern methods of sanitation, droplet infection is by far the most important route by which disease is spread. Every time a person sneezes, coughs, or even speaks loudly, he exhales a cloud of tiny droplets of saliva. Each droplet contains a little dissolved protein, as well as varying numbers of the microorganisms that inhabit the mouth and respiratory tract; the droplets quickly evaporate, leaving in the air great numbers of minute flakes of protein that bear living bacteria. A person suffering from a respiratory infection will certainly contaminate every other person in whose presence he coughs, sneezes, or speaks. The only way to prevent such spread would be to require that all individuals wear masks equipped with filters. Such an extreme measure has rarely been found practicable or enforceable. The result is that in a crowded city, a highly infective respiratory pathogen such as the influenza virus can spread from one person to several million in as short a time as 6 or 8 weeks.

Many highly important diseases involve infection of the respiratory tract and are spread by droplet exhalation. Some important bacterial respiratory diseases are described in Table 27.3.

**Diseases transmitted by direct contact**

There is a small number of pathogens for which the portal of entry is the skin or the mucous membranes and which depend on direct contact for transmission. This group includes the causative agents of the venereal diseases, syphilis and gonorrhea. The responsible organism in each case cannot survive for long outside the host and requires direct contact of mucous membranes for transmission. Sexual intercourse is thus one of the chief means of spreading these diseases, although syphilis can also be acquired before birth, and gonorrhea during birth, from an infected mother.

In the tropics, there are several diseases caused by organisms closely related to the agent of syphilis that are normally not transmitted by sexual intercourse. They all start as skin infections and require direct contact for transmission. Yaws is an example of this group.

Three other nonvenereal diseases that are transmissible by direct contact are anthrax, tularemia, and brucellosis. All are diseases of animals that can be transmitted to man. Brucellosis, a disease of goats, cattle, and swine, constitutes a severe occupational hazard for animal handlers, including veterinarians, meat packers, and dairy workers. Tularemia, a disease of wild rodents, is often contracted by hunters or butchers who handle the carcasses of wild game. The principal bacterial diseases transmitted by direct contact are described in Table 27.4.

**Diseases transmitted by animal bite**

An interesting solution to the problem of transmission has been the evolutionary adaptation by certain pathogens to existence in two or more alternative hosts. The plague bacillus, for example, can multiply in rats, fleas, and man; it is thus guaranteed a continuous supply of fresh hosts, for the flea carries it from rodent to rodent or from rodent

*Table 27.3 Some human diseases transmitted by exhalation droplets*

| Disease | Etiologic agent | Pathogenesis |
|---|---|---|
| *Diphtheria* | *Corynebacterium diphtheriae, a Gram-positive, immotile rod; tends to be club-shaped. Post-fission movements result in characteristic "palisade" arrangements. Facultative anaerobe; propionic acid fermentation. (See Chapter 21.)* | *The organism establishes itself in the throat and remains localized in the upper respiratory tract. An exotoxin is produced which is disseminated by the bloodstream to all parts of the body. Local inflammation causes an exudate that forms a "diphtheritic pseudomembrane," blocking the air passages.* |
| *Tuberculosis* | *Mycobacterium tuberculosis, a pleomorphic, immotile, acid-fast rod. Obligately aerobic. (See Chapter 21.)* | *The bacteria multiply both intracellularly and extracellularly within lesions of the lungs, called tubercles. An enlarging tubercle may discharge into a bronchus, promoting spread of the disease to other parts of the lung and (via exhalation droplets) to other individuals. Less frequently, organisms are disseminated via the bloodstream to set up secondary (metastatic) lesions in other organs. The toxins of M. tuberculosis have not been identified; at least some of the host damage results from hypersensitivity reactions of the delayed type.* |
| *Plague* | *Yersinia pestis, a Gram-negative, immotile, small rod. Facultatively anaerobic; mixed-acid fermentation. (See Chapter 18.)* | *A disease of domestic and wild rodents, plague is transmitted to man by flea bite. The disease is called bubonic plague when it is characterized by enlarged, infected lymph nodes ("buboes"). In severe cases, the organism is disseminated to other organs; when the lungs become infected, the disease becomes directly transmissible from man to man by droplet infection. This form of the disease is called pneumonic plague and is highly contagious.* |
| *Meningococcal meningitis* | *Neisseria meningitidis (meningococcus), a Gram-negative, immotile coccus forming pairs of cells. (See Chapter 19.)* | *Meningococci are carried harmlessly in the nasopharynx by 25 percent or more of the population. For unknown reasons, it occasionally invades the bloodstream and then localizes in the meninges (the membranes surrounding the spinal cord).* |
| *Streptococcal infections* | *Streptococcus pyogenes, a β-hemolytic, Gram-positive coccus growing in chains. Lactic acid homofermentation. (See Chapter 21.)* | *The organism develops first in the pharynx. Strains harboring a particular bacteriophage form erythrogenic toxin: if the individual is sensitive to the toxin a skin rash appears and the disease is called scarlet fever. Further spread of the organism may lead to mastoiditis, peritonitis, puerperal sepsis, cellulitis of the skin, or erysipelas. Rheumatic fever is a common sequel to recurrent streptococcal pharyngitis and is correlated with high titers of antibody to streptococcal antigens. No organisms can be isolated from the diseased heart tissue, which appears to be damaged by an immunological reaction.* |
| *Pneumococcal pneumonia* | *Streptococcus pneumoniae (pneumococcus), a Gram-positive coccus typically forming diplococci. Lactic acid homofermentation. See Chapter 21.)* | *Between 40 and 70 percent of the adult population carry pneumococci in their throats. The organisms reach the lungs when normal barriers are malfunctioning (e.g., during viral respiratory infections).* |
| *Other respiratory infections* | *Hemophilus influenzae; Bordetella pertussis. Small, Gram-negative, nonmotile rods. Hemophilus species require hematin and nicotinamide nucleoside for growth.* | *H. influenzae causes respiratory infections in children; it is also the most common cause of bacterial meningitis in children. B. pertussis causes whooping cough; it rarely penetrates the mucosa of the respiratory tract and does not spread to other parts of the body.* |

*Table 27.4 Some human diseases transmitted by direct contact*

| Disease | Etiologic agent | Pathogenesis |
|---|---|---|
| *Anthrax* | *Bacillus anthracis, a Gram-positive, spore-forming, immotile rod. Obligately aerobic. (See Chapter 20.)* | *Anthrax is a natural disease of domestic and wild animals, including mammals, birds, and reptiles. Man acquires the disease primarily by direct contact, the organism gaining entry through a minor abrasion of the skin. A "malignant pustule" forms at the site of infection and may be followed by a fatal septicemia.* |
| *Tularemia* | *Pasteurella tularensis, a short, immotile, Gram-negative rod.* | *Tularemia (discovered in Tulare County, California) is a natural disease of wild rodents. Man usually acquires the disease through the handling of infected carcasses and skins, although it can also be transmitted by arthropod bite (ticks, deer flies). The organism spreads throughout the body via the lymphatics and blood stream; lesions develop in the lungs, liver, spleen, and brain. Growth is primarily intracellular.* |
| *Brucellosis* | *Brucella melitensis, B. abortus, B. suis: small, Gram-negative, immotile rods.* | *All three species are capable of infecting a wide range of mammals, although each has a preferred host: B. melitensis, goats; B. abortus, cattle; B. suis, swine. In cattle, localization of the organism in the pregnant uterus (which often causes abortion) results from a specific growth stimulation by erythritol, which is present only in the vulnerable host tissues. In man, the organism is widely disseminated in the body and multiplies primarily within host phagocytic cells. The disease may undergo periodic remissions, in which case it is called undulant fever.* |
| *Gonorrhea* | *Neisseria gonorrhoeae (gonococcus), a Gram-negative coccus, forming pairs of cells. (See Chapter 19.)* | *The organism is transmitted by sexual contact; newborn infants may acquire serious eye infections from passage through an infected birth canal. Following sexual contact, the organism penetrates the mucous membranes of the genitourinary tract; infection usually is restricted to the reproductive organs, although septicemia may occur.* |
| *Syphilis* | *Treponema pallidum, a spirochete. (See Chapter 5.)* | *The organism is transmitted by sexual contact; it may also be transmitted to the fetus during pregnancy. Following sexual contact, the organism penetrates the mucous membranes of the sexual organ, forming a local primary lesion, or "chancre." Secondary lesions develop several weeks later in the eyes, bones, joints, or central nervous system. If untreated, the disease may progress several years later with the formation of tertiary lesions of the heart valves, central nervous system, eyes, bones, or skin. The toxins responsible for the disease are unknown; delayed hypersensitivity is believed to account for part or all of the damage occurring in the tertiary stage.* |

to man, and it never has to survive in environments unsuitable for growth.

The agents of plague and tularemia are the only true bacteria in this category, but there are many viral, rickettsial, protozoan, and spirochetal diseases transmitted by animal bite. Epidemics of malaria, yellow fever, rabies, typhus, and plague spread in this fashion have radically altered the course of human history. The very property of having

Table 27.5 *Some human diseases transmitted by animal bite*

| Ecology, with respect to man | Disease | Type of micro-organism | Vector | Reservoir |
|---|---|---|---|---|
| *Man serves as accidental host, not as a reservoir* | *Plague* | *Bacterium* | *Flea* | *Rat, other rodents* |
| | *Tularemia* | *Bacterium* | *Tick* | *Wild rodents, tick*[a] |
| | *Rabies* | *Virus* | *Dog, jackal, bat, etc.* | *Same as vector* |
| | *Endemic typhus*[b] | *Rickettsia* | *Flea* | *Rat* |
| | *Rocky Mountain spotted fever* | *Rickettsia* | *Tick* | *Wild rodents* |
| *Man is one of two or more reservoirs* | *African sleeping sickness* | *Protozoon* | *Tsetse fly* | *Man, wild mammals* |
| | *Yellow fever* | *Virus* | *Mosquito* | *Man, monkeys* |
| | *Relapsing fever* | *Spirochete* | *Louse, tick* | *Man, tick,*[a] *rodents (?)* |
| *Man is sole reservoir* | *Malaria* | *Protozoon* | *Mosquito* | *Man* |
| | *Epidemic typhus*[b] | *Rickettsia* | *Louse* | *Man* |
| | *Dengue fever* | *Virus* | *Mosquito* | *Man* |

[a]Ticks act both as a vector and as a reservoir, because microorganisms multiply in the body of the tick and are transmitted through ovary and egg from one tick generation to the next.
[b]The two types of typhus fever are caused by closely related strains of rickettsiae.

alternative hosts, which is such an advantage to parasites, has also led to their eventual control by man. By eliminating the *vector* (the species that transmits the pathogen), or the *reservoir of infection* (the species from which the vector derives the infection), man has been able to eradicate such diseases from large areas.

The chief diseases transmitted by animal bite, together with their vectors and reservoirs, are listed in Table 27.5.

**Wound infections**

Whenever unsterilized foreign material penetrates a wound, microorganisms will certainly be introduced. If conditions in the wound are suitable for the growth of one or more contaminating microbes, an infection results that may eventually spread through the tissues or circulatory system.

Introduction into wounds cannot be considered a "natural" route of transmission, being too irregular and infrequent to ensure perpetuation of a parasitic species. Most often infected wounds are found to harbor ordinary soil-dwelling saprophytic bacteria, such as the clostridia. The clostridia are obligate anaerobes that do not grow in healthy tissues. Deep wounds, however, form an ideal environment, since dead (necrotic) tissues are present, air is excluded, and tissue oxygenation is reduced as the result of impaired circulation. Clostridia are spore-formers and are so ubiquitous in nature that any deep wound into which clothing or soil is introduced has a high probability of being contaminated with one or another species of *Clostridium*. Many of these organisms produce potent exotoxins that kill the surrounding host tissues. One species, *C. tetani*, produces a toxin that affects the nerves and causes muscle spasms; infection, if not treated, is almost invariably fatal. This disease is called tetanus, or "lockjaw." Other clostridia cause severe local damage (gangrene) at the site of infection.

Although the clostridia are the most dangerous wound pathogens, many other bacteria may become established in wounds. Common wound contaminants include staphylococci, streptococci, enterobacteria, and pseudomonads. *Pseudomonas aeruginosa* is a common wound contaminant that liberates a blue pigment and causes the formation of blue pus.

Before Lister introduced antiseptic surgery, the surgeon's knife was itself an outstanding cause of gangrene and other infections. Modern surgical practices, combined with the application of chemotherapeutic

*Table 27.6 Some common wound infections*

| Disease | Etiologic agent | Pathogenesis |
|---|---|---|
| *Tetanus* | *Clostridium tetani, a Gram-positive, spore-forming, peritrichously flagellated rod. Obligately anaerobic. (See Chapter 20.)* | *C. tetani is a common soil organism; it is also of common occurrence in the feces of animals (but not of humans). It produces disease when accidentally introduced into wounds. Spore germination and growth require anaerobic conditions, which obtain when the wound is necrotic and vascular damage is severe. A neurotoxin is produced which is transported via the bloodstream and the peripheral nerves to the spinal cord, where it acts principally on the anterior horn cells.* |
| *Gas gangrene* | *Clostridium perfringens, C. novyi, C. septicum: Gram-positive, spore-forming, peritrichously flagellated rods. Obligately anaerobic. (See Chapter 20.)* | *Development of the organism in anaerobic wounds is accompanied by accumulation of hydrogen gas produced by fermentation. A variety of soluble toxins are produced, including the α toxin of C. perfringens, a lecithinase.* |
| *Leptospirosis* | *Leptospira icterohaemorrhagiae, L. canicola, L. pomona: spirochetes. (See Chapter 5.)* | *The leptospirae cause mild, chronic infections of rodents and domestic animals, which shed them continuously in the urine. Man acquires the disease through contact with urine-contaminated water; the portal of entry is the skin. A disseminated infection results, during which the organisms can be cultured from the blood.* |

Table 27.7 *Some representative rickettsial diseases of man*

| Disease | Reservoirs, vectors, and transmission to man | Remarks |
|---|---|---|
| *Epidemic typhus* | *The rickettsia is carried from man to man by the body louse.* | *High mortality rates in man.* |
| *Endemic typhus* | *The rickettsia is normally a parasite of rats and fleas. It is maintained in nature by rat–flea–rat chains of transmission. Mild epidemic, involving man–louse–man chain of transmission, may follow bite by rat–flea.* | *Milder disease than epidemic typhus. Epidemic and endemic typhus are caused by different, but closely related, rickettsiae.* |
| *Scrub typhus (tsutsugamushi fever)* | *The true reservoir is the mite; the rickettsiae are passed from one mite generation to the next via the eggs. They are also transmitted to rats by arthropod bite, and the rat is thus a secondary reservoir. Man is infected by bite of a mite or rat–flea.* | *Confined to the Far East.* |
| *Spotted fevers* | *The rickettsiae are tick parasites and are passed from one tick generation to the next via the egg. They are also transmitted to a variety of mammalian hosts, including rodents and domestic animals, which thus form secondary reservoirs. Man is infected by tick bite.* | *There are several closely related diseases of this type, with differing geographical distributions (e.g., Rocky Mountain spotted fever, Mediterranean fever, South African tick-bite fever).* |
| *Rickettsialpox* | *The reservoirs are the house mouse and its mites; man is infected by mite bite.* | *The rickettsia is antigenically related to the spotted fever group. Rickettsialpox has only been observed in urban areas.* |
| *Q fever* | *The rickettsia is a parasite of numerous wild animals and ticks; the latter transmit it to goats, sheep, and cattle. Man acquires the disease by inhalation of infected dust, drinking infected milk, or by direct contact with animals or animal products.* | *This is a unique rickettsial disease, in that it is transmissible to man by means other than arthropod bite.* |

agents to wounds, have all but eliminated wound diseases in civilized countries.

Leptospirosis, which begins as a wound infection, is an occupational disease among workers who are exposed to frequent contact with polluted water. The leptospirae are parasites of pigs, dogs, and rodents and are excreted in the urine of infected animals. They can infect man only through minor wounds or breaks in the skin, and the disease is thus most common among men who work in wet places, such as sewers, fish markets, wet fields, or canals. The common wound infections are described in Table 27.6.

**Diseases acquired by ingestion of bacterial toxins**

Two serious diseases are caused by the ingestion of food containing the toxin of *Clostridium botulinum* or of *Staphylococcus aureus*. Although the disease is not further transmitted by the victim, an outbreak affecting many people can occur when a common food source becomes contaminated. Before the introduction of a strict canning code, the food canning industry was responsible for many deaths each year from botulism (ingestion of *C. botulinum* toxin). Food handlers who have open staphylococcal skin lesions are still a common cause of outbreaks of staphylococcal food poisoning.

**Rickettsial diseases of man**

The rickettsiae are extremely small, obligately parasitic bacteria. Their average dimensions are 300 by 600 nm; all are *intracellular* parasites.

One of the outstanding features of the rickettsiae is their parasitic relationship to arthropods (lice, fleas, ticks, and mites). These are their natural hosts, in which they usually live without producing disease. The rickettsiae have also become adapted to mammalian hosts, to which they are transmitted by arthropod bite. Thus, arthropod–mammal–arthropod chains of transmission are common. In most cases, man is only an accidental host, not forming a part of a transmission chain; the one exception is louse-borne typhus. The principal rickettsial diseases of man are described in Table 27.7.

## *Further reading*

**Books**


Dubos, R., and J. Hirsch, *Bacterial and Mycotic Infections of Man*, 4th ed. Philadelphia: Lippincott, 1965.

Horsfall, F., and I. Tamm., *Viral and Rickettsial Invections of Man*, 4th ed. Philadelphia: Lippincott, 1965.

Jawetz, E., J. Melnick, and E. Adelberg, *Review of Medical Microbiology*, 9th ed. Los Altos, Calif.: Lange, 1970.

Smith, H., and J. Taylor (editors), *Microbial Behavior in Vivo and in Vitro: Fourteenth Symposium of the Society for General Microbiology*. New York: Cambridge University Press, 1964.


**Reviews and original articles**


Braun, W., and D. Siva Sankar (editors), "Biochemical Aspects of Microbial Pathogenicity." *Ann. N.Y. Acad. Sci.* **88,** 1021–1318 (1960).

Panos, C., and S. Ajl, "Metabolism of Microorganisms as Related to Their Pathogenicity." *Ann. Rev. Microbiol.* **17,** 297 (1963).

Smith, H., "Biochemical Challenge of Microbial Pathogenicity." *Bacteriol. Rev.* **32,** 164 (1968).

van Heyningen, W., and S. Arseculeratne, "Exotoxins." *Ann. Rev. Microbiol.* **18,** 195 (1964).

# Chapter twenty-eight
# Mechanisms of inducible host resistance

*During the Middle Ages,* Europe was repeatedly swept by epidemics of smallpox. Many people died; of those who recovered, many were disfigured for life. When the inhabitants of a town were struck by smallpox twice within a generation, the pock-marked survivors of the first attack were spared during the second; their dramatic immunity was clearly attributable to their recovery from the earlier infection.

Acquired immunity was thus a well-recognized phenomenon as early as the sixteenth century. This led to deliberate attempts at immunization, although neither the nature of the disease nor the nature of the immunization process was understood. At considerable risk, some individuals allowed themselves to be inoculated with pustular material from mild cases of smallpox, hoping to undergo mild infections themselves and thus acquire protection against a later, fatal attack of the disease; this practice, known as "variolation," was too dangerous to be widely accepted. A safe means of immunization was not devised until Edward Jenner turned his attention to the problem in 1796.

Jenner, aware of the popular belief that individuals who had recovered from the mild disease, cowpox, were immune to smallpox, compared the reactions of "cowpoxed" and normal people to variolation and found that the former group were indeed protected against infection. He was thus led to inoculate an individual with pustular material from an infected cow. Some months later he inoculated this subject with virulent material from a true smallpox pustule and found him to be completely immune.

Jenner's process of vaccination (Latin *vacca*, cow) eventually became widely accepted and was greeted as a great achievement. The basis for his success necessarily remained completely obscure, however, until the role of microorganisms in disease was established by Robert Koch and Louis Pasteur. Not until 1879 was Jenner's device of artificial immunization applied to other diseases.

In 1879 Pasteur and his collaborators were engaged in studies on chicken cholera. They had just isolated the causative bacterium in pure culture, when work was interrupted by the summer vacation. Pasteur found on his return that the cultures that had been stored over the summer had lost their virulence; chickens infected with these cultures remained healthy. Pasteur saved these chickens for use in later experiments. He proceeded to reisolate a virulent strain of the chicken cholera organism from a fresh outbreak of the disease and inoculated these chickens with the new strain. The birds that had previously been inoculated with the avirulent culture proved to be immune, although the new strain was fully virulent for chickens that had not been so treated.

Pasteur had long believed that the agent of cowpox was an avirulent form of the agent smallpox, and he now perceived a parallel example of the same relationship in his avirulent and virulent strains of the chicken cholera bacterium. Thus, he had discovered the rationale of vaccination together with a means of protecting both men and animals against infectious disease. It appeared necessary only to isolate the causative agent in pure culture, cultivate it under conditions that would bring about its attenuation (loss of virulence), and inoculate it into healthy individuals to protect them against the virulent form.

There was no predictable way to bring about attenuation, but by trial and error Pasteur found successful procedures for attenuating the organisms of chicken cholera, anthrax, and swine erysipelas. For example, *Pasteurella multocida*, the causal agent of chicken cholera, was attenuated by serial transfer in broth medium using long intervals between transfers, while *Bacillus anthracis*, the agent of anthrax, was attenuated by serial transfer at 42°C. Some of Pasteur's procedures are still in use today, although in some cases killed suspensions of virulent organisms are used for immunizations. In honor of Jenner's discovery, Pasteur proposed that the term "vaccination" be applied to all processes of immunization with killed or attenuated suspensions of microorganisms.

The *mechanisms* of induced immunity were gradually discovered. The finding that exotoxins induce the formation of specific antitoxins in the blood led to the early recognition of *antibodies* and the important role that they play in immunity. Antibodies are highly specific reagents; a given antibody will combine only with the substance (antigen) that has induced its formation or with substances that are structurally closely related. This fact, together with the long-held belief that induced resistance is caused solely by antibody formation, led to the erroneous conclusion that all induced resistance is specific.

In recent years, however, some instances of nonspecific inducible resistance have been described; antibodies seem to play little or no part in such immunity. Although very little is known about the mechanisms involved, it seems likely that induced nonspecific resistance may play an important role in some host–parasite relationships.

### *Inducible resistance conferred by circulating antibodies*

Antibodies were discovered in 1890, when E. von Behring and S. Kitasato demonstrated that the blood of an immune animal contains a protective agent. Serum from an animal that had received repeated small injections of diphtheria toxin was injected into the bloodstream of a susceptible animal, together with a lethal dose of toxin. The recipient survived, whereas "control" animals, which received toxin but no serum, died. Experiments showed that only serum from an immune animal was protective; serum from "normal" animals had no effect. Furthermore, the phenomenon was found to be highly specific; serum from an animal surviving injections of diphtheria toxin would protect against that toxin but not against others. Behring and Kitasato thus established that immune serum contains specific protective substances which can be transferred to another animal and can confer on the recipient a temporary immunity.

An animal receiving immune serum is said to have undergone *passive immunization* in contrast with *active immunization,* which results from direct exposure to the pathogen or to its toxin. The active principle in immune serum is called antibody, and immune serum itself is commonly referred to as antiserum. We now know that one of the primary roles of antibody is to increase tremendously the efficiency of the normal host resistance mechanisms in combating the specific agent. For example, antiserum against bacteria contains antibody molecules that coat the surface of the bacterial cells and render them much more susceptible to phagocytosis. In many cases, this antibody coating also makes the bacteria sensitive to killing or lysis by the normal blood component, complement. Antibodies protect not only against bacterial invasion but (as described above) also against the action of toxins.

Substances that can induce antibody formation and react specifically with the antibodies so formed are called *antigens.* Conversely, antibodies are defined as substances that are formed in the body in response to contact with antigens and react specifically with those antigens. The ability to form antibody arose very late in the evolution of the animal kingdom; it is absent from invertebrates and from the most primitive vertebrate, the hagfish. It exists in all other vertebrates, although some features of the antibody-forming system are weak or absent in the lamprey eel, which stands just above the hagfish on the evolutionary scale.

## *The structure of antibody*

Antibodies belong to the group of blood proteins known as globulins. On the basis of electrophoretic mobility the globulins of vertebrates fall into three broad classes, called $\alpha$, $\beta$, and $\gamma$. Early work on immunized animals suggested that all antibodies belong to the $\gamma$-globulin class, and for a long time antibodies were referred to as "gamma globulins." In recent years, however, more sensitive methods have revealed that antibody proteins exhibit a wider range of electrophoretic mobilities than do $\gamma$-globulins. The term *immunoglobulins* is consequently now used to refer to all proteins that either show antibody activity or possess antigenic determinants in common with antibody molecules. The immunoglobulins are *glycoproteins,* having carbohydrate residues attached to the polypeptide chains.

**The classes of immunoglobulins**

The human immunoglobulins discovered to date fall into five classes, based on their physiochemical and antigenic properties.* The serum of every individual contains molecules of all five classes. The most abundant antibodies in man belong to the class which has been designated IgG (for "immunoglobulin type G"). Next most abundant are the antibodies of the IgM class (so designated because they were originally called "macroglobulins") and an electrophoretically fast-moving group called IgA. The last two classes, IgD and IgE, are present in such low amounts that they escaped detection until very recently.† Antibodies of the five classes differ from one another in molecular size, charge, antigenicity, and carbohydrate content.

*Antibodies, like other proteins, can serve as antigens when injected into a foreign animal; hence they can be characterized by their ability to react with the new antibodies formed against them.

†Two terminologies for immunoglobulins are in current use. Some workers use the terms IgG, IgM, IgA, IgD, and IgE; others refer to the same classes as $\gamma$G, $\gamma$M, $\gamma$A, $\gamma$D, and $\gamma$E, respectively.

## The structure of immunoglobulin G

The IgG antibodies have a molecular weight of about 150,000. Their sedimentation constant under standard conditions is 6.6 S, and they are hence often referred to as "7-S antibodies." By a variety of physical methods it has been shown that the IgG molecule is roughly ellipsoidal in shape, with dimensions of 50 A and 240 A for the axes.

Chemical analysis reveals the presence of between 20 and 25 disulfide bridges, linking cysteine residues in different parts of the molecule. If very mild reducing conditions are applied, so that only four or five disulfide bonds are broken, the IgG molecule dissociates into four polypeptide chains: two identical *light chains* and two identical *heavy chains*. The light chains each have a molecular weight of about 25,000, and the heavy chains each have a molecular weight of about 50,000.

The IgG molecule thus consists of four polypeptide subunits, held together by disulfide bonds. The organization of the subunits has been revealed, to a limited extent, by cleavage of the molecule into fragments as a result of mild treatment with proteases. When IgG is digested with papain, for example, it is cleaved into two types of fragments. One type, termed *Fc*, is crystallizable, and has no affinity for antigens; the other type, termed *Fab*, is not crystallizable and has *antigen-binding* capacity.

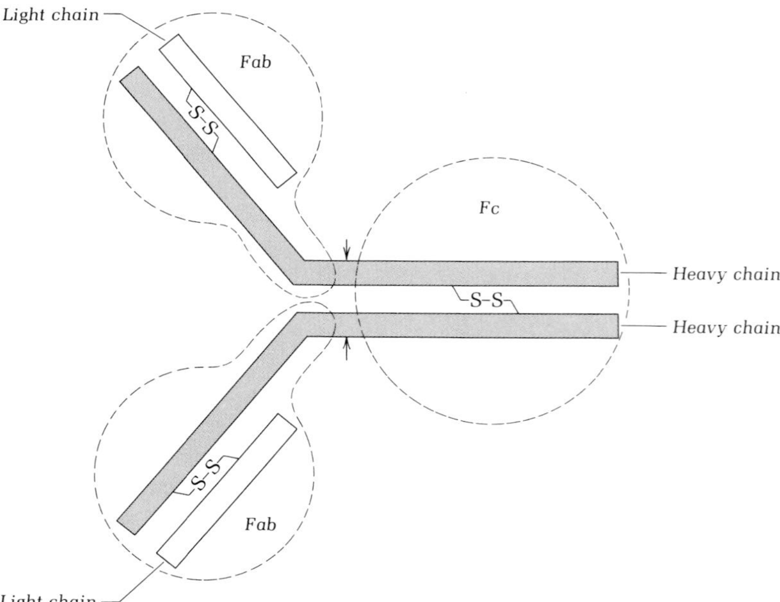


*Figure 28.1 The groundplan of IgG. The arrows (center) indicate the site of cleavage with papain, producing two Fab fragments and one Fc fragment.*

Each molecule of IgG antibody can bind two antigen molecules; it is thus said to have a *valence* of 2. When an IgG antibody molecule is cleaved with papain, each molecule gives rise to *two* Fab fragments, and each of these is univalent. Studies on the antigenicity and physicochemical properties of the fragments obtained by protease digestion have revealed the ground plan of IgG shown in Figure 28.1.

In referring to the three subregions of the molecule, it is customary to speak of the "Fc domain" and the "Fab domains." Note that each Fab domain contains an antigen binding site, so that the binding site may contain parts of both the heavy and light chains. When isolated heavy and light chains from a specific antibody are tested, only the heavy chain binds antigen; this binding, however, is stimulated by the addition of light chain from the same antibody. Thus, the binding site of the Fab domain must represent mainly amino acid residues of the heavy chain; the light chain may act indirectly by its combination with the heavy chain or may contribute a minor share of the amino acid residues of the binding site.

**The structure of other immunoglobulins**

Immunoglobulins of the IgM class have a molecular weight of about 1 million and a sedimentation constant that ranges from 18 to 23 S. They are frequently referred to as "19-S antibodies."

Under mild reducing conditions which break a small number of disulfide bridges, the IgM molecule dissociates into five identical 7-S monomers. The 7-S subunit of IgM immunoglobulin resembles the IgG molecule (which also has a sedimentation constant of approximately 7 S); it contains two identical heavy chains and two identical light chains. The IgM 7-S subunit differs from IgG, however, in the antigenic specificity (and thus in the amino acid sequence) of its heavy chains and also in carbohydrate content.

The third class of immunoglobulins, IgA, was first recognized as an electrophoretically fast-moving component of serum. Several subclasses exist, which appear to be monomers, dimers, and possibly higher multimers of 7-S subunits. The 7-S subunit contains two heavy chains and two light chains, but differs from the 7-S monomers of IgG and IgM both in the antigenic specificity of the heavy chain and in its carbohydrate content.

The last two classes of immunoglobulins, IgD and IgE, are present in extremely low amounts in human sera. IgD has the usual structure of two heavy and two light chains, the heavy chains being antigenically distinct from those of IgG, IgM, and IgA. IgE has so far been recognized only as an antigenically distinct class of antibodies responsible for certain allergic reactions; its structure is not yet known.

**Genetic variation in human immunoglobulins: allotypes**

We have said that each individual produces all five classes of immunoglobulins. Actually, each individual produces at least four types of IgG molecules ($IgG_1$, $IgG_2$, $IgG_3$, and $IgG_4$) and two types of IgA molecules ($IgA_a$ and $IgA_b$), as well as IgM, IgD, and IgE. Each type can be distinguished on the basis of the *antigenic specificity of its heavy chain*. Genetic analyses suggest that each of the nine types of heavy chain is determined by a separate locus.

Many of these loci are represented in human population by two or more different alleles: variants of the locus that differ from each other in nucleotide sequence. Thus, one individual may carry allele A of the locus for $IgG_1$, while another individual may carry allele B. These two individuals form antigenically different $IgG_1$ molecules. The different immunoglobins determined by alleles of the same locus are called *allotypes* (Greek *allos*, other).

The genetic variation in light chains is less complex. There appear to be only two loci governing the structure of light chains; the corresponding chains are called $k$ and $\lambda$. Each individual produces both $k$ and $\lambda$ light chains, which occur in all classes of immunoglobulins. A given molecule of immunoglobulin, no matter what kind of heavy chain it possesses, contains either two $k$ or two $\lambda$ light chains.

The cells of an individual thus produce nine or more types of heavy chains and two types of light chains. Heavy and light chains can associate at random (e.g., the serum of an individual will contain both $IgG_1$-$k$ and $IgG_1$-$\lambda$ immunoglobulins). Within a given immunoglobulin molecule, however, the two heavy chains are always identical and the two light chains are always identical. The mechanism that ensures this nonrandom pairing will be discussed later.

The number of distinguishable types of immunoglobulins that can be produced by an individual is thus very large. First, the nine types of heavy chains are found associated with both $k$ and $\lambda$ light chains, making possible 18 species of immunoglobulin molecules. Second, individuals are *diploid* for each of the loci concerned. If an individual were heterozygous for each locus, he might produce as many as 72 different types of immunoglobulin molecules.*

The number of different *specific antibodies* (i.e., specific in terms of antigen-binding capacity) that an individual can form is, however, several orders of magnitude greater. It has been estimated that the num-

*Nine heterozygous loci for heavy chains would produce 18 different heavy chains. Two heterozygous loci for light chains would produce four different light chains. Random association of one type of heavy chain with one type of light chain would then permit 72 different combinations.

ber may be as large as $10^6$. The basis for this tremendous diversity will be discussed later on in this chapter.

## The pathological immunoglobulins

Our understanding of the structure and diversity of antibodies has been enormously increased by studies of the immunoglobulins produced by patients suffering from *multiple myeloma* and other forms of cancer that affect lymphoid tissue. As we will discuss later, antibodies are formed exclusively by cells of the lymphoid tissues; in multiple myeloma, a neoplastic antibody-forming cell undergoes rapid and unregulated proliferation so that the serum contains excessive amounts of a single type of immunoglobulin, termed a *myeloma protein*. Furthermore, the urine of the patient contains a very high concentration of the light-chain subunit of the myeloma protein; these light chains in the urine of the multiple myeloma patients are called *Bence-Jones* proteins, after the physician who first described them over 100 years ago.

Essentially, the multiple myeloma patient is an individual whose serum contains, in addition to the usual complement of immunoglobulins, a huge excess of *one* normal immunoglobulin, as the result of the proliferation of one type of lymphoid cell. This fact tells us that *in a given lymphoid cell only one of the nine loci for heavy chains and one of the two loci for light chains is expressed, all others being repressed. Furthermore, only one of the two alleles of each locus is expressed.* Each individual suffering from multiple myeloma produces an excess of a different specific immunoglobulin.

The homogeneity and high concentration of myeloma protein and its corresponding Bence-Jones protein in each patient has made possible the complete determination of the amino acid sequences of both heavy and light chains of a large number of immunoglobulins; it has also facilitated the preparation of highly purified specific antisera against various types of heavy and light chains.

The analysis of amino acid sequences has revealed a set of striking relationships, which can be summarized as follows. The light chains of immunoglobulins contain from 214 to 219 amino acid residues. *The 107 residues constituting the C-terminal half of the molecule are identical* in all Bence-Jones proteins of a given type ($k$ or $\lambda$). In contrast, *the 107 to 112 amino acid residues constituting the N-terminal half show numerous differences.* In almost every instance, the difference between two light chains at a given position can be ascribed to a single base-pair difference in the nucleotide sequence of the genetic locus concerned.

The same pattern occurs in heavy chains. Thus, in both light and heavy chains there is a *constant region* at the C-terminal end and a *variable region* at the N-terminal end. Minor differences in amino acid residues in the constant region are responsible for antigenically different allotypes; *antibody specificity (with respect to antigen-binding capacity), on the other hand, undoubtedly reflects the major differences in amino acid sequence in the variable region of each chain.*

The analysis of amino acid sequences has also revealed striking similarities between the constant regions of light and heavy chains, and even between the first and second groups of 107 residues of heavy chains, starting at the C-terminal end. These similarities strongly suggest that the many loci for immunoglobulin synthesis have arisen during evolution by the repeated duplication of a single gene, with subsequent rearrangements. A diagram which illustrates the constant and variable regions of a typical immunoglobulin is presented here in Figure 28.2.

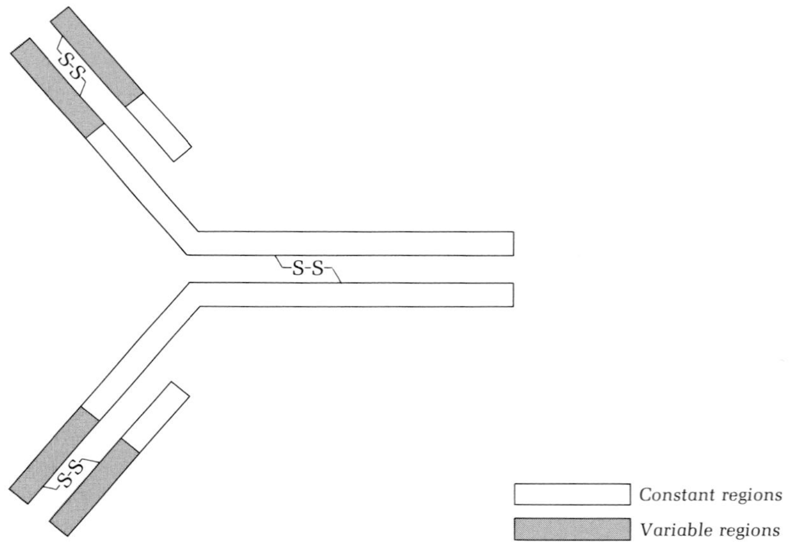


*Figure 28.2 The relative positions of the constant and variable regions of the heavy and light chains of an immunoglobulin.*

**The genetic basis of antibody diversity**

No upper limit has ever been found for the number of different antibodies (in terms of antigen-binding specificities) that can be formed by a single individual. Virtually every protein injected into an animal evokes the formation of a unique antibody, and it is commonly assumed that the individual has the potential for making $10^6$ or even more different antibodies. Antibody specificity reflects differences in amino acid sequence of the variable regions of immunoglobulin chains; this means that each individual is capable of making $10^6$ or more different amino acid sequences. Two different hypotheses have been proposed to account for this extraordinary potential.

The *germ-line* hypothesis states that all the genetic information needed for making at least $10^6$ different antibodies preexists in the genome of the individual and thus in the genome of each of his cells. If both the heavy chain and the light chain contribute to the specificity of the antigen binding site, and if any heavy chain can pair with any light chain, $10^3$ different loci of each type are required. If the specificity is determined solely by the heavy chain, $10^6$ different heavy-chain loci are required.

It is difficult to reconcile the germ-line hypothesis with the observations described above concerning allotypes. The hypothesis implies the existence of between $10^3$ and $10^6$ genes, each producing a different IgG molecule, but only four *loci* have been recognized in terms of IgG antigenic types. If the hypothesis is correct, many hundreds or thousands of genes must produce IgG molecules of identical allotype (e.g., the allotype corresponding to "allele A" of the $IgG_1$ locus). Because many alleles of the $IgG_1$ locus exist in the human population, the germ-line hypothesis requires that each "locus" for antigenic type represent a large group of different genes, *each group being inherited as a block at meiosis.* Recombination between the genes in such a block must not occur, because recombination would produce individuals forming many allotypes corresponding to a single "locus" rather than only one, as is the case.

The *somatic mutation hypothesis* states that the genome contains only a limited number of loci for immunoglobulins (nine for heavy chains and two for light chains, based on present knowledge). If this is so, extensive somatic mutation must take place in the variable region of each immunoglobulin locus during the development of the lymphoid tissue. At maturity, the animal is pictured as having over $10^6$ clones of lymphoid cells, each clone with unique nucleotide sequences in the variable region of each immunoglobulin locus. Each lymphoid clone thus produces a distinct set of specific antibodies. The difficulty with this hypothesis is that there is no known mechanism by which an extremely high rate of mutation could occur in one region of a gene (the variable region) and not in another (the constant region). Both hypotheses will be considered again in the following sections of this chapter, in connection with general theories of antibody formation in the whole animal and at the cellular level.

## *The formation of antibody*

**The primary and secondary responses of the whole animal**

Until an animal has been exposed to a particular antigen, antibodies against that antigen are not detectable in its serum. When the animal is immunized, antibodies may first be detected in the serum several days or weeks later, and the concentration of antibody increases exponentially for some time thereafter.

This observation led to the notion that there is a considerable *lag period* between the time of primary immunization and the time that antibody synthesis commences. Actually, however, the lag which is observed refers only to *detectable* antibody; when the methods used to detect antibody were improved, so that much lower levels could be measured, it was found that antibody synthesis commences much earlier than had previously been believed. In fact, there is probably no lag period at all in the primary response.

Once synthesis has begun, the amount of antibody in the serum rises exponentially to a plateau and then declines with time. The exponential increase is of particular interest, since—as we will discuss later—antigen is known to *stimulate the proliferation of antibody-forming cells.* The increase in antigen may therefore simply reflect the increase in the number of antibody-forming cells, the rate of synthesis per cell remaining constant. The doubling time for antibody concentration is similar to the doubling time of antibody-forming cells (10 to 12 hours).

When the plateau is reached, the rates of antibody synthesis and degradation are roughly equal. The concentration finally declines when the rate of synthesis falls below the rate of degradation. If at this point

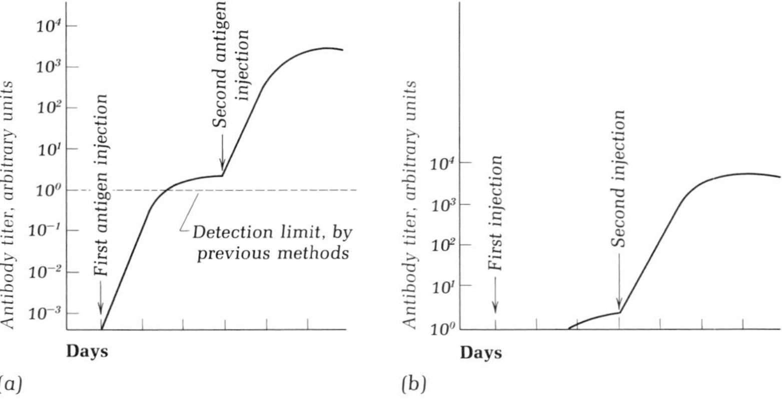


*Figure 28.3* Experimental results showing the stimulation by antigen of the production of antibody-forming plasma cells: *(a)* results obtained with highly sensitive detection methods; *(b)* results obtained with conventional detection methods.

a second injection of antigen is given, a secondary or *anamnestic response* is observed. The term "anamnestic" (Greek *anamnesis*, recall) refers to the ability of the animal to "remember" a previous immunization and produce much more antibody than it would have done if the injection had been a primary one. The secondary response occurs with little or no lag, and the rate of exponential increase in antibody concentration is the same as that observed during the primary response. These relationships are shown in Figure 28.3 and will be discussed again in connection with hypotheses concerning the mechanism of antibody formation.

## Antibody formation at the cellular level

It required several decades of research to track down the cells responsible for antibody production. Cells of the lymphoid tissue were suspected from the beginning, since the regional lymph nodes undergo rapid cellular division and may increase severalfold in weight following immunization. Antibody formation by lymphoid tissue was confirmed by showing that when such tissues, taken from immunized animals, are transferred to normal animals, specific antibodies appear in the blood of the recipients soon after the transfer. Furthermore, the antibody level in the recipient attains a level too high to be accounted for by the carryover of any residual antigen or antibody.

Lymphoid tissue contains a variety of cell types, but attention was ultimately focused on the *plasma cells,* which are large cells characterized by their high content of ribosomes. Plasma cells are relatively rare in the lymph nodes of normal animals but appear in increasing numbers after immunization. There is a good correlation between the amount of antibody that can be extracted from a given tissue and the number of plasma cells present in that tissue; furthermore, plasma cells do not appear following the injection of antigen into individuals who are congenitally unable to synthesize immunoglobulins (agammaglobulinemics).

The antibody-forming activity of plasma cells has now been fully established by several techniques. The presence of antibodies in individual cells can be detected by *immunofluorescent staining.* In this procedure, tissue sections are treated first with a specific antigen and then with fluorescein-labeled antibody to that antigen. Cells containing antibody bind the antigen, which—being multivalent with respect to antibody binding sites—then binds the fluorescent antibody. When this technique is applied to lymph node sections, antibody is found principally in plasma cells. The number of positively staining plasma cells increases in parallel with the rise of antibody titer in the serum. A few large and small lymphocytes also contain antibody.

Synthesis of antibody by isolated plasma cells has also been demonstrated. When plasma cells taken from an immunized animal are individually suspended in microdroplets, antibodies against the specific immunizing agent are liberated into the droplet and can be detected by highly sensitive techniques. For example, if the animal has been immunized with a motile strain of *Salmonella*, antiflagellar antibodies can be microscopically detected by the immobilization of bacterial cells added to the droplet. Antibodies against bacteriophage can be detected by measuring the neutralization of phage particles added to the droplet.

Finally, an extremely elegant *plaque technique* has been devised for detecting antibody production by individual plasma cells. Lymphoid cells from an animal that has been immunized against foreign red blood cells are plated in agar containing a suspension of the same red blood cells. Antibody liberated by the plated lymphoid cells diffuses into the agar and binds to the red cell surfaces; if complement is then added, the red cells lyse, and a clear plaque develops around each antibody-forming cell.

As discussed earlier, plasma cells begin to appear in lymph nodes shortly after immunization. There is evidence that they arise by differentiation from small lymphocytes; for example, animals that have been made immunologically unresponsive by massive doses of X rays can acquire the ability to form antibodies if, before immunization, they are given a population of small lymphocytes from a normal animal. Small lymphocytes are essentially resting cells. It has been suggested that exposure to antigen triggers a series of developmental steps through which the small lymphocytes differentiate into rapidly dividing large "immunoblasts." Some of these give rise to mature, nondividing plasma

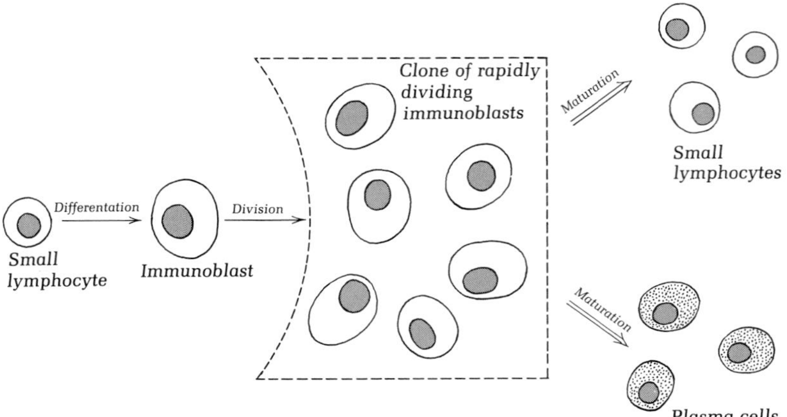


*Figure 28.4 Hypothetical scheme for the development of antibody-forming cells from small lymphocytes.*

cells; others differentiate into large lymphocytes and subsequently into small lymphocytes again (Figure 28.4). The end result is the formation of two different populations of nondividing cells: relatively short-lived plasma cells, which actively synthesize antibody, and extremely long-lived small lymphocytes. The latter group, being much more numerous than the small lymphocytes which were originally stimulated by antigen, serve as "memory cells," the significance of which we will discuss below in connection with the clonal selection hypothesis.

**One cell–one immunoglobulin**

It will be recalled that the different types of immunoglobulins can be recognized by the antigenic specificities of their heavy chains. It has thus been possible to use immunofluorescent staining to determine the type of immunoglobulin made by individual cells in lymphoid tissue. Antisera have been prepared against each type of immunoglobulin, and each antiserum has been tagged with a different-colored fluorescent substituent. When lymph node sections are stained with pairs of such stains, no individual cell is found to make more than one type of immunoglobulin—although all types are represented in the tissue. Similar tests with antisera against $k$ and $\lambda$ light chains show that a given cell produces one or the other but not both.

This discovery is surprising, in view of the fact that the genome of each cell can be inferred to contain the loci for all types of heavy chains and both types of light chains. There thus exists a mechanism whereby only one locus (and only one allele of that locus in the diploid cell) is expressed for a given type of chain, heavy or light. The locus that is "turned on" appears to be randomly determined. This finding is in perfect accord with the observation that a patient with multiple myeloma forms myeloma protein of only one antigenic type and a Bence-Jones protein of only one antigenic type. Since any given cell makes only one kind of heavy chain and only one kind of light chain, the homogeneity of heavy and light chains within a given immunoglobulin molecule can be explained by assuming that the association of chains occurs within the cell, before the molecule is released.

**The role of macrophages in the immune response**

One of the many complexities of the process of antibody formation is the fact that the cells which synthesize antibody are not those which take up antigen. By the use of radioactive antigens, as well as of immunofluorescent techniques, it can be shown that antigen is taken up exclusively by the phagocytic cells of the RE system, particularly the macrophages. Plasma cells that are actively synthesizing antibody contain no detectable antigen, even though the detection techniques are

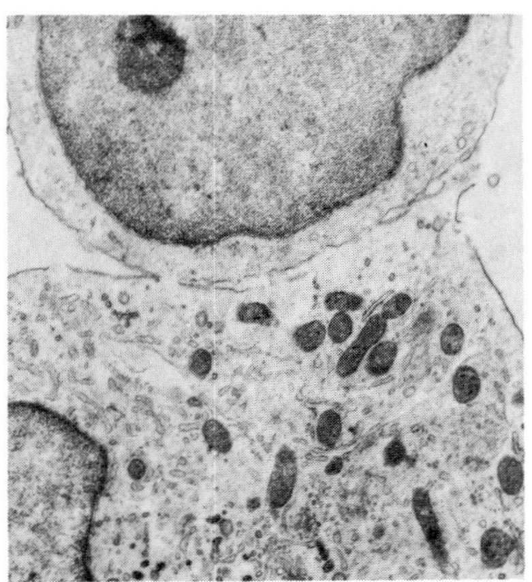

*Figure 28.5 Electron micrograph, showing direct cytoplasmic connections between a lymphocyte (top) and a macrophage. From M. D. Schoenberg, et al, "Cytoplasmic Interaction between Macrophages and Lymphocytic Cells in Antibody Synthesis." Science* **143,** *964 (1964); copyright 1964 by the American Association for the Advancement of Science.*

so sensitive that as few as 10 molecules of antigen per cell could be detected.

The first event in immunization is, accordingly, the uptake of antigen by macrophages. Some time later, cells of the lymphocyte–plasma cell series start to synthesize specific antibody. What are the steps that intervene between these two events?

One clue comes from histological studies, which show that lymphocytes cluster around macrophages that contain phagocytized material. Direct surface contacts are formed, and electron microscopic studies reveal cytoplasmic connections between lymphocytes and macrophages (Figure 28.5). These observations are strongly suggestive of a direct transfer of antigen from macrophage to lymphocyte.

A second clue comes from studies of the induction of a primary antibody response in isolated lymph node cells. Such a response cannot be induced with free antigen but can be induced with an extract made from macrophages that have taken up antigen. The extract contains RNA, and its ability to evoke antibody production is destroyed by RNase. The extract presumably also contains antigen, perhaps bound to the RNA. Although the latter point is in dispute, it has been shown that antigen fragments bound to RNA can be isolated from the liver tissue of immunized rabbits, and that this RNA-bound antigen is a better immunogen than free antigen. The best interpretation at the moment is that—to evoke a primary antibody response—antigen must first be taken up by macrophages, processed or "activated" by being bound to RNA, and then transferred to lymphocytes by direct cell-to-cell contact.

The same steps presumably take place in the secondary, or anamnestic, response, since antigen is always taken up by phagocytes rather than by lymphocytes. These steps may not be necessary for the secondary response, however, since antibody synthesis (and cell proliferation) can be stimulated by free antigen in lymphoid tissue taken from a previously immunized animal.

### Immunological unresponsiveness

Instead of evoking an antibody response, an antigen will actually *prevent a response* from occurring if it is administered to an animal under certain special conditions. These conditions are (1) administration in low doses to a fetal or neonatal animal; (2) administration in low doses to an adult animal that has had its immune mechanism blocked by "immunosuppressive drugs" or by X rays; or (3) administration in massive doses to an adult, normal animal.

The induced state of immunological unresponsiveness was originally called "tolerance" when it occurred in either condition (1) or (2) and "immune paralysis" when it occurred in condition (3); for a time, it was thought that different mechanisms were involved. It is now clear, however, that tolerance and immune paralysis are basically the same phenomenon, and we shall refer to all such states as *tolerance*. Tolerance is not a permanent change in the animal; it persists only as long as the concentration of antigen remains above a critical level. If the antigen is degraded, for example, the animal regains its ability to make a normal immune response to that antigen.

The relative ease with which tolerance is induced in the newborn animal or in the immunosuppressed adult, and the requirement for massive doses of antigen to induce tolerance in the normal adult, suggest that *tolerance is induced when the concentration of antigen is very high relative to the number of cells in the body that are actively synthesizing antibody.*

The existence of the tolerance mechanism explains why animals do not make antibodies against their own normal tissue and blood constituents. These constituents, being present as antigens during fetal life, induce tolerance which persists—as do the antigens themselves—throughout the life of the animal. As exceptions which prove the rule, those "self-antigens" which normally do not come into contact with the lymphatic system (such as eye lens tissue and nerve tissue) do *not* evoke tolerance, and will produce an active immune response if removed and then injected into the bloodstream of the same animal.

## *Theories of antibody formation*

**The phenomena for which theory must account**

Let us begin by reviewing the phenomena that we have described so far. They can be conveniently discussed under seven headings.

(1) *Induction of the primary response.* An antigen taken up by macrophages, but not free antigen, induces lymphocytes to mature into plasma cells and to liberate antibodies with specific affinity for that antigen.

(2) *The diversity of antibody specificities.* No limit has been found to the number of different antibodies (in terms of antigen-binding specificity) that an individual animal can make; the number may be $10^6$ or even higher.

(3) *The Mendelian inheritance of allotypes.* At present count, there appear to be nine loci for the heavy chains and two loci for the light chains of immunoglobulins. Many of these loci exist in the form of two or more different alleles in the human population, and these alleles segregate in Mendelian fashion during recombination.

(4) *One cell–one type of immunoglobulin.* Although the genome of the individual animal, and thus of each of his lymphoid cells, contains

all 11 loci for heavy and light chains, a given cell produces only one type of immunoglobulin.

(5) *The secondary or anamnestic response.* An animal that has been previously immunized with a given antigen responds to a second injection of the same antigen by making much more antibody than it would in a primary response to the same dose.

(6) *The phenomenon of "original antigenic sin."* A special aspect of the anamnestic response, which we have not yet mentioned, is observed when the secondary immunization is made with an antigen which closely resembles the antigen used in the primary immunization (i.e., a cross-reacting antigen: one that can react with antibody formed against another). The antibodies produced in the secondary response are directed against the antigen used in the primary immunization rather than against the one used in the secondary immunization. This phenomenon has been referred to as "original antigenic sin."

(7) *Immunological unresponsiveness.* Antigens can induce tolerance (i.e., block the production of antibodies directed against themselves) if they are given to normal adult animals in massive doses or to newborn animals or immunosuppressed adults in low doses. Tolerance persists only as long as the antigen concentration remains above a critical level.

### The clonal selection theory

The clonal selection theory, which was proposed in its original form by F. M. Burnet, comes closer than any other to explaining all the phenomena listed above. The summary that follows includes modifications which have been made since Burnet's original formulation.

(1) During the development of the animal embryo, differentiation leads to the appearance of a line of lymphoid cells which, when mature, will excrete immunoglobulin. As these cells multiply, random changes occur such that different lymphoid cells produce immunoglobulins with different antigen-binding specificities. The changes that occur may be somatic mutations in the variable regions of the loci for the heavy and light chains of immunoglobulins. Alternatively, it is possible that the genome of every cell contains a different gene for each specific antibody, in which case the random changes represent the "turning on" of one particular gene in each cell, all others remaining unexpressed.

Whichever mechanism is operating, the change is perpetuated at cell division. The result is that by the time the animal is born, its antibody-forming tissue is a mixture of thousands or even hundreds of thousands of *clones*, each clone producing a different specific antibody. The number of different types is high enough to ensure that at least one cell in the body produces an immunoglobulin with specific binding affinity for any conceivable antigen. The immunoglobulins in the serum of the animal are thus a mixture of many different antibodies, produced as a result of random changes during the differentiation of the lymphoid tissue.

(2) The mature antibody-forming cell contains a high concentration of its own type of antibody, perhaps on its surface. When the corresponding antigen (processed by macrophages) comes in contact with this cell it combines with the antibody and—for reasons unknown—stimulates the cell to proliferate. The number of cells forming the specific antibody thus increases exponentially, and this increase is reflected in the exponential increase of specific antibody in the blood. This is the *primary response* to immunization.

(3) After the antigen has disappeared, the animal still has a "memory" of the immunization in the form of a high proportion of resting small lymphocytes that have the genetic potential to form the specific antibody.

(4) If a second dose of antigen is given, the same selective stimulation takes place. This time, however, there is a much larger clone present initially, so that a much larger number of cells of the specific type is formed. The circulating antibody titer rises proportionately; this is the *anamnestic response.* For example, if at the first injection there were only 10 cells of the specific type present in the animal, and a thousand-fold multiplication occurred (about 10 doublings), $10^4$ specific cells would be formed. This population might produce barely detectable levels of antibody in the blood. A second dose of antigen, however, might bring the number up by another factor of a thousand to a total of $10^7$ cells. This population increase would be reflected by a dramatic, exponential rise in measurable antibody. It is easy to see that this feature of the clonal selection theory nicely explains the phenomenon of "original antigenic sin."

(5) If antigen comes in contact with an *immature* antibody-forming cell, that cell is killed (in contrast with the stimulation that occurs when antigen comes in contact with a mature antibody-forming cell). This feature of the theory has been added as one way of accounting for the phenomenon of immunological unresponsiveness, or *tolerance.*

**The genetic derepression theory**

The central postulate of the clonal selection theory is the preexistence in the body of clones of lymphoid cells, each clone making a different antibody. According to that theory, the role of antigen is to stimulate selectively the proliferation of that clone of cells which is already making the corresponding antibody. If, however, the diversity of antibody specificities is determined not by somatic mutation but by the existence in the genome of a different locus for each amino acid sequence, an alternative role of the antigen is conceivable. Thus, it can be postulated that each antibody serves as a specific repressor of its own locus or loci, so that antibody synthesis ceases as soon as one or a few molecules of each type have been made. The antigen then acts by *combining with its corresponding antibody and inactivating it as repressor.*

The clonal selection theory and the genetic derepression theory offer two, at present equally plausible, mechanisms to account for the specificity of the primary immune response. The clonal selection theory, however, can account for a number of phenomena for which the genetic derepression theory offers no explanations. Thus, clonal selection accounts for the anamnestic response, including the phenomenon of "original antigenic sin" and may also provide a basis for understanding tolerance. Furthermore, the clonal selection theory is compatible with either hypothesis of genetic diversity: somatic mutation or one gene–one specific antibody.

## *The combination of antibody with antigen*

**The structure of antigens**

Antigenicity is a property of many, but not all, macromolecules. Most free proteins are antigenic, as are glycoproteins, nucleoproteins, and lipoproteins. Some polysaccharides are antigenic, but many are not. Most lipids are not antigenic.

Although most antigens are large molecules, the specificity of the antibody that a given antigen evokes is determined by a very small site on its surface, called the *antigenic determinant.* An antigen as complex as a protein, which has an irregular three-dimensional shape, may have a number of different antigenic determinants; each determinant probably consists of a projection on the surface of the molecule, comprising five or six amino acid residues or a conjugated substituent of equivalent size. When such an antigen is injected into an animal, antibodies are formed that have specific combining activity for each antigenic determinant on the molecule; thus, "antibody" against a single protein is often a heterogeneous population of molecules, comprising several types with different antigen-binding specificities.

**Haptens**

The antigenic specificity of a protein can be changed if the protein is coupled chemically to a small molecule which by itself is wholly devoid of antigenicity. The antibodies that are formed against the modified protein will not combine with the original protein but are specific in their ability to combine with the modified protein. Furthermore, these antibodies can combine with the small molecule itself, unattached to any protein. For example, the molecule sulfanilic acid can be coupled to a protein, as shown in Figure 28.6, and used to immunize an animal. The ability of the resulting antibody to combine with free sulfanilic acid is shown by the fact that free sulfanilic acid will specifically interfere with the combination between antibody and the protein to which sulfanilic acid is bound. The specificity of the antibody is remarkable;

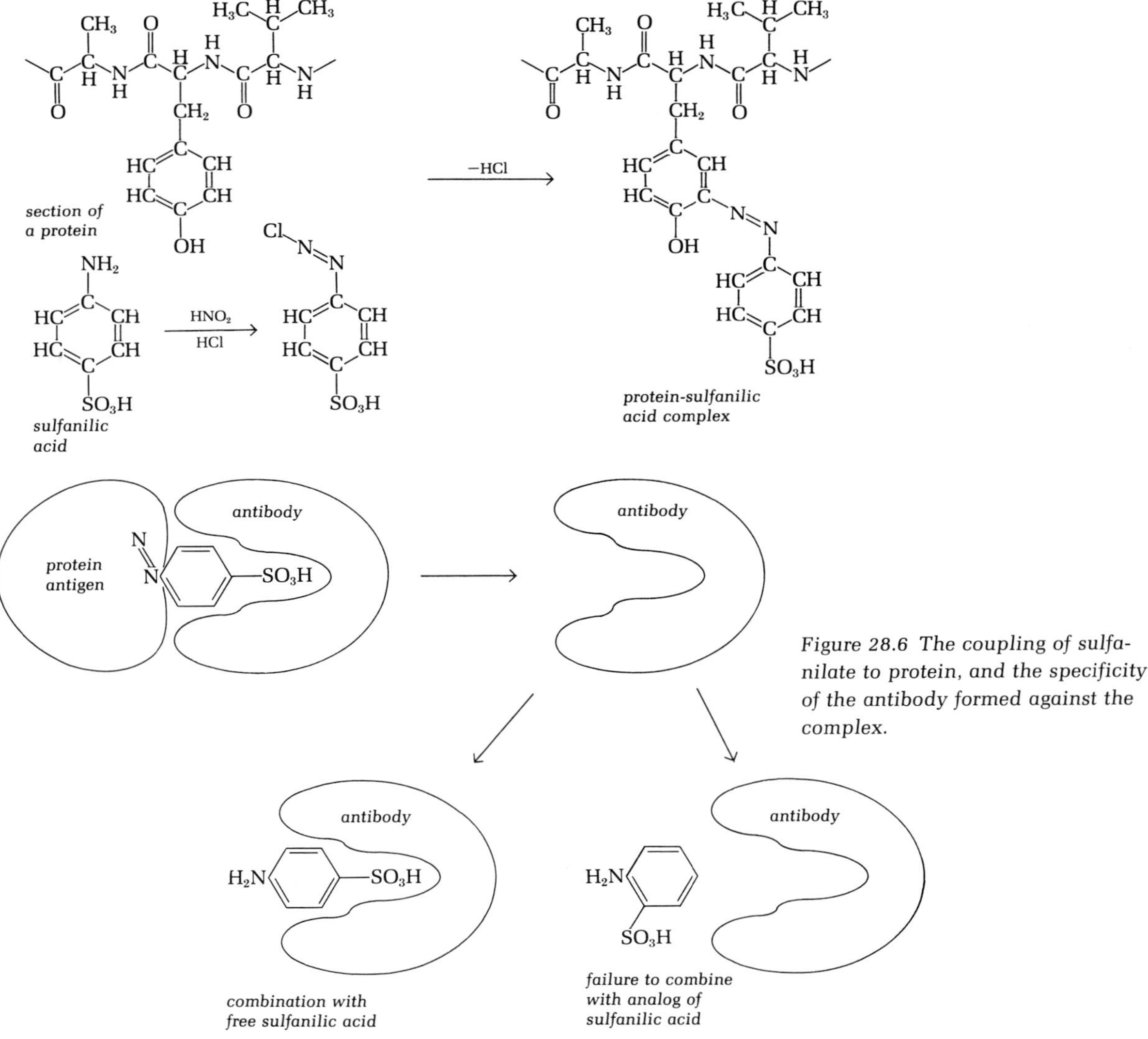


*Figure 28.6 The coupling of sulfanilate to protein, and the specificity of the antibody formed against the complex.*

for example, an isomer of sulfanilic acid, in which the —$SO_3H$ group is in the *meta* position relative to the amino group, will not combine with antibody formed against the protein–sulfanilate in which the —$SO_3H$ group is *para* to the amino group.

Sulfanilic acid behaves as a *hapten. A hapten may be defined as a nonimmunogenic molecule which, when coupled to a protein, determines the antigenic specificity of the complex; in its free form a hapten will combine with the antibody the specificity of which it has determined.*

When combined with a protein, a hapten is referred to as a *haptenic group:* in such combination it may itself constitute a complete antigenic determinant or may be part of a larger antigenic determinant that includes one or more amino acid residues of the protein.

### The nature of antigen–antibody binding

The nature and size of the antigen-binding sites on the antibody molecule have been deduced from studies of the affinity of antidextran antibody for glucose polymers of different length. The results suggest that the binding site is an invagination of the antibody molecule which can accommodate between five and six glucose residues. This means that only 15 to 30 amino acid residues of the antibody molecule, out of a total of about 1,300, are involved in each of its two active sites. Conversely, it suggests that the antigenic determinants of macromolecules represent projections which can be accommodated by binding sites of such volume.

The binding of the antigenic determinant to the amino acid residues that line the antigen-binding site of the antibody involves many types of bonds, but none of them is covalent. For example, binding can involve ionic bonds, hydrogen bonds, and hydrophobic bonds. Hydrophobic bonds are probably of major importance, because the more nonpolar the antigen (and thus the more it is repelled by water molecules), the tighter it is bound to antibody. This suggests that the amino acid residues in the antigen-binding sites are predominantly those with nonpolar side chains.

### Lattice formation and the precipitin reaction

Antibodies, as well as most antigens, are *multivalent.* Valency refers to the number of combining sites on the molecule; as we have discussed earlier, IgG antibodies have a valency of 2, and IgM antibodies a valency of 5. Antigens with more than one antigenic determinant are also multivalent and may have from 2 to 40 or more combining sites.

When a set of multivalent antibodies with specific affinity for the antigenic determinants of a multivalent antigen are mixed with a solu-

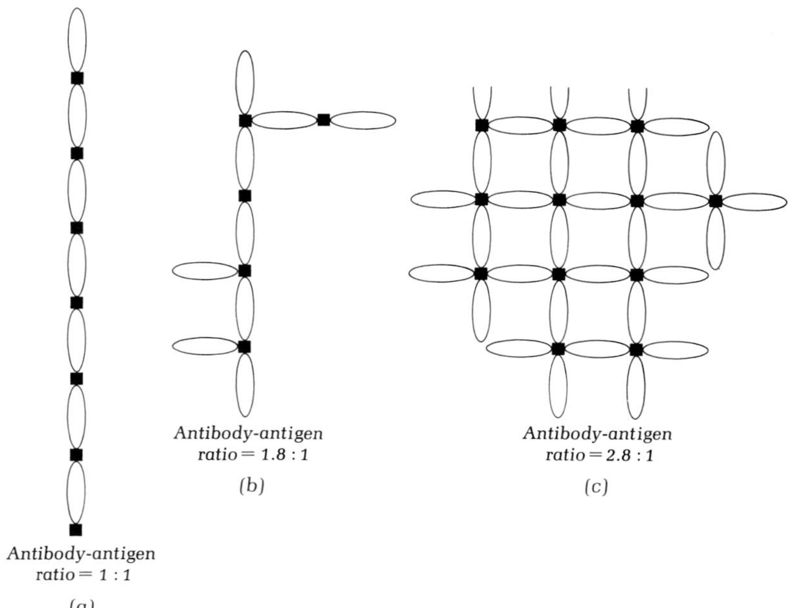


*Figure 28.7 Insoluble lattices that are formed when bivalent antibody is mixed with an antigen having a valency of four. The shaded squares represent antigen molecules, the ovals represent antibody molecules. When antigen and antibody are mixed in equal proportions (a), the ratio of antibody to antigen in the precipitate is approximately one. As the antibody is added in increasing excess (b) and (c), the ratio of antibody to antigen in the precipitate approaches the valency of the antigen (c). Modified from B. Davis et al., Microbiology. New York: Harper and Row, 1967.*

tion of that antigen, a complex or *lattice* is quickly formed and eventually precipitates from solution. Figure 28.7 shows some of the complexes that can form when a set of bivalent antibodies is mixed with an antigen having a valency of 4. As indicated in the figure, the ratio of antibody to antigen in the precipitated complex depends on their initial proportions in the mixture. As antibody is added in increasing proportion, the ratio approaches the number of combining sites on the antigen; as the antigen is added in increasing proportion, the ratio approaches 1. When the antigen is in excess, the final mixture contains not only precipitate but also *soluble complexes*, as shown in Figure 28.8.

Since a multivalent antigen may have several different antigenic determinants, the antibodies that react with it to form a lattice will be heterogeneous. This principle is illustrated in Figure 28.9, in which the different antigenic determinants have been arbitrarily numbered 1 to 4.

Precipitation requires not only the formation of lattices, but also the *aggregation* of such lattices into larger complexes. This aggregation occurs with most, but not all, types of multivalent antibodies. It is a very slow reaction, and is thought to result from the formation of ionic and hydrophobic bonds between antibody molecules in two different lattices.

*Figure 28.8 Soluble complexes that are formed in an antigen-antibody mixture when antigen is present in excess. Modified from B. Davis et al., Microbiology. New York: Harper and Row, 1967.*

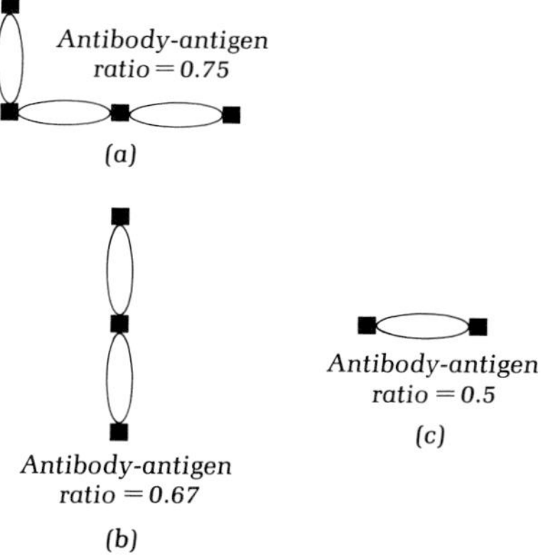

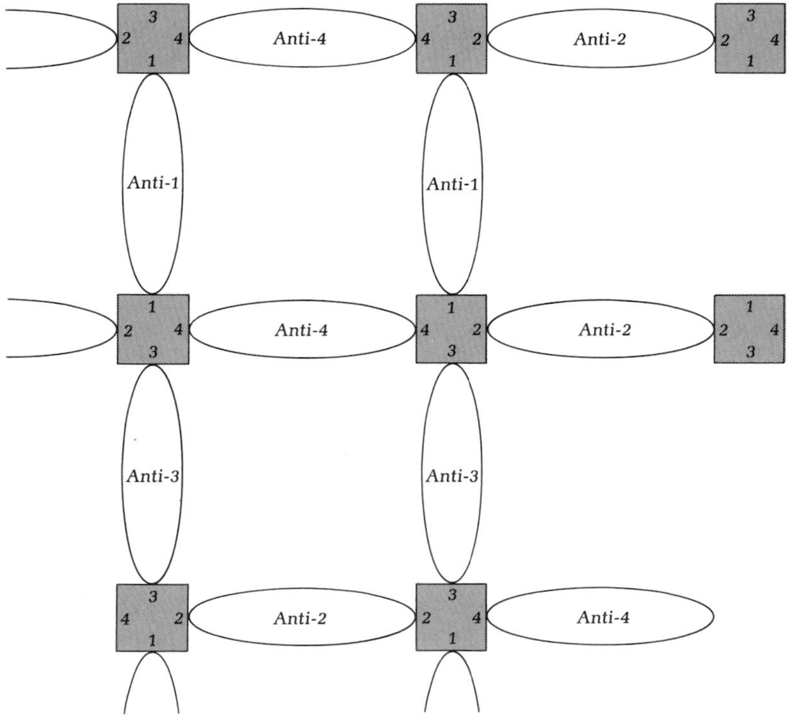


*Figure 28.9 The heterogeneity of antibodies in a lattice containing an antigen with four different antigenic determinants. Modified from B. Davis et al., Microbiology. New York: Harper and Row, 1967.*

The precipitation of antigen–antibody complexes is called the *precipitin reaction*, because it was originally thought to involve a special type of antibody called "precipitin." Now, however, it is clear that most types of antibodies are capable of precipitating with their corresponding antigens. The precipitin reaction is not known to play any role in host resistance; its importance derives from its tremendous usefulness as an immunochemical tool for the detection and identification of both antigens and antibodies.

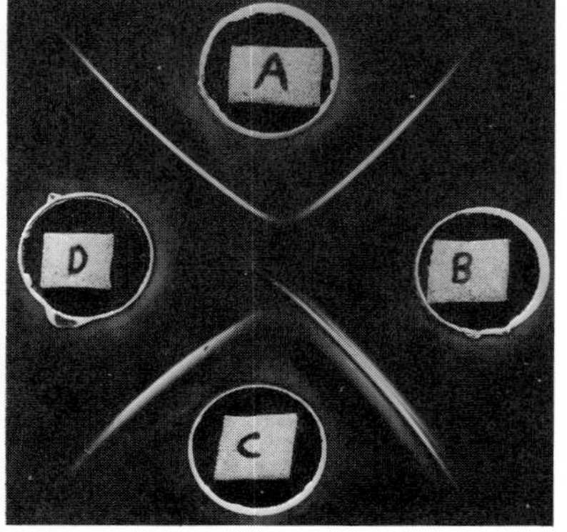


*Figure 28.10 The formation of antigen-antibody precipitates by agar diffusion. Well B contains fetal sheep serum, which comprises several antigens including the protein fetuin. Well C contains antibody against fetal sheep serum; several precipitate bands are visible between B and C. Well D contains purified fetuin, and Well A contains antibody against purified fetuin. Note the continuous band formed by the fetuin diffusing from Well D and from Well B. From F. H. Bergman, L. Levine, and R. G. Spiro, "Fetuin: Immunochemistry and Quantitative Estimation in Serum." Biochim. Biophys. Acta* **58,** *41 (1962).*

Precipitation tests for the detection of antigens are often carried out using the *agar diffusion method*, as developed by Ö. Ouchterlony. In the Ouchterlony technique, illustrated in Figure 28.10, antibody and antigen solutions are placed in wells cut in agar. As the antibody and antigen molecules diffuse outward, they meet and precipitate. Each specific antigen–antibody precipitate forms a band in the agar; since the position of the band is determined by such factors as diffusion rates and molecular proportions, there is a high probability that each precipitate will be detectable as a separate band.

The Ouchterlony technique can be combined with electrophoresis, which increases the resolution of the system even further. This procedure, known as *immunoelectrophoresis*, is illustrated in Figure 28.11.

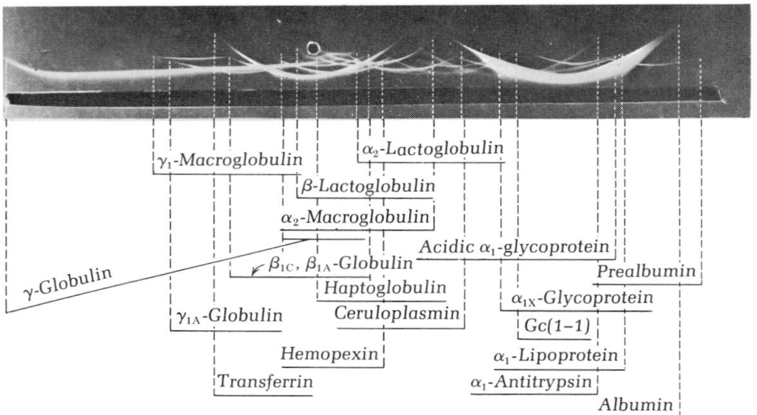


*Figure 28.11 Immunoelectrophoresis. A sample of human serum was placed in the hole visible at the center of the agar strip. An electric potential was applied to the buffer in the agar, the negative pole in this case being at the left and the positive pole at the right. The various proteins in the serum sample migrated in the electric field according to their relative charges. Following the electrophoresis process, the long trough in the agar was filled with rabbit antiserum against human serum. The antiserum diffused at right angles to the trough, and formed white bands of precipitate where it mixed with the human serum proteins diffusing from their final electrophoretic positions. From H. E. Schultze and G. Schwick, "Control of Fractionation of Plasma," in Immunoelectrophoretic Analysis (P. Grabar and P. Burtin, editors), Chapter 7. Amsterdam: Elsevier, 1964.*

**Agglutination of cells**

When bacterial cells are injected into the bloodstream of an animal, antibodies are formed against the surface antigens of the bacteria. The surface antigens include the components of the cell wall, the capsule, and the flagella. When bacteria are mixed in vitro with an antiserum directed against them, the multivalent antibodies combine with the bacterial surface antigens. Lattices, composed of bacterial cells held together by antibody bridges, are formed. Under the proper conditions, these lattices grow in size until they form large visible clumps, containing hundreds of cells.

The *agglutination reaction* is one of the most sensitive reactions known for the detection of specific antibodies. It has been extended to the detection of soluble antigens by attaching such antigens to the surface of red blood cells or to inorganic particles such as bentonite. When antibody against the attached antigen is added to a suspension of such particles, agglutination of the particles occurs. This procedure, called *passive agglutination*, can detect as little as 0.01 $\mu$g of antibody per milliliter of serum. In contrast, precipitin reactions require 20 to 60 $\mu$g of antibody per milliliter.

**Complement fixation and its mechanism**

In 1894 R. Pfeiffer and his coworkers discovered that certain Gram-negative bacteria were lysed by antiserum prepared against them. The serum lost its lytic activity when heated to 56°C for a few minutes; the activity was restored by the addition of fresh, *normal* serum. Lysis of the bacterial cells thus requires both specific antibody and a nonspecific component of normal serum; the latter, named *complement*, is now known to be a group of proteins rather than a single substance. In human serum, complement comprises 11 different proteins.

Red blood cells, which have bound antibodies directed against their surface, are also lysed by complement. The mechanism of this *complement fixation reaction* have been studied in detail and can be summarized as follows.

The first step is the binding of antibody to the surface of the red blood cell. Not all types of antibody can trigger the subsequent chain of reactions; only IgM and some types of IgG antibodies are effective. Antibody that is so bound is altered, perhaps by an allosteric transition, so that in turn it binds the component of complement called C′1.* This binding appears to involve the Fc domain of the antibody molecule, because digestion of the Fc domain with pepsin specifically destroys complement-binding activity.

The alteration in the antibody molecule which permits the binding of C′1 can be induced in several ways (e.g., by heat denaturation or by binding to large multivalent antigens). Only those conditions which cause antibody molecules to aggregate will bring about the alteration; thus, the combination of antibody with univalent antigens does not permit complement fixation.

C′1, which recently has been discovered to be a complex of three different proteins, has no enzymatic activities. When bound to the antigen–antibody complex, however, it is converted to a form that has both esterase and protease activity and reacts with C′4, modifying it so that it binds to the complex also. Activated C′1 is called C′1a; the complex now consists of cell–antibody–C′1a–C′4. This complex now binds C′2, which is cleaved by C′1a to form a bound fragment called C′2a.

*Complement is abbreviated C′. The components of complement have been numbered, the numbering system reflecting in part the historical order of their discovery.

In a further series of steps, the complex successively adds C′ components until it consists of cell–antibody–C′1a,4,2a,3,5,6,7. Finally, the reaction of this complex with C′8 and C′9 brings about lysis of the red blood cell. The cell becomes pocked with small holes, each about 100 A in diameter, through which hemoglobin escapes.

**The role of complement fixation in host defense**

The identification of the various components of complement, through the analysis of the red blood cell system, has led to a much clearer picture of the ways in which complement serves in host defense against bacterial infection. For example, bacterial cells that are coated with antibody and have bound C′1a, 4, 2a, and 3 are more readily phagocytized than cells coated with antibody alone. Antigen–antibody complexes that have bound C′1a, 4, 2a, 3, and 5 release a polypeptide fragment that causes mast cells to release *histamine*, a substance prominent in hypersensitivity reactions of the immediate type. Finally, complexes that have bound C′1a, 4, 2a, 3, 5, 6, and 7 release chemotactic substances that attract phagocytic polymorphonuclear cells.

**The complement fixation test**

The ability of antigen–antibody complexes to fix complement has led to the development of the *complement fixation test* as a means of detecting either antibodies or antigens. If, for example, one wishes to determine whether a sample of serum contains a particular antibody, specific antigen and complement are mixed with the serum to be tested ("reaction mixture A"). If the serum contains specific antibody, the complement will be fixed to the antigen–antibody complex and thus removed from solution. The test is then completed by determining whether any complement remains free in solution. For this purpose, sheep red blood cells are mixed with heated, homologous antiserum, which renders them sensitive to lysis by complement. To this test system is added a portion of reaction mixture A. If the red blood cells lyse, free complement was present in A, and the serum under test contained no antibody. If no lysis occurs, the complement must have been removed from A, demonstrating that the serum under test contained antibody. Such a test is illustrated in Figure 28.12.

**Hypersensitivity of the "immediate" type**

So far, we have spoken of antibodies only as protective or defensive agents of the immune animal. Under certain conditions, however, the combination of antibody with antigen can produce severe damage to the host. An individual who experiences such damage is said to be *hypersensitive* to the antigen.

Figure 28.12 *The complement fixation test. (See text for explanation.) The sera used both in reaction mixture A and in the test system have been treated to destroy any complement originally present.*

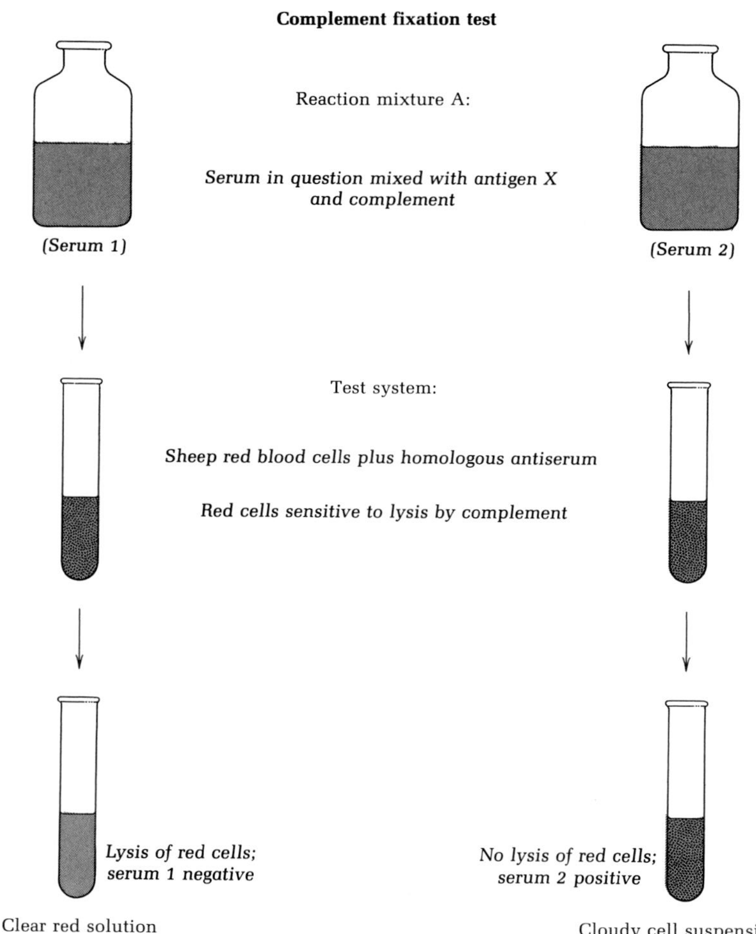


Hypersensitivity has been described under many names, the most familiar being *allergy*. For example, some individuals are unusually sensitive (allergic) to certain foodstuffs or pollens that are innocuous to others. Another instance of hypersensitivity is *allergy of infection*, in which an individual becomes allergic to some microbial substance as a consequence of infection. There are many other host reactions that can be considered manifestations of hypersensitivity, ranging from the itch that follows a mosquito bite to rheumatic fever, a serious disease that is a sequel to repeated streptococcal infections. All these host reactions have a certain pattern in common. The individual must first be sensitized by exposure to the antigen, and a certain period of time must then elapse for antibody formation to take place. After this, when the individual is again exposed to fairly large amounts of the same antigen, the ensuing antigen–antibody reaction results in damage.

Two general types of hypersensitivity occur. In one type, called *immediate hypersensitivity*, the tissue damage takes place within a few minutes after administration of the second dose of antigen. In the other type, called *delayed hypersensitivity*, the reaction to the second dose of antigen is slow and progressive, reaching a peak after 24 to 72 hours. The two types differ in a number of fundamental ways, the time lag in the appearance of damage being of only secondary importance. Hypersensitivities of the immediate type, are, as we will show, mediated by circulating antibodies. Hypersensitivities of the delayed type, on the other hand, are mediated by cells that have become sensitized in a manner not yet fully understood.

The most dramatic form of the immediate hypersensitivity response is *anaphylaxis*, in which the second exposure to the antigen is rapidly followed by violent symptoms and often death. Anaphylaxis was discovered in the course of experiments on guinea pigs. It was observed that some guinea pigs, which had received an injection of horse serum, died rapidly on receiving a second injection of the same material. The guinea pigs had become sensitized to normal horse serum protein, and the second dose of serum caused a severe antigen–antibody reaction. In guinea pigs, the anaphylactic reaction is localized in the respiratory tract and kills the animal by bringing about contraction of the bronchioles and suffocation.

Such sensitization can be *passively transferred*, just as can immunity. A guinea pig, for example, can be made hypersensitive to egg-white protein by receiving serum or purified antibodies from a second animal that has been immunized with this antigen. Furthermore, if the smooth muscles that are affected in the anaphylactic response of guinea pigs are removed from a sensitized animal and suspended in a bath, addition of specific antigen to the bath will cause violent contractions of the isolated muscle.

These and other experiments have led to the following general picture of anaphylaxis. When an animal is first exposed to an antigen, circulating antibodies are formed. Many of these leave the bloodstream and become fixed to cells, particularly mast cells and platelets. When the animal is reinjected with large amounts of antigen, the antigen is distributed to the tissues, where it combines with the fixed antibodies. This reaction causes the cells to act as they do in inflammation and pour out histamine and serotonin. These substances, in turn, bring about anaphylaxis. Those tissues that are most sensitive to histamine and serotonin are immediately affected; in the guinea pig, for example, the lungs are the site of the anaphylactic reaction, whereas in the dog the reaction takes place in the liver.

The histamine and serotonin liberated by the combination of antigen with fixed antibodies produce two characteristic responses: the blood capillaries undergo typical inflammatory changes, becoming more permeable, and the smooth muscles are contracted. The type and location of the response that predominates depend on the animal species and on the route of antigen injection. For example, injection of antigen into the skin produces only the capillary response, whereas injection of antigen into the bloodstream, permitting its systemic distribution, produces both types of response.

A number of common hypersensitivity reactions are of the immediate type. These include "serum sickness," a mild form of anaphylaxis that often follows antiserum injections, and "atopies," which include such allergic reactions as hayfever, asthma, and hives. The atopies are characteristically responses to antigens which are inhaled, ingested, or absorbed through the skin.

Very few types of immunoglobulins are capable of binding to cell surface receptors and mediating the anaphylactic response. In man, only the IgE type of immunoglobulin has this capacity. Such antibodies are called *cytotropic*, because of their affinity for cells. They usually exhibit a high degree of species specificity; the cytotropic antibodies of man, for example, can be used to sensitize the skin of man and monkeys but not of guinea pigs.

## The role of circulating antibodies in host resistance

When a microbial or viral pathogen infects a vertebrate host, the numerous antigens of the parasite induce the formation of homologous antibodies. These antibodies combined with their specific antigenic determinants, coating the surface of the microorganism or virion. Furthermore, if the parasite liberates antigenic products such as toxins and enzymes, these also elicit the formation of antibodies and complex with them.

We have summarized in the preceding sections a number of consequences of antibody–antigen combination: precipitation, agglutination, complement fixation, and hypersensitivity reactions. To what extent do these reactions constitute defense mechanisms of the host?

The precipitation of soluble antigens has no known protective function in vivo and needs no further consideration. Agglutination, on the other hand, appears to enhance the resistance of the host; agglutinated microbial cells are more readily trapped in the alveoli of the lungs or in the sinuses of the lymph nodes, which slows their spread through the body.

Complement fixation by antibodies bound to the surface of Gram-negative bacteria plays a part in several defense mechanisms of the host. It may lead to lysis or to enhanced phagocytosis of the parasite, and it may result in the chemotactic attraction of phagocytic cells. Finally, it can lead to the release of histamine from mast cells and thus to an inflammatory reaction.

Hypersensitivity reactions, which result when certain types of immunoglobulin become fixed to tissue cells, also lead to histamine release and inflammation, which, as discussed in Chapter 27, constitutes an important defense mechanism of the host. Thus, agglutination, complement fixation, and hypersensitivity are all consequences of antigen–antibody reactions that enhance host resistance.

The most important contributions that antibodies make to the defense mechanisms of the host, however, are *opsonization* and *neutralization*. Neutralization is the term given to the combination of antibody with toxins and virions, which renders them unable to react with their receptor sites on susceptible host cells. Opsonization (Greek *opson*, food) is the process by which virulent microbial cells are coated with antibody and rendered sensitive to phagocytosis. Figure 28.13 illustrates graphically the fate of virulent and avirulent strains of a pathogenic bacterium after injection into immune and nonimmune hosts. In the nonimmune animal, cells of the avirulent strain are completely cleared from the blood by phagocytosis within 80 hours, but cells of the virulent strain multiply rapidly and kill the host within 40 hours. In the immune animal, however, virulent cells are cleared from the blood as readily as are avirulent cells in the normal animal.

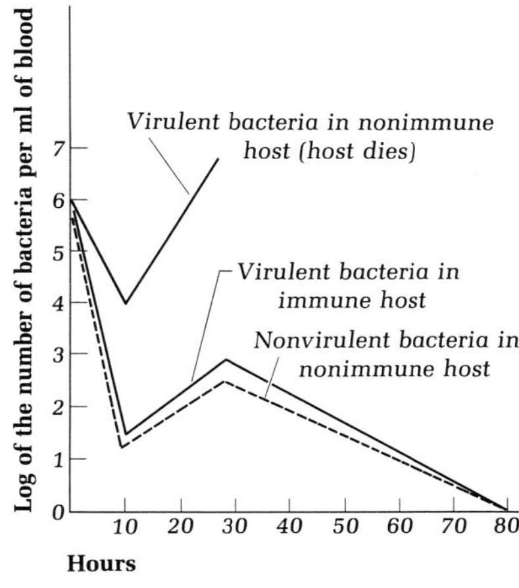


*Figure 28.13 The clearing of bacteria from the blood stream. In the nonimmune host, virulent bacteria multiply rapidly after an initial drop in number and kill the host. In the immune host, virulent bacteria are cleared from the blood, as are nonvirulent bacteria in a nonimmune host.*

## *Other mechanisms of inducible resistance*

**Hypersensitivity of the "delayed" type**

This type of hypersensitivity differs from the immediate type in several ways. The most significant difference is that *passive transfer can be accomplished with viable lymphoid cells but not with serum;* despite many attempts, it has not been possible to implicate circulating antibodies in delayed-type hypersensitivity. Furthermore, the injection of antigen into the sensitized individual does not cause the liberation of histamine or other anaphylactic mediators, and there are no special "shock tissues," such as smooth muscle.

Histologically, the delayed-type reaction is characterized by the accumulation of inflammatory cells at the site of antigen injection or —in the case of sensitization by an infectious agent such as the tubercle bacillus—at the site of localized infections. The reaction seems to involve sensitized lymphocytes, or possibly macrophages, present in the tissues. Inflammation is followed by tissue necrosis in severe cases.

The induction of delayed hypersensitivity has certain unique requirements. The inducing antigen usually must be supplied as a major fragment of a microbial cell or, if used as a chemically pure antigen, must be absorbed through the skin. The two common types of delayed hypersensitivity are thus allergy of infection and such contact allergies as the well-known responses to poison ivy and poison oak.

The most important allergy of infection occurs in tuberculosis. If an animal is infected with living tubercle bacilli, it becomes hypersensitive to a protein antigen of the bacterium called "tuberculin." Later injection of tuberculin, or the release of tuberculin by the death of bacterial cells in the tuberculous lesions, causes widespread tissue necrosis. Tuberculosis is believed to be primarily due to the delayed hypersensitivity response rather than to the action of a bacterial toxin.

The puzzling requirement for whole bacterial cells in the sensitizing process has been explained by showing that tuberculin can also induce the delayed hypersensitivity state if injected together with a waxy lipid extractable from the tubercle bacilli. Indeed, any protein can similarly be made active if injected with the waxy lipid. This finding has led to a suggested explanation of the activity of pure antigens in inducing contact allergies: the skin itself is postulated to supply a lipid that can act like the waxy lipid of tubercle bacilli to potentiate the induction of delayed hypersensitivity.

Although delayed-type hypersensitivity can be passively transferred only with cells,* and circulating antibodies may not be detectable in individuals exhibiting strong reactions, there are a number of compelling reasons for believing that antibody protein molecules—perhaps of a special type—are involved. The reasons for so believing are as follows.

(1) The reaction is specific, requiring a second injection of the antigen used for the sensitization.

(2) The sensitization period is typical of that which is required for antibody formation and is comparable to the sensitization period in immediate-type hypersensitivity.

(3) The reaction is mediated by lymphoid cells, which also have the capacity to form antibodies. (An important difference, however, is that plasma cells, which are the most active producers of *circulating* antibodies, cannot be used to transfer delayed-type hypersensitivity.)

(4) Tolerance phenomena operate in delayed-type hypersensitivity as they do in antibody formation.

(5) Antibody production and delayed-type hypersensitivity have the

*In the case of humans, delayed-type hypersensitivity can also be transferred by an *extract* of blood leucocytes. The "transfer factor" has not yet been fully characterized.

same phylogenetic distribution, occurring only in vertebrates that stand above the hagfish on the evolutionary scale.

A possible explanation for these observations is that lymphocytes, but not plasma cells, produce a special class of antibody molecules that have extremely high affinity for their antigens. The antibodies are either liberated in amounts too small to be detected or else remain bound to the surfaces of the cells that produce them. The combination of antigen with cell-bound antibody then produces cell damage, which leads to local inflammation and cell death.

Delayed-type hypersensitivity, by producing an inflammatory response, constitutes a mechanism of inducible host resistance. Although it probably represents one extreme in the spectrum of antibody reactions, it is a *cellular* response that appears not to depend on circulating antibodies.

**Induced cellular immunity of phagocytes**

Induced resistance to tuberculosis is a well-established phenomenon. The susceptibility to tuberculosis among tuberculin-negative* individuals is significantly higher than among tuberculin-positive individuals. Furthermore, vaccination with the attenuated B.C.G. strain† of *M. tuberculosis* confers partial protection against the disease. Nevertheless, little or no antibody can be detected in the serum of resistant individuals. This fact, together with the intracellular nature of the infection, led to a suspicion that immunity to tuberculosis might reflect cellular changes rather than the presence of circulating antibodies.

To test this hypothesis, animals were immunized with the B.C.G. vaccine. Monocytes were then taken from these animals, as well as from nonimmunized controls. The monocytes were thoroughly washed and allowed to ingest living, virulent tubercle bacilli. When these suspensions were maintained in normal serum for 1 week or more, a marked difference between the two batches of monocytes was observed. In the monocytes taken from nonimmunized animals, the tubercle bacilli multiplied and, in many cases, killed the phagocytes. In the monocytes taken from immunized animals, in contrast, many of the tubercle bacilli were destroyed, and the phagocytes remained healthy.

This immunity proved to be nonspecific. Immunization with either of two typical intracellular parasites, *M. tuberculosis* or *B. abortus*, led to "cellular immunity" toward both organisms. It seems unlikely that

*A tuberculin-negative individual is one who fails to exhibit a hypersensitivity reaction when tuberculin is placed in his skin. He is thus considered not to have experienced previous infection with *Mycobacterium tuberculosis*.

†The letters stand for "the bacillus of Calmette and Guérin," the latter being the workers who developed the strain.

antibodies, present within the phagocytes, can be responsible for the effect, since *Mycobacterium* and *Brucella* are not known to have antigens in common. Recently, however, it has been found that such nonspecific immunity of phagocytes is subject to secondary anamnestic response, and this response can be induced only with the agent used in the primary immunization. The phenomenon thus appears to be mediated by an antibody or antibody-like system; the unique feature of phagocytic immunity is that the specifically induced immunity, which is expressed as an enhancement of the destruction of intracellular parasites, is nonspecific in its action.

**The effect of endotoxins**

The endotoxins of enteric bacteria cause changes in nonspecific resistance to infection when injected into animals in minute doses. Figure 28.14 shows what happens when mice are injected with a purified lipopolysaccharide fraction of an endotoxin and then challenged at successive time intervals by injection with a virulent strain of *E. coli*. Immediately after the administration of the lipopolysaccharide, the mice show greatly increased susceptibility to bacterial infection. The $LD_{50}$ (the number of bacterial cells per animal required to kill 50 percent of the treated animals) is reduced a thousandfold. Within 24 hours, however, their susceptibility to infection diminishes markedly, the $LD_{50}$ rising until it reaches a value about 100,000 times greater than that of the controls. The increase in resistance lasts for 3 to 5 days and then disappears. The role of endotoxin in increasing the resistance of animals to infection is nonspecific; a purified lipopolysaccharide prepared from one species of Gram-negative bacterium confers resistance to infection by other species of Gram-negative bacteria. The mechanism of this effect is not known. It may play a very important role in nature, since all mammals (including man) are constantly exposed to the endotoxins of enteric bacteria.

*Figure 28.14 The effect of endotoxin lipopolysaccharide on resistance to infection (see text for explanation). Replotted from data of O. Westphal.*

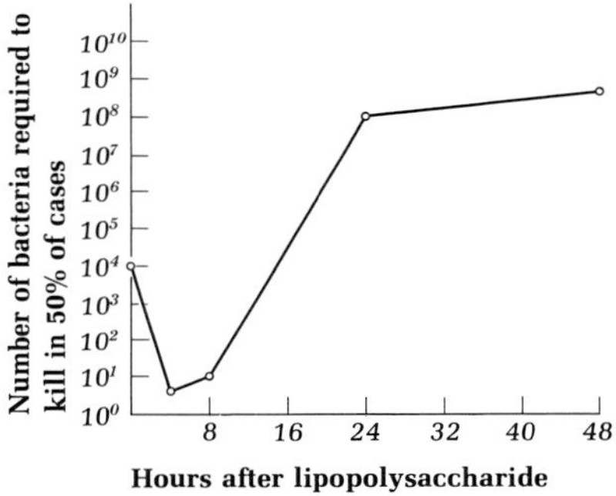


**Interferon**

When animal cells in culture are infected with a virus, they liberate a protein that inhibits the multiplication of a second kind of virus. This protein is called *interferon*.

Animal cells from different species liberate interferons of different molecular weight and antigenicity. The type of interferon liberated by cells of a given animal species is always the same, however, regardless of the virus used to induce it. Thus, interferons are *host specific, not virus specific*.

Interferons can be isolated from the fluid surrounding infected cells in culture and used to protect other cells from virus challenge. Such experiments reveal that interferon acts by inducing the cell to synthesize yet another protein, which actively inhibits the combination of viral messenger RNA with ribosomes and thus prevents the synthesis of essential viral proteins. The ability of interferon to induce the formation of the inhibitory protein is somewhat species specific: a given interferon is most effective in cells of the species that produced it.

The induction process appears to involve the derepression of a host gene. The nucleic acid of the virus plays an essential role in the induction process; empty viral cores, lacking nucleic acid, are inactive. In fact, various kinds of RNA, including heterologous cellular RNA as well as viral RNA, can induce interferon formation.

There is considerable indirect evidence that interferon plays an important role in host resistance to viral infections. The evidence consists mainly of a temporal correlation, during active infection, between the production of interferon and the decline in the multiplication of the virus.

## *Further reading*


**Books**

"Antibodies." *Cold Spring Harbor Symp. Quant. Biol.* **32** (1967).

Boyd, W. C., *Fundamentals of Immunology*, 4th ed. New York: Wiley, 1966.

Kabat, E., *Structural Concepts in Immunology and Immunochemistry*. New York: Holt, Rinehart & Winston, 1968.

Pressman, D., and A. Grossberg, *The Structural Basis of Antibody Specificity*. New York: Benjamin, 1968.

**Reviews and original articles**

Cohn, M., "The Molecular Biology of Expectation," in *Nucleic Acids in Immunology* (O. Plescia and W. Braun, eds.). New York: Springer, 1968.

Lennox, E. S., and M. Cohn, "Immunoglobulins." *Ann. Rev. Biochem.* **36,** 364 (1967).

Nossal, G., "Mechanisms of Antibody Production." *Ann. Rev. Med.* **18,** 81 (1967).

Uhr, J., "Delayed Hypersensitivity." *Physiol. Rev.* **46,** 359 (1966).

# Chapter twenty-nine
# The exploitation of microorganisms by man

*The roles of microorganisms* in the transformations of organic matter were not recognized until the middle of the nineteenth century. Nevertheless, certain microbial metabolic processes had been used by man since prehistoric times for the preparation of food, drink, and textiles; in many cases, these processes had become controlled and perfected to an astonishing degree by purely empirical methods. The outstanding examples of such *traditional microbiological processes* are those used in the production of beer and wine; the leavening of bread; the making of vinegar, cheese, and butter; and the retting of flax. The rise of microbiology, which revealed the nature of these traditional processes, led not only to great improvements in many of them, but also to the development of entirely new industries, based on the use of microorganisms not previously exploited by man.

## The use of yeast by man

The most important microorganism from the technical and industrial standpoint is yeast. Although many different yeasts exist in nature, the ones that have been principally utilized by man are varieties of *Saccharomyces cerevisiae* (e.g., var. *ellipsoideus*). The oldest known processes employing yeast are the fermentations of plant materials to produce wine and beer. The manufacture and consumption of alcoholic beverages were already established in the oldest civilizations of which records are available. Most of these civilizations had myths about the origin of wine making that attributed its discovery to divine revelation. This suggests that even in very ancient times the beginnings of the art of wine making were already shrouded in prehistoric darkness. The second main traditional use of yeast, in the leavening of bread, was a much later discovery. It seems to have originated in Egypt about 6,000 years ago and to have spread slowly from there to other parts of the western world.

The discovery that alcohol can be distilled, and so concentrated, was made relatively late, either in China or in the Arab world. Distilleries began to appear in Europe about the middle of the seventeenth century. At first, the spirits so manufactured were used primarily for direct human consumption, but with the coming of the Industrial Revolution, an increasing demand arose for alcohol as a chemical raw material, and the distilling industry grew very rapidly.

## The making of wine

Fermented beverages fall into two classes, which differ in the nature of the raw materials used and in the technique of preparation. *Wines* are made from the juice of grapes and, on a smaller scale, of other fruits. *Beers* are prepared from grain, generally barley, at least in the Western world. The principal carbohydrates in fruit juices are soluble sugars; the principal carbohydrate in grains is starch, an insoluble polysaccharide. The yeasts that bring about alcoholic fermentation can attack soluble sugars but do not produce starch-splitting enzymes. Hence, wines can be made by a direct fermentation of the raw material, whereas the production of beers requires a preliminary step, the hydrolysis of the starch to yield sugars fermentable by yeast. This process is brought about by starch-splitting enzymes from other sources.

In spite of the knowledge gained during the past century, wine making is still more of an art than a science. Given grapes of a suitable composition, the art of wine making consists largely in providing the conditions that will favor the development of a suitable microbial flora and prevent the growth of the various undesirable microorganisms present on the grapes themselves. After they have been gathered, the grapes are crushed in a press, and the juice is collected and allowed to settle. The raw juice or *must* is a strongly acid solution, containing from 10 to 25 percent soluble sugars. Its acidity and high sugar concentration make it an unfavorable medium for the growth of bacteria but highly suitable for yeasts and molds. It contains many yeasts, molds, and bacteria, derived from the surface of the ripe grapes themselves, including the so-called *true wine yeast, Saccharomyces cerevisiae* var. *ellipsoideus*. The fermentation may be allowed to proceed spontaneously, or can be "started" by inoculation with a must that has been previously successfully fermented by *S. cerevisiae* var. *ellipsoideus*. Most modern wineries on the American continent eliminate the original microbial population of the must by pasteurization or by treatment with sulfur dioxide. The must is then inoculated with a *starter culture* derived from a pure culture of a suitable strain of wine yeast. This procedure eliminates many of the uncertainties and difficulties of older methods. At the start of the fermentation, the must is aerated slightly to build up a large and vigorous yeast population; once fermentation sets in, the rapid production of carbon dioxide maintains anaerobic conditions, which prevent the growth of undesirable aerobic organisms, such as bacteria and molds. The temperature of fermentation is usually from 25 to 30°C, and the duration of the fermentation process may extend from a few days to 2 weeks. As soon as the desired degree of sugar disappearance and alcohol production has been attained, the microbiological phase of wine making is over. Thereafter, the quality and stability of the wine depend very largely on preventing further microbial activity, both during the "aging" in wooden casks and after bottling. The aging process is chemically very complex. Since some of the chemical changes that take place are oxidative, a moderate supply of oxygen is absolutely necessary, and it is for this reason that aging is carried out in wooden casks, through the walls of which a small amount of air can diffuse. A wine bottled immediately after fermentation would never develop a good flavor and aroma. Too much access of oxygen during aging brings with it the danger of aerobic microbial growth and spoilage; this is avoided as much as possible by aging wines at low temperatures and keeping the storage casks full.

At all stages during its manufacture, wine is subject to spoilage by undesirable microorganisms. The problem of the "diseases" of wines was first scientifically explored by Pasteur, whose descriptions of the organisms responsible and recommendations for overcoming them are still valid today. The most serious aerobic spoilage processes are brought about by film-forming yeasts and acetic acid bacteria, both of which grow at the expense of the alcohol, converting it to acetic acid or to carbon dioxide and water. The chief danger from these organisms arises when access of air is not carefully regulated during aging. Much more serious are the diseases caused by fermentative bacteria, particularly rod-shaped lactic acid bacteria, which grow at the expense of residual sugar and impart a mousy taste to the wine. Such wines are known as *turned* wines. Since oxygen is unnecessary for the growth of lactic acid bacteria, wine spoilage of this kind can occur even after bottling; as Pasteur showed, the risk of such infections in the finished product can be eliminated by pasteurization after bottling.

## The making of beer

Beers are manufactured from grains, the starch in which must be saccharified (converted to the fermentable sugars, maltose and glucose) by a preliminary treatment. Three principal grains were originally used for beverage production by man: barley in Europe, rice in the Orient, and corn in the Americas. In each case, a different solution to the problem of saccharifying the starch was found. In the case of barley, the starch-splitting enzymes (amylases) of the barley itself are used. Ungerminated barley contains little or no amylase, but large amounts of amylase are formed upon germination. Hence, the barley is allowed to germinate, after which the seedlings are carefully dried and stored until needed. Such germinated, dried barley is known as *malt.* The starch in the malt is attacked only to a small extent during germination because it is mechanically protected by the cellular structure of the seedling from the action of the amylases. Accordingly, the first step in brewing is the grinding of the malt and its suspension in water so as to permit the hydrolysis of the starch. Other enzymatic processes go on simultaneously, including a breakdown of malt proteins to simpler nitrogenous compounds that contribute to the final flavor of the product. After saccharification has reached the desired stage, the mixture is boiled to stop further enzyme action, and then filtered. The filtrate *(wort)* is mixed with hops, which contain resins that dissolve in the wort and contribute the characteristic bitter flavor of beer. The resins also act to some extent as preservatives against the growth of bacteria. The addition of hops is a relatively recent modification of the art of beer making, having been introduced about the middle of the sixteenth century; even today, unhopped beers are still made in some countries. After filtration, the hopped wort is ready for fermentation.

In contrast to wine making, the making of beer is never allowed to proceed by spontaneous fermentation. Instead, the wort is heavily inoculated with a special strain of yeast derived from a previous fermentation. The fermentation proceeds at low temperature for a period of 5 to 10 days. The exact conditions depend on the type of beer being manufactured. All the yeasts used in making beer belong to the species *Saccharomyces cerevisiae,* but not all strains of *S. cerevisiae* can be used to make a good beer. During the course of time, special strains with the properties desirable in brewing have been developed, and these are known as *beer yeasts.* Before Pasteur, the selection and maintenance of a good beer yeast was an empirical art. The success of a brewer depended very largely on his ability to obtain a strain suitable for his purposes and to carry it from vat to vat without getting it too heavily contaminated by undesirable microorganisms. Good brewing yeasts were developed over a period of centuries; they cannot be found in nature. The beer yeasts are a product of human art, like cultivated higher plants. In recognition of this fact, the brewer refers to other yeasts (including other strains of *S. cerevisiae*) as *wild yeasts.*

Since Pasteur's time, the recognition, testing, selection, and maintenance of good strains of beer yeasts have been placed on a scientific basis. The pioneer in this work was a Danish contemporary of Pasteur, E. C. Hansen, who worked at the Carlsberg brewery in Copenhagen. He started out by isolating and studying many strains of beer yeasts. By using these pure strains in the production of experimental beers, he was able to show that the type and quality of beer is greatly influenced by the particular yeast employed. He also devised a special yeast propagator that could be used to grow a pure yeast, under aseptic conditions, in amounts sufficient to inoculate a fermentation vat. Even in modern breweries, the actual fermentation process is never carried out aseptically, but thanks to the work of Hansen the modern brewer can start his fermentation with a pure yeast strain of known properties and can thus eliminate many of the dangers and uncertainties that beset the older practice of repeated mass transfer with mixed cultures from vat to vat.

The strains of beer yeasts fall into two principal groups, known as *top* and *bottom yeasts.* Top yeasts are vigorous fermenters, acting best at relatively high temperatures (20°C), and are used for making heavy beers with a high alcoholic content, such as the English ales. The name derives from the fact that during fermentation the cells are swept to the top of the vat by vigorous gas production. In contrast, the bottom yeasts are slow fermenters, acting best at low temperatures (12 to 15°C), and are used to produce light beers with a low alcoholic content, of the kind most commonly consumed in the United States. Since they produce gas less vigorously than do top yeasts, their cells tend to settle to the bottom of the fermentation vat.

The diseases of beer, like those of wine, were first scientifically studied by Pasteur. They occur most commonly during maturation or after bottling. One agent is a wild yeast, *Saccharomyces pasteurianus,* which gives rise to a disagreeable bitterness. Lactic acid bacteria sometimes make beer acid and cloudy. They develop principally when the temperature becomes too high during maturation and storage. Acetic acid bacteria may at times cause souring, particularly in barreled beer. The principal methods of avoiding trouble from all these sources are the use of pure yeast strains as starters and the pasteurization of the final product.

The characteristic alcoholic beverages of the Western world are wines made from grapes and beers made from barley. In the Orient, however, the most popular fermented beverages (e.g., the Japanese sake) are made from rice. The problem of hydrolyzing the starch in rice as a preliminary to alcoholic fermentation was solved in a fashion entirely different from that devised for barley. The process is conducted by means of amylases derived from molds, principally *Aspergillus oryzae*. In the manufacture of sake, which may be considered a typical rice beverage, the first step is the preparation of a culture of the mold. Mold spores, saved from a previous batch, are sown on steamed rice and allowed to grow until the mass of rice is thoroughly permeated with mycelium. This material is then used to inoculate a larger batch of steamed rice mixed with water. The amylases produced by the mold gradually transform the rice starch into sugar, and after a few days a spontaneous fermentation begins. This is not usually a pure alcoholic fermentation, for lactic acid bacteria are present in addition to yeasts, and consequently part of the sugar is decomposed to lactic acid. More rice may be added from time to time, since the mold amylases are still present and will bring about its saccharification. The oriental fermentation of grain thus differs from the occidental one in two crucial respects: both saccharification and fermentation are caused by microorganisms, and the two processes go on simultaneously, instead of being separated in time as they are in the manufacture of a Western beer.

Last, it should be mentioned that in the Americas yet a third agent of saccharification for making alcoholic beverages was discovered—human saliva, which has a high content of starch-splitting enzymes. Many Indians in Central and South America prepare a corn beer by chewing the grains and then spitting the mixture of corn and saliva into a vessel, where it is allowed to undergo a spontaneous fermentation.

**Bread making**

An alcoholic fermentation by yeast is an essential step in the production of bread; this process is known as the *leavening of bread* (after the old word for yeast, "leaven"). The moistened flour is mixed with yeast and allowed to stand for several hours in a warm place. Flour itself contains little free sugar, but there are sufficient quantities of starch-splitting enzymes in it to produce some sugar during the leavening process. In highly refined flours, such as the white flour used in the United States, the amylases have been destroyed, and free sugar must be added. The sugar is rapidly fermented by the yeast with the production of alcohol and carbon dioxide, the latter causing the rising of the bread. During the baking process, the alcohol is driven off.

Although the gas production that takes place during the leavening of bread is the most conspicuous event, the yeast produces other, more subtle changes in the physical and chemical properties of the dough.

This fact became evident when J. von Liebig, the great German chemist, invented baking powder, a mixture of chemicals that produces carbon dioxide when moistened. Liebig anticipated that baking powder would replace yeast in the manufacture of bread. However, baking powder produces a very different kind of bread from that produced by yeast, and although Liebig's invention was useful, it did not supplant yeast as a leavening agent.

The yeasts used for baking all belong to the species *Saccharomyces cerevisiae* and have been derived historically from the strains of top yeasts used in brewing. Until the nineteenth century, baking was either carried on in the home or in small local bakeries, and the yeast required was obtained from the nearest brewery. The commercial production of compressed yeast for baking purposes began about 1850. The development of the compressed yeast industry was greatly stimulated by the application of mass-production methods to baking. A large modern bakery may use many hundreds of pounds of yeast daily, for about 5 lb of yeast is needed to leaven 300 lb of flour.

**Industrial alcohol production**

Ethyl alcohol is a major industrial chemical; the annual production in the United States alone now approaches 1 billion gallons. Until recently, industrial alcohol was largely derived from the distillation of fermented plant materials, but this method of manufacture has steadily lost ground to chemical syntheses, which now account for about 75 percent of the total production.

The underlying economic problem of industrial alcohol production by fermentation is the cost of the raw material, a matter of secondary concern to the manufacturer of alcoholic beverages. A producer of wine is not so much concerned with paying the minimal possible price for grapes as with obtaining the variety of grape that will yield a particular kind of wine and that has been grown under conditions apt to give a product with the many subtle qualities that mark a good wine. A producer of industrial alcohol is always in search of the cheapest usable raw materials. Thirty years ago, most of the industrial alcohol produced in the United States was derived from the fermentation of molasses; today the scarcity and high price of this raw material has led to its virtual abandonment in favor of grains.

Special strains of yeast have been developed for industrial alcohol production. Historically derived from the top yeasts of the brewing industry, they have been specially selected since the eighteenth century for rapid growth and high product yield and now have markedly different physiological properties from those of the top yeasts still used for brewing.

When one recalls that the biological nature of yeast was recognized only about 1840, the evolution of *Saccharomyces cerevisiae* in the hands of brewers, bakers, and distillers is a very remarkable phenomenon. By selection, man succeeded in genetically modifying this microorganism to meet a variety of special needs, long before he realized that he was dealing with a biological entity.

**The glycerol fermentation**

In his studies on wine and beer, Pasteur discovered that during alcohol fermentation, glycerol ($CH_2OH—CHOH—CH_2OH$) is always formed in small amounts (2 to 3 percent of the weight of the sugar supplied). A few years before World War I, C. Neuberg, a German biochemist who was investigating the biochemical mechanism of alcoholic fermentation, discovered that the fermentation could be modified to yield large amounts of glycerol. This can be achieved in several ways. One is to add sodium bisulfite ($NaHSO_3$) to the fermenting mixture. This chemical traps the acetaldehyde, which is an intermediate in the conversion of pyruvic acid to ethyl alcohol. The trapped acetaldehyde cannot be reduced to alcohol, so there is an excess of hydrogen atoms available, and these are used to reduce triose phosphate, an earlier intermediate, to glycerol. The net result is that the sugar is broken down as follows:

$$\text{glucose} \rightarrow CO_2 + \text{acetaldehyde (trapped)} + \text{glycerol}$$

instead of in the normal fashion:

$$\text{glucose} \rightarrow 2CO_2 + 2\ \text{ethyl alcohol}$$

At the time, Neuberg's discovery was of academic interest only. Glycerol, although an important industrial chemical, could be produced cheaply and in large quantities by the hydrolysis of vegetable oils and fats, as a by-product of soap manufacture. Within a few years, the situation changed radically. After the outbreak of World War I, Germany faced an unprecedented demand for glycerol as a raw material for manufacturing explosives. At the same time, the British naval blockade of central Europe had drastically reduced the foreign supplies of vegetable oils needed to produce glycerol. In this critical situation, the production of glycerol by Neuberg's modified alcoholic fermentation was rapidly developed on an industrial scale, and by 1918 Germany was producing 1,000 tons a month from this source. Thus a discovery made incidentally to a study of the mechanism of alcoholic fermentation became, within a few years, of vital importance to the war economy of a large nation. There are few more dramatic illustrations of the difficulty of drawing a sharp line between "pure" and "applied" science.

The glycerol fermentation was also developed in England and in the United States during World War I, but it was discontinued with the return of peace, since under normal conditions the production of glycerol from fats is a cheaper process.

**Yeast as a food**

Yeasts are rich in vitamins and in most of the essential amino acids required by man and higher animals. The growth of the science of animal nutrition has led to the development of a new industry based on the cultivation of yeasts: their production for use as food, more specifically as a *food supplement*, to compensate for the dietary inadequacies of cheap food materials. The human population explosion has evoked much interest in the industrial manufacture of yeast for human consumption, particularly in the poorer regions of the world where human malnutrition is chronic. However, this industry is still largely directed to the production of yeast as a supplement for feedstuffs consumed by domestic animals.

Since the goal is to obtain cell material, the organisms are always grown under forced aeration to maximize the growth yield. The ability to produce ethanol, which lies at the base of the other industrial uses of yeast is, if anything, a disadvantage: unless the oxygen supply is always maintained at a high level, there is a risk with fermentative yeasts that part of the substrate may be diverted to alcohol. Consequently, most processes do not use yeasts of the genus *Saccharomyces*, but the strictly aerobic members of the group, belonging to the genus *Candida*.

As in the industrial production of alcohol, the cost of the raw material is a factor of paramount economic importance. Consequently, media that are prepared with cheap sources of carbohydrates (e.g., whey, molasses, sulfite waste liquor) have hitherto been used for the growth of food yeast. However, since the growth conditions are aerobic, the choice of a carbon and energy source is not restricted to carbohydrates; any organic compound that can support the respiratory metabolism of a yeast may be used. This consideration has led to the recent development in France of an exceedingly interesting and ingenious production method, based on the use of *Candida lipolytica*, a yeast that can oxidize aliphatic, unbranched hydrocarbons of chain length $C_{12}$ to $C_{18}$. These compounds form part of the complex mixture of hydrocarbons present in petroleum, and their removal is an essential and costly step in the refining of crude oil. Their elimination by biological means therefore has a potential economic value in the manufacture of petroleum products.

It has proved possible to grow *C. lipolytica* in a medium that contains an aqueous emulsion of crude petroleum. The process yields a yeast preparation which, after solvent extraction, is an excellent supplement for animal foodstuffs and at the same time removes certain hydrocarbons from the crude petroleum, thus considerably simplifying its subsequent chemical fractionation. A particular advantage of this process is that the growth yields are extremely high, since hydrocarbons are the most reduced class of organic compounds. The manufacture of yeast by this means has been put on an industrial scale and is by far the most economical method so far devised for food production from a microbial source.

## *The manufacture of vinegar*

When wine or beer is freely exposed to the air, it frequently turns sour. Souring is caused by the oxidation of the contained alcohol to acetic acid, mediated by the strictly aerobic acetic acid bacteria. The spontaneous souring of wine is the traditional method for the manufacture of vinegar, a word derived from the French "vinaigre," which literally means "sour wine."

The manufacture of vinegar still remains largely empirical. The principal modifications that have been introduced in the past century concern the mechanical rather than the microbiological aspects of manufacture. In the traditional Orleans process, which is still used in France, wooden vats or casks are partially filled with wine. The acetic bacteria develop as a gelatinous pellicle on the surface of the liquid. The conversion of ethanol takes several weeks, because its rate is limited by the rate of diffusion of air into the liquid. The survival of this slow and inefficient method is attributable to the very high quality of the vinegar that it yields, which commands a premium price.

When the taste of the product is not a factor of primary importance, vinegar is manufactured by more rapid methods from cheaper raw materials (e.g., diluted industrial alcohol, cider). These methods are also microbiologically uncontrolled and are designed primarily to accelerate the rate of the oxidation by improved aeration and control of the temperature. The oldest method, developed early in the nineteenth century, involves the use of a tank loosely filled with wooden shavings, through which the medium is circulated. The liquid is trickled into the tank, and air is blown through, countercurrent to the liquid flow. The acetic acid bacteria grow as a thin film over the wooden shavings, with the consequence that a very large area of cells is simultaneously exposed to the medium and to air. Once the bacterial population has become established on the shavings, successive batches of vinegar can be produced in a relatively short period of time: solutions containing initially 10 percent alcohol are converted to vinegar in 4 or 5 days, with a yield of acetic acid that is about 90 percent of the theoretical one. In the United States, vinegar is still largely manufactured by this method, but tank fermentors, patterned on those originally devised for the manufacture of penicillin, which allow a much stricter control of aeration and temperature, have been recently introduced.

## *The use of lactic acid bacteria by man*

**Fermented milk products**

The manufacture of such milk products as butter, cheese, and yogurt involves the action of microorganisms, among which the *lactic acid bacteria* are particularly important. The discovery of the roles played by microorganisms in these traditional processes of food preparation has led to the development of a special branch of bacteriology, known as dairy bacteriology. Here we shall indicate very briefly some aspects of the subject.

Many lactic acid bacteria occur normally in milk and are responsible for its spontaneous souring. Since they produce large amounts of acid, they inhibit the subsequent development of other microorganisms. Milk souring is thus, in a sense, a method of preserving an otherwise very unstable foodstuff, and the manufacture of cheese and other fermented milk products by primitive man was no doubt first undertaken largely as a means of preservation.

The manufacture of cheese involves two main steps: first, *curdling* of the milk proteins to form a solid material, from which much of the liquid is drained away, and second, the subsequent *ripening of this solid curd* by the action of various bacteria and fungi.

The curdling process may in itself be microbiological, because the formation of lactic acid as a result of the growth of lactic acid bacteria can lower the pH sufficiently to cause a coagulation of the milk proteins. In the manufacture of many cheeses, however, the curdling is brought about by an enzyme known as *rennin*, extracted from the stomachs of calves.

The subsequent ripening of the curd is a very complex process. In many cheeses (cheddar cheese, ordinary American cheese), it is largely brought about by the lactic acid bacteria present in the curd. As these organisms die, proteolytic and fat-splitting enzymes are released from the cells and slowly break down the milk fat and proteins with the formation of materials that impart the characteristic cheese flavor. So-called mold-ripened cheeses (blue cheeses, Camembert) are produced by inoculating the curd with special kinds of fungi that develop either throughout the curd (blue cheeses) or over its surface (Camembert). In the ripening of Swiss cheese, another group of bacteria, the propionic acid bacteria, are active. These organisms ferment the lactic acid present in the curd, converting it to propionic acid, acetic acid, and carbon dioxide. The "holes" in Swiss cheese are produced by the carbon dioxide, and the characteristic flavor is imparted by propionic acid.

Butter manufacture is also in part a microbiological process, since souring of the cream is desirable for a good subsequent separation of the butterfat in the churning process. Certain of the lactic acid streptococci, which cause the souring, produce at the same time small amounts of *acetoin,* which is spontaneously oxidized to *diacetyl,* the substance responsible for the characteristic flavor and aroma of butter. Streptococci differ considerably in their ability to produce acetoin, and in the past 20 years it has become a common practice to use *starter cultures* of proved worth in the souring of the cream. The cream is first pasteurized and then inoculated with a pure culture of the starter.

**The lactic fermentations of plant materials**

Certain lactic acid bacteria are found characteristically on plant materials. These organisms are responsible for the souring processes that occur in the preparation of pickles and sauerkraut. In these lactic acid fermentations, the sugars present in the plant tissues serve as the fermentable materials; the acid produced, in addition to imparting flavor to the product, preserves it from further spoilage by other microorganisms. The preservative value of a lactic fermentation is also taken advantage of in the *ensilaging* of green cattle fodder; after the plant materials have undergone fermentation in the silo, they can be kept indefinitely without risk of further decomposition.

**Dextran production**

Some lactic acid bacteria belonging to the genus *Leuconostoc* produce from sucrose large amounts of an extracellular polysaccharide known as *dextran.* Dextran is a polyglucose of high but variable molecular weight (15,000 to 20 million). The average molecular weight of the dextran synthesized by *Leuconostoc* varies somewhat from strain to strain. These lactic acid bacteria first came to the attention of industrial microbiologists for their nuisance value; occasionally, they develop in sugar refineries, and the large amounts of gummy polysaccharide produced may literally clog the works. About 20 years ago, the cultivation of this organism was undertaken on an industrial scale for the production of dextran, as a partial substitute, or "extender," of blood plasma; 6 percent solutions of dextrans with a molecular weight of between 50,000 and 100,000 are isosmotic with plasma and can thus be used to restore blood volume in cases of shock. More recently, dextran has acquired a use in biological research, following the discovery that dextran derivatives which have been chemically cross-linked and modified to render them insoluble in water can act as molecular sieves. Marketed under the commercial name Sephadex, these dextran derivatives are used by biochemists for the separation of proteins and for determinations of the molecular weight of large molecules.

## *The use of butyric acid bacteria by man*

**The retting processes**

*Retting* is a controlled microbial decomposition of plant materials designed to liberate certain components of the plant tissue. The oldest retting process, which has been used by the human race for several thousand years, is the retting of flax and hemp to free the bast fibers, used in the making of linen. These fibers, composed of cellulose, are held together in the plant stem by a cementing material, pectin, and their physical separation is difficult. The goal of the retting process is to bring about a decomposition of the pectin, thus freeing the bast fibers, without a simultaneous decomposition of the fibers themselves. The plant stems are immersed in tanks of water or in pits. As they become waterlogged, microbial development begins. At first, the organisms that develop are predominantly aerobes which use up the dissolved oxygen. Then a vigorous development of the anaerobic *butyric acid bacteria* begins. These organisms rapidly attack the plant pectin, loosening the stem structure and freeing the bast fibers. If retting is unduly prolonged, cellulose-fermenting bacteria will also develop and destroy the bast fibers, but when it is terminated at the right time, the fibers can be recovered undamaged. A similar retting process has sometimes been used in the preparation of potato starch, to liberate the starch-containing cells in potato tubers from the pectin in which they are embedded.

**The acetone–butanol fermentation**

In the past 50 years, certain *Clostridium* species have been used on a very large scale for the production of two industrial chemicals, acetone and butanol. Many butyric acid bacteria carry out a fermentation of sugars with the formation of carbon dioxide, hydrogen, and butyric and acetic acids. Some perform additional reactions, converting the acetic acid to ethanol and the butyric acid to acetone and butanol. It is organisms of this type (principally the species *Clostridium acetobutylicum*) that are used industrially.

Like the glycerol fermentation, the acetone–butanol fermentation became an industrial process during World War I when the British required large amounts of acetone for the manufacture of munitions. The man principally responsible for this achievement was Chaim Weizmann, later the first president of Israel, but at that time working as a chemist in England. With the entry of the United States into World War I, the acetone–butanol fermentation was also established industrially in this country, and by the end of 1918 large amounts of these two solvents were being produced by fermentation in both England and the United States. Unlike glycerol fermentation, acetone–butanol fermentation survived the return of peace, largely because butanol, which had been a more or less useless by-product during the war years, began to find increasing industrial use.

Acetone–butanol fermentation was the first large-scale microbiological process in which the exclusion of other kinds of microorganisms from the culture vessels became a factor of major importance for the success of the operation. The medium used for the cultivation of *Clostridium acetobutylicum* is also favorable for the development of lactic acid bacteria; and if these organisms begin to grow, they rapidly inhibit the further growth of the clostridia through lactic acid formation. An even more serious problem is infection with bacterial viruses, to which the clostridia are highly susceptible. Thus, acetone–butanol fermentation can be operated successfully only under conditions of the most careful microbiological control. The establishment of this industry led to the first introduction of *pure culture methods on a mass scale*, which were later improved and refined in connection with the industrial production of penicillin.

## *The microbial production of chemotherapeutic agents*

The past 25 years have seen the establishment and extremely rapid growth of a major new industry, the use of microorganisms for the synthesis of chemotherapeutic agents, particularly antibiotics and hormones. Before discussing this industry, we shall briefly review the historical origins of chemotherapy, which has revolutionized the practice of medicine in one generation, by bringing under control nearly all bacterial infectious diseases, previously a major cause of human death.

### The rise of chemotherapy

The importance of acquired immunity as a means of specific protection was recognized shortly after the discovery of the role of microorganisms as the etiological agents of infectious diseases, and for several decades thereafter, interest was centered on the control of infectious diseases

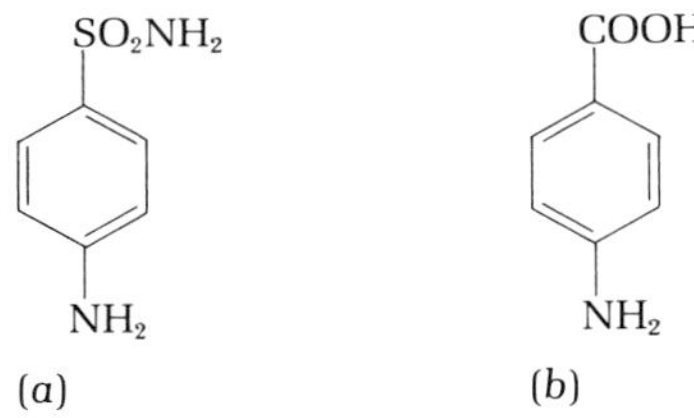


*Figure 29.1 The structures of (a) sulfanilamide and (b) p-aminobenzoic acid.*

through the development and use of vaccines and antisera. A different kind of approach to this medical problem was made by Paul Ehrlich, who first proposed that infectious diseases might be curable by the use of drugs *selectively toxic* for the parasite. This is the basic concept underlying *chemotherapy*, a name coined by Ehrlich. In 1909 Ehrlich prepared the first synthetic chemicals endowed with selective toxicity, organic compounds of arsenic that proved effective in the treatment of syphilis and other spirochetal infections.

The next advance came in 1939, when the first member of a group of chemotherapeutic agents useful in the treatment of bacterial infections, the *sulfonamides*, was discovered. This was sulfanilamide (Figure 29.1). Shortly afterward, D. D. Woods was studying its action on bacteria and observed that the inhibition of bacterial growth by sulfanilamide could be reversed by a structural analog, *p*-aminobenzoic acid (Figure 29.1). He then made a brilliant series of deductions: that *p*-aminobenzoic acid is a normal constituent of the microbial cell, that it has a coenzymatic function, and that this function is blocked by the structural analog, sulfanilamide, which competes for attachment to an enzyme. As we now know, *p*-aminobenzoic acid is not itself a coenzyme but a biosynthetic precursor of the coenzyme folic acid, and sulfanilamide blocks its conversion to this end-product. The selective toxicity of sulfanilamide results from the fact that most bacteria synthesize folic acid de novo, whereas mammals obtain this substance from dietary sources.

This work appeared to offer at last a *rational approach* to the search for chemotherapeutic agents, through the synthesis of compounds that are structural analogs of known essential metabolites. In the following years, hundreds of analogs of amino acids, vitamins, purines, and pyrimidines were synthesized and tested, but with the exception of some additional drugs of the sulfonamide group, no useful chemotherapeutic agents emerged. The next great advance in chemotherapy—the discovery of selectively toxic microbial metabolic products, or *antibiotics*—involved a return to an empirical approach and originated from a chance observation.

In 1929 Alexander Fleming observed that on a plate culture of bacteria which had become contaminated by a *Penicillium*, growth of the bacterial colonies was inhibited in the vicinity of the mold colony. Intrigued by this phenomenon, he isolated the mold and found that its culture filtrates contained a metabolic product that was bactericidal and nontoxic for animals. The active substance, which he named *penicillin*, proved to be chemically unstable; and Fleming, who was not trained as a chemist and had not received much encouragement from his colleagues, dropped the problem at this point. Ten years elapsed before another British group, headed by H. W. Florey and E. Chain, set about the isolation and purification of penicillin. Clinical trials with the first preparations of partially purified material were dramatically successful. By this time, however, England was at war and facing the prospect of a German invasion, and a rapid large-scale development of penicillin production was out of the question. The British group accordingly brought their knowledge of the problem to America, where an intensive program of research and development was begun in many laboratories. Within 3 years, penicillin was being produced on an industrial scale, an astonishing achievement in view of the many difficulties that had to be overcome. It is still one of the most effective chemotherapeutic agents for treatment of bacterial infections.

The search for other antibiotics immediately got under way, and a second clinically effective antibiotic, streptomycin, was soon dis-

*Table 29.1 Partial list of antibiotics produced industrially*

| Name | Biological source | Effective chemotherapeutically against: |
|---|---|---|
| *Antibiotics formed by fungi* | | |
| *Penicillins* | *Penicillium spp.* | *Gram-positive bacteria* |
| *Cephalosporins* | *Cephalosporium spp.* | *Gram-positive and Gram-negative bacteria* |
| *Griseofulvin* | *Penicillium griseofulvum* | *Other fungi* |
| *Antibiotics formed by unicellular bacteria* | | |
| *Gramicidin* | *Bacillus brevis* | *Other Gram-positive bacteria* |
| *Polymyxin B* | *B. polymyxa* | *Gram-negative bacteria* |
| *Bacitracin* | *B. subtilis* | *Other Gram-positive bacteria* |
| *Antibiotics formed by actinomycetes* | | |
| *Chloramphenicol* | *Streptomyces venezuelae*[a] | *Gram-positive and Gram-negative bacteria; rickettsias* |
| *Streptomycin* | *S. griseus* | *Gram-positive and Gram-negative bacteria* |
| *Tetracyclines* | *S. aureofaciens, S. rimosus* | *Gram-positive and Gram-negative bacteria; rickettsias and bedsonias* |
| *Neomycins* | *S. fradiae* | *Gram-positive and Gram-negative bacteria* |
| *Erythromycin* | *S. erythreus* | *Gram-positive bacteria* |
| *Nystatin* | *S. noursei* | *Fungi* |
| *Amphotericin B* | *S. nodosus* | *Fungi* |

[a]Now prepared commercially by chemical synthesis.

covered by A. Schatz and S. Waksman; it emerged from the first systematic screening of the actinomycetes for their ability to produce antibacterial substances, carried out by a large group in Waksman's laboratory. Since 1945, over 1,000 different antibiotics, produced by fungi, actinomycetes, or bacteria, have been isolated and characterized.

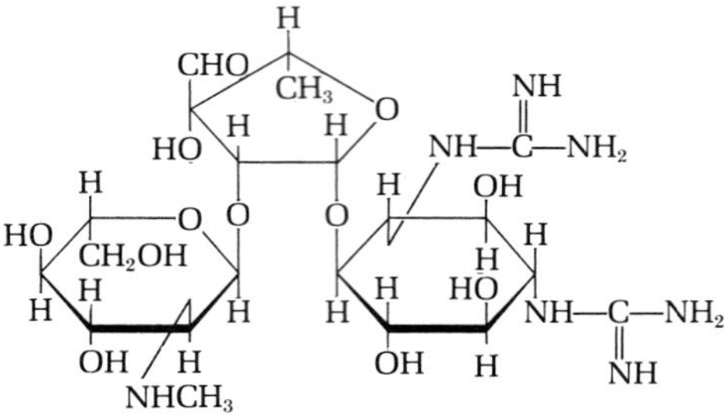

*streptomycin*

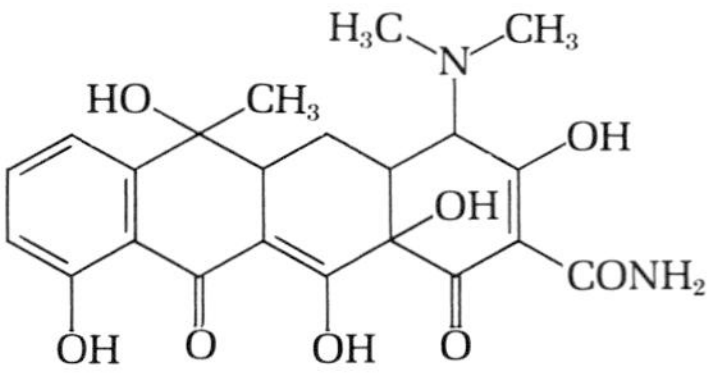

*tetracycline*

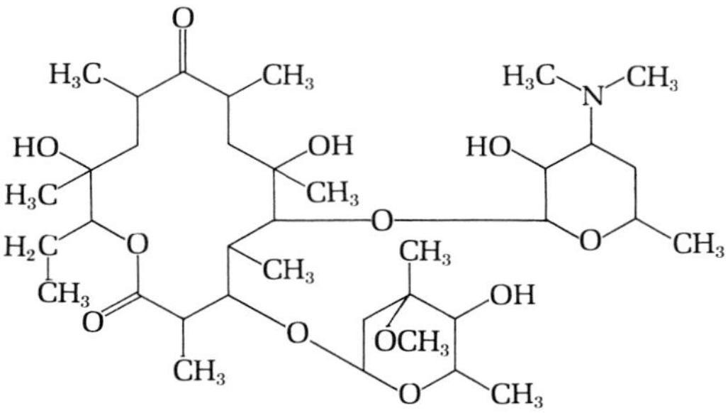

*erythromycin*

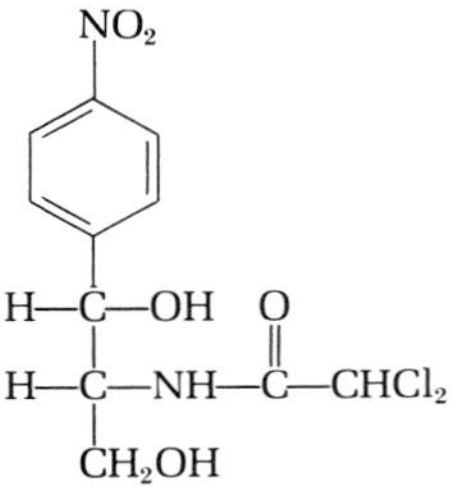

*chloramphenicol*

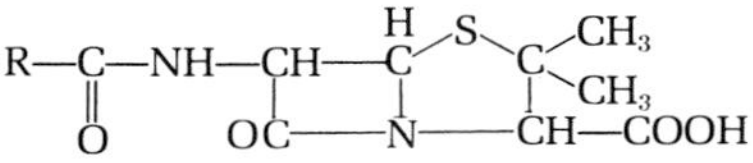

*penicillins*
*(R—group variable)*

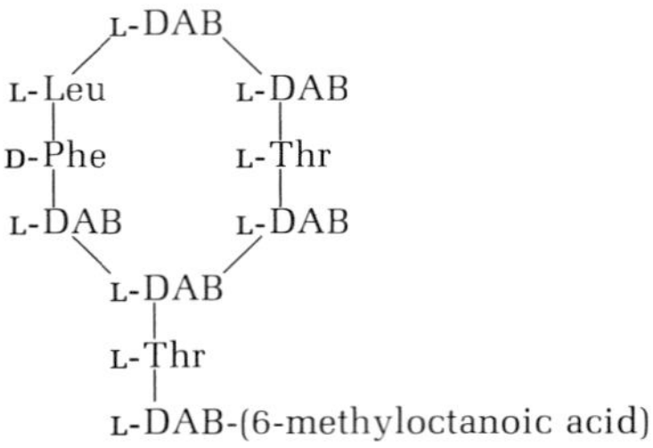

L-DAB
L-Leu
L-DAB
D-Phe
L-Thr
L-DAB
L-DAB
L-DAB
L-Thr
L-DAB-(6-methyloctanoic acid)
*polymyxin B*

*Figure 29.2 Structures of some antibiotics, illustrating the wide diversity in the chemical character of these substances. Of the antibiotics illustrated, streptomycin, tetracycline, erythromycin, and chloramphenicol are synthesized by actinomycetes; the penicillins by fungi; and polymyxin B by a bacillus. Polymyxin B is a cyclic polypeptide. Amino acid residues: leu, leucine; Phe, phenylalanine; Thr, threonine; DAB α,γ-diaminobutyric acid.*

A small fraction of these have proved useful in the treatment of disease, and about 50 are currently produced on a large scale for medical and veterinary use. The names, sources, and uses of some are shown in Table 29.1.

In terms of chemical structure, the antibiotics are exceedingly varied, as illustrated in Figure 29.2. Most are organic compounds of relatively low molecular weight but of unusual structure. Although many have been chemically synthesized on a laboratory scale, it is usually far cheaper to produce them biologically; only one, chloramphenicol, is at present manufactured entirely by chemical synthesis.

**The production of antibiotics**

The antibiotics were the first microbial metabolites to be produced industrially that were not *major* metabolic end-products. Antibiotics are typically the products of secondary and minor metabolic pathways, which come into play when normal biosynthesis and growth are almost terminated. They are synthesized and excreted by the producing organisms as cultures approach the stationary phase, and in relatively small amounts. The yields, calculated in terms of the amount of the major carbon and energy source utilized, are low and are greatly influenced by the nature of the medium and the conditions of cultivation. The problem of ascertaining the optimal environmental conditions for the formation of antibiotics is therefore complex.

*Genetic selection* has been of paramount importance in improving the yields of antibiotics. The flow of carbon through these originally minor metabolic pathways can often be greatly increased by mutation. A particular strain of *Penicillium chrysogenum*, selected initially from among a large number of wild strains for its relatively good production of penicillin, yielded 120 mg of antibiotic per liter of medium; a mutant derived from it yielded 8 g/liter under the same growth conditions, a sixtyfold improvement in yield. By a combination of genetic selection and modification of the conditions of growth, the yields of some antibiotics have been increased as much as a thousandfold over those initially obtained; in certain cases, a significant fraction of the substrate utilized is converted to the antibiotic.

The microorganisms used to produce antibiotics are all aerobes and must thus be grown in the presence of air, although the growth process is often referred to in the technical literature as a "fermentation." The provision of adequate aeration during the growth of cultures is of great importance. The production of penicillin was originally undertaken in stationary cultures, the mold being grown as a mycelial mat on the surface of shallow layers of medium in trays. It was quickly realized that growth under these conditions is severely limited by the rate of diffusion of oxygen, and so-called *submerged cultivation methods* were developed. These involve the use of deep stainless steel tanks as culture vessels, the culture being subjected to continuous forced aeration and rapid mechanical agitation. Such methods are now standard for the growth of antibiotic-producing microorganisms as well as of other aerobic microorganisms cultivated on an industrial scale. Continuous culture, although theoretically more efficient than batch cultures, has not yet been adapted to industrial use.

When microorganisms are grown in tanks with a capacity of tens of thousands of gallons, the maintenance of pure culture conditions involves a whole series of special engineering problems. The experience with penicillin production showed that the maintenance of rigorously pure cultures was essential; many bacteria destroy penicillin by means of the enzyme penicillinase, so that bacterial contamination of a batch leads to total loss of the product. The design of culture vessels with a structure that minimized the possibility of external contamination was an important part of the development of submerged culture methods. In current commercial practice, losses from contamination rarely exceed 2 percent of the batches prepared.

The manufacture of antibiotics sometimes involves secondary chemical modifications of the compound produced by the microbial agent. An interesting illustrative case history is provided by the penicillin industry. "Penicillin" is a group designation for a class of substances produced by penicillia and related molds. All naturally occurring peni-

```
            H   H
            |   |  S
acyl—NH—C—C/    \C/CH3
            |   |      |\CH3
        O=C—N———CH—COOH
```

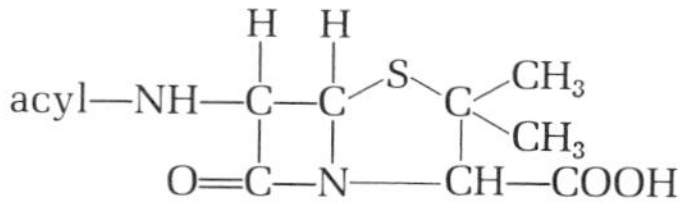

*penicillin (general structure)*

| Nature of acyl group | Name |
|---|---|
| $CH_3CH_2CH{=}CHCH_2C({=}O){-}$ | penicillin F |
| $CH_3CH_2CH_2CH_2CH_2C({=}O){-}$ | dihydropenicillin F |
| $CH_3(CH_2)_5CH_2C({=}O){-}$ | penicillin K |
| 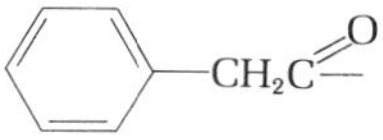 $C_6H_5{-}CH_2C({=}O){-}$ | penicillin G |
| 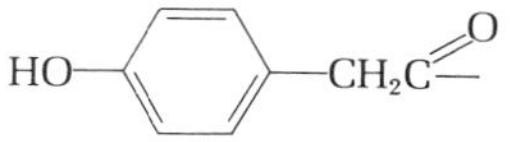 $HO{-}C_6H_4{-}CH_2C({=}O){-}$ | penicillin X |

*Figure 29.3 The structures of some naturally occurring penicillins.*

cillins have a common ring structure, to which different acyl side chains are attached, as shown in Figure 29.3. The *Penicillium* strains first used for the manufacture of penicillin produced largely penicillin F. It was found that the addition of corn steep liquor to the growth medium improves penicillin yields, and subsequently that the addition of this material causes a change in the nature of the penicillin synthesized, which shifts from penicillin F to penicillin G. Corn steep liquor contains substances that can be used by the mold as precursors of phenylacetic acid, the acyl substituent of penicillin G. This experience revealed that the chemical nature of the penicillin produced can be markedly influenced by the addition to the medium of a biosynthetic precursor and, since penicillin G is more useful clinically than penicillin F, the addition of phenylacetic acid to growing cultures became a routine part of the production process. The fungi can incorporate many other compounds into the acyl side chain of the molecule, provided that appropriate precursors are added to the medium. In this way, a very large number of new penicillins, not produced under natural conditions, have been synthesized biologically. Penicillin V, with a phenoxyacyl group, is formed upon addition of phenoxyacetic acid to the medium and has proved to be useful for oral administration, since it is not as acid-labile as the natural penicillins.

Consistent discrepancies between the results of chemical and biological assays of penicillin levels in culture filtrates led to the discovery that the mold produces small amounts of a biologically inactive biosynthetic precursor, 6-aminopenicillanic acid, together with penicillin. This compound is a penicillin without an acyl group; and it can be easily and efficiently acylated by chemical means, which makes it possible to produce an unlimited variety of *semisynthetic* penicillins. Thousands have been prepared, and a number of them have proved to be markedly superior to penicillin G in clinical use. Numerous bacteria can produce an enzyme, penicillin amidase, which hydrolyzes penicillin G with the formation of 6-aminopenicillanic acid (Figure 29.4). The semisynthetic penicillins can be prepared by the enzymatic or

```
                              S
                             / \
(C6H5)—CH2CO—NH—CH—CH      C(CH3)2    +  H2O
                |    |      |
                OC———N——————CHCOOH

                |
                v
                                       S
                                      / \
(C6H5)—CH2COOH   +   H2NCH—CH       C(CH3)2
                        |    |       |
                        OC———N———————CHCOOH
```

*Figure 29.4 The hydrolysis of penicillin G to yield phenylacetic acid and 6-aminopenicillanic acid, mediated by the enzyme penicillin amidase.*

chemical hydrolysis of penicillin G, followed by a chemical reacylation of 6-aminopenicillanic acid. The names and structures of some of the more important semisynthetic penicillins are shown in Table 29.2.

*Table 29.2 Some semisynthetic penicillins now in chemical use, showing the chemically introduced acyl substituents*

| Name | Nature of acyl group |
|---|---|
| *Propicillin* | 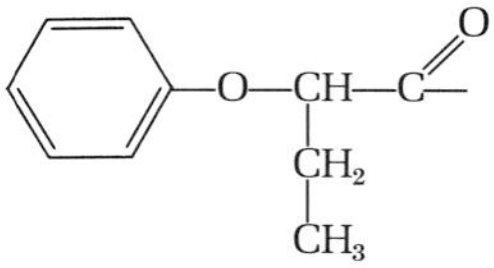 O —O—CH—C— $CH_2$ $CH_3$ |
| *Methicillin* | 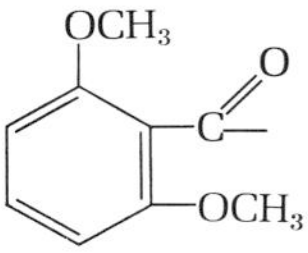 $OCH_3$ O —C— —$OCH_3$ |
| *Ampicillin* | 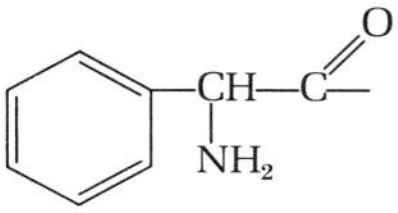 O —CH—C— $NH_2$ |
| *Oxacillin* | 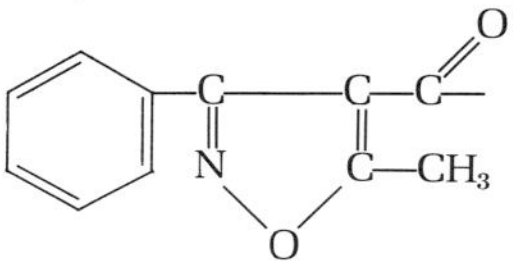 O —C——C—C— N C—$CH_3$ O |

**Microbial transformations of steroids**

Cholesterol (Figure 29.5) and chemically related steroids are structural components of eucaryotic cellular membranes and therefore universal chemical constituents of eucaryotes. During the evolution of vertebrates, special pathways were evolved for the conversion of these uni-

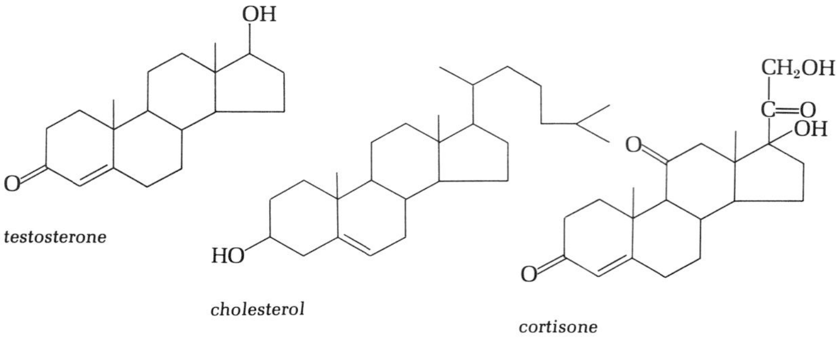

*Figure 29.5 The structure of cholesterol, a $C_{27}$ steroid, and of two mammalian steroid hormones for which cholesterol is a biosynthetic precursor—cortisone ($C_{21}$) and testosterone ($C_{19}$).*

OH
O
testosterone
HO
cholesterol
$CH_2OH$
C=O
OH
O
O
cortisone

versal cell constituents to new and functionally specialized steroids: the *steroid hormones*, which are potent regulators of animal development and metabolism. Steroid hormones are formed in specialized organs, through the secondary metabolism of cholesterol, a $C_{27}$ steroid. The adrenocortical hormones are synthesized in the adrenal gland and are all $C_{21}$ compounds, such as *cortisone* (Figure 29.5); the sex hormones are synthesized in the ovary or the testis and are $C_{18}$ or $C_{19}$ compounds (Figure 29.5). Accordingly, by relatively slight chemical modifications of the basic steroid structure, vertebrates have evolved two new subclasses of steroid molecules with highly specific physiological functions and of great potency.

The elucidation of the structures and general functions of mammalian steroid hormones was completed about 30 years ago, but it was only in 1950 that possible chemotherapeutic uses for them became apparent, with the discovery that cortisone treatment can relieve dramatically the symptoms of rheumatoid arthritis. Today cortisone and its derivatives are very widely used to treat a variety of inflammatory conditions, and additional medical uses for steroid hormones have emerged in the treatment of certain types of cancer and as oral contraceptives. The production of these compounds has now become a major industry.

Since the steroid hormones are produced by mammals in very small quantities, it was evident that their isolation from animal sources could not supply the clinical needs. Accordingly, the chemists turned their attention to the synthesis of those substances from plant sterols, which are abundant and can be cheaply prepared. One major chemical obstacle soon became apparent. All adrenocortical hormones are characterized by the insertion of an oxygen atom at position 11 on the ring system (Figure 29.6), by an organ-specific enzymatic hydroxylation of the biosynthetic precursor in the adrenal gland. Although it is easy to hydroxylate the steroid nucleus chemically, it is extremely difficult to insert a hydroxyl group at a specific position, and the specific 11-hydroxylation essential to the successful synthesis of cortisone from cheaper steroids became a major stumbling block to the development of a successful industrial process.

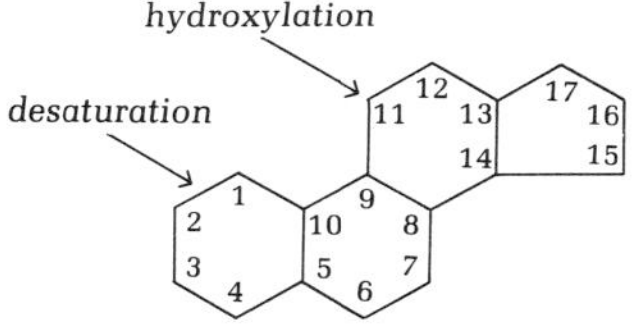


*Figure 29.6 The ring system of steroids, showing the numbering of the carbon atoms and the specific sites of two commercially important chemical modifications that are mediated by microorganisms.*

The discovery was then made that many microorganisms—fungi, actinomycetes, and bacteria—are capable of performing limited oxidations of steroids, which cause small and highly specific structural changes. The positions and nature of these changes are often characteristic for a microbial species, so that by the selection of an appropriate microorganism as an agent, it is possible to bring about any one of a large number of different modifications of the steroid molecule. Of particular practical importance is, of course, hydroxylation at the 11 position, which can be mediated by *Rhizopus* and other fungi. The introduction of a double bond by dehydrogenation between positions 1 and 2, mediated by a *Corynebacterium*, is another transformation of industrial importance, essential in the synthesis of a cortisone derivative, prednisolone.

The substrates for these microbial oxidative transformations are essentially insoluble in water. Furthermore, the limited transformation which the microorganism can effect does not provide it with either carbon or energy. The steroid substrates are, accordingly, added near the end of microbial growth, in the form of a finely dispersed suspension. The transformed products are released into the medium. In spite of the virtual insolubility of the substrates in water, many of these transformations proceed rapidly and with high yields.

## *Microbiological methods for the control of insects*

In Chapter 20 the formation of crystalline inclusions in the sporulating cells of certain *Bacillus* species was described. These bacilli (*Bacillus thuringensis* and related forms) are all pathogenic for the larvae (caterpillars) of certain insects, specifically, a very wide range of insects belonging to the Lepidoptera (butterflies and related forms). Following the isolation of the crystalline inclusions from sporulating bacterial cells, it was shown that all the primary symptoms characteristic of the natural disease of insects could be reproduced by feeding larvae on leaves coated with the purified crystals. The crystals consist of a protein which is insoluble in water under neutral or mildly acid conditions but which can be dissolved in dilute alkali. The gut contents of larvae are, in general, alkaline, and when the ingested crystals reach the gut, they are dissolved. The dissolved protein attacks the cementing substances which keep the cells of the gut wall adherent, and as a consequence, the liquid in the gut can diffuse freely into the blood of the insect. The blood of the insect becomes highly alkaline, and this change in pH induces a general paralysis of the larva. Death, which ensues much later, appears to result from bacterial invasion of the body tissues.

The protein crystals possess a highly specific toxicity for the larvae of many Lepidoptera but are wholly nontoxic for other animals (including all the vertebrates) and for plants. They thus provide an ideal agent for the control of many serious insect pests which damage plant crops. Recognition of this fact has led recently to the development of a new microbiological industry: the large-scale production of the toxic protein, for incorporation in dusting agents that can be used to protect commercial crops from the ravages of caterpillars. In industrial practice, the protein itself is not chemically isolated. Instead, the crystal-producing bacilli are grown on a large scale, harvested after the onset of sporulation with its accompanying crystal production, dried, and incorporated in a dusting powder.

## *The production of other chemicals by microorganisms*

The widespread use of microorganisms in the chemical and pharmaceutical industries has come about because of the recognition that *it is often cheaper to use a microorganism for the synthesis of a given compound than to synthesize it chemically*. Many uses of microorganisms in industry, other than those already described, have been discovered during the present century and will be briefly mentioned here.

Certain organic acids are produced commercially by the use of microorganisms. They include citric and gluconic acids, produced by *Aspergillus niger;* itaconic acid, produced by *Aspergillus itaconicus;* and lactic acid, produced by lactobacilli. In recent years, the production of specific amino acids by microorganisms has become an important industry. The production of two vitamins, riboflavin and the chemically very complex vitamin $B_{12}$, is mediated by microorganisms. Finally, it should be noted that the preparation of enzymes from microbial sources has become an important industry; the nature and sources of some enzymes now manufactured on a large scale, for a variety of industrial uses, is shown in Table 29.3.

*Table 29.3 Some microorganisms used as commercial sources of enzymes*

| Group | Organism | Enzymes prepared commercially |
|---|---|---|
| *Bacteria* | *Streptococcus hemolyticus* | *Streptokinase* |
| | *Bacillus subtilis* | *Proteases, amylases* |
| *Fungi* | *Aspergillus niger* | *Pectinases, glucose oxidase, catalase, lipase* |
| | *Aspergillus oryzae* | *Proteases, amylases* |
| | *Saccharomyces cerevisiae* | *Invertase* |

## *The use of microorganisms in bioassays*

So far, we have been considering the microorganisms as producers of chemical substances. Yet another means of exploiting microorganisms has been discovered: their use in the performance of *bioassays*.

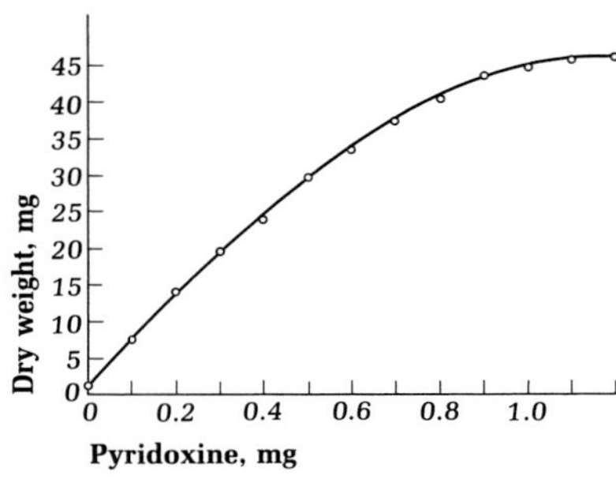


*Figure 29.7 Microbial bioassay curve for the vitamin pyridoxine. The curve expresses the relationship between the amount of pyridoxine supplied and the amount of growth (dry weight) obtained. The organism used was a pyridoxine-requiring mutant of the mold Neurospora. From J. L. Stokes, Alma Larsen, C. R. Woodward, Jr., and J. W. Foster, "A Neurospora Assay for Pyridoxine." J. Biol. Chem.* **150,** *19 (1943).*

Many microorganisms require accessory growth factors such as vitamins and amino acids for their growth and can be used to measure the quantities of these substances in foods or other materials. The principle of a bioassay is very simple. One prepares a medium that contains all the nutrients required for growth by the test microorganism, with the exception of the substance that one wishes to assay. In the absence of this substance, the test organism cannot grow. If the substance is added in limiting quantities, the growth of the test organism will be proportional to the amount added. The first step in developing a bioassay is thus to measure the relationship between the amount of growth obtained and the amount of the limiting substance added; one obtains a curve such as is shown in Figure 29.7. One can then add to the basal medium different amounts of a material that one suspects to contain the growth substance in question and measure the amount of growth that occurs. By comparison with the *standard curve,* obtained with known amounts of the growth substance, it is then possible to calculate the amount of this substance in the material that one has assayed. The lactic acid bacteria are favorite microorganisms for conducting bioassays because they have very elaborate nutritional requirements. By changing the basal medium so as to make different growth factors limiting, one can assay a large number of different amino acids and vitamins with a single test organism. In the case of lactic acid bacteria, growth is often estimated indirectly by titrating the amount of lactic acid produced (Figure 29.8). The great advantage of such microbial bioassays is that they can be used to determine quantitatively very small amounts of compounds, which cannot be easily or specifically determined by chemical analysis.

Microbial bioassays are also used for quantitative determinations of antibiotics. In this case, what one measures is not growth but the *inhibition of growth.* In the *cup plate assay,* which was developed for the estimation of penicillin, an agar plate is seeded with the test bacterium, and then a number of glass cylinders are placed on its surface, forming a series of little cups. A known dilution of a penicillin solution is added to each cup, and the plate is then incubated until bacterial growth has occurred. The penicillin diffuses out from the cup into the surrounding

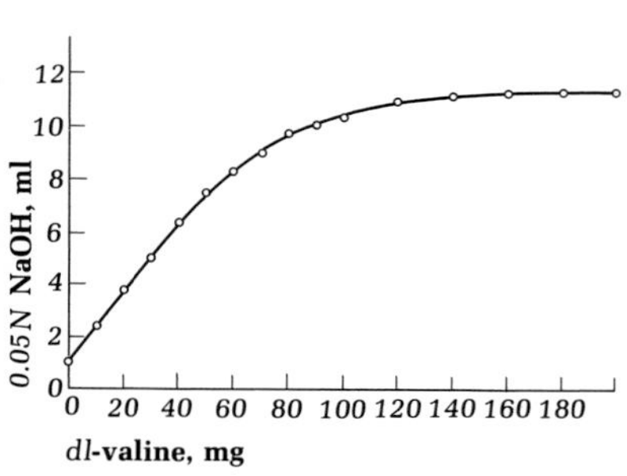


*Figure 29.8 Microbial bioassay curve for the amino acid valine. The curve expresses the relationship between the amount of valine furnished and the amount of alkali required to neutralize the acid formed by the test organism (a lactic acid bacterium) as a result of growth. From Millard J. Horn et al., "Microbiological Determination of Valine in Proteins and Foods." J. Biol. Chem.* **170,** *720 (1947).*

*Figure 29.9 A bioassay for the antibiotic penicillin. A tray of nutrient agar was seeded with the test bacterium, and differing amounts of penicillin were placed on the paper discs. The amount of penicillin increases from 10 units on the left-hand disc to 10,000 units on the right-hand disc. From the areas of the zones in which bacterial growth is inhibited, a curve relating penicillin concentration to the extent of inhibition may be derived.*

agar and produces a zone of inhibition, and the diameter of this zone is proportional to the concentration of penicillin in the cup. Thus, when one has established a standard curve relating zone diameter to penicillin concentration, it is possible to assay penicillin solutions of unknown strength. The same type of assay method has been subsequently applied to the measurement of other antibiotics. A slight modification involves the substitution of paper discs, soaked in solutions of the antibiotic, for the glass cylinders (Figure 29.9).

## Further reading


**Books**

Foster, E. M., F. E. Nelson, M. L. Speck, R. N. Doetsch, and J. C. Olson, *Dairy Microbiology*. Englewood Cliffs, N.J.: Prentice-Hall, 1957.

Peppler, H. J. (editor), *Microbial Technology*. New York: Reinhold, 1967.

Prescott, S. C., and C. G. Dunn, *Industrial Microbiology*, 3rd ed. New York: McGraw-Hill, 1959.

**Reviews**

Allgeier, R. J., and F. M. Hildebrandt. "Newer Developments in Vinegar Manufacture." *Adv. Appl. Microbiol.* **2,** 163 (1960).

Amerine, M. A., and R. C. Kunkee, "Microbiology of Winemaking." *Ann. Rev. Microbiol.* **22,** 323 (1968).

Heden, C.-G., and M. P. Starr, "Global Impacts of Applied Microbiology." *Adv. Appl. Microbiol.* **6,** 2 (1964).

Kavanagh, F., "A Commentary on Microbiological Assaying." *Adv. Appl. Microbiol.* **2,** 65 (1960).

Rogoff, M. H., "Crystal-Forming Bacteria as Insect Pathogens." *Adv. Appl. Microbiol.* **8,** 29 (1966).

Stoudt, T. H., "The Microbiological Transformation of Steroids." *Adv. Appl. Microbiol.* **2,** 183 (1960).

# Index

## A

**B**

## C

## D



## E

**F**



**G**

## N



## O



## P

## Q



## R

## S

## T



## U



## V



## W

## Y



## Z